Student Study Guide/ Solutions Manual

for use with

Organic Chemistry

Janice Gorzynski Smith
University of Hawai'i at Mānoa

and

Erin R. Smith

 **Higher Education**

Boston   Burr Ridge, IL   Dubuque, IA   Madison, WI   New York   San Francisco   St. Louis
Bangkok   Bogotá   Caracas   Kuala Lumpur   Lisbon   London   Madrid   Mexico City
Milan   Montreal   New Delhi   Santiago   Seoul   Singapore   Sydney   Taipei   Toronto

The McGraw·Hill Companies

Student Study Guide/Solutions Manual for use with
ORGANIC CHEMISTRY
JANICE GORZYNSKI SMITH AND ERIN R. SMITH

Published by McGraw-Hill Higher Education, an imprint of The McGraw-Hill Companies, Inc.,
1221 Avenue of the Americas, New York, NY 10020. Copyright © 2006 by The McGraw-Hill
Companies, Inc. All rights reserved.

This book is printed on acid-free paper.

    5 6 7 8 9 BKM BKM   0 9 8 7 6

ISBN-13: 978-0-07-239747-5

ISBN-10: 0-07-239747-0

www.mhhe.com

Table of Contents

Chapter 1: Structure and Bonding

♦ Important facts

- **The general rule of bonding.** Atoms strive to attain a complete outer shell of valence electrons (Section 1.2). H "wants" 2 electrons. Second row elements "want" 8 electrons.

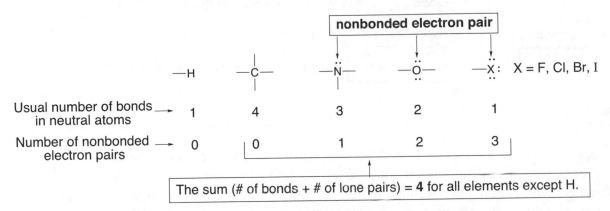

Usual number of bonds in neutral atoms →	1	4	3	2	1
Number of nonbonded electron pairs →	0	0	1	2	3

The sum (# of bonds + # of lone pairs) = **4** for all elements except H.

- **Formal charge** (FC) is the difference between the number of valence electrons of an atom and the number of electrons it "owns" (Section 1.3C). See Sample Problem 1.5 for a stepwise example.

Definition:

formal charge = number of valence electrons − number of electrons an atom "owns"

Examples:

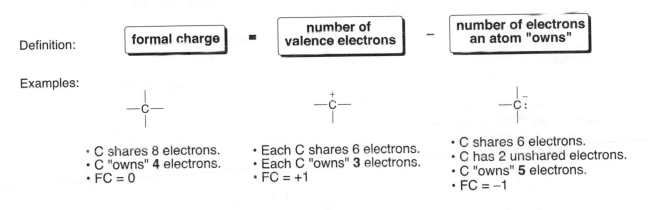

- • C shares 8 electrons.
- • C "owns" **4** electrons.
- • FC = 0

- • Each C shares 6 electrons.
- • Each C "owns" **3** electrons.
- • FC = +1

- • C shares 6 electrons.
- • C has 2 unshared electrons.
- • C "owns" **5** electrons.
- • FC = –1

- **Curved arrow notation** shows the movement of an electron pair. The tail of the arrow always begins at an electron pair, either in a bond or a lone pair. The head points to where the electron pair "moves" (Section 1.5).

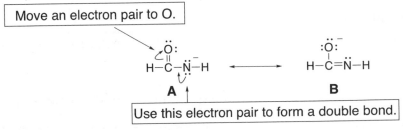

Move an electron pair to O.

Use this electron pair to form a double bond.

- **Electrostatic potential plots** are color-coded maps of electron density, indicating electron rich and electron deficient regions (Section 1.11).

♦ **The importance of Lewis structures (Section 1.3, 1.4)**

A properly drawn Lewis structure shows the number of bonds and lone pairs present around each atom in a molecule. In a valid Lewis structure, each H has 2 electrons, and each second row element has no more than 8. This is the first step needed to determine many properties of a molecule.

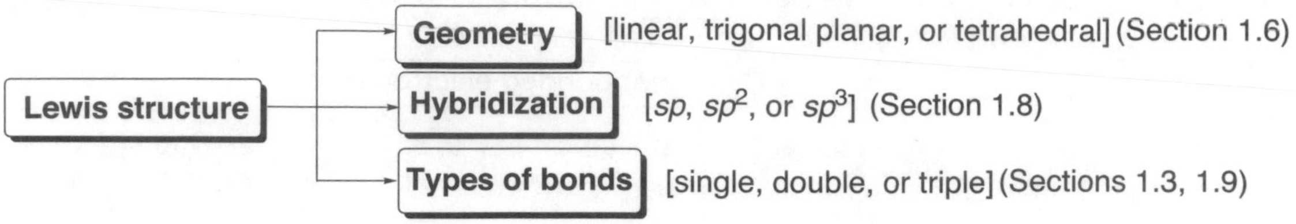

Geometry	[linear, trigonal planar, or tetrahedral] (Section 1.6)
Lewis structure → **Hybridization**	[sp, sp^2, or sp^3] (Section 1.8)
Types of bonds	[single, double, or triple] (Sections 1.3, 1.9)

♦ **Resonance (Section 1.5)**

The basic principles:

- Resonance occurs when a compound cannot be represented by a single Lewis structure.
- Two resonance structures differ *only* in the position of nonbonded electrons and π bonds.
- The resonance hybrid is the only accurate representation for a resonance-stabilized compound. A hybrid is more stable than any single resonance structure because electron density is delocalized.

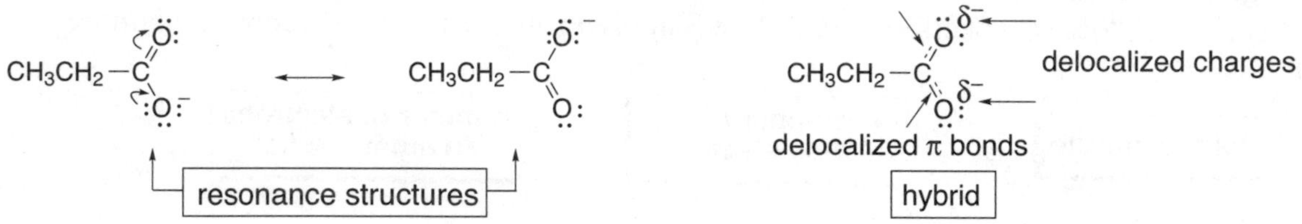

delocalized charges

delocalized π bonds

The difference between resonance structures and isomers:

- Two **isomers** differ in the arrangement of *both* atoms and electrons.
- **Resonance structures** differ *only* in the *arrangement of electrons*.

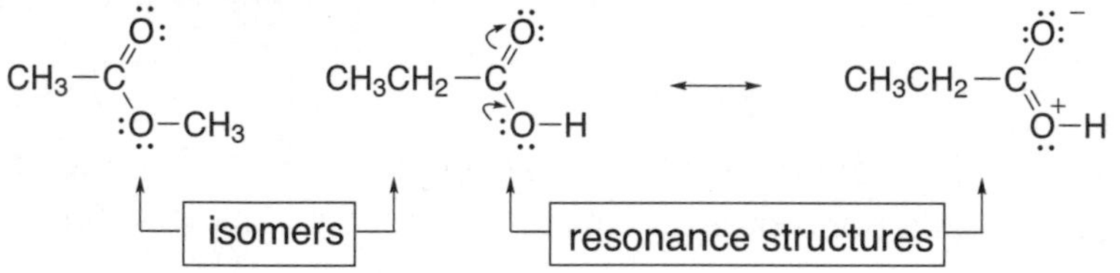

♦ **Geometry and hybridization**

The number of groups around an atom determines both its geometry (Section 1.6) and hybridization (Section 1.8).

Number of groups	Geometry	Bond angle (°)	Hybridization	Examples
2	linear	180	sp	BeH_2, $HC{\equiv}CH$
3	trigonal planar	120	sp^2	BF_3, $CH_2{=}CH_2$
4	tetrahedral	109.5	sp^3	CH_4, NH_3, H_2O

♦ **Drawing organic molecules (Section 1.7)**

- Shorthand methods are used to abbreviate the structure of organic molecules.

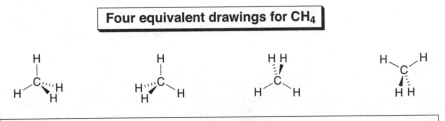

skeletal structure   =   isooctane   =   condensed structure

- A carbon bonded to four atoms is tetrahedral in shape. The best way to represent a tetrahedron is to draw two bonds in the plane, one in front, and one behind.

> **Four equivalent drawings for CH₄**

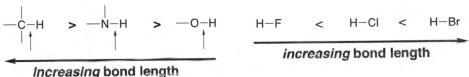

> **Each drawing has two solid lines, one wedge, and one dashed line.**

♦ **Bond length**

- Bond length decreases across a row and increases down a column of the periodic table (Section 1.6A).

$$-\overset{|}{\underset{|}{C}}-H \quad > \quad -\overset{|}{N}-H \quad > \quad -O-H \qquad H-F \quad < \quad H-Cl \quad < \quad H-Br$$

←— *Increasing* bond length   —→ *increasing* bond length

- Bond length decreases as the number of electrons between two nuclei increases (Section 1.10A).

$$CH_3-CH_3 \quad < \quad CH_2=CH_2 \quad < \quad H-C\equiv C-H$$

←— *increasing* bond length

- Bond length increases as the percent *s*-character decreases (Section 1.10B).

$$C_{sp}-H \qquad C_{sp^2}-H \qquad C_{sp^3}-H$$

—→ *increasing* bond length

- Bond length and bond strength are inversely related. Shorter bonds are stronger bonds (Section 1.10)

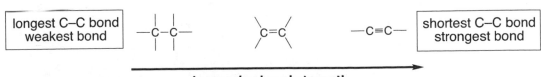

| longest C–C bond weakest bond | | | shortest C–C bond strongest bond |

—→ *increasing* bond strength

- Sigma (σ) bonds are generally stronger than π bonds (Section 1.9).

$$-\overset{|}{\underset{|}{C}}-\overset{|}{\underset{|}{C}}- \qquad \overset{\diagdown}{\diagup}C=C\overset{\diagup}{\diagdown} \qquad -C\equiv C-$$

1 strong σ bond	1 stronger σ bond	1 stronger σ bond
	1 weaker π bond	2 weaker π bonds

◆ Electronegativity and polarity (Section 1.11, 1.12)

- Electronegativity increases across a row and decreases down a column of the periodic table.
- A polar bond results when two atoms of different electronegativity are bonded together. Whenever C or H is bonded to N, O or any halogen, the bond is polar.
- A polar molecule has either one polar bond, or two or more bond dipoles that reinforce.

Chapter 1: Answers to Problems

1.1 The **mass number** is the number of protons and neutrons. The **atomic number** is the number of protons and is the same for all isotopes.

	Mass Number		
	16	17	18
a. number of protons = atomic number for O = 8	8	8	8
b. number of neutrons = mass number – atomic number	8	9	10
c. number for electrons = number of protons	8	8	8
d. The group number is the same for all isotopes.	6A	6A	6A

1.2 **Ionic bonds** form when an element on the far left side of the periodic table transfers an electron to an element on the far right side of the periodic table. **Covalent bonds** result when two atoms *share* electrons.

a. F–F
covalent

b. Li$^+$ Br$^-$
ionic

c. H–C–C–H (with H H on top and H H on bottom)
All C–H and C–C bonds are covalent.

d. Na$^+$ N–H (with H)
ionic
Both N–H bonds are covalent.

1.3 Atoms with 1, 2, or 3 valence electrons form 1, 2, or 3 bonds respectively. Atoms with four or more valence electrons form [8 – (number of valence electrons)] bonds.

a. O 8 – 6 valence e$^-$ = 2 bonds   b. Al 3 valence e$^-$ = 3 bonds   c. Br 8 –7 valence e$^-$ = 1 bond

1.4 [1] Arrange the atoms with the H's on the periphery.
[2] Count the valence electrons.
[3] Arrange the electrons around the atoms. Give the H's 2 electrons first, then fill the octets of the other atoms.
[4] Assign formal charges (Section 1.3C).

a.
[1] H H
     H C C H
     H H

[2] Count valence e$^-$.
2C x 4 e$^-$ = 8
6H x 1 e$^-$ = 6
total e$^-$ = 14

[3] H H
     H–C–C–H
     H H
All 14 e$^-$ used.
All second row elements have an octet.

b.
[1] H
     H C N H
     H H

[2] Count valence e$^-$.
1C x 4 e$^-$ = 4
5H x 1 e$^-$ = 5
1N x 5 e$^-$ = 5
total e$^-$ = 14

[3] H
     H–C–N–H
     H H
12 e$^-$ used.
N needs 2 more electrons for an octet.

→   H
    H–C–N–H
    H H

c.
[1] H
     H C
     H

[2] Count valence e$^-$.
1C x 4 e$^-$ = 4
3H x 1 e$^-$ = 3
negative charge = 1
total e$^-$ = 8

[3] H
     H–C
     H
6 e$^-$ used.
C needs 2 more electrons for an octet.

→   H
    H–C:
    H
[C has a –1 formal charge (Section 1.3C).]

1.5 Follow the directions from Answer 1.4.

a. HCN   H C N

Count valence e⁻.
$1C \times 4\ e^- = 4$
$1H \times 1\ e^- = 1$
$1N \times 5\ e^- = 5$
total e⁻   = 10

H–C–N

4 e⁻ used.

H–C≡N:

Complete N and C octets.

b. H_2CO H C O
H

Count valence e⁻.
$1C \times 4\ e^- = 4$
$2H \times 1\ e^- = 2$
$1O \times 6\ e^- = 6$
total e⁻   = 12

H–C–O
|
H

6 e⁻ used.

H–C=O:
|
H

Complete O and C octets.

c. CH_3CO^+
H
H C C O
H

Count valence e⁻.
$2C \times 4\ e^- = 8$
$3H \times 1\ e^- = 3$
$1O \times 6\ e^- = 6$
total e⁻   = 17
Subtract 1 for (+) charge = 16

H
|
H–C–C–O
|
H

10 e⁻ used.

H
|
H–C–C=O:
| +
H

or

H
|
H–C–C≡O +
|
H

Complete octets.

1.6 **Formal charge** (FC)= number of valence electrons – [number of unshared electrons + 1/2 (number of shared electrons)]

a.
$$\left[\begin{array}{c} H \\ H–N–H \\ H \end{array} \right]^+$$

$5 – [0 + 1/2(8)] = +1$

b. $CH_3–N≡C:$

$5 – [0 + 1/2(8)] = +1$

$4 – [0 + 1/2(8)] = 0$ $4 – [2 + 1/2(6)] = –1$

c. $:\ddot{O}=\ddot{O}–\ddot{O}:$

$6 – [2 + 1/2(6)] = +1$

$6 – [4 + 1/2(4)] = 0$ $6 – [6 + 1/2(2)] = –1$

1.7

a. CH_3O^-

[1] H C O
H

[2] Count valence e⁻.
$1C \times 4\ e^- = 4$
$3H \times 1\ e^- = 3$
$1O \times 6\ e^- = 6$
total e⁻   = 13
Add 1 for (–) charge = 14

[3]
H
|
H–C–O
|
H

8 e⁻ used.

H
|
H–C–Ö:
|
H

[4]
H
|
H–C–Ö:⁻
|
H

Assign charge.

b. HC_2^- [1] H C C   [2] Count valence e⁻.   [3] H—C—C ⟶ H—C≡C :   [4] H—C≡C :⁻

2C x 4 e⁻ = 8
1H x 1 e⁻ = 1
─────────────
total e⁻   = 9
Add 1 for (−) charge = 10

4 e⁻ used.

Assign charge.

c. $(CH_3NH_3)^+$ [1] H H   [2] Count valence e⁻.   [3]
H C N H
H H

1C x 4 e⁻ = 4
6H x 1 e⁻ = 6
1N x 5 e⁻ = 5
─────────────
total e⁻   = 15
Subtract 1 for (+) charge = 14

```
    H  H
    |  |
H—C—N—H
    |  |
    H  H
```
14 e⁻ used.

[4]
```
    H  H
    |  |+
H—C—N—H
    |  |
    H  H
```
Assign charge.

1.8

a. $C_2H_4Cl_2$ (two isomers)

Count valence e⁻.
2C x 4 e⁻ = 8
4H x 1 e⁻ = 4
2Cl x 7 e⁻ = 14
─────────────
total e⁻   = 26

```
   H  H
   |  |
H—C—C—Cl:
   |  ..
   H :Cl:
```

```
   H  H
   |  |
H—C—C—Cl:
   ..  |
  :Cl: H
```

b. C_3H_8O (three isomers)

Count valence e⁻.
3C x 4 e⁻ = 12
8H x 1 e⁻ = 8
1O x 6 e = 6
─────────────
total e⁻   = 26

```
   H  H  H
   |  |  |
H—C—C—C—O—H
   |  |  |
   H  H  H
```

```
       H
       |
   H  :O: H
   |  |  |
H—C—C—C—H
   |  |  |
   H  H  H
```

```
   H  H     H
   |  |     |
H—C—C—O—C—H
   |  |  ..  |
   H  H     H
```

c. C_3H_6 (two isomers)

Count valence e⁻.
3C x 4 e⁻ = 12
6H x 1 e⁻ = 6
─────────────
total e⁻   = 18

```
   H     H
   |     |
H—C—C=C
   |   |  |
   H   H  H
```

```
      H  H
       \ /
        C
       / \
   H—C—C—H
     /     \
    H       H
```

1.9 Two different definitions:

- **Isomers** have the same molecular formula and a *different* arrangement of atoms.
- **Resonance structures** have the same molecular formula and the *same* arrangement of atoms.

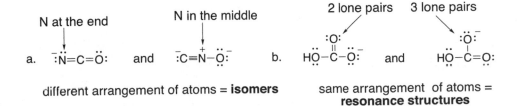

N at the end   N in the middle   2 lone pairs   3 lone pairs

a. :N̈=C=Ö: and :C≡N—Ö: b. HÖ—C—Ö:⁻ and HÖ—C=Ö:

different arrangement of atoms = **isomers**   same arrangement of atoms =
resonance structures

1.10 Curved arrow notation shows the movement of an electron pair. The tail begins at an electron pair (a bond or a lone pair) and the head points to where the electron pair moves.

a. $H-C\overset{\frown}{=}\overset{..}{O}:$ ⟷ $H-\overset{+}{C}\quad\overset{..}{\underset{..}{O}}:^{-}$
     (with H below each carbon)

b. $CH_3=\overset{..}{C}-C=CH_2$ ⟷ $CH_3-C=C-\overset{-}{C}H_2$
     (with H H below)

The net charge is the same in both resonance structures.   The net charge is the same in both resonance structures.

1.11 Compare the resonance structures to see what electrons have "moved." **Use one curved arrow to show the movement of each electron pair.**

a. $\overset{+}{C}H_2-\overset{\frown}{C}=C-CH_3$ ⟷ $CH_2=C-\overset{+}{C}-CH_3$
     (with H H below)

b. $:\overset{..}{O}=C-\overset{..}{\underset{..}{O}}:^{-}$ ⟷ $^{-}:\overset{..}{O}-C-\overset{..}{\underset{..}{O}}:$
     $\quad\quad\underset{:\overset{..}{O}:^{-}}{}$ $\quad\quad\underset{:O:}{\overset{||}{}}$

One electron pair moves: one curved arrow.   Two electron pairs move: two curved arrows.

1.12 To draw another resonance structure, **only move electrons in multiple bonds and lone pairs** and keep the number of unpaired electrons constant.

a. $CH_3-C=C-\overset{+}{C}-CH_3$ ⟷ $CH_3-\overset{+}{C}-C=C-CH_3$
     (with H H H below)

c. $H-C\overset{\frown}{=}C-\overset{..}{\underset{..}{Cl}}:$ ⟷ $H-\overset{-}{\underset{..}{C}}-C=\overset{+}{Cl}:$
     (with H H below)

b. $CH_3-\overset{+}{C}-CH_3$ ⟷ CH_3-C-CH_3
     $\quad\quad\underset{:\overset{..}{Cl}:}{}$ $\quad\quad\underset{:\overset{..}{Cl}+}{}$

1.13 To draw the resonance hybrid:
*   Use solid lines for bonds that appear in both resonance structures and dashed lines for bonds that are only in one of the resonance structures.
*   Use partial charges when the charge is on different atoms in the resonance structures.

a. $CH_2=\overset{\frown}{C}-\overset{+}{C}H_2$ ⟷ $\overset{+}{C}H_2-C=CH_2$
     (with H below each central C)

b. $H-\overset{..}{\underset{..}{C}}^{-}-\overset{:\overset{..}{O}}{\overset{||}{C}}-CH_3$ ⟷ $H-C=\overset{:\overset{..}{O}:^{-}}{\overset{|}{C}}-CH_3$
     (with H below)

$\overset{\delta+}{C}H_2=\!\!=\overset{\delta+}{\underset{H}{C}}=\!\!=CH_2$

$\overset{\delta-}{H}-C=\!\!=\overset{\delta-\ :\overset{..}{O}\ \delta-}{\underset{H}{C}}-CH_3$

1.14 **A "better" resonance structure is one that has more bonds and fewer charges.** The better structure is the major contributor and all others are minor contributors. To draw the resonance hybrid, use the rules in Answer 1.13.

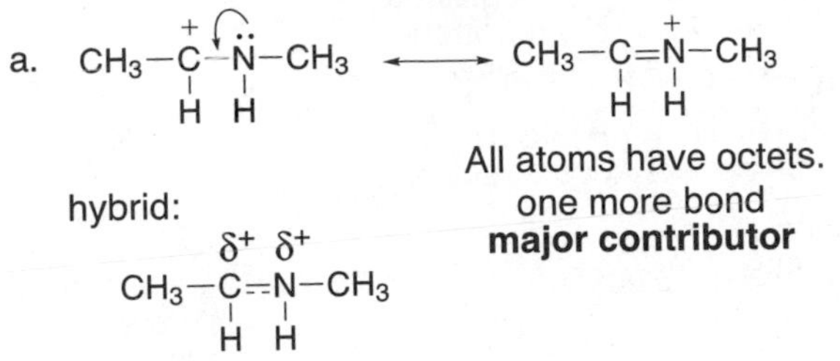

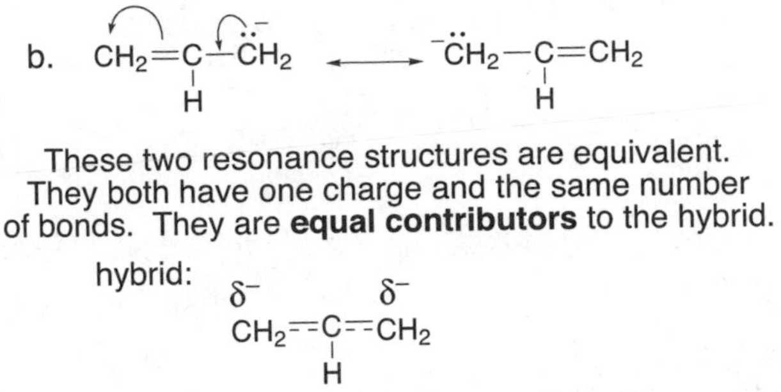

a.   CH₃–C–N–CH₃ ⟷ CH₃–C=N–CH₃   b. CH₂=C–CH₂ ⟷ CH₂–C=CH₂

All atoms have octets.
one more bond
major contributor

These two resonance structures are equivalent.
They both have one charge and the same number
of bonds. They are **equal contributors** to the hybrid.

hybrid:
δ⁺ δ⁺
CH₃–C=N–CH₃
H H

hybrid:
δ⁻   δ⁻
CH₂=C=CH₂
H

1.15 All representations have a carbon with two bonds in the plane of the page, one in front of the page (solid wedge) and one behind the page (dashed line). Four possibilities:

1.16 To predict the geometry around an atom, **count the number of groups (atoms + lone pairs),** making sure to draw in any needed lone pairs or hydrogens: 2 groups = linear, 3 groups = trigonal planar, 4 groups = tetrahedral.

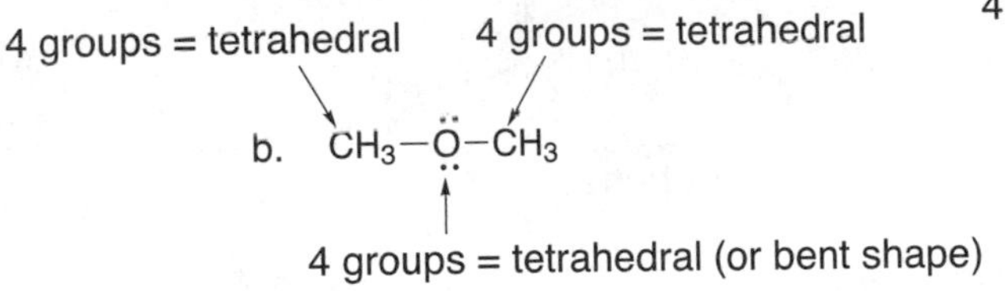

4 groups = tetrahedral

3 groups = trigonal planar
4 groups = tetrahedral

:O:
‖
a.   CH₃–C–CH₃

3 groups = trigonal planar

N has 2 atoms + 2 lone pairs
4 groups = tetrahedral (or bent shape)

c. ⁻NH₂ = :NH₂

4 groups = tetrahedral   4 groups = tetrahedral

b.   CH₃–O–CH₃

4 groups = tetrahedral (or bent shape)

4 groups = tetrahedral   2 groups = linear

d. CH₃–C≡N:

2 groups = linear

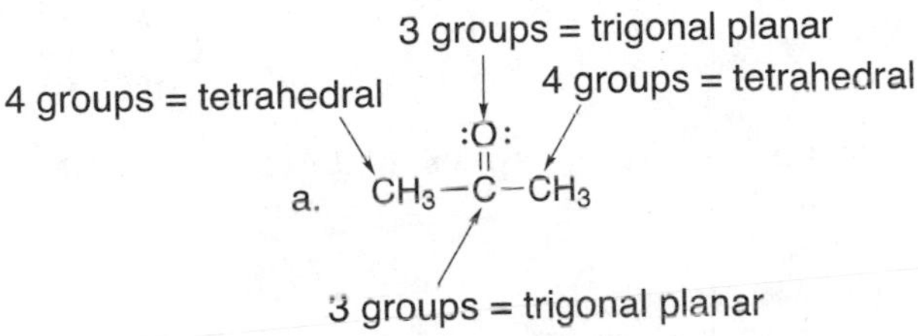

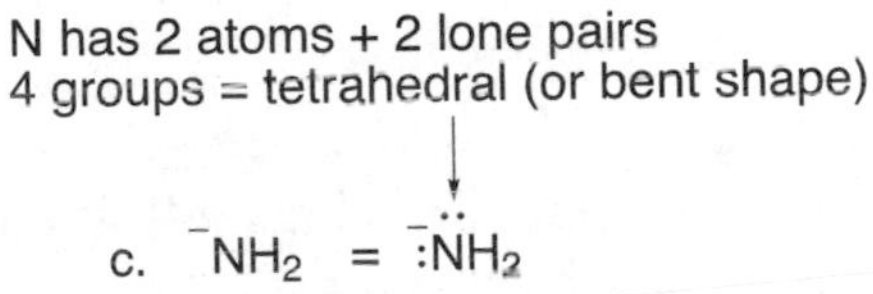

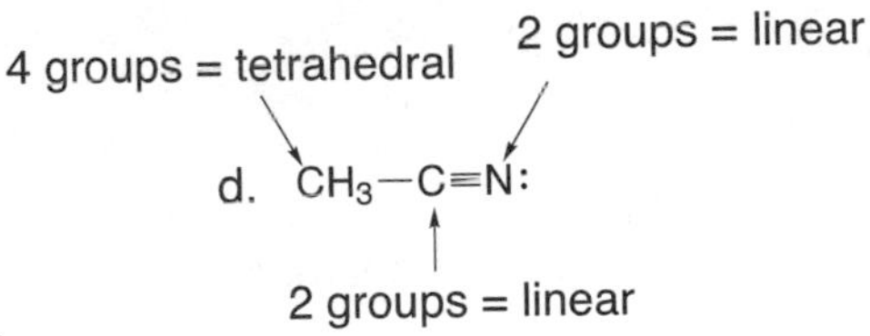

1.17 To predict the bond angle around an atom, **count the number of groups (atoms + lone pairs),** making sure to draw in any needed lone pairs or hydrogens: 2 groups = 180°, 3 groups = 120°, 4 groups = 109.5°

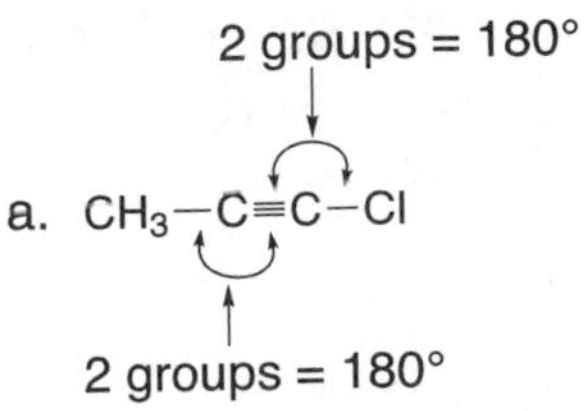

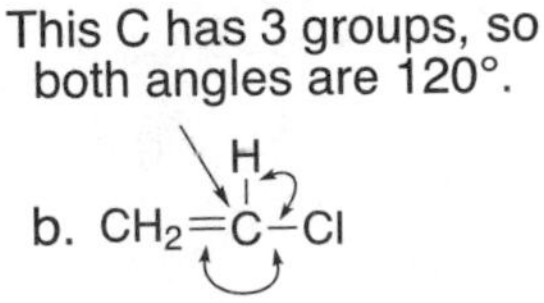

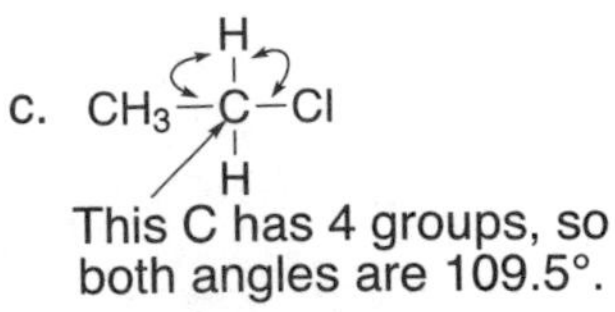

2 groups = 180°

This C has 3 groups, so
both angles are 120°.

a. CH₃–C≡C–Cl

b. CH₂=C–Cl

c. CH₃–C–Cl

2 groups = 180°

This C has 4 groups, so
both angles are 109.5°.

1.18 To predict the geometry around an atom, use the rules in Answer 1.16.

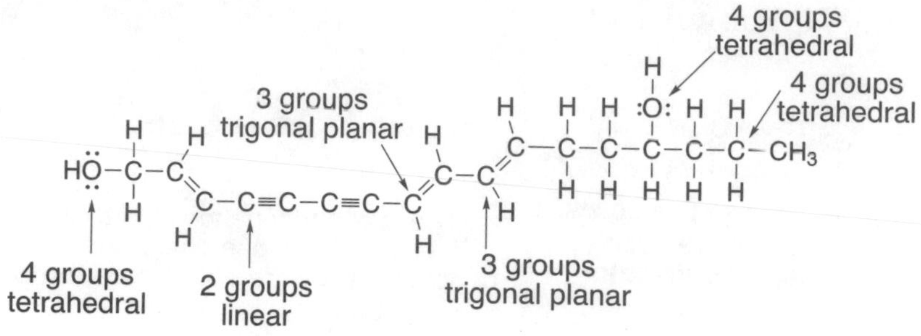

enanthotoxin

1.19 Reading from left to right, draw the molecule as a Lewis structure. Always check that carbon has four bonds and all heteroatoms have an octet by adding any needed lone pairs.

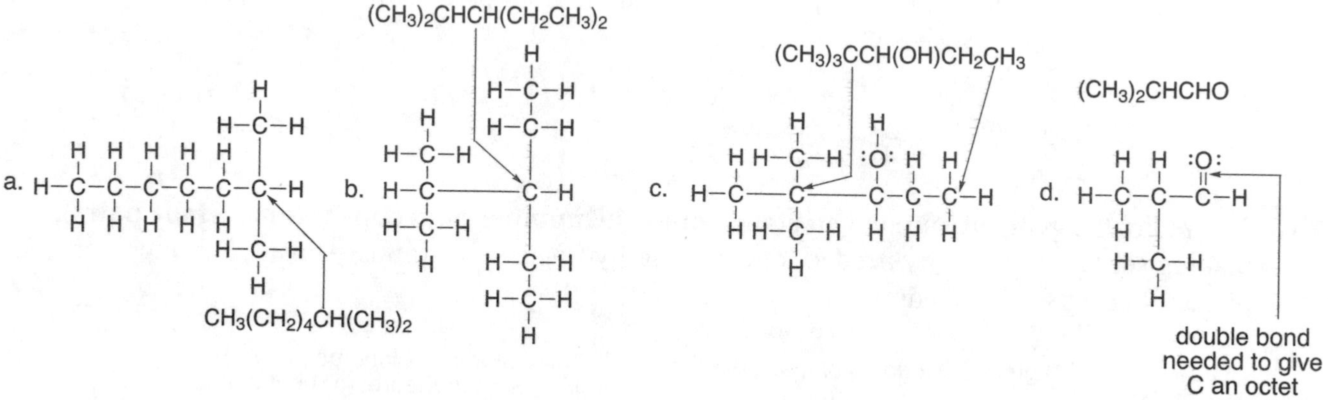

a. $(CH_3)_2CHCH(CH_2CH_3)_2$ / $CH_3(CH_2)_4CH(CH_3)_2$

b.

c. $(CH_3)_3CCH(OH)CH_2CH_3$

d. $(CH_3)_2CHCHO$

double bond
needed to give
C an octet

1.20 In shorthand or skeletal drawings, **all line junctions or ends of lines represent carbon atoms**. The carbons are all tetravalent.

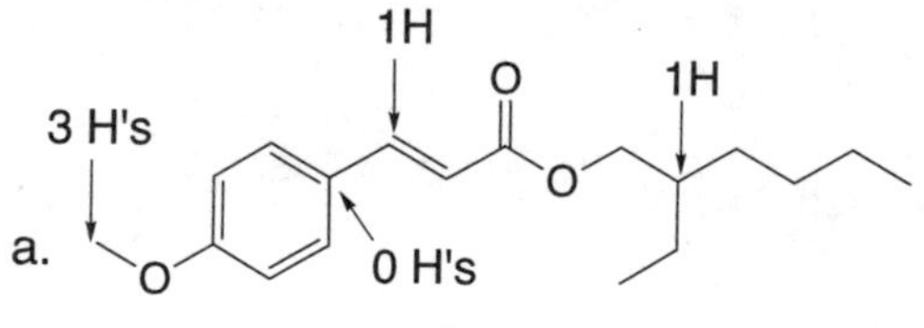

a. octinoxate
(2-ethylhexyl 4-methoxycinnamate)

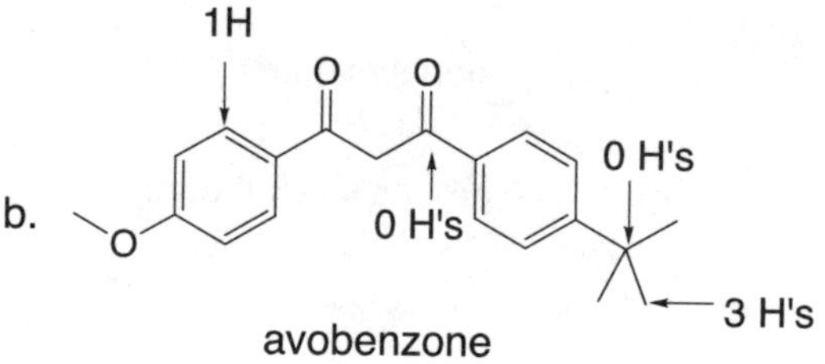

b. avobenzone

1.21 In shorthand or skeletal drawings, **all line junctions or ends of lines represent carbon atoms**. Convert by writing in all carbons, and then adding hydrogen atoms to make the carbons tetravalent.

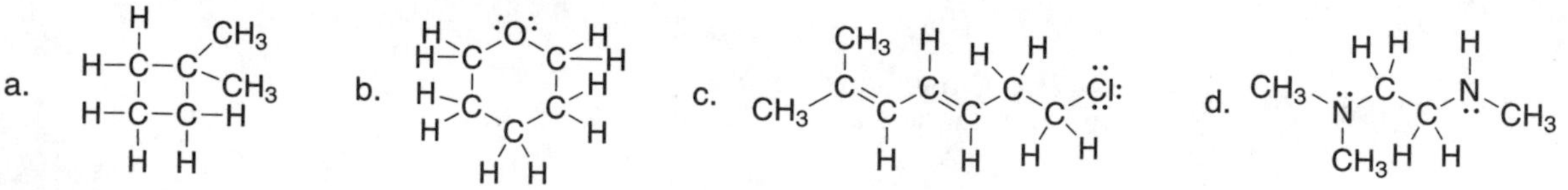

a.   b.   c.   d.

1.22 A charge on a carbon atom takes the place of one hydrogen atom. A negatively charged C has one lone pair, and a positively charged C has none.

a.

positive charge
no lone pairs
no H's needed

b.

negative charge
one lone pair
one H needed

c.

positive charge
no lone pairs
one H needed

d.

negative charge
one lone pair
one H needed

1.23

a. CO₂CH₃ =

b. OH = = CH₃CH₂C(CH₃)₂CH₂OH

c. 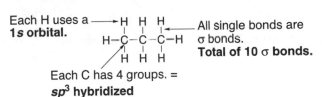 CH₃—C=C—C—C—O—C—CH₃ = (CH₃)₂C=CHCH₂COOCH₂CH₃
or
(CH₃)₂C=CHCH₂CO₂CH₂CH₃ =

1.24 To determine the orbitals used in bonding, **count the number of groups** (atoms + lone pairs): 4 groups = sp^3, 3 groups = sp^2, 2 groups = sp, H atom = $1s$ (no hybridization). All covalent single bonds are σ, and all double bonds contain one σ and one π bond.

Each H uses a ⟶ 1s orbital. | All single bonds are σ bonds. **Total of 10 σ bonds.**

Each C has 4 groups. = *sp³* hybridized

Each C–C bond is Csp^3–Csp^3.
Each C–H bond is Csp^3–H$1s$.

1.25 [1] Draw a valid Lewis structure for each molecule.
[2] **Count the number of groups** around each atom: 4 groups = sp^3, 3 groups = sp^2, 2 groups = sp, H atom = $1s$ (no hybridization).

Note: **Be and B** (Groups 2A and 3A) do not have enough valence e⁻ to form an octet, **and do not form an octet in neutral molecules**.

a. [1]

$$H-\overset{\overset{\displaystyle H}{|}}{\underset{\underset{\displaystyle H}{|}}{C}}-Be-H$$

Be has 2 bonds.

[2] Count groups around each atom:

$$H-\overset{\overset{\displaystyle H}{|}}{\underset{\underset{\displaystyle H}{|}}{C}}-Be-H$$

4 groups
sp^3 2 groups
sp

[3] All C–H bonds: C_{sp^3}–H_{1s}
C–Be bond: C_{sp^3}–Be_{sp}
Be–H bond: Be_{sp}–H_{1s}

b. [1]

$$CH_3-\overset{\overset{\displaystyle CH_3}{|}}{\underset{|}{B}}-CH_3$$

B forms 3 bonds.

[2] Count groups around each atom:

$$CH_3-\overset{\overset{\displaystyle CH_3}{|}}{\underset{|}{B}}-CH_3$$

4 groups   3 groups
sp^3 sp^2

[3] All C–H bonds: C_{sp^3}–H_{1s}
C–B bonds: C_{sp^3}–B_{sp^2}

c. [1]

$$H-\overset{\overset{\displaystyle H}{|}}{\underset{\underset{\displaystyle H}{|}}{C}}-\overset{..}{\underset{..}{O}}-\overset{\overset{\displaystyle H}{|}}{\underset{\underset{\displaystyle H}{|}}{C}}-H$$

[2] Count groups around each atom:

$$H-\overset{\overset{\displaystyle H}{|}}{\underset{\underset{\displaystyle H}{|}}{C}}-\overset{..}{\underset{..}{O}}-\overset{\overset{\displaystyle H}{|}}{\underset{\underset{\displaystyle H}{|}}{C}}-H$$

4 groups
sp^3 4 groups
sp^3

[3] All C–H bonds: C_{sp^3}–H_{1s}
C–O bonds: C_{sp^3}–O_{sp^3}

1.26 To determine the hybridization, **count the number of groups** around each atom: 4 groups = sp^3, 3 groups = sp^2, 2 groups = sp, H atom = $1s$ (no hybridization).

a. $CH_3-C\equiv CH$

4 groups   2 groups
sp^3 sp

b. [cyclohexane ring]$=\overset{..}{N}-CH_3$

3 groups   3 groups
sp^2 sp^2

c. $CH_2=C=CH_2$

3 groups   2 groups
sp^2 sp

1.27 All single bonds are σ. Multiple bonds contain one σ bond, and all others are π bonds.

All C–H bonds are σ bonds.

a.

$$\overset{\overset{\displaystyle O}{||}}{CH_3-\underset{}{C}-H}$$

one σ bond,
one π bond

σ bond   σ bond

b. $CH_3-C\equiv N$

σ bond

one σ + two π bonds

c.

$$\overset{\overset{\displaystyle O}{||}}{H-\underset{}{C}-O-CH_3}$$

σ bond — one σ bond, one π bond

σ bond

σ bond

1.28 Bond length and bond strength are inversely related: **longer bonds are weaker bonds**. Single bonds are weaker and longer than double bonds, which are weaker and longer than triple bonds.

a.

bond 1:
single bond

bond 3:
double bond

bond 2:
triple bond

increasing bond strength: 1 < 3 < 2
increasing bond length: 2 < 3 < 1

b.

bond 1:
single bond

bond 2:
double bond

bond 3:
triple bond

increasing bond strength: 1 < 2 < 3
increasing bond length: 3 < 2 < 1

1.29 Bond length and bond strength are inversely related: **longer bonds are weaker bonds**. Single bonds are weaker and longer than double bonds, which are weaker and longer than triple bonds. Increasing percent s-character increases bond strength and decreases bond length.

a. $CH_3-C\equiv C-H$ and

$C_{sp}-H_{1s}$
50% s-character
shorter bond

CH_3
$C=CH_2$
H
$C_{sp^2}-H_{1s}$
33% s-character

c. $CH_2=\overset{..}{N}-H$ and $CH_3-\overset{..}{\underset{H}{N}}-H$

$N_{sp^2}-H_{1s}$
33% s-character
shorter bond

$N_{sp^3}-H_{1s}$
25% s-character

b.

H
C=Ö
H
$C_{sp^2}-H_{1s}$
33% s-character
shorter bond

and

H
H—C—ÖH
H
$C_{sp^3}-H_{1s}$
25% s-character

1.30 Electronegativity increases across a row of the periodic table, and decreases down a column. Look at the relative position of the atoms to determine their relative electronegativity.

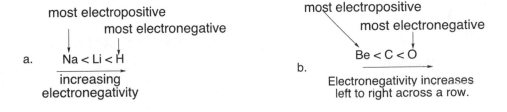

a.

most electropositive

most electronegative

Na < Li < H

increasing
electronegativity

b.

most electropositive

most electronegative

Be < C < Ö

Electronegativity increases
left to right across a row.

1.31 Dipoles result from unequal sharing of electrons in covalent bonds. More electronegative atoms "pull" electron density towards them, making a dipole. **Dipole arrows point towards the atom of higher electron density.**

a. $\overset{\delta^+}{H}-\overset{\delta^-}{F}$

b. $\overset{\delta^+}{B}-\overset{\delta^+}{C}$

c. $\overset{\delta^-}{C}-\overset{\delta^+}{Li}$

d. $\overset{\delta^+}{C}-\overset{\delta^-}{Cl}$

1.32 Polar molecules result from a net dipole. To determine polarity, draw the molecule in three dimensions around any polar bonds, draw in the dipoles, and look to see whether the dipoles cancel or reinforce.

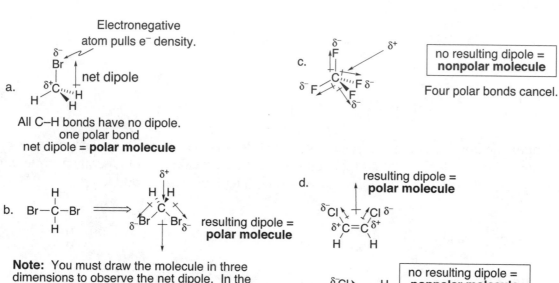

a. Electronegative atom pulls e⁻ density.
net dipole
All C–H bonds have no dipole.
one polar bond
net dipole = **polar molecule**

b. Br—C—Br ⟹ resulting dipole = **polar molecule**

Note: You must draw the molecule in three dimensions to observe the net dipole. In the Lewis structure, it appears the dipoles would cancel out, when in fact they add to make a polar molecule.

c. no resulting dipole = **nonpolar molecule**
Four polar bonds cancel.

d. resulting dipole = **polar molecule**

e. no resulting dipole = **nonpolar molecule**
Two polar bonds are **equal and opposite** and cancel.

1.33

sp^2 O
polar bond
$CH_3\ddot{O}$
$\ddot{H}\ddot{O}$
polar bond
sp^2 sp^2 sp^2
capsaicin

↑ All C with these arrows are sp^3 hybridized.

○ The two circled H's have a partial positive charge.

All C–H bonds are non polar.
All doubly bonded C's are sp^2 hybridized.

1.34 Use bonding rules in Answer 1.2.

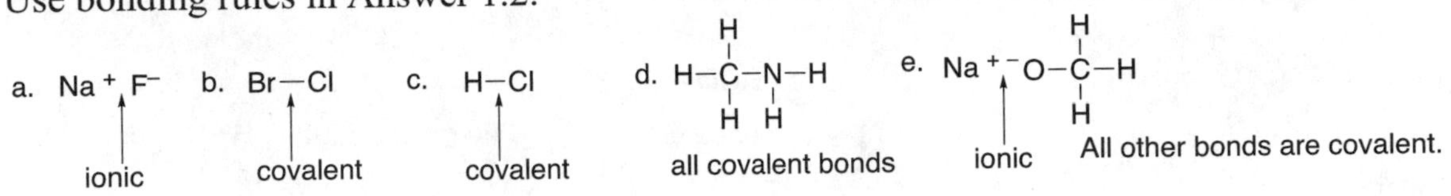

a. Na$^+$ F$^-$ b. Br—Cl   c. H—Cl   d. H—C—N—H   e. Na^{+-}O—C—H

↑ ionic   ↑ covalent   ↑ covalent   all covalent bonds   ↑ ionic   All other bonds are covalent.

1.35 Formal charge (FC) = number of valence electrons − [number of unshared electrons + 1/2 (number of shared electrons)]. C is in group 4A.

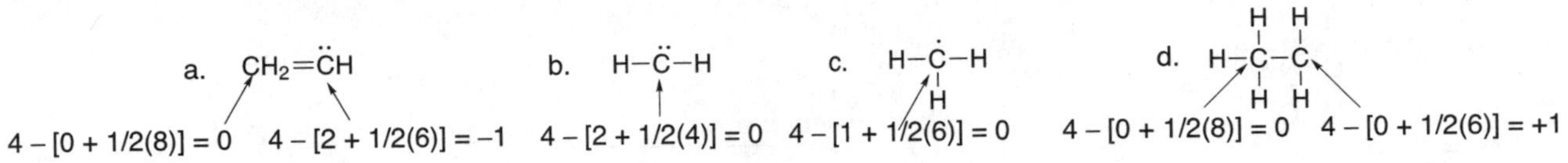

a. CH$_2$=CH   b. H—C—H   c. H—C—H   d. H—C—C

$4 - [0 + 1/2(8)] = 0$ $4 - [2 + 1/2(6)] = -1$ $4 - [2 + 1/2(4)] = 0$ $4 - [1 + 1/2(6)] = 0$ $4 - [0 + 1/2(8)] = 0$ $4 - [0 + 1/2(6)] = +1$

1.36 Formal charge (FC) = number of valence electrons − [number of unshared electrons + 1/2 (number of shared electrons)]. N is in group 5A and O is in group 6A.

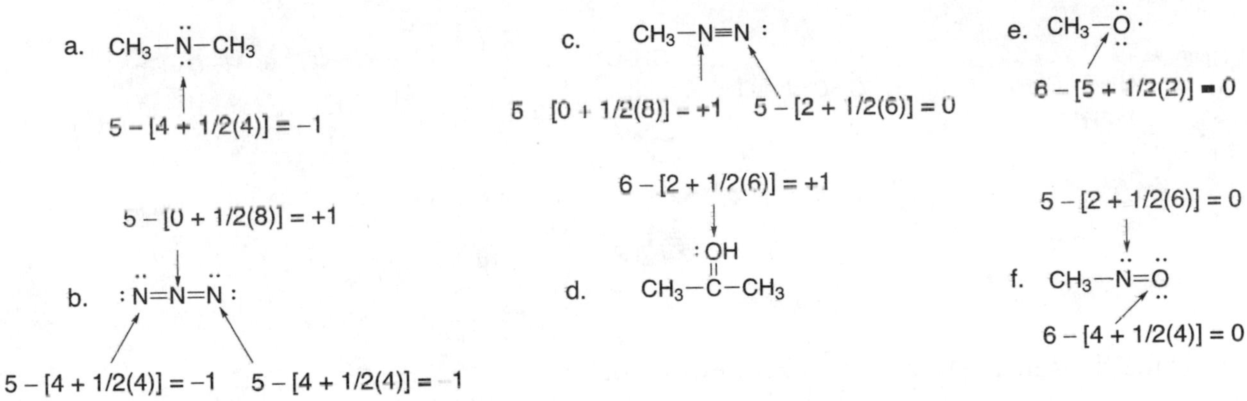

a. CH$_3$—N—CH$_3$

$5 - [4 + 1/2(4)] = -1$

$5 - [0 + 1/2(8)] = +1$

b. :N=N=N:

$5 - [4 + 1/2(4)] = -1$ $5 - [4 + 1/2(4)] = -1$

c. CH$_3$—N≡N :

$5 - [0 + 1/2(8)] = +1$ $5 - [2 + 1/2(6)] = 0$

$6 - [2 + 1/2(6)] = +1$

d. CH$_3$—C—CH$_3$ (:OH, =)

e. CH$_3$—O ·

$6 - [5 + 1/2(2)] = 0$

$5 - [2 + 1/2(6)] = 0$

f. CH$_3$—N=O

$6 - [4 + 1/2(4)] = 0$

1.37 Follow the steps in Answer 1.4 to draw Lewis structures.

a. CH_2N_2

H–C=N=N: (with + on N and – on terminal N)
|
H   or

H–C–N≡N: (with – on C and + on middle N)
|
H

valence e⁻
1C x 4 e⁻ = 4
2H x 1 e⁻ = 2
2N x 5 e⁻ = 10
total e⁻   = 16

b. CH_3NO_2

H
|
H–C–N–O:⁻   or
| ‖
H :O:

H
|
H–C–N=O:
| |
H :O:⁻

valence e⁻
1C x 4 e⁻ = 4
3H x 1 e⁻ = 3
1N x 5 e⁻ = 5
2O x 6 e⁻ = 12
total e⁻ = 24

c. CH_3CNO

H
|
H–C–C≡N–O:⁻   or
|
H

H
|
H–C–C=N=O:
|
H

or

H
|
H–C–C–N≡O: (2–, +, +)
|
H

valence e⁻
2C x 4 e⁻ = 8
3H x 1 e⁻ = 3
1N x 5 e⁻ = 5
1O x 6 e⁻ = 6
total e⁻   = 22

d. HCO_2^-

:O:⁻
‖
H–C–O: or

:O:
‖
H–C–O:⁻

valence e⁻
1C x 4 e⁻ = 4
1H x 1 e⁻ = 1
2O x 6 e⁻ = 12
1 for (–) charge = 1
total e⁻   = 18

e. HCO_3^-

:O:⁻
|
H–O–C=O: or

:O:
‖
H–O–C–O:⁻

or

:O:⁻
‖
H–O=C–O: (with + on O)

valence e⁻
1C x 4 e⁻ = 4
1H x 1 e⁻ = 1
3O x 6 e⁻ = 18
1 for (–) charge = 1
total e⁻   = 24

f. ⁻CH_2CN

H–C=C=N:⁻   or
|
H

H–C⁻–C≡N:
|
H

valence e⁻
2C x 4 e⁻ = 8
2H x 1 e⁻ = 2
1N x 5 e⁻ = 5
1 for (–) charge = 1
total e⁻   = 16

1.38 Follow the steps in Answer 1.4 to draw Lewis structures.

a. N_2

[1] N N

[2] Count valence e⁻.
2N x 5 e⁻ = 10
total e⁻ = 10

[3] N–N ⟶ :N≡N:
2 e⁻ used.   Complete N octet.

b. $(CH_3OH_2)^+$

[1]
    H
    |
H   C   O   H
    | |
    H   H

[2] Count valence e⁻.
1C x 4 e⁻ = 4
5H x 1 e⁻ = 5
1O x 6 e⁻ = 6
total e⁻ = 15
Subtract 1 for
(+) charge = 14

[3]
    H
    |
H–C–O–H
    | |
    H   H
12 e⁻ used.

[4]
    H   +
    |
H–C–O–H
    | |
    H   H
Add charge
and lone pair.

c. $(CH_3CH_2)^-$

[1]
    H
    |
H   C   C   H
    | |
    H   H

[2] Count valence e⁻.
2C x 4 e⁻ = 8
5H x 1 e⁻ = 5
total e⁻ = 13
Add 1 for
(–) charge = 14

[3]
    H   H
    | |
H–C–C–H
    | |
    H   H
12 e⁻ used.

[4]
    H   H ⁻
    | |
H–C–C–H
    | |
    H   H
Add charge
and lone pair.

d. HNNH [1] H N N H [2] Count valence e⁻. [3] H–N–N–H ⟶ H–N̈=N̈–H

2H x 1 e⁻ = 2

2N x 5 e⁻ = 10

total e⁻ = 12

6 e⁻ used. Complete N octet.

e. H₆BN [1] H H [2] Count valence e⁻. [3] H H [4] H H

H B N H

H H

1B x 3 e⁻ = 3

6H x 1 e⁻ = 6

1N x 5 e⁻ = 5

total e⁻ = 14

[3] H–B–N–H 14 e⁻ used.

[4] H–B⁻–N⁺–H Add charges.

1.39 Isomers must have a different arrangement of atoms.

a. Two isomers of molecular formula: C_3H_7Cl

b. Three isomers of molecular formula: C_2H_4O

c. Four isomers of molecular formula: C_3H_9N

1.40

Nine isomers of C_3H_6O:

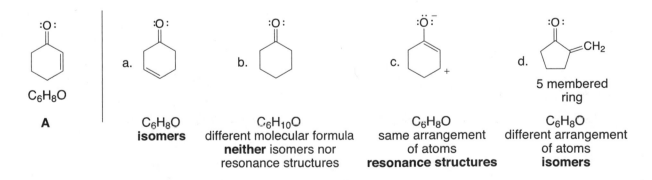

1.41 Use the definition of isomers and resonance structures in Answer 1.9.

:O:

C_6H_8O

A

a.

C_6H_8O
isomers

b.

$C_6H_{10}O$
different molecular formula
neither isomers nor
resonance structures

c.

C_6H_8O
same arrangement
of atoms
resonance structures

d.

CH₂

5 membered
ring

C_6H_8O
different arrangement
of atoms
isomers

1.42 Use the definitions of isomers and resonance structures in Answer 1.9.

a. $CH_3-\overset{..}{\underset{..}{O}}-CH_2CH_3$ and $CH_3-\overset{CH_3}{\underset{H}{\overset{|}{\underset{|}{C}}}}-\overset{..}{\underset{..}{O}}H$

two C–O bonds   one O–H bond
Different arrangement of atoms.
Both have molecular formula C_3H_8O = **isomers**

c. $CH_2=CH-\overset{..}{C}H-CH_3$ and $\overset{-}{C}H_2-CH=CH-CH_3$

Same arrangement of atoms.
Both have molecular formula $(C_4H_7)^-$.
Different arrangement of electrons = **resonance structures**

b. ☐ and $CH_3-\overset{H}{\underset{}{\overset{|}{C}}}=\overset{H}{\underset{}{\overset{|}{C}}}-CH_3$

ring   double bond
Different arrangement of atoms.
Both have molecular formula C_4H_8 = **isomers**

d. $CH_3CH_2CH_3$ and $CH_3CH_2\overset{..}{C}H_2$

molecular formula C_3H_8 molecular formula $(C_3H_7)^-$
different molecular formulas = **neither**

1.43 Compare the resonance structures to see what electrons have "moved." **Use one curved arrow to show the movement of each electron pair.**

a. $CH_3-\overset{H}{\underset{+}{\overset{|}{C}}}-\overset{..}{\underset{|}{N}}-CH_3 \longleftrightarrow CH_3-\overset{H}{\underset{}{\overset{|}{C}}}=\overset{+}{\underset{H}{\overset{|}{N}}}-CH_3$

One electron pair moves = one arrow

b. $H-\overset{\overset{..}{O}:}{\underset{}{\overset{||}{C}}}-\overset{..}{N}H_2 \longleftrightarrow H-\overset{:\overset{..}{O}:^-}{\underset{}{\overset{|}{C}}}=\overset{+}{N}H_2$

Two electron pairs move = two arrows

c.

Two electron pairs move = two arrows

1.44 Curved arrow notation shows the movement of an electron pair. The tail begins at an electron pair (a bond or a lone pair) and the head points to where the electron pair moves.

a. $CH_3-\overset{+}{N}\equiv N: \longleftrightarrow CH_3-\overset{..}{N}=\overset{+}{N}:$

b. $CH_3-\overset{:\overset{..}{O}:^-}{\underset{}{\overset{|}{C}}}=CH-\overset{+}{C}H_2 \longleftrightarrow CH_3-\overset{:O:}{\underset{}{\overset{||}{C}}}-CH=CH_2$

c. $\underset{CH_3}{\text{(ring with + charge)}} \longleftrightarrow \underset{CH_3}{\text{(ring)}}$

d. $\underset{}{\text{(ring with }\overset{+}{N}H_2)} \longleftrightarrow \underset{}{\text{(ring with }\overset{..}{N}H_2)}$

1.45 Use the rules in Answer 1.12.

a. $CH_3-\overset{:O:}{\underset{}{\overset{||}{C}}}-\overset{..}{\underset{..}{O}}:^- \longleftrightarrow CH_3-\overset{:\overset{..}{O}:^-}{\underset{}{\overset{|}{C}}}=\overset{..}{O}:$

Two electron pairs move = two arrows

b. $CH_2=\overset{+}{N}H_2 \longleftrightarrow \overset{+}{C}H_2-\overset{..}{N}H_2$

One electron pair moves = one arrow

c.

Two electron pairs move = two arrows

d. $H-\overset{+}{\overset{:OH}{\underset{}{\overset{||}{C}}}}-H \longleftrightarrow H-\overset{:\overset{..}{O}H}{\underset{}{\overset{|}{C}}}-H$

One electron pair moves = one arrow

1.46 To draw the **resonance hybrid**, use the rules in Answer 1.13.

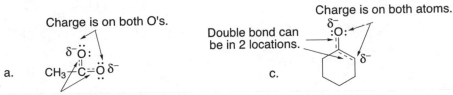

a. Charge is on both O's.

CH_3–C=O
Double bond can be in 2 locations.

b. partial double bond character

$\delta^+ \downarrow \delta^+$
CH_2==NH_2
Charge is on both atoms.

c. Double bond can be in 2 locations.
Charge is on both atoms.

d. Charge is on both atoms.

:OH
H–C–H
δ^+
C–O bond has partial double bond character.

1.47 For the compounds where the arrangement of atoms is not given, first draw a Lewis structure. Then use the rules in Answer 1.12.

a. O_3

Count valence e⁻.
$3O \times 6\ e^- = 18$
total e⁻ = 18

$:\overset{..}{O}-\overset{+}{O}=\overset{..}{O} \longleftrightarrow \overset{..}{O}=\overset{+}{O}-\overset{..}{O}:$

b. NO_3^- (a central N atom)

Count valence e⁻.
$1N \times 5\ e^- = 5$
$3O \times 6\ e^- = 18$
$(-)$ charge = 1
total e⁻ = 24

c. N_3^-

Count valence e⁻.
$3N \times 5\ e^- = 15$
$(-)$ charge = 1
total e⁻ = 16

$\overset{-}{N}=\overset{+}{N}=\overset{-}{N} \longleftrightarrow N\equiv\overset{+}{N}-\overset{2-}{N}: \longleftrightarrow :\overset{2-}{N}-\overset{+}{N}\equiv N$

d.

e.

f. $CH_2=CH-CH-CH=CH_2 \longleftrightarrow CH_2-CH=CH-CH=CH_2 \longleftrightarrow CH_2=CH-CH=CH-CH_2$

g.

1.48 To draw the **resonance hybrid**, use the rules in Answer 1.13.

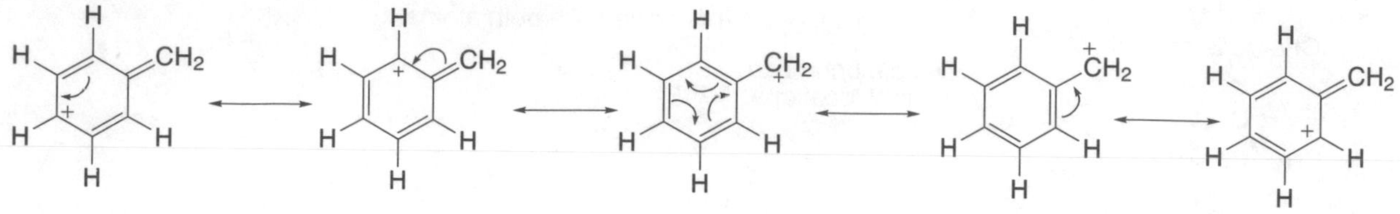

resonance hybrid

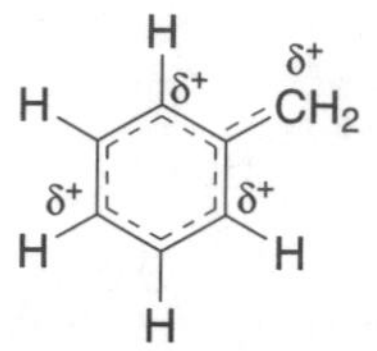

1.49 A "better" resonance structure is one that has more bonds and fewer charges. The better structure is the major contributor and all others are minor contributors.

a.

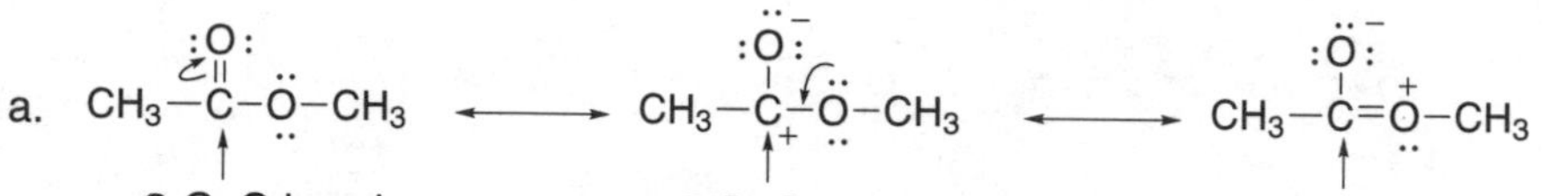

3 C–O bonds	2 C–O bonds	3 C–O bonds
no charges	2 charges	2 charges
contributes the most = **3**	contributes the least = **1**	**2**

b.

3 bonds for this N	3 bonds for this N	2 bonds for this N
no charges	2 charges	2 charges
contributes the most = **3**	**2**	contributes the least = **1**

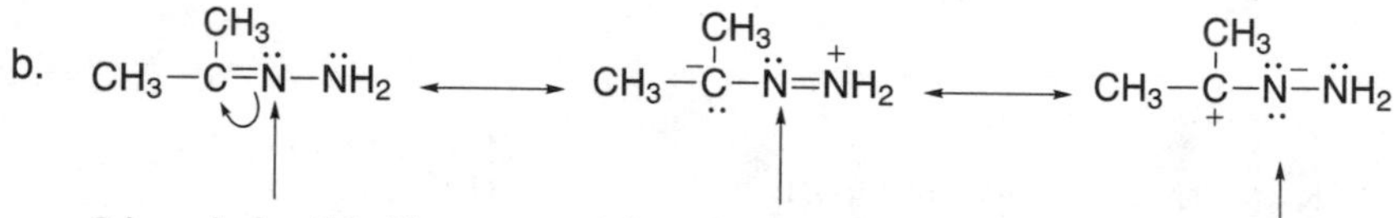

1.50 Follow the steps in Answer 1.4 to draw Lewis structures.

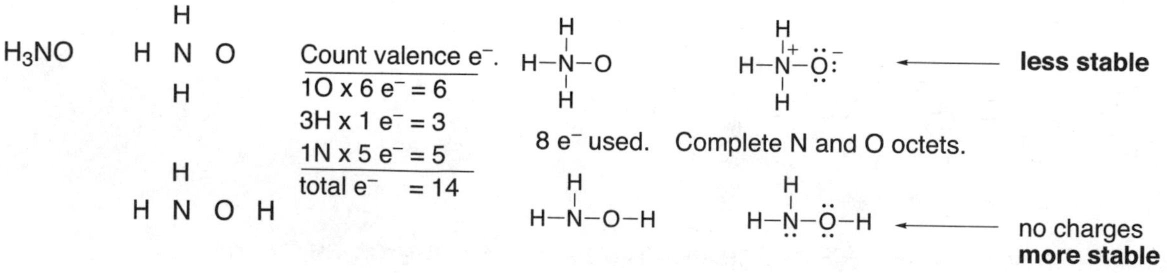

H₃NO

Count valence e⁻.

$$1 O \times 6\ e^- = 6$$
$$3 H \times 1\ e^- = 3$$
$$1 N \times 5\ e^- = 5$$
$$\text{total } e^- = 14$$

8 e⁻ used.   Complete N and O octets.

less stable

no charges
more stable

1.51 Use the rules in Answer 1.17.

a. CH₃Cl

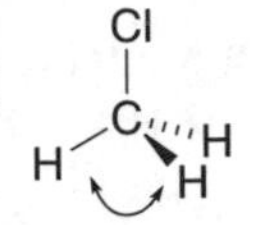

4 groups = 109.5°

c. 3 groups = 120°

120° C=N 4 groups = 109.5°

CH₃

3 groups = 120°

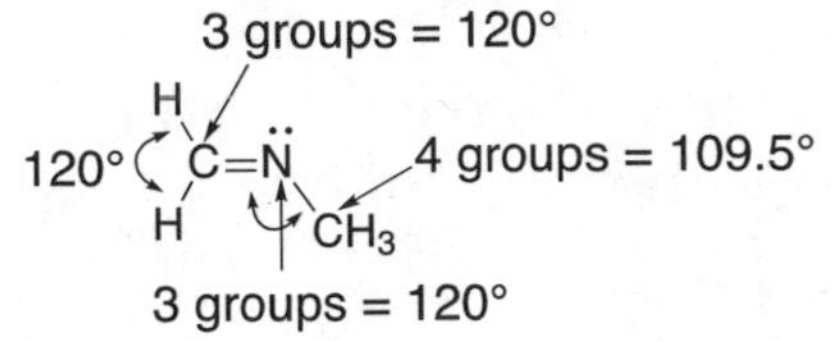

e.

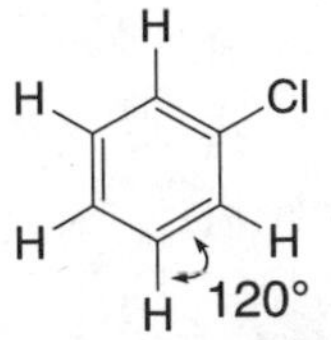

All C atoms have
3 groups. = 120°

b. 4 groups = ~109.5°

H–N–O–H

H

4 groups = ~109.5°

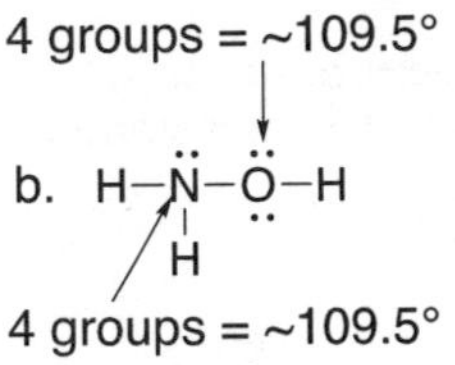

d. 4 groups = 109.5°

HC≡C–C–OH

both C's surrounded by   4 groups = ~109.5°
2 groups = 180°

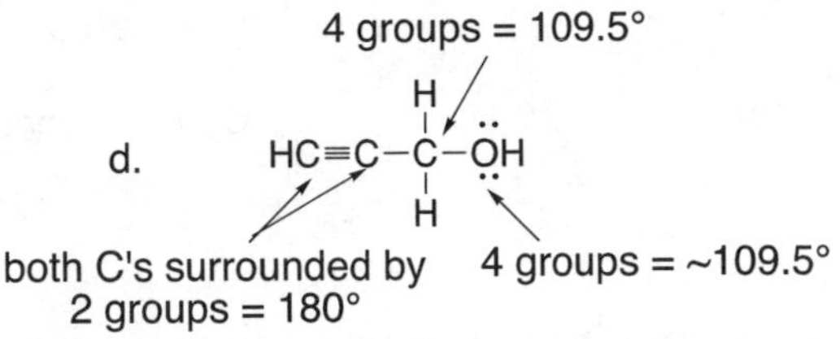

1.52 To predict the geometry around an atom, use the rules in Answer 1.16.

a. CH₃CH₂CH₂CH₃

4 groups
(4 atoms)
tetrahedral

c.

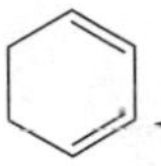

3 groups
(3 atoms)
trigonal planar

e.

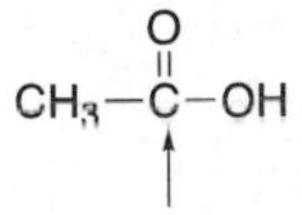

O
‖
CH₃–C–OH

3 groups
(3 atoms)
trigonal planar

b. (CH₃)₂N̈:

4 groups
(2 atoms, 2 lone pairs)
tetrahedral

d. BF₄⁻

4 groups
(4 atoms)
tetrahedral

f. (CH₃)₃N̈

4 groups
(3 atoms, 1 lone pair)
tetrahedral

1.53 Each C has two bonds in the plane of the page, one in front of the page (solid wedge) and one behind the page (dashed line).

a. CH₃CH₂CH₃

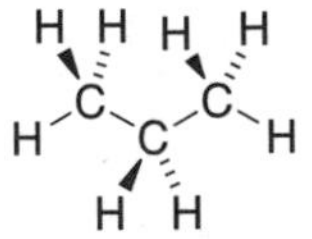

b. CH₃OH

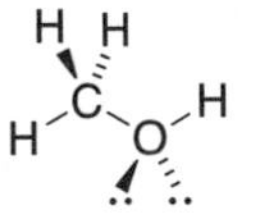

c. (CH₃)₂NH

Two sites on O have
lone pairs, not atoms.

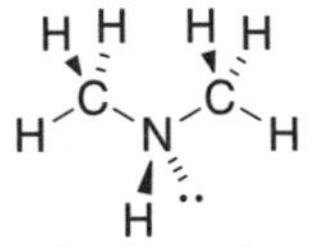

One site on N has a
lone pair, not an atom.

1.54 In shorthand or skeletal drawings, **all line junctions or ends of lines represent carbon atoms**. The C's are all tetravalent.

a.

2,4,6-undecatriene
(isolated from limu lipoa,
a common brown Hawaiian seaweed)

b.

fexofenadine
(nonsedating antihistamine)

1.55 In shorthand or skeletal drawings, **all line junctions or ends of lines represent carbon atoms**. Convert by writing in all C's, and then adding H's to make the C's tetravalent.

a.

b.

c.

d.

e.

menthol
(isolated from peppermint oil)

f.

myrcene
(isolated from bayberry)

g.

ethambutol
(drug used to treat tuberculosis)

h.

estradiol
(a female sex hormone)

1.56 In skeletal formulas, leave out all C's and H's, except H's bonded to heteroatoms.

a. (CH₃)₂CHCH₂CH₂CH(CH₃)₂

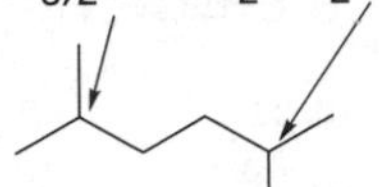

b. CH₃CH(Cl)CH(OH)CH₃

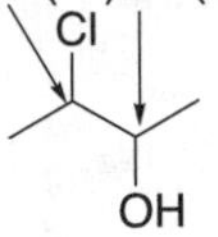

c. (CH₃)₃C(CH₂)₄CH₂CH₃

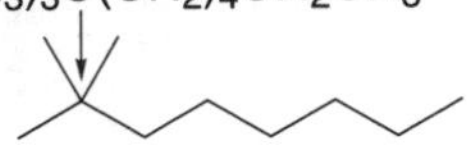

d.

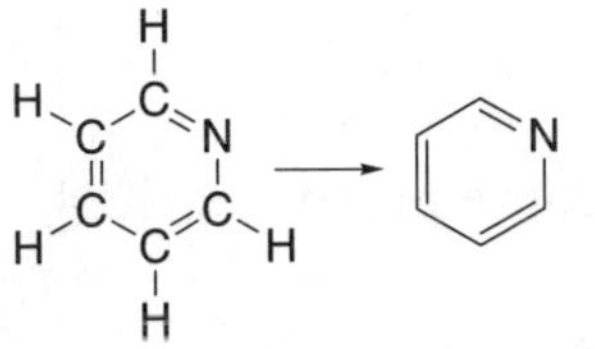

e. CH₃—C ... CH₂

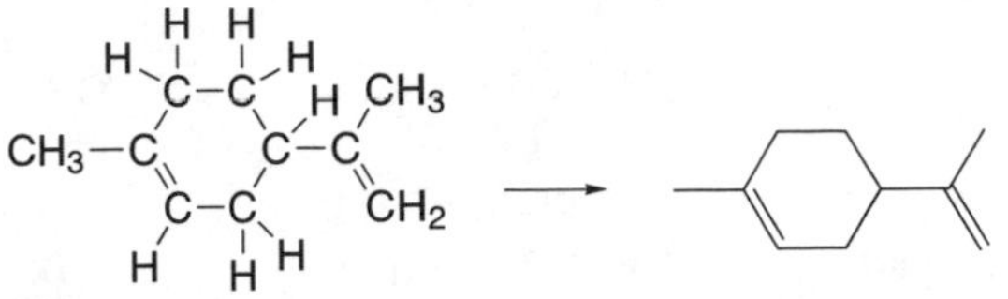

limonene
(oil of lemon)

f. CH₃(CH₂)₂C(CH₃)₂CH(CH₃)CH(CH₃)CH(Br)CH₃

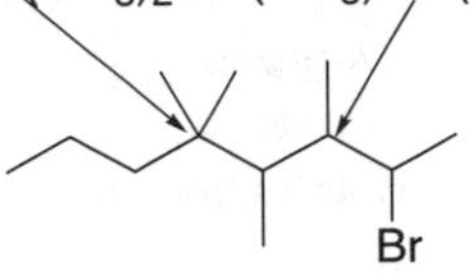

1.57 For Lewis structures, all atoms including H's and all lone pairs must be drawn in.

a. CH₃CH₂COOH   b. CH₃CONHCH₃   c. CH₃COCH₂Br   d. (CH₃)₃COH□□   e. (CH₃)₃CCHO   f. CH₃COCl

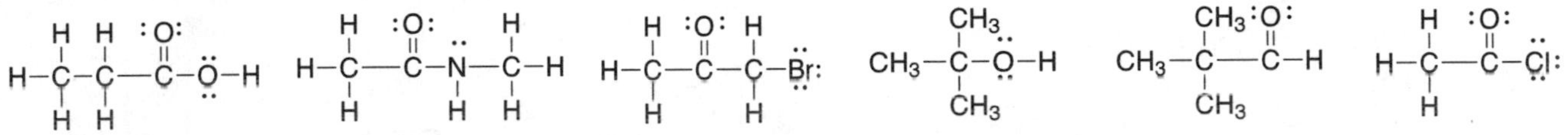

1.58 A charge on a C atom takes the place of one H atom. A negatively charged C has one lone pair, and a positively charged C has none.

a.

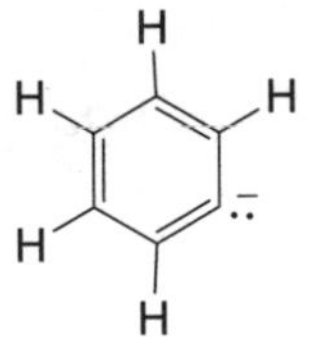

b.

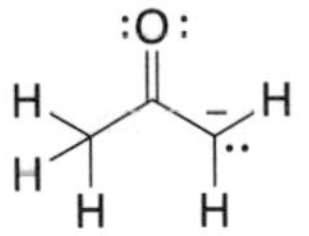

c.

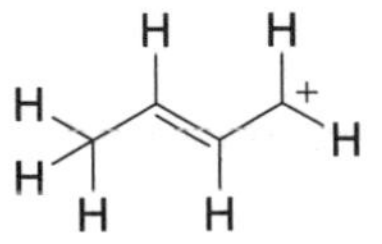

1.59 To determine the hybridization around the labeled atoms, use the procedure in Answer 1.26.

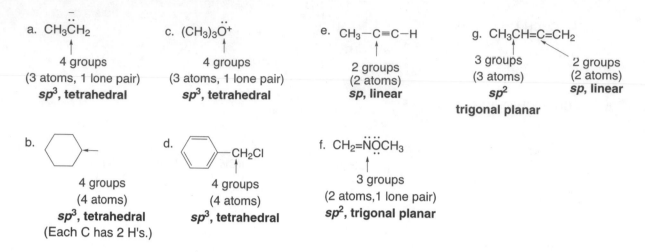

a. CH₃C̈H₂

↑

4 groups
(3 atoms, 1 lone pair)
sp³, **tetrahedral**

c. (CH₃)₃Ö⁺

↑

4 groups
(3 atoms, 1 lone pair)
sp³, **tetrahedral**

e. CH₃—C≡C—H

↑

2 groups
(2 atoms)
sp, **linear**

g. CH₃CH=C=CH₂

3 groups
(3 atoms)
sp²
trigonal planar

2 groups
(2 atoms)
sp, **linear**

b.

4 groups
(4 atoms)
sp³, **tetrahedral**
(Each C has 2 H's.)

d. —CH₂Cl

4 groups
(4 atoms)
sp³, **tetrahedral**

f. CH₂=N̈ÖCH₃

↑

3 groups
(2 atoms,1 lone pair)
sp², **trigonal planar**

1.60 To determine what orbitals are involved in bonding, use the procedure in Answer 1.24.

a.
H H
←C*sp³*–H*1s*
←C*sp³*–C*sp³*

b.
C*sp²*–H*1s* →
H
σ: C*sp²*–C*sp²*
π: C*p*–C*p*
← C*sp²*–C*sp³*

c.
C*sp²*–C*sp³*   :O:
σ: C*sp²*–O*sp²*
π: C*p*–O*p*

d.
C*sp*–C*sp²*

H–C≡C–C=N̈–CH₃
               H
C*sp*–H*1s*   σ: C*sp*–C*sp*
σ: C*sp³*–N*sp²*
π: C*p*–C*p*
π: C*p*–C*p*

1.61 To determine what orbitals are involved in bonding, use the procedure in Answer 1.24.

a.
O*sp³*–H*1s*   C*sp³*–O*sp³*
σ: C*sp²*–O*sp²*   :O: :O: H   CO₂H
π: C*p*–O*p*   :O:
HO
C*sp²*–C*sp³*
C*sp³*–C*sp³*
OH

citric acid
(responsible for the tartness
of citrus fruits)

b.
σ: C*sp³*–O*sp³*   C*sp²*–C*sp³*
:O:
CH₃   Ö
C*sp³*–C*sp³*
HO
H
C*sp²*–H*1s*

zingerone
(responsible for the pungent
taste of ginger)

1.62

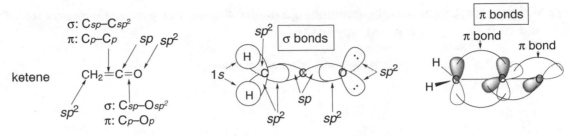

ketene

[For clarity, only the large bonding lobe of the hybrid orbitals is drawn.]

1.63

$CH_2=\overset{+}{C}H$ σ: C_{sp^2}–C_{sp}
                      π: C_p–C_p

sp^2 sp

$CH_2=\overset{..}{\overset{-}{C}}H$ σ: C_{sp^2}–C_{sp^2}
                      π: C_p–C_p

sp^2 sp^2

1.64 To determine relative bond length, use the rules in Answers 1.29.

a.

C≡C–H
triple bond
shortest

C=C–H (with H H)
double bond
middle

C–CH₃ (with H)
single bond
longest

b.   H–C≡C–CH₂–C–C–C–H (with H H H and top H)

C_{sp}–H_{1s}
highest
% s-character
shortest

C_{sp^2}–H_{1s}
middle
% s-character
middle

C_{sp^3}–H_{1s}
lowest
% s-character
longest

1.65 To determine relative bond length, use the rules in Answers 1.29.

a.   HO⤒ ⟍⟋⟍⟍ NH₂
longer
(N is to the left of O in the second row.)

b.   Br⤒ ⟍⟋⟍⟍ Cl
longer
(Br is below Cl in group 7A.)

c.   CH₃
         ⟍
          C=N⟍⟋⟍ NH₂
         ⟋
       CH₃⤒

single bond
longer

1.66 Percent s-character determines the length of a bond. **The higher percent s-character of a bond, the shorter the bond.** For the question below, determine the relative percent s-character by defining the hybridization of each atom involved in the bond. $sp = 50\%$ s-character; $sp^2 = 33\%$ s-character; $sp^3 = 25\%$ s-character

$CH_2=\overset{H}{\underset{H}{C}}-\overset{H}{\underset{H}{C}}-CH_3$

(1)   (2)

C_{sp^2}–C_{sp^3} C_{sp^3}–C_{sp^3}

Bond (1) has a *higher % s-character* due to the sp^2 hybridized carbon, making it a **shorter bond.**

1.67 Use the rule on percent s-character and bond length in Answer 1.66 and remember shorter bonds are stronger bonds. A bond formed from two sp^2 hybridized C's is stronger than a σ bond formed from two sp^3 hybridized C's because the sp^2 hybridized C bonds with an orbital having a higher percent s-character.

1.68 Percent *s*-character determines the strength of a bond. **The higher percent *s*-character of an orbital used to form a bond, the stronger the bond.**

vinyl chloride	chloroethane (ethyl chloride)
$CH_2{=}CH{-}Cl$	$CH_3{-}CH_2{-}Cl$
$C sp^2$	$C sp^3$
33% *s*-character higher percent *s*-character **stronger bond**	25% *s*-character

1.69 Dipoles result from unequal sharing of electrons in covalent bonds. More electronegative atoms "pull" electron density towards them, making a dipole. **Dipole arrows point from lower (δ^+) to higher (δ^-) electron density.**

a. $\delta^+ \, Br{-}Cl \; \delta^-$

b. $\delta^+ \quad \delta^-$
    $NH_2{-}OH$

c. $\delta^+ \quad \delta^-$
    $CH_3{-}NH_2$

d. $\delta^- {-}Li \; \delta^+$ (cyclohexyl)

1.70 Use the directions from Answer 1.32.

a. CHBr₃ — net dipole

b. CH₃CH₂OCH₂CH₃ — net dipole

c. CBr₄ — no net dipole

d. (CH₃)₂C=O — net dipole

e. (ortho-dibromobenzene) — net dipole

f. (para-dichlorobenzene) — no net dipole

1.71

The dipole of any one Cl atom opposes the resultant dipole of the other 2 Cl's, thus decreasing the net dipole on the whole molecule.

(CHCl₃ structure)

(CH₂Cl₂ structure) — net dipole

This net dipole is the result of 2 Cl atoms pulling down. They reinforce each other more to result in a **stronger net dipole.**

1.72

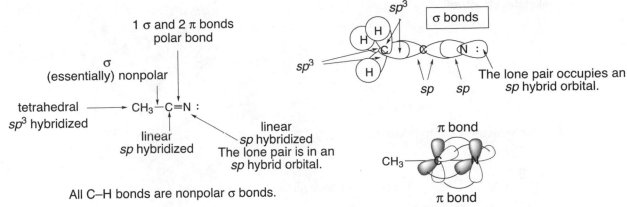

1 σ and 2 π bonds
polar bond

σ
(essentially) nonpolar

tetrahedral ———→ CH₃—C≡N :
sp^3 hybridized

linear
sp hybridized

linear
sp hybridized
The lone pair is in an
sp hybrid orbital.

All C–H bonds are nonpolar σ bonds.

sp^3

σ bonds

sp^3

sp sp

The lone pair occupies an
sp hybrid orbital.

π bond

CH₃

π bond

[For clarity, only the large bonding lobe
of the hybrid orbitals is drawn.]

1.73

a. sp^2

b. Each C is trigonal planar; the ring is flat.

c.

sp^2 hybridized C

p orbitals on C's overlap

π bonds

sp^2 hybrid orbitals on C

σ bonds

[Only the larger bonding lobe
of each orbital is drawn.]

d.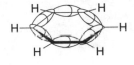

e. Benzene is stable because of its two resonance structures that contribute
equally to the hybrid. [This is only part of the story. We'll learn more about
benzene's unusual stability in Chapter 17.]

1.74

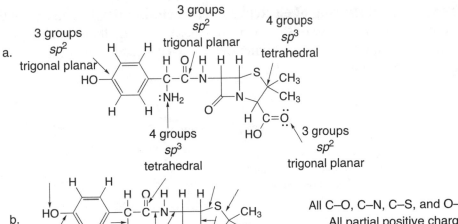

a.

3 groups
sp^2
trigonal planar

3 groups
sp^2
trigonal planar

4 groups
sp^3
tetrahedral

HO

H H H O H H H

C—C—N

:NH₂

S CH₃

CH₃

N

H C=Ö

HO

4 groups
sp^3
tetrahedral

3 groups
sp^2
trigonal planar

b.

HO

H H

H O H H H

C—C—N

NH₂

S CH₃

CH₃

N

H C=O

H—O

All C–O, C–N, C–S, and O–H bonds are polar.
All partial positive charges lie on the C.
All partial negative charges lie on the O, N or S.
In OH and NH bonds, H bears a δ⁺.

skeletal structure:

c.

d.

6 π bonds

e. **33% s-character = sp² hybridized**

These C–H bonds have 33% s-character.

1.75

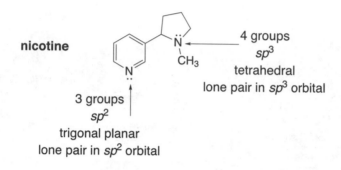

nicotine

4 groups
sp³
tetrahedral
lone pair in sp³ orbital

3 groups
sp²
trigonal planar
lone pair in sp² orbital

1.76

⁺CH₃   3 groups
sp² trigonal planar
plot **A**

The blue region is evidence of
the electron poor cation.

⁻:CH₃   4 groups
sp³ tetrahedral
(trigonal pyramidal)
plot **B**

The red region is evidence of
the electron rich anion.

1.77 Polar bonds result from unequal sharing of electrons in covalent bonds. Normally we think of more electronegative atoms "pulling" more of the electron density towards them, making a dipole. In looking at a Csp^2–Csp^3 bond, the atom with a higher percent s-character will "pull" more of the electron density towards it, creating a small dipole.

$$\overset{\delta^-}{—C}sp^2—\overset{\delta^+}{C}sp^3—$$

33% s-character
higher percent s-character
pulls more electron density
more electronegative

25% s-character

1.78

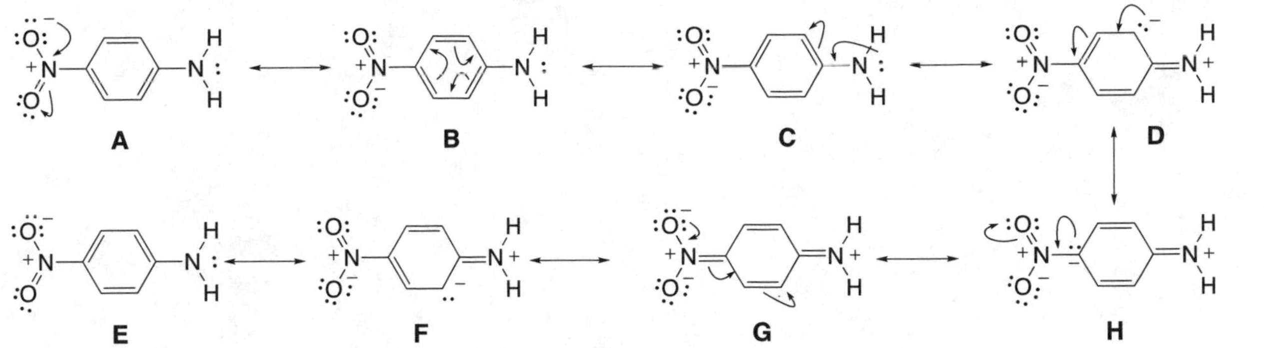

Structures **D**, **F**, and **H** have an additional resonance structure.

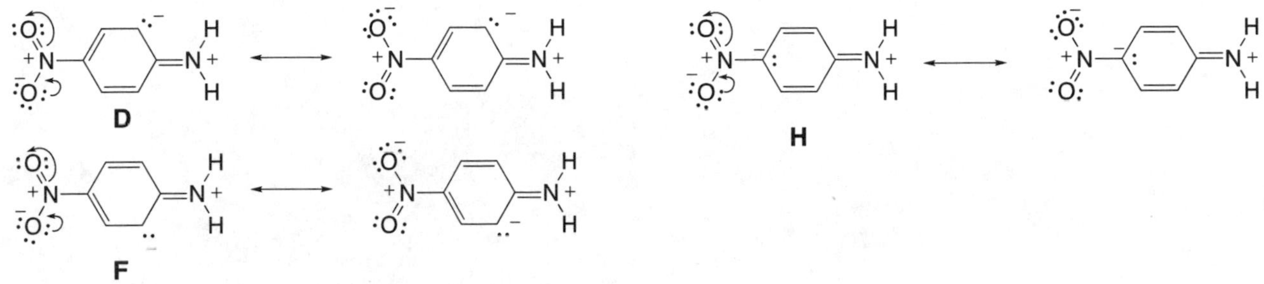

1.79

Isomers of C_4H_8:

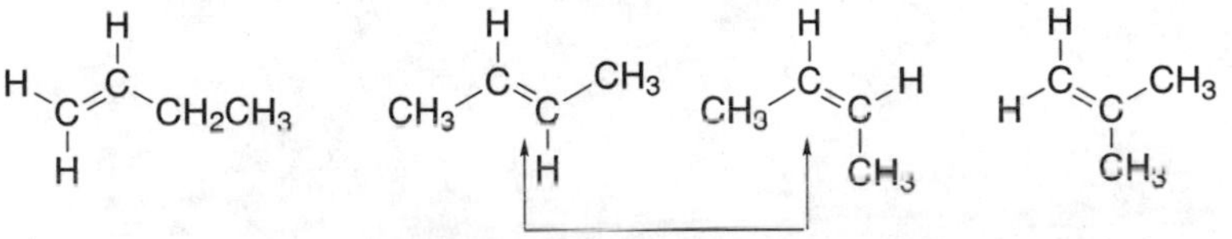

These 2 compounds are different
because of restricted rotation around
the C=C (Section 8.2B).

1.80

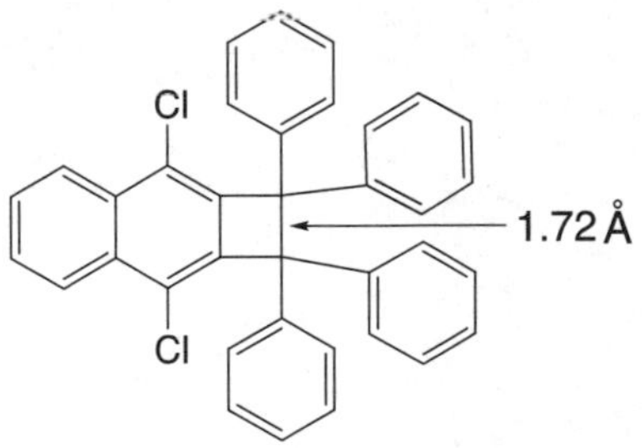

1.72Å

The four rings bonded to the C–C bond are very crowded.
With a longer C–C bond, the rings have somewhat more
space, and this adds stability to the molecule.

Chapter 2: Acids and Bases

♦ **A comparison of Brønsted–Lowry and Lewis acids and bases**

Type	Definition	Structural feature	Examples
Brønsted–Lowry acid (2.1)	proton donor	a proton	HCl, H_2SO_4, H_2O, CH_3COOH, TsOH
Brønsted–Lowry base (2.1)	proton acceptor	a lone pair *or* a π bond	^-OH, $^-OCH_3$, H^-, $^-NH_2$, $CH_2=CH_2$
Lewis acid (2.8)	electron pair acceptor	a proton, *or* an unfilled valence shell, *or* a partial (+) charge	BF_3, $AlCl_3$, HCl, CH_3COOH, H_2O
Lewis base (2.8)	electron pair donor	a lone pair *or* a π bond	^-OH, $^-OCH_3$, H^-, $^-NH_2$, $CH_2=CH_2$

♦ **Acid–base reactions**

[1] A Brønsted–Lowry acid donates a proton to a Brønsted–Lowry base (2.2).

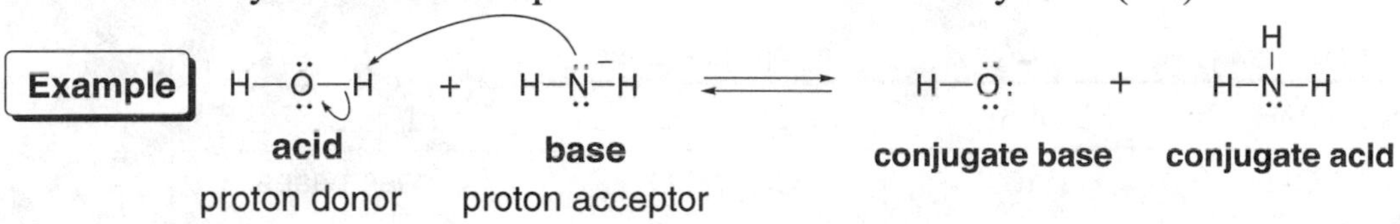

[2] A Lewis base donates an electron pair to a Lewis acid (2.8).

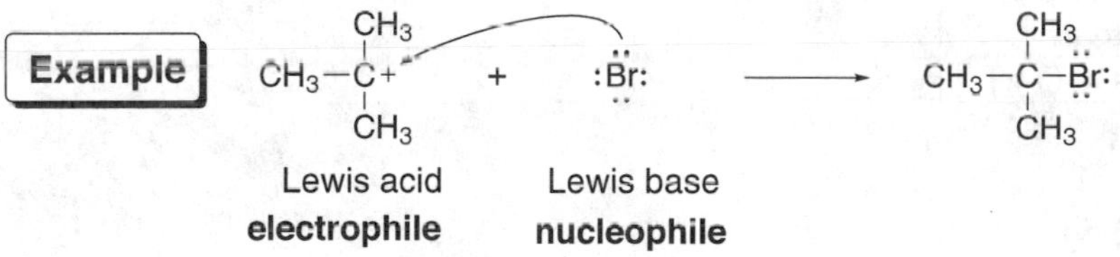

- Electron rich species react with electron poor ones.
- Nucleophiles react with electrophiles.

♦ **Important facts**

- Definition: $pK_a = -\log K_a$. The **lower the pK_a**, the **stronger** the acid (2.3).

$$NH_3 \qquad \text{versus} \qquad H_2O$$
$$pK_a = 38 \qquad\qquad\qquad pK_a = 15.7$$
$$\text{lower } pK_a = \text{stronger acid}$$

- The stronger the acid, the weaker the conjugate base (2.3).

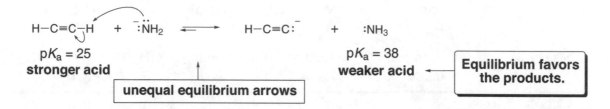

- In proton transfer reactions, equilibrium favors the weaker acid and weaker base (2.4).

$$H-C\equiv C-H \ + \ :NH_2^- \ \rightleftharpoons \ H-C\equiv C:^- \ + \ :NH_3$$

$pK_a = 25$
stronger acid

unequal equilibrium arrows

$pK_a = 38$
weaker acid

Equilibrium favors the products.

- An acid can be deprotonated by the conjugate base of any acid having a **higher pK_a** (2.4).

Acid	pK_a	Conjugate base	
$CH_3COO–H$	4.8	CH_3COO^-	
$CH_3CH_2O–H$	16	$CH_3CH_2O^-$	These bases
$HC\equiv CH$	25	$HC\equiv C^-$	can deprotonate
$H–H$	35	H^-	$CH_3COO–H$.
	higher pK_a than $CH_3COO–H$		

◆ Factors that determine acidity (2.5)

---

[1] Element effects (2.5A)   The acidity of H–A increases both across a row and down a column of the periodic table.

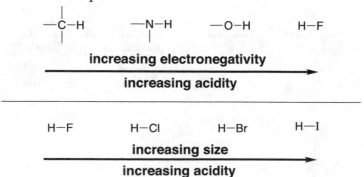

[2] **Inductive effects** (2.5B)	The acidity of H–A increases with the presence of electron withdrawing groups in A.

$$CH_3CH_2OH \longrightarrow CH_3CH_2O^-$$

weaker acid

No additional electronegative atoms stabilize the conjugate base.

$$CF_3CH_2OH \longrightarrow$$

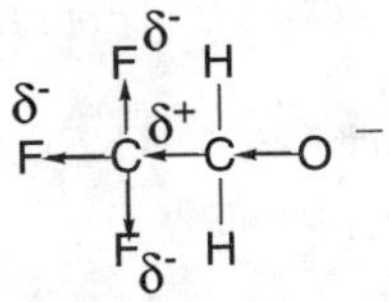

stronger acid

CF₃ withdraws electron density, stabilizing the conjugate base.

[3] **Resonance effects** (2.5C)	The acidity of H–A increases when the conjugate base A:⁻ is resonance-stabilized.

$$CH_3CH_2\ddot{O}-H \longrightarrow CH_3CH_2\ddot{O}:^-$$

ethanol

ethoxlde
conjugate base

only **one** Lewis structure

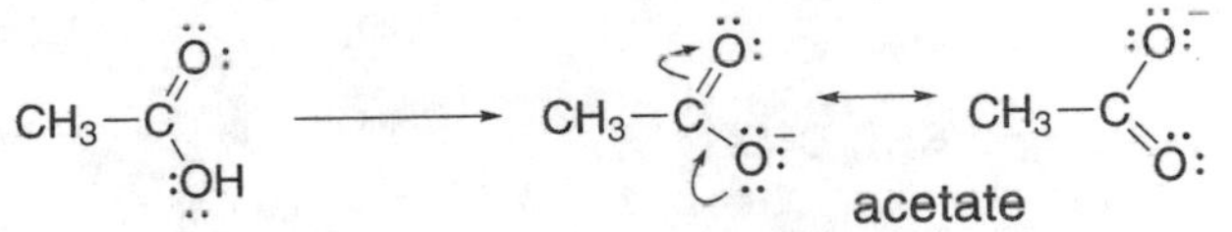

acetic acid
more acidic

acetate
conjugate base

two resonance structures

[4] **Hybridization effects** (2.5D)	The acidity of H–A increases as the percent s-character of the A:⁻ increases.

CH_3CH_3 $CH_2=CH_2$ $H-C\equiv C-H$

ethane   **ethylene**   **acetylene**

$pK_a = 50$ $pK_a = 44$ $pK_a = 25$

increasing acidity

Chapter 2: Answers to Problems

2.1 **Brønsted–Lowry acids** are **proton donors** and must contain a hydrogen atom.
Brønsted–Lowry bases are **proton acceptors** and must have an available electron pair (either a lone pair or a π bond).

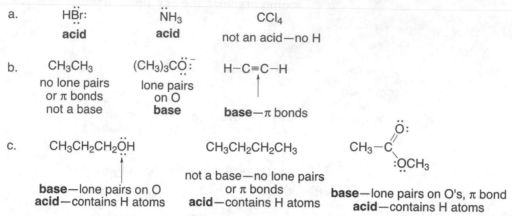

a.
HB̈r: N̈H₃   CCl₄
acid   **acid**   not an acid—no H

b.
CH₃CH₃   (CH₃)₃CÖ:⁻   H−C≡C−H
no lone pairs   lone pairs
or π bonds   on O
not a base   **base**   **base**—π bonds

c.
CH₃CH₂CH₂ÖH   CH₃CH₂CH₂CH₃

base—lone pairs on O   not a base—no lone pairs
acid—contains H atoms   or π bonds   **base**—lone pairs on O's, π bond
    acid—contains H atoms   **acid**—contains H atoms

2.2 A Brønsted–Lowry base accepts a proton to form the conjugate acid. A Brønsted–Lowry acid loses a proton to form the conjugate base.

a. NH₃ ⟶ NH₄⁺
    Cl⁻ ⟶ HCl
    (CH₃)₂C=O ⟶ (CH₃)₂C=ÖH⁺

b. HBr ⟶ Br⁻
    HSO₄⁻ ⟶ SO₄²⁻
    CH₃OH ⟶ CH₃O⁻

2.3 Use the definitions from Answer 2.2.

CH₂=CH₂ ⟶ CH₂−CH₃⁺
ethylene   accepts a proton
    conjugate acid

CH₂=CH₂ ⟶ CH₂=C̈H⁻
    loses a proton
    conjugate base

2.4 The Brønsted–Lowry base accepts a proton to form the conjugate acid. The Brønsted–Lowry acid loses a proton to form the conjugate base. Use curved arrows to show the movement of electrons (**NOT protons**). Redraw the starting materials if necessary to clarify the electron movement.

a. H−C̈l: + H₂Ö ⇌ :C̈l⁻ + H₃Ö⁺
    acid   base   conjugate base   conjugate acid

b. CH₃−C(:O:)−CH₂(H) + :ÖCH₃⁻ ⇌ CH₃−C(:O:)−C̈H₂⁻ + CH₃ÖH
    acid   base   conjugate base   conjugate acid

2.5 To draw the products:
[1] Find the acid and base.
[2] Transfer a proton from the acid to the base.
[3] Check that the charges on each side of the arrows are balanced.

a. Cl_3C-C (with O double bond, :O–H) + :Ö–CH₃ ⇌ Cl_3C-C (with O double bond, :O:⁻) + HÖ–CH₃   (–)1 charge on each side

acid   **base**

b. H–C≡C–H + :H⁻ ⇌ H–C≡C:⁻ + H₂   (–)1 charge on each side

acid   **base**

c. CH₃–N̈H₂ + H–C̈l: ⇌ CH₃–N⁺H₃ + :C̈l:⁻   net neutral on each side

base   **acid**

d. CH₃CH₂–Ö–H + H–OSO₃H ⇌ CH₃CH₂–Ö⁺–H + ⁻OSO₃H   net neutral on each side
   (with H below O)

base   **acid**

2.6 The smaller the pK_a, the stronger the acid. The larger K_a, the stronger the acid.

a. CH₃CH₂CH₃   and   CH₃CH₂OH   b.   (phenol, OH on benzene)   and   (toluene, CH₃ on benzene)

  pK_a = 50   pK_a = 16

           ↑
        smaller pK_a
        stronger acid

$K_a = 10^{-10}$ $K_a = 10^{-41}$

           ↑
        larger K_a
        stronger acid

2.7 To convert from K_a to pK_a, take (–) the log of the K_a. **pK_a = -logK_a**
To convert pK_a to K_a, take the antilog of (–) the pK_a.

a. $K_a = 10^{-10}$□ $K_a = 10^{-21}$□ $K_a = 5.2 \times 10^{-5}$ b. pK_a = 7   pK_a = 11   pK_a = 3.2

     ↓   ↓   ↓   ↓   ↓   ↓
  pK_a = 10   pK_a = 21   pK_a = 4.3 $K_a = 10^{-7}$ $K_a = 10^{-11}$ $K_a = 6.3 \times 10^{-4}$

2.8 Since **strong acids form weak conjugate bases**, the basicity of conjugate bases increases with increasing pK_a of their acids. Find the pK_a of each acid from Table 2.1 and then rank the acids in order of increasing pK_a. This will also be the order of increasing basicity of their conjugate bases.

a.   ←――――― increasing acidity   □b.   ←――――― increasing acidity

        H₂O, NH₃, CH₄   HC≡CH, CH₂=CH₂, CH₄

  pK_a = 15.7   38   50   pK_a =   25   44   50

conjugate bases: ⁻OH, ⁻NH₂, ⁻CH₃   conjugate bases: ⁻C≡CH, ⁻CH=CH₂, ⁻CH₃

        increasing basicity ――――→   increasing basicity ――――→

2.9 To estimate the pK_a of the indicated bond, find a similar bond in the pK_a table (H bonded to the same atom with the same hybridization).

a.

b.

c. $BrCH_2COO{-}H$

For NH_3, pK_a is 38.
estimated pK_a = 38

For CH_3CH_2OH,
pK_a is 16.
estimated pK_a = 16

For CH_3COOH, pK_a is 4.8.
estimated pK_a = 5

2.10 Label the acid and the base and then transfer a proton from the acid to the base. To determine if the reaction will proceed as written, compare the pK_a of the acid on the left with the conjugate acid on the right. **The equilibrium always favors the formation of the weaker acid and the weaker base.**

a. $H-C\equiv C-H$ $\quad+\quad$ $H{:}^-$ $\quad\rightleftharpoons\quad$ $H-C\equiv C{:}^-$ $\quad+\quad$ H_2

     acid   base   conjugate base   conjugate acid

    pK_a = 25 pK_a = 35

                                             weaker acid

Equilibrium favors the **products.**

b. CH_4 $\quad+\quad$ $^-\ddot{O}H$ $\quad\rightleftharpoons\quad$ $^-{:}CH_3$ $\quad+\quad$ $H_2\ddot{O}{:}$

    acid   base   conjugate base   conjugate acid

  pK_a = 50 pK_a = 15.7

weaker acid

Equilibrium favors the **starting materials.**

c. CH_3COOH $\quad+\quad$ $CH_3CH_2\ddot{O}{:}^-$ $\quad\rightleftharpoons\quad$ CH_3COO^- $\quad+\quad$ $CH_3CH_2\ddot{O}H$

     acid   base   conjugate base   conjugate acid

  pK_a = 4.8 pK_a = 16

                                             weaker acid

Equilibrium favors the **products.**

d. $^-{:}\ddot{Cl}{:}^-$ $\quad+\quad$ $CH_3CH_2\ddot{O}H$ $\quad\rightleftharpoons\quad$ $H\ddot{Cl}{:}$ $\quad+\quad$ $CH_3CH_2\ddot{O}{:}^-$

    base   acid   conjugate acid   conjugate base

               pK_a = 16 pK_a = –7

            weaker acid

Equilibrium favors the **starting materials.**

2.11 An acid can be deprotonated by the conjugate base of any acid with a higher pK_a.

CH_3COOH
pK_a = 4.8

Any base having a conjugate acid with a pK_a higher than 4.8 can deprotonate this acid.

Acid	pK_a	Conjugate base	
HCl	–7	Cl^-	not strong enough
$HC\equiv CH$	25	$HC\equiv C^-$	
H_2	35	H^-	strong enough

$HC\equiv CH$
pK_a = 25

All of these acids have a higher pK_a than $HC\equiv CH$, and a conjugate base that can deprotonate $HC\equiv CH$.

Acid	pK_a	Conjugate base
H_2	35	H^-
NH_3	38	$^-NH_2$
$CH_2=CH_2$	44	$CH_2=CH^-$
CH_4	50	CH_3^-

2.12 Acidity of H–Z **increases across a row and down a column** of the periodic table.

a. NH_3, H_2O

b. HBr, HCl

c. H_2S, HBr

O is further to the right in the periodic table.
stronger acid

Br is further down the periodic table.
stronger acid

Br is further across and down the periodic table.
stronger acid

2.13 Look at the element bonded to the acidic H and decide its acidity based on the periodic trends. **Further right and down the periodic table is more acidic.**

most acidic

most acidic

most acidic

a. $CH_3CH_2CH_2CH_2OH$

b. $HOCH_2CH_2CH_2NH_2$

c. $(CH_3)_2NCH_2CH_2CH_2NH_2$

Molecule contains C–H and O–H bonds.
O is further right; therefore, O–H hydrogen is the most acidic.

Molecule contains C–H, N–H and O–H bonds.
O is furthest right; therefore, O–H hydrogen is the most acidic.

Molecule contains C–H and N–H bonds.
N is further right; therefore, N–H hydrogen is the most acidic.

2.14 **More electronegative groups stabilize the conjugate base, making the acid stronger.** Compare the electron withdrawing groups on the acids below to decide which is a stronger acid. **(more electronegative groups = more acidic).**

a. $ClCH_2COOH$ and FCH_2COOH
more acidic

F is more electronegative than Cl making the O–H bond in the acid on the right **more acidic.**

c. CH_3COOH and O_2NCH_2COOH
more acidic

NO_2 is electron withdrawing making the O–H bond in the acid on the right **more acidic.**

b. Cl_2CHCH_2OH and $Cl_2CHCH_2CH_2OH$

Cl is closer to the acidic O–H bond.
more acidic

Cl is further from the O–H bond.

2.15 The acidity of an acid increases when the conjugate base is resonance-stabilized. Compare the conjugate bases of acetone and propane to explain why acetone is more acidic.

acetone
$pK_a = 19.2$

2 resonance structures
more stable conjugate base
Acetone is more acidic.

$CH_3CH_2CH_3$

propane
$pK_a = 50$

$CH_3CH_2CH_2$

only one Lewis structure
less stable conjugate base

(Any C–H bond in the starting material can be removed.)

2.16 The acidity of an acid increases when the conjugate base is resonance-stabilized. Acetonitrile has a resonance-stabilized conjugate base, which accounts for its acidity.

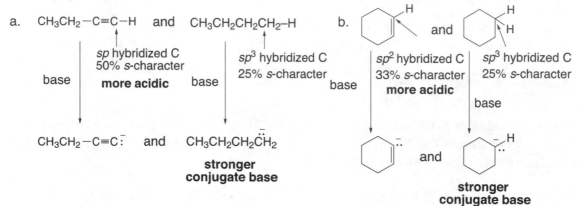

2.17 **Increasing percent *s*-character makes an acid more acidic.** Compare the percent *s*-character of the carbon atoms in each of the C–H bonds in question. A stronger acid has a weaker conjugate base.

a. $CH_3CH_2-C\equiv C-H$ and $CH_3CH_2CH_2CH_2-H$ b.

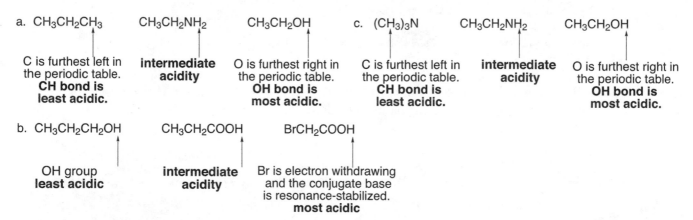

2.18 To compare the acids, first **look for element effects.** Then identify electron withdrawing groups, resonance, or hybridization differences.

a. $CH_3CH_2CH_3$ $CH_3CH_2NH_2$ CH_3CH_2OH c. $(CH_3)_3N$ $CH_3CH_2NH_2$ CH_3CH_2OH

C is furthest left in the periodic table. **CH bond is least acidic.** | **intermediate acidity** | O is furthest right in the periodic table. **OH bond is most acidic.** | C is furthest left in the periodic table. **CH bond is least acidic.** | **intermediate acidity** | O is furthest right in the periodic table. **OH bond is most acidic.**

b. $CH_3CH_2CH_2OH$ CH_3CH_2COOH $BrCH_2COOH$

OH group least acidic | **intermediate acidity** | Br is electron withdrawing and the conjugate base is resonance-stabilized. **most acidic**

2.19 Look at the element bonded to the acidic H and decide its acidity based on the periodic trends. **Further right and down the periodic table is more acidic.**

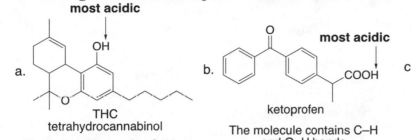

a. THC
tetrahydrocannabinol

most acidic

The molecule contains C–H and O–H bonds. O is further right; therefore, O–H hydrogen is the most acidic.

b. ketoprofen

most acidic

The molecule contains C–H and O–H bonds. O is further right; therefore, O–H hydrogen is the most acidic.

c. propranolol

most acidic

The molecule contains C–H, N–H and O–H bonds. O is furthest right; therefore, O–H hydrogen is the most acidic.

2.20 Label the acid and the base in the starting materials and then draw the products of proton transfer from the acid to the base.

a. $CH_3CH_2O–H$ + $Na^+ H^-$ ⇌ $CH_3CH_2O^- Na^+$ + H_2
   acid   base   conjugate base   conjugate acid

b. $CH_3COO–H$ + $Na^+ {}^-OCH_2CH_3$ ⇌ $CH_3COO^- Na^+$ + $HOCH_2CH_3$
   acid   base   conjugate base   conjugate acid

c. $CH_3CH_2CH_2CH_2^- Li^+$ + $H–OH$ ⇌ $CH_3CH_2CH_2CH_3$ + $Li^+ {}^-OH$
   base   acid   conjugate acid   conjugate base

d. $CH_3\text{—}$⟨⟩$\text{—}\overset{O}{\underset{O}{S}}\text{—O–H}$ + $:N(CH_2CH_3)_3$ ⇌ $CH_3\text{—}$⟨⟩$\text{—}SO_3^-$ + $HN^+(CH_2CH_3)_3$
   acid   base   conjugate base   conjugate acid

2.21 To cross a cell membrane, amphetamine must be in its neutral (not ionic) form.

amphetamine

CH_3
⟨⟩$–CH_2–C–H$
:NH_2

protonation by HCl in the stomach →

CH_3
⟨⟩$–CH_2–C–H$
$^+NH_2$
H

deprotonation in the intestines →

CH_3
⟨⟩$–CH_2–C–H$
:NH_2

↑
absorption here in the neutral form

2.22 **Lewis bases are electron pair donors**: they contain a lone pair or a π bond.

a. $\ddot{N}H_3$

yes - has lone pair

b. $CH_3CH_2CH_3$

no - no lone pair or π bond

c. $H:^-$

yes - has lone pair

d. $H–C≡C–H$

yes - has 2 π bonds

2.23 **Lewis acids are electron pair acceptors.** Most Lewis acids contain a proton or an unfilled valence shell of electrons.

a. BBr_3 | b. CH_3CH_2OH | c. $(CH_3)_3C^+$ | d. Br^-

| **yes** | **yes** | **yes** | **no** |
| unfilled valence shell on B | contains a proton | unfilled valence shell on C | no proton no unfilled valence shell |

2.24 Label the Lewis acid and Lewis base and the draw the curved arrows.

new bond

a. BF_3 + $CH_3\!-\!\ddot{O}\!-\!CH_3$ ⟶ F–B⁻–O⁺(CH₃)(CH₃) with F's

Lewis acid
unfilled valence shell on B

Lewis base
lone pairs on O

b. $(CH_3)_2CH^+$ + ^-OH ⟶ $(CH_3)_2CHOH$

Lewis acid
unfilled valence shell on C

Lewis base
lone pairs on O

2.25 A Lewis acid is also called an **electrophile**. When a Lewis base reacts with an electrophile other than a proton, it is called a **nucleophile**. Label the electrophile and nucleophile in the starting materials and then draw the products.

a. $CH_3CH_2\!-\!\ddot{O}\!-\!CH_2CH_3$ + BBr_3 ⟶ Br–B⁻–Br, $CH_3CH_2\!-\!\overset{+}{\ddot{O}}\!-\!CH_2CH_3$

Lewis base
nucleophile
lone pairs on O

Lewis acid
electrophile
unfilled valence shell on B

b. $CH_3\!-\!\overset{:O:}{C}\!-\!CH_3$ + $AlCl_3$ ⟶ Cl–Al⁻–Cl with Cl, $\overset{:O:^+}{C}$ CH_3 CH_3

Lewis base
nucleophile
lone pairs on O

Lewis acid
electrophile
unfilled valence shell on Al

2.26 Curved arrows begin at the Lewis base and point toward the Lewis acid.

new C–H bond

$CH_2\!=\!C(CH_3)(H)$ + $H_2\overset{+}{\ddot{O}}(H)$ ⟶ H–C(H)(H)–C⁺(CH₃)(H) + H_2O

Lewis base
contains a π bond

Lewis acid
contains a proton

2.27 To draw the conjugate acid of a Brønsted–Lowry base, **add a proton to the base**.

a. $H_2\overset{..}{O}:$ $\xrightarrow{H^+}$ $H_3\overset{..}{O}^+$

b. $:\overset{..}{N}H_2$ $\xrightarrow{H^+}$ $\overset{..}{N}H_3$

c. HCO_3^- $\xrightarrow{H^+}$ H_2CO_3

d. $CH_3CH_2\overset{..}{N}HCH_3$ $\xrightarrow{H^+}$ $CH_3CH_2\overset{+}{N}H_2CH_3$

e. $CH_3\overset{..}{O}CH_3$ $\xrightarrow{H^+}$ $CH_3-\overset{\overset{\displaystyle H}{|+}}{O}-CH_3$

f. CH_3COO^- $\xrightarrow{H^+}$ CH_3COOH

2.28 To draw the conjugate base of a Brønsted–Lowry acid, **remove a proton from the acid**.

a. HCN $\xrightarrow{-H^+}$ ^-CN

b. HCO_3^- $\xrightarrow{-H^+}$ CO_3^{2-}

c. $(CH_3)_2\overset{+}{N}H_2$ $\xrightarrow{-H^+}$ $(CH_3)_2NH$

d. $HC\equiv CH$ $\xrightarrow{-H^+}$ $HC\equiv C^-$

e. CH_3CH_2COOH $\xrightarrow{-H^+}$ $CH_3CH_2COO^-$

f. CH_3SO_3H $\xrightarrow{-H^+}$ $CH_3SO_3^-$

2.29 Label the Brønsted–Lowry acid and Brønsted–Lowry base in the starting materials and **transfer a proton from the acid to the base** for the products.

a. $CH_3\overset{..}{O}{-}H$ + $:\overset{..}{N}H_2$ $\rightleftharpoons$ $CH_3\overset{..}{O}:^-$ + $\overset{..}{N}H_3$
    acid   **base**   **conjugate base** **conjugate acid**

b. $CH_3CH_2-\overset{\overset{\displaystyle \overset{..}{O}:}{\|}}{C}{\underset{\underset{\displaystyle :\overset{..}{O}{-}H}{}}{}}$ + $:\overset{..}{O}-CH_3$ $\rightleftharpoons$ $CH_3CH_2-\overset{\overset{\displaystyle \overset{..}{O}:}{\|}}{C}{\underset{:\overset{..}{O}:^-}{}}$ + $H\overset{..}{O}-CH_3$
    acid   **base**   **conjugate base**   **conjugate acid**

c. $CH_3CH_2-C\equiv C{-}H$ + $:H^-$ $\rightleftharpoons$ $CH_3CH_2-C\equiv C:^-$ + H_2
    acid   **base**   **conjugate base** **conjugate acid**

d. $(CH_3CH_2)_3\overset{..}{N}$ + $H{-}\overset{..}{\underset{..}{Cl}}:$ $\rightleftharpoons$ $(CH_3CH_2)_3\overset{+}{N}H$ + $:\overset{..}{\underset{..}{Cl}}:^-$
    base   **acid**   **conjugate acid** **conjugate base**

e. $CH_3CH_2-\overset{..}{O}-H$ + $H{-}\overset{..}{\underset{..}{Br}}:$ $\rightleftharpoons$ $CH_3CH_2-\overset{\underset{\displaystyle H}{+}}{O}-H$ + $:\overset{..}{\underset{..}{Br}}:^-$
    base   **acid**   **conjugate acid**   **conjugate base**

f. $CH_3C\equiv C:^-$ + $H{-}\overset{..}{O}H$ $\rightleftharpoons$ $CH_3C\equiv CH$ + $H\overset{..}{O}:^-$
    base   **acid**   **conjugate acid** **conjugate base**

2.30 Curved arrows show the movement of electrons, **from the electron pair**.

Incorrect. Arrow must flow from attacking electron pair on O to the proton.

Incorrect. Arrow must flow from the double bond to the proton it attacks.

a. $H-\ddot{O}:^-$ + $H-\overset{\overset{H}{|}}{\underset{\underset{H}{|}}{N}}{}^+-H$ ⟶ $H-\ddot{O}-H$ + $:\overset{\overset{H}{|}}{\underset{\underset{H}{|}}{N}}-H$

b. $CH_2{=}C(CH_3)_2$ + $H{-}\ddot{B}r:$ ⟶ $H-\overset{\overset{H}{|}}{\underset{\underset{H}{|}}{C}}-\overset{\overset{CH_3}{|}}{\underset{\underset{CH_3}{|}}{C}}{}^+-CH_3$ + $:\ddot{B}r^-$

a. $H-\ddot{O}:^-$ + $H-\overset{\overset{H}{|}}{\underset{\underset{H}{|}}{N}}{}^+-H$ ⟶ $H-\ddot{O}-H$ + $:\overset{\overset{H}{|}}{\underset{\underset{H}{|}}{N}}-H$

$CH_2{=}C(CH_3)_2$ + $H{-}\ddot{B}r:$ ⟶ $H-\overset{\overset{H}{|}}{\underset{\underset{CH_3}{|}}{C}}-\overset{\overset{H}{|}}{\underset{\underset{CH_3}{|}}{C}}{}^+-CH_3$ + $:\ddot{B}r^-$

2.31 Label the acid and base in the starting materials and then draw the products of proton transfer from acid to base.

a. [structure: benzoic acid] **acid** + $\ddot{N}H_3$ **base** ⇌ [structure: benzoate] + $\overset{+}{N}H_4$

b. F_3C-C [structure with :O: and :O–H] **acid** + $Na^+\ HCO_3^-$ **base** ⇌ F_3C-C [structure with O: and $\ddot{O}:^-\ Na^+$] + H_2CO_3

c. $CH_3-C{\equiv}C{-}H$ **acid** + $Na^+\ {}^-NH_2$ **base** ⇌ $CH_3-C{\equiv}C:^-\ Na^+$ + $\ddot{N}H_3$

d. $CH_3CH_2-\ddot{O}{-}H$ **acid** + $K^+\ :\ddot{O}H^-$ **base** ⇌ $CH_3CH_2-\ddot{O}:^-\ K^+$ + $H_2\ddot{O}:$

e. $CH_3-\overset{\overset{\cdot\cdot}{|}}{\underset{\underset{CH_3}{|}}{N}}-H$ **base** + CH_3-[benzene ring]$-\overset{\overset{O}{\|}}{\underset{\underset{O}{\|}}{S}}-O{-}H$ **acid** ⇌ $CH_3-\overset{\overset{H}{|}}{\underset{\underset{CH_3}{|}}{\overset{+}{N}}}-H$ + CH_3-[benzene ring]$-SO_3^-$

f. CH_3[structure with :O:]CH_3 **base** + $H{-}OSO_3H$ **acid** ⇌ CH_3[structure with $:\overset{+}{O}{-}H$]CH_3 + HSO_4^-

2.32 To convert pK_a to K_a, take the antilog of (–) the pK_a.

a. H_2S
$pK_a = 7.0$
$K_a = 10^{-7}$

b. $ClCH_2COOH$
$pK_a = 2.8$
$K_a = 1.6 \times 10^{-3}$

c. HCN
$pK_a = 9.1$
$K_a = 7.9 \times 10^{-10}$

2.33 To convert from K_a to pK_a, take (–) the log of the K_a. $pK_a = -\log K_a$

a. ⬡—$CH_2\overset{+}{N}H_3$

$K_a = 4.7 \times 10^{-10}$
$pK_a = 9.33$

b. ⬡—$\overset{+}{N}H_3$

$K_a = 2.3 \times 10^{-5}$
$pK_a = 4.64$

c. CF_3COOH
$K_a = 5.9 \times 10^{-1}$
$pK_a = 0.23$

2.34 An acid can be deprotonated by the conjugate base of any acid with a higher pK_a.

a. H_2O
$pK_a = 15.7$ ----→
Any base with a conjugate acid having a pK_a higher than 15.7 can deprotonate it.

Acid	pK_a	Conjugate base	
CH_3CH_2OH	16	$CH_3CH_2O^-$	
$HC{\equiv}CH$	25	$HC{\equiv}C^-$	Strong enough to deprotonate H_2O.
H_2	35	H^-	
NH_3	38	$^-NH_2$	
$CH_2{=}CH_2$	44	$CH_2{=}CH^-$	
CH_4	50	CH_3^-	

c. CH_4
$pK_a = 50$
There is no base with a conjugate acid having a pK_a higher than 50 in the table.

b. NH_3
$pK_a = 38$ ---–
Any base with a conjugate acid having a pK_a higher than 38 can deprotonate it.

Acid	pK_a	Conjugate base	
$CH_2{=}CH_2$	44	$CH_2{=}CH^-$	Strong enough to deprotonate NH_3.
CH_4	50	CH_3^-	

2.35 ^-OH can deprotonate any acid with a $pK_a < 15.7$.

a. HCOOH

$pK_a = 3.8$
stronger acid
deprotonated

b. H_2S

$pK_a = 7.0$
stronger acid
deprotonated

c. ⬡—CH_3

$pK_a = 41$
weaker acid
↑

d. CH_3NH_2

$pK_a = 40$
weaker acid
↑

These acids are too weak to be
deprotonated by ^-OH.

2.36 Draw the products and then compare the pK_a of the acid on the left, and the conjugate acid on the right. **The equilibrium lies towards the side having the acid with a higher pK_a (weaker acid).**

a. $CF_3-C(\overset{\cdot\cdot}{\overset{\cdot\cdot}{O}})\overset{\cdot\cdot}{O}-H$ + $^-\overset{\cdot\cdot}{O}CH_2CH_3$ ⇌ $CF_3-C(\overset{\cdot\cdot}{\overset{\cdot\cdot}{O}})\overset{\cdot\cdot}{O}^-$ + $H\overset{\cdot\cdot}{O}CH_2CH_3$ **products favored**

$pK_a = 0.2$

$pK_a = 16$

b. $CH_3CH_2-C(\overset{\cdot\cdot}{\overset{\cdot\cdot}{O}})\overset{\cdot\cdot}{O}-H$ + $Na^+\,Cl^-$ ⇌ $CH_3CH_2-C(\overset{\cdot\cdot}{\overset{\cdot\cdot}{O}})\overset{\cdot\cdot}{O}^-\,Na^+$ + HCl   **starting material favored**

$pK_a = {\sim}5$

$pK_a = -7$

c. $(CH_3)_3C\overset{\cdot\cdot}{O}H$ + $H-\overset{\cdot\cdot}{O}SO_3H$ ⇌ $(CH_3)_3C\overset{+}{O}H_2$ + HSO_4^- **products favored**

$pK_a = -9$

$pK_a = {\sim}-3$

d.

$+ Na^+ HCO_3^-$ ⇌ $+ H_2CO_3$ **starting material favored**

$pK_a = 10$

$pK_a = 6.4$

e. $H-C≡C-H$ + $Li^+ ^-CH_2CH_3$ ⇌ $H-C≡C:^- Li^+$ + CH_3CH_3 **products favored**

$pK_a = 25$

$pK_a = 50$

f. $CH_3\ddot{N}H_2$ + $H-OSO_3H$ ⇌ $CH_3\overset{+}{N}H_3$ + HSO_4^- **products favored**

$pK_a = -9$

$pK_a = 10.7$

2.37 Compare element effects first and then, resonance, hybridization and electron withdrawing groups to determine the relative strengths of the acids.

a. Acidity increases across a row:
$NH_3 < H_2O < HF$

b. Acidity increases down a column:
$HF < HCl < HBr$

c. increasing acidity: $^-OH < H_2O < H_3O^+$

d. increasing acidity: $NH_3 < H_2O < H_2S$

Compare NH and OH bonds first:
acidity increases across a row.
OH is more acidic.

Then compare OH and SH bonds:
acidity increases down a column.
SH is more acidic.

e. Acidity increases across a row:
$CH_3CH_3 < CH_3NH_2 < CH_3OH$

f. increasing acidity: $H_2O < H_2S < HCl$

Compare HCl and SH bonds first:
acidity increases across a row.
H–Cl is more acidic.

Compare OH and SH bonds:
acidity increases down a column.
SH is more acidic.

g. $CH_3CH_2CH_3$, $ClCH_2CH_2OH$, CH_3CH_2OH

only C–H bonds
weakest acid

O–H bond and
electron withdrawing Cl
strongest acid

O–H bond

increasing acidity: $CH_3CH_2CH_3 < CH_3CH_2OH < ClCH_2CH_2OH$

h. $HC≡CCH_2CH_3$ $CH_3CH_2CH_2CH_3$ $CH_3C=CCH_3$
                                                               | |
                                                               H H

sp C–H
strongest acid

all sp^3 C–H
weakest acid

sp^2 C–H

increasing acidity: $CH_3CH_2CH_2CH_3 < CH_3CH=CHCH_3 < HC≡CCH_2CH_3$

2.38 The strongest acid has the weakest conjugate base.

a. Draw the conjugate acid.
Increasing acidity of conjugate acids:
$CH_3CH_3 < CH_3NH_2 < CH_3OH$

increasing basicity: $CH_3O^- < CH_3\overset{-}{N}H < CH_3\overset{-}{C}H_2$

b. Draw the conjugate acid.
Increasing acidity of conjugate acids:
$CH_4 < H_2O < HBr$

increasing basicity: $Br^- < HO^- < ^-CH_3$

c. Draw the conjugate acid.
Increasing acidity of conjugate acids:
$CH_3CH_2OH < CH_3COOH < ClCH_2COOH$

increasing basicity: $ClCH_2COO^- < CH_3COO^- < CH_3CH_2O^-$

d. Draw the conjugate acid.
Increasing acidity of conjugate acids:

$—CH_2CH_3 <$ $—CH=CH_2 <$ $—C≡CH$

increasing basicity:

$—C≡C^- <$ $—CH=\overset{-}{C}H <$ $—\overset{-}{C}H_2CH_2$

2.39

CH₃CH₂CH₂—H

$pK_a = 50$

conjugate base:

CH₃CH₂C̈H₂

one resonance structure
weakest acid

CH₂=C(CH₃)CH₂—H

$pK_a = \sim43$

[CH₂=C(CH₃)CH₂⁻ ⟷ ⁻:CH₂—C(CH₃)=CH₂]

two resonance structures
negative charge delocalized
on two carbons

CH₃C(=O)CH₂—H

$pK_a = 19$

The negative charge on O is good. This makes this resonance structure especially good.

[CH₃—C(:O:)—C̈H₂ ⟷ CH₃—C(:Ö:⁻)=CH₂]

two resonance structures
negative charge delocalized
on one O and one C
strongest acid

2.40 To draw the conjugate acid, look for the most basic site and protonate it. To draw the conjugate base, look for the most acidic site and remove a proton.

N̈H₂—⟨ ⟩—ÖH

most basic site **A** most acidic proton

⁺NH₃—⟨ ⟩—ÖH

conjugate acid

N̈H₂—⟨ ⟩—Ö:⁻

conjugate base

2.41 Remove the most acidic proton to form the conjugate base. Protonate the most basic electron pair to form the conjugate acid.

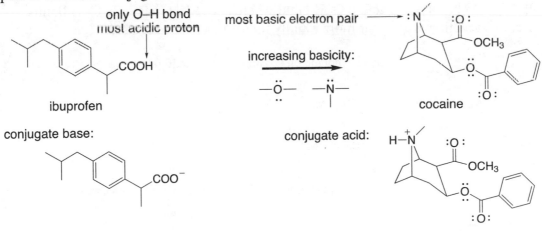

only O–H bond
most acidic proton

ibuprofen

most basic electron pair ⟶ :N

increasing basicity:
—Ö—   —N̈—

cocaine

conjugate base:

...COO⁻

conjugate acid:

H—⁺N

2.42 **A lower pK_a means a stronger acid.** The pK_a is low for the C–H bond in CH_3NO_2 due to resonance stabilization of the conjugate base.

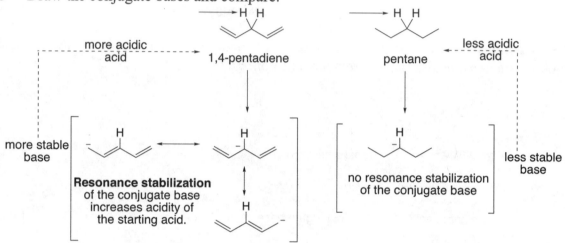

The negative charge is delocalized on the electronegative O atom. This stabilizes the conjugate base.

2.43 Draw the conjugate bases and compare.

more acidic acid → 1,4-pentadiene

less acidic acid ← pentane

more stable base

Resonance stabilization of the conjugate base increases acidity of the starting acid.

no resonance stabilization of the conjugate base

less stable base

2.44 Compare CH_3CH_2OH and CH_3CH_3. Reaction of NaH will occur more readily with a stronger acid. CH_3CH_2OH has a more acidic O–H bond whereas CH_3CH_3 has only C–H bonds. Therefore, this reaction will occur more rapidly than a reaction with CH_3CH_3.

negative charge on C—**less stable conjugate base**

CH_3CH_2–H + Na$^+$ H:$^-$ ⟶ CH_3CH_2 Na$^+$ + H$_2$

$CH_3CH_2\ddot{O}$–H + Na$^+$ H:$^-$ ⟶ $CH_3CH_2\ddot{O}$:$^-$ Na$^+$ + H$_2$

negative charge on O—**more stable conjugate base**

2.45 Compare the isomers.

dimethyl ether CH_3–O–CH_3

All H's are on C.

CH_3CH_2OH ethanol

One O–H bond
O–H bonds are more acidic than C–H bonds.
more acidic

2.46 Recall that in going from *sp* to *sp*² to *sp*³ hybridization for an atom A, H–A becomes less acidic and A:⁻ becomes more basic.

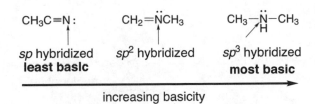

2.47 *Lewis bases* **are electron pair donors**: they contain a lone pair or a π bond. ***Brønsted–Lowry bases*** **are proton acceptors**: to accept a proton they need a lone pair or a π bond. This means all Lewis bases are Brønsted–Lowry bases.

a.
:O: ◄— lone pairs on O
‖   **both**
H—C—H

c. ⬡   **neither** = no lone pairs or π bond

d. ⬡   π bonds **both**

b. CH₃—Cl: ◄— lone pairs on Cl
          both

2.48 A *Lewis acid* **is an electron pair acceptor** and usually contains a proton or an unfilled valence shell of electrons. A ***Brønsted–Lowry acid*** **is a proton donor** and must contain a hydrogen atom. All Brønsted–Lowry acids are Lewis acids, though the reverse may not be true.

a. H_3O^+

both -
contains a H

b. Cl_3C^+

Lewis acid -
unfilled valence
shell on C

c. BCl_3

Lewis acid -
unfilled valence
shell on B

d. BF_4^-

neither -
no H or unfilled
valence shell

2.49 Label the Lewis acid and Lewis base and then draw the products.

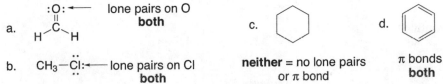

a. :Cl:⁻ + BCl₃ ⟶ Cl—B—Cl
Lewis base   Lewis acid   Cl
                          new bond

c. (acetyl chloride) + :OH⁻ ⟶ CH₃—C—Cl:
Lewis acid   **Lewis base**   :OH   new bond

b. (2,3-dimethyl-2-butene) + H—OSO₃H ⟶ product + HSO₄⁻
Lewis base   **Lewis acid**   new bond

2.50 A Lewis acid is also called an **electrophile**. When a Lewis base reacts with an electrophile other than a proton, it is called a **nucleophile**. Label the electrophile and nucleophile in the starting materials and then draw the products.

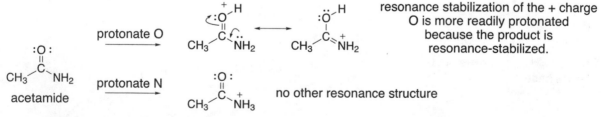

2.51

a.

b.

c.

2.52 Draw the product of protonation of either O or N and compare the conjugate acids.

protonate O

protonate N

resonance stabilization of the + charge
O is more readily protonated
because the product is
resonance-stabilized.

no other resonance structure

acetamide

When acetamide reacts with an acid, the O atom is protonated because it results in a resonance-stabilized conjugate acid.

2.53

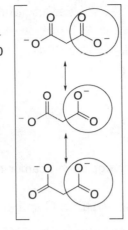

$pK_a = 2.86$ $pK_a = 5.70$

δ^+ stabilizes the $(-)$ charge
of the conjugate base.

The nearby COOH group serves
as an electron **withdrawing** group to
stabilize the negative charge. This
makes the first proton **more** acidic
than CH_3COOH.

This group destabilizes the
second negative charge.

COO^- now acts as an electron **donor** group
which destabilizes the conjugate base,
making removal of the second proton more
difficult and **less** acidic than CH_3COOH.

2.54

The COOH group of glycine gives up a proton to the basic NH_2 group to form the zwitterion.

proton transfer

a.   acts as a base $\longrightarrow$ $\overset{..}{N}H_2CH_2-C\overset{O}{\underset{OH}{}}$ $\longleftarrow$ acts as an acid

glycine

$\overset{+}{N}H_3CH_2-C\overset{O}{\underset{O^-}{}}$

zwitterion form

b. $\overset{+}{N}H_3CH_2-C\overset{O}{\underset{O^-}{}}$ $H-Cl$ $\longrightarrow$ $\overset{+}{N}H_3CH_2-C\overset{O}{\underset{OH}{}}$ $+$ Cl^-

most basic site

c. $H-\overset{+}{N}H_2-CH_2-C\overset{O}{\underset{O^-}{}}$ $Na^+ {}^-OH$ $\longrightarrow$ $NH_2CH_2-C\overset{O}{\underset{O^-}{}}$ $+$ Na^+ $+$ H_2O

most acidic site

2.55   Use curved arrows to show how the reaction occurs.

Overall reaction: $+$ $^-\overset{..}{\underset{..}{O}}H$ $\xrightarrow{H_2O}$

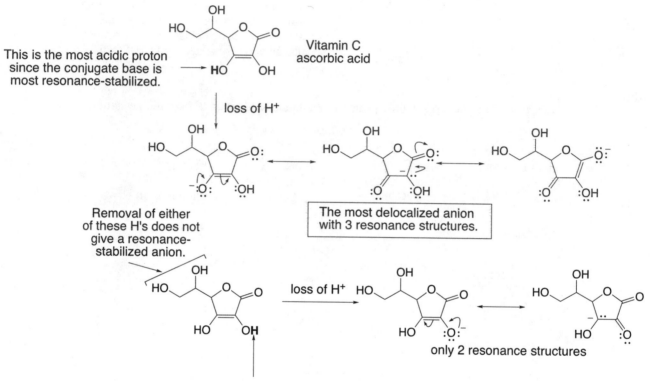

[1]

[2]

Protonate the negative charge on this carbon to form the product.

2.56   Compare the OH bonds in Vitamin C and decide which one is the most acidic.

Vitamin C
ascorbic acid

This is the most acidic proton
since the conjugate base is
most resonance-stabilized.

loss of H⁺

Removal of either
of these H's does not
give a resonance-
stabilized anion.

The most delocalized anion
with 3 resonance structures.

loss of H⁺

only 2 resonance structures

This proton is less acidic since its conjugate base is less
resonance-stabilized.

Chapter 3: Introduction to Organic Molecules and Functional Groups

♦ Types of intermolecular forces (3.4A)

	Type of force	Cause	Examples
	van der Waals (VDW)	Due to the interaction of temporary dipoles • Larger surface area, stronger forces • Larger, more polarizable atoms, stronger forces	All organic compounds
	dipole-dipole (DD)	Due to the interaction of permanent dipoles	$(CH_3)_2C=O$, H_2O
	hydrogen bonding (HB or H-bonding)	Due to the electrostatic interaction of a H atom in an O–H, N–H, or H–F bond with another N, O, or F atom.	H_2O
	ion-ion	Due to the interaction of two ions	NaCl, LiF

increasing strength (arrow pointing down along left side of table)

♦ Physical properties

Property	Observation
Boiling point (3.5A)	• For compounds of comparable molecular weight, the stronger the forces the higher the bp. $CH_3CH_2CH_2CH_2CH_3$ $CH_3CH_2CH_2CHO$ $CH_3CH_2CH_2CH_2OH$ VDW VDW, DD VDW, DD, HB MW = 72 MW = 72 MW = 74 bp = 36 $^{\circ}$C bp = 76 $^{\circ}$C bp = 118 $^{\circ}$C **increasing strength of intermolecular forces** **increasing boiling point** • For compounds with similar functional groups, the larger the surface area, the higher the bp. $CH_3CH_2CH_2CH_3$ $CH_3CH_2CH_2CH_2CH_3$ VDW VDW bp = 0 $^{\circ}$C bp = 36 $^{\circ}$C **increasing surface area** **increasing boiling point** • For compounds with similar functional groups, the more polarizable the atoms, the higher the bp. CH_3F CH_3I bp = −78 $^{\circ}$C bp = 42 $^{\circ}$C **increasing polarizability** **increasing boiling point**

Melting point (3.5B)	• For compounds of comparable molecular weight, the stronger the forces the higher the mp. $CH_3CH_2CH_2CH_2CH_3$ $CH_3CH_2CH_2CHO$ $CH_3CH_2CH_2CH_2OH$ VDW VDW, DD VDW, DD, HB MW = 72 MW = 72 MW = 74 mp = −130 °C mp = −96 °C mp = −90 °C **increasing strength of intermolecular forces** **increasing melting point** • For compounds with similar functional groups, the more symmetrical the compound, the higher the mp. $CH_3CH_2CH(CH_3)_2$ $(CH_3)_4C$ mp = −160 °C mp = −17 °C **increasing symmetry** **increasing melting point**
Solubility (3.5C)	Types of H_2O-soluble compounds: • Ionic compounds • Organic compounds having ≤ 5 C's, and an O or N atom for hydrogen bonding (for a compound with one functional group). Types of compounds soluble in organic solvents: • Organic compounds regardless of size or functional group. • Examples: 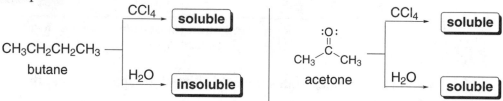
	Key: VDW - van der Waals, DD = dipole-dipole, HB = hydrogen bonding MW = molecular weight

◆ Reactivity (3.9)

• **Nucleophiles react with electrophiles.**
• Electronegative heteroatoms create electrophilic carbon atoms, which tend to react with other nucleophiles.
• Lone pairs and π bonds are nucleophilic sites that tend to react with other electrophiles.

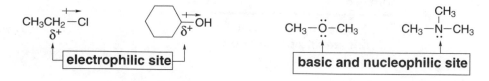

Alkenes react with electrophiles.

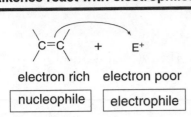

electron rich   electron poor

| nucleophile | electrophile |

Alkyl halides react with nucleophiles.

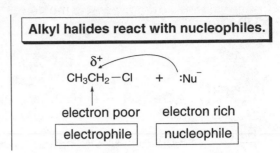

electron poor   electron rich

| electrophile | nucleophile |

Chapter 3: Answers to Problems

3.1

3.2 Identify the functional groups based on Tables 3.1, 3.2 and 3.3.

3.3 Summary of forces:
- **All compounds exhibit van der Waals forces (VDW).**
- **Polar molecules have dipole-dipole forces (DD).**
- **Hydrogen-bonding (H-bonding)** can only occur when a **H is bonded to an O, N, or F.**

a.

only nonpolar C–C
and C–H bonds
VDW only

b.

· **VDW forces**
· 2 polar C–O bonds
  and a net dipole - **DD**
· no H on O so
  no H-bonding

c. $(CH_3CH_2)_3N$

· **VDW forces**
· polar C–N bonds - **DD**
· no H on N so
  no H-bonding

d. $CH_2=CHCl$

· **VDW forces**
· polar C–Cl bond - **DD**

e. $CH_3CH_2CH_2COOH$

· **VDW forces**
· polar C–O bonds
  and a net dipole - **DD**
· H bonded to O -
  H-bonding

f. $CH_3-C\equiv C-CH_3$

only nonpolar C–H and
C–C bonds
VDW only

3.4 One principle governs boiling point:
- **Stronger intermolecular forces = higher bp.**
     Increasing intermolecular forces: van der Waals < dipole-dipole < hydrogen bonding

Two factors affect the strength of van der Waals forces, and thus affect bp:
- **Increasing surface area = increasing bp.**
     Longer molecules have a higher surface area. Any branching decreases the surface area of a molecule.
- **Increasing polarizability = increasing bp.**

a. $(CH_3)_2C=CH_2$ and $(CH_3)_2C=O$

    ↑   ↑

only VDW   VDW and DD

          polar, stronger intermolecular forces

              higher boiling point

c. $CH_3(CH_2)_4CH_3$ and $CH_3(CH_2)_5CH_3$

                       ↑

        longer molecule, more surface area

            higher boiling point

b. CH_3CH_2COOH and CH_3COOCH_3

     ↑   ↑

               no H-bonding

VDW, DD, and H-bonding

stronger intermolecular forces

    higher boiling point

d. $CH_2=CHCl$ and $CH_2=CHI$

                 ↑

        I is more polarizable.

       higher boiling point

3.5

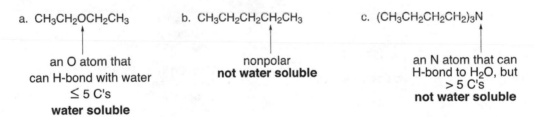

a.   and NH_2 b.   and

                    ↑

           more polar

   stronger intermolecular forces   more spherical

         (H-bonding)   packs better

         higher mp   **higher mp**

3.6 In the more ordered solid phase, molecules are much closer together than in the less ordered liquid phase. The shape of a molecule determines how close it can pack in the solid phase so symmetry is important. In the liquid phase, molecules are already further apart, so symmetry is less important and thus it doesn't affect boiling point.

3.7 A compound is water soluble if it is ionic or if it has an O or N atom and ≤ 5 C's.

a. $CH_3CH_2OCH_2CH_3$ b. $CH_3CH_2CH_2CH_2CH_3$ c. $(CH_3CH_2CH_2CH_2)_3N$

      ↑   ↑   ↑

an O atom that   nonpolar   an N atom that can

can H-bond with water   **not water soluble**   H-bond to H_2O, but

      ≤ 5 C's   > 5 C's

   water soluble   **not water soluble**

3.8 Hydrophobic portions will primarily be hydrocarbon chains. **Hydrophilic** portions will be polar.

Circled regions are **hydrophilic** because they are polar.
All other regions are **hydrophobic** since they have only C and H.

a.   b.   c.

norethindrone   arachidonic acid   benzo[a]pyrene derivative

3.9 Like dissolves like.

- To be **soluble in water**, a molecule must be ionic, or have a polar functional group capable of H-bonding for every 5 C's.
- Organic compounds are generally **soluble in organic solvents** regardless of size or functional group.

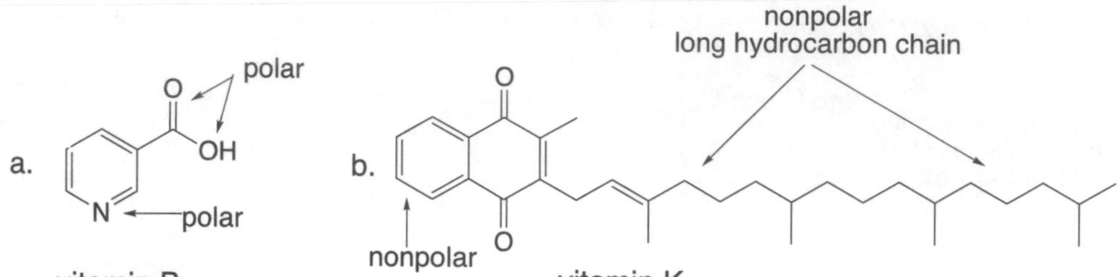

a.

vitamin B$_4$
(niacin)

soluble in water due to
two polar functional groups
and only 6 C's in the molecule

b.

vitamin K$_1$
(phylloquinone)

soluble in organic solvents
two polar C–O bonds but the
compound has > 10 C's
water insoluble

3.10 Detergents have a polar head consisting of oppositely charged ions, and a nonpolar tail consisting of C–C and C–H bonds, just like soaps do. Detergents clean by having the **hydrophobic ends of molecules surround grease**, while the **hydrophilic portion of the molecule interacts with the polar solvent** (usually water).

a detergent

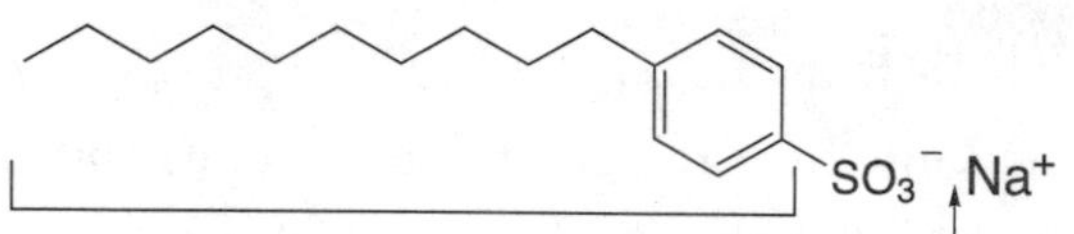

nonpolar tail
hydrophobic

This end interacts
with the grease to
dissolve it.

polar head
ionic - hydrophilic

This end interacts with
the water solvent
to maintain the micelle's
solubility in water.

3.11

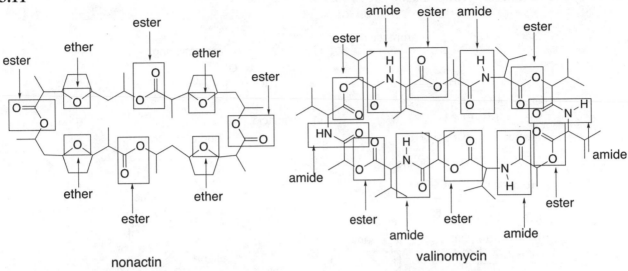

nonactin

valinomycin

3.12 Electronegative heteroatoms like N, O or X makes a carbon atom an *electrophile*.
A lone pair on a heteroatom makes it basic and nucleophilic.
π Bonds create *nucleophilic* sites and are more easily broken than σ bonds.

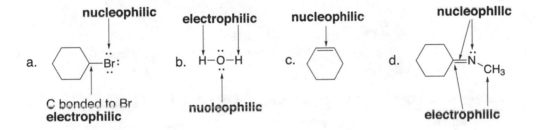

nucleophilic

a.

C bonded to Br
electrophilic

electrophilic

b. H–O–H

nucleophilic

nucleophilic

c.

nucleophilic

d.

electrophilic

3.13 Electrophiles and nucleophiles react with each other.

a. CH_3CH_2—Br + ⁻OH ⟶ **YES**

electrophile   nucleophile

b. CH_3—C≡C—CH_3 + Br⁻ ⟶ **NO**

nucleophile   nucleophile

c. $CH_3\overset{\overset{O}{\|}}{C}Cl$ + ⁻OCH_3 ⟶ **YES**

electrophile   nucleophile

d. CH_3—C≡C—CH_3 + Br⁺ ⟶ **YES**

nucleophile   electrophile

3.14 Identify the functional groups based on Tables 3.1, 3.2 and 3.3.

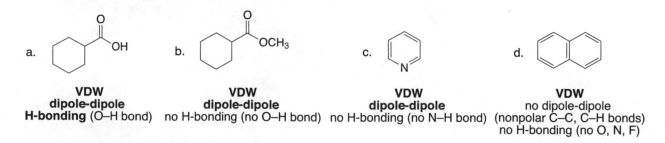

a.

CH₃CH₂CO₂—C (ester) with CH₂ (aromatic rings), CH₃—C—H, CH₂N(CH₃)₂ (amine) — **darvon**

c. HO₂C (carboxylic acid), aromatic ring — **ibuprofen**

b. aromatic ring, amine (N–H), alkene, amide, CH₃O (ether) — **melatonin**

d. amine (N), OH (alcohol), alkene, alkyne — **histrionicotoxin**

3.15

CH₃—CH(OH)—CH₂—CH₃   **alcohol**

CH₃CH₂CH₂CH₂—OH   **alcohol**

CH₃—CH(CH₂OH)—CH₃   **alcohol**

CH₃—C(CH₃)(CH₃)—OH   **alcohol**

CH₃CH₂—O—CH₃CH₂   **ether**

CH₃—O—CH(CH₃)—CH₃   **ether**

CH₃—O—CH₂CH₂CH₃   **ether**

3.16 One possible structure for each functional group:

a. aldehyde = R–CHO ⟶ CH₃CH₂CH₂–CHO

b. ketone = R–CO–R ⟶ CH₃–CO–CH₂CH₃

c. carboxylic acid = R–CO–OH ⟶ CH₃CH₂CH₂–CO–OH

d. ester = R–CO–O–R ⟶ CH₃CH₂–CO–O–CH₃

3.17 Use the rules from Answer 3.3.

a. (cyclohexane–COOH)
   VDW
   dipole-dipole
   H-bonding (O–H bond)

b. (cyclohexane–CO–OCH₃)
   VDW
   dipole-dipole
   no H-bonding (no O–H bond)

c. (pyridine)
   VDW
   dipole-dipole
   no H-bonding (no N–H bond)

d. (naphthalene)
   VDW
   no dipole-dipole
   (nonpolar C–C, C–H bonds)
   no H-bonding (no O, N, F)

3.18 **Increasing intermolecular forces**: van der Waals < dipole-dipole < H-bonding

a. **increasing intermolecular forces:**

$CH_3CH_3 < CH_3Cl < CH_3NH_2$

VDW	VDW dipole-dipole	VDW dipole-dipole H-bonding

c. **increasing intermolecular forces:**

$(CH_3)_2C=C(CH_3)_2 < (CH_3)_2CHCOCH_3 < (CH_3)_2CHCOOH$

VDW	VDW dipole-dipole	VDW dipole-dipole H-bonding

b. **increasing intermolecular forces:**

$CH_3Cl < CH_3Br < CH_3I$

increasing polarizability
stronger intermolecular forces

d. **increasing intermolecular forces:**

$CH_3Cl < CH_3OH < NaCl$

VDW dipole-dipole	VDW dipole-dipole H-bonding	ionic

3.19

CH_3-C (O- - H-O / O-H- - -O) $C-CH_3$ hydrogen bonding between 2 acetic acid molecules

3.20 **A** = VDW forces; **B** = H-bonding; **C** = ion-ion interactions; **D** = H-bonding; **E** = H-bonding; **F** = VDW forces.

3.21 Use the principles from Answer 3.4.

a. $CH_3(CH_2)_4-I$ $CH_3(CH_2)_5-I$ $CH_3(CH_2)_6-I$

increasing size, increasing surface area, increasing boiling point

b. $CH_3CH_2CH_2CH_3 < (CH_3)_3N < CH_3CH_2CH_2NH_2$

VDW	VDW dipole-dipole	VDW dipole-dipole H-bonding

increasing boiling point

c. $(CH_3)_3COC(CH_3)_3 < CH_3(CH_2)_3O(CH_2)_3CH_3 < CH_3(CH_2)_7OH$

VDW dipole-dipole smaller surface area	VDW dipole-dipole larger surface area	VDW dipole-dipole H-bonding highest bp

increasing boiling point

d. [structure] < [structure] Br < [structure] OH < [structure] OH

VDW	VDW dipole-dipole	VDW dipole-dipole H-bonding	VDW dipole-dipole H-bonding larger surface area

increasing boiling point

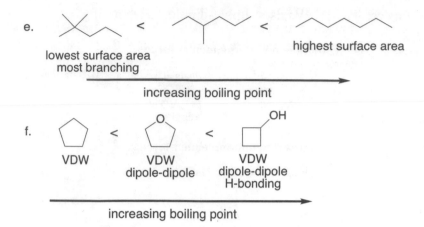

e. lowest surface area / most branching < < highest surface area

increasing boiling point

f. VDW < VDW / dipole-dipole < VDW / dipole-dipole / H-bonding

increasing boiling point

3.22 In $CH_3CH_2NHCH_3$, there is a N–H bond so the molecules exhibit intermolecular hydrogen bonding, whereas in $(CH_3)_3N$ the N is only bonded to C, so there is no hydrogen bonding. The hydrogen bonding in $CH_3CH_2NHCH_3$ makes it have much **stronger intermolecular forces** than $(CH_3)_3N$. As intermolecular forces increase, the boiling point of a molecule of the same molecular weight is higher.

3.23 Stronger forces, higher mp. More symmetrical compounds, higher mp.

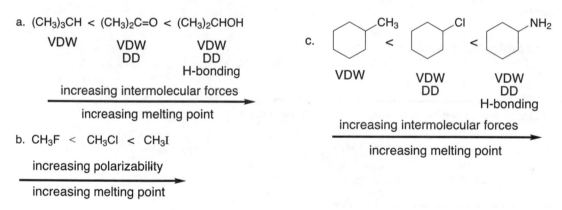

a. $(CH_3)_3CH$ < $(CH_3)_2C=O$ < $(CH_3)_2CHOH$

VDW | VDW / DD | VDW / DD / H-bonding

increasing intermolecular forces

increasing melting point

b. CH_3F < CH_3Cl < CH_3I

increasing polarizability

increasing melting point

c. (cyclohexane with CH_3) < (cyclohexane with Cl) < (cyclohexane with NH_2)

VDW | VDW / DD | VDW / DD / H-bonding

increasing intermolecular forces

increasing melting point

3.24

–119 °C	–118 °C	–91 °C	–25 °C
not symmetrical	not symmetrical	symmetrical / higher mp	most spherical / highest mp

In both compounds the CH_3 group dangling from the chain makes packing in the solid difficult, so the mp is low.

This molecule can pack somewhat better since it has no CH_3 group dangling from the chain, so the mp is somewhat higher. It also has the most surface area and this increases VDW forces compared to the first two compounds.

This compound packs the best since it is the most spherical in shape, increasing its mp.

3.25 **Boiling point is determined solely by the strength of the intermolecular forces.** Since benzene has a smaller size, it has less surface area and weaker VDW interactions and therefore a lower boiling point than toluene. The increased melting point for benzene can be explained by symmetry: benzene is much more symmetrical than toluene. More symmetrical molecules can pack more tightly together, increasing their melting point. Symmetry has no effect on boiling point.

benzene
bp = 80 °C
mp = 5 °C
very symmetrical
closer packing in solid form
higher mp

and

CH_3 toluene
bp = 111 °C
mp = –93 °C
less symmetrical
lower mp

3.26 Increasing polarity = increasing water solubility.

Neither compound is very H_2O soluble.

a.
$CH_3CH_2CH_2CH_3$ < $(CH_3)_3CH$ < $CH_3OCH_2CH_3$ < $CH_3CH_2CH_2OH$

VDW   VDW   VDW   VDW
          more spherical   DD   DD
(This nonpolar, hydrophobic   H-bonding
molecule is more compact,
making it more water soluble than its
straight chain isomer, drawn to the left.)

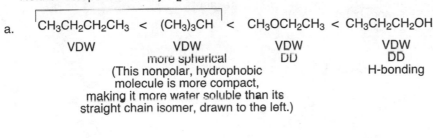

b.
polar
no H-bonding

polar
H-bonding to H_2O,
not itself

polar and
H-bonding
More opportunities
for H-bonding with its
O atom and its H on O.

3.27 Look for two things:
- To H-bond to another molecule like itself, the molecule must contain a **H bonded to O, N, or F**.
- To H-bond with water, a molecule need **only contain an O, N, or F**.

These can H-bond to another molecule like itself:
    Both compounds have N–H bonds.
    b. CH_3NH_2, e. $CH_3CH_2CH_2CONH_2$

These can H-bond with water:
    All of these molecules have an O or N atom
    b. CH_3NH_2, c. CH_3OCH_3, d. $(CH_3CH_2)_3N$,
    e. $CH_3CH_2CH_2CONH_2$, g. CH_3SOCH_3,
    h. $CH_3CH_2COOCH_3$

3.28   Draw the molecules in question and look at the intermolecular forces involved.

no H bonded to O

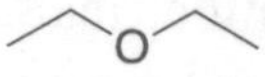

diethyl ether

1-butanol

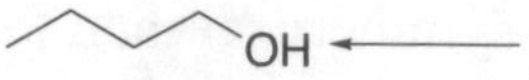

H bonded to O:
hydrogen bonding

VDW forces
dipole-dipole forces

VDW forces
dipole-dipole forces
H-bonding

• Both have ≤ 5 C's and an electronegative O atom, so they can H-bond to water,
  making them soluble in water.
  • Only 1-butanol can H-bond to another molecule life itself, and this increases its boiling point.

3.29

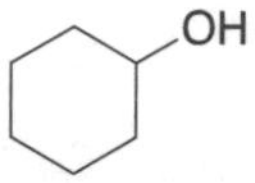

cyclohexanol
The nonpolar hydrocarbon part is more compact, so it
is easier for the OH group to solubilize it in water.

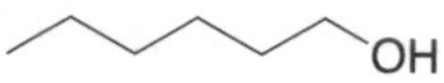

1-hexanol

3.30   Use the solubility rule from Answer 3.7.

a.

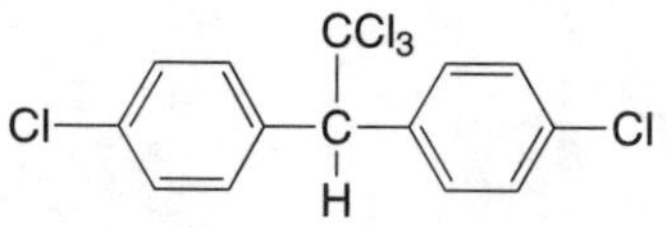

DDT
no N or O
not water soluble

c.

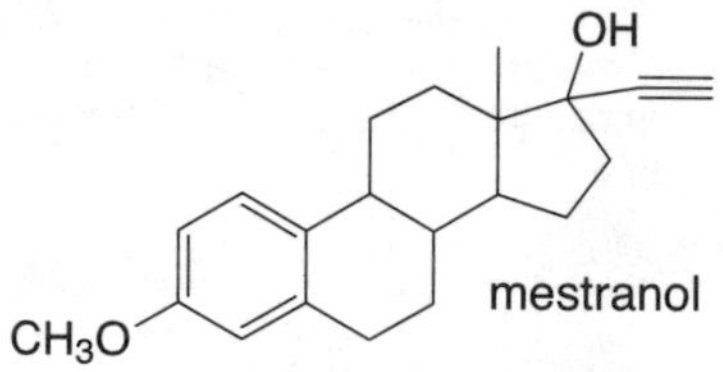

mestranol

2 polar functional groups
but > 10 C's
not water soluble

b.

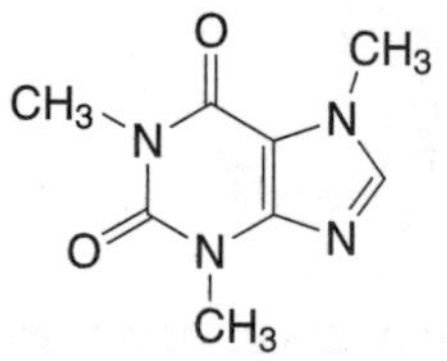

caffeine
many polar bonds with N and O atoms
Many opportunities for H-bonding.
water soluble

d.

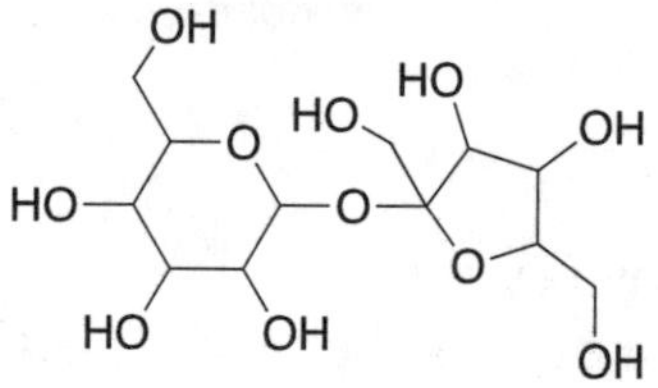

sucrose
many polar bonds with O
11 O's and 12 C's
Many opportunities for H-bonding with H₂O.
water soluble

3.31

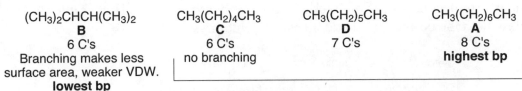

heptane
bp = 98°C

perfluoroheptane
bp = 82–84°C

molecular weight = 100 g/mol   molecular weight = 388 g/mol

F atoms are very electronegative and small, and their electron clouds
are held tightly making them very poorly polarizable. This means there
is little force of attraction between polyflourinated molecules, giving
them much lower bp's than you would expect based on their molecular
weights.

3.32

$(CH_3)_2CHCH(CH_3)_2$	$CH_3(CH_2)_4CH_3$	$CH_3(CH_2)_5CH_3$	$CH_3(CH_2)_6CH_3$
B	**C**	**D**	**A**
6 C's	6 C's	7 C's	8 C's
Branching makes less	no branching		**highest bp**
surface area, weaker VDW.			
lowest bp			

C, D, and A are all long chain hydrocarbons,
but the size increases from C to D to A, increasing
the VDW forces and increasing bp.

3.33   Water solubility is determined by polarity. Polar molecules are soluble in water, while nonpolar
molecules are soluble in organic solvents.

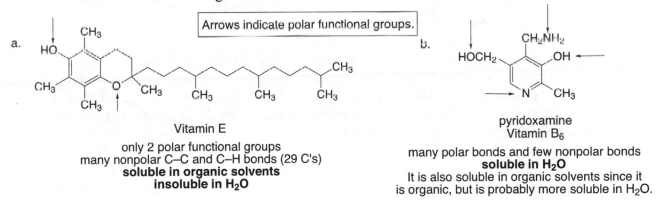

Arrows indicate polar functional groups.

a.

Vitamin E

only 2 polar functional groups
many nonpolar C–C and C–H bonds (29 C's)
soluble in organic solvents
insoluble in H₂O

b.

pyridoxamine
Vitamin B₆

many polar bonds and few nonpolar bonds
soluble in H₂O
It is also soluble in organic solvents since it
is organic, but is probably more soluble in H₂O.

3.34   Molecules that dissolve in water are readily excreted from the body in urine whereas less polar
molecules that dissolve in organic solvents are soluble in fatty tissue and are retained for longer
periods. Compare the solubility properties of THC and ethanol to determine why drug
screenings can detect THC and not ethanol weeks after introduction to the body.

Tetrahydrocannabinol
THC

ethanol

THC has relatively few polar
bonds compared to the number
of nonpolar bonds making it
soluble in organic solvents
and therefore **soluble in fatty tissue.**

Ethanol has 1 O atom and
only 2 C's making it
soluble in water.

Due to their solubilities, **THC is retained much longer in the fatty tissue of the body,** being
slowly excreted over many weeks, while ethanol is excreted rapidly in urine after ingestion.

3.35 Compare the intermolecular forces of crack and cocaine hydrochloride. Higher intermolecular
forces increase both the boiling point and the water solubility.

cocaine (crack)
neutral organic molecule

cocaine hydrochloride
a salt

The molecules are identical except for the ionic bond in cocaine hydrochloride. Ionic forces are
extremely strong forces, and therefore the cocaine hydrochloride salt has a much **higher boiling
point and is more water soluble.** Since the salt is highly water soluble, it can be injected
directly into the blood stream where it dissolves. Crack is smoked because it can dissolve in the
organic tissues of the nasal passage and lungs.

3.36 A laundry detergent must have both a highly polar end of the molecule, and a nonpolar end of
the molecule. The polar end will interact with water, while the nonpolar end surrounds the
grease/organic material.

a.

nonpolar
interacts with organic material

polar
interacts with water
by H-bonding at all O and H atoms

b.

nonpolar
interacts with
organic material

polar
interacts with water
by H-bonding
at O and H atoms

3.37 An emulsifying agent is one that dissolves a compound in a solvent in which it is not normally soluble. In this case the phospholipids can dissolve the oil in its nonpolar tails and bring it into solution in the aqueous vinegar solution. Or, the nonpolar tails dissolve in the oil, and the polar head brings the water-soluble compounds into solution. In any case, the phospholipids make a uniform medium, mayonnaise, from two insoluble layers.

vinegar
aqueous
hydrophilic

oil
organic
hydrophobic

These two ingredients will not mix. The emulsifying agent (egg yolk) has phospholipids that have both hydrophobic and hydrophilic portions, making the mayonnaise uniform.

3.38 Use the rules from Answer 3.12.

a.

nucleophilic

electrophilic

c.

nucleophilic

$\delta^+ \quad \delta^+$

electrophilic

e.

nucleophilic

$\delta^+ \quad \delta^+$
$CH_3\ddot{O}H$

electrophilic

b.

$=CH_2$

nucleophilic

d.

All the C=C's are
nucleophilic.

f.

nucleophilic
CH_3 $\overset{+}{C}l$

electrophilic
(All lone pairs on O and
Cl are nucleophilic.)

3.39

a.

+ Br⁻ ⟶ **NO**
nucleophilic

nucleophilic

d.

+ ⁻OH ⟶ **NO**
nucleophilic

nucleophilic

b.

—CH₂Cl + ⁻CN ⟶ **YES**
nucleophilic

electrophilic

e.

+ H₃O⁺ ⟶ **YES**
nucleophilic   electrophilic

c.

CH_3—C—CH_3 + ⁻CH₃ ⟶ **YES**
nucleophilic

electrophilic

3.40   More rigid cell membrane have phospholipids with *fewer* C=C's. Each C=C introduces a bend in the molecule, making the phospholipids pack less tightly. Phospholipids without C=C's can pack very tightly, making the membrane less fluid, and more rigid.

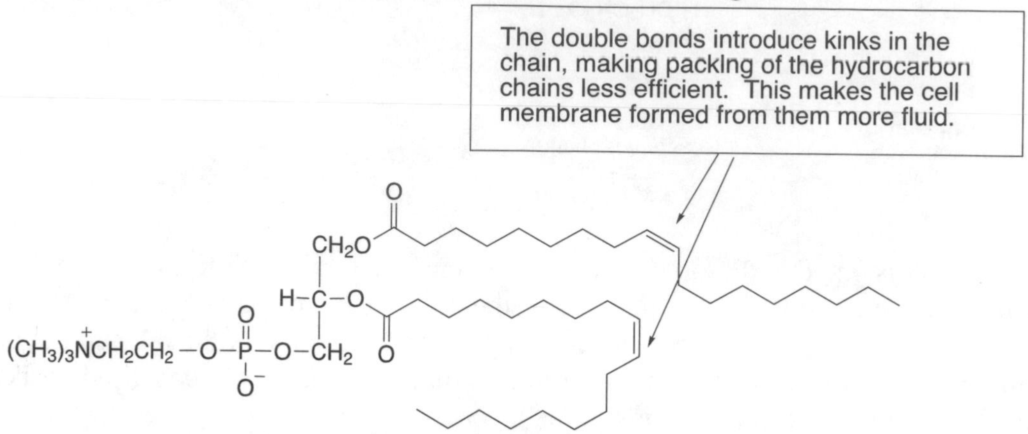

> The double bonds introduce kinks in the chain, making packing of the hydrocarbon chains less efficient. This makes the cell membrane formed from them more fluid.

The π bonds in the long hydrocarbon chains make the phospholipids pack less tightly, and the cell membrane less rigid.

3.41

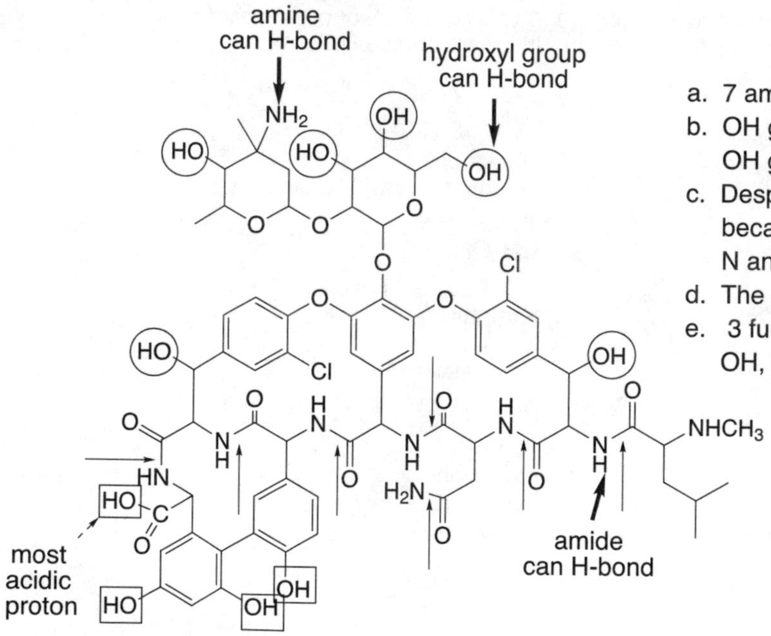

vancomycin

a. 7 amide groups (unlabeled arrows)
b. OH groups bonded to sp^3 C's are circled.
   OH groups bonded to sp^2 C's have a square.
c. Despite its size, vancomycin is water soluble because it contains many polar groups and many N and O atoms that can H-bond to H_2O.
d. The most acidic proton is labeled (COOH group).
e. 3 functional groups capable of H-bonding are OH, amides, and amines.

3.42

A

These two functional groups are close enough that they can intramolecularly H-bond to each other. Since the two polar functional groups are involved in intramolecular H-bonding, they are less available for H-bonding to H_2O. This makes **A** less H_2O soluble than **B**, whose two functional groups are both available for H-bonding to the H_2O solvent.

B

The OH and the CHO are too far apart to intramolecular H-bond to each other, leaving more opportunity to H-bond with solvent.

3.43

a. melting point

fumaric acid

Fumaric acid has its two larger COOH groups on opposite ends of the molecule, and in this way it can pack better in a lattice than maleic acid, giving it a **higher mp**.

b. solubility

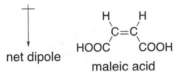

net dipole

maleic acid

Maleic acid is more polar, giving it greater **H_2O solubility.**

c. removal of the first proton (K_{a1})

loss of 1 proton

In maleic acid, intramolecular H-bonding stabilizes the conjugate base after one H is removed, making maleic acid more acidic than fumaric acid.

loss of 1 proton

Intramolecular H-bonding is not possible here.

d. removal of the second proton (K_{a2})

Now the dianion is held in close proximity in maleic acid, and this destabilizes the conjugate base. So removing the second H in maleic acid is harder, making it a weaker acid than fumaric acid for removal of the second proton.

The two negative charges are much further apart. This makes the dianion from fumaric acid more stable and thus K_{a2} is larger for fumaric acid than maleic acid.

Chapter 4 – Alkanes

◆ General facts about alkanes (4.1-4.3)

- Alkanes are composed of **tetrahedral, sp^3** hybridized C's.
- There are two types of alkanes: acyclic alkanes having molecular formula C_nH_{2n+2}, and cycloalkanes having molecular formula C_nH_{2n}.
- Alkanes have only **nonpolar C−C and C−H bonds** and no functional group so they undergo few reactions.
- Alkanes are named with the suffix **–ane.**

◆ Classifying C's and H's (4.1A)

- Carbon atoms are classified by the number of C's bonded to them; **a $1°$ C is bonded to one other C,** and so forth.

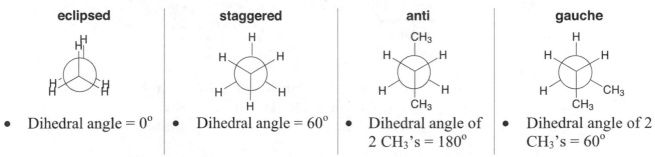

- Hydrogen atoms are classified by the type of carbon atom to which they are bonded; **a $1°$ H is bonded to a $1°$ C,** and so forth.

◆ Conformations in acyclic alkanes (4.9, 4.10)

- Alkane conformations can be classified as **staggered, eclipsed, anti,** or **gauche** depending on the relative orientation of the groups on adjacent carbons.

eclipsed	staggered	anti	gauche
• Dihedral angle = $0°$	• Dihedral angle = $60°$	• Dihedral angle of 2 CH₃'s = $180°$	• Dihedral angle of 2 CH₃'s = $60°$

- A staggered conformation is **lower in energy** than an eclipsed conformation.
- An anti conformation is **lower in energy** than a gauche conformation.

◆ Types of strain

- **Torsional strain**—an increase in energy due to eclipsing interactions (4.9).
- **Steric strain**—an increase in energy when atoms are forced too close to each other (4.10).
- **Angle strain**—an increase in energy when bond angles deviate from $109.5°$ (4.11).

◆ Two types of isomers

[1] **Constitutional isomers** – isomers that differ in the way the atoms are connected to each other (4.1A).

[2] **Stereoisomers** – isomers that differ only in the way atoms are oriented in space (4.13B).

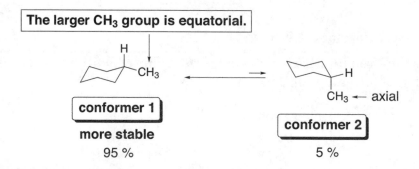

◆ Conformations in cyclohexane (4.12, 4.13)

- Cyclohexane exists as **two chair conformers** in rapid equilibrium at room temperature.
- Each carbon atom on a cyclohexane ring has **one axial** and **one equatorial hydrogen**. Ring-flipping converts axial to equatorial H's, and vice versa.

An axial H flips equatorial.

H_{ax} H_{eq} Ring-flip. H_{eq} H_{ax}

An equatorial H flips axial.

- In substituted cyclohexanes, groups larger than hydrogen are more stable in the **more roomy equatorial position.**

The larger CH₃ group is equatorial.

H   CH₃   H   CH₃ ← axial

conformer 1
more stable
95 %

conformer 2
5 %

- Disubstituted cyclohexanes with substituents on different atoms, exist as two possible stereoisomers.
  - The **cis** isomer has two groups on the **same side** of the ring, either both up or both down.
  - The **trans** isomer has two groups on **opposite sides** of the ring, one up and one down.

CH₃   CH₃   CH₃   CH₃
H   H   H   H

trans isomer   **cis** isomer

♦ Oxidation–reduction reactions (4.14)

- **Oxidation** results in an **increase in the number of C–Z bonds** or a **decrease in the number of C–H bonds.**

$$CH_3CH_2-OH \longrightarrow CH_3-\overset{\displaystyle O}{\overset{\|}{C}}-OH$$

ethanol   acetic acid

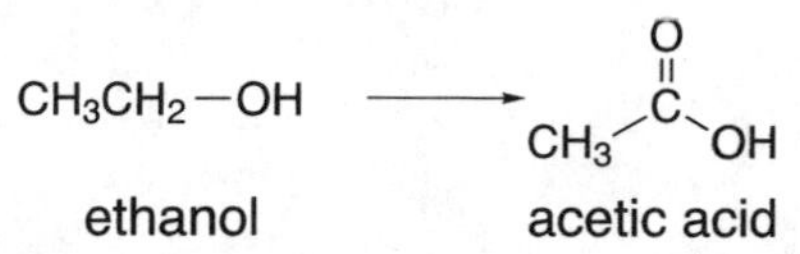

Increase in C–O bonds = **oxidation.**

- **Reduction** results in a **decrease in the number of C–Z bonds** or an **increase in the number of C–H bonds.**

$$\overset{\displaystyle H \qquad H}{\underset{\displaystyle H \qquad H}{C=C}} \longrightarrow H-\overset{\displaystyle H}{\underset{\displaystyle H}{C}}-\overset{\displaystyle H}{\underset{\displaystyle H}{C}}-H$$

ethylene   ethane

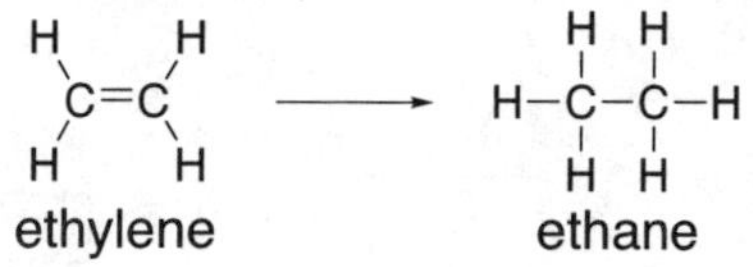

Increase in C–H bonds = **reduction.**

Chapter 4: Answers to Problems

4.1 The general molecular formula for an acyclic alkane is C_nH_{2n+2}.

a. $C_{12}H_{26}$	b. C_8H_{16}	c. $C_{30}H_{64}$
$2n + 2 = $ # H's	$2n + 2 = $ # H's	$2n + 2 = $ # H's
$2(12) + 2 = 26$	$2(8) + 2 = 18$	$2(30) + 2 = 62$
yes	**no**	**no**

4.2 Butane has 4 C's in a row. **Isobutane** has 3 C's in a row with a 1 C branch.

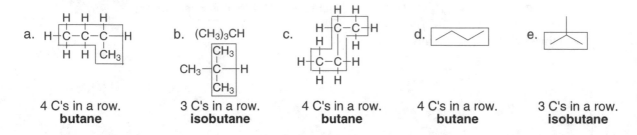

4 C's in a row.	3 C's in a row.	4 C's in a row.	4 C's in a row.	3 C's in a row.
butane	**isobutane**	**butane**	**butane**	**isobutane**

4.3 Isopentane has 4 C's in a row with a 1 C branch.

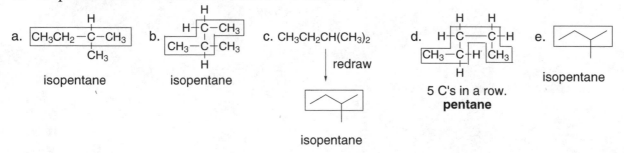

| isopentane | isopentane | c. $CH_3CH_2CH(CH_3)_2$ | 5 C's in a row. **pentane** | isopentane |

redraw → isopentane

4.4 To classify a carbon atom as 1°, 2°, 3° or 4° **determine how many carbon atoms it is bonded to** (**1° C** = bonded to **one** other C, **2° C** = bonded to **two** other C's, **3° C** = bonded to **three** other C's, **4° C** = bonded to **four** other C's). Redraw if necessary to see each carbon clearly.

a. $\overset{\text{1°C}}{\downarrow}$ $CH_3CH_2CH_2CH_3$ $\overset{\text{1°C}}{\downarrow}$
    $\underset{\text{2°C's}}{\uparrow \quad \uparrow}$

b. $(CH_3)_3CH$

1°C's ┤ $CH_3-\underset{CH_3}{\overset{CH_3}{C}}-H$ 3°C

c. **4°C's** All other C's are **1°C's.**

d. **1°C's** **4°C** **1°C** **3°C** All other C's are **2°C's.**

4.5 To classify a hydrogen atom as 1°, 2°, or 3°, **determine if it is bonded to a 1°, 2° or 3° C (a 1° H is bonded to a 1° C, a 2° H is bonded to a 2° C, a 3° H is bonded to a 3° C).** Redraw if necessary.

a. $CH_3CH_2CH_3$ 1° H, 1° H, 2° H's

b. $CH_3CH_2CH(CH_3)C(CH_3)_3$ redraw

c.   redraw   2° H's, 3° H, 3° H's, 2° H's, All other H's are 1° H's.

4.6 Constitutional isomers differ in the way the atoms are connected to each other. To draw all the constitutional isomers:

[1] Draw all of the C's in a long chain.

[2] Take off one C and use it as a substituent. (Don't add it to the end carbon: this re-makes the long chain.)

[3] Take off two C's and use these as substituents, etc.

Five **constitutional isomers** of molecular formula C_6H_{14}

[1] long chain

[2] with one C as a substituent

[3] using 2 C's as substituents

$CH_3CH_2CH_2CH_2CH_2CH_3$

4.7

Molecular formula C_8H_{18} with one CH_3 substituent.

4.8 Use the steps from Answer 4.6 to draw the constitutional isomers.

Five **constitutional isomers** of molecular formula C_5H_{10} having one ring

4.9 Cycloalkanes have molecular formula C_nH_{2n}. For a cycloalkane with 288 C's, there would be $2(288) = 576$ H's. **Molecular formula = $C_{288}H_{576}$.**

4.10 Follow these steps to name an alkane:

[1] **Name the parent chain** by finding the longest C chain.

[2] **Number the chain** so that the first substituent gets the lower number. Then **name and number all substituents**, giving like substituents a prefix (di, tri, etc.).

[3] **Combine all parts**, alphabetizing the substituents, ignoring all prefixes except *iso*.

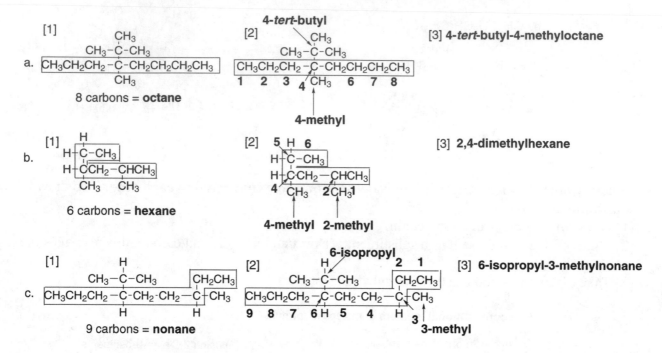

4.11 Use the steps in Answer 4.10 to name each alkane.

a. CH₃CH₂CH(CH₃)CH₂CH₃

b. (CH₃)₃CCH₂CH(CH₂CH₃)₂

c. CH₃(CH₂)₃CH(CH₂CH₂CH₃)CH(CH₃)₂

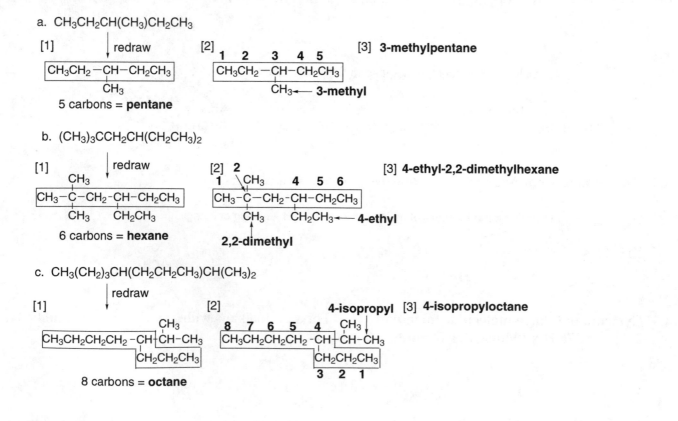

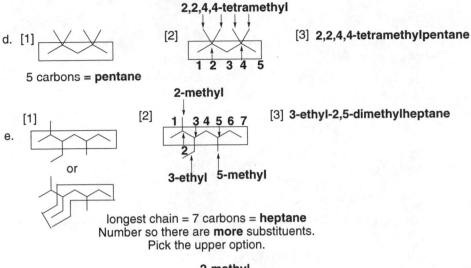

d. [1] 5 carbons = **pentane**

[2] **2,2,4,4-tetramethyl**
1 2 3 4 5

[3] **2,2,4,4-tetramethylpentane**

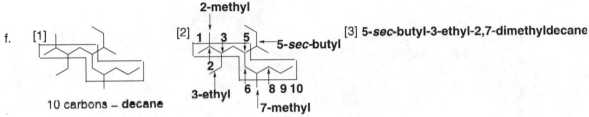

e. [1]

[2] **2-methyl**
1 3 4 5 6 7
2
3-ethyl **5-methyl**

[3] **3-ethyl-2,5-dimethylheptane**

or

longest chain = 7 carbons = **heptane**
Number so there are **more** substituents.
Pick the upper option.

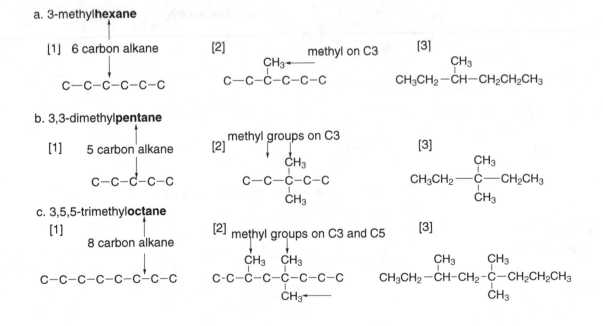

f. [1]

10 carbons – **decane**

[2] **2-methyl**
1 3 5 **5-sec-butyl**
2
3-ethyl 6 8 9 10
7-methyl

[3] **5-sec-butyl-3-ethyl-2,7-dimethyldecane**

4.12 To work backwards from a name to a structure:

[1] Find the parent name and draw that number of C's. Use the suffix to identify the functional group. (**-ane = alkane**)

[2] Arbitrarily number the C's in the chain. Add the substituents to the appropriate C's.

[3] Redraw with H's to make C's have four bonds.

a. 3-methyl**hexane**

[1] 6 carbon alkane

C—C—C—C—C—C

[2] methyl on C3

CH₃
C—C—C—C—C—C

[3]

CH₃
CH₃CH₂—CH—CH₂CH₂CH₃

b. 3,3-dimethyl**pentane**

[1] 5 carbon alkane

C—C—C—C—C

[2] methyl groups on C3

CH₃
C—C—C—C—C
CH₃

[3]

CH₃
CH₃CH₂—C—CH₂CH₃
CH₃

c. 3,5,5-trimethyl**octane**

[1] 8 carbon alkane

C—C—C—C—C—C—C—C

[2] methyl groups on C3 and C5

CH₃ CH₃
C-C—C—C—C—C—C—C
CH₃

[3]

CH₃   CH₃
CH₃CH₂—CH-CH₂-C—CH₂CH₂CH₃
CH₃

d. 3-ethyl-4-methyl**hexane**

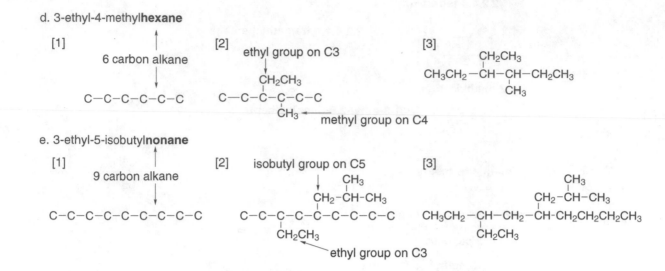

[1]

6 carbon alkane

C—C—C—C—C—C

[2] ethyl group on C3

CH₂CH₃

C—C—C—C—C—C

CH₃ ← methyl group on C4

[3]

CH₂CH₃

CH₃CH₂—CH—CH—CH₂CH₃

CH₃

e. 3-ethyl-5-isobutyl**nonane**

[1]

9 carbon alkane

C—C—C—C—C—C—C—C—C

[2] isobutyl group on C5

CH₃

CH₂—CH—CH₃

C—C—C—C—C—C—C—C—C

CH₂CH₃

ethyl group on C3

[3]

CH₃

CH₂—CH—CH₃

CH₃CH₂—CH—CH₂—CH-CH₂CH₂CH₂CH₃

CH₂CH₃

4.13 Use the steps in Answer 4.10 to name each alkane.

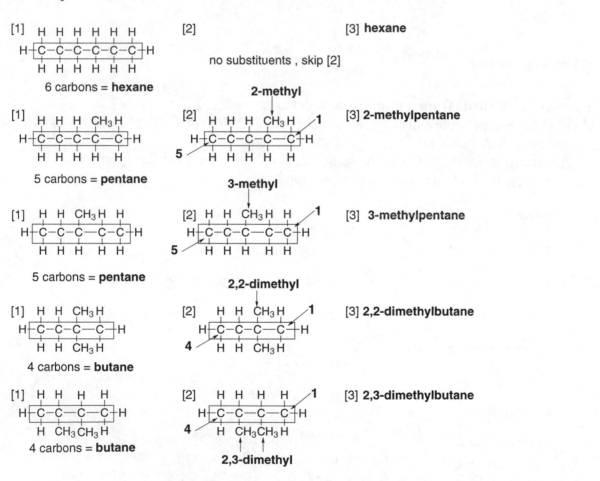

[1]
H H H H H H
H┤C–C–C–C–C–C├H
H H H H H H
6 carbons = **hexane**

[2]
no substituents , skip [2]

[3] **hexane**

[1]
H H H CH₃ H
H┤C–C–C–C—C├H
H H H H H
5 carbons = **pentane**

[2] **2-methyl**
H H H CH₃ H   1
H┤C–C–C–C—C├H
5   H H H H H

[3] **2-methylpentane**

[1]
H H CH₃ H H
H┤C–C–C—C–C├H
H H H H H
5 carbons = **pentane**

[2] **3-methyl**
H H CH₃ H H   1
H┤C–C–C—C–C├H
5 H H H H H

[3] **3-methylpentane**

[1]
H H CH₃ H
H┤C–C–C—C├H
H H CH₃ H
4 carbons = **butane**

[2] **2,2-dimethyl**
H H CH₃ H   1
H┤C–C–C—C├H
4 H H CH₃ H

[3] **2,2-dimethylbutane**

[1]
H H H H
H┤C–C–C—C├H
H CH₃ CH₃ H
4 carbons = **butane**

[2]
H H H H   1
H┤C–C–C—C├H
4 H CH₃ CH₃ H
2,3-dimethyl

[3] **2,3-dimethylbutane**

4.14 Follow these steps to name a cycloalkane:

    [1] Name the parent cycloalkane by counting the C's in the ring and adding cyclo-.

    [2] Numbering:

        [2a] Number around the ring beginning at a substituent and giving the second substituent the lower number.

        [2b] Number to assign the lower number to the substituents alphabetically.

        [2c] Name and number all substituents, giving like substituents a prefix (di, tri, etc.).

   [3] Combine all parts, alphabetizing the substituents, ignoring all prefixes except *iso*.

    (Remember: If a carbon chain has more C's than the ring, the chain is the parent, and the ring is a substituent.)

a.

[1]

6 carbons in ring =
cyclohexane

[2] **1,1-dimethyl**

Number so the
substituents are at C1.

[3] **1,1-dimethylcyclohexane**

b.

[1]

5 carbons in ring =
cyclopentane

[2] **1,2,3-trimethyl**

Number so the first substituent
is at C1, second at C2.

[3] **1,2,3-trimethylcyclopentane**

c.

[1]

6 carbons in ring =
cyclohexane

[2] **1-butyl**

4-methyl

Number so the earliest alphabetical
substituent is at C1, **b**utyl before **m**ethyl.

[3] **1-butyl-4-methylcyclohexane**

d.

[1]

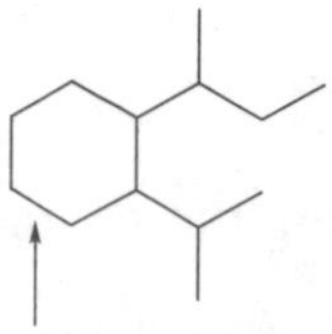

6 carbons in ring =
cyclohexane

[2]

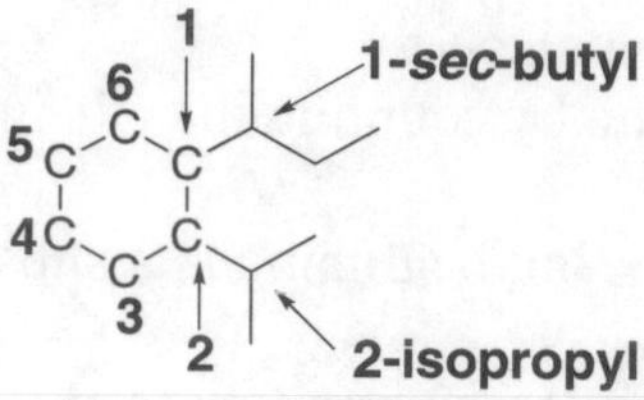

1-*sec*-butyl

2-isopropyl

Number so the earliest alphabetical
substituent is at C1, **butyl** before **isopropyl**.

[3] **1-*sec*-butyl-2-isopropylcyclohexane**

e.

[1]

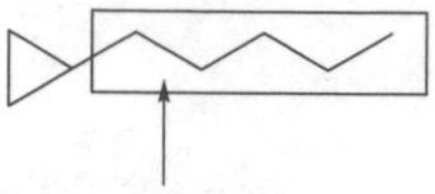

longest chain =
5 carbons =
pentane

[2]

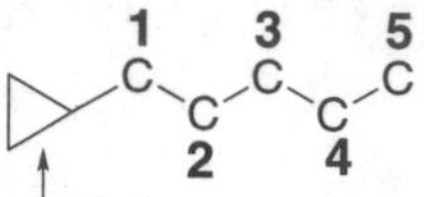

1-cyclopropyl

Number so the
cyclopropyl is at C1.

[3] **1-cyclopropylpentane**

4.15 To draw the structures, use the steps in Answer 4.12.

a. 1,2-dimethyl**cyclobutane**

[1] 4 carbon cycloalkane

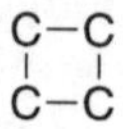

C—C
C—C

[2]

methyl groups
on C1 and C2

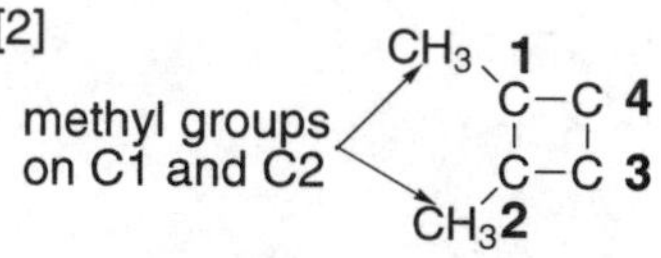

[3]

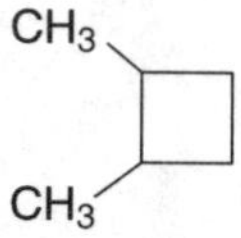

b. 1,1,2-trimethyl**cyclopropane**

[1] 3 carbon cycloalkane

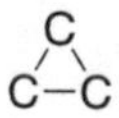

[2]

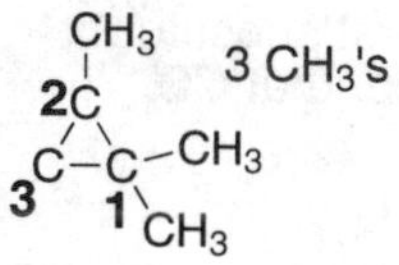

3 CH₃'s

[3]

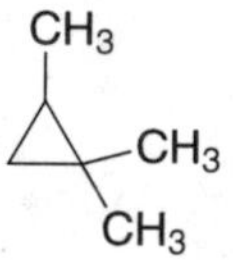

c. 4-ethyl-1,2-dimethyl**cyclohexane**

[1] 6 carbon cycloalkane

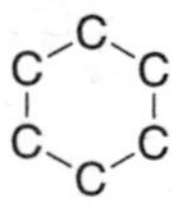

[2]

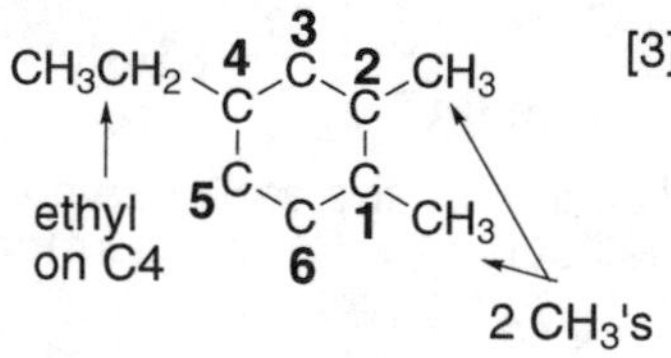

ethyl
on C4

2 CH₃'s

[3]

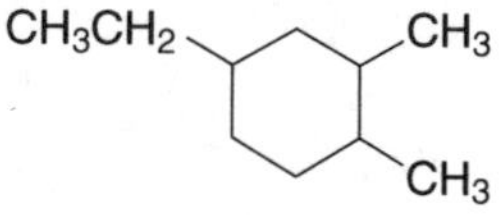

d. 1-*sec*-butyl-3-isopropyl**cyclopentane**

[1] 5 carbon cycloalkane

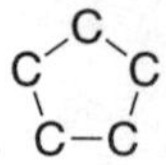

[2]

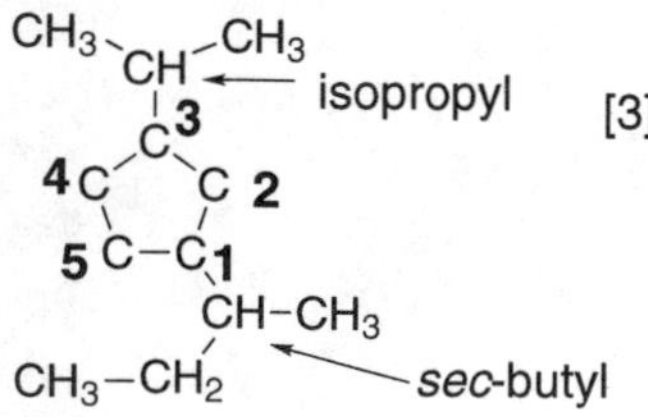

isopropyl

sec-butyl

[3]

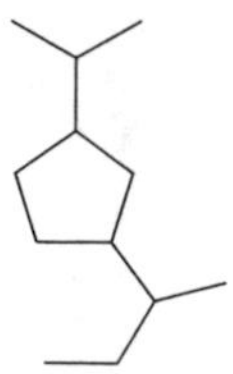

e. 1,1,2,3,4-pentamethyl**lcycloheptane**

[1] 7 carbon cycloalkane   [2] 5 CH₃'s

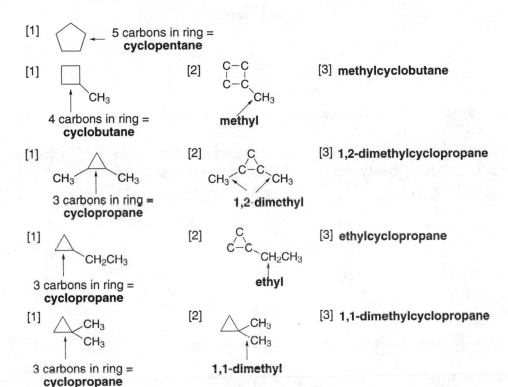

4.16 To name the cycloalkanes, use the steps from Answer 4.14.

[1]   5 carbons in ring = **cyclopentane**

[1]   4 carbons in ring = **cyclobutane**   [2] **methyl**   [3] **methylcyclobutane**

[1]   3 carbons in ring = **cyclopropane**   [2] **1,2-dimethyl**   [3] **1,2-dimethylcyclopropane**

[1]   3 carbons in ring = **cyclopropane**   [2] **ethyl**   [3] **ethylcyclopropane**

[1]   3 carbons in ring = **cyclopropane**   [2] **1,1-dimethyl**   [3] **1,1-dimethylcyclopropane**

4.17 Compare the molecular weights to determine relative boiling points.

gasoline: C_5H_{12} - $C_{12}H_{26}$

lowest molecular weight: **lowest boiling point**

kerosene: $C_{12}H_{26}$ - $C_{16}H_{34}$

middle molecular weight: **intermediate boiling point**

diesel fuel: $C_{15}H_{32}$ - $C_{18}H_{38}$

highest molecular weight: **highest boiling point**

4.18 **Compare the number of C's and surface area to determine relative boiling points**. Rules:
   [1] Increasing number of C's = increasing boiling point.
   [2] Increasing surface area = increasing boiling point (branching decreases surface area).

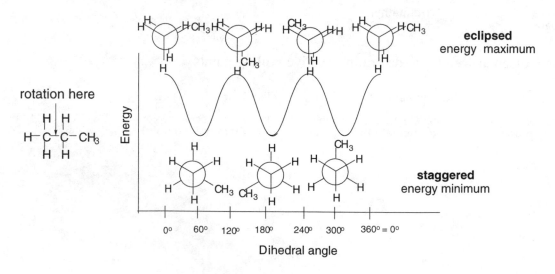

CH₃(CH₂)₆CH₃

8 C's
linear
largest number of C's
no branching
highest bp

CH₃(CH₂)₅CH₃

7 C's
linear

CH₃CH₂CH₂CH₂CH(CH₃)₂

7 C's
one branch

(CH₃)₃CCH(CH₃)₂

7 C's
three branches

increasing branching
decreasing surface area
decreasing bp

increasing boiling point: (CH₃)₃CCH(CH₃)₂ < CH₃CH₂CH₂CH₂CH(CH₃)₂ < CH₃(CH₂)₅CH₃ < CH₃(CH₂)₆CH₃

4.19 To draw a Newman projection, visualize the carbons as one in front and one in back of each other. The C–C bond is not drawn. There is only one staggered and one eclipsed conformation.

rotation here

H–C–C–Br

C in front   C behind

Br

60°

1
staggered

2
eclipsed

4.20 Staggered conformations are more stable than eclipsed conformations.

rotation here

H–C–C–CH₃

eclipsed
energy maximum

Energy

staggered
energy minimum

0°   60°   120°   180°   240°   300°   360° = 0°

Dihedral angle

4.21

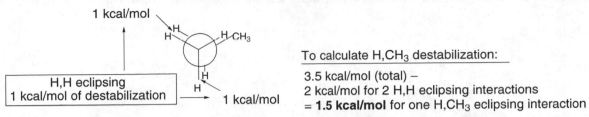

To calculate H,CH$_3$ destabilization:

3.5 kcal/mol (total) −
2 kcal/mol for 2 H,H eclipsing interactions
= **1.5 kcal/mol** for one H,CH$_3$ eclipsing interaction

4.22 To determine the energy of conformers keep two things in mind:
[1] Staggered conformers are more stable than eclipsed.
[2] Minimize steric interactions: keep large groups away from each other.
The highest energy conformation is the eclipsed conformation in which the two largest groups are eclipsed. The lowest energy conformation is the staggered conformation in which the two largest groups are anti.

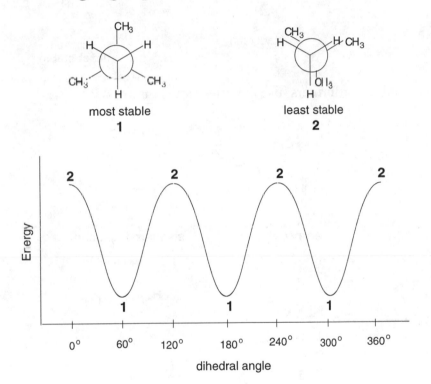

4.23 To determine the most and least stable conformers, use the rules from Answer 4.22.

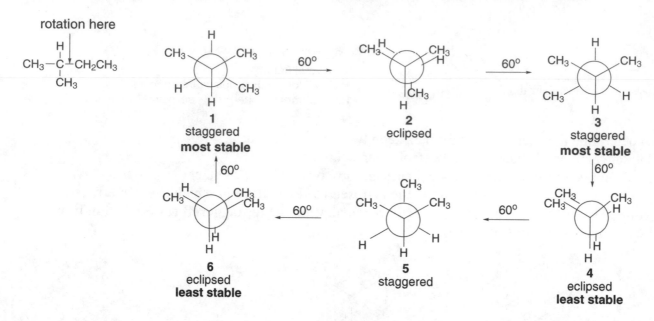

rotation here

CH₃—C—CH₂CH₃
|
CH₃

1
staggered
most stable

2
eclipsed

3
staggered
most stable

6
eclipsed
least stable

5
staggered

4
eclipsed
least stable

4.24 To determine the most and least stable conformers, use the rules from Answer 4.22.

1,2-dichloroethane

ClCH₂—CH₂Cl

rotation here

1
staggered, anti

2
eclipsed

3
staggered, gauche

6
eclipsed

5
staggered, gauche

4
eclipsed

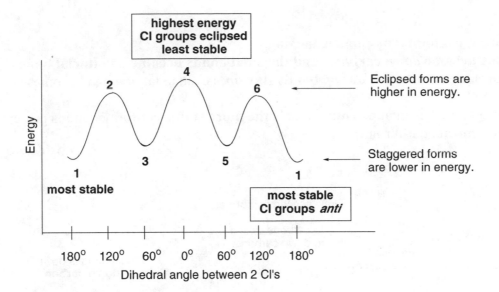

4.25 Add the energy increase for each eclipsing interaction to determine the destabilization.

a.

1 H,H interaction =	1 kcal/mol
2 H,CH₃ interactions	
(2 x 1.5 kcal/mol) =	3 kcal/mol

Total destabilization = 4 kcal/mol

b.

3 H,CH₃ interactions
(3 x 1.5 kcal/mol) =   **4.5 kcal/mol**

Total destabilization

4.26 Two points:
- Axial bonds point up or down, while equatorial bonds point out.
- An *up* carbon has an axial *up* bond, and a *down* carbon has an axial *down* bond.

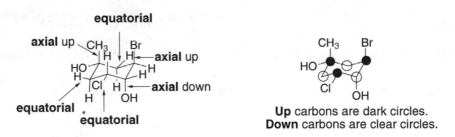

Up carbons are dark circles.
Down carbons are clear circles.

4.27 Draw the second chair conformer by flipping the ring.

- **The *up* carbons become *down* carbons, and the axial bonds become equatorial bonds.**
- **Axial bonds become equatorial, but *up* bonds stay *up*;** i.e. an axial *up* bond becomes an equatorial *up* bond.
- The conformer with **larger groups equatorial is the more stable** conformer and is present in higher concentration at equilibrium.

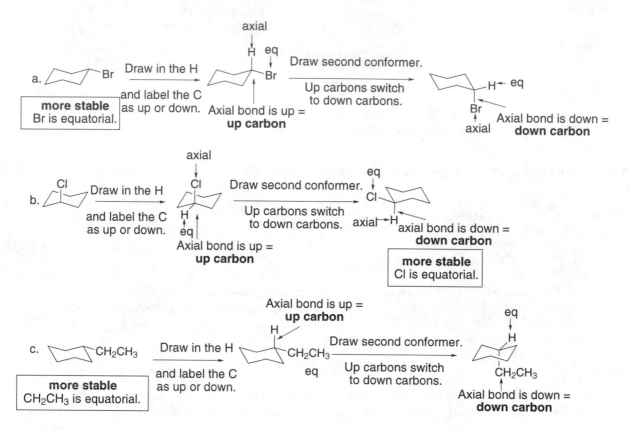

4.28 Wedges represent "up" groups in front of the page, and dashes are "down" groups in back of the page.

Cis groups are on the same side of the ring, and trans groups are on opposite sides of the ring.

cis-1,2-dimethylcyclopropane

CH₃   CH₃   or   CH₃   ''CH₃

cis = same side of the ring
both groups on wedges or
both on dashes

trans-1-ethyl-2-methylcyclopentane

CH₃CH₂   CH₃   or   CH₃CH₂   CH₃

trans = opposite sides of the ring
one group on a wedge,
one group on a dash

4.29 To classify a compound as a cis or trans isomer, **classify each non-hydrogen group as up or down. Groups on the same side = cis isomer, groups on opposite sides = trans isomer.**

a.

down bond (equatorial)

down bond (equatorial)

both groups down =
cis isomer

b.

down bond (equatorial)

up bond (equatorial)

one group up, one down =
trans isomer

c.

up bond (axial)

down bond (equatorial)

one group up, one down =
trans isomer

4.30

both groups equatorial
more stable

4.31

a.

groups on same side
cis isomer

groups on opposite sides
trans isomer

b. cis:

two chair conformers for the **cis isomer**

Same stability since they are identical groups with one equatorial, one axial.

c. trans:

both groups equatorial
more stable
two chair conformers for the **trans isomer**

d. The **trans isomer is more stable** because it can have both methyl groups in the more roomy **equatorial** position.

4.32 *Oxidation* results in an *increase* in the number of C–Z bonds, or a *decrease* in the number of C–H bonds.

Reduction results in a *decrease* in the number of C–Z bonds, or an *increase* in the number of C–H bonds.

a.

Decrease in the number of C–H bonds.
Increase in the number of C–O bonds.
Oxidation

b.

Decrease in the number of C–O bonds.
Increase in the number of C–H bonds.
Reduction

c.

No change in the number of C–O or C–H bonds. **Neither**

d.

Decrease in the number of C–O bonds.
Increase in the number of C–H bonds.
Reduction

4.33 The products of a combustion reaction of a hydrocarbon are always the same: **CO_2 and H_2O.**

a. $CH_3CH_2CH_3$ + $5 O_2$ $\xrightarrow{\text{flame}}$ $3 CO_2$ + $4 H_2O$ + heat

b.   ⬡   + $9 O_2$ $\xrightarrow{\text{flame}}$ $6 CO_2$ + $6 H_2O$ + heat

4.34 **Lipids contain many nonpolar C–C and C–H bonds and few polar functional groups.**

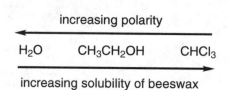

a. $CH_3(CH_2)_7CH=CH(CH_2)_7COOH$

oleic acid

only one polar functional group
18 carbons
a lipid

b.

aspartame

many polar functional groups
only 14 carbons
not a lipid

c.

tristearin

three polar functional groups
57 carbons
a lipid

4.35 "Like dissolves like." Beeswax is a lipid, and therefore, it will be more soluble in nonpolar solvents. H_2O is very polar, ethanol is slightly less polar, and chloroform is least polar. Beeswax is most soluble in the least polar solvent.

increasing polarity
⟵————————————

H_2O CH_3CH_2OH $CHCl_3$

————————————⟶
increasing solubility of beeswax

4.36 Use the rules from Answers 4.4 and 4.5.

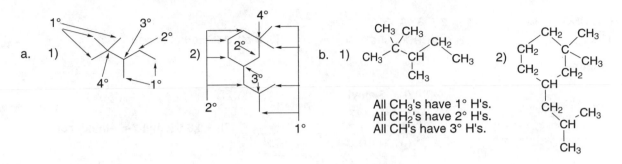

b. 1)

CH₃'s CH₃

CH₃—C—CH with CH₂ CH₃ / CH₃

All CH₃'s have 1° H's.
All CH₂'s have 2° H's.
All CH's have 3° H's.

4.37

One possibility:

a.
$$CH_3-\underset{\underset{CH_3}{|}}{\overset{\overset{CH_3}{|}}{C}}-CH_3$$

b. (hexagon)

c.
H H H H
H—C—C—C—H
    H H

d.
CH₃ CH₃
   CH
   CH₃

4.38

a. Five consitutional isomers of molecular formula C_4H_8:

(square) (methylcyclopropane with CH₃) $CH_3CH=CHCH_3$ $CH_2=CHCH_2CH_3$

CH₂
‖
$CH_3-\overset{}{C}-CH_3$

b. Nine constitutional isomers of molecular formula C_7H_{16}.

$CH_3CH_2CH_2CH_2CH_2CH_2CH_3$

H
$CH_3-\overset{|}{C}-CH_2CH_2CH_2CH_3$
CH₃

H
$CH_3CH_2-\overset{|}{C}-CH_2CH_2CH_3$
CH₃

CH₃
$CH_3-\overset{|}{\underset{|}{C}}-CH_2CH_2CH_3$
CH₃

CH₃
$CH_3CH_2-\overset{|}{C}-CH_2CH_3$
CH₃

H H
$CH_3-\overset{|}{\underset{CH_3}{C}}-\overset{|}{\underset{CH_3}{C}}-CH_2CH_3$

H H
$CH_3-\overset{|}{\underset{CH_3}{C}}-CH_2-\overset{|}{\underset{CH_3}{C}}-CH_3$

H
$CH_3CH_2-\overset{|}{\underset{CH_2CH_3}{C}}-CH_2CH_3$

H CH₃
$CH_3-\overset{|}{\underset{CH_3}{C}}-\overset{|}{\underset{CH_3}{C}}-CH_3$

c. Twelve constitutional isomers of molecular formula C_6H_{12} containing one ring.

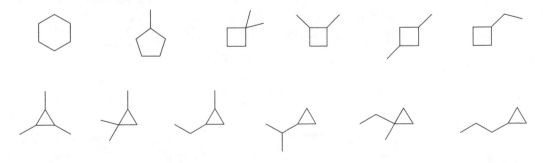

4.39 Use the steps in Answers 4.10 and 4.14 to name the alkanes.

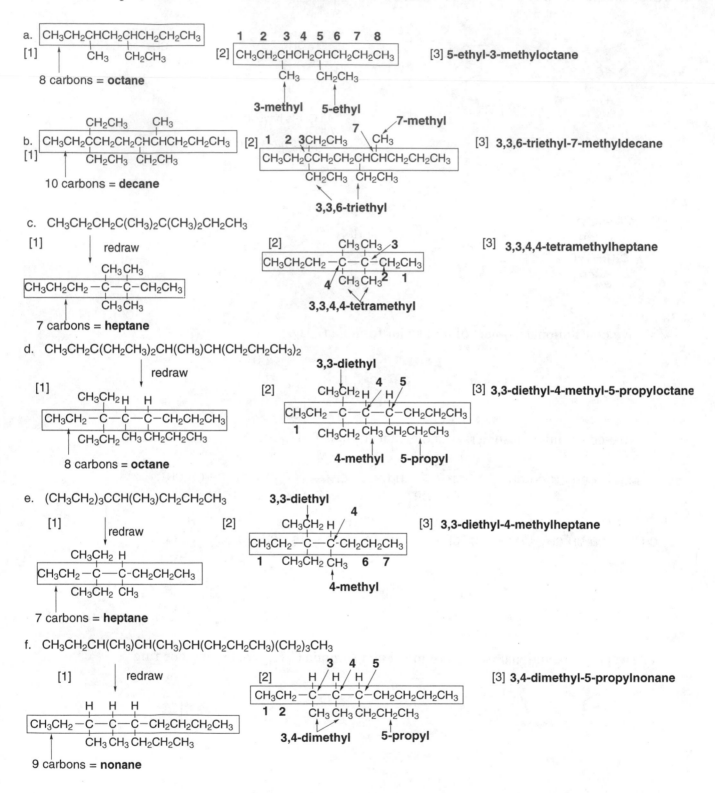

g. (CH$_3$CH$_2$CH$_2$)$_4$C

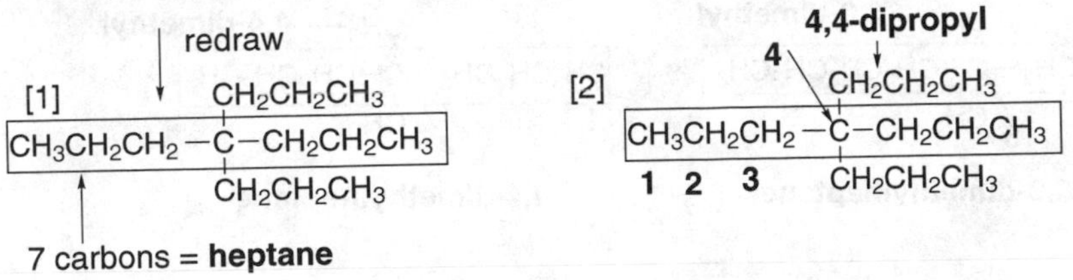

↓ redraw

4,4-dipropyl

[1] CH$_2$CH$_2$CH$_3$
CH$_3$CH$_2$CH$_2$—C—CH$_2$CH$_2$CH$_3$
CH$_2$CH$_2$CH$_3$

7 carbons = **heptane**

[2] 4 CH$_2$CH$_2$CH$_3$
CH$_3$CH$_2$CH$_2$—C—CH$_2$CH$_2$CH$_3$
1   2   3   CH$_2$CH$_2$CH$_3$

[3] **4,4-dipropylheptane**

h. [1]

[2] 1 2 6
3 5 7
3-methyl   **6-isopropyl**

10 carbons =**decane**

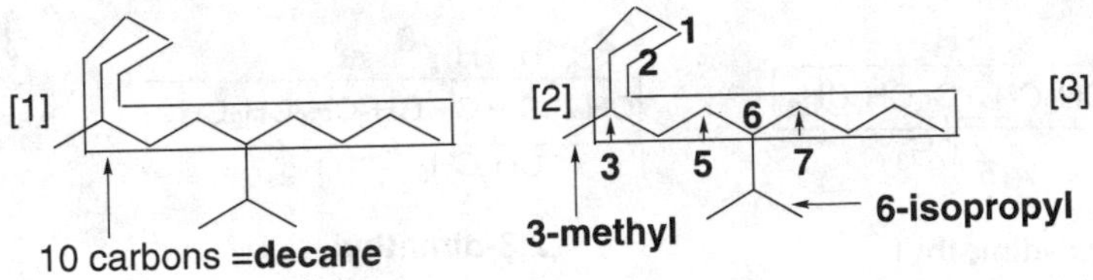

[3] **6-isopropyl-3-methyldecane**

i. [1]

10 carbons = **decane**

[2] 8 6 4 ← **4-isopropyl**
8-ethyl   1
2,6-dimethyl

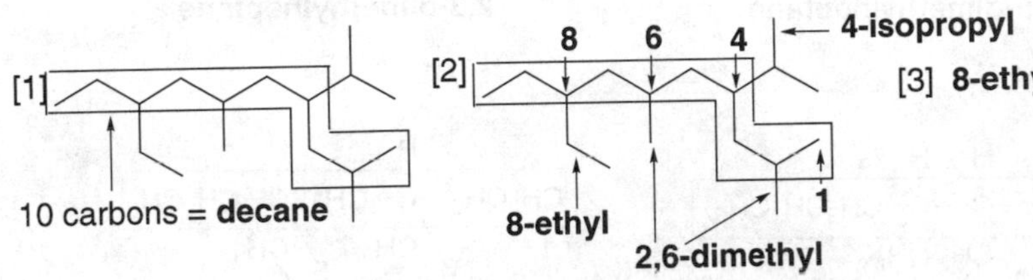

[3] **8-ethyl-4-isopropyl-2,6-dimethyldecane**

j. [1]

8 carbons = **octane**

[2] 4 **4-isopropyl**
1

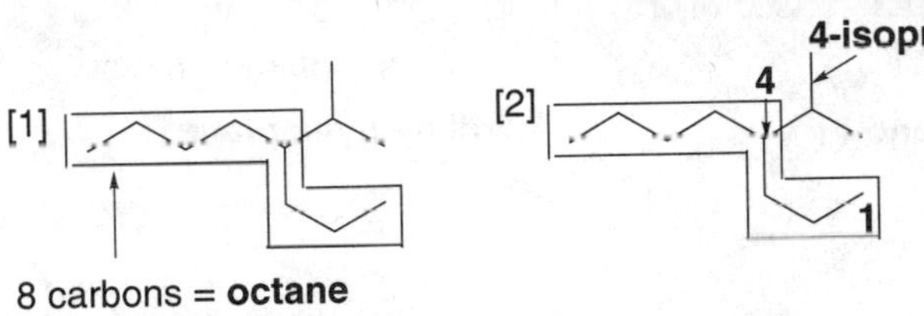

[3] **4-isopropyloctane**

2,2,5-trimethyl

k.

1 2

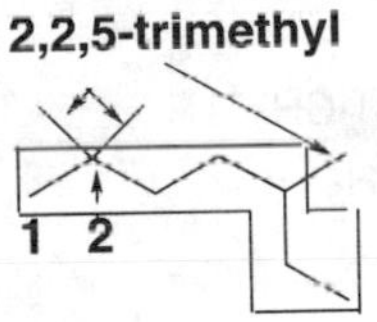

2,2,5-trimethylheptane

l. ◻—CH(CH$_2$CH$_3$)$_2$ =

1
2
3
4
5

3-cyclobutyl

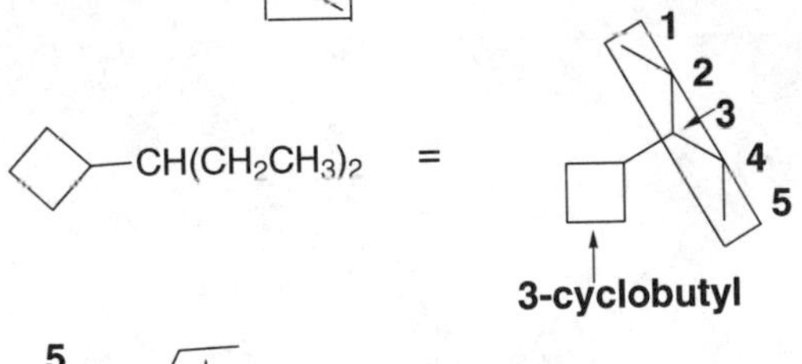

3-cyclobutylpentane

m. 5 1
4   **1-sec-butyl**
2
3
2-isopropyl

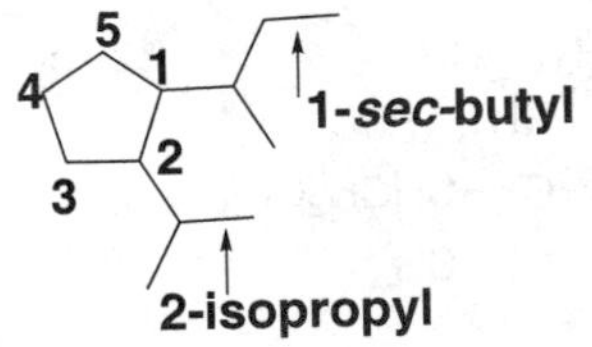

1-sec-butyl-2-isopropylcyclopentane

n. 5
6   4
1   3
2

1-isobutyl   **3-isopropyl**

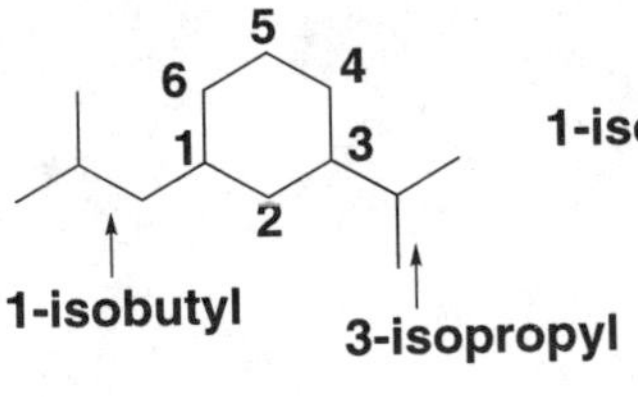

1-isobutyl-3-isopropylcyclohexane

4.40

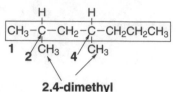

CH₃ — **2,2-dimethyl**
CH₃—C—CH₂CH₂CH₂CH₂CH₃
1 | CH₃
2
2,2-dimethylheptane

CH₃ — **3,3-dimethyl**
CH₃CH₂—C—CH₂CH₂CH₂CH₃
1 3 | CH₃
3,3-dimethylheptane

CH₃ — **4,4-dimethyl**
CH₃CH₂CH₂—C—CH₂CH₂CH₃
1 4 | CH₃
4,4-dimethylheptane

H   H
CH₃—C—CH₂—C—CH₂CH₂CH₃
1 2 CH₃ 4 CH₃
2,4-dimethyl
2,4-dimethylheptane

H   H
CH₃—C—CH₂CH₂—C—CH₂CH₃
1 CH₃ CH₃
2   5
2,5-dimethyl
2,5-dimethylheptane

2 H   H 3
CH₃—C—C—CH₂CH₂CH₂CH₃
1 CH₃CH₃
2,3-dimethyl
2,3-dimethylheptane

H   H
CH₃—C—CH₂CH₂CH₂—C—CH₃
1 2 CH₃ 6 CH₃
2,6-dimethyl
2,6-dimethylheptane

H H
CH₃CH₂—C—C—CH₂CH₂CH₃
1 3 CH₃ CH₃ 4
3,4-dimethyl
3,4-dimethylheptane

H   H
CH₃CH₂—C—CH₂—C—CH₂CH₃
1 3 CH₃ 5 CH₃
3,5-dimethyl
3,5-dimethylheptane

4.41 Use the steps in Answer 4.12 to draw the structures.

a. 3-ethyl-2-methyl**hexane**

[1] 6 C chain
C—C—C—C—C—C

[2] C—C—C—C—C—C
            CH₃ CH₂CH₃
     methyl   ethyl on C3
     on C2

[3] CH₃—C—C—CH₂CH₂CH₃
          | |
          CH₃ CH₂CH₃
     (with H H above the two central carbons)

b. *sec*-butyl**cyclopentane**

[1] 5 C ring

[2]

c. 4-isopropyl-2,4,5-trimethyl**heptane**

[1] 7 C chain
C—C—C—C—C—C—C

[2] isopropyl on C4
     CH₃   CH₃
        \ /
         CH
C—C—C—C—C—C—C
     CH₃ CH₃ CH₃
     methyls on C2, C4, and C5

[3] CH₃ CH₃
        \ /
         CH
CH₃—CH—CH₂—C—CHCH₂CH₃
        CH₃   CH₃ CH₃

d. cyclobutyl**cycloheptane**

[1] 7 C cycloalkane

[2]

e. 3-ethyl-1,1-dimethyl**cyclohexane**

[1] 6 C cycloalkane

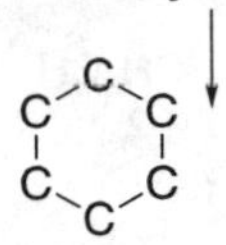

[2]

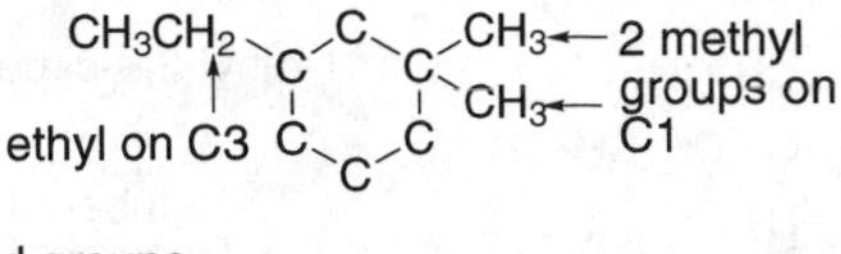

CH₃CH₂—C—C—CH₃ ← 2 methyl
⟍ ⟋   CH₃ ← groups on
ethyl on C3 C—C—C   C1

[3]

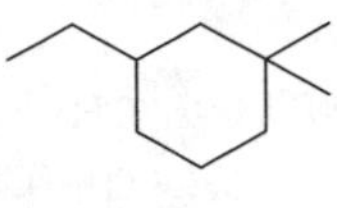

f. 4-butyl-1,1-diethyl**cyclooctane**

[1] 8 C cycloalkane

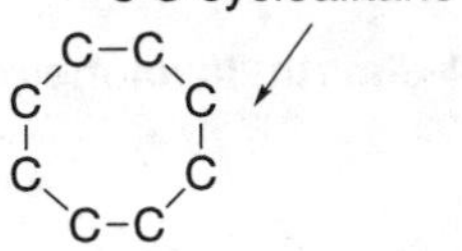

2 ethyl groups

[2]

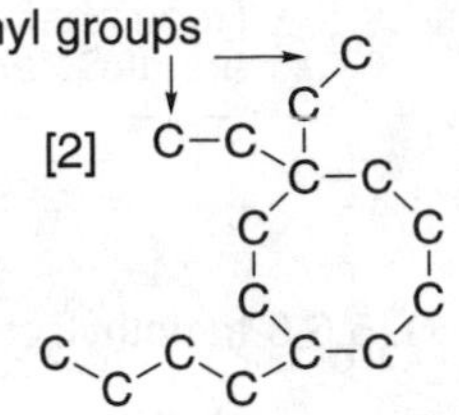

[3]

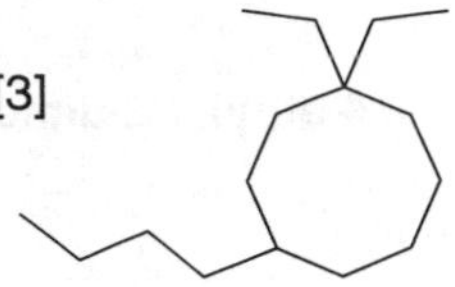

g. 6-isopropyl-2,3-dimethyl**nonane**

[1] 9 C alkane

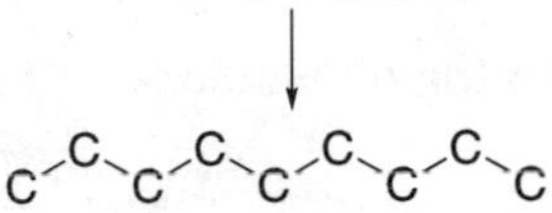

[2]

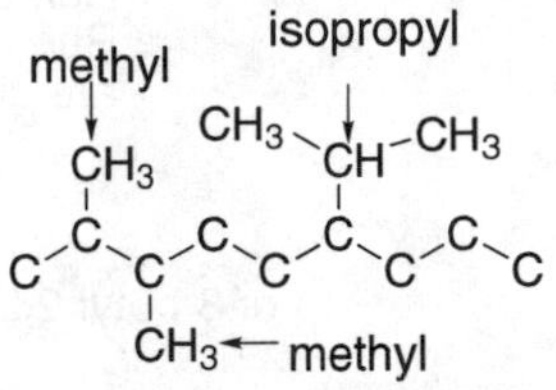

methyl   isopropyl
CH₃   CH₃—CH—CH₃
C—C—C—C—C—C—C
CH₃ ← methyl

5 methyl groups

[3]

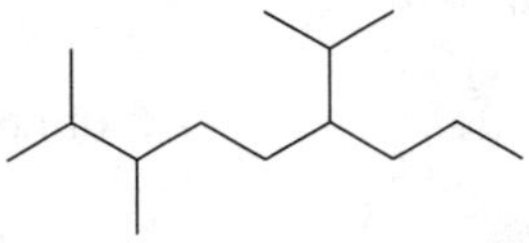

h. 2,2,6,6,7-pentamethyl**octane**

[1] 8 C alkane

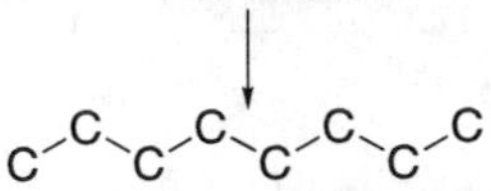

[2] CH₃ CH₃   CH₃ CH₃
C—C—C—C—C—C—C
CH₃

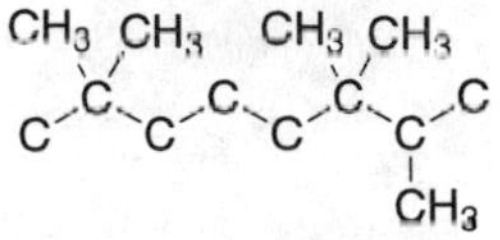

[3]

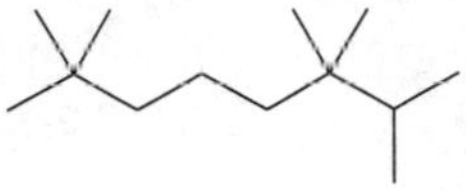

i. *cis*-1-ethyl-3-methyl**cyclopentane**

[1] 5 C ring

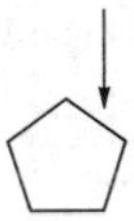

[2]

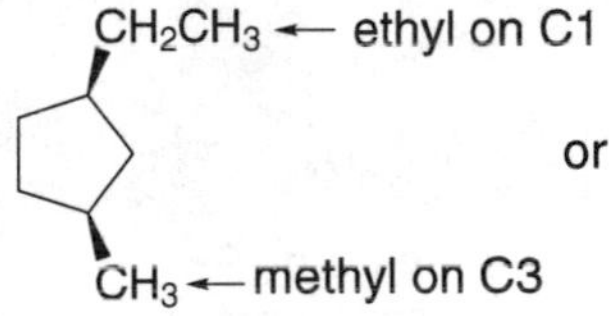

CH₂CH₃ ← ethyl on C1

CH₃ ← methyl on C3

or

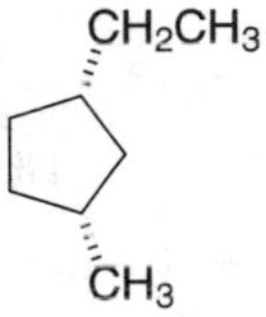

CH₂CH₃

CH₃

j. *trans*-1-*tert*-butyl-4-ethyl**cyclohexane**

[1] 6 C ring

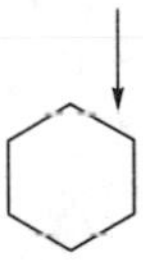

[2]

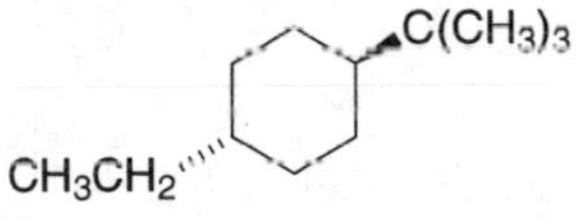

C(CH₃)₃

CH₃CH₂

4.42 Draw the compounds.

a. 2,2-dimethyl-4-ethylheptane

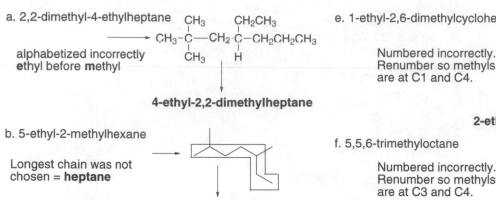

alphabetized incorrectly
ethyl before **m**ethyl

4-ethyl-2,2-dimethylheptane

b. 5-ethyl-2-methylhexane

Longest chain was not
chosen = **heptane**

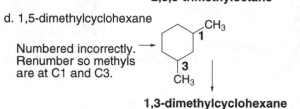

2,5-dimethylheptane

c. 2-methyl-2-isopropylheptane

longest chain was not
chosen = **octane**

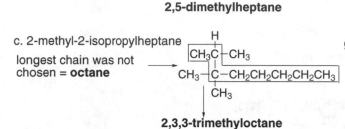

2,3,3-trimethyloctane

d. 1,5-dimethylcyclohexane

Numbered incorrectly.
Renumber so methyls
are at C1 and C3.

1,3-dimethylcyclohexane

e. 1-ethyl-2,6-dimethylcycloheptane

Numbered incorrectly.
Renumber so methyls
are at C1 and C4.

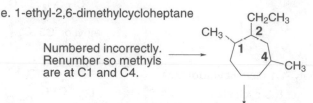

2-ethyl-1,4-dimethylcycloheptane

f. 5,5,6-trimethyloctane

Numbered incorrectly.
Renumber so methyls
are at C3 and C4.

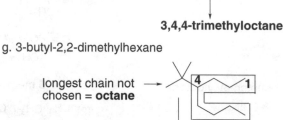

3,4,4-trimethyloctane

g. 3-butyl-2,2-dimethylhexane

longest chain not
chosen = **octane**

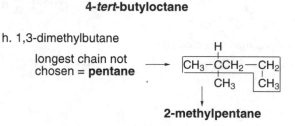

4-*tert*-butyloctane

h. 1,3-dimethylbutane

longest chain not
chosen = **pentane**

2-methylpentane

4.43 Use the rules from Answer 4.18.

a. $CH_3CH_2CH_3$, $CH_3CH_2CH_2CH_3$, $CH_3CH_2CH_2CH_2CH_3$
    3 C's   4 C's   5 C's
lowest boiling point   **highest boiling point**

b. $(CH_3)_2CHCH(CH_3)_2$, $CH_3CH_2CH_2CH(CH_3)_2$, $CH_3(CH_2)_4CH_3$
    most branching   least branching
lowest boiling point   **highest boiling point**

4.44

$CH_3(CH_2)_6CH_3$
no branching = higher surface area
higher boiling point

$(CH_3)_3C(CH_3)_3$
branching = lower surface area
lower boiling point
more spherical, better packing =
higher melting point

4.45

a.   and

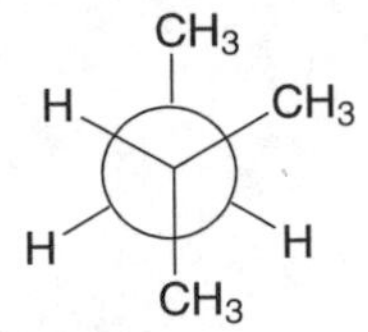

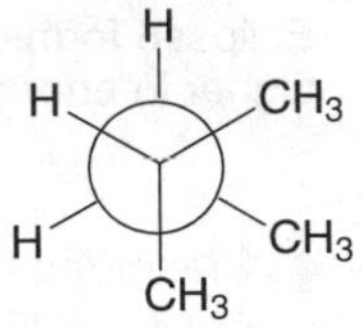

1 gauche CH_3,CH_3
= 0.9 kcal/mol
of destabilization

higher energy
2 gauche CH_3,CH_3
0.9 kcal/mol x 2 = 1.8 kcal/mol
of destabilization

Energy difference =
1.8 kcal/mol – 0.9 kcal/mol = | **0.9 kcal/mol** |

b.   and

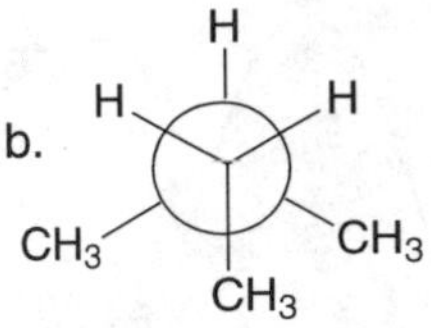

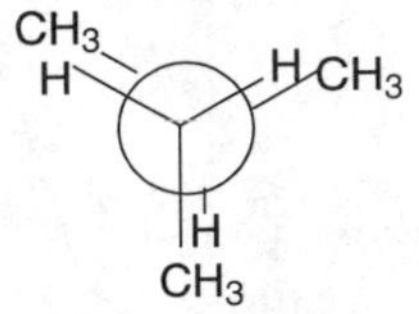

2 gauche CH_3,CH_3
0.9 kcal/mol x 2 =
1.8 kcal/mol
of destabilization

higher energy
3 eclipsed H,CH_3
1.5 kcal/mol x 3 = 4.5 kcal/mol
of destabilization

Energy difference =
4.5 kcal/mol – 1.8 kcal/mol = | **2.7 kcal/mol** |

4.46 Use the rules from Answer 4.22 to determine the most and least stable conformers.

a. $CH_3-CH_2CH_2CH_2CH_3$

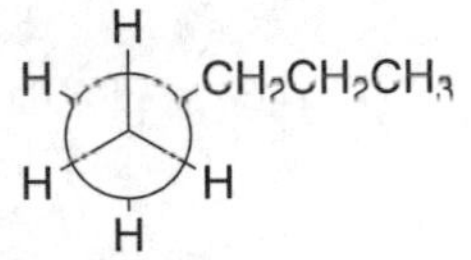

staggered
most stable

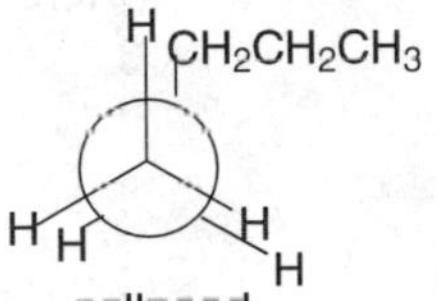

eclipsed
least stable

All staggered conformers are equal in energy.
All eclipsed conformers are equal in energy.

b. $CH_3CH_2CH_2-CH_2CH_2CH_3$

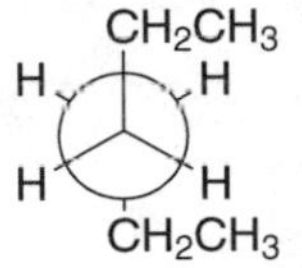

staggered
ethyl groups anti
most stable

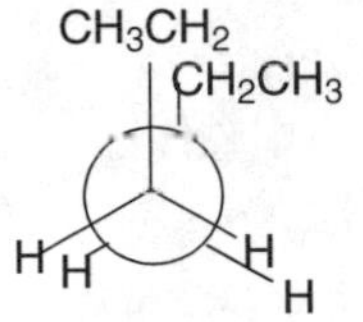

eclipsed
ethyl groups eclipsed
least stable

4.47

(1) $CH_3CH_2-CH_2CH_2CH_3$

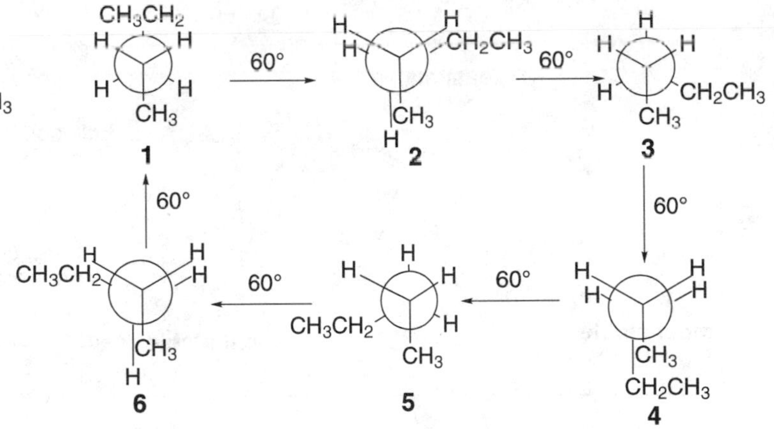

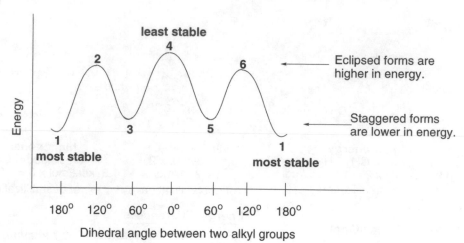

(2) $CH_3CH_2-\underset{\underset{CH_3}{|}}{C}HCH_2CH_3$

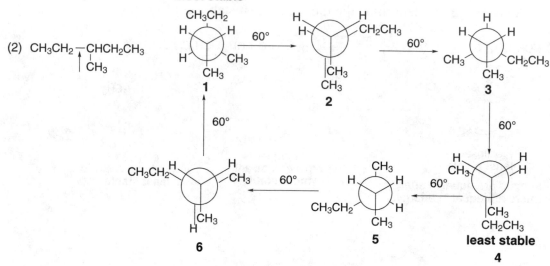

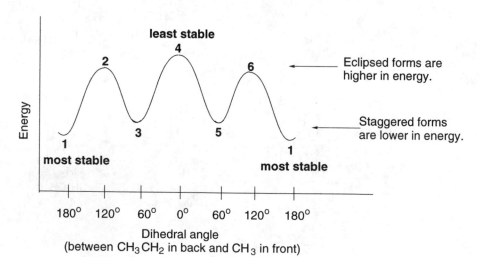

4.48 Two types of strain:
- *Torsional strain* is due to eclipsed groups on adjacent carbon atoms.
- *Steric strain* is due to overlapping electron clouds of large groups (ex: gauche interactions).

a.

two sites
three bulky methyl groups close =
steric strain

b.

eclisped conformation=
torsional strain

c.

two bulky ethyl groups close =
steric strain
eclipsed conformation =
torsional strain

4.49 The barrier to rotation is equal to the difference in energy between the highest energy eclipsed and lowest energy staggered conformations of the molecule.

a. $CH_3\!-\!CH(CH_3)_2$

most stable   least stable

Destabilization energy =

2 H,CH$_3$ eclipsing interactions
$\qquad$ 2(1.5 kcal/mol) = $\quad$ 3 kcal/mol
1 H,H eclipsing interaction = $\quad$ 1 kcal/mol

Total destabilization = **4 kcal/mol**

| 4 kcal/mol = rotation barrier |

b. $CH_3\!-\!C(CH_3)_3$

most stable   least stable

Destabilization energy =

3 H,CH$_3$ eclipsing interactions
$\qquad$ 3(1.5 kcal/mol) = $\quad$ 4.5 kcal/mol

Total destabilization = **4.5 kcal/mol**

| 4.5 kcal/mol = rotation barrier |

4.50

most stable   least stable

2 H,H eclipsing interactions = 2(1 kcal/mol) = 2 kcal/mol

Since the barrier to rotation is 3.7 kcal/mol, the difference between this value and the destabilization due to H,H eclipsing is the destabilization due to H,Cl eclipsing.

| 3.7 kcal/mol - 2 kcal/mol = **1.7 kcal/mol** |
| **destabilization due to H,Cl eclipsing** |

4.51 The gauche conformer can intramolecularly hydrogen bond, making it the more stable conformer.

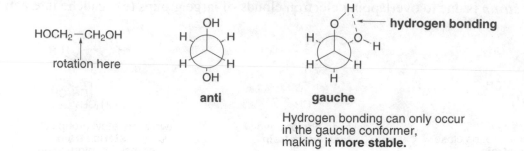

$HOCH_2—CH_2OH$

rotation here

anti

gauche

← **hydrogen bonding**

Hydrogen bonding can only occur
in the gauche conformer,
making it **more stable.**

4.52

(1)

[a] **axial** H OH **axial** HO **eq** H **eq**

[b] H **up** OH **down** HO H

one up, one down =
trans

[c] HO,,, OH

[d] ax H ax OH eq HO H eq ⇌ H eq eq HO OH ax H ax

(2)

[a] **axial** Br H **eq** CH₃ **eq** H **axial**

[b] **up** Br H CH₃ **up** H

both up =
cis

[c] Br CH₃

[d] ax Br H eq CH₃ eq H ax ⇌ ax CH₃ eq Br H eq H ax

(3)

[a] **axial** H HO OH **eq** **eq** H **axial**

[b] up Br HO **up** H OH **down** H

one up, one down =
trans

[c] OH HO

[d] ax H eq HO OHeq H ax ⇌ ax OH eq H H eq OH ax

4.53 A **cis isomer** has two groups on the **same side** of the ring. The two groups can be drawn both up or both down. Only one possibility is drawn. A **trans isomer** has one group on one side of the ring and one group on the other side. Either group can be drawn on either side. Only one possibility is drawn.

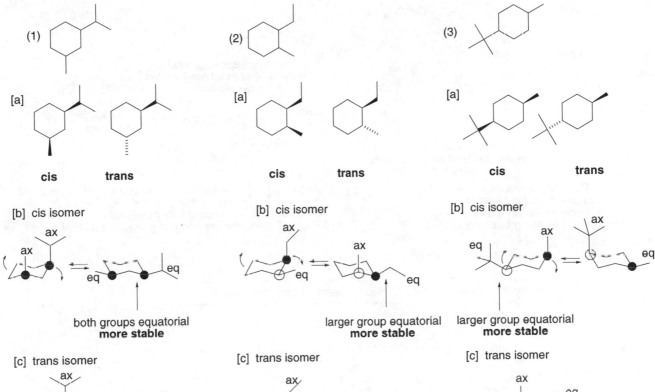

(1)

(2)

(3)

[a]

cis   trans

cis   trans

cis   trans

[b] cis isomer

ax

ax

eq

both groups equatorial
more stable

[b] cis isomer

ax

ax

eq

eq

larger group equatorial
more stable

[b] cis isomer

eq

ax

ax

eq

larger group equatorial
more stable

[c] trans isomer

ax

eq

ax

eq

larger group equatorial
more stable

[c] trans isomer

ax

eq

eq

ax

both groups equatorial
more stable

[c] trans isomer

ax

eq

eq

ax

both groups equatorial
more stable

[d]

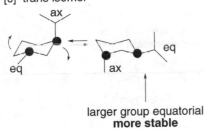

The cis isomer is more
stable than the trans
since one conformer has
both groups equatorial.

[d]

The trans isomer is more
stable than the cis
since one conformer has
both groups equatorial.

[d]

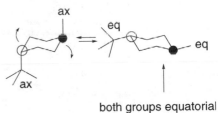

The trans isomer is more
stable than the cis
since one conformer has
both groups equatorial.

4.54 Compare the isomers by drawing them in chair conformations. Equatorial substituents are more stable. See the definitions in Problem 4.53.

(a) 1,2-diethylcyclohexane

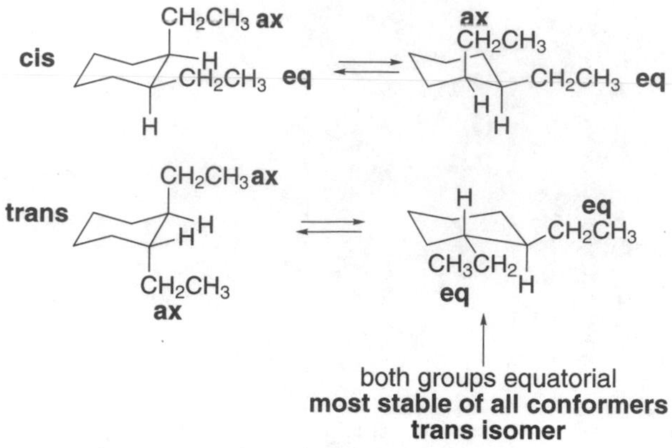

The trans isomer is more stable than the cis isomer because its more stable conformer has two groups equatorial.

(b) 1-ethyl-3-isopropylcyclohexane

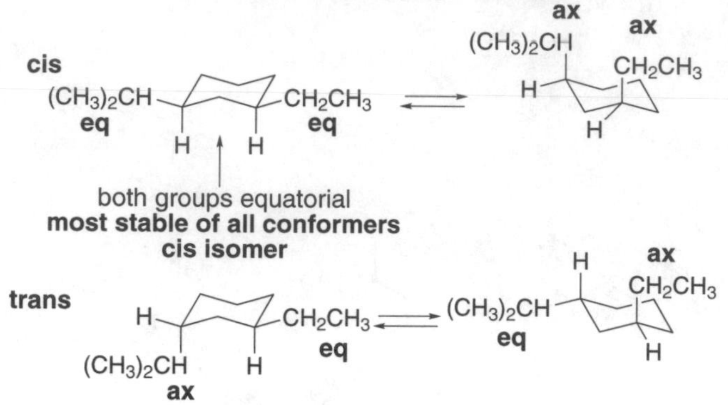

both groups equatorial
**most stable of all conformers
cis isomer**

The cis isomer is more stable than the trans isomer because its more stable conformer has two groups equatorial.

4.55

> Only the more stable conformer of each compound is drawn.

a. or

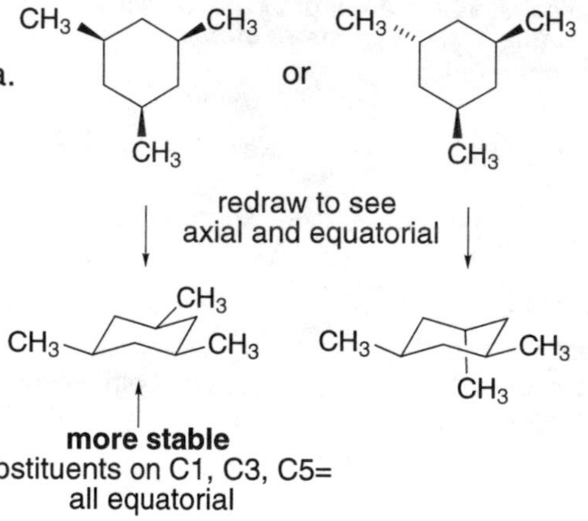

b. or

redraw to see axial and equatorial

redraw to see axial and equatorial

more stable
substituents on C1, C3, C5=
all equatorial

more stable
can all be equatorial

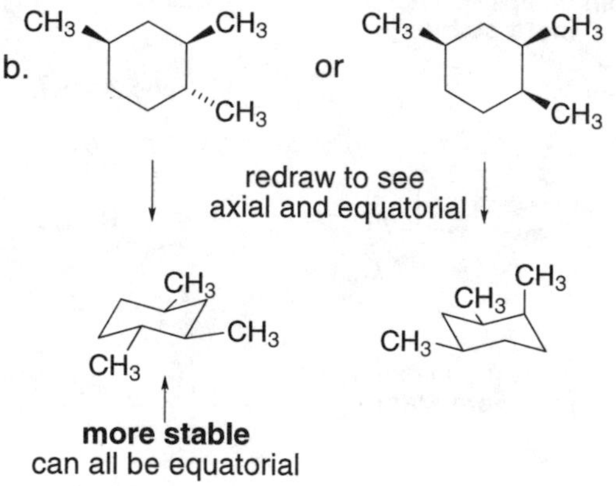

4.56

a.

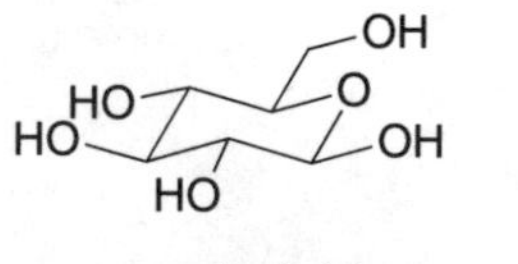

most stable
all groups are equatorial

b.

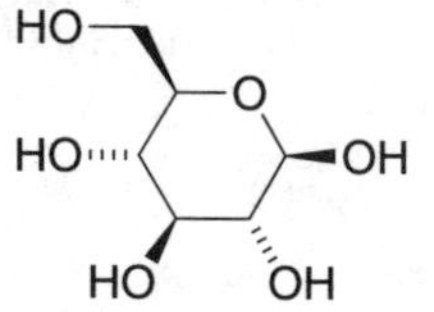

4.57

a.   ∕∕∕∕ and ☐

same molecular formula C₄H₈
different connectivity
constitutional isomers

b. and

different arrangement in three dimensions
stereoisomers

c. CH₃ ... and ... CH₃ = CH₃

1 down, 1 up =
trans

1 down, 1 up =
trans

same arrangement in three dimensions
identical

d, CH₂CH₃   and   CH₂CH₃ / CH₂CH₃
CH₂CH₃

same molecular formula C₁₀H₂₀
different connectivity
constitutional isomers

e. and —CH₃

molecular formula: C₆H₁₀   **molecular formula: C₆H₁₂**

different molecular formulas
not isomers

f. CH₃ CH CH₃
H   H
H   CH₂CH₃
CH₃

and

CH₃CH₂   CH₃   H
H   CH₃
CH₂CH₃

redraw

CH₃
CH₃–CH–CH–CH₂–CH₃
CH₂CH₃
3-ethyl-2-methylpentane

CH₃ CH₂CH₃
CH₃–CH–CH–CH₂CH₃
3-ethyl-2-methylpentane

same molecular formula
same name
identical molecules

g.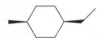

ax
CH₃ ← up
eq
CH₃CH₂
H
↑
up

and

ax
CH₂CH₃ up
eq
CH₃
H
↑
up

both up = **cis**   both up = **cis**

same arrangement in three dimensions
identical

h. and

3,4-dimethylhexane   **2,4-dimethylhexane**

same molecular formula C₈H₁₈
different IUPAC names
constitutional isomers

4.58

One possibility: **constitutional isomer**   **stereoisomer**

a.

cis   **trans**

b. H   H
OH
HO   **cis**

H
OH
H
OH

H
OH   OH
H   **trans**

c. **cis**
Cl   Cl

Cl
Cl

Cl   Cl
trans

4.59

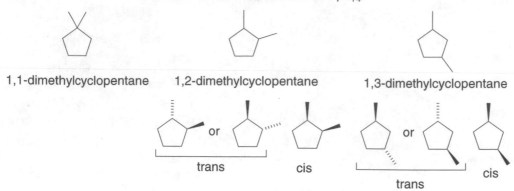

Three constitutional isomers of C_7H_{14}:

1,1-dimethylcyclopentane   1,2-dimethylcyclopentane   1,3-dimethylcyclopentane

trans   cis   trans   cis

4.60 Use the definitions from Answer 4.32 to classify the reactions.

a. CH_3CHO = (structure) $\longrightarrow$ CH_3CH_2OH

Decrease in the number of C–O
bonds. **Reduction**

d. $CH_2{=}CH_2$ $\longrightarrow$ $H{-}C{\equiv}C{-}H$

Decrease in the number of C–H
bonds. **Oxidation**

b.   (cyclohexane) $\longrightarrow$ (cyclohexanone)

Increase in the number of C–O
bonds. **Oxidation**

e.   (toluene) $\longrightarrow$ (benzyl bromide)

Increase in the number of C–Z
bonds. **Oxidation**

c. $CH_2{=}CH_2$ $\longrightarrow$ $HOCH_2CH_2OH$

Two new C–O
bonds. **Oxidation**

f. CH_3CH_2OH $\longrightarrow$ $CH_2{=}CH_2$

Loss of one C–O
bond *and* one C–H
bond. **Neither**

4.61 Use the rule from Answer 4.33.

a. $CH_3CH_2CH_2CH_2CH(CH_3)_2$ $\xrightarrow[\text{11 } O_2]{\text{flame}}$ $7\ CO_2 + 8\ H_2O + heat$

b. (structure) $\xrightarrow[\text{(13/2) } O_2]{\text{flame}}$ $4\ CO_2 + 5\ H_2O + heat$

4.62

a.   2 C–O bonds

(benzene) $\longrightarrow$ (an arene oxide) 2 C–H bonds $\longrightarrow$ (phenol) 1 C–O bond, OH, 1 C–H bond, H

benzene   **an arene oxide**   **phenol**

increase in C–O bonds
oxidation reaction

loss of 1 C–O bond,
loss of 1 C–H bond
neither

b.   Phenol is more water soluble than benzene because it is **polar (contains an O–H group)
and can hydrogen bond with water,** whereas benzene is nonpolar and cannot hydrogen bond

4.63 Lipids contain many nonpolar C–C and C–H bonds and few polar functional groups.

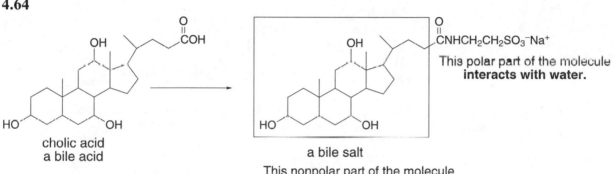

a. mevalonic acid

many polar functional groups
not a lipid

c. estradiol

few polar functional groups
a lipid

d. sucrose

many polar functional groups
not a lipid

b. squalene

no polar functional groups
a lipid

4.64

cholic acid
a bile acid

a bile salt

This polar part of the molecule
interacts with water.

This nonpolar part of the molecule
can **interact with lipids** to create
micelles that allow for transport
of lipids through aqueous environments.

4.65 The amide in the four-membered ring has 90° bond angles giving it angle strain, and therefore making it more reactive.

penicillin G

amide

strained amide
more reactive

4.66

trans-1,4-dimethylcyclohexane
more symmetrical
better packing
higher melting point

cis-1,4-dimethylcyclohexane

4.67

Example:

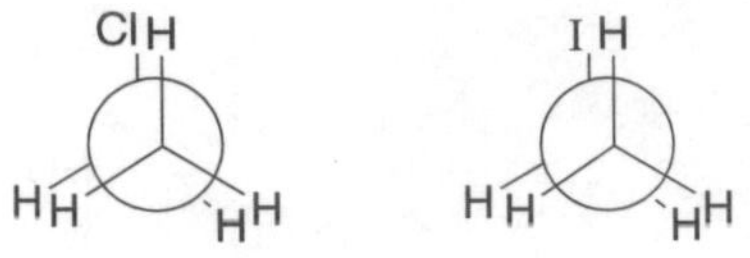

Although I is a much bigger atom than Cl, the C–I bond is also much longer than the C–Cl bond. As a result the eclipsing interaction of the H and I atoms is not very much different from the H,Cl eclipsing interaction in magnitude.

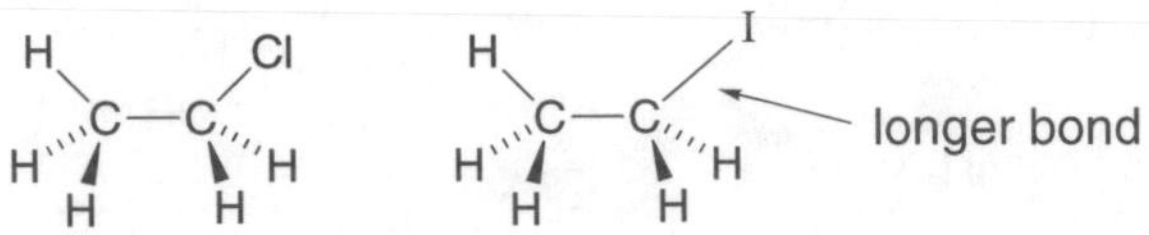

longer bond

4.68

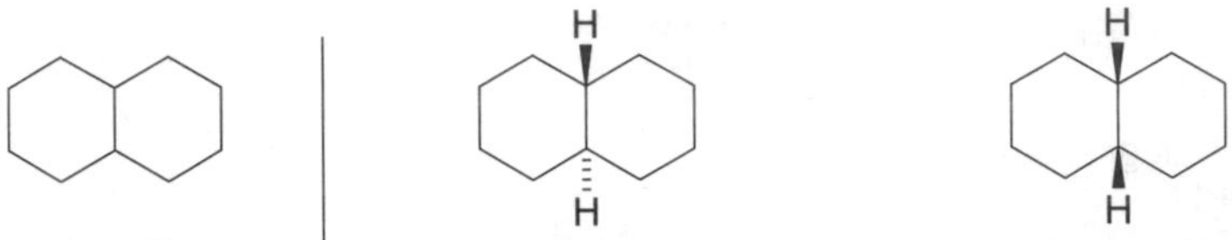

decalin   *trans*-decalin   *cis*-decalin

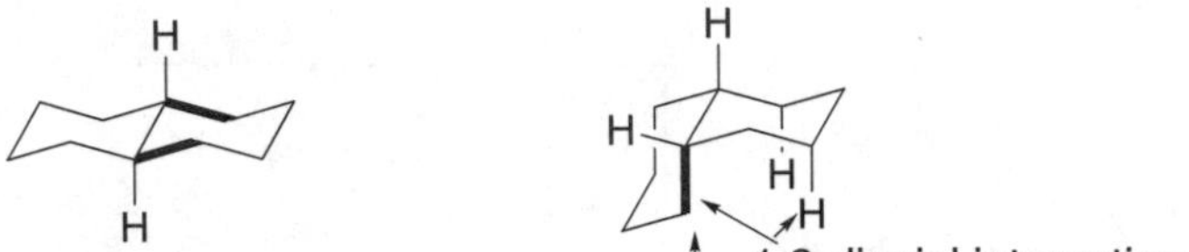

trans

The trans isomer is more stable since the carbon groups at the ring junction are both in the favorable equatorial position.

1,3-diaxial interaction

cis

This bond is axial, creating unfavorable 1,3-diaxial interactions.

Chapter 5: Stereochemistry

♦ **Isomers are different compounds with the same molecular formula. (5.2, 5.11)**

---

[1] Constitutional isomers - isomers that differ in the way the atoms are connected to each other. They have:
- different IUPAC names;
- the same or different functional groups;
- different physical and chemical properties.

[2] Stereoisomers - isomers that differ only in the way atoms are oriented in space. They have the same functional group and the same IUPAC name except for prefixes such as cis, trans, R, and S.
- **Enantiomers** – stereoisomers that are mirror images of each other (5.4).
- **Diastereomers** – stereoisomers that are not mirror images of each other (5.7).

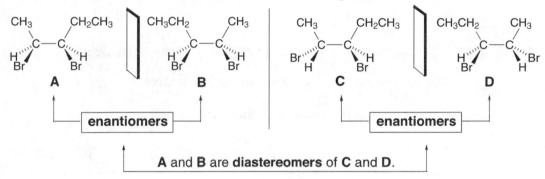

♦ **Assigning priority**

- Assign priorities (1, 2, 3, or 4) to the atoms directly bonded to the stereogenic center in order of decreasing atomic number. The atom of *highest* atomic number gets the *highest* priority (1).
- If two atoms on a stereogenic center are the *same*, assign priority based on the atomic number of the atoms bonded to these atoms. *One* atom of higher atomic number determines a higher priority.
- If two isotopes are bonded to the stereogenic center, assign priorities in order of decreasing *mass* number.
- To assign a priority to an atom that is part of a multiple bond, consider a multiply bonded atom as an equivalent number of singly bonded atoms.

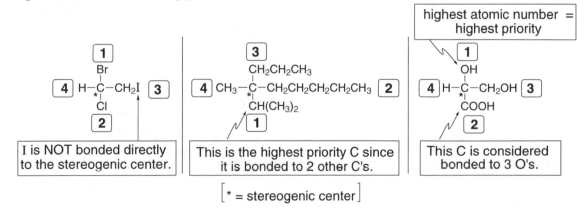

♦ **Some basic principles**

- When a compound and its mirror image are **superimposable**, they are **identical achiral compounds.** An achiral compound has a plane of symmetry in one conformation (5.3).
- When a compound and its mirror image are **not superimposable**, they are **different chiral compounds** called **enantiomers.** A chiral compound has no plane of symmetry in any conformation (5.3).
- A **tetrahedral stereogenic center** is a carbon atom bonded to four different groups (5.4, 5.5).
- For n **stereogenic centers**, the maximum number of stereoisomers is 2^n (5.7).

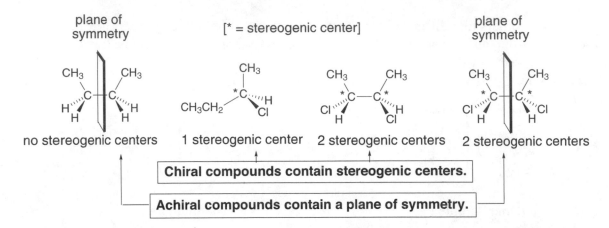

♦ **Optical activity is the ability of a compound to rotate plane-polarized light (5.12).**

- An optically active solution contains a chiral compound.
- An optically inactive solution contains one of the following:
  - An achiral compound with no stereogenic centers.
  - A meso compound – an achiral compound with two or more stereogenic centers.
  - A racemic mixture – an equal amount of two enantiomers.

♦ **The prefixes R and S compared with d and l**

The prefixes R and S are labels used in nomenclature. Rules on assigning R,S are found in Section 5.6.
- An enantiomer has every stereogenic center opposite in configuration. If a compound with two stereogenic centers has the R,R configuration, its enantiomer has the S,S configuration.
- A diastereomer of this same compound has either the R,S or S,R configuration; at least one stereogenic center has the same configuration and at least one is opposite.

The prefixes d (or +) and l (or –) tell the direction a compound rotates plane-polarized light (5.12).
- d (or +) stands for dextrorotatory, rotating polarized light clockwise.
- l (or –) stands for levorotatory, rotating polarized light counterclockwise.

♦ **The physical properties of isomers compared (5.12)**

Type of isomer	Physical properties
Constitutional isomers	Different
Enantiomers	Identical except the direction of rotation of polarized light
Diastereomers	Different
Racemic mixture	Possibly different from either enantiomer

◆ Equations

- Specific rotation (5.12C):

$$\text{specific rotation} = [\alpha] = \frac{\alpha}{l \times c}$$

α = observed rotation (°)
l = length of sample tube (dm)
c = concentration (g/mL)

$\begin{bmatrix} \text{dm = decimeter} \\ \text{1 dm = 10 cm} \end{bmatrix}$

- Enantiomeric excess (5.12D):

$$\text{ee} = \% \text{ of one enantiomer - } \% \text{ other enantiomer}$$

$$= \frac{[\alpha] \text{ mixture}}{[\alpha] \text{ pure enantiomer}} \times 100\%$$

Chapter 5: Answers to Problems

5.1 Cellulose consists of long chains held together by intermolecular hydrogen bonds forming sheets that stack in extensive three-dimensional arrays. Most of the OH groups in cellulose are in the interior of this three-dimensional network, unavailable for hydrogen bonding to water. Thus, even though cellulose has many OH groups, its three-dimensional structure prevents many of the OH groups from hydrogen bonding with the solvent and this makes it water insoluble.

5.2 **Constitutional isomers** have atoms bonded to different atoms.
Stereoisomers differ only in the three-dimensional arrangement of atoms.

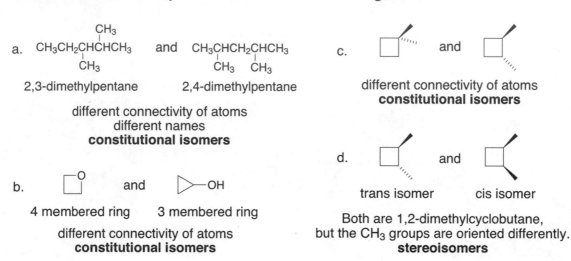

5.3 Draw the mirror image of each molecule by drawing a mirror plane and then drawing the molecule's reflection. **A chiral molecule is one that is not superimposable on its mirror image**. A molecule with one stereogenic center is always chiral. A molecule with zero stereogenic centers is not chiral (in general).

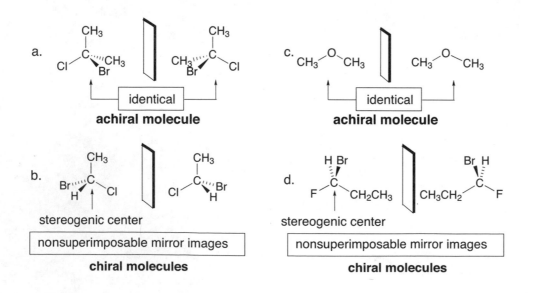

5.4 The plane of symmetry cuts the molecule into **two identical halves**.

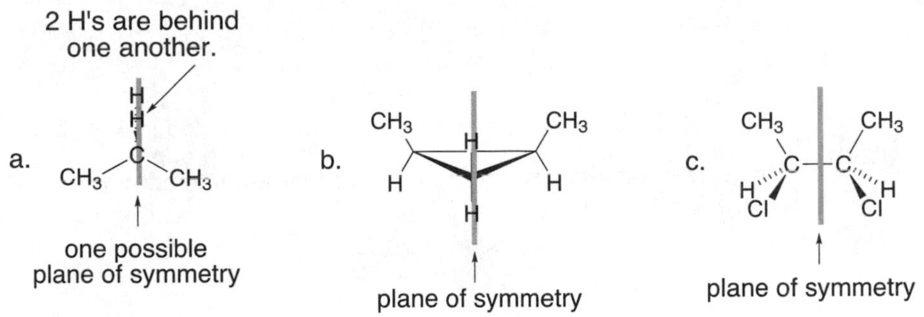

a.

2 H's are behind
one another.

one possible
plane of symmetry

b.

CH_3 CH_3

H   H

H

plane of symmetry

c.

CH_3 CH_3

H   Cl   Cl   H

plane of symmetry

5.5 Rotate around the middle C–C bond so that the Br groups are eclipsed.

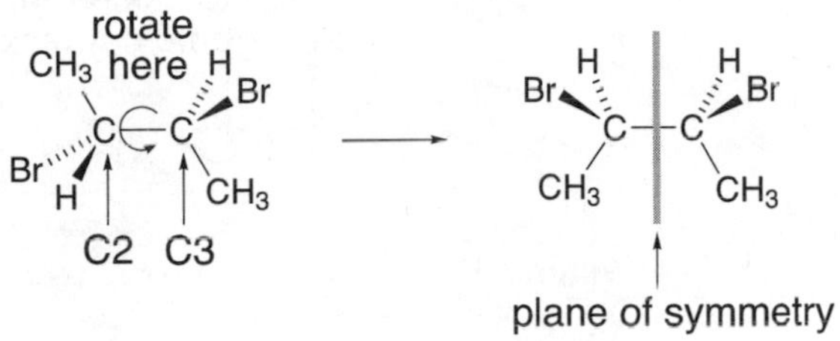

rotate
CH_3 here H
Br

Br CH_3
H

C2 C3

⟶

H   H

Br   Br

CH_3 CH_3

plane of symmetry

5.6 To locate a stereogenic center, omit:

All C's with 2 or more H's, all *sp* and sp^2 hybridized atoms, and all heteroatoms.
Then evaluate any remaining atoms: a tetrahedral stereogenic center has a carbon bonded to **four
different groups**.

a.

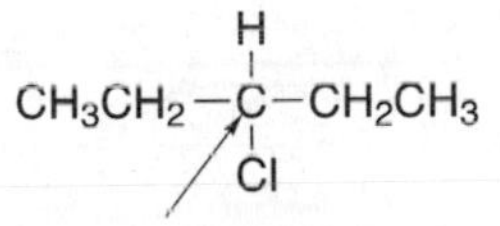

CH_3CH_2 — C — CH_2CH_3
with H above
and Cl below

bonded to 2 identical
ethyl groups
0 stereogenic centers

b. $(CH_3)_3CH$

0 stereogenic centers

c.

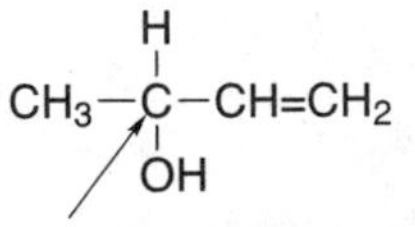

CH_3 — C — $CH=CH_2$
with H above
and OH below

This C is bonded to
4 different groups.
1 stereogenic center

d. $CH_3CH_2CH_2OH$

0 stereogenic centers

e. 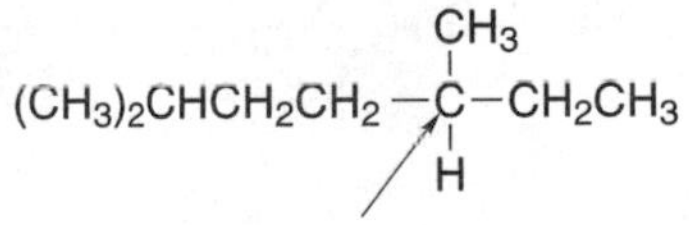 $(CH_3)_2CHCH_2CH_2$ — C — CH_2CH_3
with CH_3 above
and H below

This C is bonded to
4 different groups.
1 stereogenic center

f.

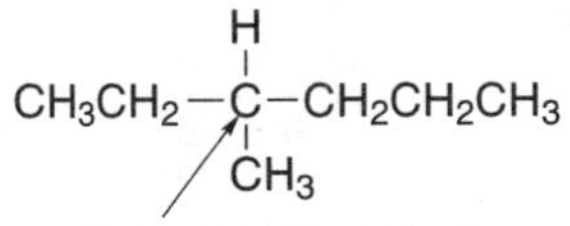

CH_3CH_2 — C — $CH_2CH_2CH_3$
with H above
and CH_3 below

This C is bonded to
4 different groups.
1 stereogenic center

5.7 Use the directions from Answer 5.6 to label the stereogenic centers.

a. CH₃CH₂CH₂—C—CH₃ with H above, OH below
stereogenic center

b. (CH₃)₂CHCH₂—C—COOH with H above, NH₂ below
stereogenic center

c.
Both C's bonded to 4
different groups.
2 stereogenic centers

d.
3 C's bonded to 4
different groups.
3 stereogenic centers

5.8 Use the directions from Answer 5.6 to label the stereogenic centers.

a.
CHO
HO–C–H
HO–C–H
H–C–OH
H–C–OH
CH₂OH
mannose
4 C's bonded to
4 different groups:
4 stereogenic centers.

b.
vitamin K₁
Both C's bonded to 4 different groups:
2 stereogenic centers.

5.9 Find the C bonded to 4 different groups in each molecule. At the stereogenic center, draw two bonds in the plane of the page, one in front (on a wedge) and one behind (on a dash). Then draw the mirror image (enantiomer).

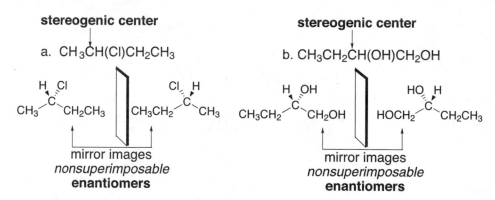

stereogenic center

a. CH₃CH(Cl)CH₂CH₃

stereogenic center

b. CH₃CH₂CH(OH)CH₂OH

mirror images
nonsuperimposable
enantiomers

mirror images
nonsuperimposable
enantiomers

5.10 Use the directions from Answer 5.6 to label the stereogenic centers.

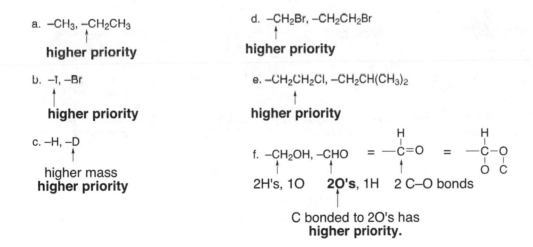

a. C bonded to
H, OH, 2 different C's:
1 stereogenic center

c. All ring C's are sp^2.
Other C's have ≥ 2 H's.
no stereogenic centers

e. 4 C's bonded to 4
different groups:
4 stereogenic centers

b. C bonded to
H and 3 different C's:
1 stereogenic center

d. Each labeled C
is bonded to:
H, Cl, CH₂, CHCl:
2 stereogenic centers

f. 3 C's bonded to 4
different groups:
3 stereogenic centers

5.11

All stereogenic C's are circled. Each C is sp^3
hybridized and bonded to 4 different groups.

cholesterol

HO

5.12 Assign priority based on atomic number: atoms with a higher atomic number get a higher priority.
If two atoms are the same, look at what they are bonded to and assign priority based on the atomic
number of these atoms.

a. –CH₃, –CH₂CH₃
↑
higher priority

b. –I, –Br
↑
higher priority

c. –H, –D
↑
higher mass
higher priority

d. –CH₂Br, –CH₂CH₂Br
↑
higher priority

e. –CH₂CH₂Cl, –CH₂CH(CH₃)₂
↑
higher priority

f. –CH₂OH, –CHO = —C=O = —C–O
2H's, 1O   2O's, 1H   2 C–O bonds
↑
C bonded to 2O's has
higher priority.

5.13 Rank by decreasing priority. Lower atomic number = lower priority.

Lowest priority = 4, Highest priority = 1

a. –COOH C = second lowest atomic number — priority **3**

 –H, H = lowest atomic number — **4**

 –NH₂, N = second highest atomic number — **2**

 –OH O = highest atomic number — **1**

decreasing priority: –OH, –NH₂, –COOH, –H

c. –CH₂CH₃, C bonded to 2H's + **1C** — priority **2**

 –CH₃, C bonded to 3H's — **3**

 –H, H = lowest atomic number — **4**

 –CH(CH₃)₂ C bonded to 1H + **2C's** — **1**

decreasing priority: –CH(CH₃)₂, –CH₂CH₃, –CH₃, –H

b. –H, H = lowest atomic number — priority **4**

 –CH₃, C bonded to 3H's — **3**

 –Cl, Cl = highest atomic number — **1**

 –CH₂Cl C bonded to 2H's + **1 Cl** — **2**

decreasing priority: –Cl, –CH₂Cl, –CH₃, –H

d. –CH=CH₂, C bonded to 1H + **2C's** — priority **2**

 –CH₃, C bonded to 3H's — **3**

 –C≡CH, C bonded to **3C's** — **1**

 –H H = lowest atomic number — **4**

decreasing priority: –C≡CH, –CH=CH₂, –CH₃, –H

5.14 To assign *R* or *S* to the molecule, first rank the groups. The lowest priority group must be oriented behind the page. If tracing a circle from (1) → (2) → (3), proceeds in the clockwise direction, the stereogenic center is labeled *R*; if the circle is counterclockwise, it is labeled *S*.

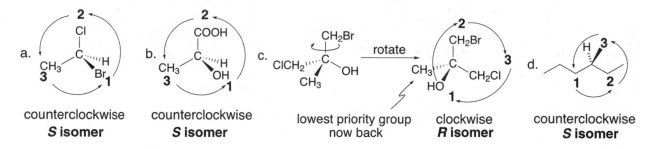

a. counterclockwise **S isomer**   b. counterclockwise **S isomer**   c. lowest priority group now back / clockwise **R isomer**   d. counterclockwise **S isomer**

5.15

a. CH₃–C(H)(Cl)–CH₂CH₃

b. CH₃O–C(H)(CH₂CH₃)–CHO

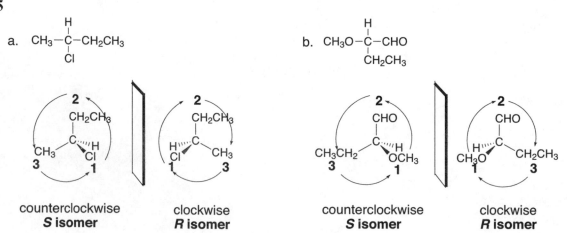

counterclockwise **S isomer**   clockwise **R isomer**   counterclockwise **S isomer**   clockwise **R isomer**

5.16

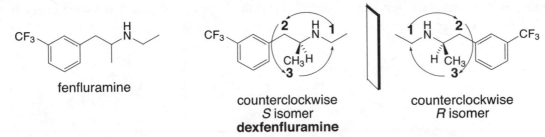

fenfluramine

counterclockwise
S isomer
dexfenfluramine

counterclockwise
R isomer

5.17 The maximum number of stereoisomers $= 2^n$ where $n =$ the number of stereogenic centers.

3 stereogenic centers
$2^3 = 8$ stereoisomers

8 stereogenic centers
$2^8 = 256$ stereoisomers

5.18

a. $CH_3CH_2CH(Cl)CH(OH)CH_2CH_3$

2 stereogenic centers = 4 stereoisomers

b. $CH_3CH(Br)CH_2CH(Cl)CH_3$.

2 stereogenic centers = 4 stereoisomers

A **B**

C **D**

A **B**

C **D**

5.19

a. $CH_3CH(OH)CH(OH)CH_3$

2 stereogenic centers = 4 possible stereoisomers

b. $CH_3CH(OH)CH(Cl)CH_3$.

2 stereogenic centers = 4 possible stereoisomers

A **B**

C **C**

identical

C is a meso compound.

A and **B** are enantiomers.
Pairs of diastereomers: **A** and **C**, **B** and **C**.

A **B**

C **D**

Pairs of enantiomers: **A** and **B**, **C** and **D**.
Pairs of diastereomers: **A** and **C**, **A** and **D**,
B and **C**, **B** and **D**.

5.20 An **enantiomer** is a stereoisomer that is a nonsuperimposable mirror image. A **diastereomer** is a stereoisomer that is not a mirror image.

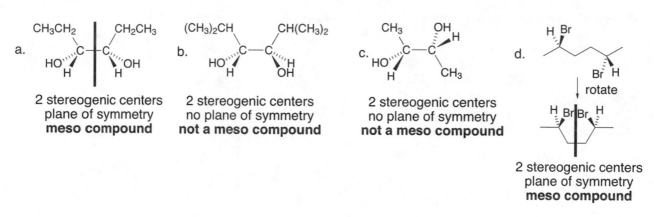

5.21 A meso compound must have at least 2 stereogenic centers and a plane of symmetry. You may have to rotate around a C–C bond to see the plane of symmetry clearly.

5.22 The enantiomer must have the exact opposite *R,S* designations. Diastereomers with two stereogenic centers have one center the same and one different.

If a compound is **R,S:**

Its enantiomer is: **S,R** ←——————— Exact opposite: *R* and *S* interchanged.

Its diastereomers are: **R,R and S,S** ←——————— One designation remains the same, the other changes.

5.23 The enantiomer must have the exact opposite *R,S* designations. For diastereomers, at least one of the *R,S* designations is the same, but not all of them.

a. (2*R*,3*S*)-2,3-hexanediol and (2*R*,3*R*)-2,3-hexanediol

One changes; one remains the same:
diastereomers

b. (2*R*,3*R*)-2,3-hexanediol and (2*S*,3*S*)-2,3-hexanediol

Both *R*'s change to *S*'s:
enantiomers

c. (2*R*,3*S*,4*R*)-2,3,4-hexanetriol and (2*S*,3*R*,4*R*)-2,3,4-hexanetriol

Two change; one remains the same:
diastereomers

5.24 To decide how the compounds are related, label all stereogenic centers.
- **Identical** molecules will have the same *R,S* designations.
- **Enantiomers** have opposite *R,S* designations.
- **Diastereomers** have at least one *R,S* designation the same, but not all of them.

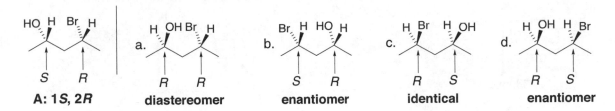

A: 1*S*, 2*R*   a. **diastereomer**   b. **enantiomer**   c. **identical**   d. **enantiomer**

5.25 All **meso compounds have a plane of symmetry**. They cannot have just one stereogenic center.

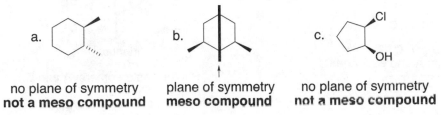

a.   b.   c.

no plane of symmetry
not a meso compound

plane of symmetry
meso compound

no plane of symmetry
not a meso compound

5.26

a. 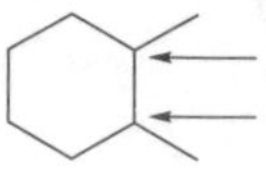 2 stereogenic centers =
4 stereoisomers maximum

Draw the cis and trans isomers:

cis identical

trans

B   **C**

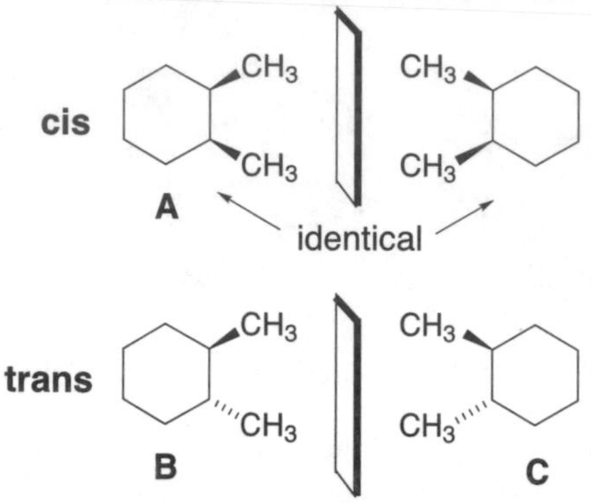

Pair of enantiomers: **B** and **C**.
Pairs of diastereomers: **A** and **B**, **A** and **C**.

Only 3 stereoisomers exist.

c.

Draw the cis and trans isomers:

identical

B identical

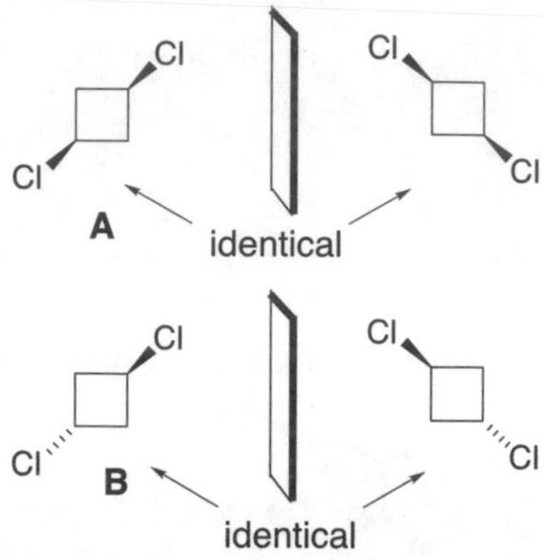

Pair of diastereomers: **A** and **B**.

Only 2 stereoisomers exist.

b. 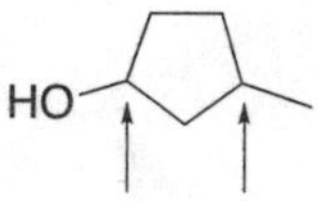 2 stereogenic centers =
4 stereoisomers maximum

Draw the cis and trans isomers:

cis

A   **B**

trans

C   **D**

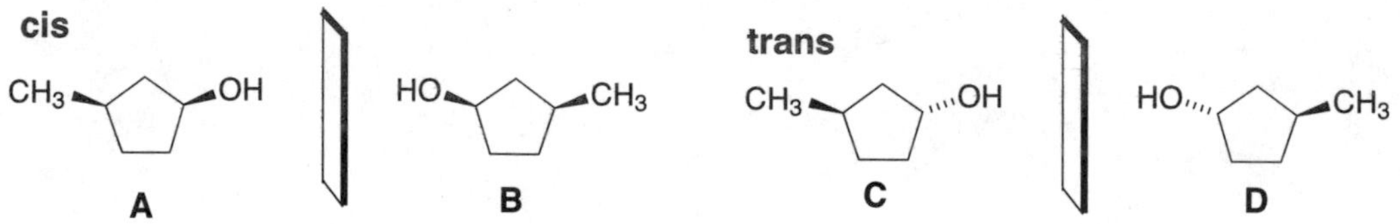

Pairs of enantiomers: **A** and **B**, **C** and **D**.
Pairs of diastereomers: **A** and **C**, **A** and **D**, **B** and **C**, **B** and **D**.

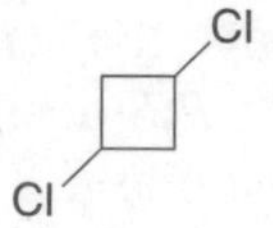

All 4 stereoisomers exist.

5.27 Four facts:

- **Enantiomers** are mirror image isomers.
- **Diastereomers** are stereoisomers that are not mirror images.
- **Constitutional isomers** have the same molecular formula but the atoms are bonded to different atoms.
- **Cis and trans isomers** are always diastereomers.

a.
same molecular formula
same *R,S* designation:
identical

c.
1,4- isomer   1,3-isomer
constitutional isomers

b.
same molecular formula,
opposite configuration at one
stereogenic center
enantiomers

d.
trans   **cis**
Both 1,3 isomers,
cis and trans:
diastereomers

5.28

(*S*)-alanine
$[\alpha] = +8.5°$
mp = 297 °C

a. Mp = same as the *S* isomer
b. The mp of a racemic mixture is often different from the melting point of the enantiomers.
c. –8.5°, same as *S* but opposite sign
d. 0°
e. Solution of pure (*S*)-alanine: **optically active**
    Equal mixture of (*R*) and (*S*)-alanine: **optically inactive**
    75% (*S*) and 25% (*R*)-alanine: **optically active**

5.29

$$[\alpha] = \frac{\alpha}{l \times c}$$

α = observed rotation
l = length of tube (dm)
c = concentration (g/mL)

$$[\alpha] = \frac{10°}{1\,dm \times (1g/10mL)} = +100° = \text{specific rotation}$$

5.30 Enantiomeric excess = ee = % of one enantiomer − % of other enantiomer.

a. 95 − 5 = **90% ee**   b. 85 − 15 = **70% ee**

5.31

    a. 90% ee means 90% excess of **A**, and 10% racemic mixture of **A** and **B** (5% each). Therefore, **95% A and 5% B**

    b. 99% ee means 99% excess of **A**, and 1% racemic mixture of **A** and **B** (0.5% each). Therefore, **99.5% A and 0.5% B.**

    c. 60% ee means 60% excess of **A**, and 40% racemic mixture of **A** and **B** (20%each). Therefore, **80% A and 20% B**

5.32

$$ee = \frac{[\alpha]\ \text{mixture}}{[\alpha]\ \text{pure enantiomer}} \times 100\%$$

    a. $\dfrac{15}{25} \times 100\% = 60\%$ ee   b. $80\% = \dfrac{[\alpha]}{25} \times 100\%$ $[\alpha] = +20°$

5.33 • **Enantiomers have the same physical properties** (mp, bp, solubility), and rotate the plane of polarized light to an equal but opposite extent.
    • **Diastereomers have different physical properties.**
    • **A racemic mixture is optically inactive.**

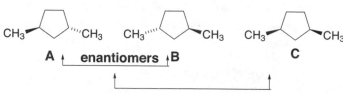

A and B are diastereomers of C.
three stereoisomers of 1,3-dimethylcyclopentane

    a. The bp's of **A** and **B** are the same. The bp's of **A** and **C** are different.
    b. Pure **A**: optically active
       Pure **B**: optically active
       Pure **C**: optically inactive
       Equal mixture of **A** and **B**: optically inactive
       Equal mixture of **A** and **C**: optically active
    c. There would be two fractions: one containing **A** and **B** (optically inactive), and one containing **C** (optically inactive).

5.34 Use the definitions from Answer 5.2.

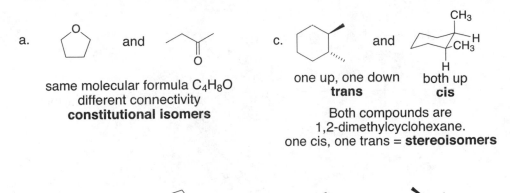

a.   same molecular formula C_4H_8O
different connectivity
constitutional isomers

c.   one up, one down   both up
trans   **cis**

Both compounds are
1,2-dimethylcyclohexane.
one cis, one trans = **stereoisomers**

b. $C_5H_{10}O$ C_5H_8O
different molecular formulas
not isomers

d.   same molecular formula C_7H_{14}
different connectivity
constitutional isomers

5.35 Use the definitions from Answer 5.3.

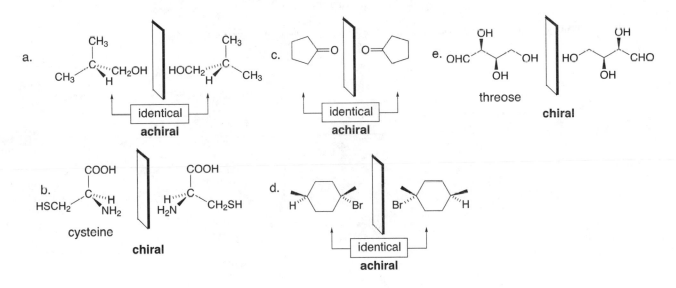

a. |identical|
achiral

c. |identical|
achiral

e.   threose   **chiral**

b.   cysteine   **chiral**

d. |identical|
achiral

5.36

A
R isomer

a.   *S*
enantiomer

b.   *R*
identical

c.   *S*
enantiomer

5.37 The plane of symmetry cuts the molecule into **two identical halves**.

a.

2 H's are aligned.

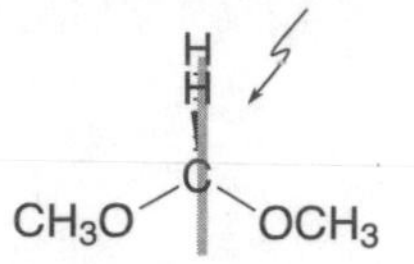

The plane of symmetry is
drawn through the
hydrogen atoms.

b.

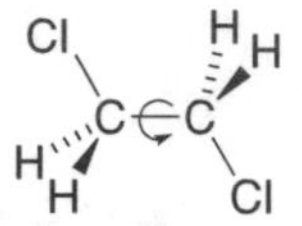

A plane of symmetry is
present in the plane
of the page; or

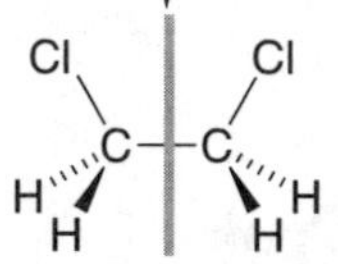

c.

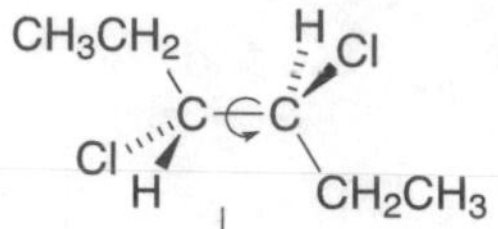

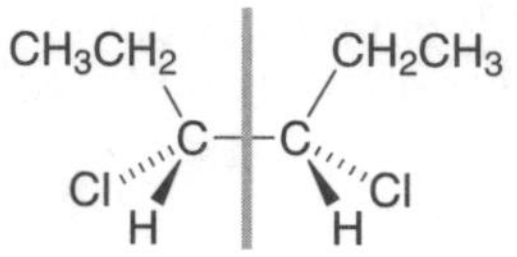

plane of symmetry

d.

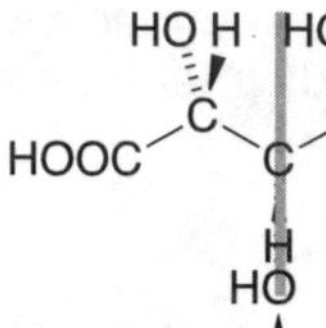

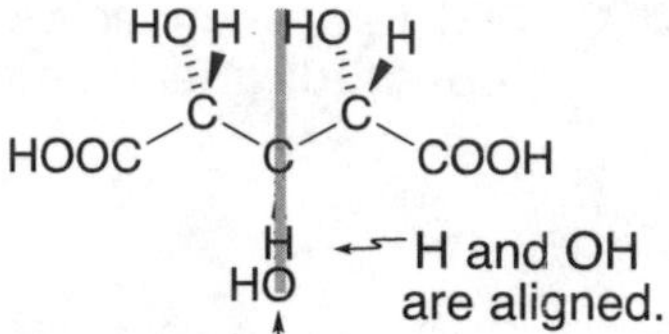

H and OH
are aligned.

plane of symmetry

e.

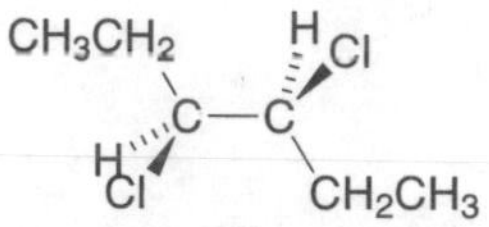

no plane of symmetry

f.

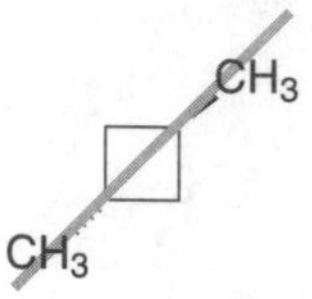

The plane of symmetry
bisects the molecule.

g.

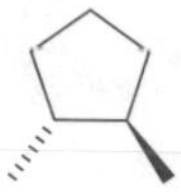

no plane of symmetry

h.

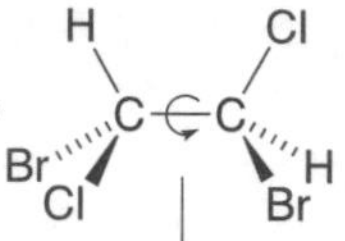

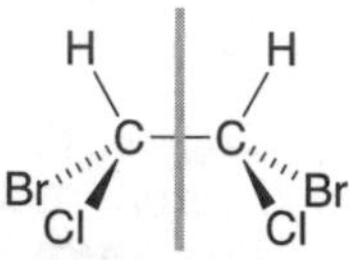

plane of symmetry

5.38 Use the directions from Answer 5.6 to label the stereogenic centers.

a. $CH_3CH_2CH_2CH_2CH_2CH_3$
   All C's have 2 or more H's.
   0 stereogenic centers

b.

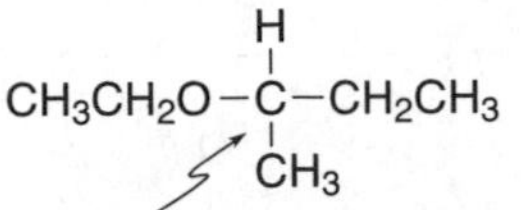

$$CH_3CH_2O-\underset{\underset{CH_3}{|}}{\overset{\overset{H}{|}}{C}}-CH_2CH_3$$

   1 stereogenic center

c. $(CH_3)_2CHCH(OH)CH(CH_3)_2$
   0 stereogenic centers

d. $(CH_3)_2CHCH_2-\overset{\overset{H}{|}}{\underset{\underset{CH_3}{|}}{C}}-CH_2-\overset{\overset{H}{|}}{\underset{\underset{CH_3}{|}}{C}}-\overset{\overset{H}{|}}{\underset{\underset{CH_3}{|}}{C}}-CH_2CH_3$

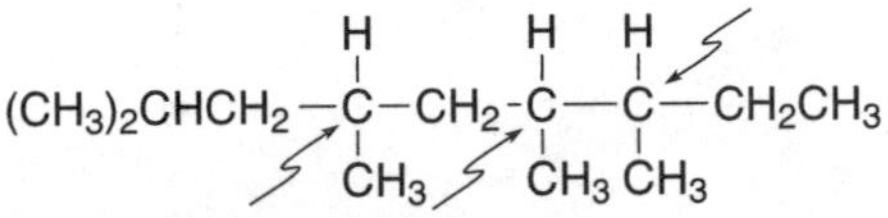

   3 stereogenic centers.

e. $CH_3-\overset{\overset{H}{|}}{\underset{\underset{D}{|}}{C}}-CH_2CH_3$

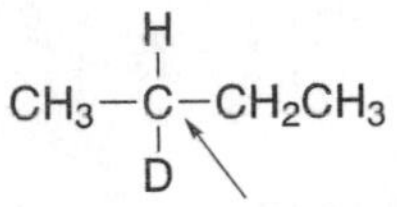

   bonded to 4 different groups
   1 stereogenic center

f.

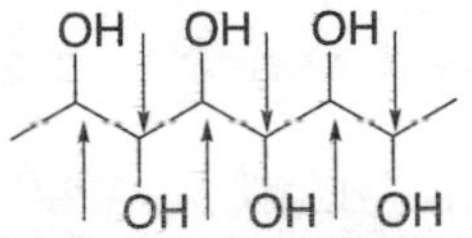

   OH   OH   OH

   OH   OH   OH

   Each labeled C bonded to 4 different groups =
   6 stereogenic centers.

g.

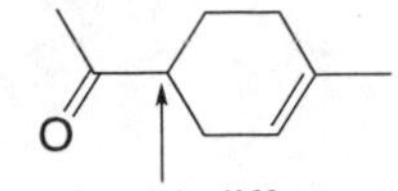

bonded to 4 different groups
1 stereogenic center

h.

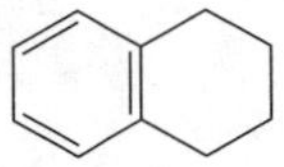

All C's have 2 or more H's, or
are sp^2 hybridized.
0 stereogenic centers

i.

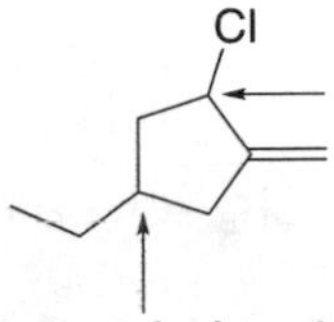

Each bonded to 4 different groups =
2 stereogenic centers.

j.

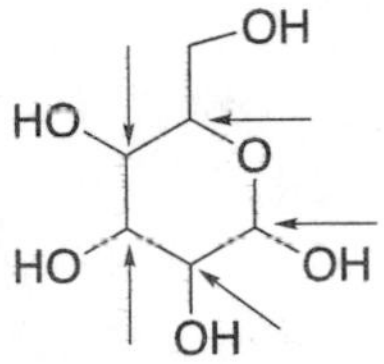

Each labeled C bonded to 4 different groups =
5 stereogenic centers.

5.39 Stereogenic centers are circled.

Eight constitutional isomers:

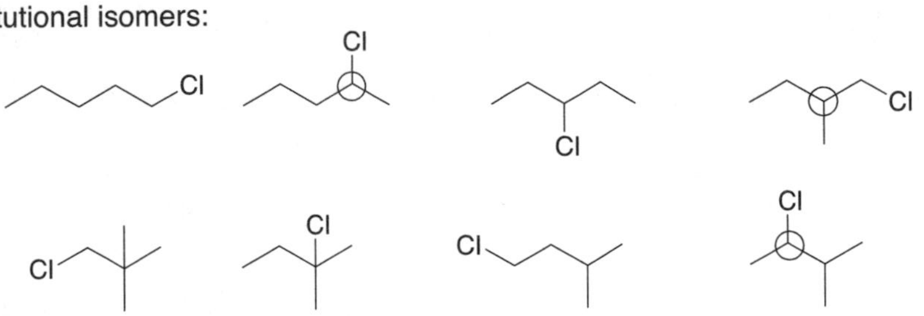

5.40

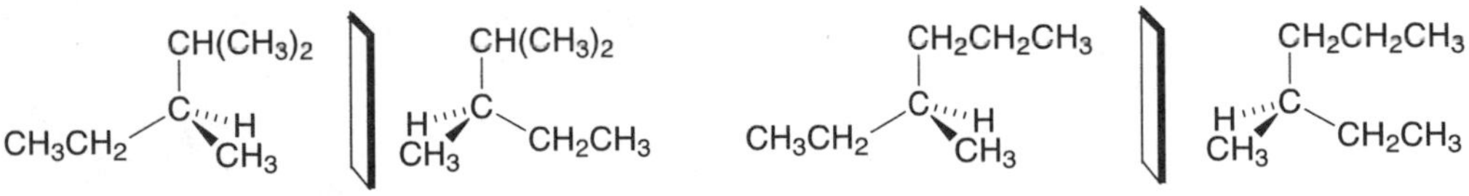

$$CH_3CH_2\underset{\underset{CH_3}{}}{\overset{\overset{CH(CH_3)_2}{|}}{C}}{\cdots}H$$
$$H{\cdots}\underset{\underset{CH_3}{}}{\overset{\overset{CH(CH_3)_2}{|}}{C}}CH_2CH_3$$
$$CH_3CH_2\underset{\underset{CH_3}{}}{\overset{\overset{CH_2CH_2CH_3}{|}}{C}}{\cdots}H$$
$$H{\cdots}\underset{\underset{CH_3}{}}{\overset{\overset{CH_2CH_2CH_3}{|}}{C}}CH_2CH_3$$

5.41

a.

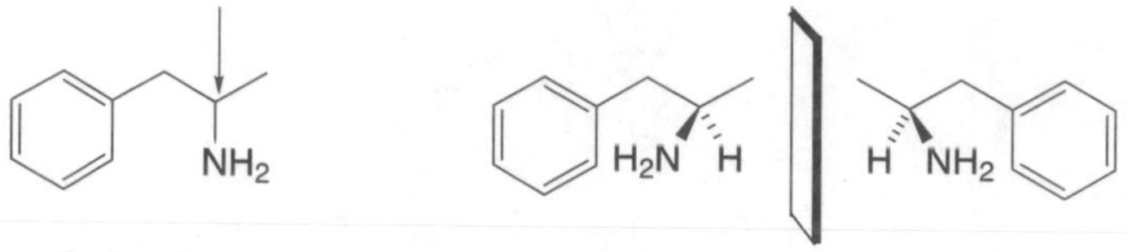

amphetamine

b.

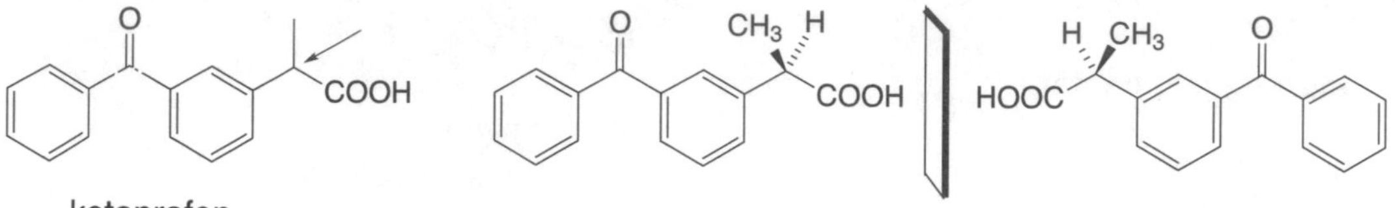

ketoprofen

5.42

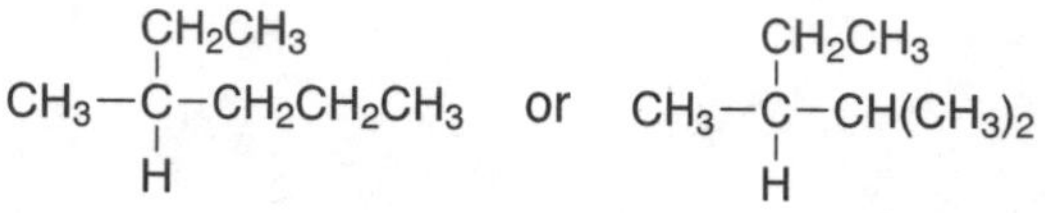

$$CH_3-\underset{\underset{H}{|}}{\overset{\overset{CH_2CH_3}{|}}{C}}-CH_2CH_2CH_3 \quad or \quad CH_3-\underset{\underset{H}{|}}{\overset{\overset{CH_2CH_3}{|}}{C}}-CH(CH_3)_2$$

5.43 Assign priority based in the rules in Answer 5.12.

a. –OH, –NH₂

↑
higher atomic number
higher priority

b. –CD₃, –CH₃

↑
D higher mass than H
higher priority

c. –CH(CH₃)₂, –CH₂OH

↑
C bonded to O
higher priority

d. –CH₂Cl, –CH₂CH₂CH₂Br

↑
C bonded to Cl
higher priority

e. –CHO, –COOH

↑
C has 3 bonds to O
higher priority

f. –CH₂NH₂, –NHCH₃

↑
higher atomic number
higher priority

5.44 Assign priority based in the rules in Answer 5.12.

a. –F > –OH > –NH₂ > –CH₃

b. –(CH₂)₃CH₃ > –CH₂CH₂CH₃ > –CH₂CH₃ > –CH₃

c. –NH₂ > –CH₂NHCH₃ > –CH₂NH₂ > –CH₃

d. –COOH > –CHO > –CH₂OH > –H

e. –Cl > –SH > –OH > –CH₃

f. –C≡CH > –CH=CH₂ > –CH(CH₃)₂ > –CH₂CH₃

5.45 Use the rules in Answer 5.14 to assign *R* or *S* to each stereogenic center.

a.

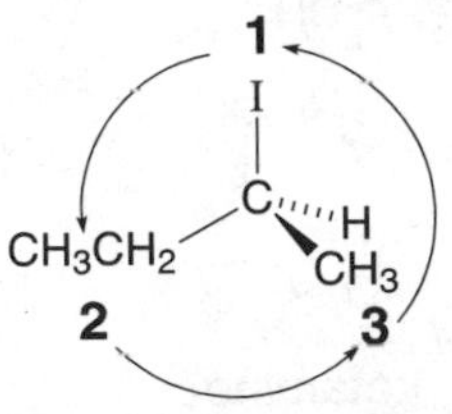

counterclockwise
S isomer

c.

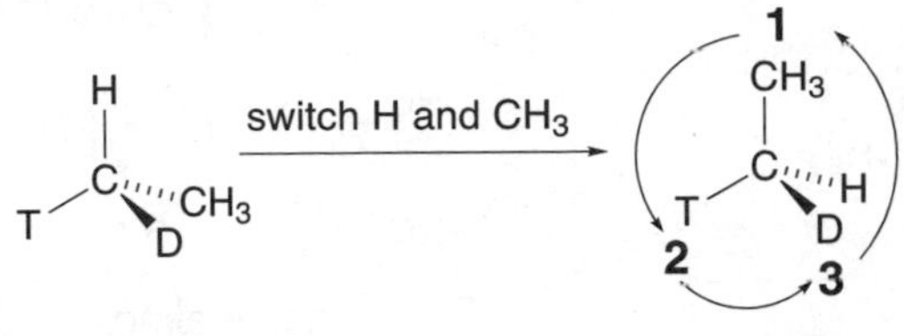

switch H and CH₃

counterclockwise
It looks like an *S* isomer, but we
must reverse the answer, *S* to *R*.
R isomer

b.

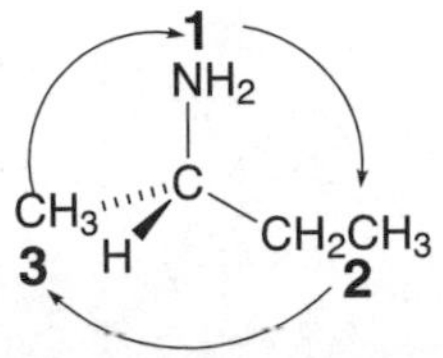

clockwise, but H in front
S isomer

d.

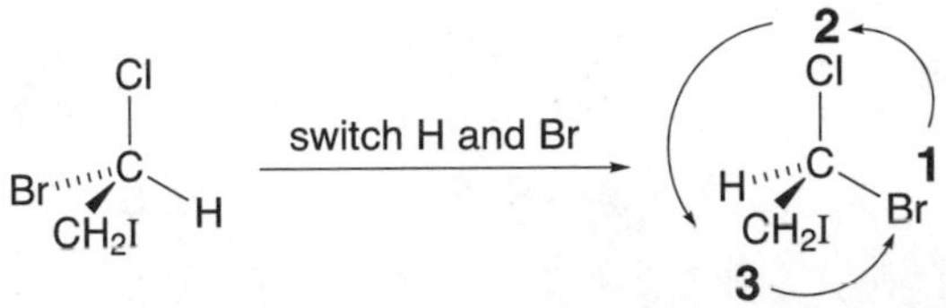

switch H and Br

counterclockwise
It looks like an *S* isomer, but we
must reverse the answer, *S* to *R*.
R isomer

e.

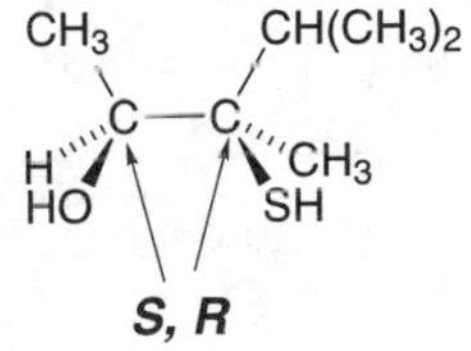

S, R

f.

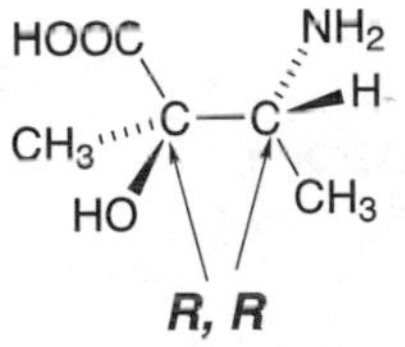

R, R

g.

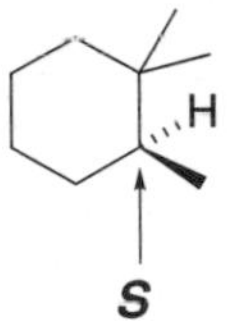

S

h.

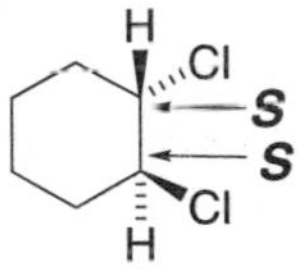

5.46

a. (3*R*)-3-methylhexane ☐☐

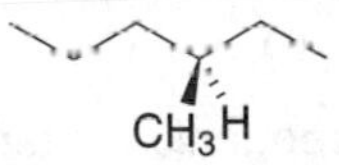

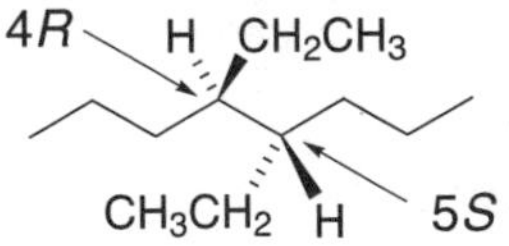

c. (3*R*,5*S*,6*R*)-5-ethyl-3,6-dimethylnonane

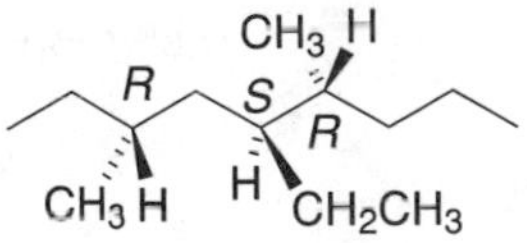

b. (4*R*,5*S*)-4,5-diethyloctane ☐☐

d. (3*S*,6*S*)-6-isopropyl-3-methyldecane

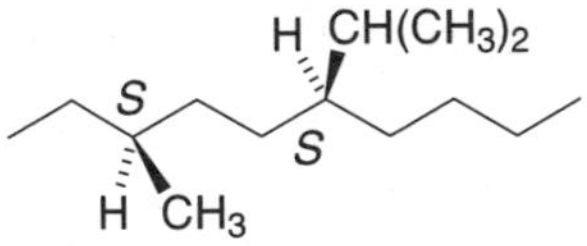

5.47 Two enantiomers of the amino acid leucine.

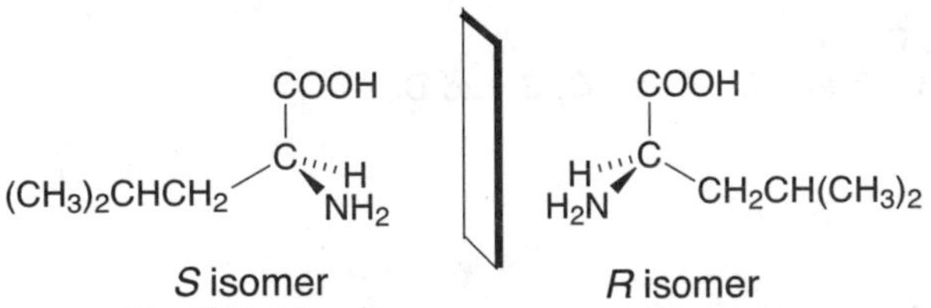

(CH₃)₂CHCH₂

S isomer
naturally occurring

R isomer

5.48

a. L-dopa

b. adrenaline

c. ketamine

5.49

methylphenidate **R, R**   **S, S**

5.50

a. amoxicillin

b. norethindrone

c. heroin

5.51

a. $CH_3CH(OH)CH(OH)CH_2CH_3$

2 stereogenic centers
$2^2 = 4$ possible stereoisomers

b. $CH_3CH_2CH_2CH(CH_3)_2$

0 stereogenic centers

c.

4 stereogenic centers
$2^4 = 16$ possible stereoisomers

5.52

a. $CH_3CH(OH)CH(OH)CH_2CH_3$

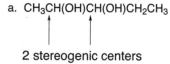

A   **B**   **C**   **D**

Pairs of enantiomers: **A** and **B**, **C** and **D**.
Pairs of diastereomers: **A** and **C**, **A** and **D**, **B** and **C**, **B** and **D**.

b. CH₃CH(OH)CH₂CH₂CH(OH)CH₃

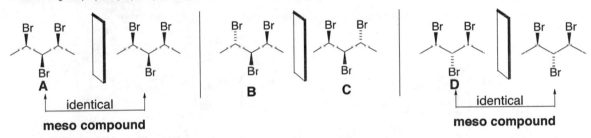

A   **B**   **C**   ↑ identical ↑
meso compound

Pair of enantiomers: **A** and **B**.
Pairs of diastereomers: **A** and **C**, **B** and **C**.

c. CH₃CH(Cl)CH₂CH(Br)CH₃

A   **B**   **C**   **D**

Pairs of enantiomers: **A** and **B**, **C** and **D**.
Pairs of diastereomers: **A** and **C**, **A** and **D**, **B** and **C**, **B** and **D**.

□□d. CH₃CH(Br)CH(Br)CH(Br)CH₃

A   **B**   **C**   **D**
↑ identical ↑   ↑ identical ↑
meso compound   **meso compound**

Pair of enantiomers: **B** and **C**.
Pairs of diastereomers: **A** and **B**, **A** and **C**, **A** and **D**, **B** and **D**, **C** and **D**

5.53

a.

COOH ——— 2
NH₂—C—H ——— 3
H—C—OH
CH₃

threonine

b. and c.

2S,3R
naturally occurring

2S,3S

2R,3R

2R,3S

5.54

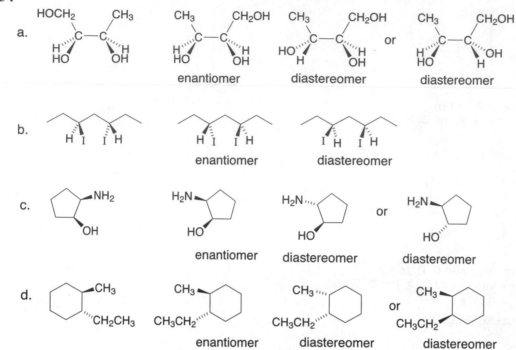

a.

enantiomer   diastereomer   or   diastereomer

b.

enantiomer   diastereomer

c.

enantiomer   diastereomer   or   diastereomer

d.

enantiomer   diastereomer   or   diastereomer

5.55

a.

A   identical   **B**   **C**

meso compound

Pair of enantiomers: **B** and **C**.
Pairs of diastereomers: **A** and **B**, **A** and **C**.

b.

A   identical   **B**   identical

Pair of diastereomers: **A** and **B**.

c.

A   **B**   **C**   **D**

Pairs of enantiomers: **A** and **B**, **C** and **D**.
Pairs of diastereomers: **A** and **C**, **A** and **D**, **B** and **C**, **B** and **D**.

5.56

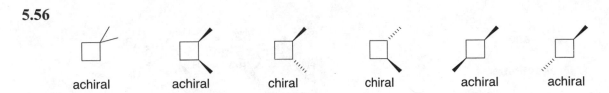

achiral   achiral   chiral   chiral   achiral   achiral

5.57 A has two stereogenic centers and a plane of symmetry, making it an achiral meso compound. Since it superimposable on its mirror image it has no enantiomer. **B** has only one stereogenic center. Its two possible stereoisomers consist of a pair of enantiomers, but no diastereomer.

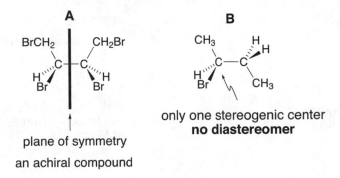

A

B

plane of symmetry

an achiral compound

only one stereogenic center
no diastereomer

5.58

C2 C3

D-erythrose

2R,3R

a.

2S,3R
diastereomer

b.

2R,3R
identical

c.

2S,3S
enantiomer

d.

2R,3S
diastereomer

5.59

a. 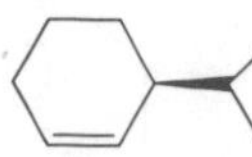 and 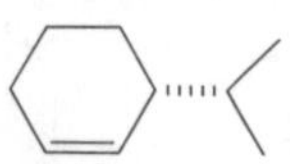

enantiomers

b. 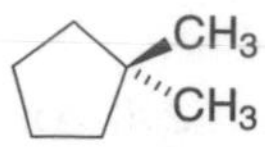 and 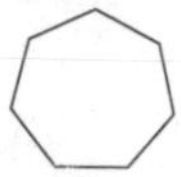

same molecular formula
different connectivity
constitutional isomers

c. 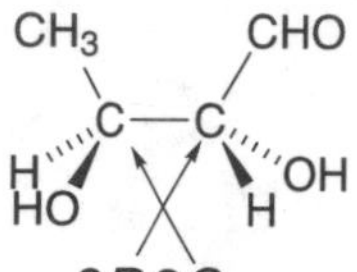 and 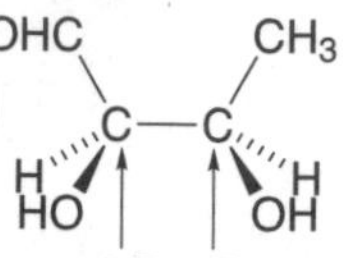

2*R*,3*S* 2*R*,3*R*

one different configuration
diastereomers

d. 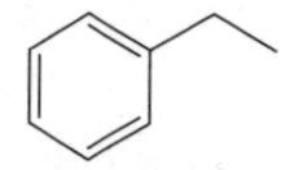 and 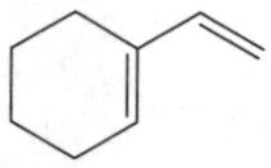

different molecular formulas
not isomers

e. 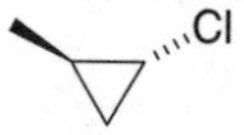 and 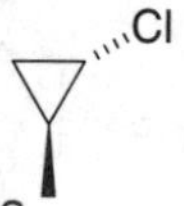

mirror images
not superimposable
enantiomers

f. 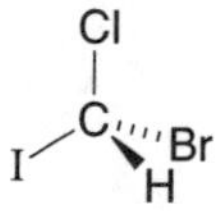 and 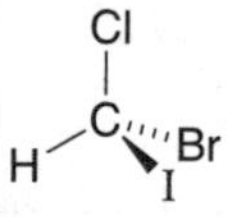

enantiomers

g. 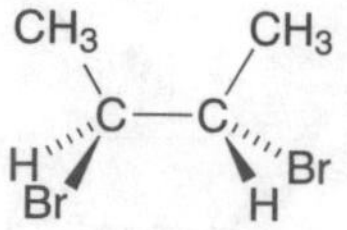 and 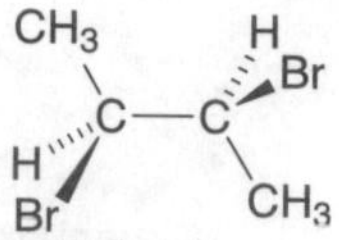

2*S*,3*S* 2*S*,3*S*
identical

h. 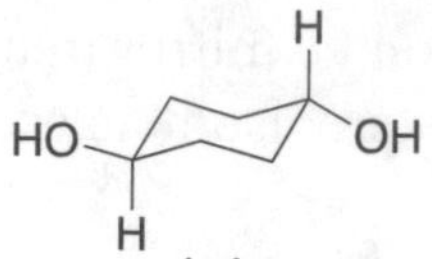 and 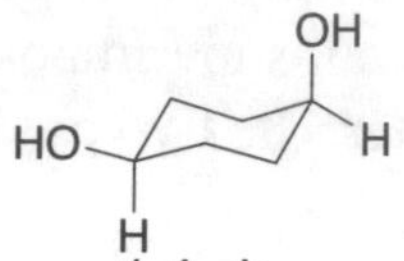

1,4-trans 1,4-cis
diastereomers

i. 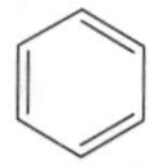 and 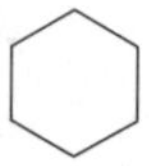

6 H's 12 H's
different molecular formulas
not isomers

j. 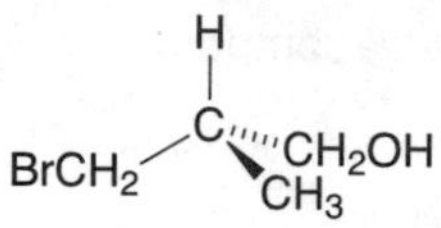 and 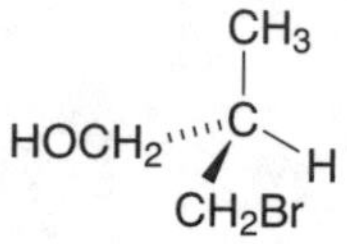

enantiomers

k. 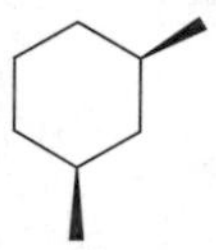 and 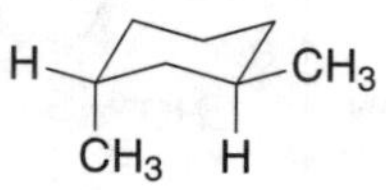

1,3-cis 1,3-trans
diastereomers

Br on end

l. 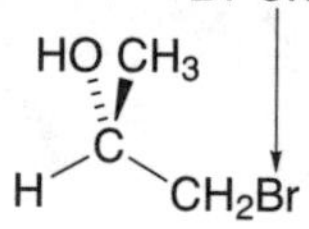 and 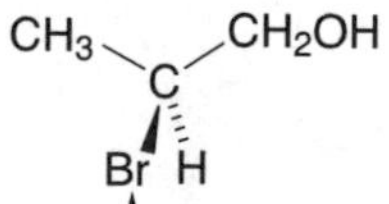

Br in middle

different connectivity
constitutional isomers

5.60

a. **A** and **B** are constitutional isomers.
   A and **C** are constitutional isomers.
   B and **C** are diastereomers (cis and trans).
   C and **D** are enantiomers.

b.

plane of symmetry

A

A plane of symmetry
in the plane of the page
achiral

B

achiral

C

chiral

D

chiral

mirror images and not
superimposable
enantiomers

c. Alone, **C** and **D** would be optically active.
d. **A** and **B** have a plane of symmetry.
e. **A** and **B** have different boiling points.
   B and **C** have different boiling points.
   C and **D** have the same boiling point.
f. **B** is a meso compound.
g. An equal mixture of **C** and **D** is optically inactive because it is a racemic mixture.
   An equal mixture of **B** and **C** would be optically active.

5.61

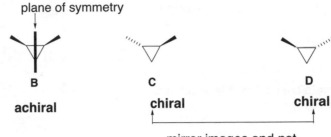

quinine

$$ee = \frac{[\alpha]\ \text{mixture}}{[\alpha]\ \text{pure enantiomer}} \times 100\%$$

quinine = **A**
quinine's enantiomer = **B**

a.

$$\frac{-50}{-165} \times 100\% = 30\%\ ee$$

b. 30% ee = 30% excess one compound (**A**)
   remaining 70% = mixture of 2 compounds (35% each **A** and **B**)
   Amount of **A** = 30 + 35 = **65%**
   Amount of **B** = **35%**

$$\frac{-83}{-165} \times 100\% = 50\%\ ee$$

50% ee = 50% the excess one compound (**A**)
remaining 50% = mixture of 2 compounds (25% each **A** and **B**)
Amount of **A** = 50 + 25 = **75%**
Amount of **B = 25%**

$$\frac{-120}{-165} \times 100\% = 73\%\ ee$$

73% ee = 73% excess of one compound (**A**)
remaining 27% = mixture of 2 compounds (13.5% each **A** and **B**)
Amount of **A** 73 + 13.5 = **86.5%**
Amount of **B** = **13.5%**

c. $[\alpha] = +165°$
d. 80% – 20% = 60% ee

e. $60\% = \dfrac{[\alpha]\ \text{mixture}}{-165°} \times 100\%$

$[\alpha]$ mixture = −99°

5.62

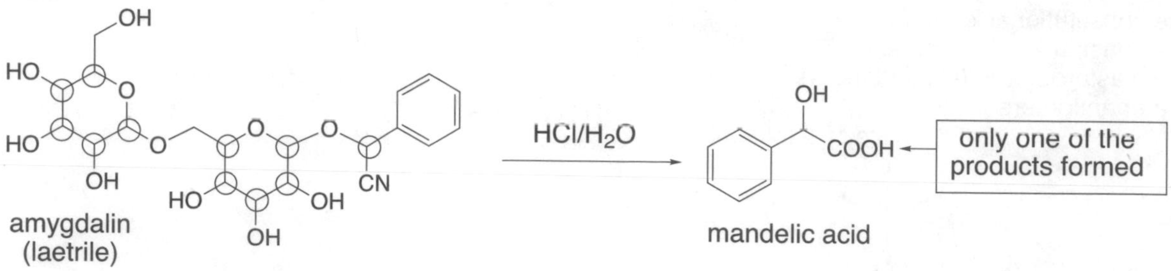

amygdalin
(laetrile)

mandelic acid

only one of the products formed

a. The 11 stereogenic centers are circled. Maximum number of stereoisomers = 2^{11} = 2048
b. Enantiomers of mandelic acid:

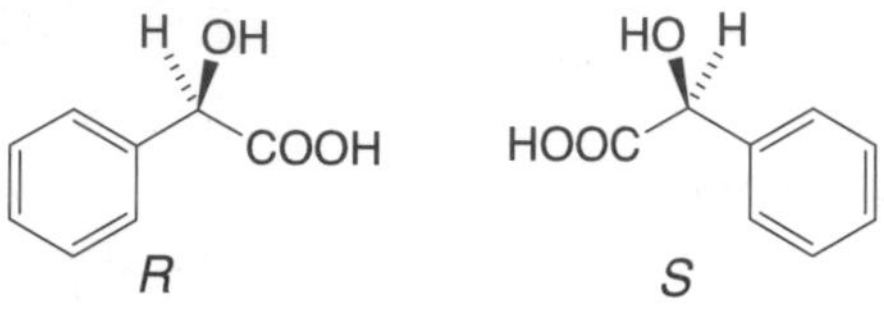

c. 60% – 40% = 20% ee
   20% = [α] mixture/–154° x 100%
   [α] mixture = –31°

d. ee = $\dfrac{+50°}{+154°}$ x 100% = 32% ee

   32% excess of the *S* enantiomer
   68% of racemic *R* and *S* = 34% *S* and 34% *R*

   [α] for (*S*)-mandelic acid = +154°

   S enantiomer: 32% + 34% = 66%
   R enantiomer = 34%

5.63 Allenes contain an *sp* hybridized carbon atom doubly bonded to two other carbons. This makes the double bonds of an allene perpendicular to each other. When each end of the allene has two like substituents, the allene contains two planes of symmetry and it is achiral. When each end of the allene has two different groups, the allene has no plane of symmetry and it becomes chiral.

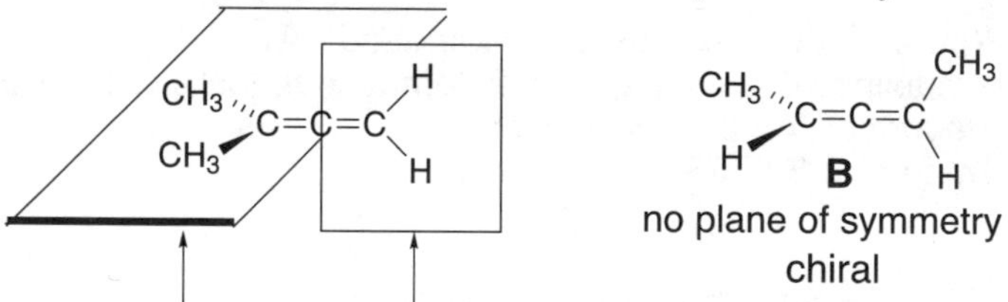

These two substituents are at 90° to these two substituents.
Allene **A** contains two planes of symmetry, making it **achiral.**

no plane of symmetry
chiral

5.64

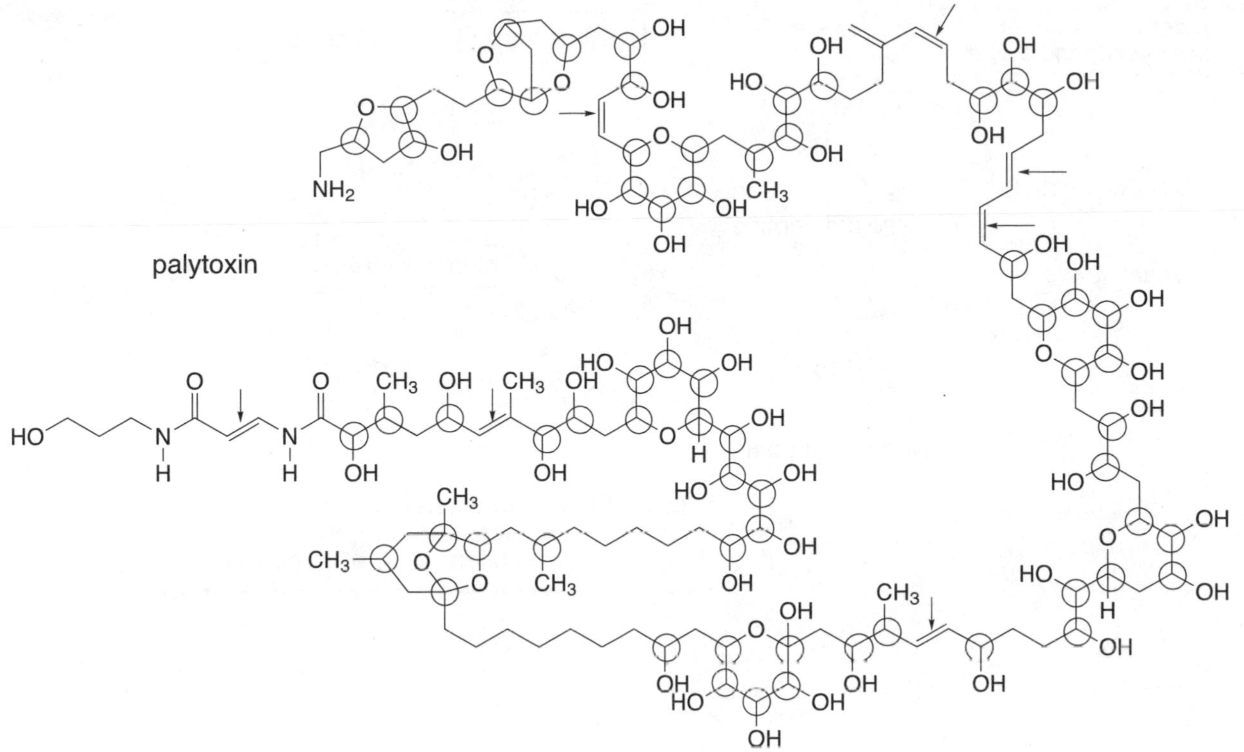

palytoxin

a. The 64 tetrahedral stereogenic centers are circled.

b. Because there is restricted rotation around a C–C double bond, groups on the end of the double bond cannot interconvert. Whenever the substituents on each end of the double bond are different from each other, the double bond is a stereogenic site. Thus, the following two double bonds are isomers:

These compounds are isomers.

Double bonds in palytoxin that are substituted by one carbon atom at each end are stereogenic. There are seven stereogenic double bonds in palytoxin, labeled with arrows.

c. The maximum number of stereoisomers for palytoxin must include the 64 tetrahedral stereogenic centers and the seven double bonds. Maximum number of stereoisomers = 2^{71} = 2.4 x 10^{21}.

5.65

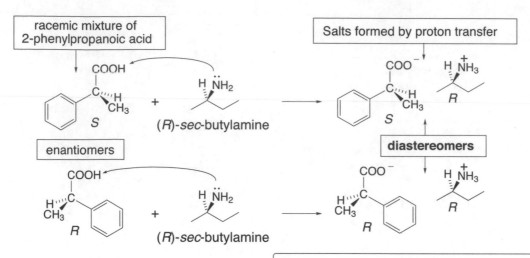

racemic mixture of 2-phenylpropanoic acid	Salts formed by proton transfer

enantiomers

diastereomers

(*R*)-*sec*-butylamine

(*R*)-*sec*-butylamine

These salts are now **diastereomers,**
and they are now separable by physical methods
since they have different physical properties.

Chapter 6: Understanding Organic Reactions

♦ Writing organic reactions (6.1)

- Use curved arrows to show the movement of electrons. Full-headed arrows are used for electron pairs and half-headed arrows are used for single electrons.

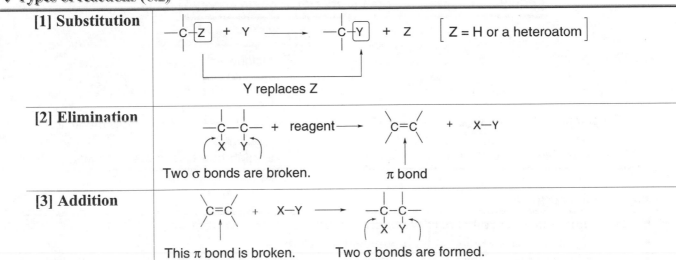

half-headed arrows   full-headed arrow

- Reagents can be drawn either on the left side of an equation or over an arrow. Catalysts are drawn over or under an arrow.

♦ Types of reactions (6.2)

[1] Substitution	$-\overset{\mid}{\underset{\mid}{C}}-Z$ + Y ⟶ $-\overset{\mid}{\underset{\mid}{C}}-Y$ + Z $\left[\text{Z = H or a heteroatom}\right]$ Y replaces Z
[2] Elimination	$-\overset{\mid}{\underset{\underset{X}{\mid}}{C}}-\overset{\mid}{\underset{\underset{Y}{\mid}}{C}}-$ + reagent ⟶ $\overset{\diagdown}{\diagup}C=C\overset{\diagup}{\diagdown}$ + X—Y Two σ bonds are broken. π bond
[3] Addition	$\overset{\diagdown}{\diagup}C=C\overset{\diagup}{\diagdown}$ + X—Y ⟶ $-\overset{\mid}{\underset{\underset{X}{\mid}}{C}}-\overset{\mid}{\underset{\underset{Y}{\mid}}{C}}-$ This π bond is broken. Two σ bonds are formed.

♦ Important trends

Values compared	Trend
bond dissociation energy and bond strength	The **higher** the bond dissociation energy, the **stronger** the bond (6.4). *increasing* size of the halogen ⟶ CH_3—F CH_3—Cl CH_3—Br CH_3—I ΔH^o = 109 kcal/mol 84 kcal/mol 70 kcal/mol 56 kcal/mol ⟵ *increasing* bond strength

E_a and **reaction rate**	The *larger* the energy of activation, the *slower* the reaction (6.9A). 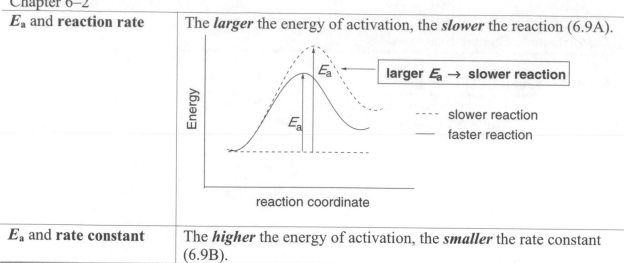
E_a and **rate constant**	The *higher* the energy of activation, the *smaller* the rate constant (6.9B).

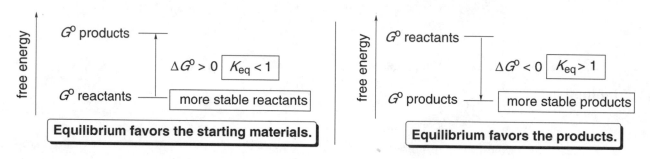

♦ Reactive intermediates (6.3)

- Breaking bonds generates reactive intermediates.
- Homolysis generates radicals with unpaired electrons.
- Heterolysis generates ions.

Reactive intermediate	General structure	Reactive feature	Reactivity
free radical	—Ċ·	unpaired electron	electrophilic
carbocation	—C+	positive charge; only 6 electrons around C	electrophilic
carbanion	—Ċ:⁻	net negative charge; lone electron pair on C	nucleophilic

♦ **Energy diagrams (6.7, 6.8)**

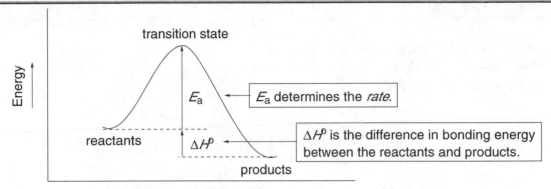

♦ **Conditions favoring product formation (6.5, 6.6)**

Variable	Value	Meaning
K_{eq}	$K_{eq} > 1$	More product than starting material is present at equilibrium.
ΔG°	$\Delta G^\circ < 0$	The energy of the products is **lower** than the energy of the reactants.
ΔH°	$\Delta H^\circ < 0$	Bonds in the products are **stronger** than bonds in the reactants.
ΔS°	$\Delta S^\circ > 0$	The product is **more disordered** than the reactant.

♦ **Equations (6.5, 6.6)**

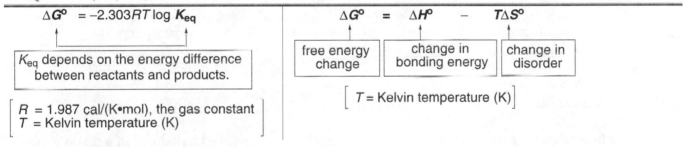

$$\Delta G^\circ = -2.303RT \log K_{eq}$$

K_{eq} depends on the energy difference between reactants and products.

R = 1.987 cal/(K•mol), the gas constant
T = Kelvin temperature (K)

$$\Delta G^\circ = \Delta H^\circ - T\Delta S^\circ$$

free energy change | change in bonding energy | change in disorder

$$\left[T = \text{Kelvin temperature (K)} \right]$$

♦ **Factors affecting reaction rate (6.9)**

Factor	Effect
energy of activation	higher E_a → slower reaction.
concentration	higher concentration → faster reaction.
temperature	higher temperature → faster reaction

6.1 [1] In a **substitution reaction**, one group replaces another.
[2] In an **elimination reaction**, elements of the starting material are lost and a π bond is formed.
[3] In an **addition reaction**, elements are added to the starting material.

a.
Br replaces OH =
substitution reaction

c.
Cl replaces H =
substitution reaction

b.
addition of 2 H's
addition reaction

d. CH₃CH₂CH(OH)CH₃ → CH₃CH=CHCH₃

elements lost
(H + OH)

π bond formed

elimination reaction

6.2 **Heterolysis** means one atom gets both of the electrons when a bond is broken. A carbocation is a C
with a positive charge, and a carbanion is a C with a negative charge.

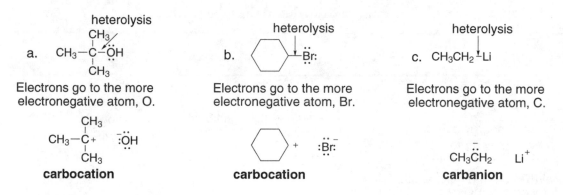

a. CH₃−C−ÖH

Electrons go to the more
electronegative atom, O.

CH₃−C+   ⁻:ÖH

carbocation

b.

Electrons go to the more
electronegative atom, Br.

+   :Br:⁻

carbocation

c. CH₃CH₂−Li

Electrons go to the more
electronegative atom, C.

CH₃CH₂⁻   Li⁺

carbanion

6.3 Use **full-headed arrows** to show the movement of electron pairs, and **half-headed arrows** to show
the movement of single electrons.

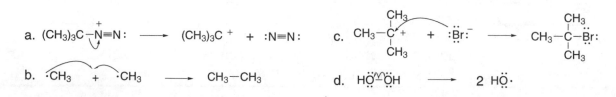

a. (CH₃)₃C−N≡N: → (CH₃)₃C⁺ + :N≡N:

c. CH₃−C⁺ + :Br:⁻ → CH₃−C−Br:
   (with CH₃ groups)

b. ·CH₃ + ·CH₃ → CH₃−CH₃

d. HÖ–ÖH → 2 HÖ·

6.4 Increasing number of electrons between atoms = increasing bond strength = increasing bond
dissociation energy = decreasing bond length.
Increasing size of an atom = increasing bond length = decreasing bond strength.

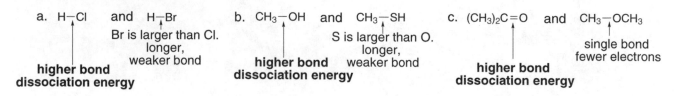

a. H−Cl   and   H−Br

Br is larger than Cl.
longer,
weaker bond

**higher bond
dissociation energy**

b. CH₃−OH   and   CH₃−SH

S is larger than O.
longer,
weaker bond

**higher bond
dissociation energy**

c. (CH₃)₂C=O   and   CH₃−OCH₃

single bond
fewer electrons

**higher bond
dissociation energy**

6.5 To determine ΔH° for a reaction:

[1] Add the bond dissociation energies for all bonds *broken* in the equation (+ values).

[2] Add the bond dissociation energies for all of the bonds *formed* in the equation (– values).

[3] *Add the energies together* to get the ΔH° for the reaction.

A positive ΔH° means the reaction is **endothermic**. **A negative ΔH°** means the reaction is *exothermic*.

a. $CH_3CH_2-Br + H_2O \longrightarrow CH_3CH_2-OH + HBr$

[1] Bonds broken

	ΔH° (kcal/mol)
CH_3CH_2-Br	+ 68
$H-OH$	+ 119
Total	+ 187 kcal/mol

[2] Bonds formed

	ΔH° (kcal/mol)
CH_3CH_2-OH	– 91
$H-Br$	– 88
Total	– 179 kcal/mol

[3] Overall ΔH° =

> sum in Step [1]
> +
> sum in Step [2]

+ 187 kcal/mol
– 179 kcal/mol

ANSWER: + 8 kcal/mol
endothermic

b. $CH_4 + Cl_2 \longrightarrow CH_3Cl + HCl$

[1] Bonds broken

	ΔH° (kcal/mol)
CH_3-H	+ 104
$Cl-Cl$	+ 58
Total	+ 162 kcal/mol

[2] Bonds formed

	ΔH° (kcal/mol)
CH_3-Cl	– 84
$H-Cl$	– 103
Total	– 187 kcal/mol

[3] Overall ΔH° =

> sum in Step [1]
> +
> sum in Step [2]

+ 162 kcal/mol
– 187 kcal/mol

ANSWER: – 25 kcal/mol
exothermic

6.6 Use the directions from Answer 6.5. In determining the number of bonds broken or formed, you must take into account the coefficients needed to balance an equation.

a. CH_4 + $2O_2$ $\longrightarrow$ CO_2 + $2H_2O$

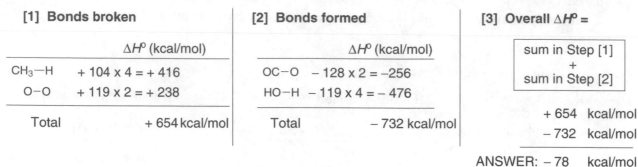

[1] Bonds broken	[2] Bonds formed	[3] Overall ΔH^o =
ΔH^o (kcal/mol)	ΔH^o (kcal/mol)	sum in Step [1] + sum in Step [2]
CH_3-H + 104 x 4 = + 416	$OC-O$ – 128 x 2 = –256	
$O-O$ + 119 x 2 = + 238	$HO-H$ – 119 x 4 = – 476	+ 654 kcal/mol – 732 kcal/mol
Total + 654 kcal/mol	Total – 732 kcal/mol	

ANSWER: – 78   kcal/mol

b. CH_3CH_3 + $(7/2)O_2$ $\longrightarrow$ $2CO_2$ + $3H_2O$

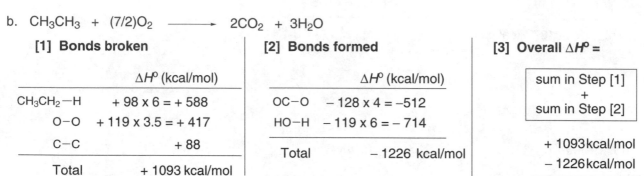

[1] Bonds broken	[2] Bonds formed	[3] Overall ΔH^o =
ΔH^o (kcal/mol)	ΔH^o (kcal/mol)	sum in Step [1] + sum in Step [2]
CH_3CH_2-H + 98 x 6 = + 588	$OC-O$ – 128 x 4 = –512	
$O-O$ + 119 x 3.5 = + 417	$HO-H$ – 119 x 6 = – 714	+ 1093 kcal/mol – 1226 kcal/mol
$C-C$ + 88	Total – 1226 kcal/mol	
Total + 1093 kcal/mol		

ANSWER: – 133 kcal/mol

6.7 Use the following relationships to answer the questions:
K_{eq} = 1 then ΔG^o = 0; K_{eq} > 1 then ΔG^o < 0; K_{eq} < 1 then ΔG^o > 0

a. A negative value of ΔG^o means the equilibrium favors the product and K_{eq} is > 1. Therefore K_{eq} = 1000 is the answer.
b. A lower value of ΔG^o means a larger value of K_{eq}, and the products are more favored. K_{eq} = 10^{-2} is larger than K_{eq} = 10^{-5}, so ΔG^o is lower.

6.8 Use the relationships from Answer 6.7.

a. K_{eq} = 5.5. K_{eq} > 1 means that the equilibrium favors the **product**.
b. ΔG^o = 10.3 kcal. A positive ΔG^o means the equilibrium favors the **starting material**.

6.9 When the product is lower in energy than the starting material, the equilibrium favors the product. When the starting material is lower in energy than the product, the equilibrium favors the starting material.

a. ΔG^o **is positive** so the equilibrium favors the starting material. Therefore the *starting material is lower in energy than the product.*
b. K_{eq} **is > 1** so the equilibrium fa rs the product. Therefore the *product is lower in energy than the starting material.*
c. ΔG^o **is negative** so the equilibrium favors the products. Therefore the *products are lower in energy than the starting material.*

d. K_{eq} is < 1 so the equilibrium favors the starting material. Therefore *the starting material is lower in energy than the product.*

6.10

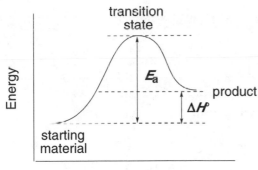

$K_{eq} = 2.7$

a. The K_{eq} is > 1 and therefore the **product** (the conformer on the right) is favored at equilibrium.
b. The $\Delta G°$ for this process must be **negative** since the product is favored.
c. $\Delta G°$ is somewhere between 0 and -1.4 kcal/mol.

6.11 A positive $\Delta H°$ favors the starting material. A negative $\Delta H°$ favors the product.

a. $\Delta H°$ is positive (20 kcal). The starting material is favored.
b. $\Delta H°$ is negative (-10 kcal). The product is favored.

6.12
a. **false**. The reaction is endothermic.
b. **true**. This assumes that $\Delta G°$ is approximately equal to $\Delta H°$.
c. **false**. $K_{eq} < 1$
d. **true**
e. **false**. The starting material is favored at equilibrium.

6.13

transition
state

Energy

E_a

product

$\Delta H°$

starting
material

reaction coordinate

6.14 A transition state is drawn with dashed lines to indicate the partially broken and partially formed bonds. Any atom that gains or loses a charge contains a partial charge in the transition state.

a. $CH_3-\overset{\overset{\displaystyle CH_3}{|}}{\underset{\underset{\displaystyle CH_3}{|}}{\overset{+}{C}}}-OH_2 \longrightarrow CH_3-\overset{\overset{\displaystyle CH_3}{|}}{\underset{\underset{\displaystyle CH_3}{|}}{C}}{+} \ + \ H_2O$

b. $CH_3O-H \ + \ {}^-OH \longrightarrow CH_3O^- \ + \ H_2O$

transition state: $\left[CH_3 \overset{\overset{\displaystyle CH_3}{|}}{\underset{\underset{\displaystyle CH_3}{|}}{\overset{\delta +}{\overset{H}{C}}}} - - - \overset{\delta +}{OH_2} \right]^{\ddagger}$

transition state: $\left[CH_3O - - - \overset{\delta -}{H} - - - \overset{\delta -}{OH} \right]^{\ddagger}$

6.15

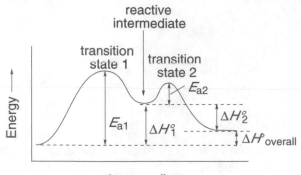

reactive
intermediate

transition
state 1 | transition
state 2

E_{a2}

Energy →

E_{a1} ΔH_1°

ΔH_2°

$\Delta H_{overall}^{\circ}$

reaction coordinate

a. Two steps since there are two energy barriers.
b. See labels.
c. See labels.
d. One reactive intermediate is formed (see label).
e. The first step is rate-determining since its transition state is at higher energy.
f. The overall reaction is endothermic since the energy of the products is higher than the energy of the reactants.

6.16

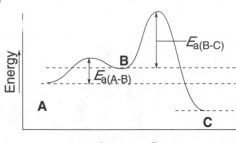

Energy

$E_{a(B-C)}$

B

$E_{a(A-B)}$

A

C

reaction coordinate

relative energies: C < A < B
B → C is rate limiting.

6.17 E_a, **concentration, and temperature affect reaction rate.** $\Delta H°$, $\Delta G°$, and K_{eq} do not affect reaction rate.

a. E_a = **1 kcal** corresponds to a faster reaction rate.
b. A temperature of **25 °C** will have a faster reaction rate since a higher temperature corresponds to a faster reaction.
c. **No change**: K_{eq} does not affect reaction rate.
d. **No change**: $\Delta H°$ does not affect reaction rate.

6.18 All reactants in the rate equation determine the rate of the reaction.

(1) rate = $k[CH_3CH_2Br][^-OH]$
a. Tripling the concentration of CH_3CH_2Br only → **The rate is tripled.**
b. Tripling the concentration of ^-OH only → **The rate is tripled.**
c. Tripling the concentration of both $CH_3CH_2CH_2Br$ and ^-OH → **The rate increases by a factor of 9 (3 x 3 = 9).**

(2) rate = $k[(CH_3)_3COH]$
a. Doubling the concentration of $(CH_3)_3COH$ → **The rate is doubled.**
b. Increasing the concentration of $(CH_3)_3COH$ by a factor of 10 → **The rate increases by a factor of 10.**

6.19 The rate equation is determined by the rate-determining step.

one step

a. CH_3CH_2-Br + ^-OH $\longrightarrow$ $CH_2=CH_2$ + H_2O + Br^- **rate = k[CH$_3$CH$_2$Br][$^-$OH]**

b. $(CH_3)_3C-Br$ $\xrightarrow{slow}$ $(CH_3)_3C +$ $\xrightarrow[fast]{^-OH}$ $(CH_3)_2C=CH_2$ + H_2O The slow step determines the rate equation.
$+$ Br^-

two steps

rate = k[(CH$_3$)$_3$CBr]

6.20 A catalyst is not used up or changed in the reaction. It only speeds up the reaction rate.

OH and H are added to
the starting material.

a. $CH_2=CH_2$ $\xrightarrow[H_2SO_4]{H_2O}$ CH_3CH_2OH

H_2SO_4 is not used up = **catalyst.**

b. CH_3Cl $\xrightarrow[OH^-]{I^-}$ CH_3OH

I^- not used up = **catalyst.**

^-OH substitutes for Cl^-.

c. $\xrightarrow[Pt]{H_2}$

H_2 adds to the starting material.

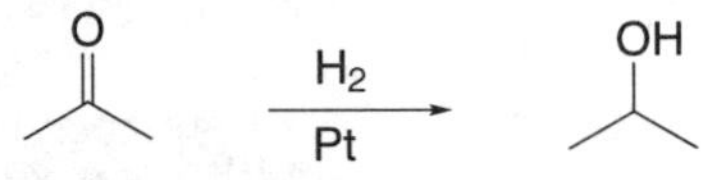

Pt not used up = **catalyst.**

6.21 Use the directions from Answer 6.1.

a.

elements lost
(H + OH)   π bond formed
elimination reaction

c.

addition of 2 H's
addition reaction

b.

Cl replaces H =
substitution reaction

d.

H replaces Cl =
substitution reaction

6.22

a. homolysis of $CH_3-\overset{\overset{\displaystyle H}{|}}{\underset{\underset{\displaystyle H}{|}}{C}}-H$

$CH_3-\overset{\overset{\displaystyle H}{|}}{\underset{\underset{\displaystyle H}{|}}{C}}\cdot \; + \; \cdot H$

radical

b. heterolysis of CH_3-O-H

$CH_3-\ddot{\underset{\cdot\cdot}{O}}:^- \; + \; H^+$

c. heterolysis of CH_3-MgBr

$^-\ddot{C}H_3 \; + \; ^+MgBr$

carbanion

6.23 Use the rules in Answer 6.3 to draw the arrows.

a.   ⟶ $^+$ + $:\ddot{\underset{\cdot\cdot}{Br}}:^-$

d.   + $:\ddot{\underset{\cdot\cdot}{Br}}-\ddot{\underset{\cdot\cdot}{Br}}:$ ⟶   + $:\ddot{\underset{\cdot\cdot}{Br}}\cdot$

b. $CH_3-\overset{\overset{\displaystyle :\ddot{O}:^-}{|}}{\underset{\underset{\displaystyle :\ddot{Cl}:}{|}}{C}}-CH_3$ ⟶ $\underset{CH_3}{\overset{:O:}{\overset{||}{C}}}CH_3$ + $:\ddot{\underset{\cdot\cdot}{Cl}}:^-$

e. $CH_3CH_2-\ddot{\underset{\cdot\cdot}{Br}}:$ + $^-\ddot{\underset{\cdot\cdot}{O}}H$ ⟶ CH_3CH_2OH + $:\ddot{\underset{\cdot\cdot}{Br}}:^-$

c. $\cdot CH_3$ + $\cdot \ddot{\underset{\cdot\cdot}{Cl}}:$ ⟶ $CH_3-\ddot{\underset{\cdot\cdot}{Cl}}:$

f. $CH_3-\overset{\overset{\displaystyle CH_3}{|}}{\underset{\underset{\displaystyle +}{|}}{C}}-\overset{\overset{\displaystyle H}{|}}{\underset{\underset{\displaystyle H}{|}}{C}}-H$ + $^-\ddot{\underset{\cdot\cdot}{O}}H$ ⟶ $\underset{CH_3}{\overset{CH_3}{C}}=C\overset{H}{\underset{H}{}}$ + $H_2\ddot{\underset{\cdot\cdot}{O}}:$

6.24

a. $-\ddot{\underset{\cdot\cdot}{I}}:$ + $^-\ddot{\underset{\cdot\cdot}{O}}H$ ⟶ $-\ddot{\underset{\cdot\cdot}{O}}H$ + $:\ddot{\underset{\cdot\cdot}{I}}:^-$

b. $CH_3-\overset{\overset{\displaystyle :\ddot{O}:}{|}}{\underset{\underset{\displaystyle :OCH_2CH_3}{|}}{C}}-CH_2CH_2CH_3$ ⟶ $\underset{CH_3}{\overset{:O:}{\overset{||}{C}}}CH_2CH_2CH_3$

+ $^-\ddot{\underset{\cdot\cdot}{O}}CH_2CH_3$

c. $^-\ddot{\underset{\cdot\cdot}{O}}H$ $H-\overset{\overset{\displaystyle H}{|}}{\underset{\underset{\displaystyle H}{|}}{C}}-\overset{\overset{\displaystyle H}{|}}{\underset{\underset{\displaystyle :Br:}{|}}{C}}-H$ ⟶ $\overset{H}{\underset{H}{}}C=C\overset{H}{\underset{H}{}}$ + $H_2\ddot{\underset{\cdot\cdot}{O}}:$ + $:\ddot{\underset{\cdot\cdot}{Br}}:^-$

d. $\overset{\overset{\displaystyle H}{|}}{\underset{\underset{\displaystyle H}{|}}{C}}-H$ + $\cdot \ddot{\underset{\cdot\cdot}{Cl}}:$ ⟶ $\overset{\overset{\displaystyle H}{|}}{\underset{\underset{\displaystyle H}{|}}{C}}\cdot$ + $H\ddot{\underset{\cdot\cdot}{Cl}}:$

6.25 Use the rules from Answer 6.4.

a. I—CCl$_3$ Br—CCl$_3$ Cl—CCl$_3$

largest halogen   **intermediate**   smallest halogen
weakest bond   **bond strength**   **strongest bond**

b. H$_2$N—NH$_2$ HN=NH   N≡N

single bond   double bond   triple bond
weakest bond   **intermediate**   **strongest bond**
                     bond strength

6.26 Use the directions from Answer 6.5.

a. CH$_3$CH$_2$—H + Br$_2$ ⟶ CH$_3$CH$_2$—Br + HBr

[1] Bonds broken

	ΔH^o (kcal/mol)
CH$_3$CH$_2$—H	+ 98
Br—Br	+ 46
Total	+ 144 kcal/mol

[2] Bonds formed

	ΔH^o (kcal/mol)
CH$_3$CH$_2$—Br	− 68
H—Br	− 88
Total	− 156 kcal/mol

[3] Overall ΔH^o =

+ 144 kcal/mol
− 156 kcal/mol

ANSWER: − 12 kcal/mol

b. ·OH + CH$_4$ ⟶ ·CH$_3$ + H$_2$O

[1] Bonds broken

	ΔH^o (kcal/mol)
CH$_3$—H	+ 104 kcal/mol

[2] Bonds formed

	ΔH^o (kcal/mol)
H—OH	− 119 kcal/mol

[3] Overall ΔH^o =

+ 104 kcal/mol
− 119 kcal/mol

ANSWER: − 15 kcal/mol

c. CH$_3$—OH + HBr ⟶ CH$_3$—Br + H$_2$O

[1] Bonds broken

	ΔH^o (kcal/mol)
CH$_3$—OH	+ 91
H—Br	+ 88
Total	+ 179 kcal/mol

[2] Bonds formed

	ΔH^o (kcal/mol)
CH$_3$—Br	− 70
H—OH	− 119
Total	− 189 kcal/mol

[3] Overall ΔH^o =

+ 179 kcal/mol
− 189 kcal/mol

ANSWER: − 10 kcal/mol

d. ·Br + CH$_4$ ⟶ ·H + CH$_3$Br

[1] Bonds broken

	ΔH^o (kcal/mol)
CH$_3$—H	+ 104 kcal/mol

[2] Bonds formed

	ΔH^o (kcal/mol)
CH$_3$—Br	− 70 kcal/mol

[3] Overall ΔH^o =

+ 104 kcal/mol
− 70   kcal/mol

ANSWER: + 34 kcal/mol

6.27

propane

$CH_3-CH_2CH_3 \longrightarrow \cdot CH_3 + \cdot CH_2CH_3$

$\Delta H^\circ = 85$ kcal/mol

propene

$CH_3-CH=CH_2 \longrightarrow \cdot CH_3 + \cdot CH=CH_2$

$\Delta H^\circ = 92$ kcal/mol

$CH_2=CH\cdot$ has an unpaired electron on a C that is part of the double bond. This C must have a higher percent s-character in its hybrid orbitals than does the C with the unpaired electron in $\cdot CH_2CH_3$, so the unpaired electron is held closer to the nucleus in $\cdot CH=CH_2$, making it less stable. [Note: The hybridization of radicals is not discussed until Chapter 13. But, since $\cdot CH_2CH_3$ has the unpaired electron on a carbon bonded to three other atoms, and $\cdot CH=CH_2$ has the unpaired electron on a carbon bonded to only two other atoms, the latter C must have a higher percent s-character in its hybrid orbitals.]

6.28

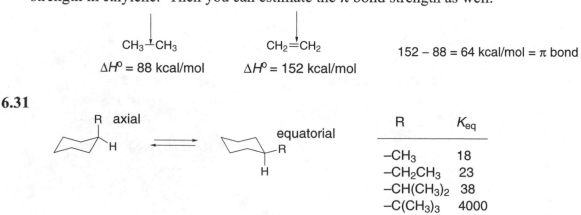

6.29

The more stable radical is formed by a reaction with a smaller ΔH°.

$CH_3-CH_2-\overset{H}{\underset{H}{C}}-H \longrightarrow CH_3-CH_2-\overset{H}{\underset{H}{\overset{\cdot}{C}}}-H$ $\Delta H^\circ = 98$ kcal/mol = less stable radical

This C–H bond is stronger.   **A**

$CH_3-\overset{H}{\underset{H}{C}}-CH_3 \longrightarrow CH_3-\overset{\cdot}{\underset{H}{C}}-CH_3$ $\Delta H^\circ = 95$ kcal/mol = more stable radical

This C–H bond is weaker.   **B**

Since the bond dissociation for cleavage of the C–H bond to form radical **A** is higher, more energy must be added to form it. This makes **A** higher in energy and therefore less stable than **B**.

6.30 Use the bond dissociation energy for the C–C σ bond in ethane as an estimate of the σ bond strength in ethylene. Then you can estimate the π bond strength as well.

CH_3-CH_3

$\Delta H^\circ = 88$ kcal/mol

$CH_2=CH_2$

$\Delta H^\circ = 152$ kcal/mol

$152 - 88 = 64$ kcal/mol = π bond

6.31

R axial

equatorial

R	K_{eq}
$-CH_3$	18
$-CH_2CH_3$	23
$-CH(CH_3)_2$	38
$-C(CH_3)_3$	4000

a. The equatorial conformer is always present in the larger amount at equilibrium since the K_{eq} for all R groups is greater than 1.

b. The cyclohexane with the –$C(CH_3)_3$ group will have the greatest amount of equatorial conformer at equilibrium since this group has the highest K_{eq}.

c. The cyclohexane with the –CH_3 group will have the greatest amount of axial conformer at equilibrium since this group has the lowest K_{eq}.

d. The cyclohexane with the –$C(CH_3)_3$ group will have the most negative $\Delta G°$ since it has the largest K_{eq}.

e. The larger the R group, the more favored the equatorial conformer.

f. The K_{eq} for *tert*-butylcyclohexane is much higher because it is bulkier than the other groups. With a *tert*-butyl group, a CH_3 group is always oriented over the ring when the group is axial, creating severe 1,3-diaxial interactions. With all other substituents, the larger CH_3 groups can be oriented away from the ring, placing a H over the ring, making the 1,3-diaxial interactions less severe. Compare:

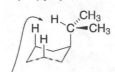

tert-butylcyclohexane	isopropylcyclohexane
severe 1,3-diaxial interactions with the CH_3 groups and the axial H	less severe 1,3-diaxial interactions

6.32 Use the rules from Answer 6.9.

a. $K_{eq} = 0.5$. K_{eq} is less than one so the **starting material** is favored.

b. $\Delta G° = -25$ kcal/mol. $\Delta G°$ is less than 0 so the **product** is favored.

c. $\Delta H° = 2.0$ kcal/mol. $\Delta H°$ is positive, so the **starting material** is favored.

d. $K_{eq} = 16$. K_{eq} is greater than one so the **product** is favored.

e. $\Delta G° = 0.5$ kcal/mol. $\Delta G°$ is greater than zero so the **starting material** is favored.

f. $\Delta H° = 100$ kcal/mol. $\Delta H°$ is positive so the **starting material** is favored.

g. $\Delta S° = 2$ cal/K•mol. $\Delta S°$ is greater than zero so the **product** is more disordered and favored.

h. $\Delta S° = -2$ cal/K•mol. $\Delta S°$ is less than zero so the **starting material** is more disordered and favored.

6.33

a. A negative $\Delta G°$ must have $K_{eq} > 1$. $K_{eq} = 10^2$.

b. $K_{eq} = $ [products]/[reactants] = [1]/[5] = 0.2 = K_{eq}. $\Delta G°$ is positive.

c. A negative $\Delta G°$ has $K_{eq} > 1$, and a positive $\Delta G°$ has $K_{eq} < 1$. $\Delta G° = -2$ kcal/mol will have a larger K_{eq}.

6.34 Reactions resulting in an increase in entropy are favored. When a single molecule forms two molecules, there is an increase in entropy.

a.   [structure] ⟶ [structure] + [structure]   increased number of molecules
$\Delta S°$ is positive.
products favored.

b. $CH_3 \cdot$ + $CH_3 \cdot$ ⟶ CH_3CH_3 decreased number of molecules
$\Delta S°$ is negative.
starting material favored.

c. $(CH_3)_2C(OH)_2$ ⟶ $(CH_3)_2C{=}O$ + H_2O increased number of molecules
$\Delta S°$ is positive.
products favored.

d. CH_3COOCH_3 + H_2O ⟶ CH_3COOH + CH_3OH no change in the number of molecules
neither favored

6.35 Use the directions in Answer 6.14 to draw the transition state.

a. [cyclohexane–Br] ⟶ [cyclohexyl cation]⁺ + Br⁻
transition state: $\left[\,[cyclohexyl]\ \delta^{+}Br\ {}^{\delta^{-}}\,\right]^{\ddagger}$

c. [cyclohexanol] + ⁻NH_2 ⟶ [cyclohexanolate] O⁻ + NH_3
transition state: $\left[\,[cyclohexane]{-}O{\cdots}H{\cdots}NH_2\ {}^{\delta^{-}\ \ \delta^{-}}\,\right]^{\ddagger}$

b. BF_3 + Cl^- ⟶ $F{-}\overset{\overset{F}{|}}{\underset{\underset{F}{|}}{B}}{-}Cl$
transition state: $\left[\,F{-}\overset{\overset{F}{|}}{\underset{\underset{F}{|}}{B}}{\cdot\cdot}Cl\ {}_{\delta^{-}\ \ \delta^{-}}\,\right]^{\ddagger}$

d. $CH_3{-}\overset{CH_3}{\underset{+}{C}}{-}\overset{H}{\underset{H}{C}}{-}H$ + H_2O ⟶ $\overset{CH_3}{\underset{CH_3}{C}}{=}\overset{H}{\underset{H}{C}}$ + H_3O^+
transition state: $\left[\,CH_3{-}\overset{CH_3\ \ \ \ H}{\underset{\delta^{+}\ \ \ \ \ H}{C{=\!=\!=}C}}{-}H{\cdots}OH_2\ {}^{\delta^{+}}\,\right]^{\ddagger}$

6.36

a.
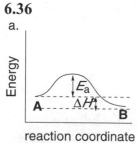
reaction coordinate

• one step
• exothermic since **B** lower than **A**
• low energy of activation (small energy barrier)

b.
Energy | E_a | $\Delta H°$ | **B** | **A**
reaction coordinate

• one step
• endothermic since **B** higher than **A**
• high energy of activation (large energy barrier)

c.
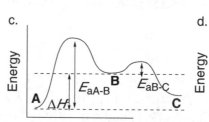
reaction coordinate

• two steps
• **A** lowest energy
• **B** highest energy
• $E_{a(A\text{-}B)}$ larger than $E_{a(B\text{-}C)}$; $E_{a(A\text{-}B)}$ **is rate-determining**

d.
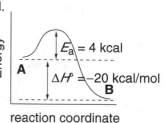
reaction coordinate

• one step
• exothermic since **B** lower than **A**

6.37

a. 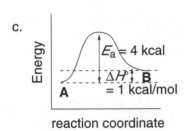 $CH_3\!-\!H$ $+$ $\cdot Cl$ $\longrightarrow$ $\cdot CH_3$ $+$ HCl

b. $\cdot Cl + CH_4$ $\longrightarrow$ $\cdot CH_3$ $+$ HCl

[1] Bonds broken		**[2] Bonds formed**		**[3] Overall $\Delta H^\circ =$**
	ΔH° (kcal/mol)		ΔH° (kcal/mol)	$+ 104$ kcal/mol
$CH_3\!-\!H$	$+ 104$ kcal/mol	$H\!-\!Cl$	$- 103$ kcal/mol	$- 103$ kcal/mol
				ANSWER: $+ 1$ kcal/mol

c.

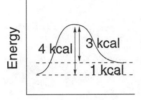

d. The E_a for the reverse reaction is the difference in energy between the products and the transition state, 3 kcal.

6.38

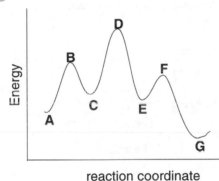

reaction coordinate

a. **B, D** and **F** are transition states.
b. **C** and **E** are reactive intermediates.
c. The overall reaction has **3 steps.**

6.39

a. Step [1] breaks one π bond and the H–Cl bond, and one C–H bond is formed. The ΔH° for this step should be positive since more bonds are broken than formed.
b. Step [2] forms one bond. The ΔH° for this step should be negative since one bond is formed and none are broken.
c. Step [1] is rate-determining since it is more difficult.

d. Transition state for Step [1]:   Transition state for Step [2]:

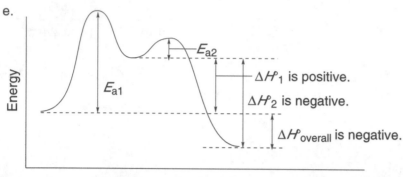

e.

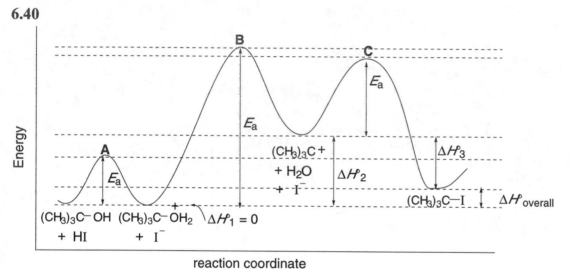

ΔH^o_1 is positive.

ΔH^o_2 is negative.

$\Delta H^o_{overall}$ is negative.

reaction coordinate

6.40

(CH$_3$)$_3$C+ + H$_2$O + I$^-$ ΔH^o_2

(CH$_3$)$_3$C—I ΔH^o_3 $\Delta H^o_{overall}$

(CH$_3$)$_3$C—OH (CH$_3$)$_3$C—OH$_2$ $\Delta H^o_1 = 0$
+ HI   + I$^-$

reaction coordinate

a. The reaction has three steps, since there are three energy barriers.
b. See above.
c. Transition state **A** (see graph for location):   Transition state **B**:   Transition state **C**:

$$\left[(CH_3)_3C-\overset{\delta+}{O}H \atop \underset{H---I}{\delta-} \right]^{\ddagger}$$ $$\left[(CH_3)_3\overset{}{\underset{\delta+}{C}}---\overset{}{\underset{\delta+}{O}}H_2 \right]^{\ddagger}$$ $$\left[(CH_3)_3\overset{\delta+}{C}---\overset{\delta-}{I} \right]^{\ddagger}$$

d. Step 2 is rate-determining since this step has the highest energy transition state.

6.41 E_a, concentration, catalysts, rate constant and temperature affect reaction rate so c, d, e, g and h affect rate.

6.42

    a. **rate = k[CH$_3$Br][NaCN]**

    b. Double [CH$_3$Br] = **rate doubles.**

    c. Halve [NaCN] = **rate halved.**

    d. Increase both [CH$_3$Br] and [NaCN] by factor of 5 = [5][5] = **rate increases by a factor of 25.**

6.43

    a. Only the slow step is included in the rate equation: **Rate = k[CH$_3$O$^-$][CH$_3$COCl]**

    b. CH$_3$O$^-$ is in the rate equation. Increasing its concentration by 10 times would increase the rate by **10 times.**

    c. When both reactant concentrations are increased by 10 times, the rate increases by **100 times (10 x 10 = 100).**

    d. This is a **substitution reaction** (OCH$_3$ substitutes for Cl).

6.44

    a. **True**: Increasing temperature increases reaction rate.

    b. **True**: If a reaction is fast, it has a large rate constant.

    c. **False: Corrected** - There is no relationship between $\Delta G°$ and reaction rate.

    d. **False: Corrected** - When the E_a is large, *the rate constant is small.*

    e. **False: Corrected** - There is no relationship between K_{eq} and reaction rate.

    f. **False: Corrected** - Increasing the concentration of a reactant increases the rate of a reaction *only if the reactant appears in the rate equation.*

6.45

    a. The first mechanism is one step: **Rate = k[(CH$_3$)$_3$CI][$^-$OH]**

    b. The second mechanism is two steps, but only the first step would be in the rate equation since it is slow and therefore rate-determining: **Rate = k[(CH$_3$)$_3$CI]**

    c. Possibility [1] is second order; possibility [2] is first order.

    d. These rate equations can be used to show which mechanism is correct by changing the concentration of [$^-$OH]. If this affects the rate, possibility [1] is correct. If it does not affect the rate, possibility [2] is correct.

6.46 The difference in both acidity and bond dissociation of CH$_3$CH$_3$ versus HC≡CH is due to the same factor: percent *s*-character. The difference results because one process is based on homolysis and one is based on heterolysis

Bond dissociation energy:

Acidity. To compare acidity, we must compare the stability of the conjugate bases:

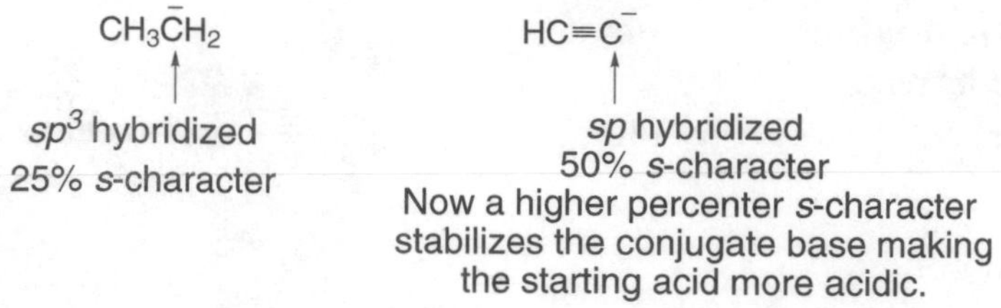

$CH_3\bar{C}H_2$ $HC{\equiv}C^-$

sp^3 hybridized sp hybridized
25% s-character   50% s-character
Now a higher percenter s-character
stabilizes the conjugate base making
the starting acid more acidic.

6.47 In Reaction [1], the number of molecules of reactants and products stays the same, so entropy is not a factor. In Reaction [2], a single molecule of starting material forms two molecules of products, so entropy increases. This makes $\Delta G°$ more favorable, thus increasing K_{eq}.

6.48

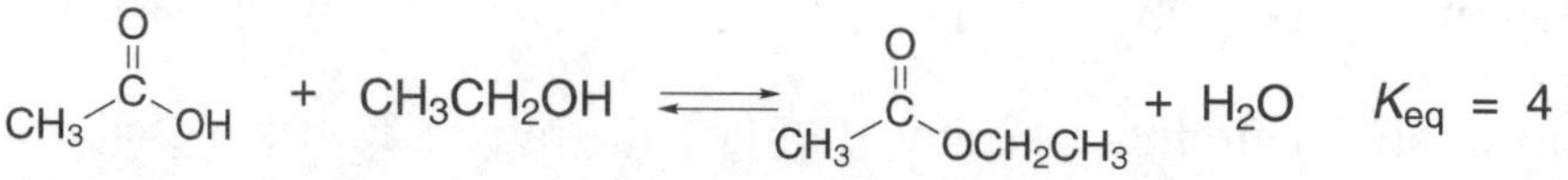

ethyl acetate

To increase the yield of ethyl acetate, H_2O can be removed from the reaction mixture, or there can be a large excess of one of the starting materials.

6.49 Since O atoms are more electronegative than all other atoms except F, each O atom in the O–O bond pulls electron density towards itself. The O atoms are less "willing" to share electron density in a two-electron bond, and this weakens the bond.

Chapter 7: Alkyl Halides and Nucleophilic Substitution

◆ General facts about alkyl halides

- Alkyl halides contain a halogen atom X bonded to an sp^3 hybridized carbon (7.1).
- Alkyl halides are named as halo alkanes, with the halogen as a substituent (7.2).
- Alkyl halides have a polar C–X bond, so they exhibit dipole-dipole interactions but are incapable of intermolecular hydrogen bonding (7.3).
- The polar C–X bond containing an electrophilic carbon makes alkyl halides reactive towards nucleophiles and bases (7.5).

◆ The central theme (7.6)

- Nucleophilic substitution is one of the two main reactions of alkyl halides. A nucleophile replaces a leaving group on an sp^3 hybridized carbon.

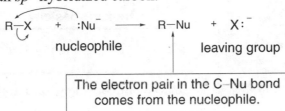

- One σ bond is broken and one σ bond is formed.
- There are two possible mechanisms: S_N1 and S_N2.

◆ S_N1 and S_N2 mechanisms compared

	S_N2 mechanism	**S_N1 mechanism**
[1] Mechanism	• One step (7.11B)	• Two steps (7.13B)
[2] Alkyl halide	• Order of reactivity: $CH_3X >$ $RCH_2X > R_2CHX > R_3CX$ (7.11D)	• Order of reactivity: $R_3CX >$ $R_2CHX > RCH_2X > CH_3X$ (7.13D)
[3] Rate equation	• rate = $k[RX][:Nu^-]$ • second order kinetics (7.11A)	• rate = $k[RX]$ • first order kinetics (7.13A)
[4] Stereochemistry	• backside attack of the nucleophile (7.11C) • inversion of configuration at a stereogenic center	• trigonal planar carbocation intermediate (7.13C) • racemization at a stereogenic center
[5] Nucleophile	• favored by stronger nucleophiles (7.17B)	• favored by weaker nucleophiles (7.17B)
[6] Leaving group	• better leaving group → faster reaction (7.17C)	• better leaving group → faster reaction (7.17C)
[7] Solvent	• favored by polar aprotic solvents (7.17D)	• favored by polar protic solvents (7.17D)

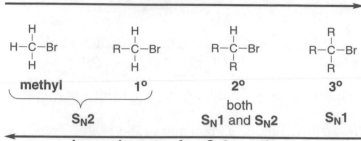

increasing rate of the S$_N$1 reaction

◆ **Important trends**

- The best leaving group is the weakest base. Leaving group ability increases across a row and down a column of the periodic table (7.7).

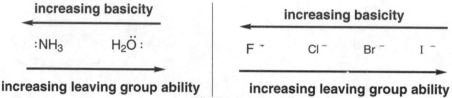

- Nucleophilicity decreases across a row of the periodic table (7.8A).

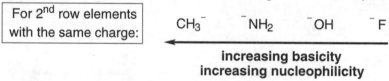

- Nucleophilicity decreases down a column of the periodic table in polar aprotic solvents (7.8C).

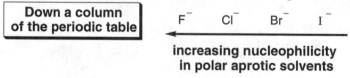

- Nucleophilicity increases down a column of the periodic table in polar protic solvents (7.8C).

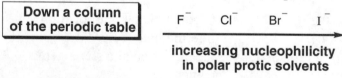

- The stability of a carbocation increases as the number of R groups bonded to the positively charged carbon increases (7.14).

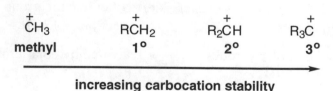

♦ Important principles

Principle	Example
• Electron donating groups (such as R groups) stabilize a positive charge (7.14A).	• 3° Carbocations (R_3C^+) are more stable than 2° carbocations (R_2CH^+), which are more stable than 1° carbocations (RCH_2^+).
• Steric hindrance decreases nucleophilicity but not basicity (7.8B).	• $(CH_3)_3CO^-$ is a stronger base but a weaker nucleophile than $CH_3CH_2O^-$.
• Hammond postulate: In an endothermic reaction, the more stable product is formed faster. In an exothermic reaction, this fact is not necessarily true (7.15).	• S_N1 reactions are faster when more stable (more substituted) carbocations are formed, because the rate-determining step is endothermic.
• Planar, sp^2 hybridized atoms react with reagents from both sides of the plane (7.13C).	• A trigonal planar carbocation reacts with nucleophiles from both sides of the plane.

7.1 Classify the alkyl halide as 1°, 2° or 3° **by counting the number of carbons directly bonded to the carbon bonded to the halogen.**

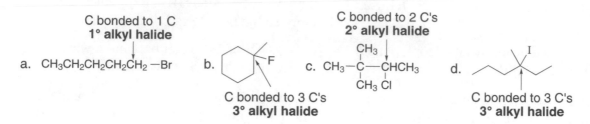

7.2 To name a compound with the IUPAC system:

[1] **Name the parent** chain by finding the longest carbon chain.

[2] **Number the chain** so the first substituent gets the lower number. Then **name and number all substituents**, giving like substituents a prefix (di, tri, etc.). **To name the halogen substituent, change the *–ine* ending to *–o*.**

[3] **Combine all parts**, alphabetizing substituents, and ignoring all prefixes except iso.

a. $(CH_3)_2CHCH(Cl)CH_2CH_3$

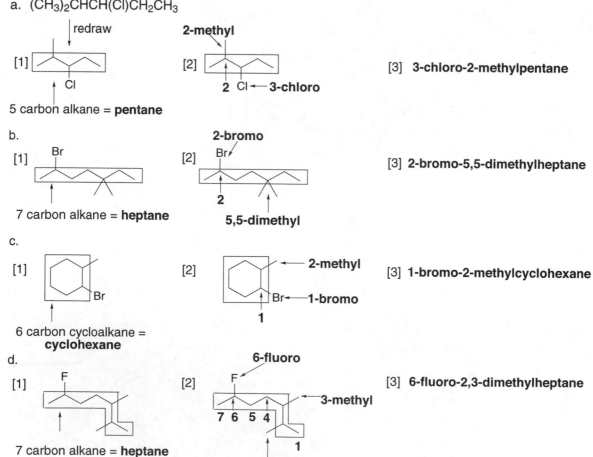

7.3 To work backwards from a name to a structure:
   [1] Find the parent name and draw that number of carbons. Use the suffix to identify the functional group. **(–ane = alkane)**
   [2] Arbitrarily number the carbons in the chain. Add the substituents to the appropriate carbon.

a. 3-chloro-2-methyl**hexane**☐☐☐

   [1]   6 carbon alkane   [2]   **methyl at C2**

   1 2 3 4 5 6   **chloro at C3**

b. 4-ethyl-5-iodo-2,2-dimethyl**octane**

   [1]   8 carbon alkane   [2]   ethyl at C4

   1 2 3 4 5 6 7 8

   2 methyls at C2   I ← iodo at C5

c. *cis*-1,3-dichloro**cyclopentane**

   [1]   5 carbon cycloalkane   [2]   chloro groups at C1 and C3, both on the same side

   Cl   Cl
   C1   C3

d. 1,1,3-tribromo**cyclohexane**

   [1]   6 carbon cycloalkane   [2]   3 Br groups
                                           Br
                                       C3 →   Br
                                                   Br
                                              C1

e. **propyl** chloride

   [1] 3 carbon alkyl group   [2] **chloride on end**

   CH₃CH₂CH₂ —   CH₃CH₂CH₂ —Cl

f. *sec*-**butyl** bromide

   [1] 4 carbon alkyl group   [2] **bromide**

   CH₃—CHCH₂CH₃   CH₃—CHCH₂CH₃
        | |
                                       Br

7.4 Boiling points of alkyl halides increase as the size (and polarizability) of X increases. Remember: **stronger intermolecular forces = higher boiling point.**

a.   CH₃CH₂CH₂F   CH₃CH₃CH₂Cl   CH₃CH₂CH₂I
         ↑   ↑   ↑
    smallest halogen   middle size halogen   largest halogen
    least polarizable   **intermediate**   most polarizable
   lowest boiling point **boiling point**   **highest boiling point**

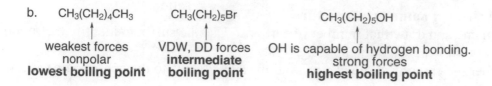

b. $CH_3(CH_2)_4CH_3$ $CH_3(CH_2)_5Br$ $CH_3(CH_2)_5OH$

weakest forces   VDW, DD forces   OH is capable of hydrogen bonding.
nonpolar   **intermediate**   strong forces
lowest boiling point   **boiling point**   **highest boiling point**

7.5 Since more polar molecules are more water soluble, look for polarity differences between methoxychlor and DDT.

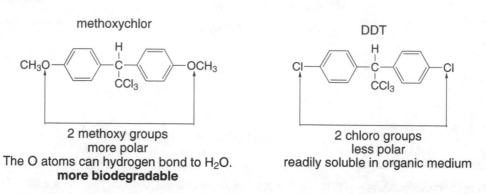

methoxychlor   DDT

2 methoxy groups   2 chloro groups
more polar   less polar
The O atoms can hydrogen bond to H_2O.   readily soluble in organic medium
more biodegradable

7.6 To draw the products of a nucleophilic substitution reaction:
[1] **Find the sp^3 hybridized electrophilic carbon** with a leaving group.
[2] **Find the nucleophile** with lone pairs or electrons in π bonds.
[3] **Substitute the nucleophile for the leaving group** on the electrophilic carbon.

a.

Br OCH_2CH_3 + Br⁻
leaving group   nonbonded e⁻ pairs
                      nucleophile

b.

Cl   OH
leaving group   Na⁺ ⁻OH   + Na⁺Cl⁻
                      nonbonded e⁻ pairs
                      nucleophile

c.

I +   N₃⁻   N₃   + I⁻
leaving group   nonbonded e⁻ pairs
                    nucleophile

d.

Br   CN
             +   Na⁺ ⁻CN   + Na⁺Br⁻
leaving group   nonbonded e⁻ pairs
                    nucleophile

7.7 Use the steps from Answer 7.6 and then draw the proton transfer reaction.

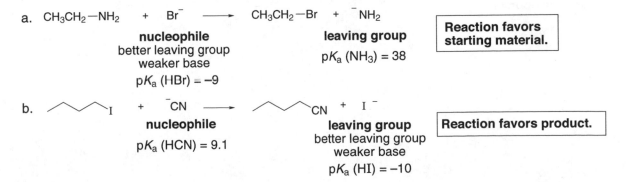

a. [structure with Br, **leaving group**] + :N(CH₂CH₃)₃ **nucleophile** →(substitution)→ [structure] ⁺N(CH₂CH₃)₃ + ⁻Br

b. (CH₃)₃C—Cl **leaving group** + H₂Ö: **nucleophile** →(substitution)→ (CH₃)₃C—⁺Ö—H (with H below) + Cl⁻ →(proton transfer)→ (CH₃)₃C—Ö—H + HCl

c. [cyclohexane with CH₃ and Br, **leaving group**] + CH₃—Ö—H **nucleophile** →(substitution)→ [cyclohexane with CH₃ and Ö⁺—CH₃ with H] + Br⁻ →(proton transfer)→ [cyclohexane with CH₃ and Ö—CH₃] + HBr

7.8 Compare the compounds based on these leaving group trends:
- Better leaving groups are weaker bases.
- A neutral leaving group is always better than its conjugate base.

a. Cl⁻, I⁻

further down a column
of the periodic table
less basic
better leaving group

b. NH₃, NH₂⁻

neutral compound
less basic
better leaving group

c. H₂O, H₂S

further down a column
of the periodic table
less basic
better leaving group

7.9 Good leaving groups include Cl⁻, Br⁻, I⁻, H₂O.

a. CH₃CH₂CH₂—Br

Br⁻ is a **good leaving group.**

b. CH₃CH₂CH₂OH

No good leaving group. ⁻OH is too strong a base.

c. CH₃CH₂CH₂—⁺OH₂

H₂O is a **good leaving group.**

d. CH₃CH₃

No good leaving group. H⁻ is too strong a base.

7.10 To decide whether the equilibrium favors the starting material or the products, **compare the nucleophile and the leaving group.** The reaction proceeds towards the weaker base.

a. CH₃CH₂—NH₂ + Br⁻ → CH₃CH₂—Br + ⁻NH₂

nucleophile
better leaving group
weaker base
pKₐ (HBr) = –9

leaving group
pKₐ (NH₃) = 38

Reaction favors starting material.

b. [structure]—I + ⁻CN → [structure]—CN + I⁻

nucleophile
pKₐ (HCN) = 9.1

leaving group
better leaving group
weaker base
pKₐ (HI) = –10

Reaction favors product.

7.11 Use these three rules to find the stronger nucleophile in each pair:

    [1] Comparing two nucleophiles having the *same attacking atom*, **the stronger base is a stronger nucleophile**.

    [2] **Negatively charged nucleophiles** are always **stronger than their conjugate acids**.

    [3] **Across a row of the periodic table, nucleophilicity decreases** when comparing a species of similar charge.

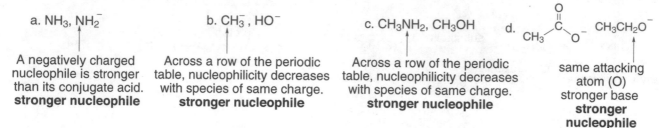

a. NH_3, NH_2^-

A negatively charged
nucleophile is stronger
than its conjugate acid.
stronger nucleophile

b. CH_3^-, HO^-

Across a row of the periodic
table, nucleophilicity decreases
with species of same charge.
stronger nucleophile

c. CH_3NH_2, CH_3OH

Across a row of the periodic
table, nucleophilicity decreases
with species of same charge.
stronger nucleophile

d.

same attacking
atom (O)
stronger base
**stronger
nucleophile**

7.12 *Polar protic solvents* are capable of **H-bonding**, and therefore must contain a **H bonded to an electronegative O or N**. *Polar aprotic solvents* are incapable of **H-bonding**, and therefore do not contain any O–H or N–H bonds.

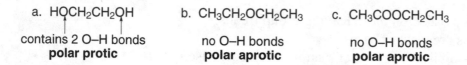

a. $HOCH_2CH_2OH$

contains 2 O–H bonds
polar protic

b. $CH_3CH_2OCH_2CH_3$

no O–H bonds
polar aprotic

c. $CH_3COOCH_2CH_3$

no O–H bonds
polar aprotic

7.13   • In *polar protic solvents*, **the trend in nucleophilicity is opposite to the trend in basicity** down a column of the periodic table so that nucleophilicity increases.

     • In *polar aprotic solvents*, **the trend is identical to basicity** so that nucleophilicity decreases down a column.

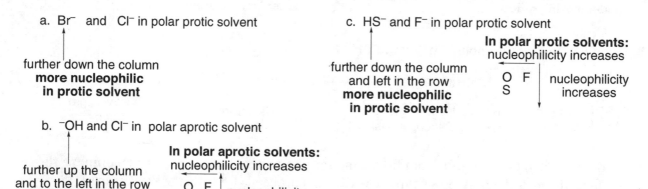

a. Br^- and Cl^- in polar protic solvent

further down the column
**more nucleophilic
in protic solvent**

c. HS^- and F^- in polar protic solvent

further down the column
and left in the row
**more nucleophilic
in protic solvent**

In polar protic solvents:
nucleophilicity increases

O F   nucleophilicity
S   increases

b. ^-OH and Cl^- in polar aprotic solvent

further up the column
and to the left in the row
more basic
more nucleophilic

In polar aprotic solvents:
nucleophilicity increases

O F   nucleophilicity
   Cl   increases

7.14 The stronger base is the stronger nucleophile except in polar protic solvents when nucleophilicity increases down a column. For other rules, see Answers 7.11 and 7.13.

a.
H_2O	⁻OH	⁻NH₂
no charge	negatively charged	negatively charged
weakest nucleophile	**intermediate nucleophile**	further left on periodic table
		strongest nucleophile

b.
Br⁻	F⁻	⁻OH
Basicity decreases down a column in polar aprotic solvents.	Basicity decreases across a row.	**strongest nucleophile**
weakest nucleophile	**intermediate nucleophile**	

c.
H_2O	CH_3COO⁻	⁻OH
weakest nucleophile	weaker base than ⁻OH	**strongest nucleophile**
	intermediate nucleophile	

7.15 To determine what nucleophile is needed to carry out each reaction, look at the product to see what has replaced the leaving group.

a. $CH_3CH_2CH_2$—Br $\longrightarrow$ $CH_3CH_2CH_2$—SH

SH replaces Br.
HS⁻ is needed.

b. [cyclohexyl–I] $\longrightarrow$ [cyclohexyl–O₂CCH₃]⁻

CH₃COO replaces I.
CH₃COO⁻ is needed.

c. [branched chain]–Cl $\longrightarrow$ [branched chain]–OCH₂CH₃

CH_3CH_2O replaces Cl.
CH_3CH_2O⁻ is needed.

d. [cyclopentyl]–Br $\longrightarrow$ [cyclopentyl]–C≡C–H

HC≡C replaces Br.
HC≡C⁻ is needed.

7.16 The general rate equation for an S$_N$2 reaction is rate = k[RX][:Nu⁻]

a. [RX] is tripled, and [:Nu⁻] stays the same: **rate triples.**
b. Both [RX] and [:Nu⁻] are tripled: **rate increases by a factor of 9 (3 x 3 = 9).**
c. [RX] is halved, and [:Nu⁻] stays the same: **rate halved.**
d. [RX] is halved, and [:Nu⁻] is doubled: **rate stays the same (1/2 x 2 = 1).**

7.17 The transition state in an S$_N$2 reaction has **dashed bonds to both the leaving group and the nucleophile,** and must contain partial charges.

a. $CH_3CH_2CH_2$—Cl + ⁻OCH₃ $\longrightarrow$ $CH_3CH_2CH_2$—OCH₃ + Cl⁻ $\left[\begin{array}{c} CH_3CH_2CH_2\text{ - - -}Cl\ \delta^- \\ | \\ CH_3O\ \delta^- \end{array} \right]^{\ddagger}$

b. [propyl]–Br + ⁻SH $\longrightarrow$ [propyl]–SH + Br⁻ $\left[\begin{array}{c} \delta^-SH \\ | \\ Br\ \delta^- \end{array} \right]^{\ddagger}$

7.18 All S_N2 reactions have one step.

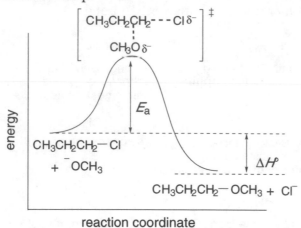

reaction coordinate

7.19 To draw the products of S_N2 reactions, **replace the leaving group by the nucleophile, and then draw the stereochemistry with *inversion* at the stereogenic center**.

a.

b.

c.

7.20 *Increasing* the number of R groups *increases* crowding of the transition state and *decreases* the rate of an S_N2 reaction.

a. CH_3CH_2—Cl or CH_3—Cl

1° alkyl halide   methyl halide
faster reaction

b.

2° alkyl halide   1° alkyl halide
faster reaction

c.

2° alkyl halide   3° alkyl halide
faster reaction

7.21

These 3 methyl groups make the alkyl halide sterically hindered. This slows the rate of an S_N2 reaction even though it is a 1° alkyl halide.

7.22

+ SR_2

loss of a proton

nicotine

7.23 In a first order reaction, **the rate changes with any change in [RX]**. The rate is independent of any change in [nucleophile].

    a. [RX] is tripled, and [:Nu⁻] stays the same: **rate triples.**
    b. Both [RX] and [:Nu⁻] are tripled: **rate triples.**
    c. [RX] is halved, and [:Nu⁻] stays the same: **rate halved.**
    d. [RX] is halved, and [:Nu⁻] is doubled: **rate halved.**

7.24 The two steps for an S_N1 reaction are:
    [1] **Breaking the C–Z bond to form a carbocation (Z = leaving group).**
    [2] **Forming the C–Nu bond.**

7.25 In S_N1 reactions, **racemization always occurs at a stereogenic center**. Draw two products, with the two possible configurations at the stereogenic center.

7.26 Carbocations are classified by the number of R groups bonded to the carbon: 0 R groups = methyl, 1 R group = 1°, 2 R groups = 2°, and 3 R groups = 3°.

a. $CH_3\overset{+}{C}HCH_2CH_3$

    2 R groups
    2° carbocation

b.

    2 R groups
    2° carbocation

c. $(CH_3)_3C\overset{+}{C}H_2$

    1 R group
    1° carbocation

d.

    3 R groups
    3° carbocation

e.

    2 R groups
    2° carbocation

7.27 For carbocations: **Increasing number of R groups = Increasing stability**.

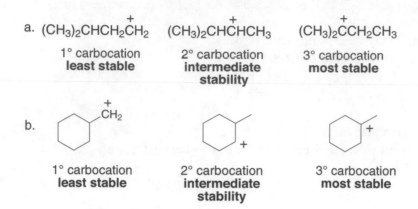

a. $(CH_3)_2CHCH_2\overset{+}{C}H_2$ $(CH_3)_2CH\overset{+}{C}HCH_3$ $(CH_3)_2\overset{+}{C}CH_2CH_3$

1° carbocation	2° carbocation	3° carbocation
least stable	**intermediate stability**	**most stable**

b.

1° carbocation	2° carbocation	3° carbocation
least stable	**intermediate stability**	**most stable**

7.28

3 Cl groups - **electron *withdrawing*** destabilizing **less stable**

methyl group without added Cl's **more stable**

In $Cl_3CCH_2^+$, the three electron withdrawing Cl atoms place a partial positive charge on the carbon adjacent to the carbocation, destabilizing it.

7.29 **The rate of S$_N$1 reaction increases with increasing alkyl substitution.**

a. $(CH_3)_3CBr$ and $(CH_3)_3CCH_2Br$

3° alkyl halide	1° alkyl halide
faster S$_N$1 reaction	**slower S$_N$1 reaction**

c.

3° alkyl halide	2° alkyl halide
faster S$_N$1 reaction	**slower S$_N$1 reaction**

b.

2° alkyl halide	3° alkyl halide
slower S$_N$1 reaction	**faster S$_N$1 reaction**

7.30 • For **methyl and 1° alkyl halides**, only S$_N$2 will occur.
• For **2° alkyl halides**, S$_N$1 and S$_N$2 will occur.
• For **3° alkyl halides**, only S$_N$1 will occur.

a.

2° alkyl halide
S$_N$1 and S$_N$2

b.

1° alkyl halide
S$_N$2

c.

2° alkyl halide
S$_N$1 and S$_N$2

d.

3° alkyl halide
S$_N$1

7.31 • For **methyl and 1° alkyl halides**, only S_N2 will occur.

  • For **2° alkyl halides**, S_N1 and S_N2 will occur and other factors determine which mechanism operates.

  • For **3° alkyl halides**, only S_N1 will occur.

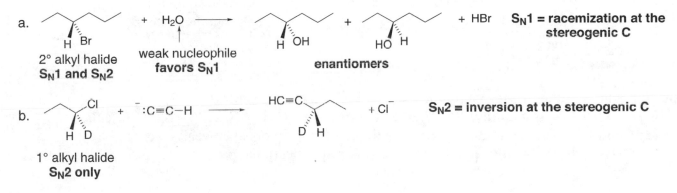

7.32 First decide whether the reaction will proceed via an S_N1 or S_N2 mechanism. Then draw the product with stereochemistry.

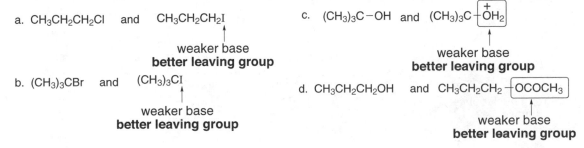

7.33 Compounds with better leaving groups react faster. Weaker bases are better leaving groups.

a. $CH_3CH_2CH_2Cl$ and $CH_3CH_2CH_2I$
↑
weaker base
better leaving group

b. $(CH_3)_3CBr$ and $(CH_3)_3CI$
↑
weaker base
better leaving group

c. $(CH_3)_3C-OH$ and $(CH_3)_3C\boxed{-\overset{+}{O}H_2}$
↑
weaker base
better leaving group

d. $CH_3CH_2CH_2OH$ and $CH_3CH_2CH_2\boxed{-OCOCH_3}$
↑
weaker base
better leaving group

7.34 • **Polar protic solvents** favor the S_N1 mechanism by solvating the intermediate carbocation and halide.
  • **Polar aprotic solvents** favor the S_N2 mechanism by making the nucleophile stronger.

a. CH_3CH_2OH	b. CH_3CN	c. CH_3COOH	d. $CH_3CH_2OCH_2CH_3$
polar protic solvent	*polar aprotic solvent*	*polar protic solvent*	*polar aprotic solvent*
contains an O–H bond	no O–H or N–H bond	contains an O–H bond	no O–H or N–H bond
favors S_N1	**favors S_N2**	**favors S_N1**	**favors S_N2**

7.35 Compare the solvents in the reactions below. **For the solvent to increase the reaction rate of an S_N1 reaction, the solvent must be *polar protic*.**

a. $(CH_3)_3CBr$ + H_2O $\xrightarrow[\text{or}]{H_2O}$ $(CH_3)_3COH$ + HBr

$3°$ RX - S_N1 reaction $(CH_3)_2C{=}O$

H_2O
Polar protic solvent increases the rate of an S_N1 reaction.

b. (structure) + CH_3OH $\xrightarrow[\text{DMSO}]{CH_3OH \text{ or}}$ (structure) + HCl

$3°$ RX - S_N1 reaction

CH_3OH
Polar protic solvent increases the rate of an S_N1 reaction.

c. (structure) Br + $^-$OH $\xrightarrow[\text{DMF}]{H_2O \text{ or}}$ (structure) OH + Br$^-$

$1°$ RX - S_N2 reaction

DMF [$HCON(CH_3)_2$]
Polar aprotic solvent increases the rate of an S_N2 reaction.

d. (structure) H Cl + CH_3O^- $\xrightarrow[\text{HMPA}]{CH_3OH \text{ or}}$ (structure) H OCH$_3$ + Cl$^-$

$2°$ RX   strong nucleophile
S_N2 reaction

HMPA [$(CH_3)_2N]_3P{=}O$
Polar aprotic solvent increases the rate of an S_N2 reaction.

7.36 To predict whether the reaction follows an S_N1 or S_N2 mechanism:
  [1] **Classify RX as a methyl, 1°, 2°, or 3° halide.** (methyl, 1° = S_N2; 3° = S_N1; 2° = either)
  [2] **Classify the nucleophile as strong or weak.** (strong favors S_N2, weak favors S_N1)
  [3] **Classify the solvent as polar protic or polar aprotic.** (polar protic favors S_N1, polar aprotic favors S_N2).

a. (cyclopentane)–CH_2Br + $CH_3CH_2O^-$ $\longrightarrow$ (cyclopentane)–$CH_2OCH_2CH_3$ + Br$^-$ S_N2 reaction

1° alkyl halide
S_N2

b. (cyclopentane structure)–Br + N$_3^-$ $\longrightarrow$ (cyclopentane structure)‧‧‧N$_3$ + Br$^-$

2° alkyl halide
S_N1 or S_N2

Strong nucleophile
favors S_N2.

S_N2 reaction = *inversion* at the stereogenic center
The leaving group was "up."
The nucleophile attacks from below.

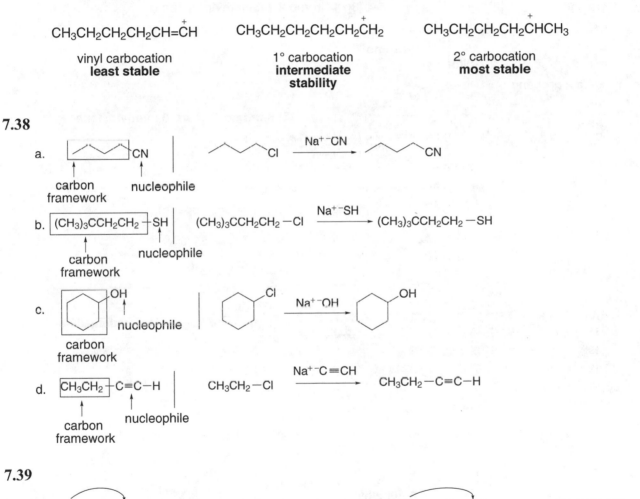

c.

3° alkyl halide
S_N1

+ CH₃OH →

Weak nucleophile
favors S_N1.

+ HI

S_N1 reaction

d.

3° alkyl halide
S_N1

+ H₂O →

Weak nucleophile
favors S_N1.

+

+ HCl

S_N1 reaction
forms **two enantiomers.**

7.37 Vinyl carbocations are even less stable than 1° carbocations.

$$CH_3CH_2CH_2CH_2CH{=}\overset{+}{C}H$$

vinyl carbocation
least stable

$$CH_3CH_2CH_2CH_2CH_2\overset{+}{C}H_2$$

1° carbocation
**intermediate
stability**

$$CH_3CH_2CH_2CH_2\overset{+}{C}HCH_3$$

2° carbocation
most stable

7.38

a.

carbon
framework

nucleophile

Na⁺ ⁻CN →

b. (CH₃)₃CCH₂CH₂ —SH

carbon
framework

nucleophile

(CH₃)₃CCH₂CH₂ —Cl

Na⁺ ⁻SH →

(CH₃)₃CCH₂CH₂ —SH

c.

carbon
framework

nucleophile

Na⁺ ⁻OH →

d. CH₃CH₂ —C≡C—H

carbon
framework

nucleophile

CH₃CH₂ —Cl

Na⁺ ⁻C≡CH →

CH₃CH₂ —C≡C—H

7.39

CH₃O⁻ + Cl—CH₂CH₃ ——→ CH₃OCH₂CH₃   CH₃CH₂O⁻ + Cl—CH₃ ——→ CH₃OCH₂CH₃

7.40 Use the directions from Answer 7.2 to name the compounds.

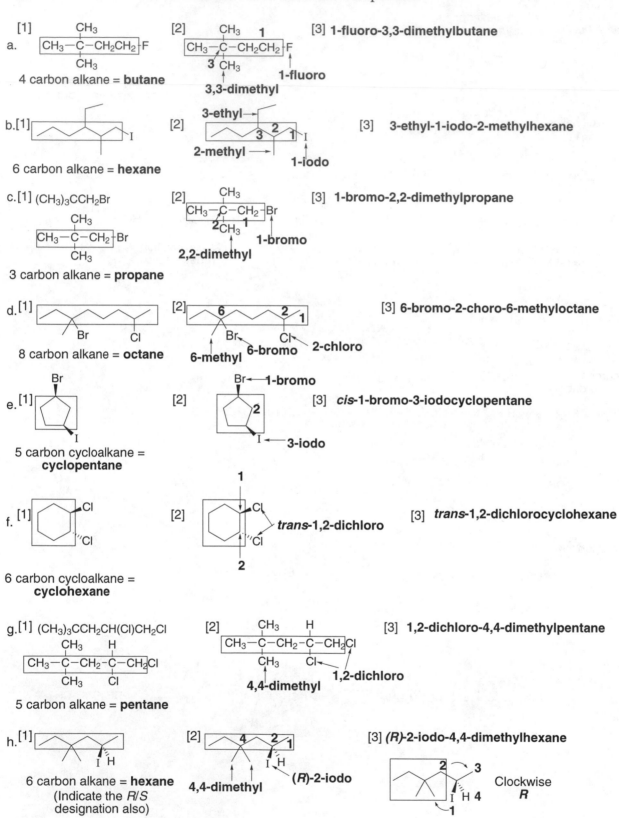

a. [1]

CH₃
|
CH₃—C—CH₂CH₂—F
|
CH₃

4 carbon alkane = **butane**

[2]

CH₃ 1
|
CH₃—C—CH₂CH₂—F
3 | CH₃
3,3-dimethyl   **1-fluoro**

[3] **1-fluoro-3,3-dimethylbutane**

b. [1]

I

6 carbon alkane = **hexane**

[2]

3-ethyl
2-methyl
3 2 1 I
1-iodo

[3] **3-ethyl-1-iodo-2-methylhexane**

c. [1] (CH₃)₃CCH₂Br

CH₃
|
CH₃—C—CH₂—Br
|
CH₃

3 carbon alkane = **propane**

[2]

CH₃
|
CH₃—C—CH₂—Br
2 CH₃ 1
2,2-dimethyl   **1-bromo**

[3] **1-bromo-2,2-dimethylpropane**

d. [1]

Br   Cl

8 carbon alkane = **octane**

[2]

6   2 1
Br   Cl
6-methyl **6-bromo** **2-chloro**

[3] **6-bromo-2-choro-6-methyloctane**

e. [1]

Br

I

5 carbon cycloalkane =
cyclopentane

[2]

Br **1-bromo**

2

I **3-iodo**

[3] *cis*-**1-bromo-3-iodocyclopentane**

f. [1]

Cl

Cl

6 carbon cycloalkane =
cyclohexane

[2]

1

Cl
trans-1,2-dichloro
Cl

2

[3] *trans*-**1,2-dichlorocyclohexane**

g. [1] (CH₃)₃CCH₂CH(Cl)CH₂Cl

CH₃   H
| |
CH₃—C—CH₂—C—CH₂Cl
| |
CH₃   Cl

5 carbon alkane = **pentane**

[2]

CH₃   H
| |
CH₃—C—CH₂—C—CH₂Cl
| |
CH₃   Cl
4,4-dimethyl   **1,2-dichloro**

[3] **1,2-dichloro-4,4-dimethylpentane**

h. [1]

I H

6 carbon alkane = **hexane**
(Indicate the *R/S*
designation also)

[2]

4   2 1
I H
4,4-dimethyl   **(R)-2-iodo**

[3] **(R)-2-iodo-4,4-dimethylhexane**

2   3

I H 4
1

Clockwise
R

7.41 To work backwards to a structure, use the directions in Answer 7.3.

a.   **isopropyl** bromide

Br ◄── Bromine on middle C
CH₃─CHCH₃   makes it an isopropyl group.

b.   3-bromo-4-ethyl**heptane**

4-ethyl

3 **4**

Br ◄── 3-bromo

c.   1,1-dichloro-2-methyl**cyclohexane**

1 Cl ◄── 1,1-dichloro
Cl ◄──

2 ◄── 2-methyl

d.   *trans*-1-chloro-3-iodo**cyclobutane**

3 Cl ── 1-chloro
1

I ◄── 3-iodo

e.   1-bromo-4-ethyl-3-fluoro**octane**

4-ethyl

Br   **3** **4**
1

1-bromo   F ◄── 3-fluoro

7.42

a.   CH₃
      CH₃─C─CH₂CH₂F
           CH₃
                1° halide

b.   I
                1° halide

c.   (CH₃)₃CCH₂Br
           1° halide

d.   Br   Cl
      3° halide   **2° halide**

e.   Br ◄── **2° halide**

      I ◄── **2° halide**

f.   Cl ◄──
                Both are
                2° halides.
      Cl ◄──

g.   H H
      (CH₃)₂CCH₂─C─C─Cl ◄── **1° halide**
                Cl H

      2° halide

h.   I H

      2° halide

7.43

1-chloro
1
Cl
1-chloropentane

1-chloro
Cl
2 **1**
1-chloro-2,2-dimethylpropane

3-chloro ── Cl
3
3-chloropentane

2-chloro
2-methyl ──
Cl
2
2-chloro-2-methylbutane

3-methyl ──
Cl **1** **3**
1-chloro
1-chloro-3-methylbutane

Two stereoisomers

2-chloro → Cl

2-chloropentane

[* denotes stereogenic center]

Clockwise
"4" in back =
R

Clockwise
"4" in *front* =
S

Two stereoisomers

3-methyl →

2-chloro → Cl

2-chloro-3-methylbutane

[* denotes stereogenic center]

Counterclockwise
"4" in front =
R

Counterclockwise
"4" in *back* =
S

Two stereoisomers

1-chloro

2-methyl →

1-chloro-2-methylbutane

[* denotes stereogenic center]

Clockwise
"4" in *front* =
S

Clockwise
"4" in *back* =
R

7.44 Use the directions from Answer 7.4.

a.   (CH₃)₃CBr   and   CH₃CH₂CH₂CH₂Br

larger surface area =
stronger intermolecular forces =
higher boiling point

c.   and   Br

nonpolar
only VDW forces

more polar =
higher boiling point

b.   I   and   Br

larger halide = more polarizable =
higher boiling point

7.45

a.   CH₃CH₂CH₂CH₂ —Br + ⁻OH ⟶ CH₃CH₂CH₂CH₂OH + Br⁻

b.   CH₃CH₂CH₂CH₂ —Br + ⁻SH ⟶ CH₃CH₂CH₂CH₂SH + Br⁻

c.   CH₃CH₂CH₂CH₂ —Br + ⁻CN ⟶ CH₃CH₂CH₂CH₂CN + Br⁻

d. $CH_3CH_2CH_2CH_2 -Br + ^-OCH(CH_3)_2 \longrightarrow CH_3CH_2CH_2CH_2OCH(CH_3)_2 + Br^-$

e. $CH_3CH_2CH_2CH_2 -Br + ^-C\equiv CH \longrightarrow CH_3CH_2CH_2CH_2C\equiv CH + Br^-$

f. $CH_3CH_2CH_2CH_2 -Br + H_2\ddot{O}: \longrightarrow CH_3CH_2CH_2CH_2\overset{+}{O}H_2 + Br^- \longrightarrow CH_3CH_2CH_2CH_2OH + HBr$

g. $CH_3CH_2CH_2CH_2 -Br + \ddot{N}H_3 \longrightarrow CH_3CH_2CH_2CH_2\overset{+}{N}H_3 + Br^- \longrightarrow CH_3CH_2CH_2CH_2NH_2 + HBr$

h. $CH_3CH_2CH_2CH_2 -Br + Na^+I^- \longrightarrow CH_3CH_2CH_2CH_2I + Na^+Br^-$

i. $CH_3CH_2CH_2CH_2 -Br + Na^+N_3^- \longrightarrow CH_3CH_2CH_2CH_2N_3 + Na^+Br^-$

7.46 Use the steps from Answer 7.6 and then draw the proton transfer reaction, when necessary.

a. [structure] + [structure] CH$_3$—C(=O)—O$^-$ (nucleophile) → [structure] + Cl$^-$

leaving group   nucleophile

b. [structure]—I + Na$^+$ $^-$CN → [structure]—CN + NaI

leaving group   nucleophile

c. [structure]—I + H$_2\ddot{O}$: → [structure]—OH + HI

leaving group   nucleophile

d. [structure]—Cl + CH$_3$CH$_2\ddot{O}$H → [structure]—OCH$_2$CH$_3$ + HCl

leaving group   nucleophile

e. [structure]—I + $^-$N$_3$ → [structure]—N$_3$ + I$^-$

leaving group   nucleophile

f. [structure]—Br + Na$^+$ $^-$OCH$_3$ → [structure]—OCH$_3$ + NaBr

leaving group   nucleophile

g. $(CH_3)_3CBr$ + CH_3COOH → $(CH_3)_3COOCCH_3$ + HBr

leaving group   nucleophile

h. [structure]—Cl + CH$_3\ddot{S}$CH$_3$ → [structure]—$\overset{+}{S}$(CH$_3$)CH$_3$ + Cl$^-$

leaving group   nucleophile

7.47 A good leaving group is a weak base.

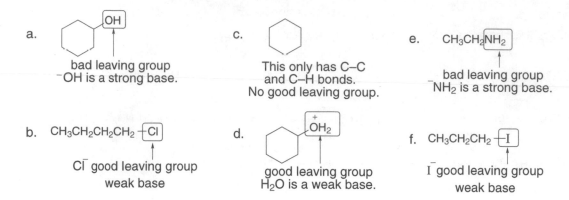

a.
OH
bad leaving group
⁻OH is a strong base.

c.
This only has C–C
and C–H bonds.
No good leaving group.

e. CH₃CH₂NH₂
bad leaving group
⁻NH₂ is a strong base.

b. CH₃CH₂CH₂CH₂ —Cl
Cl⁻ good leaving group
weak base

d.
⁺OH₂
good leaving group
H₂O is a weak base.

f. CH₃CH₂CH₂ —I
I⁻ good leaving group
weak base

7.48 Use the rules from Answer 7.8.

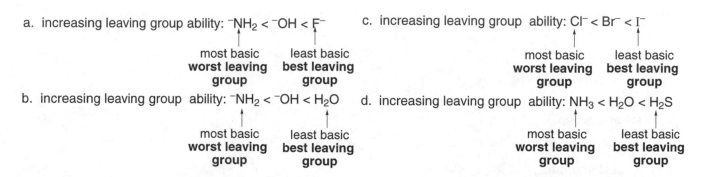

a. increasing leaving group ability: ⁻NH₂ < ⁻OH < F⁻

most basic — worst leaving group

least basic — best leaving group

c. increasing leaving group ability: Cl⁻ < Br⁻ < I⁻

most basic — worst leaving group

least basic — best leaving group

b. increasing leaving group ability: ⁻NH₂ < ⁻OH < H₂O

most basic — worst leaving group

least basic — best leaving group

d. increasing leaving group ability: NH₃ < H₂O < H₂S

most basic — worst leaving group

least basic — best leaving group

7.49 Compare the nucleophile and the leaving group in each of the following reactions. The reaction will occur if it proceeds towards the weaker base. Remember that the stronger the acid (lower pK_a), the weaker the conjugate base.

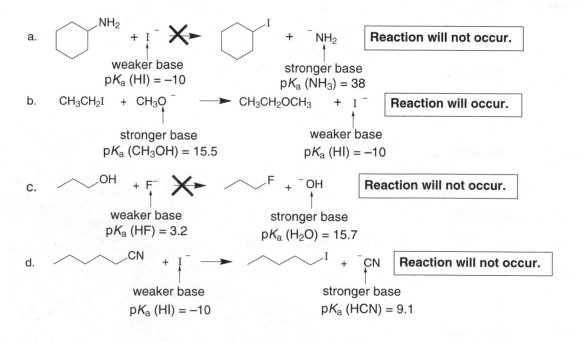

a.
NH₂   + I⁻ ✗→   I   + ⁻NH₂   **Reaction will not occur.**

weaker base
pK_a (HI) = −10

stronger base
pK_a (NH₃) = 38

b. CH₃CH₂I + CH₃O⁻ ⟶ CH₃CH₂OCH₃ + I⁻   **Reaction will occur.**

stronger base
pK_a (CH₃OH) = 15.5

weaker base
pK_a (HI) = −10

c.
OH + F⁻ ✗→   F + ⁻OH   **Reaction will not occur.**

weaker base
pK_a (HF) = 3.2

stronger base
pK_a (H₂O) = 15.7

d.
CN + I⁻ ⟶   I + ⁻CN   **Reaction will not occur.**

weaker base
pK_a (HI) = −10

stronger base
pK_a (HCN) = 9.1

7.50

a. $CH_3CH_2CH_2CH_2Br \longrightarrow CH_3CH_2CH_2CH_2OH$ OH replaces Br.
   $^-$**OH is needed.**

b.   SCH$_3$ replaces I.
   $^-$**SCH$_3$ is needed.**

c.   I replaces Br.
   I$^-$ is needed.

d. $CH_3CH_2Cl \longrightarrow CH_3CH_2OCH_2CH_3$ OCH$_2$CH$_3$ replaces Cl.
   $^-$**OCH$_2$CH$_3$ is needed.**

e.   C≡CCH$_3$ replaces Cl.
   $^-$**C≡CCH$_3$ is needed.**

f. $CH_3CH_2Br \longrightarrow CH_3CH_2\overset{+}{N}(CH_3)_3 \ Br^-$ N(CH$_3$)$_3$ replaces Br.
   N(CH$_3$)$_3$ is needed.

7.51 Use the directions in Answer 7.14.

a. Across a row of the periodic table nucleophilicity decreases.
$$^-OH < {}^-NH_2 < CH_3{}^-$$

b. • In a **polar protic solvent** (CH$_3$OH), nucleophilicity *increases down a column* of the periodic table, so: $^-$SH is more nucleophilic than $^-$OH.
   • *Negatively charged species* are more nucleophilic than neutral species so $^-$OH is more nucleophilic than H$_2$O.
$$H_2O < {}^-OH < {}^-SH$$

c. • In a **polar protic solvent** (CH$_3$OH), nucleophilicity *increases down a column* of the periodic table, so: CH$_3$CH$_2$S$^-$ is more nucleophilic than CH$_3$CH$_2$O$^-$.
   • For two species with same attacking atom the more basic is more nucleophilic so CH$_3$CH$_2$O$^-$ is more nucleophilic than CH$_3$COO$^-$.
$$CH_3COO^- < CH_3CH_2O^- < CH_3CH_2S^-$$

d. Compare the nucleophilicity of N, S and O. In a polar aprotic solvent (acetone), nucleophilicity parallels basicity.
$$CH_3SH < CH_3OH < CH_3NH_2$$

e. In a **polar aprotic solvent** (acetone), nucleophilicity parallels basicity. Across a row and down a column of the periodic table nucleophilicity decreases.
$$Cl^- < F^- < {}^-OH$$

f. Nucleophilicity decreases across a row so $^-$SH is more nucleophilic than Cl$^-$.
In a **polar protic solvent** (CH$_3$OH), nucleophilicity increases down a column so Cl$^-$ is more nucleophilic than F$^-$.
$$F^- < Cl^- < {}^-SH$$

7.52 Halide ions in the gas phase would experience no solvent effects. Therefore, the trends would be the same as basicity trends – a stronger base is a stronger nucleophile. Thus **F$^-$ > Cl$^-$ > Br$^-$ > I$^-$.**

7.53 *Polar protic solvents* **are capable of hydrogen bonding**, and therefore must contain a H bonded to an electronegative O or N. *Polar aprotic solvents* **are incapable of hydrogen bonding**, and therefore do not contain any O–H or N–H bonds.

a. $(CH_3)_2CHOH$

    contains O–H bond
    protic

b. CH_3NO_2

    no O–H or N–H bond
    aprotic

c. CH_2Cl_2

    no O–H or N–H bond
    aprotic

d. NH_3

    contains N–H bond
    protic

e. $N(CH_3)_3$

    no O–H or N–H bond
    aprotic

f. $HCONH_2$

    contains an N–H bond
    protic

7.54

a. Mechanism:

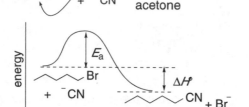

1° alkyl halide
S_N2 reaction

b. Energy diagram:

c. Transition state:

reaction coordinate

d. Rate equation: one step reaction with both nucleophile and alkyl halide in the only step:

    rate = k[R-Br][̄CN]

e.   [1] The leaving group is changed from Br ̄ to I ̄:

        Leaving group becomes less basic → a better leaving group → faster reaction.

    [2] The solvent is changed from acetone to CH_3CH_2OH:

        Solvent changed to polar protic → decreases reaction rate.

    [3] The alkyl halide is changed from $CH_3(CH_2)_4Br$ to $CH_3CH_2CH_2CH(Br)CH_3$:

        Changed from 1° to 2° alkyl halide → the alkyl halide gets more crowded and the reaction rate decreases.

    [4] The concentration of ̄CN is increased by a factor of 5.

        Reaction rate will increase by a factor of five.

    [5] The concentration of both the alkyl halide and ̄CN are increased by a factor of 5:

        Reaction rate will increase by a factor of 25 (5 x 5 = 25).

7.55 Use the directions for Answer 7.20.

a.

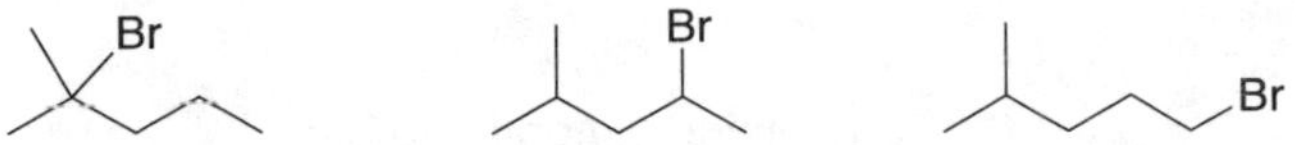

3° alkyl halide
least reactive

2° alkyl halide
intermediate reactivity

1° alkyl halide
most reactive

b.

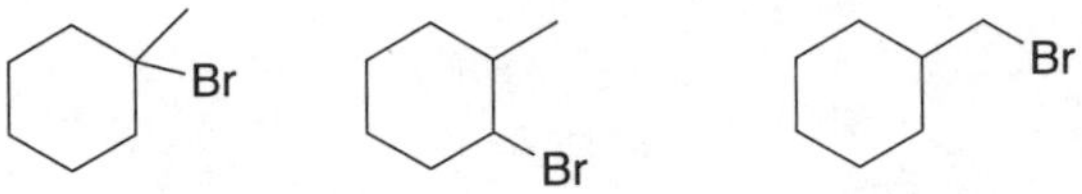

3° alkyl halide
least reactive

2° alkyl halide
intermediate reactivity

1° alkyl halide
most reactive

c.

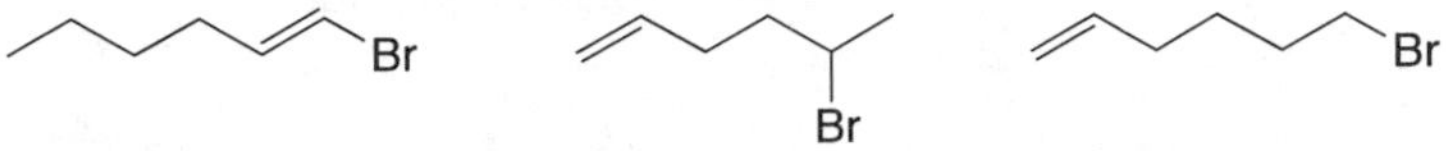

vinyl halide
least reactivity

2° alkyl halide
intermediate reactivity

1° alkyl halide
most reactive

7.56

a.

better leaving group

CH_3CH_2Br + ^-OH ⟶ **faster reaction**

CH_3CH_2Cl + ^-OH ⟶

b.

stronger nucleophile

Br + ^-OH ⟶ **faster reaction**

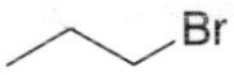

Br + H_2O ⟶

c.

stronger nucleophile

Cl + NaOH ⟶ **faster reaction**

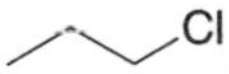

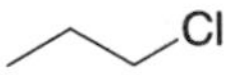

Cl + $NaOCOCH_3$ ⟶

d.

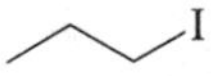

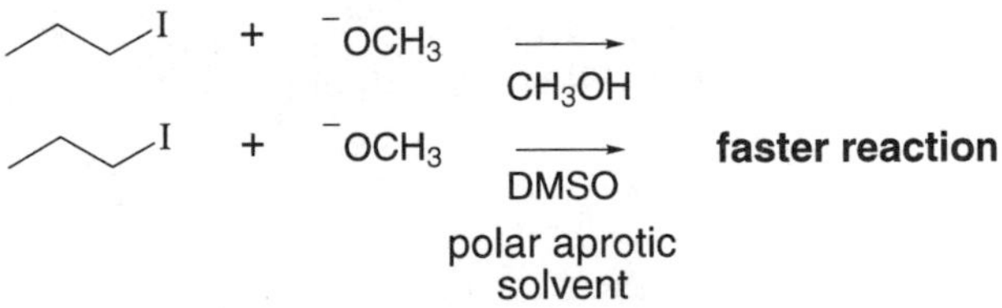

I + $^-OCH_3$ ⟶
CH₃OH

I + $^-OCH_3$ ⟶ **faster reaction**
DMSO
polar aprotic
solvent

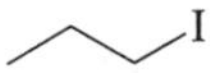

e.

less steric hinderance

Br + $^-OCH_2CH_3$ ⟶ **faster reaction**

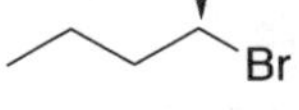

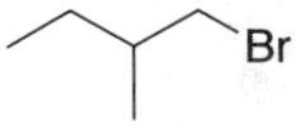

Br + $^-OCH_2CH_3$ ⟶

7.57 All S_N2 reactions proceed with backside attack of the nucleophile. When nucleophilic attack occurs at a stereogenic center, inversion of configuration occurs.

a. 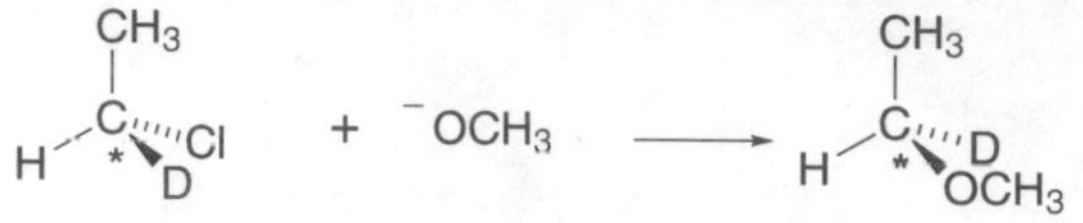 inversion of configuration

b.

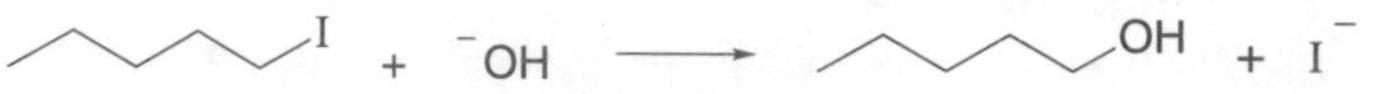

c.

No bond to the stereogenic center is broken, since the leaving group is not bonded to the stereogenic center.

d.  inversion of configuration

[* denotes a stereogenic center]

7.58 Follow the definitions from Answer 7.26.

a. $CH_3CH_2\overset{+}{C}HCH_2CH_3$

2° carbocation

c. $(CH_3)_2CHCH_2\overset{+}{C}H_2$

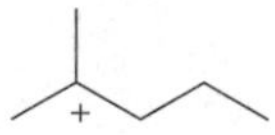

1° carbocation

e.

2° carbocation

b.

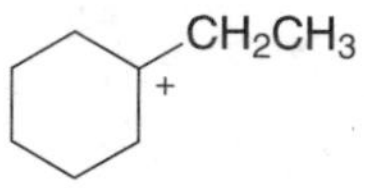

3° carbocation

d.

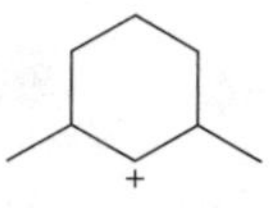

3° carbocation

f.

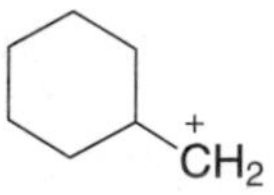

1° carbocation

7.59 For carbocations: **Increasing number of R groups = Increasing stability**.

a.

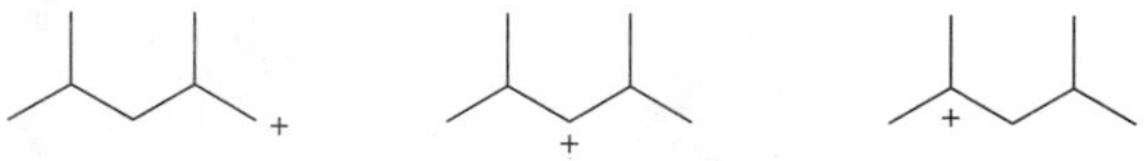

1° carbocation
least stable

2° carbocation
intermediate stablity

3° carbocation
most stable

b.

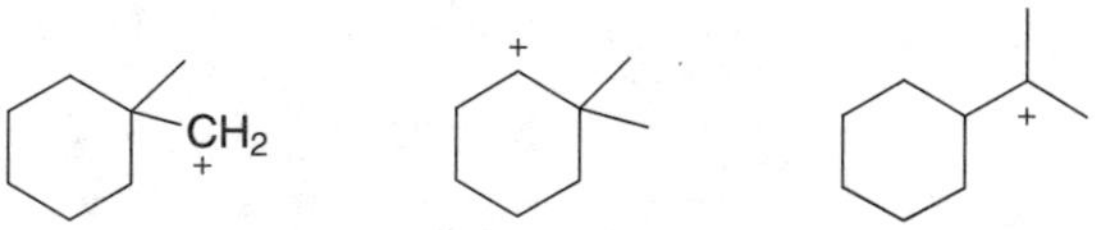

1° carbocation
least stable

2° carbocation
intermediate stablity

3° carbocation
most stable

7.60

a. Mechanism: S_N1 only

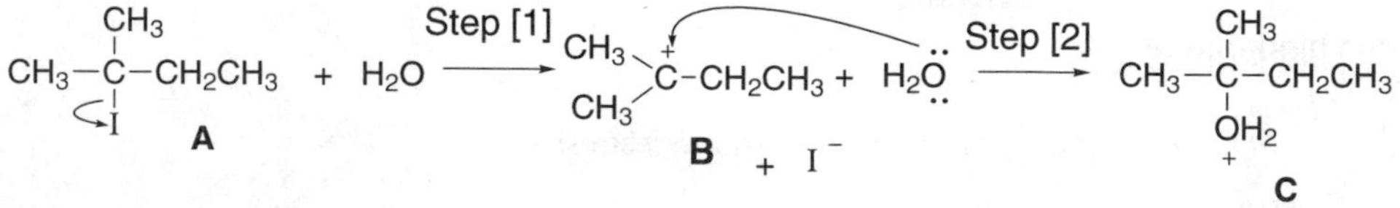

b. Energy diagram:

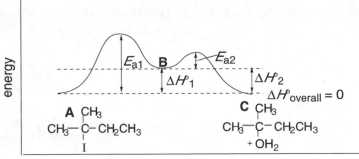

c. Transition states:

$$\left[\begin{array}{c} CH_3 \\ \delta^+ \\ CH_3 \end{array} C\text{-}CH_2CH_3 \\ \vdots \\ I\ \delta^- \right]^{\ddagger} \qquad \left[\begin{array}{c} CH_3 \\ \delta^+ \\ CH_3 \end{array} C\text{-}CH_2CH_3 \\ \vdots \\ \underset{\delta^+}{OH_2} \right]^{\ddagger}$$

d. rate equation: **rate = $k[(CH_3)_2CICH_2CH_3]$**

e. [1] Leaving group changed from I⁻ to Cl⁻: **rate decreases** since I⁻ is a better leaving group.
[2] Solvent changed from H_2O (polar protic) to DMF(polar aprotic):
   rate decreases since polar protic solvent favors S_N1.
[3] Alkyl halide changed from 3° to 2°: **rate decreases** since 2° carbocations are less stable.
[4] [H_2O] increased by factor of five: **no change in rate** since H_2O is not in rate equation.
[5] [R-X] and [H_2O] increased by factor of five: **rate increases** by a factor of five. (Only the concentration of R-X affects the rate.)

7.61 The rate of S_N1 reaction increases with increasing alkyl substitution.

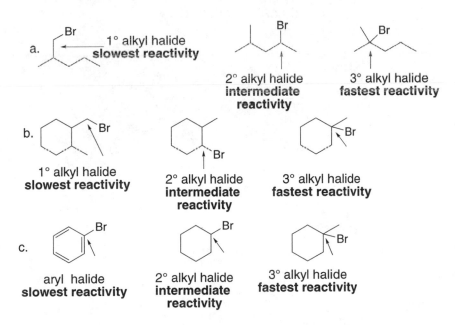

7.62 The rate of S_N1 reaction increases with increasing alkyl substitution, polar protic solvents, and better leaving groups.

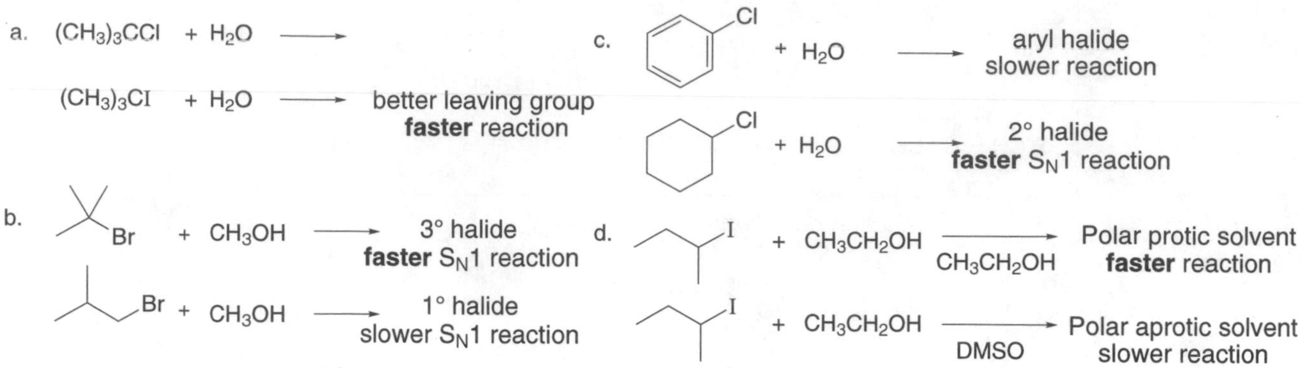
a. (CH3)3CCl + H2O
(CH3)3CI + H2O
better leaving group
faster reaction
c. + H2O
aryl halide
slower reaction
+ H2O
2° halide
faster SN1 reaction
b. Br + CH3OH
3° halide
faster SN1 reaction
Br + CH3OH
1° halide
slower SN1 reaction
d. + CH3CH2OH
CH3CH2OH
Polar protic solvent
faster reaction
+ CH3CH2OH
DMSO
Polar aprotic solvent
slower reaction

7.63

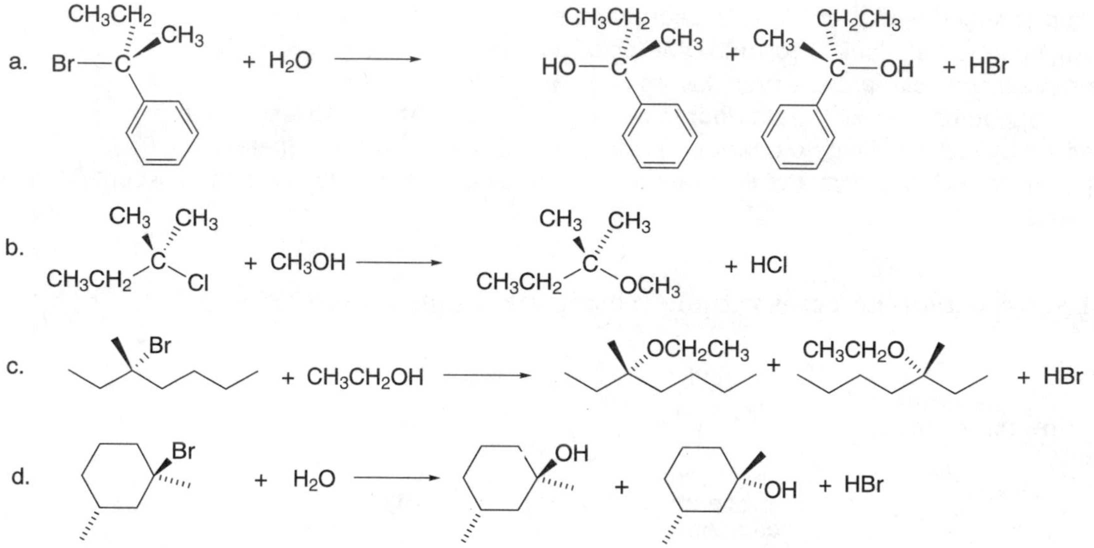
a. + H2O
+ HBr
b. + CH3OH
+ HCl
c. + CH3CH2OH
+ HBr
d. + H2O
+ HBr

7.64

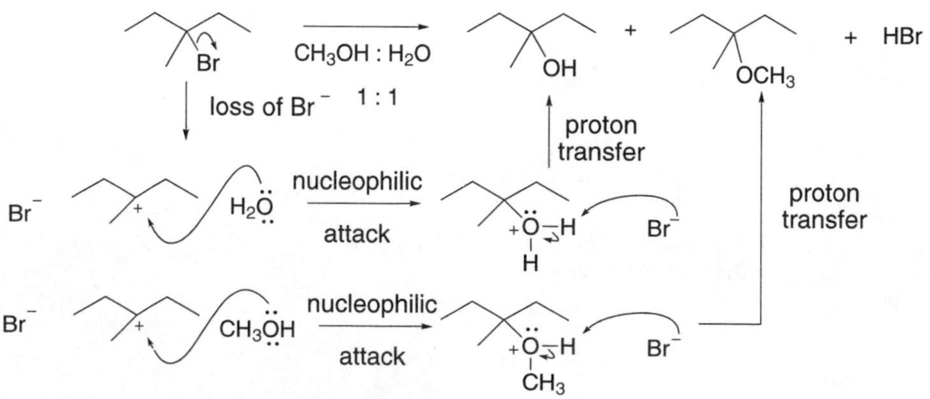
CH3OH : H2O
1 : 1
loss of Br⁻
+ HBr
proton
transfer
nucleophilic
attack
proton
transfer
nucleophilic
attack

7.65 More polar solvents favor S_N1 reactions by stabilizing the carbocation intermediate. Decreasing the polarity of a solvent would decrease the rate of an S_N1 reaction by making the E_a higher.

7.66 First decide whether the reaction will proceed via an S_N1 or S_N2 mechanism (Answer 7.36), and then draw the mechanism.

a. 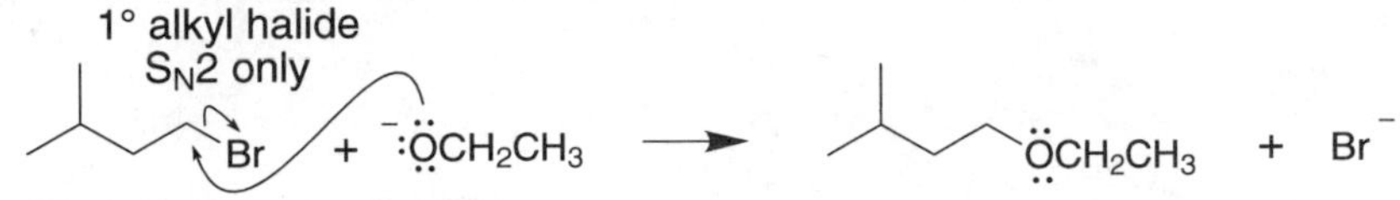
1° alkyl halide
SN2 only
Br
:ÖCH2CH3
ÖCH2CH3
Br⁻

b. 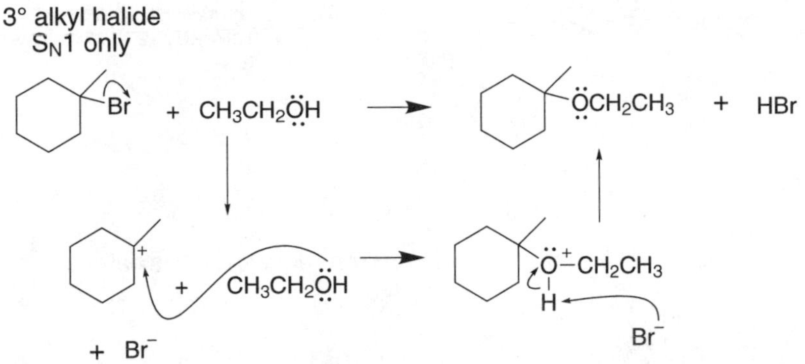
3° alkyl halide
SN1 only
Br
CH3CH2ÖH
ÖCH2CH3
HBr
CH3CH2ÖH
Ö–CH2CH3
H
Br⁻
+ Br⁻

c. 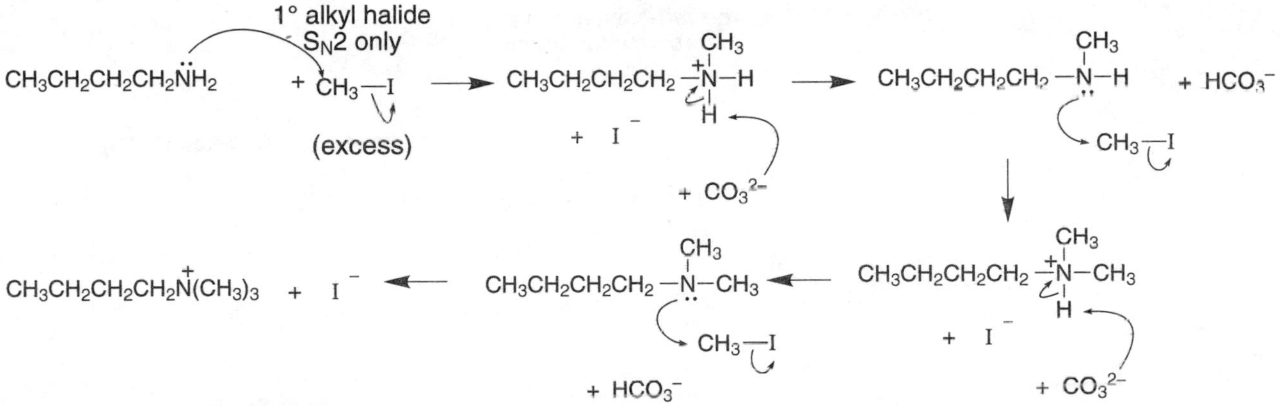
1° alkyl halide
SN2 only
CH3CH2CH2CH2N̈H2
+ CH3–I
(excess)
CH3CH2CH2CH2–N–H
CH3
H
+ I⁻
+ CO3 2–
CH3CH2CH2CH2–N–H
+ HCO3⁻
CH3–I
CH3CH2CH2CH2–N–CH3
+ CO3 2–
CH3CH2CH2CH2–N–CH3
CH3–I
+ HCO3⁻
CH3CH2CH2CH2N(CH3)3 + I⁻

7.67

a. 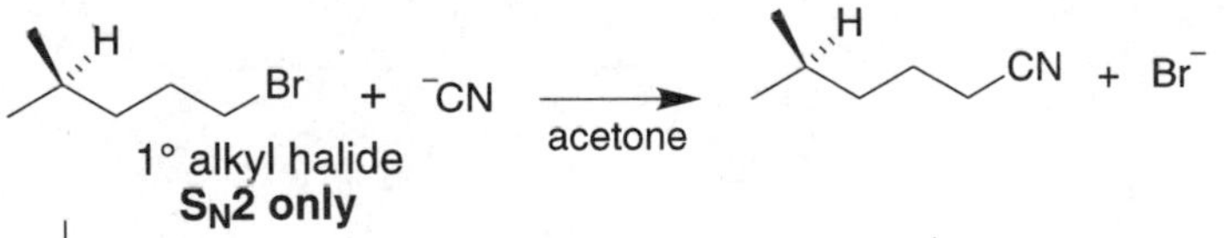
H
Br
+ ⁻CN
acetone
CN + Br⁻
1° alkyl halide
SN2 only

b. 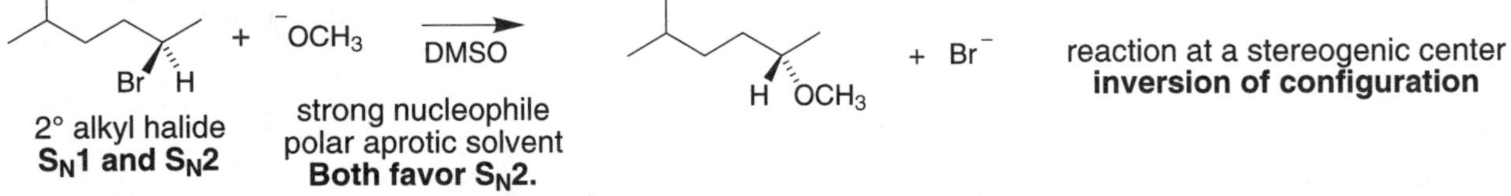
+ ⁻OCH3
DMSO
Br H
H OCH3
+ Br⁻
2° alkyl halide
SN1 and SN2
strong nucleophile
polar aprotic solvent
Both favor SN2.
reaction at a stereogenic center
inversion of configuration

c. 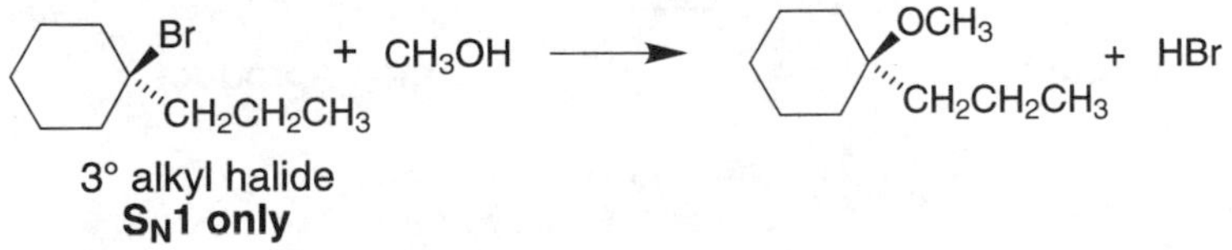
Br
CH2CH2CH3
+ CH3OH
OCH3
CH2CH2CH3
+ HBr
3° alkyl halide
SN1 only

d. 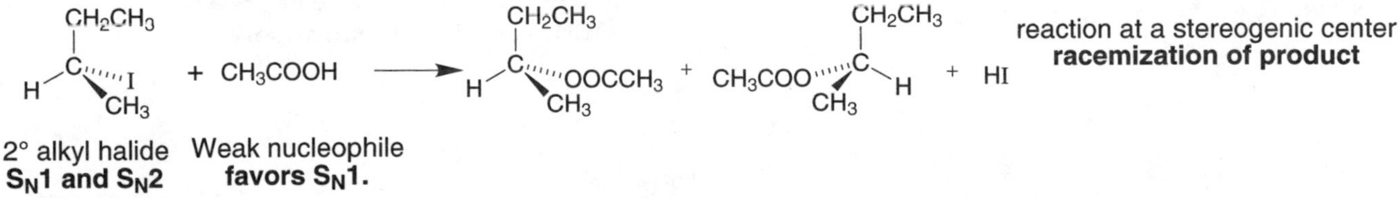
CH2CH3
H
C
I
CH3
+ CH3COOH
OOCCH3
CH3COO
+ HI
2° alkyl halide
SN1 and SN2
Weak nucleophile
favors SN1.
reaction at a stereogenic center
racemization of product

e.

2° alkyl halide
S_N1 and S_N2

strong nucleophile
polar aprotic solvent
Both favor S_N2.

reaction at a stereogenic center
inversion of configuration

f.

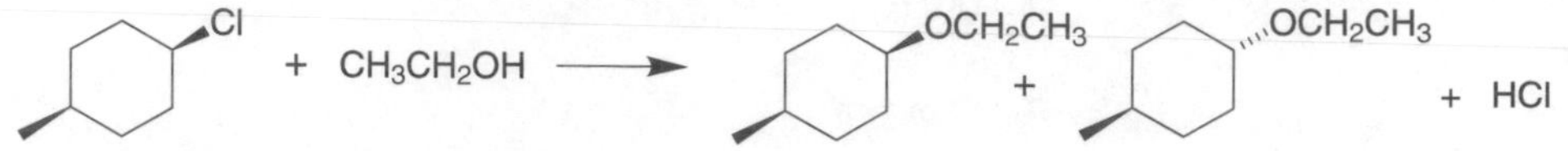

2° alkyl halide
S_N1 and S_N2

Weak nucleophile
favors S_N1.

two products - **diastereomers**
Nucleophile attacks
from above and below.

7.68

a.

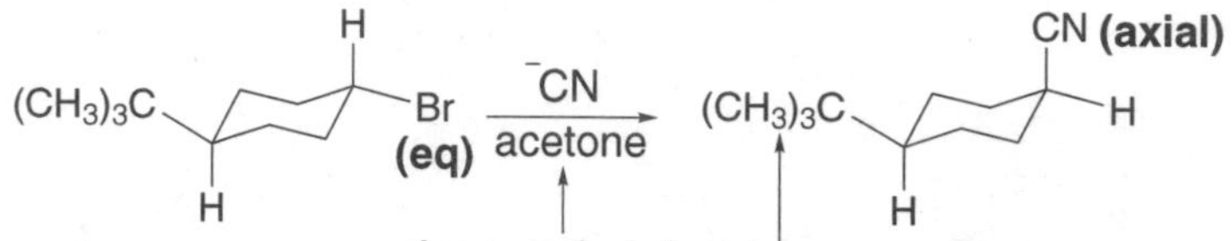

inversion (equatorial to axial)

polar aprotic solvent
S_N2 reaction

Large *tert*-butyl group in
more roomy equatorial
position.

b.

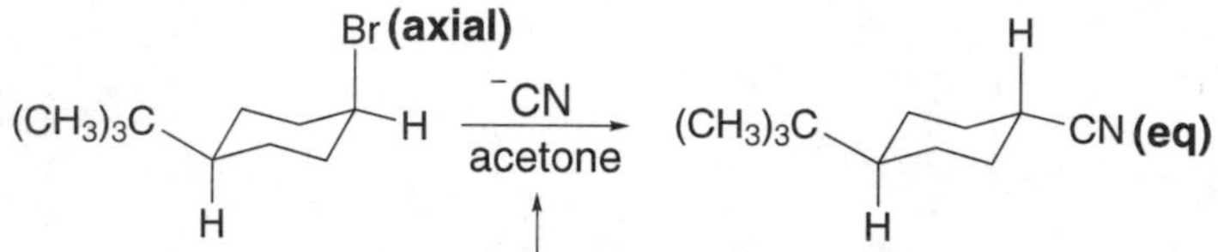

inversion (axial to equatorial)

polar aprotic solvent
S_N2 reaction

7.69

a.

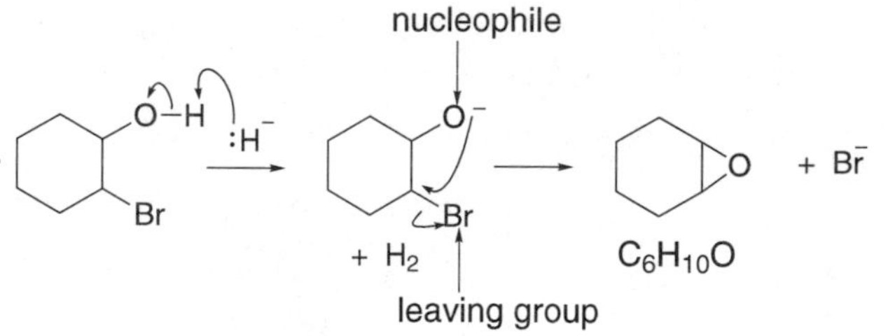

nucleophile

+ H_2

leaving group

$C_6H_{10}O$

b.

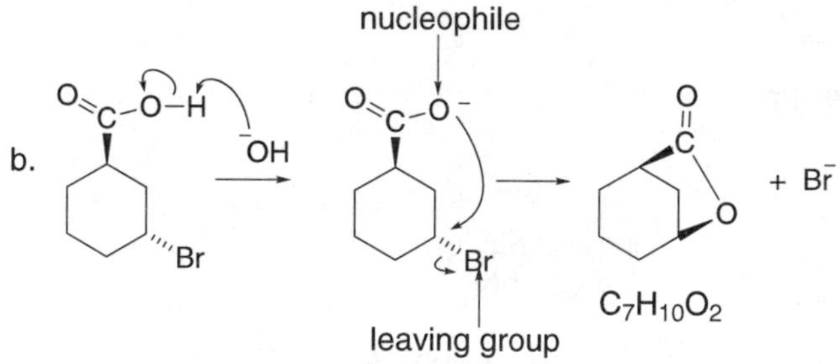

nucleophile

leaving group

$C_7H_{10}O_2$

7.70

CH_3—I
NaOCH$_3$
+ NaSCH$_3$
$\xrightarrow{CH_3OH}$
CH_3—OCH_3 + CH_3—SCH_3
major product

polar protic solvent
Nucleophilicity increases
down a column as anion size increases.
$^-$**SCH$_3$ is the better nucleophile.**

CH_3—I
NaOCH$_3$
+ NaSCH$_3$
$\xrightarrow{(CH_3)_2S=O}$
CH_3—OCH_3 + CH_3—SCH_3
major product

polar aprotic solvent
Nucleophilicity decreases
down a column as anion size increases.
$^-$**OCH$_3$ is the better nucleophile.**

7.71

a. Hexane is nonpolar and therefore few nucleophiles will dissolve in it.

b. $(CH_3)_3CO^-$ is a stronger base than $CH_3CH_2O^-$:

The three electron donating CH_3 groups add electron density to the negative charge
of the conjugate base, destabilizing it and making it a stronger base.

c. By the Hammond postulate, the S_N1 reaction is faster with RX that form more stable carbocations.

$(CH_3)_3C +$

3° Carbocation is stabilized by
three electron donor CH_3 groups.

$(CH_3)_2C +$
CF_3 Although this carbocation is also 3°,
the three electron withdrawing F atoms
destabilize the positive charge. Since
the carbocation is less stable, the
reaction to form it is slower.

d. The identifty of the nucleophile does not affect the rate of S_N1 reactions since the nucleophile
does not appear in the rate-determining step.

e.

Polar aprotic solvent
favors S_N2 reaction.

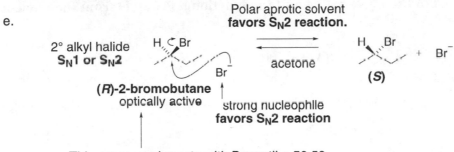

2° alkyl halide
S_N1 or S_N2

acetone

(R)-2-bromobutane
optically active

(S)   + Br^-

strong nucleophile
favors S_N2 reaction

This compound reacts with Br^- until a 50:50
mixture results, making the mixture optically inactive.
Then either compound can react with Br^- and the
mixture remains optically inactive.

7.72

1° alkyl halides react by
S_N2 reactions.
H_2O is a weak nucleophile
and favors **S_N1** reactions.
This makes the reactions slow.

$$CH_3Cl \xrightarrow[\text{slow}]{H_2O} CH_3OH$$

$$CH_2Cl_2 \xrightarrow[\text{extremely slow}]{H_2O} CH_3OH$$

H_2O is a weak nucleophile. Since CH_3Cl must
react by S_N2, the weak nucleophile means a
slower reaction. Adding more Cl's adds steric
hindrance, decreasing the rate even more.

7.73

a. ～Br → ～CN   The nucleophile has replaced the leaving group.
Missing reagent: $^-$**CN**

b. The nucleophile has replaced the leaving group.
Missing reagent:

c. The nucleophile has replaced the leaving group.
Missing reagent:
$^-C\equiv CH$

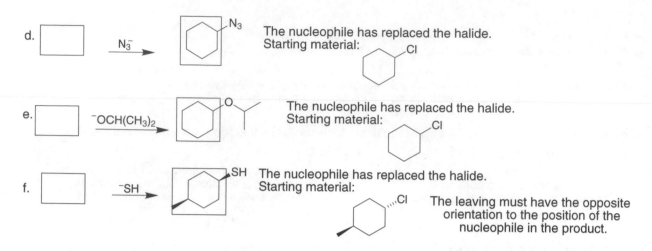

d. The nucleophile has replaced the halide.
Starting material:

e. The nucleophile has replaced the halide.
Starting material:

f. The nucleophile has replaced the halide.
Starting material:

The leaving must have the opposite
orientation to the position of the
nucleophile in the product.

7.74 To devise a synthesis, look for the carbon framework and the functional group in the product. **The carbon framework is from the alkyl halide and the functional group is from the nucleophile.**

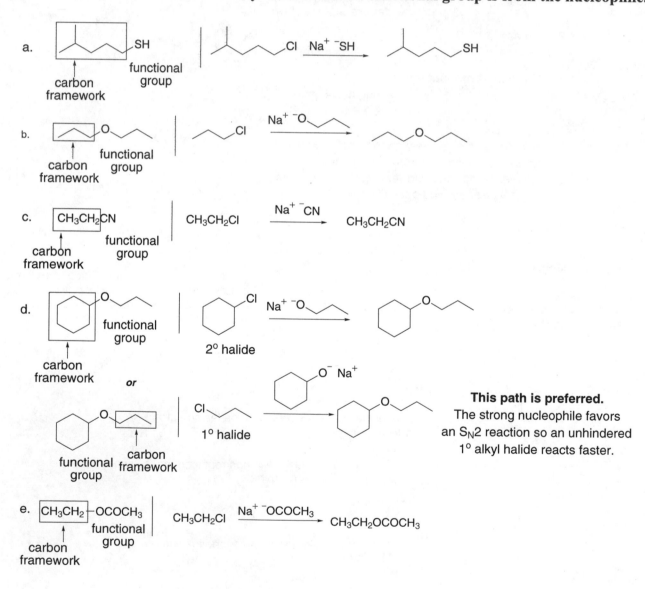

a. SH
functional group
carbon framework

Cl Na⁺ ⁻SH → SH

b. O
functional group
carbon framework

Cl Na⁺ ⁻O →

c. CH₃CH₂CN
functional group
carbon framework

CH₃CH₂Cl Na⁺ ⁻CN → CH₃CH₂CN

d. O
functional group
carbon framework

Cl Na⁺ ⁻O →
2° halide

or

O
functional group carbon framework

Cl
1° halide
O⁻ Na⁺ →

This path is preferred.
The strong nucleophile favors
an S$_N$2 reaction so an unhindered
1° alkyl halide reacts faster.

e. CH₃CH₂—OCOCH₃
functional group
carbon framework

CH₃CH₂Cl Na⁺ ⁻OCOCH₃ → CH₃CH₂OCOCH₃

7.75

B
very crowded 3° halide

Na$^+$ $^-$OCH$_3$

C

E

D

$^-$O$^-$Na$^+$ CH$_3$I

A
unhindered methyl halide

OCH$_3$

E

preferred method
The strong nucleophile favors S$_N$2 reaction so the alkyl halide should be unhindered for a faster reaction.

7.76

A and **B** can't react by an S$_N$2 mechanism because the backside attack of the nucleophile is blocked:

:Nu$^-$

Cl

The S$_N$1 reaction would require a planar carbocation, and geometry doesn't allow this to occur. The resulting carbocation cannot adopt the needed trigonal planar geometry and thus it does not form.

Br

This carbocation cannot adopt a trigonal planar geometry.

7.77 Steric hindrance decreases nucleophilicity.

quinuclidine

triethylamine

This electron pair is more hindered by the three CH$_2$CH$_3$ groups.

The three alkyl groups are "tied back" in a ring, making the electron pair more available.

CH$_3$CH$_2$ N   CH$_2$CH$_3$
CH$_2$CH$_3$

These bulky groups around the N cause steric hindrance and this decreases nucleophilicity.

This electron pair on quinuclidine is much more available than the one on triethylamine.

less steric hindrance
more nucleophilic

7.78

O

H

H:$^-$

[1]

O$^-$

+ H$_2$

O$^-$

CH$_3$—Br

[2]

O—CH$_3$

minor product

+ NaBr

O

CH$_3$—Br   [2]

O

CH$_3$

major product

7.79

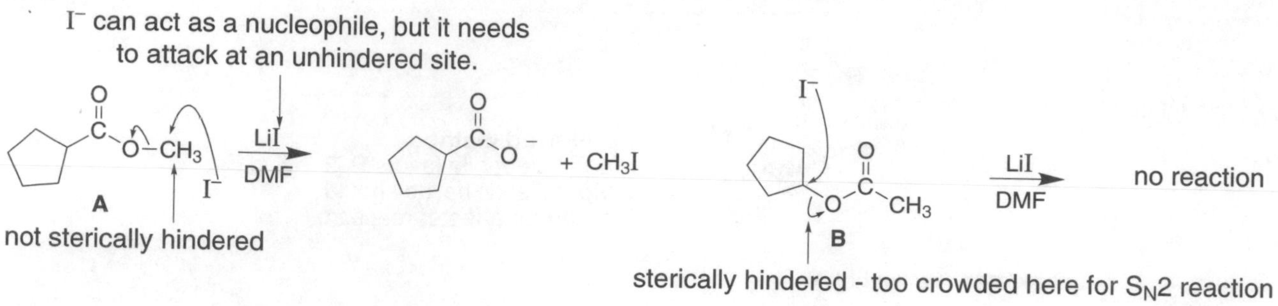

I⁻ can act as a nucleophile, but it needs to attack at an unhindered site.
LiI
DMF
A
+ CH3I
not sterically hindered
LiI
DMF
no reaction
B
sterically hindered - too crowded here for SN2 reaction

7.80 $CH_3CH_2OCH_2Cl$ affords a resonance-stabilized carbocation, making an S_N1 reaction possible even though the alkyl halide is 1°.

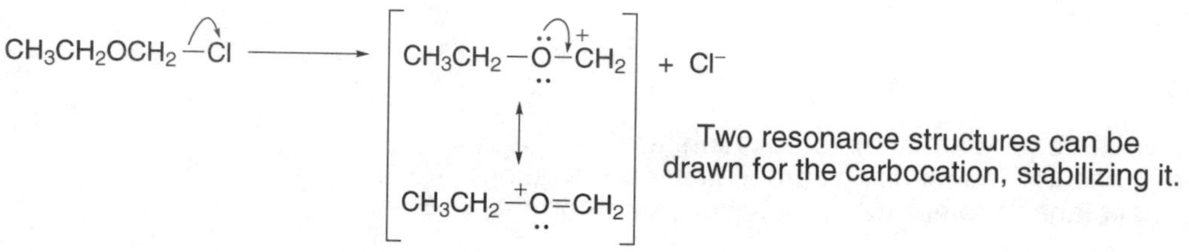

CH3CH2OCH2–Cl
CH3CH2–O–CH2
+ Cl⁻
CH3CH2–O=CH2
Two resonance structures can be drawn for the carbocation, stabilizing it.

The carbocation can then continue the reaction:

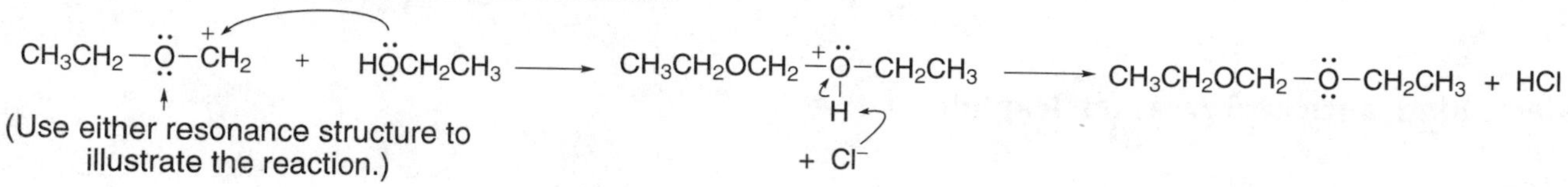

CH3CH2–O–CH2
+
HOCH2CH3
CH3CH2OCH2–O–CH2CH3
H
+ Cl⁻
CH3CH2OCH2–O–CH2CH3 + HCl
(Use either resonance structure to illustrate the reaction.)

Chapter 8: Alkyl Halides and Elimination Reactions

◆ A comparison between nucleophilic substitution and β-elimination

Nucleophilic substitution—A nucleophile attacks a carbon atom (7.6).

substitution
product

good
leaving group

β-Elimination—A base attacks a proton (8.1).

elimination
product

good
leaving group

Similarities	Differences
• In both reactions RX acts as an electrophile, reacting with an electron rich reagent. • Both reactions require a **good leaving group X:⁻** willing to accept the electron density in the C–X bond.	• In substitution, a nucleophile attacks a single carbon atom. • In elimination, a Brønsted–Lowry base removes a proton to form a π bond, and two carbons are involved in the reaction.

◆ The importance of the base in E2 and E1 reactions (8.9)

The strength of the base determines the mechanism of elimination.
- Strong bases favor E2 reactions.
- Weak bases favor E1 reactions.

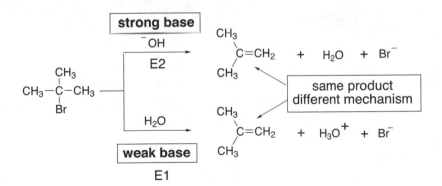

♦ E1 and E2 mechanisms compared

	E2 mechanism	E1 mechanism
[1] Mechanism	• One step (8.4B)	• Two steps (8.6B)
[2] Alkyl halide	• rate: $R_3CX > R_2CHX >$ RCH_2X (8.4C)	• rate: $R_3CX > R_2CHX >$ RCH_2X (8.6C)
[3] Rate equation	• rate = k[RX][B:] • second order kinetics (8.4A)	• rate = k[RX] • first order kinetics (8.6A)
[4] Stereochemistry	• anti periplanar arrangement of H and X (8.8)	• trigonal planar carbocation intermediate (8.6B)
[5] Base	• favored by strong bases (8.4B)	• favored by weak bases (8.6C)
[6] Leaving group	• better leaving group → faster reaction (8.4B)	• better leaving group → faster reaction (Table 8.4)
[7] Solvents	• favored by polar aprotic solvents (8.4B)	• favored by polar protic solvents (Table 8.4)
[8] Product	• more substituted alkene favored (Zaitsev Rule, 8.5)	• more substituted alkene favored (Zaitsev Rule, 8.6C)

♦ Summary chart on the four mechanisms: S_N1, S_N2, E1 or E2

Alkyl halide type	Conditions	Mechanism
1° RCH_2X	strong nucleophile	S_N2
	strong bulky base	E2
2° R_2CHX	strong base and nucleophile	S_N2 + E2
	strong bulky base	E2
	weak base and nucleophile	S_N1 + E1
3° R_3CX	weak base and nucleophile	S_N1 + E1
	strong base	E2

♦ Zaitsev Rule

- β-Elimination affords the more stable product having the more substituted double bond.
- Zaitsev products predominate in E2 reactions except when a cyclohexane ring prevents trans diaxial arrangement.

Chapter 8: Answers to Problems

8.1 • The carbon bonded to the leaving group is the **α carbon**. Any carbon bonded to it is a **β carbon**.
  • **To draw the products of an elimination reaction:** Remove the leaving group from the α carbon and a H from the β carbon and form a π bond.

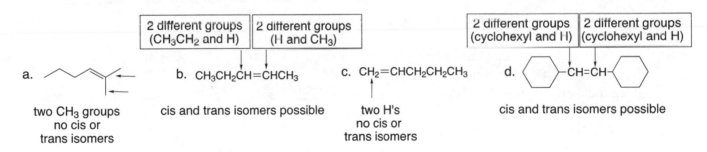

8.2 Alkenes are classified by the number of carbon atoms bonded to the double bond. A monosubstituted alkene has one carbon atom bonded to the double bond, a disubstituted alkene has two carbon atoms bonded to the double bond, etc. (Boldface atoms or bonds are bonded to the C=C.)

a. $CH_3CH=CHCH_2CH_3$

2 C's bonded to C=C
disubstituted

b.

3 C's bonded to C=C
trisubstituted

c.

CH_3
CH_3

4 C's bonded to C=C
tetrasubstituted

d.

$=CH_2$

2 C's bonded to C=C
disubstituted

8.3 To have a cis and trans isomer, the two groups on each end of the double bond must be different from each other.

a.

two CH₃ groups
no cis or
trans isomers

2 different groups (CH₃CH₂ and H)	2 different groups (H and CH₃)

b. $CH_3CH_2CH=CHCH_3$

cis and trans isomers possible

c. $CH_2=CHCH_2CH_2CH_3$

two H's
no cis or
trans isomers

2 different groups (cyclohexyl and H)	2 different groups (cyclohexyl and H)

d.

$-CH=CH-$

cis and trans isomers possible

8.4 Two definitions:
  • **Constitutional isomers** differ in the connectivity of the atoms.
  • **Stereoisomers** differ only in the 3-D arrangement of atoms in space.

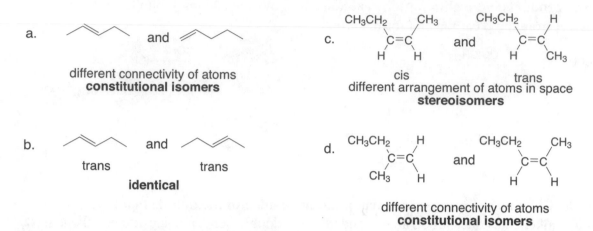

a. [structure] and [structure]

different connectivity of atoms
constitutional isomers

c. CH_3CH_2 CH_3 CH_3CH_2 H
        C=C   and   C=C
       H   H   H CH_3

cis   trans
different arrangement of atoms in space
stereoisomers

b. [structure] and [structure]

trans   trans
identical

d. CH_3CH_2 H CH_3CH_2 CH_3
        C=C   and   C=C
       CH_3 H   H   H

different connectivity of atoms
constitutional isomers

8.5 Two rules to predict the relative stability of alkenes:
  [1] Trans alkenes are more stable than cis alkenes.
  [2] The stability of an alkene increases as the number of R groups on the C=C bond increases.

a. [structure] and [structure]

monosubstituted   disubstituted
                   more stable

c. [structure] and [structure]

trisubstituted   disubstituted
more stable

b. CH_3CH_2 CH_2CH_3 CH_3CH_2 H
        C=C   and   C=C
       H   H   H CH_2CH_3

cis   trans
                        more stable

8.6 In an E2 mechanism, four bonds are involved in the single step. Use curved arrows to show these simultaneous actions:
  [1] The base attacks a hydrogen on a β carbon.
  [2] A π bond forms.
  [3] The leaving group comes off.

β carbon
CH_3CH_2

$CH_3CH_2-C-CHCH_3$ ⟶ $(CH_3CH_2)_2C=CHCH_3$ + $HOCH_2CH_3$ + Br^-
        Br H
                                    new π bond
        $^-OCH_2CH_3$

transition state:

$$\left[\begin{array}{c} CH_3CH_2 \\ CH_3CH_2-C=CHCH_3 \\ \delta^-Br \quad H--OCH_2CH_3 \\ \delta^- \end{array} \right]^{\ddagger}$$

8.7 For E2 elimination to occur there must be at least one hydrogen on a β carbon.

β carbon

CH₃—C(CH₃)(CH₃)—C(H)(Br)—H

no H's on β carbon
inert to E2 elimination

8.8

a. CH_3CH_2—Br + ⁻OH ⟶

CH_3CH_2—Br + ⁻OC(CH₃)₃ ⟶

stronger base
faster reaction

better leaving group
faster reaction
↓
b. CH_3CH_2—Br + ⁻OC(CH₃)₃ ⟶

CH_3CH_2—Cl + ⁻OC(CH₃)₃ ⟶

8.9 As the number of R groups on the carbon with the leaving group increases, the rate of an E2 reaction increases.

a. (CH₃)₂CHCH₂CH₂CH₂Br   (CH₃)₂CHCH₂CH(Br)CH₃   (CH₃)₂C(Br)CH₂CH₂CH₃

1° alkyl halide   2° alkyl halide   3° alkyl halide
least reactive   **intermediate reactivity**   **most reactive**

b.
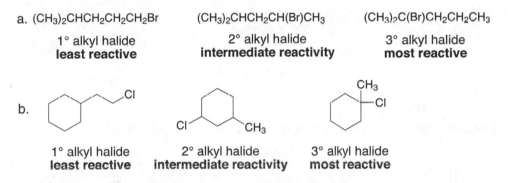

1° alkyl halide   2° alkyl halide   3° alkyl halide
least reactive   **intermediate reactivity**   **most reactive**

8.10 Use the following characteristics of an E2 reaction to answer the questions:
[1] E2 reactions are second order and one step.
[2] More substituted halides react faster.
[3] Reactions with strong bases or better leaving groups are faster.
[4] Reactions with polar aprotic solvents are faster.

Rate equation: rate = k[RX][Base]
  a. tripling the concentration of the alkyl halide = **rate triples**
  b. halving the concentration of the base = **rate halved**
  c. changing the solvent from CH₃OH to DMSO = **rate increases** (Polar aprotic solvent is better for E2.)
  d. changing the leaving group from I⁻ to Br⁻ = **rate decreases** (I⁻ is a better leaving group.)
  e. changing the base from ⁻OH to H₂O = **rate decreases** (weaker base)
  f. changing the alkyl halide from CH₃CH₂Br to (CH₃)₂CHBr = **rate increases** (More substituted halide reacts faster.)

8.11 The Zaitsev Rule states: In a β-elimination reaction, the major product has the more substituted double bond.

a.

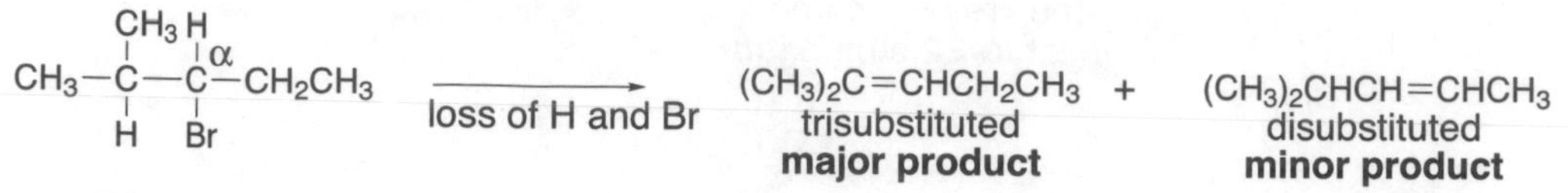

$CH_3-\underset{\underset{Br}{|}}{\overset{\overset{CH_3\ H}{|\ \ \ |\alpha}}{C}}-CH_2CH_3$ →(loss of H and Br)→ $(CH_3)_2C{=}CHCH_2CH_3$ + $(CH_3)_2CHCH{=}CHCH_3$

trisubstituted
major product

disubstituted
minor product

b.

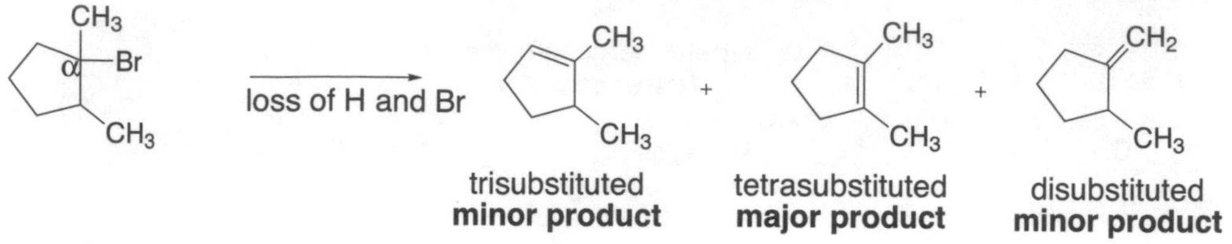

loss of H and Br

trisubstituted
minor product

tetrasubstituted
major product

disubstituted
minor product

c.

loss of H and Cl

monosubstituted
minor product

+ $CH_3CH_2CH_2CH_2CH{=}CHCH_3$

disubstituted
major product

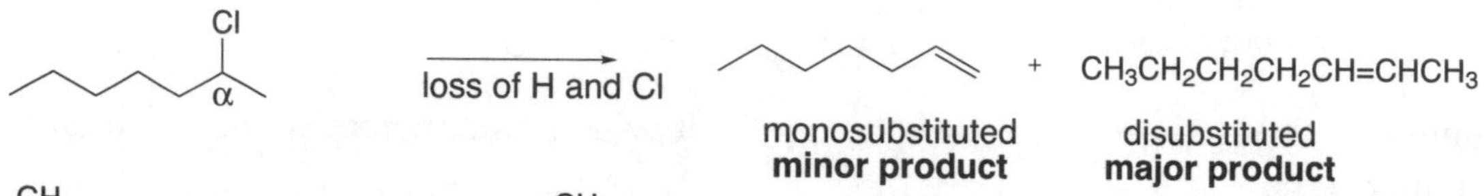

d.

loss of H and Cl

trisubstituted
ONLY product

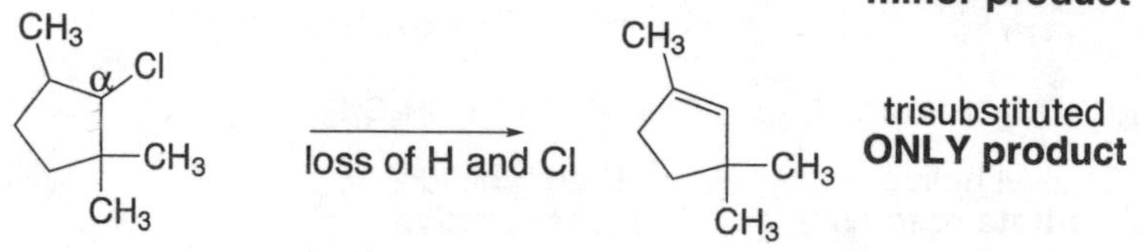

8.12 An E1 mechanism has two steps:
[1] The leaving group comes off, creating a carbocation.
[2] A base pulls off a proton from a β carbon, and a π bond forms.

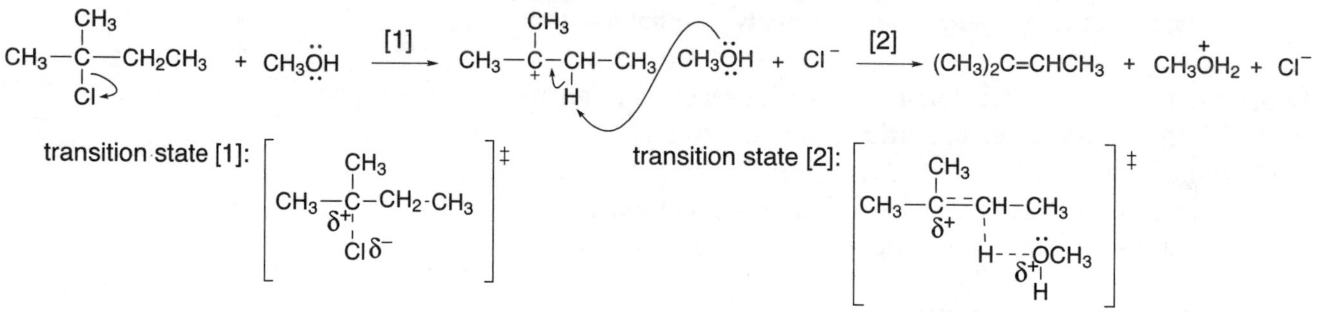

transition state [1]:

transition state [2]:

8.13 The Zaitsev Rule states: In a β-elimination reaction, the major product has the more substituted double bond.

a.

trisubstituted
major product

+

disubstituted

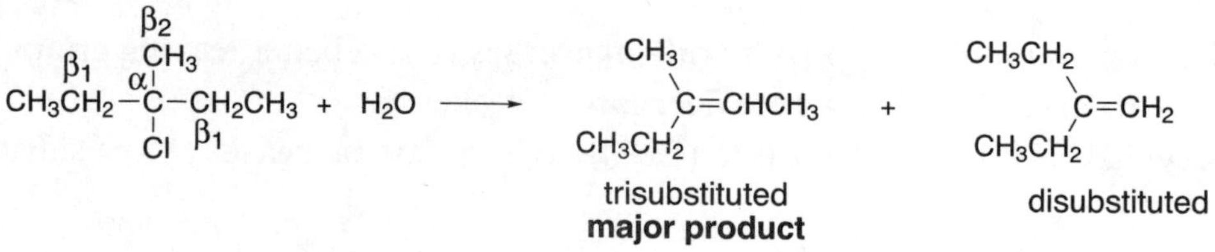

tetrasubstituted
major product

disubstituted

trisubstituted

8.14 Use the following characteristics of an **E1 reaction** to answer the questions:
[1] E1 reactions are first order and two steps.
[2] More substituted halides react faster.
[3] Weaker bases are preferred.
[4] Reactions with better leaving groups are faster.
[5] Reactions in polar protic solvents are faster.

Rate equation: rate = k[RX] The base doesn't affect rate.
a. doubling the concentration of the alkyl halide = **rate doubles**
b. doubling the concentration of the base = **no change** (Base is not in the rate equation.)
c. changing the alkyl halide from $(CH_3)_3CBr$ to $CH_3CH_2CH_2Br$ = **rate decreases** (More substituted halides react faster.)
d. changing the leaving group from Cl^- to Br^- = **rate increases** (better leaving group)
e. changing the solvent from DMSO to CH_3OH = **rate increases** (Polar protic solvent favors E1.)

8.15 Both S_N1 and E1 reactions occur by forming a carbocation. To draw the products:
[1] **For the S_N1 reaction,** substitute the nucleophile for the leaving group.
[2] **For the E1 reaction,** remove a proton from a β carbon and create a new π bond.

8.16 E2 reactions occur with anti periplanar geometry. **The anti periplanar arrangement uses a *staggered* conformation and has the H and X on *opposite sides* of the C–C bond.**

H and Br are on opposite sides =
anti periplanar

8.17 The E2 elimination reactions will occur in the anti periplanar orientation as drawn. To draw the product of elimination, maintain the orientation of the remaining groups around the C=C.

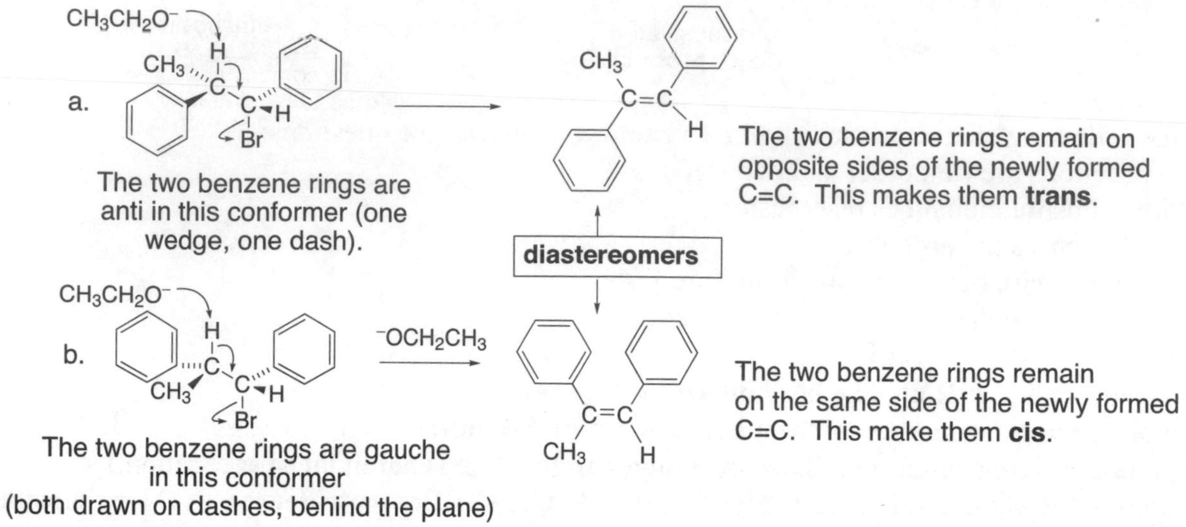
a.
CH3CH2O−
The two benzene rings are anti in this conformer (one wedge, one dash).
The two benzene rings remain on opposite sides of the newly formed C=C. This makes them trans.
diastereomers
b.
CH3CH2O−
−OCH2CH3
The two benzene rings are gauche in this conformer (both drawn on dashes, behind the plane)
The two benzene rings remain on the same side of the newly formed C=C. This make them cis.

8.18 Note: The Zaitsev products predominate in E2 elimination *except* when substituents on a cyclohexane ring prevent a **trans diaxial** arrangement of H and X.

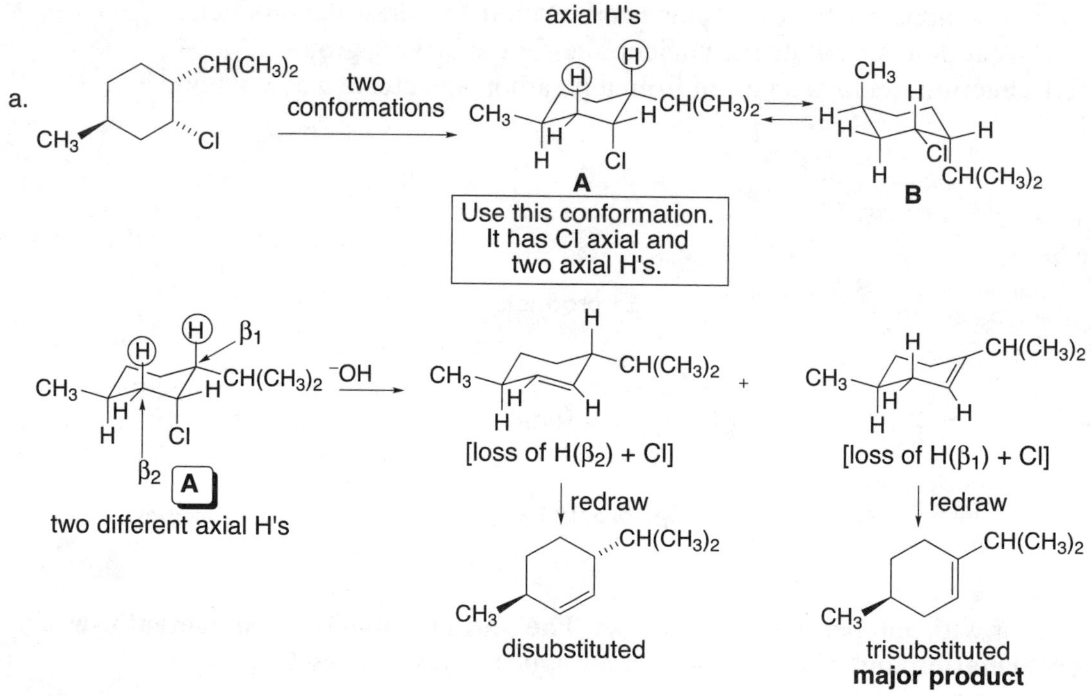
a.
two conformations
axial H's
A
B
Use this conformation. It has Cl axial and two axial H's.
β1
β2
A
two different axial H's
−OH
[loss of H(β2) + Cl]
redraw
disubstituted
[loss of H(β1) + Cl]
redraw
trisubstituted
major product

b.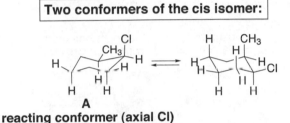

A

B

Use this conformation.
It has Cl axial and
one axial H.

β₂

⁻OH

B

only one axial H
on a β carbon

[loss of H(β₁) + Cl]

=

disubstituted
only product

8.19 Draw the chair conformation of *cis*-1-chloro-2-methylcyclohexane and its trans isomer. For E2 elimination reactions to occur, **there must be an H and X trans diaxial to each other.**

Two conformers of the cis isomer:

A
reacting conformer (axial Cl)

This reacting conformer has only one
group axial, making it more stable than **B**.
This makes a **faster elimination reaction
with the cis isomer.**

Two conformers of the trans isomer:

B
reacting conformer (axial Cl)

This conformer is unstable
since both CH₃ and Cl are axial.
Since this reacting conformer is less
stable than **A**, **this slows the rate of
elimination from the trans isomer.**

8.20 **E2 reactions are favored by strong negatively charged bases** and occur with 1°, 2° and 3° halides, with 3° being the most reactive.

E1 reactions are favored by weaker neutral bases and do not occur with 1° halides since they form highly unstable carbocations.

a. $CH_3-\underset{\underset{Cl}{|}}{\overset{\overset{CH_3}{|}}{C}}-CH_3$ + ⁻OCH₃ ⟶
strong negatively
charged base
E2

c.
+ CH₃OH ⟶
weak neutral
base
E1

b.
+ H₂O ⟶
weak neutral
base
E1

d. CH₃CH₂Br + ⁻OC(CH₃)₃ ⟶
strong negatively
charged base
E2

8.21 Draw the alkynes that result from removal of two equivalents of HX.

a. cyclohexyl–C(Cl)(H)–C(Cl)(H)–CH$_2$CH$_3$ $\xrightarrow{\ ^-NH_2\ }$ cyclohexyl–C≡C–CH$_2$CH$_3$

c. CH$_3$–C(Br)(Br)–CH$_2$CH$_3$ $\xrightarrow{\ ^-NH_2\ }$ CH$_3$C≡CCH$_3$

+ HC≡CCH$_2$CH$_3$

b. CH$_3$CH$_2$CH$_2$CHCl$_2$ $\xrightarrow[\text{DMSO}]{\text{KOC(CH}_3)_3}$ CH$_3$CH$_2$C≡CH

d. (Ph)CH(Br)–CH(Br)(Ph) $\xrightarrow{\ ^-NH_2\ }$ (Ph)C≡C(Ph)

8.22

a. CH$_3$CH$_2$CH$_2$CH$_2$CH$_2$CH$_2$–Cl $\xrightarrow{\text{K}^+\ ^-\text{OC(CH}_3)_3}$ (heptene)

  1° halide
  S$_N$2 or E2
  strong sterically
  hindered base
  E2

b. CH$_3$–C(H)(Cl)–CH$_2$CH$_3$ $\xrightarrow{\ ^-OH\ }$ CH$_3$–C(H)(OH)–CH$_2$CH$_3$ + CH$_3$–CH=CHCH$_3$ + CH$_2$=CH–CH$_2$CH$_3$

  2° halide   strong base   S$_N$2 product   disubstituted   monosubstituted
  any mechanism   S$_N$2 and E2   **major E2 product**   **minor E2 product**

c. cyclohexane with CH$_2$CH$_3$ and I $\xrightarrow[\text{weak base}]{\text{CH}_3\text{CH}_2\text{OH}}$ cyclohexane with CH$_2$CH$_3$ and OCH$_2$CH$_3$ + cyclohexane=CHCH$_3$ with CH$_2$CH$_3$ + cyclohexene with CH$_2$CH$_3$

  3° halide   S$_N$1 and E1   **S$_N$1 product**   + **E1 product**   + **E1 product**
  no S$_N$2

d. CH$_3$CH$_2$CH$_2$–C(CH$_3$)(Cl)–CH$_3$ type $\xrightarrow[\text{CH}_3\text{CH}_2\text{OH}]{\text{strong base} \ \text{CH}_3\text{CH}_2\text{O}^-}$ (alkene) + (alkene)

  3° halide   **E2**   **major E2 product**   **minor E2 product**
  no S$_N$2

8.23

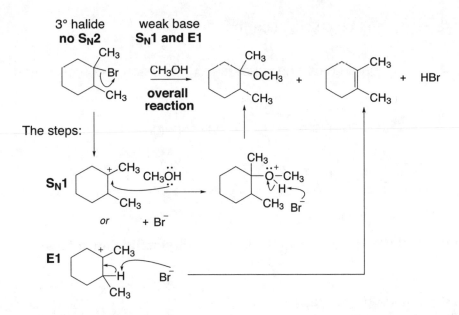

The steps:

S_N1

or   + Br⁻

E1

8.24

a. $CH_3CH_2CH_2CH_2CH_2CH_2Br \longrightarrow CH_3CH_2CH_2CH_2CH{=}CH_2$

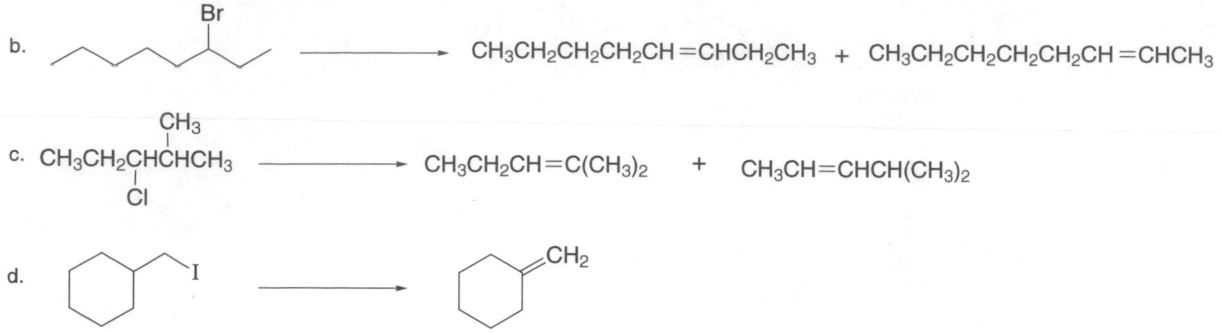
b.
Br
CH3CH2CH2CH2CH=CHCH2CH3 + CH3CH2CH2CH2CH2CH=CHCH3
c. CH3CH2CHCHCH3
CH3
Cl
CH3CH2CH=C(CH3)2 + CH3CH=CHCH(CH3)2
d.
I
CH2

8.25 To give only one product in an elimination reaction, **the starting alkyl halide must have only one type of β carbon.**

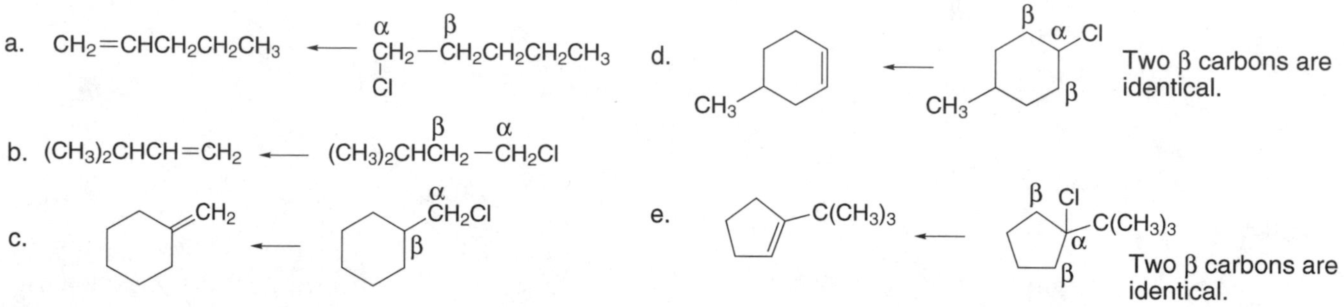
a. CH2=CHCH2CH2CH3 ← CH2—CH2CH2CH2CH3
α β
Cl
b. (CH3)2CHCH=CH2 ← (CH3)2CHCH2—CH2Cl
β α
c.
CH2
CH2Cl
α
β
d.
CH3
Cl
α
β
Two β carbons are identical.
e.
C(CH3)3
β Cl
α
Two β carbons are identical.

8.26 To have cis and trans isomers, the two groups on each end of the double bond must be different from each other.

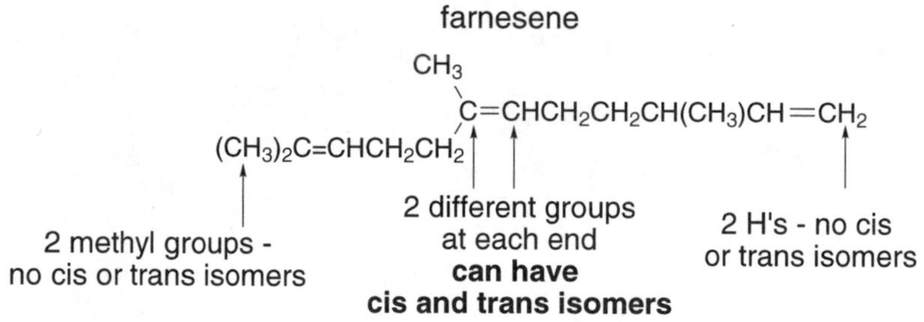
farnesene
CH3
C=CHCH2CH2CH(CH3)CH=CH2
(CH3)2C=CHCH2CH2
2 methyl groups - no cis or trans isomers
2 different groups at each end
can have cis and trans isomers
2 H's - no cis or trans isomers

8.27 Use the definitions in Answer 8.4.

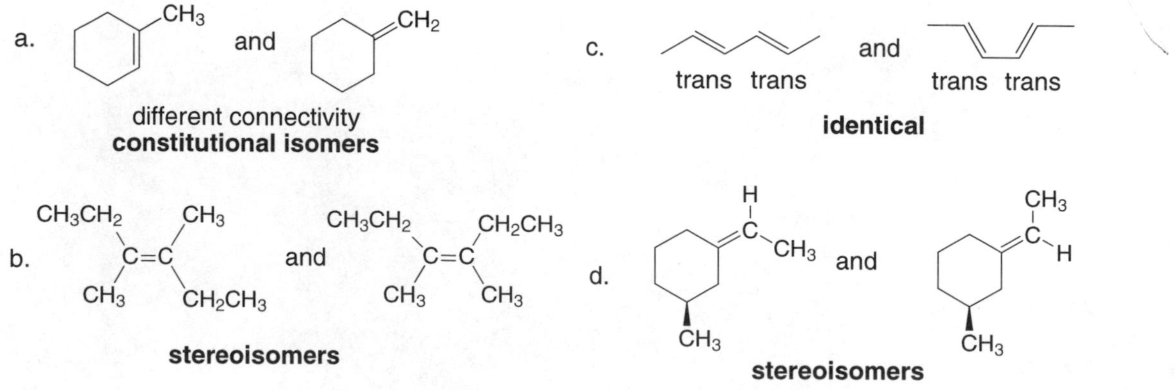
a.
CH3
and
CH2
different connectivity
constitutional isomers
c.
and
trans trans
trans trans
identical
b.
CH3CH2
CH3
C=C
CH3
CH2CH3
and
CH3CH2
CH2CH3
C=C
CH3
CH3
stereoisomers
d.
H
C
CH3
and
CH3
C
H
CH3
CH3
stereoisomers

8.28 There are three different isomers. Cis and trans isomers are diastereomers.

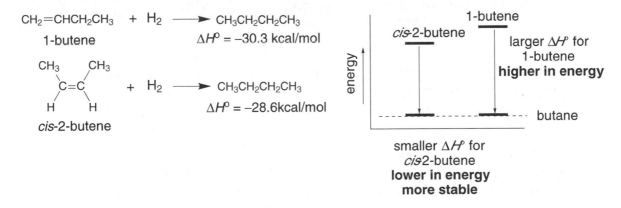

A
**constitutional isomer
of B and C**

B   **C**
diastereomers

8.29

Double bond can be cis or trans.

PGF$_{2\alpha}$

$CH_2CH=CH(CH_2)_3COOH$

$CH=CHCH(OH)(CH_2)_4CH_3$
sp³ stereogenic center
Double bond can be cis or trans.

a. 5 *sp³* stereogenic centers (4 circled, one labeled)
b. 2 double bonds can both be cis or trans.
c. 2^7 = 128 stereoisomers possible

8.30 Use the rules from Answer 8.5 to rank the alkenes.

a. $CH_2=CHCH_2CH_2CH_3$

monosubstituted

least stable

CH_3CH_2 CH_3
  C=C
 H   H

disubstituted
cis

**intermediate
stability**

CH_3CH_2 H
  C=C
 H CH_3

disubstituted
trans

most stable

b. $CH_2=CHCH(CH_3)_2$ $CH_2=C(CH_3)CH_2CH_3$ $(CH_3)_2C=CHCH_3$

monosubstituted
least stable

disubstituted
**intermediate
stability**

trisubstituted
most stable

8.31 **A more negative value for $\Delta H°$ means the reaction is more exothermic.** Since both 1-butene and *cis*-2-butene form the same product (butane) this data shows that 1-butene was higher in energy to begin with, **since more energy is released in the hydrogenation reaction.**

$CH_2=CHCH_2CH_3$ + H$_2$ $\longrightarrow$ $CH_3CH_2CH_2CH_3$
1-butene
$\Delta H° = -30.3$ kcal/mol

CH_3 CH_3
  C=C   + H$_2$ $\longrightarrow$ $CH_3CH_2CH_2CH_3$
 H   H
cis-2-butene
$\Delta H° = -28.6$ kcal/mol

energy

cis-2-butene

1-butene

larger $\Delta H°$ for
1-butene
higher in energy

butane

smaller $\Delta H°$ for
*cis*2-butene
**lower in energy
more stable**

8.32

a.

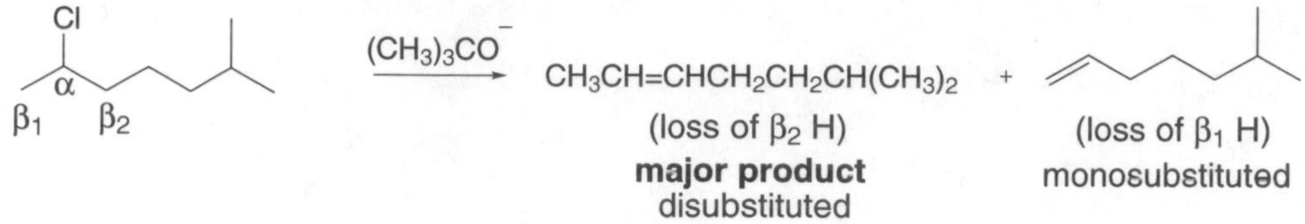
Cl
α
β1 β2
(CH3)3CO⁻
CH3CH=CHCH2CH2CH(CH3)2
(loss of β2 H)
major product
disubstituted
+
(loss of β1 H)
monosubstituted

b.

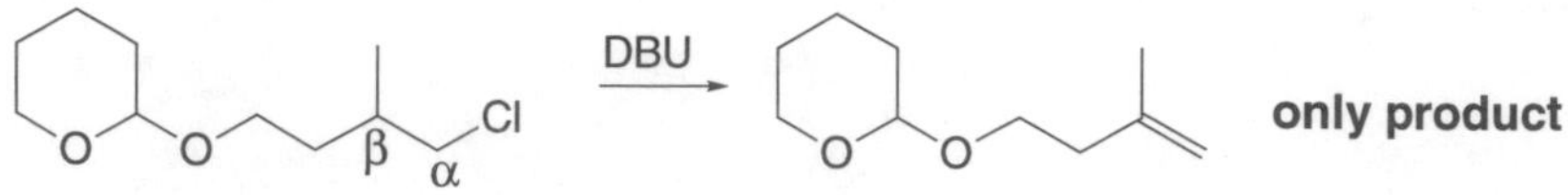
O O
β α
Cl
DBU
only product

c.

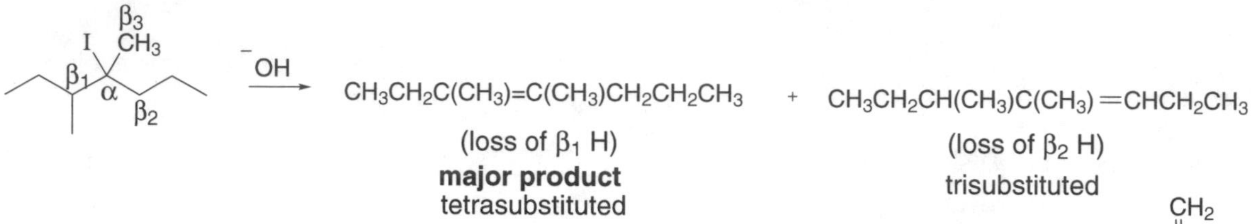
β3
I CH3
β1 α
β2
⁻OH
CH3CH2C(CH3)=C(CH3)CH2CH2CH3
(loss of β1 H)
major product
tetrasubstituted
+
CH3CH2CH(CH3)C(CH3)=CHCH2CH3
(loss of β2 H)
trisubstituted

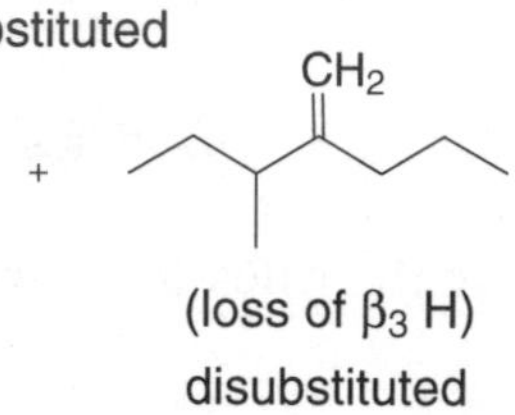
+
CH2
(loss of β3 H)
disubstituted

d.

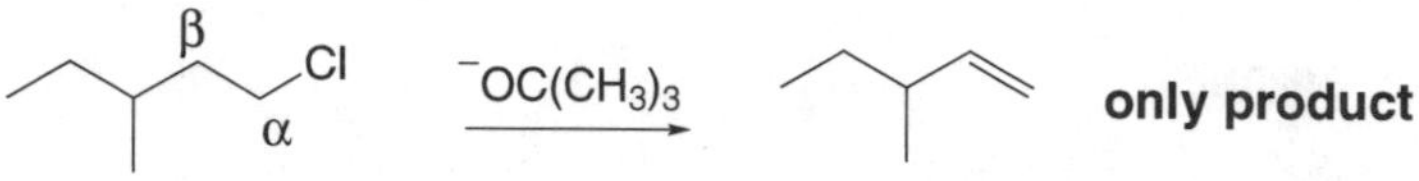
β
Cl
α
⁻OC(CH3)3
only product

e.

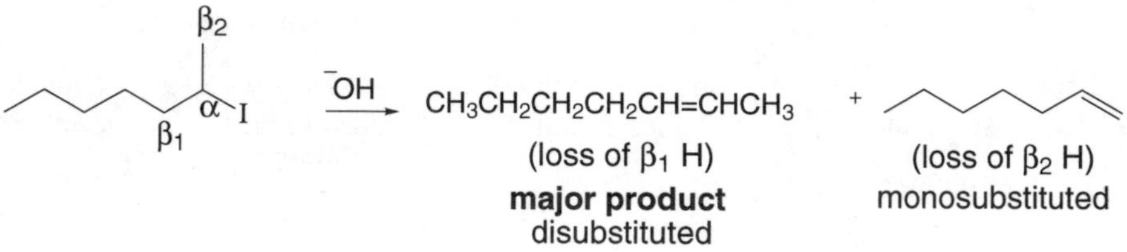
β2
α I
β1
⁻OH
CH3CH2CH2CH2CH=CHCH3
(loss of β1 H)
major product
disubstituted
+
(loss of β2 H)
monosubstituted

f.

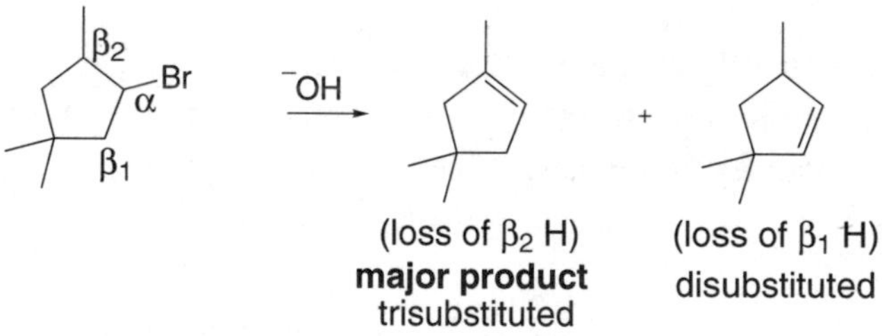
β2
Br
α
β1
⁻OH
+
(loss of β2 H)
major product
trisubstituted
(loss of β1 H)
disubstituted

8.33 To give only one alkene as the product of elimination, the alkyl halide must have either:

- Only one β carbon with a hydrogen atom.
- All identical β carbons so the resulting elimination products are identical.

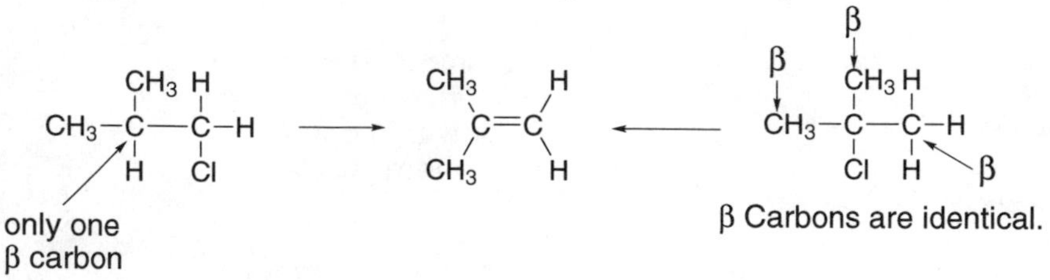
CH3 H
CH3–C—C–H
H Cl
only one
β carbon
CH3 H
C=C
CH3 H
β
β
CH3 H
CH3–C—C–H
Cl H
β
β Carbons are identical.

8.34 Draw the products of the E2 reaction and compare the number of C's bonded to the C=C.

A yields a trisubstituted alkene as the major product and a disubstituted alkene as minor product. **B** yields a disubstituted alkene as the major product and a monosubstituted alkene as minor product. Since the major and minor products formed from **A** have more alkyl groups (making them more stable) than those formed from **B**, **A** reacts faster in an elimination reaction.

8.35

a. Mechanism:

By-products

b. Rate = k[R-Br][⁻OC(CH₃)₃]

Rate = k[R-Br][$^-$OC(CH$_3$)$_3$]

1. Solvent changed to DMF (polar aprotic) = **rate increases**
2. [$^-$OC(CH$_3$)$_3$] decreased = **rate decreases**
3. Base changed to $^-$OH = **rate decreases** (weaker base)
4. Halide changed to 2° = **rate increases** (More substituted RX reacts faster.)
5. Leaving group changed to I$^-$ = **rate increases** (better leaving group)

8.36

CH₃CH₂O⁻	21%	79%
(CH₃)₃CO⁻	73%	27%

With a less sterically hindered base, more of the more stable product is formed.

As the base gets bigger, the more accessible proton is removed more easily.

Removal of the less accessible 2° H gives the more substituted, more stable alkene.

This pathway is usually **favored**, as is the case with CH₃CH₂O⁻ as base.

Removal of the more accessible 1° H gives the less substituted, less stable alkene.

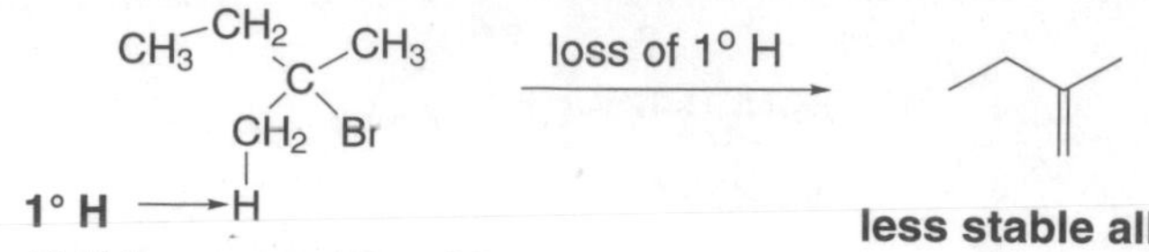
CH3
CH2
CH3
C
CH2
Br
1° H
H
loss of 1° H

less stable alkene

The 1° H is more accessible,
less sterically hindered.
With a bulkier base,
this proton is more readily removed.

As the base gets **bulkier**, the more accessible proton is removed faster; thus, the 1° H reacts faster than the 2° H, and the less stable alkene predominates.

Explanation: 1° H's are more easily removed than 2° H's with sterically hindered bases.

8.37

a. 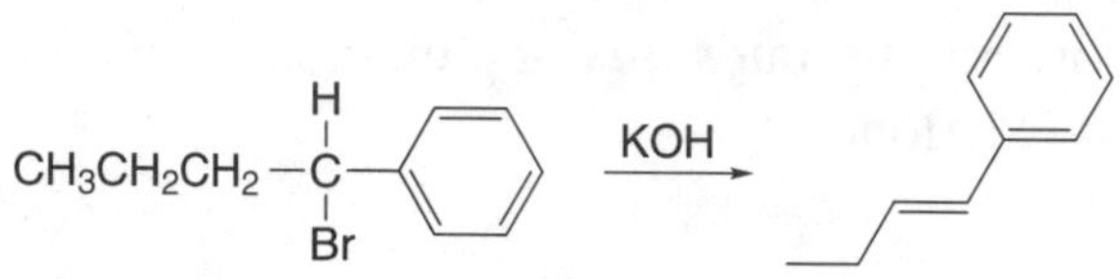
CH3CH2CH2
H
C
Br
KOH

trans isomer more stable
major product

b. 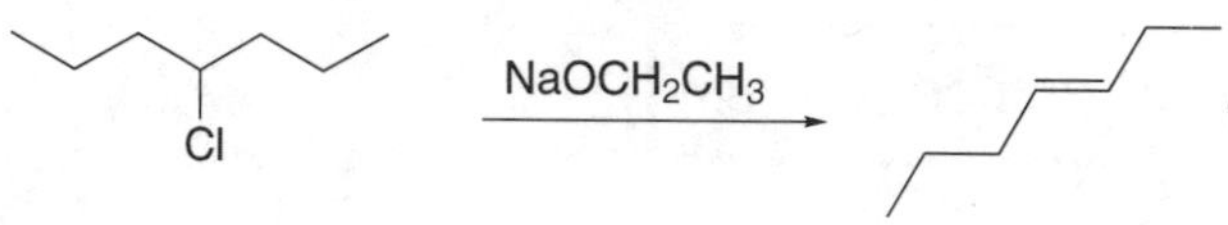
Cl
NaOCH2CH3

trans isomer more stable
major product

8.38

a. 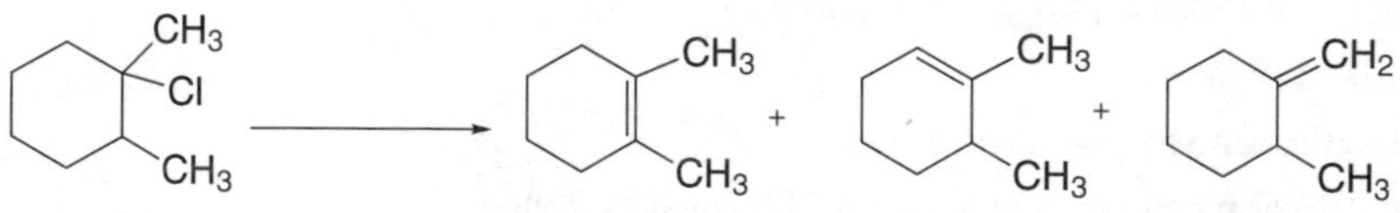
CH3
Cl
CH3
CH3
CH3
+
CH3
CH3
+
CH2
CH3

tetrasubstituted **major product** | trisubstituted | disubstituted

b. 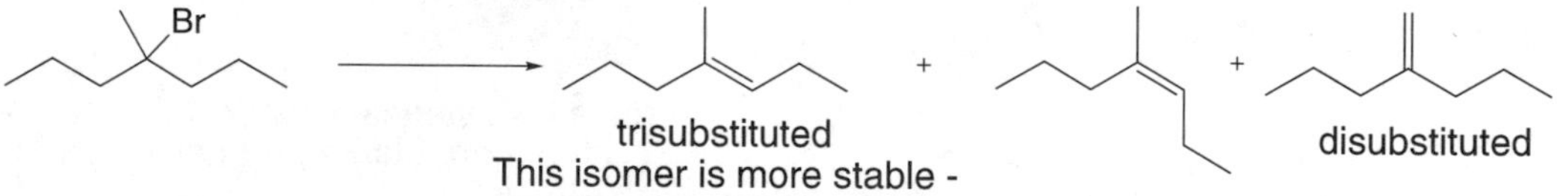
Br
+
+

trisubstituted
This isomer is more stable - large groups further away.
major product

trisubstituted

disubstituted

c. 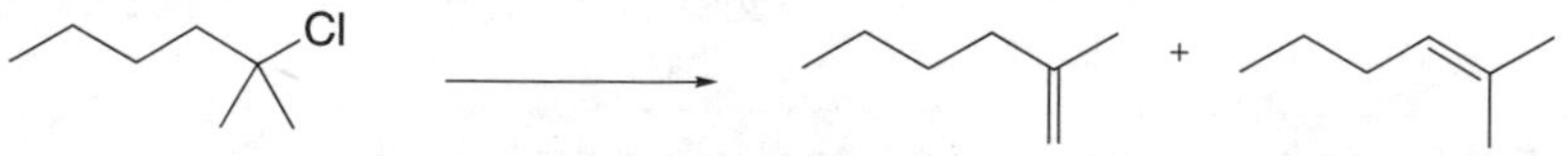
Cl
+

disubstituted

trisubstituted
major product

8.39 Use the rules from Answer 8.20.

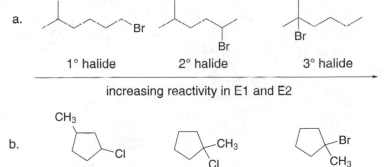

a. [2° halide, Br] $\xrightarrow[\text{strong base}]{^-\text{OCH}_3}$ E2 → $CH_3CH=CHCH_3$ (cis and trans) + $CH_3CH_2CH=CH_2$

b. [2° halide, Br] $\xrightarrow[\text{weak base}]{CH_3OH}$ E1 → $CH_3CH=CHCH_3$ (cis and trans) + $CH_3CH_2CH=CH_2$

c. [1° halide, I] $\xrightarrow[\text{strong base}]{^-\text{OC(CH}_3)_3}$ E2 → [pentene]

d. [3° halide, $CH_2CH_2CH_3$, Cl, CH_3] $\xrightarrow[\text{weak base}]{H_2O}$ E1 → [$=CHCH_2CH_3$, CH_3] + [$-CH_2CH_2CH_3$, CH_3] + [$-CH_2CH_2CH_3$, CH_3]

e. [2° halide, Cl] $\xrightarrow[\text{strong base}]{^-\text{OH}}$ E2 → [cyclohexene product]

f. [2° halide, Cl] $\xrightarrow[\text{strong base}]{^-\text{OH}}$ E2 → [product] + [product]

8.40 The order of reactivity is the same for both E2 and E1: $1° < 2° < 3°$

a. [1° halide, Br]   [2° halide, Br]   [3° halide, Br]

increasing reactivity in E1 and E2 →

b. [CH_3, Cl, 2° halide]   [CH_3, Cl, 3° halide]   [Br, CH_3, 3° halide + better leaving group]

increasing reactivity in E1 and E2 →

8.41

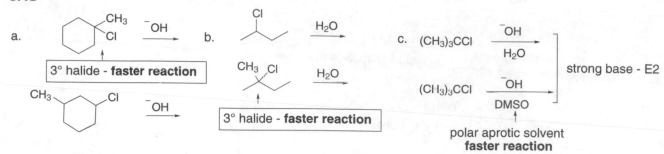

a.

3° halide - **faster reaction**

b.

H_2O

H_2O

3° halide - **faster reaction**

c. $(CH_3)_3CCl$ ^-OH / H_2O

$(CH_3)_3CCl$ ^-OH / DMSO

strong base - E2

polar aprotic solvent
faster reaction

8.42

a.

two chair conformations

$(CH_3)_2CH$ A

$(CH_3)_2CH$ **axial** Cl B CH₃ (H)

Choose this conformer.
axial Cl

$(CH_3)_2CH$ B CH₃ (H)
one axial H

$(CH_3)_2CH$ CH₃ H

= only product

$^{\prime\prime}CH_3$
$CH(CH_3)_2$

b.

two chair conformations

$(CH_3)_2CH$ A

$(CH_3)_2CH$ CH₃ **axial** Cl B (H) (H)

Choose this conformer.
axial Cl

$(CH_3)_2CH$ CH₃ Cl β₂ H β₁ (H) (H)
B two axial H's

$(CH_3)_2CH$ CH₃ H
(loss of β₁ H)
major product
trisubstituted

$(CH_3)_2CH$ CH₃ H H
(loss of β₂ H)

↓ redraw

CH₃
$CH(CH_3)_2$

↓ redraw

$^{\prime\prime}CH_3$
$CH(CH_3)_2$

c.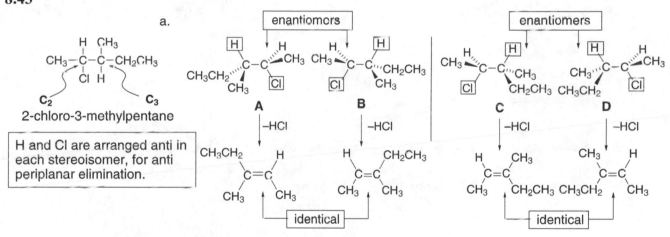

(loss of β₂ H) rendered in image

This conformer reacts.

d.

(loss of β₂ D)

This conformer reacts.

(loss of β₁ D)

8.43

a.

CH₃—C—C—CH₂CH₃

C₂   **C₃**

2-chloro-3-methylpentane

| H and Cl are arranged anti in each stereoisomer, for anti periplanar elimination. |

enantiomers

A   **B**

enantiomers

C   **D**

–HCl

identical

identical

b. Two different alkenes are formed as products.

c. The products are diastereomers: Two enantiomers (**A** and **B**) give identical products.

A and B are diastereomers of **C** and **D**. Each pair of enantiomers gives a single alkene. Thus diastereomers give diastereomeric products.

8.44

a. (cyclohexyl)CH_2CHCl_2 $\xrightarrow[\text{(2 equiv)}]{^-NH_2}$ (cyclohexyl)$C{\equiv}CH$

b. $CH_3CH_2-\overset{\overset{\displaystyle CH_3}{|}}{\underset{\underset{\displaystyle CH_3}{|}}{C}}-\overset{}{\underset{\underset{\displaystyle Br}{|}}{C}}HCH_2Br$ $\xrightarrow[\text{(2 equiv)}]{^-NH_2}$ $CH_3CH_2-\overset{\overset{\displaystyle CH_3}{|}}{\underset{\underset{\displaystyle CH_3}{|}}{C}}-C{\equiv}CH$

c. $CH_3-\overset{\overset{\displaystyle Cl}{|}}{\underset{\underset{\displaystyle Cl}{|}}{C}}-CH_2CH_3$ $\xrightarrow[\text{(excess)}]{^-NH_2}$ $HC{\equiv}C-CH_2CH_3$ + $CH_3-C{\equiv}C-CH_3$

d. (cyclohexyl)$-\overset{\overset{\displaystyle H}{|}}{\underset{\underset{\displaystyle Cl}{|}}{C}}-\overset{\overset{\displaystyle H}{|}}{\underset{\underset{\displaystyle Cl}{|}}{C}}-$(cyclohexyl) $\xrightarrow[\text{(2 equiv)}]{^-NH_2}$ (cyclohexyl)$-C{\equiv}C-$(cyclohexyl)

8.45

a. $CH_3C{\equiv}CCH_3$ | $CH_3-\overset{\overset{\displaystyle Br}{|}}{\underset{\underset{\displaystyle Br}{|}}{C}}-CH_2CH_3$ or $CH_3-\overset{\overset{\displaystyle H}{|}}{\underset{\underset{\displaystyle Br}{|}}{C}}-\overset{\overset{\displaystyle H}{|}}{\underset{\underset{\displaystyle Br}{|}}{C}}-CH_3$

b. $CH_3-\overset{\overset{\displaystyle CH_3}{|}}{\underset{\underset{\displaystyle CH_3}{|}}{C}}-C{\equiv}CH$ | $CH_3-\overset{\overset{\displaystyle CH_3}{|}}{\underset{\underset{\displaystyle CH_3}{|}}{C}}-\overset{\overset{\displaystyle Br}{|}}{\underset{\underset{\displaystyle Br}{|}}{C}}-CH_3$ or $CH_3-\overset{\overset{\displaystyle CH_3}{|}}{\underset{\underset{\displaystyle CH_3}{|}}{C}}-\overset{\overset{\displaystyle Br}{|}}{\underset{}{C}}H-CH_2Br$ or $CH_3-\overset{\overset{\displaystyle CH_3}{|}}{\underset{\underset{\displaystyle CH_3}{|}}{C}}-CH_2CHBr_2$

c. (phenyl)$-C{\equiv}C-$(phenyl) | (phenyl)$-\overset{\overset{\displaystyle Br}{|}}{\underset{\underset{\displaystyle Br}{|}}{C}}-\overset{\overset{\displaystyle H}{|}}{\underset{\underset{\displaystyle H}{|}}{C}}-$(phenyl) or (phenyl)$-\overset{\overset{\displaystyle Br}{|}}{\underset{\underset{\displaystyle H}{|}}{C}}-\overset{\overset{\displaystyle Br}{|}}{\underset{\underset{\displaystyle H}{|}}{C}}-$(phenyl)

8.46

$CH_3-\overset{\overset{\displaystyle H}{|}}{\underset{\underset{\displaystyle Br}{|}}{C}}-\overset{\overset{\displaystyle H}{|}}{\underset{\underset{\displaystyle Br}{|}}{C}}-CH_3$ $\longrightarrow$ $CH_3-C{\equiv}C-CH_3$ $CH_3-CH{=}C{=}CH_2$ $CH_2{=}CH-CH{=}CH_2$
2,3-dibromobutane   ↑   ↑   ↑   **C**
                                                sp sp   **sp**
                                                   A   **B**

8.47

a. (structure)$-Br$ $\xrightarrow[\text{sterically hindered base}]{^-OC(CH_3)_3}$ (structure) **E2**
   1° halide
   S_N2 or E2

b. (structure)$-I$ $\xrightarrow[\text{strong nucleophile}]{^-OCH_2CH_3}$ (structure)$-OCH_2CH_3$ **S_N2**
   1° halide
   S_N2 or E2

c. $CH_3-\overset{\overset{\displaystyle Cl}{|}}{\underset{\underset{\displaystyle Cl}{|}}{C}}-CH_3$ $\xrightarrow[\substack{\text{(2 equiv)}\\\text{strong base}}]{^-NH_2}$ $HC{\equiv}C-CH_3$
   dihalide

d.

1° halide
S$_N$2 or E2

DBU
sterically
hindered
base

E2

e.

2° halide
S$_N$1, S$_N$2, E1, E2

$^-$OC(CH$_3$)$_3$
sterically
hindered
base

CH$_2$CH$_3$ + CH$_2$CH$_3$ **E2**

major product

f.

Br
CH$_2$CH$_3$
3° halide
no S$_N$2

CH$_3$CH$_2$OH
weak base

OCH$_2$CH$_3$
CH$_2$CH$_3$
S$_N$1 product

+ CH$_2$CH$_3$ + CHCH$_3$

E1 products

g.   (CH$_3$)$_2$CH—CHCH$_2$Br
dihalide Br

2NaNH$_2$
strong base

(CH$_3$)$_2$CH—C≡CH

h.

Cl Cl

dihalide

KOC(CH$_3$)$_3$
(2 equiv)
DMSO
strong base

CH$_3$
CH$_3$—C—C≡CH
CH$_3$

i.

I

2° halide
S$_N$1, S$_N$2, E1, E2

CH$_3$CH$_2$OH
weak base

OCH$_2$CH$_3$

S$_N$1 product

+

E1 product

+ CH$_3$CH=CHCH$_3$
(cis and trans)
E1 product

j.

Cl

3° halide
no S$_N$2

H$_2$O
weak base

OH

S$_N$1 product

+ CH$_3$CH$_2$C(CH$_3$)=CHCH$_3$
(cis and trans)
E1 product

+

E1 product

8.48

a.

Cl H

2° halide
S$_N$1, S$_N$2, E1, E2

$^-$OH
strong base
S$_N$2 and E2

H OH

S$_N$2 product
inversion at
stereogenic center

+ CH$_3$CH=CHCH$_2$CH$_2$CH$_3$
(cis and trans)
E2 product

+

E2 product

b.

Cl H

2° halide
S$_N$1, S$_N$2, E1, E2

H$_2$O
weak base
S$_N$1 and E1

HO H

+

H OH

+

S$_N$1 products

+

major E1 product   **minor E1 product**   **minor E1 product**

c.

CH$_3$
Cl
CH$_3$

3° halide
no S$_N$2

CH$_3$OH
weak base
S$_N$1 and E1

CH$_3$
OCH$_3$
CH$_3$

+

CH$_3$
CH$_3$
OCH$_3$

S$_N$1 products

+

CH$_3$
CH$_3$

major E1 product

+

CH$_3$
CH$_3$

+

CH$_3$
CH$_2$

minor E1 products

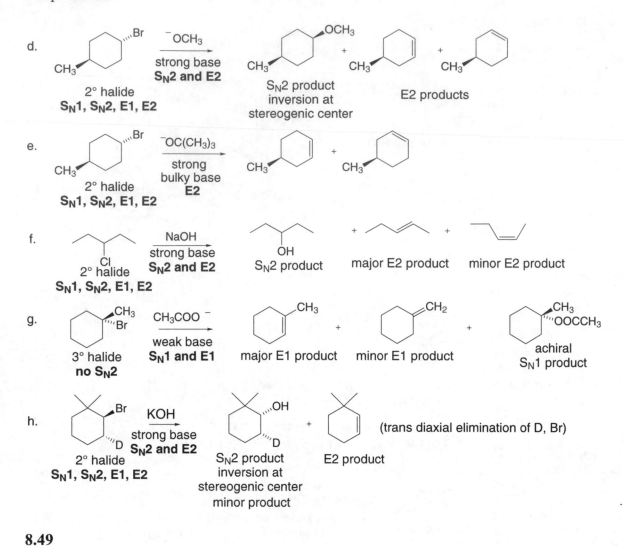

8.49

a.

No substitution occurs with a strong bulky base and a 3° RX. The C with the leaving group is too crowded for an S_N2 substitution to occur. Elimination occurs instead by an E2 mechanism.

b.

All elimination reactions are slow with 1° halides.
The strong nucleophile reacts by an S_N2 mechanism instead.

c.

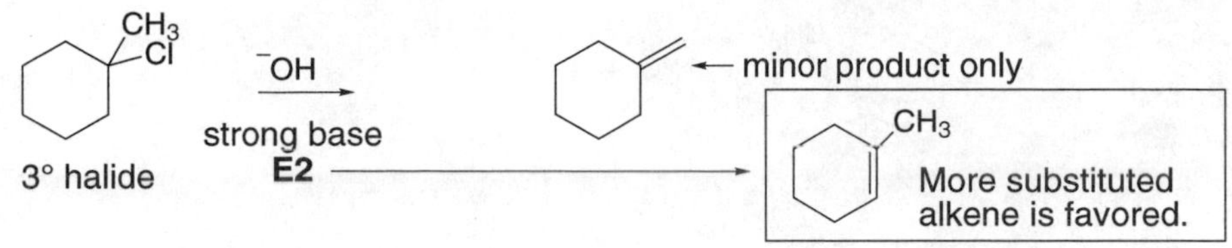

3° halide

⁻OH

strong base
E2

← minor product only

More substituted
alkene is favored.

d.

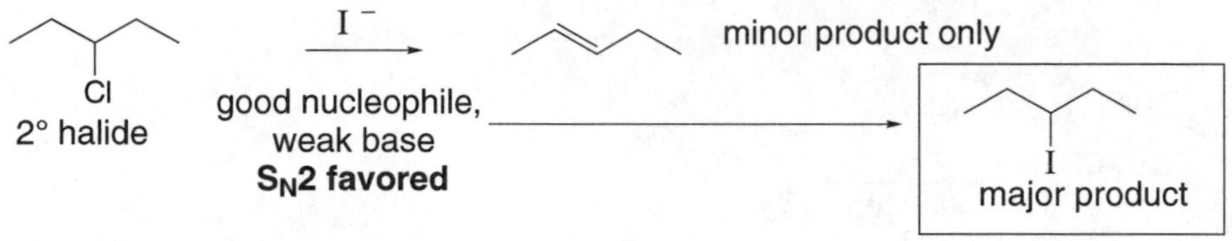

I⁻

2° halide

good nucleophile,
weak base
S$_N$2 favored

minor product only

I
major product

The 2° halide can react by an E2 or S$_N$2 reaction with a negatively charged nucleophile or base. Since I⁻ is a weak base, substitution by an S$_N$2 mechanism is favored.

8.50

2° halide, weak base:
S$_N$1 and E1

a.

CH$_3$CH$_2$OH

**overall
reaction**

OCH$_2$CH$_3$

+ + + HCl

The steps:

S$_N$1

CH$_3$CH$_2$ÖH + Cl⁻

or

+ :OCH$_2$CH$_3$

CH$_3$CH$_2$ÖH

+ CH$_3$CH$_2$ÖH$_2$

E1 H

CH$_3$CH$_2$ÖH

or

+ CH$_3$CH$_2$ÖH$_2$

E1

H

CH$_3$CH$_2$ÖH

+ CH$_3$CH$_2$ÖH$_2$

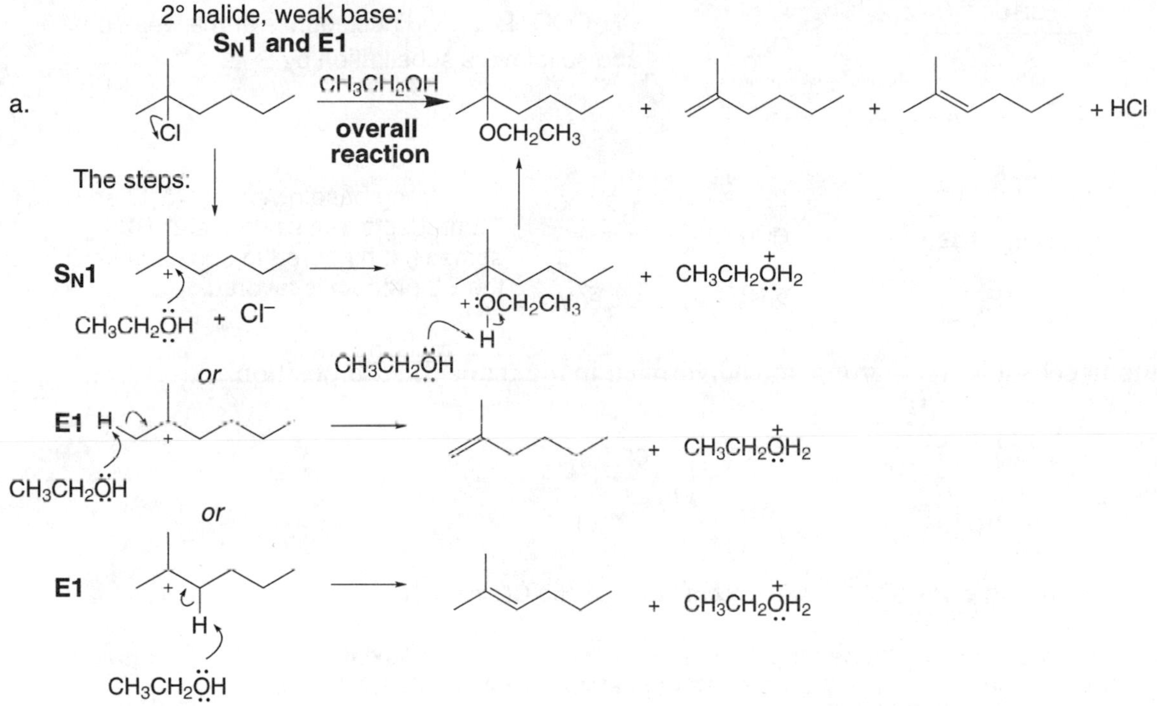

Any base (such as CH$_3$CH$_2$OH or Cl⁻) can be used to remove a proton to form an alkene. If Cl⁻ is used, HCl is formed as a reaction by-product. If CH$_3$CH$_2$OH is used, (CH$_3$CH$_2$OH$_2$)⁺ is formed instead.

b.

3° halide
strong base
E2

Each product:

or

one step

+ H₂O + Cl⁻

one step

8.51

good nucleophile

CH₃COO⁻ is a good nucleophile and a weak base and so it favors substitution by S_N2.

(only)

CH₃CH₂O⁻

strong base

CH_3-CHCH_3 + $CH_3CH=CH_2$
      |
   OCH₂CH₃

20%   80%

The strong base gives both S_N2 and E2 products, but since the 2° RX is somewhat hindered to substitution, the E2 product is favored.

8.52 E2 elimination needs a leaving group and a hydrogen in the **trans diaxial** position.

Two different
conformers:

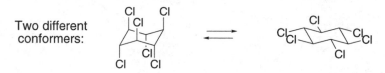

This conformer has Cl's
axial, but no H's axial.

This conformer has no Cl's axial.

For elimination to occur a cyclohexane must have a H and Cl in the trans diaxial arrangement. Neither conformer of this isomer has both atoms—H and Cl—axial; thus, this isomer only slowly loses HCl by elimination.

8.53

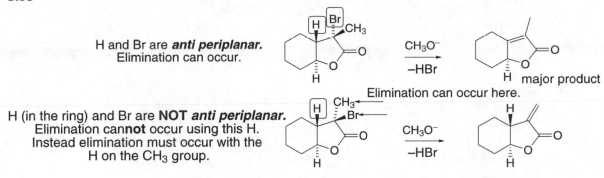

H and Br are *anti periplanar.*
Elimination can occur.

Elimination can occur here.

H (in the ring) and Br are **NOT** *anti periplanar.*
Elimination can**not** occur using this H.
Instead elimination must occur with the
H on the CH₃ group.

Elimination cannot occur in the ring
because the required anti periplanar geometry is not present.

8.54

leaving group

DBN
overall reaction

E2

A sequence of two reactions forms the
final product: E2 elimination opens the
five-membered ring. Then the sulfur
nucleophile displaces the Cl⁻ leaving group
to form the six-membered ring.

S_N2

Chapter 9: Alcohols, Ethers, and Epoxides

◆ General facts about ROH, ROR, and epoxides

- All three compounds contain an O atom that is sp^3 hybridized and tetrahedral (9.2).

$$CH_3 \overset{\cdot\cdot\overset{\cdot\cdot}{O}\cdot\cdot}{\diagdown} H \qquad CH_3 \overset{\cdot\cdot\overset{\cdot\cdot}{O}\cdot\cdot}{\diagdown} CH_3 \qquad \overset{60°}{\overset{\cdot\cdot\overset{\cdot\cdot}{O}\cdot\cdot}{\triangle}}$$

  109°   111°   H H   H H

  An alcohol   **An ether**   **An epoxide**

- All three compounds have polar C–O bonds, but only alcohols have an O–H bond for intermolecular hydrogen bonding (9.4).

  hydrogen bond

- Alcohols and ethers do not contain a good leaving group. Nucleophilic substitution can occur only after the OH (or OR) group is converted to a better leaving group (9.7A).

$$R-\overset{\cdot\cdot}{\underset{\cdot\cdot}{O}}H \;+\; H-Cl \;\rightleftharpoons\; R-\overset{+}{\underset{\cdot\cdot}{O}}H_2 \;+\; Cl^-$$

  strong acid

  weak base
  good leaving group

- Epoxides have a leaving group located in a strained three-membered ring, making them reactive to strong nucleophiles and acids HZ that contain a nucleophilic atom Z (9.15).

  leaving group

  With strong nucleophiles,
  :Nu⁻

  [1]   Nu   [2]   Nu   + ⁻OH

◆ A new reaction of carbocations (9.9)

- Less stable carbocations rearrange to more stable carbocations by shift of a hydrogen atom or an alkyl group. Besides rearrangement, carbocations also react with nucleophiles (7.13) and bases (8.6).

$$-\overset{|}{\underset{|}{C}}-\overset{|}{\underset{R}{\overset{+}{C}}}- \quad \xrightarrow{\text{1,2-shift}} \quad -\overset{|}{\underset{+}{C}}-\overset{|}{\underset{R}{C}}-$$

  (or H)   (or H)

◆ Preparation of alcohols, ethers, and epoxides (9.6)

[1] Preparation of alcohols

$$R-X \;+\; \boxed{^-OH} \longrightarrow R-\boxed{OH} \;+\; X^-$$

- The mechanism is S_N2.
- The reaction works best for CH_3X and 1° RX.

[2] Preparation of alkoxides (a Brønsted–Lowry acid–base reaction)

$$R-O-H \ + \ Na^+H^- \ \longrightarrow \ \boxed{R-O^-} Na^+ \ + \ H_2$$

alkoxide

[3] Preparation of ethers (Williamson ether synthesis)

$$R-X \ + \ \boxed{^-OR'} \longrightarrow \ R-\boxed{OR'} \ + \ X^-$$

- The mechanism is S_N2.
- The reaction works best for CH_3X and $1°$ RX.

[4] Preparation of epoxides (Intramolecular S_N2 reaction)

halohydrin

- A two step-reaction sequence:
  [1] Removal of a proton with base forms an alkoxide;
  [2] Intramolecular **S_N2** reaction forms the epoxide.

◆ Reactions of alcohols

[1] Dehydration to form alkenes

[a] Using strong acid (9.8, 9.9)

- Order of reactivity: $R_3COH > R_2CHOH > RCH_2OH$.
- The mechanism for $2°$ and $3°$ ROH is E1; carbocations are intermediates and rearrangements occur.
- The mechanism for $1°$ ROH is E2.
- The Zaitsev rule is followed.

[b] Using POCl$_3$ and pyridine (9.10)

- The mechanism is E2.
- No carbocation rearrangements occur.

[2] Reaction with HX to form RX (9.11)

$$R-OH \ + \ H-X \ \longrightarrow \ \boxed{R-X} \ + \ H_2O$$

- Order of reactivity: $R_3COH > R_2CHOH > RCH_2OH$.
- The mechanism for $2°$ and $3°$ ROH is S_N1; carbocations are intermediates and rearrangements occur.
- The mechanism for CH_3OH and $1°$ ROH is S_N2.

[3] Reaction with other reagents to form RX (9.12)

R—OH + SOCl$_2$ $\xrightarrow{\text{pyridine}}$ R—Cl

R—OH + PBr$_3$ $\longrightarrow$ R—Br

- Reactions occur with CH$_3$OH and 1° and 2° ROH.
- The reactions follow an S$_N$2 mechanism.

[4] Reaction with tosyl chloride to form tosylates (9.13A)

R—OH + Cl—S(=O)(=O)—⟨benzene⟩—CH$_3$ $\xrightarrow{\text{pyridine}}$ R—O—S(=O)(=O)—⟨benzene⟩—CH$_3$

R—OTs

- The C—O bond is not broken so the configuration at a stereogenic center is retained.

♦ **Reactions of tosylates**

Tosylates undergo either substitution or elimination depending on the reagent (9.13B).

:Nu$^-$ → product with H Nu + $^-$OTs

B: → C=C + TsOH

- Substitution is carried out with strong :Nu$^-$ so the mechanism is S$_N$2.

- Elimination is carried out with strong bases so the mechanism is E2.

♦ **Reactions of ethers**

Only one reaction is useful: Cleavage with strong acids (9.14)

R—O—R' + H—X $\longrightarrow$ R—X + R'—X + H$_2$O

[X = Br or I]

- With 2° and 3° R groups, the mechanism is S$_N$1.
- With 1° R groups the mechanism is S$_N$2.

♦ **Reactions of epoxides**

Epoxides are ring-opened with nucleophiles :Nu$^-$ and acids HZ (9.15).

C—C (epoxide) $\xrightarrow[\text{or HZ}]{\text{[1] :Nu$^-$ [2] H$_2$O}}$ Nu—C—C—OH (Z)

- The reaction occurs with backside attack, resulting in trans or anti products.
- With :Nu$^-$, the mechanism is S$_N$2, and nucleophilic attack occurs at the less substituted C.
- With HZ, the mechanism is between S$_N$1 and S$_N$2, and attack of Z$^-$ occurs at the more substituted C.

Chapter 9: Answers to Problems

9.1 • **Alcohols** are classified as 1°, 2° or 3°, depending on the number of carbon atoms bonded to the carbon with the OH group.
   • **Symmetrical ethers** have two identical R groups, and **unsymmetrical ethers** have R groups that are different.

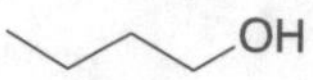

1° alcohol	**2° alcohol**	**symmetrical ether**	**unsymmetrical ether**

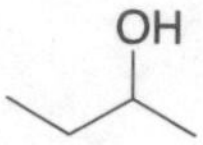

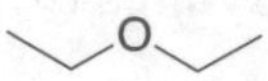

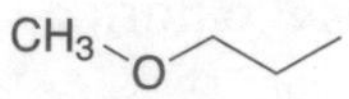

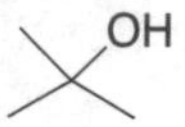

3° alcohol	**1° alcohol**	**unsymmetrical ether**

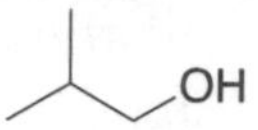

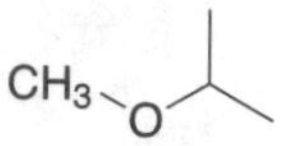

9.2 To name an alcohol:

[1] **Find the longest chain that has the OH group as a substituent.** Name the molecule as a derivative of that number of carbons by changing the –e ending of the alkane to the suffix **–ol**.

[2] **Number the carbon chain to give the OH group the lower number.** When the OH group is bonded to a ring, the ring is numbered beginning with the OH group, and the "1" is usually omitted.

[3] Apply the other rules of nomenclature to complete the name.

a. [1] 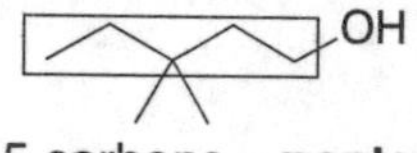 5 carbons = **pentanol**   [2] 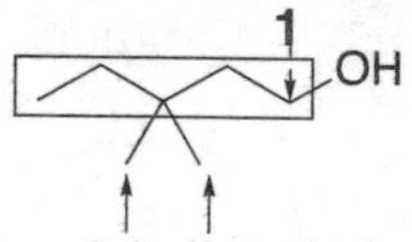 **3,3-dimethyl**   [3] **3,3-dimethyl-1-pentanol**

b. (CH₃CH₂)₂CHCH(OH)CH₂CH₃
   ↓ redraw
   [1] CH₃CH₂–C–C–CH₂CH₃ ... 6 carbons = **hexanol**   [2] 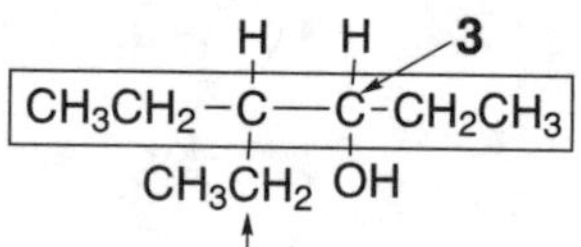 **4-ethyl**   [3] **4-ethyl-3-hexanol**

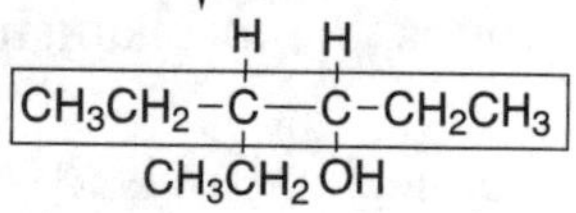

c. [1] 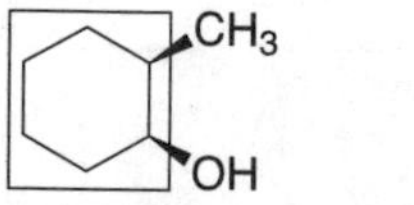 6 carbon ring = **cyclohexanol**   [2] 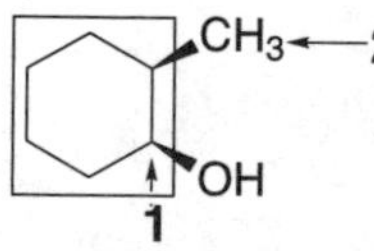 **2-methyl**, **1**   [3] *cis*-**2-methylcyclohexanol**

d. [1] 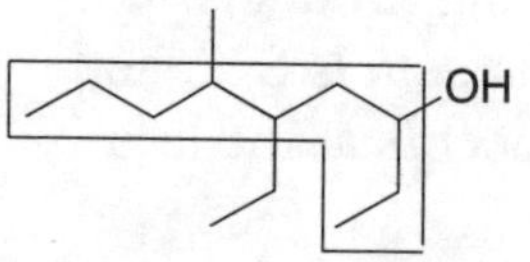 9 carbons = **nonanol**   [2] 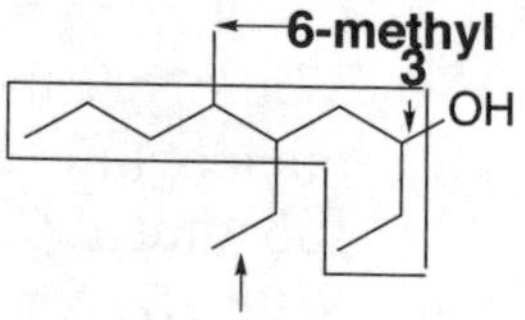 **6-methyl**, **5-ethyl**   [3] **5-ethyl-6-methyl-3-nonanol**

9.3 To work backwards from a structure to a name:
[1] Find the parent name and draw its structure.
[2] Add the substituents to the long chain.

a. 7,7-dimethyl-4-**octanol**

c. 2-*tert*-butyl-3-methyl**cyclohexanol**

b. 5-methyl-4-propyl-3-**heptanol**

d. *trans*-1,2-**cyclohexanediol**

9.4 To name simple ethers:
[1] Name both alkyl groups bonded to the oxygen.
[2] Arrange these names alphabetically and add the word ***ether***. For symmetrical ethers, name the alkyl group and add the prefix ***di***.

To name ethers using the IUPAC system:
[1] Find the two alkyl groups bonded to the ether oxygen. The smaller chain becomes the substituent, named as an alkoxy group.
[2] Number the chain to give the lower number to the first substituent.

a. **common name:**

$CH_3-O-CH_2CH_2CH_2CH_3$

methyl   butyl

butyl methyl ether

IUPAC name:

$CH_3-O|CH_2CH_2CH_2CH_3$ ← larger group - 4 C's
                                                    butane

substituent:
methoxy

1-methoxybutane

b. **common name:**

OCH_3

methyl

cyclohexyl

cyclohexyl methyl ether

IUPAC name:

OCH_3 ← substituent - methoxy

larger group - 6 C's
cyclohexane

methoxycyclohexane

c. **common name:**

$CH_3CH_2CH_2-O-CH_2CH_2CH_3$

propyl   propyl

dipropyl ether

IUPAC name:

$CH_3CH_2CH_2-O|CH_2CH_2CH_3$

propoxy   propane

1-propoxypropane

9.5 Three ways to name epoxides:
  [1] Epoxides are named as derivations of oxirane, the simplest epoxide.
  [2] Epoxides can be named by considering the oxygen a substituent called an **epoxy** group, bonded to a hydrocarbon chain. Use two numbers to designate which two atoms the oxygen is bonded to.
  [3] Epoxides can be named as **alkene oxides** by mentally replacing the epoxide oxygen by a double bond. Name the alkene (Chapter 10) and add the word *oxide*.

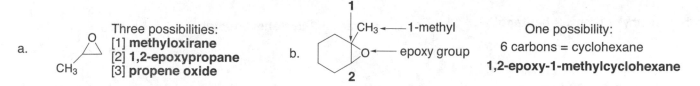

a.

Three possibilities:
[1] **methyloxirane**
[2] **1,2-epoxypropane**
[3] **propene oxide**

b.

CH₃ ← 1-methyl
O ← epoxy group

One possibility:
6 carbons = cyclohexane
1,2-epoxy-1-methylcyclohexane

9.6 Two rules for boiling point:
  [1] **The stronger the forces the higher the bp.**
  [2] **Bp increases as the extent of the hydrogen bonding increases.** For alcohols with the same number of carbon atoms: hydrogen bonding and bp's increase: $3° ROH < 2° ROH < 1° ROH$.

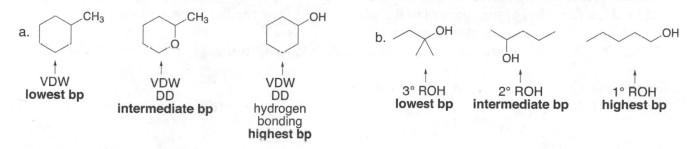

a.

VDW
lowest bp

VDW
DD
intermediate bp

VDW
DD
hydrogen
bonding
highest bp

b.

3° ROH
lowest bp

2° ROH
intermediate bp

1° ROH
highest bp

9.7 Draw dimethyl ether and ethanol and analyze their intermolecular forces to explain the observed trend.

dimethyl ether

CH₃ ‒O‒ CH₃

VDW
DD
no HB
much lower bp

ethanol

CH₃CH₂OH

VDW
DD
HB
Two molecules of CH₃CH₂OH
can hydrogen bond to each other.
stronger forces =
much higher bp

Both molecules contain an O atom
and can hydrogen bond with water. They
have fewer than 5 C's and are
therefore **water soluble.**

H
O
H
CH₃ O CH₃

H
O
H
CH₃CH₂ O H

9.8 Strong nucleophiles (like ⁻CN) favor S_N2 reactions. The use of crown ethers in nonpolar solvents increases the nucleophilicity of the anion, and this increases the rate of the S_N2 reaction. The nucleophile does not appear in the rate equation for the S_N1 reaction. Nonpolar solvents cannot solvate carbocations so this disfavors S_N1 reactions as well.

9.9 To carry out the reaction below, crown ethers can be used.

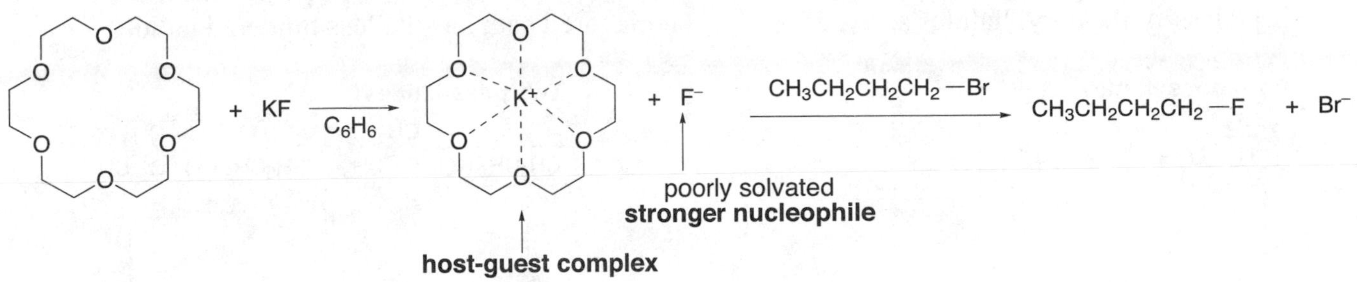

+ KF $\xrightarrow{C_6H_6}$ host-guest complex + F⁻ poorly solvated **stronger nucleophile** $\xrightarrow{CH_3CH_2CH_2CH_2-Br}$ $CH_3CH_2CH_2CH_2-F$ + Br⁻

9.10 Although epothilone B contains 27 C's, each of the labeled functional groups can hydrogen bond to H_2O, thus making it somewhat water soluble.

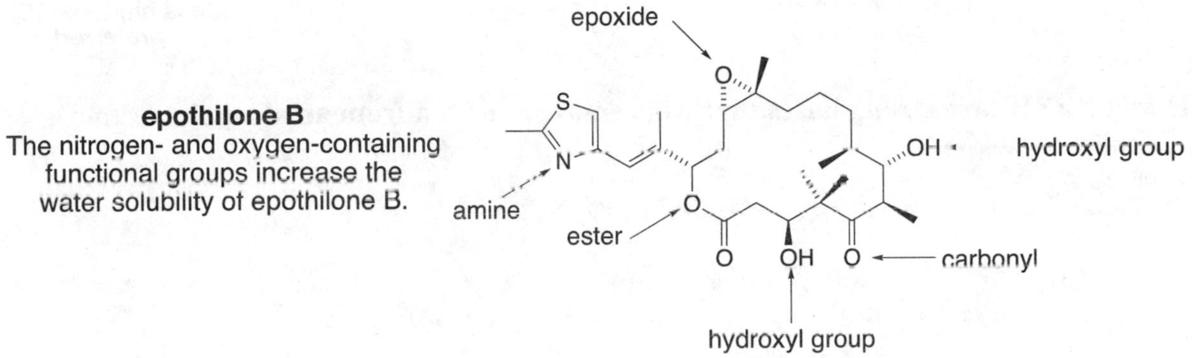

epothilone B
The nitrogen- and oxygen-containing functional groups increase the water solubility of epothilone B.

epoxide

amine

ester

hydroxyl group

OH · hydroxyl group

carbonyl

9.11 Draw the products of substitution in the following reactions by substituting ⁻OH or ⁻OR for X in the starting material.

a.  $CH_3CH_2CH_2CH_2-Br$ + ⁻OH ⟶ $CH_3CH_2CH_2CH_2-OH$ + Br⁻ **alcohol**

b. 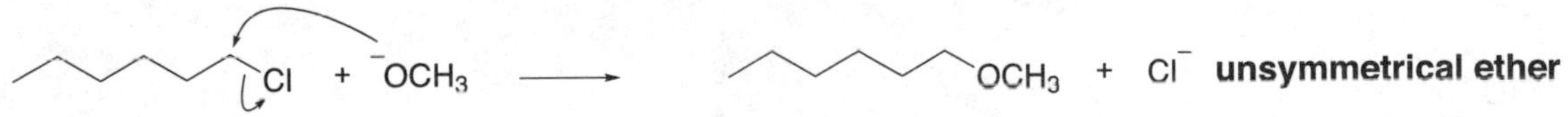 ⟋⟍⟋⟍CI + ⁻OCH₃ ⟶ ⟋⟍⟋⟍OCH₃ + Cl⁻ **unsymmetrical ether**

c. 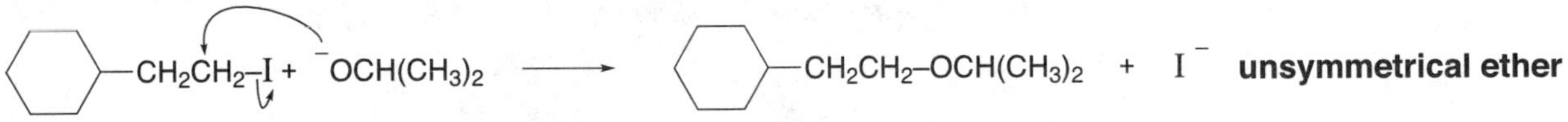 ◯—CH₂CH₂—I + ⁻OCH(CH₃)₂ ⟶ ◯—CH₂CH₂—OCH(CH₃)₂ + I⁻ **unsymmetrical ether**

d. 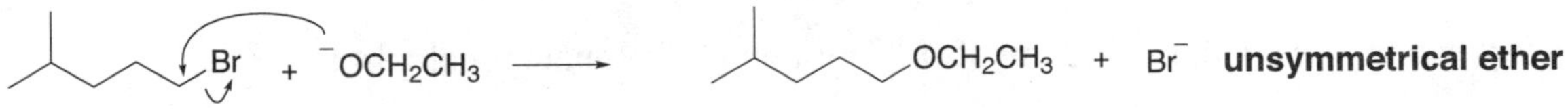 ⟍⟋⟍Br + ⁻OCH₂CH₃ ⟶ ⟍⟋⟍OCH₂CH₃ + Br⁻ **unsymmetrical ether**

9.12 To synthesize an ether using a Williamson ether synthesis:
   [1] First find the two possible alkoxides and alkyl halides needed for nucleophilic substitution.
   [2] Classify the alkyl halides as 1°, 2°, or 3°. The favored path has the less hindered halide.

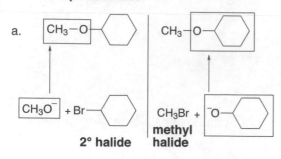

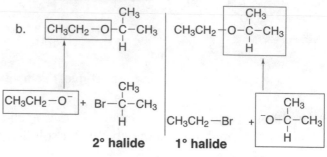

9.13 NaH and NaNH₂ are strong bases that will remove a proton from an alcohol, creating a nucleophile.

a. $CH_3CH_2CH_2-O-H$ + Na^+H^- ⟶ $CH_3CH_2CH_2-O^-$ Na^+ + H_2

b. [structure: cyclopentyl–C(CH₃)(H)–O–H] + $Na^+ \, ^-NH_2$ ⟶ [structure: cyclopentyl–C(CH₃)(H)–O⁻] + Na^+ + NH_3

c. [structure] Na^+H^- ⟶ [structure] $CH_3CH_2CH_2-Br$ + Na^+ + H_2

   ⟶ [structure] $O-CH_2CH_2CH_3$ + Br^-

d. [structure with OH and Br] Na^+H^- ⟶ [structure with O⁻ and Br] + Na^+ + H_2 ⟶ [epoxide structure] O + Br^-

   $C_6H_{10}O$

9.14 Only acids having pK_a's below –2 are strong enough to protonate an alcohol.

a. HF (pK_a = 3) = above –2; **not strong enough**

b. HClO₄ (pK_a = –10) = below –2; **strong enough to protonate an alcohol.**

c. $C_6H_5SO_3H$ (pK_a = –7) = below –2; **strong enough to protonate an alcohol.**

d. CH_3COOH (pK_a = 5) = above –2; **not strong enough**

9.15 **Dehydration follows the Zaitsev rule**, so the more stable, more substituted alkene is the major product.

a. $CH_3-\underset{\underset{OH}{|}}{\overset{\overset{H}{|}}{C}}-CH_3 \xrightarrow{H_2SO_4} CH_2=CH-CH_3 + H_2O$

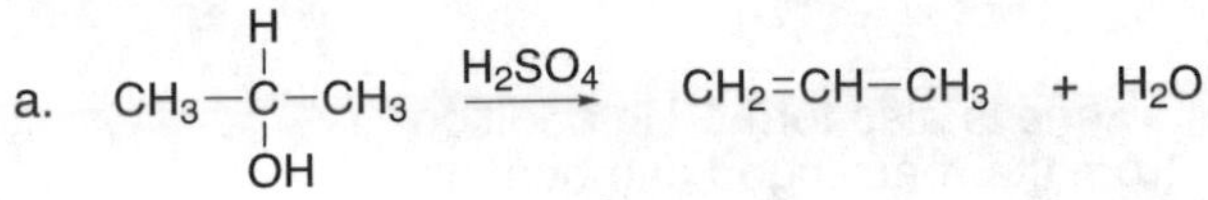

b.

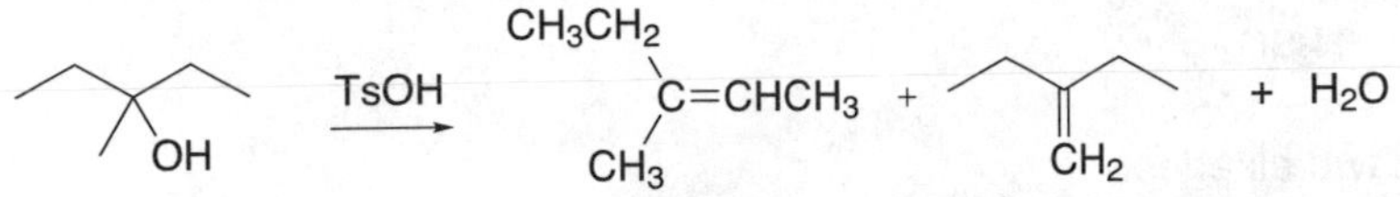

$\xrightarrow{TsOH}$

$\underset{\underset{CH_3}{|}}{\overset{\overset{CH_3CH_2}{|}}{C}}=CHCH_3$ +

$+ H_2O$

trisubstituted   disubstituted
major product   **minor product**

c.

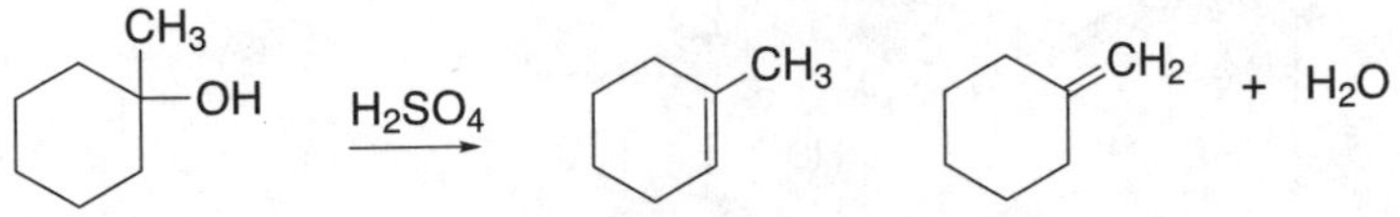

$\xrightarrow{H_2SO_4}$

$+ H_2O$

trisubstituted   disubstituted
major product   **minor product**

9.16 The rate of dehydration increases as the number of R groups increases.

a. $(CH_3)_2CHCH_2CH_2CH_2OH$ $(CH_3)_2CHCH_2CH(OH)CH_3$ $(CH_3)_2C(OH)CH_2CH_2CH_3$
         1° alcohol   2° alcohol   3° alcohol
     slowest reaction   **intermediate**   **fastest reaction**
                              reactivity

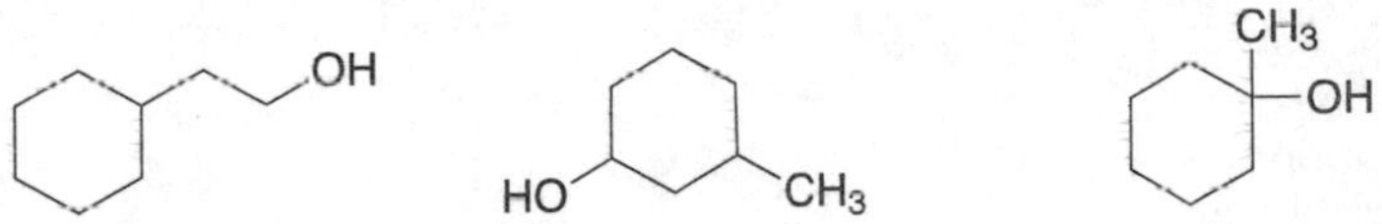

b.

     1° alcohol   2° alcohol   3° alcohol
slowest reaction   **intermediate**   **fastest reaction**
                         reactivity

9.17 There are three steps in the E1 mechanism for dehydration of alcohols, and three transition states.

transition state [1]:   **transition state [2]:** 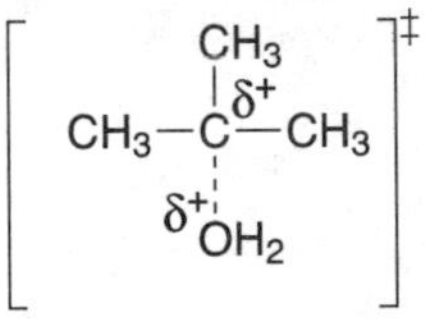   **transition state [3]:**

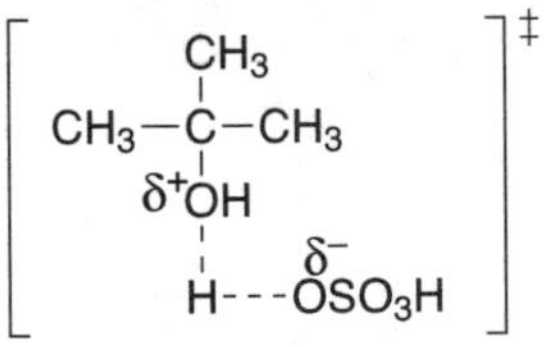

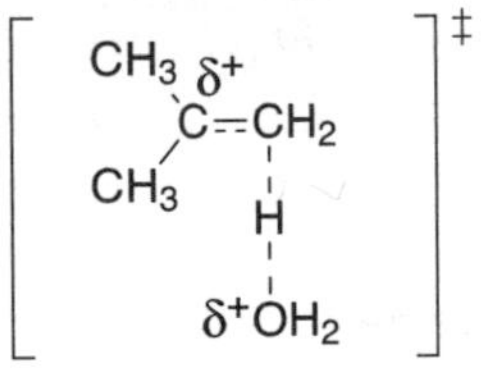

9.18

     transition state [1]:   **transition state [2]:**

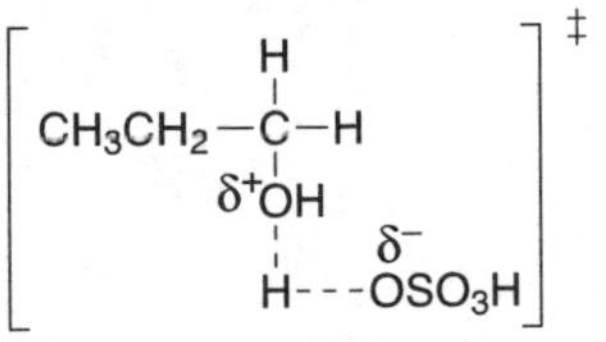

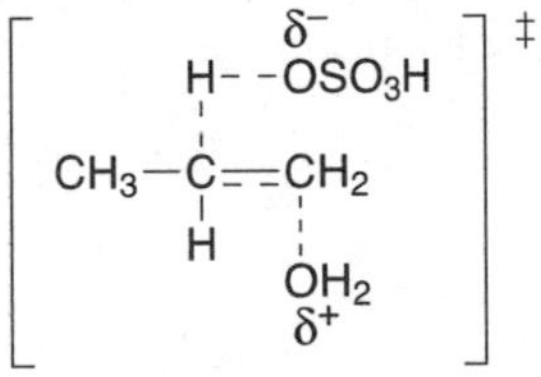

9.19

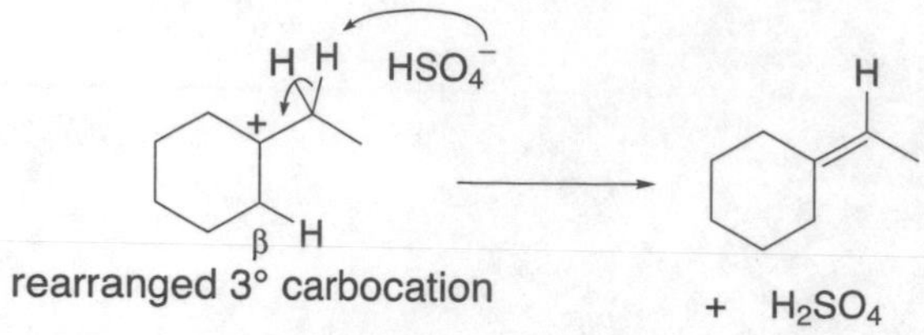

rearranged 3° carbocation

+ H₂SO₄

This alkene is also formed in addition to **Y** from the rearranged carbocation.

The initially formed 2° carbocation gives two alkenes:

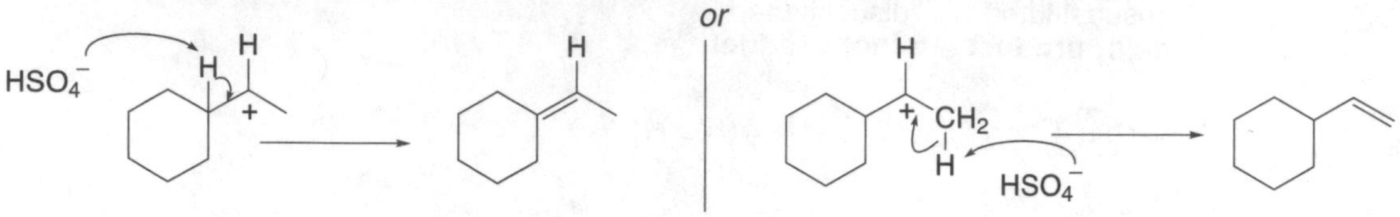

9.20

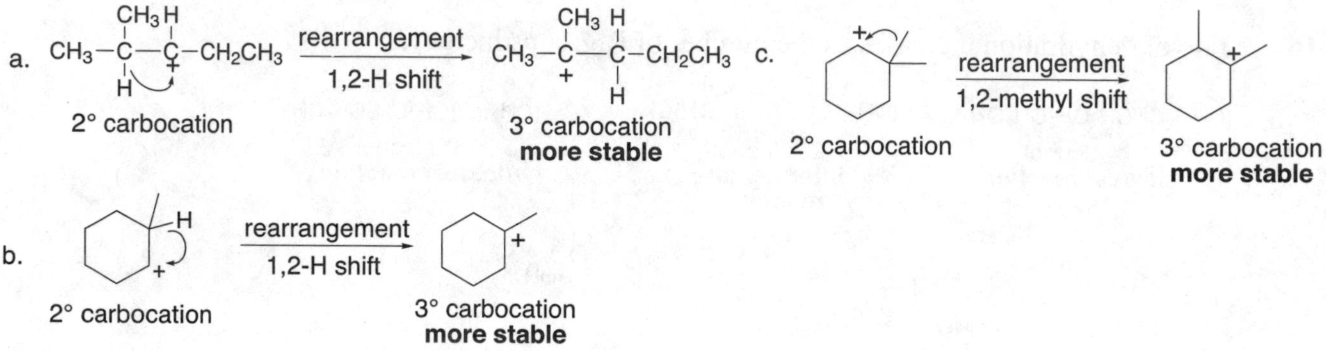

a. $CH_3-\overset{CH_3}{\underset{H}{C}}-\overset{H}{\underset{+}{C}}-CH_2CH_3$ →(rearrangement, 1,2-H shift)→ $CH_3-\overset{CH_3}{\underset{+}{C}}-\overset{H}{\underset{H}{C}}-CH_2CH_3$

2° carbocation   3° carbocation **more stable**

c. →(rearrangement 1,2-methyl shift)→

2° carbocation   3° carbocation **more stable**

b. →(rearrangement 1,2-H shift)→

2° carbocation   3° carbocation **more stable**

9.21

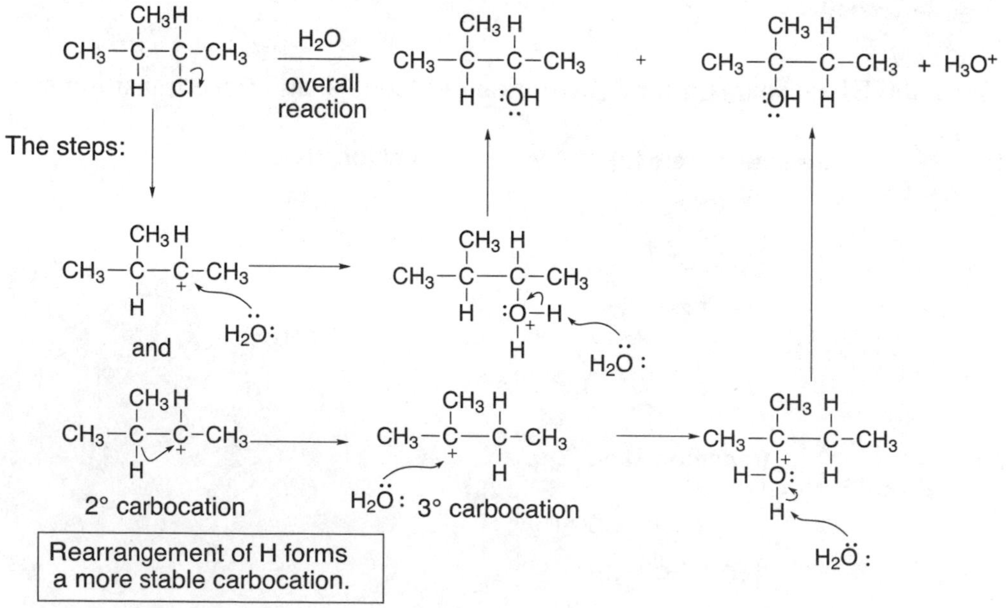

The steps:

and

2° carbocation   H₂Ö: 3° carbocation

Rearrangement of H forms a more stable carbocation.

9.22

a.

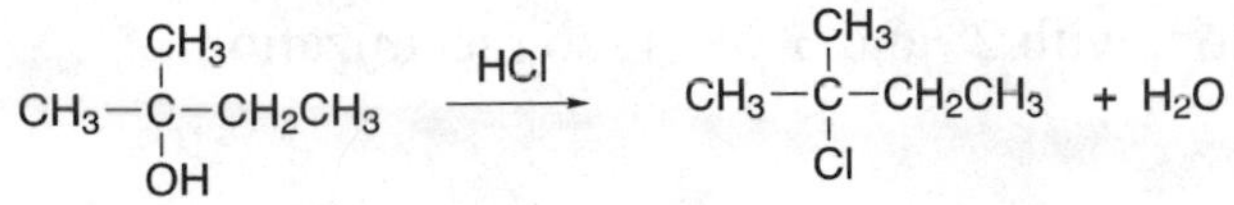

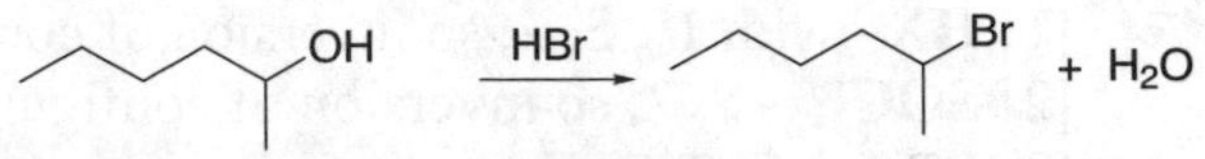

b.

c.

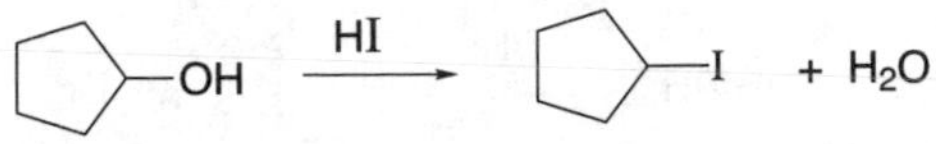

9.23 • **CH_3OH and 1° alcohols** follow an S_N2 mechanism and result in inversion of configuration.
   • **2° and 3° alcohols** follow an S_N1 mechanism and result in racemization at a stereogenic center.

a. 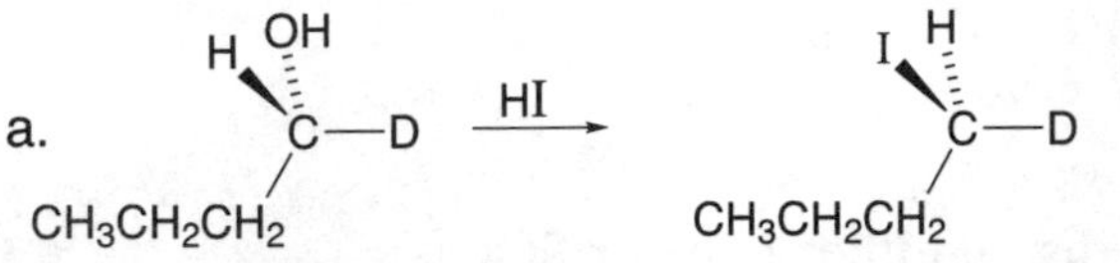   1° alcohol so **inversion of configuration**

b. 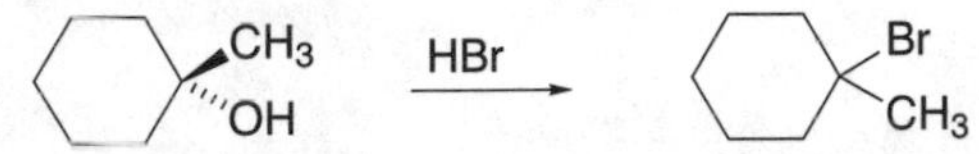   3° alcohol, so Br⁻ attacks from above and below.
   The product is achiral.

achiral starting material   achiral product

c. 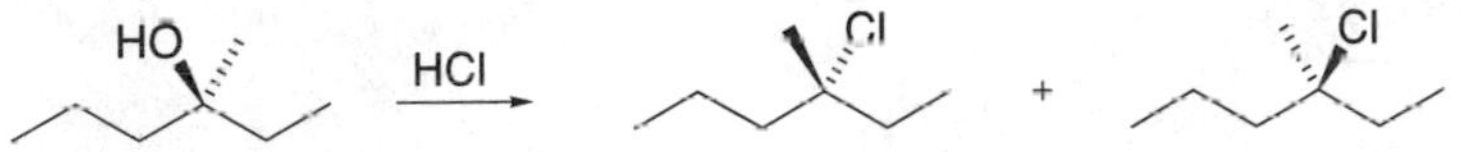   3° alcohol – **racemization**

9.24

a.

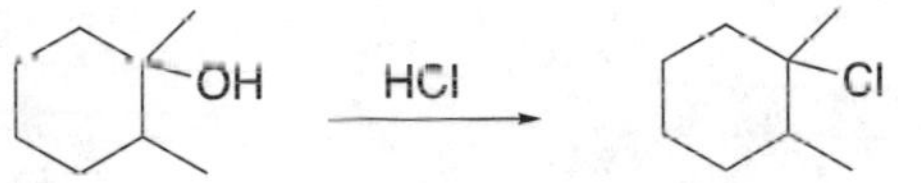

c. 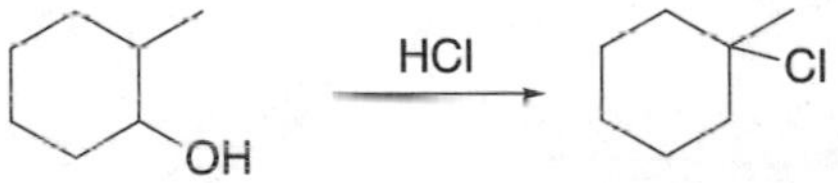   (product formed after a 1,2–H shift)

b. 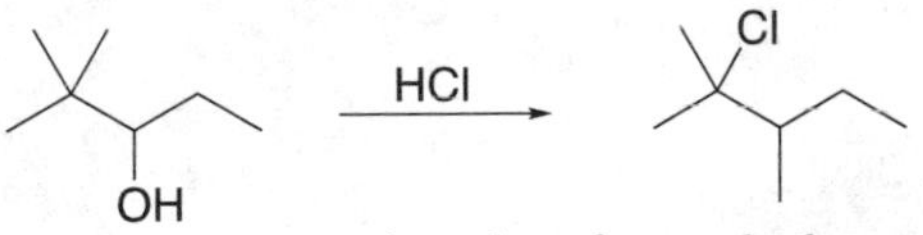   (product formed after a 1,2–CH_3 shift)

9.25 Substitution reactions of alcohols using $SOCl_2$ proceed by an S_N2 mechanism. Therefore, there is **inversion of configuration** at a stereogenic center.

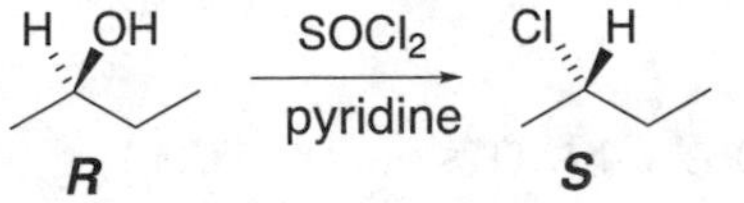

Reactions using $SOCl_2$
proceed by an S_N2 mechanism =
inversion of configuration.

9.26 Substitution reactions of alcohols using PBr_3 proceed by an S_N2 mechanism. Therefore, there is inversion of configuration at a stereogenic center.

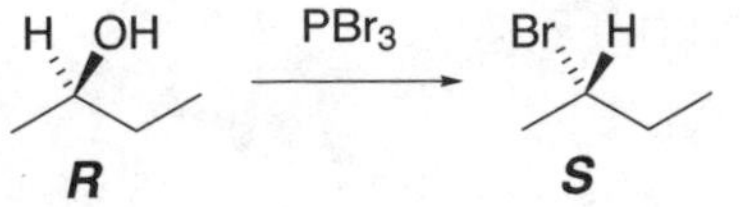

Reactions using PBr_3
proceed by an S_N2 mechanism =
inversion of configuration.

9.27 Stereochemistry for conversion of ROH to RX by reagent:

[1] **HX**– with 1°, S_N2, so inversion of configuration; with 2° and 3°, S_N1, so racemization.

[2] **SOCl$_2$** – S_N2, so inversion of configuration.

[3] **PBr$_3$** – S_N2, so inversion of configuration.

a.

$\sim\!\!\sim\!\!\sim$OH $\xrightarrow[\text{pyridine}]{\text{SOCl}_2}$ $\sim\!\!\sim\!\!\sim$Cl

c.

(cyclopentane ring with CH$_3$ and OH) $\xrightarrow{\text{PBr}_3}$ (cyclopentane ring with CH$_3$ and ''''Br)

$S_N2 =$
inversion

b.

(structure with OH) $\xrightarrow{\text{HI}}$ (structure with ''''I) + (structure with '''' I)

3° alcohol, $S_N1 =$
racemization

9.28 To do a two-step synthesis with this starting material:

[1] Convert the OH group into a good leaving group (by using either PBr$_3$ or SOCl$_2$).

[2] Add the nucleophile for the S_N2 reaction.

$$CH_3-\overset{\overset{H}{|}}{\underset{\underset{CH_3}{|}}{C}}-OH \xrightarrow{PBr_3} CH_3-\overset{\overset{H}{|}}{\underset{\underset{CH_3}{|}}{C}}-Br$$

bad leaving group   **good leaving group**

$\xrightarrow{\overline{N}_3}$ $CH_3-\overset{\overset{H}{|}}{\underset{\underset{CH_3}{|}}{C}}-N_3$

$\xrightarrow{CH_3CH_2O^-}$ $CH_3-\overset{\overset{H}{|}}{\underset{\underset{CH_3}{|}}{C}}-OCH_2CH_3$

9.29

a. $CH_3CH_2CH_2CH_2-OH$ + $CH_3-\!\!\left\langle\!\!\bigcirc\!\!\right\rangle\!\!-SO_2Cl$ $\xrightarrow{\text{pyridine}}$ $CH_3CH_2CH_2CH_2-O-\overset{\overset{O}{\|}}{\underset{\underset{O}{\|}}{S}}-\!\!\left\langle\!\!\bigcirc\!\!\right\rangle\!\!-CH_3$ + Cl^-

b. $CH_3CH_2CH_2-\overset{\overset{H}{\,}\,\,\overset{OH}{\,}}{\underset{\underset{CH_3}{\,}}{C}}$ $\xrightarrow[\text{pyridine}]{\text{TsCl}}$ $CH_3CH_2CH_2-\overset{\overset{H}{\,}\,\,\overset{OTs}{\,}}{\underset{\underset{CH_3}{\,}}{C}}$ + Cl^-

9.30

a. $\sim\!\!\sim$OTs + $^-$CN $\xrightarrow{S_N2}$ $\sim\!\!\sim$CN + $^-$OTs

1° tosylate   strong nucleophile

b. $CH_3CH_2CH_2-OTs$ + $K^+\ ^-OC(CH_3)_3$ $\xrightarrow{E2}$ $CH_3CH=CH_2$ + $K^+\ ^-OTs$ + $HOC(CH_3)_3$

1° tosylate   strong bulky base

c. $CH_3-\overset{\overset{H}{\,}\,\,\overset{OTs}{\,}}{\underset{\underset{CH_2CH_2CH_3}{\,}}{C}}$ + ^-SH $\xrightarrow{S_N2}$ $CH_3-\overset{\overset{HS}{\,}\,\,\overset{H}{\,}}{\underset{\underset{CH_2CH_2CH_3}{\,}}{C}}$

S_N2 product
(inversion of configuration)

2° tosylate   strong nucleophile

(Substitution is favored over elimination.)

9.31

$$HO \quad H \xrightarrow[\text{pyridine}]{\text{TsCl}} TsO \quad H \xrightarrow[\text{S}_N 2]{\text{NaOH}} H \quad OH$$

S

retention
S

inversion
R

One inversion from starting material to product.

enantiomers

9.32 These reagents can be classified as:

[1] $SOCl_2$, PBr_3, HCl, and HBr replace OH with X by a substitution reaction.

[2] **Tosyl chloride** (TsCl) makes OH a better leaving group by converting it to OTs.

[3] **Strong acids** (H_2SO_4) result in elimination by dehydration.

a.
$$CH_3-\overset{\underset{|}{CH_3}}{\overset{|}{\underset{}{C}}}-OH \xrightarrow[\text{pyridine}]{SOCl_2} CH_3-\overset{\underset{|}{CH_3}}{\overset{|}{\underset{}{C}}}-Cl$$

b.
$$CH_3-\overset{\underset{|}{CH_3}}{\overset{|}{\underset{}{C}}}-OH \xrightarrow[\text{pyridine}]{TsCl} CH_3-\overset{\underset{|}{CH_3}}{\overset{|}{\underset{}{C}}}-OTs$$

c.
$$CH_3-\overset{\underset{|}{CH_3}}{\overset{|}{\underset{}{C}}}-OH \xrightarrow{H_2SO_4} CH_2=CHCH_3$$

d.
$$CH_3-\overset{\underset{|}{CH_3}}{\overset{|}{\underset{}{C}}}-OH \xrightarrow{HBr} CH_3-\overset{\underset{|}{CH_3}}{\overset{|}{\underset{}{C}}}-Br$$

e.
$$CH_3-\overset{\underset{|}{CH_3}}{\overset{|}{\underset{}{C}}}-OH \xrightarrow[\text{[2] NaCN}]{\text{[1] PBr}_3} CH_3-\overset{\underset{|}{CH_3}}{\overset{|}{\underset{}{C}}}-CN$$

f.
$$CH_3-\overset{\underset{|}{CH_3}}{\overset{|}{\underset{}{C}}}-OH \xrightarrow[\text{pyridine}]{POCl_3} CH_2=CHCH_3$$

9.33

a. $CH_3CH_2-O-CH_2CH_3 \xrightarrow{HBr} 2\ CH_3CH_2-Br + H_2O$

c. ⬡$-O-CH_3 \xrightarrow{HBr}$ ⬡$-Br + CH_3Br + H_2O$

b.
$$CH_3-\overset{\underset{|}{H}}{\overset{|}{\underset{}{C}}}-O-CH_2CH_3 \xrightarrow{HBr} CH_3-\overset{\underset{|}{H}}{\overset{|}{\underset{}{C}}}-Br + CH_3CH_2Br + H_2O$$
(with CH_3 on top)

9.34 Compare epoxides and cyclopropane. For a compound to be reactive towards nucleophiles, it must be electrophilic.

epoxide

O is electronegative and pulls electron density away from C's. This makes them electrophilic and reactive with nucleophiles.

cyclopropane

Cyclopropane has all C's and H's, so all nonpolar bonds. There are no electrophilic C's so it will not react with nucleophiles.

9.35 Two rules for reaction of an epoxide:

[1] Nucleophiles attack from the **back side** of the epoxide.

[2] Negatively charged nucleophiles attack at the **less substituted carbon**.

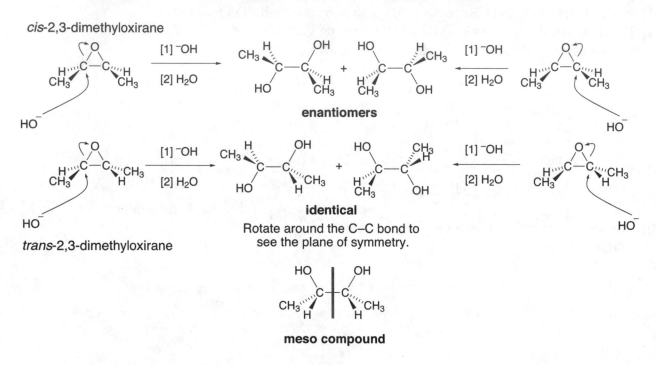

a.

Attack here:
less substituted C
backside attack

b.

Attack here:
less substituted C
backside attack

9.36 In both isomers, ⁻OH attacks from the back side at either C–O bond.

cis-2,3-dimethyloxirane

enantiomers

trans-2,3-dimethyloxirane

identical

Rotate around the C–C bond to
see the plane of symmetry.

meso compound

9.37 Remember the difference between negatively charged nucleophiles and neutral nucleophiles:

- **Negatively charged nucleophiles attack first**, followed by protonation and the nucleophile attacks at the **less substituted carbon**.
- **Neutral nucleophiles have protonation first**, followed by nucleophilic attack at the **more substituted carbon**.

BUT – trans or anti products are always formed regardless of the nucleophile.

a. HBr

neutral nucleophile:
attack at **more**
substituted C

b. [1] ⁻CN [2] H₂O

negatively charged
nucleophile:
attack at **less**
substituted C

c. CH_3CH_2OH / H_2SO_4

neutral nucleophile:
attack at **more**
substituted C

d. [1] CH_3O^- [2] CH_3OH

negatively charged
nucleophile:
attack at **less**
substituted C

9.38 Use the directions from Answer 9.2.

a. (CH₃)₂CHCH₂CH₂CH₂OH

[1]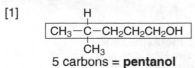
5 carbons = **pentanol**

[2]
CH₃—C—CH₂CH₂CH₂OH with H above, **1** label, CH₃ ← **4-methyl**

[3] **4-methyl-1-pentanol**

b. (CH₃)₂CHCH₂CH(CH₂CH₃)CH(OH)CH₂CH₃

[1]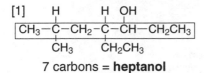
CH₃—C—CH₂—C—CH—CH₂CH₃ with H, H, OH; CH₃, CH₂CH₃
7 carbons = **heptanol**

[2]
CH₃—C—CH₂—C—CH—CH₂CH₃ with H, H, OH, **3**; CH₃ (**6-methyl**), CH₂CH₃ (**4-ethyl**)

[3] **4-ethyl-6-methyl-3-heptanol**

c.

[1]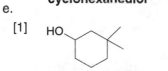
8 carbons = **octanol**

[2]
5-methyl, **4-ethyl**, OH, **3**

[3] **4-ethyl-5-methyl-3-octanol**

d.

[1] HO—⬡—OH
cyclohexanediol

[2] **4** HO—⬡—OH **1**
cis

[3] *cis*-1,4-cyclohexanediol

e.

[1]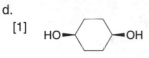
6 carbons = **cyclohexanol**

[2] HO on cyclohexane ring with **3,3-dimethyl**

[3] **3,3-dimethylcyclohexanol**

f.

[1]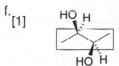
HO H / HO H
4 carbons = **butanediol**

[2] HO H / **2 3** / HO H
2R, 3R

[3] **(2R,3R)-2,3-butanediol**

g.

[1]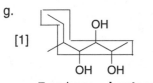
OH / OH OH
7 carbons = **heptanetriol**

[2] OH / **4 3 2** / OH OH / **5-methyl**

[3] **5-methyl-2,3,4-heptanetriol**

h.

[1] HO‖‖‖—⬠—CH(CH₃)₂
5 carbons = **cyclopentanol**

[2] HO‖‖‖—⬠(**1**)—CH(CH₃)₂ / **3-isopropyl**

[3] **3-isopropylcyclopentanol**

9.39 Use the rules from Answers 9.4 and 9.5.

a.

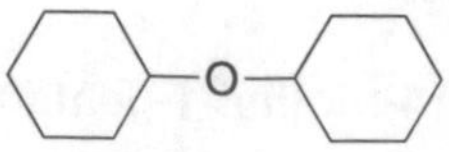

common name: **dicyclohexyl ether**

b. 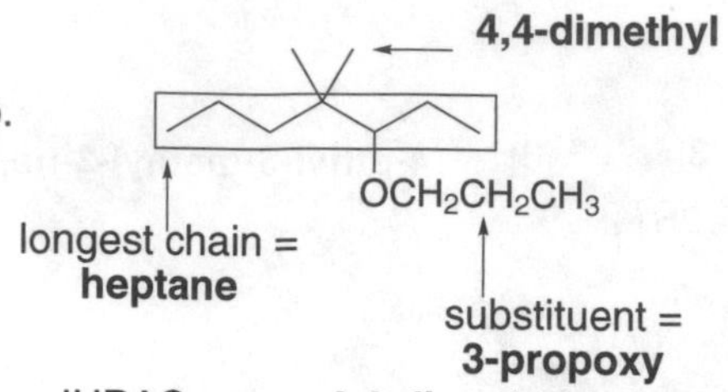
4,4-dimethyl

OCH₂CH₂CH₃

longest chain =
heptane

substituent =
3-propoxy

IUPAC name: **4,4-dimethyl-3-propoxyheptane**

c.

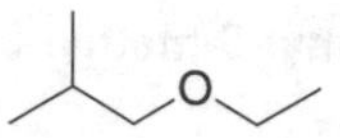

common name: **ethyl isobutyl ether**

IUPAC name: **1-ethoxy-2-methylpropane**

d.

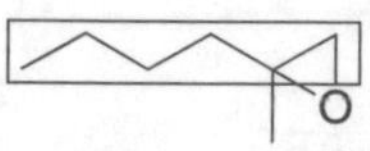

1,2-epoxy-2-methylhexane
or **1-butyl-1-methyloxirane**
or **2-methylhexene oxide**

e. 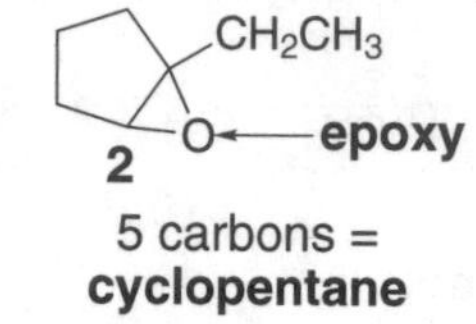
CH₂CH₃

epoxy

2

5 carbons =
cyclopentane

IUPAC name: **1,2-epoxy-1-ethylcyclopentane**

f. 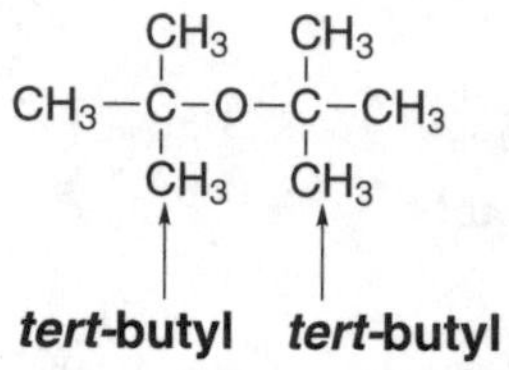
CH₃ CH₃

CH₃—C—O—C—CH₃

CH₃ CH₃

tert-butyl   *tert*-butyl

common name: **di-*tert*-butyl ether**

9.40 Use the directions from Answer 9.3.

a. 4-ethyl-3-**heptanol**

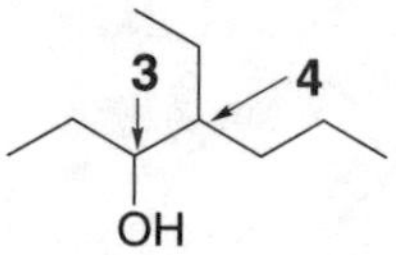

b. *trans*-2-methyl**cyclohexanol**

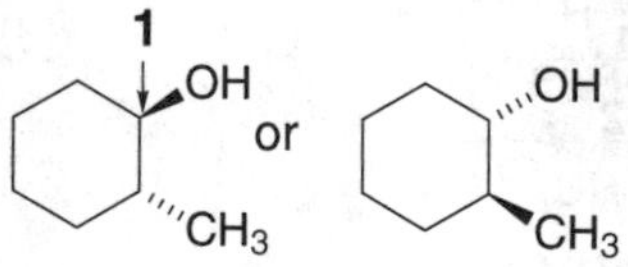

or

c. 2,3,3-trimethyl-2-**butanol**

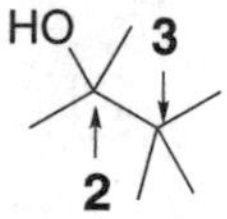

d. 6-*sec*-butyl-7,7-diethyl-4-**decanol**

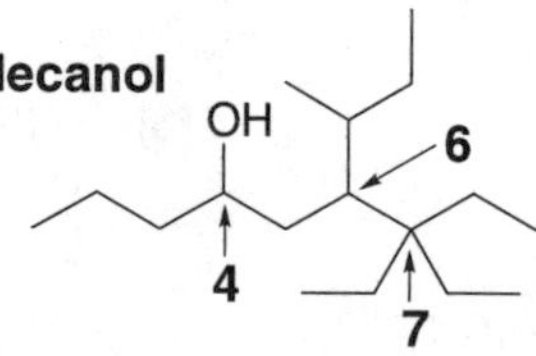

e. 3-chloro-1,2-**propanediol**

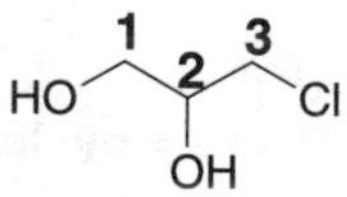

f. diisobutyl ether

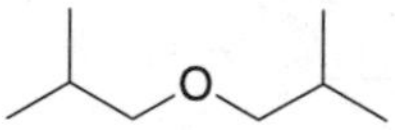

g. 1,2-epoxy-1,3,3-trimethyl**cyclohexane**

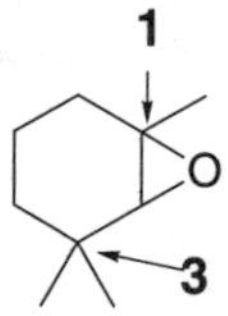

h. 1-ethoxy-3-ethyl**heptane**

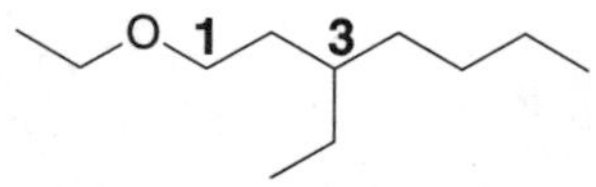

9.41

Eight constitutional isomers of molecular formula $C_5H_{12}O$ containing an OH group:

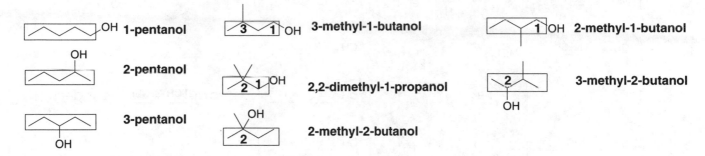

9.42 Use the boiling point rules from Answer 9.6.

a.

$CH_3CH_2OCH_3$	$(CH_3)_2CHOH$	$CH_3CH_2CH_2OH$
↑	↑	↑
ether	2° alcohol	1° alcohol
no hydrogen bonding	hydrogen bonding	hydrogen bonding
lowest bp	**intermediate bp**	**highest bp**

b.

$CH_3CH_2CH_2CH_2CH_2CH_3$	$CH_3CH_2CH_2CH_2CH_2CH_2OH$	$HOCH_2CH_2CH_2CH_2CH_2CH_2OH$
no OH group	one OH group	two OH groups
lowest water solubility	**intermediate water solubility**	**highest water solubility**

9.43

a. $CH_3CH_2CH_2OH$ $\xrightarrow{H_2SO_4}$ $CH_3CH=CH_2$ + H_2O

b. $CH_3CH_2CH_2OH$ $\xrightarrow{NaH}$ $CH_3CH_2CH_2O^- \ Na^+$ + H_2

c. $CH_3CH_2CH_2OH$ $\xrightarrow[ZnCl_2]{HCl}$ $CH_3CH_2CH_2Cl$ + H_2O

d. $CH_3CH_2CH_2OH$ $\xrightarrow{HBr}$ $CH_3CH_2CH_2Br$ + H_2O

e. $CH_3CH_2CH_2OH$ $\xrightarrow[pyridine]{SOCl_2}$ $CH_3CH_2CH_2Cl$

f. $CH_3CH_2CH_2OH$ $\xrightarrow{PBr_3}$ $CH_3CH_2CH_2Br$

g. $CH_3CH_2CH_2OH$ $\xrightarrow[pyridine]{TsCl}$ $CH_3CH_2CH_2OTs$

h. $CH_3CH_2CH_2OH$ $\xrightarrow{[1]\ NaH}$ $CH_3CH_2CH_2O^- \ Na^+$ $\xrightarrow{[2]\ CH_3CH_2Br}$ $CH_3CH_2CH_2OCH_2CH_3$

i. $CH_3CH_2CH_2OH$ $\xrightarrow{[1]\ TsCl}$ $CH_3CH_2CH_2OTs$ $\xrightarrow{[2]\ NaSH}$ $CH_3CH_2CH_2SH$

9.44

a.

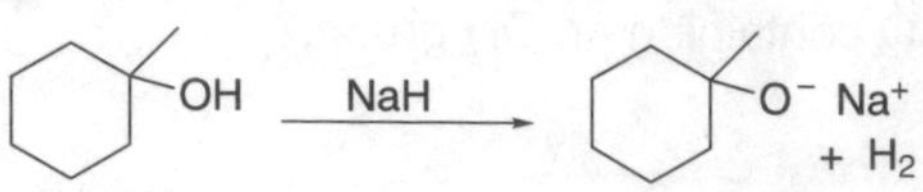

b.

c.

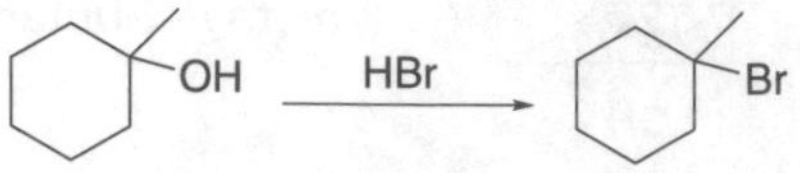

d.

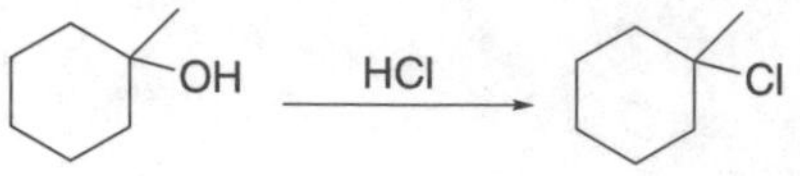

e.

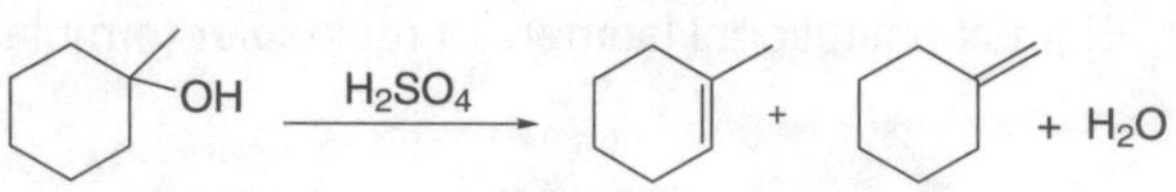

f.

g.

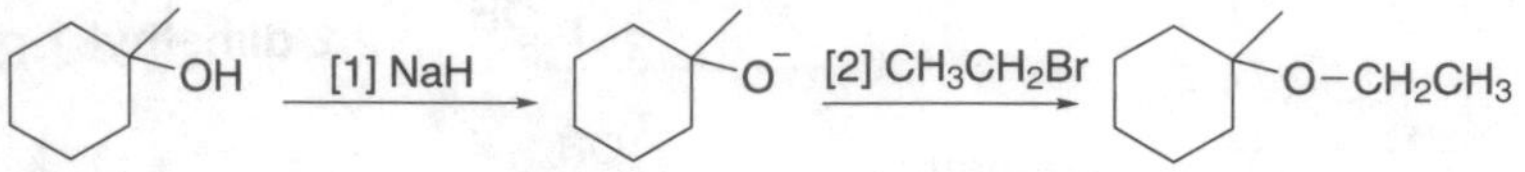

h.

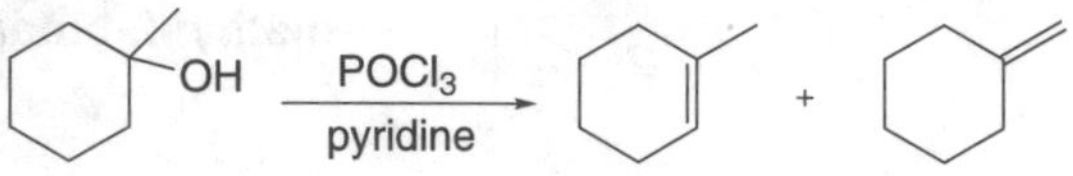

9.45 Dehydration follows the Zaitsev rule, so the more stable, more substituted alkene is the major product.

a.

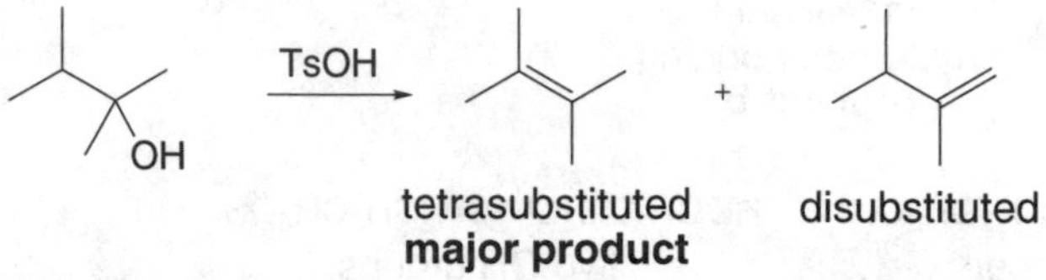

tetrasubstituted
major product   disubstituted

b.

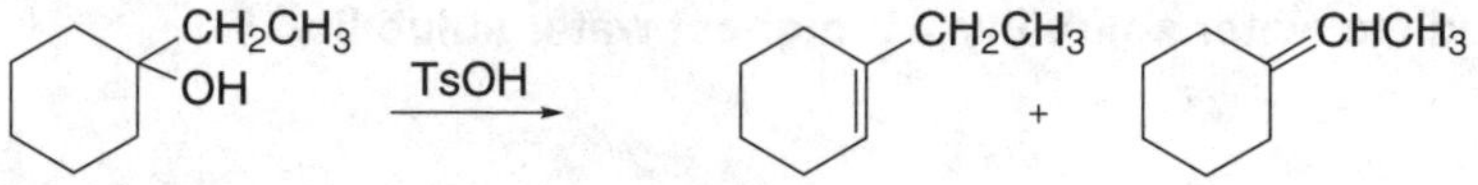

c.

trisubstituted
major product   disubstituted

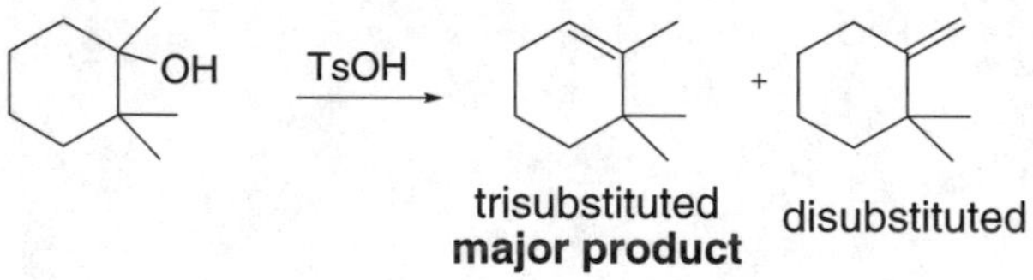

d. CH₃CH₂CH₂CH₂OH $\xrightarrow{\text{TsOH}}$ CH₃CH₂CH=CH₂

e.

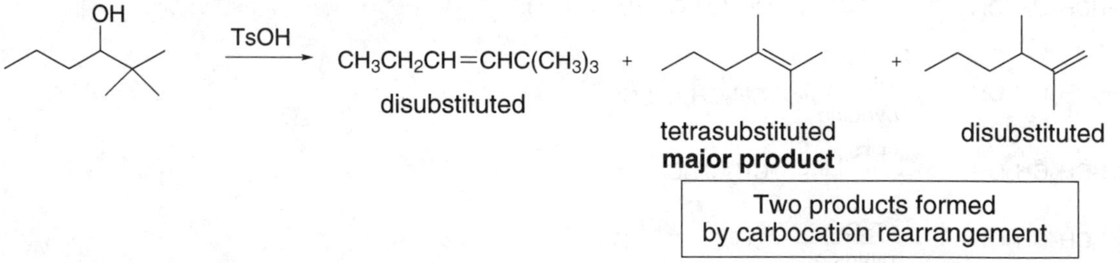

CH₃CH₂CH=CHC(CH₃)₃ +

disubstituted

tetrasubstituted
major product   disubstituted

Two products formed
by carbocation rearrangement

9.46 The more stable alkene is the major product.

trans and disubstituted   monosubstituted   cis and disubstituted
major product

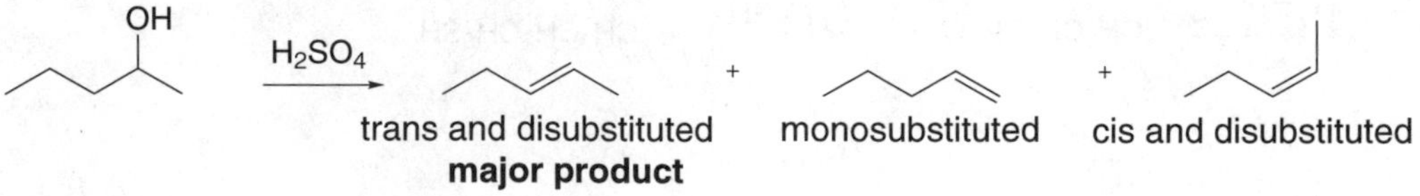

9.47 OTs is a good leaving group and will easily be replaced by a nucleophile. Draw the products by substituting the nucleophile in the reagent for OTs in the starting material.

a. $CH_3CH_2CH_2CH_2-OTs$ $\xrightarrow[\text{S}_\text{N}2]{CH_3SH}$ $CH_3CH_2CH_2CH_2-SCH_3$ + HOTs

b. $CH_3CH_2CH_2CH_2-OTs$ $\xrightarrow[\text{S}_\text{N}2]{NaOCH_2CH_3}$ $CH_3CH_2CH_2CH_2-OCH_2CH_3$ + Na^+ ^-OTs

c. $CH_3CH_2CH_2CH_2-OTs$ $\xrightarrow[\text{S}_\text{N}2]{NaOH}$ $CH_3CH_2CH_2CH_2-OH$ + Na^+ ^-OTs

d. $CH_3CH_2CH_2CH_2-OTs$ $\xrightarrow[\text{E2}]{K^+ \ ^-OC(CH_3)_3}$ $CH_3CH_2CH=CH_2$ + $(CH_3)_3COH$ + K^+ ^-OTs

9.48

a. 2° Alcohol will undergo S_N1. **racemization**

b. 1° Alcohol will undergo S_N2. **inversion**

c. $SOCl_2$ always implies S_N2. **Inversion**

d. Configuration is maintained. C–O bond is not broken.

9.49

a.

b. **B** and **D** are enantiomers.

c. **B** and **F** are identical.

9.50 Acid-catalyzed dehydration follows an E1 mechanism for 2° and 3° ROH with an added step to make a good leaving group. The three steps are:
[1] Protonate the oxygen to make a good leaving group.
[2] Break the C–O bond to form a carbocation.
[3] Remove a β hydrogen to form the π bond.

a.

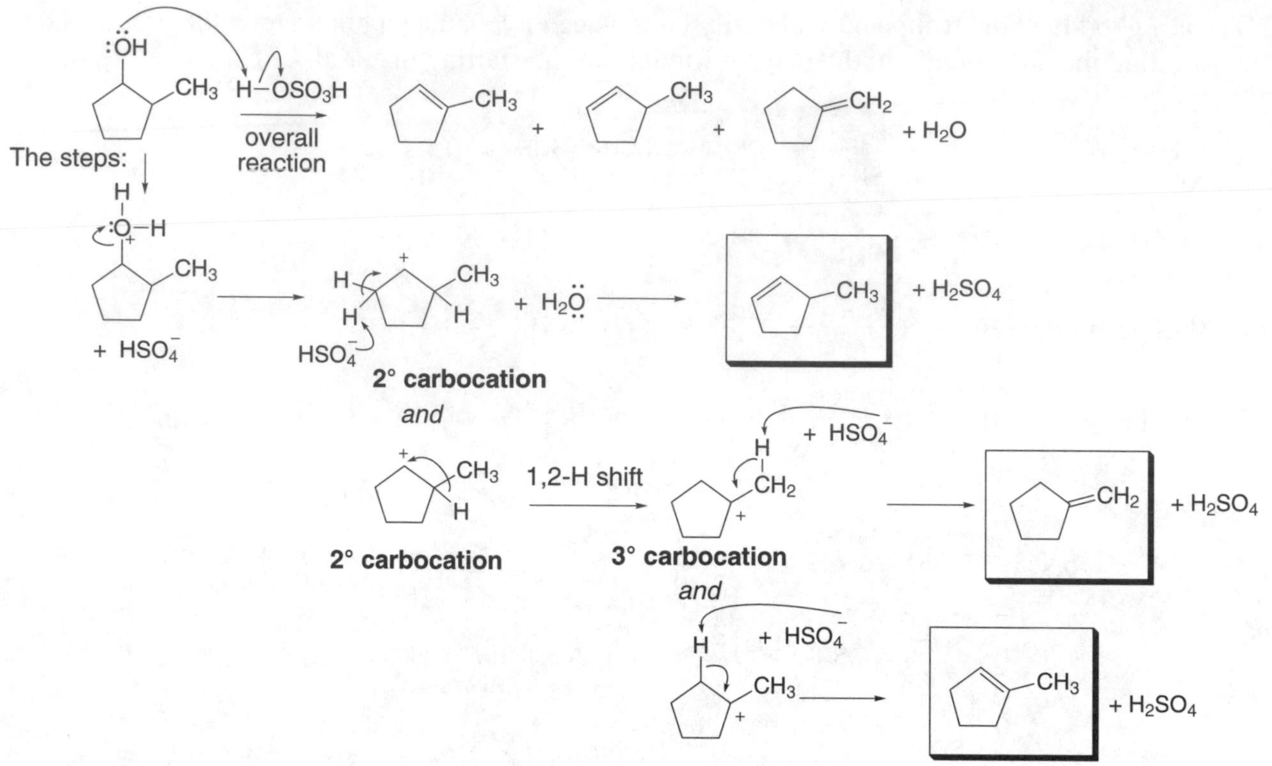

The steps:

2° carbocation

and

2° carbocation → 1,2-H shift → **3° carbocation**

and

b.

The steps:

2° carbocation → 1,2-CH₃ shift → **3° carbocation**

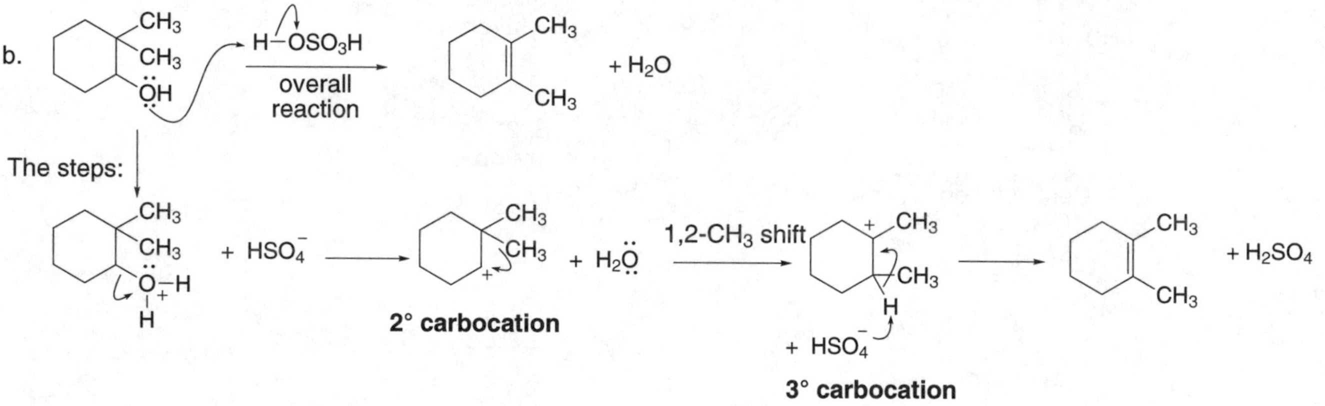

9.51 To draw the mechanisms:

   [1] Protonate the oxygen to make a good leaving group.
   [2] Break the C–O bond to form a carbocation.
   [3] Look for possible rearrangements to make a more stable carbocation.
   [4] Remove a β hydrogen to form the π bond.

Dark and light circle are meant to show where the carbons in the starting material appear in the product.

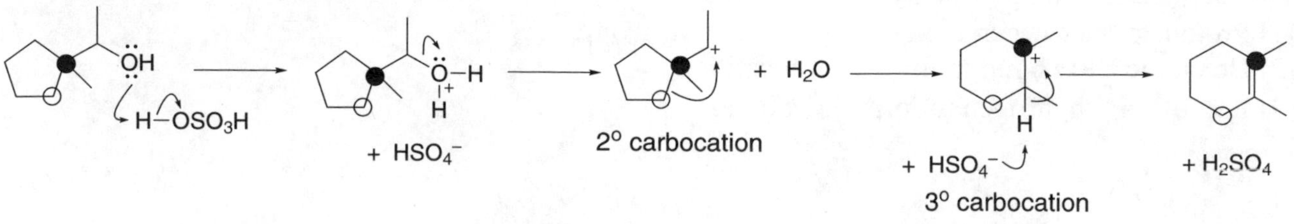

2° carbocation

3° carbocation

9.52

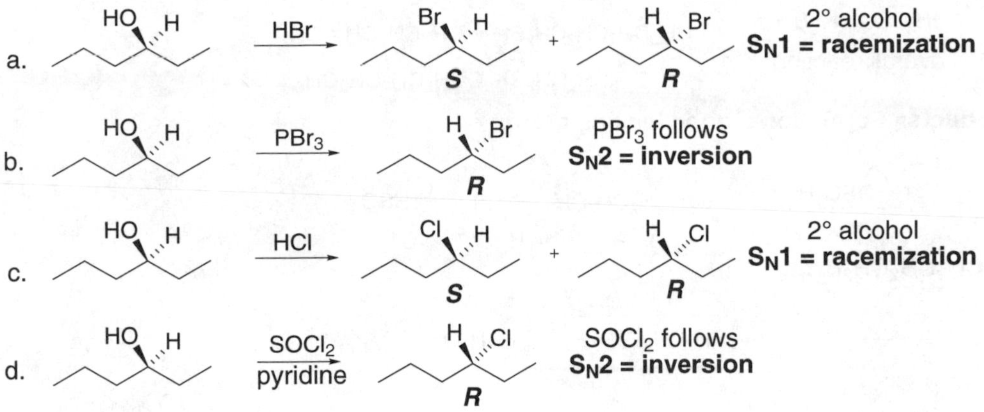
a.
HBr
S
R
2° alcohol
SN1 = racemization
b.
PBr3
R
PBr3 follows
SN2 = inversion
c.
HCl
S
R
2° alcohol
SN1 = racemization
d.
SOCl2
pyridine
R
SOCl2 follows
SN2 = inversion

9.53

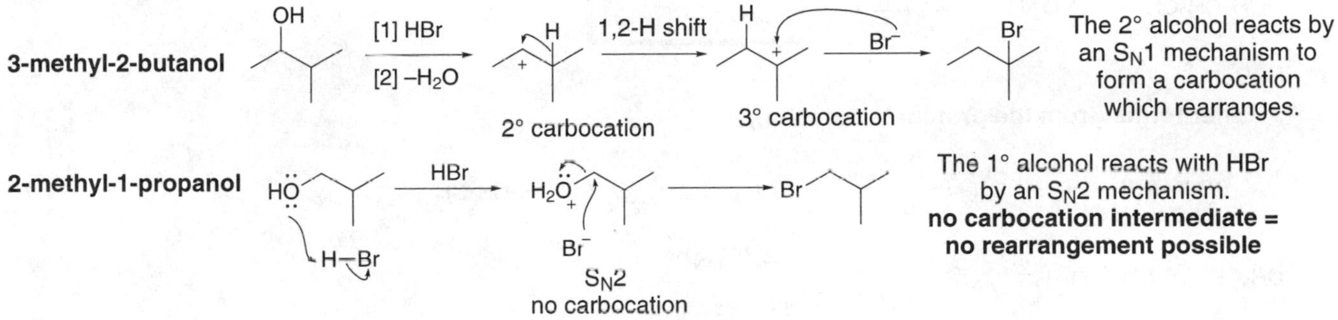
3-methyl-2-butanol
[1] HBr
[2] –H2O
1,2-H shift
2° carbocation
3° carbocation
Br–
The 2° alcohol reacts by an SN1 mechanism to form a carbocation which rearranges.
2-methyl-1-propanol
HBr
H–Br
SN2
no carbocation
The 1° alcohol reacts with HBr by an SN2 mechanism.
no carbocation intermediate = no rearrangement possible

9.54 Conversion of a 1° alcohol into a 1° alkyl chloride occurs by an S_N2 mechanism. S_N2 mechanisms occur more readily in polar aprotic solvents by making the nucleophile stronger. No added $ZnCl_2$ is necessary.

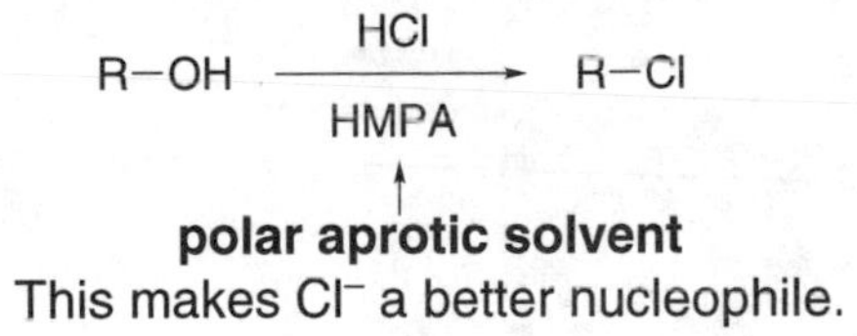
R–OH
HCl
HMPA
R–Cl
polar aprotic solvent
This makes Cl– a better nucleophile.

9.55

$$CH_3CH_2CH_2CH_2OH \xrightarrow[\text{overall reaction}]{H_2SO_4, \ NaBr} CH_3CH_2CH_2CH_2Br + CH_3CH_2CH=CH_2$$

$$+ \ CH_3CH_2CH_2CH_2OCH_2CH_2CH_2CH_3$$

Step [1] for all products: Formation a good leaving group

$$CH_3CH_2CH_2CH_2-\ddot{O}H \xrightarrow{H-OSO_3H} CH_3CH_2CH_2CH_2-\overset{+}{\underset{H}{\ddot{O}}}-H + HSO_4^-$$

Formation of $CH_3CH_2CH_2CH_2Br$:

$$CH_3CH_2CH_2CH_2-\overset{+}{\underset{H}{\ddot{O}}}-H \longrightarrow \boxed{CH_3CH_2CH_2CH_2Br} + H_2\ddot{O}$$

$$Na^+ \ Br^-$$

Formation of $CH_3CH_2CH=CH_2$:

$$CH_3CH_2CH-CH_2-\overset{+}{\underset{H}{\ddot{O}}}-H \longrightarrow \boxed{CH_3CH_2CH=CH_2} + H_2\ddot{O}$$

$$HSO_4^-$$

Ether forms (from the protonated alcohol):

$$CH_3CH_2CH_2CH_2-\overset{+}{\ddot{O}}H_2 \longrightarrow CH_3CH_2CH_2CH_2-\overset{+}{\underset{H}{\ddot{O}}}-CH_2CH_2CH_2CH_3 + H_2O$$

$$CH_3CH_2CH_2CH_2-\ddot{O}-H \qquad HSO_4^-$$

$$\boxed{CH_3CH_2CH_2CH_2-\ddot{O}-CH_2CH_2CH_2CH_3} + H_2SO_4$$

9.56

9.57

a.

axial OTs group

2 anti H's
2 elimination products
This conformer reacts.

two axial β hydrogens
two possible products

elimination
–H, OTs

major product
trisubstituted

disubstituted

b.

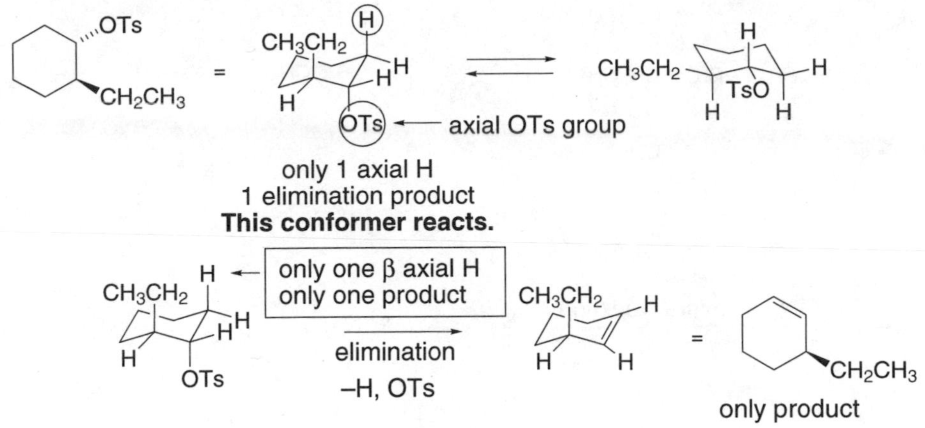

only 1 axial H
1 elimination product
This conformer reacts.

only one β axial H
only one product

elimination

−H, OTs

only product

9.58

a.

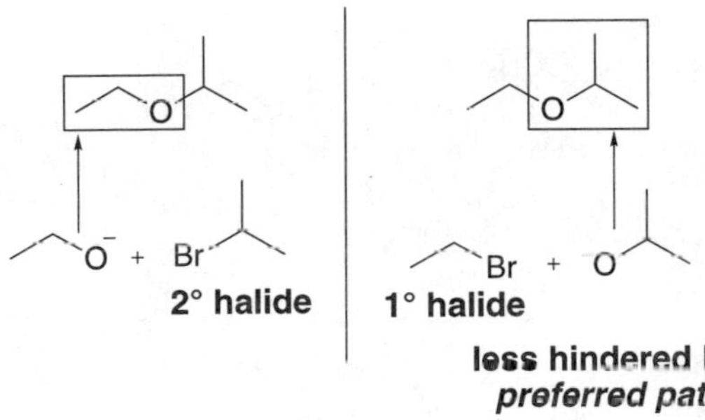

2° halide | 1° halide

**less hindered RX
preferred path**

c.

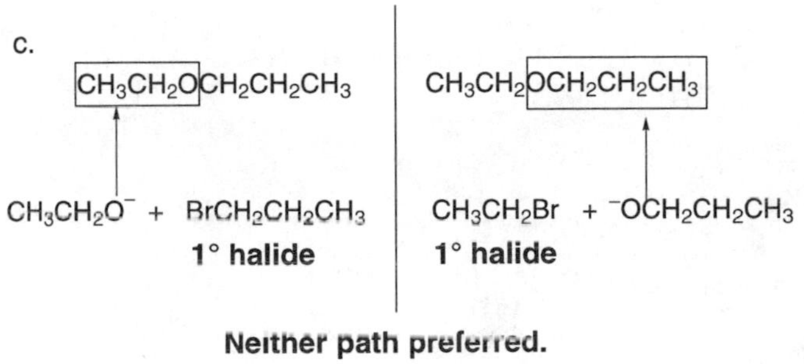

$CH_3CH_2OCH_2CH_2CH_3$ $CH_3CH_2OCH_2CH_2CH_3$

$CH_3CH_2O^-$ + $BrCH_2CH_2CH_3$ CH_3CH_2Br + $^-OCH_2CH_2CH_3$
1° halide   1° halide

Neither path preferred.

b.

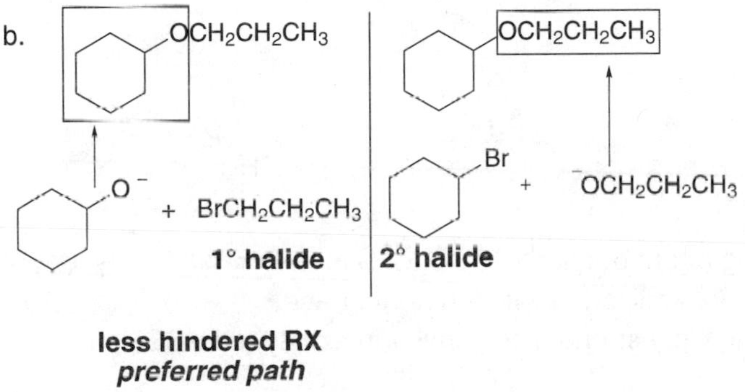

1° halide | 2° halide

**less hindered RX
preferred path**

9.59 A tertiary halide is too hindered and an aryl halide too unreactive to undergo a Willamson ether synthesis.

Two possible starting materials:

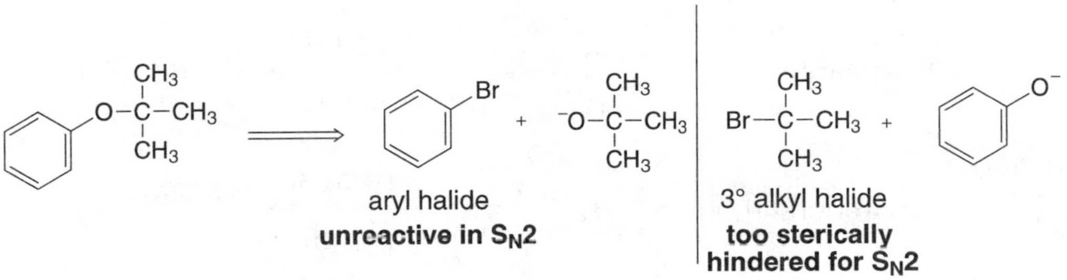

aryl halide
unreactive in S_N2

3° alkyl halide
**too sterically
hindered for S_N2**

9.60

a. $(CH_3)_3COCH_2CH_2CH_3$ $\xrightarrow{HBr}$ $(CH_3)_3CBr$ + $BrCH_2CH_2CH_3$ + H_2O

c. 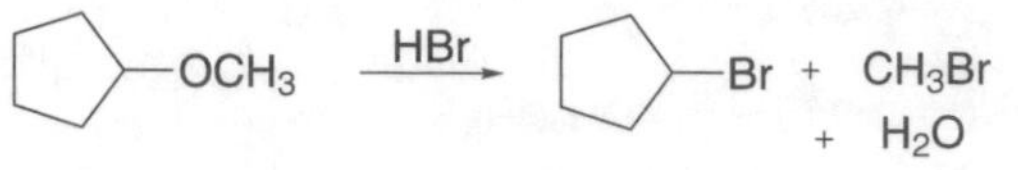 —OCH_3 $\xrightarrow{HBr}$ —Br + CH_3Br + H_2O

b. 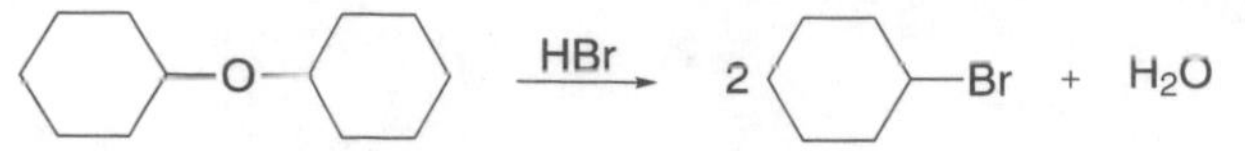 $\xrightarrow{HBr}$ 2 —Br + H_2O

9.61

a.

The steps:

b.

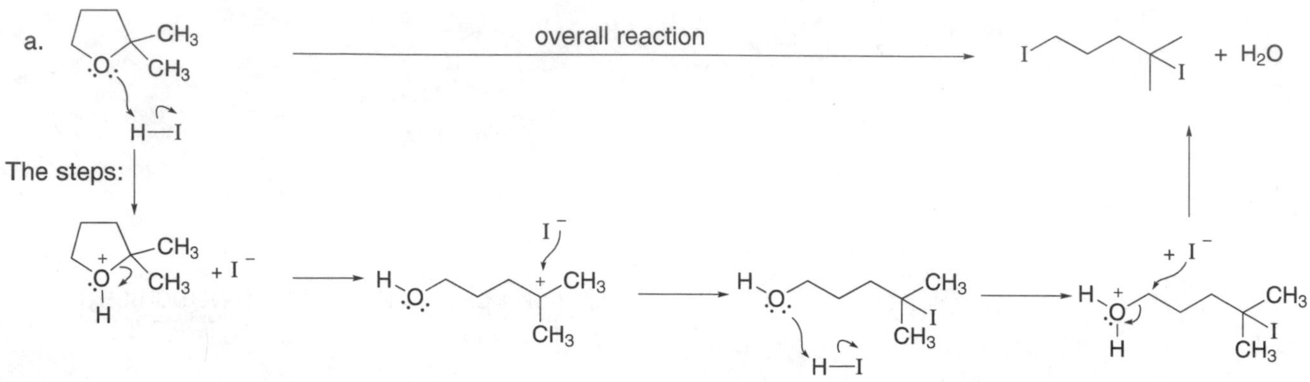

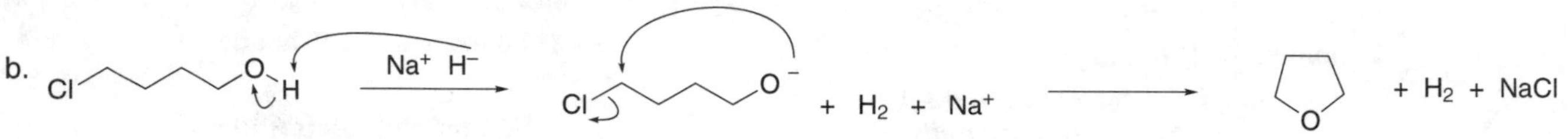

9.62

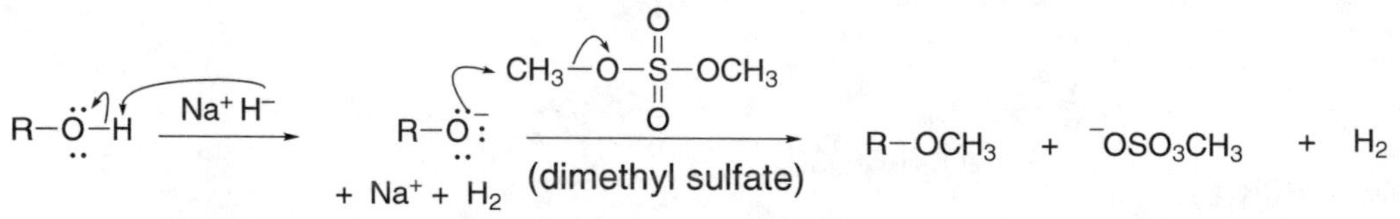

Dimethyl sulfate is a reactive methylating agent because $^-OSO_3CH_3$ is a
very good leaving group; it is a resonance-stabilized, weak conjugate base.
The conjugate acid of $^-OSO_3CH_3$ is $HOSO_3CH_3$, which is a strong acid, similar in acidity to H_2SO_4.

9.63

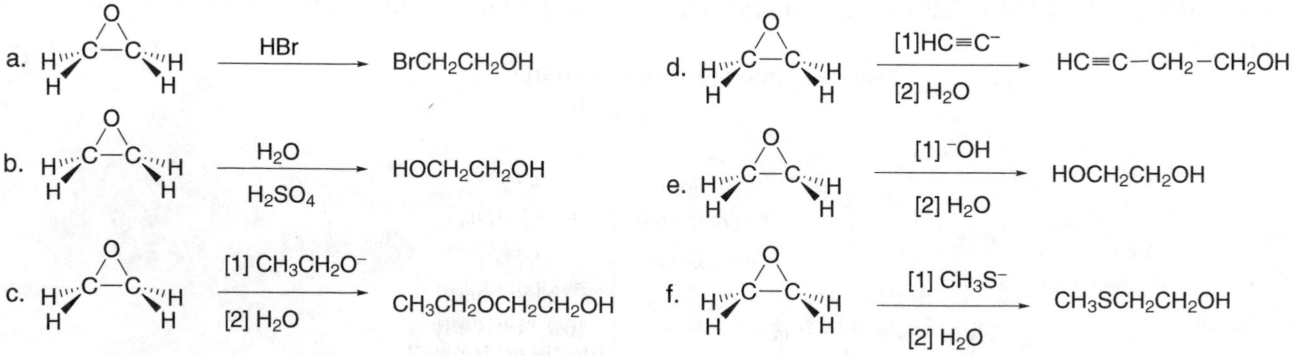

a. $\xrightarrow{HBr}$ $BrCH_2CH_2OH$

d. $\xrightarrow[\text{[2] }H_2O]{\text{[1]}HC\equiv C^-}$ $HC\equiv C-CH_2-CH_2OH$

b. $\xrightarrow[H_2SO_4]{H_2O}$ $HOCH_2CH_2OH$

e. $\xrightarrow[\text{[2] }H_2O]{\text{[1] }^-OH}$ $HOCH_2CH_2OH$

c. $\xrightarrow[\text{[2] }H_2O]{\text{[1] }CH_3CH_2O^-}$ $CH_3CH_2OCH_2CH_2OH$

f. $\xrightarrow[\text{[2] }H_2O]{\text{[1] }CH_3S^-}$ $CH_3SCH_2CH_2OH$

9.64

a.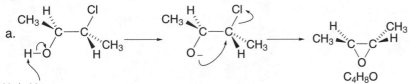

CH$_3$ epoxide + CH$_3$CH$_2$OH / H$_2$SO$_4$ →

CH$_3$ OH
CH$_3$–C–C–H
CH$_3$CH$_2$O H

c. spiro epoxide + HBr → cyclohexane with OH and Br

b. methylcyclohexene oxide + [1] CH$_3$CH$_2$O$^-$ Na$^+$ [2] H$_2$O →

CH$_3$
OH
OCH$_2$CH$_3$

d. spiro epoxide + [1] NaCN [2] H$_2$O →

OH
CN

9.65

a.

CH$_3$ H, Cl
C–C
H–O H CH$_3$
Na$^+$ H$^-$

→

CH$_3$ H, Cl
C–C
O$^-$ H CH$_3$

→

H CH$_3$
C–C
CH$_3$ O H
C$_4$H$_8$O

The 2 CH$_3$ groups are anti in the starting material, making them trans in the product.

b.

CH$_3$ Cl
H–C–C
H–O H CH$_3$
Na$^+$ H$^-$

→

CH$_3$ Cl
H–C–C
O$^-$ H CH$_3$

→

CH$_3$ CH$_3$
H–C–C–H
O
C$_4$H$_8$O

The 2 CH$_3$ groups are gauche in the starting material, making them cis in the product.

9.66

a. cyclopentyl–OTs + KOC(CH$_3$)$_3$ → cyclopentene Bulky base favors E2.

b.

chain with OH, H, CH$_3$ + HBr → chain with Br, H, CH$_3$ Keep the stereochemistry at the stereogenic center [*] the same here since no bond is broken to it.

c.

CH$_3$CH$_2$–C–C–H epoxide with CH$_2$CH$_3$ + [1] $^-$CN [2] H$_2$O →

CH$_3$CH$_2$ OH
H–C–C–H
CN CH$_2$CH$_3$

+

HO H
C–C–CH$_2$CH$_3$
CH$_3$CH$_2$ CN

identical

d. (CH$_3$)$_3$C– cyclohexane –OTs, H + KCN / S$_N$2 inversion → (CH$_3$)$_3$C– cyclohexane with H and CN

e. cyclohexane with OH, CH$_3$ + PBr$_3$ → cyclohexane with Br, CH$_3$ S$_N$2 inversion

f. chain with OH, H, D + TsCl / pyridine → chain with OTs, H, D + CH$_3$CO$_2^-$ → CH$_3$CO$_2$– chain with H, D

g. $\xrightarrow{\text{HBr}}$ +

h. $\xrightarrow[\text{[2] H}_2\text{O}]{\text{[1] NaOCH}_3}$ +

i. $\xrightarrow{\text{NaH}}$ $\xrightarrow{\text{CH}_3\text{CH}_2\text{I}}$

j.

$$\text{CH}_3\text{CH}_2-\underset{\underset{\text{CH}_3}{|}}{\overset{\overset{\text{CH}_3}{|}}{C}}-\text{O}-\text{CH}_3 \xrightarrow{\text{HI}} \text{CH}_3\text{CH}_2-\underset{\underset{\text{CH}_3}{|}}{\overset{\overset{\text{CH}_3}{|}}{C}}-\text{I} \ + \ \text{I}-\text{CH}_3 \ + \ \text{H}_2\text{O}$$

9.67

a. $\xrightarrow[\substack{\text{or} \\ \text{PBr}_3}]{\text{HBr}}$

b. $\xrightarrow[\substack{\text{or} \\ \text{SOCl}_2}]{\text{HCl}}$

c. $\xrightarrow{\text{[1] Na}^+\text{H}^-}$ $\xrightarrow{\text{[2] CH}_3\text{CH}_2\text{Cl}}$

d. $\xrightarrow{\text{[1] TsCl/pyridine}}$ $\xrightarrow{\text{[2] N}_3^-}$

Make OH a good leaving
group (use TsCl), then add N_3^-.

9.68

a. $\xrightarrow[\substack{\text{or} \\ \text{SOCl}_2}]{\text{HCl}}$

b. $\xrightarrow{\text{H}_2\text{SO}_4}$

c. $\xrightarrow{\text{[1] Na}^+\text{H}^-}$ $\xrightarrow{\text{[2] CH}_3\text{Cl}}$

d. $\xrightarrow{\text{[1] TsCl/pyridine}}$ $\xrightarrow{\text{[2] }^-\text{CN}}$

Make OH a good
leaving group (use TsCl),
then add $^-$CN.

9.69

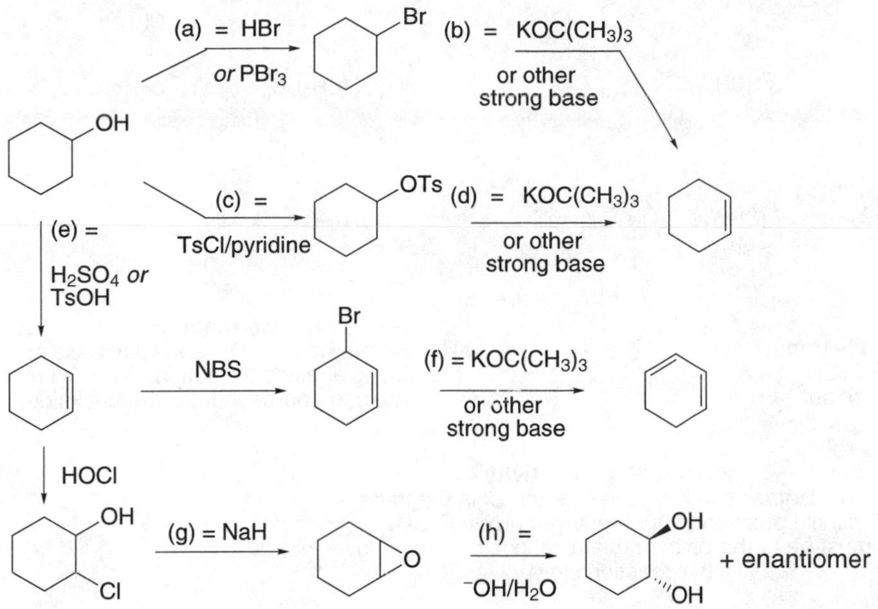
(a) = HBr
or PBr3
Br
(b) = KOC(CH3)3
or other
strong base
OH
(c) =
TsCl/pyridine
OTs
(d) = KOC(CH3)3
or other
strong base
(e) =
H2SO4 or
TsOH
NBS
Br
(f) = KOC(CH3)3
or other
strong base
HOCl
OH
Cl
(g) = NaH
O
(h) =
-OH/H2O
OH
OH
+ enantiomer

9.70

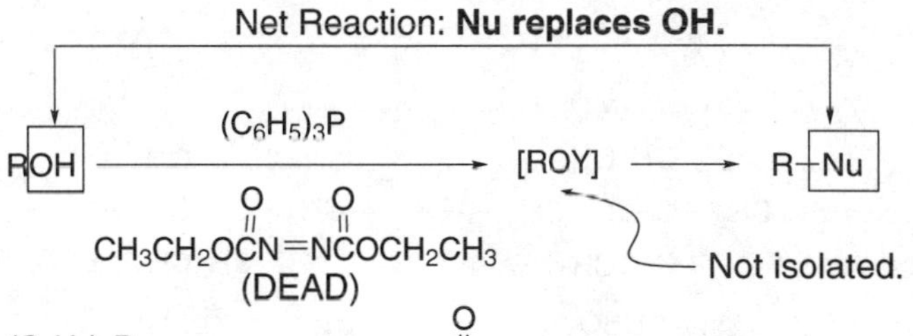
Net Reaction: Nu replaces OH.
ROH
(C6H5)3P
CH3CH2OCN=NCOCH2CH3
(DEAD)
[ROY]
R–Nu
Not isolated.

a. 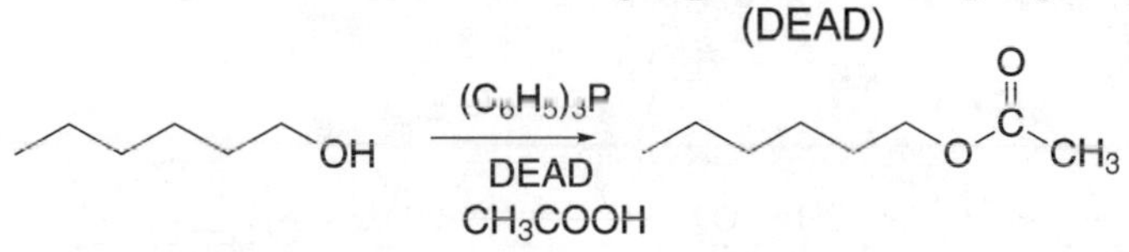
OH
(C6H5)3P
DEAD
CH3COOH
O
CH3

b. 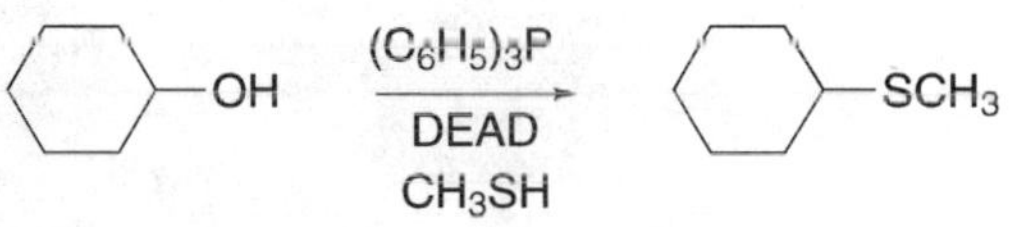
OH
(C6H5)3P
DEAD
CH3SH
SCH3

9.71

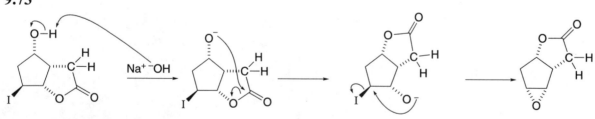

A

B

C

rapid reaction
This isomer can have the OH and Cl in the trans diaxial positions, while the larger group C(CH₃)₃ can be in the favorable equatorial position.

intermediate reactivity
This isomer can have the OH and Cl in the trans diaxial positions. But the larger group C(CH₃)₃ must be in the unfavorable axial position, making this reaction slower.

no reaction
This isomer has the OH and Cl in a cis position, prohibiting epoxide formation. The OH must be able to approach from the backside.

OH is in a favorable arrangement for backside attack of the nucleophile on the leaving group.

9.72

pinacol

+ HSO₄⁻

+ H₂O

1,2-CH₃ shift

pinacolone

9.73

Na⁺ ⁻OH

9.74 You must draw the product that places the nucleophile and leaving group (O⁻) trans and diaxial.

a. (CH₃)₃C⌬epoxide

This bond was above the six-membered ring
in the epoxide and it stays above it in the product.

(CH₃)₃C⌬ ⟶ (CH₃)₃C⌬ $\xrightarrow{H_2O}$ (CH₃)₃C⌬ = (CH₃)₃C⌬—OH

attack from below

OH (axial)

OCH₃

OCH₃ (axial)

OCH₃
only

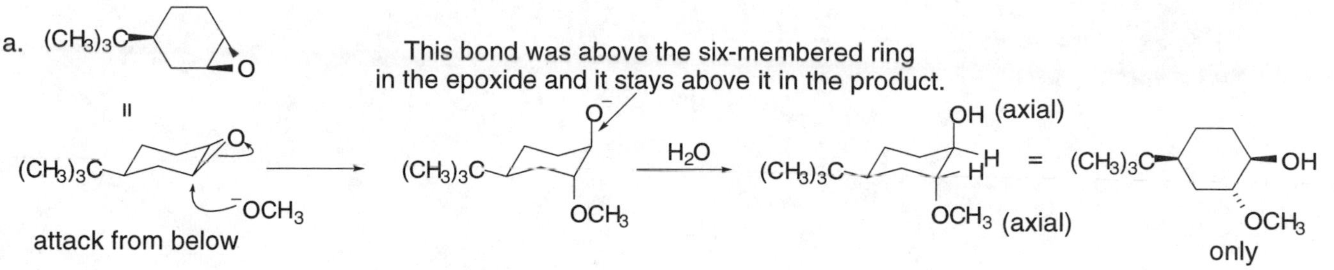

b. (CH₃)₃C⌬epoxide

(CH₃)₃C⌬ ⟶ (CH₃)₃C⌬ $\xrightarrow{H_2O}$ (CH₃)₃C⌬ = (CH₃)₃C⌬—OCH₃

attack from above

This bond was below the six-membered ring
in the epoxide and it stays below it in the product.

OCH₃

OCH₃ (axial)

OH (axial)

OH
only

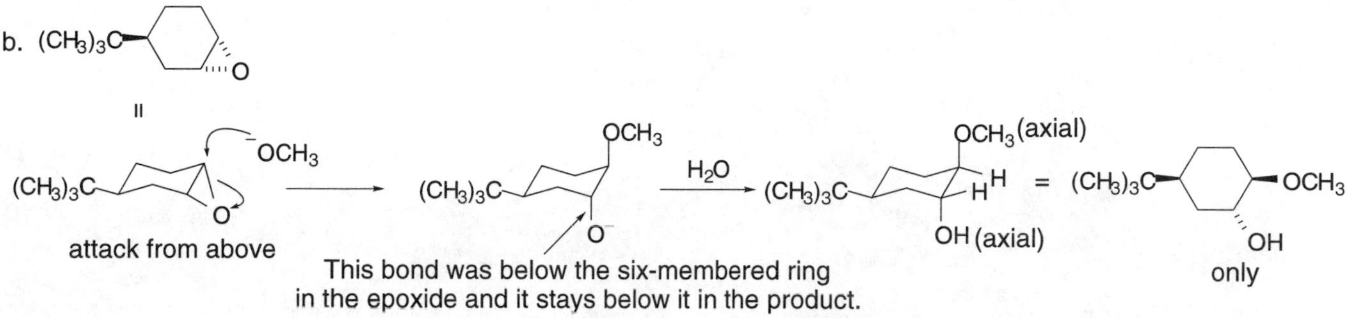

The nucleophile must always approach by backside attack; i.e. if the epoxide is drawn "up" it must attack from below. Even though both ends of the epoxide are equally substituted, nucleophilic attack occurs at only one C–O bond, the one that gives trans diaxial products, as drawn.

Chapter 10: Alkenes

◆ General facts about alkenes

- Alkenes contain a carbon–carbon double bond consisting of a stronger σ bond and a weaker π bond. Each carbon is sp^2 hybridized and trigonal planar (10.1).
- Alkenes are named using the suffix –*ene* (10.3).
- Alkenes with different groups on each end of the double bond exist as a pair of diastereomers, identified by the prefixes *E* and *Z* (10.3B).

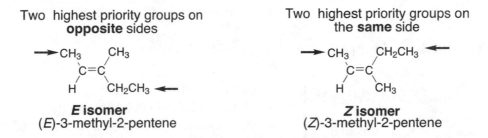

- Alkenes have weak intermolecular forces, giving them low mp's and bp's, and making them water insoluble. A cis alkene is more polar than a trans alkene, giving it a slightly higher boiling point (10.4).

cis-2-butene   *trans*-2-butene

more polar isomer ⟶   CH₃ C=C CH₃   H H   CH₃ C=C H   H CH₃   ⟵ **less polar isomer**

a small net dipole   **no net dipole**
higher bp   **lower bp**

- Since a π bond is electron rich and much weaker than a σ bond, alkenes undergo addition reactions with electrophiles (10.8)

◆ Stereochemistry of alkene addition reactions (10.8)

A reagent XY adds to a double bond in one of three different ways:

- **Syn addition** – X and Y add from the same side.

  $$C=C \xrightarrow{\text{H–BH}_2} \text{H, BH}_2\ C\text{–}C$$

  - Syn addition occurs in **hydroboration.**

- **Anti addition** – X and Y add from opposite sides.

  $$C=C \xrightarrow[\text{X}_2/\text{H}_2\text{O}]{\text{X}_2 \text{ or}} \text{X, X(OH)}\ C\text{–}C$$

  - Anti addition occurs in **halogenation** and **halohydrin formation.**

- **Both syn and anti addition** occur when carbocations are intermediates.

  $$C=C \xrightarrow[\text{H}_2\text{O/H}^+]{\text{H–X or}} \text{H, X(OH)}\ C\text{–}C \text{ and } \text{H, X(OH)}\ C\text{–}C$$

  - Syn and anti addition occur in **hydrohalogenation** and **hydration.**

♦ **Addition reactions of alkenes**

[1] Hydrohalogenation – Addition of HX (X = Cl, Br, I) (10.9–10.11)

$RCH=CH_2$ + H—X ⟶

R—CH—CH$_2$
| |
X H
alkyl halide

- The mechanism has two steps.
- Carbocations are formed as intermediates.
- Carbocation rearrangements are possible.
- Markovnikov's rule is followed. H bonds to the less substituted C to form the more stable carbocation.
- Syn and anti addition occur.

[2] Hydration and related reactions (Addition of H$_2$O or ROH) (10.12)

$RCH=CH_2$ + H—OH $\xrightarrow{H_2SO_4}$

R—CH—CH$_2$
| |
OH H
alcohol

$RCH=CH_2$ + H—OR $\xrightarrow{H_2SO_4}$

R—CH—CH$_2$
| |
OR H
ether

For both reactions:
- The mechanism has three steps.
- Carbocations are formed as intermediates.
- Carbocation rearrangements are possible.
- Markovnikov's rule is followed. H bonds to the less substituted C to form the more stable carbocation.
- Syn and anti addition occur.

[3] Halogenation (Addition of X$_2$; X = Cl or Br) (10.13–10.14)

$RCH=CH_2$ + X—X ⟶

R—CH—CH$_2$
| |
X X
vicinal dihalide

- The mechanism has two steps.
- Bridged halonium ions are formed as intermediates.
- No rearrangements occur.
- Anti addition occurs.

[4] Halohydrin formation (Addition of OH and X; X = Cl, Br) (10.15)

$RCH=CH_2$ + X—X $\xrightarrow{H_2O}$

R—CH—CH$_2$
| |
OH X
halohydrin

- The mechanism has three steps.
- Bridged halonium ions are formed as intermediates.
- No rearrangements occur.
- X bonds to the less substituted C.
- Anti addition occurs.
- NBS in DMSO and H$_2$O adds Br and OH in the same fashion.

[5] Hydroboration–oxidation (Addition of H$_2$O) (10.16)

$RCH=CH_2$ $\xrightarrow[\text{[2] } H_2O_2/HO^-]{\text{[1] } BH_3 \text{ or 9-BBN}}$

R—CH—CH$_2$
| |
H OH
alcohol

- Hydroboration has a one-step mechanism.
- No rearrangements occur.
- OH bonds to the less substituted C.
- Syn addition of H$_2$O results.

Chapter 10: Answers to Problems

10.1

Six alkenes of molecular formula C_5H_{10}.

$CH_2{=}CHCH_2CH_2CH_3$

trans   cis

diastereomers

10.2 To determine the number of degrees of unsaturation:
  [1] Calculate the maximum number of H's ($2n + 2$).
  [2] Subtract the actual number of H's from the maximum number.
  [3] Divide by two.

a. C_2H_2
  [1] maximum number of H's = $2n + 2 = 2(2) + 2 = 6$
  [2] subtract actual from maximum = $6 - 2 = 4$
  [3] divide by two – 4/2 = **2 degrees of unsaturation**

b. C_6H_6
  [1] maximum number of H's – $2n + 2 = 2(6) + 2 = 14$
  [2] subtract actual from maximum = $14 - 6 = 8$
  [3] divide by two = 8/2 = **4 degrees of unsaturation**

c. C_8H_{18}
  [1] maximum number of H's = $2n + 2 = 2(8) + 2 = 18$
  [2] subtract actual from maximum = $18 - 18 = 0$
  [3] divide by two – 0/2 – **0 degrees of unsaturation**

d. C_7H_8O
  Ignore the O.
  [1] maximum number of H's = $2n + 2 = 2(7) + 2 = 16$
  [2] subtract actual from maximum = $16 - 8 = 8$
  [3] divide by two = 8/2 = **4 degrees of unsaturation**

e. $C_7H_{11}Br$
  Because of Br, add one more H (11 + 1 H's – 12 H's)
  [1] maximum number of H's = $2n + 2 = 2(7) + 2 = 16$
  [2] subtract actual from maximum = $16 - 12 = 4$
  [3] divide by two = 4/2 = **2 degrees of unsaturation**

10.3 First determine the number of degrees of unsaturation in the compound. Then decide which combinations of rings and π bonds could exist.

$C_{10}H_{14}$
  [1] maximum number of H's = $2n + 2 = 2(10) + 2 = 22$
  [2] subtract actual from maximum = $22 - 14 = 8$
  [3] divide by two = 8/2 = **4 degrees of unsaturation**

possibilities:
  4 π bonds
  3 π bonds + 1 ring
  2 π bonds + 2 rings
  1 π bond + 3 rings
  4 rings

10.4 To name an alkene:
  [1] Find the longest chain that contains the double bond. Change the ending from *–ane* to *–ene*.
  [2] Number the chain to give the double bond the lower number. The alkene is named by the first number.
  [3] Apply all other rules of nomenclature.

To name a cycloalkene:
  [1] When a double bond is located in a ring, it is always located between C1 and C2. Omit the "1" in the name. Change the ending from *–ane* to *–ene*.
  [2] Number the ring clockwise or counterclockwise to give the first substituent the lower number.
  [3] Apply all other rules of nomenclature.

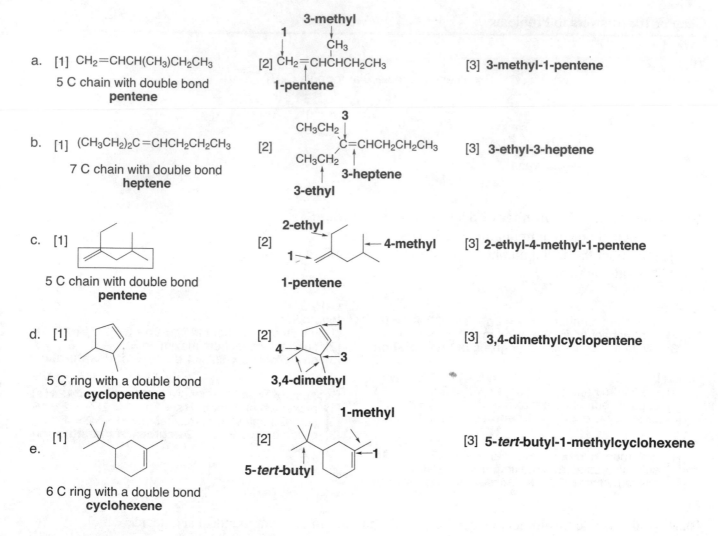

10.5 To label an alkene as *E* or *Z*:

[1] **Assign priorities** to the two substituents *on each end* using the rules for *R,S* nomenclature.

[2] **Assign E or Z** depending on the location of the two highest priority groups.

- The *E* prefix is used when the two higher priority groups are on **opposite sides**.
- The *Z* prefix is used when the two higher priority groups are on the **same side** of the double bond.

a.

higher priority →CH₃ | Cl
C=C
H | Br ← higher priority
Two higher priority groups are
on opposite sides: *E* isomer

b.

higher priority → CH₃CH₂ | CH₂CH₃ ← higher priority
C=C
H | CH₃
Two higher priority groups are
on the same side: *Z* isomer

c.

higher priority → ⟨benzene⟩ | ⟨benzene⟩ OCH₂CH₂N(CH₃)₂ ← higher priority
C=C
CH₃CH₂ | ⟨benzene⟩
tamoxifen
Two higher priority groups are
on the same side: *Z* isomer

10.6 To name an alkene: First follow the rules from Answer 10.4. Then, when necessary, assign an *E* or *Z* prefix based on priority, as in 10.5.

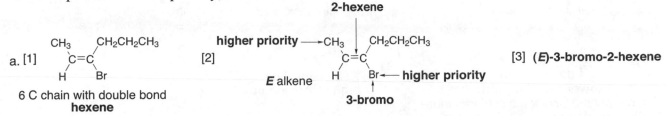

a. [1]

CH₃   CH₂CH₂CH₃
　　C=C
H　　Br

6 C chain with double bond
hexene

[2]

2-hexene

higher priority ──►CH₃ | CH₂CH₂CH₃
　　　　　　　C=C
　　　　　H　　Br ◄── **higher priority**

E alkene

3-bromo

[3] **(*E*)-3-bromo-2-hexene**

b. [1]

CH₃CH₂   CH₂CH₂CH₂CH₂C(CH₃)₃
　　C=C
H　　CH₂CH₃

10 C chain with double bond
decene

[2]

higher priority ──►CH₃CH₂

Z alkene

3-decene

CH₃CH₂ | CH₂CH₂CH₂CH₂CCH₃
　　C=C
H　　CH₂CH₃

CH₃◄── **9,9-dimethyl**
CH₃◄── **higher priority**
CH₃◄── **9,9-dimethyl**

4-ethyl

[3] **(*Z*)-4-ethyl-9,9-dimethyl-3-decene**

10.7 To work backwards from a name to a structure:
　[1] Find the parent name and functional group and draw, remembering that the double bond is between C1 and C2 for cycloalkenes.
　[2] Add the substituents to the appropriate carbon.

a. (*Z*)-4-ethyl-3-heptene☐☐ The higher priority groups are on the same side = **Z**.

7 carbons

4-ethyl

The double bond is between C3 and C4.

c. (*Z*)-2-bromo-1-iodo-1-hexene　The double bond is between C1 and C2.

6 carbons

I
Br

The higher priority groups are on the same side = **Z**.

b. (*E*)-3,5,6-trimethyl-2-octene☐　The double bond is between C2 and C3.

8 carbons

The higher priority groups are on opposite sides = **E**.

3,5,6-trimethyl

10.8 To rank the isomers by increasing boiling point:
　Look for polarity differences: *small net dipoles* make an alkene more polar giving it a higher boiling point than an alkene with *no net dipole*. Cis isomers have a higher boiling point than their trans isomers.

CH₃　　CH₃
　　C=C
CH₃　　CH₃

All dipoles cancel.
smallest surface area
no net dipole
lowest bp

CH₃CH₂　　H
　　C=C
H　　CH₂CH₃

Two dipoles cancel.
no net dipole
trans isomer
intermediate bp

CH₃CH₂　　CH₂CH₃
　　C=C
H　　H

Two dipoles reinforce.
net dipole
cis isomer
highest bp

10.9 Increasing number of double bonds = decreasing melting point

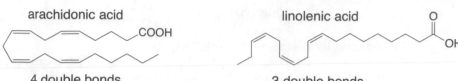

arachidonic acid .COOH   linolenic acid

4 double bonds   3 double bonds
lower melting point   **higher melting point**
Even though arachidonic acid has more
C's, it also has more C=C's and this
decreases the melting point.

10.10

a. [structure] OH →(H₂SO₄)→ [cyclopentene] + [methylenecyclopentane]

b. [structure with Br] →(NaOCH₂CH₃)→ $CH_2=CHCH_2CH_2CH_2CH_3$
+
$CH_3CH=CHCH_2CH_2CH_3$

10.11 To draw the products of an addition reaction:
[1] Locate the two bonds that will be broken in the reaction. Always break the π bond.
[2] Draw the product by forming two new σ bonds.

a. [cyclopentene] + HI → [structure with H, I] two new σ bonds

b. [cyclohexene with CH₃, CH₃] + HCl → [structure with H, CH₃, CH₃, Cl] two new σ bonds

10.12 **Addition reactions of HX occur in two steps:**
[1] The double bond attacks the H atom of HX to form a carbocation.
[2] X⁻ attacks the carbocation to form a C–X bond.

[mechanism: cyclohexene + H–Cl → carbocation with Cl⁻ → chlorocyclohexane]

10.13 Addition to alkenes follows Markovnikov's rule: When HX adds to an unsymmetrical alkene, the
H bonds to the C that has more H's to begin with.

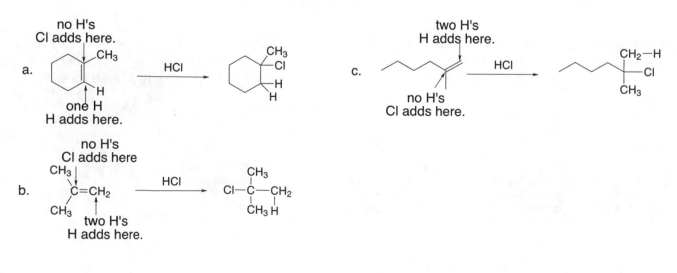

a.
no H's
Cl adds here.
[cyclohexene structure with CH₃, H]
one H
H adds here.
→(HCl)→ [product with CH₃, Cl, H, H]

c.
two H's
H adds here.
[alkene structure]
no H's
Cl adds here.
→(HCl)→ [product with CH₂–H, Cl, CH₃]

b.
no H's
Cl adds here
CH_3 $C=CH_2$ CH_3
two H's
H adds here.
→(HCl)→ $Cl–C–CH_2$ with CH_3, CH_3 H

10.14 To determine which alkene in each pair will react faster, draw the carbocation that forms in the rate-determining step. The more stable, more substituted carbocation, the lower the E_a to form it and the faster the reaction.

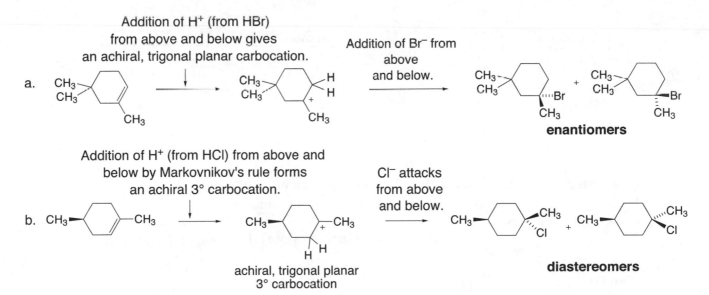

10.15 Look for rearrangements of a carbocation intermediate to explain these results.

10.16 To draw the products, remember that addition of HBr proceeds via a carbocation intermediate.

Addition of H⁺ (from HBr) from above and below gives an achiral, trigonal planar carbocation.

Addition of Br⁻ from above and below.

a.

enantiomers

Addition of H⁺ (from HCl) from above and below by Markovnikov's rule forms an achiral 3° carbocation.

Cl⁻ attacks from above and below.

b.

achiral, trigonal planar
3° carbocation

diastereomers

10.17 The product of syn addition will have H and Cl both up or down (both on wedges or both dashes), while the product of anti addition will have one up and one down (one wedge, one dash).

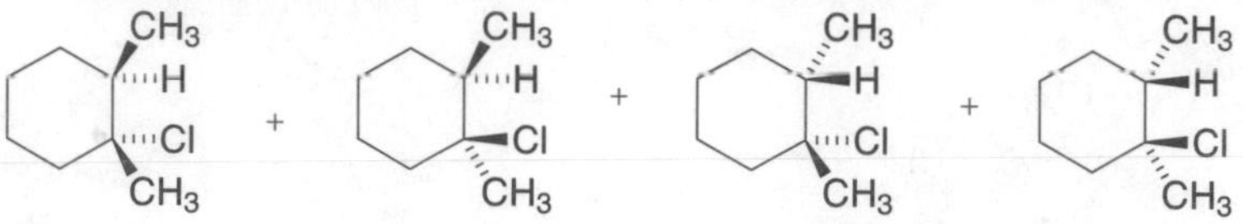

A	B	C	D
syn addition	anti addition	anti addition	syn addition

10.18

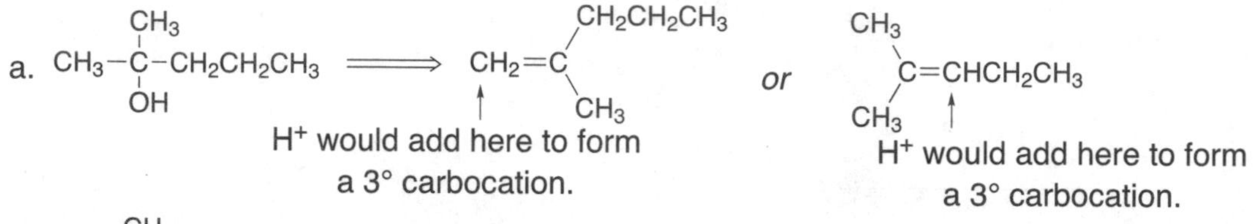

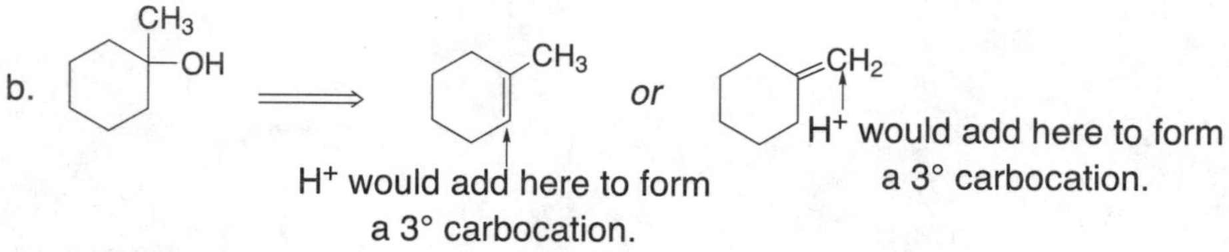

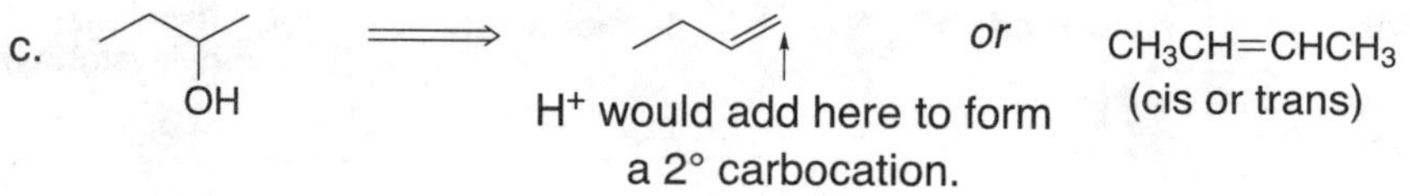

a.

H⁺ would add here to form a 3° carbocation.

or

H⁺ would add here to form a 3° carbocation.

b.

H⁺ would add here to form a 3° carbocation.

or

H⁺ would add here to form a 3° carbocation.

c.

H⁺ would add here to form a 2° carbocation.

or

$CH_3CH=CHCH_3$
(cis or trans)

10.19

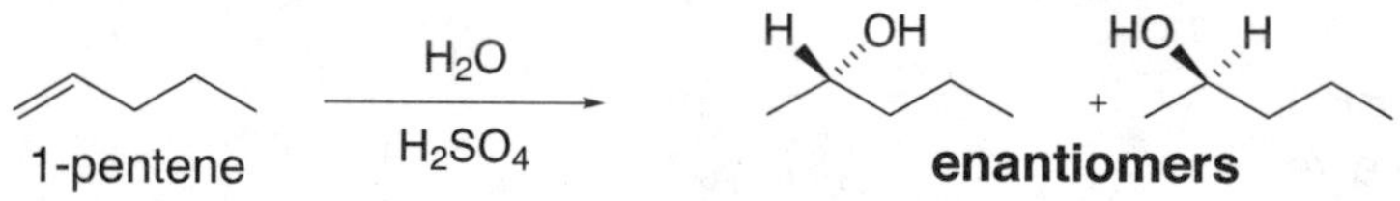

1-pentene H_2O H_2SO_4 **enantiomers**

10.20

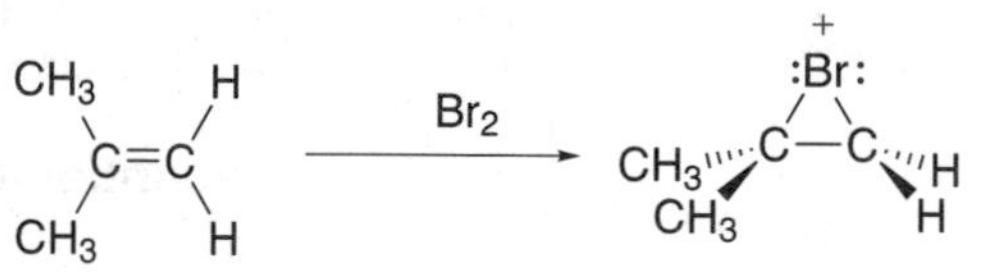

CH_3 H
   C=C Br_2
CH_3 H

10.21 **The two steps in the mechanism for the halogenation of an alkene are:**

[1] Addition of X⁺ to the alkene to form a bridged halonium ion.
[2] Nucleophilic attack by X⁻.

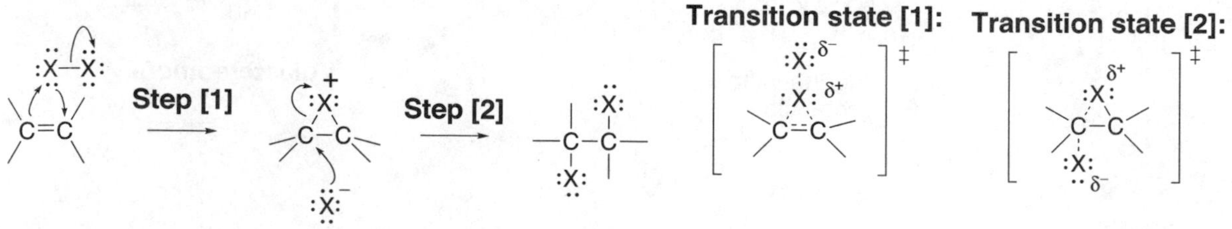

Step [1]   Step [2]

Transition state [1]:   **Transition state [2]:**

10.22 Halogenation of an alkene adds two elements of X in an anti fashion.

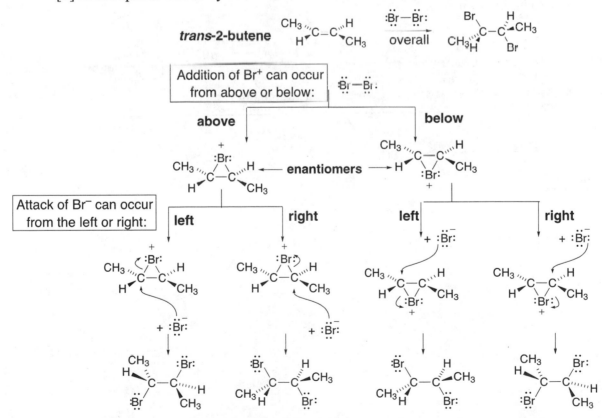

a. [square] $\xrightarrow{Br_2}$ [cyclobutane with Br, Br] + [cyclobutane with Br, Br]

b. [methylcyclohexene] $\xrightarrow{Cl_2}$ [cyclohexane with CH₃, Cl, Cl] + [cyclohexane with CH₃, Cl, Cl]

10.23 To draw the products of halogenation of an alkene, remember that the halogen adds to both ends of the double bond but only anti addition occurs.

a. [methylcyclopentene] $\xrightarrow{Cl_2}$ [product with Cl, Cl, CH₃] + [product with Cl, Cl, CH₃]

enantiomers

c. CH_3-[cyclohexene] $\xrightarrow{Br_2}$ CH_3-[product with Br, Br] + CH_3-[product with Br, Br]

diastereomers

b. [1,2-diphenylethene with H, H] $\xrightarrow{Br_2}$ [product with C–C, Br, H, H, Br, two phenyls]

achiral meso compound

10.24 **The two steps in the mechanism for the halogenation of an alkene are:**
[1] Addition of X^+ to the alkene to form a bridged halonium ion.
[2] Nucleophilic attack by X^-.

trans-2-butene $CH_3\overset{}{\underset{H}{C}}-\overset{H}{\underset{CH_3}{C}}$ $\overset{:\ddot{Br}-\ddot{Br}:}{\xrightarrow{\hspace{1cm}}}$ $CH_3\overset{Br}{\underset{H}{C}}-\overset{H}{\underset{Br}{C}}CH_3$
overall

Addition of Br⁺ can occur from above or below: $\ddot{Br}-\ddot{Br}:$

above ← | → **below**

$CH_3\overset{+:\ddot{Br}:}{\underset{H}{C}}-\overset{H}{\underset{CH_3}{C}}$ ← **enantiomers** → $CH_3\overset{H}{\underset{:\ddot{Br}:+}{C}}-\overset{H}{\underset{CH_3}{C}}$

Attack of Br⁻ can occur from the left or right: | **left** | **right** | **left** | **right**

$CH_3\overset{+:\ddot{Br}:}{\underset{H}{C}}-\overset{H}{\underset{CH_3}{C}}$ $CH_3\overset{+:\ddot{Br}:}{\underset{H}{C}}-\overset{H}{\underset{CH_3}{C}}$ $CH_3\overset{H}{\underset{:\ddot{Br}:+}{C}}-\overset{H}{\underset{CH_3}{C}}$ $CH_3\overset{H}{\underset{:\ddot{Br}:+}{C}}-\overset{H}{\underset{CH_3}{C}}$

$+:\ddot{Br}:^-$ $+:\ddot{Br}:^-$ $+:\ddot{Br}:^-$ $+:\ddot{Br}:^-$

$H\overset{CH_3}{\underset{:\ddot{Br}:}{C}}-\overset{:\ddot{Br}:}{\underset{CH_3}{C}}H$ $\overset{:\ddot{Br}}{\underset{CH_3}{C}}H-C\overset{H}{\underset{:\ddot{Br}:}{}}CH_3$ $\overset{:\ddot{Br}}{\underset{CH_3}{C}}H-C\overset{H}{\underset{:\ddot{Br}:}{}}CH_3$ $H\overset{CH_3}{\underset{:\ddot{Br}:}{C}}-\overset{:\ddot{Br}:}{\underset{CH_3}{C}}H$

All four compounds are identical—an achiral meso compound.

10.25 Halohydrin formation adds the elements of X and OH across the double bond in an anti fashion. The reaction is regioselective so X ends up on the carbon that had more H's to begin with.

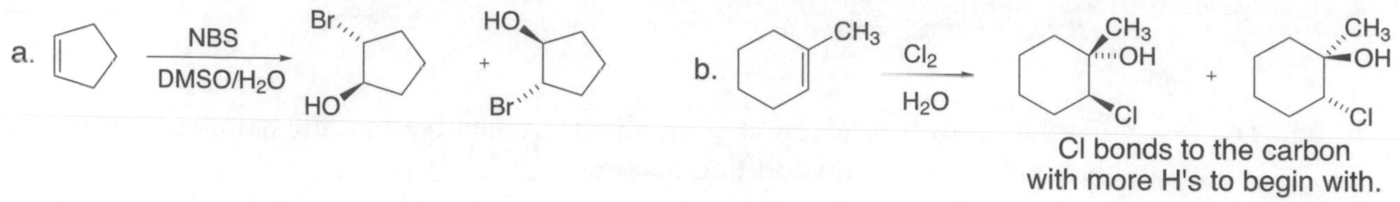
a.
NBS
DMSO/H2O
Br
HO
+
HO
Br
b.
CH3
Cl2
H2O
CH3
OH
Cl
+
CH3
OH
Cl
Cl bonds to the carbon
with more H's to begin with.

10.26

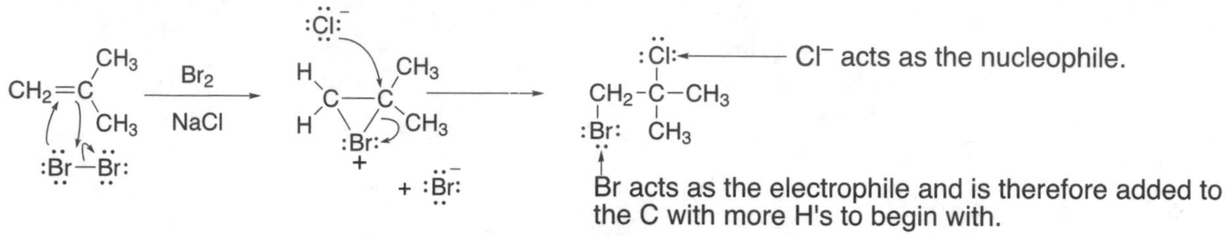
CH3
CH2=C
CH3
:Br—Br:
Br2
NaCl
:Cl:
H
H
CH3
CH3
:Br:
+
+ :Br:
:Cl:
Cl⁻ acts as the nucleophile.
CH2-C-CH3
:Br: CH3
Br acts as the electrophile and is therefore added to
the C with more H's to begin with.

10.27

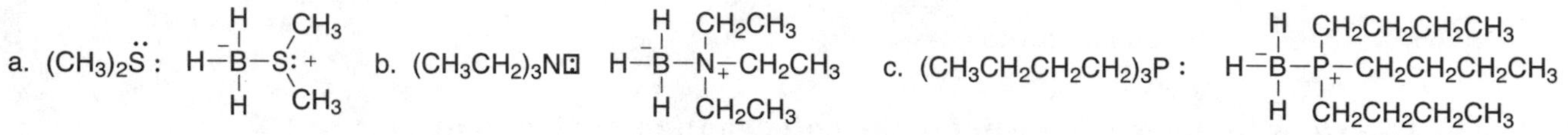
a. (CH3)2S :
H—B—S: +
H
CH3
CH3
b. (CH3CH2)3N
H—B—N—CH2CH3
H CH2CH3
H CH2CH3
c. (CH3CH2CH2CH2)3P :
H—B—P—CH2CH2CH2CH3
H CH2CH2CH2CH3
H CH2CH2CH2CH3

10.28 In hydroboration the boron atom is the electrophile and becomes bonded to the carbon atom that had more H's to begin with.

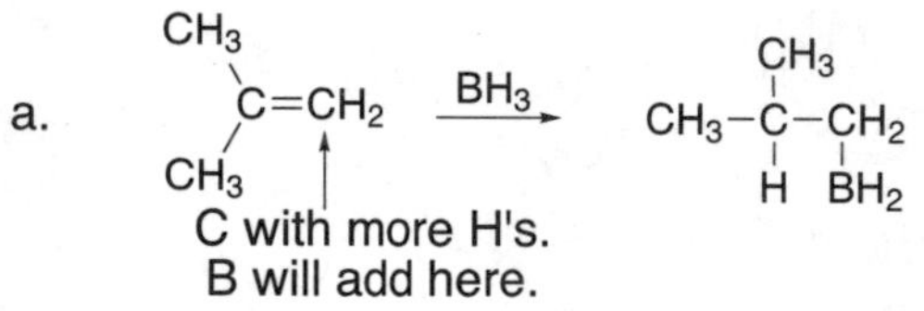
a.
CH3
C=CH2
CH3
BH3
C with more H's.
B will add here.
CH3
CH3—C—CH2
H BH2

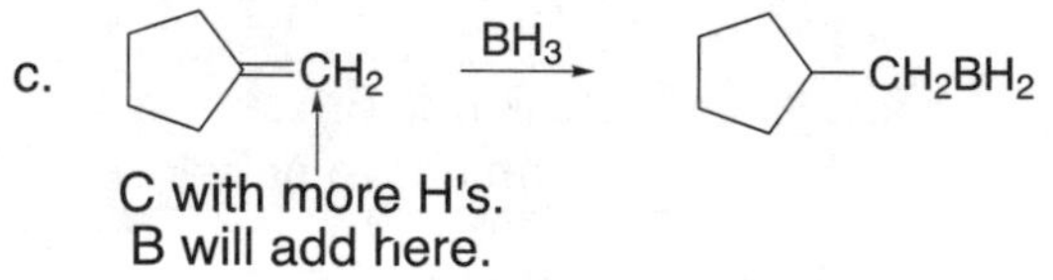
c.
=CH2
BH3
—CH2BH2
C with more H's.
B will add here.

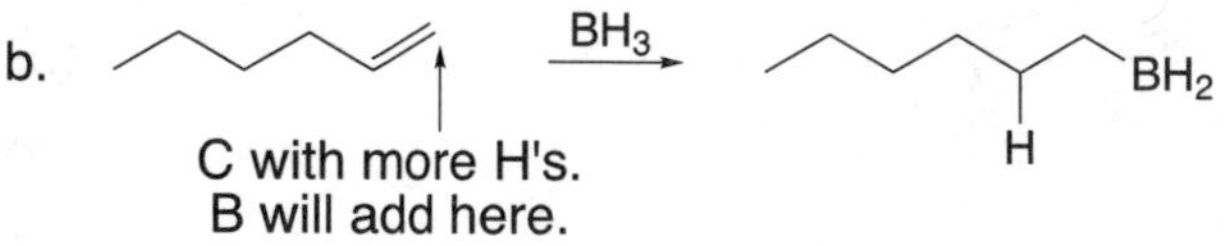
b.
BH3
BH2
H
C with more H's.
B will add here.

10.29 The hydroboration–oxidation reaction occurs in two steps:
[1] Syn addition of BH_3, with the borane on the less substituted carbon atom.
[2] OH replaces the BH_2 with retention of configuration.

a. $CH_3CH_2CH=CH_2$ $\xrightarrow{BH_3}$ $CH_3CH_2CH-CH_2$ $\xrightarrow{H_2O_2/^-OH}$ $CH_3CH_2CH-CH_2$
$\quad\quad\quad\quad\quad\quad\quad\quad\quad\quad\quad\quad\quad\quad\quad$ $\overset{|}{H}\quad\overset{|}{BH_2}$ $\quad\quad\quad\quad\quad\quad\quad\quad$ $\overset{|}{H}\quad\overset{|}{OH}$

b.

c.

10.30

a. Hydration places the OH on the more substituted carbon.

b. Hydroboration–oxidation places the OH on the less substituted carbon.

c. Hydration places the OH on the more substituted carbon.

d. Hydroboration–oxidation places the OH on the less substituted carbon.

10.31 There are always two steps in this kind of question:
[1] **Identify the functional group and decide what types of reactions it undergoes** (for example, substitution, elimination, or addition).
[2] **Look at the reagent and determine if it is an electrophile, nucleophile, acid, or base.**

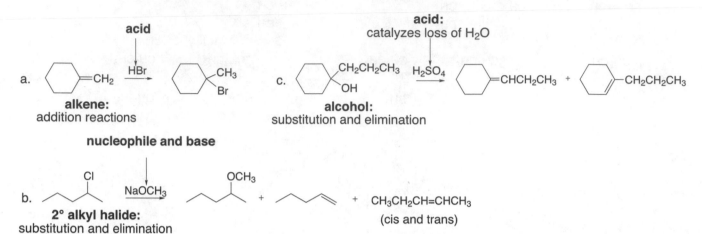

10.32 To devise a synthesis:
[1] Look at the starting material and decide what reactions it can undergo.
[2] Look at the product and decide what reactions could make it.

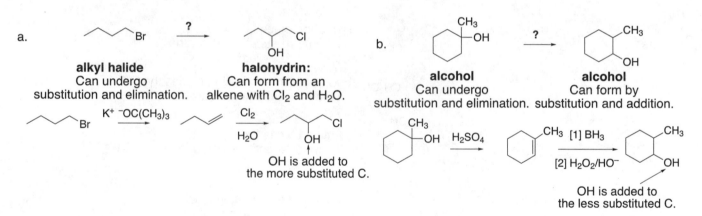

10.33 Name the alkenes using the rules in Answer 10.4 and 10.5.

a. $CH_2=CHCH_2CH(CH_3)CH_2CH_3$

    6 C chain with a double bond =
        hexene

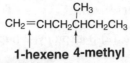

$CH_2=CHCH_2CHCH_2CH_3$ with CH_3 branch

1-hexene   **4-methyl**

4-methyl-1-hexene

b.

    8 C chain with a double bond =
        octene

2-methyl →

5-ethyl

2-octene

5-ethyl-2-methyl-2-octene

c.

    5 C chain with a double bond =
        pentene

2-isopropyl

← **1-pentene**

4-methyl

2-isopropyl-4-methyl-1-pentene

d.

$CH_3 \quad CH_3$
    $C=C$
$H \quad\quad CH_2CH(CH_3)_2$

    6 C chain with a double bond =
        hexene

$CH_3 \quad\quad CH_3$ ← **3-methyl**
   $C=C$ CH_3 ← **5-methyl**
$H \quad\quad CH_2CHCH_3$

2-hexene

higher priority → $CH_3 \quad\quad CH_3$ **(E)-3,5-dimethyl-2-hexene**
        $C=C$
$H \quad\quad\quad CH_2CH(CH_3)_2$ ← higher priority

Higher priority groups
are on opposite sides =
_E_ alkene.

e.

    6 C ring with a double bond =
        cyclohexene

← **1-ethyl**

5-isopropyl

1-ethyl-5-isopropylcyclohexene

f.

    5 C ring with a double bond =
        cyclopentene

2-methyl →

← **1-_sec_-butyl**

1-_sec_-butyl-2-methylcyclopentene

10.34 Use the directions from Answer 10.7.

a. (*E*)-4-ethyl-3-**heptene**

      7 carbons

⟵ **4-ethyl**

3

Higher priority groups on
opposite sides = *E*.

b. 3,3-dimethyl**cyclopentene**

      5 carbon ring

⟵ **3,3-dimethyl**

⟵ **1**

c. *cis*-4-**octene**

   8 carbons

4

Higher priority groups on
the same side = cis.

d. 4-vinyl**cyclopentene**

     5 carbon ring

2 **1**

4⟶

e. (*Z*)-3-isopropyl-2-**heptene**

       7 carbons

2

⟵ **3-isopropyl**

Higher priority groups on
the same side = *Z*.

f. *cis*-3,4-dimethyl**cyclopentene**

      5 carbon ring

1

4⟶   ⟵**3**

3,4-dimethyl

g. *trans*-2-**heptene**

      7 carbons

2   Higher priority groups on
opposite sides = trans.

4-propyl

h. 1-isopropyl-4-propyl**cyclohexene**

      6 carbon ring

⟵**1-isopropyl**

10.35

a.   2-butyl-3-methyl-1-pentene

As written, this is the parent chain,
but there is another longer chain
containing the double bond.

⟵ **2-*sec*-butyl**

new name:
2-*sec*-butyl-1-hexene

b.   (*Z*)-2-methyl-2-hexene

Two groups on one end of the C=C
are the same (2 CH₃'s), so no *E* and *Z* isomers are possible.

new name:
2-methyl-2-hexene

c.   (*E*)-1-isopropyl-1-butene

As written, this is the parent chain,
but there is another longer chain
containing the double bond.

1

new name:
(*E*)-2-methyl-3-hexene

d. 5-methylcyclohexene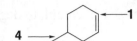

As written the methyl
is at C5. Renumber
to put it at C4.

new name:
4-methylcyclohexene

e. 4-isobutyl-2-methylcylohexene

As written this methyl
is at C2. Renumber
to put it at C1.

new name:
5-isobutyl-1-methylcyclohexene

f. 1-*sec*-butyl-2-cyclopentene

This has the double bond between
C2 and C3. Cycloalkenes must
have the double bond between
C1 and C2. Renumber.

new name:
3-*sec*-butylcyclopentene

10.36

a and b.

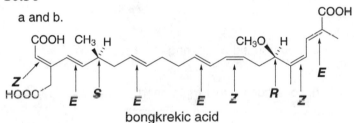

bongkrekic acid

c. Since there are 7 double bonds
and 2 tetrahedral stereogenic
centers, $2^9 = 512$ possible stereoisomers.

10.37

2-methyl-1-pentene   (*E*)-4-methyl-2-pentene   4-methyl-1-pentene   (*E*)-3-methyl-2-pentene   (*R*)-3-methyl-1-pentene

2-methyl-2-pentene   (*Z*)-4-methyl-2-pentene   2-ethyl-1-butene   (*Z*)-3-methyl-2-pentene   (*S*)-3-methyl-1-pentene

10.38 Use the directions from Answer 10.2 to calculate degrees of unsaturation.

a. C_3H_4
  [1] maximum number of H's = $2n + 2 = 2(3) + 2 = 8$
  [2] subtract actual from maximum – $8 – 4 = 4$
  [3] divide by 2 = $4/2$ = **2 degrees of unsaturation**

b. C_6H_8
  [1] maximum number of H's = $2n + 2 = 2(6) + 2 = 14$
  [2] subtract actual from maximum = $14 – 8 = 6$
  [3] divide by 2 = $6/2$ = **3 degrees of unsaturation**

c. $C_{40}H_{56}$
  [1] maximum number of H's = $2n + 2 = 2(40) + 2 = 82$
  [2] subtract actual from maximum = $82 – 56 = 26$
  [3] divide by 2 = $26/2$ = **13 degrees of unsaturation**

d. C_8H_8O
  Ignore the O
  [1] maximum number of H's = $2n + 2 = 2(8) + 2 = 18$
  [2] subtract actual from maximum = $18 – 8 = 10$
  [3] divide by 2 = $10/2$ = **5 degrees of unsaturation**

e. $C_{10}H_{16}O_2$
  Ignore both O's
  [1] maximum number of H's = $2n + 2 = 2(10) + 2 = 22$
  [2] subtract actual from maximum = $22 – 16 = 6$
  [3] divide by 2 = $6/2$ = **3 degrees of unsaturation**

f. C_8H_9Br
  Because of Br, add one H ($9 + 1 = 10$ H's)
  [1] maximum number of H's = $2n + 2 = 2(8) + 2 = 18$
  [2] subtract actual from maximum = $18 – 10 = 8$
  [3] divide by 2 = $8/2$ = **4 degrees of unsaturation**

g. C_8H_9ClO
  Ignore the O, count Cl as one more H ($9 + 1 = 10$ H's)
  [1] maximum number of H's = $2n + 2 = 2(8) + 2 = 18$
  [2] subtract actual from maximum = $18 – 10 = 8$
  [3] divide by 2 = $8/2$ = **4 degrees of unsaturation**

h. C_7H_9Br
  Because of Br, add one H ($9 + 1 = 10$ H's)
  [1] maximum number of H's = $2n + 2 = 2(7) + 2 = 16$
  [2] subtract actual from maximum = $16 – 10 = 6$
  [3] divide by 2 = $6/2$ = **3 degrees of unsaturation**

10.39

One possibility for C_6H_{10}:

a. a compound that has 2 π bonds

c. a compound with 2 rings

b. a compound that has 1 ring and 1 π bond

d. compound with 1 triple bond

10.40

steric acid — **highest melting point** — no double bonds

elaidic acid — **intermediate melting point** — one *E* double bond

oleic acid — **lowest melting point** — one *Z* double bond

10.41

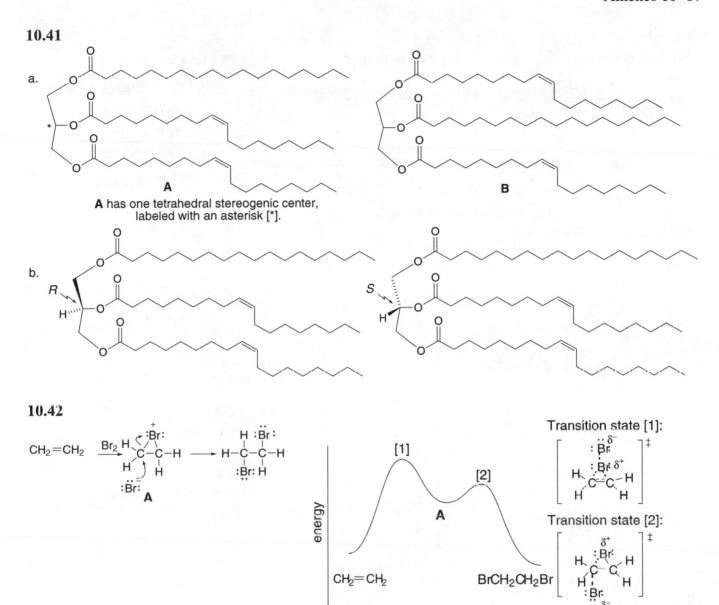

A has one tetrahedral stereogenic center, labeled with an asterisk [*].

10.42

10.43 The more negative the $\Delta H°$, the larger the K_{eq} assuming entropy changes are comparable. Calculate the $\Delta H°$ for each reaction and compare.

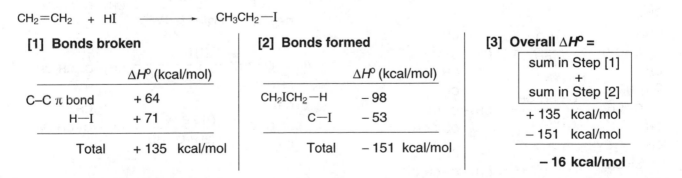

$$CH_2=CH_2 + HI \longrightarrow CH_3CH_2-I$$

[1] Bonds broken

	$\Delta H°$ (kcal/mol)
C–C π bond	+ 64
H–I	+ 71
Total	+ 135 kcal/mol

[2] Bonds formed

	$\Delta H°$ (kcal/mol)
CH₂ICH₂–H	– 98
C–I	– 53
Total	– 151 kcal/mol

[3] Overall $\Delta H°$ =

sum in Step [1] + sum in Step [2]
+ 135 kcal/mol
– 151 kcal/mol
– 16 kcal/mol

$CH_2{=}CH_2$ + HCl $\longrightarrow$ $CH_3CH_2{-}Cl$

[1] **Bonds broken**		[2] **Bonds formed**		[3] **Overall** $\Delta H° =$

	$\Delta H°$ (kcal/mol)		$\Delta H°$ (kcal/mol)	sum in Step [1] + sum in Step [2]
C–C π bond	+ 64	$CH_2ClCH_2{-}H$	– 98	+ 167 kcal/mol
H–Cl	+ 103	C–Cl	– 81	– 179 kcal/mol
Total	+ 167 kcal/mol	Total	– 179 kcal/mol	– 12 kcal/mol

Compare the $\Delta H°$s :
Addition of HI: **–16 kcal/mol** more negative $\Delta H°$, larger K_{eq}.
Addition of HCl: **–12 kcal/mol**

10.44

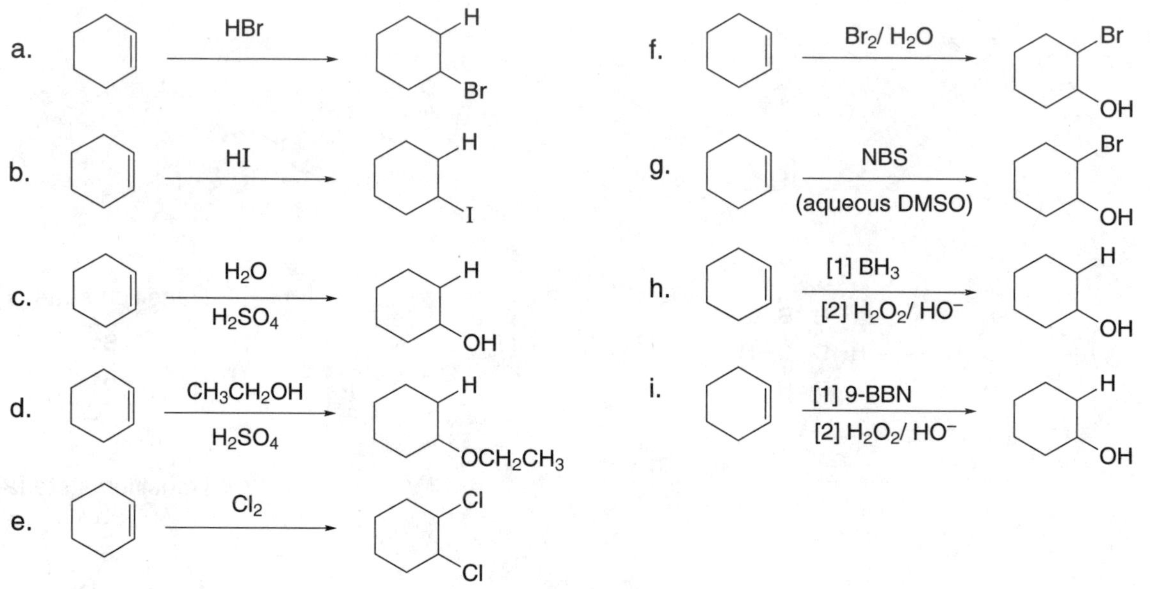

a. HBr

b. HI

c. H_2O / H_2SO_4

d. CH_3CH_2OH / H_2SO_4

e. Cl_2

f. Br_2/ H_2O

g. NBS (aqueous DMSO)

h. [1] BH_3 [2] H_2O_2/ HO^-

i. [1] 9-BBN [2] H_2O_2/ HO^-

10.45

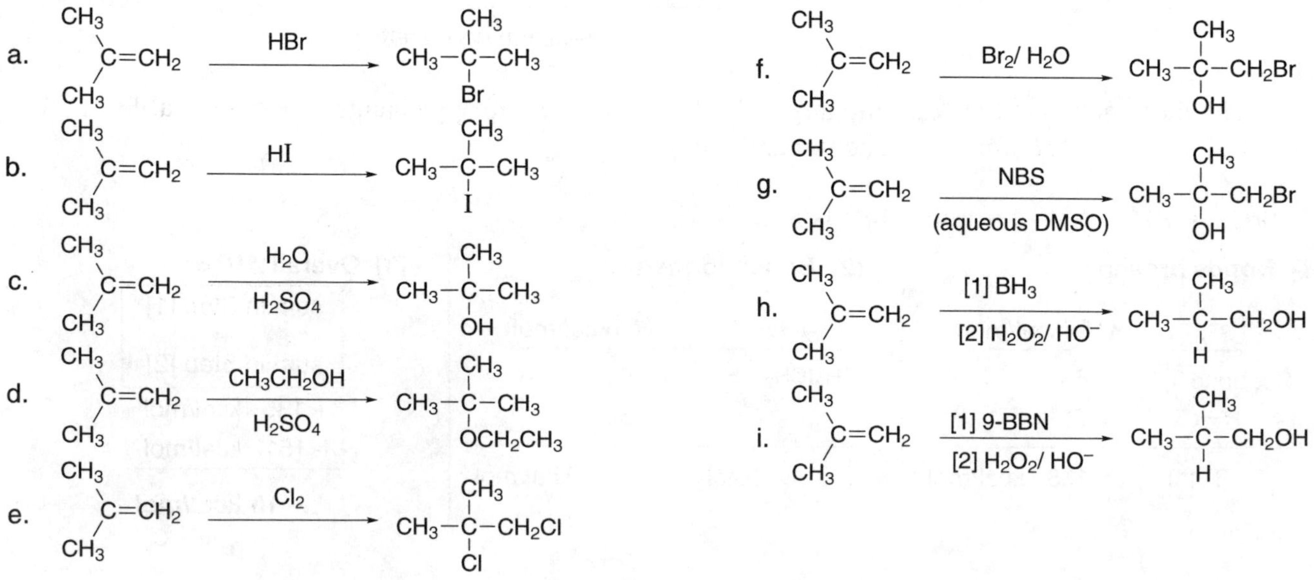

a. HBr

b. HI

c. H_2O / H_2SO_4

d. CH_3CH_2OH / H_2SO_4

e. Cl_2

f. Br_2/ H_2O

g. NBS (aqueous DMSO)

h. [1] BH_3 [2] H_2O_2/ HO^-

i. [1] 9-BBN [2] H_2O_2/ HO^-

10.46

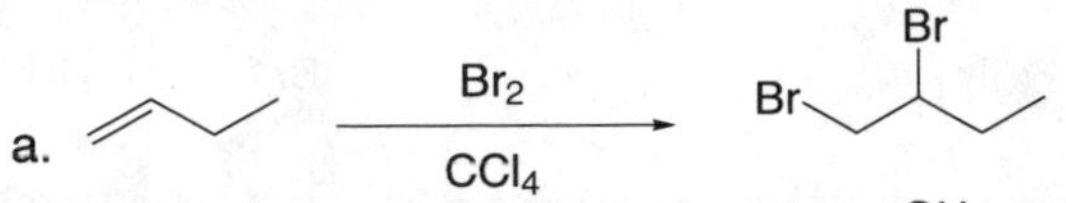
a.
Br2
CCl4
Br
Br

Halogenation

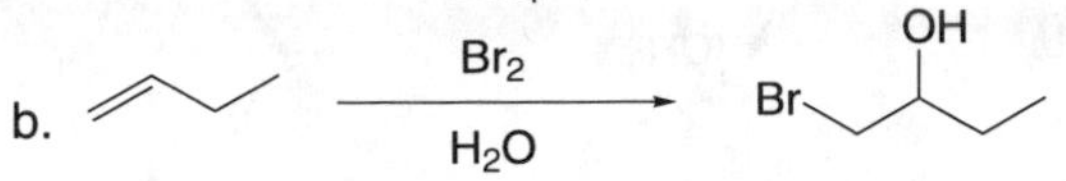
b.
Br2
H2O
OH
Br

Halohydrin formation: Br adds to the C that had more H's to begin with.

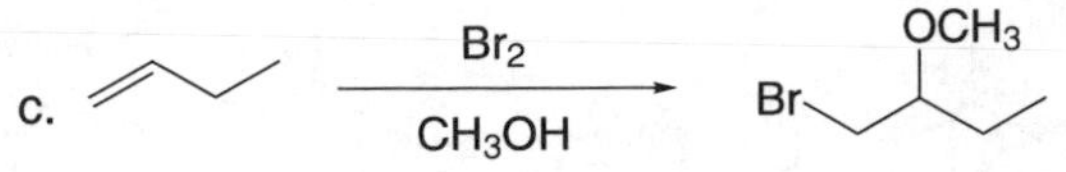
c.
Br2
CH3OH
OCH3
Br

Same as halohydrin formation, except CH_3OH in place of H_2O.

10.47

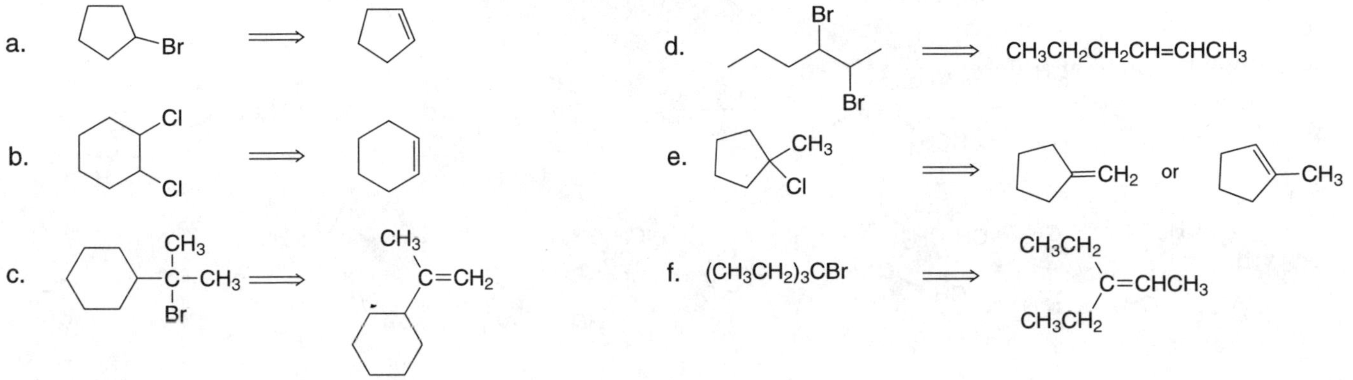
a.
Br
b.
Cl
Cl
c.
CH3
C–CH3
Br
CH3
C=CH2
d.
Br
Br
CH3CH2CH2CH=CHCH3
e.
CH3
Cl
=CH2
or
–CH3
f. (CH3CH2)3CBr
CH3CH2
C=CHCH3
CH3CH2

10.48

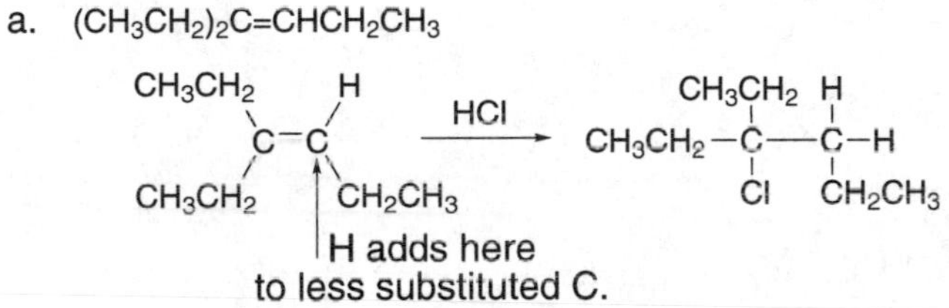
a. (CH3CH2)2C=CHCH2CH3
CH3CH2
H
C=C
CH3CH2
CH2CH3
HCl
CH3CH2 H
CH3CH2–C—C–H
Cl CH2CH3
H adds here
to less substituted C.

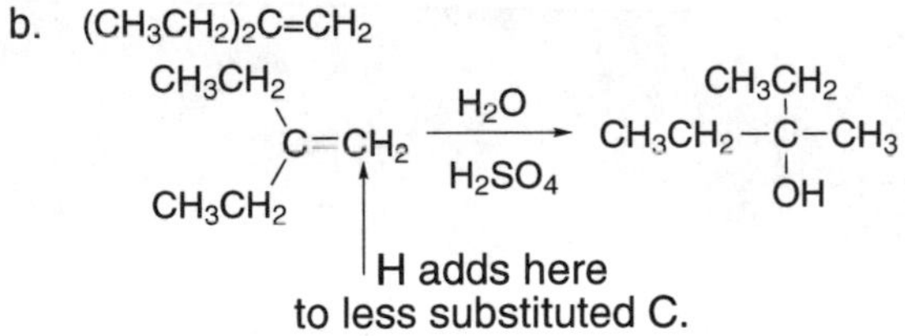
b. (CH3CH2)2C=CH2
CH3CH2
C=CH2
CH3CH2
H2O
H2SO4
CH3CH2
CH3CH2–C–CH3
OH
H adds here
to less substituted C.

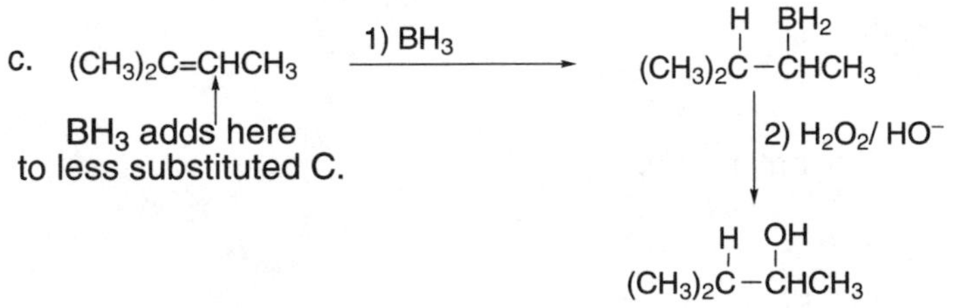
c. (CH3)2C=CHCH3
1) BH3
H BH2
(CH3)2C–CHCH3
2) H2O2/ HO–
H OH
(CH3)2C–CHCH3
BH3 adds here
to less substituted C.

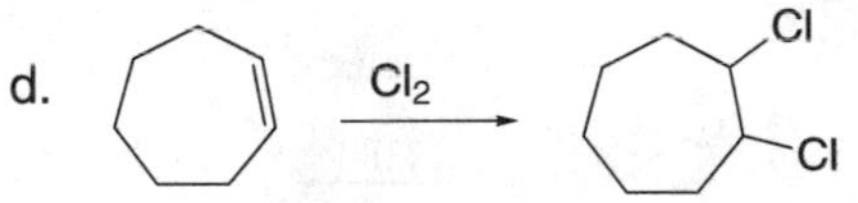
d.
Cl2
Cl
Cl

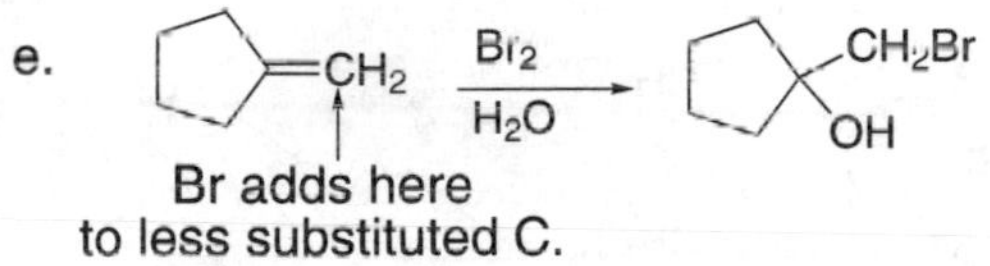
e.
=CH2
Br2
H2O
CH2Br
OH
Br adds here
to less substituted C.

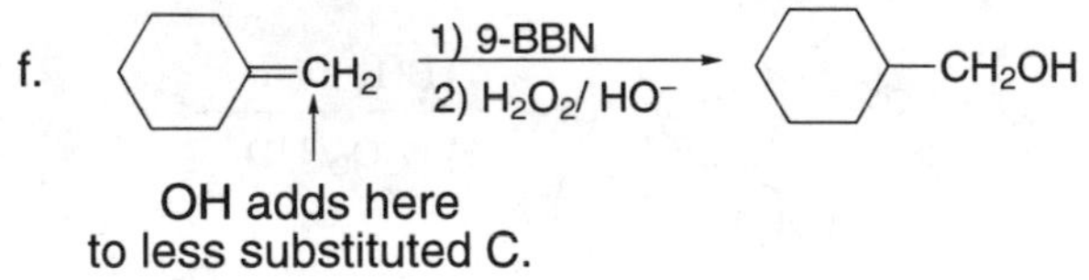
f.
=CH2
1) 9-BBN
2) H2O2/ HO–
–CH2OH
OH adds here
to less substituted C.

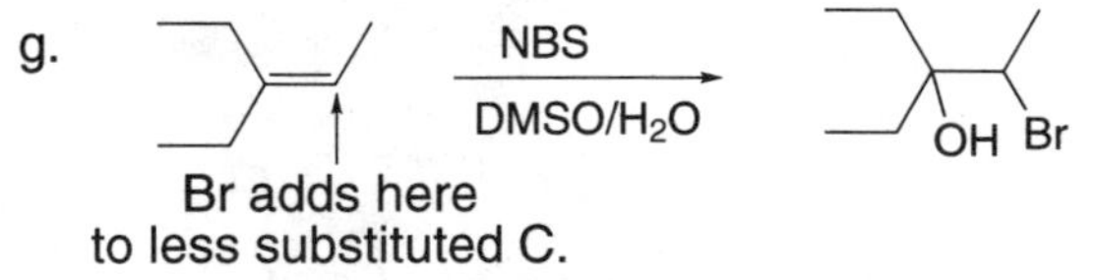
g.
NBS
DMSO/H2O
OH Br
Br adds here
to less substituted C.

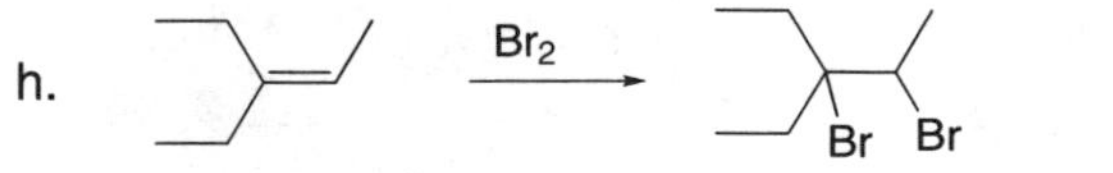
h.
Br2
Br Br

10.49

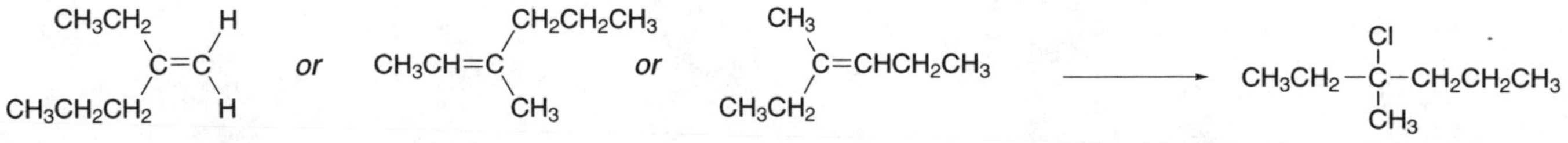
CH3CH2 H
C=C
CH3CH2CH2 H
or
CH2CH2CH3
CH3CH=C
CH3
or
CH3
C=CHCH2CH3
CH3CH2
Cl
CH3CH2–C–CH2CH2CH3
CH3

10.50

a.

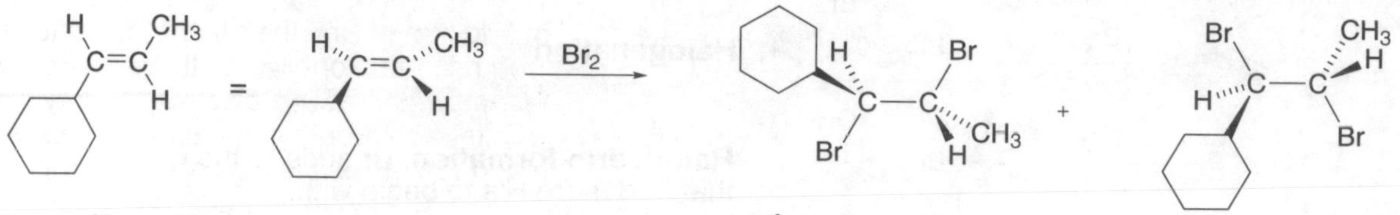

b.

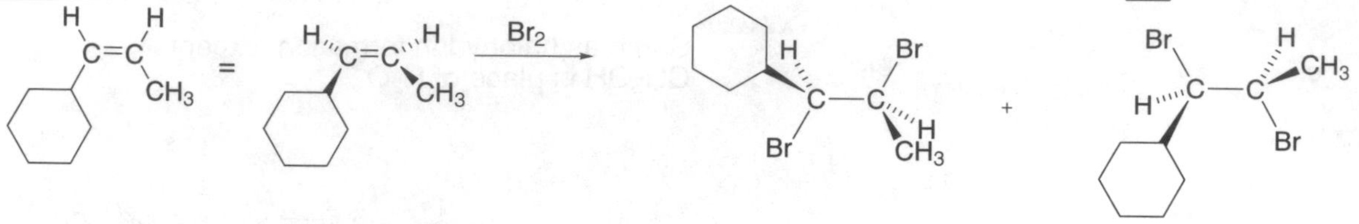

c.

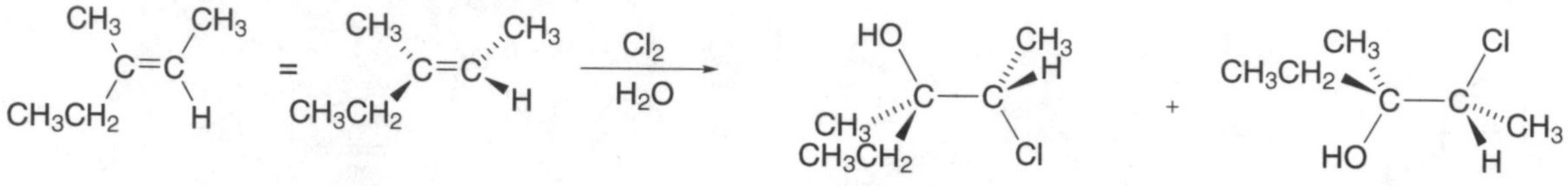

d.

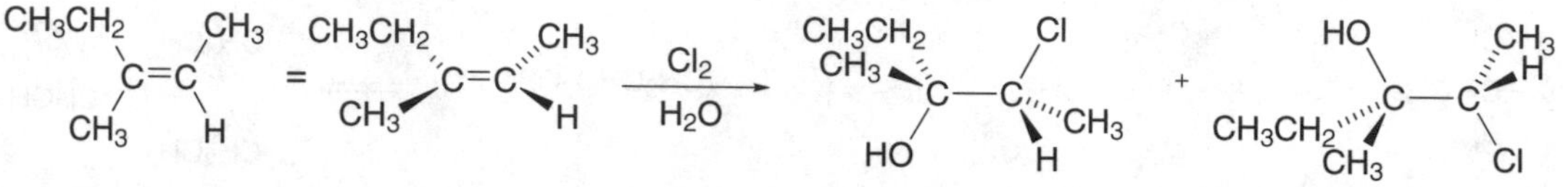

10.51

a. 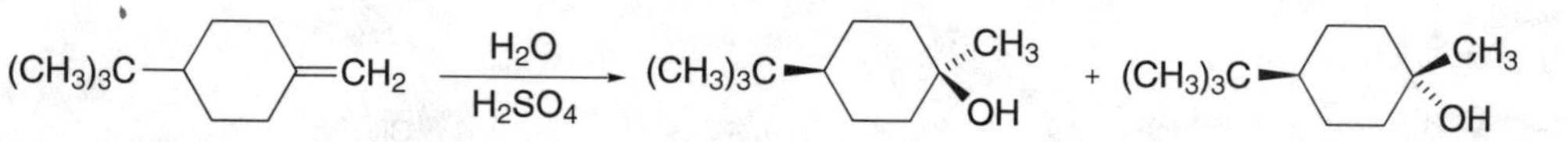 $(CH_3)_3C$—〈ring〉=CH_2 $\xrightarrow[H_2SO_4]{H_2O}$ (structures) + (structures)

b. 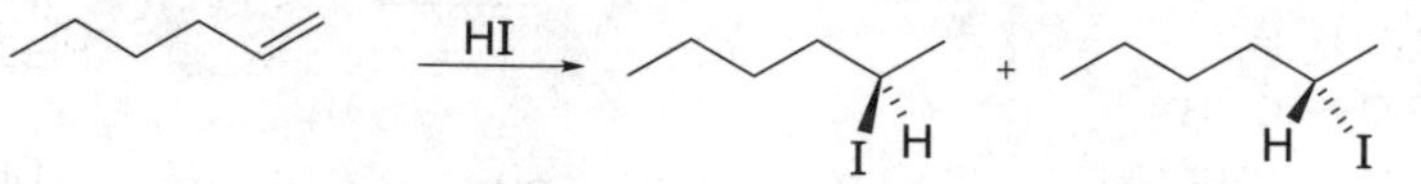 $\xrightarrow{HI}$ (structures) + (structures)

c. 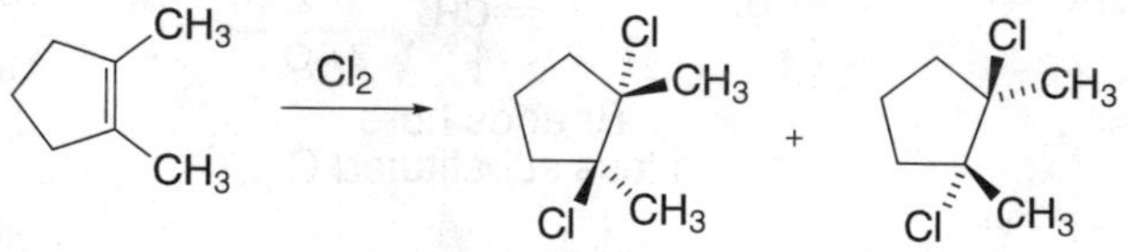 $\xrightarrow{Cl_2}$ (structures) + (structures)   Only anti addition.

d. 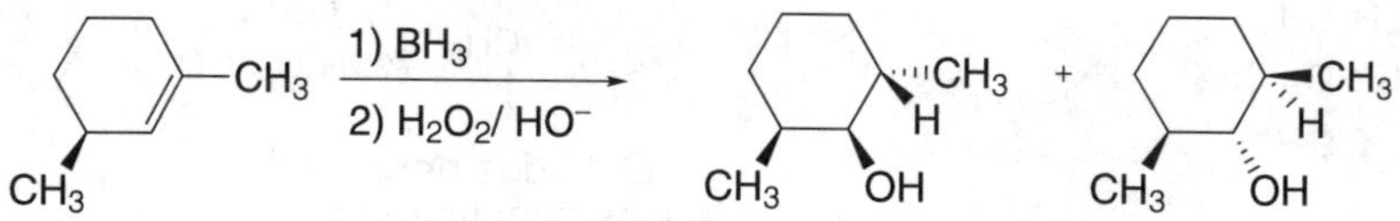 $\xrightarrow[\text{2) } H_2O_2/\ HO^-]{\text{1) } BH_3}$ (structures) + (structures)

e. 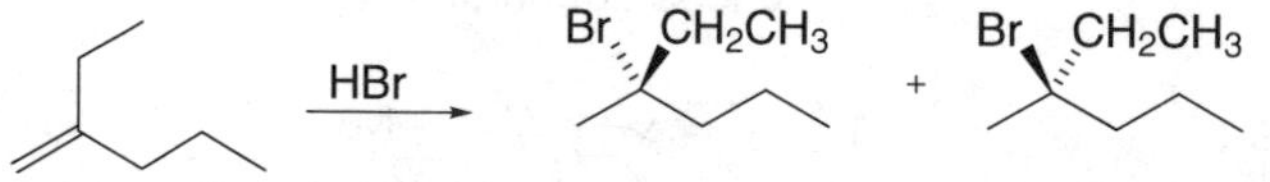 $\xrightarrow{HBr}$ (structures) + (structures)

f. 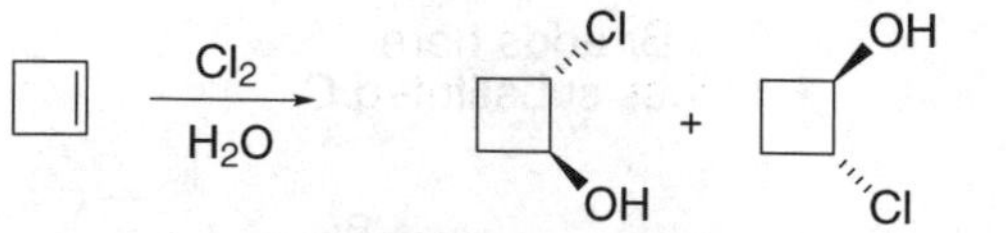 $\xrightarrow[H_2O]{Cl_2}$ (structures) + (structures)   Only anti addition.

g. 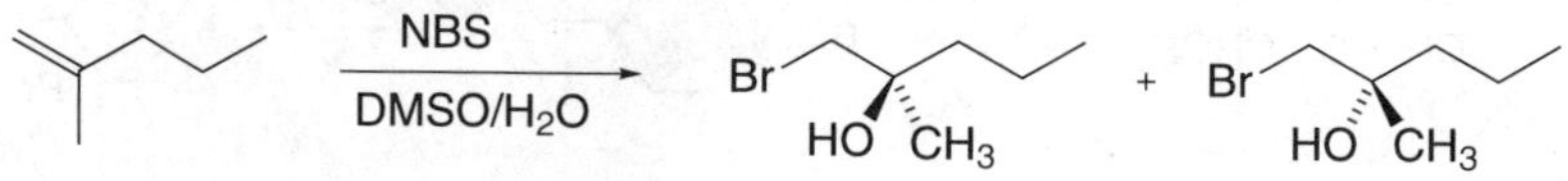 $\xrightarrow[DMSO/H_2O]{NBS}$ (structures) + (structures)

h. $CH_3CH{=}CHCH_2CH_3$ $\xrightarrow[H_2SO_4]{H_2O}$ (structures) + (structures) + (structures)

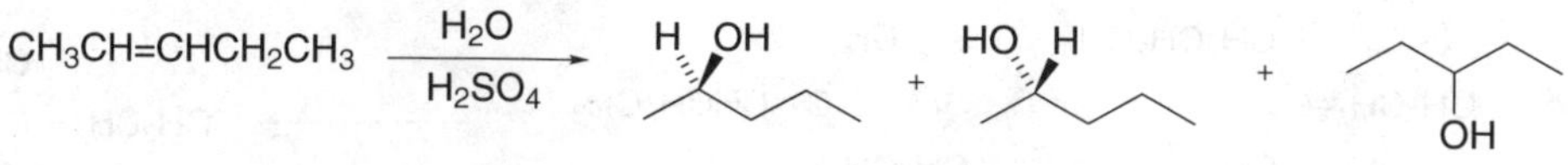

10.52

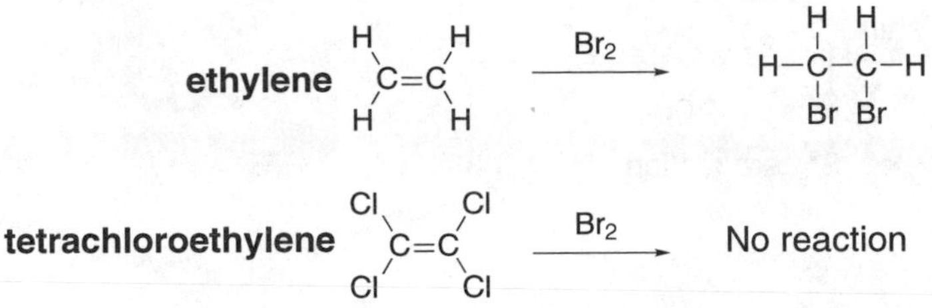

ethylene

tetrachloroethylene

In addition of Br_2, the alkene must act as an electron rich nucleophile. With four electronegative Cl atoms withdrawing electron density from the C=C, tetrachloroethylene is much less electron rich, and therefore it is a poorer nucleophile. This makes it less reactive with Br_2.

10.53

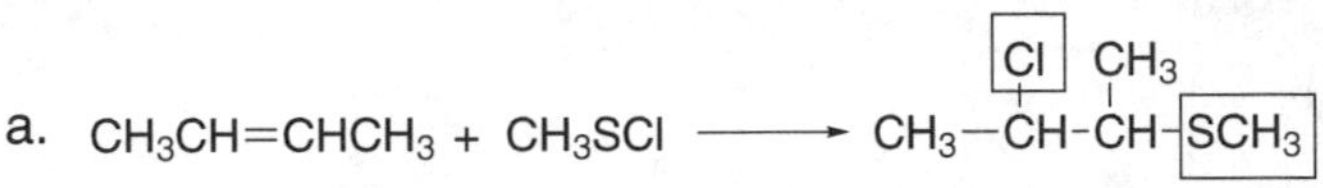

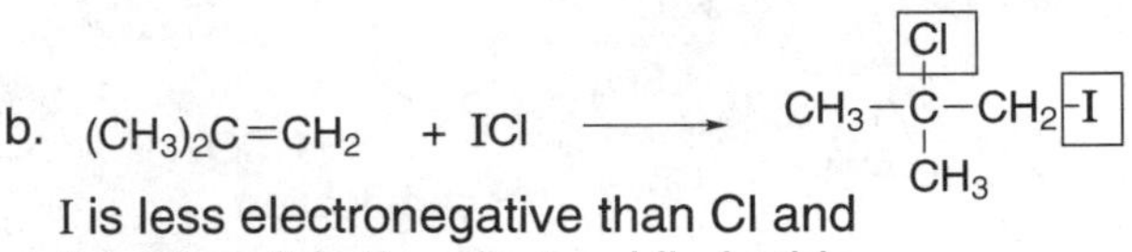

a. $CH_3CH=CHCH_3 + CH_3SCl \longrightarrow CH_3-CH-CH-SCH_3$ (with Cl and CH_3)

b. $(CH_3)_2C=CH_2 + ICl \longrightarrow CH_3-C-CH_2-I$ (with Cl and CH_3)

I is less electronegative than Cl and therefore it is the electrophile in this reaction. It ends up on the less substituted C.

10.54

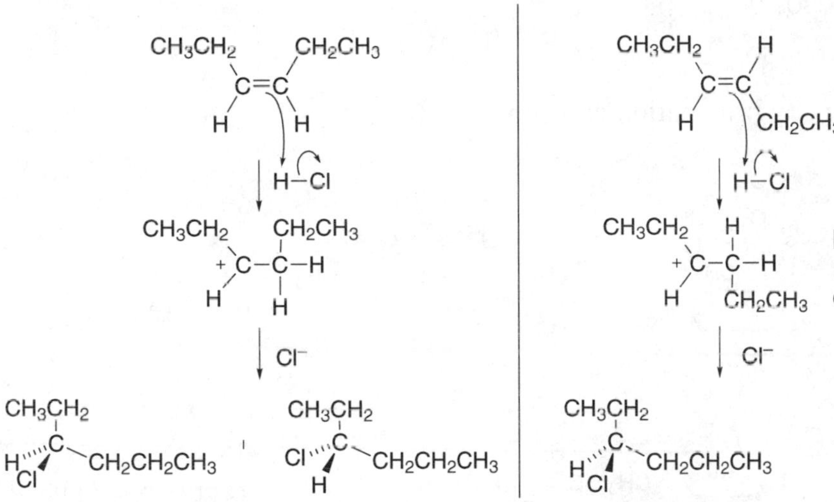

By protonation of the alkene, the *cis* and *trans* isomers produce **identical** carbocation intermediates.

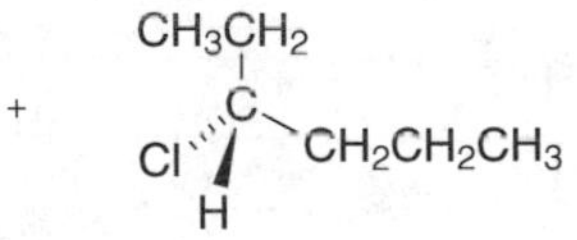

Both *cis-* and *trans*-3-hexene give the same racemic mixture of products, so the reaction is not stereospecific.

10.55

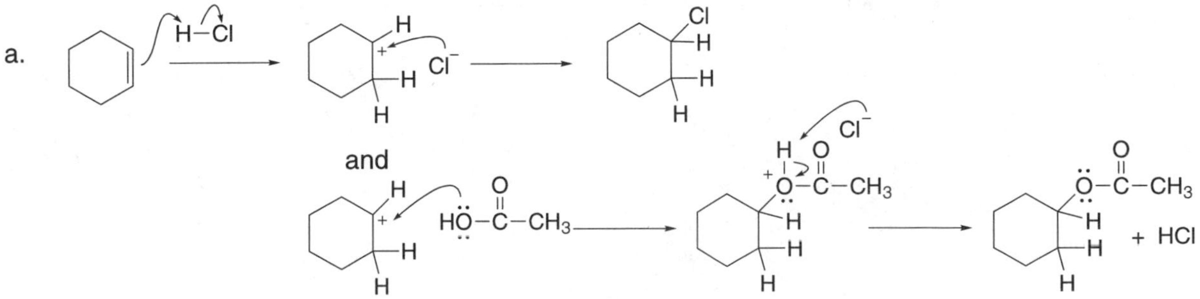

a.

and

b. 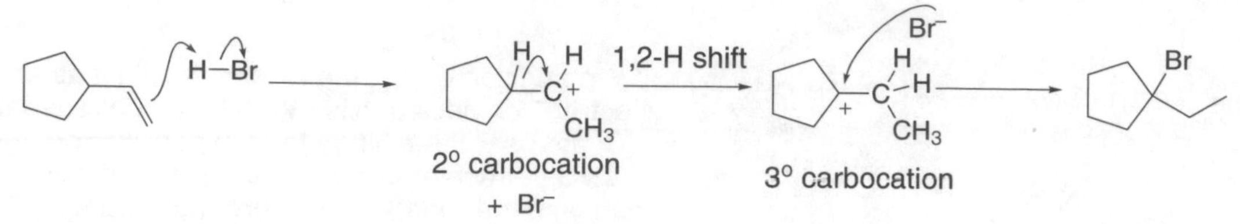
H—Br
H H
1,2-H shift
C+
CH3
2° carbocation
+ Br⁻
Br⁻
H
C—H
CH3
3° carbocation
Br

10.56

a. 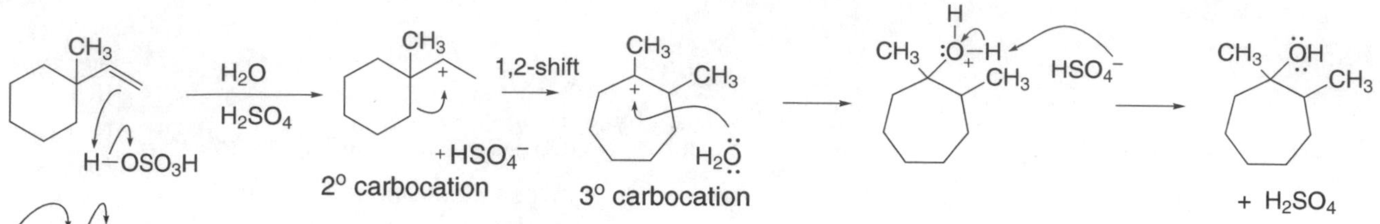
CH3
H—OSO3H
H2O
H2SO4
CH3
2° carbocation
+HSO4⁻
1,2-shift
CH3
CH3
3° carbocation
H2O
H
CH3 :O—H
CH3
HSO4⁻
CH3 OH
CH3
+ H2SO4

b. 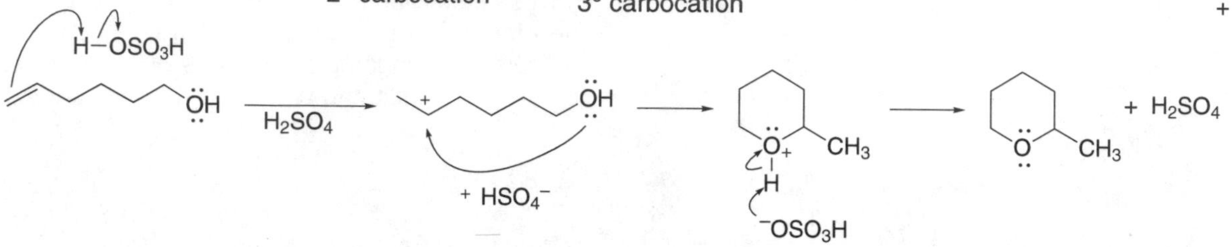
H—OSO3H
OH
H2SO4
OH
+ HSO4⁻
O+
CH3
H
⁻OSO3H
O
CH3
+ H2SO4

10.57 The isomerization reaction occurs by protonation and deprotonation.

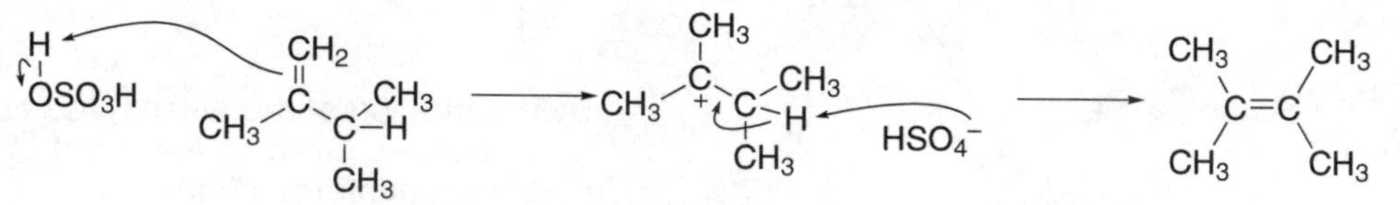
H
OSO3H
CH2
C
CH3
CH3
C—H
CH3
CH3
C
CH3
CH3
C—H
CH3
HSO4⁻
CH3 CH3
C=C
CH3 CH3

10.58

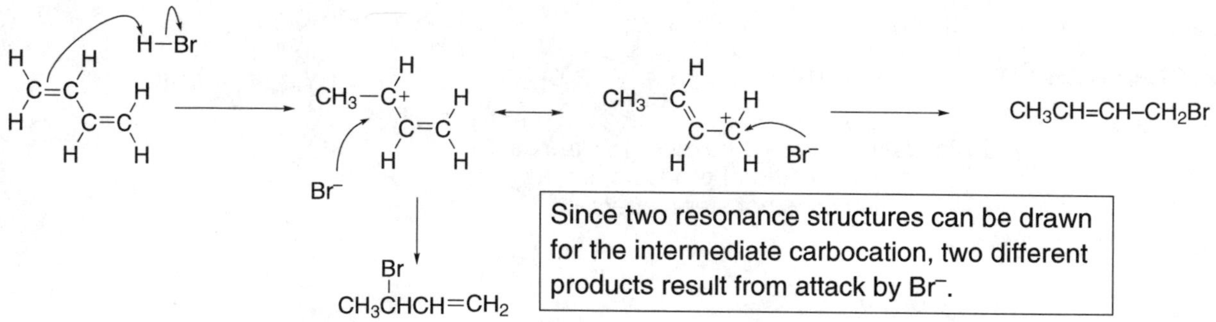
H—Br
H H
C=C
H H
C=C
H H
CH3—C+ H
C=C
H H
Br⁻
Br
CH3CHCH=CH2
CH3—C H
C—C
H H
Br⁻
CH3CH=CH–CH2Br
Since two resonance structures can be drawn for the intermediate carbocation, two different products result from attack by Br⁻.

10.59

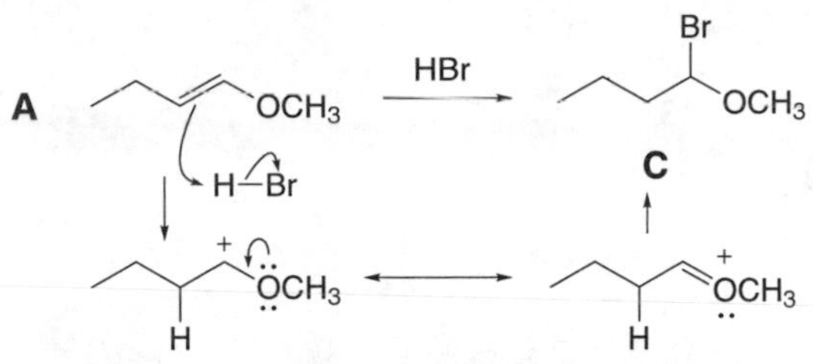

This carbocation is resonance stabilized by the O atom, and therefore preferentially forms and results in **C**.

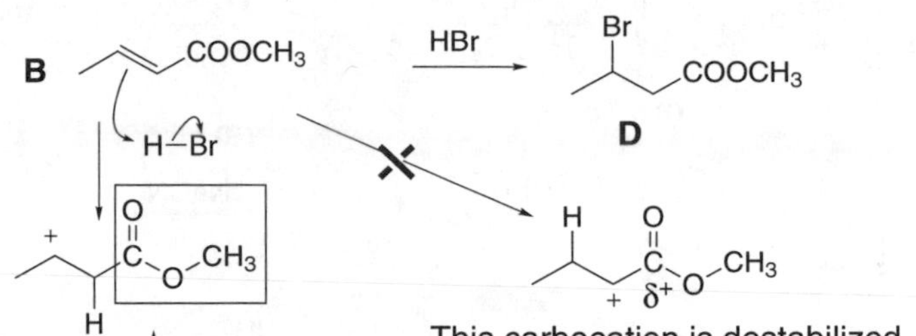

This carbocation is formed preferentially and results in product **D**. It is not destabilized by an adjacent electron withdrawing COOCH$_3$ group.

This carbocation is destabilized by the δ$^+$ on the adjacent C, so it does not form.

10.60

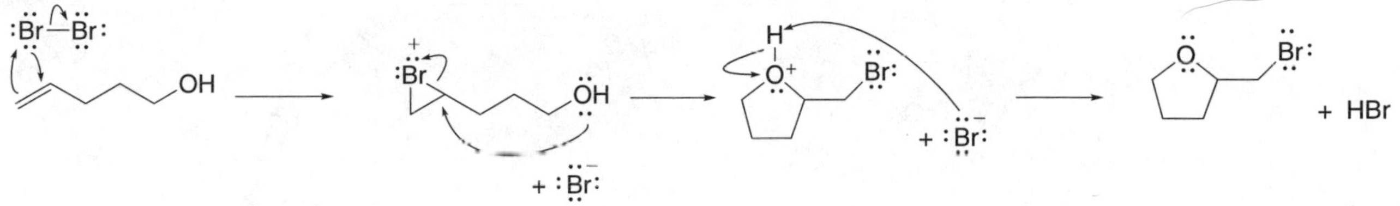

+ HBr

10.61

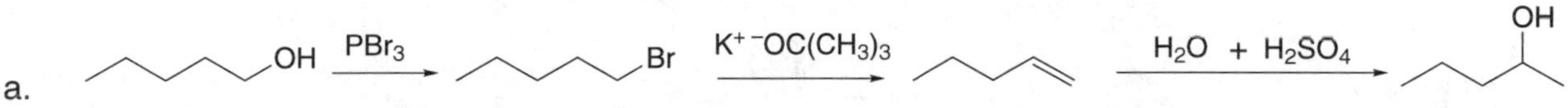

a.

OH adds to **more** substituted C.

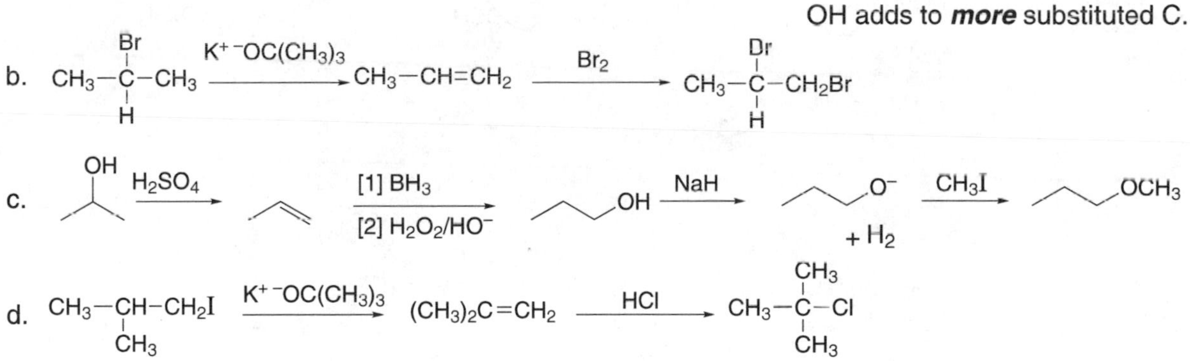

b.

c.

+ H$_2$

d.

10.62

a.

b.

c.

+ H$_2$

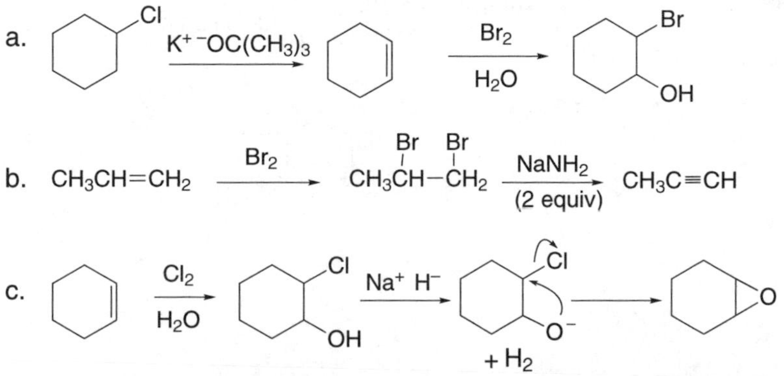

d.

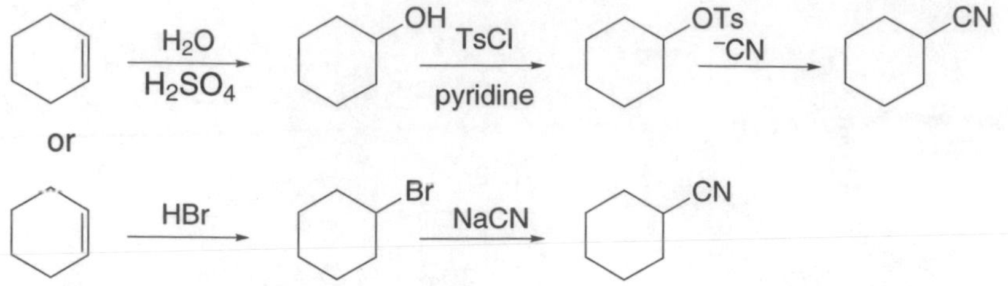

or

e.

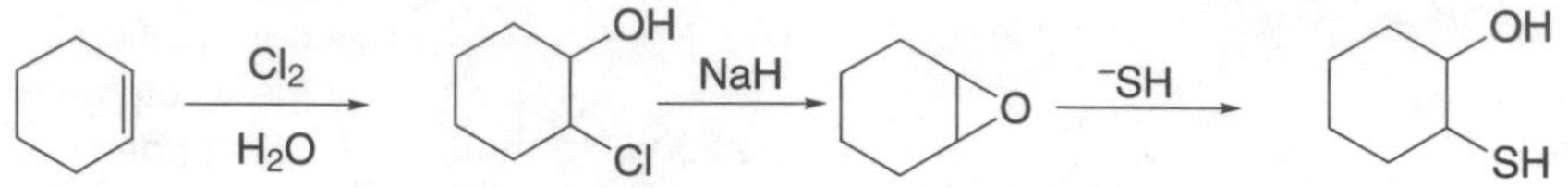

10.63

A + TsO⁻ isocomene

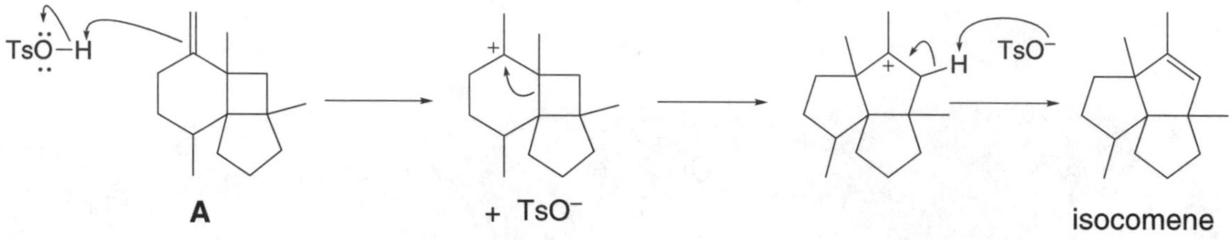

10.64

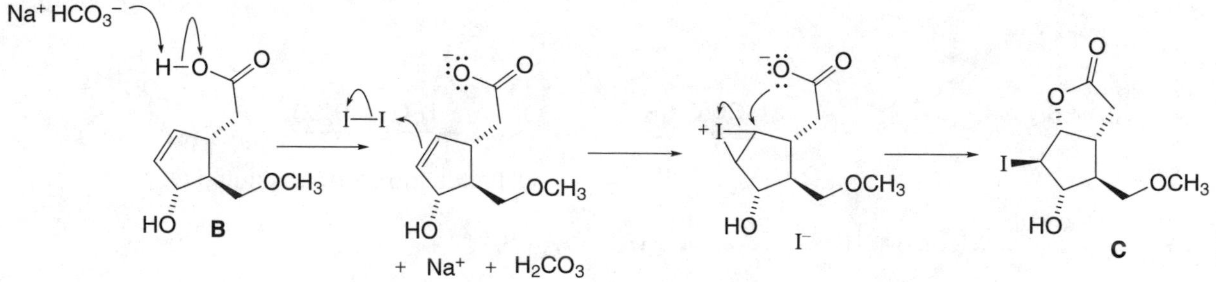

10.65

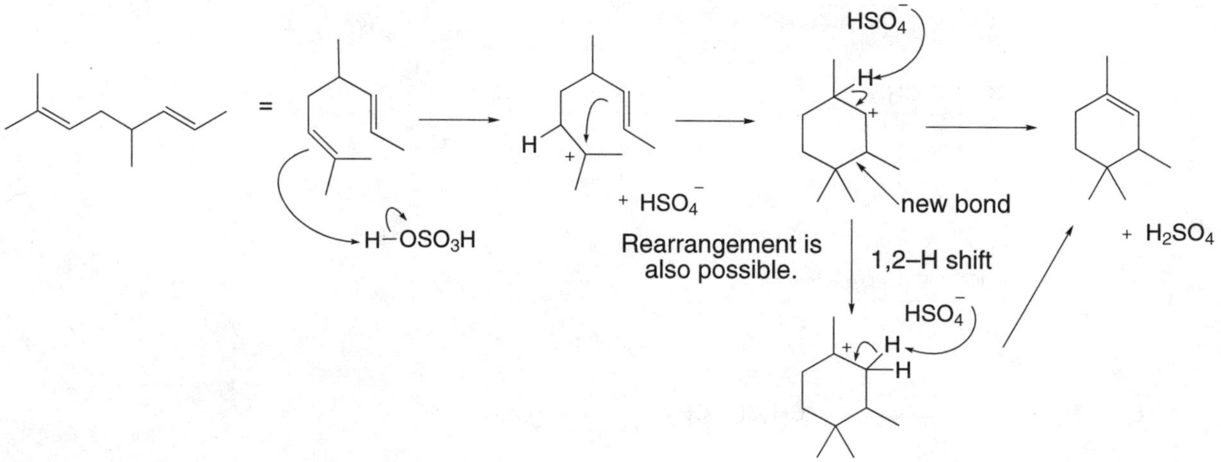

Chapter 11: Alkynes

◆ General facts about alkynes

- Alkynes contain a carbon–carbon triple bond consisting of a strong σ bond and two weak π bonds. Each carbon is *sp* hybridized and linear (11.1).

$$H-C\equiv C-H$$
acetylene

180°

sp hybridized

- Alkynes are named using the suffix *–yne* (11.2).
- Alkynes have weak intermolecular forces, giving them low mp's and low bp's, and making them water insoluble (11.3).
- Since its weaker π bonds make an alkyne electron rich, alkynes undergo addition reactions with electrophiles (11.6).

◆ Addition reactions of alkynes

[1] Hydrohalogenation – Addition of HX (X = Cl, Br, I) (11.7)

R–C≡C–H $\xrightarrow[\text{(2 equiv)}]{\text{H—X}}$

X H
R–C–C–H
X H
geminal dihalide

- Markovnikov's rule is followed. H bonds to the less substituted C in order to form the more stable carbocation.

[2] Halogenation (Addition of X$_2$, X = Cl or Br) (11.8)

R–C≡C–H $\xrightarrow[\text{(2 equiv)}]{\text{X—X}}$

X X
R–C–C–H
X X
tetrahalide

- Bridged halonium ions are formed as intermediates.
- Anti addition of X$_2$ occurs.

[3] Hydration (Addition of H$_2$O) (11.9)

R–C≡C–H $\xrightarrow[\substack{H_2SO_4 \\ HgSO_4}]{H_2O}$

$$\left[\begin{array}{c} R \quad\quad H \\ C=C \\ HO \quad\quad H \end{array} \right]$$
enol

$\longleftarrow$

O
‖
R–C–CH$_3$
ketone

- Markovnikov's rule is followed. H bonds to the less substituted C in order to form the more stable carbocation.
- An unstable enol is first formed, which rearranges to a carbonyl group.

[4] Hydroboration–oxidation (Addition of H_2O) (11.10)

- The unstable enol, first formed after oxidation, rearranges to a carbonyl group.

♦ Reactions involving acetylide anions

[1] Formation of acetylide anions from terminal alkynes (11.6B)

- Typical bases used for the reaction are $NaNH_2$ and NaH.

[2] Reaction of acetylide anions with alkyl halides (11.11A)

- The reaction follows an S_N2 mechanism.
- The reaction works best with CH_3X and RCH_2X.

[3] Reaction of acetylide anions with epoxides (11.11B)

- The reaction follows an S_N2 mechanism.
- Ring opening occurs from the back side at the less substituted end of the epoxide.

Chapter 11: Answers to Problems

11.1 • An *internal alkyne* has the triple bond somewhere in the *middle* of the carbon chain.
  • A *terminal alkyne* has the triple bond at the *end* of the carbon chain.

$$HC\equiv C-CH_2CH_2CH_3 \qquad CH_3-C\equiv C-CH_2CH_3 \qquad HC\equiv C-\underset{\underset{CH_3}{|}}{CH}-CH_3$$

  terminal alkyne   internal alkyne   terminal alkyne

11.2

11.3 Like alkenes, the larger the number of alkyl groups bonded to the *sp* hybridized C, the more stable the alkyne. This makes internal alkynes more stable than terminal alkynes.

11.4 To name an alkyne:
  [1] Find the longest chain that contains both atoms of the triple bond, and number the chain to give the first carbon of the triple bond the lower number.
  [2] Name all substituents following the other rules of nomenclature.

  1-hexyne   3-hexyne   3-methyl-1-pentyne   3,3-dimethyl-1-butyne

      2-hexyne   4-methyl-1-pentyne   4-methyl-2-pentyne

11.5 Use the directions from Answer 11.4 to name the alkyne.

a. $H-C\equiv C-CH_2C(CH_2CH_2CH_3)_3$ =

b. $CH_3C\equiv CC(CH_3)ClCH_2CH_3$ =

11.6 To work backwards from a name to a structure:
[1] Find the parent name and the functional group.
[2] Add the substituents to the appropriate carbon.

a. *trans*-2-ethynyl**cyclopentanol**
   5 C ring with OH at C1

b. 4-*tert*-butyl-5-**decyne**
   10 C chain with
   a triple bond

c. 3-methyl**cyclononyne**
   9 C ring with a
   triple bond at C1

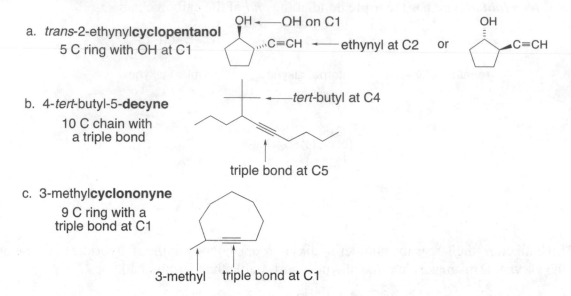

11.7 Two factors cause the boiling point increase. The linear *sp* hybridized C's of the alkyne allow for more van der Waals attraction between alkyne molecules. Also, since a triple bond is more polarizable than a double bond, this increases the van der Waals forces between two molecules as well.

11.8 To convert an alkene to an alkyne:
[1] Make a vicinal dihalide from the alkene by addition of X_2.
[2] Add base to remove two equivalents of HX and form the alkyne.

a. $Br_2CH(CH_2)_4CH_3$ $\xrightarrow{Na^+ \ ^-NH_2}$ [$BrCH=CHCH_2CH_2CH_2CH_3$] $\xrightarrow{Na^+ \ ^-NH_2}$ $HC\equiv CCH_2CH_2CH_2CH_3$
   not isolated

b. $CH_2=CCl(CH_2)_3CH_3$ $\xrightarrow{Na^+ \ ^-NH_2}$ $HC\equiv CCH_2CH_2CH_2CH_3$

□□c. $CH_2=CH(CH_2)_3CH_3$ $\xrightarrow{Cl_2}$ $CH_2CHCH_2CH_2CH_2CH_3$ $\xrightarrow[\text{(2 equiv)}]{Na^+ \ ^-NH_2}$ $HC\equiv CCH_2CH_2CH_2CH_3$
   $\overset{|}{Cl} \ \overset{|}{Cl}$

11.9

11.10 Acetylene has a pK_a of 25, so **bases having a conjugate acid with a pK_a *above* 25 will be able** to deprotonate it.

a. CO_3^{2-} [pK_a (HCO_3^-)= 10.2]
   pK_a < 25 = **Cannot deprotonate acetylene.**

b. $CH_2=CH^-$ [pK_a ($CH_2=CH_2$) = 44]
   pK_a > 25 = **Can deprotonate acetylene.**

c. $(CH_3)_3CO^-$ {pK_a [$(CH_3)_3COH$]= 18}
   pK_a < 25 = **Cannot deprotonate acetylene.**

11.11 To draw the products of reactions with HX:
   • Add two moles of HX to the triple bond, following Markovnikov's rule.
   • Both X's end up on the more substituted C.

a. $CH_3CH_2CH_2CH_2$ $C{\equiv}C-H$ $\xrightarrow{\text{2 HI}}$ $CH_3CH_2CH_2CH_2-\overset{\overset{\displaystyle I}{|}}{\underset{\underset{\displaystyle I}{|}}{C}}-CH_3$

b. $CH_3-C{\equiv}C-CH_2CH_3$ $\xrightarrow{\text{2 HBr}}$ $CH_3-CH_2-\overset{\overset{\displaystyle Br}{|}}{\underset{\underset{\displaystyle Br}{|}}{C}}-CH_2CH_3$ + $CH_3-\overset{\overset{\displaystyle Br}{|}}{\underset{\underset{\displaystyle Br}{|}}{C}}-CH_2-CH_2CH_3$

11.12

11.13 Addition of one equivalent of X_2 to alkynes forms a trans dihalide.
   Addition of two equivalents of X_2 to alkynes forms tetrahalides.

$CH_3CH_2-C{\equiv}C-CH_2CH_3$ $\xrightarrow{\text{2Br}_2}$ $CH_3CH_2-\overset{\overset{\displaystyle Br}{|}}{\underset{\underset{\displaystyle Br}{|}}{C}}-\overset{\overset{\displaystyle Br}{|}}{\underset{\underset{\displaystyle Br}{|}}{C}}-CH_2CH_3$

$CH_3CH_2-C{\equiv}C-CH_2CH_3$ $\xrightarrow{\text{Cl}_2}$

***trans* dihalide**

11.14

CH$_3$—C≡C—CH$_3$ $\xrightarrow{\text{Cl}_2}$

The two Cl atoms are electron withdrawing, making the π bond less electron rich and therefore less reactive with an electrophile.

11.15 **To draw the keto form of each enol:**
[1] Change the C–OH to a C=O at one end of the double bond.
[2] At the other end of the double bond, add a proton.

a.

new C–H bond

c.

new C–H bond

b.

new C–H bond

11.16 Treatment of alkynes with H$_2$O, H$_2$SO$_4$, and HgSO$_4$ yields ketones.

CH$_3$—C≡C—CH$_2$CH$_3$ $\xrightarrow[\text{H}_2\text{SO}_4,\ \text{HgSO}_4]{\text{H}_2\text{O}}$

CH$_3$CH=C(OH)CH$_2$CH$_3$
+
CH$_3$C(OH)=CHCH$_2$CH$_3$
Two enols form.

**two ketones after
tautomerization**

11.17 To determine what two alkynes could yield the given ketone, work backwards by drawing the enols and then the alkynes.

HC≡C—CH$_2$CH$_3$

2-butanone

CH$_3$—C≡C—CH$_3$

11.18 Reaction with H$_2$O, H$_2$SO$_4$, and HgSO$_4$ adds the oxygen to the *more* substituted carbon.
Reaction with [1] BH$_3$, [2] H$_2$O$_2$/⁻OH adds the oxygen to the *less* substituted carbon.

a. (CH$_3$)$_2$CHCH$_2$—C≡C—H $\xrightarrow[\text{H}_2\text{SO}_4/\text{HgSO}_4]{\text{H}_2\text{O}}$

Forms a **ketone.** H$_2$O is added with the O atom on the *more* substituted carbon.

b. (CH$_3$)$_2$CHCH$_2$—C≡C—H $\xrightarrow[\text{[2] H}_2\text{O}_2/\text{HO}^-]{\text{[1] BH}_3}$

Forms an **aldehyde.** H$_2$O is added with the O atom on the *less* substituted carbon.

c. $\xrightarrow[\text{H}_2\text{SO}_4/\text{HgSO}_4]{\text{H}_2\text{O}}$

Forms a **ketone.** H$_2$O is added with the O atom on the *more* substituted carbon.

d. $\xrightarrow[\text{[2] H}_2\text{O}_2/\text{HO}^-]{\text{[1] BH}_3}$

Forms an **aldehyde.** H$_2$O is added with the O atom on the *less* substituted carbon.

11.19

a. $H-C\equiv C-H$ $\xrightarrow{\text{[1] NaH}}$ $H-C\equiv C^-$ $+ H_2$ $\xrightarrow{\text{[2] }(CH_3)_2CHCH_2-Cl}$ $(CH_3)_2CHCH_2-C\equiv C-H$ $+$ NaCl

b. cyclopentyl$-C\equiv CH$

$\xrightarrow{\text{[1] NaH}}$ cyclopentyl$-C\equiv C^-$ $+ H_2$ $\xrightarrow{\text{[2] }CH_3CH_2-Br}$ cyclopentyl$-C\equiv C-CH_2CH_3$ $+$ NaBr

1° alkyl halide **substitution product**

$\xrightarrow{\text{[1] }^-NH_2}$ cyclopentyl$-C\equiv C^-$ $+ NH_3$ $\xrightarrow{\text{[2] }(CH_3)_3CCl}$
$$\begin{matrix} CH_3 \\ | \\ C=CH_2 \\ | \\ CH_3 \end{matrix}$$ $+$ NaCl

3° alkyl halide **elimination product**

$+$ cyclopentyl$-C\equiv CH$

11.20

a. $(CH_3)_2CHCH_2C\equiv CH$

$$\begin{matrix} CH_3 \\ | \\ CH_3-C-CH_2-\{-C\equiv C-H \\ | \\ H \end{matrix} \Longrightarrow \begin{matrix} CH_3 \\ | \\ CH_3-C-CH_2Cl \\ | \\ H \end{matrix} + \;\;^-C\equiv C-H$$

1° RX

terminal alkyne only one possibility

b. $CH_3C\equiv CCH_2CH_2CH_2CH_3$

$CH_3-\{-C\equiv C-\}-CH_2CH_2CH_2CH_3$
     [1]   [2]

$\Longrightarrow$ [1] CH_3Cl $+$ $^-C\equiv C-CH_2CH_2CH_2CH_3$

[2] $CH_3-C\equiv C^-$ $+$ $Cl-CH_2CH_2CH_2CH_3$

1° RX

internal alkyne two possibilities

c. $(CH_3)_3CC\equiv CCH_2CH_3$

$$\begin{matrix} CH_3 \\ | \\ CH_3-C-C\equiv C-\{-CH_2CH_3 \\ | \\ CH_3 \end{matrix} \Longrightarrow \begin{matrix} CH_3 \\ | \\ CH_3-C-C\equiv C^- \\ | \\ CH_3 \end{matrix} + \;Cl-CH_2CH_3$$

1° RX

internal alkyne only one possibility

$$\xrightarrow{\quad\not\quad} \begin{matrix} CH_3 \\ | \\ CH_3-C-Cl \\ | \\ CH_3 \end{matrix} + \;\;^-C\equiv CCH_2CH_3$$

The 3° alkyl halide would undergo elimination.

3° RX
too crowded for S$_N$2 reaction

11.21

$HC\equiv C-H$ $\xrightarrow{Na^+ H^-}$ $HC\equiv C^-$ $\xrightarrow{CH_3CH_2-Br}$ $H-C\equiv C-CH_2CH_3$ $\xrightarrow{Na^+ H^-}$ $^-C\equiv C-CH_2CH_3 + H_2$

$+ H_2$
$+ Na^+$

$+ Na^+Br^-$

$(CH_3)_2CHCH_2\overset{\displaystyle H}{\underset{\displaystyle H}{C}}-Br$

$(CH_3)_2CHCH_2CH_2-C\equiv C-CH_2CH_3 + Br^-$

11.22

$$\begin{matrix} CH_3 & & CH_3 \\ | & & | \\ CH_3-C-C\equiv C-C-CH_3 \\ | & & | \\ CH_3 & & CH_3 \end{matrix} \Longrightarrow \begin{matrix} CH_3 \\ | \\ CH_3-C-C\equiv C^- \\ | \\ CH_3 \end{matrix} + \begin{matrix} CH_3 \\ | \\ X-C-CH_3 \\ | \\ CH_3 \end{matrix}$$

The 3° alkyl halide is too crowded for nucleophilic substitution. Instead, it would undergo elimination with the acetylide anion.

2,2,5,5-tetramethyl-3-hexyne

11.23

a.

[1] ⁻:C≡C—H
[2] H₂O

Epoxide is drawn up, so the
**acetylide anion attacks from *below*
at *less* substituted C.**

b.

[1] ⁻:C≡C—H
[2] H₂O

+

enantiomers

Backside attack of the nucleophile
(⁻C≡CH) at either C since both
ends are equally substituted.

11.24 To use a retrosynthetic analysis:

[1] **Count the number of carbon atoms** in the starting material and product.

[2] **Look at the functional groups** in the starting material and product.

  Determine what types of reactions can form the product.

  Determine what types of reactions the starting material can undergo.

[3] **Work backwards** from the product to make the starting material.

[4] Write out the synthesis in synthetic direction.

$$CH_3CH_2C≡CCH_2CH_3 \xRightarrow{?} HC≡CH$$

6 C's   2 C's

$$CH_3CH_2C≡CCH_2CH_3 \Longrightarrow CH_3CH_2C≡C^- + CH_3CH_2Br \Longrightarrow HC≡C^- + CH_3CH_2Br$$

$$HC≡C-H \xrightarrow{Na^+ H^-} HC≡C^- \xrightarrow{CH_3CH_2-Br} CH_3CH_2C≡C-H \xrightarrow{Na^+ H^-} CH_3CH_2C≡C^- \xrightarrow{CH_3CH_2-Br} CH_3CH_2C≡CCH_2CH_3$$

11.25

product:
4 carbons, aldehyde functional group
(can be made by hydroboration–oxidation of
a terminal alkyne)

HC≡CH

starting material:
2 carbons, C≡C functional group
(can form an acetylide
anion by reaction with NaH)

Retrosynthetic: $CH_3CH_2CH_2-C\overset{O}{\underset{H}{}} \Longrightarrow CH_3CH_2C≡C-H \Longrightarrow H-C≡C-H$

Forward direction: $H-C≡C-H \xrightarrow{Na^+ H^-} {}^-C≡C-H \xrightarrow{CH_3CH_2-Br} CH_3CH_2C≡C-H \xrightarrow[{[2] H_2O_2/HO^-}]{[1] BH_3} CH_3CH_2CH_2-C\overset{O}{\underset{H}{}}$

11.26 Use the rules from Answer 11.4 to name the alkynes.

a. CH₃CH₂CH(CH₃)C≡CCH₂CH₃   **5-methyl-3-heptyne**

5-methyl   3-heptyne

3-hexyne

b.   CH₃CHC≡CCHCH₃   **2,5-dimethyl-3-hexyne**
         | |
        CH₃   CH₃

2,5-dimethyl

4-nonyne
c.   CH₃CH₂CHC≡CCHCHCH₂CH₃   CH₃← 7-methyl
         | |
        CH₃CH₂   CH₂CH₃
3,6-diethyl
                    3,6-diethyl-7-methyl-4-nonyne

d.   HC≡C–CH(CH₂CH₃)CH₂CH₂CH₃   **3-ethyl-1-hexyne**
      1-hexyne   3-ethyl

e.   CH₃CH₂–C–C≡CH   CH₃← 3-methyl
              | **3-ethyl-3-methyl-1-hexyne**
           CH₂CH₂CH₃
      3-ethyl

f.   CH₃CH₂C≡CCH₂C≡CCH₃   **2,5-octadiyne**
              5   2

11.27 Use the directions from Answer 11.6 to draw each structure.

a. 5,6-dimethyl-2-**heptyne**

2-heptyne

b. 5-*tert*-butyl-6,6-dimethyl-3-**nonyne**

6,6-dimethyl

3-nonyne

5-*tert*-butyl

c. (S)-4-chloro-2-**pentyne**

Cl ← 4-chloro
ₙₙH   S stereochemistry

2-pentyne

d. □*cis*-1-ethynyl-2-methyl**cyclopentane**

1-ethynyl

C≡CH   or   C≡CH

2-methyl

11.28 Keto-enol tautomers are constitutional isomers in equilibrium that differ in the location of a double bond and a hydrogen. The OH in an enol must be bonded to a C=C.

a.
O   OH
|| |
CH₃–C–CH₃   and   CH₂=C–CH₃

• C=O   • C=C
• one more CH bond • OH on C=C
keto-enol tautomers

b.
O   and   OH

OH is not bonded
to the C=C.
NOT keto-enol tautomers

c.
OH   and   H
                    ||
                    O
• C=C   • C=O
• OH on C=C   • one more CH bond
keto-enol tautomers

d.
O   and   OH← OH is not bonded
|| to the C=C.

NOT keto-enol tautomers

11.29 **To draw the enol form of each keto form:**
[1] Change the C=O to a C–OH.
[2] Change one single C–C bond to a double bond, making sure the OH group is bonded to the C=C.

a. $\begin{bmatrix} E/Z \text{ isomers} \\ \text{possible} \end{bmatrix}$ c.

b. $CH_3CH_2CHO =$ $\begin{bmatrix} E/Z \text{ isomers} \\ \text{possible} \end{bmatrix}$

11.30 Use the directions from Answer 11.15 to draw each keto form.

a.   c.

b.

11.31

11.32 The equilibrium always favors the formation of the weaker acid and the weaker base.

a. $HC\equiv C^- + CH_3OH \rightleftharpoons HC\equiv CH + CH_3O^-$
  $pK_a = 15.5$ $pK_a = 25$
              weaker acid
       Equilibrium favors products.

c. $HC\equiv CH + Na^+Br^- \rightleftharpoons HC\equiv C^- Na^+ + HBr$
   $pK_a = 25$ $pK_a = -9$
   weaker acid
**Equilibrium favors
starting materials.**

b. $CH_3C\equiv CH + CH_3^- \rightleftharpoons CH_3C\equiv C^- + CH_4$
   $pK_a = 25$ $pK_a = 50$
                              weaker acid
              Equilibrium favors products.

d. $CH_3CH_2C\equiv C^- + CH_3COOH \rightleftharpoons CH_3CH_2C\equiv CH + CH_3COO^-$
          $pK_a = 4.8$ $pK_a = 25$
                                weaker acid
              Equilibrium favors products.

11.33

HC≡CCH₂CH₂CH₂CH₃

a. $\xrightarrow[\text{(2 equiv)}]{\text{HCl}}$

$CH_3-\underset{\underset{\displaystyle Cl}{|}}{\overset{\overset{\displaystyle Cl}{|}}{C}}-CH_2CH_2CH_2CH_3$

b. $\xrightarrow[\text{(2 equiv)}]{\text{HBr}}$

$CH_3-\underset{\underset{\displaystyle Br}{|}}{\overset{\overset{\displaystyle Br}{|}}{C}}CH_2CH_2CH_2CH_3$

c. $\xrightarrow[\text{(2 equiv)}]{\text{Cl}_2}$

$\underset{\underset{\displaystyle Cl}{|}}{\overset{\overset{\displaystyle Cl}{|}}{HC}}-\underset{\underset{\displaystyle Cl}{|}}{\overset{\overset{\displaystyle Cl}{|}}{C}}CH_2CH_2CH_2CH_3$

d. $\xrightarrow[\text{H}_2\text{SO}_4,\ \text{HgSO}_4]{\text{H}_2\text{O}}$

$\underset{H}{\overset{H}{>}}C=C\underset{CH_2CH_2CH_2CH_3}{\overset{OH}{<}}$ ⟶ $CH_3-\overset{O}{\overset{||}{C}}-CH_2CH_2CH_2CH_3$

e. $\xrightarrow[\text{[2] H}_2\text{O}_2/\ \text{HO}^-]{\text{[1] BH}_3}$ $\overset{O}{\overset{||}{HC}}-CH_2CH_2CH_2CH_2CH_3$

f. $\xrightarrow{\text{NaH}}$ $^-C≡CCH_2CH_2CH_2CH_3$

g. $\xrightarrow[\text{[2] CH}_3\text{CH}_2\text{Br}]{\text{[1] }^-\text{NH}_2}$ $CH_3CH_2C≡CCH_2CH_2CH_2CH_3$

h. $\xrightarrow[\text{[2] }\triangle\!\!\!\!O]{\text{[1] }^-\text{NH}_2}$ $HOCH_2CH_2-C≡CCH_2CH_2CH_2CH_3$

11.34

a. $\xrightarrow[\text{(2 equiv)}]{\text{HBr}}$

b. $\xrightarrow[\text{(2 equiv)}]{\text{Br}_2}$

c. $\xrightarrow[\text{H}_2\text{SO}_4]{\text{H}_2\text{O}}$

d. $\xrightarrow[\text{[2] H}_2\text{O}_2/\ \text{HO}^-]{\text{[1] BH}_3}$

11.35

a. $(CH_3CH_2)_3C-C≡CH$

$\xrightarrow[\text{H}_2\text{SO}_4,\ \text{HgSO}_4]{\text{H}_2\text{O}}$ $(CH_3CH_2)_3C-\overset{O}{\overset{||}{C}}-CH_3$

$\xrightarrow[\text{[2] H}_2\text{O}_2/\text{HO}^-]{\text{[1] BH}_3}$ $(CH_3CH_2)_3C-CH_2CHO$

$\xrightarrow[\text{[2] CH}_3\text{Br}]{\text{[1] NaH}}$ $(CH_3CH_2)_3C-C≡CCH_3$

b. $CH_3CH_2CH_2CH(CH_3)CH_2CHBr_2$ $\xrightarrow{2\ ^-\text{NH}_2}$ $CH_3CH_2CH_2CH(CH_3)C≡CH$

c. $CH_3CH_2CH=CHCH_2CH_3$ $\xrightarrow{\text{Br}_2}$ $CH_3CH_2\underset{\underset{\displaystyle Br}{|}}{CH}-\underset{\underset{\displaystyle Br}{|}}{CH}CH_2CH_3$ $\xrightarrow{2\ ^-\text{NH}_2}$ $CH_3CH_2C≡CCH_2CH_3$

11.36

a. ⟹ ⟹

b. $CH_3-\overset{O}{\overset{||}{C}}-CH_3$ ⟹ $CH_3-\underset{}{\overset{OH}{C}}=CH_2$ ⟹ $CH_3-C≡CH$

c.

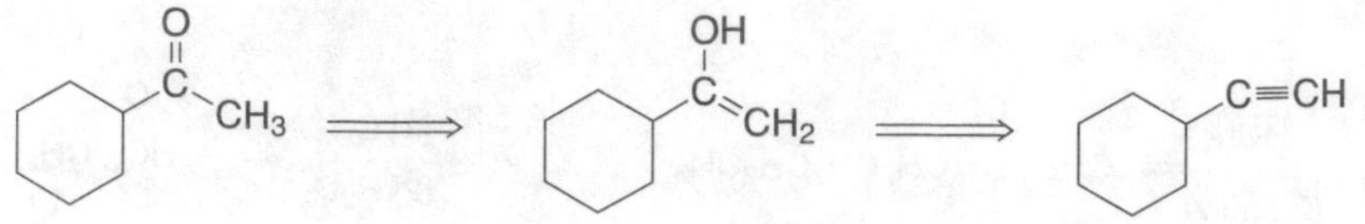

d.

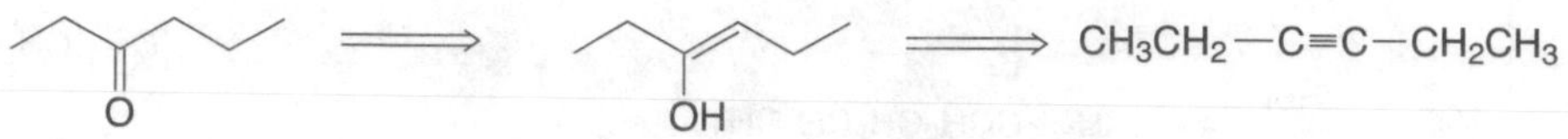

11.37

a.

b. 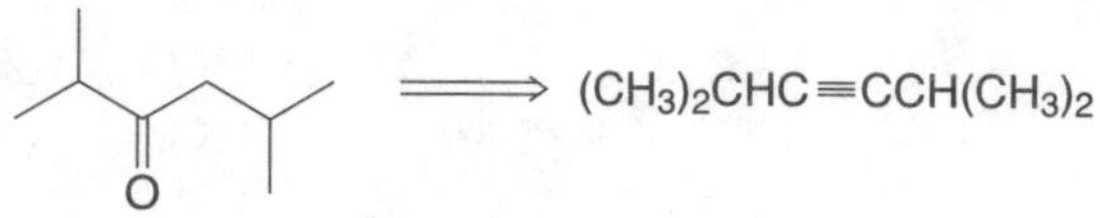 $\Longrightarrow$ $(CH_3)_2CHC\equiv CCH(CH_3)_2$

11.38 The alkyne that yields the same aldehyde on treatment with H_2O, H_2SO_4, $HgSO_4$ and BH_3, then H_2O_2/HO^- must not have any internal carbons. It must be $HC\equiv CH$.

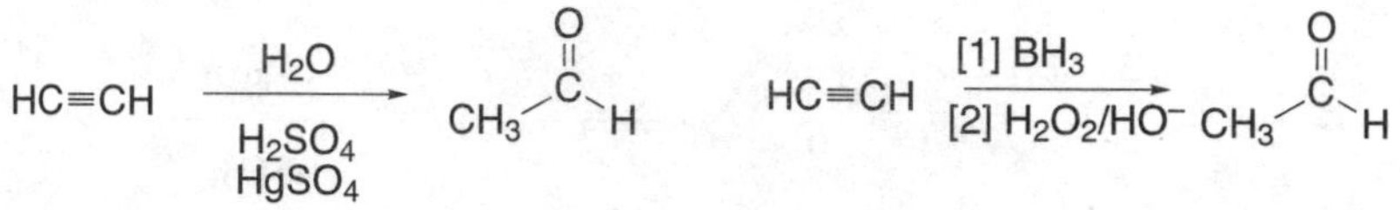

11.39

a.

b. 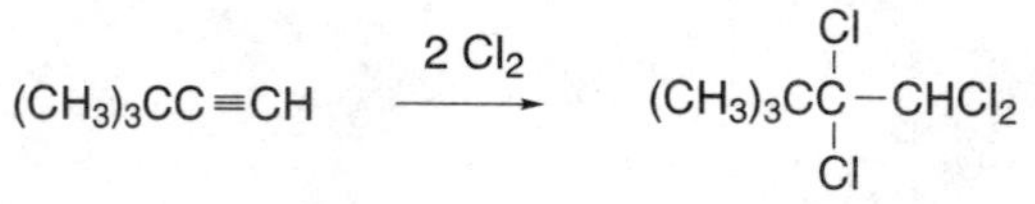 $(CH_3)_3CC\equiv CH$ $\xrightarrow{2\ Cl_2}$

c.

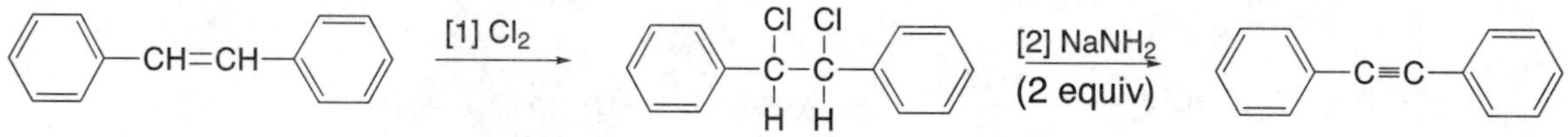

d.

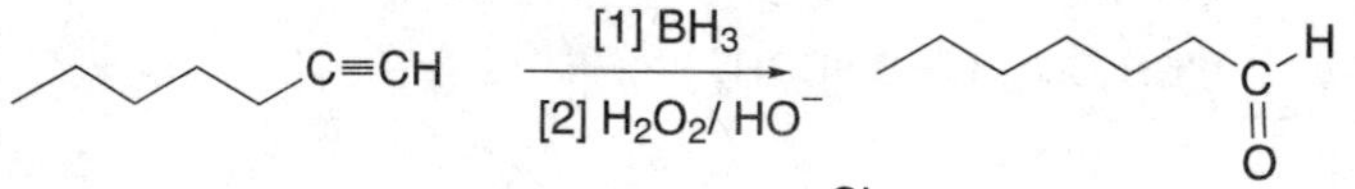

e.

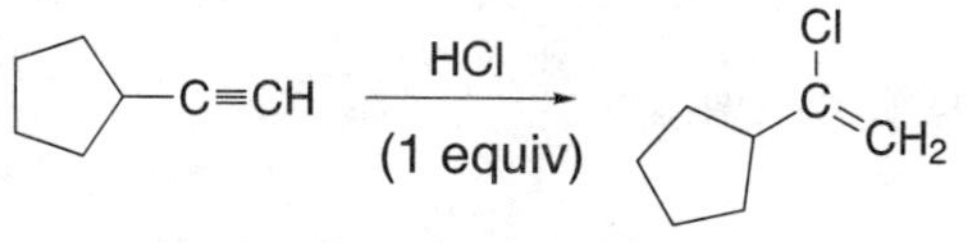

f. $HC\equiv C^- + D_2O \longrightarrow HC\equiv CD + DO^-$

g.

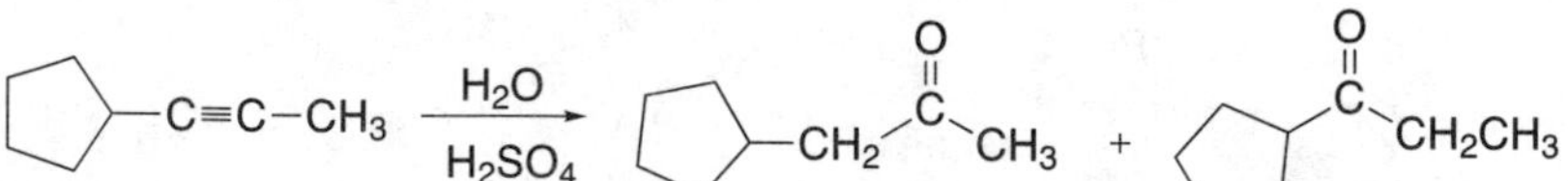

h.

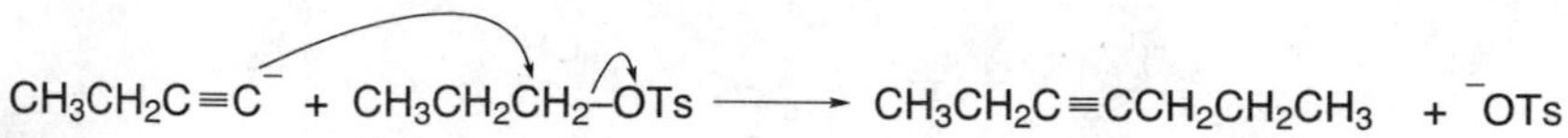

i.

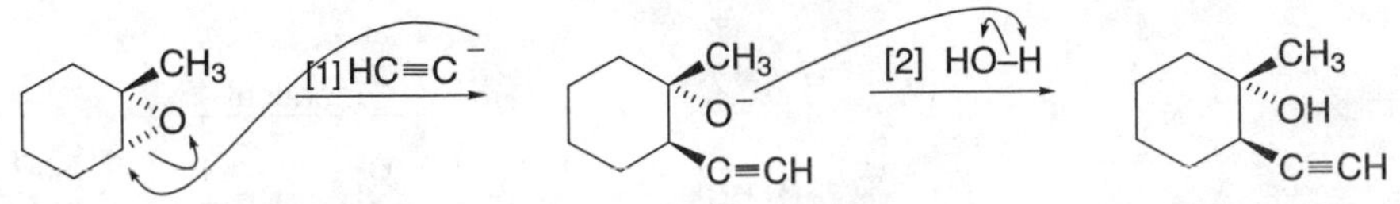

j. $CH_3C\equiv C-H$ $\xrightarrow{[1]\ Na^+\ H^-}$ $CH_3C\equiv C^-$ $\xrightarrow[\substack{2°\ halide\\E2}]{[2]}$ $+$ $CH_3C\equiv C-H$

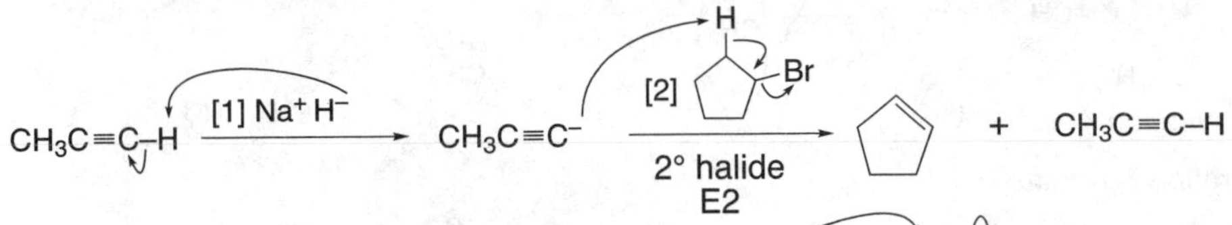

k. $CH_3CH_2C\equiv C-H$ $\xrightarrow{[1]\ Na^+\ NH_2}$ $CH_3CH_2C\equiv C^-$ $\xrightarrow{[2]}$ $CH_3CH_2C\equiv C-CH_2$⬡

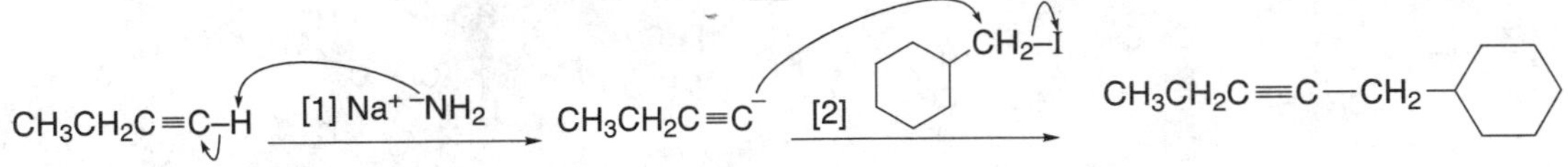

l. $\xrightarrow{[1]\ Na^+\ H^-}$ ⬡$C\equiv C^-$ $\xrightarrow{[2]}$ ⬡$C\equiv CCH_2CH_2O^-$ $\xrightarrow{[3]\ HO-H}$ ⬡$C\equiv C-CH_2CH_2OH$

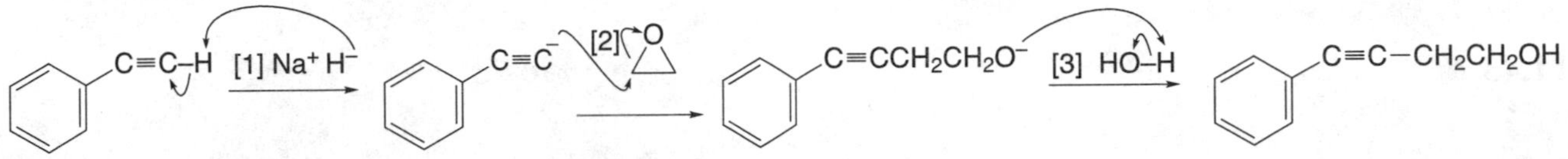

11.40

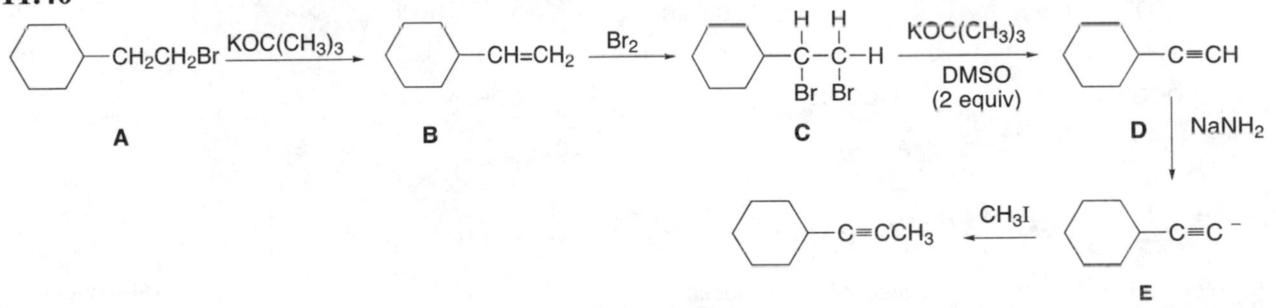

$⬡-CH_2CH_2Br$ $\xrightarrow{KOC(CH_3)_3}$ $⬡-CH=CH_2$ $\xrightarrow{Br_2}$ C $\xrightarrow[\substack{DMSO\\(2\ equiv)}]{KOC(CH_3)_3}$ $⬡-C\equiv CH$

A   B   C   D $\downarrow NaNH_2$

$⬡-C\equiv CCH_3$ $\xleftarrow{CH_3I}$ $⬡-C\equiv C^-$

E

11.41

most acidic H

$H-C\equiv C-CH_2CH_2CH_2OH$ $\xrightarrow[2)\ CH_3I]{1)\ NaNH_2}$ ✗ $CH_3-C\equiv C-CH_2CH_2CH_2OH$

A

NaNH$_2$ will remove the
proton from the
OH since it is more acidic.

$H-C\equiv C-CH_2CH_2CH_2O^-$ $\longrightarrow$ $H-C\equiv C-CH_2CH_2CH_2OCH_3$ = **B**

CH_3-I

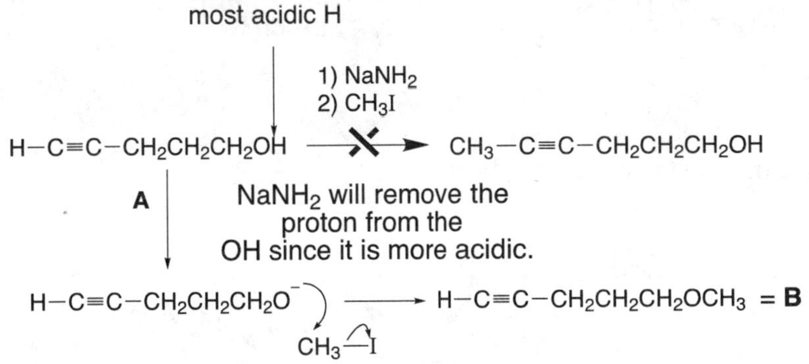

11.42

Stereogenic center at
the site of reaction.

a.

Cl   HC≡C⁻ →

H D

C=CH

D H

inversion

Stereogenic center NOT
at the site of reaction.   *Configuration is retained.*

b.

Cl   HC≡C⁻ →

CH₃ H

C≡CH

CH₃ H

c.

O

H''''CH₃

CH₃ H

[1] HC≡C⁻
[2] H₂O
→

identical

HO   CH₃

H   H

C—C   +   CH₃   OH

H   CH₃   C—C

CH₃   CH₃

CH   HC

d.

O

H''''H

CH₃ CH₃

[1] HC≡C⁻
[2] H₂O
→

HO   H

H''''   CH₃

C—C   +   CH₃   OH

H   CH₃   C—C

CH₃   H

CH   HC

enantiomers

11.43

a.

OH

OCOR

PBr₃ →

H–C≡CCH₂OR'

CH₃⁻ Li⁺

⁻C≡CCH₂OR'

B

Br

OCOR

A

→

new C–C bond

C≡CCH₂OR'

OCOR

C

b.

O

OR

[1] **D** = HC≡C⁻
[2] H₂O
→

OH

H–C≡C   OR

These 2 C's are added.

→

OH

H–C≡C   OH

E =
TsCl/pyridine

↓

OH

H–C≡C   OTs

←

OH

H–C≡C   OH

2) **F** = CH₃CH₂Br

[1] NaH

←

OCH₂CH₃

H–C≡C

[1] NaH
[2] CH₃I
→

OCH₂CH₃

CH₃–C≡C

G

new C–C bond

11.44   Draw two diagrams to show σ and π bonds.

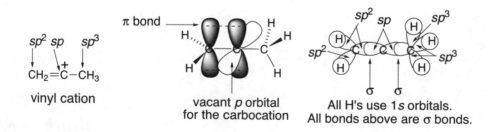

sp^2 sp sp^3

$CH_2 = \overset{+}{C} - CH_3$

vinyl cation

π bond

H   H

C

H   H

H   H

vacant *p* orbital
for the carbocation

sp^2 sp sp^3

H   H

sp^2 C   C   C   H

H   H

σ   σ

All H's use 1*s* orbitals.
All bonds above are σ bonds.

The positive charge in a vinyl carbocation resides on a carbon that is *sp* hybridized, while in $(CH_3)_2CH^+$, the positive charge is located on an sp^2 hybridized carbon. The higher percent *s*-character on carbon destabilizes the positive charge in the vinyl cation. Moreover, the positively charged carbocation is now bonded to an sp^2 hybridized carbon, which donates electrons less readily than an sp^3 hybridized carbon.

11.45

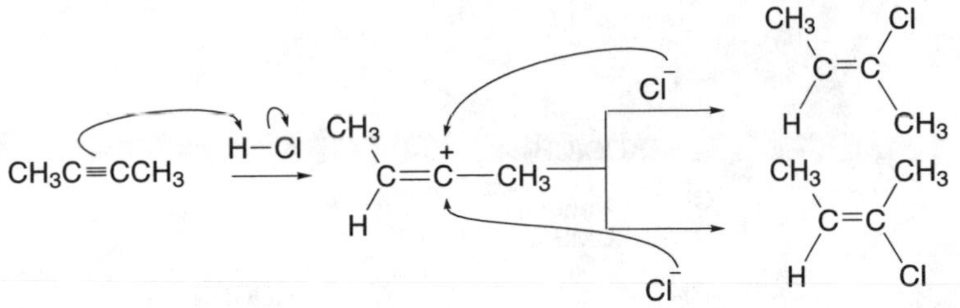

CH₃C≡CCH₃   H–Cl

Cl⁻ attack on the opposite side to the H yields the **Z** isomer.

Cl⁻ attack on the same side as the H yields the **E** isomer.

11.46

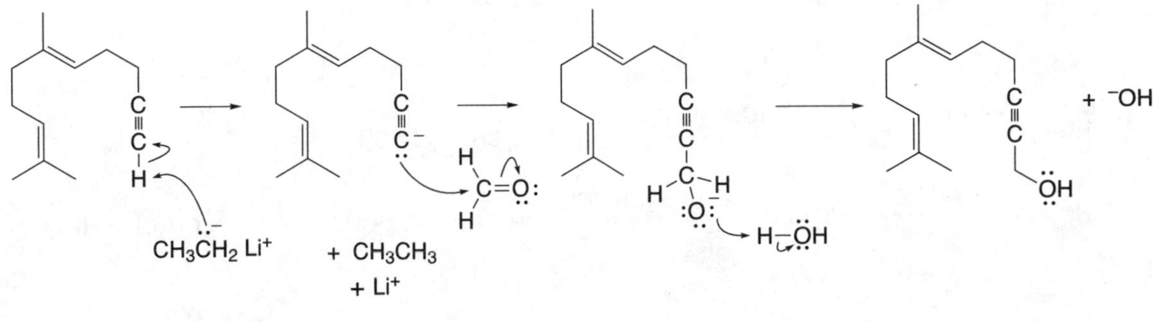

CH₃ĊH₂ Li⁺   + CH₃CH₃   + Li⁺

+ ⁻OH

11.47

a.

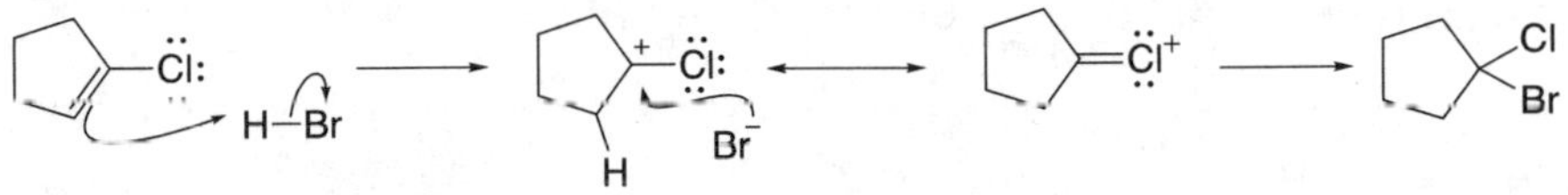

b.   CH₃–C̈–C≡CH ⟶ CH₃–C⁺–C≡CH ⟶ [CH₃–C⁺–C=CH ⟷ CH₃–C=C=CH] ⟶ CH₃–C=C=C–H

+ HSO₄⁻   resonance structures

CH₃–CH=CH C=Ö: ⟵ [CH₃–CH=C–C ⟷ CH₃–CH=C–C⁺] ⟵ CH₃–CH=C=C–H

+ H₂SO₄

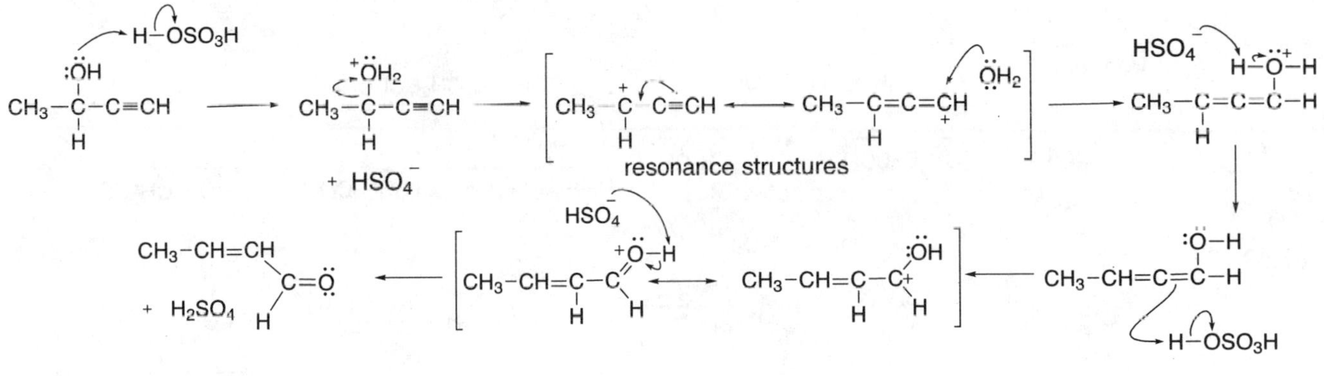

11.48   The alkyl halides must be methyl or 1°.

a.   HC≡C–CH₂CH₂CH(CH₃)₂ ⟹ HC≡C:⁻ + Cl–CH₂CH₂CH(CH₃)₂   **1° RX**

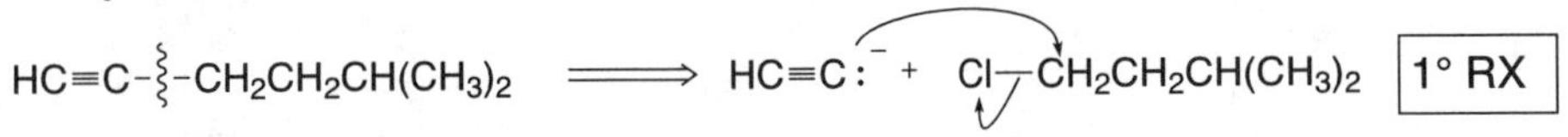

b.   CH₃–C≡C–C(CH₃)₂–CH₂CH₃ ⟹ CH₃–Cl + ⁻:C≡C–C(CH₃)₂–CH₂CH₃

c.   (cyclohexyl)–C≡C–CH₂CH₂CH₃ ⟹ (cyclohexyl)–C≡C:⁻ + Cl–CH₂CH₂CH₃   **1° RX**

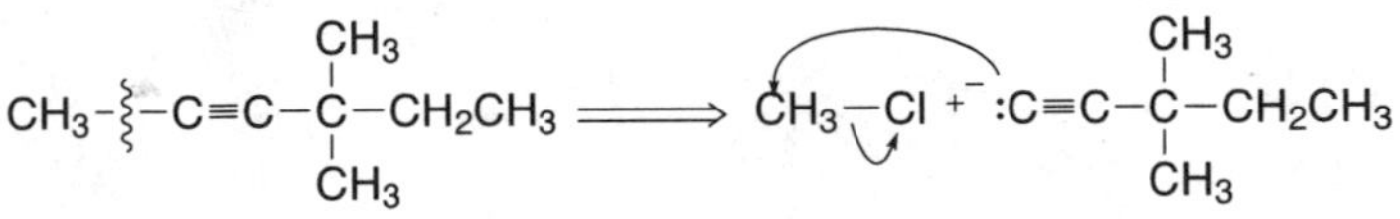

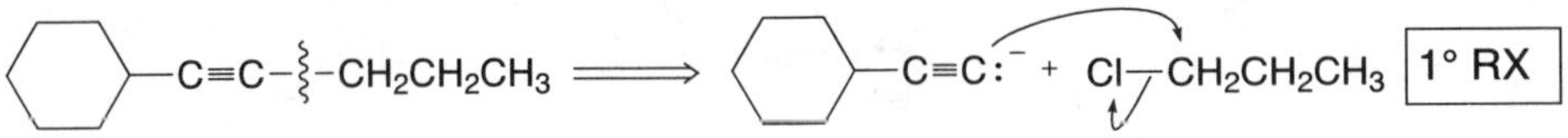

11.49

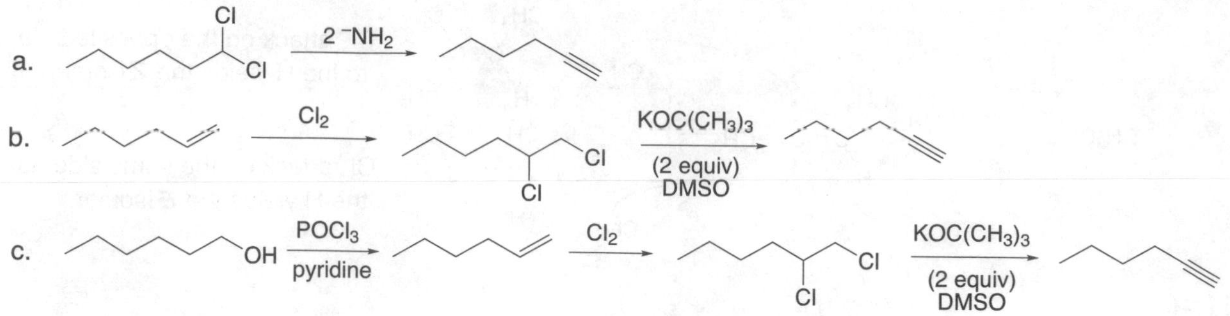
a.
Cl
Cl
2 ⁻NH₂
b.
Cl₂
Cl
Cl
KOC(CH₃)₃
(2 equiv)
DMSO
c.
OH
POCl₃
pyridine
Cl₂
Cl
Cl
KOC(CH₃)₃
(2 equiv)
DMSO

11.50

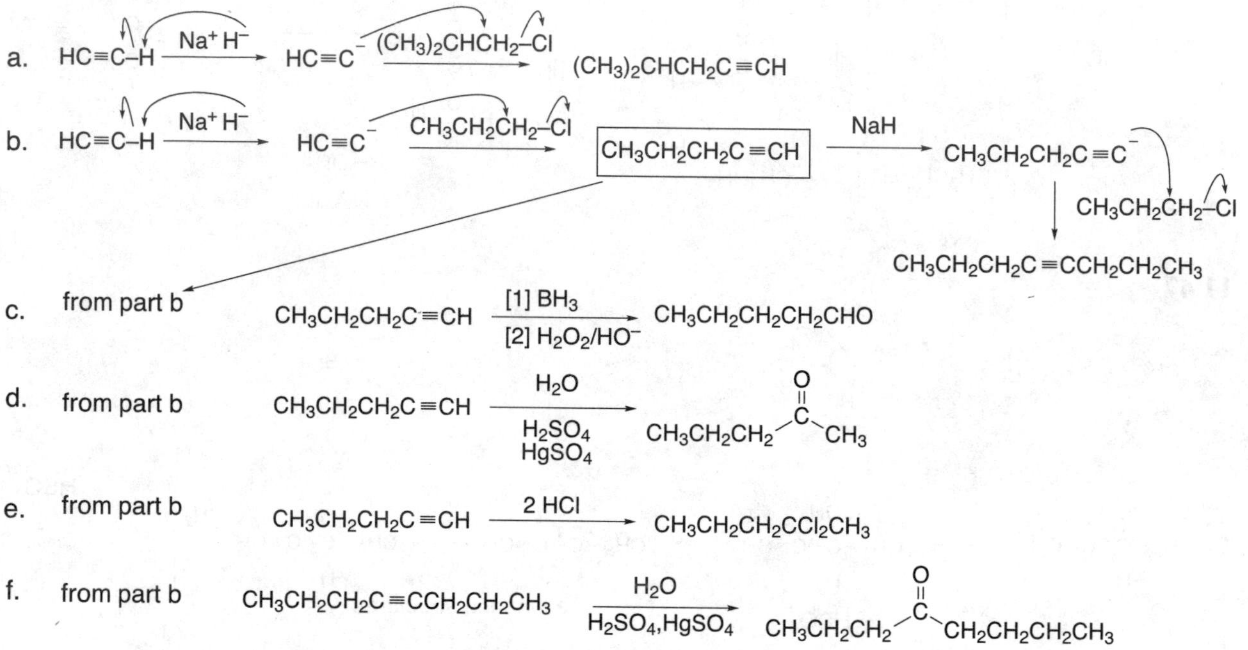
a.
HC≡C–H
Na⁺ H⁻
HC≡C⁻
(CH₃)₂CHCH₂–Cl
(CH₃)₂CHCH₂C≡CH
b.
HC≡C–H
Na⁺ H⁻
HC≡C⁻
CH₃CH₂CH₂–Cl
CH₃CH₂CH₂C≡CH
NaH
CH₃CH₂CH₂C≡C⁻
CH₃CH₂CH₂–Cl
CH₃CH₂CH₂C≡CCH₂CH₂CH₃
c.
from part b
CH₃CH₂CH₂C≡CH
[1] BH₃
[2] H₂O₂/HO⁻
CH₃CH₂CH₂CH₂CHO
d.
from part b
CH₃CH₂CH₂C≡CH
H₂O
H₂SO₄
HgSO₄
CH₃CH₂CH₂COCH₃
e.
from part b
CH₃CH₂CH₂C≡CH
2 HCl
CH₃CH₂CH₂CCl₂CH₃
f.
from part b
CH₃CH₂CH₂C≡CCH₂CH₂CH₃
H₂O
H₂SO₄,HgSO₄
CH₃CH₂CH₂COCH₂CH₂CH₂CH₃

11.51

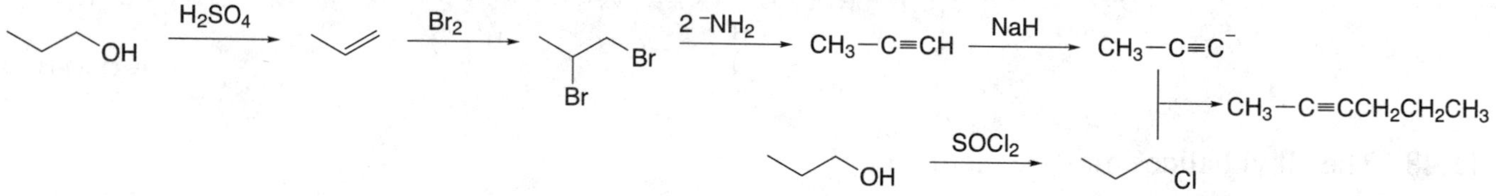
OH
H₂SO₄
Br₂
Br
Br
2 ⁻NH₂
CH₃–C≡CH
NaH
CH₃–C≡C⁻
CH₃–C≡CCH₂CH₂CH₃
OH
SOCl₂
Cl

11.52

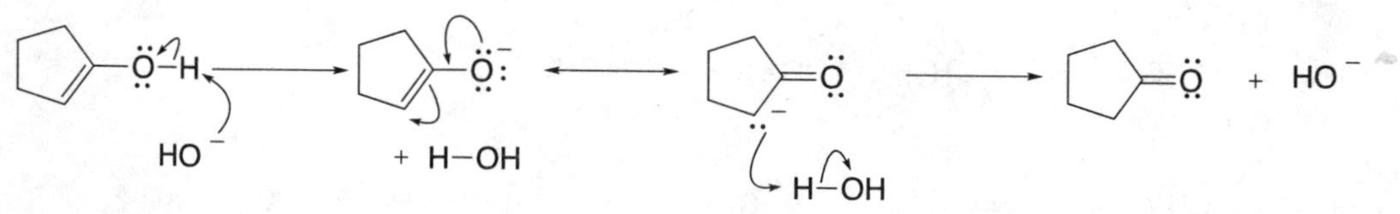
HO⁻
+ H–OH
H–OH
+ HO⁻

11.53

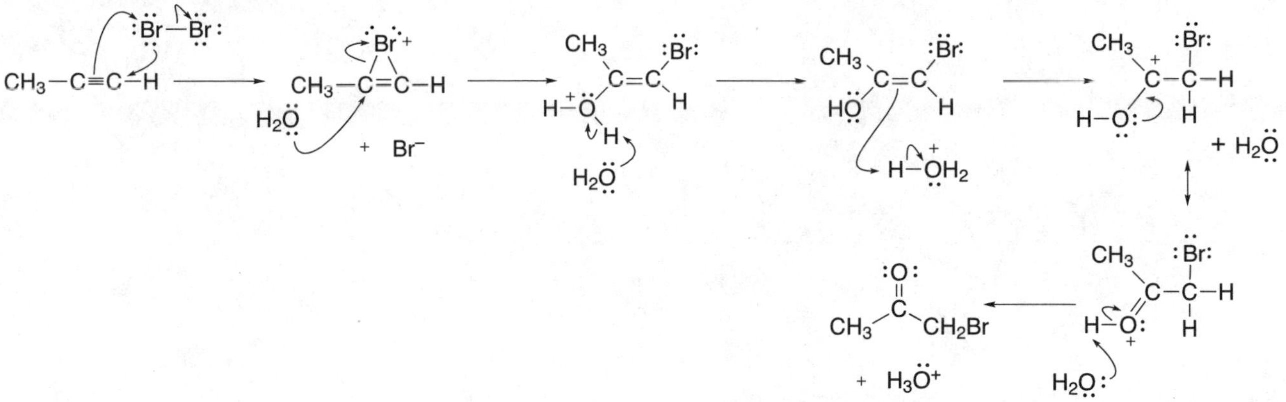

11.54

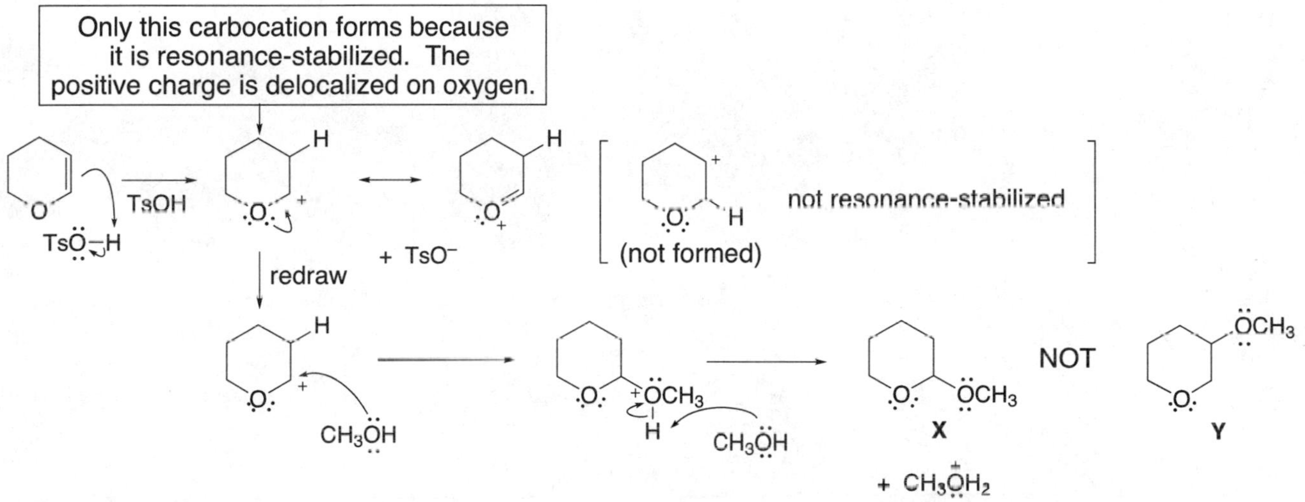
Only this carbocation forms because it is resonance-stabilized. The positive charge is delocalized on oxygen.
TsOH
redraw
+ TsO⁻
(not formed)
not resonance-stabilized
NOT
X
Y

11.55

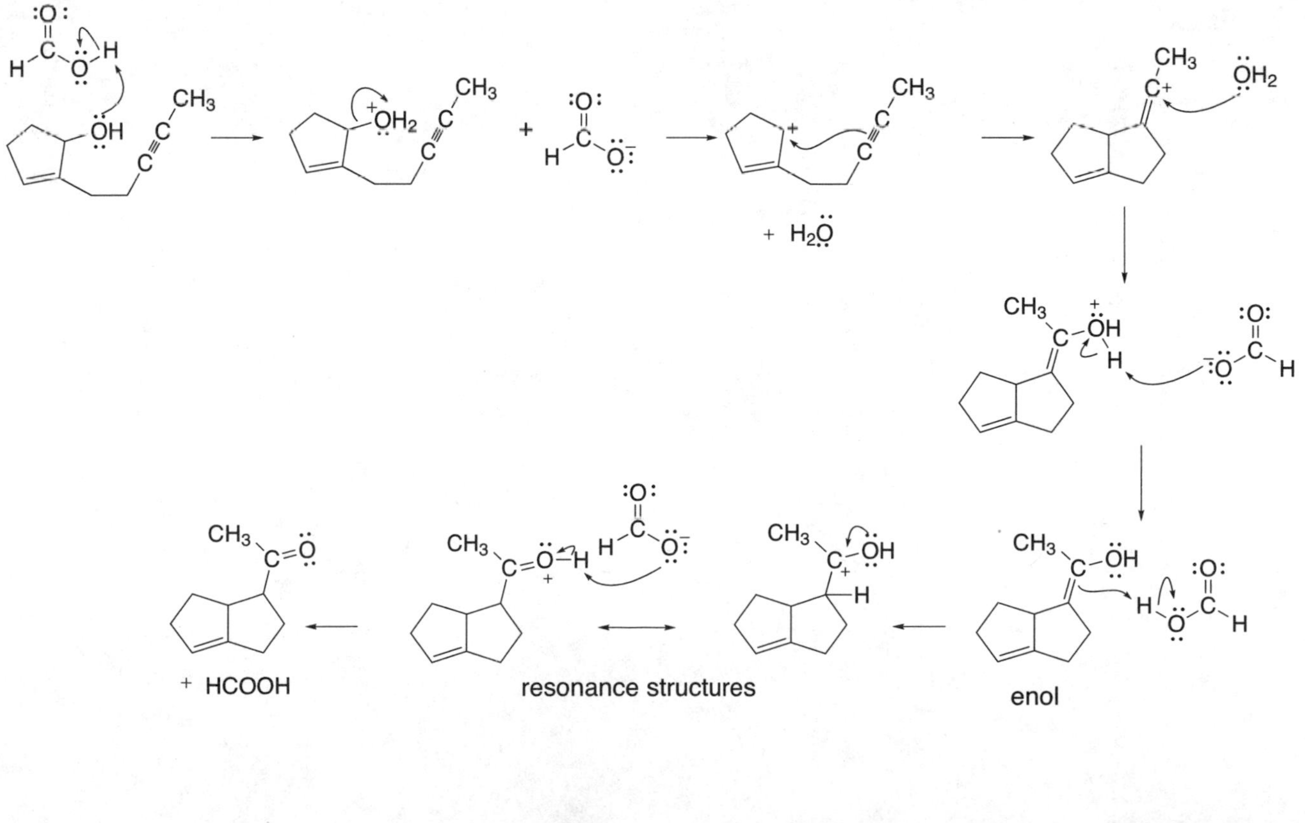
enol
resonance structures
+ HCOOH

Chapter 12: Oxidation and Reduction

♦ Summary: Terms that describe reaction selectivity

- A **regioselective reaction** forms predominately or exclusively one constitutional isomer (Section 8.5).

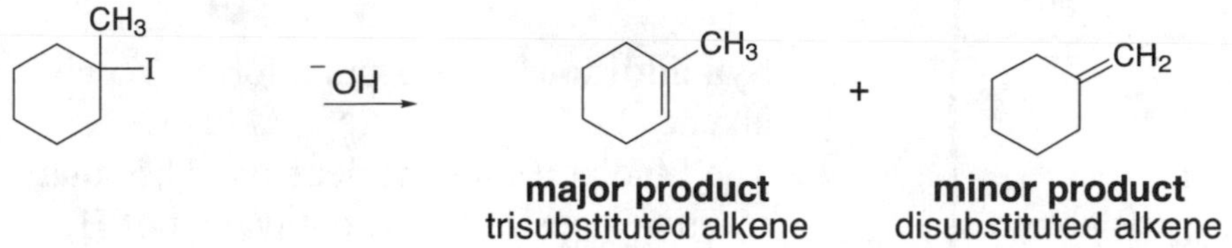

major product
trisubstituted alkene

minor product
disubstituted alkene

- A **stereoselective reaction** forms predominately or exclusively one stereoisomer (Section 8.5).

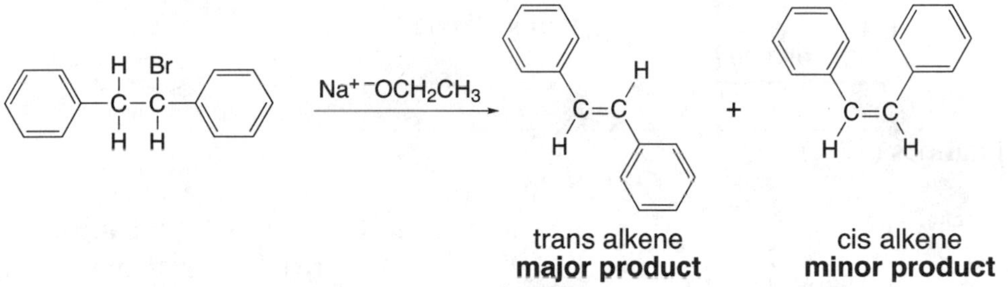

trans alkene
major product

cis alkene
minor product

- An **enantioselective reaction** forms predominately or exclusively one enantiomer (Section 12.14).

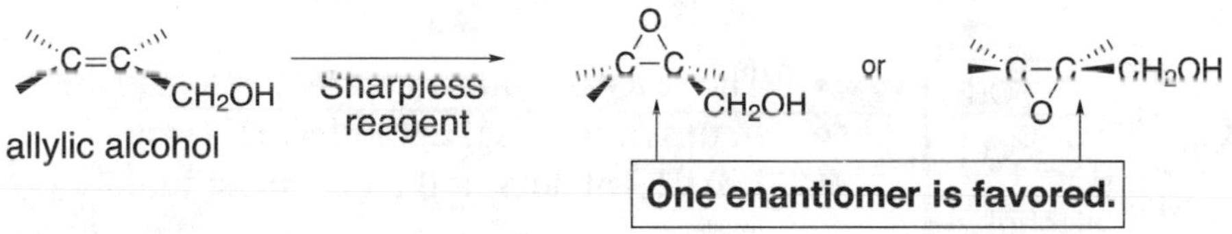

allylic alcohol

or

One enantiomer is favored.

♦ Definitions of oxidation and reduction

Oxidation reactions result in:
- an increase in the number of C–Z bonds, *or*
- a decrease in the number of C–H bonds.

Reduction reactions result in:
- a decrease in the number of C–Z bonds, *or*
- an increase in the number of C–H bonds.

[Z = an element more electronegative than C]

♦ Reduction reactions

[1] Reduction of alkenes – Catalytic hydrogenation (12.3)

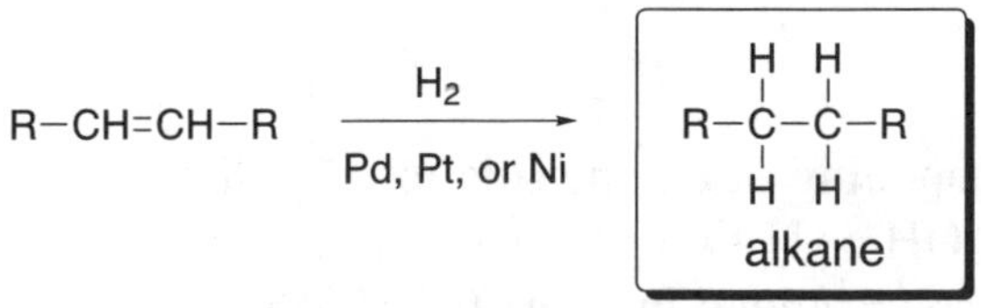

alkane

- **Syn addition** of H_2 occurs.
- Increasing alkyl substitution on the C=C decreases the rate of reaction.

[2] Reduction of alkynes

$$R-C \equiv C-R \xrightarrow[\text{Pd/C}]{2 H_2} \quad \underset{\text{alkane}}{R-\overset{\overset{\text{H}}{|}}{C}-\overset{\overset{\text{H}}{|}}{\underset{\underset{\text{H}}{|}}{C}}-R}$$

- Two equivalents of H_2 are added and four new C–H bonds are formed (12.5A).

$$R-C \equiv C-R \xrightarrow[\text{catalyst}]{\underset{\text{Lindlar}}{H_2}} \quad \underset{\text{cis alkene}}{\underset{\text{H}}{\overset{\text{R}}{\diagdown}} C = C \underset{\text{H}}{\overset{\text{R}}{\diagup}}}$$

- **Syn addition** of H_2 occurs, forming a **cis** alkene (12.5B).
- The Lindlar catalyst is deactivated so that reaction stops after one equivalent of H_2 has added.

$$R-C \equiv C-R \xrightarrow[\text{NH}_3]{\text{Na}} \quad \underset{\text{trans alkene}}{\underset{\text{H}}{\overset{\text{R}}{\diagdown}} C = C \underset{\text{R}}{\overset{\text{H}}{\diagup}}}$$

- **Anti addition** of H_2 occurs, forming a **trans** alkene (12.5C).

---

[3] Reduction of alkyl halides (12.6)

$$R-X \xrightarrow[\text{[2] } H_2O]{\text{[1] LiAlH}_4} \quad \underset{\text{alkane}}{R-H}$$

- The reaction follows an S_N2 mechanism.
- CH_3X and RCH_2X react faster than more substituted RX.

---

[4] Reduction of epoxides (12.6)

$$\underset{}{\overset{\overset{\text{O}}{\diagdown\diagup}}{C-C}} \xrightarrow[\text{[2] } H_2O]{\text{[1] LiAlH}_4} \quad \underset{\text{alcohol}}{\underset{\text{H}}{\overset{}{C}}-\overset{\text{OH}}{C}}$$

- The reaction follows an S_N2 mechanism.
- In unsymmetrical epoxides, H^- (from $LiAlH_4$) attacks at the less substituted carbon.

♦ Oxidation reactions

[1] Oxidation of alkenes

[a] Epoxidation (12.8)

$$\overset{}{\underset{}{C=C}} + RCO_3H \longrightarrow \underset{\text{epoxide}}{\overset{\overset{\text{O}}{\diagdown\diagup}}{C-C}}$$

- The mechanism has **one step**.
- **Syn addition** of an O atom occurs.
- The reaction is stereospecific.

---

[b] Anti dihydroxylation (12.9A)

$$\overset{}{\underset{}{C=C}} \xrightarrow[\text{[2] } H_2O \text{ (}H^+ \text{ or } HO^-\text{)}]{\text{[1] RCO}_3H} \quad \underset{\text{1,2-diol}}{\overset{\text{HO}}{\underset{\text{OH}}{C-C}}}$$

- Ring opening of an epoxide intermediate with ^-OH or H_2O forms a 1,2-diol with two OH groups added in an **anti** fashion.

[c] Syn dihydroxylation (12.9B)

$$\begin{array}{c} \text{C=C} \xrightarrow[\substack{\text{or} \\ \text{[1] OsO}_4\text{, NMO; [2] NaHSO}_3 \\ \text{or} \\ \text{KMnO}_4/\text{H}_2\text{O}/\text{HO}^-}]{\text{[1] OsO}_4\text{; [2] NaHSO}_3} \underset{\text{1,2-diol}}{\overset{\text{HO}\quad\text{OH}}{\text{C—C}}} \end{array}$$

- Each reagent adds two new C–O bonds to the C=C in a **syn** fashion.

[d] Oxidative cleavage (12.10)

$$\underset{\text{R}}{\overset{\text{R}}{}}\text{C=C}\underset{\text{H}}{\overset{\text{R'}}{}} \xrightarrow[\substack{\text{[2] Zn/ H}_2\text{O or} \\ \text{CH}_3\text{SCH}_3}]{\text{[1] O}_3} \underset{\text{ketone}}{\overset{\text{R}}{\text{C=O}}} + \underset{\text{aldehyde}}{\overset{\text{R'}}{\text{O=C}}}$$

- Both the σ and π bond of the alkene are cleaved to form two carbonyl groups.

[2] Oxidative cleavage of alkynes (12.11)

$$\underset{\text{Internal alkyne}}{\text{R—C≡C—R'}} \xrightarrow[\text{[2] H}_2\text{O}]{\text{[1] O}_3} \underset{\text{carboxylic acids}}{\underset{\text{HO}}{\overset{\text{R}}{\text{C=O}}} + \underset{\text{OH}}{\overset{\text{R'}}{\text{O=C}}}}$$

- The σ bond and both π bonds of the alkyne are cleaved.

$$\underset{\text{terminal alkyne}}{\text{R—C≡C—H}} \xrightarrow[\text{[2] H}_2\text{O}]{\text{[1] O}_3} \underset{\text{HO}}{\overset{\text{R}}{\text{C=O}}} + \text{CO}_2$$

[3] Oxidation of alcohols (12.12)

$$\underset{\text{1° alcohol}}{\text{R—}\overset{\overset{\text{H}}{|}}{\underset{\underset{\text{H}}{|}}{\text{C}}}\text{—OH}} \xrightarrow{\text{PCC}} \underset{\text{aldehyde}}{\overset{\text{R}}{\underset{\text{H}}{\text{C=O}}}}$$

- Oxidation of a 1° alcohol with PCC stops at the aldehyde stage. Only one C–H bond is replaced by a C–O bond.

$$\underset{\text{1° alcohol}}{\text{R—}\overset{\overset{\text{H}}{|}}{\underset{\underset{\text{H}}{|}}{\text{C}}}\text{—OH}} \xrightarrow[\text{H}_2\text{SO}_4/\text{H}_2\text{O}]{\text{CrO}_3} \underset{\text{carboxylic acid}}{\overset{\text{R}}{\underset{\text{HO}}{\text{C=O}}}}$$

- Oxidation of a 1° alcohol under harsher reaction conditions—CrO$_3$ (or Na$_2$Cr$_2$O$_7$ or K$_2$Cr$_2$O$_7$) + H$_2$O + H$_2$SO$_4$—affords a RCOOH. Two C–H bonds are replaced by two C–O bonds.

$$\underset{\text{2° alcohol}}{\text{R—}\overset{\overset{\text{H}}{|}}{\underset{\underset{\text{R}}{|}}{\text{C}}}\text{—OH}} \xrightarrow{\text{PCC or CrO}_3} \underset{\text{ketone}}{\overset{\text{R}}{\underset{\text{R}}{\text{C=O}}}}$$

- Since a 2° alcohol has only one C–H bond on the carbon bearing the OH group, all Cr^{6+} reagents—PCC, CrO$_3$, Na$_2$Cr$_2$O$_7$, or K$_2$Cr$_2$O$_7$—oxidize a 2° alcohol to a ketone.

[4] Asymmetric epoxidation of allylic alcohols (12.14)

$$\underset{\text{H}}{\overset{\text{R}}{}}\text{C=C}\underset{\text{OH}}{\overset{\text{H}}{}} \xrightarrow[\text{Ti[OCH(CH}_3)_2]_4]{(\text{CH}_3)_3\text{C—OOH}} \underset{\text{with (–)-DET}}{\overset{\text{O}}{\underset{\text{H}}{\text{R—}\triangle\text{—OH}}}} \quad\text{or}\quad \underset{\text{with (+)-DET}}{\overset{\text{H}}{\underset{\text{O}}{\text{R—}\triangle\text{—OH}}}}$$

Chapter 12: Answers to Problems

12.1 *Oxidation* results in an *increase* in the number of C–Z bonds (usually C–O bonds) *or* a *decrease* in the number of C–H bonds.
Reduction results in a *decrease* in the number of C–Z bonds (usually C–O bonds) *or* an *increase* in the number of C–H bonds.

a. **oxidation**

c. **oxidation**

b. **reduction**

d. $CH_2=CH_2 \longrightarrow CH_3CH_2Cl$ **neither**

1 new C–H bond
and 1 new C–Cl bond

12.2 Hydrogenation is the addition of hydrogen. When alkenes are hydrogenated, they are *reduced* by the addition of H_2 to the π bond. To draw the alkane product, add a H to each C of the double bond.

a.

c.

b.

12.3 Cis alkenes are less stable than trans alkenes, so they have larger heats of hydrogenation. Increasing alkyl substitution increases the stability of a C=C, decreasing the heat of hydrogenation.

a.

cis alkane
less stable
larger **heat of hydrogenation**

trans alkane

b.

trisubstituted

disubstituted
less stable
larger **heat of hydrogenation**

12.4 Hydrogenation products must be identical to use hydrogenation data to evaluate the relative stability of the starting materials.

2-methyl-2-pentene

Different products are formed. Hydrogenation data can't be used to determine the relative stability of the starting materials.

3-methyl-1-pentene

12.5 Increasing alkyl substitution on the C=C decreases the rate of hydrogenation.

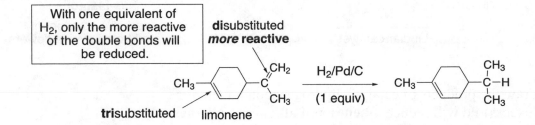

12.6

a. new stereogenic center

Two enantiomers are formed in equal amounts:

b.

diastereomers

12.7

Compound	Molecular formula before hydrogenation	Molecular formula after hydrogenation	Number of rings	Number of π bonds
A	$C_{10}H_{12}$	$C_{10}H_{16}$	3	2
B	C_4H_8	C_4H_{10}	0	1
C	C_6H_8	C_6H_{12}	1	2

12.8

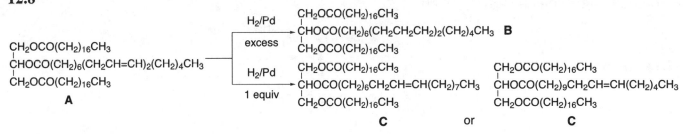

A has 2 double bonds. **lowest melting point**

C has 1 double bond. **intermediate melting point**

B has 0 double bonds. **highest melting point**

12.9

$$HC{\equiv}CCH_2CH_2CH_3$$
or
$$CH_3C{\equiv}CCH_2CH_3 \quad \xrightarrow[\text{Pd}]{H_2} \quad CH_3CH_2CH_2CH_2CH_3$$

12.10

cis-jasmone
(perfume component
isolated from jasmine flowers)

12.11 To draw the products of catalytic hydrogenation remember:
- H_2 (excess)/Pd will reduce **alkenes and alkynes to alkanes**.
- H_2 (excess)/Lindlar catalyst will reduce *only* **alkynes to cis alkenes**.

a. $CH_2=CHCH_2CH_2-C\equiv C-CH_3$ $\xrightarrow[\text{Pd}]{\text{H}_2\text{ (excess)}}$ $CH_3CH_2CH_2CH_2CH_2CH_2CH_3$

b. $CH_2=CHCH_2CH_2-C\equiv C-CH_3$ $\xrightarrow[\substack{\text{Lindlar}\\\text{catalyst}}]{\text{H}_2\text{ (excess)}}$

12.12 Use the directions from Answer 12.11.

a. $CH_3OCH_2CH_2C\equiv CCH_2CH(CH_3)_2$ $\xrightarrow[\text{Pd/C}]{\text{H}_2\text{ (excess)}}$ $CH_3OCH_2CH_2CH_2CH_2CH_2CH(CH_3)_2$

b. $CH_3OCH_2CH_2C\equiv CCH_2CH(CH_3)_2$ $\xrightarrow[\text{Lindlar catalyst}]{\text{H}_2\text{ (1 equiv)}}$

c. $CH_3OCH_2CH_2C\equiv CCH_2CH(CH_3)_2$ $\xrightarrow[\text{Lindlar catalyst}]{\text{H}_2\text{ (excess)}}$

d. $CH_3OCH_2CH_2C\equiv CCH_2CH(CH_3)_2$ $\xrightarrow{\text{Na/NH}_3}$

12.13

$CH_3-C\equiv C-CH_2CH_2CH_3$

$\xrightarrow[\text{Pd}]{\text{D}_2}$ $CH_3CD_2CD_2CH_2CH_2CH_3$

$\xrightarrow[\substack{\text{Lindlar}\\\text{catalyst}}]{\text{D}_2}$

$\xrightarrow[\text{ND}_3]{\text{Na}}$

12.14 LiAlH$_4$ reduces alkyl halides to alkanes and epoxides to alcohols.

a. (structure with CI) $\xrightarrow[\text{[2] H}_2\text{O}]{\text{[1] LiAlH}_4}$ (structure with H)

H replaces Cl.

b. (cyclopentane epoxide with CH$_3$) $\xrightarrow[\text{[2] H}_2\text{O}]{\text{[1] LiAlH}_4}$ (cyclopentane with CH$_3$, OH, H)

12.15 To draw the product, add an O atom across the π bond of the C=C.

a. $(CH_3)_2C=CH_2$ $\xrightarrow{\text{mCPBA}}$ $(CH_3)_2\overset{O}{\overset{\triangle}{C}}-CH_2$

b. $(CH_3)_2C=C(CH_3)_2$ $\xrightarrow{\text{mCPBA}}$ $(CH_3)_2\overset{O}{\overset{\triangle}{C}}-C(CH_3)_2$

c. (cyclohexane)$=CH_2$ $\xrightarrow{\text{mCPBA}}$ (cyclohexane epoxide)CH_2

12.16 For epoxidation reactions:
- There are two possible products: O adds from above and below the double bond.
- Substituents on the C=C retain their original configuration in the products.

a. (cis-2-butene structure) $\xrightarrow{\text{mCPBA}}$ (epoxide) + (epoxide enantiomer) enantiomers

b. (3,4-hexene structure) $\xrightarrow{\text{mCPBA}}$ (epoxide) + (epoxide) identical

c. (1-methylcyclohexene) $\xrightarrow{\text{mCPBA}}$ (epoxide with CH$_3$, H) + (epoxide with CH$_3$, H) enantiomers

12.17 Treatment of an alkene with a peroxyacid followed by H$_2$O/HO⁻ adds two hydroxyl groups in an **anti** fashion. *cis*-2-Butene and *trans*-2-butene yield different products of dihydroxylation. *cis*-2-Butene gives a mixture of two enantiomers and *trans*-2-butene gives a meso compound. The reaction is stereospecific because two stereoisomeric starting materials give different products that are also stereoisomers of each other.

(cis-2-butene structure) $\xrightarrow[\text{[2] H}_2\text{O/HO}^-]{\text{[1] RCO}_3\text{H}}$ (diol structure) + (diol structure)

cis-2-butene   **enantiomers**

(trans-2-butene structure) $\xrightarrow[\text{[2] H}_2\text{O/HO}^-]{\text{[1] RCO}_3\text{H}}$ (diol structure) + (diol structure)

trans-2-butene   **identical**
meso compound

12.18 Treatment of an alkene with OsO₄ adds two hydroxyl groups in a **syn** fashion. *cis*-2-Butene and *trans*-2-butene yield different stereoisomers in this dihydroxylation, so the reaction is stereospecific.

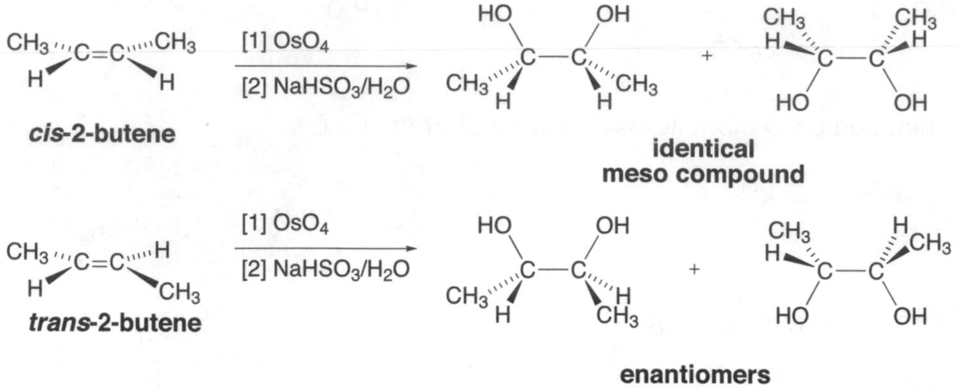

identical meso compound

enantiomers

12.19

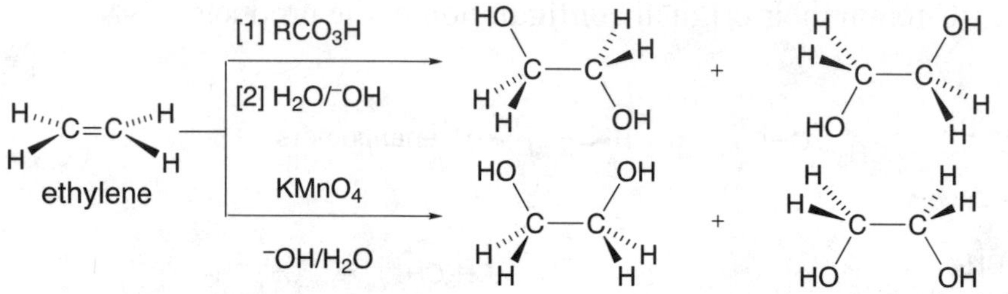

There are no stereogenic centers, so all the products are identical ($HOCH_2CH_2OH$).

12.20 To draw the oxidative cleavage products:
- **Locate all the π bonds** in the molecule.
- **Replace all C=C's with *two* C=O's.**

Replace this π bond with two C=O's.

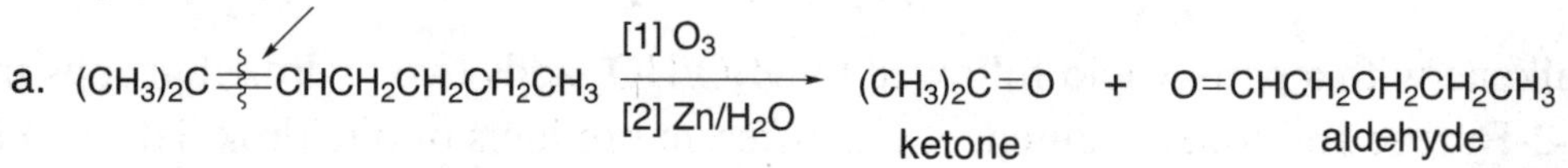

a.

One **ketone** and one **aldehyde** are formed.

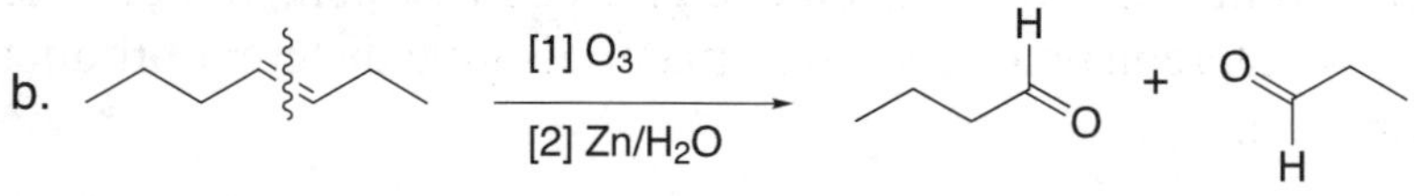

b.

Two **aldehydes** are formed.

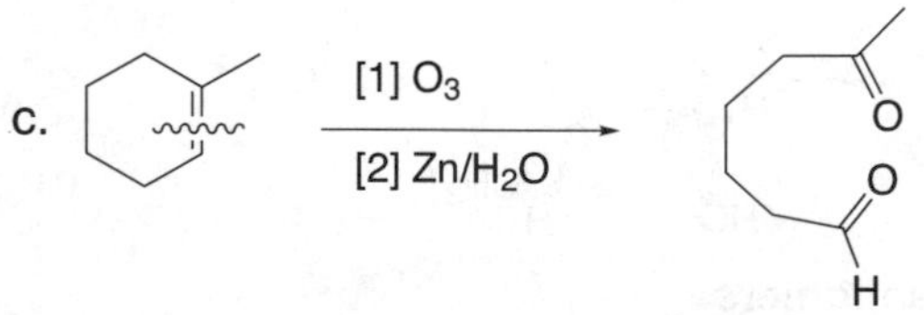

c.

A **dicarbonyl** compound is formed.

12.21 To find the alkene that yields the oxidative cleavage products:
- **Find the two carbonyl groups** in the products.
- **Join the two carbonyl carbons** together with a double bond. This is the double bond that was broken during ozonolysis.

a. $(CH_3)_2C=O$ + $(CH_3CH_2)_2C=O$ $\Longrightarrow$ $(CH_3)_2C=C(CH_2CH_3)_2$

Join these two C's.

c.
$$\begin{array}{c} CH_3 \\ | \\ CH_3 \end{array} C=O \text{ only} \Longrightarrow \begin{array}{cc} CH_3 & CH_3 \\ C=C \\ CH_3 & CH_3 \end{array}$$

With only one product, the alkene must be symmetrical around the double bond. Join this C to the same C in another identical molecule.

b. (cyclohexanone) + CH_3CHO $\Longrightarrow$ (cyclohexane with =CHCH_3)

Join these two C's.

12.22 To draw the products of oxidative cleavage of alkynes:
- **Locate the triple bond**.
- For internal alkynes, **convert the *sp* hybridized C to COOH.**
- For terminal alkynes, the *sp* hybridized C becomes CO_2.

a. $CH_3CH_2-C\equiv C$ $CH_2CH_2CH_3$

internal alkyne

$\xrightarrow[\text{[2] } H_2O]{\text{[1] } O_3}$ $CH_3CH_2\overset{O}{\underset{}{C}}OH$ + $HO\overset{O}{\underset{}{C}}CH_2CH_2CH_3$

h. (phenyl)$-C\equiv C-$(phenyl)

internal alkyne

$\xrightarrow[\text{[2] } H_2O]{\text{[1] } O_3}$ (benzoic acid) + (benzoic acid)

identical compounds

c.
terminal alkyne   internal alkyne

$H-C\equiv C-CH_2-CH_2-C\equiv C-CH_3$ $\xrightarrow[\text{[2] } H_2O]{\text{[1] } O_3}$ CO_2 + $HO\overset{O}{\underset{O}{C}}\cdots\overset{O}{\underset{}{C}}OH$ + $HO\overset{O}{\underset{}{C}}CH_3$

12.23 For **oxidation of alcohols**, remember:
- **1° alcohols** are oxidized to aldehydes with PCC.
- **1° alcohols** are oxidized to carboxylic acids with oxidizing agents like CrO_3 or $Na_2Cr_2O_7$.
- **2° alcohols** are oxidized to ketones with all Cr^{6+} reagents.

a. (pentanol chain)OH $\xrightarrow{PCC}$ (aldehyde)

b. (butanol with OH) $\xrightarrow{PCC}$ (ketone)

c. (cyclohexylmethanol)OH $\xrightarrow[\text{H}_2\text{SO}_4/\text{H}_2\text{O}]{CrO_3}$ (carboxylic acid)

d. (1-cyclohexylethanol, OH) $\xrightarrow[\text{H}_2\text{SO}_4/\text{H}_2\text{O}]{CrO_3}$ (ketone)

12.24

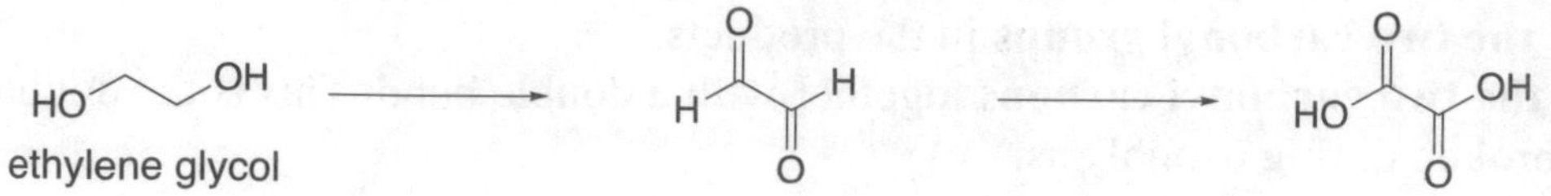
HO
OH
ethylene glycol
O
H
H
O
O
HO
OH
O

12.25 To draw the products of a **Sharpless epoxidation**:

- With the C=C vertical, draw the allylic alcohol with the OH on the **bottom right** of the alkene.
- Add the new oxygen **above** the plane if (–)-DET is used and **below** the plane if (+)-DET is used.

a.

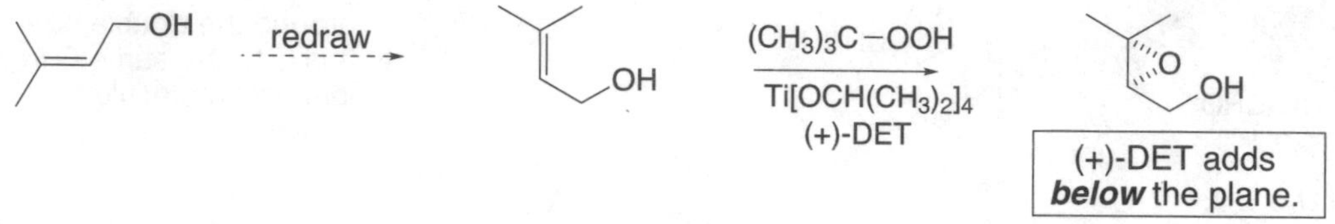
OH
redraw
OH
(CH3)3C–OOH
Ti[OCH(CH3)2]4
(+)-DET
O
OH
(+)-DET adds
below the plane.

b.

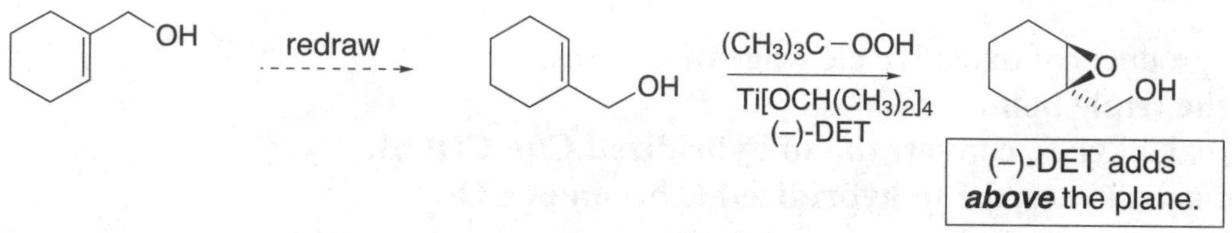
OH
redraw
OH
(CH3)3C–OOH
Ti[OCH(CH3)2]4
(–)-DET
O
OH
(–)-DET adds
above the plane.

12.26 Sharpless epoxidation needs an ***allylic alcohol*** as the starting material. Alkenes with no allylic OH group will not undergo reaction with the Sharpless reagent.

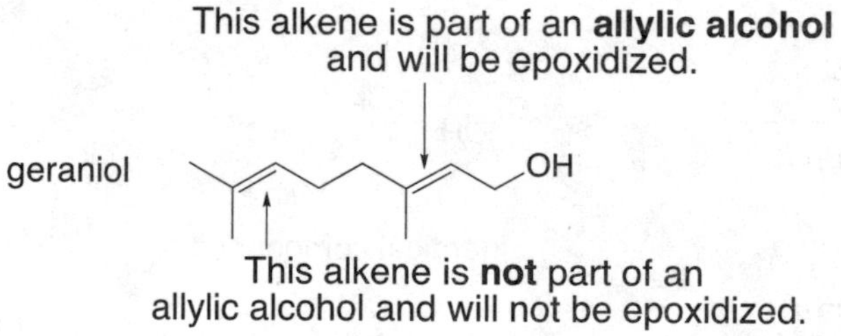
This alkene is part of an allylic alcohol
and will be epoxidized.
geraniol
OH
This alkene is not part of an
allylic alcohol and will not be epoxidized.

12.27 Use the rules from Answer 12.1.

a. (cyclohexyl)–C≡CH ⟶ (cyclohexyl)–CH₂CH₃ **reduction**

b. (structure with OH) ⟶ (structure with C=O and OH) **oxidation**

c. CH_3CH_2Br ⟶ $CH_2=CH_2$ **neither**
1 C–H and 1 C–Br
bond are removed.

d. (epoxide on cyclohexane) ⟶ (cyclohexanol) **reduction**

e. $CH_2=CH_2$ ⟶ $ClCH_2CH_2Cl$ **oxidation**
(2 new C–Cl bonds)

f. HO–(benzene ring)–OH ⟶ $O=$(ring)$=O$ **oxidation**

12.28 Use the principles from Answer 12.2 and draw the products of syn addition of H_2 from above and below the C=C.

a. (pentene structure) $\xrightarrow{\text{H}_2 \; \text{Pd/C}}$ (pentane structure)

b. (dimethyl cyclohexene with CH₃) $\xrightarrow{\text{H}_2 \; \text{Pd/C}}$ (product with CH₃, CH₃) + (product with CH₃, CH₃)

c. (cyclohexene with CH₂CH₃ and CH₃) $\xrightarrow{\text{H}_2 \; \text{Pd/C}}$ (product CH₂CH₃, CH₃) + (product CH₂CH₃, CH₃)

d. CH_3CH_2 / $(CH_3)_2CH$ $C=CH_2$ $\xrightarrow{\text{H}_2 \; \text{Pd/C}}$ (CH₂CH₃ product, H–C–CH₃, (CH₃)₂CH) + (CH₂CH₃ product, (CH₃)₂CH–C–CH₃, H)

12.29 Increasing alkyl substitution increases alkene stability, decreasing the heat of hydrogenation.

2-methyl-2-butene | **2-methyl-1-butene** | **3-methyl-1-butene**

trisubstituted | **di**substituted | **mono**substituted

smallest $\Delta H^\circ = -26.9$ kcal/mol | **intermediate** $\Delta H^\circ = -28.5$ kcal/mol | **largest** $\Delta H^\circ = -30.3$ kcal/mol

12.30

A possible structure:

a. Compound **A**: molecular formula C_5H_8: hydrogenated to C_5H_{10}.
2 degrees of unsaturation, 1 is hydrogenated.
1 ring and 1 π bond ⟶ (methylenecyclobutane structure)

b. Compound **B**: molecular formula $C_{10}H_{16}$: hydrogenated to $C_{10}H_{18}$.
3 degrees of unsaturation, 1 is hydrogenated.
2 rings and 1 π bond ⟶ (bicyclic structure)

c. Compound **C**: molecular formula C_8H_8: hydrogenated to C_8H_{16}.
5 degrees of unsaturation, 4 are hydrogenated.
1 ring and 4 π bonds ⟶ (cyclooctatetraene structure)

12.31

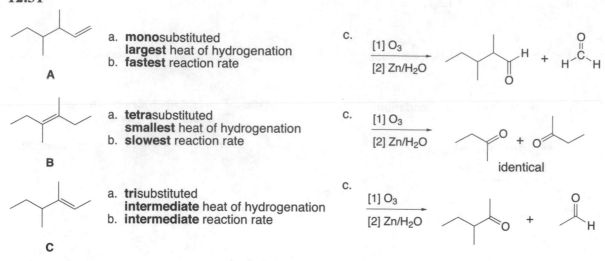

A
- a. **mono**substituted
     largest heat of hydrogenation
- b. **fastest** reaction rate

B
- a. **tetra**substituted
     smallest heat of hydrogenation
- b. **slowest** reaction rate

C
- a. **tri**substituted
     intermediate heat of hydrogenation
- b. **intermediate** reaction rate

12.32 Work backwards to find the alkene that will be hydrogenated to form 2-methylpentane.

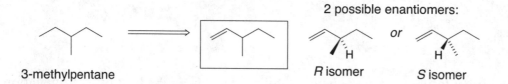

2-methylpentane

12.33 Work backwards to find the alkene that will be hydrogenated to form 3-methylpentane.

3-methylpentane

2 possible enantiomers:

R isomer   *S* isomer

12.34 Hydrogenation without a catalyst has a large E_a, making it a kinetically slow reaction, so hydrogenation only occurs in the presence of a catalyst. The mechanism with the catalyst is multistep, but no step has a transition state as high in energy as the uncatalyzed reaction. It is thermodynamically favorable because energy is released during hydrogenation: the starting materials are at a higher energy than the products, giving hydrogenation a negative ΔH°.

12.35

a. [cyclopentene] $\xrightarrow[\text{Pd/C}]{\text{H}_2}$ [cyclopentane]

b. [cyclopentene] $\xrightarrow[\text{Lindlar catalyst}]{\text{H}_2}$ no reaction

c. [cyclopentene] $\xrightarrow[\text{NH}_3]{\text{Na}}$ no reaction

d. [cyclopentene] $\xrightarrow{\text{CH}_3\text{CO}_3\text{H}}$ [epoxide O]

e. [cyclopentene] $\xrightarrow[\text{[2] H}_2\text{O/HO}^-]{\text{[1] CH}_3\text{CO}_3\text{H}}$ [trans-diol] + [trans-diol] anti addition

f. [cyclopentene] $\xrightarrow[\text{[2] NaHSO}_3]{\text{[1] OsO}_4 + \text{NMO}}$ [cis-diol] syn addition

g. [cyclopentene] $\xrightarrow[\text{H}_2\text{O/HO}^-]{\text{KMnO}_4}$ [cis-diol] syn addition

h. [cyclopentene] $\xrightarrow[\text{[2] H}_2\text{O}]{\text{[1] LiAlH}_4}$ no reaction

i. [cyclopentene] $\xrightarrow[\text{[2] CH}_3\text{SCH}_3]{\text{[1] O}_3}$ [dialdehyde: H–C(=O)–CH$_2$CH$_2$CH$_2$–C(=O)–H]

j. [cyclopentene] $\xrightarrow[\substack{\text{Ti[OCH(CH}_3)_2]_4 \\ (-)\text{-DET}}]{\text{(CH}_3)_3\text{COOH}}$ no reaction

k. [cyclopentene] $\xrightarrow{\text{mCPBA}}$ [epoxide O]

l. [epoxide] $\xrightarrow[\text{[2] H}_2\text{O}]{\text{[1] LiAlH}_4}$ [cyclopentanol OH]

12.36

a. CH$_3$CH$_2$CH$_2$—C≡C—CH$_2$CH$_2$CH$_3$ $\xrightarrow[\text{Pd/C}]{\text{H}_2 \text{ (excess)}}$ [octane chain]

b. CH$_3$CH$_2$CH$_2$—C≡C—CH$_2$CH$_2$CH$_3$ $\xrightarrow[\text{Lindlar catalyst}]{\text{H}_2}$

$\begin{array}{c} \text{CH}_3\text{CH}_2\text{CH}_2 \qquad \text{CH}_2\text{CH}_2\text{CH}_3 \\ \text{C=C} \\ \text{H} \qquad\qquad \text{H} \end{array}$ cis alkene

c. CH$_3$CH$_2$CH$_2$—C≡C—CH$_2$CH$_2$CH$_3$ $\xrightarrow[\text{NH}_3]{\text{Na}}$

$\begin{array}{c} \text{H} \qquad\qquad \text{CH}_2\text{CH}_2\text{CH}_3 \\ \text{C=C} \\ \text{CH}_3\text{CH}_2\text{CH}_2 \qquad \text{H} \end{array}$ trans alkene

d. CH$_3$CH$_2$CH$_2$—C≡C—CH$_2$CH$_2$CH$_3$ $\xrightarrow[\text{[2] H}_2\text{O}]{\text{[1] O}_3}$

$\text{CH}_3\text{CH}_2\text{CH}_2–\overset{\text{O}}{\overset{\|}{\text{C}}}–\text{OH}$ + $\text{CH}_3\text{CH}_2\text{CH}_2–\overset{\text{O}}{\overset{\|}{\text{C}}}–\text{OH}$

identical

12.37

a. $\begin{array}{c} \text{CH}_3\text{CH}_2 \qquad \text{H} \\ \text{C=C} \\ \text{CH}_3 \qquad \text{CH}_2\text{OH} \end{array}$ $\xrightarrow[\text{Pd/C}]{\text{H}_2}$ $\begin{array}{c} \text{CH}_3\text{CH}_2 \ \text{H} \\ \text{H–}\overset{*}{\text{C}}–\text{C–CH}_2\text{OH} \\ \text{CH}_3 \ \text{H} \end{array}$

[* = new stereogenic center.]

$\text{CH}_3\text{CH}_2–\overset{\text{CH}_3}{\underset{\text{H}}{\text{C}}}\cdots\text{CH}_2\text{CH}_2\text{OH}$ + $\text{HOCH}_2\text{CH}_2\cdots\overset{\text{CH}_3}{\underset{\text{H}}{\text{C}}}–\text{CH}_2\text{CH}_3$

Two enantiomers are formed.

b. $\begin{array}{c} \text{CH}_3\text{CH}_2 \qquad \text{H} \\ \text{C=C} \\ \text{CH}_3 \qquad \text{CH}_2\text{OH} \end{array}$ $\xrightarrow{\text{mCPBA}}$ $\text{CH}_3\text{CH}_2\cdots\overset{\text{H}}{\underset{\text{CH}_3}{\text{epoxide}}}\text{CH}_2\text{OH}$

+

$\text{CH}_3\text{CH}_2\cdots\overset{\text{H}}{\underset{\text{CH}_3}{\text{epoxide}}}\text{CH}_2\text{OH}$

e. $\text{CH}_3\cdots\overset{}{\text{C=C}}\cdots\text{CH}_2\text{OH}$ (CH_3CH_2, H) $\xrightarrow[\substack{\text{Ti[OCH(CH}_3)_2]_4 \\ (+)\text{-DET}}]{\text{(CH}_3)_3\text{COOH}}$ $\text{CH}_3\cdots\overset{\text{CH}_2\text{OH}}{\underset{\text{H}}{\text{epoxide}}}$ (CH_3CH_2)

f. $\text{CH}_3\cdots\overset{}{\text{C=C}}\cdots\text{CH}_2\text{OH}$ (CH_3CH_2, H) $\xrightarrow[\substack{\text{Ti[OCH(CH}_3)_2]_4 \\ (-)\text{-DET}}]{\text{(CH}_3)_3\text{COOH}}$ $\text{CH}_3\cdots\overset{}{\underset{\text{H}}{\text{epoxide}}}\cdots\text{CH}_2\text{OH}$ (CH_3CH_2)

c. 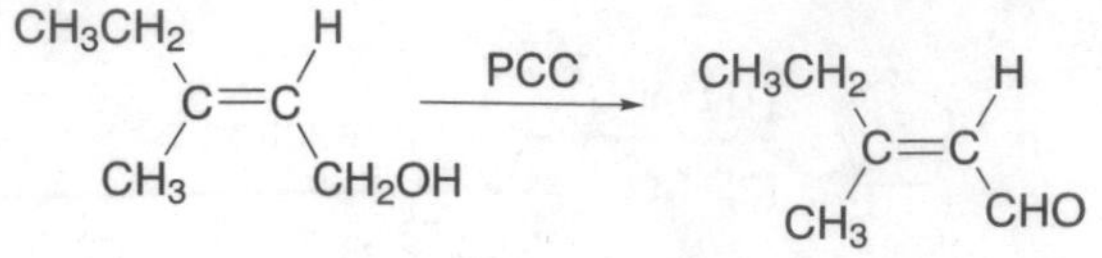
CH3CH2
H
C=C
CH3
CH2OH
PCC
CH3CH2
H
C=C
CH3
CHO

d. 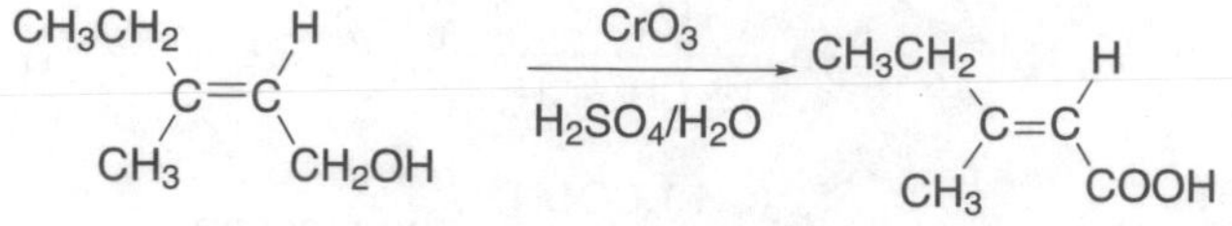
CH3CH2
H
C=C
CH3
CH2OH
CrO3
H2SO4/H2O
CH3CH2
H
C=C
CH3
COOH

g. 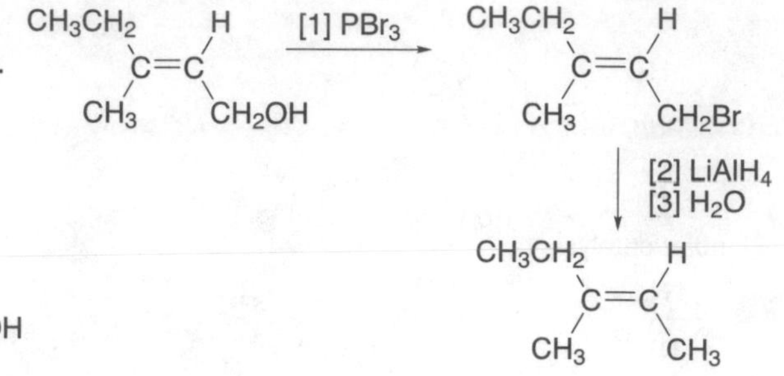
CH3CH2
H
C=C
CH3
CH2OH
[1] PBr3
CH3CH2
H
C=C
CH3
CH2Br
[2] LiAlH4
[3] H2O
CH3CH2
H
C=C
CH3
CH3

12.38

a. 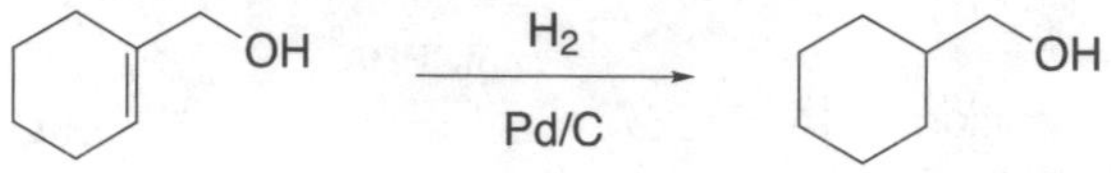
OH
H2
Pd/C
OH

b. 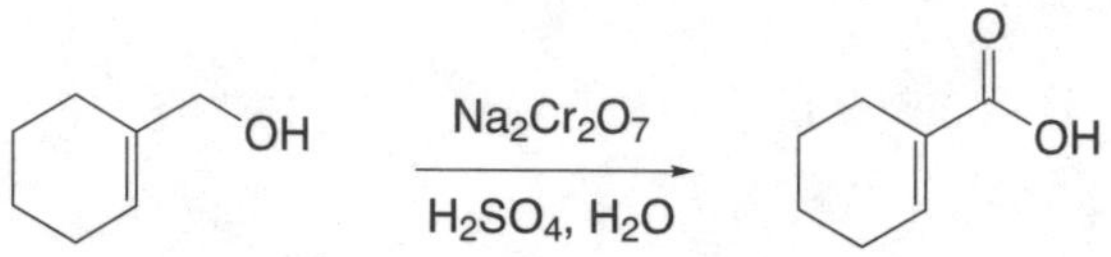
OH
Na2Cr2O7
H2SO4, H2O
O
OH

c. 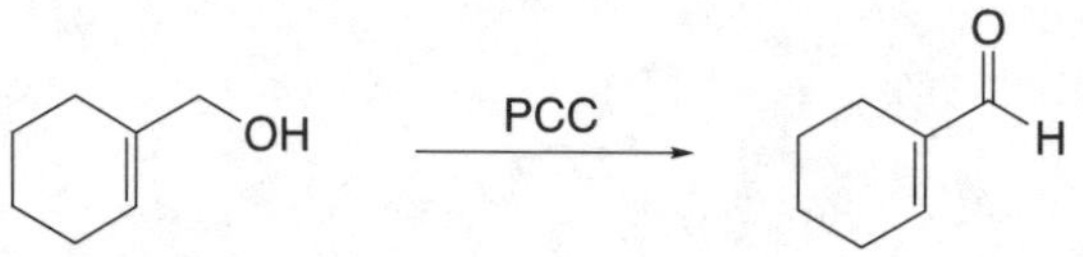
OH
PCC
O
H

d. 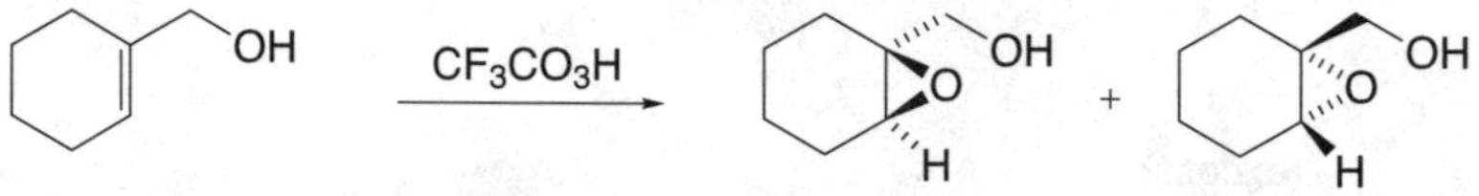
OH
CF3CO3H
OH
O
H
+
OH
O
H

e. 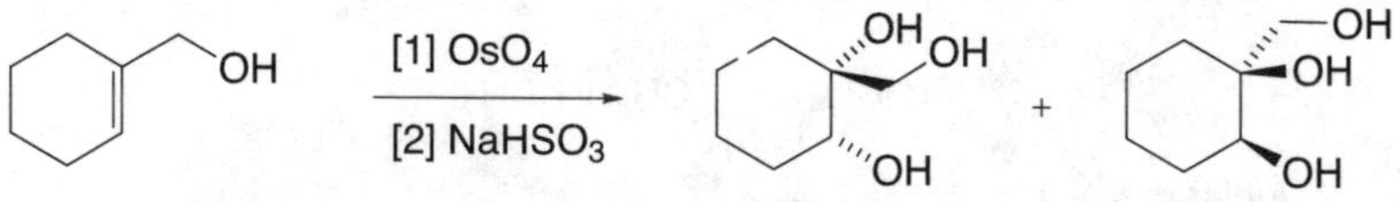
OH
[1] OsO4
[2] NaHSO3
OH
OH
OH
+
OH
OH
OH

f. 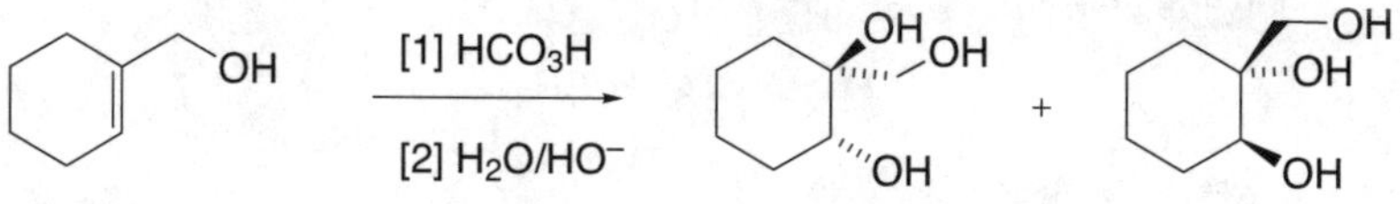
OH
[1] HCO3H
[2] H2O/HO⁻
OH
OH
OH
+
OH
OH
OH

g. 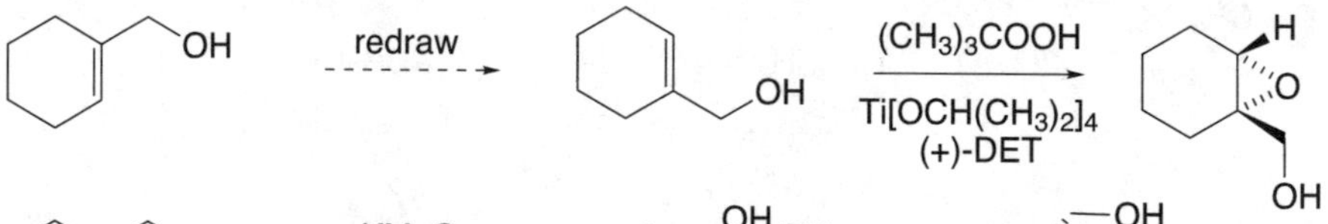
OH
redraw
OH
(CH3)3COOH
Ti[OCH(CH3)2]4
(+)-DET
H
O
OH

h. 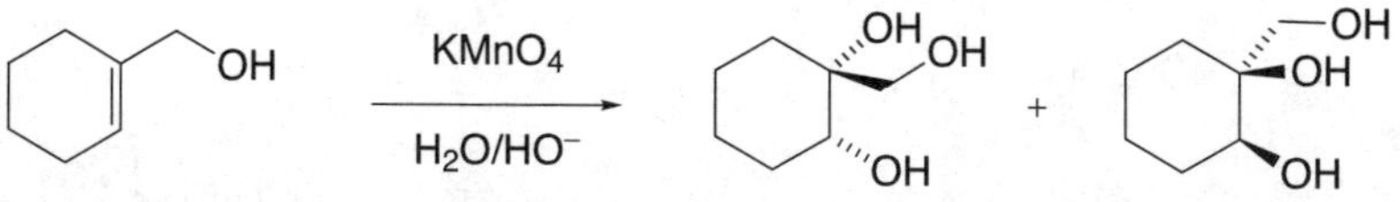
OH
KMnO4
H2O/HO⁻
OH
OH
OH
+
OH
OH
OH

12.39

a. 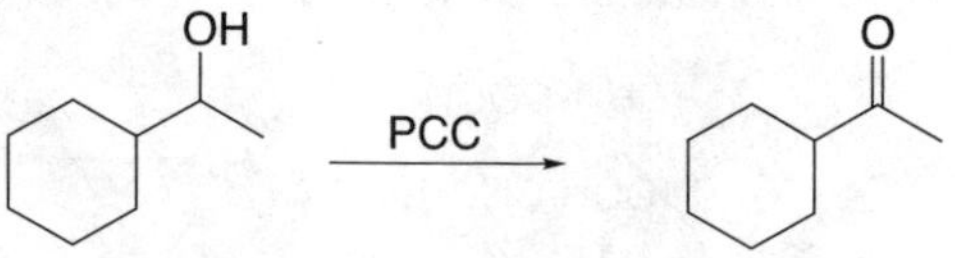
OH
PCC
O

b. 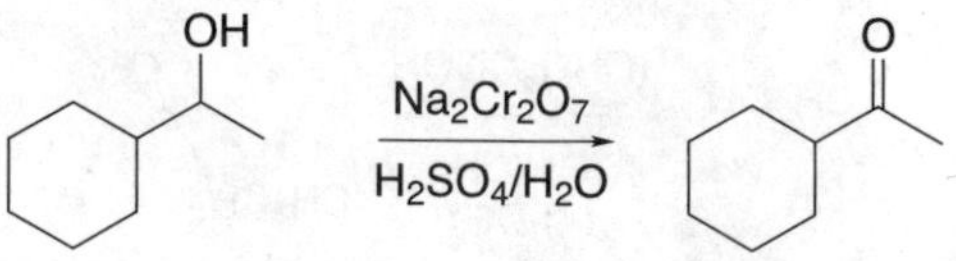
OH
Na2Cr2O7
H2SO4/H2O
O

c. 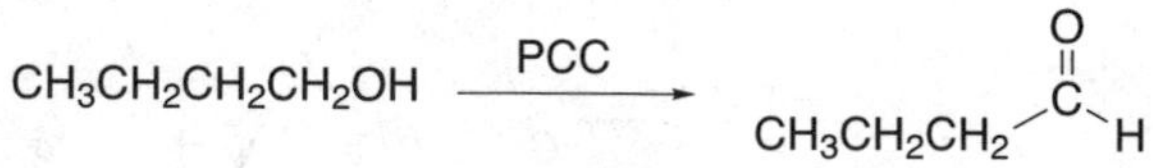
CH3CH2CH2CH2OH
PCC
O
CH3CH2CH2
C
H

d. 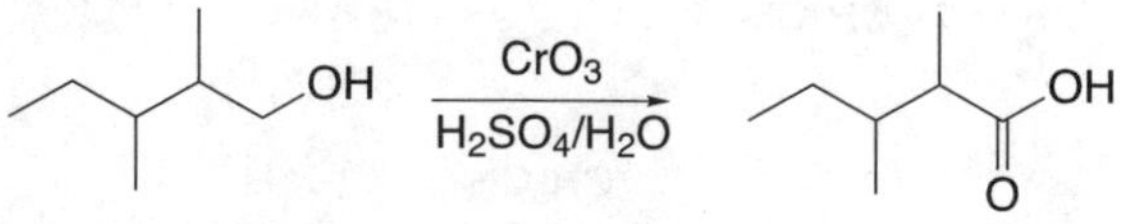
OH
CrO3
H2SO4/H2O
OH
O

12.40

a. [1] $SOCl_2$ → [2] $LiAlH_4$ / [3] H_2O

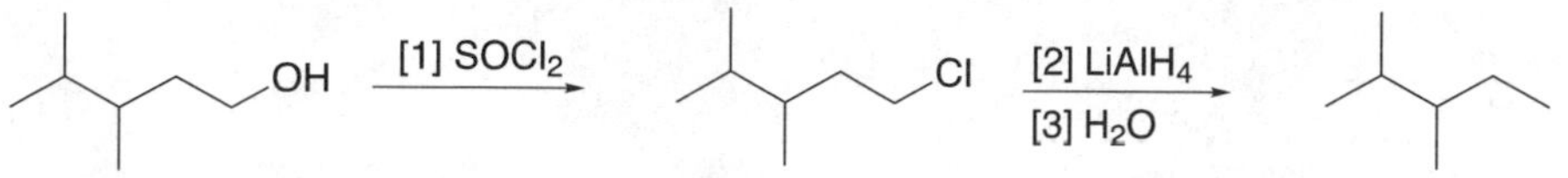

b. $=CH_2$ [1] OsO_4 / [2] $NaHSO_3$ → OH CH_2OH

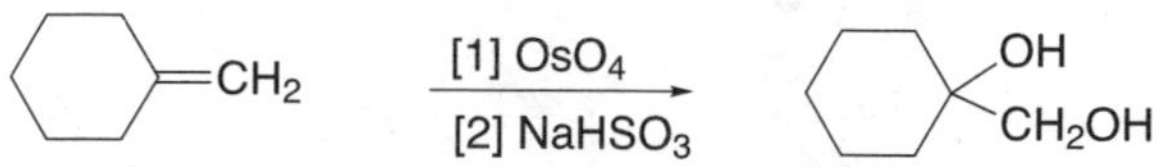

c. $=CH_2$ [1] mCPBA → O [2] $LiAlH_4$ / [3] H_2O → OH CH_3

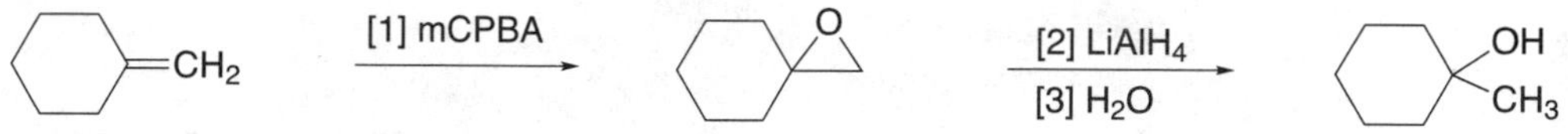

d. H_2 / Lindlar catalyst

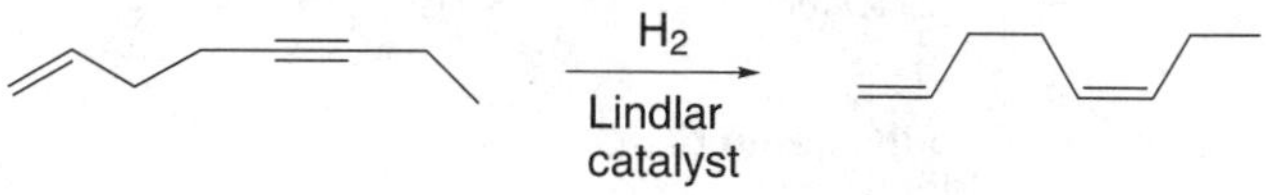

12.41

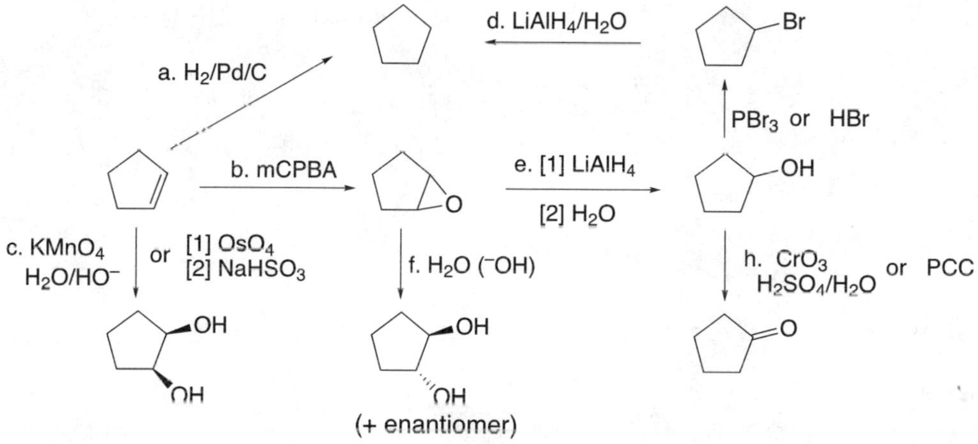

a. H_2/Pd/C

d. $LiAlH_4/H_2O$ ← Br

PBr$_3$ or HBr

b. mCPBA

e. [1] $LiAlH_4$ / [2] H_2O ← OH

c. $KMnO_4$ H_2O/HO$^-$ or [1] OsO_4 [2] $NaHSO_3$

f. H_2O ($^-$OH)

h. CrO_3 H_2SO_4/H_2O or PCC

OH OH

OH OH (+ enantiomer)

O

12.42

a. **A** C_8H_{12} [1] O_3 / [2] CH_2SCH_3 → (product: H–CO–CH$_2$CH$_2$–CO–H)

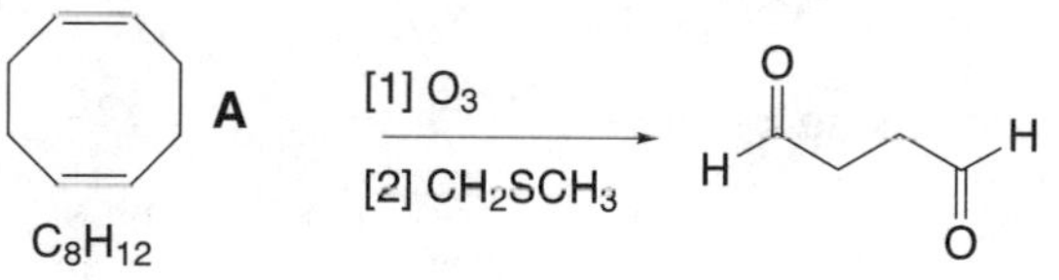

b. C_6H_{10} **B** H_2 (excess) / Pd → → NaNH$_2$ / CH$_3$I → C_7H_{12} **C**

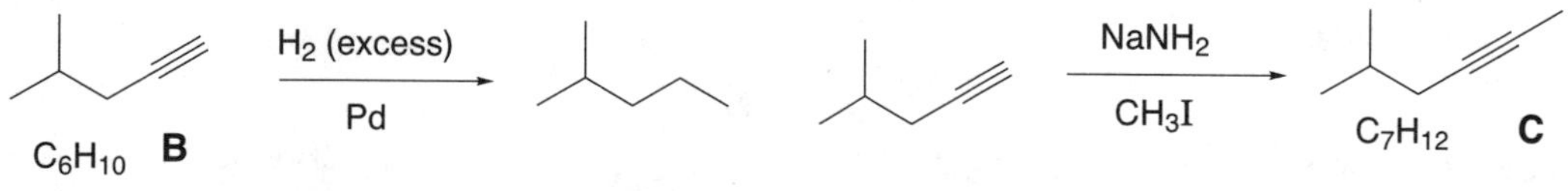

12.43 Use the directions from Answer 12.20.

a. $(CH_3CH_2)_2C=CHCH_2CH_3$ 1) O_3 / 2) CH_3SCH_3 → $(CH_3CH_2)_2C=O$ + $O=CHCH_2CH_3$

b. 1) O_3 / 2) Zn/H_2O → O + $O=C(CH_3)_2$ (with CH_3 groups)

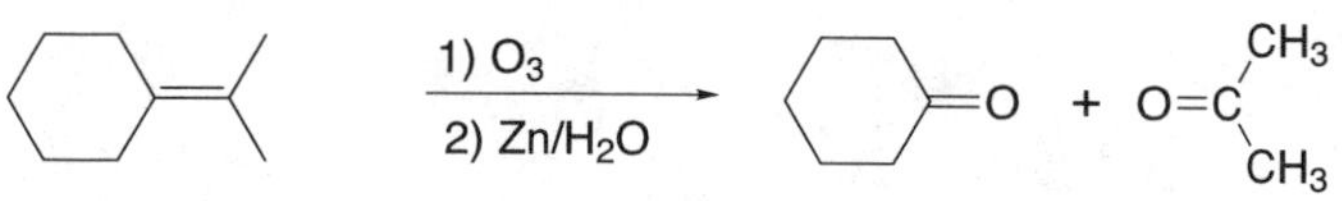

c.

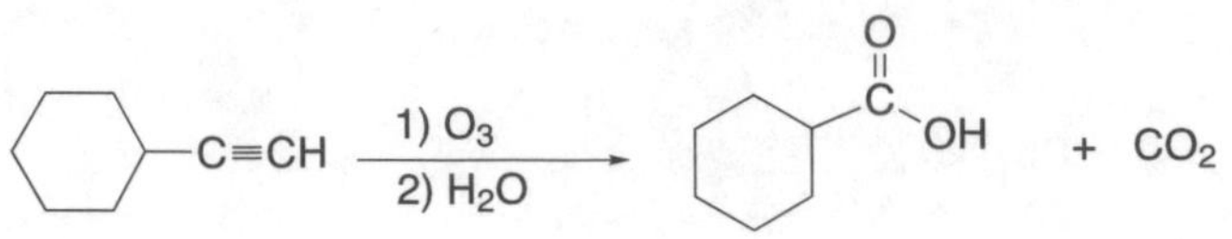

d.

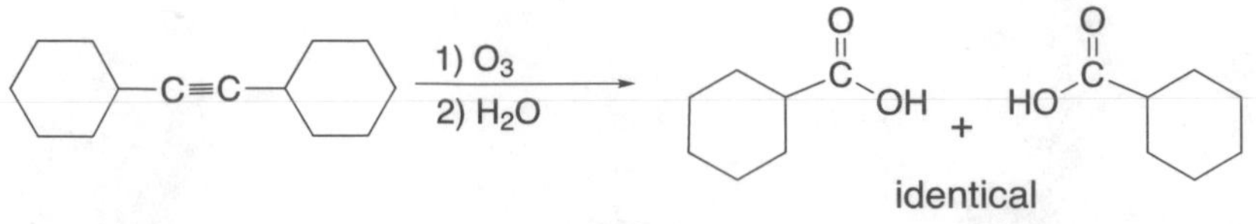

identical

12.44

a. $(CH_3)_2C=O$ and $CH_2=O$ $\Longrightarrow$ $(CH_3)_2C=CH_2$

b. 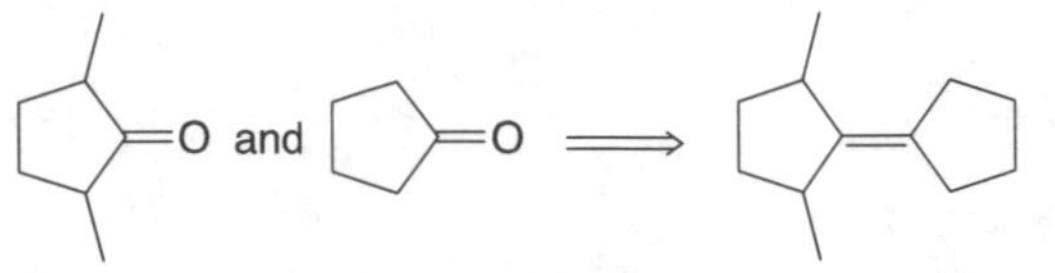 =O and =O $\Longrightarrow$

c. $CH_3CH_2CH_2CHO$ only $\Longrightarrow$ $CH_3CH_2CH_2CH=CHCH_2CH_2CH_3$

Join this C to the same C
in another identical molecule.

d.

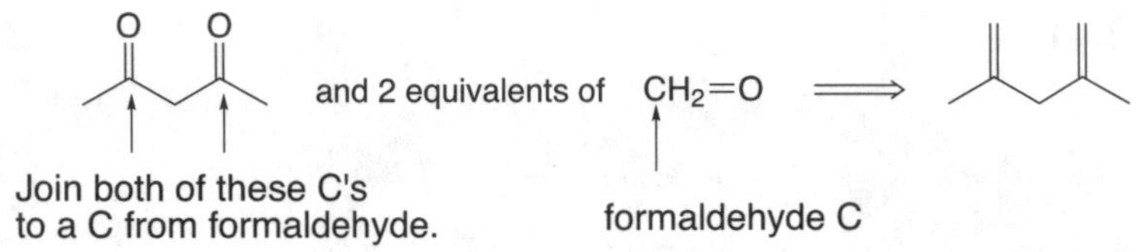

and 2 equivalents of $CH_2=O$ $\Longrightarrow$

Join both of these C's
to a C from formaldehyde.

formaldehyde C

12.45 Use the directions from Answer 12.21.

Join these two C's.

a. 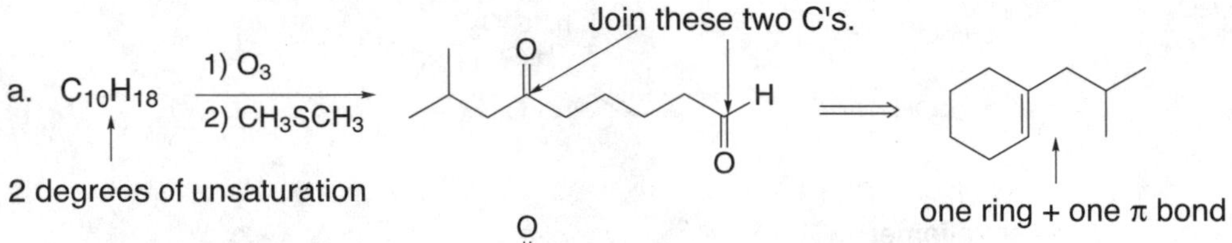 $C_{10}H_{18}$ $\xrightarrow[\text{2) }CH_3SCH_3]{\text{1) }O_3}$ $\Longrightarrow$

2 degrees of unsaturation

one ring + one π bond

b. 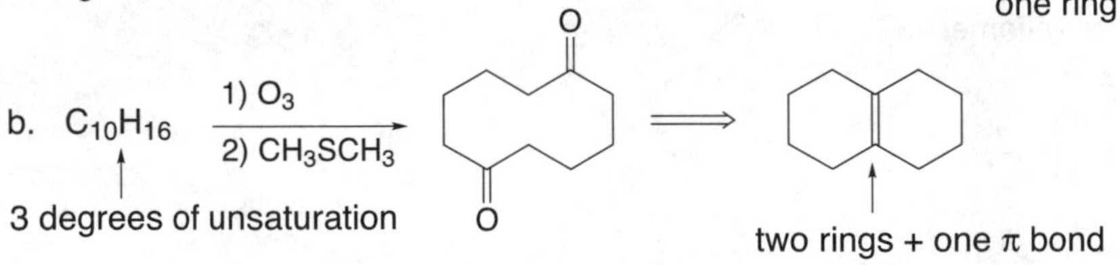 $C_{10}H_{16}$ $\xrightarrow[\text{2) }CH_3SCH_3]{\text{1) }O_3}$ $\Longrightarrow$

3 degrees of unsaturation

two rings + one π bond

12.46

Join these two C's.

a. $CH_3CH_2CH_2CH_2COOH$ and CO_2 $\Longrightarrow$ $CH_3CH_2CH_2CH_2C\equiv CH$

c. Join these two C's.

and CH_3COOH $\Longrightarrow$

Join these two C's.

b. CH_3CH_2COOH and $CH_3CH_2CH_2COOH$ $\Longrightarrow$ $CH_3CH_2C\equiv CCH_2CH_2CH_3$

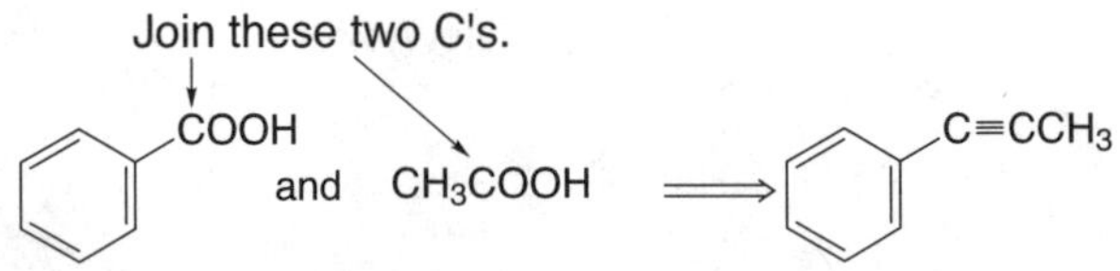

12.47

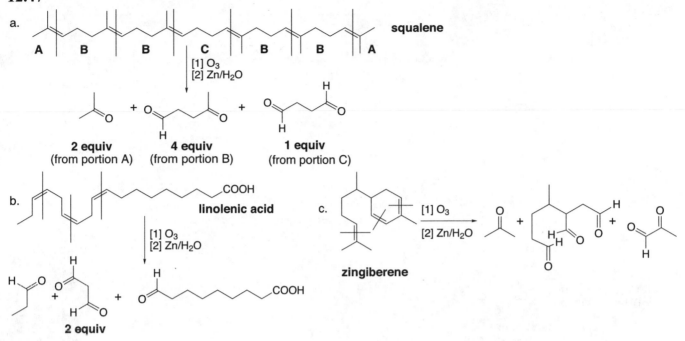

a.

A B B C B B A **squalene**

[1] O_3
[2] Zn/H_2O

2 equiv
(from portion A)

4 equiv
(from portion B)

1 equiv
(from portion C)

b.

linolenic acid

[1] O_3
[2] Zn/H_2O

2 equiv

c.

zingiberene

[1] O_3
[2] Zn/H_2O

12.48

$C_{10}H_{16}$ $\xrightarrow{\text{H}_2}{\text{Pd/C}}$ **2,6-dimethyloctane**

3 degrees of unsaturation

The hydrogenation reaction tells you that both oximene and myrcene have 3 π bonds (and no rings). Use this carbon backbone and add in the double bonds based on the oxidative cleavage products.

Oximene: $(CH_3)_2C=O$ $CH_2=O$ $CH_2(CHO)_2$ $CH_3-\overset{O}{\underset{}{C}}-CHO$ $\Longrightarrow$

Myrcene: $(CH_3)_2C-O$ CH_2-O $H-\overset{O}{\underset{}{C}}-CH_2CH_2-\overset{O}{\underset{}{C}}-CHO$ $\longrightarrow$

(2 equiv)

12.49

a. $\xrightarrow{\text{redraw}}$ $\xrightarrow[\text{Ti[OC(CH}_3)_2]_4 \\ (-)\text{-DET}]{(CH_3)_3COOH}$

b. $\underset{(CH_3)_3C}{\overset{H}{\diagdown}}C=C\underset{H}{\overset{CH_2OH}{\diagup}}$ $\xrightarrow[\text{Ti[OC(CH}_3)_2]_4 \\ (+)\text{-DET}]{(CH_3)_3COOH}$ $(CH_3)_3C\overset{H}{\underset{}{\diagdown}}C\overset{}{\underset{O}{\diagup}}C\underset{H}{\overset{CH_2OH}{\diagup}}$

12.50

enantiomeric excess =
% one enantiomer – % second enantiomer

ee = 87% – 13% = **74%**

major product
87%

minor product
13%

12.51

a. Replace this O to make an alkene.

(–)-DET

c. Replace this O to make an alkene.

(+)-DET

b.

(–)-DET

Replace this O to make an alkene.

12.52 Use retrosynthetic analysis to devise a synthesis of each hydrocarbon from acetylene.

a. $CH_3CH_2CH=CH_2 \Longrightarrow CH_3CH_2CH\equiv CH \Longrightarrow {}^-C\equiv CH \Longrightarrow HC\equiv CH$

$HC\equiv CH \xrightarrow{NaH} {}^-C\equiv CH \xrightarrow{CH_3CH_2Cl} CH_3CH_2C\equiv CH \xrightarrow[\text{Lindlar catalyst}]{H_2} CH_3CH_2CH=CH_2$

b.

$\Longrightarrow CH_3-C\equiv C-CH_3 \Longrightarrow CH_3-C\equiv CH \Longrightarrow {}^-C\equiv CH \Longrightarrow HC\equiv CH$

$HC\equiv CH \xrightarrow{NaH} {}^-C\equiv CH \xrightarrow{CH_3Cl} CH_3-C\equiv CH \xrightarrow{NaH} CH_3-C\equiv C^- \xrightarrow{CH_3Cl} CH_3-C\equiv C-CH_3 \xrightarrow[\text{Lindlar catalyst}]{H_2}$

c.

$\Longrightarrow CH_3-C\equiv C-CH_3 \Longrightarrow CH_3-C\equiv CH \Longrightarrow {}^-C\equiv CH \Longrightarrow HC\equiv CH$

$HC\equiv CH \xrightarrow{NaH} {}^-C\equiv CH \xrightarrow{CH_3Cl} CH_3-C\equiv CH \xrightarrow{NaH} CH_3-C\equiv C^- \xrightarrow{CH_3Cl} CH_3-C\equiv C-CH_3 \xrightarrow[NH_3]{Na}$

d. $(CH_3)_2CHCH_2CH_2CH_2CH_2CH(CH_3)_2 \Longrightarrow HC\equiv CCH_2CH(CH_3)_2 \Longrightarrow {}^-C\equiv CH \Longrightarrow HC\equiv CH$

$HC\equiv CH \xrightarrow{NaH} {}^-C\equiv CH \xrightarrow{Cl} $ $\xrightarrow{NaH} {}^-C\equiv C-$ $\xrightarrow{Cl}$ $\xrightarrow{H_2}{Pd/C}$

12.53

Retrosynthetic analysis:

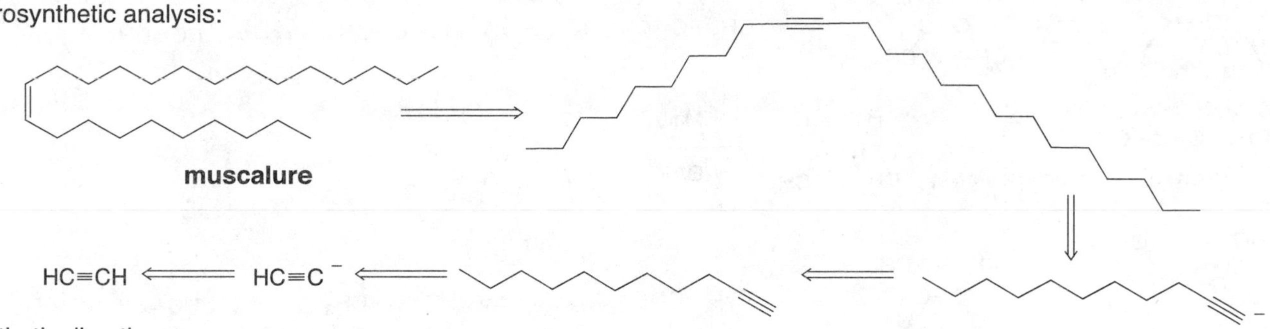

muscalure

$$HC\equiv CH \Longleftarrow HC\equiv C^- \Longleftarrow$$ $$\Longleftarrow$$

Synthetic direction:

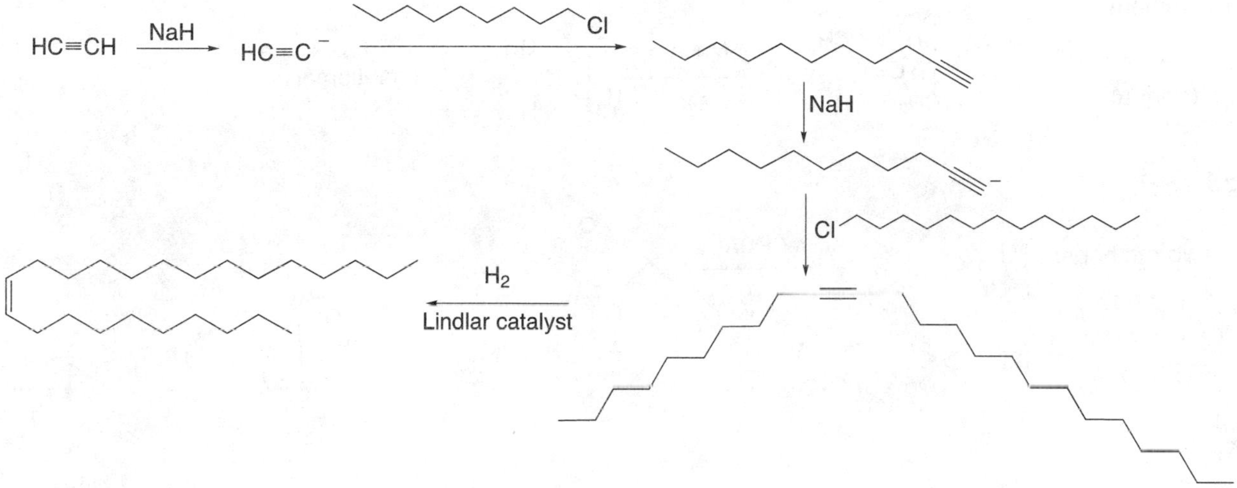

$$HC\equiv CH \xrightarrow{\text{NaH}} HC\equiv C^-$$

12.54

a.

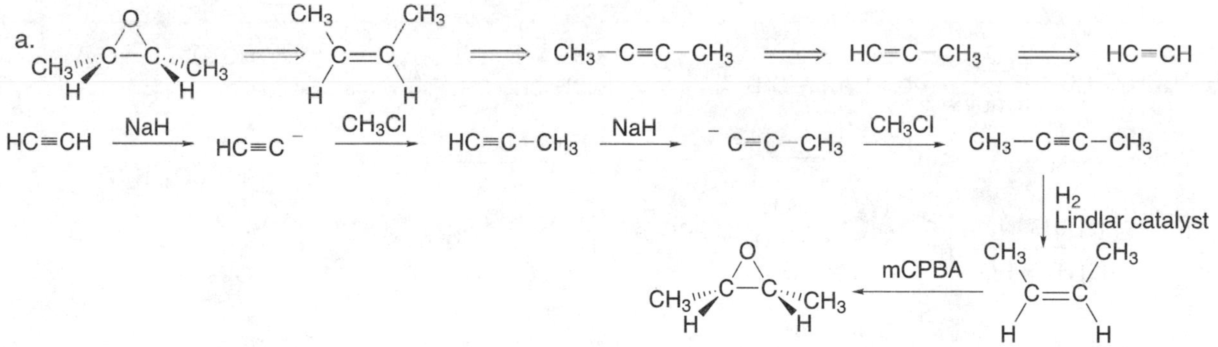

$$HC\equiv CH \xrightarrow{\text{NaH}} HC\equiv C^- \xrightarrow{CH_3Cl} HC\equiv C-CH_3 \xrightarrow{\text{NaH}} {}^-C\equiv C-CH_3 \xrightarrow{CH_3Cl} CH_3-C\equiv C-CH_3$$

b.

(+ enantiomer)

$$CH_3-C\equiv C-CH_3 \xrightarrow{\text{Na/NH}_3}$$ $$\xrightarrow{\text{mCPBA}}$$ + enantiomer

(from a)

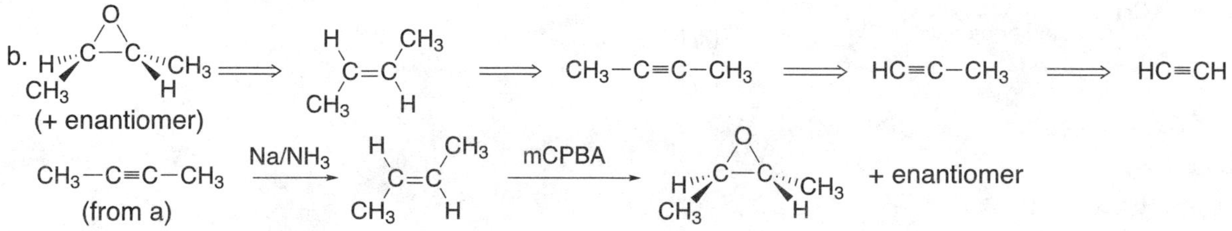

c.

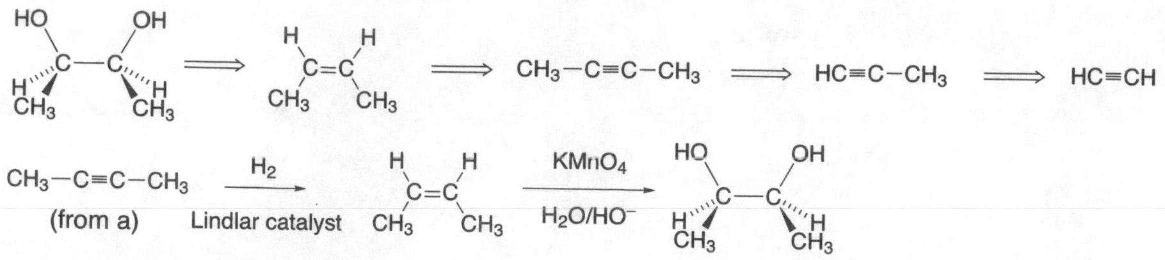

CH$_3$—C≡C—CH$_3$ $\xrightarrow[\text{Lindlar catalyst}]{H_2}$ (from a)

d.

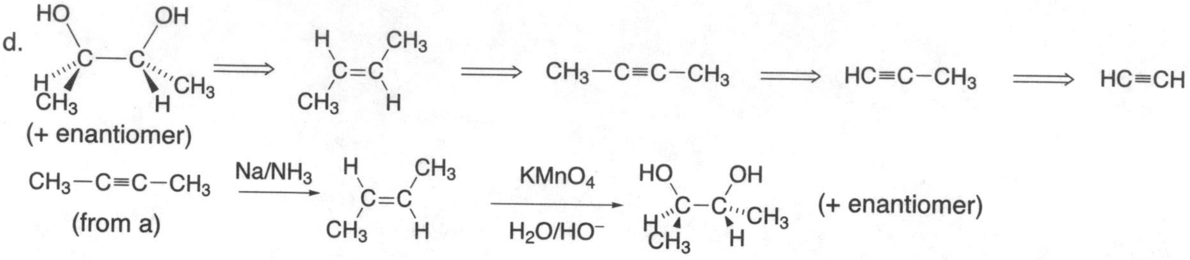

(+ enantiomer)

CH$_3$—C≡C—CH$_3$ $\xrightarrow{Na/NH_3}$ (from a)

(+ enantiomer)

12.55

Two methods: [1]

$\xrightarrow{mCPBA}$

[2] $\xrightarrow{Br_2/H_2O}$ $\xrightarrow{NaH}$

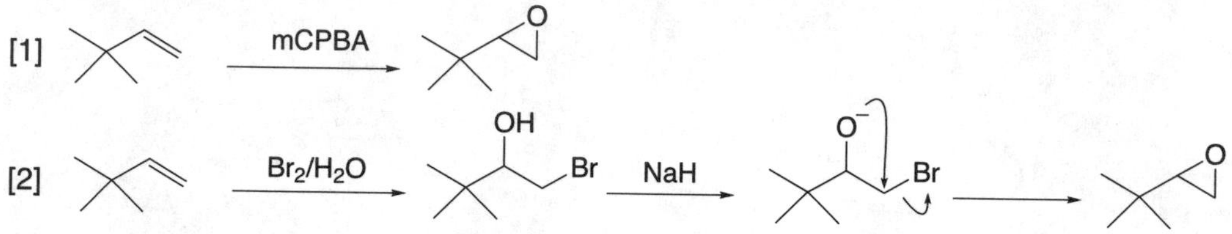

12.56

a. CH$_3$CH$_2$CH=CH$_2$ $\xrightarrow[\text{[2] } H_2O_2/HO^-]{\text{[1] 9-BBN}}$ CH$_3$CH$_2$CH$_2$CH$_2$OH $\xrightarrow[\text{H}_2\text{SO}_4, \text{H}_2\text{O}]{CrO_3}$ CH$_3$CH$_2$CH$_2$COOH

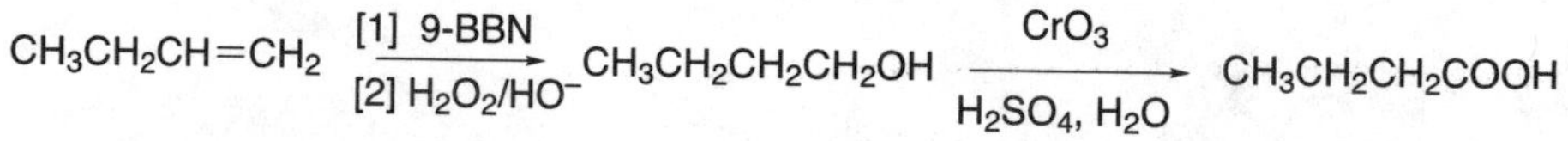

b. CH$_3$CH$_2$CH$_2$CH$_2$OH $\xrightarrow[\text{pyridine}]{POCl_3}$ CH$_3$CH$_2$CH=CH$_2$ $\xrightarrow{mCPBA}$ CH$_3$CH$_2$CH—CH$_2$ $\xrightarrow{LiAlH_4}$ CH$_3$CH$_2$CH—CH$_3$

H$_2$O/H$_2$SO$_4$

CH$_3$CH$_2$—C—CH$_3$ $\xrightarrow[\text{H}_2\text{SO}_4, \text{H}_2\text{O}]{CrO_3}$

c. $\xrightarrow[\text{[2] } H_2O_2/HO^-]{\text{[1] 9-BBN}}$ $\xrightarrow{PCC}$

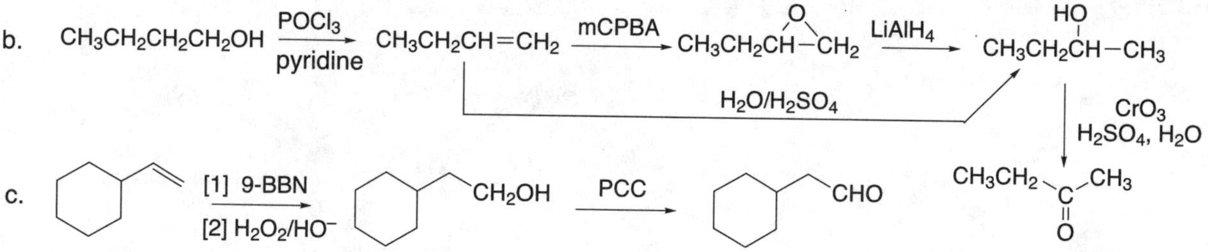

d. $\xrightarrow[\text{Lindlar catalyst}]{H_2}$ $\xrightarrow[\text{[2] } H_2O_2/HO^-]{\text{[1] 9-BBN}}$

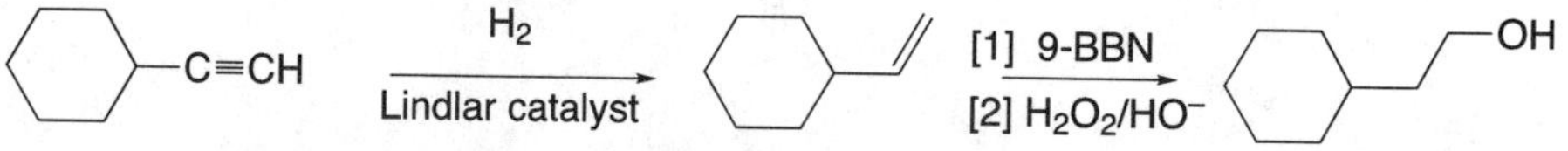

12.57

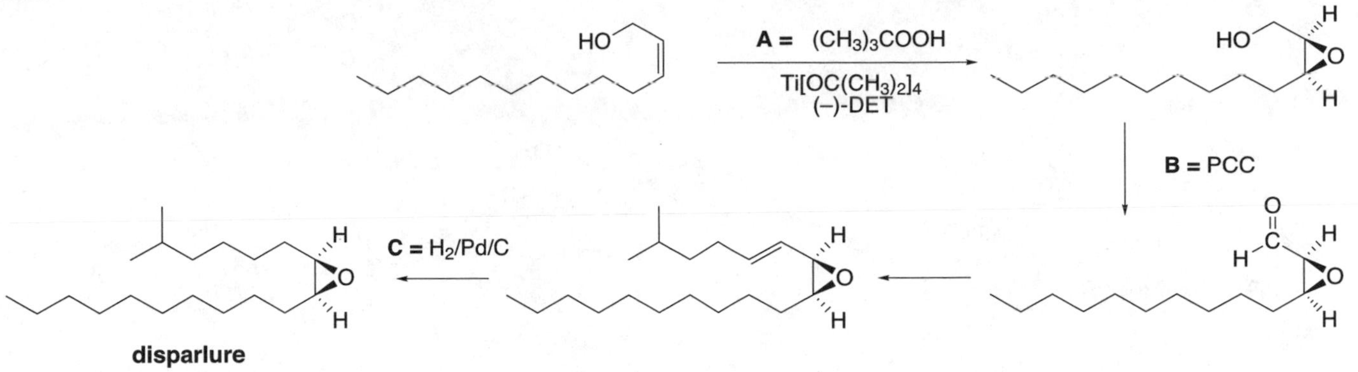

disparlure

12.58

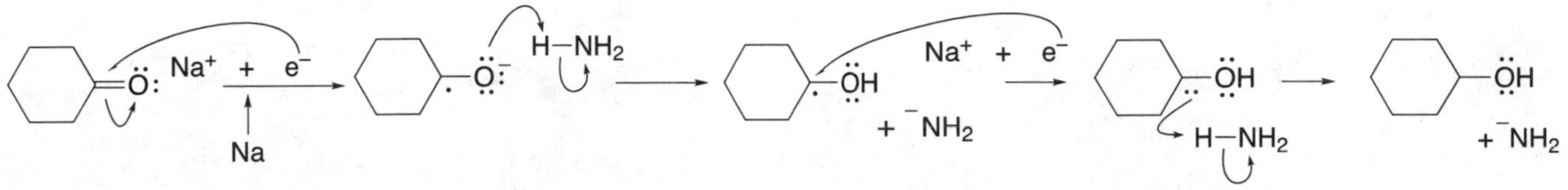

12.59

The favored conformation
for both molecules places
the *tert*-butyl group equatorial.

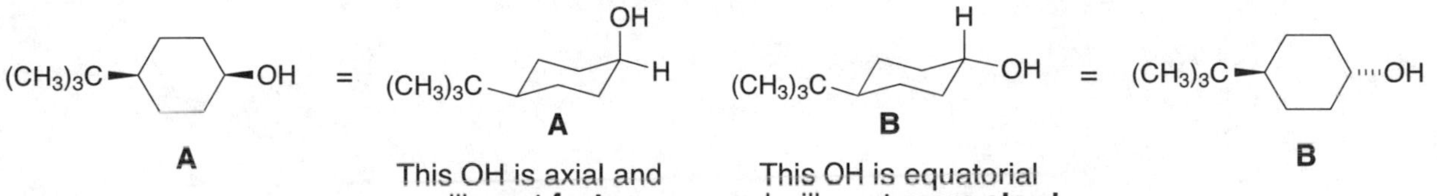

This OH is axial and
will react **faster**.

This OH is equatorial
and will react **more slowly**.

12.60

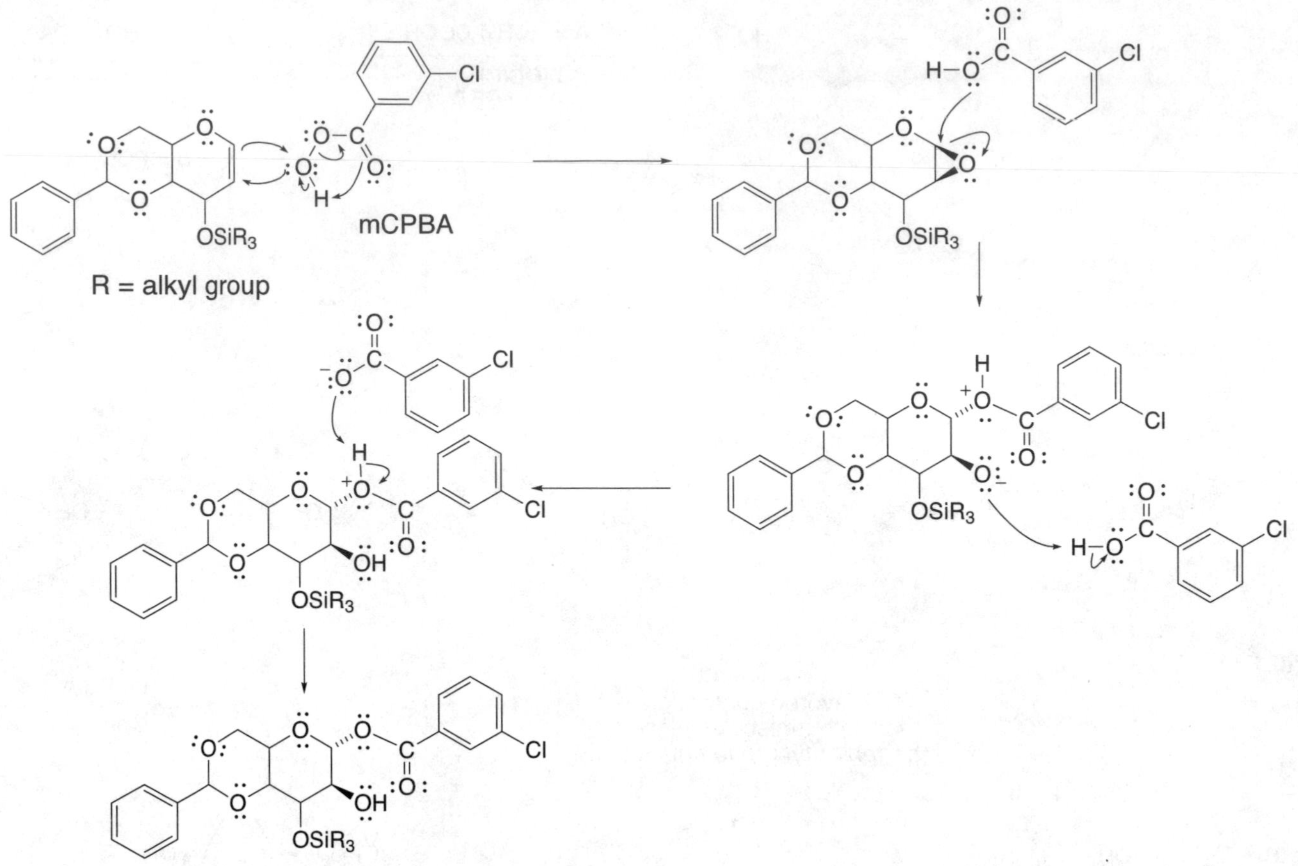

R = alkyl group

mCPBA

Chapter 13: Radical Reactions

♦ General features of radicals

- A radical is a reactive intermediate with an unpaired electron (13.1)
- A carbon radical is sp^2 hybridized and trigonal planar (13.1)
- The stability of a radical increases as the number of C's bonded to the radical carbon increases (13.1).

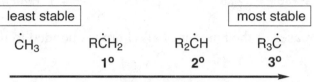

least stable			most stable
$\overset{\displaystyle \cdot}{C}H_3$	$R\overset{\displaystyle \cdot}{C}H_2$	$R_2\overset{\displaystyle \cdot}{C}H$	$R_3\overset{\displaystyle \cdot}{C}$
	1°	2°	3°

increasing alkyl substitution
increasing radical stability

- Allylic radicals are stabilized by resonance, making them more stable than 3° radicals (13.10).

$$CH_2=CH-\overset{\displaystyle \cdot}{C}H_2 \longleftrightarrow \overset{\displaystyle \cdot}{C}H_2-CH=CH_2$$

two resonance structures for the allyl radical

♦ Radical reactions

[1] Halogenation of alkanes (13.4)

$$R-H \xrightarrow[\text{hv or } \Delta]{X_2} \boxed{\begin{array}{c} R-X \\ \text{alkyl halide} \end{array}}$$

X = Cl or Br

- The reaction follows a radical chain mechanism.
- The weaker the C–H bond, the more readily it is replaced by X.
- Chlorination is faster and less selective than bromination (13.6).
- Radical substitution results in racemization at a stereogenic center (13.8)

[2] Allylic halogenation (13.10)

$$CH_2=CH-CH_3 \xrightarrow[\text{hv or ROOR}]{NBS} \boxed{\begin{array}{c} CH_2=CHCH_2Br \\ \text{allylic halide} \end{array}}$$

- The reaction follows a radical chain mechanism.

[3] Radical addition of HBr to an alkene (13.13)

$$RCH=CH_2 \xrightarrow[\substack{\text{hv, }\Delta\text{, or} \\ \text{ROOR}}]{HBr} \boxed{\begin{array}{c} \text{H H} \\ R-\overset{|}{C}-\overset{|}{C}-H \\ \text{H Br} \\ \text{alkyl bromide} \end{array}}$$

- A radical addition mechanism is followed.
- Br bonds to the less substituted carbon atom to form the more substituted, more stable radical.

[4] Radical polymerization of alkenes (13.14)

$$CH_2=CHZ \xrightarrow{ROOR} \boxed{\begin{array}{c} \text{polymer} \end{array}}$$

- A radical addition mechanism is followed.

Chapter 13: Answers to Problems

13.1 1° radicals are on C's bonded to one other C; 2° radicals are on C's bonded to two other C's; 3° radicals are on C's bonded to three other C's.

a. $CH_3CH_2-\overset{\bullet}{C}HCH_2CH_3$ b.   c.   d.

    2° radical   **3° radical**   **2° radical**   **1° radical**

13.2 The stability of a radical increases as the number of alkyl groups bonded to the radical carbon increases.

a. $(CH_3)_2CH\overset{\bullet}{C}H_2$ $CH_3CH_2\overset{\bullet}{C}HCH_3$ $(CH_3)_3C\cdot$ b.

   1° radical   **2° radical**   **3° radical**   **1° radical**   **2° radical**   **3° radical**
   least stable   **intermediate stability**   **most stable**   **least stable**   **intermediate stability**   **most stable**

13.3 Reaction of a radical with:
- an alkane abstracts a hydrogen atom and creates a new carbon radical.
- an alkene generates a new bond to one carbon, and a new carbon radical.
- another radical forms a bond.

a. $CH_3-CH_3 \xrightarrow{\ :\overset{..}{\underset{..}{C}l}\cdot\ } CH_3-\overset{\bullet}{C}H_2 + H-\overset{..}{\underset{..}{C}l}:$ c. $:\overset{..}{\underset{..}{C}l}\cdot \xrightarrow{\ :\overset{..}{\underset{..}{C}l}\cdot\ } :\overset{..}{\underset{..}{C}l}-\overset{..}{\underset{..}{C}l}:$

b. $CH_2{=}CH_2 \xrightarrow{\ :\overset{..}{\underset{..}{C}l}\cdot\ } \overset{\bullet}{C}H_2-CH_2-\overset{..}{\underset{..}{C}l}:$

13.4 **Monochlorination** is a radical substitution reaction in which a Cl replaces a H generating an alkyl halide.

a.

b. $CH_3CH_2CH_2CH_2CH_2CH_3 \xrightarrow{Cl_2} ClCH_2CH_2CH_2CH_2CH_2CH_3 + CH_3-\overset{\overset{\displaystyle H}{|}}{\underset{\underset{\displaystyle Cl}{|}}{C}}-CH_2CH_2CH_2CH_3 + CH_3CH_2-\overset{\overset{\displaystyle H}{|}}{\underset{\underset{\displaystyle Cl}{|}}{C}}-CH_2CH_2CH_3$

c. $(CH_3)_3CH \xrightarrow{Cl_2} CH_3-\overset{\overset{\displaystyle Cl}{|}}{\underset{\underset{\displaystyle CH_3}{|}}{C}}-CH_3 + CH_3-\overset{\overset{\displaystyle H}{|}}{\underset{\underset{\displaystyle CH_3}{|}}{C}}-CH_2Cl$

13.5 If an alkane yields only one product on monohalogenation with Cl_2, all of the hydrogens must be identical in the starting material. Only one compound of molecular formula C_5H_{12} has all identical H's: $(CH_3)_4C$.

13.6

Initiation: $:\ddot{B}r \overset{\frown}{\underset{\smile}{}} \ddot{B}r: \quad \xrightarrow[\text{or } \Delta]{h\nu} \quad :\ddot{B}r\cdot \ + \ \cdot\ddot{B}r:$

Propagation: $CH_3 \overset{\frown}{\underset{\smile}{}} H \ + \ \cdot\ddot{B}r: \quad \longrightarrow \quad \dot{C}H_3 \ + \ H{-}\ddot{B}r:$

$\dot{C}H_3 \ + \ :\ddot{B}r \overset{\frown}{\underset{\smile}{}} \ddot{B}r: \quad \longrightarrow \quad CH_3{-}Br \ + \ \cdot\ddot{B}r:$

Termination: $:\ddot{B}r\cdot \ + \ \cdot\ddot{B}r: \quad \longrightarrow \quad :\ddot{B}r{-}\ddot{B}r:$

or

$\dot{C}H_3 \ + \ \dot{C}H_3 \quad \longrightarrow \quad CH_3{-}CH_3$

or

$\dot{C}H_3 \ + \ \cdot\ddot{B}r: \quad \longrightarrow \quad CH_3{-}\ddot{B}r:$

13.7

Initiation: $:\ddot{C}l \overset{\frown}{\underset{\smile}{}} \ddot{C}l: \quad \xrightarrow[\text{or } \Delta]{h\nu} \quad :\ddot{C}l\cdot \ + \ \cdot\ddot{C}l:$

Propagation: $CH_3 \overset{\frown}{\underset{\smile}{}} H \ + \ \cdot\ddot{C}l: \quad \longrightarrow \quad \dot{C}H_3 \ + \ H{-}\ddot{C}l:$

$\dot{C}H_3 \ + \ :\ddot{C}l{-}\ddot{C}l: \quad \longrightarrow \quad CH_3{-}Cl \ + \ \cdot\ddot{C}l:$

One possibility for termination: $\dot{C}H_3 \ + \ \dot{C}H_3 \quad \longrightarrow \quad \boxed{CH_3{-}CH_3}$

$\downarrow$

$CH_3{-}CH_2 \overset{\frown}{\underset{\smile}{}} H \ + \ \cdot\ddot{C}l: \quad \longrightarrow \quad CH_3\dot{C}H_2 \ + \ H{-}\ddot{C}l:$

$CH_3\dot{C}H_2 \ + \ :\ddot{C}l{-}\ddot{C}l: \quad \longrightarrow \quad \boxed{CH_3CH_2{-}Cl} \ + \ \cdot\ddot{C}l:$

13.8 Transition states are hypothetical structures at the "peaks" on an energy diagram. They are drawn with dashed lines to indicate partially broken and partially formed bonds.

Transition state 1:

$$\left[:\overset{\delta\cdot}{\ddot{C}l}\text{----}H\text{---}CH_2CH_3 \right]^{\ddagger}$$

Transition state 2:

$$\left[:\overset{\delta\cdot}{\ddot{C}l}\text{----}\overset{}{\ddot{C}l}\text{---}\overset{\delta\cdot}{C}H_2CH_3 \right]^{\ddagger}$$

13.9

Step 1:

$$CH_3-H \;+\; \cdot \ddot{B}r: \;\longrightarrow\; \dot{C}H_3 \;+\; H-\ddot{B}r: \qquad \Delta H° = +16 \text{ kcal/mol}$$

1 bond broken   1 bond formed
+ 104 kcal/mol   −88 kcal/mol

Step 2:

$$\dot{C}H_3 \;+\; :\ddot{B}r-\ddot{B}r: \;\longrightarrow\; CH_3-Br \;+\; \cdot \ddot{B}r: \qquad \Delta H° = -24 \text{ kcal/mol}$$

1 bond broken   1 bond formed
+ 46 kcal/mol   −70 kcal/mol

13.10 The rate determining step for halogenation reactions is formation of $CH_3\cdot + HX$.

$$CH_3-H \;+\; \cdot \ddot{I}: \;\longrightarrow\; \cdot CH_3 \;+\; H-\ddot{I}: \qquad \Delta H° = +33 \text{ kcal/mol}$$

1 bond broken   1 bond formed
+ 104 kcal/mol   −71 kcal/mol

This reaction is more endothermic and has a higher E_a than a similar reaction with Cl_2 or Br_2.

13.11 The **weakest C–H bond** in each alkane is the **most readily cleaved** during radical halogenation.

a.

3°
most reactive

b.

3°
most reactive

c. $CH_3CHCH_2CH_3$

H

2°
most reactive

13.12 To draw the product of bromination:
- Draw out the starting material and find the most reactive C–H bond (on the most substituted C).
- The major product is formed by **cleavage of the** *weakest* **C–H bond**.

$$\xrightarrow[hv]{Br_2}$$

This 3° C has the most reactive C–H bond.

13.13

$$\xrightarrow[hv]{Cl_2}$$

This is the desired product,
1-chloro-1-methylcyclohexane,
but many other products are formed.

13.14 If 1° C–H and 3° C–H bonds were equally reactive there would be nine times as much $(CH_3)_2CHCH_2Cl$ as $(CH_3)_3CCl$ since the ratio of 1° to 3° H's is 9:1. The fact that the ratio is only 63:37 shows that the 1° C–H bond is less reactive than the 3° C–H bond. $(CH_3)_2CHCH_2Cl$ is still the major product, though, because there are nine 1° C–H bonds and only one 3° C–H bond.

13.15

a. $CH_3\text{-}\underset{\underset{CH_3}{|}}{\overset{\overset{CH_3}{|}}{C}}\text{-}H \xrightarrow[\Delta]{Br_2} CH_3\text{-}\underset{\underset{CH_3}{|}}{\overset{\overset{CH_3}{|}}{C}}\text{-}Br$

c. $\underset{CH_3}{\overset{CH_3}{}}C{=}CH_2 \xrightarrow[H_2SO_4]{H_2O} CH_3\text{-}\underset{\underset{CH_3}{|}}{\overset{\overset{CH_3}{|}}{C}}\text{-}OH$ [from (b)]

b. $CH_3\text{-}\underset{\underset{CH_3}{|}}{\overset{\overset{CH_3}{|}}{C}}\text{-}Br \xrightarrow{(CH_3)_3CO^-K^+} \underset{CH_3}{\overset{CH_3}{}}C{=}CH_2$ [from (a)]

d. $\underset{CH_3}{\overset{CH_3}{}}C{=}CH_2 \xrightarrow[CCl_4]{Cl_2} CH_3\text{-}\underset{\underset{CH_3}{|}}{\overset{\overset{CH_2Cl}{|}}{C}}\text{-}Cl$ [from (b)]

13.16 Since the reaction does not occur at the stereogenic center, leave it as is.

(R)-2-bromobutane $\xrightarrow[h\nu]{Cl_2}$ S + R

13.17

a. $CH_3CH_2CH_2CH_2CH_3 \xrightarrow{Cl_2 \; \Delta} CH_3CH_2CH_2CH_2CH_2Cl + CH_3CH_2\overset{\overset{Cl}{|}}{C}HCH_2CH_3 + CH_3CH_2CH_2\underset{CH_3}{\overset{Cl}{}}{C}H + CH_3CH_2CH_2\underset{H}{\overset{Cl}{}}{C}CH_3$

b. ▷—$CH_3 \xrightarrow{Cl_2 \; \Delta}$ ▷—CH_2Cl + ▷⟨$^{Cl}_{CH_3}$ + (4 cyclopropane stereoisomers) —CH_3

c. $CH_3CH_2\text{-}\underset{\underset{CH_2CH_3}{|}}{\overset{\overset{H}{|}}{C}}\text{-}CH_2CH_3 \xrightarrow[h\nu]{Cl_2} CH_3CH_2\text{-}\underset{\underset{CH_2CH_3}{|}}{\overset{\overset{Cl}{|}}{C}}\text{-}CH_2CH_3 + ClCH_2CH_2CH(CH_2CH_3)_2 + CH_3\underset{CH(CH_2CH_3)_2}{\overset{Cl\;H}{}}C$

d. $\xrightarrow[h\nu]{Cl_2}$ (Consider attack at C2 and C3 only.)

13.18

Chain propagation:

$:\ddot{O}{=}\ddot{N}\cdot + O_3 \longrightarrow :\ddot{O}{=}\ddot{N}\text{-}\ddot{O}\cdot + O_2$

$:\ddot{O}{=}\ddot{N}\text{-}\ddot{O}\cdot + \cdot\ddot{O}\cdot \longrightarrow :\ddot{O}{=}\ddot{N}\cdot + O_2$

The radical is re-formed.

13.19 Draw the resonance structure by moving the π bond and the unpaired electron. The hybrid is drawn with dashed lines for bonds that are in one resonance structure but not another. The symbol δ• is used on any atom that has an unpaired electron in any resonance structure.

a. 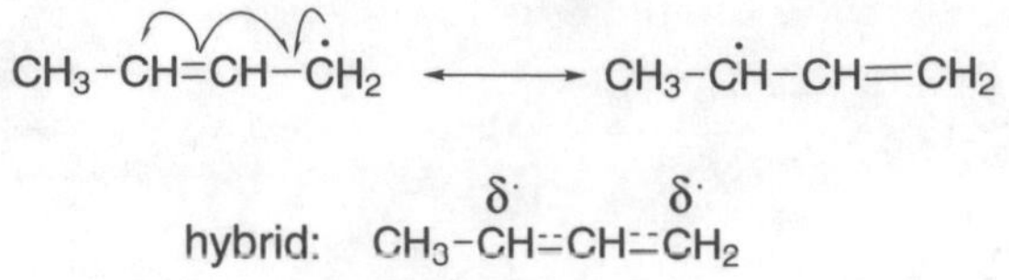 $CH_3-CH=CH-\overset{\cdot}{C}H_2$ ⟷ $CH_3-\overset{\cdot}{C}H-CH=CH_2$ **b.**

hybrid: $CH_3-CH\overset{\delta\cdot}{=}CH\overset{\delta\cdot}{=}CH_2$

hybrid:

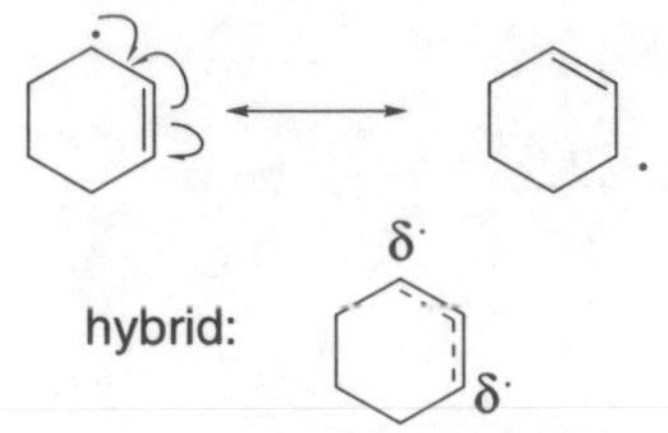

13.20   Reaction of an alkene with NBS or $Br_2 + h\nu$ yields allylic substitution products.

a. 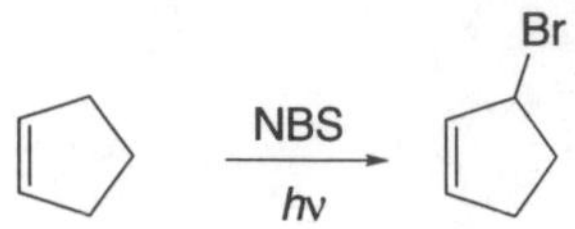 ⟶ (NBS, $h\nu$) Br-cyclopentene

c. $CH_2=CH-CH_3$ ⟶ (Br_2) $\underset{\underset{Br\ \ Br}{|\ \ |}}{CH_2CHCH_3}$

b. $CH_2=CH-CH_3$ ⟶ (NBS, $h\nu$) $CH_2=CH-CH_2Br$

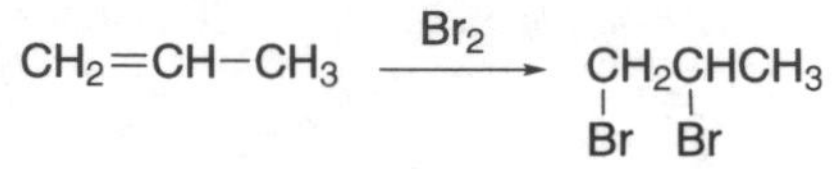

13.21

a. $CH_3-CH=CH-CH_3$ ⟶ (NBS, $h\nu$) $CH_3-CH=CH-CH_2Br$
+
$CH_3CH(Br)CH=CH_2$

c. 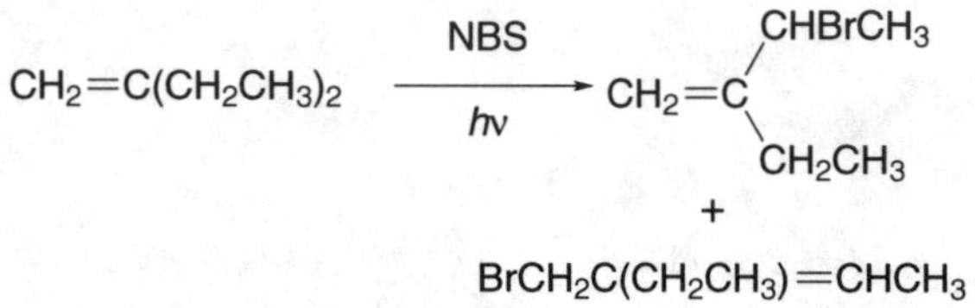 $CH_2=C(CH_2CH_3)_2$ ⟶ (NBS, $h\nu$) $CH_2=C\underset{CH_2CH_3}{\overset{CHBrCH_3}{<}}$
+
$BrCH_2C(CH_2CH_3)=CHCH_3$

b. 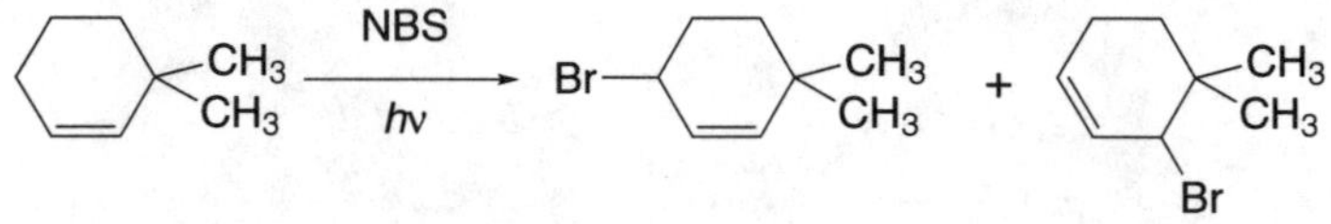 ⟶ (NBS, $h\nu$) Br-cyclohexene with CH_3, CH_3 + cyclohexene with CH_3, CH_3 and Br

13.22

methylcyclohexene ⟶ (NBS, $h\nu$) products with CH_3 and Br +

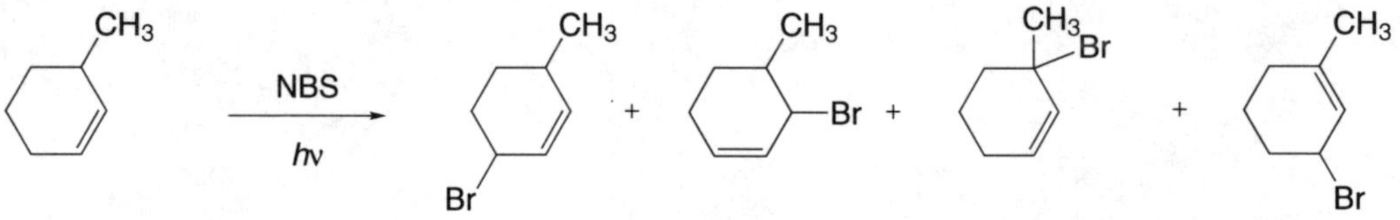

13.23

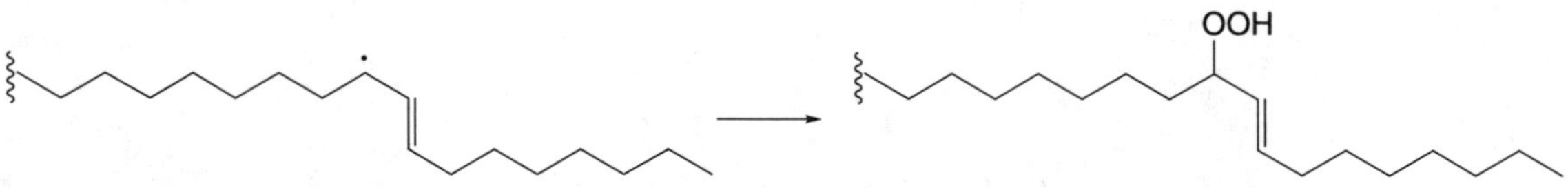

second resonance structure ⟶ (OOH product)

13.24 The weakest C–H bond is most readily cleaved. To draw the hydroperoxide products, add OOH to each carbon that bears a radical in one of the resonance structures.

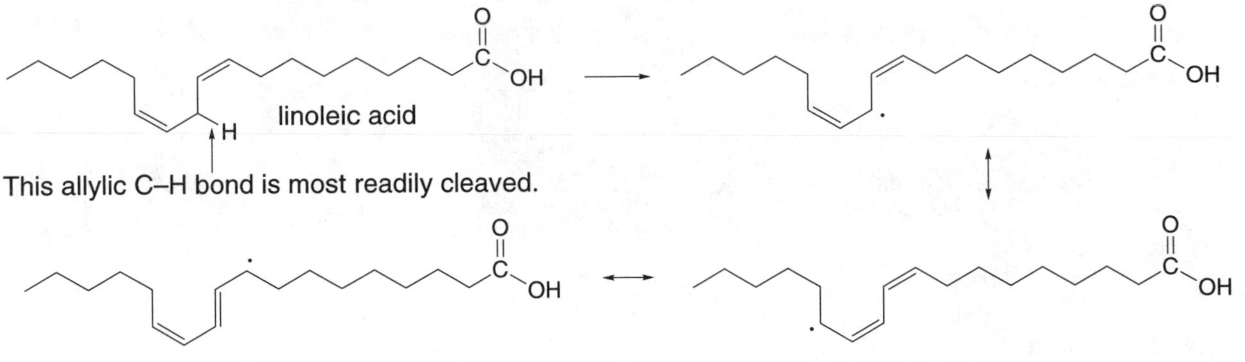

linoleic acid

This allylic C–H bond is most readily cleaved.

hydroperoxide products:

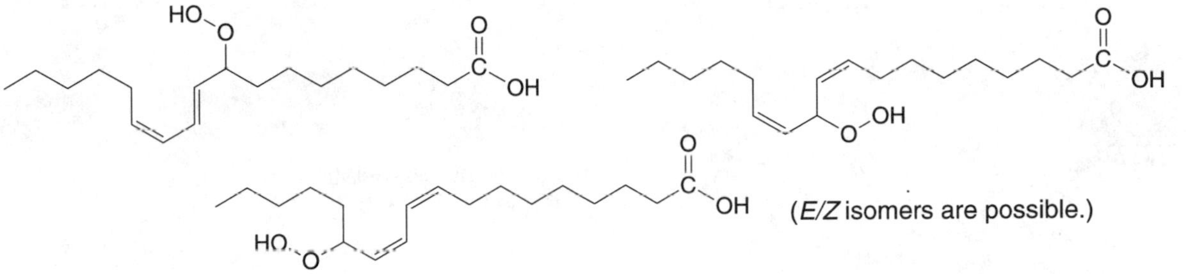

(*E/Z* isomers are possible.)

13.25

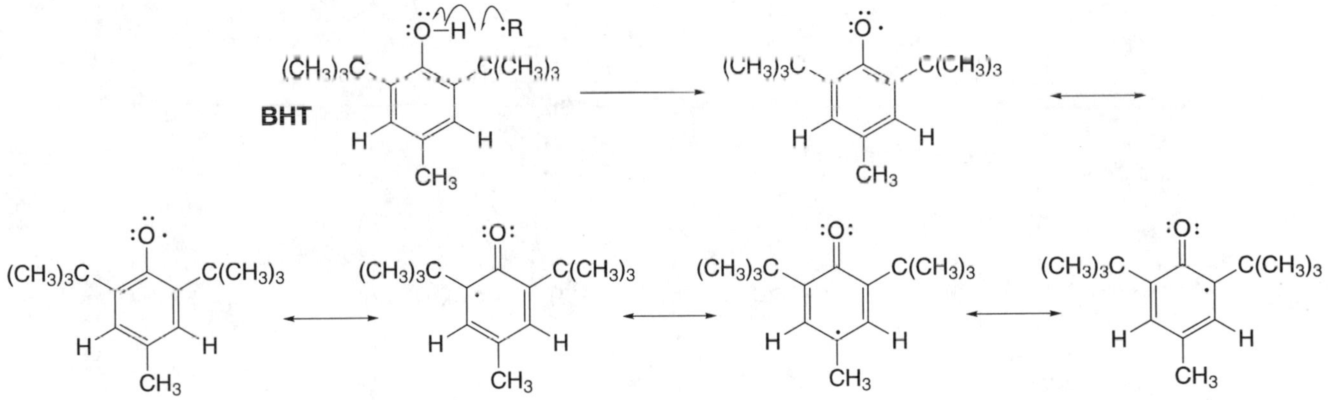

BHT

13.26

a. $CH_2{=}CHCH_2CH_2CH_2CH_3$ $\xrightarrow{\text{HBr}}$ $CH_3\underset{\underset{Br}{|}}{C}HCH_2CH_2CH_2CH_3$

b. 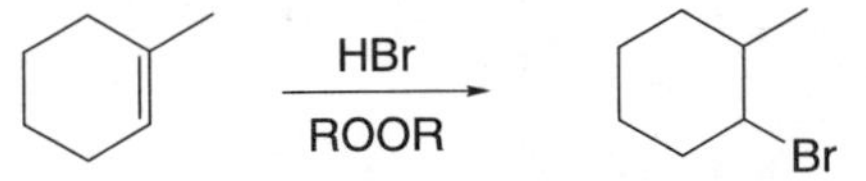 $\xrightarrow[\text{ROOR}]{\text{HBr}}$

c. $(CH_3)_2C{=}CHCH_3$ $\xrightarrow{\text{HBr}}$ $(CH_3)_2\overset{\overset{Br}{|}}{C}{-}CH_2CH_3$

d. $CH_3CH{=}CHCH_2CH_2CH_3$ $\xrightarrow[\text{ROOR}]{\text{HBr}}$ $CH_3\underset{\underset{Br}{|}}{C}H{-}CH_2CH_2CH_2CH_3$ $+$ $CH_3CH_2{-}\underset{\underset{Br}{|}}{C}HCH_2CH_2CH_3$

13.27 In addition of HBr under radical conditions:
- Br• adds first to form the more stable radical.
- Then H• is added to the carbon radical.

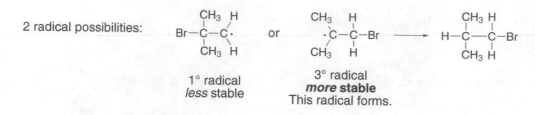

2 radical possibilities:

1° radical
less stable

3° radical
more **stable**
This radical forms.

13.28

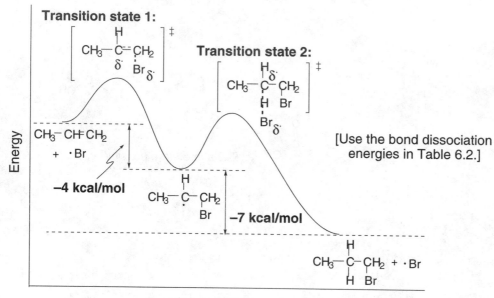

[Use the bond dissociation energies in Table 6.2.]

reaction coordinate

13.29

Step 1: CH$_3$–CH=CH$_2$ + •Cl ⟶ CH$_3$–C–CH$_2$ $\Delta H_1° = -17$ kcal/mol

1 bond broken
+ 64 kcal/mol

1 bond formed
−81 kcal/mol

Step 2: CH$_3$–C–CH$_2$ + H–Cl ⟶ CH$_3$–C–CH$_2$ $\Delta H_2° = +8$ kcal/mol
This step of propagation is endothermic. It prohibits chain propagation from occurring over and over.

1 bond broken
+ 103 kcal/mol

1 bond formed
−95 kcal/mol

13.30

a. CH$_2$=C ⟶ {–CH$_2$–C–CH$_2$–C–CH$_2$–C–}

b. ⟹ CH$_2$=CH–OCOCH$_3$

poly(vinyl acetate)

13.31

Initiation:

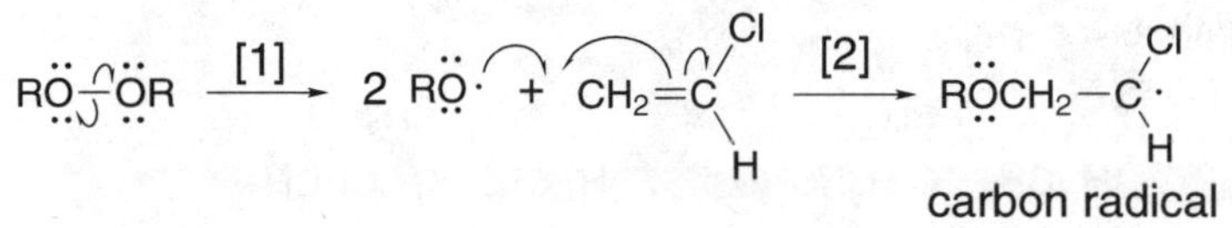

$$RO \cdot \overset{\frown}{\cdot} OR \xrightarrow{[1]} 2 \ RO \cdot \ + \ CH_2=C \overset{Cl}{\underset{H}{}} \xrightarrow{[2]} ROCH_2-C \overset{Cl}{\underset{H}{\cdot}}$$

carbon radical

Propagation:

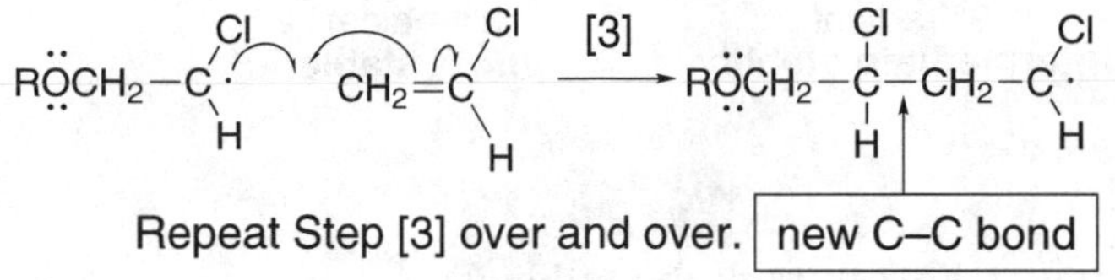

$$ROCH_2-C \overset{Cl}{\underset{H}{\cdot}} \quad CH_2=C \overset{Cl}{\underset{H}{}} \xrightarrow{[3]} ROCH_2-C \overset{Cl}{\underset{H}{}}-CH_2-C \overset{Cl}{\underset{H}{\cdot}}$$

Repeat Step [3] over and over. | new C–C bond |

Termination:

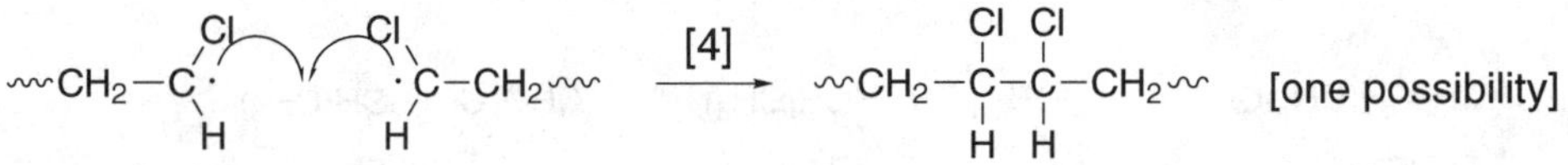

$$\sim\sim CH_2-C\overset{Cl}{\underset{H}{\cdot}} \quad \cdot C\overset{Cl}{\underset{H}{}}-CH_2\sim\sim \xrightarrow{[4]} \sim CH_2-\overset{Cl}{\underset{H}{C}}-\overset{Cl}{\underset{H}{C}}-CH_2\sim \quad \text{[one possibility]}$$

13.32

  a. increasing bond strength: b < c < a

  b and c.

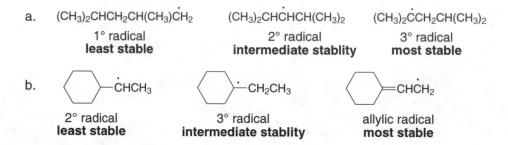

1° radical	2° radical	3° radical
least stable	**intermediate stablity**	**most stable**

  d. increasing H abstraction: a < c < b

13.33 Use the directions from Answer 13.2 to rank the radicals.

  a. $(CH_3)_2CHCH_2CH(CH_3)\dot{C}H_2$ $(CH_3)_2CH\dot{C}HCH(CH_3)_2$ $(CH_3)_2\dot{C}CH_2CH(CH_3)_2$

  1° radical   2° radical   3° radical
  least stable   **intermediate stablity**   **most stable**

  b.

  2° radical   3° radical   allylic radical
  least stable   **intermediate stablity**   **most stable**

13.34 Draw the radical formed by cleavage of the benzylic C–H bond. Then draw all of the resonance structures. Having more resonance structures (five in this case) makes the radical more stable, and the benzylic C–H bond weaker.

  benzylic C–H bond
  bond dissociation energy = 85 kcal/mol

13.35

$H_b \rightarrow H \quad H \leftarrow H_d$
$CH_2{=}CHCHCHC(CH_3)CH_2\text{-}H{\leftarrow}H_c$
$\quad H_a{\rightarrow}\dot{H}$

H_a = bonded to an sp^3 3° carbon
H_b = bonded to an allylic carbon
H_c = bonded to an sp^3 1° carbon
H_d = bonded to an sp^3 2° carbon

Increasing ease of abstraction:
$\mathbf{H_c < H_d < H_a < H_b}$

13.36 Use the directions from Answer 13.4.

a.

b. $(CH_3)_3CCH_2CH_2CH_2CH_3 \longrightarrow (CH_3)_3CCH_2CH_2CH_2CH_2Cl + (CH_3)_3CCH_2CH_2\overset{Cl}{\underset{}{C}}HCH_3 + (CH_3)_3CCH_2\overset{Cl}{\underset{}{C}}HCH_2CH_3$

$(CH_3)_2\overset{CH_2Cl}{\underset{}{C}}CH_2CH_2CH_2CH_3 + (CH_3)_3C\overset{Cl}{\underset{}{C}}HCH_2CH_2CH_3$

c.

d.

13.37 To draw the product of bromination:
- Draw out the starting material and find the most reactive C–H bond (on the most substituted C).
- The major product is formed by **cleavage of the *weakest* C–H bond**.

a.

b. $(CH_3)_3CCH_2CH(CH_3)_2 \longrightarrow (CH_3)_3CCH_2\overset{}{\underset{Br}{C}}(CH_3)_2$

c.

d. $(CH_3)_3CCH_2CH_3 \longrightarrow (CH_3)_3C\overset{Br}{\underset{}{C}}HCH_3$

13.38 Draw all of the alkane isomers of C_6H_{14} and their products on chlorination. Then determine which letter corresponds to which alkane.

[* = stereogenic center]

13.39 Halogenation replaces a C–H bond with a C–X bond. To find the alkane needed to make each of the alkyl halides, replace the X with a H.

a.

c.

b.

d. $(CH_3)_3CCH_2Cl \implies (CH_3)_3CCH_3$

13.40 For an alkane to yield one major product on monohalogenation with Cl_2, all of the hydrogens must be identical in the starting material. For an alkane to yield one major product on bromination, it must have a more substituted carbon in the starting material.

a.

b.

These two compounds can be formed in high yield from an alkane.

c.

many different
C–H bonds

d.

Br on 2° carbon
The product with Br on 3° carbon
will form predominantly.

These two compounds cannot be formed in high yield from an alkane.

13.41 Chlorination with two equivalents of Cl_2 yields a variety of products.

The desired product is only one of four products formed.

13.42 Draw the resonance structures by moving the π bonds and the radical.

a.

b.

13.43 Reaction of an alkene with NBS + $h\nu$ yields allylic substitution products.

a.

b. $CH_3CH_2CH=CHCH_2CH_3 \xrightarrow[h\nu]{NBS} CH_3CH_2CH=CHCHCH_3 \overset{Br}{|} + CH_3CH_2CHCH=CHCH_3 \overset{Br}{|}$

c. $(CH_3)_2C=CHCH_3 \xrightarrow[h\nu]{NBS} (CH_3)_2C=CHCH_2Br + BrCH_2C(CH_3)=CHCH_3 + CH_2=C(CH_3)CH(Br)CH_3$

$+$

$(CH_3)_2C(Br)CH=CH_2$

d.

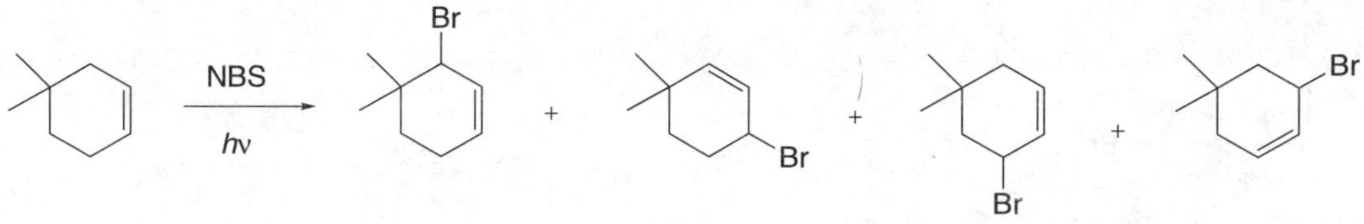

13.44 Reaction of an alkene with NBS + hv yields allylic substitution products.

a. one possible product: **high yield**

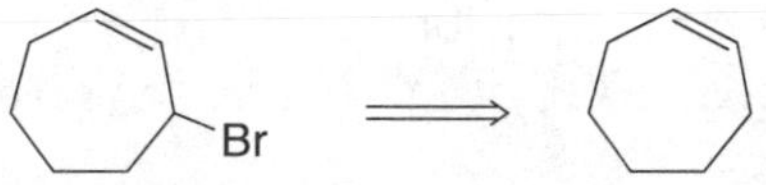

b. $CH_3CH_2CH=CHCH_2Br$ c.

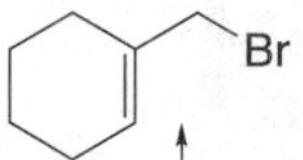

> Cannot be made in high yield by allylic halogenation.
> Any alkene starting material would yield a mixture of allylic halides.

13.45

a.

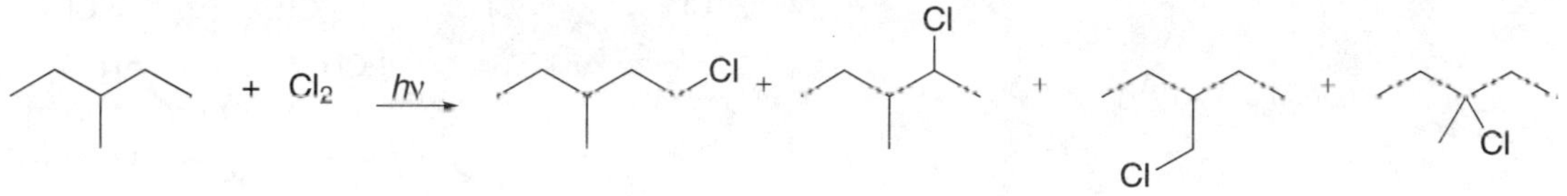

b. 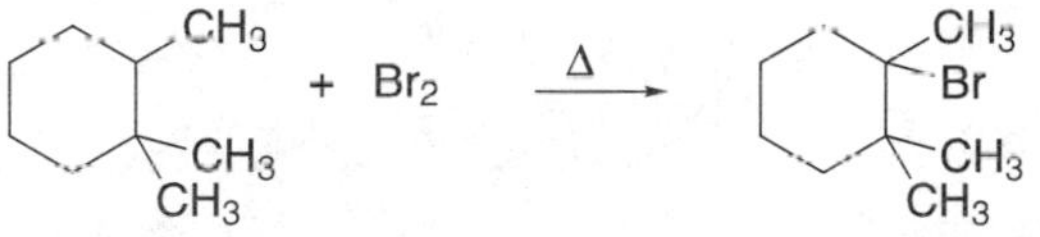 (major product)

c. 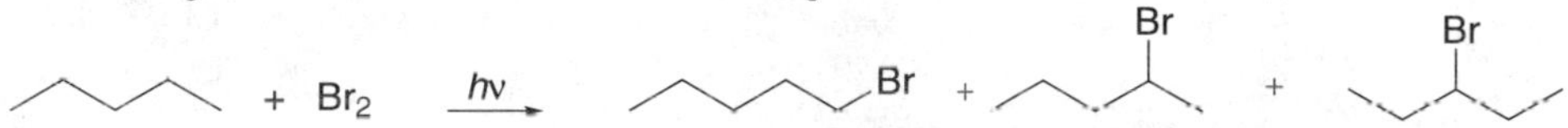
   (minor product)   (two major products)

d.

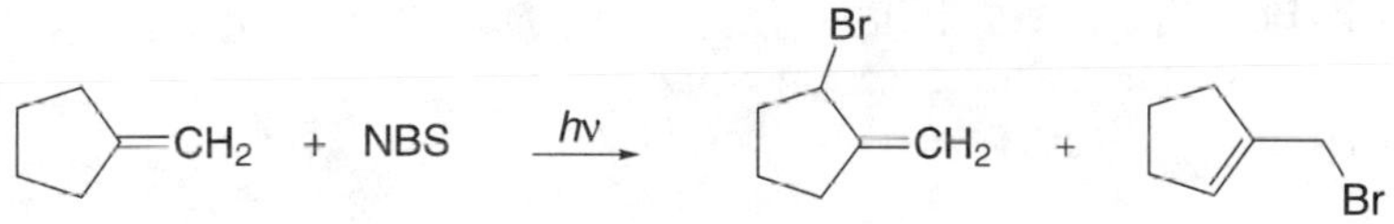

e.

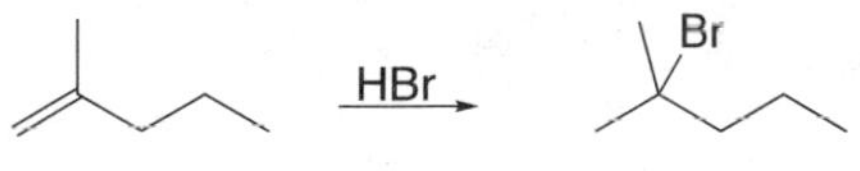

f.

g.

h.

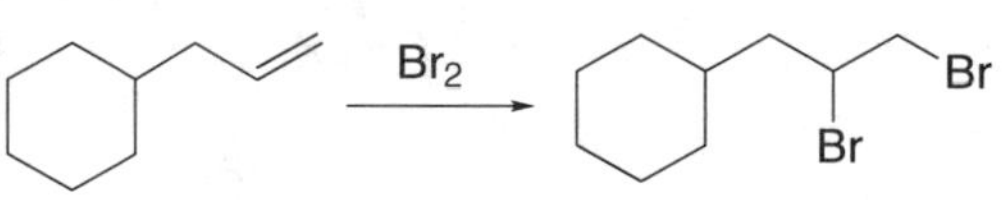

i.

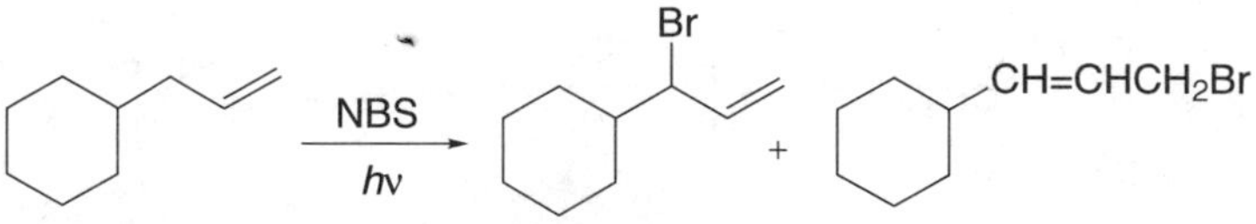

13.46

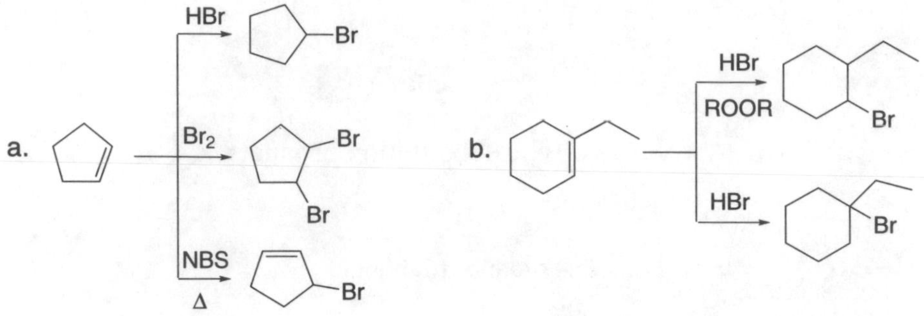

a.

b.

13.47

a.

b.

c.

d.

e.

f.

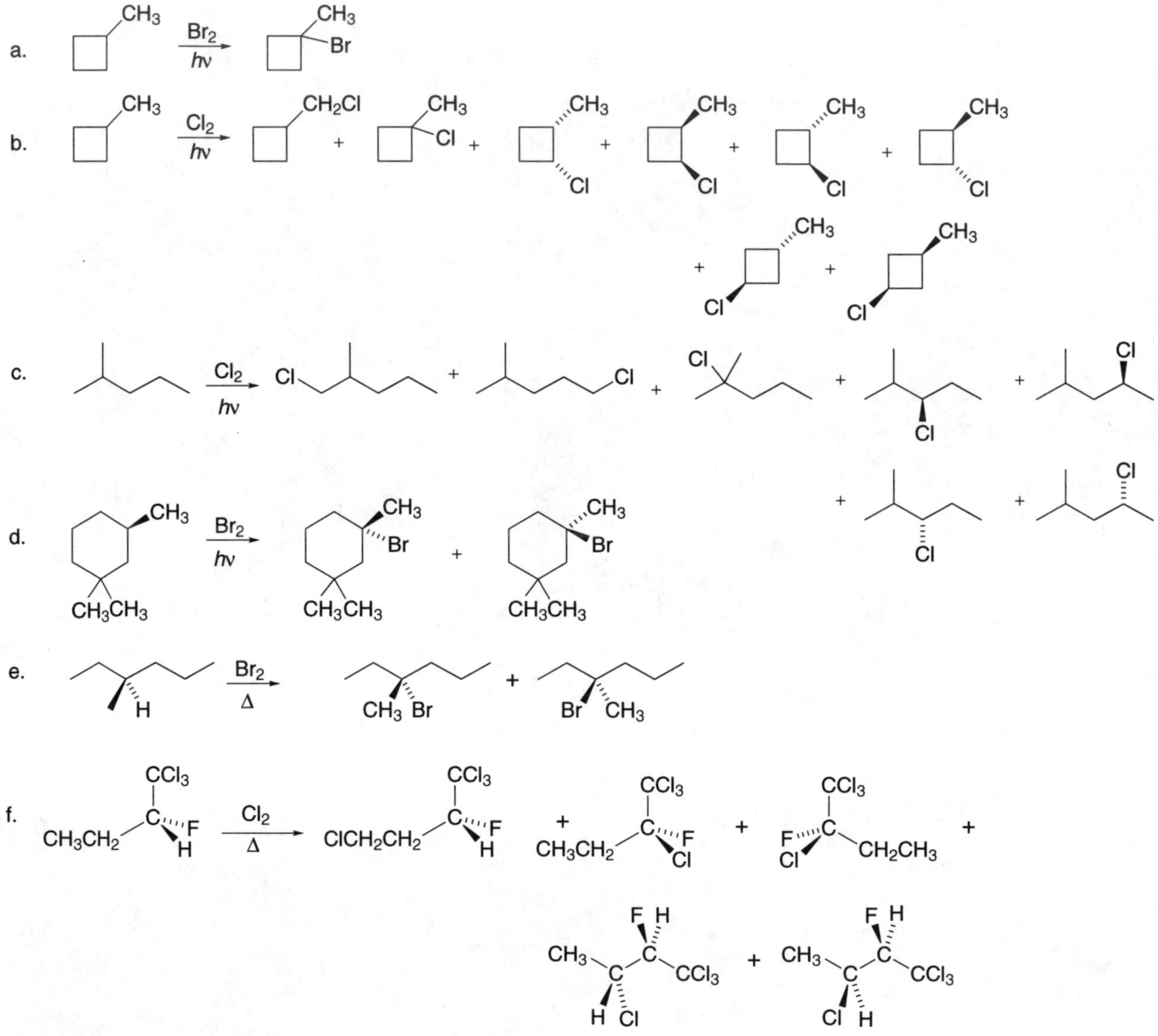

13.48

a.

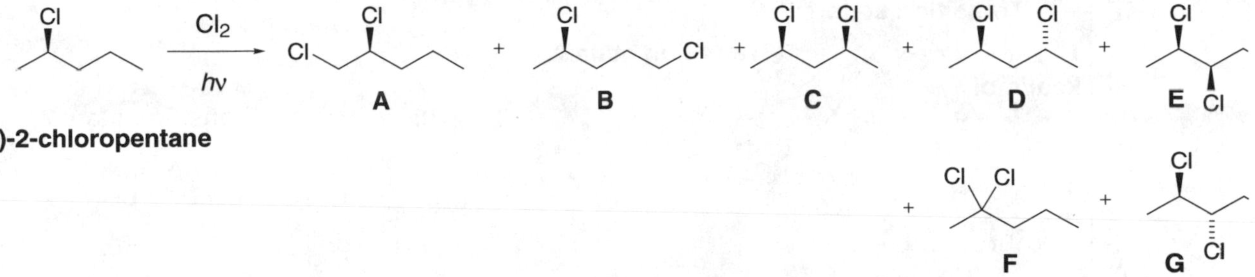

(*R*)-2-chloropentane

b. There would be seven fractions, since each molecule drawn has different physical properties.
c. Fractions **A**, **B**, **D**, **E**, and **G** would show optical activity.

13.49

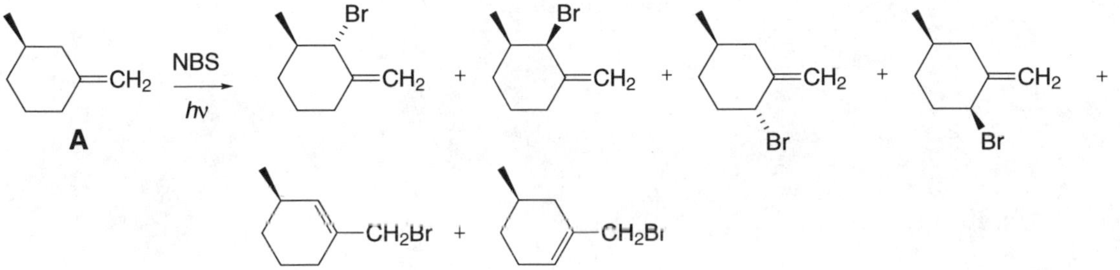

13.50

a. $CH_3-C(CH_3)(H)-CH_3 + Br_2 \xrightarrow{h\nu} CH_3-C(CH_3)(Br)-CH_3 + HBr$

C–H bond broken	Br–Br bond broken	C–Br bond formed	H–Br bond formed
+91 kcal/mol	+46 kcal/mol	–65 kcal/mol	–88 kcal/mol

total bonds broken = +137 kcal/mol   total bonds formed = –153 kcal/mol $\Delta H° = -16$ kcal/mol

b. Initiation:

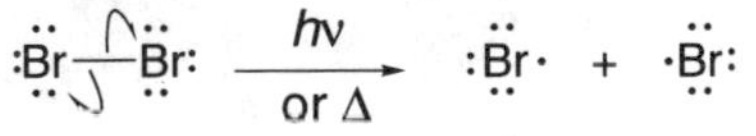

Propagation: $(CH_3)_3C-H + \cdot Br: \longrightarrow (CH_3)_3C\cdot + H-Br:$

$(CH_3)_3C\cdot + :Br-Br: \longrightarrow (CH_3)_3C-Br + \cdot Br:$

c. $\Delta H°$ = (bonds broken) – (bonds formed)
= (+91 kcal/mol) + (–88 kcal/mol)
= +3 kcal/mol

$\Delta H°$ = (bonds broken) – (bonds formed)
= (+46 kcal/mol) + (–65 kcal/mol)
= –19 kcal/mol

Termination:
(one possibility)

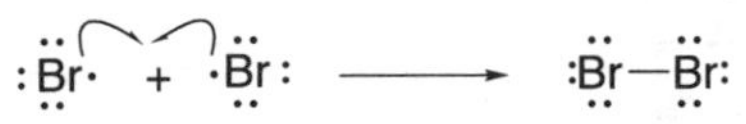

d and e.

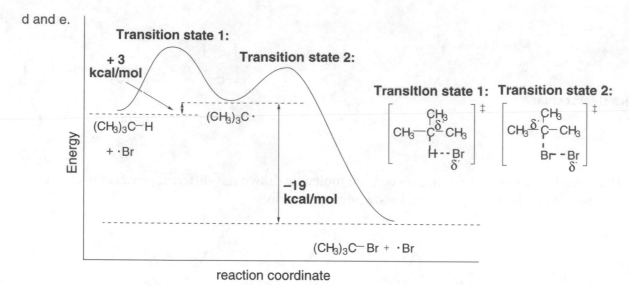

13.51

Initiation:

NBS
$\xrightarrow{h\nu}$
+ ·Br:

Propagation:

+ ·Br: $\longrightarrow$ ⟷ ⟷

:Br—Br: (from NBS)   :Br—Br: (from NBS)   + H—Br:

Br + Br + ·Br:

Termination: :Br· + ·Br $\longrightarrow$:Br—Br:
(one possibility)

13.52 Calculate the $\Delta H°$ for the propagation steps of the reaction of CH_4 with I_2 to show why it does not occur at an appreciable rate.

CH_3—H + ·I: $\longrightarrow$ ·CH_3 + H—I: $\Delta H° = +33$ kcal/mol

+104 kcal/mol   −71 kcal/mol

> This step is highly endothermic, making it difficult for chain propagation to occur over and over again.

·CH_3 + :I—I: $\longrightarrow$ CH_3—I + ·I: $\Delta H° = -20$ kcal/mol

+36 kcal/mol   −56 kcal/mol

13.53 Calculate $\Delta H°$ for each of these steps, and use these values to explain why this alternate mechanism is unlikely.

[1] CH_4 + $:\ddot{C}l\cdot$ ⟶ CH_3Cl + $H\cdot$

C–H bond broken   C–Cl bond formed $\Delta H° = +20$ kcal/mol
+104 kcal/mol   −84 kcal/mol

> The $\Delta H°$ for Step [1] is very endothermic, making this mechanism unlikely.

[2] $H\cdot$ + Cl_2 ⟶ HCl + $:\ddot{C}l\cdot$

Cl–Cl bond broken   H–Cl bond formed $\Delta H° = -45$ kcal/mol
+58 kcal/mol   −103 kcal/mol

13.54

3,3-dimethyl-1-butene   2° carbocation   3° carbocation   2-bromo-2,3-dimethylbutane

3,3-dimethyl-1-butene   HBr peroxide   The 2° radical does NOT rearrange.   1-bromo-3,3-dimethylbutane

Addition of HBr without added peroxide occurs by an ionic mechanism and forms a 2° carbocation, which rearranges to a more stable 3° carbocation. The addition of H^+ occurs first, followed by Br^-. Addition of HBr with added peroxide occurs by a radical mechanism and forms a 2° radical that does not rearrange. In the radical mechanism Br• adds first, followed by H•.

13.55

a. [Cl₂, Δ]

b. [from (a)] K⁺ ⁻OC(CH₃)₃

c. [from (b)] NBS / ROOR

d. [from (b)] H₂O / H₂SO₄

e. [from (b)] Cl₂

f. [from (c)] Br₂

g. [from (c)] ⁻OH

h. [from (g)] mCPBA

13.56

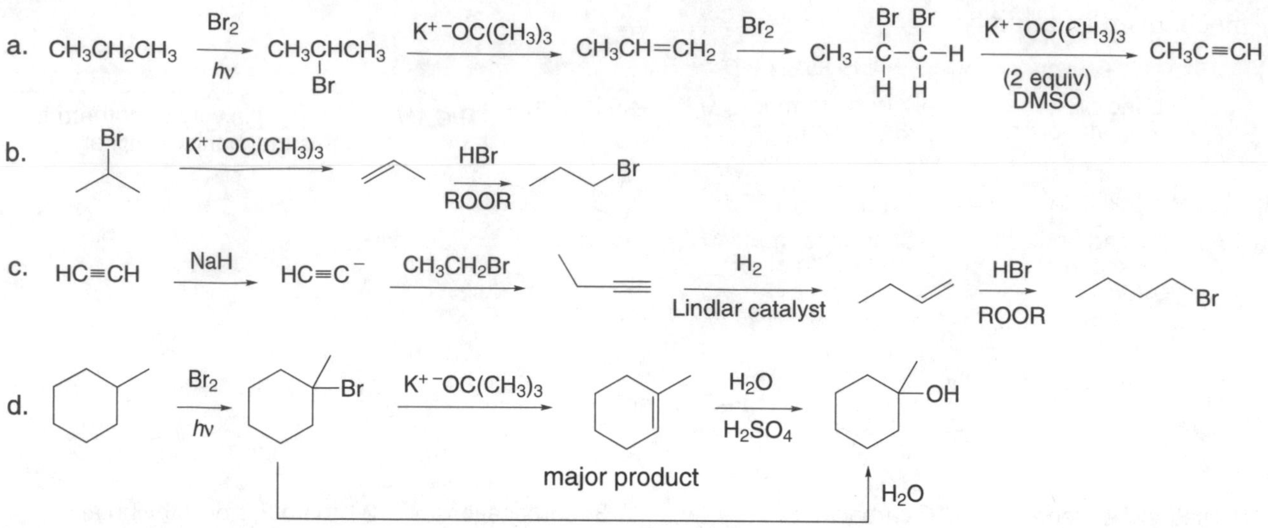

a. $CH_3CH_2CH_3$ $\xrightarrow{Br_2, h\nu}$ $CH_3\underset{Br}{CHCH_3}$ $\xrightarrow{K^+ \, ^-OC(CH_3)_3}$ $CH_3CH{=}CH_2$ $\xrightarrow{Br_2}$ $CH_3\overset{Br}{\underset{H}{C}}{-}\overset{Br}{\underset{H}{C}}{-}H$ $\xrightarrow[\substack{(2\ equiv) \\ DMSO}]{K^+ \, ^-OC(CH_3)_3}$ $CH_3C{\equiv}CH$

b.

c. $HC{\equiv}CH$ $\xrightarrow{NaH}$ $HC{\equiv}C^-$ $\xrightarrow{CH_3CH_2Br}$ $\xrightarrow{\substack{H_2 \\ \text{Lindlar catalyst}}}$ $\xrightarrow{\substack{HBr \\ ROOR}}$

d.

major product

13.57

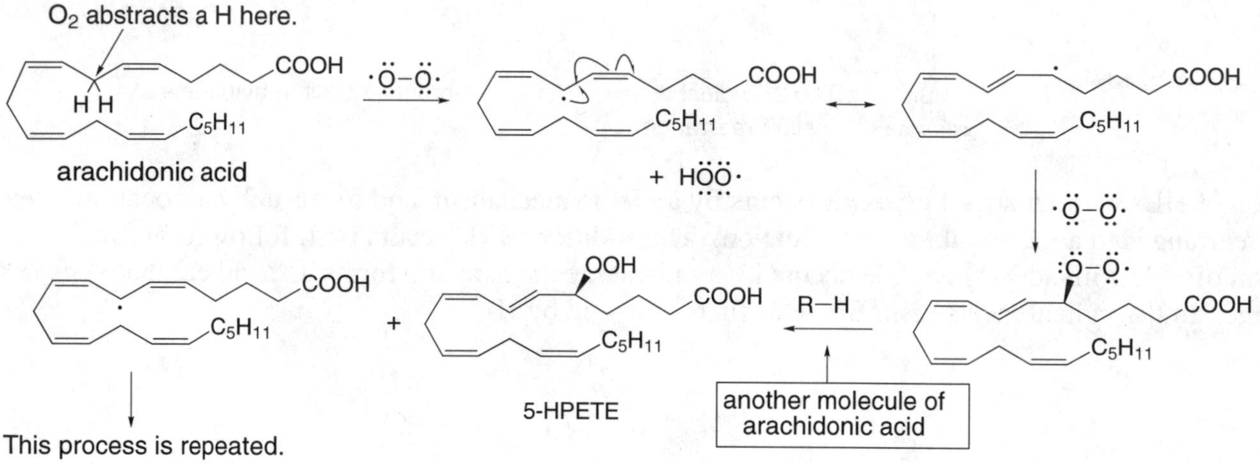

O₂ abstracts a H here.

arachidonic acid

+ HOO·

5-HPETE

another molecule of
arachidonic acid

This process is repeated.

13.58

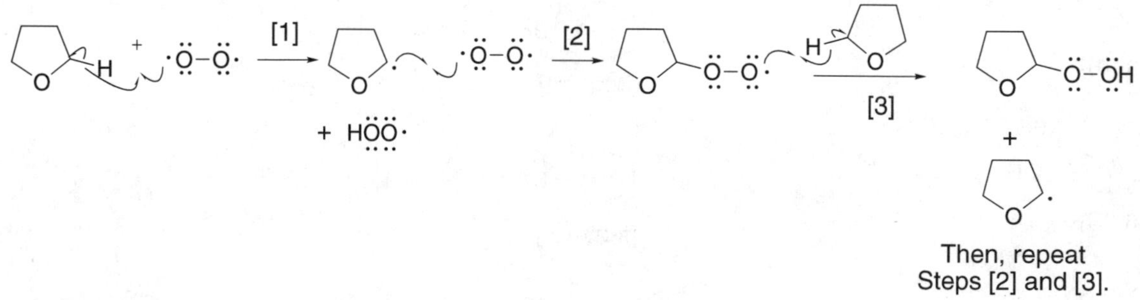

+ HOO·

[1] [2] [3]

Then, repeat
Steps [2] and [3].

13.59

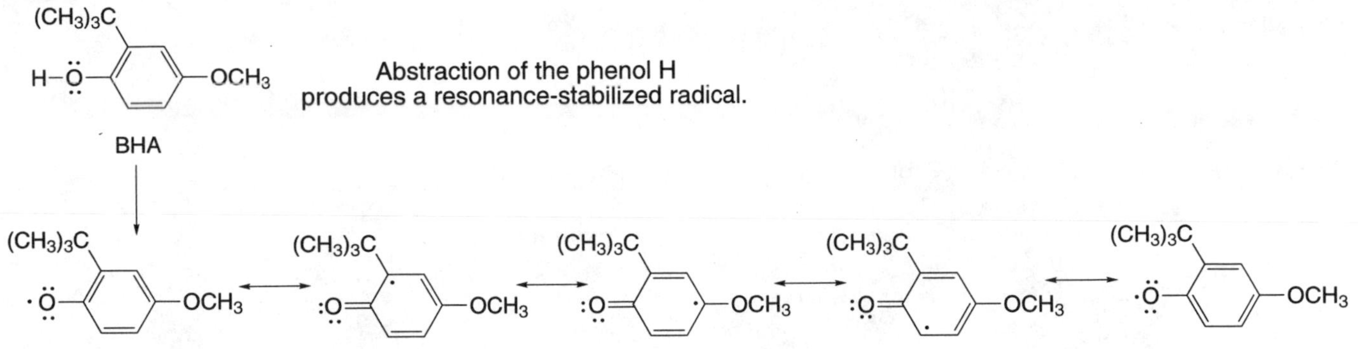

Abstraction of the phenol H produces a resonance-stabilized radical.

BHA

13.60 Abstraction of the labeled H forms a highly resonance-stabilized radical. Four of the possible resonance structures are drawn.

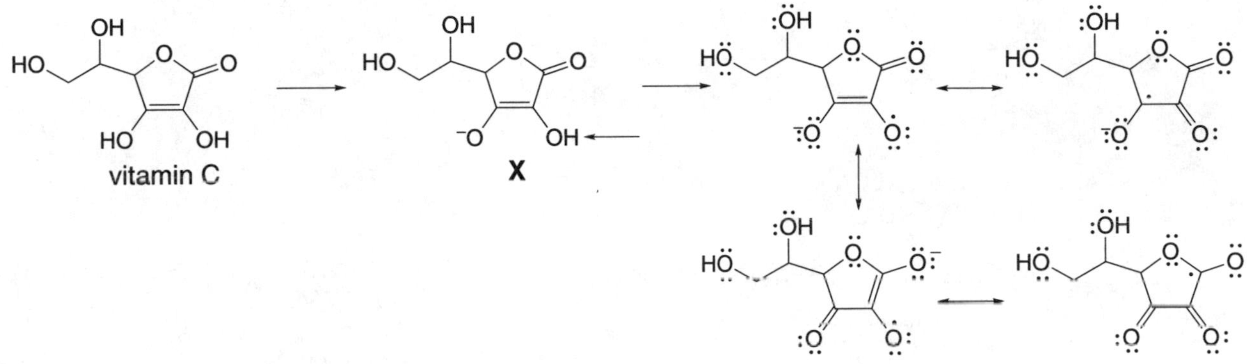

vitamin C

X

13.61 The monomers used in polymerization always contain double bonds.

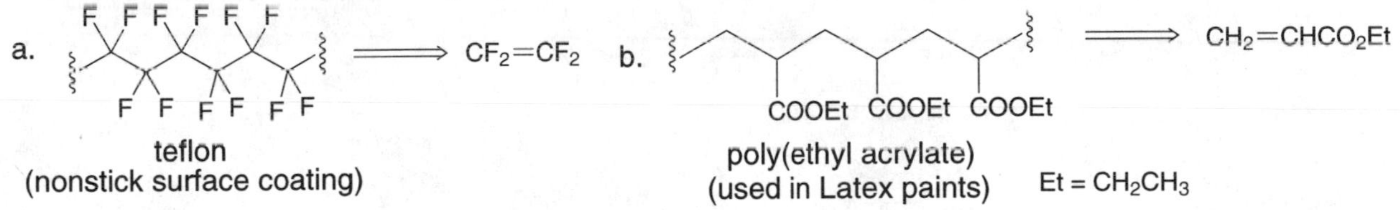

a.

$\Longrightarrow$ $CF_2{=}CF_2$ b. $\Longrightarrow$ $CH_2{=}CHCO_2Et$

teflon
(nonstick surface coating)

poly(ethyl acrylate)
(used in Latex paints)   Et = CH_2CH_3

13.62

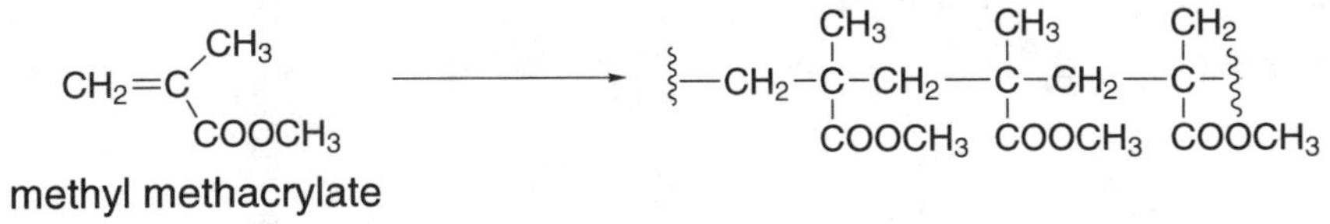

methyl methacrylate

13.63

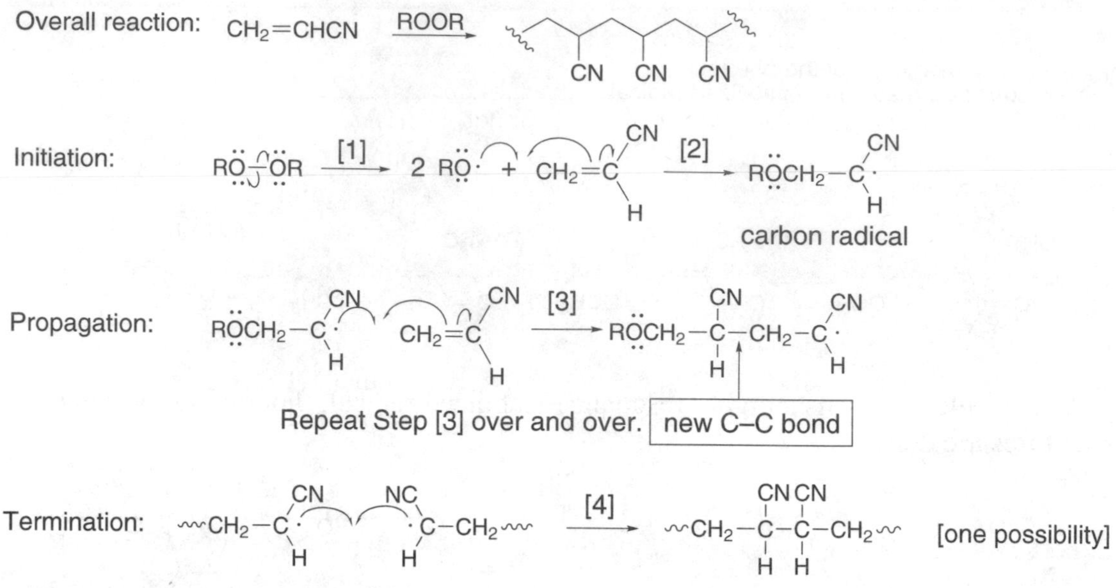
Overall reaction:
CH2=CHCN
ROOR
CN CN CN
Initiation:
[1]
[2]
carbon radical
Propagation:
[3]
Repeat Step [3] over and over.
new C–C bond
Termination:
[4]
[one possibility]

13.64

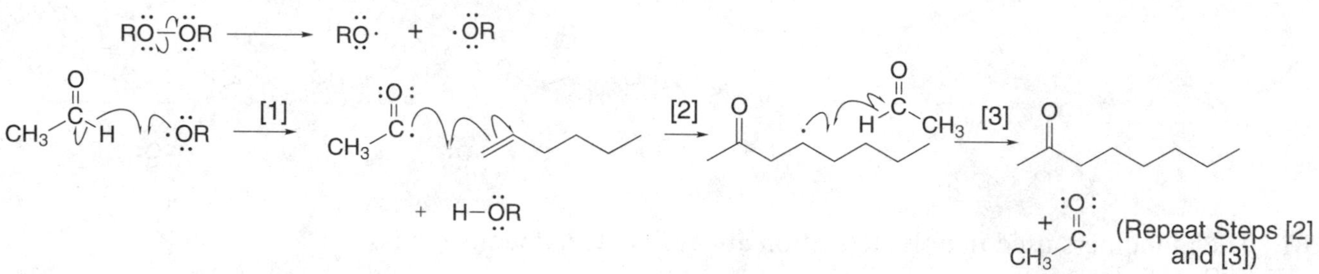
[1]
[2]
[3]
(Repeat Steps [2] and [3])

13.65

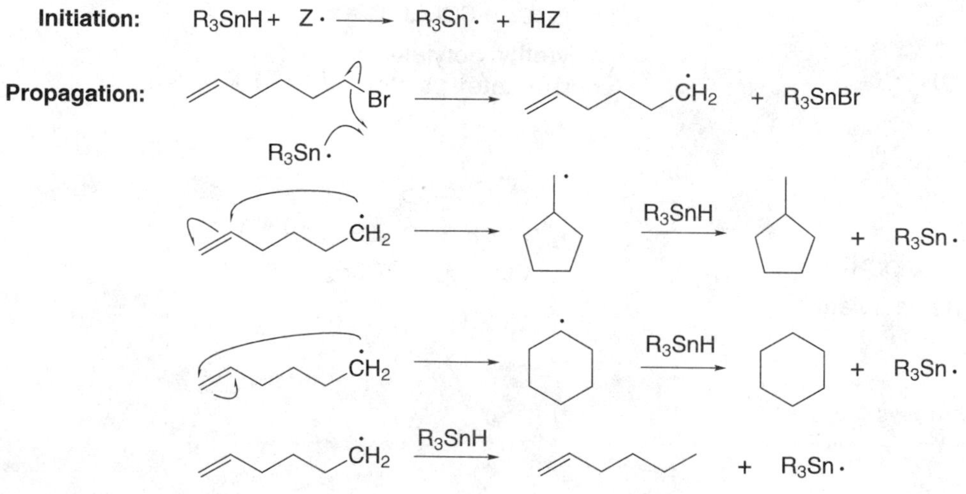
Initiation:
R3SnH + Z· → R3Sn· + HZ
Propagation:
R3SnBr
R3SnH
R3Sn·

Chapter 14: Mass Spectrometry and Infrared Spectroscopy

♦ Mass spectrometry (MS)

- Mass spectrometry measures the molecular weight of a compound (14.1A).
- The mass of the molecular ion (**M**) = the molecular weight of a compound. Except for isotope peaks at M + 1 and M + 2, the molecular ion has the highest mass in a mass spectrum (14.1A).
- The base peak is the tallest peak in a mass spectrum (14.1A).
- A compound with an odd number of N atoms gives an odd molecular ion. A compound with an even number of N atoms (including zero) gives an even molecular ion (14.1B).
- Organic chlorides show two peaks for the molecular ion (M and M + 2) in a 3:1 ratio (14.2).
- Organic bromides show two peaks for the molecular ion (M and M + 2) in a 1:1 ratio (14.2).
- High-resolution mass spectrometry gives the molecular formula of a compound (14.3A).

♦ Electromagnetic radiation

- The wavelength and frequency of electromagnetic radiation are *inversely* related, by the following equations: $\lambda = c/\nu$ or $\nu = c/\lambda$ (14.4).
- The energy of a photon is proportional to its frequency; the higher the frequency the higher the energy: $E = h\nu$ (14.4).

♦ Infrared spectroscopy (IR, 14.5 and 14.6)

- Infrared spectroscopy identifies functional groups.
- IR absorptions are reported in wave numbers $\tilde{\nu} = 1/\lambda$.
- The functional group region from **4000–1500 cm^{-1}** is the most useful region of an IR spectrum.
- C–H, O–H, and N–H bonds absorb at high frequency, ≥ 2500 cm^{-1}.
- As bond strength increases, the $\tilde{\nu}$ of absorption increases; thus triple bonds absorb at higher $\tilde{\nu}$ than double bonds.

$$C=C \qquad C{\equiv}C$$
$$\sim 1650 \text{ cm}^{-1} \quad \sim 2250 \text{ cm}^{-1}$$

increasing bond strength
increasing $\tilde{\nu}$

- The higher the percent *s*-character, the stronger the bond, and the higher the $\tilde{\nu}$ of an IR absorption.

Csp^3–H	Csp^2–H	Csp–H
25% *s*-character	33% *s*-character	50% *s*-character
3000–2850 cm^{-1}	3150–3000 cm^{-1}	3300 cm^{-1}

increasing percent *s*-character
increasing $\tilde{\nu}$

Chapter 14: Answers to Problems

14.1 The molecular ion formed from each compound is equal to its molecular weight.

a. C_3H_6O
molecular weight = **58**
molecular ion (m/z) = **58**

b. $C_{10}H_{20}$
molecular weight = **140**
molecular ion (m/z) = **140**

c. $C_8H_8O_2$
molecular weight = **136**
molecular ion (m/z) = **136**

d. $C_{10}H_{15}N$
molecular weight = **149**
molecular ion (m/z) = **149**

14.2 If the molecular ion is 88, the molecular weight must also be 88.

a.

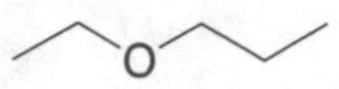

molecular formula = $C_5H_{12}O$
molecular weight = **88**
This could be **X**.

b.

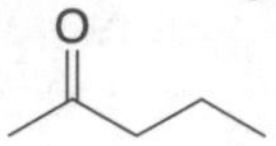

molecular formula = $C_5H_{10}O$
molecular weight = **86**

c.

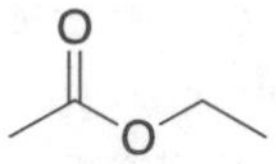

molecular formula = $C_4H_8O_2$
molecular weight = **88**
This could be **X**.

14.3 Some possible formulas for each molecular ion:

a.   Molecular ion at 72: C_5H_{12}, C_4H_8O, $C_3H_4O_2$
b.   Molecular ion at 100: C_8H_4, C_7H_{16}, $C_6H_{12}O$, $C_5H_8O_2$
c.   Molecular ion at 73: $C_4H_{11}N$, $C_2H_7N_3$

14.4 To calculate the molecular ions you would expect for compounds with Cl, calculate the molecular weight using each of the two most common isotopes of Cl (^{35}Cl and ^{37}Cl). Do the same for Br, using ^{79}Br and ^{81}Br.

a. $C_4H_9{}^{35}Cl$ = **92**
$C_4H_9{}^{37}Cl$ = **94**
Two peaks in 3:1 ratio at m/z 92 and 94.

c. $C_6H_{11}{}^{79}Br$ = **162**
$C_6H_{11}{}^{81}Br$ = **164**
Two peaks in a 1:1 ratio at m/z 162 and 164.

b. C_3H_7F = **62**
One peak at m/z 62.

d. $C_4H_{11}N$ = **73**
One peak at m/z 73.

14.5 Use the exact values given in Table 14.1 to calculate the exact mass of each compound.

$C_7H_5NO_3$

mass: 151.0270

$C_8H_9NO_2$

mass: 151.0634
compound **X**

$C_{10}H_{17}N$

mass: 151.1362

14.6

benzene
C_6H_6 $m/z = 78$

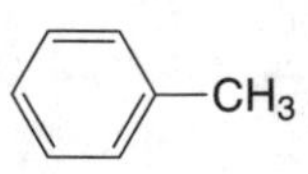

toluene
C_7H_8 $m/z = 92$

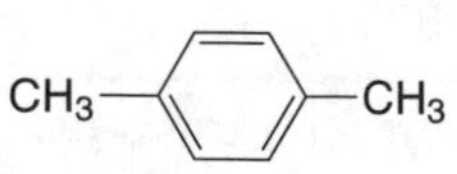

p-xylene
C_8H_{10} $m/z = 106$

GC–MS analysis:
3 peaks in the gas chromatogram.
Order of peaks: benzene, toluene, p-xylene, in order of increasing bp.
Molecular ions observed in the three mass spectra: 78, 92, 106.

14.7 **Wavelength and frequency are inversely proportional**. The higher frequency light will have a shorter wavelength.
   a.   Light having a λ of 10^2 nm has a higher ν than light with a λ of 10^4 nm.
   b.   Light having a λ of 100 nm has a higher ν than light with a λ of 100 μm.
   c.   Blue light has a higher ν than red light.

14.8 The **energy of a photon** is *proportional* to its **frequency**, and inversely proportional to its wavelength.
   a.   Light having a ν of 10^8 Hz is of higher energy than light having a ν of 10^4 Hz.
   b.   Light having a λ of 10 nm is of higher energy than light having a λ of 1000 nm.
   c.   Blue light is of higher energy than red light.

14.9 The larger the energy difference between two states, the higher the frequency needed for absorption. The 100 kcal/mol transition requires a higher ν of radiation than a 5 kcal/mol transition.

14.10   Higher wave numbers are proportional to higher frequencies and higher energies.
   a.   IR light with a wave number of 3000 cm^{-1} is higher in energy than IR light with a wave number of 1500 cm^{-1}.
   b.   IR light having a λ of 10 μm is higher in energy than IR light having a λ of 20 μm.

14.11   Stronger bonds absorb at a higher wave number. Bonds to lighter atoms (H versus D) absorb at higher wave number.

   a. $CH_3-C{\equiv}C-CH_2CH_3$ or $CH_2{=}C(CH_3)_2$ b. CH_3CH_2-H or $CH_2{=}CH-H$ c. CH_3-H or CH_3-D

        stronger bond   higher % *s*-character   lighter atom H
        higher wave number   higher wave number   higher wave number

14.12   Cyclopentane and 1-pentene are both composed of C–C and C–H bonds, but 1-pentene also has a C=C bond. This difference will give the IR of 1-pentene an additional peak at 1650 cm^{-1} (for the C=C). 1-Pentene will also show C–H absorptions for sp^2 hybridized C–H bonds at 3150–3000 cm^{-1}.

14.13   Look at the functional groups in each compound below to explain how each IR is different.

   $CH_3-\overset{\overset{\displaystyle O}{\|}}{C}-CH_3$ $CH_3OCH{=}CH_2$ ▷—OH
          A   **B**   **C**

   C=O peak at ~1700 cm^{-1} C=C peak at 1650 cm^{-1} O–H peak at 3200–3600cm^{-1}
                                    Csp^2–H at 3150–3000 cm^{-1}

14.14
   a.   Compound **A** has peaks at ~3150 (sp^2 hybridized C–H), 3000–2850 (sp^3 hybridized C–H), and 1650 (C=C) cm^{-1}.
   b.   Compound **B** has a peak at 3000–2850 (sp^3 hybridized C–H) cm^{-1}.

14.15 All compounds show an absorption at 3000–2850 cm^{-1} due to the sp^3 hybridized C–H bonds. Additional peaks in the functional group region for each compound are shown.

a.

no additional peaks

b. —OH

O–H bond at 3600–3200 cm^{-1}

c.

Csp^2–H at 3150–3000 cm^{-1}
C=C bond at 1650 cm^{-1}

d. =O

C=O bond at ~1700 cm^{-1}

14.16 Possible structures are (a) $CH_3COOCH_2CH_3$ and (c) $CH_3CH_2COOCH_3$. Compounds (b) and (d) also have an OH group that would give a strong absorption at ~3600–3200 cm^{-1}, which is absent in the IR spectrum of **X**, thus excluding them as possibilities.

14.17

a. Hydrocarbon with a molecular ion at $m/z = 68$

IR absorptions at 3310 cm^{-1} = Csp–H bond
3000–2850 cm^{-1} = Csp^3–H bonds
2120 cm^{-1} = C≡C bond

Molecular formula: C_5H_8

H–C≡C–CH$_2$CH$_2$CH$_3$ or   H–C≡C–CHCH$_3$
                                              |
                                             CH$_3$

b. Compound with C, H, and O with a molecular ion at $m/z = 60$

IR absorptions at 3600–3200 cm^{-1} = O–H bond
3000–2850 cm^{-1} = Csp^3–H bonds

Molecular formula: C_3H_8O

CH$_3$CH$_2$CH$_2$—O–H   or   CH$_3$CH–O–H
                                          |
                                         CH$_3$

14.18

a.

molecular formula: C_6H_6
molecular ion (m/z): **78**

c.

molecular formula: $C_5H_{10}O$
molecular ion (m/z): **86**

e. $(CH_3)_3CCH(Br)CH(CH_3)_2$

molecular formula: $C_8H_{17}Br$
molecular ions (m/z): **192, 194**

b.

molecular formula: $C_{10}H_{16}$
molecular ion (m/z): **136**

d.

Cl

molecular formula: $C_5H_{11}Cl$
molecular ions (m/z): **106, 108**

14.19

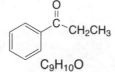
$CH_2CH_2CH_3$

C_9H_{12}
molecular weight = 120

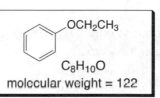
O
‖
C—CH_2CH_3

$C_9H_{10}O$
molecular weight = 134

OCH_2CH_3

$C_8H_{10}O$
molecular weight = 122

14.20 Examples are given for each molecular ion.
a. molecular ion 102: C_8H_6, $C_6H_{14}O$, $C_5H_{10}O_2$, $C_5H_{14}N_2$
b. molecular ion 98: C_8H_2, C_7H_{14}, $C_6H_{10}O$, $C_5H_6O_2$
c. molecular ion 119: C_8H_9N, $C_6H_5N_3$
d. molecular ion 74: C_6H_2, $C_4H_{10}O$, $C_3H_6O_2$

14.21 Likely molecular formula, C_8H_{16} (one degree of unsaturation—one ring or one π bond).

Four structures with m/z = 112

14.22

CH_3—CH—CO_2CH_3
|
Cl
B

$C_4H_7O_2Cl$
molecular weight: **122, 124**
should show 2 peaks for the
molecular ion with a **3:1 ratio**

Mass spectrum [1]

CH_3 OCH_3
C

$C_8H_{10}O$
molecular weight: **122**

Mass spectrum [2]

$CH_3CH_2CH_2Br$
A

C_3H_7Br
molecular weight: **122, 124**
should show 2 peaks for the
molecular ion with a **1:1 ratio**

Mass spectrum [3]

14.23

Possible structures
C_7H_{12}
(exact mass 96.0940)

14.24 One possible structure is drawn for each set of data:

a. a compound that contains a benzene ring and has a molecular ion at $m/z = 107$.

C_7H_9N

b. a hydrocarbon that contains only sp^3 hybridized carbons and a molecular ion at $m/z = 84$.

C_6H_{12}

c. a compound that contains a carbonyl group and gives a molecular ion at $m/z = 114$.

CH_3 — $C(=O)$ — $CH_2CH_2CH_2CH_2CH_3$

$C_7H_{14}O$

d. A compound that contains C, H, N, and O and has an exact mass for the molecular ion at 101.0841.

CH_3 — $C(=O)$ — $NHCH_2CH_2CH_3$

$C_5H_{11}NO$

14.25 Use the exact values given in Table 14.1 to calculate the exact mass of each compound as in Answer 14.5. $C_8H_{11}NO_2$ (exact mass 153.0790) is the correct molecular formula.

14.26 Molecules with an odd number of N's have an odd number of H's, making the molecular ion odd as well.

14.27 Two isomers such as $CH_2=CHCH_2CH_2CH_2CH_3$ and $(CH_3)_2C=CHCH_2CH_3$ have the same molecular formulas and therefore give the same exact mass, so they are not distinguishable by their exact mass spectra.

14.28

a. $(CH_3)_2C=O$ or $(CH_3)_2CH-OH$

stronger bond
higher $\tilde{\nu}$ absorption

b. $(CH_3)_2C=NCH_3$ or $(CH_3)_2CH-NCH_3$

stronger bond
higher $\tilde{\nu}$ absorption

c.

stronger bond
higher $\tilde{\nu}$ absorption

14.29 Locate the functional groups in each compound. Use Table 14.2 to determine what IR absorptions each would have.

a. (cyclopentane) Csp^3–H at 2850–3000 cm^{-1}

f. (cyclohexene–OH)   O–H at 3200–3600 cm^{-1}
Csp^2–H at 3000–3150 cm^{-1}
Csp^3–H at 2850–3000 cm^{-1}
C=C at 1650 cm^{-1}

b. (cyclohexyl–C≡CH) Csp–H at 3300 cm^{-1}
Csp^3–H at 2850–3000 cm^{-1}
C–C triple bond at 2250 cm^{-1}

g. (cyclohexene ketone) Csp^2–H at 3000–3150 cm^{-1}
Csp^3–H at 2850–3000 cm^{-1}
C=O at ~1700 cm^{-1}
C=C at 1650 cm^{-1}

c. (alcohol chain OH)   O–H at 3200–3600 cm^{-1}
Csp^3–H at 2850–3000 cm^{-1}

d. (ketone chain) Csp^3–H at 2850–3000 cm^{-1}
C=O at 1700 cm^{-1}

h. (benzoic acid)   O–H at ~3200–3600 cm^{-1}
Csp^2–H at 3000–3150 cm^{-1}
C=O at ~1700 cm^{-1}
phenyl group at 1600, 1500 cm^{-1}

e. (anisole –OCH$_3$) Csp^2–H at 3000–3150 cm^{-1}
Csp^3–H at 2850–3000 cm^{-1}
phenyl group at 1600, 1500 cm^{-1}

i. (Br alkyl chain)   Br Csp^3–H at 2850–3000 cm^{-1}

14.30

a. (cyclopentene)   and   HC≡CCH$_2$CH$_2$CH$_3$
C=C bond   C≡C bond
1650 cm^{-1} 2250 cm^{-1}
Csp^2–H at 3150–3000 cm^{-1} Csp–H at 3300 cm^{-1}

d. (cyclohexane OCH$_3$ OCH$_3$)   and   CH$_3$(CH$_2$)$_5$–C(=O)–OCH$_3$
no C=O bond   C=O bond
~1700 cm^{-1}

b.   CH$_3$CH$_2$–C(=O)–OH   and   CH$_3$–C(=O)–OCH$_3$
O–H bond   no O–H bond
~3200–3600 cm^{-1}

e.   CH$_3$C≡CCH$_3$ and   CH$_3$CH$_2$C=CH
no C≡C absorption Csp–H bond
due to symmetry   3300 cm^{-1}
C≡C bond at ~2250 cm^{-1}

c.   CH$_3$CH$_2$–C(=O)–CH$_3$ and   CH$_3$CH=CHCH$_2$OH
C=O bond   O–H bond
1700 cm^{-1} 3200–3600 cm^{-1}
Csp^2–H at 3150–3000 cm^{-1}
C=C bond at 1650 cm^{-1}

f.   HC≡CCH$_2$N(CH$_2$CH$_3$)$_2$ and   CH$_3$(CH$_2$)$_5$C≡N
Csp–H bond
3300 cm^{-1}

14.31 Look for a **change in functional groups** from starting material to product to see how IR could be used to determine when the reaction is complete.

a. $\xrightarrow[\text{Pd}]{\text{H}_2}$ Loss of the C=C will be visible in the IR by disappearance of the peak at 1650 cm $^{-1}$.

b. $\xrightarrow{\text{PCC}}$ Loss of the O–H group will be visible in the IR by disappearance of the peak at 3200–3600 cm^{-1} and appearance of the C=O at ~1700 cm^{-1}.

c. $\xrightarrow[\text{2) CH}_3\text{SCH}_3]{\text{1) O}_3}$ + O=C(CH$_3$)(CH$_3$)   Loss of the C=C will be visible in the IR by disappearance of the peak at 1650 cm^{-1} and appearance of the C=O at ~1700 cm^{-1}.

14.32 In addition to Csp^3–H at ~3000–2850 cm^{-1}:

Spectrum [1]:

CH$_2$=C(CH$_3$)CH$_2$CH$_2$CH$_2$CH$_3$ (**B**)
    C=C peak at 1650 cm^{-1}
    Csp^2–H at ~3150 cm^{-1}

Spectrum [2]:

(CH$_3$CH$_2$)$_3$COH (**F**)
    OH at 3600–3200 cm^{-1}

Spectrum [3]:

(CH$_3$)$_2$CHOCH(CH$_3$)$_2$ (**D**)
    No other peaks above 1500 cm^{-1}.

Spectrum [4]:

⬡—CH(CH$_3$)$_2$
C
    Csp^2–H at ~3150 cm^{-1}
    Phenyl peaks at 1600 and 1500 cm^{-1}

Spectrum [5]:

CH$_3$CH$_2$CH$_2$CH$_2$COOH (**A**)
    OH at ~3500–2500 cm^{-1}
    C=O at ~1700 cm^{-1}

Spectrum [6]:

CH$_3$COOC(CH$_3$)$_3$ (**E**)
    C=O at ~1700 cm^{-1}

14.33 In addition to Csp^3–H at ~3000–2850 cm^{-1}:

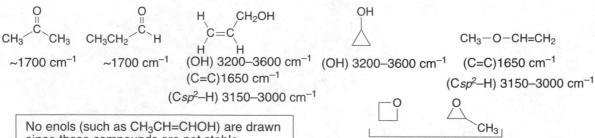

CH$_3$—C(=O)—CH$_3$ CH$_3$CH$_2$—C(=O)—H
~1700 cm^{-1} ~1700 cm^{-1}

H(CH$_2$OH)C=C(H)(H)
(OH) 3200–3600 cm^{-1}
(C=C)1650 cm^{-1}
(Csp^2–H) 3150–3000 cm^{-1}

(OH) 3200–3600 cm^{-1}

CH$_3$–O–CH=CH$_2$
(C=C)1650 cm^{-1}
(Csp^2–H) 3150–3000 cm^{-1}

No enols (such as CH$_3$CH=CHOH) are drawn since these compounds are not stable.

No additional peaks above 1500 cm^{-1}.

14.34 Stronger bonds absorb at a higher frequency. Bonds to heavier atoms tend to absorb at lower frequency.

$CH_3 \overset{\uparrow}{-} I$

weakest bond
I—highest mass
lowest IR frequency

$CH_3 \overset{\uparrow}{-} Br$

medium bond strength
intermediate IR frequency

$CH_3 \overset{\uparrow}{-} Cl$

strongest bond
Cl—lowest mass
highest IR frequency

14.35

a. Compound with a molecular ion at $m/z = 72$
   IR absorption at 1725 cm^{-1} = C=O bond
   Molecular formula: C_4H_8O

b. Compound with a molecular ion at $m/z = 55$
   The odd molecular ion means an odd number of N's present. Molecular formula: C_3H_5N
   IR absorption at 2250 cm^{-1} = C≡N bond

   $CH_3CH_2C≡N$

c. Compound with a molecular ion at $m/z = 74$
   IR absorption at 3600–3200 cm^{-1} = O–H bond
   Molecular formula: $C_4H_{10}O$

14.36

Chiral hydrocarbon with a molecular ion at $m/z = 82$
Molecular formula: C_6H_{10}
IR absorptions at 3300 cm^{-1} = C_{sp}–H bond
3000–2850 cm^{-1} = C_{sp^3}–H bonds
2250 cm^{-1} = C≡C bond

HC≡C\overset{*}{C}HCH_2CH_3
|
CH_3

stereogenic center

14.37 The chiral compound **Y** has a strong absorption at 2970-2840 cm^{-1} in its IR spectrum due to sp^3 hybridized C–H bonds. The two peaks of equal intensity at 136 and 138 indicate the presence of a Br atom. The molecular formula is C_4H_9Br. Only one alkyl bromide of this molecular formula has a stereogenic center:

Y =   [* = stereogenic center]

14.38

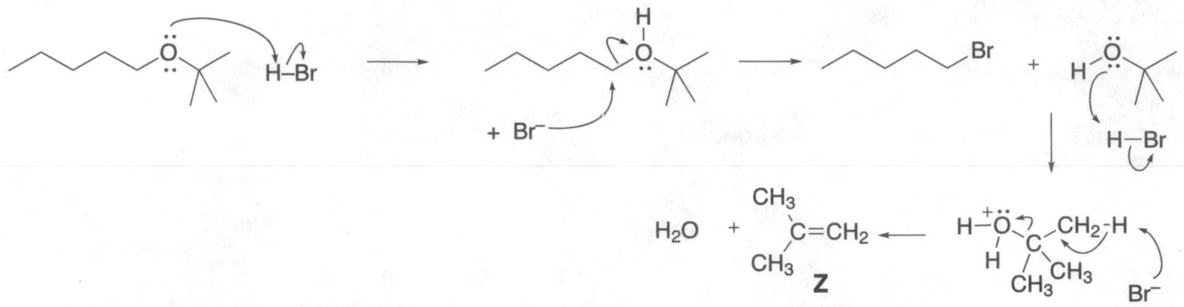

14.39 The mass spectrum has a molecular ion at 71. The odd mass suggests the presence of an odd number of N atoms; likely formula, C_4H_9N. The IR absorption at ~3300 cm^{-1} is due to N–H and the 3000–2850 cm^{-1} is due to sp^3 hybridized C–H bonds.

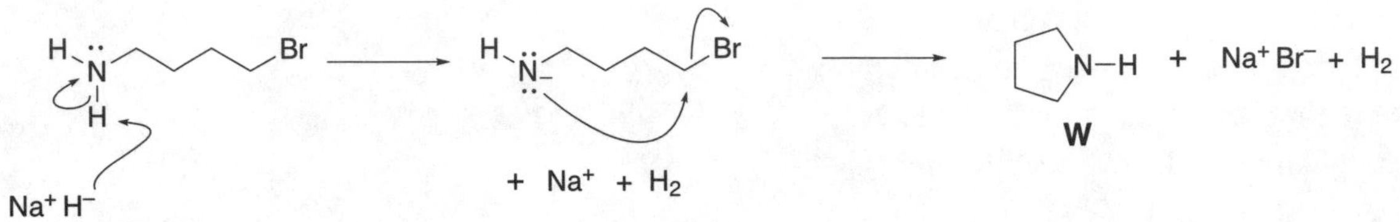

14.40 The α,β-unsaturated carbonyl compound has three resonance structures, two of which place a single bond between the C and O atoms. This means that the C–O bond has partial single bond character, making it weaker than a regular C=O bond, and moving the absorption to lower wave number.

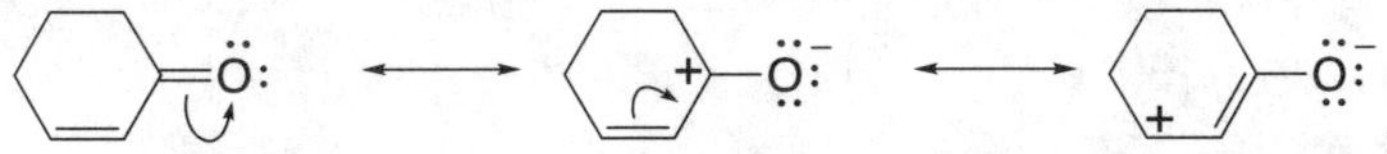

three resonance structures for 2-cyclohexenone

14.41

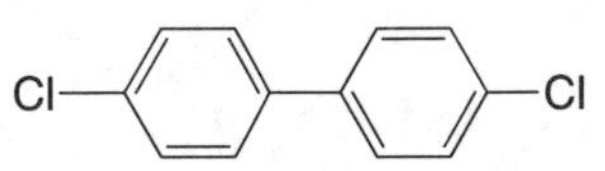

4,4'-dichlorobiphenyl
(a common PCB)

Three peaks would be seen for the molecular ions:

m/z

$C_{12}H_8Cl_2^{35}$: 222 (Both ^{35}Cl isotopes)
$C_{12}H_8Cl_2^{35,37}$: 224 (One ^{35}Cl and one ^{37}Cl isotope)
$C_{12}H_8Cl_2^{37}$: 226 (Both ^{37}Cl isotopes)

14.42

a and b.

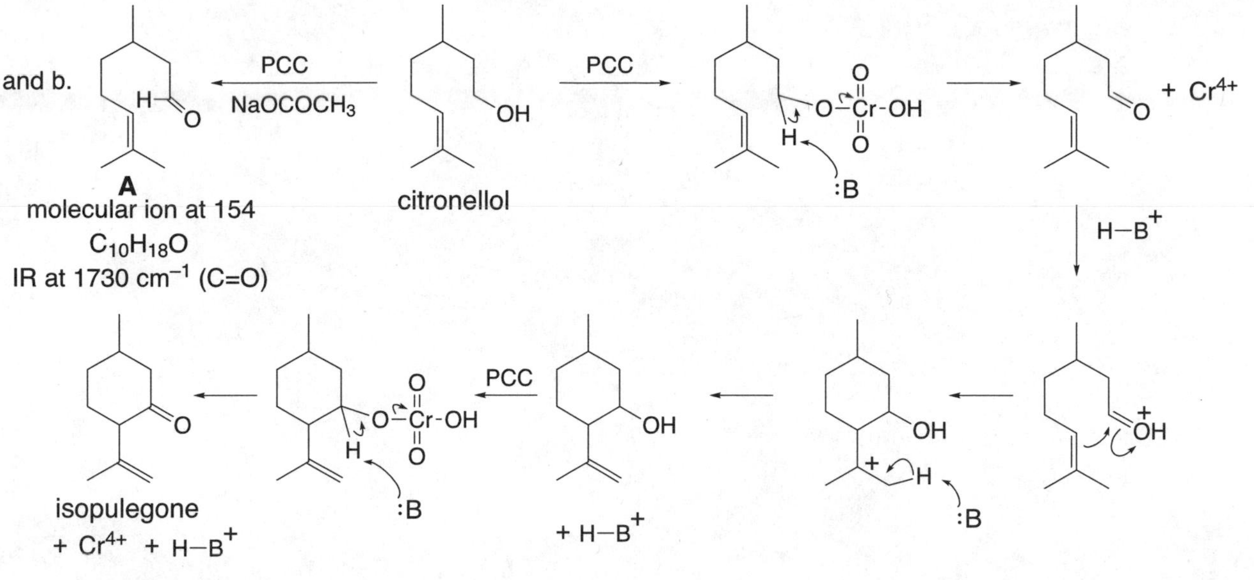
PCC
NaOCOCH3
PCC
OH
citronellol
A
molecular ion at 154
C10H18O
IR at 1730 cm−1 (C=O)
O-Cr-OH
:B
+ Cr4+
H–B+
PCC
isopulegone
+ Cr4+ + H–B+
+ H–B+
OH
:B

a and c.

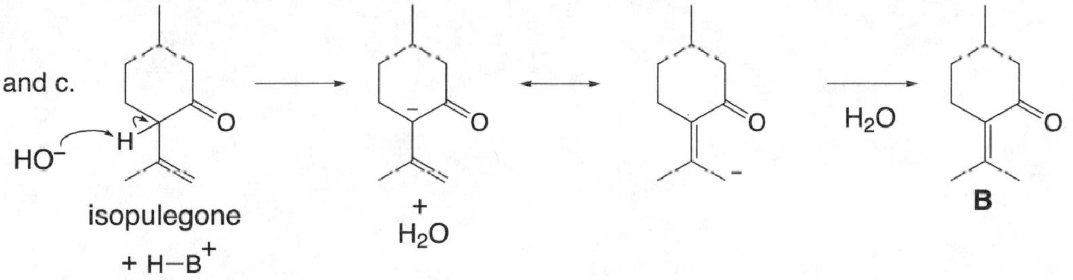
HO−
isopulegone
+ H–B+
+
H2O
H2O
B

Chapter 15: Nuclear Magnetic Resonance Spectroscopy

◆ ^{1}H NMR spectroscopy

[1] The **number of signals** equals the number of different types of protons (15.2).

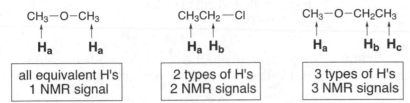

[2] The **position of a signal** (its chemical shift) is determined by shielding and deshielding effects.
- Shielding shifts an absorption upfield; deshielding shifts an absorption downfield.
- Electronegative atoms withdraw electron density, deshield a nucleus, and shift an absorption downfield (15.3).

—C—H ←	This proton is shielded. Its absorption is upfield, 0.9–2 ppm.

—C—H ←	This proton is deshielded. Its absorption is further downfield, 2.5–4 ppm.
X	

- Loosely held π electrons can either shield or deshield a nucleus. Protons on benzene rings and double bonds are deshielded and absorb downfield, whereas protons on triple bonds are shielded and absorb upfield (15.4).

deshielded H downfield absorption	shielded H upfield absorption

[3] The **area under an NMR signal** is proportional to the number of absorbing protons (15.5).

[4] **Spin-spin splitting** tells about nearby nonequivalent protons (15.6–15.8).
- Equivalent protons do not split each other's signals.
- A set of n nonequivalent protons on the same carbon or adjacent carbons split an NMR signal into $n + 1$ peaks.
- OH and NH protons do not cause splitting (15.9).
- When an absorbing proton has two sets of nearby nonequivalent protons that are equivalent to each other, use the $n + 1$ rule to determine splitting.
- When an absorbing proton has two sets of nearby nonequivalent protons that are not equivalent to each other, the number of peaks in the NMR signal $= (n + 1)(m + 1)$.

◆ ^{13}C NMR spectroscopy (15.11)

[1] The number of signals equals the number of different types of carbon atoms. All signals are single peaks.

[2] The relative position of ^{13}C signals is determined by shielding and deshielding effects.
- Carbons that are sp^3 hybridized are shielded and absorb upfield.
- Electronegative elements (N, O and X) shift absorptions downfield.
- The carbons of alkenes and benzene rings absorb downfield.
- Carbonyl carbons are highly deshielded, and absorb further downfield than other carbon types.

Chapter 15: Answers to Problems

15.1 To calculate chemical shift, use the following formula:

$$\delta = ppm = [\text{observed chemical shift (Hz)}] / \nu \text{ of the NMR (MHz)}$$

a. $\delta = [60 Hz] / [60 MHz]$
   $= \mathbf{1\ ppm}$

b. $\delta = [1600\ Hz] / [200\ MHz]$
   $= \mathbf{8\ ppm}$

15.2 To determine if two H's are equivalent replace each by an atom X. If this yields the same compound or mirror images, the two H's are equivalent. Each kind of H will give one NMR signal.

a. CH_3CH_3

   1 kind of H
   1 NMR signal

c. $CH_3CH_2CH_2CH_3$

   2 kinds of H's
   2 NMR signals

e. $CH_3CH_2CO_2CH_2CH_3$

   4 kinds of H's
   4 NMR signals

g. $CH_3CH_2OCH_2CH_3$

   2 kinds of H's
   2 NMR signals

b. $CH_3CH_2CH_3$

   2 kinds of H's
   2 NMR signals

d. $(CH_3)_2CHCH(CH_3)_2$

   2 kinds of H's
   2 NMR signals

f. $CH_3OCH_2CH(CH_3)_2$

   4 kinds of H's
   4 NMR signals

h. $CH_3CH_2CH_2OH$

   4 kinds of H's
   4 NMR signals

15.3

$CH_3CH_2CH_2CH_2CH_2CH_2CH_2CH_2Cl$

Each C is a different distance from the Cl. This makes each C different, and each set of H's different. There are 8 different kinds of protons.

15.4 Compare the H's in the three isomeric ethylenes by looking at the cis and trans groups. If two H's are cis and trans to the same group, they are equivalent.

Both H's are
identical.
1 NMR signal

Both H's are cis
to another H
and trans to Br.
1 NMR signal

Both H's are cis
to Br and trans
another H.
1 NMR signal

15.5 Draw in all of the H's and compare them. If two H's are cis and trans to the same group, they are equivalent.

a. 4 identical H's

2 NMR signals

b.

4 NMR signals

c.

3 NMR signals

15.6 The two protons of a CH_2 group are usually different from each other if a compound has one stereogenic center. Replace one of the protons of one enantiomer with X and compare the products.

2-chlorobutane   diastereomers   **5 NMR signals altogether**

This means the two H's on the CH_2 group are diastereotopic protons. They are not equivalent and give different signals.

15.7 If replacement of H with X yields enantiomers, the protons are **enantiotopic**.
If replacement of H with X yields diastereomers, the protons are **diastereotopic**. In general, if the compound has **one stereogenic center**, the protons in a CH_2 group are **diastereotopic**.

a. $CH_3CH_2CH_2CH_2CH_2CH_3$

replacement of H with X

enantiomers =
enantiotopic H's

b. $CH_3CH_2CH_2CH_2CH_3$

replacement of H with X

$CH_3CH_2CHCH_2CH_3$ no stereogenic center
         |
         X   **neither enantiotopic nor diastereotopic**

c. $CH_3CH(OH)CH_2CH_2CH_3$

Pick one configuration at the existing stereogenic center.

replacement of H with X

diastereomers =
diastereotopic H's

15.8 The two protons of a CH_2 group are usually different from each other if the compound has one stereogenic center. Replace one proton with an X and compare the products.

a. The stereogenic center makes the H's in the CH_2 group diastereotopic and therefore different from each other.

5 NMR signals

b.

7 NMR signals

15.9 Decreased electron density deshields a nucleus and the absorption goes downfield. Absorption also shifts downfield with increasing alkyl substitution.

a. FCH$_2$CH$_2$CH$_2$Cl
   F is more electronegative than Cl. The CH$_2$ group adjacent to the F is more deshielded and the H's will absorb further downfield.

c. CH$_3$OC(CH$_3$)$_3$
   The CH$_3$ group bonded to the O atom will absorb further downfield.

b. CH$_3$CH$_2$CH$_2$CH$_2$OCH$_3$
   The CH$_2$ group adjacent to the O will absorb further downfield because it is closer to the electronegative O atom.

15.10

a. ClCH$_2$CH$_2$CH$_2$Br

   H$_a$ H$_b$ H$_c$

   3 types of protons:
   H$_b$ < H$_c$ < H$_a$

b. CH$_3$OCH$_2$OC(CH$_3$)$_3$

   H$_a$ H$_b$ H$_c$

   3 types of protons:
   H$_c$ < H$_a$ < H$_b$

c. CH$_3$—C(=O)—CH$_2$CH$_3$

   H$_a$ H$_b$ H$_c$

   3 types of protons:
   H$_c$ < H$_a$ < H$_b$

15.11

a. CH$_3$—C≡C—H   CH$_3$CH=CH$_2$ CH$_3$CH$_2$CH$_3$

   H$_a$ H$_b$ H$_c$

   H$_c$ protons are shielded because they are bonded to an *sp^3* C.

   H$_a$ is shielded because it is bonded to an *sp* C.

   H$_b$ protons are deshielded because they are bonded to an *sp^2* C.

   H$_c$ < H$_a$ < H$_b$

b. CH$_3$—C(=O)—OCH$_2$CH$_3$

   H$_a$ H$_b$ H$_c$

   H$_c$ protons are shielded because they are bonded to an *sp^3* C.

   H$_a$ protons are deshielded slightly because the CH$_3$ group is bonded to a C=O.

   H$_b$ protons are deshielded because the CH$_2$ group is bonded to an O atom.

   H$_c$ < H$_a$ < H$_b$

15.12 An integration ratio of 2:3 means that there are two types of hydrogens in the compound, and that the ratio of one type to another type is 2:3.

a. CH$_3$CH$_2$Cl
   2 types of H's
   3:2 - YES

b. CH$_3$CH$_2$CH$_3$
   2 types of H's
   6:2 or 3:1 - no

c. CH$_3$CH$_2$OCH$_2$CH$_3$
   2 types of H's
   6:4 or 3:2 - YES

d. CH$_3$OCH$_2$CH$_2$OCH$_3$
   2 types of H's
   6:4 or 3:2 - YES

15.13 To determine how many protons give rise to each signal:
- Divide the total number of integration units by the total number of protons to find the number of units per H.
- Divide each integration value by this value and round to the nearest whole number.

   C$_8$H$_{14}$O$_2$

   total number of integration units = 14 + 12 + 44 = 70 units
   total number of protons = 14H's
   70 units/14H's = **5 units per H**

   Signal [A] = 14/5 = **3H**
   Signal [B] = 12/5 = **2H**
   Signal [C] = 44/5 = **9H**

15.14 To determine the **splitting pattern** for a molecule:
- Determine the number of different kinds of protons.
- Nonequivalent protons on the same C or adjacent C's split each other
- Apply the $n + 1$ rule.

a.

$$CH_3CH_2-\overset{\overset{\displaystyle O}{\|}}{C}\cdots Cl$$
$\uparrow\quad\uparrow$
$H_a\ H_b$

H_a: 3 peaks - triplet
H_b: 4 peaks - quartet

c.

$$CH_3-\overset{\overset{\displaystyle O}{\|}}{C}-CH_2CH_2Br$$
$\uparrow\qquad\uparrow\ \uparrow$
$H_a\qquad H_b\ H_c$

H_a: 1 peak - singlet
H_b: 3 peaks - triplet
H_c: 3 peaks - triplet

e.

$$\overset{\displaystyle CH_3CH_2}{\underset{\displaystyle CH_3}{}}C=C\overset{\displaystyle H \leftarrow H_a}{\underset{\displaystyle H \leftarrow H_b}{}}$$

H_a: 2 peaks - doublet
H_b: 2 peaks - doublet

b.

$$\overset{\displaystyle H\leftarrow H_b}{|}$$
$$CH_3-\overset{|}{\underset{|}{C}}-Br$$
$$\underset{Br}{}$$
$\uparrow$
H_a

H_a: 2 peaks - doublet
H_b: 4 peaks - quartet

d.

$$\overset{\displaystyle H_a \rightarrow H\quad Cl}{}C=C\overset{\displaystyle}{}$$
$$\underset{\displaystyle Br\qquad H \leftarrow H_b}{}$$

H_a: 2 peaks - doublet
H_b: 2 peaks - doublet

15.15 Use the directions from Answer 15.14.

a. $Cl_2CHCHCl_2$

Both H's are equivalent.
1 NMR signal
no splitting

c.

$$\overset{\displaystyle H_a\qquad H_a}{\downarrow\qquad\downarrow}$$
$$\underset{\displaystyle O}{}$$
$$H_b\qquad\qquad H_b$$

H_a: quartet
H_b: triplet
2 NMR signals

e. $CH_3-\overset{\overset{\displaystyle O}{\|}}{C}-H\!\!-\!\!H_b$
$\quad H_a$

H_a: doublet
H_b: quartet
2 NMR signals

b. $Cl_2CHCHBr_2$

The H's are not equivalent.
2 NMR signals
both doublets

d. $CH_3-\overset{\overset{\displaystyle O}{\|}}{C}-OCH_2CH_2OCH_3$
$\quad\uparrow\qquad\qquad\uparrow\quad\uparrow\quad\uparrow$
$\quad H_a\qquad\qquad H_b\ H_c\ H_d$

H_a and H_d are both singlets.
H_b: triplet
H_c: triplet
4 NMR signals

15.16

CH_3CH_2Cl

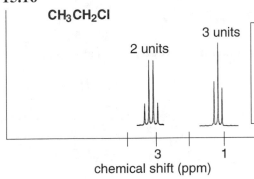

2 units

3 units

3   1

chemical shift (ppm)

There are two kinds of protons, and they can split each other.
The CH_3 protons will be split by the CH_2 protons into 2+1 = 3 peaks.
It will be upfield from the CH_2 protons since it is further from the Cl.
The CH_2 protons will be split by the CH_3 protons into 3+1 = 4 peaks.
It will be downfield from the CH_3 protons since they are closer to the Cl.
The ratio of intergration units will be 3:2.

15.17

a. $(CH_3)_2CHCO_2CH_3$

split by 6 equivalent H's
(6+1) = **7 peaks**

b. $CH_3CH_2CH_2CH_2CH_3$
   $H_a\ H_b\ H_c$

H_a: split by 2 H's
3 peaks
H_b: spilt by 2 sets of H's
(3+1)(2+1) = **12 peaks**
H_c: split by 4 equivalent H's
5 peaks

c.

Cl   CH₂Br
  \   /
   C=C
  /   \
$H_a\!\!\rightarrow\!\!H$ $H\!\leftarrow\! H_b$

H_a: split by 1 H
2 peaks
H_b: spilt by 2 sets of H's
(1+1)(2+1) = **6 peaks**

d.

$H_a\!\!\rightarrow\!\!H$ $H\!\leftarrow\! H_b$
       \   /
        C=C   (all H's)
       /   \
     Br $H\!\leftarrow\! H_c$

H_a: split by 2 different H's
(1+1)(1+1) = **4 peaks**
H_b: spilt by 2 different H's
(1+1)(1+1) = **4 peaks**
H_c: split by 2 different H's
(1+1)(1+1) = **4 peaks**

15.18

$\boxed{A}$

$\begin{array}{ccc} Cl & Cl & H \\ H-C-C-C-H \\ H_a\!\!\rightarrow\!\!H & Cl & H \end{array} \Big\}\ H_b$

H_a: singlet at 4.04
H_b: singlet at 2.23

$\boxed{B}$

$\begin{array}{cc} Cl & Cl \\ H-C-C-CH_3 \leftarrow H_c \\ Cl & H \leftarrow H_b \\ H_a \end{array}$

H_a: doublet at 5.85
H_b: multiplet at 4.34
H_c: doublet at 1.69

15.19

a. $CH_3OCH_2CH_3$
   $H_a\ \ H_b\ H_c$

H_a: singlet at ~3 ppm
H_b: quartet at ~3.5 ppm
H_c: triplet at ~1 ppm

b.

$\begin{array}{c} O \\ \| \\ CH_3CH_2-C-OCH(CH_3)_2 \\ H_a\ H_bH_c\ H_d \end{array}$

H_a: triplet at ~1 ppm
H_b: quartet at ~2 ppm
H_c: septet at ~3.5 ppm
H_d: doublet at ~1 ppm

c. $CH_3OCH_2CH_2CH_2OCH_3$
   $H_a\ \ H_b\ \ H_c\ \ H_b\ \ H_a$

H_a: singlet at ~3 ppm
H_b: triplet at ~3.5 ppm
H_c: quintet at ~1.5 ppm

d.

$\begin{array}{c} H_b \\ CH_3CH_2 \diagdown\ \diagup CH_2CH_3 \\ H_a\quad C=C\quad H_a \\ H\diagup\ \ \diagdown H \\ H_c \end{array}$

H_a: triplet at ~1 ppm
H_b: multiplet (8 peaks) at ~2.5 ppm
H_c: triplet at ~5 ppm

15.20

$\begin{array}{c} H_b\diagdown\quad \diagup H_a \\ C=C \\ H_c\diagup\quad \diagdown CN \end{array}$

J_{ab} = 11.8 Hz
J_{bc} = 0.9 Hz
J_{ac} = 18 Hz

H_a: doublet of doublets at 5.7 ppm. Two large J values are seen for the H's cis (J_{ab} = 11.8 Hz) and trans (J_{ac} = 18 Hz) to H_a.

H_b: doublet of doublets at ~6.2 ppm. One large J value is seen for the cis H (J_{ab} = 11.8 Hz). The geminal coupling (J_{bc} = 0.9 Hz) is hard to see.

H_c: doublet of doublets at ~6.6 ppm. One large J value is seen for the trans H (J_{ac} = 18 Hz). The geminal coupling (J_{bc} = 0.9 Hz) is hard to see.

Splitting diagram for H_a

1 **trans** H_c proton splits H_a into
1 + 1 = 2 peaks
a doublet

1 **cis** H_b proton splits H_a into
1 + 1 = 2 peaks
Now it's a doublet of doublets.

H_a

J_{ac} = the coupling constant between H_a and H_c

J_{ab} = the coupling constant between H_a and H_b

15.21 Remember that **OH (or NH) protons** do not split other signals, and are not split by adjacent protons.

singlet

a. $(CH_3)_3CCH_2OH$

singlet   singlet

3 NMR signals

triplet   triplet

b. $CH_3CH_2CH_2OH$

12 peaks   singlet

4 NMR signals

15.22

H_d {
5 H's on benzene ring

$H \leftarrow H_c$
$C-CH_3 \leftarrow H_a$
$OH \leftarrow H_b$

A

H_a: doublet at ~1.4 due to the CH_3 group, split into two peaks by one adjacent nonequivalent H (H_c).

H_b: singlet at ~2.7 due to the OH group. OH protons are not split by nor do they split adjacent protons.

H_c: quartet at ~4.7 due to the CH group, split into four peaks by the adjacent CH_3 group.

H_d: Five protons on the benzene ring.

15.23 Use these steps to propose a structure consistent with the molecular formula, IR, and NMR data.

- Calculate the **degrees of unsaturation**.
- Use the IR data to determine what types of **functional groups** are present.
- Determine the number of different **types of protons**.
- Calculate the **number of H's** giving rise to each signal.
- Analyze the **splitting pattern** and put together a molecule.
- Use the **chemical shift** information to check the structure.

- Molecular formula $C_7H_{14}O_2$

  $2n + 2 = 2(7) + 2 = 16$
  $16 - 14 = 2/2 = $ **1 degree of unsaturation**
  1 π bond or one ring

- IR peak at 1740 cm^{-1}

  C=O absorption is around 1700 cm^{-1} (causes the degree of unsaturation)
  no signal at 3200–3600 cm^{-1} means there is no O–H bond

- NMR data:

absorption	ppm	integration	
singlet	1.2	26	$\rightarrow$ 26/3 = **9 H's**
triplet	1.3	10	$\rightarrow$ 10/3 = **3 H's** (probably a CH_3 group)
quartet	4.1	6	$\rightarrow$ 6/3 = **2 H's** (probably a CH_2 group)

- 3 kinds of H's
- number of H's per signal
  total integration units: 26 + 10 + 6 = 42 units
  42 units / 14 H's = 3 units per H
- look at the splitting pattern
  the singlet (9H) is likely from a *tert*-butyl group:

  CH_3
  |
  $-C-CH_3$
  |
  CH_3

  the CH_3 and CH_2 groups split each other: CH_3-CH_2-

• join the pieces together

$$CH_3CH_2O-\overset{\overset{\displaystyle O}{\parallel}}{C}\overset{\overset{\displaystyle CH_3}{|}}{\underset{\underset{\displaystyle CH_3}{|}}{C}}-CH_3 \quad or \quad CH_3CH_2-\overset{\overset{\displaystyle O}{\parallel}}{C}-O-\overset{\overset{\displaystyle CH_3}{|}}{\underset{\underset{\displaystyle CH_3}{|}}{C}}-CH_3$$

Pick this structure due to the chemical shift data.
The CH₂ group is shifted downfield (4 ppm), so it
is close to the electron withdrawing O.

15.24

• Molecular formula: C_3H_8O

➤ Calculate degrees of unsaturation
$$2n + 2 = 2(3) + 2 = 8$$
$$8 - 8 = \textbf{0 degrees of unsaturation}$$

• IR peak at 3200–3600 cm^{-1}

➤ peak at 3200–3600 cm^{-1} is due to an **O–H bond**

• NMR data:
  • Doublet at ~1.2 (6H)
  • Singlet at ~2.2 (1H)
  • Septet at ~4 (1H)

3 types of H's
septet from 1 H ◄———— split by 6 H's
→ **singlet** from 1 H
doublet from 6 H's ◄———— split by 1 H
from the O–H proton

➤ Put information together:

$$HO-\overset{\overset{\displaystyle CH_3}{|}}{\underset{\underset{\displaystyle H}{|}}{C}}-CH_3$$

15.25

a.

Absorption [A]:	Singlet at ~3.8 ppm	CH₃O–
Absorption [B]:	Triplet at ~3.6 ppm	CH₂N
Absorption [C]:	Triplet at ~2.9 ppm	CH₂ adjacent to five-membered ring
Absorption [D]:	Singlet at ~1.9 ppm	CH₃C=O

b.

split by 2 adjacent
nonequivalent H's
into a triplet

15.26 Each different kind of carbon atom will give a different ^{13}C NMR signal.

a. $CH_3CH_2CH_2CH_3$
     C_a C_b C_b C_a
2 kinds of C's
2 ^{13}C NMR signals

b.
$$CH_3CH_2\overset{\overset{\displaystyle O}{\parallel}}{C}OCH_3$$
each C is different
4 kinds of C's
4 ^{13}C NMR signals

c. $CH_3CH_2CH_2-O-CH_2CH_2CH_3$
    C_a C_b C_c C_c C_b C_a
same groups on both sides of O
3 kinds of C's
3 ^{13}C NMR signals

d.
$$\overset{CH_3CH_2}{}\underset{H}{\overset{}{}}C=C\underset{H}{\overset{H}{}}$$
each C is different
4 kinds of C's
4 ^{13}C NMR signals

15.27 To give only one peak in both its ^{1}H and ^{13}C NMR spectra, a compound must have only one type of C and H.

$$CH_3O-\overset{\overset{O}{\|}}{C}-OCH_3 \qquad \triangle \qquad CH_3-\overset{\overset{CH_3}{|}}{\underset{\underset{CH_3}{|}}{C}}-O-\overset{\overset{CH_3}{|}}{\underset{\underset{CH_3}{|}}{C}}-CH_3 \qquad Br_2CHCHBr_2$$

1 ^{1}H NMR signal	**1 ^{1}H NMR signal**	1 ^{1}H NMR signal	**1 ^{1}H NMR signal**
2 ^{13}C signals	**1 ^{13}C signal**	2 ^{13}C signals	**1 ^{13}C signal**

15.28 Electronegative elements shift absorptions downfield. Carbons in alkenes and benzene rings, and carbonyl carbons are also shifted downfield.

a. $CH_3CH_2OCH_2CH_3$

The CH_2 group is closer to the electronegative O and will be further downfield.

b. $BrCH_2CHBr_2$

The C of the $CHBr_2$ group has two bonds to electronegative Br atoms and will be further downfield.

c. $H-\overset{\overset{O}{\|}}{C}-OCH_3$

The carbonyl carbon is highly deshielded and will be further downfield.

d. $CH_3CH=CH_2$

The CH_2 group is part of a double bond and will be further downfield.

15.29

a. In order of lowest to highest chemical shift:

$$\overset{C_a \quad C_b \quad C_c \quad C_d}{CH_3CHCH_2CH_3}$$
$$\underset{OH}{|}$$

$$C_d < C_a < C_c < C_b$$

b. In order of lowest to highest chemical shift:

$$(CH_3CH_2)_2C=O$$
$$C_a \ C_b \quad \ C_c$$

$$C_a < C_b < C_c$$

15.30

- molecular formula $C_4H_8O_2$
    $2n + 2 = 2(4) + 2 = 10$
    $10 - 8 = 2/2 = $ **1 degree of unsaturation**

- no IR peaks at 3200–3600 or 1700 cm^{-1}
    no O–H or C=O

- ^{1}H NMR spectrum at 3.69 ppm
    only one kind of proton

- ^{13}C NMR spectrum at 67 ppm
    only one kind of carbon

This structure satisifies all the data. One ring is one degree of unsaturation. All carbons and protons are identical.

15.31 Use the directions from Answer 15.2.

a. $(CH_3)_3CH$

2 kinds of H's

b. $(CH_3)_3CC(CH_3)_3$
1 kind of H

c. $CH_3CH_2OCH_2CH_2CH_2CH_3$
7 kinds of H's

d.
$H_a \rightarrow CH_3 \quad H \leftarrow H_c$
$C=C$
$H_b \rightarrow H \quad H \leftarrow H_d$
4 kinds of H's

e.
5 kinds of H's

f.
$CH_3CH_2 \quad H$
$C=C$
$H \quad CH_2CH_3$
3 kinds of H's

g. $CH_3CH_2CH_2OCH_2CH_2CH_3$
3 kinds of H's

h.
$H_a \rightarrow CH_3 \quad CH_2CH_3 \leftarrow H_d$
$C=C$
$H_b \rightarrow CH_3 \quad Br \quad H_c$
4 kinds of H's

i.
$H_b \rightarrow H \quad H \leftarrow H_d$
$CH_3 - C - C - CH_3 \leftarrow H_f$
$H_a \uparrow \quad HO \quad H \leftarrow H_e$
H_c
6 kinds of H's

j.
$H_a \rightarrow H \quad CH_3 \leftarrow H_c$
$H_b \rightarrow H \quad O \quad H \leftarrow H_d$
4 kinds of H's

15.32

$Cl \quad \quad \quad Cl$
2 kinds of protons

$H_a \rightarrow H \quad H \leftarrow H_c$
$Cl - C - C - CH_3 \leftarrow H_d$
$H_b \rightarrow H \quad Cl$
4 kinds of protons

Cl
Cl
3 kinds of protons

$Cl \quad Cl$
1 kind of proton

15.33

a.
$H_b \{ \quad CH_3 \} H_a$
$H \quad CH_3$
H_c
H_b
3 kinds of protons

b.
$H_e \quad H_d \quad H_b$
$H_d \rightarrow H \quad CH_2CH_3 \leftarrow H_a$
$H_f \rightarrow H \quad H_g \quad H_f \quad H \leftarrow H_c$
7 kinds of protons

c.
$H_c \rightarrow H \quad H \leftarrow H_c$
$H_a \rightarrow CH_3 \quad CH_3 \leftarrow H_a$
$H \quad H$
$H_b \rightarrow H \quad H_d \quad H_d \quad H \leftarrow H_b$
4 kinds of protons

d.
$H_d \rightarrow H \quad H \leftarrow H_c$
$H_a \rightarrow CH_3 \quad H \leftarrow H_b$
$H \quad H$
$H_b \rightarrow H \quad H_c \quad H_d \quad CH_3 \leftarrow H_a$
4 kinds of protons

15.34

caffeine
4 NMR signals

vanillin
6 NMR signals

15.35

$$\delta = [\text{observed chemical shift (Hz)}] / \nu \text{ of the NMR (MHz)}]$$

  a. $2.5 = $ observed chemical shift/300 MHz
      observed chemical shift = 750 Hz
  b. ppm = 1200 Hz/300 Hz
      4 ppm
  c. $2.0 = $ Hz/300 MHz
      600 Hz

15.36 Use the directions from Answer 15.9.

  a. $CH_3CH_2CH_2CH_2CH_3$ or $CH_3CH_2CH_2OCH_3$

      Adjacent O deshields the H's.
      further downfield

  c. $CH_3OCH_2CH_3$
      or
      Increasing alkyl substitution
      further downfield

  b. $CH_3CH_2CH_2I$ or $CH_3CH_2CH_2F$

      More electronegative F
      deshields the H's.
      further downfield

  d. $CH_3CH_2CHBr_2$ or $CH_3CH_2CH_2Br$

      Two electronegative
      Br's deshield the H.
      further downfield

15.37 Use the directions from Answer 15.13.

[total number of integration units] / [total number of protons]   signal of 13 units is from **1 H**
$[13 + 33 + 73] / 10 = \sim12$ units per proton   signal of 33 units is from **3 H's**
                                                                    signal of 73 units is from **6 H's**

15.38 The following compounds give one singlet in a ^1H NMR spectrum:

CH_3CH_3 $CH_3-C\equiv C-CH_3$

15.39

$CH_3CH_2CH_2CH_2CH_2CH_3$
  H_a H_b H_c H_c H_b H_a

3 signals:
H_a: split by 2 H_b protons - triplet
H_b: split by 3 H_a + 2 H_c protons - 12 peaks
H_c: split by 2 H_b protons - triplet

$CH_3 \leftarrow H_a$
$CH_3CHCH_2CH_2CH_3$
  H_a H_b H_c H_d H_e

5 signals:
H_a: split by 1 H_b proton - doublet
H_b: split by 6 H_a + 2 H_c protons - 21 peaks
H_c: split by 1 H_b + 2 H_d protons - 6 peaks
H_d: split by 2 H_c + 3 H_e protons - 12 peaks
H_e: split by 2 H_d protons - triplet

H_c
$CH_3CH_2CHCH_2CH_3$
H_a H_b CH_3 H_b H_a
        H_d

4 signals:
H_a: split by 2 H_b protons - triplet
H_b: split by 3 H_a + 1 H_c protons - 8 peaks
H_c: split by 4 H_b + 3 H_d protons - 20 peaks
H_d: split by 1 H_c proton - doublet

H_a H_a
CH_3 CH_3
$CH_3CH-CHCH_3$
H_a H_b H_b H_a

2 signals:
H_a: split by 1 H_b proton - doublet
H_b: split by 6 H_a protons - septet

$CH_3 \leftarrow H_c$
$CH_3CH_2-C-CH_3 \leftarrow H_c$
  H_a H_b $CH_3 \leftarrow H_c$

3 signals:
H_a: split by 2 H_b protons - triplet
H_b: split by 3 H_a protons - quartet
H_c: no splitting - singlet

15.40

a. $CH_3CH(OCH_3)_2$

CH_3 protons split by 1 H = **doublet**
CH proton split by 3 H's = **quartet**

b. $CH_3OCH_2CH_2\overset{\overset{\displaystyle O}{\|}}{C}OCH_3$

both CH_2 groups split
each other = **triplets**

c. [benzene ring]$-CH_2CH_3$

CH_3 protons split by 2 H's = **triplet**
CH_2 protons split by 3 H's = **quartet**

d. $CH_3OCH_2CHCl_2$

CH_2 protons split by 1 H = **doublet**
CH proton split by 2 H's = **triplet**

e. $(CH_3)_2CH\overset{\overset{\displaystyle O}{\|}}{C}OCH_2CH_3$
  H_a H_b H_c H_d

H_a protons split by 1 H = **doublet**
H_b proton split by 6 H's = **septet**
H_c protons split by 3 H's = **quartet**
H_d protons split by 2 H's = **triplet**

f. $HOCH_2CH_2CH_2OH$
        H_a H_b

H_a protons split by 2 CH_2 groups =
quintet
H_b protons split by 2 H's = **triplet**

g. $CH_3CH_2CH_2CH_2OH$
    H_a H_b H_c H_d

H_a protons split by 2 H's = **triplet**
H_b protons split by CH_3 + CH_2 protons = **12 peaks**
H_c protons spilt by 2 different CH_2 groups = **9 peaks**
H_d protons split by 2 H's = **triplet**

h. $CH_3CH_2CH_2\overset{\overset{\displaystyle O}{\|}}{C}OH$
    H_a H_b H_c

H_a protons split by 2 H's = **triplet**
H_b protons split by CH_3 + CH_2 protons = **12 peaks**
H_c protons spilt by 2 H's = **triplet**

i. $CH_3CH_2\overset{\overset{\displaystyle O}{\|}}{C}H$
       H_a H_b

H_a: split by CH_3 group + H_b =
8 peaks
H_b: split by 2 H's= **triplet**

j. CH_3, CH_3CH_2 C=C $H—H_a$, $H—H_b$

H_a: split by 1 H = **doublet**
H_b: split by 1 H = **doublet**

k. CH_3, Br C=C $H—H_a$, $H—H_b$

H_a: split by 1 H = **doublet**
H_b: split by 1 H = **doublet**

l. CH_3, $H_c—H$ C=C $H—H_a$, $H—H_b$

H_a: split by H_b + H_c -
doublet of doublets (4 peaks)
H_b: split by H_a + H_c -
doublet of doublets (4 peaks)
H_c: split by CH_3, H_a + H_b - **16 peaks**

15.41 Use the directions from Answer 15.23.

a. $C_4H_8Br_2$:0 degrees of unsaturation
   IR peak at 3000–2850 cm^{-1}: **Csp^3–H bonds**
   NMR: singlet at 1.87 ppm (6H) (2 CH$_3$ groups)
   singlet at 3.86 ppm (2H) (CH$_2$ group)

$$CH_3-\underset{\underset{Br}{|}}{\overset{\overset{CH_3}{|}}{C}}-CH_2Br$$

b. **$C_9H_{18}O$: 1 degree of unsaturation**
   IR peak at 1710 cm^{-1}: **C=O**
   NMR: singlet at 1.2 ppm (all identical H's)

$$\underset{CH_3\ CH_3}{\overset{CH_3}{CH_3}}\underset{C}{\overset{O}{\underset{\|}{C}}}\underset{CH_3}{\overset{CH_3}{C}}$$

c. $C_2H_4Cl_2$: 0 degrees of unsaturation
   IR peak at 3000–2850 cm^{-1}: **Csp^3–H bonds**
   NMR: doublet at 2.1 ppm (split by 1 H)
   quartet at 5.9 ppm (split by 3 H's)

$$CH_3-CHCl_2$$

d. $C_3H_6Br_2$: 0 degrees of unsaturation
   IR peak at 3000–2850 cm^{-1}: **Csp^3–H bonds**
   NMR: quintet at 2.4 ppm (split by 2 CH$_2$ groups)
   triplet at 3.5 ppm (split by 2 H's)

$$Br\diagup\diagdown\diagup Br$$

e. **$C_5H_{10}O_2$: 1 degree of unsaturation**
   IR peak at 1740 cm^{-1}: **C=O**
   NMR: triplet at 1.15 ppm (3H) (CH$_3$ split by 2 H's)
   triplet at 1.25 ppm (3H) (CH$_3$ split by 2 H's)
   quartet at 2.30 ppm (2H) (CH$_2$ split by 3 H's)
   quartet at 4.72 ppm (2H) (CH$_2$ split by 3 H's)

$$CH_3CH_2-\overset{\overset{O}{\|}}{C}-O-CH_2CH_3$$

f. $C_6H_{14}O$:0 degrees of unsaturation
   IR peak at 3600–3200 cm^{-1}: **O–H**
   NMR: triplet at 0.8 ppm (6H) (2 CH$_3$ groups
   split by CH$_2$ groups)
   singlet at 1.0 ppm (3H) (CH$_3$)
   quartet at 1.5 ppm (4H) (2 CH$_2$ groups split
   by CH$_3$ groups)
   singlet at 1.6 ppm (1H) (O–H proton)

$$CH_3CH_2-\underset{\underset{OH}{|}}{\overset{\overset{CH_3}{|}}{C}}-CH_2CH_3$$

g. $C_6H_{14}O$: 0 degrees of unsaturation
   IR peak at 3000–2850 cm^{-1}: **Csp^3–H bonds**
   NMR: doublet at 1.10 ppm (integration = 30 un
   (from 12 H's)
   septet at 3.60 ppm (integration = 5 units)
   (from 2 H's)

$$CH_3-\underset{\underset{CH_3}{|}}{\overset{\overset{H}{|}}{C}}-O-\underset{\underset{CH_3}{|}}{\overset{\overset{H}{|}}{C}}-CH_3$$

h. **C_3H_6O: 1 degree of unsaturation**
   IR peak at 1730 cm^{-1}: **C=O**
   NMR: triplet at 1.11 ppm
   multiplet at 2.46 ppm
   triplet at 9.79 ppm

$$CH_3CH_2-\overset{\overset{O}{\|}}{C}\diagdown H$$

15.42
Two isomers of $C_9H_{10}O$: 5 degrees of unsaturation (benzene ring likely)

Compound A:
  IR absorption at 1742 cm^{-1}: **C=O**
  NMR data:
    Absorptions:
    singlet at 2.15 (3H) (CH_3 group)
    singlet at 3.70 (2H) (CH_2 group)
    broad singlet at 7.20 (5H)
      (likely a monosubstituted benzene ring)

Compound B:
  IR absorption at 1688 cm^{-1}: **C=O**
  NMR data:
    Absorptions:
    triplet at 1.22 (3H) (CH_3 group split by 2 H's)
    quartet at 2.98 (2H) (CH_2 group split by 3 H's
    multiplet at 7.28–7.95 (5H)
      (likely a monosubstituted benzene ring)

15.43

In addition to protons for the benzene rings, the following absorptions are observed:

a.
and

one singlet for the CH_3 protons   **two singlets** for the O–H proton and CH_2 protons

b.
and

The CH_2 group is adjacent to the benzene ring, but not an O atom, and so absorbs around 2.5 ppm.

The CH_2 group is adjacent to an O atom, and so it is more deshielded and absorbs around 3.5 ppm.

15.44

Compound [C]:
$C_4H_8O_2$:
   1 degree of unsaturation
IR absorption at 1743 cm^{-1}: **C=O**
NMR data:
   total integration units/# H's
   (23 + 29 + 30)/8 = ~10 units per H

   H_a: quartet at 4.1 (22 units - **2H**)
   H_b: singlet at 2.0 (29 units - **3H**)
   H_c: triplet at 1.4 (30 units - **3H**)

Compound [D]:
$C_4H_8O_2$:
   1 degree of unsaturation
IR absorption at 1730 cm^{-1}: **C=O**
NMR data:
   total integration units/# H's
   (18 + 30 + 31)/8 = ~10 units per H

   H_a: singlet at 4.1 (18 units - **2H**)
   H_b: singlet at 3.4 (30 units - **3H**)
   H_c: singlet at 2.1 (31 units - **3H**)

15.45

a. $C_9H_{10}O_2$:

5 degrees of unsaturation

IR absorption at 1718 cm^{-1}: **C=O**

NMR data:

multiplet at 7.8 ppm, **5H** on a benzene ring

quartet at 4.4 ppm, **2H**, split by 3 H's

triplet at 1.3 ppm, **3H**, split by 2 H's

downfield due to the O atom

b. C_9H_{12}:

4 degrees of unsaturation

IR absorption at 2850–3150 cm^{-1}:

C–H bonds

NMR data:

singlet at 7.2 ppm, **5H**, benzene

septet at 2.8 ppm, **1H**, split by 6 H's

doublet at 1.3 ppm, **6H**, split by 1 H

15.46

a. Compound **A** has a molecular ion at 72: molecular formula C_4H_8O

1 degree of unsaturation

IR spectrum at 1710 cm^{-1}: **C=O**

^{1}H NMR data (ppm):

1.0 (triplet, 3H), split by 2 H's

2.1 (singlet, 3H)

2.4 (quartet, 2H), split by 3 H's

b. Compound **B** has a molecular ion at 88: molecular formula $C_5H_{12}O$

0 degrees if unsaturation

IR spectrum at 3600–3200 cm^{-1}: **O–H bond**

^{1}H NMR data (ppm):

0.9 (triplet, 3H), split by 2 H's

1.2 (singlet, 6H), due to 2 CH$_3$ groups

1.5 (quartet, 2H), split by 3 H's

1.6 (singlet, 1H), due to the OH proton

15.47

$C_7H_{16}O_2$: 0 degrees of unsaturation

IR: 3000 cm^{-1}: **C–H bonds**

NMR data (ppm):

H_a: quartet at 3.5 (**4H**), split by 3 H's

H_b: singlet at 1.4 (**6H**)

H_c: triplet at 1.2 (**6H**), split by 2 H's

15.48

Compound [W]: Molecular ion at 86.

Molecular formula: $C_5H_{10}O$:

1 degree of unsaturation

IR absorption at ~1700 cm^{-1}: **C=O**

NMR data:

H_a: doublet at 1.1 ppm, 2 CH$_3$ groups split by 1 H

H_b: singlet at 2.1 ppm, CH$_3$ group

H_c: septet at 2.6 ppm, 1 H split by 6 H's

15.49

a. Compound **A**, the odor of banana: $C_7H_{14}O_2$
   1 degree of unsaturation
   ^{1}H NMR (ppm):

   H_a: 0.93 (doublet, 6H)
   H_b: 1.52 (multiplet, 2H)
   H_c: 1.69 (multiplet, 1H)
   H_d: 2.04 (singlet, 3H)
   H_e: 4.10 (triplet, 2H)

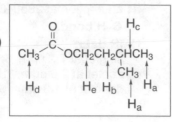

b. Compound **B**, the odor of rum: $C_7H_{14}O_2$
   1 degree of unsaturation
   ^{1}H NMR (ppm):

   H_a: 0.94 (doublet, 6H)
   H_b: 1.15 (triplet, 3H)
   H_c: 1.91 (multiplet, 1H)
   H_d: 2.33 (quartet, 2H)
   H_e: 3.86 (doublet, 2H)

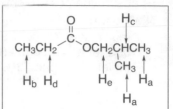

15.50

C_6H_{12}:
1 degree of unsaturation

$\xrightarrow{K^+ \; {}^-OC(CH_3)_3}$ **A** ^{1}H NMR of **A** (ppm):

   H_a: 1.01 (singlet, 9H)
   H_b: 4.82 (doublet of doublets, 1H, $J = 10, 1.7$ Hz)
   H_c: 4.93 (doublet of doublets, 1H, $J = 18, 1.7$ Hz)
   H_d: 5.83 (doublet of doublets, 1H, $J = 18, 10$ Hz)

$\xrightarrow{H^+}$ **B** ^{1}H NMR of **B**: 1.60 (singlet) ppm.

All H's are identical, so there is only one singlet in the NMR.

15.51   Both **A** and **B** have the same molecular ion—since they are isomers— and show a C=O peak in their IR spectra. ^{1}H NMR spectroscopy is the best way to distinguish the two compounds.

CH_3O—C(=O)—$C(CH_3)_3$ or CH_3—C(=O)—$OC(CH_3)_3$
   A   **B**

Both **A** and **B** have two singlets in a 3:1 ratio in their ^{1}H NMR spectra. But **A** has a peak at ~3 ppm due to the deshielded CH_3 group bonded to the O atom. **B** has no proton that is so deshielded. Both of its singlets are in the 1–2.5 ppm region.

15.52

Four constitutional isomers of C_4H_9Br:

4 different C's   4 different C's   2 different C's   3 different C's

15.53   Only two compounds in Problem 15.38 give one signal in their ^{13}C NMR spectrum:

CH_3CH_3

15.54

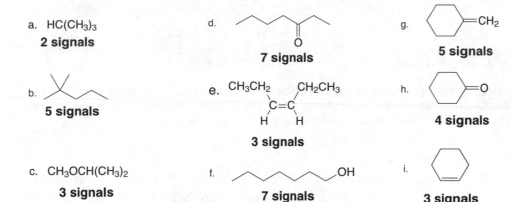

a. HC(CH₃)₃ — $HC(CH_3)_3$
   2 signals

d. **7 signals**

g. **5 signals**

b. **5 signals**

e. $\underset{H}{\overset{CH_3CH_2}{C}}=\underset{H}{\overset{CH_2CH_3}{C}}$
   3 signals

h. **4 signals**

c. CH₃OCH(CH₃)₂ — $CH_3OCH(CH_3)_2$
   3 signals

f. **7 signals**

i. **3 signals**

15.55

a. $CH_3CH_2\overset{\overset{O}{\|}}{C}OH$

   C_a C_b C_c

   $C_a < C_b < C_c$

b. $CH_3CH_2\overset{OH}{CH}CH_2CH_3$

   C_a C_b C_c

   $C_a < C_b < C_c$

c. $Ph\overset{\overset{O}{\|}}{C}CH_2CH_3$

   C_a C_b C_c

   $C_c < C_b < C_a$

d. $CH_2=CHCH_2CH_2CH_2Br$

   C_a C_b C_c

   $C_b < C_c < C_a$

15.56

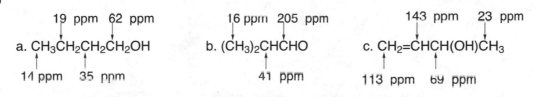

a. $\underset{\underset{14\ ppm}{\uparrow}\ \ \ \ \underset{35\ ppm}{\uparrow}}{\overset{\overset{19\ ppm}{\downarrow}\ \ \ \ \overset{62\ ppm}{\downarrow}}{CH_3CH_2CH_2CH_2OH}}$

b. $\underset{\underset{41\ ppm}{\uparrow}}{\overset{\overset{16\ ppm\ \ \ 205\ ppm}{\downarrow\ \ \ \ \ \ \downarrow}}{(CH_3)_2CHCHO}}$

c. $\underset{\underset{113\ ppm}{\uparrow}\ \ \ \underset{69\ ppm}{\uparrow}}{\overset{\overset{143\ ppm\ \ \ \ 23\ ppm}{\downarrow\ \ \ \ \ \ \downarrow}}{CH_2=CHCH(OH)CH_3}}$

15.57

a.

   $C_6H_{12}O_2$:
   1 degree of unsaturation
   IR peak at 1740 cm⁻¹: **C=O**
   ¹H NMR two signals: two types of H's
   ¹³C NMR: 4 signals: 4 kinds of C's,
   including one at ~170 ppm due a C=O

   $CH_3\underset{\underset{CH_3\ CH_3}{|}}{\overset{|}{C}}\overset{\overset{O}{\|}}{C}O{-}CH_3$

b.

   C_6H_{10}:
   2 degrees of unsaturation
   IR peak at 3000 cm⁻¹: **Csp³–H bonds**
      peak at 3300 cm⁻¹: **Csp–H bond**
      peak at ~2150 cm⁻¹: **C≡C bond**
   ¹³C NMR: 4 signals: 4 kinds of C's

   $HC{\equiv}C{-}\underset{\underset{CH_3}{|}}{\overset{\overset{CH_3}{|}}{C}}{-}CH_3$

15.58

N,N-dimethylformamide

cis to the O atom

CH$_3$ ← cis to the H atom

A second resonance structure for *N,N*-dimethylformamide places the two CH$_3$ groups in different environments. One CH$_3$ group is cis to the O atom, and one is cis to the H atom. This gives rise to two different absorptions for the CH$_3$ groups.

15.59

18-Annulene has 18 π electrons that create an **induced magnetic field** similar to the 6 π electrons of benzene. 18-Annulene has 12 protons that are oriented on the outside of the ring (labeled H$_o$), and 6 protons that are oriented inside the ring (labeled H$_i$). The induced magnetic field reinforces the external field in the vicinity of the protons on the outside of the ring. These protons are deshielded and so they absorb downfield (8.9 ppm). In contrast, the induced magnetic filed is opposite in direction to the applied magnetic field in the vicinity of the protons on the inside of the ring. This shields the protons and the absorption is therefore very far upfield, even higher than TMS (−1.8 ppm).

15.60

3-methyl-2-butanol

replace a CH$_3$ group with an X

Replace C$_a$.   or   Replace C$_b$.

The CH$_3$ groups are not equivalent to each other, since replacement of each by X forms two diastereomers.

Thus, every C in this compound is different and there are five ^{13}C signals.

Chapter 16: Conjugation, Resonance, and Dienes

◆ Conjugation and delocalization of electron density

- Having p orbitals on three or more adjacent atoms allows electron density to delocalize, thus adding stability (16.1).
- An allyl carbocation ($CH_2=CHCH_2^+$) is more stable than a 1° carbocation because of p orbital overlap (16.2).
- In any system $X=Y–Z:$, Z is sp^2 hybridized to allow the lone pair to occupy a p orbital, making the system conjugated (16.5).

◆ Four common examples of resonance (16.3)

[1] The three atom "allyl" system:

$$X=Y-\overset{*}{Z} \longleftrightarrow X-Y=\overset{*}{Z} \qquad \boxed{* = +, -, \bullet, \text{ or } \bullet\bullet}$$

[2] Conjugated double bonds:

[3] Cations having a positive charge adjacent to a lone pair:

$$\overset{..}{X}-\overset{+}{Y} \longleftrightarrow \overset{+}{X}=Y$$

[4] Double bonds having one atom more electronegative than the other:

$$X=Y \longleftrightarrow \overset{+}{X}-\overset{-}{Y}: \qquad \boxed{\text{Electronegativity of Y > X}}$$

◆ Rules on evaluating the relative "stability" of resonance structures (16.4)

[1] Structures with more bonds and fewer charges are more stable.

$$\boxed{\begin{array}{c}\textbf{more stable}\\\textbf{resonance structure}\end{array}}$$

all neutral atoms
one more bond

charge separation

[2] Structures in which every atom has an octet are more stable.

$$CH_3-\overset{..}{\underset{..}{O}}-\overset{+}{C}H_2 \longleftrightarrow CH_3-\overset{+}{O}=CH_2 \qquad \boxed{\begin{array}{c}\textbf{more stable}\\\textbf{resonance structure}\end{array}}$$

All 2nd row elements have an octet.

[3] Structures that place a negative charge on a more electronegative element are more stable.

The (–) charge is on the more electronegative O atom.

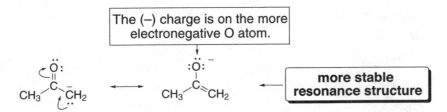

$$\boxed{\begin{array}{c}\textbf{more stable}\\\textbf{resonance structure}\end{array}}$$

♦ The unusual properties of conjugated dienes

[1] The C–C σ bond joining the two double bonds is unusually short (16.8).

[2] Conjugated dienes are more stable than similar isolated dienes. $\Delta H°$ of hydrogenation is smaller for a conjugated diene than for an isolated diene converted to the same product (16.9).

[3] The reactions are unusual:

- Electrophilic addition affords products of 1,2-addition and 1,4-addition (16.10, 16.11).
- Conjugated dienes undergo the Diels–Alder reaction, a reaction that does not occur with isolated dienes (16.12–16.14).

[4] Conjugated dienes absorb UV light in the 200–400 nm region. As the number of conjugated π bonds increases, the absorption shifts to longer wavelength (16.15).

♦ Reactions of conjugated dienes

[1] Electrophilic addition of HX (X = halogen) (16.10–16.11)

$$CH_2{=}CH{-}CH{=}CH_2 \xrightarrow[\text{(1 equiv)}]{HX}$$

$CH_2{-}CH{-}CH{=}CH_2$
     | |
     H   X
1,2-product
kinetic product
	$CH_2{-}CH{=}CH{-}CH_2$
	     | |
	     H X
	1,4-product
	thermodynamic product

- The mechanism has two steps.
- Markovnikov's rule is followed. Addition of H^+ forms the more stable allylic carbocation.
- The 1,2-product is the kinetic product. When H^+ adds to the double bond, X^- adds to the end of the allylic carbocation to which it is closer (C2 not C4). The kinetic product is formed faster at low temperature.
- The thermodynamic product has the more substituted, more stable double bond. The thermodynamic product predominates at equilibrium. With 1,3-butadiene, the thermodynamic product is the 1,4-product.

[2] Diels-Alder reaction (16.12–16.14)

1,3-diene   dienophile

| The three new bonds are labeled in **bold**. |

- The reaction forms two σ and one π bond in a six-membered ring.
- The reaction is initiated by heat.
- The mechanism is concerted: all bonds are broken and formed in a single step.
- The diene must react in the *s*-cis conformation (16.13A).
- Electron withdrawing groups in the dienophile increase the reaction rate (16.13B).
- The stereochemistry of the dienophile is retained in the product (16.13C).
- Endo products are preferred (16.13D).

16.1 **Isolated dienes** have two double bonds separated by two or more σ bonds.
Conjugated dienes have two double bonds separated by only one σ bond.

a.

one σ bond separates
two double bonds =
conjugated diene

b.

two σ bonds separate
two double bonds =
isolated diene

c.

one σ bond separates
two double bonds =
conjugated diene

d.

four σ bonds separate
two double bonds =
isolated diene

16.2 Conjugation occurs when there are *p* orbitals on three or more adjacent atoms. Double bonds separated by 2 σ bonds are not conjugated.

a. $CH_2=CH-CH=CH-CH=CH_2$

All of the carbon atoms are *sp²*
hybridized. Each π bond is
separated by only 1 σ bond.
conjugated

b.

The two π bonds are
separated by three σ bonds.
NOT conjugated

c.

The two π bonds are
separated by only 1 σ bond.
conjugated

d.

Three adjacent carbon atoms are *sp²* hybridized
and have an unhybridized *p* orbital.
conjugated

e.

This carbon is not *sp²*
hybridized.
NOT conjugated

16.3 Two resonance structures differ only in the placement of electrons. All σ bonds stay in the same place. Nonbonded electrons and π bonds can be moved. To draw the hybrid:

- Use a dashed line between atoms that have a π bond in one resonance structure and not the other.
- Use a δ symbol for atoms with a charge or radical in one structure but not the other.

a.

resonance hybrid:

δ^+ δ^+

The + charge is delocalized on two carbons.

b.

resonance hybrid:

δ^+
δ^+

The + charge is delocalized on two carbons.

c.

resonance hybrid:

δ^+ δ^+

The + charge is delocalized on two carbons.

16.4 Each different kind of carbon atom will give a different ^{13}C signal. When a carbocation is delocalized as in structure **B**, carbons become equivalent.

The two δ^+ carbons are identical.

A
5 different kinds of C
5 ^{13}C NMR signals

B
4 different kinds of C
4 ^{13}C NMR signals

16.5

a. $CH_2=CH-CH-CH=CH_2$ ⟷ $CH_2-CH=CH-CH=CH_2$ ⟷ $CH_2=CH-CH=CH-CH_2$

Move the charge
and the double bond.

b. $CH_3CH_2-\overset{\overset{\displaystyle :O:}{\|}}{C}-\overset{\overset{\displaystyle}{|}}{\underset{H}{C}}-CH_3$ ⟷ $CH_3CH_2-\overset{\overset{\displaystyle :O:^-}{}}{C}=\underset{H}{C}-CH_3$

c. $CH_3-\overset{+}{CH}-\ddot{C}l:$ ⟷ $CH_3-CH=\overset{+}{C}l:$

Move the lone pair.

Move the charge
and the double bond.

16.6 To compare the resonance structures remember:

- Resonance structures with **more bonds** are better.
- Resonance structures in which **every atom has an octet** are better.
- Resonance structures with **neutral atoms** are better than those with charge separation.
- Resonance structures that place a **negative charge on a more electronegative atom** are better.

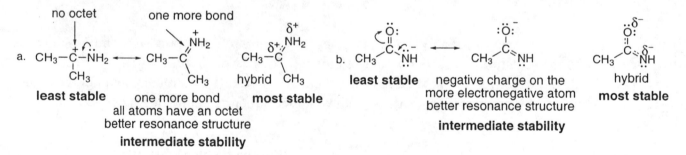

a.

no octet

$CH_3-\overset{\overset{\displaystyle +}{|}}{\underset{CH_3}{C}}-\ddot{N}H_2$ ⟷ $CH_3-\overset{\overset{\displaystyle \overset{+}{N}H_2}{}}{\underset{CH_3}{C}}$

least stable

one more bond

one more bond
all atoms have an octet
better resonance structure
intermediate stability

$CH_3-\overset{\overset{\displaystyle \overset{\delta^+}{N}H_2}{}}{\underset{CH_3}{\overset{\displaystyle \delta^+}{C}}}$

hybrid
most stable

b. $CH_3-\overset{\overset{\displaystyle :\ddot{O}:}{\|}}{C}-\ddot{N}H$ ⟷ $CH_3-\overset{\overset{\displaystyle :\ddot{O}:^-}{}}{C}=\overset{+}{N}H$

least stable   negative charge on the
more electronegative atom
better resonance structure
intermediate stability

$CH_3-\overset{\overset{\displaystyle \overset{\delta^-}{\ddot{O}}:}{}}{\underset{\overset{\displaystyle \delta^-}{N}H}{C}}$

hybrid
most stable

16.7 Remember that in any allyl system, there must be p orbitals to delocalize the lone pair.

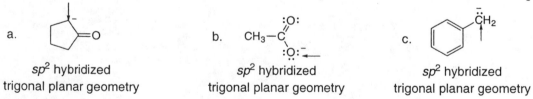

a.

sp^2 hybridized
trigonal planar geometry

b. $CH_3-\overset{\overset{\displaystyle :O:}{}}{\underset{:O:^-}{C}}$

sp^2 hybridized
trigonal planar geometry

c.

sp^2 hybridized
trigonal planar geometry

16.8 The *s-cis* conformer has two double bonds on the **same side** of the single bond. The *s-trans* conformer has two double bonds on **opposite sides** of the single bond.

a. (2*E*,4*E*)-2,4-**octadiene** in the *s-trans* conformation

double bonds on opposite sides
s-trans

c. (3*Z*,5*Z*)-4,5-dimethyl-3,5-**decadiene** in both the *s-cis* and *s-trans* conformers.

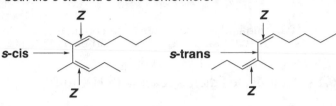

b. (3*E*,5*Z*)- 3,5-**nonadiene** in the *s-cis* conformation

double bonds on the same side
s-cis

16.9 Bond length depends on hybridization and percent *s*-character. Bonds with a higher percent *s*-character have smaller orbitals and are shorter.

HC≡C–C≡CH

sp hybridized carbons
50% *s*-character
shortest bond

CH₂=CH–CH=CH₂

sp² hybridized carbons
33% *s*-character
intermediate length bond

CH₃–CH₃

sp³ hybridized carbons
25% *s*-character
longest bond

16.10 Bond length depends on hybridization and percent *s*-character. Bonds with a higher percent *s*-character have smaller orbitals and are shorter. Compare the hybridization of the carbons in the bonds to rank the bonds by increasing length.

CH₃CH₂–CH₃

two *sp³* hybridized carbons
25% *s*-character
longest bond

CH₂=CH–CH₃

one *sp²* hybridized carbon
one *sp³* hybridized carbon
intermediate length bond

CH₂=CH–CH=CH₂

two *sp²* hybridized carbons
33% *s*-character
shortest bond

16.11 The **less stable** (higher energy) **diene** has the **higher heat of hydrogenation.** Isolated dienes are higher in energy than conjugated dienes, so they will have a higher heat of hydrogenation.

a.

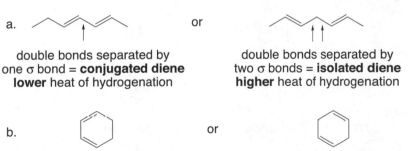

or

double bonds separated by
one σ bond = **conjugated diene**
lower heat of hydrogenation

double bonds separated by
two σ bonds = **isolated diene**
higher heat of hydrogenation

b.

or

double bonds separated by
one σ bond = **conjugated diene**
lower heat of hydrogenation

double bonds separated by
two σ bonds = **isolated diene**
higher heat of hydrogenation

16.12 Isolated dienes are higher in energy than conjugated dienes. Compare the location of the double bonds in the compounds below.

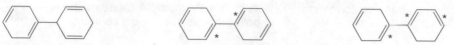

0 conjugated double bonds
least stable

2 conjugated double bonds
intermediate stability

3 conjugated double bonds
most stable

16.13 Conjugated dienes react with HX to form 1,2- and 1,4-products.

a. $CH_3-CH=CH-CH=CH-CH_3$ $\xrightarrow{HCl}$

$\underset{\underset{H}{|} \ \underset{Cl}{|}}{CH_3-CH-CH-CH=CH-CH_3}$ + $\underset{\underset{H}{|} \qquad \underset{Cl}{|}}{CH_3-CH-CH=CH-CH-CH_3}$

1,2-product **1,4-product**

b. [structure] $\xrightarrow{HCl}$ [structure with Cl]

isolated diene

c. [structure] $\xrightarrow{HCl}$ [structure with Cl]

d. [structure] $\xrightarrow{HCl}$ [structure **A**] + [structure **B**] + [structure **C**]

A   **B**   **C**

This double bond is more reactive, so **C** is probably a minor product because it results from HCl addition to the less reactive double bond.

16.14 The mechanism for addition of DCl has two steps:
[1] **Addition of D⁺** forms a resonance-stabilized carbocation.
[2] **Nucleophilic attack of Cl⁻** forms 1,2- and 1,4-products.

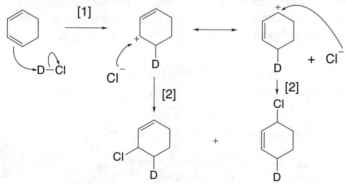

16.15 Label the products as 1,2- or 1,4-products. The 1,2-product is the kinetic product, and the 1,4-product, which has the more substituted double bond, is the thermodynamic product.

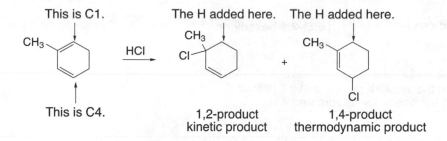

16.16 To draw the products of a Diels-Alder reaction:
[1] Find the 1,3-diene and the dienophile.
[2] Arrange them so the diene is on the left and the dienophile is on the right.
[3] Cleave three bonds and use arrows to show where the new bonds will be formed.

a.

diene   **dienophile**

b.

diene   **dienophile**
rotate to
make it *s*-cis

c.

dienophile   **diene**
rotate to
make it *s*-cis

16.17 For a diene to be reactive in a Diels-Alder reaction, **a diene must be able to adopt an *s*-cis conformation.**

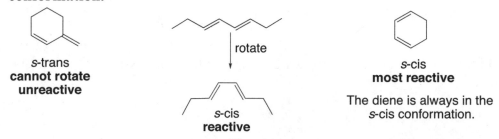

s-trans
**cannot rotate
unreactive**

rotate

s-cis
reactive

s-cis
most reactive

The diene is always in the
s-cis conformation.

16.18

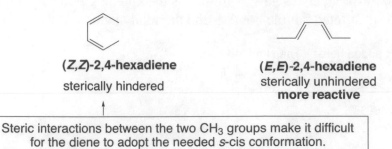

(Z,Z)-2,4-hexadiene	**(E,E)-2,4-hexadiene**
sterically hindered	sterically unhindered
	more reactive

Steric interactions between the two CH₃ groups make it difficult
for the diene to adopt the needed *s*-cis conformation.

16.19 Electron withdrawing substituents in the dienophile increase the reaction rate.

$CH_2=CH_2$ $CH_2=C\overset{H}{\underset{COOH}{}}$ $\overset{H\quad H}{\underset{HOOC\quad COOH}{C=C}}$

no electron withdrawing groups   one electron withdrawing group   two electron withdrawing groups
        least reactive   **intermediate reactivity**   **most reactive**

16.20 A cis dienophile forms a cis-substituted cyclohexene.
A trans dienophile forms a trans-substituted cyclohexene.

a.

$\overset{CH_3OOC\qquad COOCH_3}{\underset{H\qquad\quad H}{C=C}}$ → cis-substituted products

cis-dienophile

b.

$\overset{CH_3OOC\qquad H}{\underset{H\qquad\quad COOCH_3}{C=C}}$ → trans-substituted products

trans-dienophile

c.

cis-dienophile → cis-substituted product   + identical

16.21 The **endo product** (with the substituents under the plane of the new six-membered ring) is the
preferred product.

a.   + $CH_2=CHCOOCH_3$ → COOCH₃ **endo** substituent

b.   + $\overset{COOCH_3}{\underset{COOCH_3}{}}$ → COOCH₃ **both groups endo**
                                                              COOCH₃

16.22 To find the diene and dienophile needed to make each of the products:
  [1] Find the six-membered ring with a C–C double bond.
  [2] Draw three arrows to work backwards.
  [3] Follow the arrows to show the diene and dienophile.

a.

b.

c.

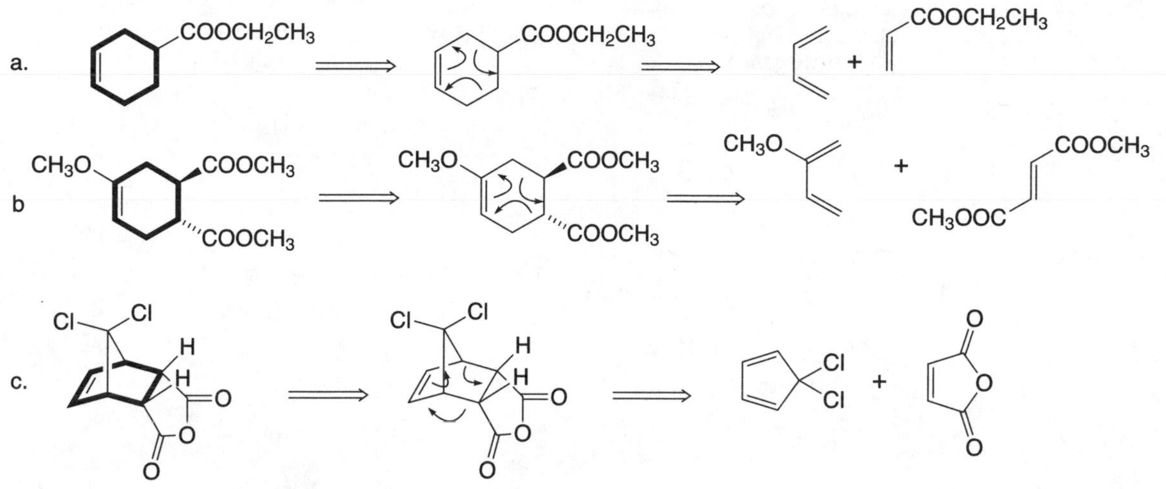

16.23

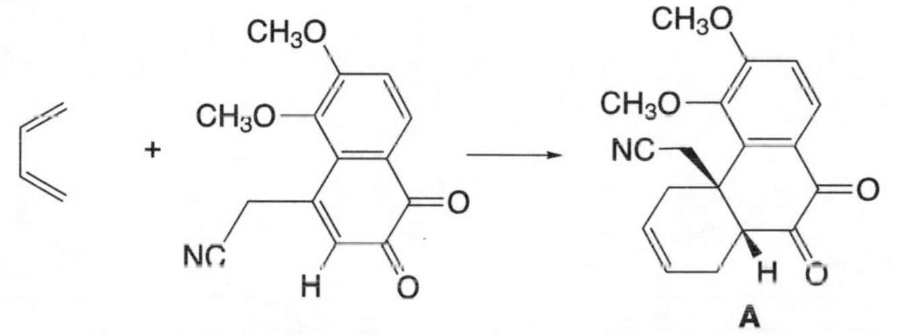

16.24 Conjugated molecules absorb light at a longer wavelength than molecules that are not conjugated.

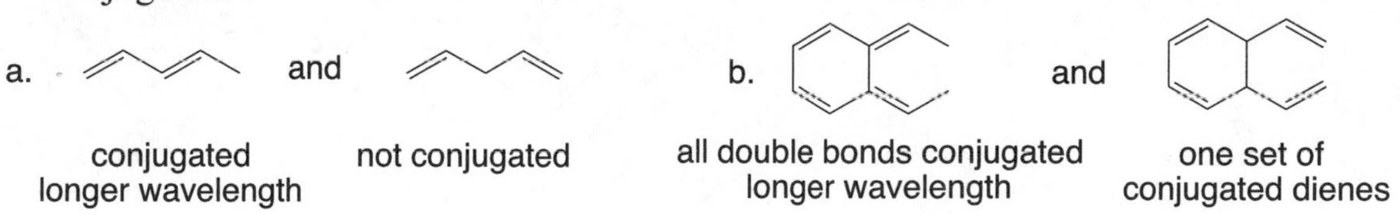

a.

conjugated
longer wavelength

not conjugated

b.

all double bonds conjugated
longer wavelength

one set of
conjugated dienes

16.25 Sunscreens contain conjugated systems to absorb UV radiation from sunlight. Look for conjugated systems in the compounds below.

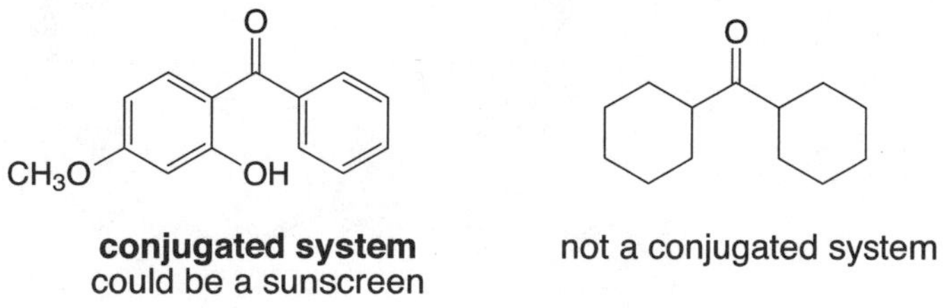

conjugated system
could be a sunscreen

not a conjugated system

16.26 Use the definition from Answer 16.1.

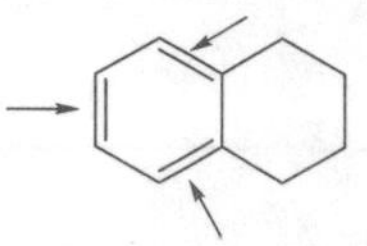

3 π bonds with only
1 σ bond between
conjugated

CH₂=CHC≡N

2 multiple bonds with only
1 σ bond between
conjugated

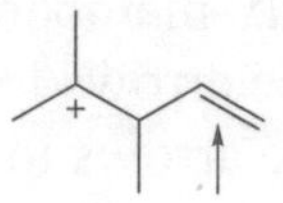

1 π bond with
no adjacent *sp²* hybridized atoms
NOT conjugated

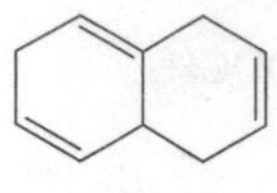

3 π bonds with 2 or
more σ bonds between
NOT conjugated

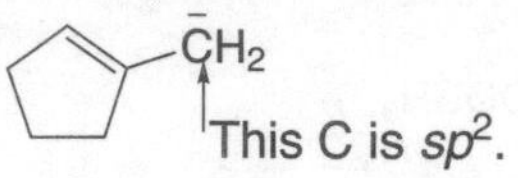

ĊH₂

This C is *sp²*.

1 π bond with
an adjacent *sp²* hybridized atom

The lone pair occupies a *p* orbital,
so there are *p* orbitals on
three adjacent atoms.

conjugated

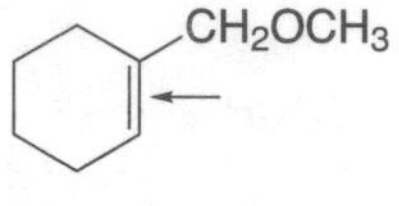

CH₂OCH₃

1 π bond with
no adjacent *sp²* hybridized atoms
NOT conjugated

16.27

a. $(CH_3)_2\overset{+}{C}-CH=CH_2 \longleftrightarrow (CH_3)_2C=CH-\overset{+}{C}H_2$

b.

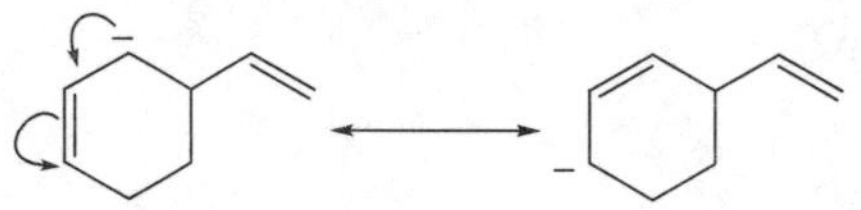

c.

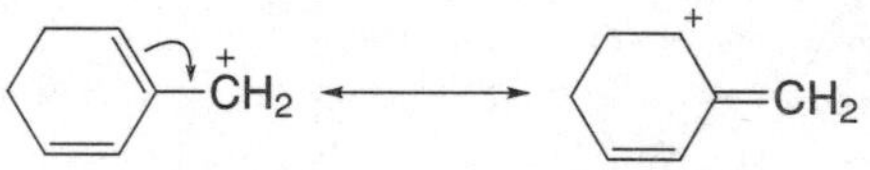

d.

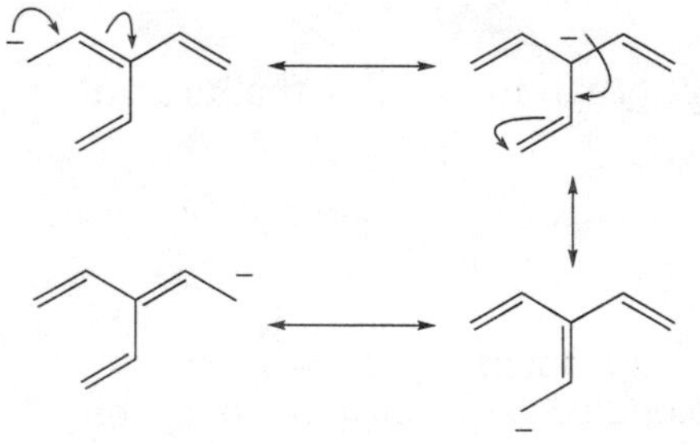

e. $CH_3\overset{..}{O}-CH=CH-\overset{+}{C}H_2 \longleftrightarrow CH_3\overset{..}{\underset{..}{O}}-\overset{+}{C}H-CH=CH_2$

$CH_3\overset{+}{\underset{..}{O}}=CH-CH=CH_2$

f.

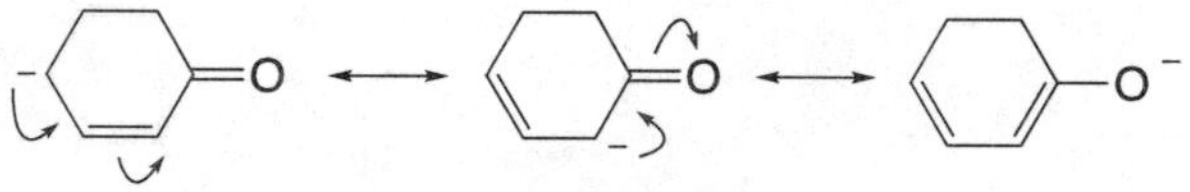

g. 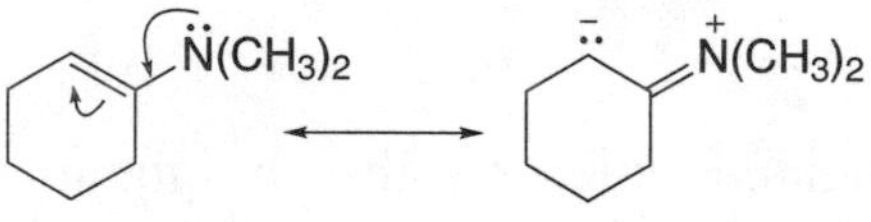 $\dot{N}(CH_3)_2$ $\ddot{N}(CH_3)_2$

h.

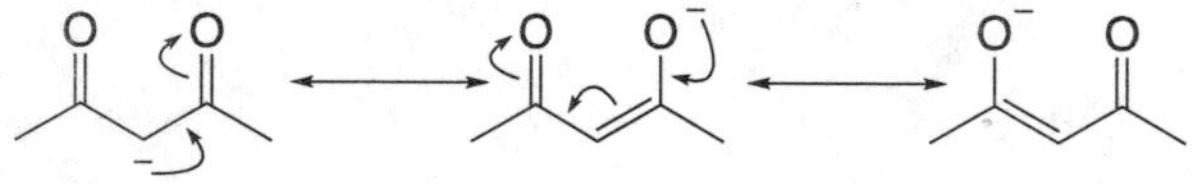

16.28

a. 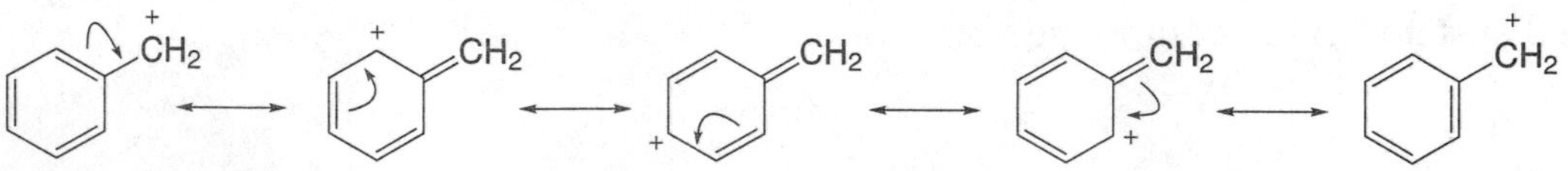 CH₂ CH₂ CH₂ CH₂ CH₂

b. 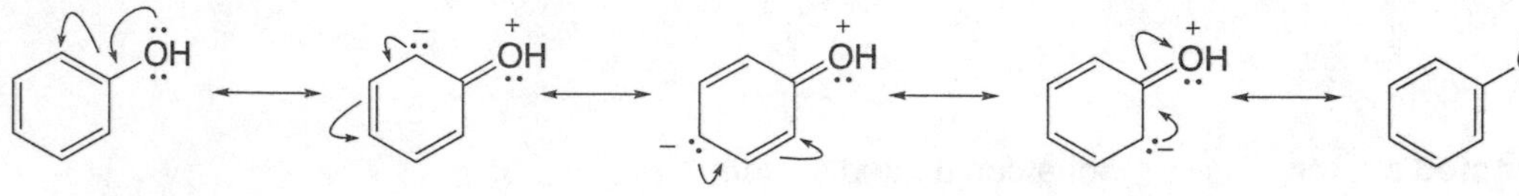 ÖH ÖH ÖH ÖH ÖH

16.29 Use the directions from Answer 16.3.

A	**B**	**C**	**D**
no charges	2 charges	2 charges	resonance hybrid
all atoms have an octet	C does not have an octet	all atoms have an octet	
4 bonds (C–C + C–O)	3 bonds (C–C + C–O)	4 bonds (C–C + C–O)	
most stable	**least stable**	**intermediate stability**	

16.30

resonance hybrid:

Five resonance structures delocalize the negative charge on five C's making them all equivalent.

All of the carbons are identical in the anion.

16.31

a.

The benzylic C is sp^2 hybridized.
trigonal planar

b.

The benzylic C is sp^2 hybridized.
trigonal planar

c.

This carbon is NOT part of a conjugated system.
4 groups = sp^3 hybridized
tetrahedral

16.32

a.

Two equivalent resonance structures delocalize the π bond and the negative charge.

hybrid:

These bond lengths are equal because they are identical.

b.

$CH_2=CH-CH_2-H$
| more acidic

$\bar{C}H_2-CH=CH_2 \longleftrightarrow CH_2=CH-\bar{C}H_2$

Resonance stabilization delocalizes the negative charge on 2 C's after loss of a proton. This makes propene more acidic than propane.

$CH_3CH_2CH_2-H$
| less acidic

$CH_3CH_2\bar{C}H_2$

only one Lewis structure

c. S_N1 reactions proceed via a carbocation intermediate. Draw the carbocation formed on loss of Cl and compare.

3-chloro-1-propene $CH_2=CHCH_2Cl$
more reactive

$$\left[\overset{+}{C}H_2-CH=CH_2 \longleftrightarrow CH_2=CH-\overset{+}{C}H_2 \right]$$

resonance-stabilized carbocation

Two resonance structures delocalize the positive charge on 2 C's making the 3-chloro-1-propene more reactive.

$CH_3CH_2CH_2Cl$ is a 1° halide, which does not react by an S_N1 reaction because cleavage of the C–Cl bond forms a highly unstable 1° carbocation.

1-chloropropane $CH_3CH_2CH_2Cl$
less reactive

$$CH_3CH_2\overset{+}{C}H_2$$

only one Lewis structure
very unstable

d. Draw the products of cleavage of the bond.

ethane CH_3-CH_3

$\cdot CH_3$ + $\cdot CH_3$

Two unstable radicals form.

1-butene $CH_3-CH_2CH=CH_2$

$\cdot CH_3$ + $\overset{\cdot}{C}H_2-CH=CH_2 \longleftrightarrow CH_2=CH-\overset{\cdot}{C}H_2$

One resonance-stabilized radical forms.
This makes the bond dissociation
energy lower because a more stable radical is formed.

16.33

$CH_3CH=CHCH_2\overset{..}{O}H \longrightarrow CH_3CH=CHCH_2-\overset{+}{\overset{..}{O}}H_2$ + $Br^- \longrightarrow CH_3CH=CH\overset{+}{C}H_2$ + $H_2O \longrightarrow CH_3CH=CHCH_2Br$

H–Br

two resonance
structures:

+ Br^-

$CH_3\overset{+}{C}HCH=CH_2 \longrightarrow CH_3CHCH=CH_2$
|
Br

+ Br^-

16.34 Use the directions from Answer 16.8.

a. (3Z)-1,3-pentadiene in the s-trans conformation

double bonds on opposide sides
s-trans

b. (2E,4Z)-1-bromo-3-methyl-2,4-hexadiene

Br

c. (2E,4E,6E)-2,4,6-octatriene

d. (E,E)-3-methyl-2,4-hexadiene in the s-cis conformation

s-cis

16.35

2E,4E 2E,4Z

2Z,4E 2Z,4Z

16.36

a. 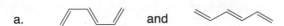 and

(3*E*)-1,3,5-hexatriene   (3*E*)-1,3,5-hexatriene
both *s*-cis   both *s*-trans

different conformers

c. 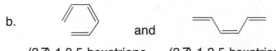 and

(3*E*)-1,3,5-hexatriene   (3*Z*)-1,3,5-hexatriene

different stereoisomers

b. 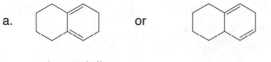 and

(3*Z*)-1,3,5-hexatriene   (3*Z*)-1,3,5-hexatriene
both *s*-cis   both *s*-trans

different conformers

16.37 Use the directions from Answer 16.11.

a.   or

conjugated diene   isolated diene
less stable
higher heat of hydrogenation

b.   or

isolated diene   conjugated diene
less stable
higher heat of hydrogenation

16.38 Conjugated dienes react with HX to form 1,2- and 1,4-products.

a. isolated diene → HBr (1 equiv)

major product, formed by addition of HBr to the more substituted C=C

b. → HCl (1 equiv)

1,2-product   +   1,4-product   (*E* and *Z* isomers can form.)

c. → DCl (1 equiv)

1,2-product   +   1,4-product

d. → HBr (1 equiv)

1,2-product   +   1,4-product   +   1,2-product   +   (*E* and *Z* isomers)
1,4-product

16.39

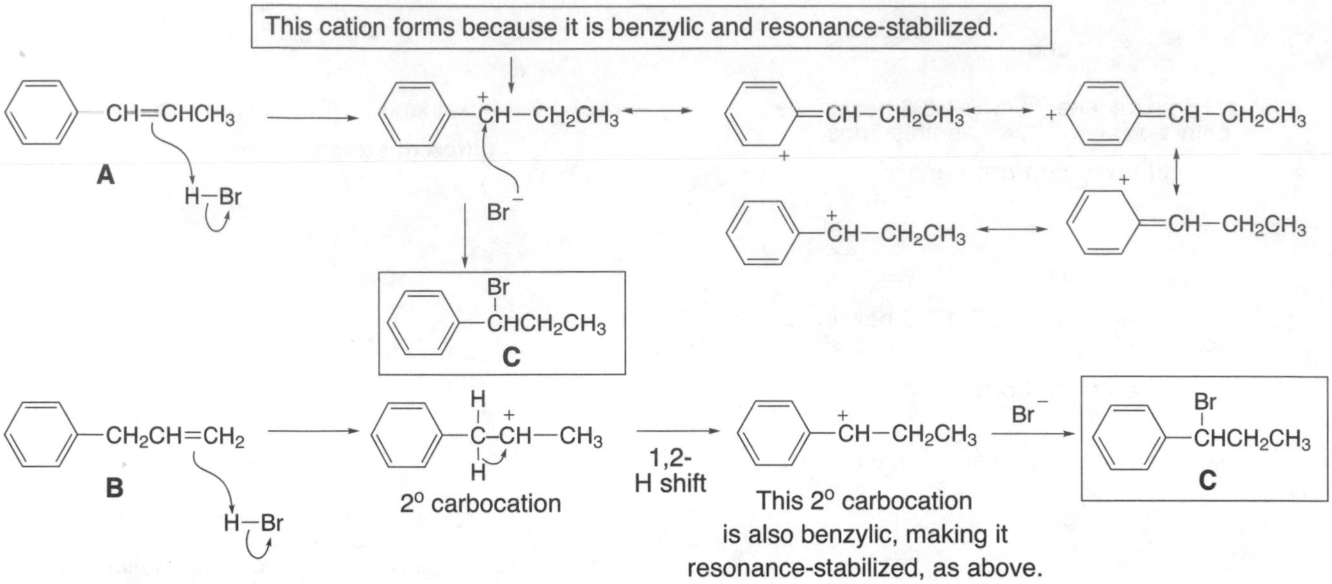
This cation forms because it is benzylic and resonance-stabilized.
A
B
C
2° carbocation
1,2-
H shift
This 2° carbocation
is also benzylic, making it
resonance-stabilized, as above.

16.40 To draw the mechanism for reaction of a diene with HBr/ROOR, recall from Chapter 13 that when an alkene is treated with HBr under these radical conditions, the Br ends up on the carbon with more H's to begin with.

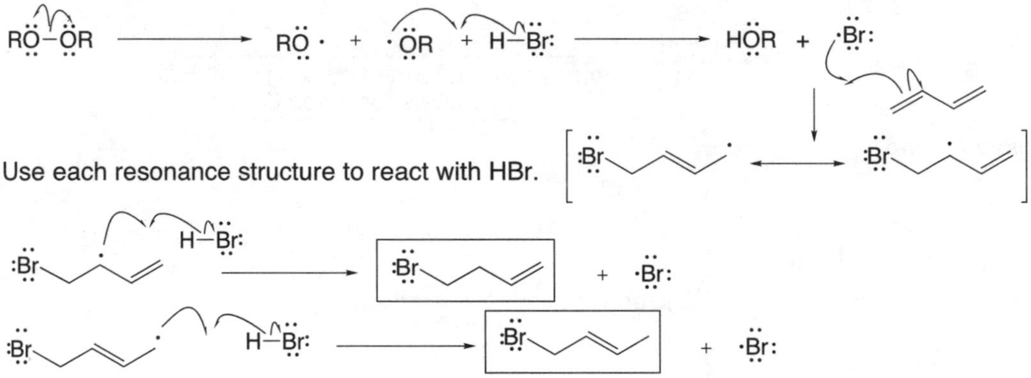
Use each resonance structure to react with HBr.

16.41

a. and b.

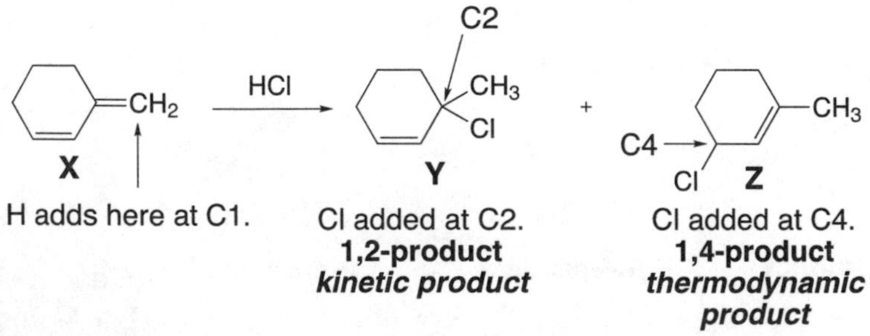
HCl
X
Y
Z
H adds here at C1.
Cl added at C2.
1,2-product
kinetic product
Cl added at C4.
1,4-product
thermodynamic product

Y is the kinetic product because of the proximity effect. H and Cl add across two adjacent atoms.
Z is the thermodynamic product because it has a more stable trisubstituted double bond.

Addition occurs at the labeled double bond due to the stability of the carbocation intermediate.

c.

The two resonance structures for this allylic cation are 3° and 2° carbocations.

| more stable intermediate |
| Addition occurs here. |

If addition occurred at the other C=C, the following allylic carbocation would form:

The two resonance structures for this allylic cation are 1° and 2° carbocations.

less stable

16.42

Addition of HCl at the terminal double bond forms a carbocation that is highly resonance-stabilized since it is both allylic and benzylic. Such stabilization does not occur when HCl is added to the other double bond. This gives rise to two products of electrophilic addtion.

1,2-product

1,4-product

(+ four more resonance structures that delocalize the positive charge onto the benzene ring)

16.43 The electron pairs on O can be donated to the double bond through resonance. This increases the electron density of the double bond, making it less electrophilic and therefore less reactive in a Diels–Alder reaction.

$$CH_2-CH\ddot{O}CH_3 \longleftrightarrow \ddot{C}H_2-CH=\overset{+}{\ddot{O}}CH_3$$

methyl vinyl ether

This C now bears a net negative charge.

16.44 Use the directions from Answer 16.16.

a.

diene   **dienophile**   re-draw   Δ

b.

diene   trans-dienophile   Δ   **trans-substituted products**

c.

diene   cis-dienophile   Δ   **cis-substituted products**

d. 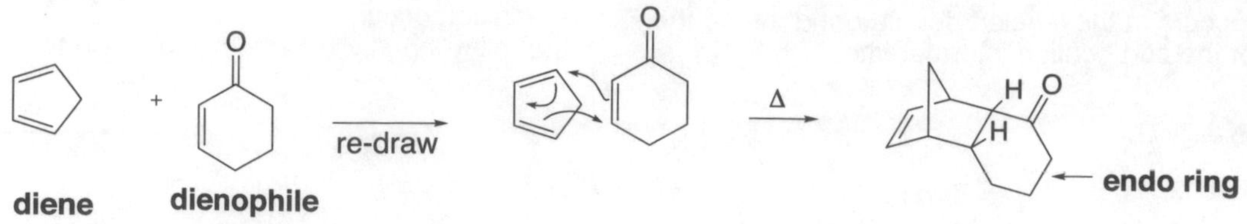
re-draw
Δ
H
O
H
endo ring
diene
dienophile

e. 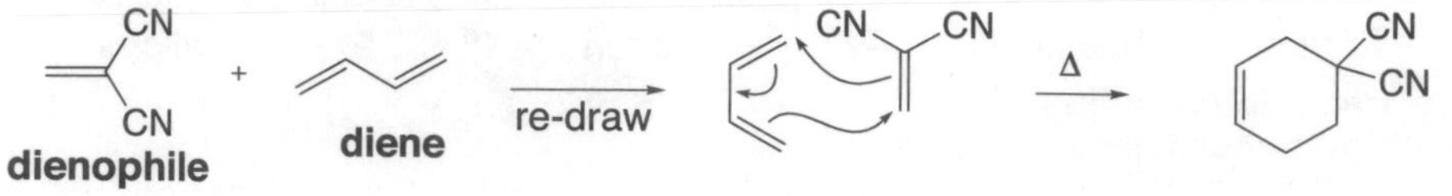
CN
CN
dienophile
diene
re-draw
CN
CN
Δ
CN
CN

f. 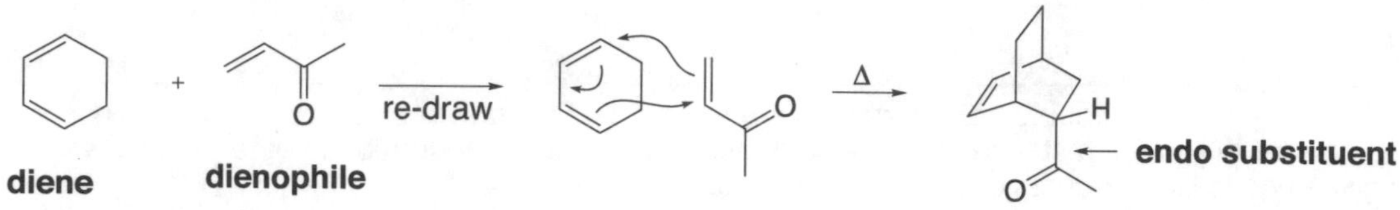
O
re-draw
O
Δ
H
endo substituent
O
diene
dienophile

16.45 Use the directions from 16.22.

a. 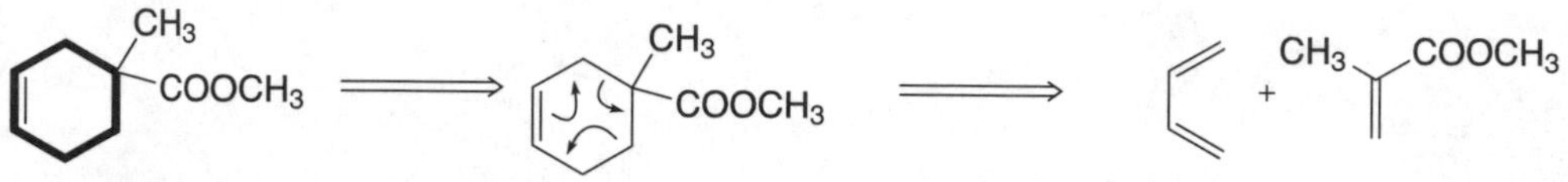
CH3
COOCH3
CH3
COOCH3
CH3
COOCH3

b. 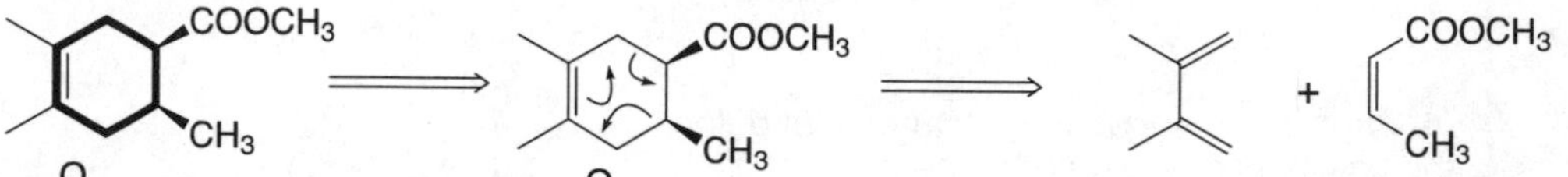
COOCH3
CH3
COOCH3
CH3
COOCH3
CH3

c. 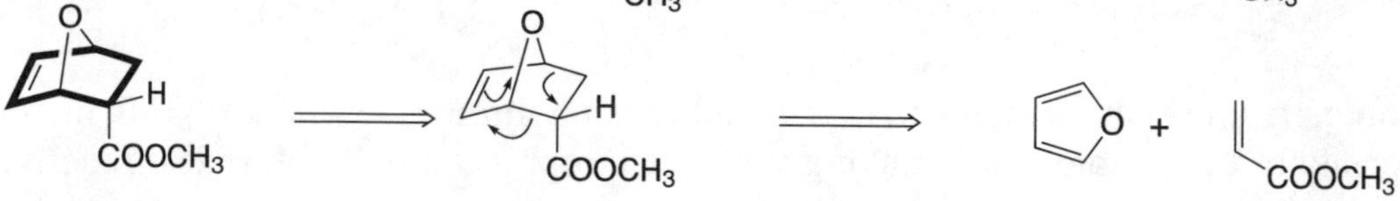
O
H
COOCH3
O
H
COOCH3
O
COOCH3

d. 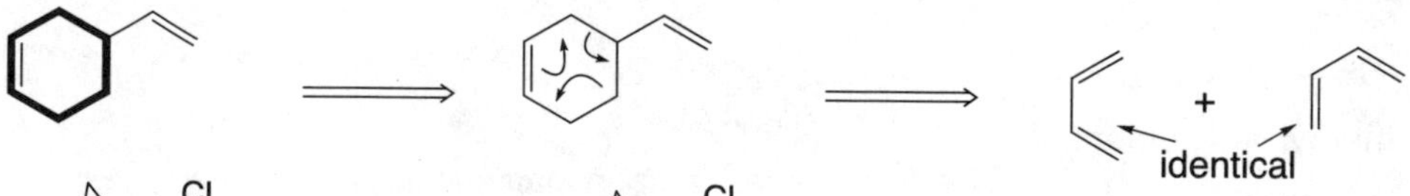
identical

e. 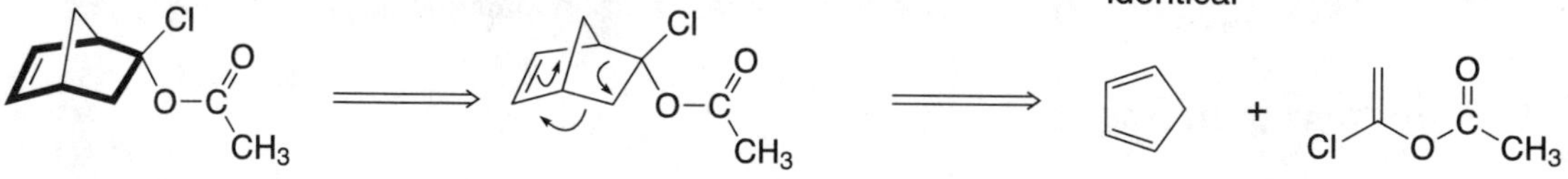
Cl
O
O–C
CH3
Cl
O
O–C
CH3
O
Cl
O
C
CH3

f. 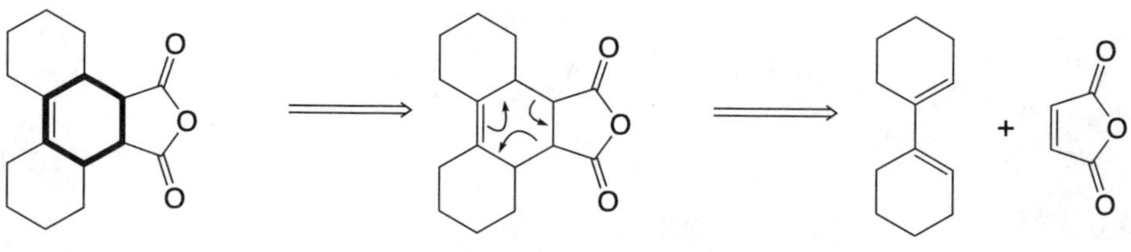
O
O
O
O
O
O
O
O
O

16.46

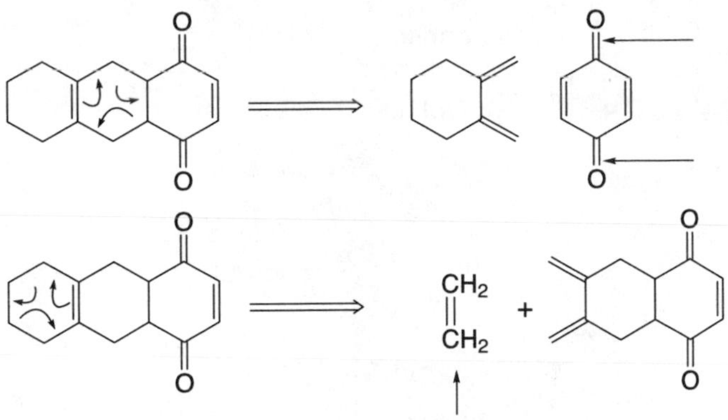

This pathway is **prefered** because the dienophile has electron-withdrawing C=O groups which make it more reactive.

no electron-withdrawing groups
less reactive

16.47

a.

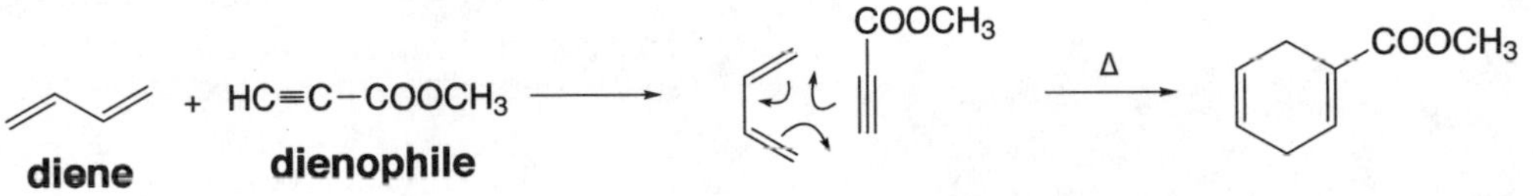

diene   dienophile

b.

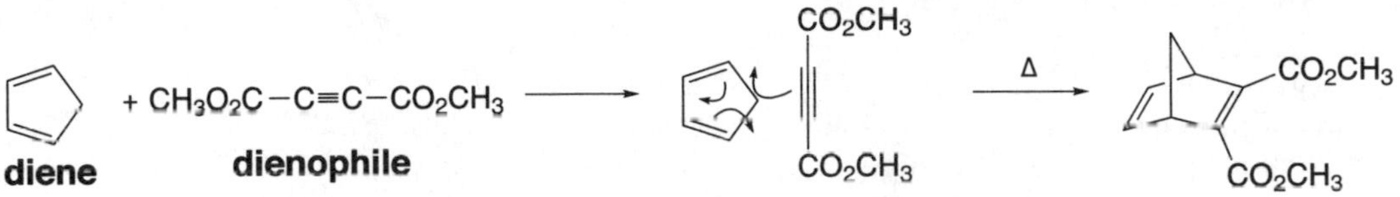

diene   dienophile

16.48

or

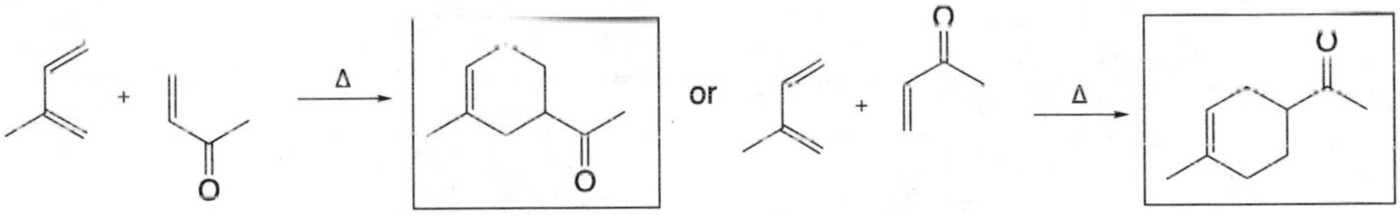

16.49

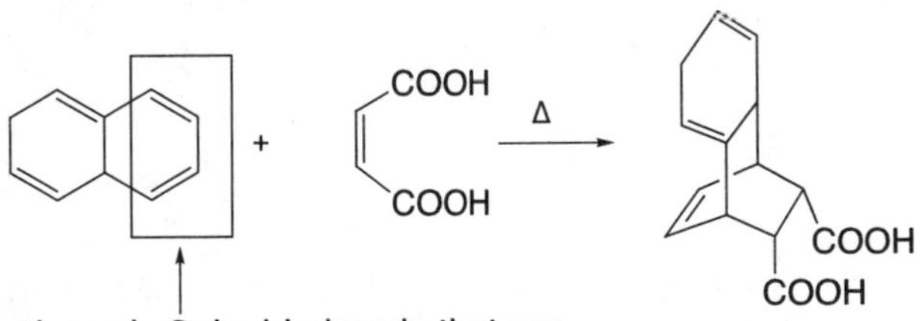

These are the only 2 double bonds that are conjugated and have the *s*-cis conformation needed for a Diels–Alder reaction.

16.50 Benzoquinone has two double bonds that can undergo reaction with excess 1,3-butadiene.

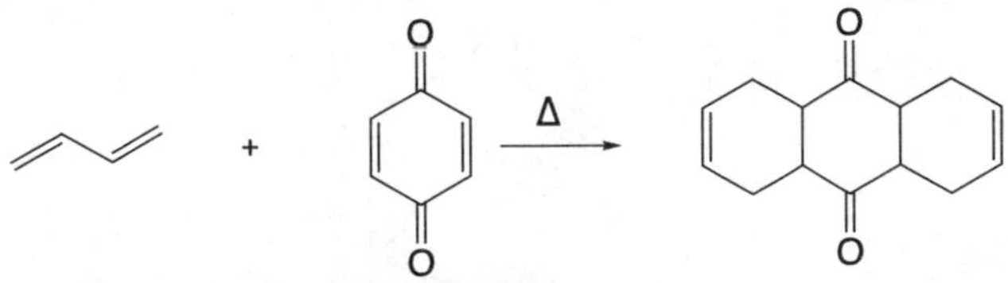

excess   benzoquinone-
two dienophiles $C_{14}H_{16}O_2$

16.51

16.52

16.53

16.54

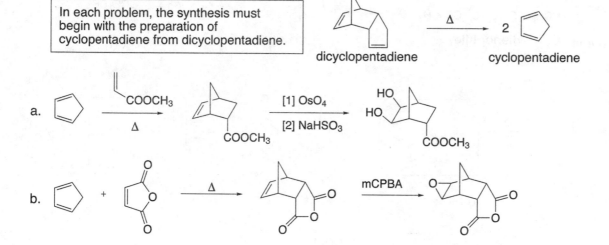

16.55

a. (CH₃)₂C=CHCH₂CH₂CH=CH₂ →(HCl) (CH₃)₂C(Cl)CH₂CH₂CH₂CH=CH₂ + (CH₃)₂C=CHCH₂CH₂CHClCH₃
    isolated diene   **major product**   **minor product**

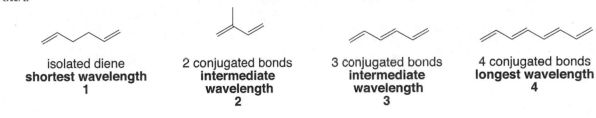

16.56

isolated diene — **shortest wavelength 1**; 2 conjugated bonds — **intermediate wavelength 2**; 3 conjugated bonds — **intermediate wavelength 3**; 4 conjugated bonds — **longest wavelength 4**

16.57

16.58

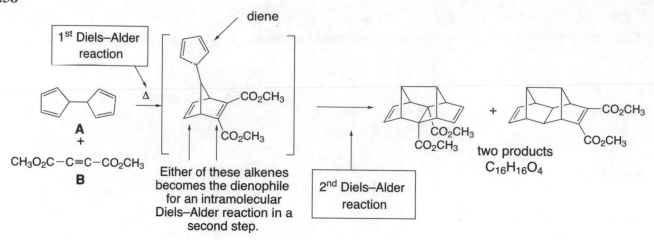

two products
$C_{16}H_{16}O_4$

Chapter 17: Benzene and Aromatic Compounds

◆ Comparing aromatic, antiaromatic, and nonaromatic compounds (17.7)

- **Aromatic compound**
  - A cyclic, planar, completely conjugated compound that contains $4n + 2$ π electrons ($n = 0, 1, 2, 3$, and so forth).
  - An aromatic compound is more stable than a similar acyclic compound having the same number of π electrons.

- **Antiaromatic compound**
  - A cyclic, planar, completely conjugated compound that contains $4n$ π electrons ($n = 0, 1, 2, 3$, and so forth).
  - An antiaromatic compound is less stable than a similar acyclic compound having the same number of π electrons.

- **A compound that is not aromatic**
  - A compound that is either not cyclic or not planar or not completely conjugated.

◆ Properties of aromatic compounds

- Every carbon has a p orbital to delocalize electron density (17.2).
- They are unusually stable. $\Delta H°$ for hydrogenation is much less than expected, given the number of degrees of unsaturation (17.6).
- They do not undergo the usual addition reactions of alkenes (17.6).
- ^{1}H NMR spectra show highly deshielded protons because of ring currents (17.4).

◆ Examples of aromatic compounds with 6π electrons (17.8)

benzene   pyridine   pyrrole   cyclopentadienyl anion   tropylium cation

◆ Examples of compounds that are not aromatic (17.8)

not cyclic   not planar   not completely conjugated

Chapter 17: Answers to Problems

17.1 Move the electrons in the π bonds to draw all major resonance structures.

17.2 Look at the hybridization of the atoms involved in each bond. Carbons in a benzene ring are surrounded by three groups and are sp^2 hybridized.

a.

C_{sp^2}–C_{sp^3}

CH$_3$

C_{sp^2}–H$_{1s}$

H

b.

C_{sp^2}–C_{sp^2}

C_{sp^2}–C_{sp^2}
C_p–C_p

17.3

- To name a benzene ring with **one substituent**, name the substituent and add the word *benzene*.
- To name a **disubstituted ring**, select the correct prefix (ortho = 1,2; meta = 1,3; para = 1,4) and alphabetize the substituents. Use a common name if it is a derivative of that monosubstituted benzene.
- To name a **polysubstituted ring**, number the ring to give the lowest possible numbers and then follow other rules of nomenclature.

a.

isopropyl group

PhCH(CH$_3$)$_2$

isopropylbenzene

c.

OH

2 groups are 1,3 = meta

butyl group

phenol

m-butylphenol

b.

CH$_2$CH$_3$ ←— ethyl

iodo   2 groups are 1,4 = para
p-ethyliodobenzene

d.

CH$_3$

Br ——→2-bromo

5-chloro

→toluene   (CH$_3$ group must be at the "1" position, if the molecule is named as a toluene derivative.)

2-bromo-5-chlorotoluene

17.4 Work backwards to draw the structure from the names.

a. isobutylbenzene

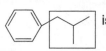

isobutyl group

c. *cis*-1,2-diphenylcyclohexane

d. *m*-bromoaniline

Br

NH₂ → aniline

b. *o*-dichlorobenzene

Cl
Cl

e. 4-chloro-1,2-diethylbenzene

Cl

17.5

Molecular formula $C_{10}H_{14}O_2$: 4 degrees of unsaturation
IR absorption at 3150–2850 cm⁻¹: sp^2 and sp^3 hybridized C–H bonds
NMR absorptions (ppm):
    1.4 (triplet, 6H)
    4.0 (quartet, 4H)
    6.8 (singlet, 4H)

O—⟨ ⟩—O

17.6 Count the different types of carbons to determine the number of ¹³C NMR signals.

a.
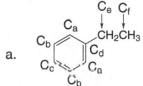
Cₐ
C_b
C_c
C_d
Cₐ
C_b
Cₑ C_f
CH₂CH₃

4 types of C in the benzene ring
6 peaks

b.
CH₃
Cl

all C's are different
7 peaks

c.
C_c C_b   C_b C_c
C_d   C_d
C_c C_b C_a C_b C_c

4 peaks

17.7 The less stable compound has a higher heat of hydrogenation.

CH₃

A

benzene ring, more stable
lower Δ*H*°

CH₂

B

no benzene ring, less stable
higher Δ*H*°

17.8 The protons on sp^2 hybridized carbons in aromatic hydrocarbons are highly deshielded and absorb at 6.5–8 ppm whereas hydrocarbons that are not aromatic show an absorption at 4.5–6 ppm, typical of protons bonded to the C=C of an alkene.

a.
H H H H H

aromatic ring
H's ~ 6.5–8 ppm

b.

not aromatic
alkene H's ~ 4.5–6 ppm

c.
H H H H H H H H

aromatic ring
H's ~ 6.5–8 ppm

17.9 To be aromatic, a ring must have $4n + 2$ π electrons.

16 π e⁻	20 π e⁻	22 π e⁻
$4n$	$4n$	$4n + 2$
$4(4) = 16$	$4(5) = 20$	$4(5) + 2 = 22$
antiaromatic	**antiaromatic**	**aromatic**

17.10 Annulenes have alternating double and single bonds. An odd number of carbon atoms in the ring would mean there would be two adjacent single bonds. Therefore an annulene having an odd number of carbon atoms cannot exist.

17.11

17.12 In determining if a heterocycle is aromatic, count a nonbonded electron pair if it makes the ring aromatic in calculating $4n + 2$.

a.

count 2 of O's
nonbonded electrons
$4n + 2 = 4(1) + 2 = 6$
aromatic

b.

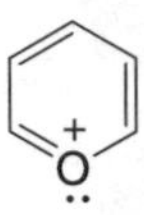

no lone pair from O
$4n + 2 = 4(1) + 2 = 6$
aromatic

c.

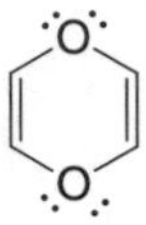

with one lone
pair from each O
there are 8 electrons
not aromatic

17.13

quinine
(antimalarial drug)

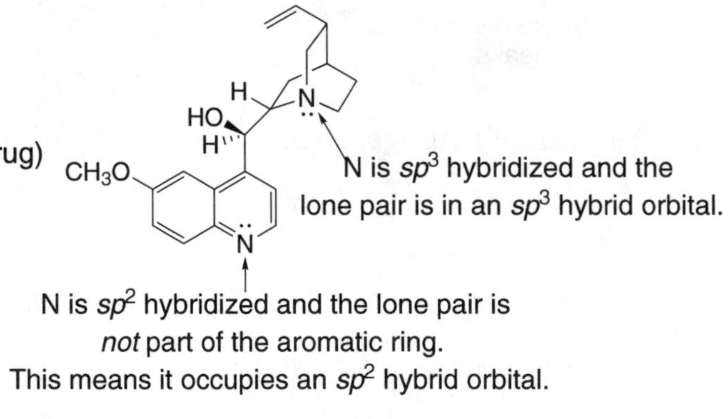

N is sp^3 hybridized and the
lone pair is in an sp^3 hybrid orbital.

N is sp^2 hybridized and the lone pair is
not part of the aromatic ring.
This means it occupies an sp^2 hybrid orbital.

17.14

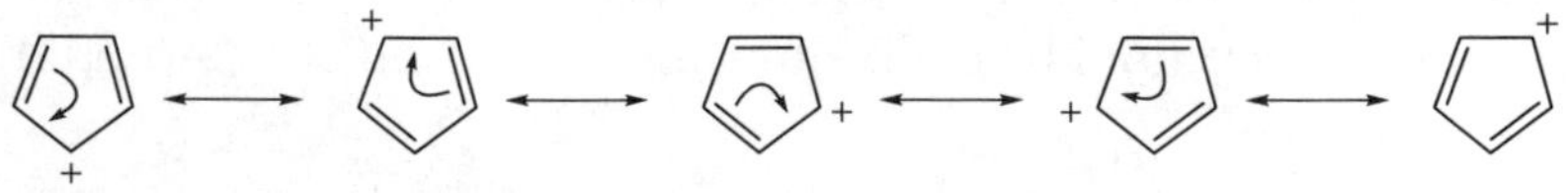

17.15 Compare the reaction of 1,3,5-cycloheptatriene with base, with that of cyclopentadiene with base. Remember that the compound with the more stable conjugate base will have a lower pK_a.

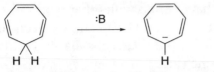 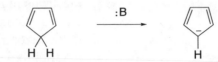

1,3,5-cycloheptatriene pK_a = 39	8 π Electrons make this conjugate base especially unstable (**antiaromatic**).

Since the conjugate base is unstable,
the pK_a of 1,3,5-cycloheptatriene is **high**.

cyclopentadiene

6 π electrons
aromatic conjugate base
very stable anion

Since the conjugate base is very stable, the
pK_a of cyclopentadiene is much **lower**.

17.16 The compound with the most stable conjugate base is the most acidic.

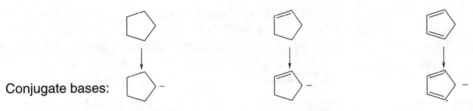

Conjugate bases:

no resonance
delocalization

Most unstable base so
least acidic acid.

2 resonance
structures

The acid is
intermediate in acidity.

aromatic conjugate base
most stable

The acid is the
most acidic.

17.17

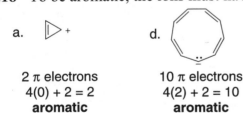

17.18 To be aromatic, the ions must have $4n + 2$ π electrons.

a.

2 π electrons
4(0) + 2 = 2
aromatic

d.

10 π electrons
4(2) + 2 = 10
aromatic

17.19

absorbs at 7.6 ppm

A =

The NMR indicates that **A** is aromatic. The C's of the triple bond are *sp* hybridized. Each triple bond has one set of electrons in *p* orbitals that overlap with other *p* orbitals on adjacent atoms in the ring. This overlap allows electrons to delocalize. Each C of the triple bonds also has a *p* orbital parallel to the plane of the ring. The electrons in these *p* orbitals are localized between the C's of the triple bond, and not delocalized in the ring. Although **A** has 24 π e$^-$ electrons total, only 18 e$^-$ are delocalized around the ring.

17.20 In using the inscribed polygon method, always draw the vertex pointing down.

2 antibonding MOs

1 bonding MO

2 π electrons
All bonding MOs are filled.
aromatic

17.21 Draw the inscribed pentagons with the vertex pointing down. Then draw the molecular orbitals (MOs) and add the electrons.

2 antibonding MOs

3 bonding MOs

Cation:

4 π electrons
Not all bonding MOs are filled.
not aromatic

Radical:

5 π electrons
Not all bonding MOs are filled.
not aromatic

17.22 C_{60} would exhibit only one ^{13}C NMR signal because all the carbons are identical.

17.23

a. If benzene could be described by a single Kekulé structure, only one product would form in Reaction [1], but there would be four (not three) dibromobenzenes (**A–D**), because adjacent C–C bonds are different—one is single and one is double. Thus, compounds **A** and **B** would *not* be identical. **A** has two Br's bonded to the same double bond, but **B** has two Br's on different double bonds.

b. In the resonance description, only one product would form in Reaction [1], since all C's are identical, but only three dibromobenzenes (ortho, meta, and para isomers) are possible. **A** and **B** are identical because each C–C bond is identical and intermediate in bond length between a C–C single and C–C double bond.

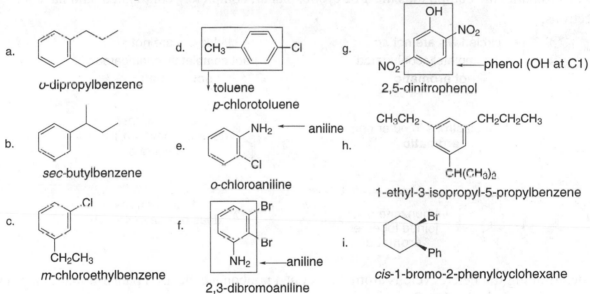

17.24 To name the compounds use the directions from Answer 17.3.

a. *o*-dipropylbenzene

b. *sec*-butylbenzene

c. *m*-chloroethylbenzene

d. toluene
p-chlorotoluene

e. aniline
o-chloroaniline

f. aniline
2,3-dibromoaniline

g. phenol (OH at C1)
2,5-dinitrophenol

h. 1-ethyl-3-isopropyl-5-propylbenzene

i. cis-1-bromo-2-phenylcyclohexane

17.25

a. *p*-dichlorobenzene

b. *m*-chlorophenol

c. *p*-iodoaniline

d. *o*-bromonitrobenzene

e. 2,6-dimethoxytoluene

f. 2-phenyl-1-butene

g. 2-phenyl-2-propen-1-ol

h. *trans*-1-benzyl-3-phenylcyclopentane

or

17.26

1,2,3-trichlorobenzene   1,2,4-trichlorobenzene   1,3,5-trichlorobenzene

17.27 Count the electrons in the π bonds. Each π bond holds two electrons.

a.   b.   c.

   10 π electrons   7 π electrons   10 π electrons

17.28 To be aromatic, the compounds must be cyclic, planar, completely conjugated, and have 4*n* + 2 π electrons.

a.   circled C's are not *sp²*   d.   circled C's are not *sp²*
            not completely conjugated   not completely conjugated
                not aromatic   **not aromatic**

b.   14 π electrons in outer ring   e.   12 π electrons
                 aromatic   does **not** have 4*n* + 2
                                         π electrons

c.   4 benzene rings   f.   12 π electrons
            joined together   does **not** have 4*n* + 2
               aromatic   π electrons

17.29 In determining if a heterocycle is aromatic, count a nonbonded electron pair if it makes the ring aromatic in calculating 4*n* +2.

a.   c.   e.   g.

   6 π electrons
counting a lone pair from S   **not aromatic**   **not aromatic**   N is not *sp²* (no *p* orbital)
       4(1) + 2 = 6   **not aromatic**
        aromatic

b.   d.   f.   h.

   6 π electrons
counting a lone pair from O   10 π electrons   6 π electrons,   10 π electrons
       4(1) + 2 = 6   4(2) + 2 = 10   counting a lone pair from O   4(2) + 2 = 10
       aromatic   **aromatic**   4(1) + 2 = 6   **aromatic**
                                           aromatic

17.30

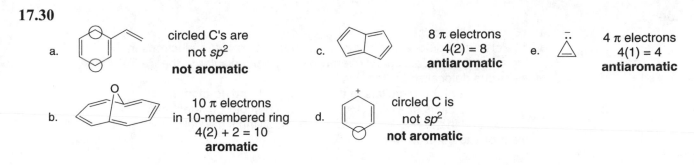

a. circled C's are
    not sp^2
    not aromatic

c. 8 π electrons
    4(2) = 8
    antiaromatic

e. 4 π electrons
    4(1) = 4
    antiaromatic

b. 10 π electrons
    in 10-membered ring
    4(2) + 2 = 10
    aromatic

d. circled C is
    not sp^2
    not aromatic

17.31

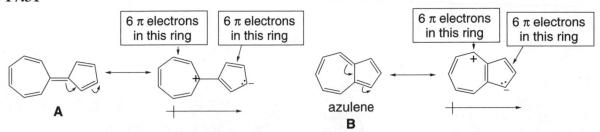

6 π electrons
in this ring

6 π electrons
in this ring

6 π electrons
in this ring

6 π electrons
in this ring

A

azulene
B

In both **A** and **B**, resonance structures can be drawn that place a negative charge in the five-membered ring and a positive charge in the seven-membered ring. These resonance structures show that each ring has 6 π electrons, making it aromatic. Each molecule possesses a dipole such that the seven-membered ring is electron deficient and the five-membered ring is electron rich.

17.32 Benzene has C–C bonds of equal length, intermediate between a C–C double and single bond. Cyclooctatetraene is not planar and not aromatic so its double bonds are localized.

cyclooctatetraene

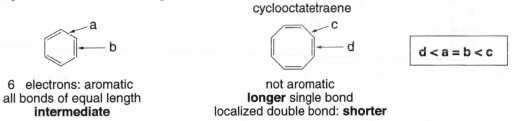

a
b

c
d

d < a = b < c

6 electrons: aromatic
all bonds of equal length
intermediate

not aromatic
longer single bond
localized double bond: **shorter**

17.33

:N≡N
N–N
H
purine

sp^2 hybridized but
with lone pair in p orbital.

a. Each N atom is sp^2 hybridized.

b. The three unlabeled N atoms are sp^2 hybridized with lone pairs in one of the sp^2 hybrid orbitals. The labeled N has its lone pair in a p orbital.

c. 10 π electrons

d. Purine is cyclic, planar, completely conjugated, and has 10 π electrons [4(2) + 2] so it is aromatic.

17.34

C

a. 16 total electrons
b. 14 electrons delocalized in the ring. [Note: Two of the electrons in the triple bond are localized between two C's, perpendicular to the π electrons delocalized in the ring.]
c. By having 2 of the *p* orbitals of the C–C triple bond co-planar with the *p* orbitals of all the C=C's, the total number of electrons delocalized in the ring is 14. 4(3) + 2 = 14, giving it the right number of electrons to be **aromatic.**

17.35 The rate of an S_N1 reaction increases with increasing stability of the intermediate carbocation.

increasing reactivity

4 electrons
antiaromatic
very *unstable* intermediate

2° carbocation

6 electrons
aromatic
very *stable* intermediate

The aromatic carbocation is delocalized over the whole ring making it a very stable intermediate and most easily formed in an S_N1 reaction.

increasing stability

17.36

17.37 α-Pyrone reacts like benzene because it is aromatic. A second resonance structure can be drawn showing how the ring has six π electrons. Thus, α-pyrone undergoes reactions characteristic of aromatic compounds; that is, substitution rather than addition.

α-pyrone

6 π electrons

17.38

a.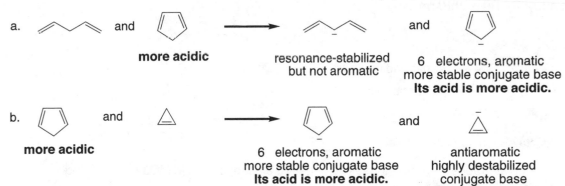

cyclopropenyl radical

b.

pyrrole

c.

phenanthrene

17.39

Naphthalene can be drawn as three resonance structures:

In two of the resonance structures bond (a) is a double bond, and bond (b) is a single bond. Therefore, bond (b) has more single bond character, making it longer.

17.40 The compound with the more stable conjugate base is the stronger acid. Draw and compare the conjugate bases of each pair of compounds.

conjugate bases

a.   and   **more acidic**   ⟶   and

resonance-stabilized but not aromatic

6 electrons, aromatic more stable conjugate base **Its acid is more acidic.**

b.   and   **more acidic**   ⟶   and

6 electrons, aromatic more stable conjugate base **Its acid is more acidic.**

antiaromatic highly destabilized conjugate base

17.41

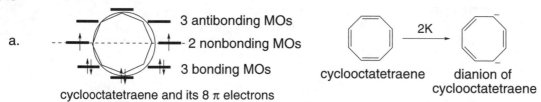

indene   + NaNH₂ ——→   + NH₃

Na⁺

The conjugate base of indene has 10 electrons making it aromatic
and very stable. Therefore, indene is more acidic than many hydrocarbons.

17.42

A		pyrrole
conjugate acid		
more acidic		

B		pyridine
conjugate acid		
less acidic		

Loss of a proton from **A** (which is not aromatic)
gives two electrons to N, so pyrrole has six π
electrons that can then delocalize in the five-
membered ring, making it aromatic. This makes
deprotonation a highly favorable process.

Both **B** and its conjugate base (pyridine) are
aromatic. Since **B** has six π electrons, it is already
aromatic to begin with, so there is less to be gained
by deprotonation, and **B** is thus less acidic than **A**.

17.43

a.

3 antibonding MOs
2 nonbonding MOs
3 bonding MOs

cyclooctatetraene and its 8 π electrons

cyclooctatetraene   2K ——→   dianion of
cyclooctatetraene

b. Even if cyclooctatetraene were flat, it has two unpaired electrons in it HOMOs (nonbonding
MOs) so it cannot be aromatic.

c. The dianion has 10 electrons.

d. The two additional electrons fill the nonbonding MOs; that is, all the bonding and
nonbonding MOs are filled with electrons in the dianion.

e. The dianion is aromatic since its HOMOs are completely filled, and it has no electrons in
antibonding MOs.

17.44

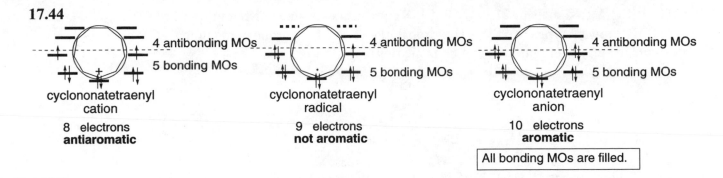

cyclononatetraenyl cation	cyclononatetraenyl radical	cyclononatetraenyl anion
8 electrons	9 electrons	10 electrons
antiaromatic	**not aromatic**	**aromatic**

All bonding MOs are filled.

17.45 Use the directions from Answer 17.6.

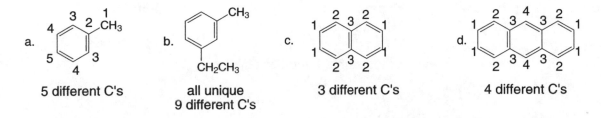

a. 5 different C's

b. all unique 9 different C's

c. 3 different C's

d. 4 different C's

17.46 Draw the three isomers and count the different types of carbon in each. Then match the structures with the data.

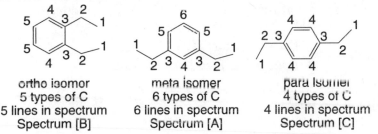

ortho isomer	meta isomer	para isomer
5 types of C	6 types of C	4 types of C
5 lines in spectrum	6 lines in spectrum	4 lines in spectrum
Spectrum [B]	Spectrum [A]	Spectrum [C]

17.47

a. $C_{10}H_{14}$: IR absorptions at 3150–2850 (sp^2 and sp^3 hybridized C–H), 1600, and 1500 (due to a benzene ring) cm^{-1}.

^{1}H NMR data:

Absorption	ppm	# of H's	Explanation
doublet	1.2	6	6H's adjacent to 1H
singlet	2.3	3	CH$_3$
septet	3.1	1	1H adjacent to 6H's
multiplet	7–7.4	4	a disubstituted benzene ring

$(CH_3)_2CH$ group

You can't tell from these data where the two groups are on the benzene ring. They are not para. That usually gives two sets of distinct peaks (resembling two doublets) so there are two possible structures—ortho and meta isomers.

or

b. C_9H_{12}: ^{13}C NMR signals at 21, 127, and 138 ppm → means three different types of C's.
^{1}H NMR shows 2 types of H's: 9H's probably means 3 CH_3 groups; the other 3 H's are very deshielded so they are bonded to a benzene ring.
Only one possible structure fits:

c. C_8H_{10}:IR absorptions at 3108–2875 (sp^2 and sp^3 hybridized C–H), 1606, and 1496 (due to a benzene ring) cm^{-1}.
^{1}H NMR data:

Absorption	ppm	# of H's	Explanation	Structure:
triplet	1.3	3	3H's adjacent to 2H's	
quartet	2.7	2	2H's adjacent to 3H's	
multiplet	7.3	5	a monosubstituted benzene ring	

17.48

a. Compound **A**: Molecular formula $C_8H_{10}O$.
IR absorption at 3150–2850 (sp^2 and sp^3 hybridized C–H) cm^{-1}.
^{1}H NMR data:

Absorption	ppm	# of H's	Explanation	Structure:
triplet	1.4	3	3H's adjacent to 2H's	
quartet	3.95	2	2H's adjacent to 3H's	
multiplet	6.8–7.3	5	a monosubstituted benzene ring	

b. Compound **B**: Molecular formula $C_9H_{10}O_2$.
IR absorption at 1669 (C=O) cm^{-1}.
^{1}H NMR data:

Absorption	ppm	# of H's	Explanation	Structure:
singlet	2.5	3	CH_3 group	
singlet	3.8	3	CH_3 group	
doublet	6.9	2	2H's on a benzene ring	
doublet	7.9	2	2H's on a benzene ring	

17.49

The induced magnetic field by the circulating π electrons opposes the applied field in this vicinity, shifting the absorption upfield to a lower chemical shift than other sp^3 C–H protons.
In this case the protons absorb upfield from TMS, an unusual phenomenon for C–H protons.

Induced magnetic field

H H

2.84 ppm ⟶ H

H

–0.6 ppm
(upfield from TMS)

Induced magnetic field

B_0

The induced magnetic field by the circulating π electrons reinforces the applied field in this vicinity, shifting the absorption downfield to a higher chemical shift.

17.50 A second resonance structure for **A** shows that the ring is completely conjugated and has 6 π electrons, making it aromatic and especially stable. A similar charge-separated resonance structure for **B** makes the ring completely conjugated, but gives the ring 4 π electrons, making it antiaromatic and especially unstable.

A   6 electrons   **B**   4 electrons
                aromatic   antiaromatic
                stable   not stable

17.51

$H_2\ddot{N}$←N1

N2 ⟶ :N   N3

N2: The electron pair occupies an sp^3 hybrid orbital on the N atom.
N2: The electron pair is contained in an sp^2 hybrid orbital and is not delocalized over the five-membered ring.
N3: The electron pair is in a p orbital, delocalized on the five-membered ring to make it aromatic.

N1: The electron pair occupies an sp^3 hybrid orbital on the N atom.

Basicity depends on the ability of an atom to donate an electron pair. Electron pairs that are delocalized in a π system to make it aromatic are less available for electron donation, making them less basic. Basicity is also affected by percent s-character. The higher the percent s-character, the weaker the base. Electrons are held closer to the nucleus so they are less available for electron donation. Thus, in order of increasing basicity: N3 < N2 < N1.

17.52 Resonance structures for triphenylene:

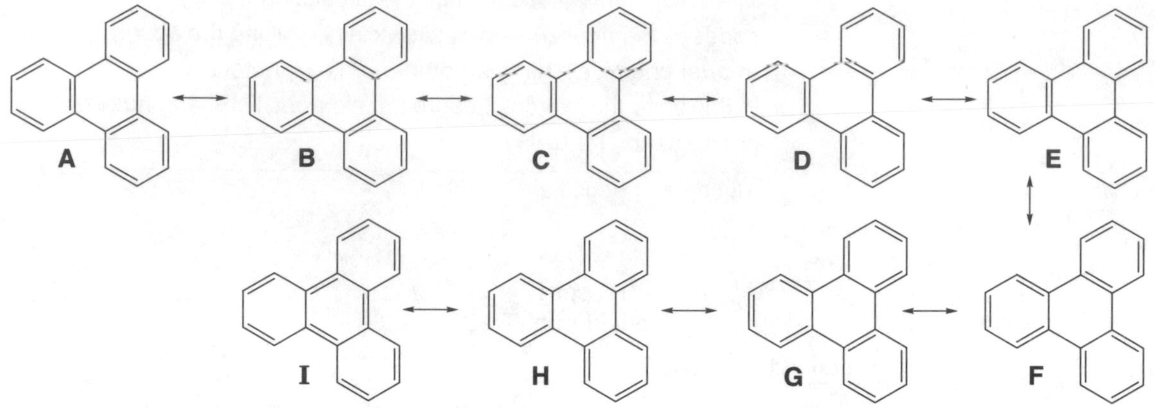

Resonance structures **A–H** all keep three double and three single bonds in the three six-membered rings on the periphery of the molecule. This means that each ring behaves like an isolated benzene ring undergoing substitution rather than addition because the π electron density is delocalized within each six-membered ring. Only resonance structure **I** does not have this form. Each C–C bond of triphenylene has four (or five) resonance structures in which it is a single bond and four (or five) resonance structures in which it is a double bond.

Resonance structures for phenanthrene:

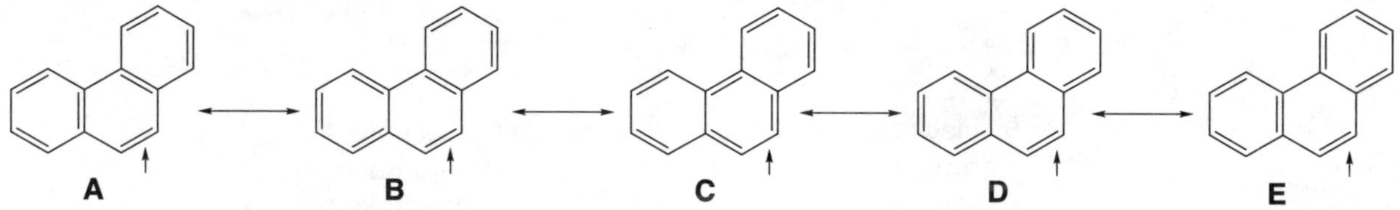

With phenanthrene, however, four of the five resonance structures keep a double bond at the labeled C's. (Only **C** does not.) This means that these two C's have more double bond character than other C–C bonds in phenanthrene, making them more susceptible to addition rather than substitution.

17.53

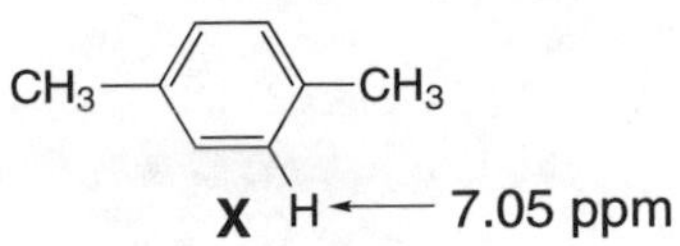

CH$_3$— (ring) —CH$_3$

X H ←— 7.05 ppm

CH$_3$ groups are electron donating. Increasing the electron density of the ring shields the protons and shifts the absorption slightly upfield.

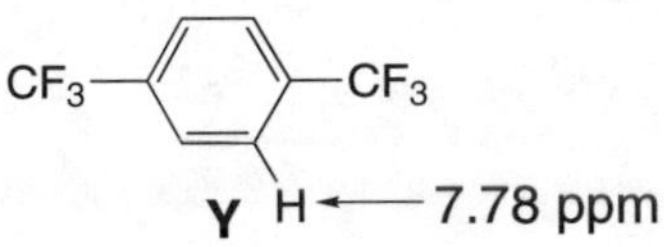

CF$_3$— (ring) —CF$_3$

Y H ←—7.78 ppm

The 3 electronegative F's make CF$_3$ an electron withdrawing group. Decreasing the electron density of the ring deshields the protons and the absorption goes slightly downfield.

When F's are directly bonded to the benzene ring, two conflicting factors come into play. Since F is very electronegative, it withdraws electron density from the ring. But, F atoms also contain lone pairs of electrons that can be donated to the ring by resonance. For example:

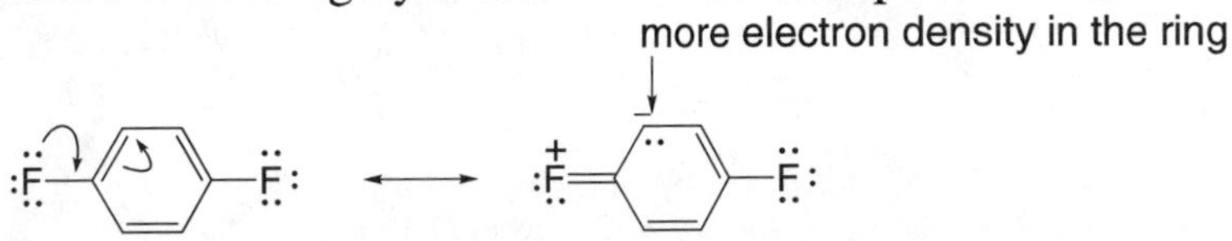

more electron density in the ring

:F̈—(ring)—F̈: ←——→ :F̈=(ring)—F̈:

This increases the electron density of the ring. On balance, these factors just about cancel, so the absorption occurs at ~7 ppm.

Chapter 18: Electrophilic aromatic substitution

♦ Mechanism of electrophilic aromatic substitution (18.2)

- Electrophilic aromatic substitution follows a two-step mechanism. Reaction of the aromatic ring with an electrophile forms a carbocation, and loss of a proton regenerates the aromatic ring.
- The first step is rate-determining.
- The intermediate carbocation is stabilized by resonance; a minimum of three resonance structures can be drawn. The positive charge is always located ortho or para to the new C–E bond.

| (+) *ortho* to E | (+) *para* to E | (+) *ortho* to E |

♦ Three rules describing the reactivity and directing effects of common substituents (18.7–18.9)

[1] All ortho, para directors except the halogens activate the benzene ring.
[2] All meta directors deactivate the benzene ring.
[3] The halogens deactivate the benzene ring.

♦ Summary of substituent effects in electrophilic aromatic substitution (18.6–18.9)

	Substituent	Inductive effect	Resonance effect	Reactivity	Directing effect
[1]	R = alkyl	donating	none	activating	ortho, para
[2]	Z = N or O	withdrawing	donating	activating	ortho, para
[3]	X = halogen	withdrawing	donating	deactivating	ortho, para
[4]	Y (δ⁺ or +)	withdrawing	withdrawing	deactivating	meta

♦ Five examples of electrophilic aromatic substitution

[1] Halogenation–Replacement of H by Cl or Br (18.3)

- Polyhalogenation occurs on benzene rings substituted by OH and NH$_2$ (and related substituents) (18.10A).

[2] Nitration–Replacement of H by NO$_2$ (18.4)

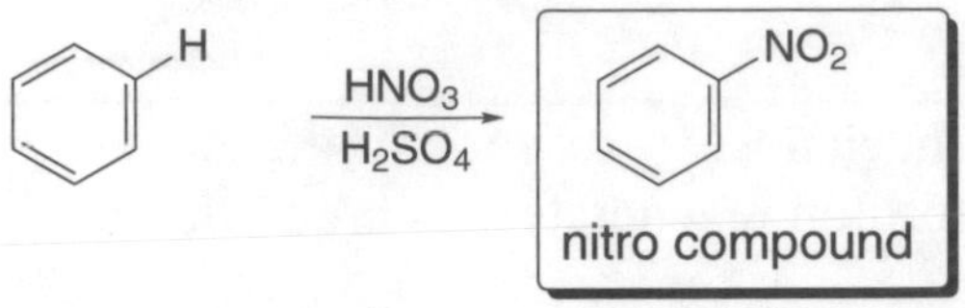

[3] Sulfonation–Replacement of H by SO$_3$H (18.4)

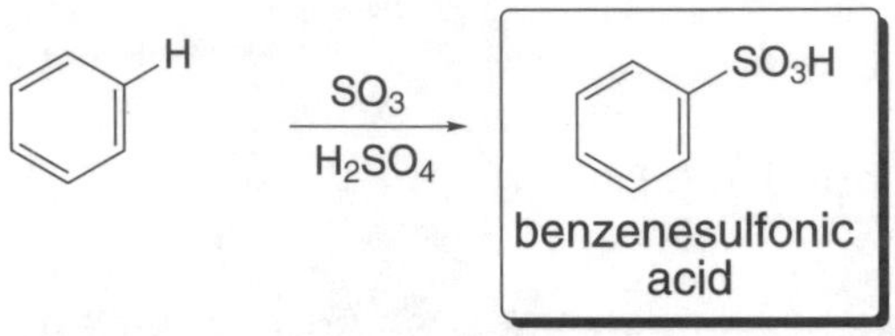

[4] Friedel–Crafts alkylation–Replacement of H by R (18.5)

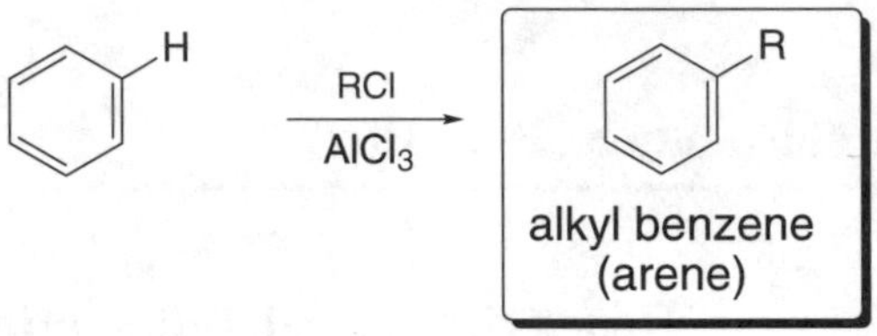

- Rearrangements can occur.
- Vinyl halides and aryl halides are unreactive.
- The reaction does not occur on benzene rings substituted by meta deactivating groups or NH$_2$ groups (18.10B).
- Polyalkylation can occur.

Variations:

[1] with alcohols

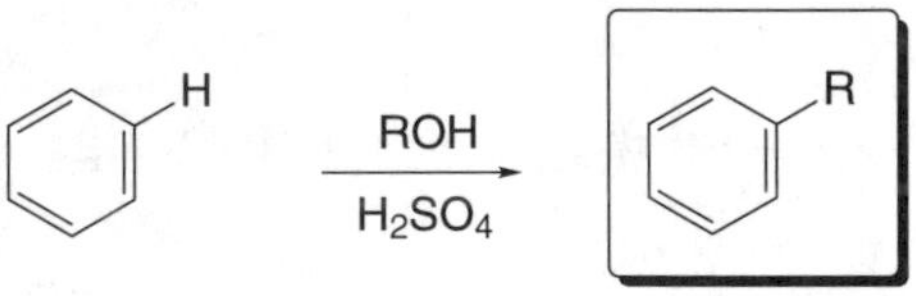

[2] with alkenes

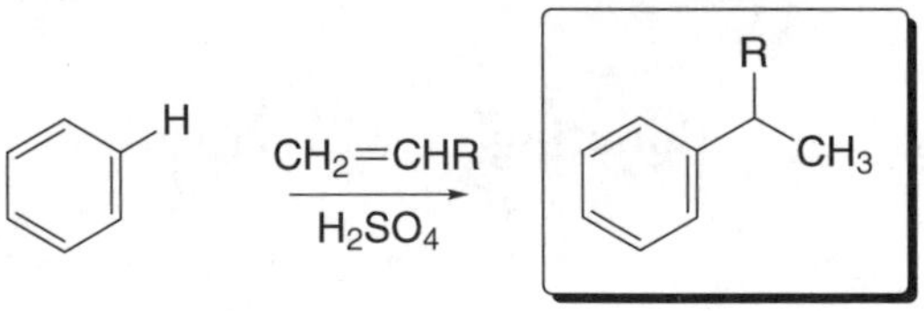

[5] Friedel–Crafts acylation–Replacement of H by RCO (18.5)

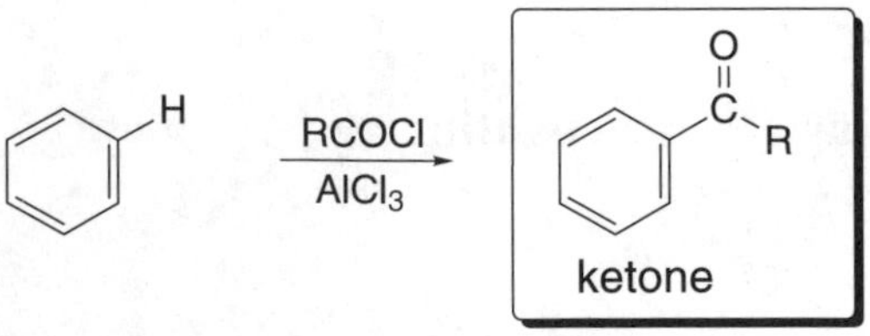

- The reaction does not occur on benzene rings substituted by meta deactivating groups or NH$_2$ groups (18.10B).

◆ Other reactions of benzene derivatives

[1] Benzylic halogenation (18.13)

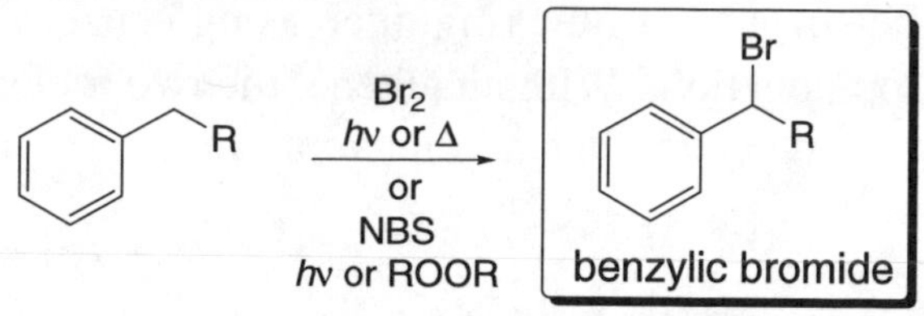
R
Br_2
hν or Δ
or
NBS
hν or ROOR
Br
R
benzylic bromide

[2] Oxidation of alkyl benzenes (18.14A)

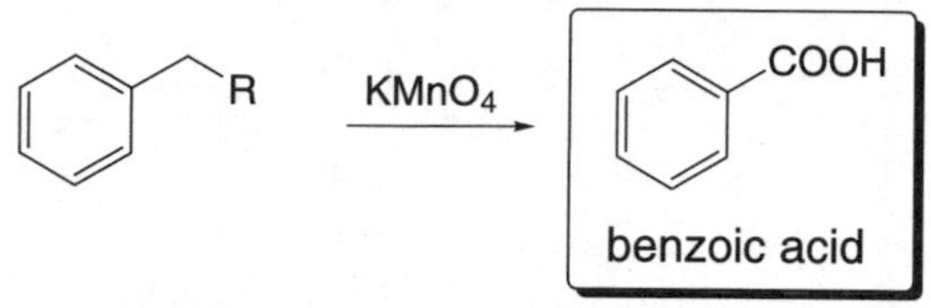
R
$KMnO_4$
COOH
benzoic acid

- A benzylic C–H bond is needed for reaction.

[3] Reduction of ketones to alkyl benzenes (18.14B)

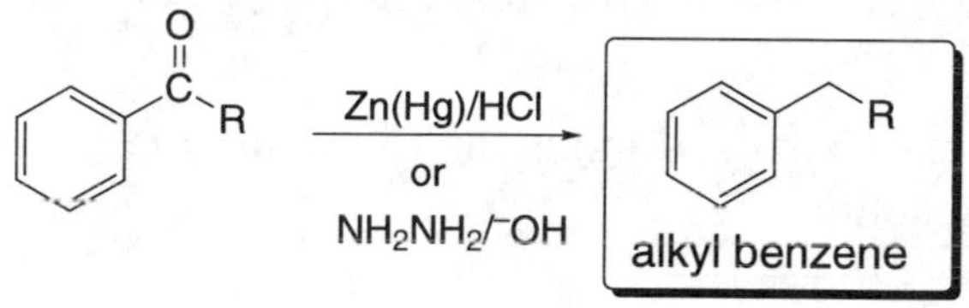
O
C
R
Zn(Hg)/HCl
or
$NH_2NH_2/^{-}OH$
R
alkyl benzene

[4] Reduction of nitro groups to amino groups (18.14C)

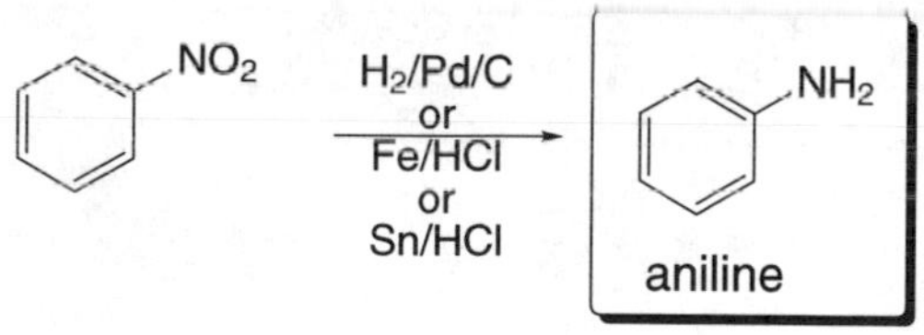
NO_2
$H_2/Pd/C$
or
Fe/HCl
or
Sn/HCl
NH_2
aniline

18.1 The π electrons of benzene are delocalized over the six atoms of the ring, increasing benzene's stability and making them less available for electron donation. With an alkene, the two π electrons are localized between the two C's making them more nucleophilic and thus more reactive with an electrophile than the delocalized electrons in benzene.

18.2 In drawing the resonance structures, the charge is always located ortho or para to the one sp^3 hybridized atom on the ring.

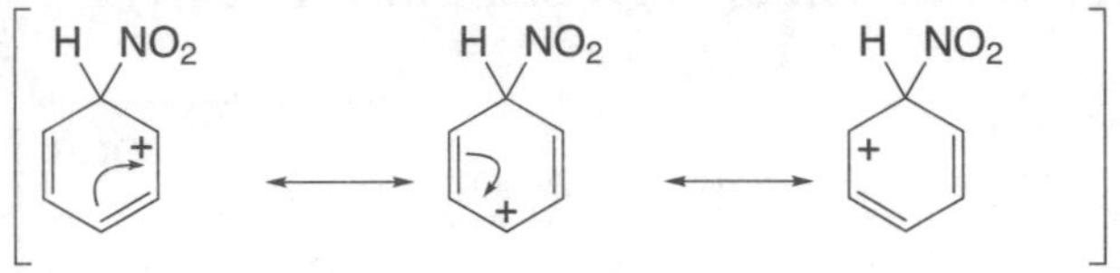

18.3

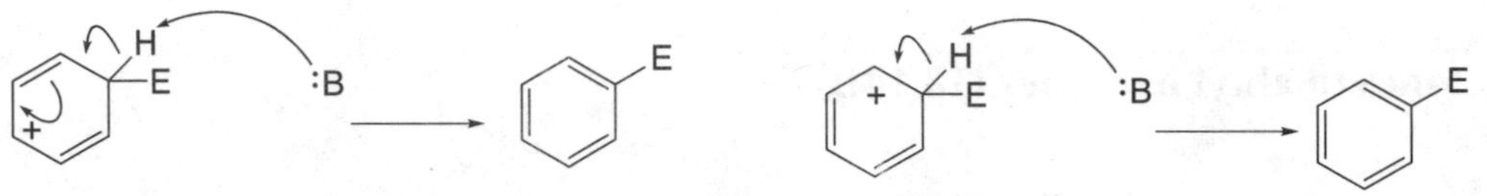

18.4 Addition of Cl_2 with $FeCl_3$ as the catalyst occurs in two parts. First is the formation of an electrophile, followed by a two-step substitution reaction.

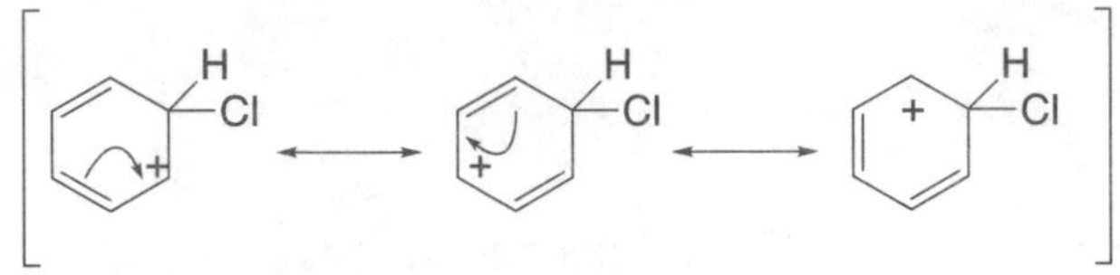

resonance-stabilized carbocation

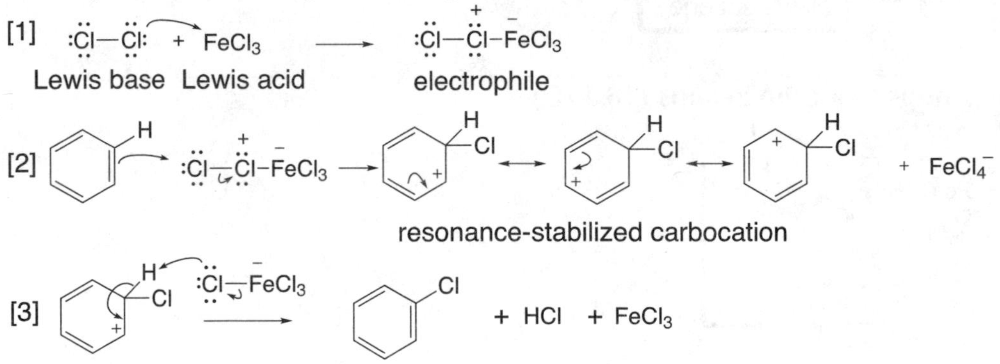

18.5 There are two parts in the mechanism. The first part is formation of an electrophile. The second part is a two-step substitution reaction.

electrophile

resonance-stabilized carbocation

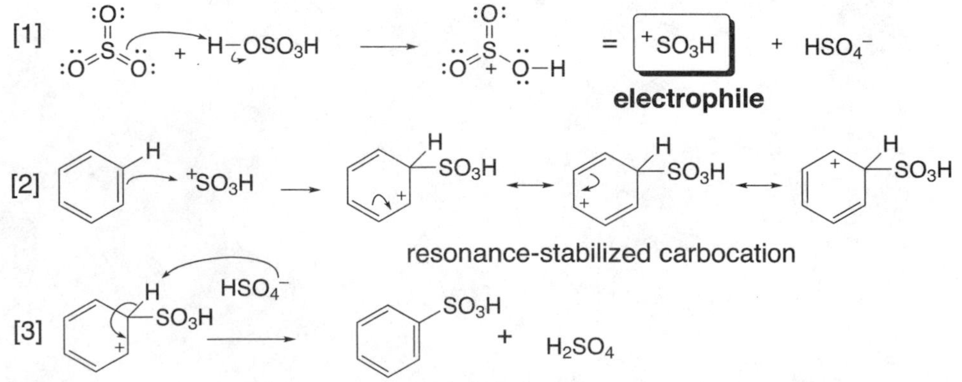

18.6 Friedel–Crafts alkylation results in the transfer of an alkyl group from a halogen to a benzene ring. In Friedel–Crafts acylation an acyl group is transferred from a halogen to a benzene ring.

a. benzene + $(CH_3)_2CHCl$ $\xrightarrow{AlCl_3}$ isopropylbenzene [$CH(CH_3)_2$]

b. benzene + cyclohexyl chloride (Cl) $\xrightarrow{AlCl_3}$ cyclohexylbenzene

c. benzene + CH_3CH_2—C(=O)—Cl $\xrightarrow{AlCl_3}$ phenyl ethyl ketone (C(=O)CH$_2$CH$_3$)

18.7 Remember that an acyl group is transferred from a Cl atom to a benzene ring. To draw the acid chloride, substitute a Cl for the benzene ring.

a. Ph—C(=O)—$CH_2CH_2CH(CH_3)_2$ $\Longrightarrow$ Cl—C(=O)—$CH_2CH_2CH(CH_3)_2$

b. Ph—C(=O)—Ph $\Longrightarrow$ Cl—C(=O)—Ph

c. cyclopentyl—C(=O)—Ph $\Longrightarrow$ cyclopentyl—C(=O)—Cl

18.8

[1] CH_3CH_2—$\overset{..}{\underset{..}{Cl}}:$ + AlCl$_3$ $\longrightarrow$ CH_3CH_2—$\overset{+}{\underset{..}{Cl}}$—$\overset{-}{AlCl_3}$
electrophile

[2] benzene (H) + CH_3CH_2—$\overset{+}{\underset{..}{Cl}}$—$\overset{-}{AlCl_3}$ $\longrightarrow$ [arenium ion with H and CH$_2$CH$_3$] $\longleftrightarrow$ [arenium ion] $\longleftrightarrow$ [arenium ion]
resonance-stabilized carbocation   +   AlCl$_4^-$

[3] [arenium ion with H and CH$_2$CH$_3$] :$\overset{-}{Cl}$—AlCl$_3$ $\longrightarrow$ ethylbenzene (CH$_2$CH$_3$) + HCl + AlCl$_3$

18.9 To be reactive in a Friedel–Crafts alkylation reaction, the X must be bonded to an sp^3 hybridized carbon atom.

a. cyclohexenyl–Br, sp^2 — **unreactive**

b. cyclohexenyl–CH$_2$Br, sp^3 — **reactive**

c. phenyl–Br, sp^2 — **unreactive**

d. benzyl–CH$_2$Br, sp^3 — **reactive**

18.10 The product has an "unexpected" carbon skeleton, so rearrangement must have occurred.

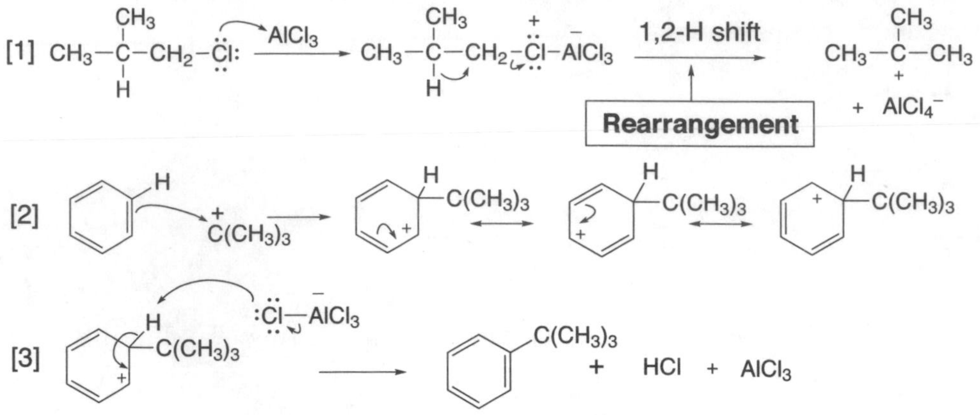

18.11 Both alkenes and alcohols can form carbocations for Friedel–Crafts alkylation reactions.

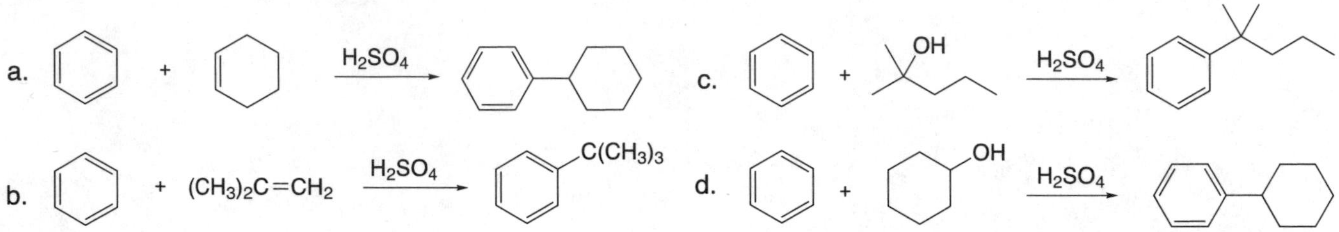

18.12

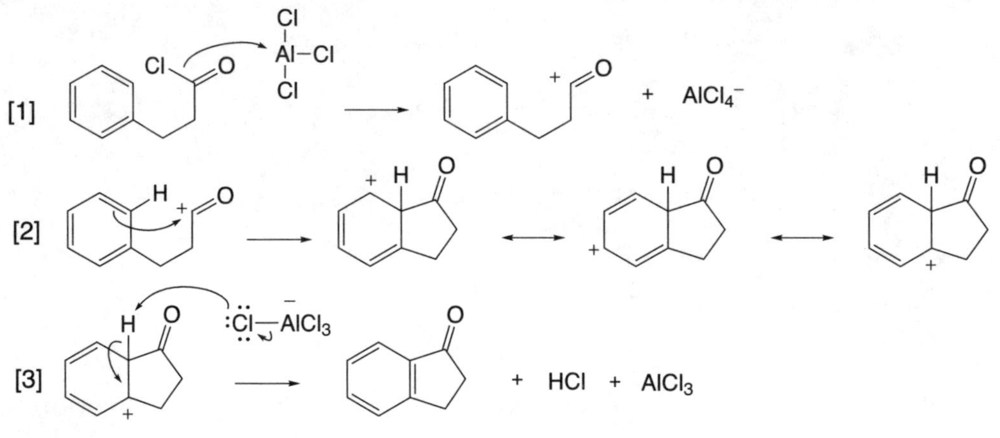

18.13

a. —CH$_2$CH$_2$CH$_2$CH$_3$
alkyl group
electron donating

b. —Br
halide
electron withdrawing

c. —OCH$_2$CH$_3$
electronegative O
electron withdrawing

18.14 Electron donating groups place a negative charge in the benzene ring. Draw the resonance structures to show how –OCH$_3$ puts a negative charge in the ring. Electron withdrawing groups place a positive charge in the benzene ring. Draw the resonance structures to show how –COCH$_3$ puts a positive charge in the ring.

a.

b.

18.15 To classify each substituent, look at the atom directly bonded to the benzene ring. All R groups and Z groups (except halogens) are electron donating. All groups with a positive charge, δ^+, or halogens are electron withdrawing.

a.

  lone pair on O
  electron donating

b.

  halogen
  electron withdrawing

c.

  R groups
  electron donating

18.16 First classify the starting material as: ortho, para activating, ortho, para deactivating, or meta deactivating. Then draw the products.

a.

  lone pair on O
  o,p activating

  ortho product   **para product**

b.

  halogen
  o,p deactivating

  ortho product   **para product**

c.

  meta deactivating   **meta product**

18.17 **Electron donating groups** make the compound *react faster* than benzene in electrophilic aromatic substitution. **Electron withdrawing groups** make the compound *react more slowly* than benzene in electrophilic aromatic substitution.

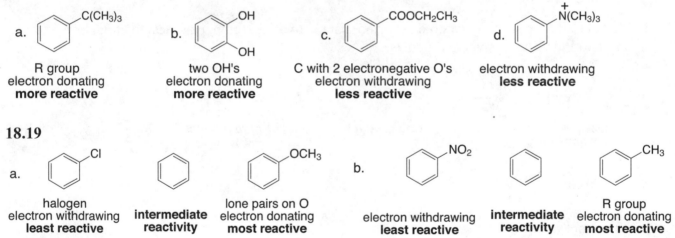

a.

electron withdrawing
reacts slower

b.

electron withdrawing
reacts slower

c.

lone pairs on O
electron donating
reacts faster

d.

halogen
electron withdrawing
reacts slower

e.

R group
electron donating
reacts faster

18.18 **Electron donating groups** make the compound *more reactive* than benzene in electrophilic aromatic substitution. **Electron withdrawing groups** make the compound *less reactive* than benzene in electrophilic aromatic substitution.

a.

C(CH₃)₃

R group
electron donating
more reactive

b.

OH
OH

two OH's
electron donating
more reactive

c.

COOCH₂CH₃

C with 2 electronegative O's
electron withdrawing
less reactive

d.

⁺N(CH₃)₃

electron withdrawing
less reactive

18.19

a.

Cl

halogen
electron withdrawing
least reactive

intermediate
reactivity

OCH₃

lone pairs on O
electron donating
most reactive

b.

NO₂

electron withdrawing
least reactive

intermediate
reactivity

CH₃

R group
electron donating
most reactive

18.20 Especially stable resonance structures have all atoms with an octet. Carbocations with additional electron donor R groups are also more stable structures. Especially unstable resonance structures have adjacent like charges.

a.

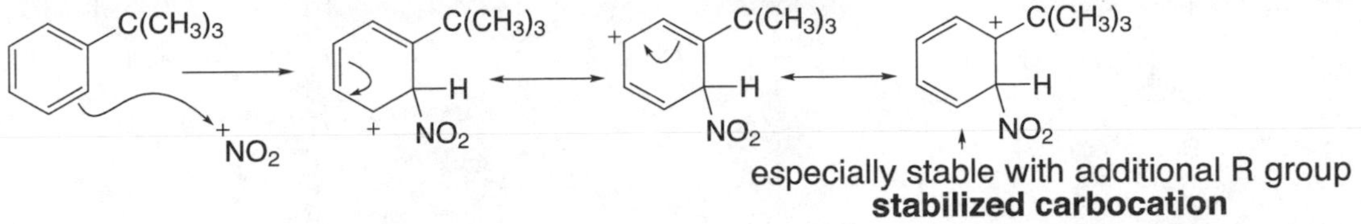

especially stable with additional R group
stabilized carbocation

b.

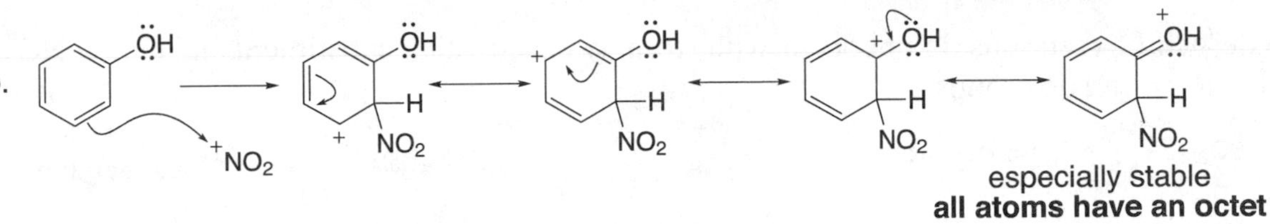

especially stable
all atoms have an octet

c.

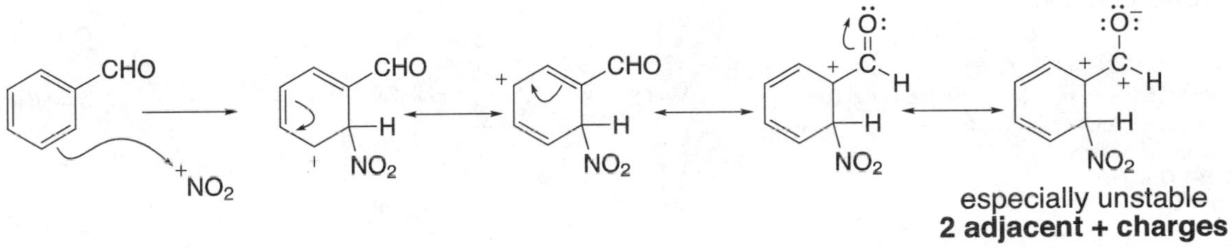

especially unstable
2 adjacent + charges

18.21

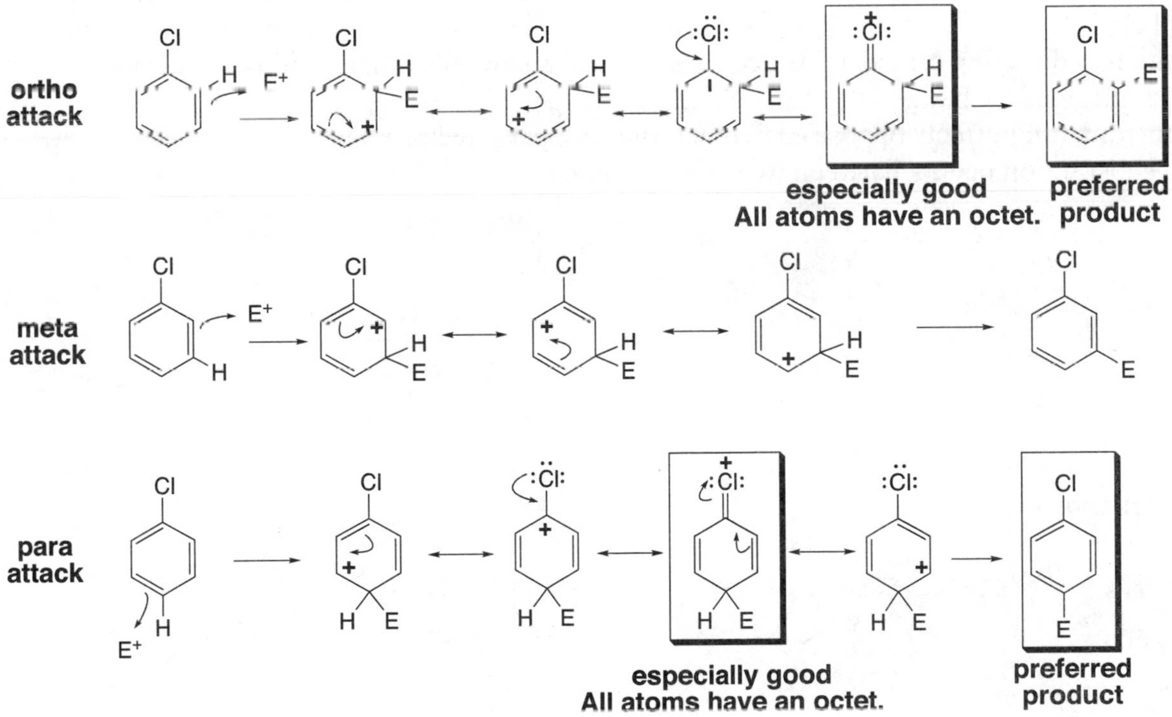

ortho attack

especially good
All atoms have an octet. **preferred product**

meta attack

para attack

especially good
All atoms have an octet.

preferred product

18.22 Draw the products of each reaction. Polyhalogenation occurs with highly activated benzene rings containing OH, NH$_2$, and related groups with a catalyst.

a.
$\dfrac{Cl_2}{FeCl_3}$

b.
$\xrightarrow{Cl_2}$
+

c.
$\dfrac{Cl_2}{FeCl_3}$
+

18.23 Friedel-Crafts reactions do not occur with strongly deactivating substituents including NO$_2$, or with NR$_2$, or NHR groups.

a.
—SO$_3$H $\dfrac{CH_3Cl}{AlCl_3}$ **no reaction**

strongly deactivating

c.
—N(CH$_3$)$_2$ $\dfrac{CH_3Cl}{AlCl_3}$ **no reaction**

b.
—Cl $\dfrac{CH_3Cl}{AlCl_3}$ + CH$_3$——Cl

Cl is an o,p
director.

d.
—NHCOCH$_3$ $\dfrac{CH_3Cl}{AlCl_3}$ —NHCOCH$_3$

+

CH$_3$——NHCOCH$_3$

18.24 To draw the product of reaction with HNO$_3$/H$_2$SO$_4$ and these disubstituted benzene derivatives remember:

- If the two directing effects reinforce each other, the new substituent will be on the position reinforced by both.
- If the directing effects oppose each other, the stronger director wins.
- No substitution occurs between two meta directors.

a.
o,p
$\dfrac{HNO_3}{H_2SO_4}$
meta

b.
o,p (strong)
$\dfrac{HNO_3}{H_2SO_4}$
o,p
oppose
products due to
OCH$_3$ directing effects
+

c.
o,p
meta
$\dfrac{HNO_3}{H_2SO_4}$
+

d.
o,p
$\dfrac{HNO_3}{H_2SO_4}$
o,p
+

18.25

a.

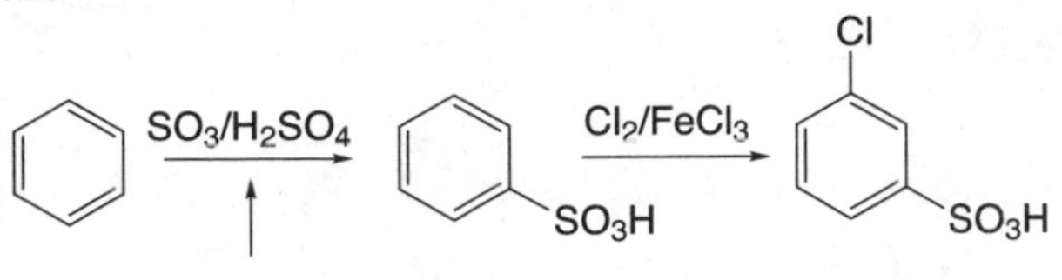

Put meta director on first.

c.

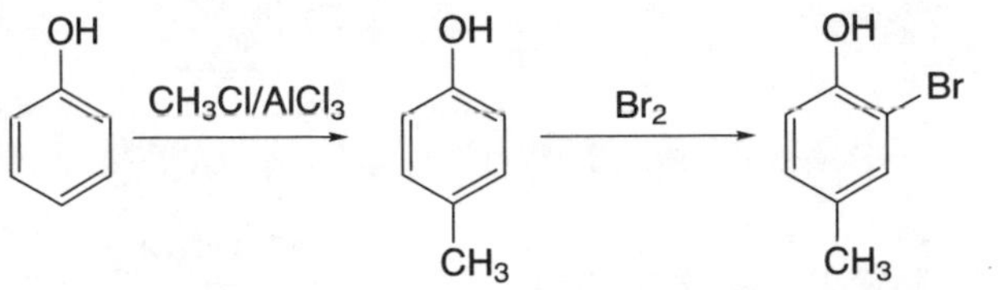

(+ ortho isomer) Br goes ortho to the stronger activator.

b.

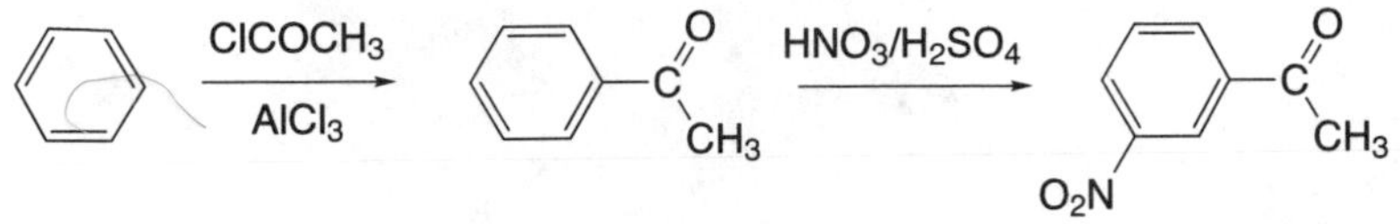

18.26 This reaction proceeds via a radical bromination mechanism and two radicals are possible: **A** (2° and benzylic) and **B** (1°). Since **B** (which leads to $C_6H_5CH_2CH_2Br$) is much less stable, this radical is not formed so only $C_6H_5CH(Br)CH_3$ is formed as product.

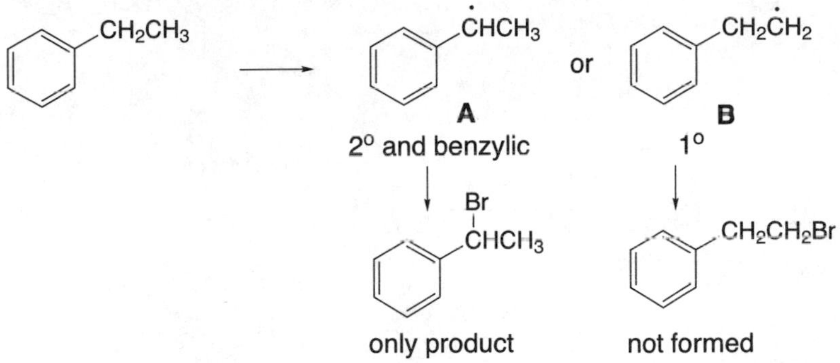

A
2° and benzylic

or

B
1°

only product   not formed

18.27 **Radical substitution** occurs at the carbon adjacent to the benzene ring (at the **benzylic position**).

a.

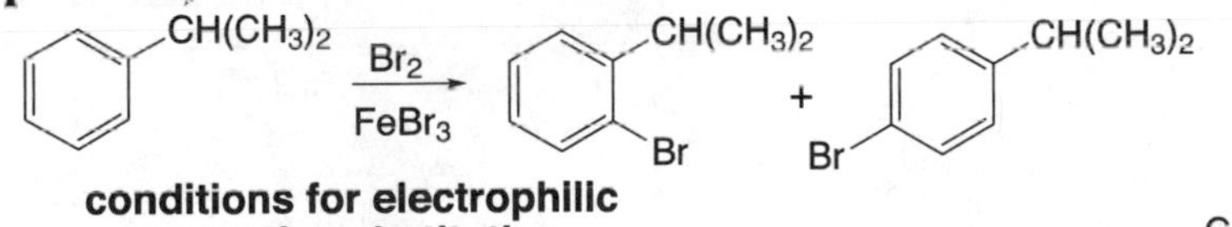

conditions for electrophilic aromatic substitution

b.

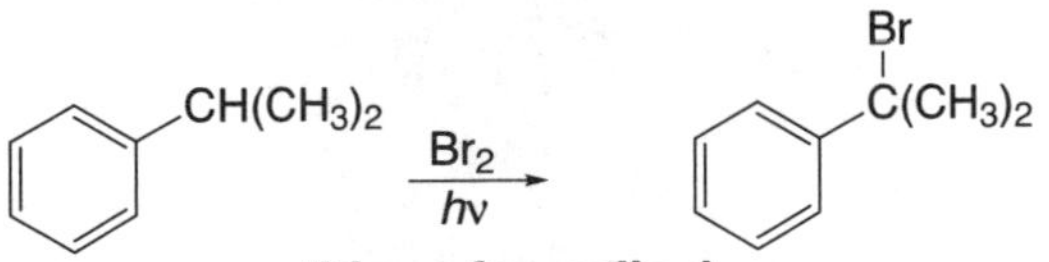

conditions for radical substitution

c.

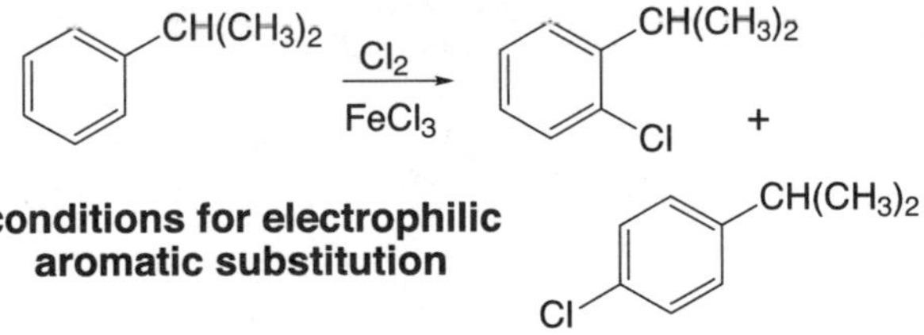

conditions for electrophilic aromatic substitution

18.28

a.

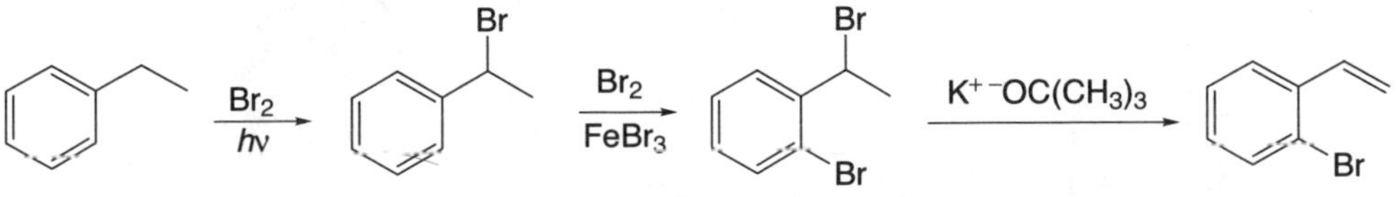

(+ para isomer)

b.

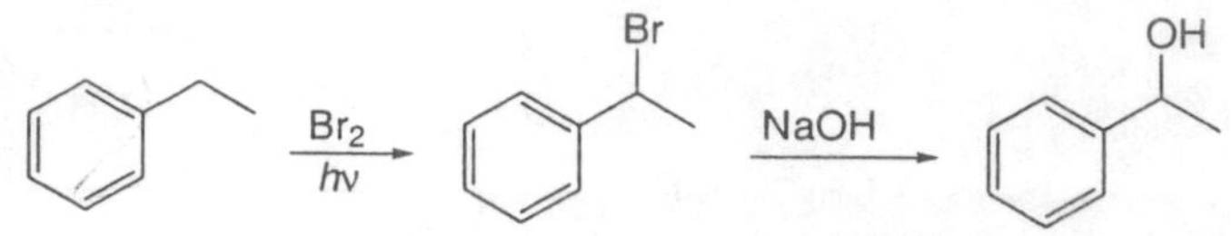

c.

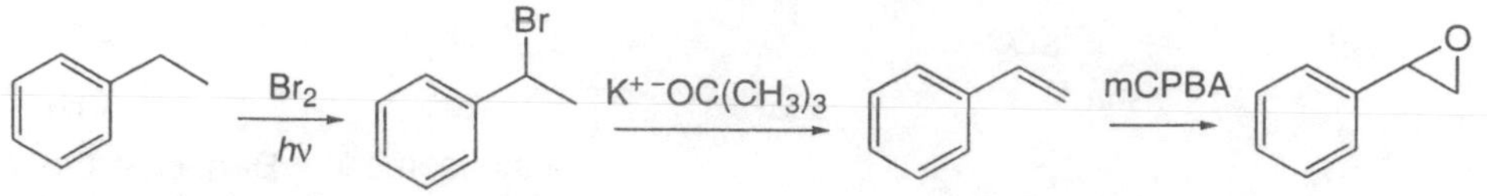

18.29

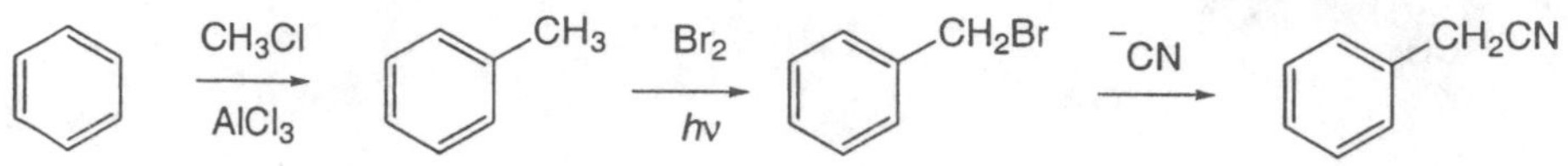

18.30 First use an acylation reaction, and the reduce the carbonyl group to form the alkyl benzenes.

a.

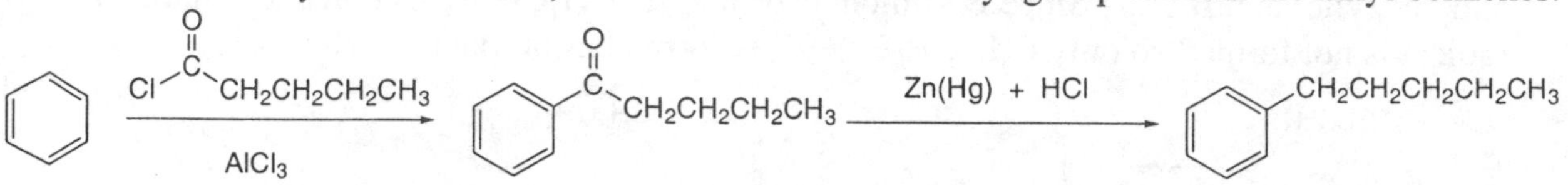

b.

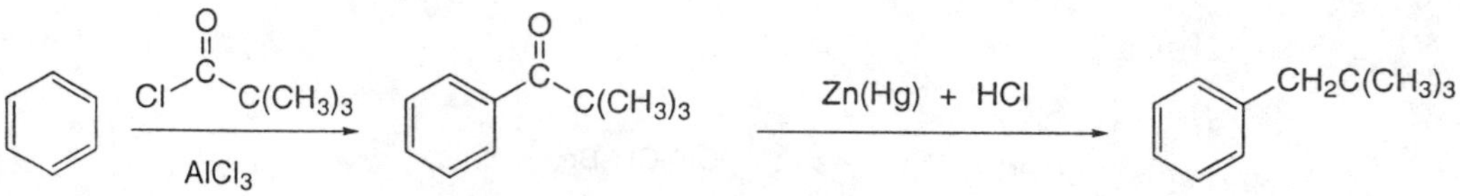

18.31

$C_6H_5C(CH_3)_3$ cannot be made by Friedel–Crafts acylation because there are no benzylic hydrogens. All products of Friedel–Crafts acylation followed by reduction have two benzylic hydrogens where the C=O was reduced. $C_6H_5C(CH_3)_3$ can be made by Friedel–Crafts alkylation using $(CH_3)_3CCl$ and $AlCl_3$.

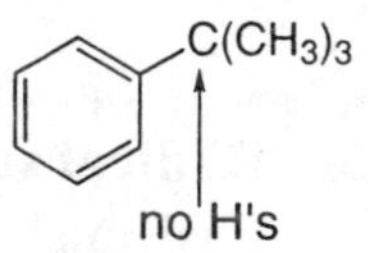

no H's

18.32

a.

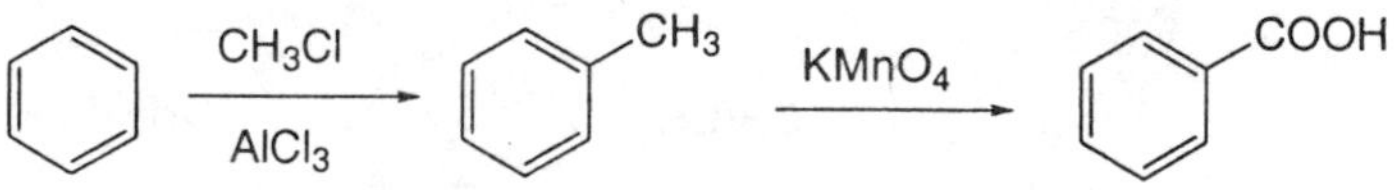

b.

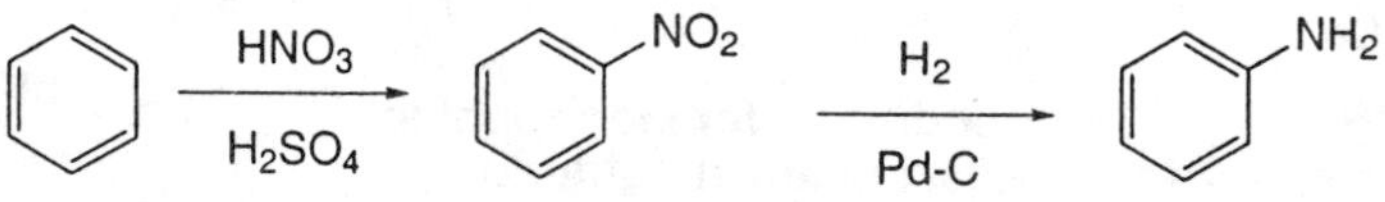

c.

(+ para isomer)

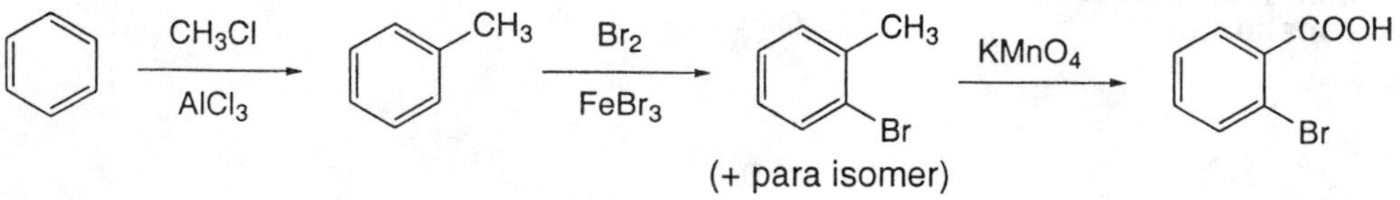

18.33

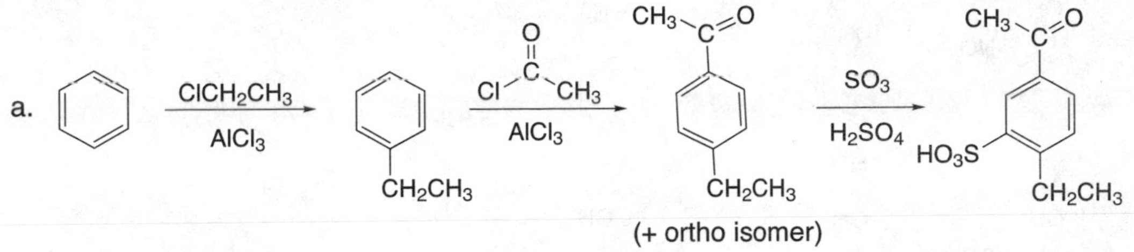

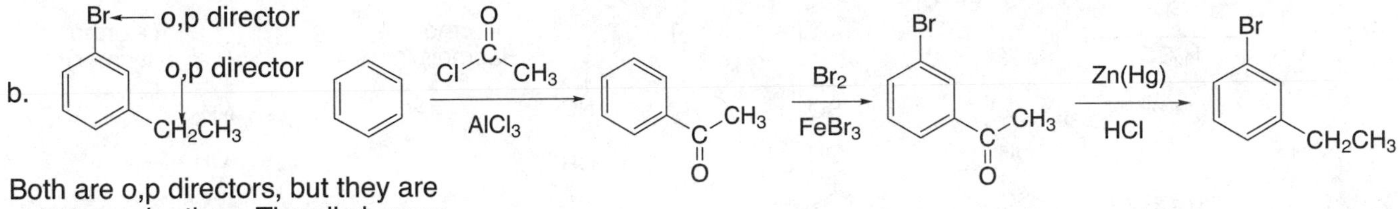

a.

(+ ortho isomer)

b.

Both are o,p directors, but they are
meta to each other. The alkyl group
must be obtained by reduction
of a carbonyl.

18.34 OH is an ortho, para director.

a.

g.

b.

h.

c.

i.

d.

j.

k.

e.

l.

f.

18.35 CN is a meta director that deactivates the benzene ring.

a.

c.

b.

d.

e.

18.36

a. $CH(CH_3)_2$ CH_3CH_2COCl / $AlCl_3$ → [2-isopropyl phenyl with $CH(CH_3)_2$ and $COCH_2CH_3$] + CH_3CH_2—CO—[para $CH(CH_3)_2$ phenyl]

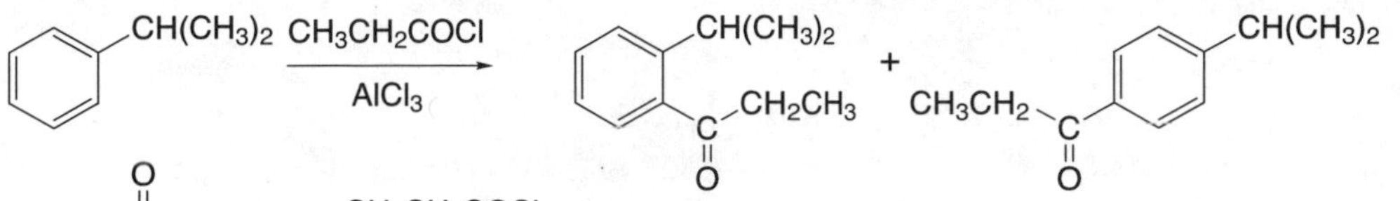

b. [phenyl—CO—$CH(CH_3)_2$] CH_3CH_2COCl / $AlCl_3$ → **no reaction**

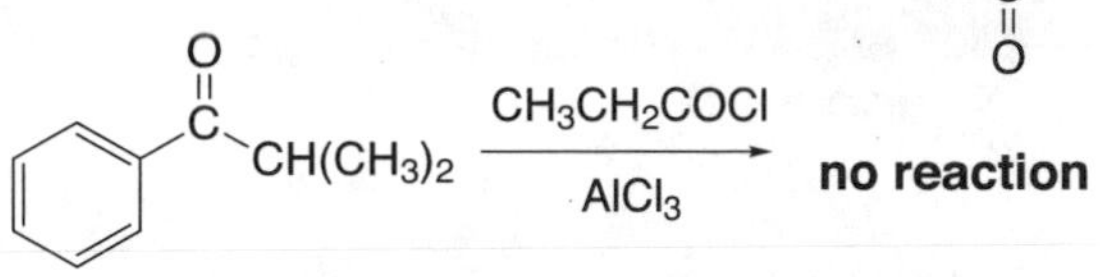

c. [phenyl—$N(CH_3)_2$] CH_3CH_2COCl / $AlCl_3$ → **no reaction**

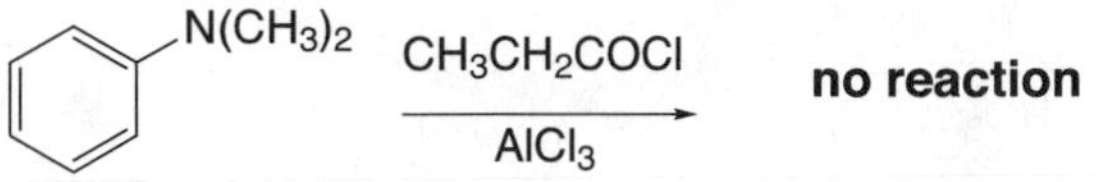

d. [phenyl—Br] CH_3CH_2COCl / $AlCl_3$ → [2-bromophenyl CO—CH_2CH_3] + CH_3CH_2—CO—[para Br phenyl]

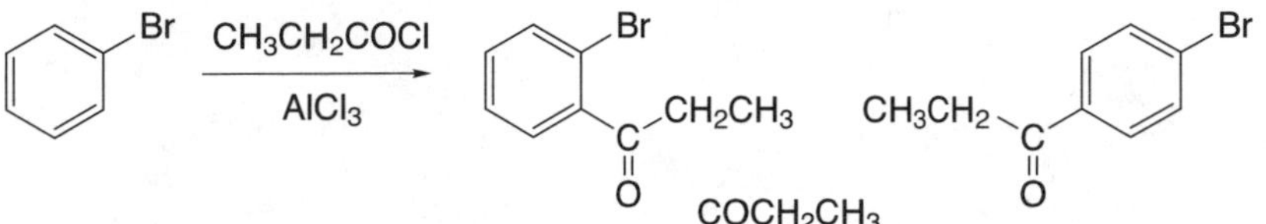

e. [phenyl—NH—CO—CH_3] CH_3CH_2COCl / $AlCl_3$ → [ortho $COCH_2CH_3$, NH—CO—CH_3 phenyl] + CH_3CH_2—CO—[para NH—CO—CH_3 phenyl]

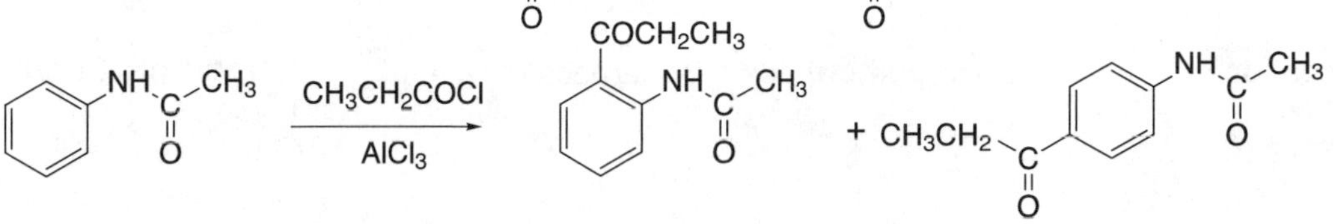

18.37

a. [HO, NO_2 phenyl] HNO_3 / H_2SO_4 → O_2N—[HO, NO_2 phenyl]—NO_2 + [HO, NO_2, NO_2 phenyl]

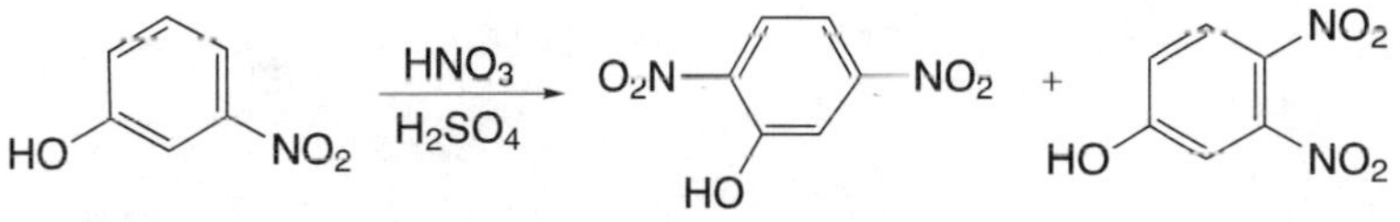

b. [CH_3, OH phenyl] SO_3 / H_2SO_4 → [CH_3, OH, SO_3H phenyl] + HO_3S—[CH_3, OH phenyl]

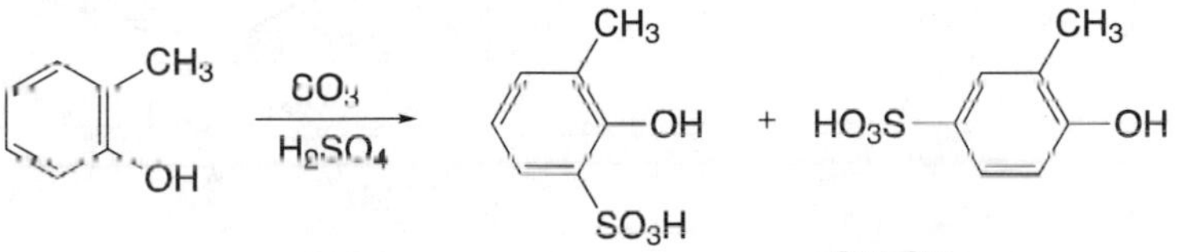

c. Cl—[phenyl]—$OCOCH_3$ CH_3CH_2Cl / $AlCl_3$ → Cl—[CH_2CH_3, $OCOCH_3$ phenyl]

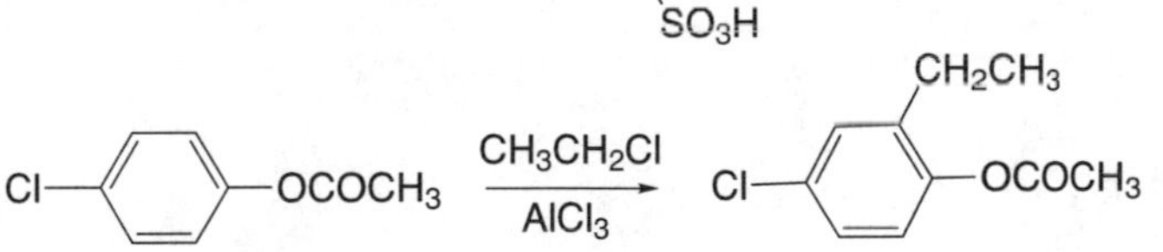

d. [CHO, CH_3 phenyl] Br_2 / $FeBr_3$ → [CHO, CH_3, Br phenyl] + Br—[CHO, CH_3 phenyl]

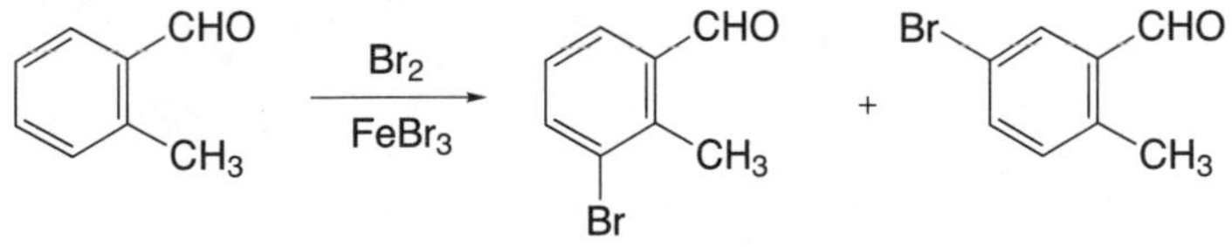

e. CH_3OCO—[phenyl]—$NHCOCH_3$ CH_3COCl / $AlCl_3$ → CH_3OCO—[$COCH_3$, $NHCOCH_3$ phenyl]

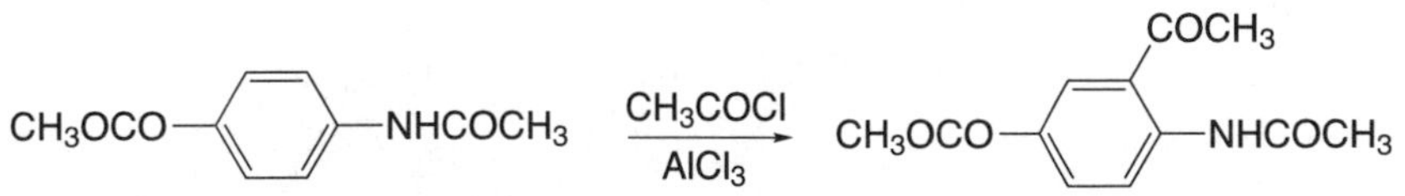

f.

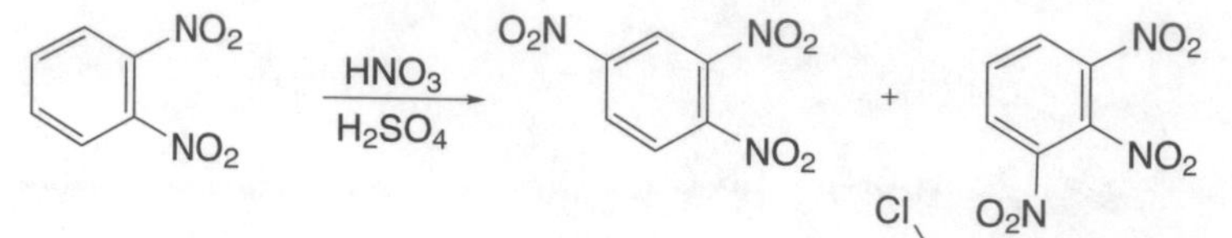

HNO₃ → H₂SO₄

$$HNO_3 / H_2SO_4$$

g.

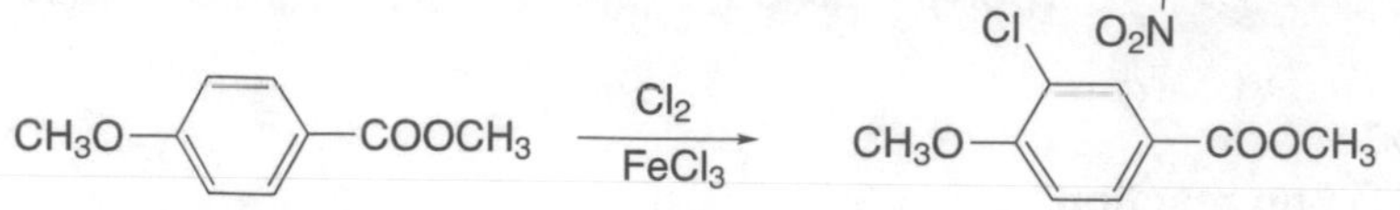

CH₃O—⟨ ⟩—COOCH₃ $\xrightarrow{Cl_2 / FeCl_3}$ CH₃O—⟨ ⟩—COOCH₃

h.

Br—⟨ ⟩—OCH₃ $\xrightarrow{SO_3 / H_2SO_4}$ Br—⟨ ⟩—OCH₃ (SO₃H) + Br—⟨ ⟩—OCH₃ (HO₃S)

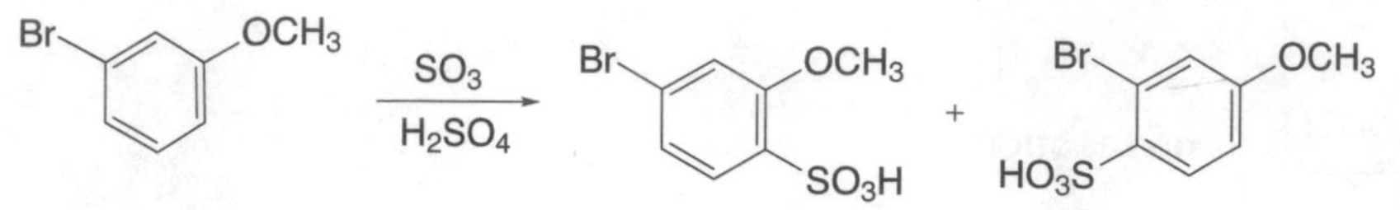

18.38 Watch out for rearrangements.

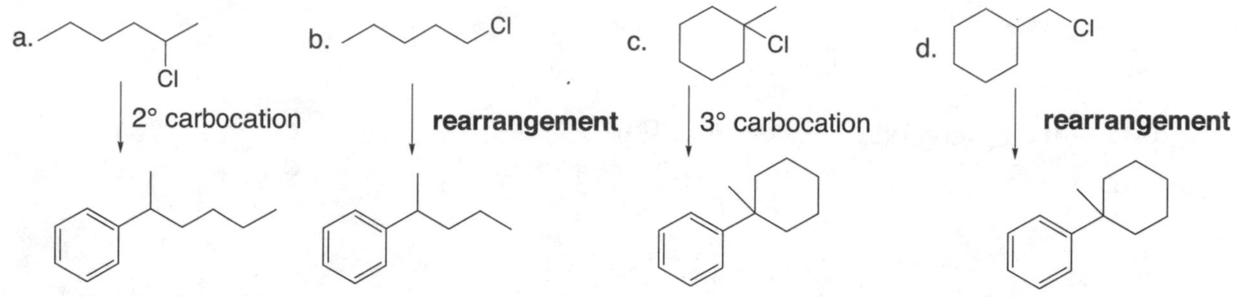

a. 2° carbocation

b. **rearrangement**

c. 3° carbocation

d. **rearrangement**

18.39

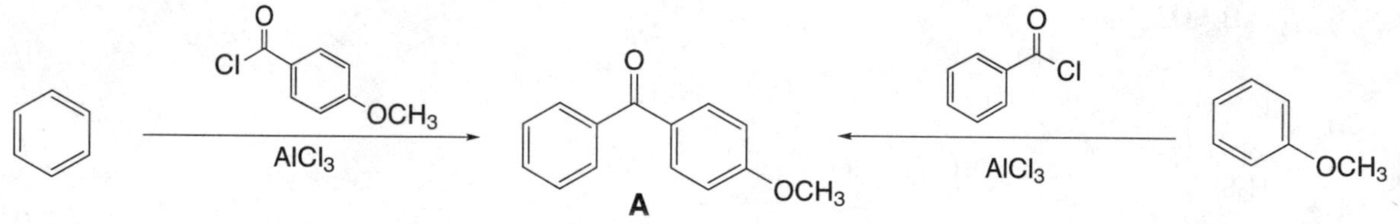

⟨ ⟩ $\xrightarrow[\text{AlCl}_3]{}$ **A** $\xleftarrow[\text{AlCl}_3]{}$ ⟨ ⟩—OCH₃

18.40

a. CH₃—⟨ ⟩—C(CH₃)₃ $\xrightarrow{KMnO_4}$ HOOC—⟨ ⟩—C(CH₃)₃

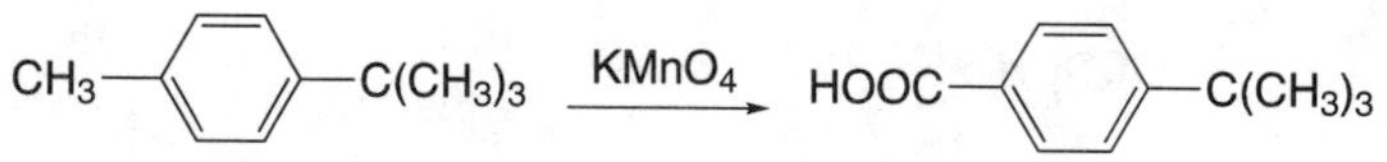

b. ⟨ ⟩—CH₂CH₂CH₂CH₃ $\xrightarrow[h\nu]{Br_2}$ ⟨ ⟩—CHBr—CH₂CH₃

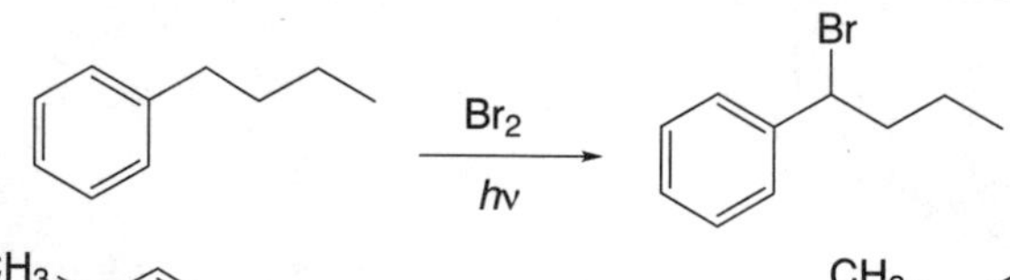

c. CH₃—⟨ ⟩—C(=O)—CH₂CH₂CH₃ $\xrightarrow{Zn(Hg)/HCl}$ CH₃—⟨ ⟩—CH₂—CH₂CH₂CH₃

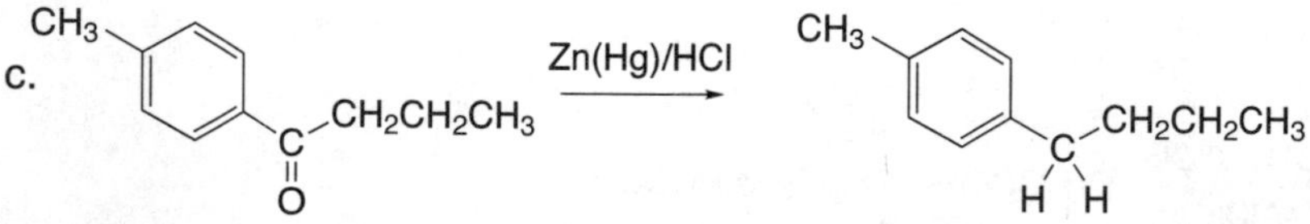

d. ⟨ ⟩—CH₂CH₂CH₂CH₃ $\xrightarrow[FeBr_3]{Br_2}$ (ortho-Br) + (para-Br)

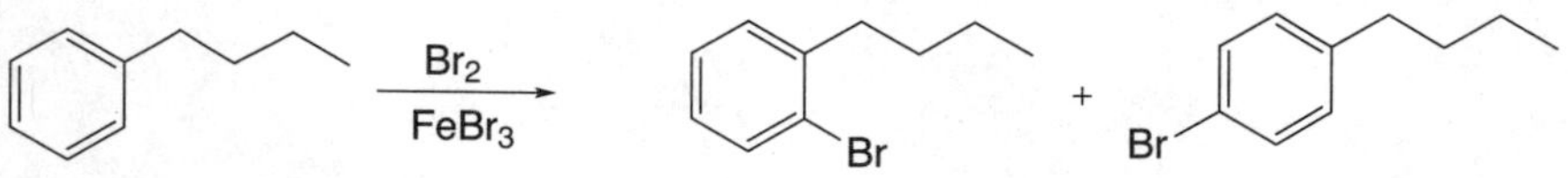

e.

NH$_2$NH$_2$
$^-$OH

f.

OCH$_2$CH$_3$ Br$_2$ ——→   FeBr$_3$

18.41

C bonded to 2 H's
must use acylation
followed by reduction

a.

$$\text{Cl} \overset{O}{\underset{}{\text{C}}} \quad \xrightarrow{\text{AlCl}_3} \quad \quad \xrightarrow[\text{HCl}]{\text{Zn(Hg)}}$$

C bonded to 1H
can be added directly
by alkylation

b.

Cl ——→ AlCl$_3$

c.

CH$_2$CH$_3$

$$\text{Cl}\overset{O}{\overset{\|}{\text{C}}}\text{CH}_3 \xrightarrow{\text{AlCl}_3} \quad \overset{O}{\overset{\|}{\text{C}}}\text{CH}_3 \xrightarrow[\text{HCl}]{\text{Zn(Hg)}} \quad \text{CH}_2\text{CH}_3 \quad \xleftarrow[\text{AlCl}_3]{\text{CH}_3\text{CH}_2\text{Cl}}$$

Method [1]   Method [2]

Ethyl group can be introduced
by two methods.

18.42

a.

SO$_3$H   1) CH$_3$COCl/AlCl$_3$ ——→   2) Cl$_2$/FeCl$_3$

SO$_3$H

Cl

O   CH$_3$

= **A**

Step [1] won't work because a Friedel–Crafts reaction can't be done on a deactivated benzene ring, as is the case with the SO$_3$H substituent. Even if Step [1] did work, the second step would introduce Cl meta to SO$_3$H, not para as drawn.

Alternate synthesis:

Cl$_2$
FeCl$_3$

Cl

CH$_3$COCl
AlCl$_3$

Cl

O   CH$_3$

(+ para isomer)

SO$_3$
H$_2$SO$_4$

Cl

SO$_3$H

O   CH$_3$

(+ isomer)

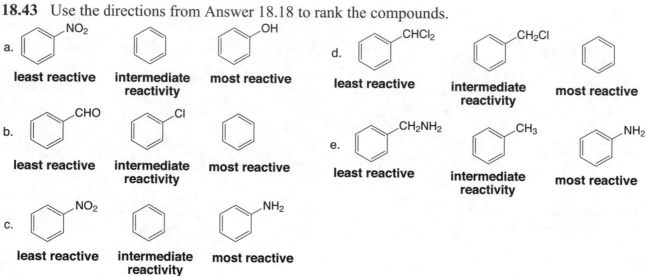

18.43 Use the directions from Answer 18.18 to rank the compounds.

a.

least reactive   intermediate reactivity   most reactive

d.

least reactive   intermediate reactivity   most reactive

b.

least reactive   intermediate reactivity   most reactive

e.

least reactive   intermediate reactivity   most reactive

c.

least reactive   intermediate reactivity   most reactive

18.44 Electron withdrawing groups place a positive charge in the benzene ring. Draw the resonance structures to show how NO_2 puts a positive charge in the ring, giving it an electron withdrawing resonance effect. Electron donating groups place a negative charge in the benzene ring. Draw the resonance structures to show how F puts a negative charge in the ring, giving it an electron donating resonance effect.

a.

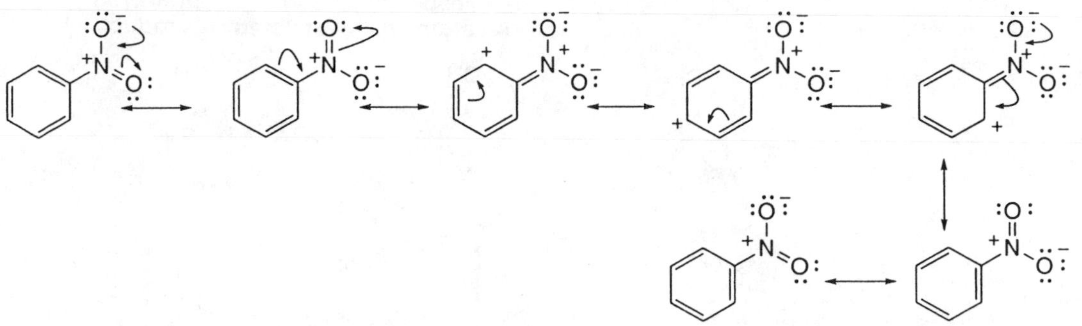

b.

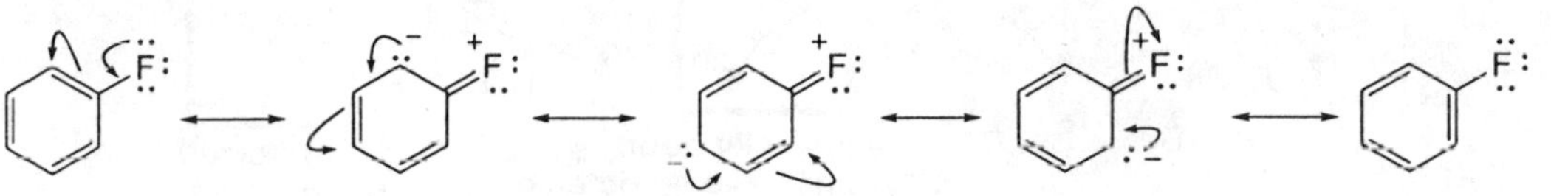

18.45

[1]   Br

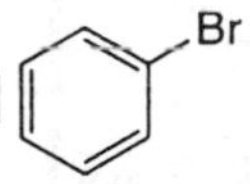

a. withdraw
b. donate
c. less
d. deactivate

[2]   C≡N

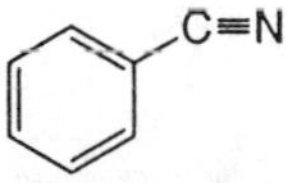

a. withdraw
b. wtihdraw
c. less
d. deactivate

[3]   O—C—CH₃ (C=O)

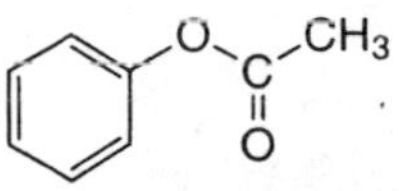

a. withdraw
b. donate
c. more
d. activate

18.46

a.

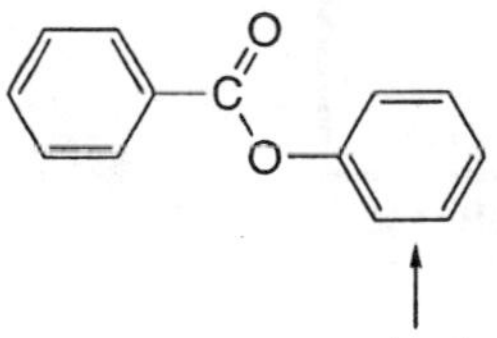

more electron rich
due to O atom
more reactive

b.   —CH₃

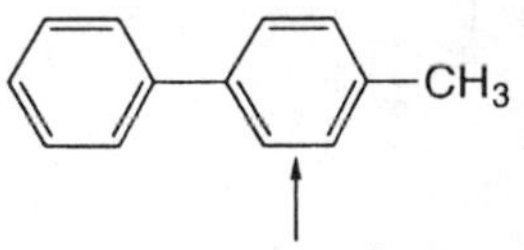

more electron rich
due to CH₃ group
more reactive

18.47

a.

ortho attack

**especially good
All atoms have an octet.**

preferred product

meta attack

para attack

**especially good
All atoms have an octet.**

preferred product

b.

ortho attack

**destabilized
two adjacent like charges**

meta attack

preferred product

para attack

**destabilized
two adjacent like charges**

18.48

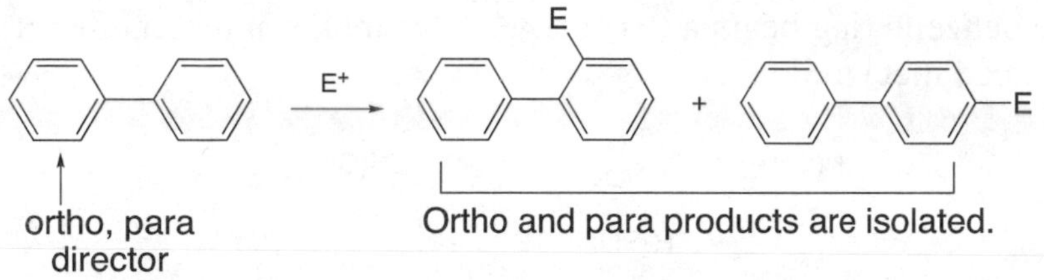

ortho, para
director

Ortho and para products are isolated.

A benzene ring is an electron rich substituent that stabilizes an intermediate positive charge by an electron donating resonance effect. As a result, it activates a benzene ring toward reaction with electrophiles.

With ortho and para attack there is additional resonance stabilization that delocalizes the positive charge onto the second benzene ring. Such additional stabilization is not possible with meta attack.

Ortho attack:

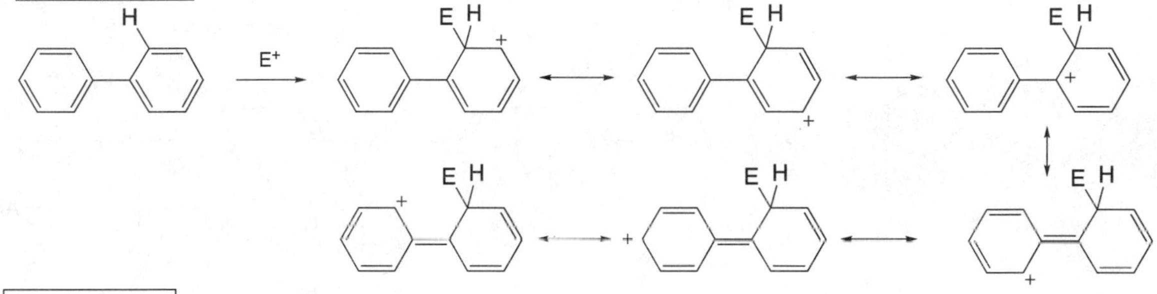

Meta attack:

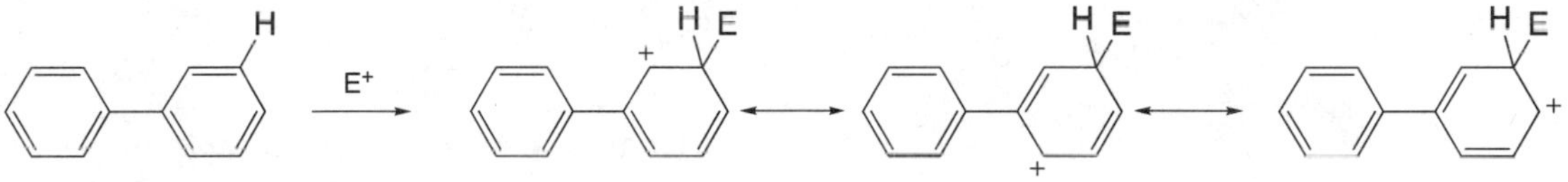

Para attack:

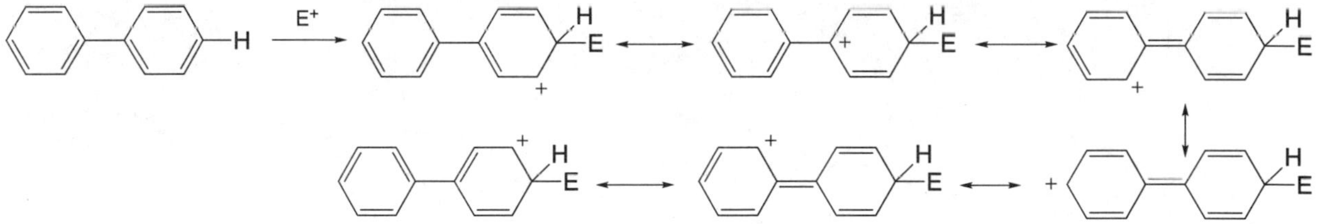

18.49

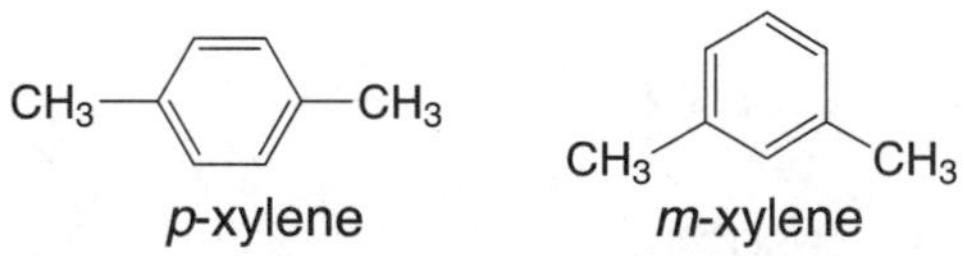

p-xylene
slower

m-xylene
faster

CH_3 is an ortho, para director. In m-xylene, the directing effects of the two CH_3 groups reinforce each other, making it react faster. In p-xylene, the two groups activate different C's, so the reaction is slower.

18.50 Under the acidic conditions of nitration, the N atom of the starting material gets protonated, so the atom directly bonded to the benzene ring bears a (+) charge. This makes it a meta director, so the new NO_2 group is introduced meta to it.

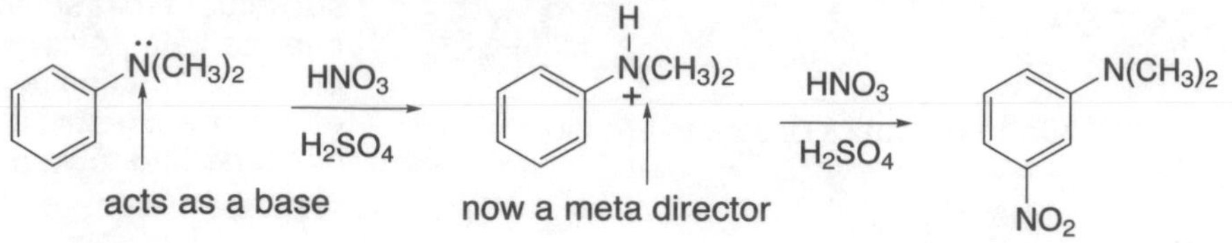
N(CH3)2
HNO3
H2SO4
acts as a base
H
N(CH3)2
+
now a meta director
HNO3
H2SO4
N(CH3)2
NO2

18.51

a.

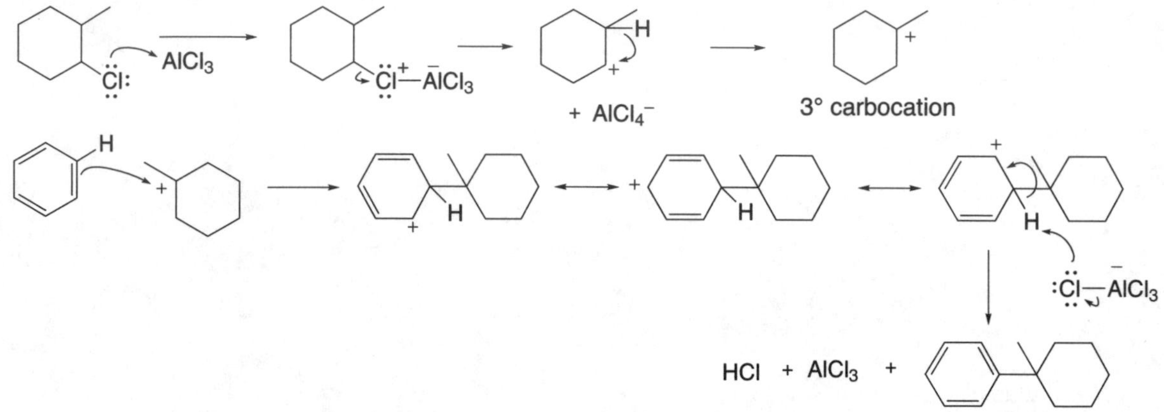
Cl:
AlCl3
Cl—AlCl3
H
+ AlCl4−
3° carbocation
H
H
H
H
:Cl—AlCl3
HCl + AlCl3 +

b.

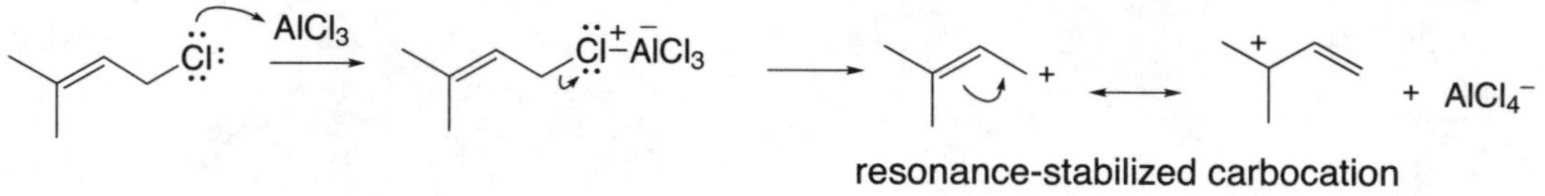
Cl:
AlCl3
Cl—AlCl3
+ AlCl4−
resonance-stabilized carbocation

Use both resonance forms to show how two products are formed.

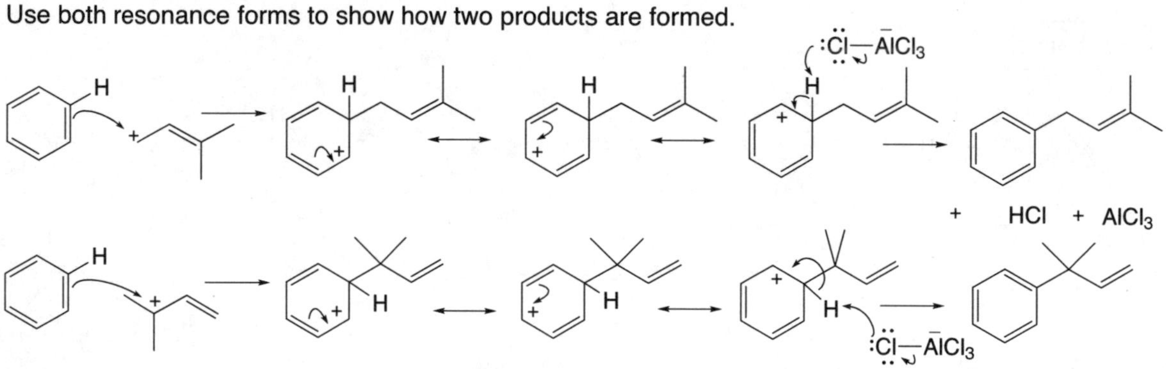
H
H
H
:Cl—AlCl3
H
+ HCl + AlCl3
H
H
H
H
:Cl—AlCl3

c.

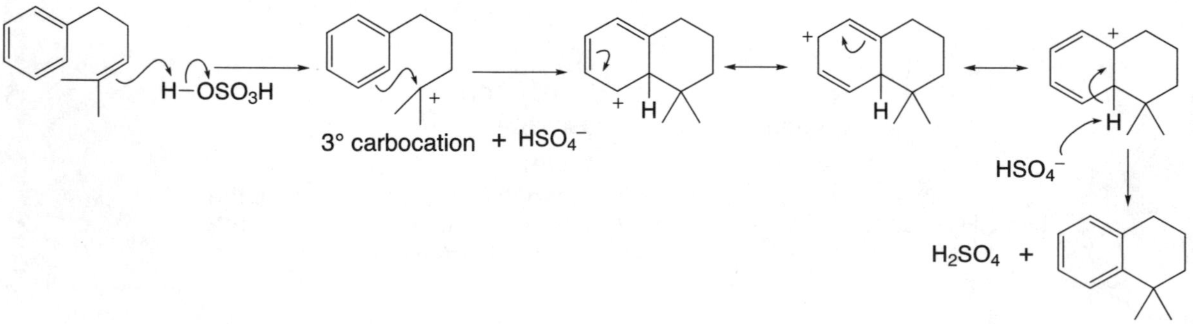
H—OSO3H
3° carbocation + HSO4−
H
H
H
HSO4−
H2SO4 +

18.52

a. The product has one stereogenic center.

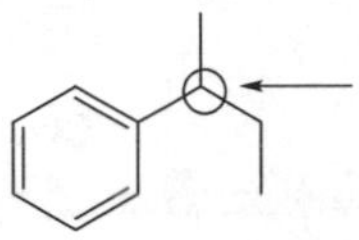

stereogenic center

b. The mechanism for Friedel–Crafts alkylation with this 2° halide involves formation of a trigonal planar carbocation. Since the carbocation is achiral, it reacts with benzene with equal probability from two possible directions (above and below) to afford an optically inactive, racemic mixture of two products.

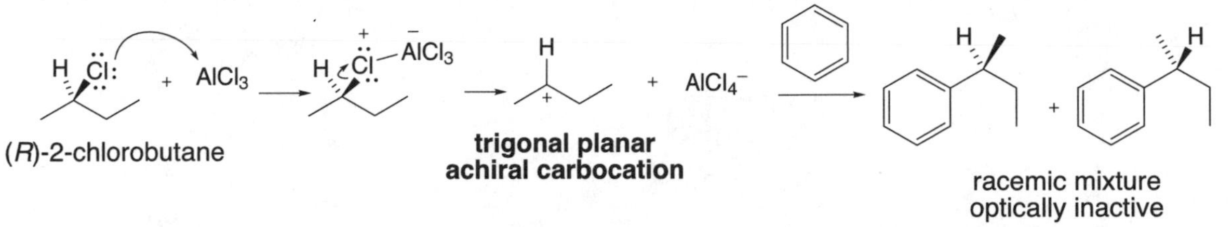

(R)-2-chlorobutane   **trigonal planar achiral carbocation**

racemic mixture
optically inactive

18.53

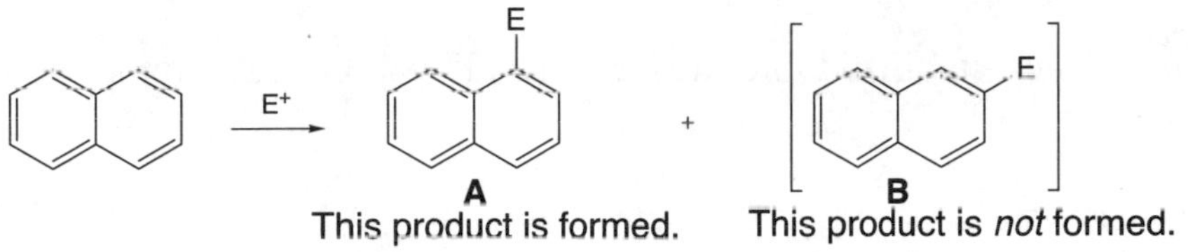

E⁺

A
This product is formed.

B
This product is *not* formed.

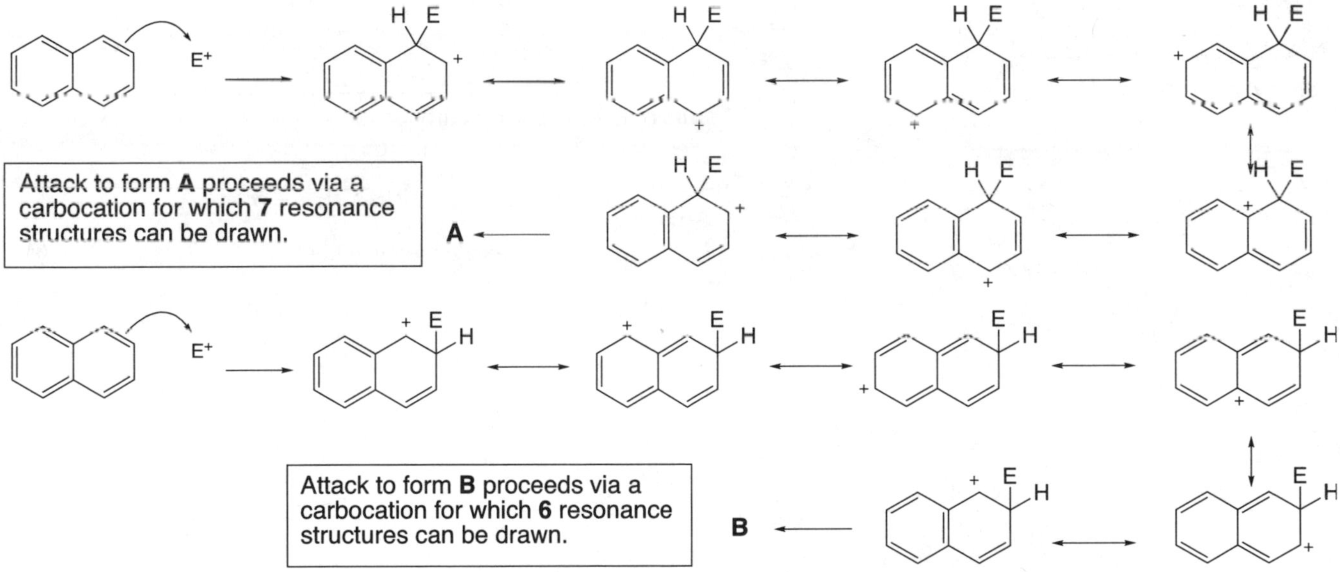

E⁺

Attack to form **A** proceeds via a carbocation for which **7** resonance structures can be drawn.

A

E⁺

Attack to form **B** proceeds via a carbocation for which **6** resonance structures can be drawn.

B

A reaction that occurs by way of the more stable carbocation is preferred so product **A** is formed.

18.54

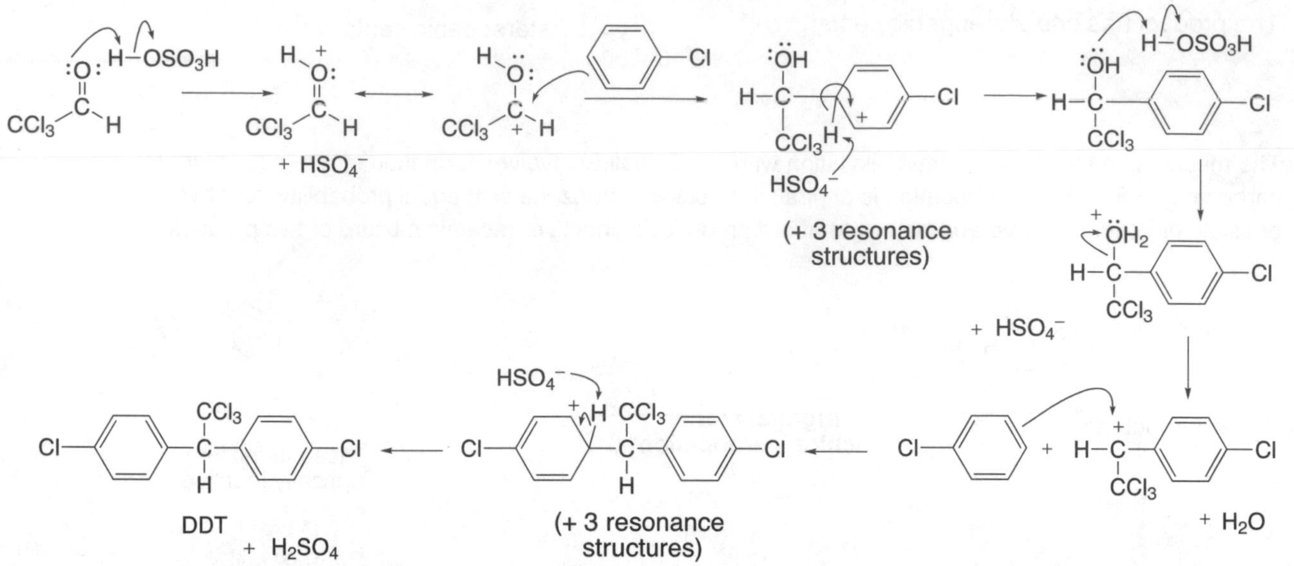

18.55 Benzyl bromide forms a resonance-stabilized intermediate that allows it to react rapidly under S_N1 conditions.

Formation of a resonance-stabilized carbocation:

benzyl bromide

resonance-stabilized carbocation

benzyl methyl ether

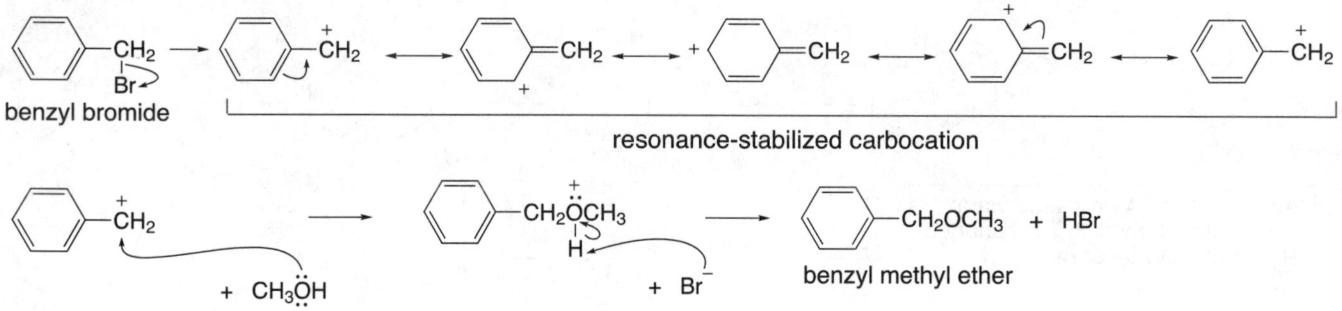

18.56 Addition of HBr will only afford one alkyl bromide because the intermediate carbocation leading to its formation is resonance-stabilized.

18.57

g.

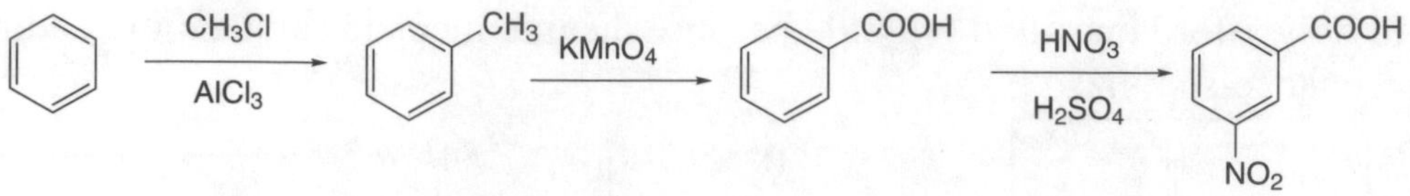

h.

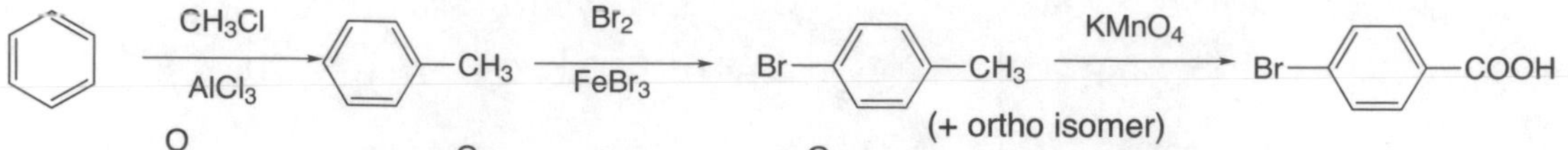

(+ ortho isomer)

i.

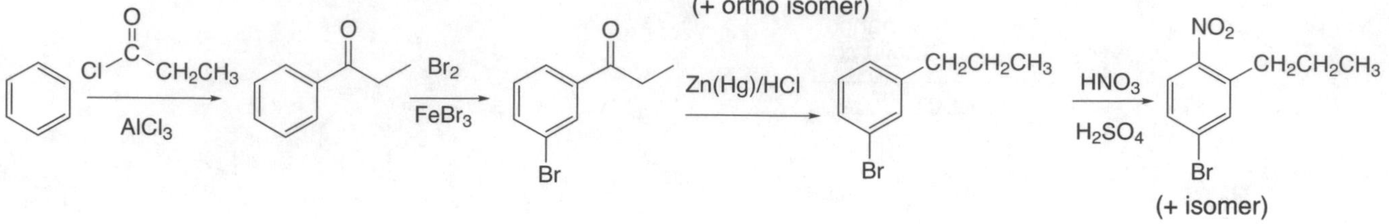

(+ isomer)

18.58

a.

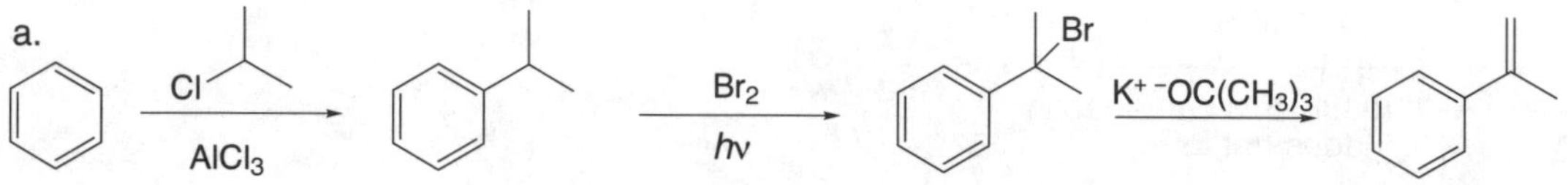

b.

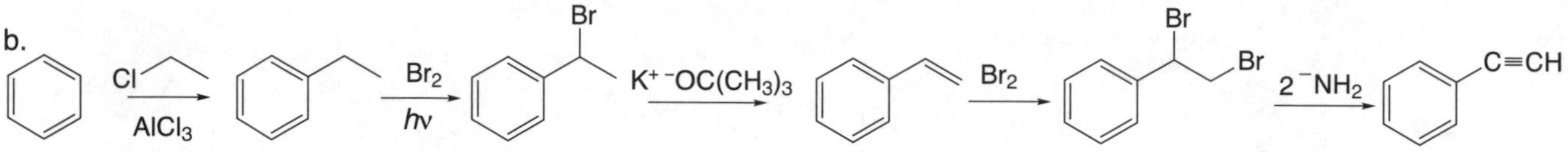

18.59

a.

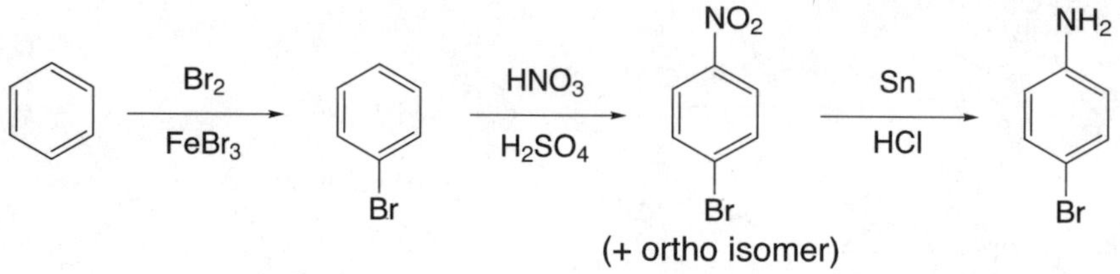

(+ ortho isomer)

b.

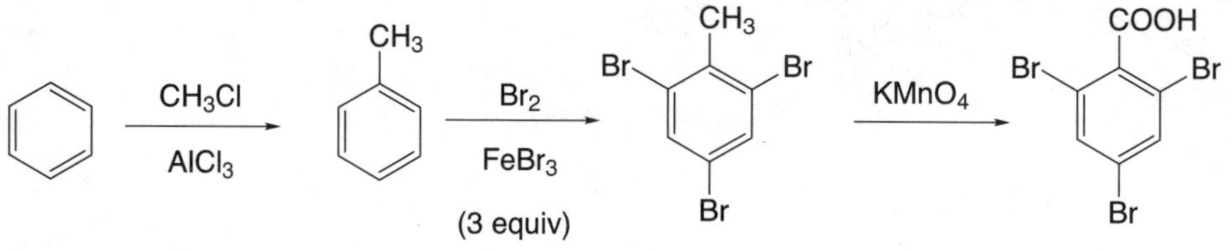

(3 equiv)

c.

(+ ortho isomer)

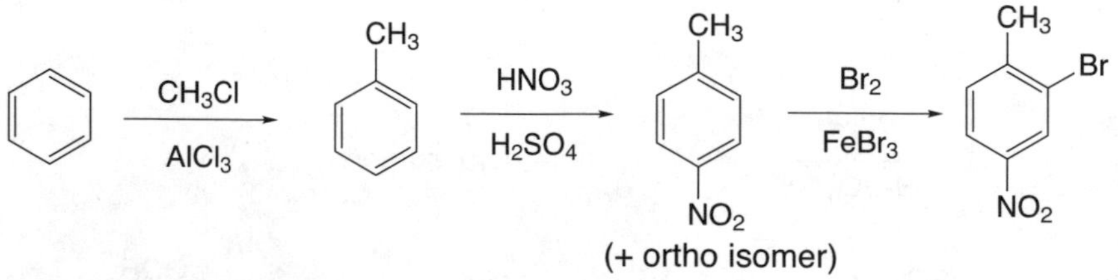

d.

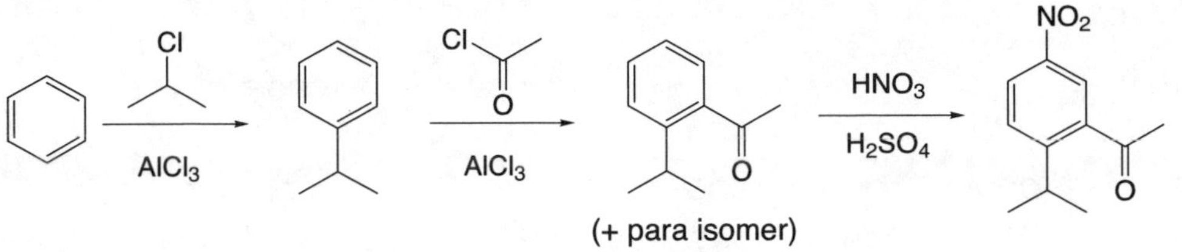

(+ para isomer)

e.

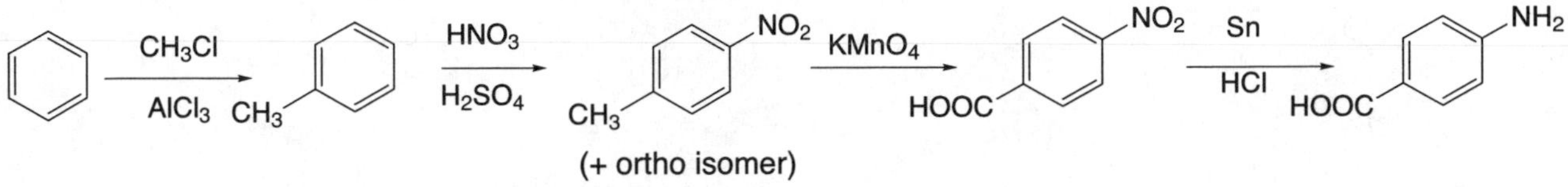

(+ ortho isomer)

18.60

a.

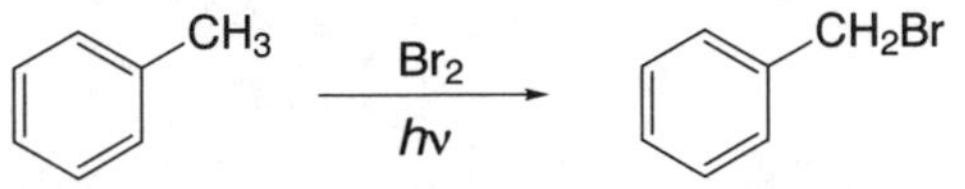

b. 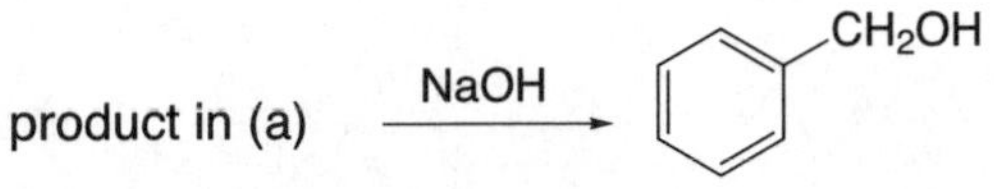 product in (a) → NaOH → CH₂OH

c. 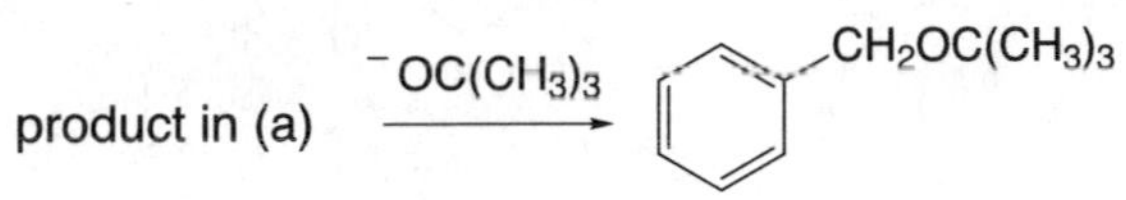 product in (a) → ⁻OC(CH₃)₃ → CH₂OC(CH₃)₃

d. 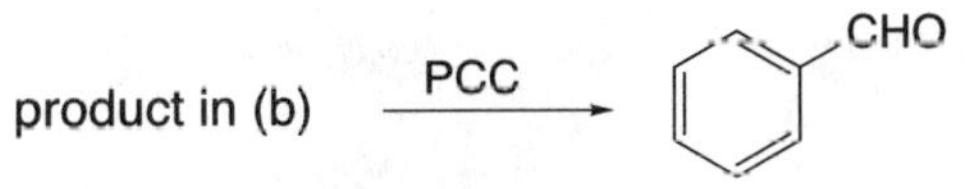 product in (b) → PCC → CHO

e.

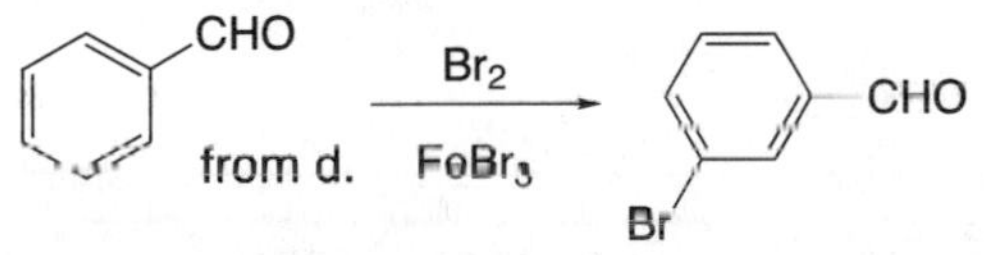

f.

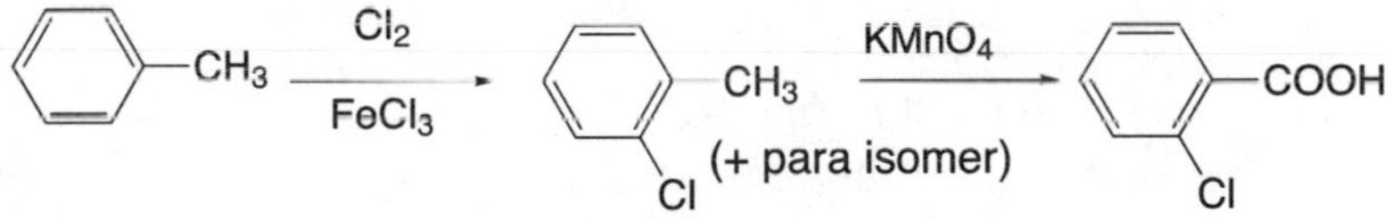

(+ para isomer)

g.

(+ ortho isomer)

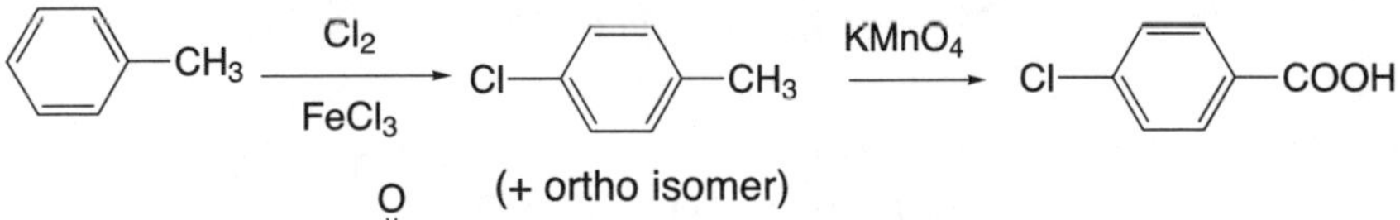

h.

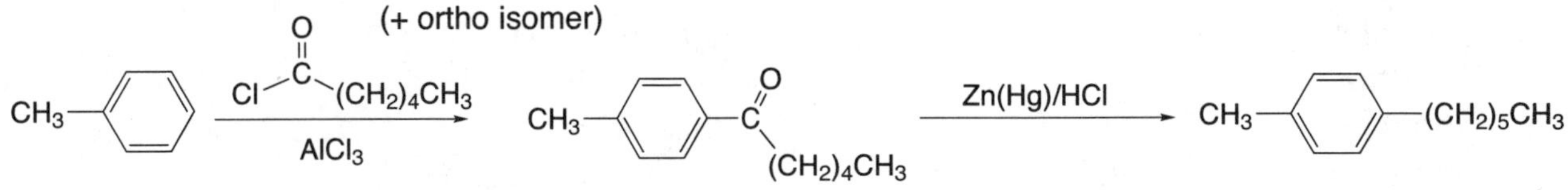

i.

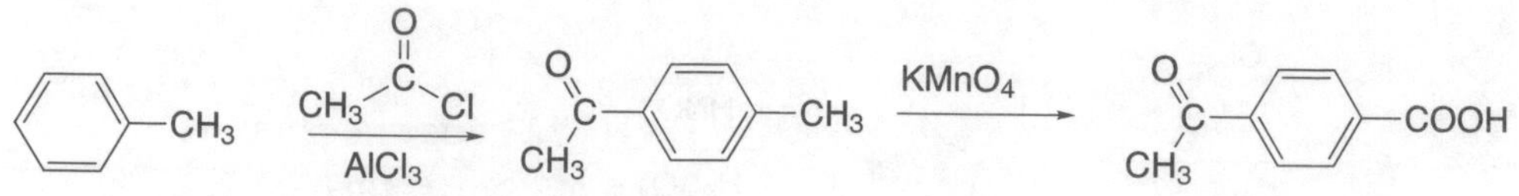

j.

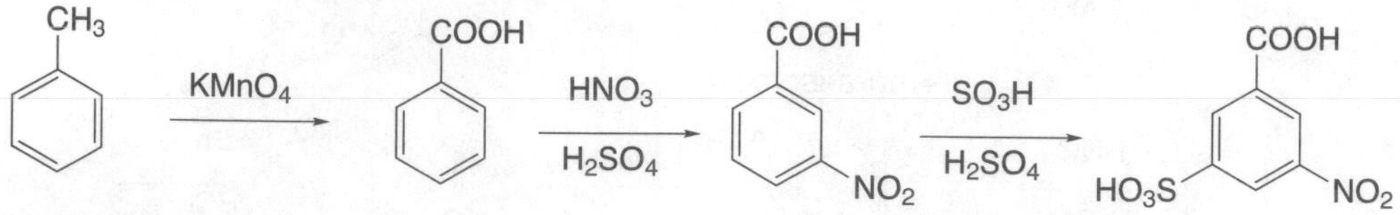

k.

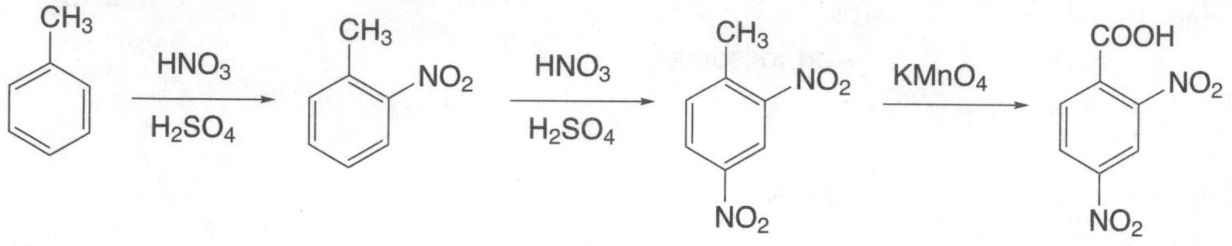

18.61

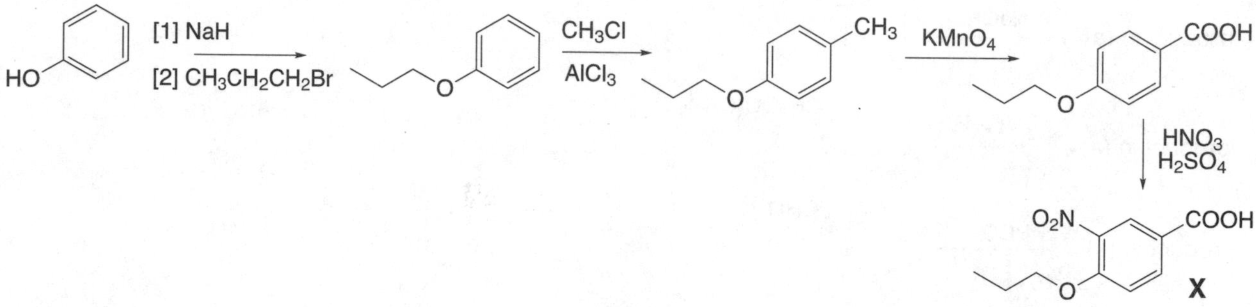

18.62

^{1}H NMR data Compound **X** ($C_{10}H_{12}O$):

absorption	ppm	# of H's	Explanation	Structure:
doublet	1.3	6	6H's adjacent to 1H	
septet	3.5	1	1H adjacent to 6H's	
multiplet	7.4–8.1	5	monosubstituted benzene	

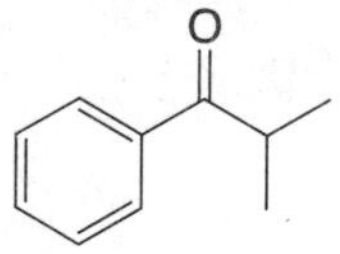

^{1}H NMR data Compound **Y** ($C_{10}H_{14}$):

absorption	ppm	# of H's	Explanation	Structure:
doublet	0.9	6	6H's adjacent to 1H	
multiplet	1.8	1	1H adjacent to many H's	
doublet	2.5	2	2H's adjacent to 1H	
multiplet	7.1–7.3	5	monosubstituted benzene	

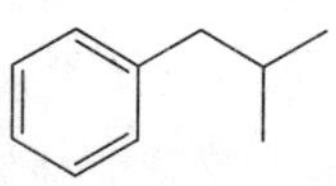

18.63

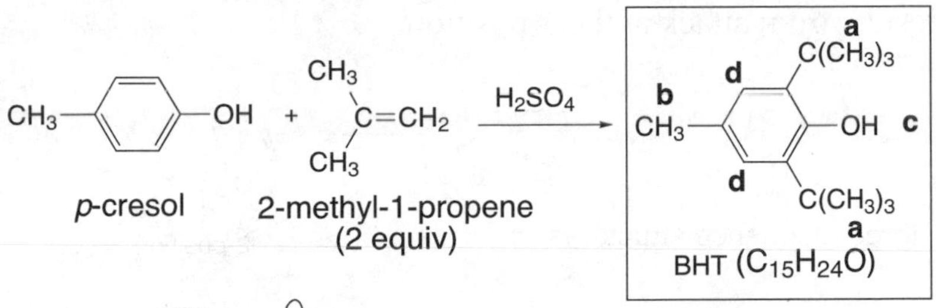

^{1}H NMR spectral data:

1.4 (singlet, 18H) (a)
2.27 (singlet, 3H) (b)
5.0 (singlet, 1H) (c)
7.0 (singlet, 2H) (d) ppm

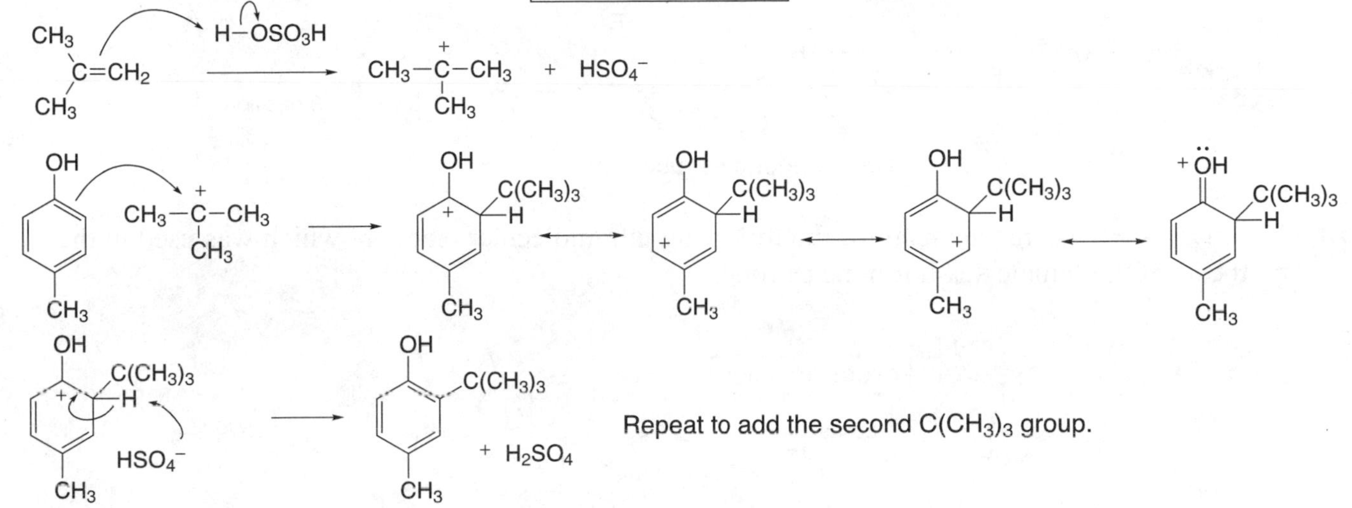

Repeat to add the second $C(CH_3)_3$ group.

18.64

Molecular formula (**Z**): C_9H_9ClO
IR absorption at 1683 cm^{-1}: C=O
^{1}H NMR spectral data:

absorption	ppm	# of H's	Explanation
triplet	1.2	3	3H's adjacent to 2H's
quartet	2.9	2	2H's adjacent to 3H's
multiplet	7.2–8.0	4	disubstituted benzene

Structure:

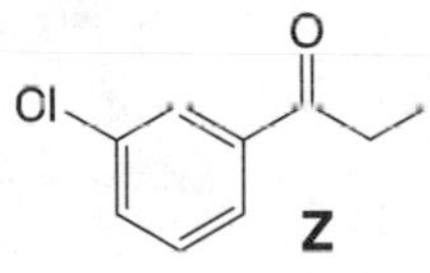

18.65 Five resonance structures can be drawn for phenol, three of which place a negative charge on the ortho and para carbons. These illustrate that the electron density at these positions is increased, thus shielding the protons at these positions, and shifting the absorptions to lower chemical shift. Similar resonance structures cannot be drawn with a negative charge at the meta position, so it is more deshielded and absorbs further downfield, at higher chemical shift.

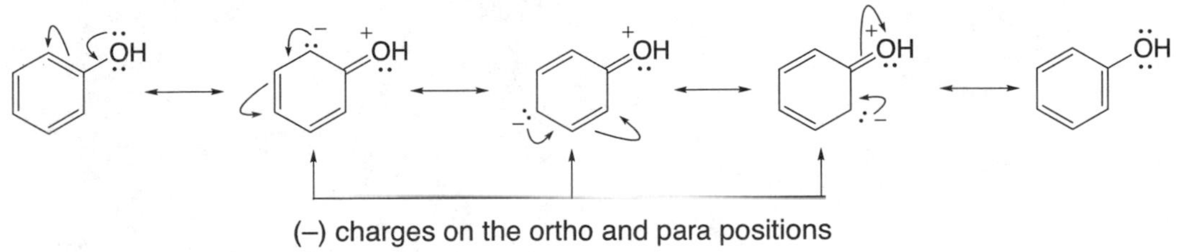

(–) charges on the ortho and para positions

18.66 Attack at the 2-position is favored because the resulting carbocation is more highly resonance-stabilized than the carbocation that results from attack at the 3-position.

18.67 Draw a stepwise mechanism for the following intramolecular reaction, which was used in the synthesis of the female sex hormone estrone.

Chapter 19: Carboxylic Acids and the Acidity of the O–H Bond

♦ General facts

- Carboxylic acids contain a carboxy group (COOH). The central carbon is sp^2 hybridized and trigonal planar (19.1).
- Carboxylic acids are identified by the suffixes *–oic acid*, *carboxylic acid*, or *–ic acid* (19.2).
- Carboxylic acids are polar compounds that exhibit hydrogen bonding interactions (19.3).

♦ Summary of spectroscopic absorptions (19.4)

IR absorptions	C=O	~1710 cm^{-1}
	O–H	3500–2500 cm^{-1} (very broad and strong)
^{1}H NMR absorptions	O–H	10–12 ppm (highly deshielded proton)
	C–H α to COOH	2–2.5 ppm (somewhat deshielded Csp^3–H)
^{13}C NMR absorption	C=O	170–210 ppm (highly deshielded carbon)

♦ General acid-base reaction of carboxylic acids (19.9)

$pK_a \approx 5$ carboxylate anion

- Carboxylic acids are especially acidic because carboxylate anions are resonance-stabilized.
- For equilibrium to favor the products, the base must have a conjugate acid with a $pK_a > 5$. Common bases are listed in Table 19.3.

♦ Factors that affect acidity

Resonance effects. A carboxylic acid is more acidic than an alcohol or phenol because its conjugate base is more effectively stabilized by resonance (19.9)

ROH   pK_a = 16–18   pK_a = 10   pK_a ≈ 5

increasing acidity →

Inductive effects. Acidity increases with the presence of electron withdrawing groups (like the electronegative halogens) and decreases with the presence of electron donating groups (like polarizable alkyl groups) (19.10).

Substituted benzoic acids.
- Electron donor groups (D) make a substituted benzoic acid less acidic than benzoic acid.
- Electron withdrawing groups (W) make a substituted benzoic acid more acidic than benzoic acid.

less acidic
higher pK_a
pK_a > 4.2

pK_a = 4.2

more acidic
lower pK_a
pK_a < 4.2

increasing acidity

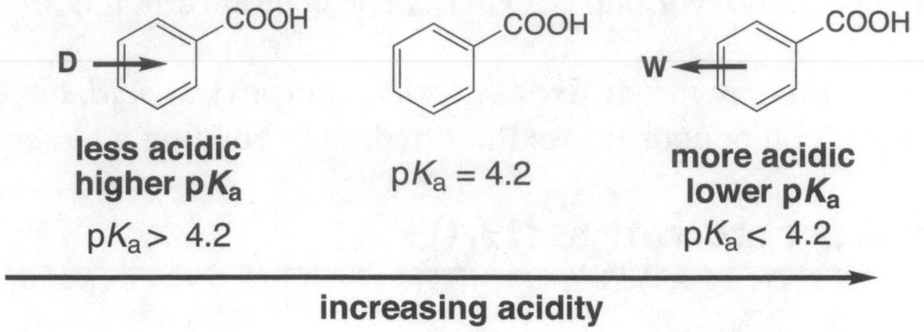

◆ Other facts

- Extraction is a useful technique for separating compounds having different solubility properties. Carboxylic acids can be separated from other organic compounds by extraction, because aqueous base converts a carboxylic acid into a water-soluble carboxylate anion (19.12).
- A sulfonic acid (RSO_3H) is a strong acid because it forms a weak, resonance-stabilized conjugate base on deprotonation (19.13).
- Amino acids have an amino group on the α carbon to the carboxyl group [$RCH(NH_2)COOH$]. Amino acids exist as zwitterions at pH $\approx$ 7. Adding acid forms a species with a net (+1) charge [$RCH(NH_3)COOH$]$^+$. Adding base forms a species with a net (–1) charge [$RCH(NH_2)COO$]$^-$ (19.14).

19.1 To name a carboxylic acid:
    [1] Find the longest chain containing the COOH group and change the –*e* ending to –*oic acid*.
    [2] Number the chain to put the COOH carbon at C1, but omit the number from the name.
    [3] Follow all other rules of nomenclature.

a. $CH_3CH_2CH_2\overset{3}{\underset{CH_3}{\overset{CH_3}{C}}}-CH_2COOH$
    Number the chain to put COOH at C1.
    6 carbon chain = **hexanoic acid**
    3,3-dimethylhexanoic acid

c. $CH_3CH_2\overset{4}{\underset{CH_3CH_2}{\overset{H}{C}}}-CH_2\overset{2}{\underset{CH_2CH_3}{\overset{H}{C}}}-COOH$
    Number the chain to put COOH at C1.
    6 carbon chain = **hexanoic acid**
    2,4-diethylhexanoic acid

b. $CH_3-\overset{4}{\underset{Cl}{\overset{H}{C}}}-CH_2CH_2COOH$
    Number the chain to put COOH at C1.
    5 carbon chain = **pentanoic acid**
    4-chloropentanoic acid

d.   Number the chain to put COOH at C1.
    9 carbon chain = **nonanoic acid**
    4-isopropyl-6,8-dimethylnonanoic acid

19.2
a. 2-bromo**butanoic acid**   c. 3,3,4-trimethyl**heptanoic acid**   e. 3,4-diethyl**cyclohexanecarboxylic aci**

b. 2,3-dimethyl**pentanoic acid**   d. 2-sec-butyl-4,4-diethyl**nonanoic acid**   f. 1-isopropyl**cyclobutane-**
                                                                 carboxylic acid

19.3
a. α-methoxy**valeric acid**   c. α,β-dimethyl**caproic acid**

b. β-phenyl**propionic acid**   d. α-chloro-β-methyl**butyric acid**

19.4
a.   COOH   6 carbon chain = **hexanoic acid** *or*
                      caproic acid
    Cl Br

           4-bromo-4-chlorohexanoic acid
                     or
           γ-bromo-γ-chlorocaproic acid

b.   COOH   4 carbon chain = **butanoic acid** *or*
                      butyric acid
    I

           2-iodo-2-phenylbutanoic acid
                     or
           α-iodo-α-phenylbutyric acid

19.5

sodium benzoate

19.6 More polar molecules have a higher boiling point and are more soluble.

COOCH₃	CH₂CH₂CH₂OH	CH₂COOH
least polar	intermediate polarity	most polar
lowest boiling point	**intermediate boiling point**	**highest boiling point**
least H₂O soluble		**most H₂O soluble**

19.7 Look for functional group differences to distinguish the compounds by IR. Besides sp^3 hybridized C–H bonds at 3000–2850 cm^{-1} (which all three compounds have), the following functional group absorptions are seen:

$CH_3CH_2CH_2CH_2$—C(=O)—OH	$CH_3CH_2CH_2CH_2$—C(=O)—OCH₃	(tetrahydropyran)—OH
carboxylic acid	**ester**	**alcohol**
2 strong absorptions	1 strong absorption	1 strong absorption
~1710 (C=O)	~1700 (C=O) cm^{-1}	~3600–3200 (OH) cm^{-1}
~2500–3500 (OH) cm^{-1}		

19.8

Molecular formula: $C_4H_8O_2$
one degree of unsaturation

^{1}H NMR data (ppm):
  0.95 (triplet, 3H)
  1.65 (multiplet, 2H)
  2.30 (triplet, 2H)
  11.8 (singlet, 1H)

19.9

PGF$_{2\alpha}$
a prostaglandin

enantiomer

There are 5 tetrahedral stereogenic centers. Both double bonds can exhibit cis–trans isomerism. Therefore, there are $2^7 = 128$ stereoisomers.

19.10 1° alcohols are converted to carboxylic acids by oxidation reactions.

a.

c.

b.

19.11

a. $\xrightarrow[\text{H}_2\text{SO}_4/\text{H}_2\text{O}]{\text{Na}_2\text{Cr}_2\text{O}_7}$ benzene-COOH   c. O_2N-benzene-CH_3 $\xrightarrow{\text{KMnO}_4}$ O_2N-benzene-COOH

A

(Any R group with benzylic H's can be present para to NO_2.)

b. $CH_3C\equiv CCH_3$ $\xrightarrow[\text{[2] H}_2\text{O}]{\text{[1] O}_3}$ CH_3COOH (2 equiv)

B

C

19.12 Carboxylic acids react with all the listed reagents except CrO_3. They are already highly oxidized, so they cannot be readily oxidized to other compounds with oxidizing agents like CrO_3.

a. NaOH   b. HCl   c. $LiAlH_4$ d. CrO_3 **NO REACTION**   e. $NaOCH_3$

19.13

a. cyclohexyl-COOH $\xrightarrow{\text{NaOH}}$ cyclohexyl-COO$^-$ Na$^+$ + H_2O

c. $CH_3-\overset{CH_3}{\underset{CH_3}{C}}-OH$ $\xrightarrow{\text{NaH}}$ $CH_3-\overset{CH_3}{\underset{CH_3}{C}}-O^-$ Na$^+$ + H_2

b. CH_3-benzene-OH $\xrightarrow{\text{NaOCH}_3}$ CH_3-benzene-O$^-$ Na$^+$ + $HOCH_3$

d. benzene-COOH $\xrightarrow{\text{NaHCO}_3}$ benzene-COO$^-$ Na$^+$ + H_2CO_3

19.14 CH_3COOH has a pK_a of 4.8. Any base with a conjugate acid with a pK_a higher than 4.8 can deprotonate it.

a. F$^-$ pK_a (HF) = 3.2 **not strong enough**
b. $(CH_3)_3CO^-$ pK_a [$(CH_3)_3COH$]= 18 **strong enough**
c. CH_3^- pK_a (CH$_4$) = 50 **strong enough**
d. $^-NH_2$ pK_a (NH$_3$) = 38 **strong enough**
e. Cl$^-$ pK_a (HCl) = −7.0 **not strong enough**

19.15 The stronger acid has a weaker conjugate base. Remember the periodic trends that determine acidity.

benzene-OH → benzene-O$^-$	benzene-NH$_2$ → benzene-$^-$NH
acid ... conjugate base	acid ... conjugate base
weaker, more stable conjugate base	stronger, less stable conjugate base
O is further right in the periodic table	N is further left in the periodic table
O is more electronegative.	**weaker acid**
stronger acid	

19.16 Electron withdrawing groups make an acid more acidic, lowering its pK_a.

CH_3CH_2-COOH
least acidic
pK_a = 4.9

ICH_2-COOH
one electron withdrawing group
intermediate acidity
pK_a = 3.2

CF_3-COOH
three electron withdrawing F's
most acidic
pK_a = 0.2

19.17 Acetic acid has an electron donating methyl group adjacent to the carboxyl group. The CH_3 group both stabilizes the acid and destabilizes the nearby negative charge on the conjugate base, making CH_3COOH less acidic (with a higher pK_a) than HCOOH.

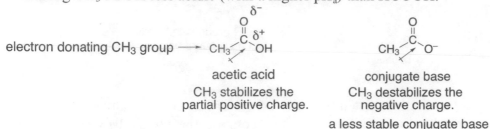

electron donating CH_3 group ⟶

acetic acid
CH_3 stabilizes the
partial positive charge.

conjugate base
CH_3 destabilizes the
negative charge.
a less stable conjugate base

19.18

a. CH_3COOH $HSCH_2COOH$ $HOCH_2COOH$ b. ICH_2CH_2COOH ICH_2COOH $I_2CHCOOH$

least acidic   **intermediate acidity**   **most acidic**   **least acidic**   **intermediate acidity**   **most acidic**

19.19

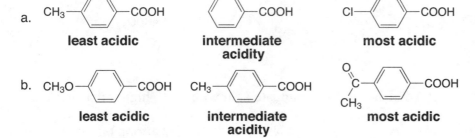

a.

least acidic   **intermediate acidity**   **most acidic**

b.

least acidic   **intermediate acidity**   **most acidic**

19.20

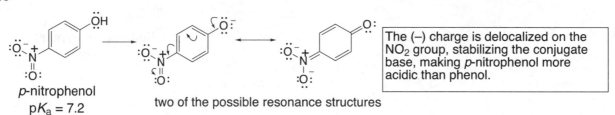

p-nitrophenol
pK_a = 7.2

two of the possible resonance structures

The (−) charge is delocalized on the NO_2 group, stabilizing the conjugate base, making *p*-nitrophenol more acidic than phenol.

19.21 To separate compounds by an extraction procedure, they must have different solubility properties.

a. $CH_3(CH_2)_6COOH$ and $CH_3CH_2CH_2CH_2CH=CH_2$: **YES.** The acid can be extracted into aqueous base, while the alkene will remain in an organic layer.

b. $CH_3CH_2CH_2CH_2CH=CH_2$ and $(CH_3CH_2CH_2)_2O$: **NO.** Both compounds are soluble in organic solvents and insoluble in water. Neither is acidic enough to be extracted into aqueous base.

c. $CH_3(CH_2)_6COOH$ and NaCl: one carboxylic, one salt: **YES.** The carboxylic acid is soluble in an organic solvent while the salt is soluble in water.

d. NaCl and KCl: two salts: **NO.**

19.22

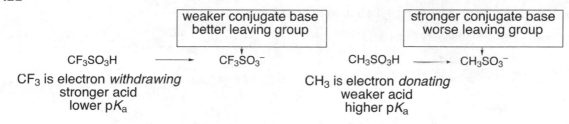

CF$_3$SO$_3$H $\longrightarrow$ CF$_3$SO$_3^-$

CF$_3$ is electron *withdrawing*
stronger acid
lower pK_a

CH$_3$SO$_3$H $\longrightarrow$ CH$_3$SO$_3^-$

CH$_3$ is electron *donating*
weaker acid
higher pK_a

19.23

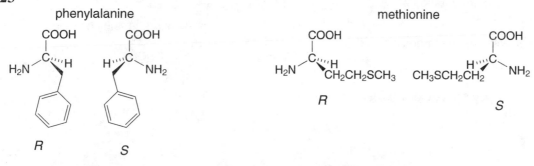

phenylalanine

R   S

methionine

R   S

19.24 Since amino acids exist as zwitterions (i.e. salts), they are too polar to be soluble in organic solvents like diethyl ether. Thus, they are soluble in water.

19.25

COOH
H$_3$N$^+$–C–H
H

pH = 1

COO$^-$
H$_3$N$^+$–C–H
H

glycine
neutral form

COO$^-$
H$_2$N–C–H
H

pH = 11

19.26

$$p I = \frac{pK_a(COOH) + pK_a(NH_3^+)}{2} = \frac{(2.58) + (9.24)}{2} = 5.91$$

COO$^-$
H$_3$N$^+$–C–H
CH$_2$

19.27

H$_3$N$^+$–CH–COOH $\longrightarrow$ H$_3$N$^+$–CH–COO$^-$
H   H

electron withdrawing group

The nearby (+) stabilizes the conjugate base by an electron withdrawing inductive effect, thus making the starting acid more acidic.

19.28 Use the directions from Answer 19.1 to name the compounds.

a. (CH₃)₂CHCH₂CH₂CO₂H   4-methylpentanoic acid

b. BrCH₂COOH   2-bromoacetic acid

c.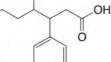

4,4,5,5-tetramethyloctanoic acid

d. CH₃CH₂CH₂COO⁻Li⁺   lithium butanoate

e. 1-ethylcyclopentanecarboxylic acid

f. 2,4-dimethylcyclohexanecarboxylic acid

g. o-bromobenzoic acid

h. CH₃CH₂——COOH p-ethylbenzoic acid

i. sodium 2-methylhexanoate

j. 7-ethyl-5-isopropyl-3-methyldecanoic acid

19.29

a. 3,3-dimethylpentanoic acid

b. 4-chloro-3-phenylheptanoic acid

c. (R)-2-chloropropanoic acid

d. β,β-dichloropropionic acid

e. m-hydroxybenzoic acid

f. o-chlorobenzoic acid

g. potassium acetate

h. sodium α-bromobutyrate

19.30

pentanoic acid   3-methylbutanoic acid   2-methylbutanoic acid   2,2-dimethylpropanoic acid

sodium pentanoate   sodium 3-methylbutanoate   sodium 2-methylbutanoate   sodium 2,2-dimethylpropanoate

19.31

a.

lowest boiling point   intermediate boiling point   highest boiling point

b.

lowest boiling point   intermediate boiling point   highest boiling point

19.32

a.

$$\xrightarrow[\text{H}_2\text{SO}_4/\text{H}_2\text{O}]{\text{CrO}_3}$$

c.

$$\xrightarrow[\text{2) H}_2\text{O}]{\text{1) O}_3} \quad \text{—COOH} \; + \; \text{CO}_2$$

b. $(CH_3)_2CH$—

—$CH_3 \xrightarrow{\text{KMnO}_4}$ HOOC—

—COOH   d. $CH_3(CH_2)_6CH_2OH \xrightarrow[\text{H}_2\text{SO}_4/\text{H}_2\text{O}]{\text{Na}_2\text{Cr}_2\text{O}_7} CH_3(CH_2)_6COOH$

19.33

a.

$$\xrightarrow[\text{2) H}_2\text{O}_2/^-\text{OH}]{\text{1) BH}_3}$$

—$CH_2OH \xrightarrow[\text{H}_2\text{SO}_4/\text{H}_2\text{O}]{\text{CrO}_3}$ —COOH

A   B

b. $HC\equiv CH \xrightarrow[\text{2) CH}_3\text{I}]{\text{1) NaNH}_2} HC\equiv CCH_3 \xrightarrow[\text{2) CH}_3\text{CH}_2\text{I}]{\text{1) NaNH}_2} CH_3CH_2C\equiv CCH_3 \xrightarrow[\text{2) H}_2\text{O}]{\text{1) O}_3} CH_3CH_2COOH \; + \; CH_3COOH$

C   D   E + F

c.

$$\xrightarrow[\text{AlCl}_3]{(CH_3)_2CHCl}$$

—$CH(CH_3)_2 \xrightarrow{\text{KMnO}_4}$ —COOH

G   H

19.34

Bases: [1] $^-$OH pK_a (H_2O) = 15.7; [2] $CH_3CH_2^-$ pK_a (CH_3CH_3) = 50; [3] $^-$NH$_2$ pK_a (NH_3) = 38; [4] NH_3 pK_a (NH_4^+) = 9.4; [5] $HC\equiv C^-$ pK_a ($HC\equiv CH$) = 25,

a. CH_3—

—COOH

pK_a = 4.3
All of the bases
can deprotonate this.

b. Cl—

—OH

pK_a = 9.4
$^-$OH, $CH_3CH_2^-$, $^-$NH$_2$ and $HC\equiv C^-$
can deprotonate this.

c. $(CH_3)_3COH$

pK_a = 18
$CH_3CH_2^-$, $^-$NH$_2$ and $HC\equiv C^-$
can deprotonate this.

19.35

a.

—COOH + K^+ $^-OC(CH_3)_3 \longrightarrow$ —COO$^-$ K^+ + HOC(CH$_3$)$_3$

pK_a = 4.2 pK_a = 18 | Reaction favors products. |

b.

OH + $NH_3 \longrightarrow$

O$^-$ + NH_4^+

$pK_a \approx 16$ pK_a = 9.4 | Reaction favors reactants. |

c.

—OH + $NH_2^- \longrightarrow$

—O$^-$ + NH_3

pK_a = 10 pK_a = 38 | Reaction favors products. |

d.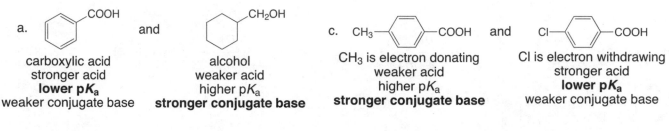

Reaction favors products.

$pK_a = 50$

$pK_a \approx 4$

e.

+ Na$^+$H$^-$ ⟶

Reaction favors products.

$pK_a \approx 16$ $pK_a = 35$

f. CH$_3$—⟨ ⟩—OH + Na$_2$CO$_3$ ⟶ CH$_3$—⟨ ⟩—O$^-$Na$^+$ + Na$^+$ HCO$_3^-$

$pK_a = 10.2$ $pK_a = 10.2$

With the same pK_a for the starting acid and the conjugate acid, an equal amount of starting materials and products is present.

19.36   The stronger acid has a lower pK_a and a weaker conjugate base.

a.

and

c.   CH$_3$—⟨ ⟩—COOH   and   Cl—⟨ ⟩—COOH

carboxylic acid
stronger acid
lower pK_a
weaker conjugate base

alcohol
weaker acid
higher pK_a
stronger conjugate base

CH$_3$ is electron donating
weaker acid
higher pK_a
stronger conjugate base

Cl is electron withdrawing
stronger acid
lower pK_a
weaker conjugate base

b.   ClCH$_2$COOH   and   FCH$_2$COOH

weaker acid
higher pK_a
stronger conjugate base

F is more electronegative
stronger acid
lower pK_a
weaker conjugate base

d.   NCCH$_2$COOH   and   CH$_3$COOH

CN is electron withdrawing
stronger acid
lower pK_a
weaker conjugate base

weaker acid
higher pK_a
stronger conjugate base

19.37

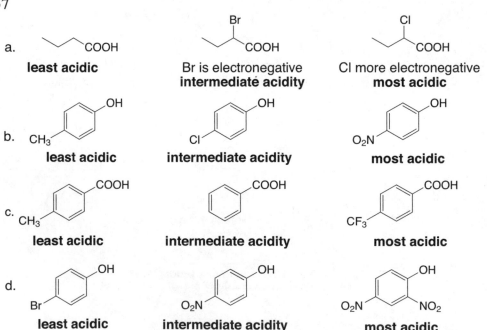

a.

least acidic

Br is electronegative
intermediate acidity

Cl more electronegative
most acidic

b.   CH$_3$

least acidic

Cl

intermediate acidity

O$_2$N

most acidic

c.   CH$_3$

least acidic

intermediate acidity

CF$_3$

most acidic

d.   Br

least acidic

O$_2$N

intermediate acidity

O$_2$N   NO$_2$

most acidic

19.38

a. $BrCH_2COO^-$ $BrCH_2CH_2COO^-$ $(CH_3)_3CCOO^-$

    weakest base   **intermediate basicity**   **strongest base**

c.

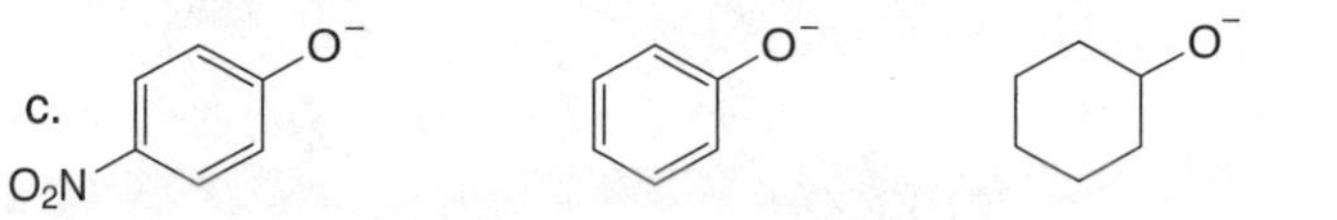

    weakest base   **intermediate basicity**   **strongest base**

b.

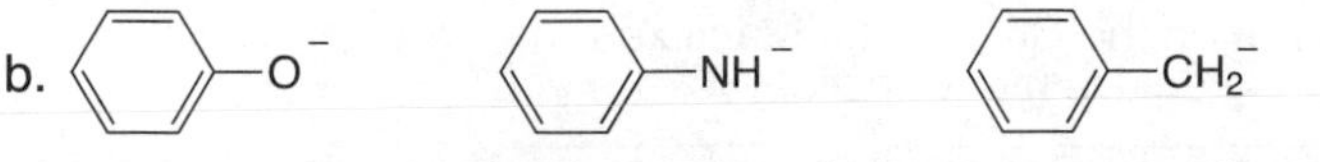

    weakest base   **intermediate basicity**   **strongest base**

19.39

	increasing acidity →				
	ICH_2COOH	$BrCH_2COOH$	FCH_2COOH	$F_2CHCOOH$	F_3CCOOH
pK_a values	least acidic 3.12	2.86	2.66	1.24	most acidic 0.28

19.40 In the para isomer, the negative charge of the conjugate base is delocalized over both the benzene ring and onto the NO_2 group, whereas in the meta isomer it cannot be delocalized onto the NO_2 group. This makes the conjugate base from the para isomer more highly resonance-stabilized, and the para substituted phenol more acidic than its meta isomer.

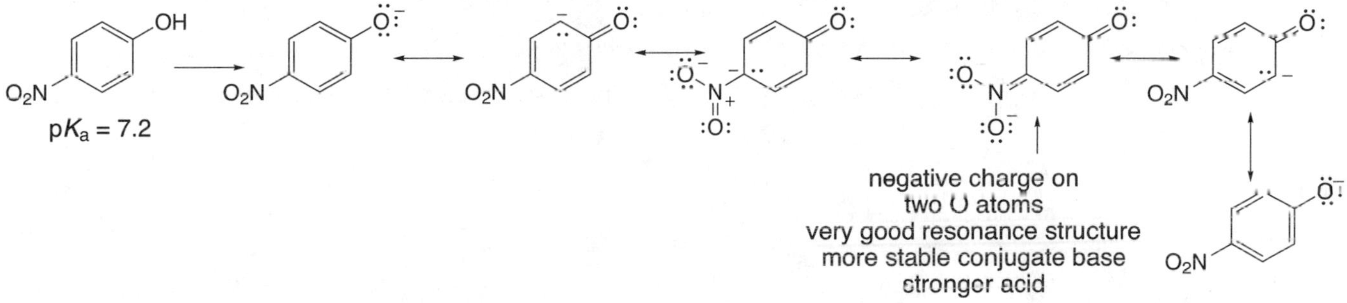

$pK_a = 7.2$

negative charge on two O atoms
very good resonance structure
more stable conjugate base
stronger acid

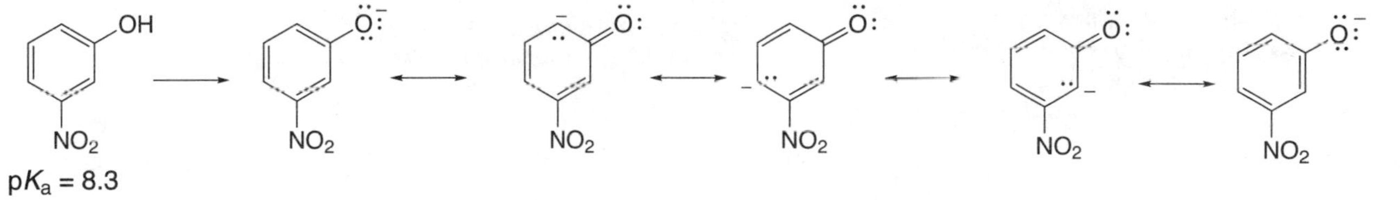

$pK_a = 8.3$

19.41 Both **A** and **B** have stabilizing groups for the conjugate base, making them more acidic than CH_3CH_2COOH.

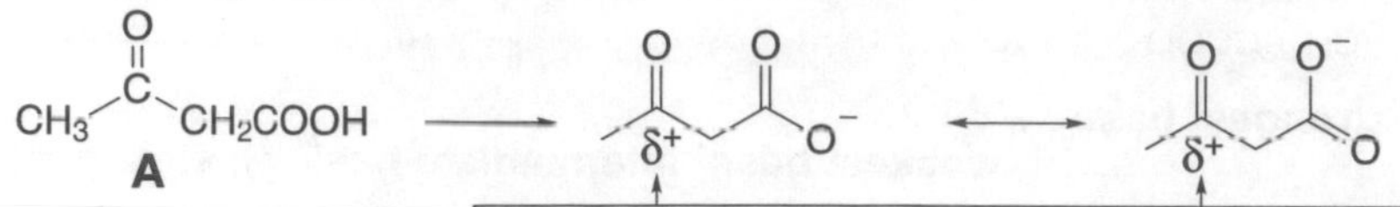

The nearby partial positive charge on the C=O group stabilizes the conjugate base, so the starting acid is more acidic.

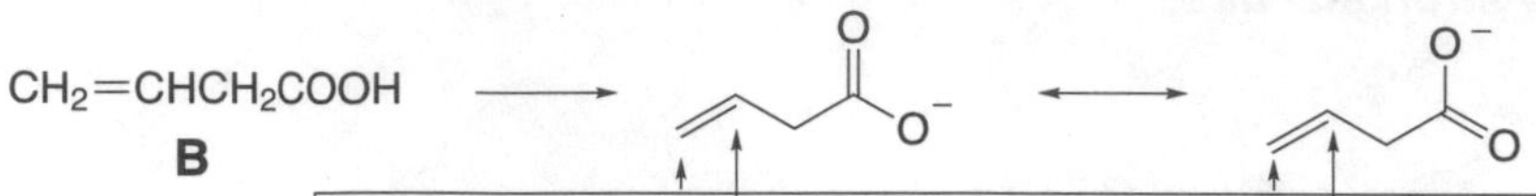

The sp^2 hybridized C's of the double bond have a higher percent s-character than an sp^3 hybridized C, so they pull more electron density toward them, stabilizing the conjugate base. This makes **B** more acidic than CH_3CH_2COOH.

19.42

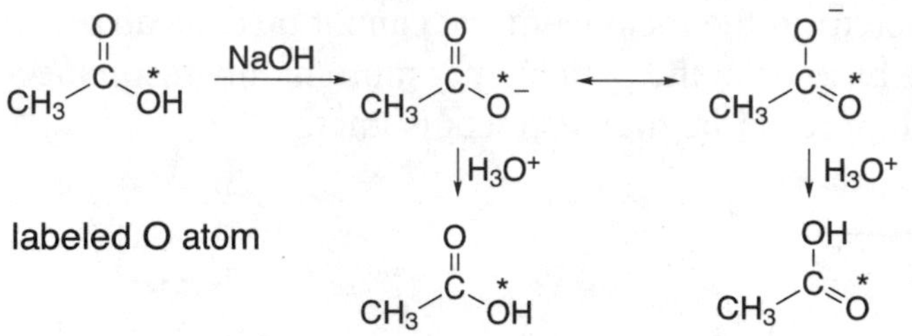

The resonance-stabilized carboxylate anion can now be protonated on either O atom, the one with the label and the one without the label.

labeled O atom

The label is now in two different locations.

19.43

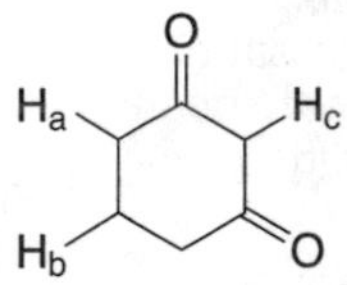

1,3-cyclohexanedione
increasing acidity: $H_b < H_a < H_c$

loss of H_b:

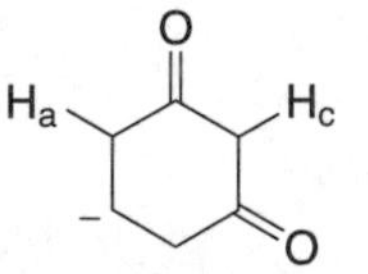

The most acidic proton forms the most stable conjugate base.

one Lewis structure
least stable conjugate base

loss of H_a:

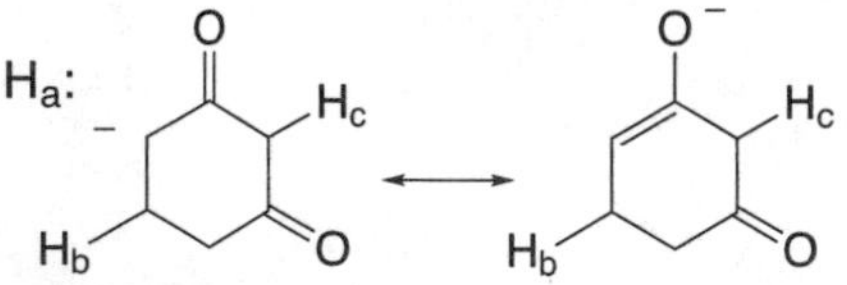

2 resonance structures
intermediate stability

loss of H_c:

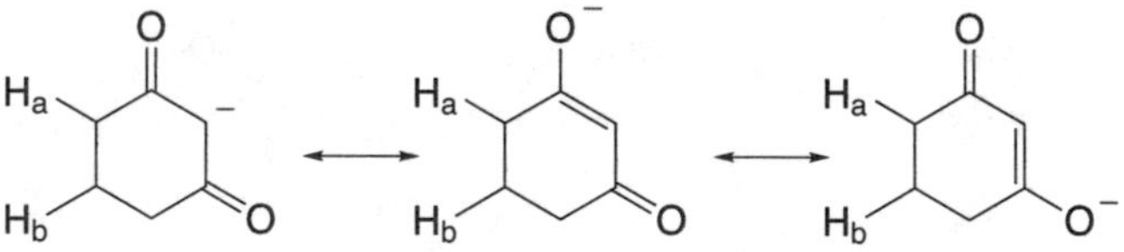

3 resonance structures
most stable conjugate base

19.44 As usual, compare the stability of the conjugate bases. With RSO_3H, loss of a proton forms a conjugate base that has three resonance structures, all of which are equivalent and place a negative charge on a more electronegative O atom. With the conjugate base of RCOOH, there are only two of these such resonance structures. Thus, the conjugate base RSO_3^- is more highly resonance-stabilized than $RCOO^-$, so RSO_3H is a stronger acid than RCOOH.

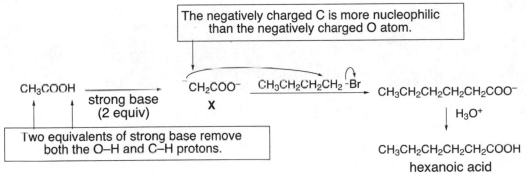

three resonance structures
for the conjugate base

two resonance structures
for the conjugate base

19.45

The negatively charged C is more nucleophilic than the negatively charged O atom.

CH_3COOH ⟶ (strong base (2 equiv)) $^-CH_2COO^-$ **X** $CH_3CH_2CH_2CH_2\text{-}Br$ ⟶ $CH_3CH_2CH_2CH_2CH_2COO^-$

Two equivalents of strong base remove both the O–H and C–H protons.

$\downarrow H_3O^+$

$CH_3CH_2CH_2CH_2CH_2COOH$
hexanoic acid

19.46

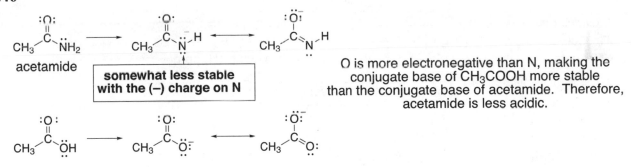

acetamide

somewhat less stable
with the (−) charge on N

O is more electronegative than N, making the conjugate base of CH_3COOH more stable than the conjugate base of acetamide. Therefore, acetamide is less acidic.

19.47

A   **B** (COOH)

- Dissolve both compounds in CH_2Cl_2.
- Add 10% $NaHCO_3$ solution. This makes a carboxylate anion ($C_{10}H_7COO^-$) from **B**, which dissolves in the aqueous layer. The other compound (**A**) remains in the CH_2Cl_2.
- Separate the layers.

19.48

and

- Dissolve both compounds in CH_2Cl_2.
- Add 10% NaOH solution. This converts C_6H_5OH into a phenoxide anion, $C_6H_5O^-$, which dissolves in the aqueous solution. The alcohol remains in the organic layer (neutral) since it is not acidic enough to be deprotonated to any significant extent by NaOH.
- Separate the layers.

19.49 To separate two compounds in an aqueous extraction, one must be water soluble (or be able to be converted into a water-soluble ionic compound by an acid–base reaction), and the other insoluble. 1-Octanol has greater than 5 C's, making it insoluble in water. Octane is an alkane, also insoluble in water. Neither compound is acidic enough to be deprotonated by a base in aqueous solution. Since their solubility properties are similar, they cannot be separated by an extraction procedure.

19.50

a. Molecular formula: $C_3H_5ClO_2$ ———— one double bond or ring
   IR: 3500–2500 cm^{-1}; 1714 cm^{-1} ———— C=O and O–H
   NMR data: 2.87 (triplet, 2H); 3.76 (triplet, 2H); 11.8 (singlet, 1H) ppm

b. Molecular formula: $C_8H_8O_3$ ———— 5 double bonds or rings
   IR: 3500–2500 cm^{-1}; 1688 cm^{-1} ———— C=O and O–H
   NMR data: 3.8 (singlet, 3H); 7.0 (doublet, 2H), 7.9 (doublet, 2H); 12.7 (singlet, 1H) ppm

para disubstituted benzene ring

c. Molecular formula: $C_8H_8O_3$ ———— 5 double bonds or rings
   IR: 3500–2500 cm^{-1}; 1710 cm^{-1} ———— C=O and O–H
   NMR data: 4.7 (singlet, 2H); 6.9–7.3 (multiplet, 5H), 11.3 (singlet, 1H) ppm

monosubstituted benzene ring

19.51

Compound A: Molecular formula $C_4H_8O_2$ (one degree of unsaturation)
IR absorptions at 3600–3200 (O–H), 3000–2800 (C–H), and 1700 (C=O) cm^{-1}.
^{1}H NMR data:

absorption	ppm	# of H's	Explanation	Structure:
singlet	2.2	3	a CH$_3$ group	
singlet	2.55	1	1H adjacent to none or OH	
triplet	2.7	2	2H's adjacent to 2H's	
triplet	3.9	2	2H's adjacent to 2H's	

Compound **B**: Molecular formula $C_4H_8O_2$ (one degree of unsaturation)
IR absorptions at 3500–2500 (O–H) and 1700 (C=O) cm^{-1}.
^{1}H NMR data:

absorption	ppm	# of H's	Explanation	Structure:
doublet	1.6	6	6H's adjacent to 1H	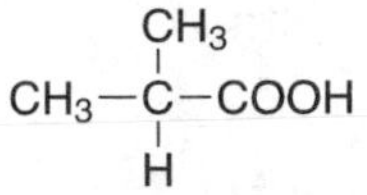
septet	2.3	1	1H adjacent to 6H's	$CH_3-\overset{\displaystyle CH_3}{\underset{\displaystyle H}{C}}-COOH$
singlet (very broad)	10.7	1	OH of RCOOH	

19.52

Molecular formula $C_6H_{12}O_2$ (1 double bond due to COOH)

^{1}H NMR: 1.1 δ (singlet), 2.2 δ (singlet), 11.9 δ (singlet)

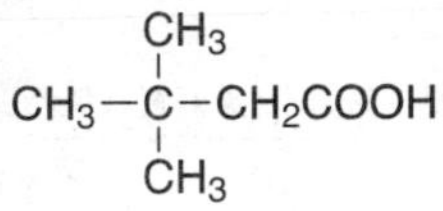

$CH_3-\overset{\displaystyle CH_3}{\underset{\displaystyle CH_3}{C}}-CH_2COOH$

19.53

Molecular formula: $C_8H_6O_4$: 6 degrees of unsaturation
IR 1692 cm^{-1} (C=O)
^{1}H NMR 8.2 and 10.0 ppm (singlets)

↑ aromatic H   ↑ COOH

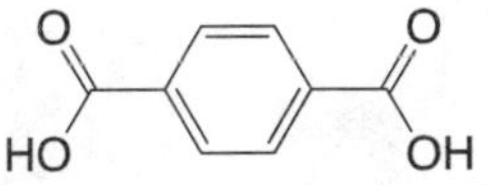

19.54

A 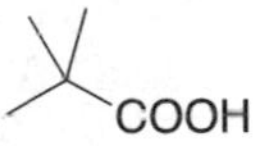   3 different C's
Spectrum [2]: peaks at 27, 39, 186 ppm

B 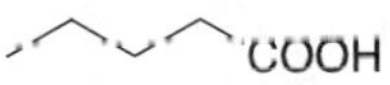   5 different C's
Spectrum [1]: peaks at 14, 22, 27, 34, 181 ppm

C 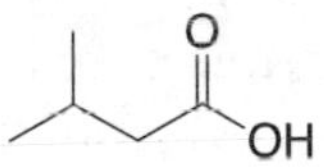   4 different C's
Spectrum [3]: peaks at 22, 26, 43, 180 ppm

19.55

threonine

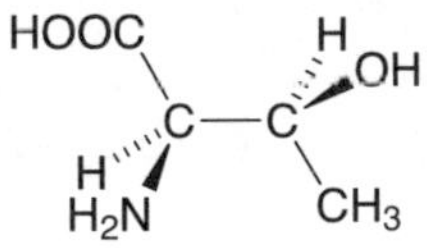

2R,3S   2S,3S   2R,3R   2S,3R
naturally
occurring

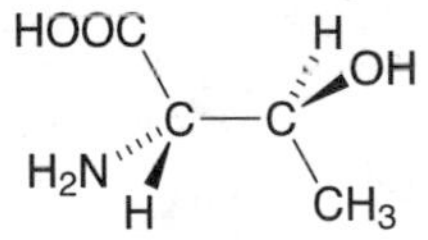

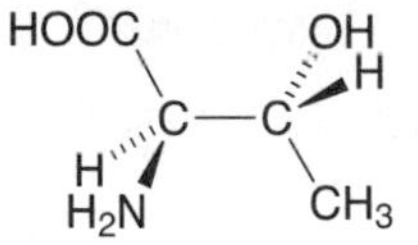

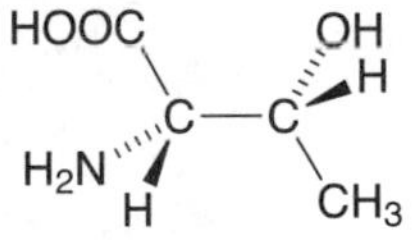

19.56

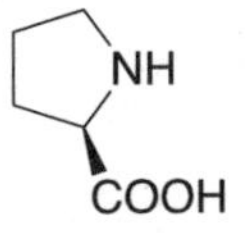

proline 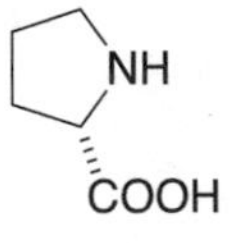   enantiomer 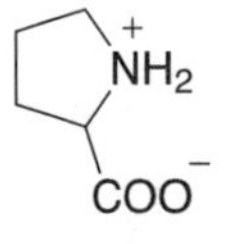   zwitterion

19.57

a. methionine

b. serine

$$\underset{\substack{\text{pH = 1}}}{\overset{\displaystyle \text{O} \\ \text{H}_3\overset{+}{\text{N}}-\text{CH}-\overset{\parallel}{\text{C}}-\text{OH} \\ \overset{|}{\text{CH}_2} \\ \overset{|}{\text{CH}_2\text{SCH}_3}}{}}$$

$$\underset{\substack{\text{pH = 7} \\ \text{form at isoelectric point}}}{\overset{\displaystyle \text{O} \\ \text{H}_3\overset{+}{\text{N}}-\text{CH}-\overset{\parallel}{\text{C}}-\text{O}^- \\ \overset{|}{\text{CH}_2} \\ \overset{|}{\text{CH}_2\text{SCH}_3}}{}}$$

$$\underset{\substack{\text{pH = 11}}}{\overset{\displaystyle \text{O} \\ \text{H}_2\text{N}-\text{CH}-\overset{\parallel}{\text{C}}-\text{O}^- \\ \overset{|}{\text{CH}_2} \\ \overset{|}{\text{CH}_2\text{SCH}_3}}{}}$$

$$\underset{\substack{\text{pH = 1}}}{\overset{\displaystyle \text{O} \\ \text{H}_3\overset{+}{\text{N}}-\text{CH}-\overset{\parallel}{\text{C}}-\text{OH} \\ \overset{|}{\text{CH}_2} \\ \overset{|}{\text{OH}}}{}}$$

$$\underset{\substack{\text{pH = 7} \\ \text{form at isoelectric point}}}{\overset{\displaystyle \text{O} \\ \text{H}_3\overset{+}{\text{N}}-\text{CH}-\overset{\parallel}{\text{C}}-\text{O}^- \\ \overset{|}{\text{CH}_2} \\ \overset{|}{\text{OH}}}{}}$$

$$\underset{\substack{\text{pH = 11}}}{\overset{\displaystyle \text{O} \\ \text{H}_2\text{N}-\text{CH}-\overset{\parallel}{\text{C}}-\text{O}^- \\ \overset{|}{\text{CH}_2} \\ \overset{|}{\text{OH}}}{}}$$

19.58

a. cysteine $pI = \dfrac{pK_a(\text{COOH}) + pK_a(\text{NH}_3^+)}{2} = (2.05) + (10.25) / 2 = \mathbf{6.15}$

b. methionine $pI = \dfrac{pK_a(\text{COOH}) + pK_a(\text{NH}_3^+)}{2} = (2.28) + (9.21) / 2 = \mathbf{5.75}$

19.59

The first equivalent of NH_3 acts as a base to remove a proton from the carboxylic acid. A second equivalent then acts as a nucleophile to displace X to form the ammonium salt of the amino acid.

19.60 The first equivalent of NaH removes the most acidic proton; that is, the OH proton on the phenol. The resulting phenoxide can then act as a nucleophile to displace I to form a substitution product. With two equivalents, both OH protons are removed. In this case the more nucleophilic O atom is the strongest base; that is, the alkoxide derived from the alcohol (not the phenoxide) so this negatively charged O atom reacts first in a nucleophilic substitution reaction.

19.61

p-hydroxybenzoic acid
less acidic than benzoic acid

like charges on nearby atoms
destabilizing

The OH group donates electron density by its resonance effect and this destabilizes the conjugate base, making the acid less acidic than benzoic acid.

Intramolecular hydrogen bonding stabilizes the conjugate base, making the acid more acidic than benzoic acid.

o-hydroxybenzoic acid
more acidic than benzoic acid

19.62 By periodic trends, H_d is least acidic since it is bonded to N (not O). H_c is least acidic of the OH protons since it is not bonded to the benzene ring, and thus its conjugate base is not resonance-stabilized. H_a and H_b are thus the most acidic H's, although their relative acidity is difficult to determine. Usually alkyl groups on benzene rings decrease the acidity of phenolic OH's more in the para position than the meta position. For example, *p*-cresol is somewhat less acidic than *m*-cresol:

m-cresol
$pK_a = 10.01$
more acidic

p-cresol
$pK_a = 10.17$
less acidic

If the alkyl group on the benzene ring has a similar electron donating effect, that would make H_a more acidic than H_b. In increasing acidity: $H_d < H_c < H_b < H_a$.

adrenaline

Chapter 20: Introduction to Carbonyl Chemistry

♦ Reduction reactions

[1] Reduction of aldehydes and ketones to 1° and 2° alcohols (20.4)

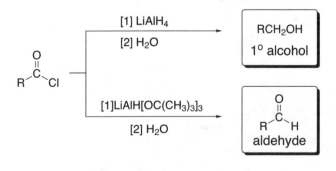

NaBH$_4$/CH$_3$OH
or
[1] LiAlH$_4$ [2] H$_2$O
or
H$_2$/Pd/C

1° or 2° alcohol

[2] Reduction of α,β-unsaturated aldehydes and ketones (20.4C)

NaBH$_4$ / CH$_3$OH
• reduction of the C=O only

H$_2$ (1 equiv) / Pd/C
• reduction of the C=C only

H$_2$ (excess) / Pd/C
• reduction of both π bonds

[3] Enantioselective ketone reduction (20.6)

[1] (R)- or (S)- CBS reagent
[2] H$_2$O

(S) 2° alcohol or (R) 2° alcohol

• A single enantiomer is formed.

[4] Reduction of acid chlorides (20.7A)

[1] LiAlH$_4$
[2] H$_2$O

RCH$_2$OH
1° alcohol

[1] LiAlH[OC(CH$_3$)$_3$]$_3$
[2] H$_2$O

aldehyde

• LiAlH$_4$, a strong reducing agent, reduces an acid chloride all the way to a 1° alcohol.

• With LiAlH[OC(CH$_3$)$_3$]$_3$, a milder reducing agent, reduction stops at the aldehyde stage.

[5] Reduction of esters (20.7A)

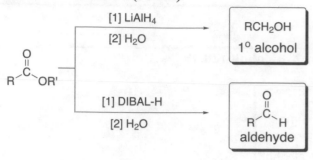

$\xrightarrow{\begin{array}{c}\text{[1] LiAlH}_4\\\text{[2] H}_2\text{O}\end{array}}$ RCH₂OH 1° alcohol

- LiAlH₄, a strong reducing agent, reduces an ester all the way to a 1° alcohol.

$\xrightarrow{\begin{array}{c}\text{[1] DIBAL-H}\\\text{[2] H}_2\text{O}\end{array}}$ aldehyde

- With DIBAL-H, a milder reducing agent, reduction stops at the aldehyde stage.

[6] Reduction of carboxylic acids to 1° alcohols (20.7B)

$\xrightarrow{\begin{array}{c}\text{[1] LiAlH}_4\\\text{[2] H}_2\text{O}\end{array}}$ RCH₂OH 1° alcohol

[7] Reduction of amides to amines (20.7B)

$\xrightarrow{\begin{array}{c}\text{[1] LiAlH}_4\\\text{[2] H}_2\text{O}\end{array}}$ RCH₂—N— amine

♦ Oxidation reactions

Oxidation of aldehydes to carboxylic acids (20.8)

$\xrightarrow[\begin{array}{c}\text{or}\\\text{Ag}_2\text{O/NH}_4\text{OH}\end{array}]{\text{CrO}_3,\ \text{Na}_2\text{Cr}_2\text{O}_7,\ \text{K}_2\text{Cr}_2\text{O}_7,\ \text{KMnO}_4}$ carboxylic acid

♦ Preparation of organometallic reagents (20.9)

[1] Organolithium reagents:

R—X + 2 Li ⟶ R—Li + LiX

[2] Grignard reagents:

R—X + Mg $\xrightarrow{\text{(CH}_3\text{CH}_2)_2\text{O}}$ R—Mg-X

[3] Organocuprate reagents:

R—X + 2 Li ⟶ R—Li + LiX

2 R—Li + CuI ⟶ R—Cu⁻ Li⁺ + LiI

[4] Lithium and sodium acetylides:

R—C≡C—H $\xrightarrow{\text{Na}^+\ ^-\text{NH}_2}$ R—C≡C⁻ Na⁺ + NH₃
a sodium acetylide

R—C≡C—H $\xrightarrow{\text{R—Li}}$ R—C≡C—Li + R—H
a lithium acetylide

♦ Reactions with organometallic reagents

[1] Reaction as a base (20.9C)

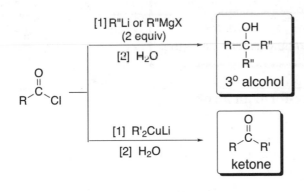

$R-M + H-\ddot{O}-R \longrightarrow \boxed{R-H} + M^+ \; \ddot{:}\ddot{O}-R$

- $RM = RLi, RMgX, R_2CuLi$
- This acid-base reaction occurs with: H_2O, ROH, RNH_2, R_2NH, RSH, $RCOOH$, $RCONH_2$, $RCONHR$.

[2] Reaction with aldehydes and ketones to form 1°, 2°, and 3° alcohols (20.10)

$$R-\overset{O}{\overset{\|}{C}}-H\,(R') \xrightarrow[\text{[2] } H_2O]{\text{[1] } R''MgX \text{ or } R''Li} \quad R-\overset{OH}{\underset{R''}{\overset{|}{C}}}-H\,(R')$$

1°, 2°, or 3° alcohol

[3] Reaction with esters to form 3° alcohols (20.13A)

$$R-\overset{O}{\overset{\|}{C}}-OR' \xrightarrow[\text{[2] } H_2O]{\substack{\text{[1]} R''Li \text{ or } R''MgX \\ \text{(2 equiv)}}} \quad R-\overset{OH}{\underset{R''}{\overset{|}{C}}}-R''$$

3° alcohol

[4] Reaction with acid chlorides (20.13)

$$R-\overset{O}{\overset{\|}{C}}-Cl$$

$\xrightarrow[\text{[2] } H_2O]{\substack{\text{[1]} R''Li \text{ or } R''MgX \\ \text{(2 equiv)}}}$ $R-\overset{OH}{\underset{R''}{\overset{|}{C}}}-R''$

3° alcohol

$\xrightarrow[\text{[2] } H_2O]{\text{[1] } R'_2CuLi}$ $R-\overset{O}{\overset{\|}{C}}-R'$

ketone

- More reactive organometallic reagents—R''Li and R''MgX—add two equivalents of R'' to an acid chloride to form a 3° alcohol with two identical R'' groups.

- Less reactive organometallic reagents—R'$_2$CuLi—add only one equivalent of R' to an acid chloride to form a ketone.

[5] Reaction with carbon dioxide—Carboxylation (20.14A)

$$R-MgX \xrightarrow[\text{[2] } H_2O]{\text{[1] } CO_2} R-\overset{O}{\overset{\|}{C}}-OH$$

carboxylic acid

[6] Reaction with epoxides (20.14B)

[7] Reaction with α,β-unsaturated aldehydes and ketones (20.15B)

- More reactive organometallic reagents—R'Li and R'MgX—react with α,β–unsaturated carbonyls by 1,2-addition.

- Less reactive organometallic reagents—R'$_2$CuLi— react with α,β–unsaturated carbonyls by 1,4-addition.

♦ **Protecting groups (20.12)**

[1] Protecting an alcohol as a *tert*-butyldimethylsilyl ether

[2] Deprotecting a *tert*-butyldimethylsilyl ether to re-form an alcohol

Chapter 20: Answers to Problems

20.1

a.

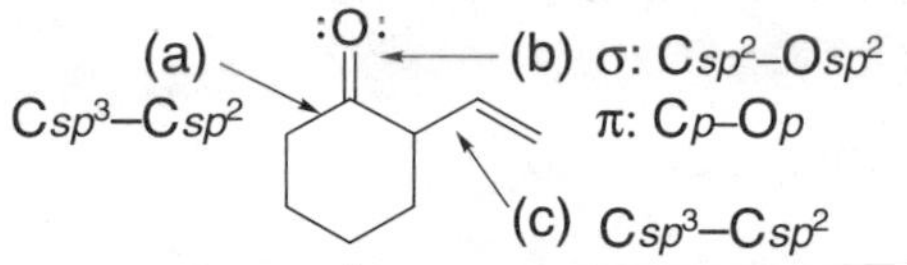

(a) C_{sp^3}–C_{sp^2}

(b) σ: C_{sp^2}–O_{sp^2}
π: C_p–O_p

(c) C_{sp^3}–C_{sp^2}

A

b. The O is sp^2 hybridized.
Both lone pairs occupy sp^2 hybrid orbitals.

20.2 A carbonyl compound with a reasonable leaving group undergoes substitution reactions. Those without good leaving groups undergo addition.

a.

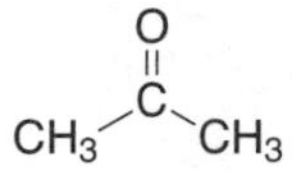

no good leaving group
addition reactions

b.

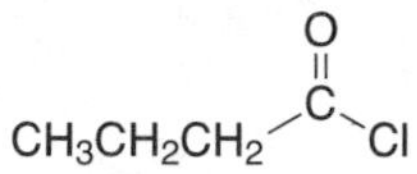

Cl–good leaving group
substitution reactions

c.

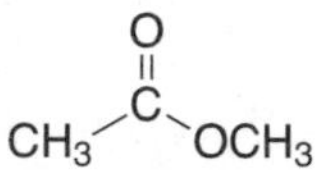

OCH_3–reasonable leaving group
substitution reactions

d.

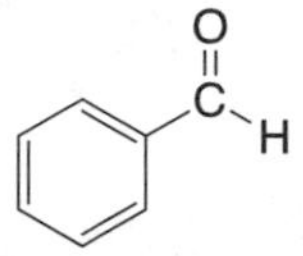

no good leaving group
addition reactions

20.3 Aldehydes are more reactive than ketones. In carbonyl compounds with leaving groups, the better the leaving group, the more reactive the carbonyl compound.

a. and

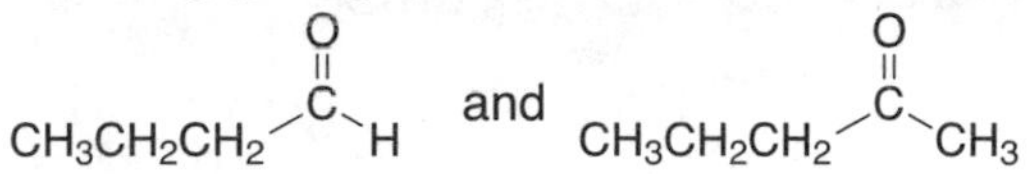

less hindered carbonyl
more reactive

c. and

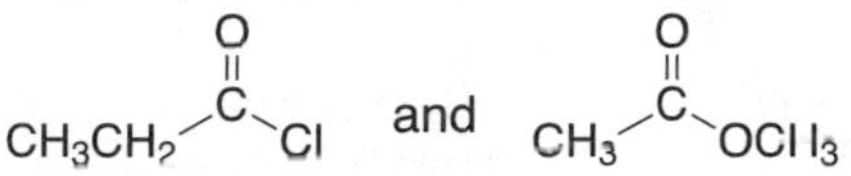

better leaving group
more reactive

b. and

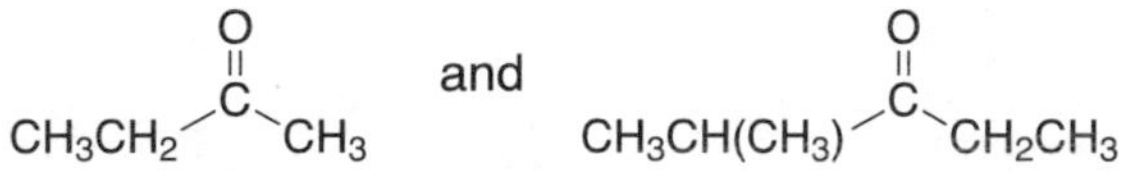

less hindered carbonyl
more reactive

d. and

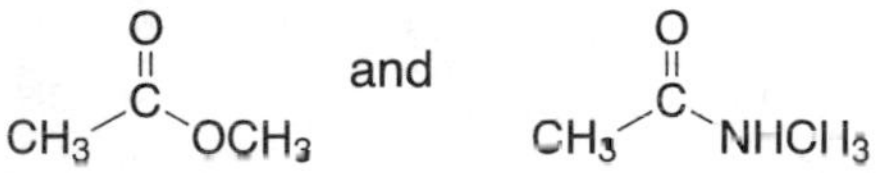

better leaving group
more reactive

20.4 $NaBH_4$ reduces aldehydes to 1° alcohols, and ketones to 2° alcohols.

a.

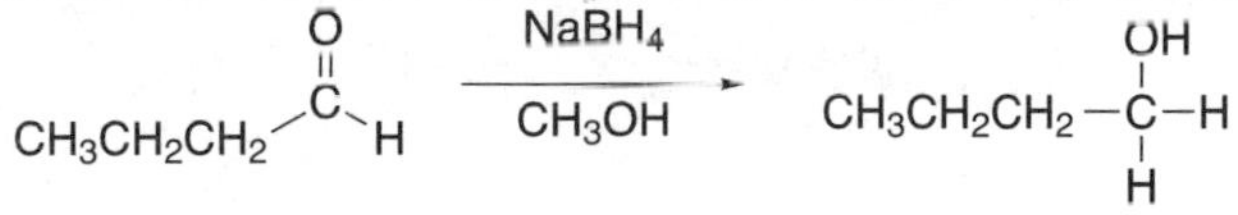

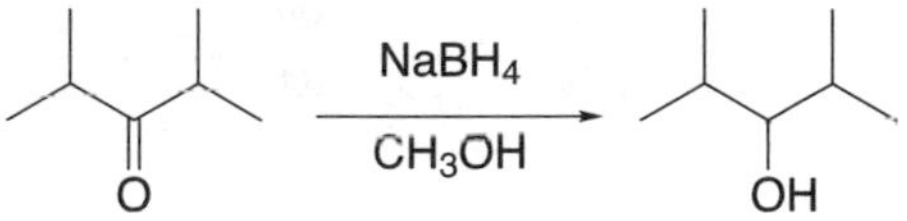

b.

c.

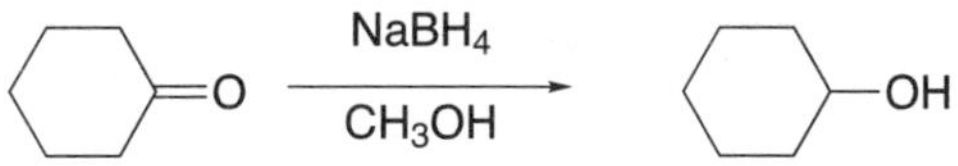

20.5 1° alcohols are prepared from aldehydes and 2° alcohols are from ketones.

a.

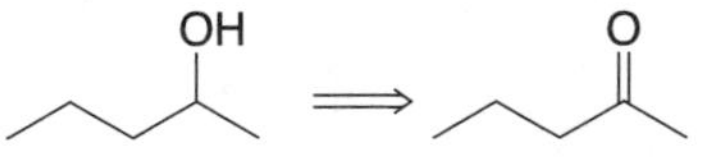

b.

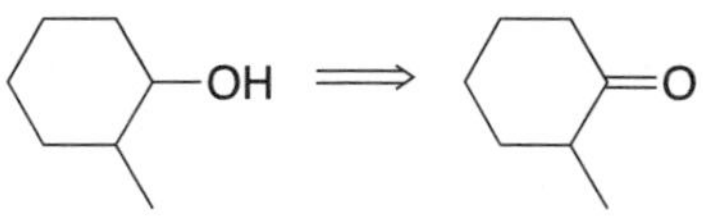

c.

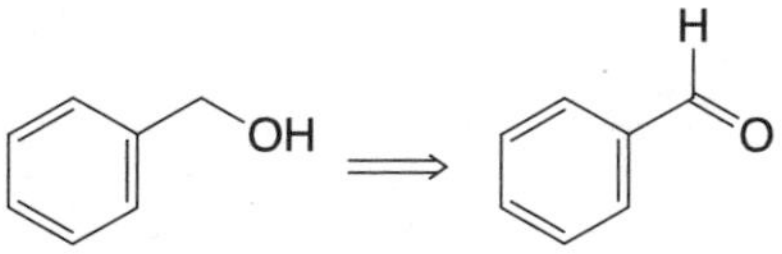

20.6

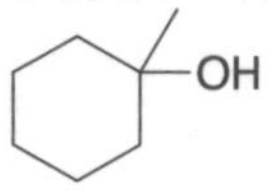
OH

3° Alcohols cannot be made by reduction of a carbonyl group, because they do not contain a H on the C with the OH.

1-methylcyclohexanol

20.7

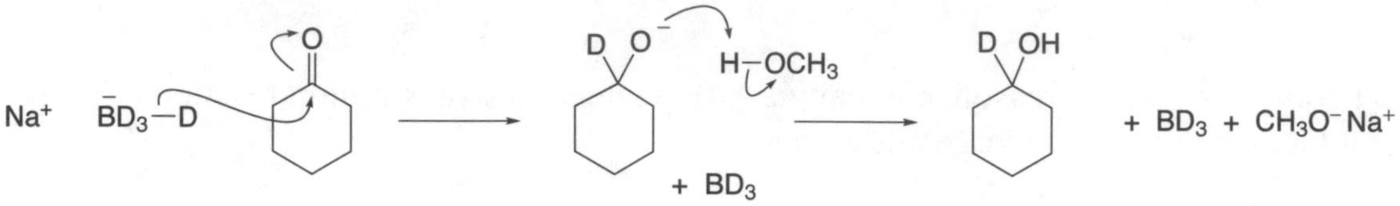
Na+
BD3—D
O
D O−
H—OCH3
+ BD3
D OH
+ BD3 + CH3O− Na+

20.8

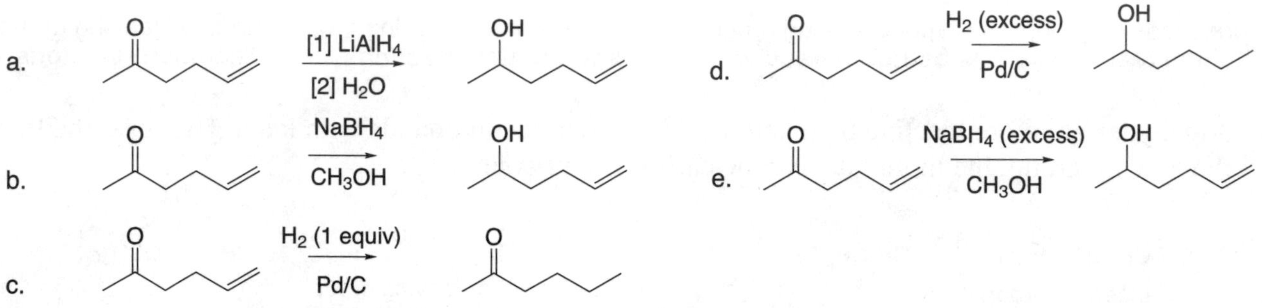
a.
O
[1] LiAlH4
[2] H2O
OH
b.
O
NaBH4
CH3OH
OH
c.
O
H2 (1 equiv)
Pd/C
O
d.
O
H2 (excess)
Pd/C
OH
e.
O
NaBH4 (excess)
CH3OH
OH

20.9

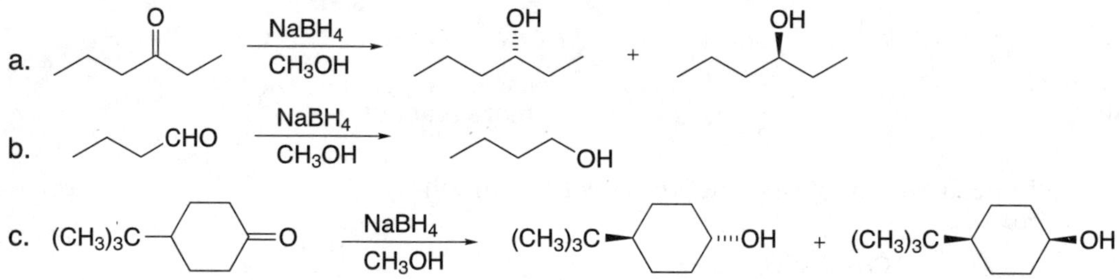
a.
O
NaBH4
CH3OH
OH
+
OH
b.
CHO
NaBH4
CH3OH
OH
c. (CH3)3C
O
NaBH4
CH3OH
(CH3)3C
OH
+
(CH3)3C
OH

20.10

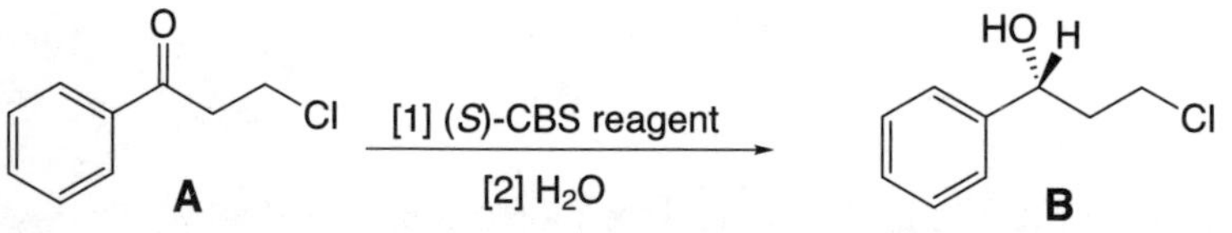
O
Cl
A
[1] (S)-CBS reagent
[2] H2O
HO H
Cl
B

20.11

Part [1]: Nucleophilic substitution of H for Cl

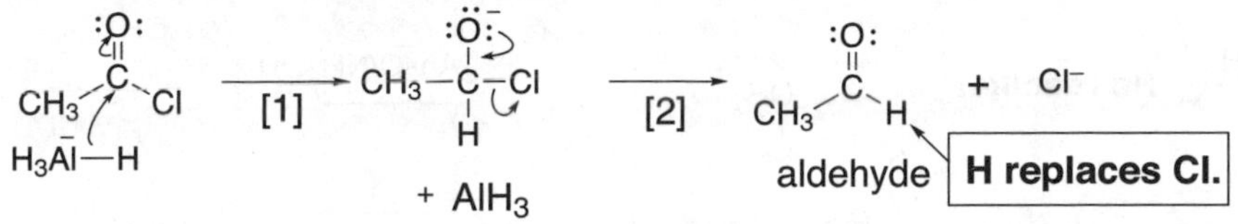

aldehyde | H replaces Cl. |

+ AlH₃

Part [2]: Nucleophilic addition of H⁻ to form an alcohol

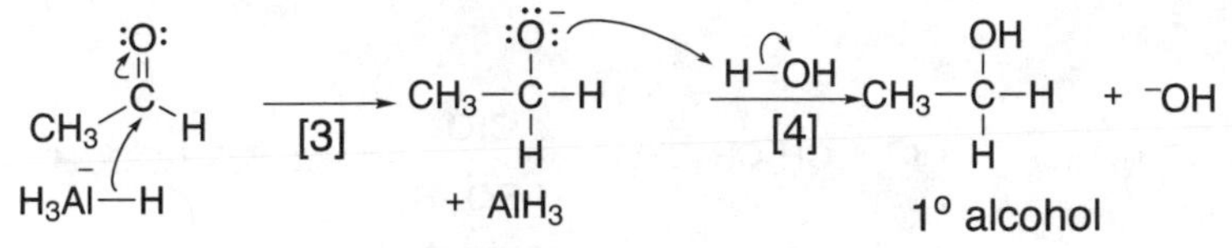

1° alcohol

20.12 Acid chlorides and esters can be reduced to 1° alcohols. Keep the carbon skeleton the same in drawing an ester and acid chloride precursor.

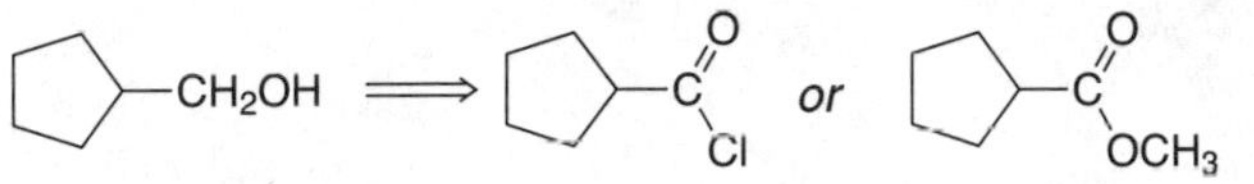

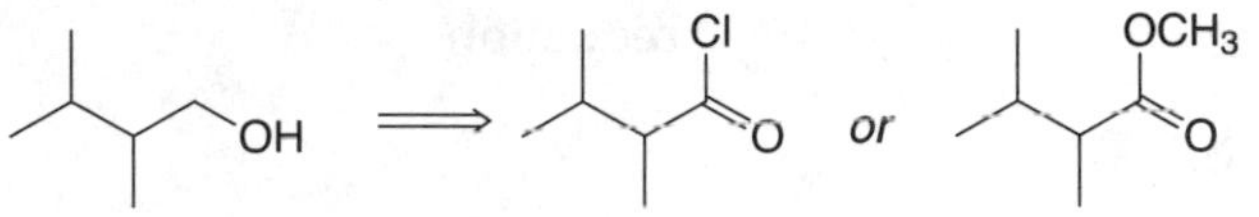

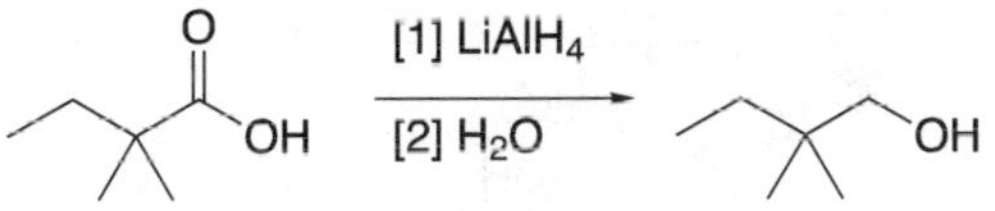

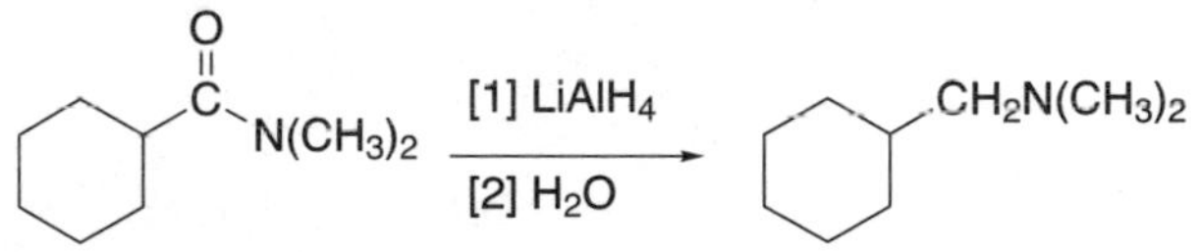

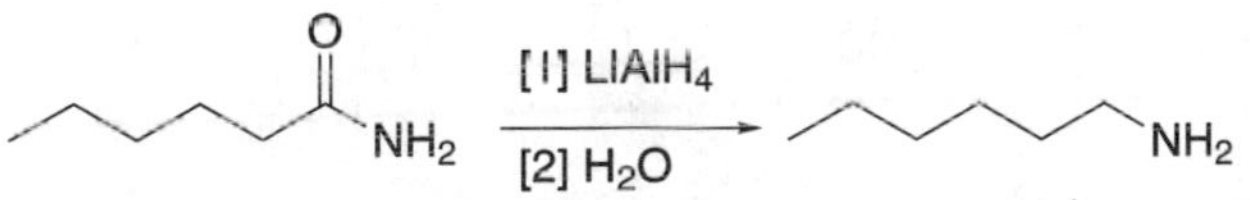

20.13

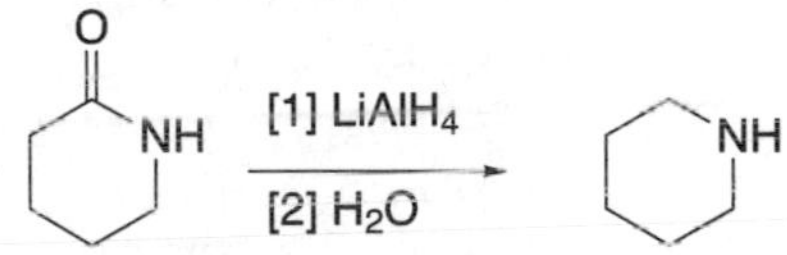

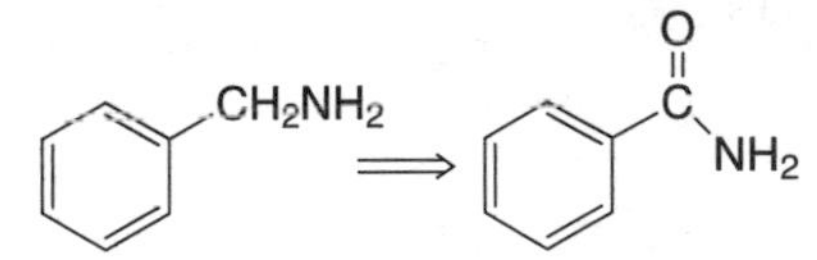

20.14

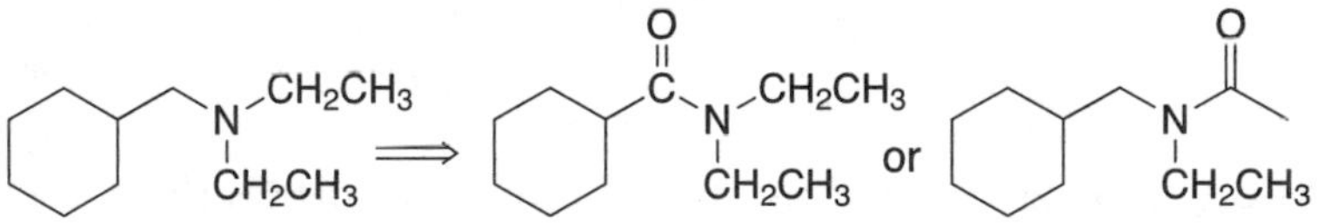

20.15

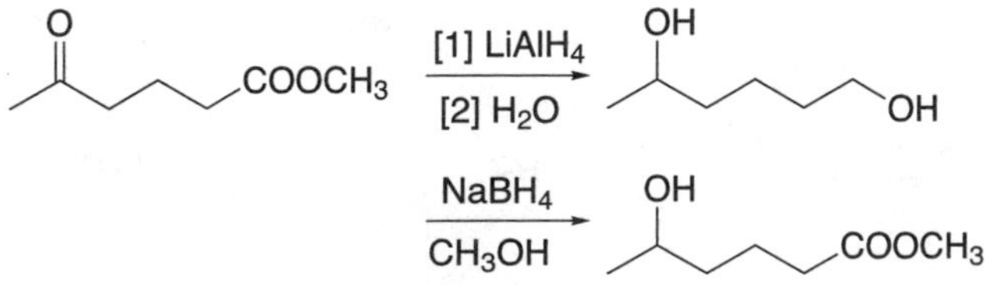

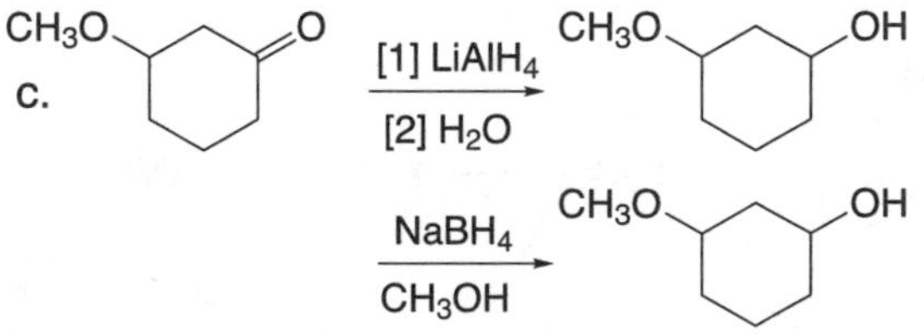

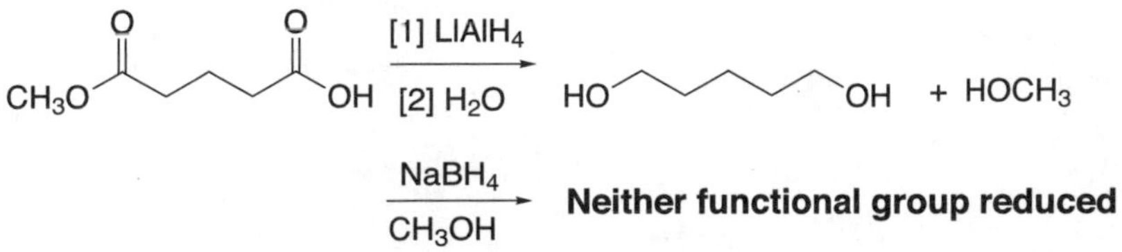

20.16 Tollens reagent reacts only with aldehydes.

a.

$$C_6H_5-CH_2OH \xrightarrow{Ag_2O/NH_4OH} \textbf{No reaction}$$

$$CH_2OH \xrightarrow{Na_2Cr_2O_7\ H_2SO_4/H_2O} C_6H_5-COOH$$

b.

$$\xrightarrow{Ag_2O/NH_4OH}$$

$$\xrightarrow{Na_2Cr_2O_7\ H_2SO_4/H_2O}$$

20.17 One possible reagent is shown:

a. $CH_3CH_2CHO \xrightarrow[\text{[2] } H_2O]{\text{[1] } LiAlH_4} CH_3CH_2CH_2OH$

reduction

c. $CH_3CH_2CHO \xrightarrow[H_2SO_4/H_2O]{CrO_3} CH_3CH_2COOH$

oxidation

b. $\xrightarrow[\text{[2] } H_2O]{\text{[1] } LiAlH_4}$

reduction

d. $C_6H_5 \xrightarrow[HCl]{Zn(Hg)} C_6H_5$

reduction

20.18

B

c. **B** $\xrightarrow{PCC}$

a. **B** $\xrightarrow[CH_3OH]{NaBH_4}$

d. **B** $\xrightarrow{Ag_2O/NH_4OH}$

b. **B** $\xrightarrow[\text{[2] } H_2O]{\text{[1] } LiAlH_4}$

e. **B** $\xrightarrow[H_2SO_4/H_2O]{CrO_3}$

20.19

a. $CH_3CH_2Br + 2Li \longrightarrow CH_3CH_2Li + LiBr$

b. $CH_3CH_2Br + Mg \longrightarrow CH_3CH_2MgBr$

c. $CH_3CH_2Br + 2Li \longrightarrow CH_3CH_2Li + LiBr$

$2CH_3CH_2Li + CuI \longrightarrow LiCu(CH_2CH_3)_2 + LiI$

20.20

a. $\overset{}{\bigcirc}-Li + H_2O \longrightarrow \bigcirc + LiOH$

c. $\xrightarrow{}$ $+ H_2O \longrightarrow$ $+ HOMgBr$

b. $CH_3-\overset{CH_3}{\underset{CH_3}{\overset{|}{\underset{|}{C}}}}-MgBr + H_2O \longrightarrow CH_3-\overset{CH_3}{\underset{CH_3}{\overset{|}{\underset{|}{C}}}}-H + HOMgBr$

d. $CH_3CH_2C\equiv C-Li + H_2O \longrightarrow CH_3CH_2C\equiv CH + LiOH$

20.21 To draw the product, add the benzene ring to the carbonyl carbon and protonate the oxygen.

a.

b.

c.

d.

20.22 To draw the products, add the alkyl group to the carbonyl carbon and protonate the oxygen.

a.

b.

c.

d.

20.23 Addition of RM always occurs from above and below the place of the molecule, so if a new stereogenic center is formed, there is a mixture of enantiomers.

a.

b.

20.24

a. $CH_3-\overset{OH}{\underset{H}{C}}-CH_3 \implies CH_3MgBr + \underset{CH_3}{\overset{O}{H}}$

b. (cyclohexyl)$-CH_2OH \implies$ (cyclohexyl)$-MgBr + \underset{H}{\overset{O}{H}}$

c. (cyclohexyl)$-\overset{OH}{\underset{H}{C}}-CH_2CH_3 \implies$ (cyclohexyl)$-MgBr + \underset{CH_2CH_3}{\overset{O}{H}}$

or

(cyclohexyl)$-\overset{OH}{\underset{H}{C}}-CH_2CH_3 \implies$ (cyclohexyl)$-\overset{O}{C}H + CH_3CH_2MgBr$

d. $\implies \sim\sim\sim MgBr + \underset{CH_3}{\overset{O}{H}}$

or

$\implies \sim\sim\sim\overset{O}{C}H + CH_3MgBr$

20.25

a. $CH_3-\overset{OH}{\underset{CH_2CH_2CH_3}{C}}-CH_2CH_3 \implies CH_3-\overset{O}{C}-CH_2CH_3 + CH_3CH_2CH_2MgBr$

$CH_3-\overset{OH}{\underset{CH_2CH_2CH_3}{C}}-CH_2CH_3 \implies CH_3-\overset{O}{C}-CH_2CH_2CH_3 + CH_3CH_2MgBr$

$CH_3-\overset{OH}{\underset{CH_2CH_2CH_3}{C}}-CH_2CH_3 \implies CH_3CH_2CH_2-\overset{O}{C}-CH_2CH_3 + CH_3MgBr$

b. $CH_3-\overset{OH}{\underset{\text{(phenyl)}}{C}}-\text{(cyclohexyl)} \implies CH_3-\overset{O}{C}-\text{(cyclohexyl)} + \text{(phenyl)}-MgBr$

$CH_3-\overset{OH}{\underset{\text{(phenyl)}}{C}}-\text{(cyclohexyl)} \implies CH_3-\overset{O}{C}-\text{(phenyl)} + \text{(cyclohexyl)}-MgBr$

$CH_3-\overset{OH}{\underset{\text{(phenyl)}}{C}}-\text{(cyclohexyl)} \implies \text{(phenyl)}-\overset{O}{C}-\text{(cyclohexyl)} + CH_3MgBr$

20.26

$HO-\text{(cyclohexanone)} \xrightarrow[\underset{N\diagdown NH}{\text{TBDMS}-Cl}]{} TBDMSO-\text{(cyclohexanone)} \xrightarrow[\text{[2] } H_2O]{\text{[1] } BrMgCH_2CH_2CH_2CH_3} TBDMSO-\text{(cyclohexane)}\overset{CH_2CH_2CH_2CH_3}{\underset{OH}{}}$

$\xrightarrow{(CH_3CH_2CH_2CH_2)_4N^+ F^-} HO-\text{(cyclohexane)}\overset{CH_2CH_2CH_2CH_3}{\underset{OH}{}}$

20.27

a. CH$_3$CH$_2$$-$C($=$O)$-$Cl $\xrightarrow[\text{[2] H}_2\text{O}]{\text{[1] CH}_3\text{CH}_2\text{CH}_2\text{CH}_2\text{MgBr}}$ CH$_3$CH$_2$$-$C(OH)($-CH_2CH_2CH_2CH_3$)$-CH_2CH_2CH_2CH_3$

b. Ph$-$C($=$O)$-$OCH$_3$ $\xrightarrow[\text{[2] H}_2\text{O}]{\text{[1] CH}_3\text{CH}_2\text{CH}_2\text{CH}_2\text{MgBr}}$ CH$_3$CH$_2$CH$_2$CH$_2$$-$C(OH)(Ph)$-CH_2CH_2CH_2CH_3$

c. (pentanoate) $-$OCH$_2$CH$_3$ $\xrightarrow[\text{[2] H}_2\text{O}]{\text{[1] CH}_3\text{CH}_2\text{CH}_2\text{CH}_2\text{MgBr}}$ CH$_3$CH$_2$CH$_2$CH$_2$$-$C(OH)($-CH_2CH_2CH_2CH_3$)$-CH_2CH_2CH_2CH_3$

20.28

a. (dicyclohexyl, C(OH)(CH$_3$)) $\Longrightarrow$ CH$_3$O$-$C($=$O)$-$CH$_3$ + cyclohexyl$-$MgBr

2 equiv

b. (CH$_3$CH$_2$CH$_2$)$_3$COH $\Longrightarrow$ CH$_3$O$-$C($=$O)$-$CH$_2$CH$_2$CH$_3$ + CH$_3$CH$_2$CH$_2$MgBr

2 equiv

c. CH$_3$$-$C(OH)(CH$_3$)$-CH_2$CH(CH$_3$)$_2$ $\Longrightarrow$ CH$_3$O$-$C($=$O)$-$CH$_2$CH(CH$_3$)$_2$ + CH$_3$MgBr

2 equiv

20.29 The R group of the organocuprate has replaced the Cl on the acid chloride.

a. CH$_3$CH$_2$$-$C($=$O)$-$Cl $\xrightarrow[\text{[2] H}_2\text{O}]{\text{[1] (CH}_3\text{)}_2\text{CuLi}}$ CH$_3$CH$_2$$-$C($=$O)$-$$\boxed{\text{CH}_3}$

c. CH$_3$CH$_2$$-$C($=$O)$-$Cl $\xrightarrow[\text{[2] H}_2\text{O}]{\text{[1] (Ph)}_2\text{CuLi}}$ $-$C($=$O)$-$$\boxed{\text{Ph}}$

b. $-$C($=$O)$-$Cl $\xrightarrow[\text{[2] H}_2\text{O}]{\text{[1] [CH}_3\text{CH}_2\text{CH(CH}_3\text{)]}_2\text{CuLi}}$ $\boxed{\text{(product)}}$

20.30

a. (CH$_3$)$_2$CHCH$_2$$-$C($=$O)$-$Cl $\xrightarrow[\text{[2] H}_2\text{O}]{\text{[1] LiAlH[OC(CH}_3\text{)}_3\text{]}_3}$ (CH$_3$)$_2$CHCH$_2$$-$C($=$O)$-$H

c. (CH$_3$)$_2$CHCH$_2$$-$C($=$O)$-$Cl $\xrightarrow[\text{[2] H}_2\text{O}]{\text{[1] (CH}_3\text{CH}_2\text{)}_2\text{CuLi}}$ $\boxed{\text{(product)}}$

b. (CH$_3$)$_2$CHCH$_2$$-$C($=$O)$-$Cl $\xrightarrow[\text{[2] H}_2\text{O}]{\text{[1] CH}_3\text{CH}_2\text{Li (2 equiv)}}$ $\boxed{\text{HO (product)}}$

d. (CH$_3$)$_2$CHCH$_2$$-$C($=$O)$-$Cl $\xrightarrow[\text{[2] H}_2\text{O}]{\text{[1] LiAlH}_4}$ (CH$_3$)$_2$CHCH$_2$CH$_2$OH

20.31

a.

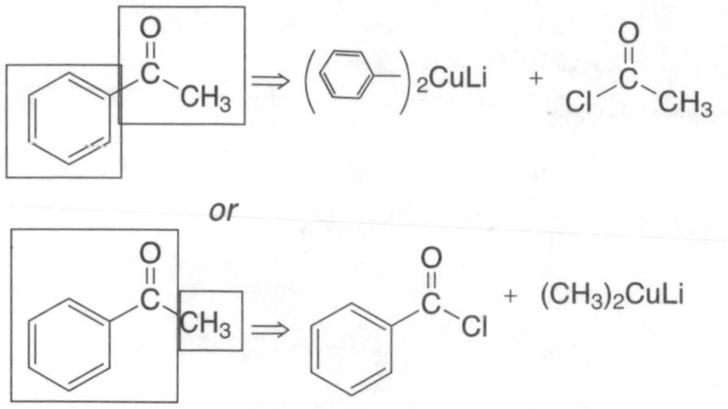

or

b.

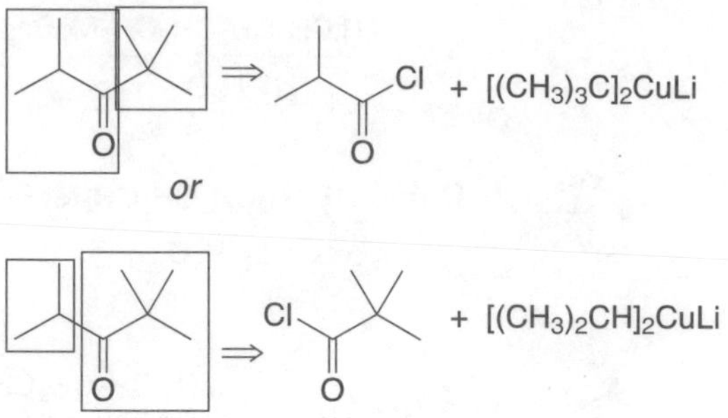

or

20.32

a. 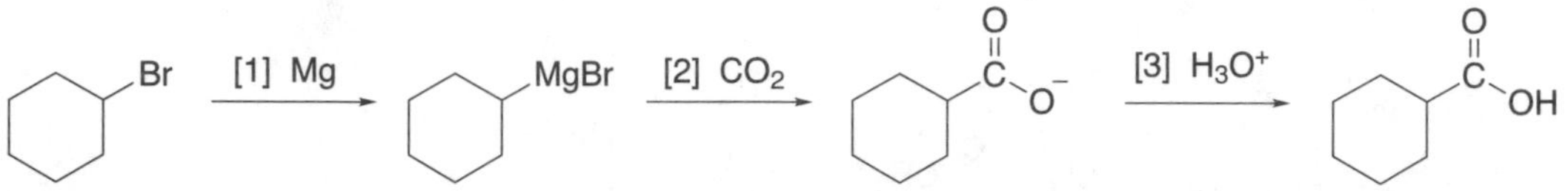 [1] Mg → MgBr [2] CO_2 → [3] H_3O^+

b. 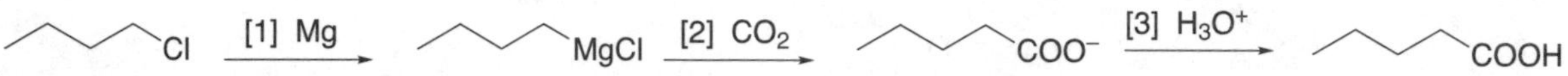 [1] Mg → MgCl [2] CO_2 → COO⁻ [3] H_3O^+ → COOH

c. 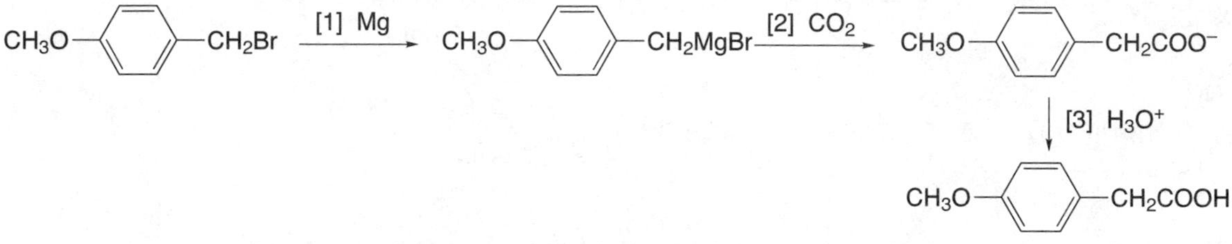 CH_3O—⟨⟩—CH_2Br [1] Mg → CH_3O—⟨⟩—CH_2MgBr [2] CO_2 → CH_3O—⟨⟩—CH_2COO^-

[3] H_3O^+

CH_3O—⟨⟩—CH_2COOH

20.33

+ HOMgBr

+ HOMgBr

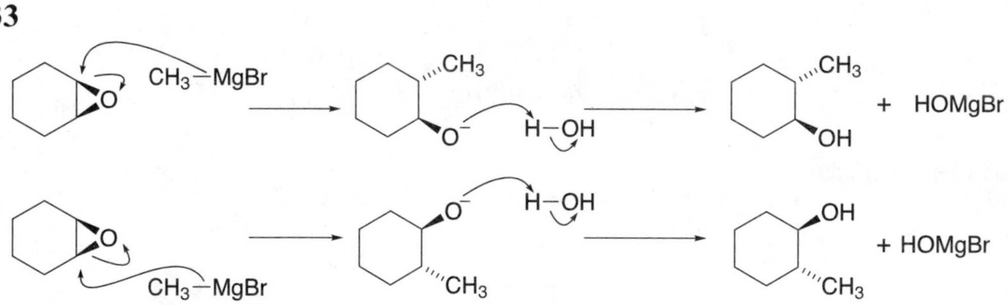

20.34 The characteristic reaction of α,β-unsaturated carbonyl compounds is nucleophilic addition. Grignard and organolithium reagents undergo 1,2-addition and organocuprate reagents undergo 1,4-addition.

a. 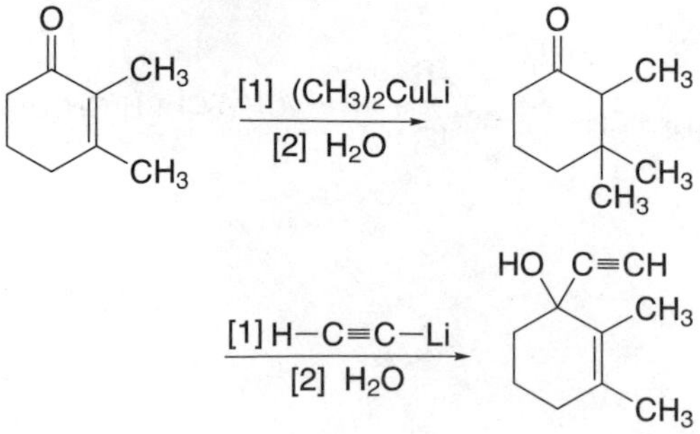 [1] $(CH_3)_2CuLi$ [2] H_2O

b. 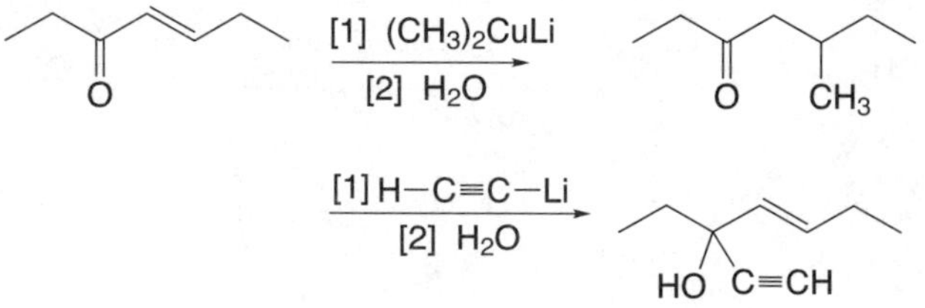 [1] $(CH_3)_2CuLi$ [2] H_2O

[1] H—C≡C—Li [2] H_2O

[1] H—C≡C—Li [2] H_2O

20.35

a. 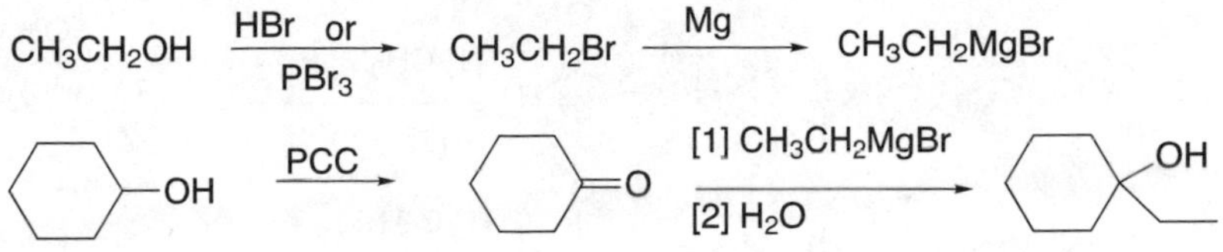 CH_3CH_2OH $\xrightarrow[\text{PBr}_3]{\text{HBr or}}$ CH_3CH_2Br $\xrightarrow{\text{Mg}}$ CH_3CH_2MgBr

b. 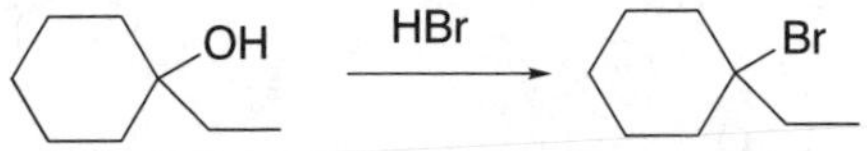 [image: cyclohexanol with ethyl] $\xrightarrow{\text{HBr}}$ [image: cyclohexyl bromide with ethyl]

[from (a)]

c. 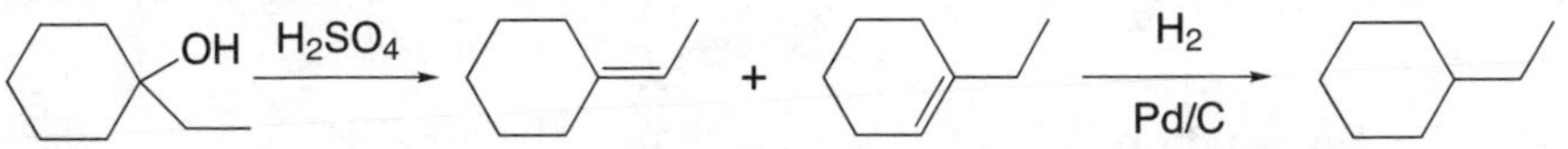 [image: 1-ethylcyclohexanol] $\xrightarrow{\text{H}_2\text{SO}_4}$ [image: ethylidenecyclohexane] + [image: ethylcyclohexene] $\xrightarrow[\text{Pd/C}]{\text{H}_2}$ [image: ethylcyclohexane]

[from (a)]

d. 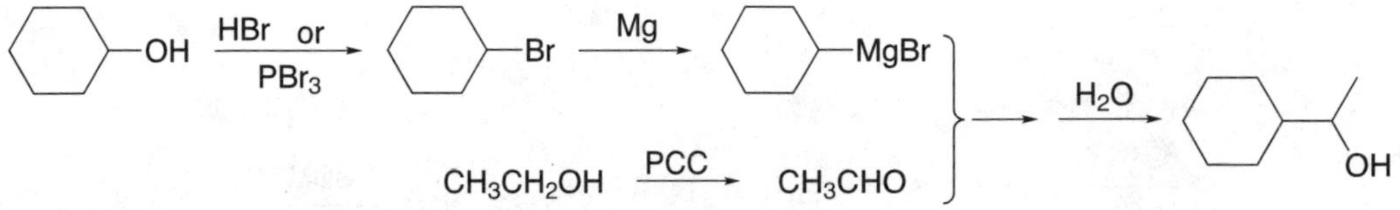 [image: cyclohexanol] $\xrightarrow[\text{PBr}_3]{\text{HBr or}}$ [image: cyclohexyl bromide] $\xrightarrow{\text{Mg}}$ [image: cyclohexyl–MgBr] $\Bigg\}$ $\xrightarrow{\text{H}_2\text{O}}$ [image: 1-cyclohexylethanol]

CH_3CH_2OH $\xrightarrow{\text{PCC}}$ CH_3CHO

e. 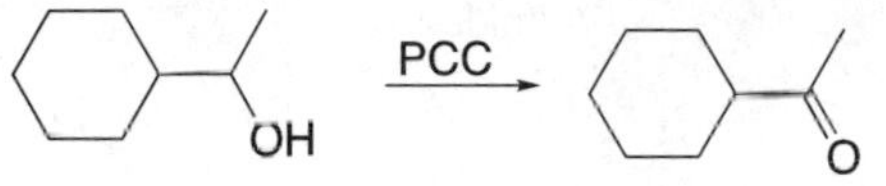 [image: 1-cyclohexylethanol] $\xrightarrow{\text{PCC}}$ [image: cyclohexyl methyl ketone]

[from (d)]

20.36

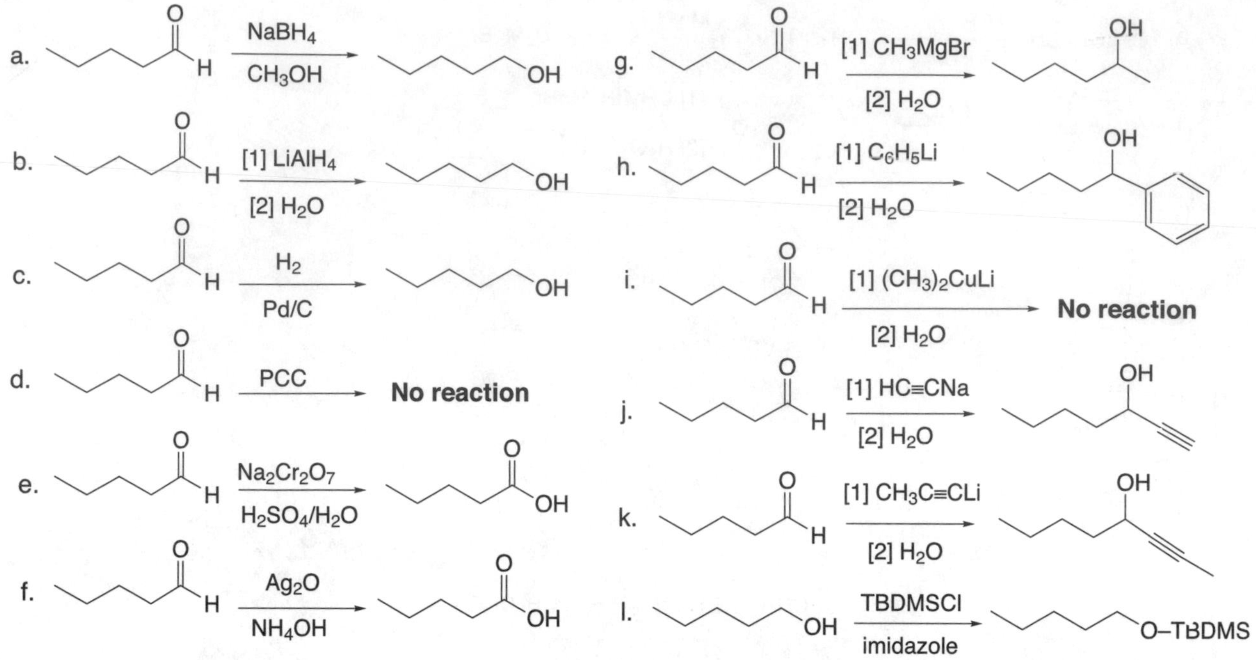

a. NaBH$_4$ / CH$_3$OH

b. [1] LiAlH$_4$ / [2] H$_2$O

c. H$_2$ / Pd/C

d. PCC → **No reaction**

e. Na$_2$Cr$_2$O$_7$ / H$_2$SO$_4$/H$_2$O

f. Ag$_2$O / NH$_4$OH

g. [1] CH$_3$MgBr / [2] H$_2$O

h. [1] C$_6$H$_5$Li / [2] H$_2$O

i. [1] (CH$_3$)$_2$CuLi / [2] H$_2$O → **No reaction**

j. [1] HC≡CNa / [2] H$_2$O

k. [1] CH$_3$C≡CLi / [2] H$_2$O

l. TBDMSCl / imidazole

20.37

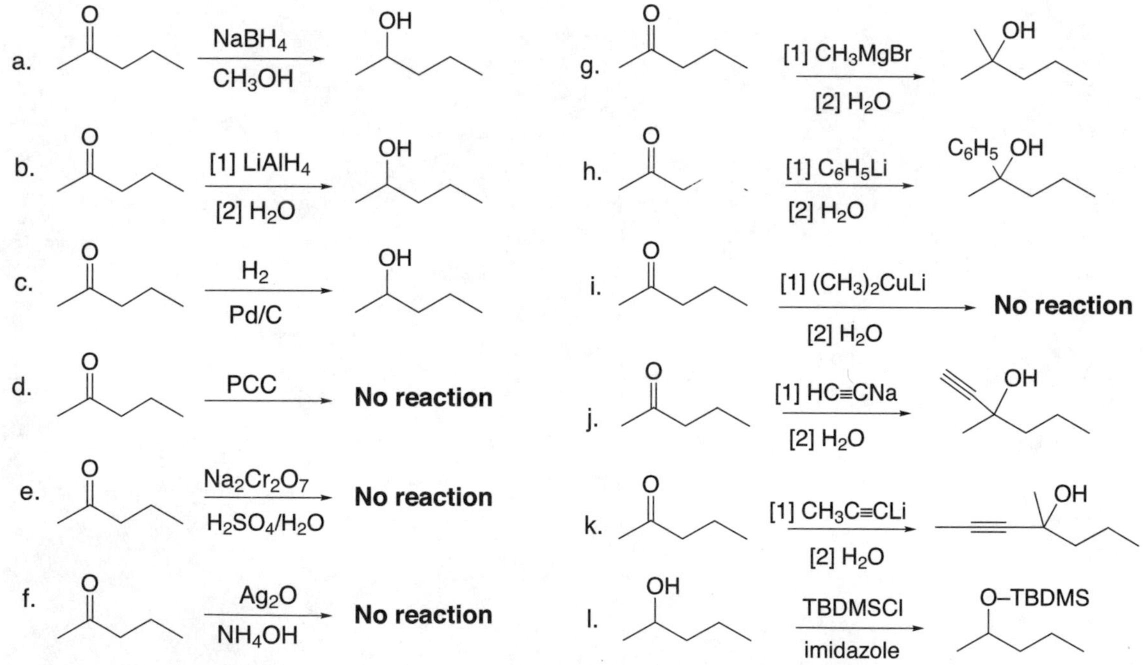

a. NaBH$_4$ / CH$_3$OH

b. [1] LiAlH$_4$ / [2] H$_2$O

c. H$_2$ / Pd/C

d. PCC → **No reaction**

e. Na$_2$Cr$_2$O$_7$ / H$_2$SO$_4$/H$_2$O → **No reaction**

f. Ag$_2$O / NH$_4$OH → **No reaction**

g. [1] CH$_3$MgBr / [2] H$_2$O

h. [1] C$_6$H$_5$Li / [2] H$_2$O

i. [1] (CH$_3$)$_2$CuLi / [2] H$_2$O → **No reaction**

j. [1] HC≡CNa / [2] H$_2$O

k. [1] CH$_3$C≡CLi / [2] H$_2$O

l. TBDMSCl / imidazole

20.38

a. CH₃CH₂CH₂CH₂Br $\xrightarrow{\text{Li (2 equiv)}}$ CH₃CH₂CH₂CH₂Li + LiBr

d. CH₃CH₂CH₂CH₂Li $\xrightarrow{H_2O}$ + LiOH

b. CH₃CH₂CH₂CH₂Br $\xrightarrow{Mg}$ MgBr

e. MgBr $\xrightarrow{D_2O}$ D + DOMgBr

c. Br $\xrightarrow[\text{[2] CuI (0.5 equiv)}]{\text{[1] Li (2 equiv)}}$ (CH₃CH₂CH₂CH₂)₂CuLi

f. Li $\xrightarrow{CH_3C\equiv CH}$ + LiC≡CCH₃

20.39

a. MgBr $\xrightarrow{CH_2=O \quad H_2O}$ OH

g. MgBr $\xrightarrow{CH_3COOH}$ + CH₃COO⁻

b. MgBr $\xrightarrow{H_2O}$ HO

h. MgBr $\xrightarrow{HC\equiv CH}$ + HC≡C⁻

c. MgBr $\xrightarrow{CH_3CH_2COCl \quad H_2O}$ OH

i. MgBr $\xrightarrow{CO_2 \quad H_3O^+}$ OH (carboxylic acid)

d. MgBr $\xrightarrow{CH_3CH_2COOCH_3 \quad H_2O}$ OH

j. MgBr $\xrightarrow{\quad \quad} \xrightarrow{H_2O}$ OH

e. MgBr $\xrightarrow{H_2O}$ + ⁻OH

k. MgBr $\xrightarrow{D_2O}$ D + ⁻OD

f. MgBr $\xrightarrow{CH_3CH_2OH}$ + CH₃CH₂O⁻

l. MgBr $\xrightarrow{\quad \quad} \xrightarrow{H_2O}$ OH

20.40

a. (benzoyl chloride, C₆H₅COCl) $\xrightarrow{\text{(CH}_3\text{CH}_2\text{CH}_2\text{CH}_2)_2\text{CuLi}}$ C₆H₅CO·CH₂CH₂CH₂CH₃

b. (methyl benzoate, C₆H₅COOCH₃) $\xrightarrow{\text{(CH}_3\text{CH}_2\text{CH}_2\text{CH}_2)_2\text{CuLi}}$ **No reaction**

c. (2-methylcyclohex-2-enone) $\xrightarrow[\text{[2] H}_2\text{O}]{\text{[1] (CH}_3\text{CH}_2\text{CH}_2\text{CH}_2)_2\text{CuLi}}$ (2-methyl-3-butylcyclohexanone)

d. (2,2-dimethyloxirane) $\xrightarrow[\text{[2] H}_2\text{O}]{\text{[1] (CH}_3\text{CH}_2\text{CH}_2\text{CH}_2)_2\text{CuLi}}$ HO-product

20.41

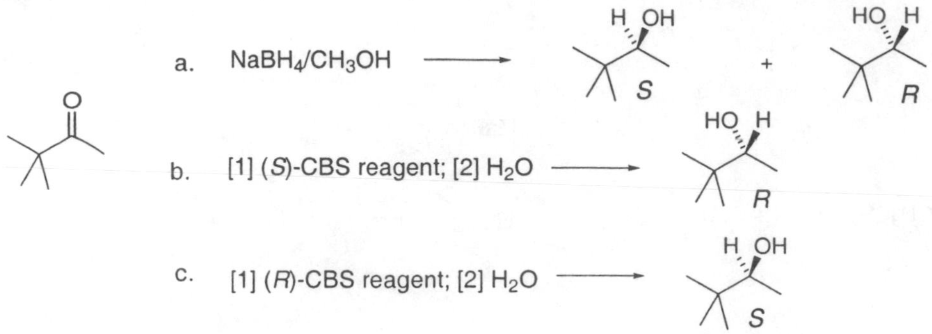

a. NaBH$_4$/CH$_3$OH ⟶

b. [1] (S)-CBS reagent; [2] H$_2$O ⟶

c. [1] (R)-CBS reagent; [2] H$_2$O ⟶

20.42

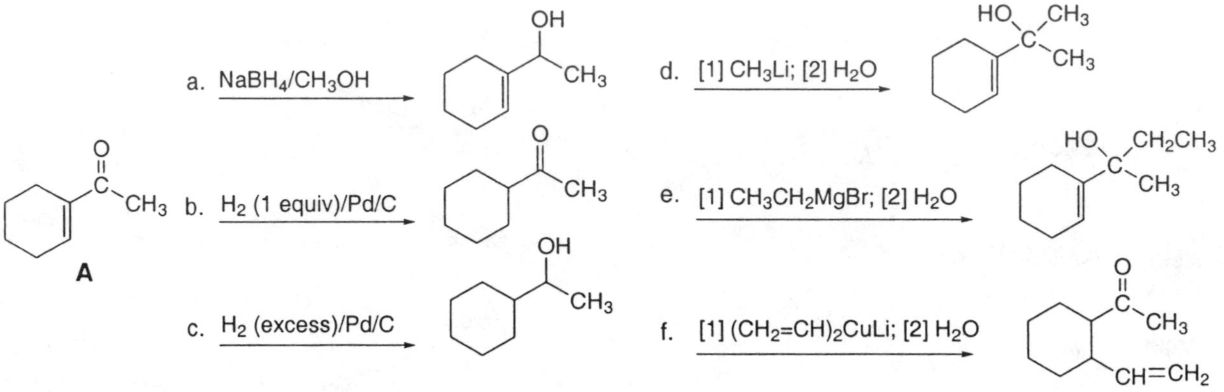

A

a. NaBH$_4$/CH$_3$OH

b. H$_2$ (1 equiv)/Pd/C

c. H$_2$ (excess)/Pd/C

d. [1] CH$_3$Li; [2] H$_2$O

e. [1] CH$_3$CH$_2$MgBr; [2] H$_2$O

f. [1] (CH$_2$=CH)$_2$CuLi; [2] H$_2$O

20.43

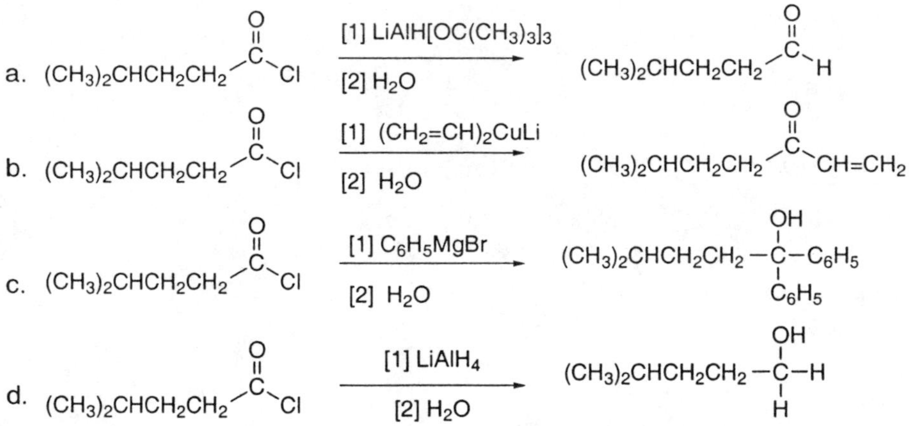

a. (CH$_3$)$_2$CHCH$_2$CH$_2$C(=O)Cl   [1] LiAlH[OC(CH$_3$)$_3$]$_3$ / [2] H$_2$O   (CH$_3$)$_2$CHCH$_2$CH$_2$C(=O)H

b. (CH$_3$)$_2$CHCH$_2$CH$_2$C(=O)Cl   [1] (CH$_2$=CH)$_2$CuLi / [2] H$_2$O   (CH$_3$)$_2$CHCH$_2$CH$_2$C(=O)CH=CH$_2$

c. (CH$_3$)$_2$CHCH$_2$CH$_2$C(=O)Cl   [1] C$_6$H$_5$MgBr / [2] H$_2$O   (CH$_3$)$_2$CHCH$_2$CH$_2$—C(OH)(C$_6$H$_5$)—C$_6$H$_5$

d. (CH$_3$)$_2$CHCH$_2$CH$_2$C(=O)Cl   [1] LiAlH$_4$ / [2] H$_2$O   (CH$_3$)$_2$CHCH$_2$CH$_2$—CH(OH)—H

20.44

a. CH$_3$CH$_2$C(=O)OCH$_2$CH$_2$CH$_3$ [1] LiAlH$_4$ / [2] H$_2$O   CH$_3$CH$_2$CH$_2$OH

b. CH$_3$CH$_2$C(=O)OCH$_2$CH$_2$CH$_3$ [1] CH$_3$CH$_2$CH$_2$MgCl / [2] H$_2$O   CH$_3$CH$_2$—C(OH)(CH$_2$CH$_2$CH$_3$)—CH$_2$CH$_2$CH$_3$

c. CH$_3$CH$_2$C(=O)OCH$_2$CH$_2$CH$_3$ [1] DIBAL-H / [2] H$_2$O   CH$_3$CH$_2$C(=O)H

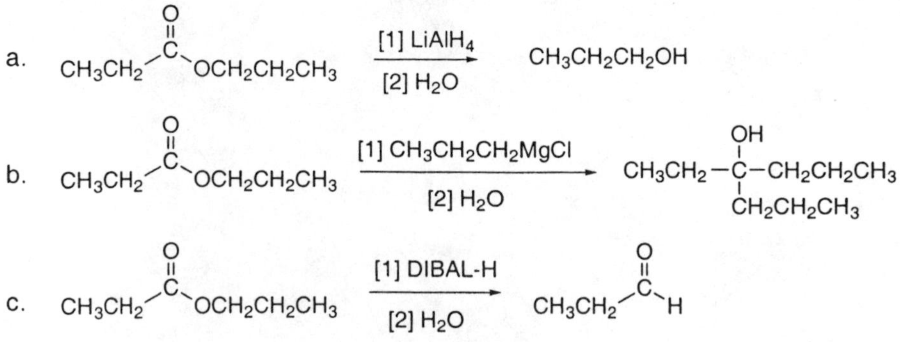

20.45

a. HO–...–CHO $\xrightarrow[\text{H}_2\text{SO}_4/\text{H}_2\text{O}]{\text{CrO}_3}$ HOOC–...–COOH

c. HO–...–CHO $\xrightarrow[\text{NH}_4\text{OH}]{\text{Ag}_2\text{O}}$ HO–...–COOH

b. HO–...–CHO $\xrightarrow{\text{PCC}}$ OHC–...–CHO

d. HO–...–CHO $\xrightarrow[\text{H}_2\text{SO}_4/\text{H}_2\text{O}]{\text{Na}_2\text{Cr}_2\text{O}_7}$ HOOC–...–COOH

20.46

a. $\xrightarrow[\text{CH}_3\text{OH}]{\text{NaBH}_4}$

c. $(\text{CH}_3)_2\text{N}$–...–OH $\xrightarrow[\text{[2] H}_2\text{O}]{\text{[1] LiAlH}_4}$ $(\text{CH}_3)_2\text{N}$–...–OH

b. $\xrightarrow[\text{[2] H}_2\text{O}]{\text{[1] LiAlH}_4}$

d. $\xrightarrow[\text{[2] H}_2\text{O}]{\text{[1] LiAlH[OC(CH}_3)_3]_3}$

20.47

a. $\xrightarrow[\text{[2] H}_3\text{O}^+]{\text{[1] CO}_2}$

f. $\xrightarrow[\text{[2] H}_2\text{O}]{\text{[1] CH}_2\text{=O}}$ CH$_2$OH

b. $\xrightarrow[\text{[2] H}_2\text{O}]{\text{[1] CH}_3\text{CH}_2\text{MgBr}}$

g. $\xrightarrow[\text{[2] H}_2\text{O}]{\text{[1] (CH}_3)_2\text{CuLi}}$

c. $\xrightarrow[\text{[2] H}_2\text{O}]{\text{[1] C}_6\text{H}_5\text{Li}}$

h. $\xrightarrow[\text{[2] H}_2\text{O}]{\text{[1] CH}_3\text{MgBr}}$

d. $\xrightarrow[\text{[2] H}_2\text{O}]{\text{[1] C}_6\text{H}_5\text{MgBr (excess)}}$

i. $\xrightarrow[\text{[2] H}_2\text{O}]{\text{[1] C}_6\text{H}_5\text{Li}}$

j. $\xrightarrow[\text{[2] H}_2\text{O}]{\text{[1] (CH}_3)_2\text{CuLi}}$

e. $\xrightarrow[\text{[2] H}_2\text{O}]{\text{[1] CH}_3\text{MgCl (excess)}}$

20.48

a. $\xrightarrow[\text{[2] H}_2\text{O}]{\text{[1] C}_6\text{H}_5\text{MgBr}}$

b. $(\text{CH}_3)_3\text{C}$–...–O $\xrightarrow[\text{[2] H}_2\text{O}]{\text{[1] CH}_3\text{Li}}$ $(\text{CH}_3)_3\text{C}$– ... –OH + $(\text{CH}_3)_3\text{C}$– ... –OH

c.

$$\xrightarrow[\text{[2] H}_2\text{O}]{\text{[1] CH}_3\text{CH}_2\text{MgBr}}$$

d.

$$\xrightarrow[\text{[2] H}_2\text{O}]{\text{[1] (CH}_2=\text{CH)}_2\text{CuLi}}$$

e.

$$\xrightarrow[\substack{\text{[2] CO}_2 \\ \text{[3] H}_3\text{O}^+}]{\text{[1] Mg}}$$

f.

$$\xrightarrow[\text{[2] H}_2\text{O}]{\text{[1] (S)-CBS reagent}}$$

g.

$$\xrightarrow[\text{[2] H}_2\text{O}]{\text{[1] (R)-CBS reagent}}$$

h.

$$\xrightarrow[\text{[2] H}_2\text{O}]{\text{[1] LiAlH}_4}$$

CH$_2$OH + HOCH$_2$CH$_3$

20.49 Since a Grignard reagent contains a carbon atom with a partial negative charge, it acts as a base and reacts with the OH of the starting halide, BrCH$_2$CH$_2$CH$_2$CH$_2$OH. This acid–base reaction destroys the Grignard reagent so that addition cannot occur. To get around this problem, the OH group can be protected as a *tert*-butyldimethylsilyl ether, from which a Grignard reagent can be made.

INSTEAD: Use a protecting group.

20.50

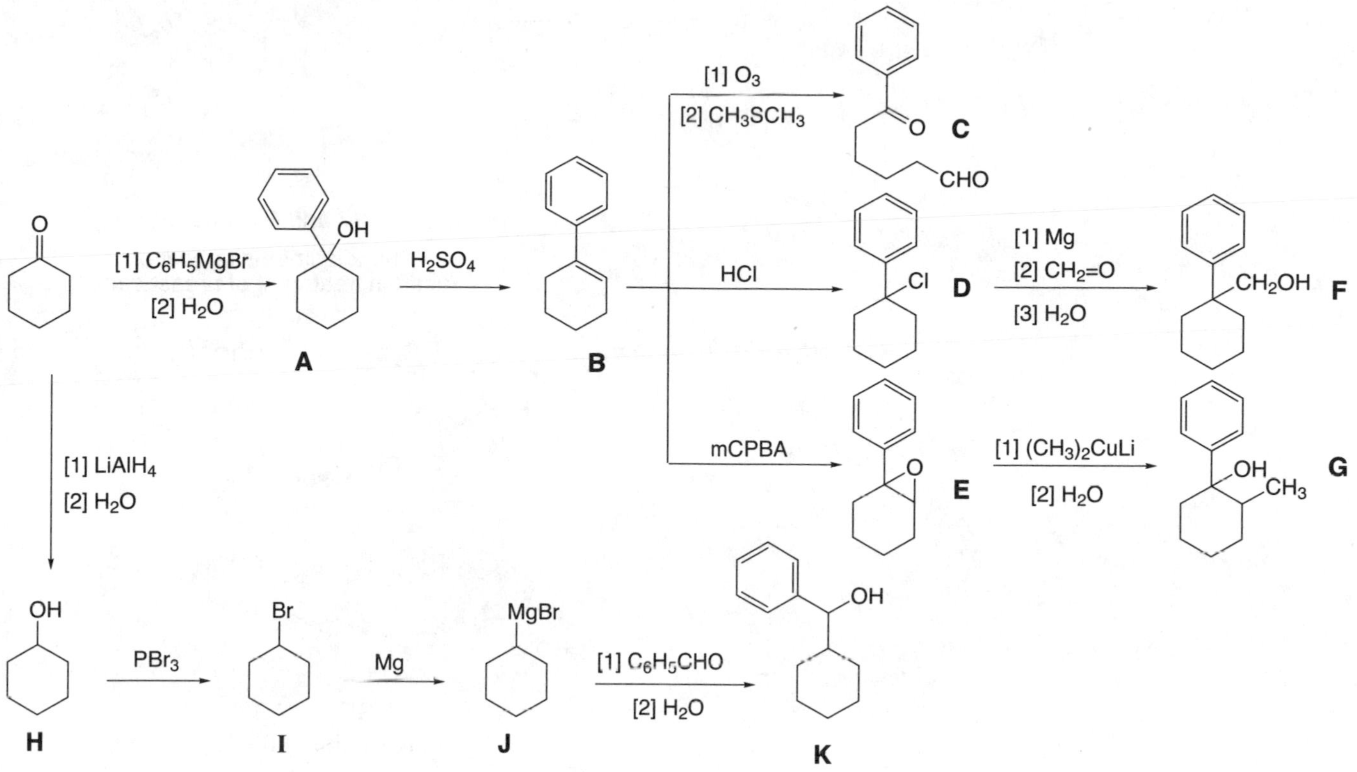

Compounds **F**, **G**, and **K** are all alcohols with aromatic rings so there will be many similarities in their proton NMR spectra. These compounds will, however, show differences in absorptions due to the CH protons on the carbon bearing the OH group. **F** has a CH$_2$OH group, which will give a singlet in the 3–4 ppm region of the spectrum. **G** is a 3° alcohol that has no protons on the C bonded to the OH group so it will have no peak in the 3–4 ppm region of the spectrum. **K** is a 2° alcohol that will give a doublet in the 3–4 ppm region of the spectrum for the CH proton on the carbon with the OH group.

20.51

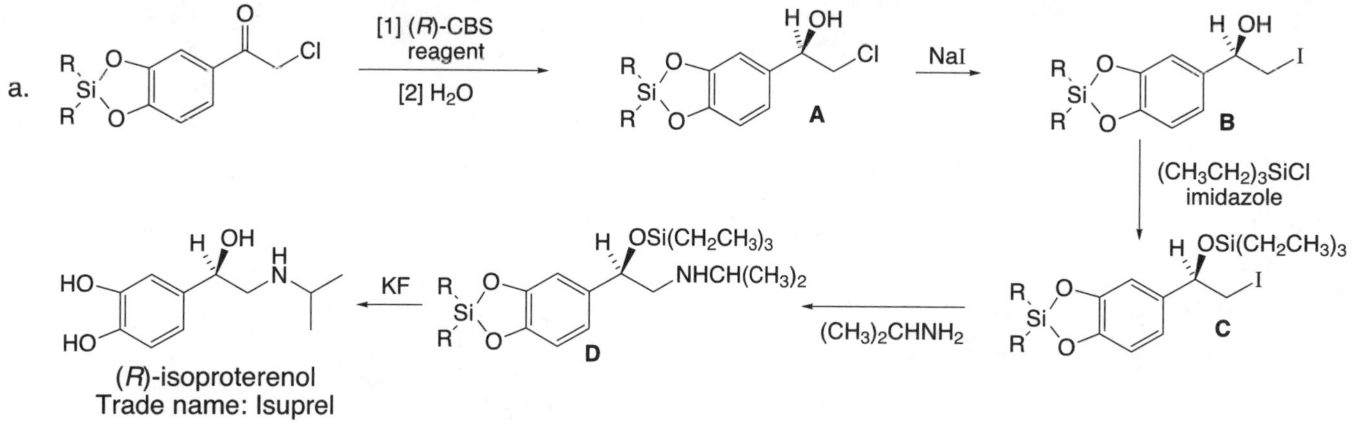

b.

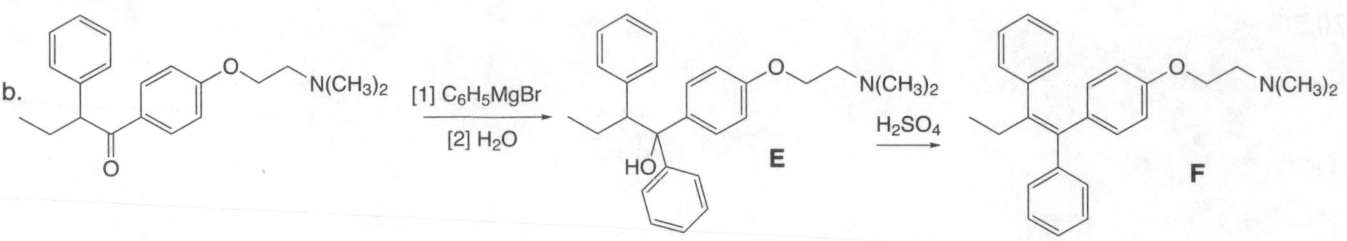

(Z and E isomers)
The Z isomer is tamoxifen.
(used in treatment of breast cancer)

c.

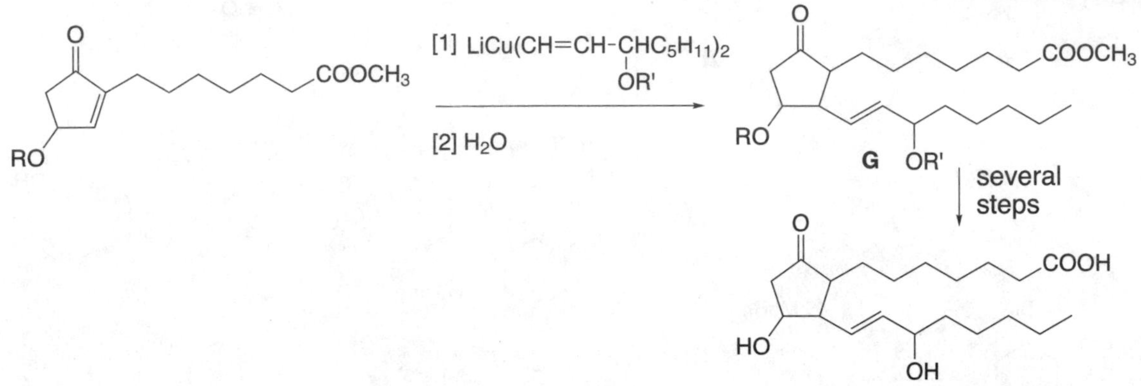

PGE$_1$
(a prostaglandin)

20.52

a.

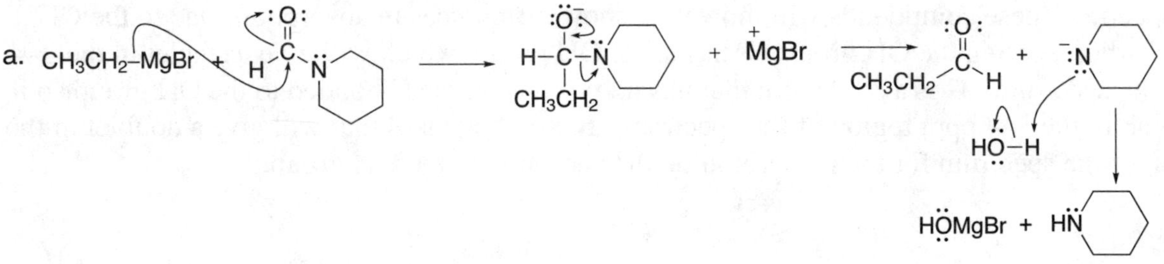

b.

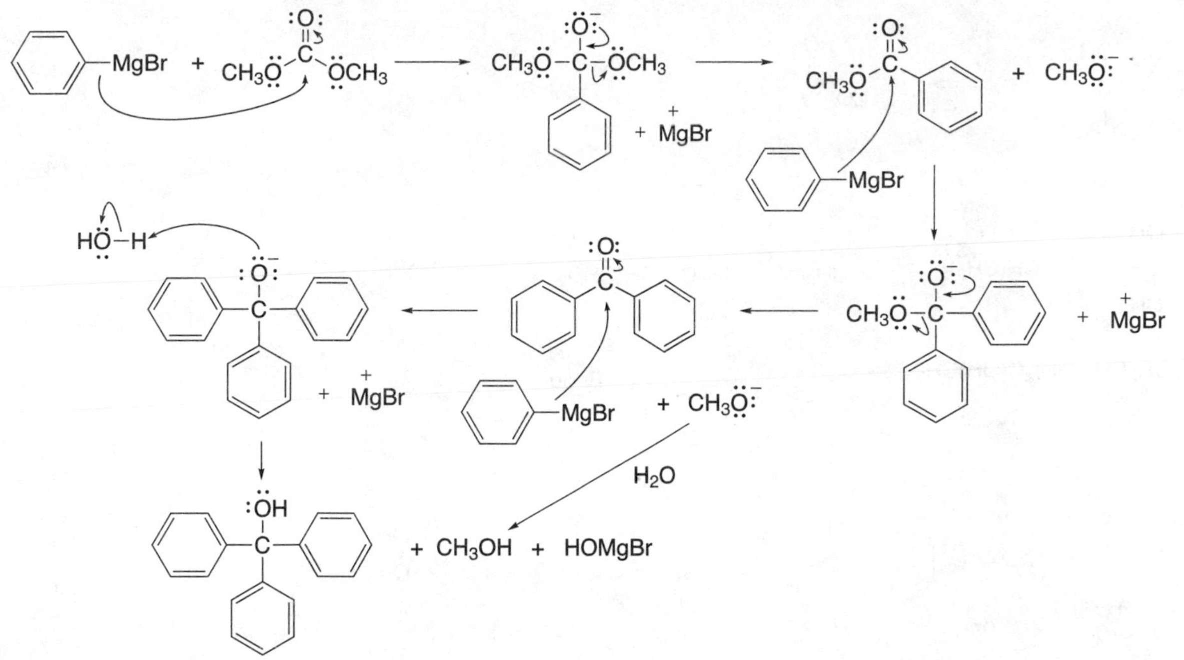

20.53

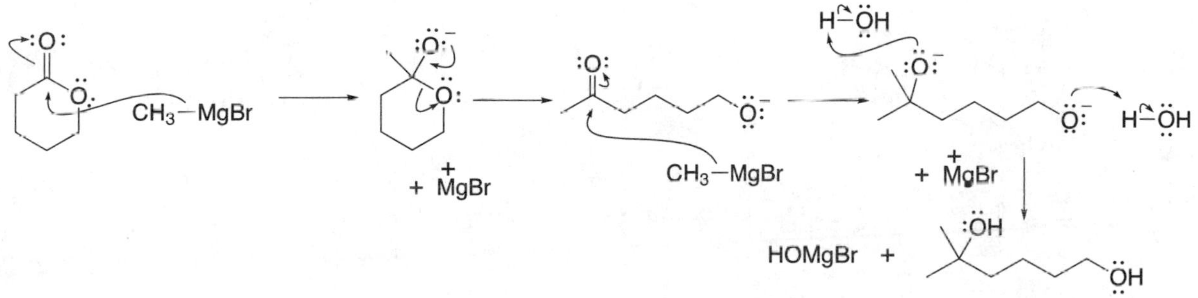

20.54

a.

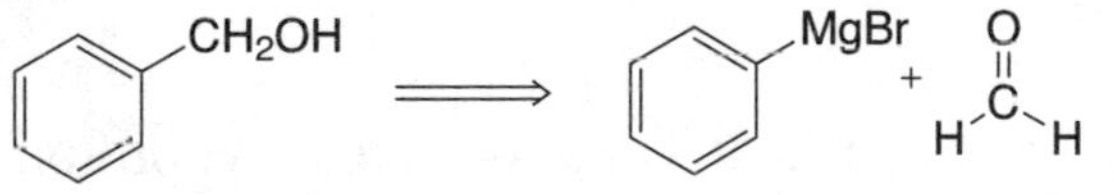

b.

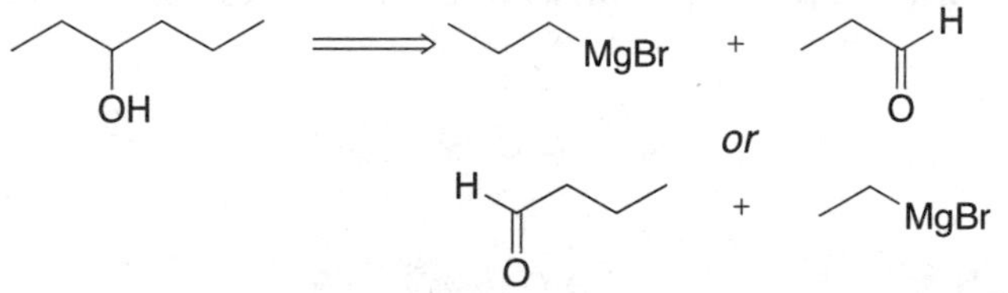

or

c. $(C_6H_5)_3COH \implies (C_6H_5)_2C=O +$ BrMg—

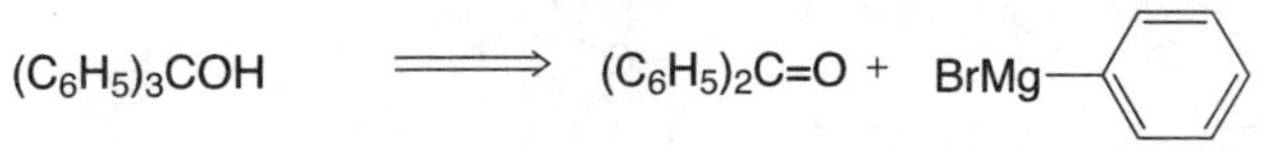

d.

or

or

CH$_3$MgBr +

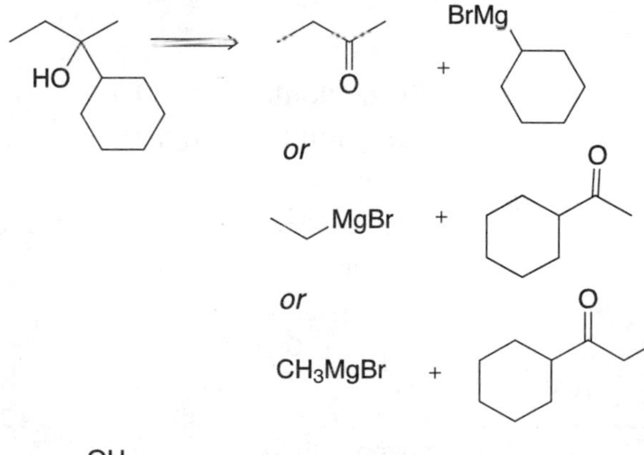

e.

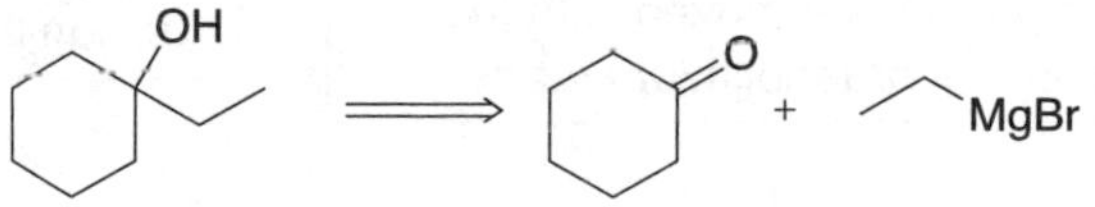

20.55

a.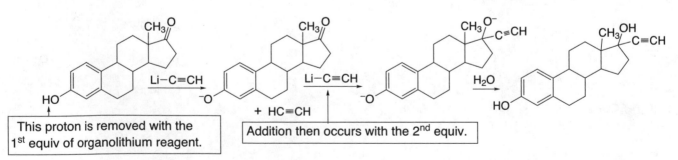

b. CH$_3$—C(OH)(CH$_3$)—CH$_2$CH$_2$CH(CH$_3$)$_2$ $\Longrightarrow$ CH$_3$O—C(=O)—CH$_2$CH$_2$CH(CH$_3$)$_2$ + CH$_3$—MgBr

c. (CH$_3$CH$_2$CH$_2$CH$_2$)$_2$C(OH)CH$_3$ $\Longrightarrow$ CH$_3$O—C(=O)—CH$_3$ + BrMg—

20.56

a. (2 ways)

b. (3 ways)

c. (3 ways)

20.57 a. Since estrone contains an OH group, the first equivalent of LiC≡CH removes the OH proton by an acid–base reaction to form a phenoxide. Then, the second equivalent adds to the carbonyl group.

This proton is removed with the 1st equiv of organolithium reagent.

+ HC≡CH

Addition then occurs with the 2nd equiv.

b. **Alternate synthesis:**

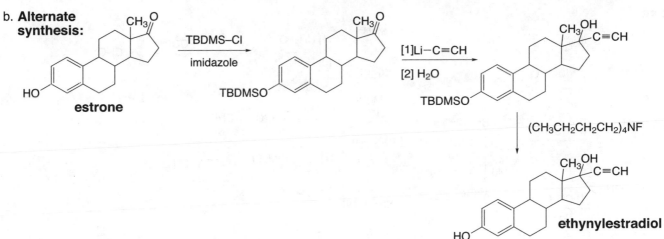

20.58

a.

cyclohexanol $\xrightarrow[\text{H}_2\text{SO}_4/\text{H}_2\text{O}]{\text{CrO}_3}$ cyclohexanone

b. cyclohexanol $\xrightarrow{\text{PBr}_3}$ bromocyclohexane $\xrightarrow{\text{Mg}}$ cyclohexyl MgBr $\xrightarrow[\text{[2] H}_2\text{O}]{\text{[1] CH}_2=\text{O}}$ cyclohexyl-CH$_2$OH

c. cyclohexyl MgBr $\xrightarrow[\text{[2] H}_3\text{O}^+]{\text{[1] CO}_2}$ cyclohexyl-COOH

[from (b)]

d. cyclohexyl MgBr $\xrightarrow[\text{[2] H}_2\text{O}]{\text{[1] epoxide}}$ cyclohexyl-CH$_2$CH$_2$OH

[from (b)]

e. cyclohexanone $\xrightarrow[\text{[2] H}_2\text{O}]{\text{[1] CH}_3\text{MgBr}}$ 1-methylcyclohexanol $\xrightarrow{\text{H}_2\text{SO}_4}$ methylcyclohexene $\xrightarrow{\text{mCPBA}}$ epoxide

[from (a)]

f. cyclohexanol $\xrightarrow{\text{H}_2\text{SO}_4}$ cyclohexene $\xrightarrow{\text{mCPBA}}$ epoxide $\xrightarrow[\text{[2] H}_2\text{O}]{\text{[1] C}_6\text{H}_5\text{MgBr}}$ 2-phenylcyclohexanol $\xrightarrow{\text{PCC}}$ 2-phenylcyclohexanone

g. cyclohexanone $\xrightarrow[\text{[2] H}_2\text{O}]{\text{[1] CH}_3\text{CH}_2\text{CH}_2\text{MgBr}}$ 1-propylcyclohexanol $\xrightarrow{\text{H}_2\text{SO}_4}$ propylcyclohexene $\xrightarrow[\text{Pd/C}]{\text{H}_2}$ propylcyclohexane

+ (propylidenecyclohexane)

[from (a)]

h. cyclohexanone $\xrightarrow[\text{[2] H}_2\text{O}]{\text{[1] HC≡CLi}}$ 1-ethynylcyclohexanol $\xrightarrow[\text{Hg}_2\text{SO}_4]{\text{H}_2\text{O, H}_2\text{SO}_4}$ 1-acetylcyclohexanol

[from (a)]

20.59

a. OH $\xrightarrow{\text{SOCl}_2}$ Cl

b. OH $\xrightarrow{\text{PCC}}$ O

c. OH $\xrightarrow{\text{PBr}_3}$ Br $\xrightarrow{\text{Mg}}$ MgBr

OH $\xleftarrow[\text{[2] H}_2\text{O}]{\text{[1] CH}_3\text{CHO}}$ MgBr

d. MgBr [from (c)] $\xrightarrow[\text{[2] H}_2\text{O}]{\text{[1] (epoxide)}}$ OH

e. MgBr [from (c)] $\xrightarrow[\text{[2] H}_3\text{O}^+]{\text{[1] CO}_2}$ COOH

f. MgBr [from (c)] $\xrightarrow[\text{[2] H}_2\text{O}]{\text{[1] CH}_2=\text{O}}$ OH $\xrightarrow{\text{PCC}}$ CHO

g. MgBr [from (c)] $\xrightarrow{\text{D}_2\text{O}}$ (CH$_3$)$_2$CHD

h. O (acetone) [from (b)] $\xrightarrow[\text{[2] H}_2\text{O}]{\text{[1] cyclohexyl-MgBr}}$ (cyclohexyl C(CH$_3$)$_2$OH)

i. MgBr [from (c)] $\xrightarrow[\text{[2] H}_2\text{O}]{\text{[1] cyclohexanone}}$ (1-isopropylcyclohexanol)

j. (CH$_3$)$_2$CHBr [from (c)] $\xrightarrow[\text{[2] CuI (0.5 equiv)}]{\text{[1] 2Li}}$ [(CH$_3$)$_2$CH]$_2$CuLi

$\xrightarrow[\text{[2] H}_2\text{O}]{\text{[1] 2-cyclopentenone}}$ (3-isopropylcyclopentanone)

20.60

a. (benzene) $\xrightarrow[\text{FeBr}_3]{\text{Br}_2}$ (bromobenzene) $\xrightarrow{\text{Mg}}$ (phenyl-MgBr) $\xrightarrow[\text{[2] H}_3\text{O}^+]{\text{[1] CO}_2}$ COOH (benzoic acid)

b. MgBr [from (a)] $\xrightarrow[\text{[2] H}_2\text{O}]{\text{[1] CH}_2=\text{O}}$ CH$_2$OH $\xrightarrow{\text{PCC}}$ CHO | CH$_3$OH $\xrightarrow{\text{PCC}}$ CH$_2$=O

c. MgBr [from (a)] $\xrightarrow[\text{[2] H}_2\text{O}]{\text{[1] (epoxide)}}$ CH$_2$CH$_2$OH | CH$_3$CH$_2$OH $\xrightarrow{\text{H}_2\text{SO}_4}$ CH$_2$=CH$_2$ $\xrightarrow{\text{mCPBA}}$ (epoxide)

d.

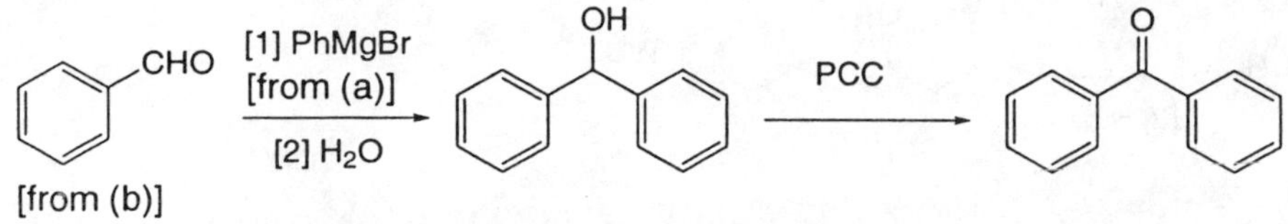

[from (b)]

[1] PhMgBr [from (a)]

[2] H₂O

PCC

e. 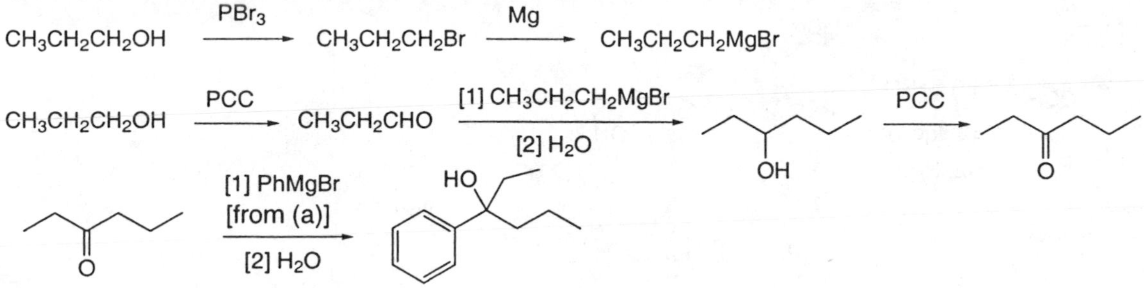 $CH_3CH_2CH_2OH$ →(PBr₃) $CH_3CH_2CH_2Br$ →(Mg) $CH_3CH_2CH_2MgBr$

$CH_3CH_2CH_2OH$ →(PCC) CH_3CH_2CHO →([1] $CH_3CH_2CH_2MgBr$ / [2] H₂O)

→(PCC)

[1] PhMgBr [from (a)]

[2] H₂O

20.61

a. 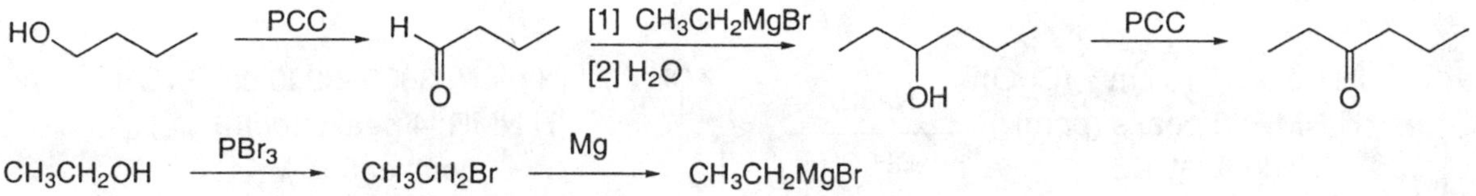 HO →(PCC) →([1] CH_3CH_2MgBr / [2] H₂O) →(PCC)

CH_3CH_2OH →(PBr₃) CH_3CH_2Br →(Mg) CH_3CH_2MgBr

b.

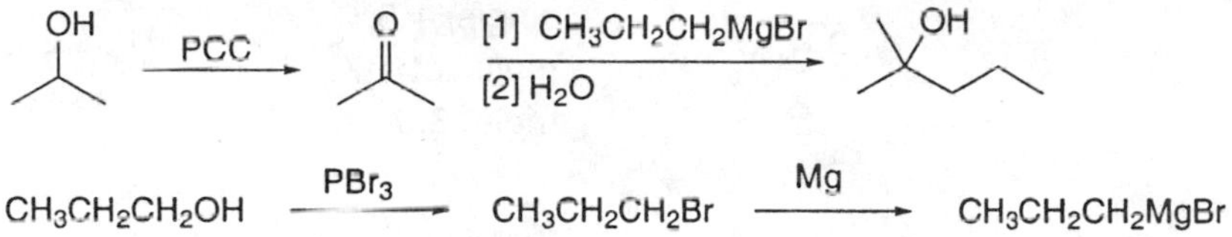

→(PCC) →([1] $CH_3CH_2CH_2MgBr$ / [2] H₂O)

$CH_3CH_2CH_2OH$ →(PBr₃) $CH_3CH_2CH_2Br$ →(Mg) $CH_3CH_2CH_2MgBr$

c. 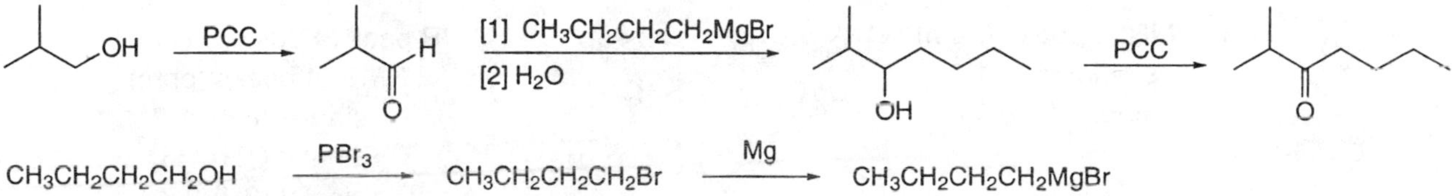 →(PCC) →([1] $CH_3CH_2CH_2CH_2MgBr$ / [2] H₂O) →(PCC)

$CH_3CH_2CH_2CH_2OH$ →(PBr₃) $CH_3CH_2CH_2CH_2Br$ →(Mg) $CH_3CH_2CH_2CH_2MgBr$

d.

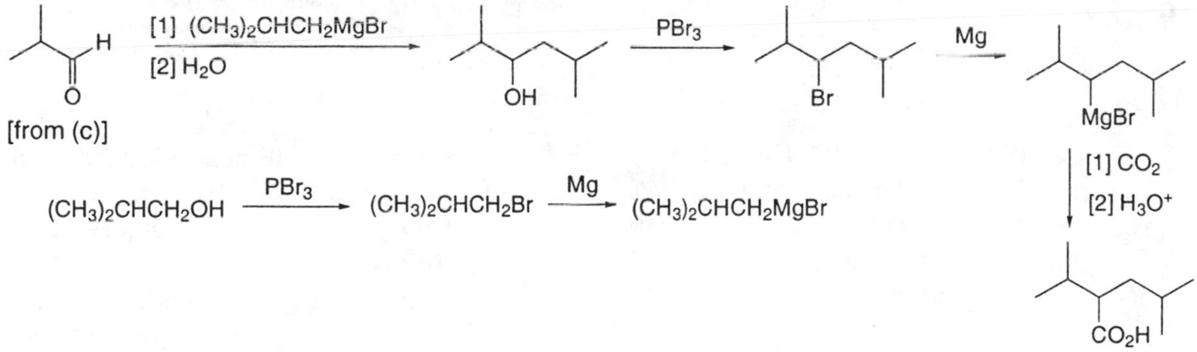

[from (c)]

[1] $(CH_3)_2CHCH_2MgBr$

[2] H₂O

→(PBr₃) →(Mg)

[1] CO₂

[2] H₃O⁺

$(CH_3)_2CHCH_2OH$ →(PBr₃) $(CH_3)_2CHCH_2Br$ →(Mg) $(CH_3)_2CHCH_2MgBr$

20.62

a.

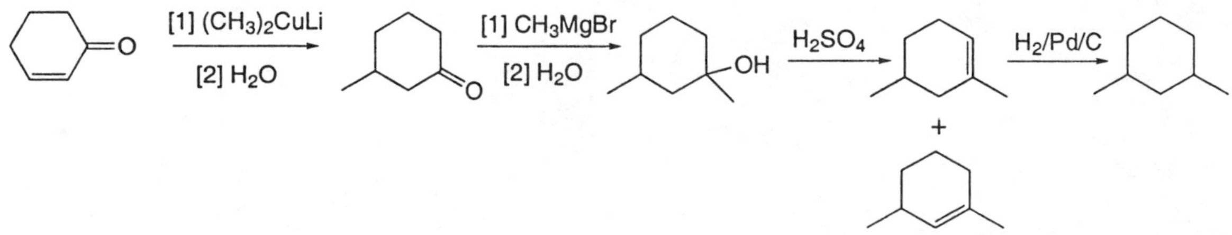

[1] $(CH_3)_2CuLi$

[2] H₂O

[1] CH_3MgBr

[2] H₂O

H_2SO_4

$H_2/Pd/C$

+

b.

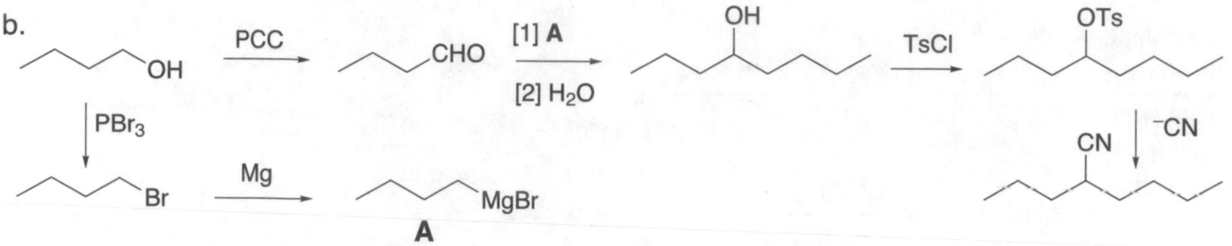

c.

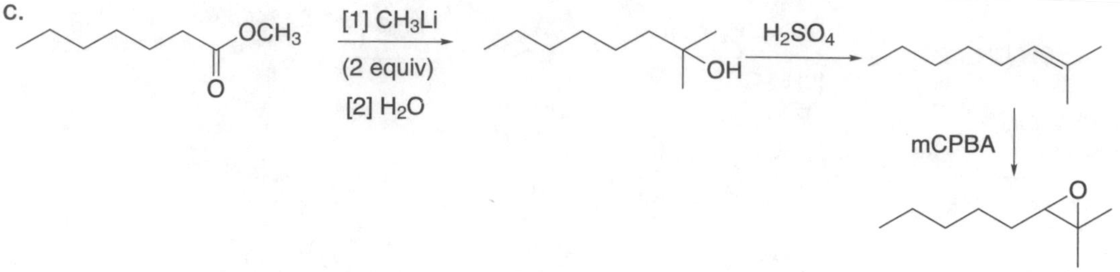

20.63

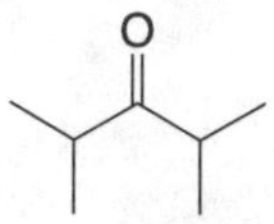

IR peak: 1716 cm^{-1} (C=O)
^{1}H NMR: 2 peaks (ppm)
    doublet 1.2
    septet 2.7

$C_7H_{14}O$

A

NaBH$_4$ / CH$_3$OH →

IR peak: 3600–3200 cm^{-1} (OH)
^{1}H NMR: 4 peaks (ppm)
    doublet 0.9
    singlet 1.5
    multiplet 1.7
    triplet 3.0

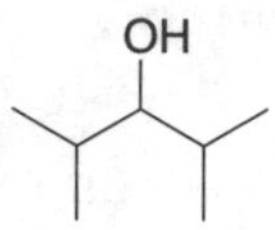

$C_7H_{16}O$

B

20.64

^{1}H NMR: 2 peaks (ppm)
    singlet (6H) 1.3
    singlet (2H) 2.4

C_4H_8O

C

[1] C$_6$H$_5$MgBr
[2] H$_2$O →

IR peak 3600–3200 cm^{-1} (OH)
^{1}H NMR: 4 peaks (ppm)
    singlet (6H) 1.2
    singlet (1H) 1.6
    singlet (2H) 2.7
    multiplet (5H) 7.2

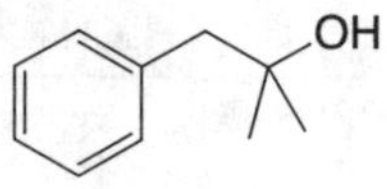

$C_{10}H_{14}O$

D

20.65

CH$_3$–C(=O)–OCH$_2$CH$_3$

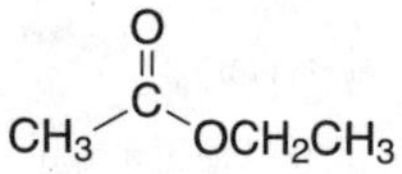

$C_4H_8O_2$

E

IR peak 1743 cm^{-1} (C=O)
^{1}H NMR: 2 peaks (ppm)
    triplet (3H) 1.2
    singlet (3H) 2.0
    quartet (2H) 4.1

[1] CH$_3$CH$_2$MgBr (excess)
[2] H$_2$O →

$$CH_3CH_2-\underset{\underset{CH_3}{|}}{\overset{\overset{OH}{|}}{C}}-CH_2CH_3$$

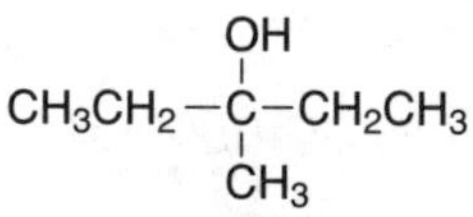

$C_6H_{14}O$

F

IR peak 3600-3200 cm^{-1} (OH)
^{1}H NMR: 4 peaks (ppm)
    triplet (6H) 0.9
    singlet (3H) 1.1
    quartet (4H) 1.5
    singlet (1H) 1.55

20.66

The reaction scheme shows:

A ketone (2-bromo-1-(4-hydroxy-3-(methoxycarbonyl)phenyl)ethanone) with a bromomethyl ketone, HO on the ring, and COOCH₃ substituent reacts with:

[1] (R)-CBS reagent
[2] H₂O

to give the secondary alcohol (with OH and Br), which reacts with:

TBDMSCl, imidazole (2 equiv)

to give **A** (OTBDMS, Br, TBDMS—O, COOCH₃).

Second row:

HO—(chain)—C₆H₅ reacts with:
[1] NaH
[2] Br—(chain)—Br

to give Br—(chain)—O—(chain)—C₆H₅, which reacts with NH₃ to give:

H₂N—(chain)—O—(chain)—C₆H₅ **B**

Third row:

A + **B** → (OTBDMS, N—H, chain, O, chain, C₆H₅; TBDMS—O, COOCH₃)

[1] LiAlH₄
[2] H₂O

→ (OTBDMS, N—H, chain, O, chain, C₆H₅; TBDMS—O, CH₂OH)

(CH₃CH₂CH₂CH₂)₄NF

→ (R) salmeterol

20.67

4-*tert*-butylcyclohexanone   *cis*-4-*tert*-butylcyclohexanol
major product

L-selectride adds H⁻ to a C=O group. There are two possible reduction products—cis and trans isomers—but the cis isomer is favored. The key element is that the three *sec*-butyl groups make L-selectride a large, bulky reducing agent that attacks the carbonyl group from the less hindered direction.

> When H⁻ adds from the equatorial direction, the product has an axial OH and a new
> equatorial H. Since the equatorial direction is less hindered, this mode of attack is favored
> with large bulky reducing agents like L-selectride. In this case, the product is cis.

> The axial H's hinder H⁻ attack from the axial direction. As a result, this mode of attack is more difficult
> with larger reducing agents. In this case the product is trans. This product is not formed to any
> appreciable extent.

Chapter 21: Aldehydes and Ketones—Nucleophilic addition

◆ General facts

- Aldehydes and ketones contain a carbonyl group bonded only to H atoms or R groups. The carbonyl carbon is sp^2 hybridized and trigonal planar (21.1).
- Aldehydes are identified by the suffix *–al*, while ketones are identified by the suffix *–one* (21.2).
- Aldehydes and ketones are polar compounds that exhibit dipole–dipole interactions (21.3).

◆ Summary of spectroscopic absorptions of RCHO and R$_2$CO (21.4)

IR absorptions	C=O	~1715 cm^{-1} for ketones
		• increasing frequency with decreasing ring size
		~1730 cm^{-1} for aldehydes
		• For both RCHO and R$_2$CO, the frequency decreases with conjugation.
	Csp^2–H of CHO	~2700–2830 cm^{-1} (one or two peaks)
^{1}H NMR absorptions	CHO	9–10 ppm (highly deshielded proton)
	C–H α to C=O	2–2.5 ppm (somewhat deshielded Csp^3–H)
^{13}C NMR absorption	C=O	190–215 ppm

◆ Nucleophilic addition reactions

[1] Addition of hydride (H$^-$) (21.8)

$$R\overset{\overset{O}{\|}}{C}H(R') \xrightarrow[\text{[1] LiAlH}_4\text{ [2] H}_2\text{O}]{\overset{\text{NaBH}_4/\text{CH}_3\text{OH}}{\text{or}}} \underset{\substack{\text{H} \\ 1° \text{ or } 2° \text{ alcohol}}}{R-\overset{\overset{\text{OH}}{|}}{C}-H(R')}$$

- The mechanism has two steps.
- H:$^-$ adds to the planar C=O from both sides.

[2] Addition of organometallic reagents (R$^-$) (21.8)

$$R\overset{\overset{O}{\|}}{C}H(R') \xrightarrow[\text{[2] H}_2\text{O}]{\text{[1] R''MgX or R''Li}} \underset{\substack{\text{R''} \\ 1°, 2°, \text{ or } 3° \text{ alcohol}}}{R-\overset{\overset{\text{OH}}{|}}{C}-H(R')}$$

- The mechanism has two steps.
- R:$^-$ adds to the planar C=O from both sides.

[3] Addition of cyanide ($^-$CN) (21.9)

$$R\overset{\overset{O}{\|}}{C}H(R') \xrightarrow[\text{HCl}]{\text{NaCN}} \underset{\substack{\text{CN} \\ \text{cyanohydrin}}}{R-\overset{\overset{\text{OH}}{|}}{C}-H(R')}$$

- The mechanism has two steps.
- $^-$CN adds to the planar C=O from both sides.

[4] Wittig reaction (21.10)

- The reaction forms a new C–C σ bond and a new C–C π bond.
- Ph$_3$P=O is formed as by-product.

[5] Addition of 1° amines (21.11)

- The reaction is fastest at pH 4–5.
- The intermediate carbinolamine is unstable, and loses H$_2$O to form the C=N.

[6] Addition of 2° amines (21.12)

- The reaction is fastest at pH 4–5.
- The intermediate carbinolamine is unstable, and loses H$_2$O to form the C=C.

[7] Addition of H₂O—Hydration (21.13)

- The reaction is reversible. Equilibrium favors the product only with less stable carbonyl compounds (e.g. H$_2$CO and Cl$_3$CCHO).
- The reaction is catalyzed with either H$^+$ or $^-$OH.

[8] Addition of alcohols (21.14)

- The reaction is reversible.
- The reaction is catalyzed with acid.
- Removal of H$_2$O drives the equilibrium to favor the products.

♦ Other reactions

[1] Synthesis of Wittig reagents (21.10A)

- Step [1] is best with CH$_3$X and RCH$_2$X since the reaction follows an S$_N$2 mechanism.
- A strong base is needed for proton removal in Step [2].

[2] Conversion of cyanohydrins to aldehydes and ketones (21.9)

- This reaction is the reverse of cyanohydrin formation.

[3] Hydrolysis of nitriles (21.9)

[4] Hydrolysis of imines and enamines (21.12)

[5] Hydrolysis of acetals (21.14)

- The reaction is acid-catalyzed and is the reverse of acetal synthesis.
- A large excess of H_2O drives the equilibrium to favor the products.

Chapter 21: Answers to Problems

21.1 As the number of R groups bonded to the carbonyl C increases, reactivity towards nucleophilic attack decreases.

a. $(CH_3)_2C=O$ $CH_3CH=O$ $CH_2=O$

      2 R groups   1 R group 0 R groups

             increasing reactivity
         decreasing alkyl substitution

b.

               increasing reactivity
            decreasing steric hindrance

21.2 More stable aldehydes are less reactive towards nucleophilic attack.

                 CHO   CHO

        benzaldehyde   cyclohexanecarbaldehyde
Several resonance structures delocalize the partial   This aldehyde has no added resonance stabilization.
positive charge on the carbonyl carbon making
it more stable, and less reactive towards
nucleophilic attack.

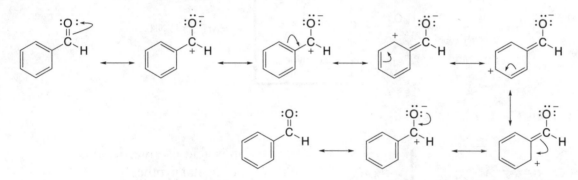

21.3
- To name an aldehyde with a chain of atoms: [1] Find the longest chain with the CHO group and change the –e ending to –al. [2] Number the carbon chain to put the CHO at C1, but omit this number from the name. Apply all other nomenclature rules.
- To name an aldehyde with the CHO bonded to a ring: [1] Name the ring and add the suffix –carbaldehyde. [2] Number the ring to put the CHO group at C1, but omit this number from the name. Apply all other nomenclature rules.

a. $(CH_3)_3CC(CH_3)_2CH_2CHO$

                  5 C chain = pentanal   **3,3,4,4-tetramethylpentanal**

c.

             4 C ring =   **3,3-dichlorocyclobutane-**
      cyclobutanecarbaldehyde   **carbaldehyde**

b.

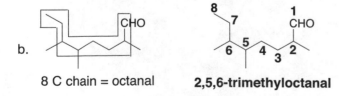

    8 C chain = octanal   **2,5,6-trimethyloctanal**

21.4 Work backwards from the name to the structure, referring to the nomenclature rules in Answer 21.3.

a. 2-isobutyl-3-isopropyl**hexanal**

6 C chain

c. 1-methyl**cyclopropanecarbaldehyde**

3 carbon ring

b. *trans*-3-methyl**cyclopentanecarbaldehyde**

5 carbon ring

or

d. 3,6-diethyl**nonanal**

9 C chain

21.5 • To name an acyclic ketone: [1] Find the longest chain with the carbonyl group and change the –e ending to –one. [2] Number the carbon chain to give the carbonyl C the lower number. Apply all other nomenclature rules.

• To name a cyclic ketone: [1] Name the ring and change the –e ending to –one. [2] Number the C's to put the carbonyl C at C1 and the next substituent the lower number. Apply all other nomenclature rules.

a.

8 C chain = octanone

5-ethyl-4-methyl-3-octanone

c. (CH₃)₃CCOC(CH₃)₃

5 C chain = pentanone

2,2,4,4-tetramethyl-3-pentanone

b.

5 C ring = cyclopentanone

3-*tert*-butyl-2-methylcyclopentanone

21.6 Most common names are formed by naming both alkyl groups on the carbonyl C, arranging them alphabetically, and adding the word ketone.

a. *sec*-butyl ethyl ketone   c. *p*-ethylacetophenone

b. methyl vinyl ketone

d. 2-benzyl-3-benzoylcyclopentanone.

benzyl group: benzoyl group: C1 ketone

21.7 Even though both compounds have polar C–O bonds, the electron pairs around the sp^3 hybridized O atom of diethyl ether are more crowded and less able to interact with electron-deficient sites in other diethyl ether molecules. Because the carbonyl group of 2-butanone contains an sp^2 hybridized C and O atom, it is less crowded. The lone pairs of electrons on the O atom can more readily interact with the electron deficient sites in the other molecules, resulting in stronger forces.

2-butanone   diethyl ether

21.8 For cyclic ketones, the carbonyl absorption shifts to higher wave number as the size of the ring decreases and the ring strain increases. Conjugation of the carbonyl group with a C=C or a benzene ring shifts the absorption to lower wavenumber.

a. CHO and CHO

C=O conjugated with   **higher wave number**
a benzene ring
lower wave number

c. =O and =O

smaller ring
higher wave number

b. CHO and CHO

conjugated C=O   **higher wave number**
lower wave number

21.9

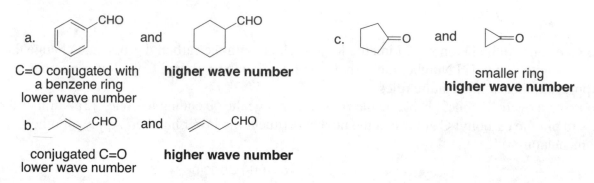

C_4H_8O

3 types of C   4 types of C   4 types of C
aldehyde   aldehyde   methyl ketone
A   **B**   **C**

Compound **A** shows three peaks in its ^{13}C NMR spectrum.

Compound **B** has a peak at 9.8 ppm in its ^1H NMR spectrum, due to the H of an aldehyde.

Compound **C** has a singlet at 2.1 ppm in its ^1H NMR spectrum, due to the protons on the α C, a methyl ketone.

21.10

a. $CH_3CH_2CH_2COOCH_3$ $\xrightarrow[\text{[2] H}_2\text{O}]{\text{[1] DIBAL–H}}$ $CH_3CH_2CH_2CHO$

c. $HC\equiv CCH_2CH_3$ $\xrightarrow[\text{[2] H}_2\text{O}_2/\text{HO}^-]{\text{[1] BH}_3}$ $CH_3CH_2CH_2CHO$

b. $CH_3CH_2CH_2CH_2OH$ $\xrightarrow{\text{PCC}}$ $CH_3CH_2CH_2CHO$

d. $CH_3CH_2CH_2CH=CHCH_2CH_2CH_3$ $\xrightarrow[\text{[2] Zn/H}_2\text{O}]{\text{[1] O}_3}$ $CH_3CH_2CH_2CHO$

21.11

a.

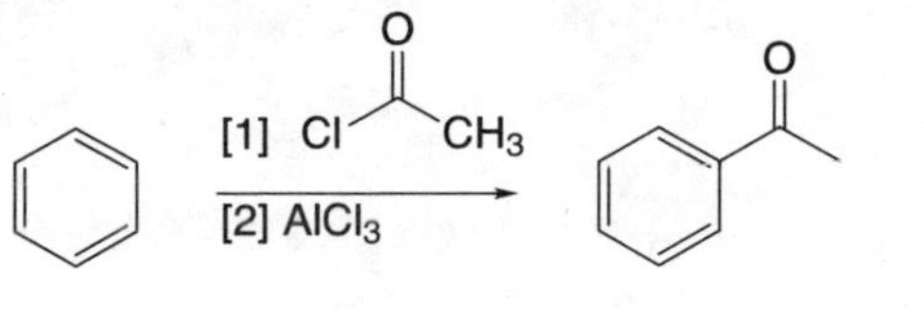

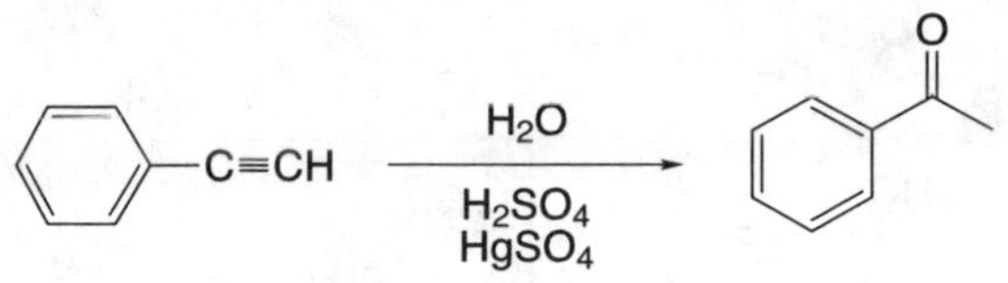

b.

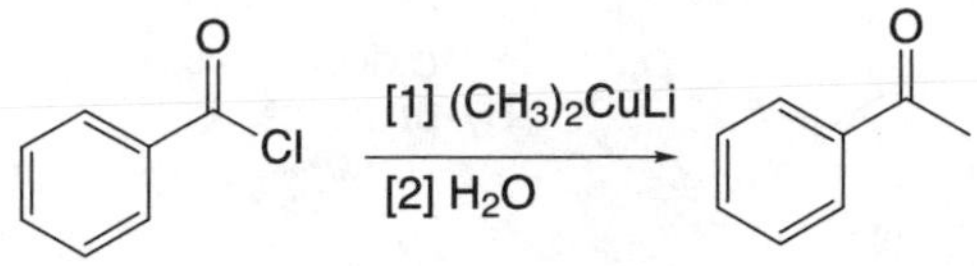

c.

21.12

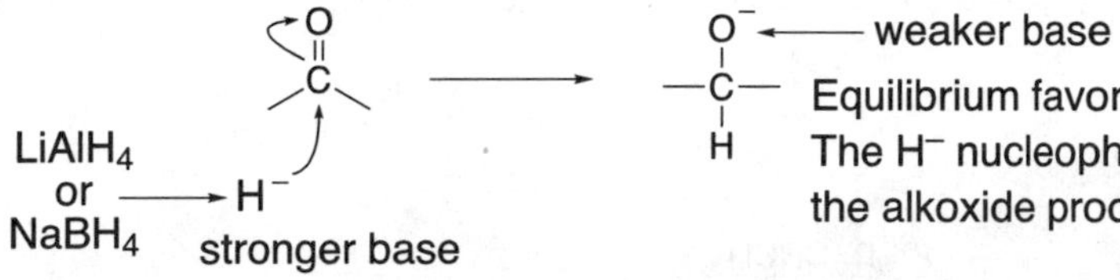

LiAlH$_4$
or
NaBH$_4$ stronger base

O$^-$ ← weaker base
Equilibrium favors the weaker base.
The H$^-$ nucleophile is a much stronger base than
the alkoxide product.

21.13 Addition of hydride occurs at a planar carbonyl C, so two different configurations at a new stereogenic center are possible.

a.

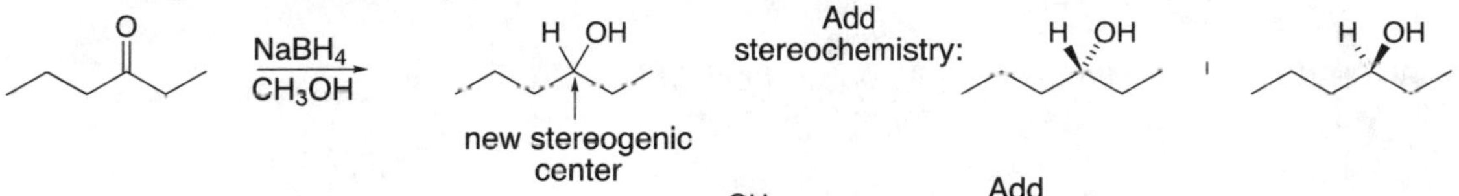

new stereogenic center

Add stereochemistry:

b.

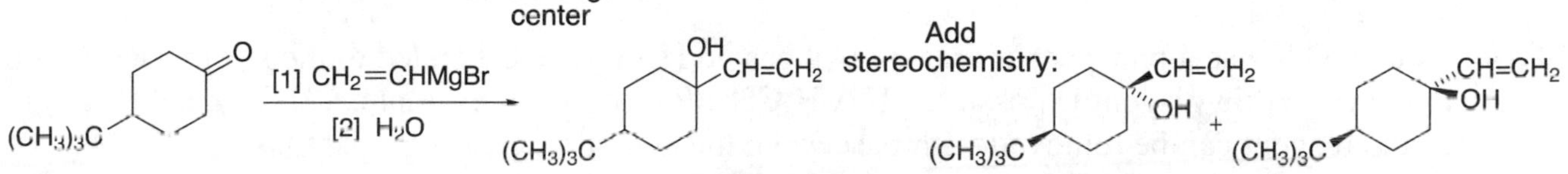

Add stereochemistry:

c.

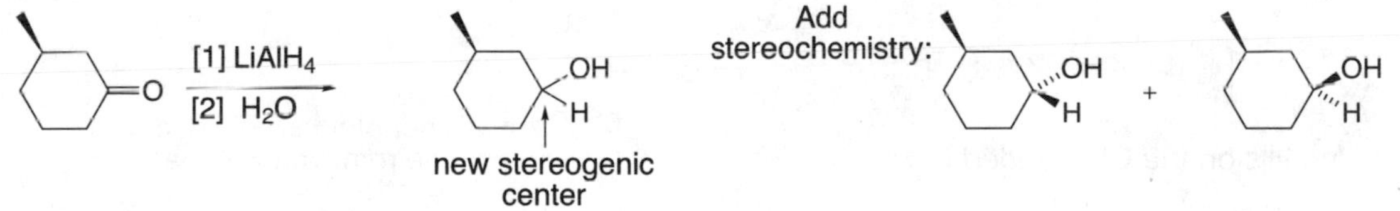

new stereogenic center

Add stereochemistry:

21.14

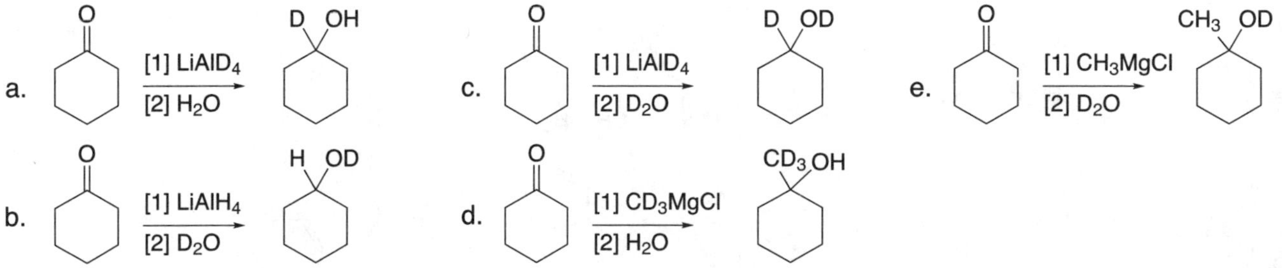

a.

b.

c.

d.

e.

21.15 Treatment of an aldehyde or ketone with NaCN/HCl adds HCN across the double bond. Cyano groups are hydrolyzed by H$_3$O$^+$ to replace the 3 C–N bonds with 3 C–O bonds.

a.

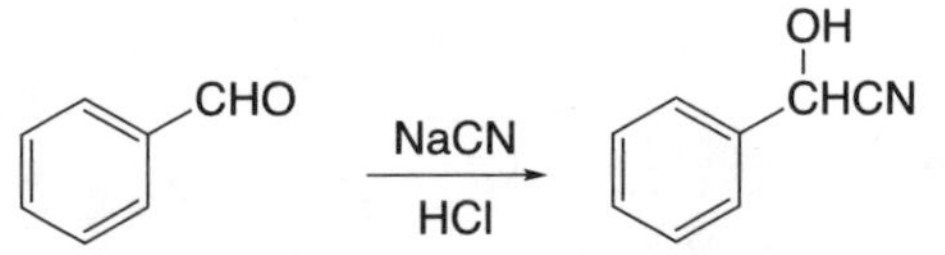

b.

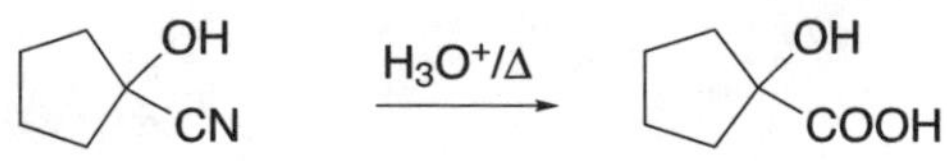

21.16

amygdalin → enzyme → HO–C–CN / H → enzyme → (benzaldehyde) + HCN
toxic by-product

21.17

a. $\begin{array}{c} CH_3 \\ C=O \\ CH_3 \end{array}$ + $Ph_3P=CH_2$ ⟶ $\begin{array}{c} CH_3 \\ C=CH_2 \\ CH_3 \end{array}$

b. (cyclopentanone)=O + $Ph_3P=CHCH_2CH_2CH_2CH_3$ ⟶ (cyclopentylidene)=$CHCH_2CH_2CH_2CH_3$

21.18

a. $Ph_3P:$ + $Br–CH_2CH_3$ ⟶ $Ph_3\overset{+}{P}–CH_2CH_3$ Br^- $\xrightarrow{BuLi}$ $Ph_3P=CHCH_3$

b. $Ph_3P:$ + $Br–CH(CH_3)_2$ ⟶ $Ph_3\overset{+}{P}–CH(CH_3)_2$ Br^- $\xrightarrow{BuLi}$ $Ph_3P=C(CH_3)_2$

c. $Ph_3P:$ + $Br–CH_2C_6H_5$ ⟶ $Ph_3\overset{+}{P}–CH_2C_6H_5$ Br^- $\xrightarrow{BuLi}$ $Ph_3P=CHC_6H_5$

21.19 You need to use a phosphorus reagent that has no H's on the C bonded to the P, otherwise more than one Wittig reagent is possible. If $(CH_3CH_2)_3P$ is used, for example, a H from a CH_2 group of the reagent can be removed with base when the Wittig reagent is prepared.

No H's on the C's bonded to P

$(CH_3CH_2)_3P:$ + RCH_2X ⟶ $(CH_3CH_2)_2\overset{+}{P}–CH_2R$ + X^-
CH_3CH_2 ← Either of these H's can now be removed with base.

21.20

a. (benzaldehyde with CHO) + $Ph_3P=CHCH_2CH_3$ ⟶ (Z-alkene with CH_2CH_3) + (E-alkene with CH_2CH_3)

b. (benzaldehyde with CHO) + $Ph_3P=CHC_6H_5$ ⟶ (trans-stilbene) + (cis-stilbene)

c. (benzaldehyde with CHO) + $Ph_3P=CHCOOCH_3$ ⟶ (E-cinnamate with $COOCH_3$) + (Z-cinnamate with $COOCH_3$)

21.21 To draw the starting materials of the Wittig reactions, find the C=C and cleave it. Replace it with a C=O in one half of the molecule and a C=PPh₃ in the other half. The preferred pathway uses a Wittig reagent derived from a less hindered alkyl halide.

a.

$$CH_3, CH_2CH_3 / CH_3, H \quad C=C \implies CH_3, CH_3 \quad C=PPh_3 \; + \; O=C \; CH_2CH_3, H$$

2° halide precursor
(CH₃)₂CHX

or

$$CH_3, CH_3 \quad C=O \; + \; Ph_3P=C \; CH_2CH_3, H$$

1° halide precursor
XCH₂CH₂CH₃
preferred pathway

b.

$$CH_3CH_2, H / CH_2CH_3, H \quad C=C \implies CH_3CH_2, H \quad C=PPh_3 \; + \; O=C \; CH_2CH_3, H$$

(*cis or trans*)

(only one route possible)

21.22

a.   Two-step sequence:

[1] CH₃MgBr
[2] H₂O

HO CH₃

H₂SO₄

minor product   **tetrasubstituted major product**   (*E* and *Z* isomers)

One-step sequence:

Ph₃P=CH₂

only product

b.   Two-step sequence:

[1] C₆H₅CH₂MgBr
[2] H₂O

OH
CH₂C₆H₅

H₂SO₄

—CHC₆H₅
trisubstituted conjugated C=C

+

—CH₂C₆H₅
trisubstituted

One-step sequence:

Ph₃P=CHC₆H₅

=CHC₆H₅ **only product**

21.23 When a 1° amine reacts with an aldehyde or ketone, the C=O is replaced by C=NR.

a.   —CHO   CH₃CH₂CH₂CH₂NH₂ →   —CH=NCH₂CH₂CH₂CH₃

b.   CH₃CH₂CH₂CH₂NH₂ →   NCH₂CH₂CH₂CH₃

c.   =O   CH₃CH₂CH₂CH₂NH₂ →   =NCH₂CH₂CH₂CH₃

21.24 Remember that the C=NR is formed from a C=O and an NH₂ group of a 1° amine.

a. $CH_3, H \quad C=NCH_2CH_2CH_3 \implies CH_3, H \quad C=O \; + \; NH_2CH_2CH_2CH_3$

b.   CH₃—⟨ ⟩=N—⟨ ⟩ ⟹ CH₃—⟨ ⟩=O + NH₂—⟨ ⟩

21.25

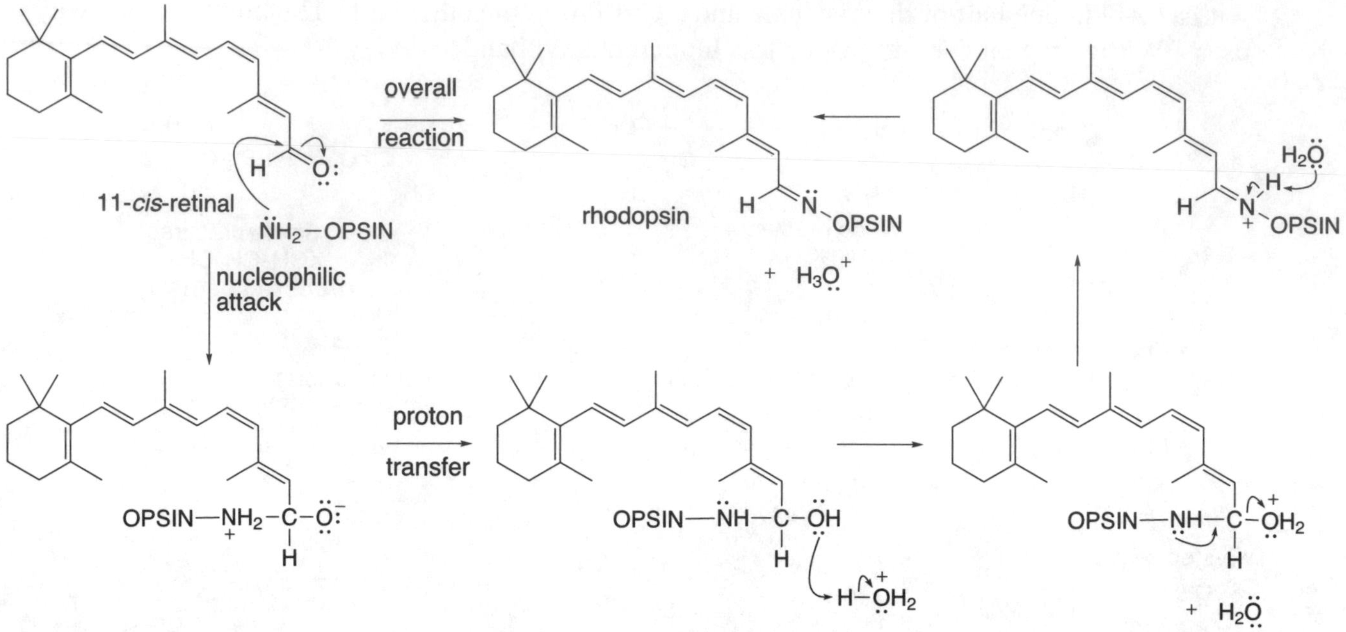

11-*cis*-retinal

NH_2—OPSIN

nucleophilic
attack

overall
reaction

rhodopsin

H $\ddot{N}$ OPSIN

+ $H_3\overset{+}{O}$

H₂$\ddot{O}$

H

$\overset{+}{N}$ OPSIN

proton
transfer

OPSIN—$\overset{+}{NH_2}$—$\overset{|}{\underset{H}{C}}$—$\ddot{\underset{..}{O}}$:

OPSIN—$\ddot{N}H$—$\overset{|}{\underset{H}{C}}$—$\ddot{O}H$

H—$\overset{+}{OH_2}$

OPSIN—$\ddot{N}H$—$\overset{|}{\underset{H}{C}}$—$\overset{+}{O}H_2$

+ $H_2\ddot{O}$

21.26

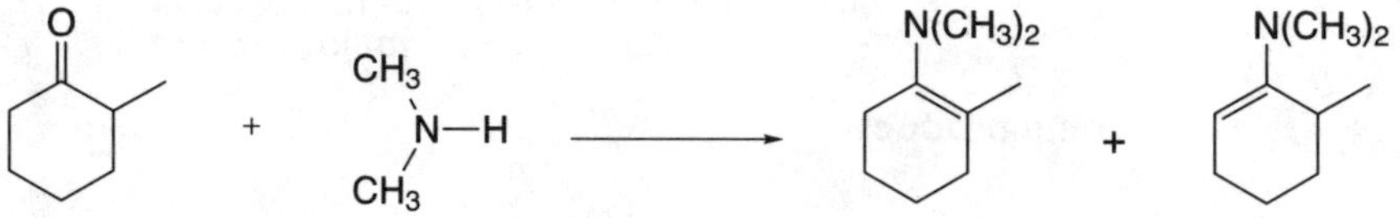

21.27

This carbon has 4 bonds to C's. To make an enamine, it needs
a H atom, which is lost as H_2O when the enamine is formed.

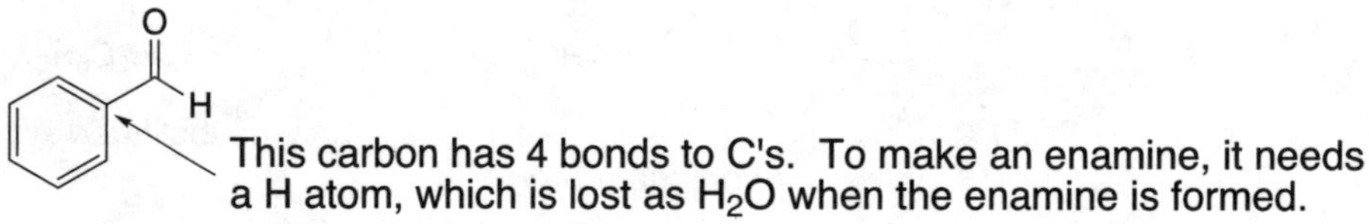

21.28 • Imines are hydrolyzed to 1° amines and a carbonyl compound.
   • Enamines are hydrolyzed to 2° amines and a carbonyl compound.

a.

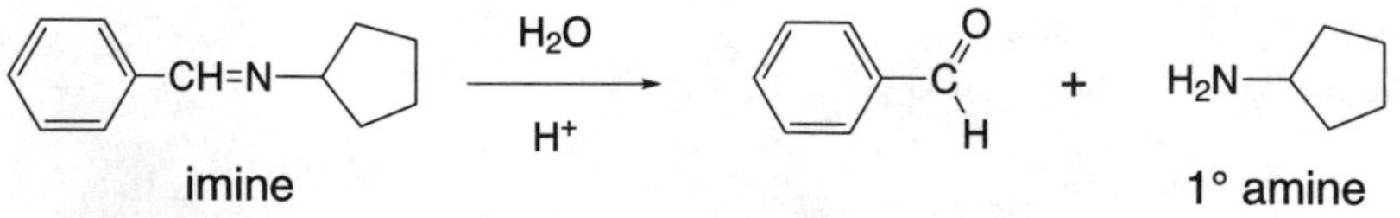

imine

H_2O / H^+

+ H_2N—

1° amine

b.

enamine

H_2O / H^+

+

2° amine

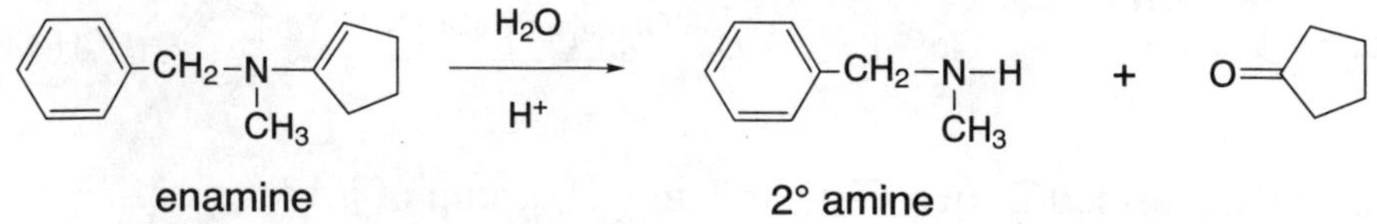

21.29 • A substituent that **donates** electron density to the carbonyl C stabilizes it, **decreasing** the percentage of hydrate at equilibrium.
• A substituent that **withdraws** electron density from the carbonyl C destabilizes it, **increasing** the percentage of hydrate at equilibrium.

a. $CH_3CH_2CH_2CHO$ and $CH_3CH_2COCH_3$

one R group on C=O
higher percentage of hydrate

2 R groups
on C=O

c.

and

NO_2 is electron withdrawing.
higher percentage of hydrate

electron donating

b. CH_3CF_2CHO and CH_3CH_2CHO

F atoms are electron withdrawing.
higher percentage of hydrate

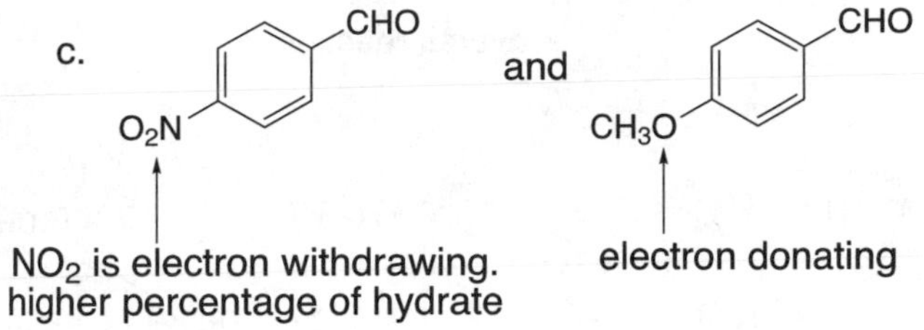

21.30

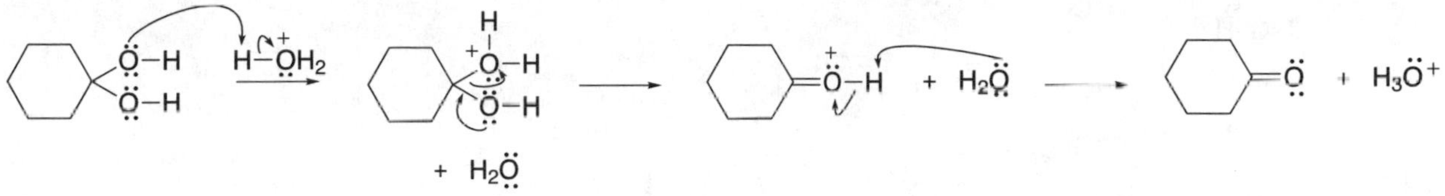

21.31 Treatment of an aldehyde or ketone with two equivalents of alcohol results in the formation of acetals (a C bonded to 2 OR groups).

a.

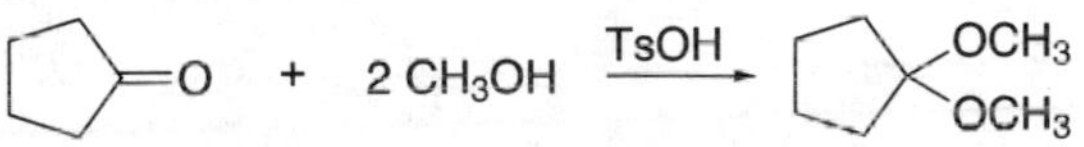

b.

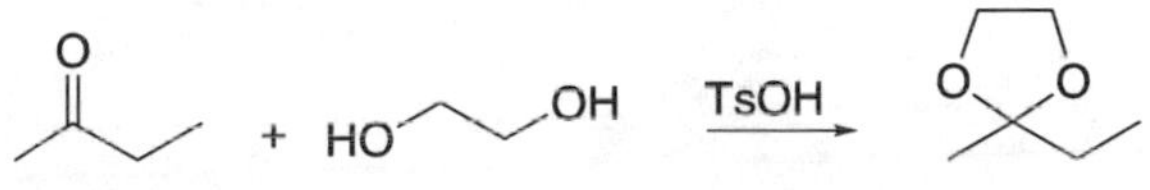

21.32

a.

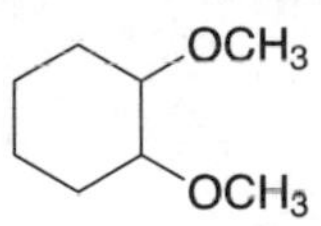

2 OR groups
on different C's.
2 ethers

b.

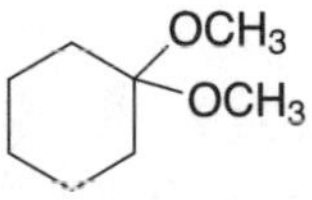

2 OR groups
on same C.
acetal

c.

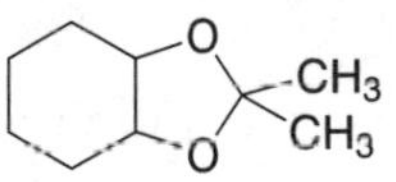

2 OR groups
on same C.
acetal

d.

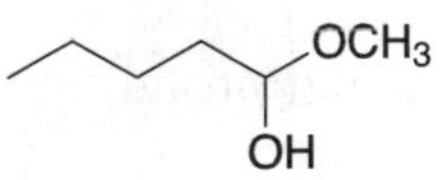

1 OR group and
1 OH group
on same C.
hemiacetal

21.33 The mechanism has 2 parts: [1] Nucleophilic addition of ROH to form a hemiacetal; [2] Conversion of the hemiacetal to an acetal.

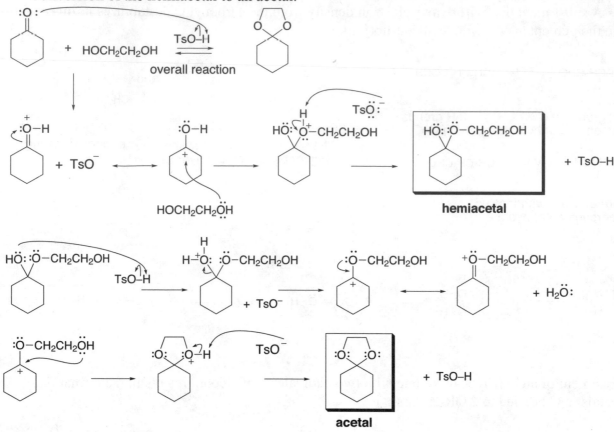

21.34

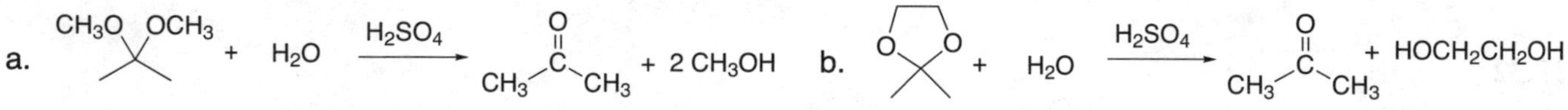

a. <image: CH₃O OCH₃ structure> + H₂O →(H₂SO₄)→ CH₃–C(=O)–CH₃ + 2 CH₃OH   b. <image: dioxolane structure> + H₂O →(H₂SO₄)→ CH₃–C(=O)–CH₃ + HOCH₂CH₂OH

21.35 Use an acetal protecting group to carry out the reaction.

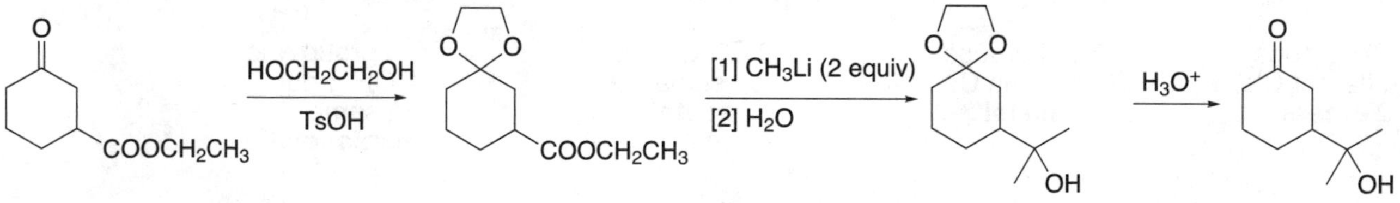

<image: reaction scheme showing cyclohexanone ester → acetal protected → [1] CH₃Li (2 equiv) [2] H₂O → H₃O⁺ → product>

HOCH₂CH₂OH / TsOH   [1] CH₃Li (2 equiv) / [2] H₂O   H₃O⁺

21.36

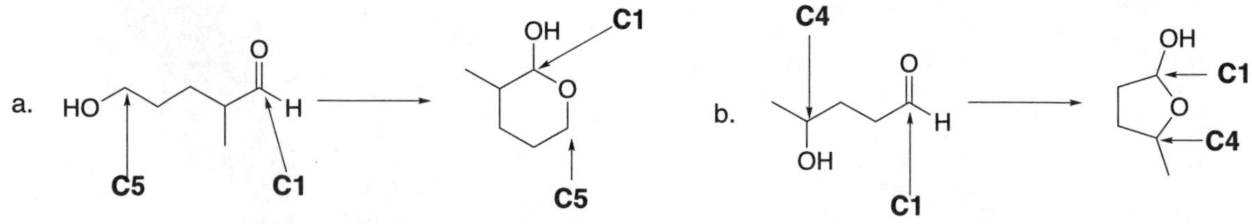

a. <image: HO...CHO structure with C5 and C1 labels> → <image: cyclic product with OH, C1, O, C5 labels>

b. <image: diol aldehyde structure with C4, OH, C1 labels> → <image: cyclic product with OH, C1, O, C4 labels>

21.37 The hemiacetal OH is replaced by an OR group to form an acetal.

a. 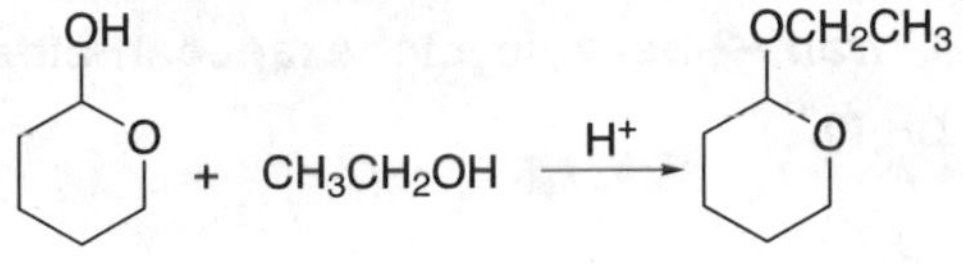 + CH₃CH₂OH $\xrightarrow{H^+}$

b. 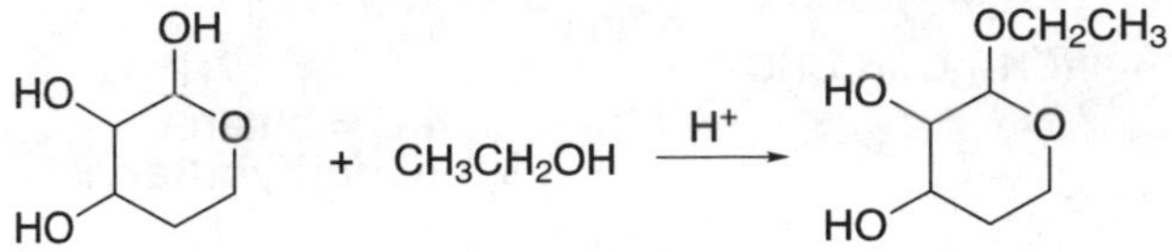 + CH₃CH₂OH $\xrightarrow{H^+}$

21.38

monensin

paeoniflorin

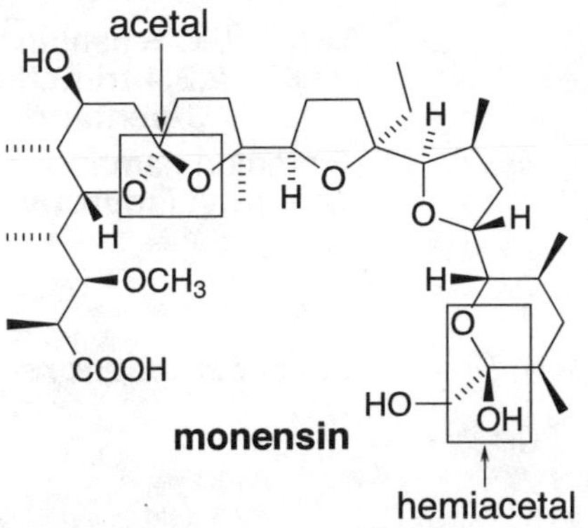

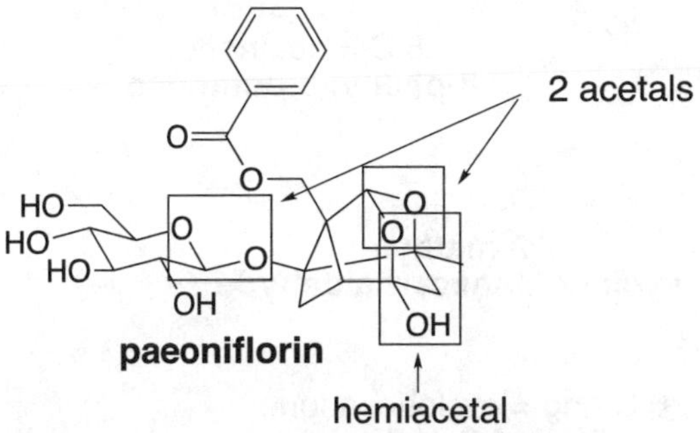

21.39 β-D-glucose has five stereogenic centers. α- and β-D-glucose are stereoisomers (diastereomers). A and β-D-glucose are constitutional isomers.

21.40

a. 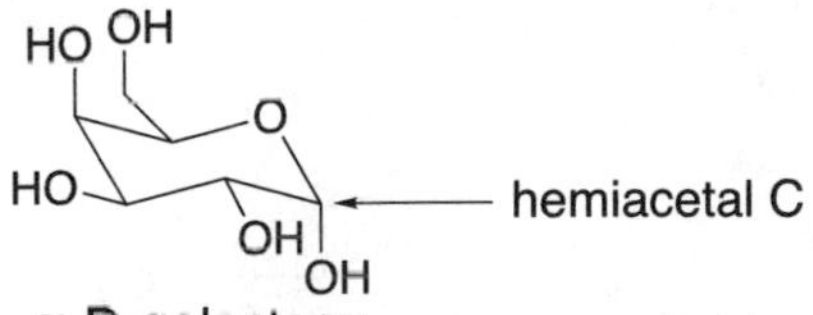 ← hemiacetal C

α-D-galactose

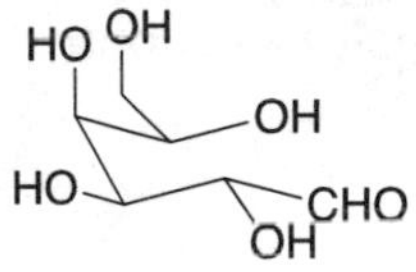

b.

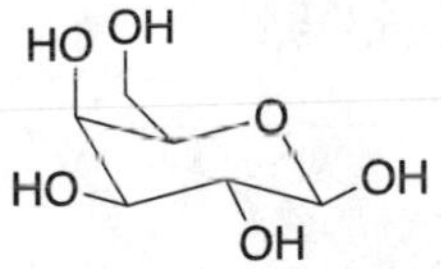

β-D-galactose

c.

d.

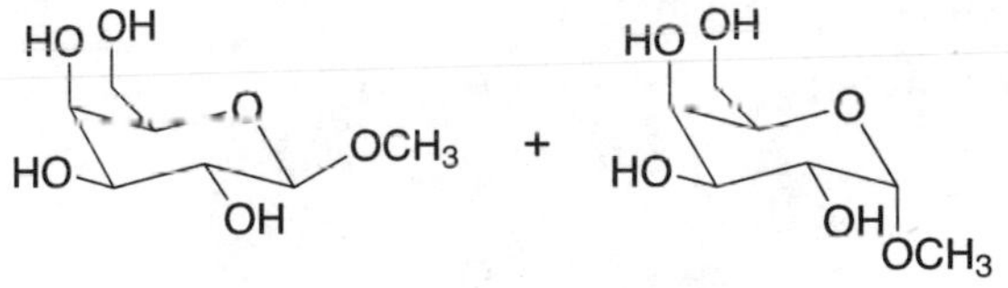

21.41 Use the rules from Answer 21.3 and 21.5 to name the aldehydes and ketones.

a. $(CH_3)_3CCH_2CHO$ —redraw→

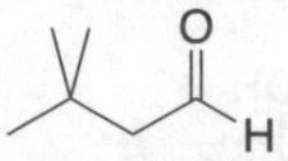

4 C = butanal
3,3-dimethybutanal

g. *trans*-2-benzylcyclohexanecarbaldehyde

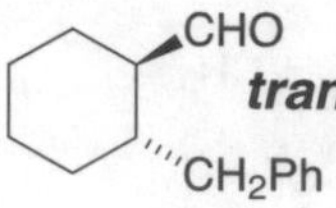

b.

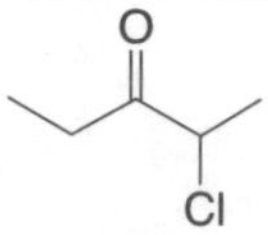

5 C = pentanone
2-chloro-3-pentanone

h. $(CH_3)_3C$ —redraw→

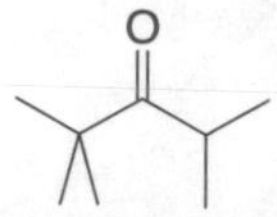

5 C = pentanone
2,2,4-trimethyl-3-pentanone
(common name:
***tert*-butyl isopropyl ketone**)

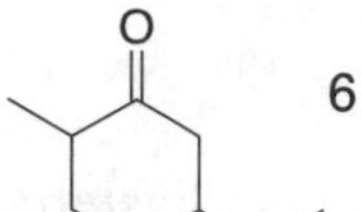

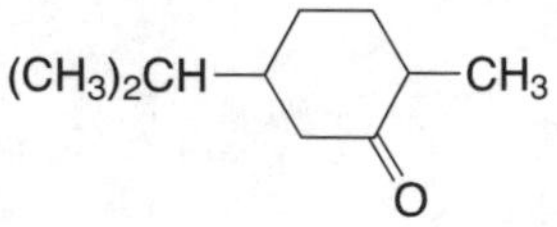

c. Ph

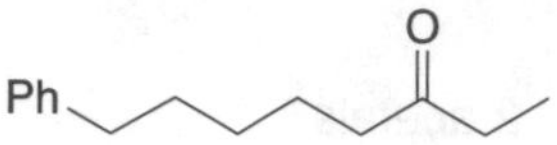

8 C = octanone
8-phenyl-3-octanone

d.

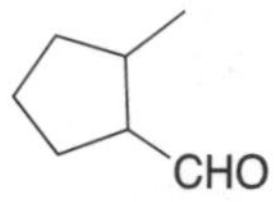

5 C ring
2-methyl-cyclopentanecarbaldehyde

i. 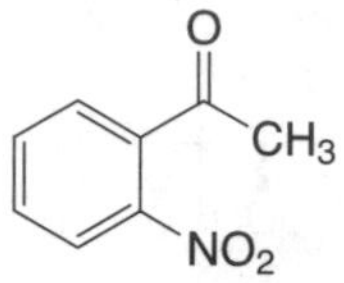 *o*-nitroacetophenone

e.

6 C ring = cyclohexanone
5-ethyl-2-methyl-cyclohexanone

j.

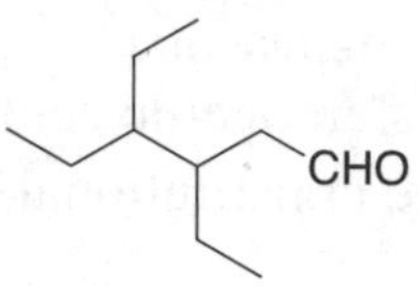

6 C = hexanal
3,4-diethylhexanal

f. $(CH_3)_2CH$—CH_3

6 C ring = cyclohexanone
5-isopropyl-2-methyl-cyclohexanone

21.42

a. 2-methyl-3-phenylbutanal

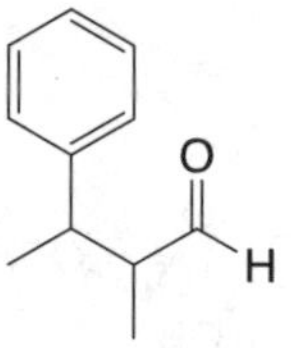

b. dipropyl ketone

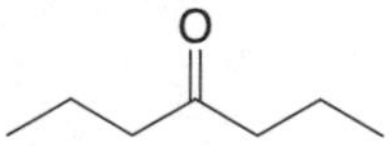

c. 3,3-dimethylcyclohexanecarbaldehyde

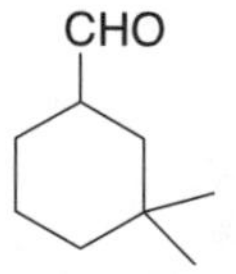

d. α-methoxypropionaldehyde

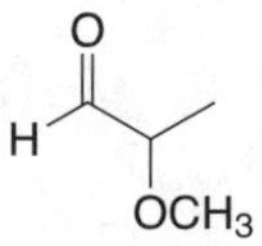

e. 3-benzoylcyclopentanone

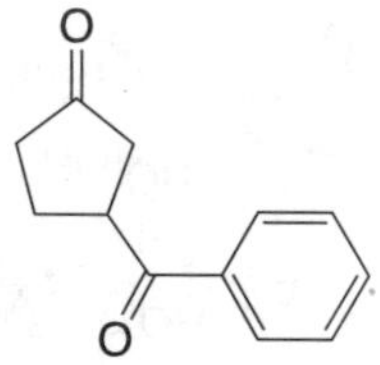

f. 2-formylcyclopentanone

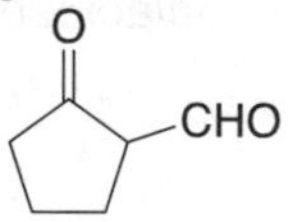

g. (*R*)-3-methyl-2-heptanone

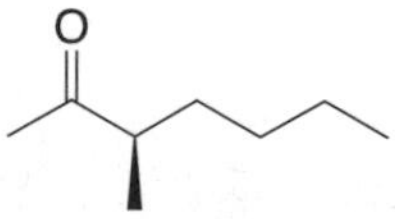

h. *m*-acetylbenzaldehyde

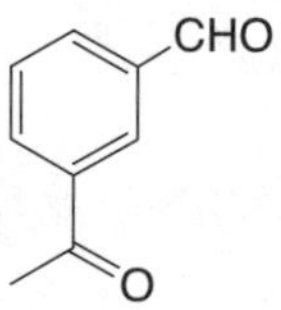

21.43

hexanal   2-methylpentanal   3-methylpentanal   4-methylpentanal   3,3-dimethylbutanal

2-ethylbutanal   2,2-dimethylbutanal   2,3-dimethylbutanal   3-hexanone   3,3-dimethyl-2-butanone

2-hexanone   3-methyl-2-pentanone   4-methyl-2-pentanone   2-methyl-3-pentanone

21.44

$= C_6H_5CH_2CHO$ phenylacetaldehyde

a. $\underrightarrow{\text{NaBH}_4/\text{CH}_3\text{OH}}$ $C_6H_5CH_2CH_2OH$

b. $\underrightarrow{[1]\ \text{LiAlH}_4;\ [2]\ \text{H}_2\text{O}}$ $C_6H_5CH_2CH_2OH$

c. $\underrightarrow{[1]\ \text{CH}_3\text{MgBr};\ [2]\ \text{H}_2\text{O}}$ $C_6H_5CH_2CH(OH)CH_3$

d. $\underrightarrow{\text{NaCN/HCl}}$ $C_6H_5CH_2CH(OH)CN$

e. $\underrightarrow{\text{Ph}_3\text{P}=\text{CHCH}_3}$ $C_6H_5CH_2CH=CHCH_3$
(E and Z isomers)

f. $\underrightarrow{\substack{(\text{CH}_3)_2\text{CHNH}_2 \\ \text{mild acid}}}$

g. $\underrightarrow{(\text{CH}_3\text{CH}_2)_2\text{NH/mild acid}}$ (E and Z isomers)

h. $\underrightarrow{\text{CH}_3\text{CH}_2\text{OH (excess)/H}^+}$

i. $\underrightarrow{\text{NH + mild acid}}$ (E and Z isomers)

j. $\underrightarrow{\text{HO}\diagup\diagdown\text{OH /H}^+}$

21.45

2-butanone

a. NaBH₄/CH₃OH → (2-butanol with OH)

b. [1] LiAlH₄; [2] H₂O → (2-butanol with OH)

c. [1] CH₃MgBr; [2] H₂O → (tertiary alcohol OH)

d. NaCN/HCl → NC—C(OH) cyanohydrin

e. Ph₃P=CHCH₃ → =CHCH₃ alkene

f. (CH₃)₂CHNH₂/mild H⁺ → =NCH(CH₃)₂ imine

g. (CH₃CH₂)₂NH/mild H⁺ → N(CH₂CH₃)₂ + N(CH₂CH₃)₂ (*E* and *Z* isomers)

h. CH₃CH₂OH (excess)/H⁺ → CH₃CH₂O OCH₂CH₃ acetal

i. ⬡NH + mild H⁺ → enamine + enamine (*E* and *Z* isomers)

j. HO—CH₂CH₂—OH /H⁺ → dioxolane

21.46

a. (cyclopentanone) =O $\xrightarrow{Ph_3P=CHCH_2CH_3}$ =CHCH₂CH₃

b. (cyclopentane)—CHO $\xrightarrow{Ph_3P=\text{(cyclohexylidene)}}$ product

c. (cyclopentane)—CHO $\xrightarrow{Ph_3P=CHCOOCH_3}$ (H / COOCH₃ alkene) + (COOCH₃ alkene)

d. (cyclopentanone) =O $\xrightarrow{Ph_3P=CH(CH_2)_5COOCH_3}$ =CH(CH₂)₅COOCH₃

21.47

a. CH₃CH₂Cl $\xrightarrow[\substack{[2]\ BuLi \\ [3]\ (CH_3)_2C=O}]{[1]\ Ph_3P}$ CH₃CH=C(CH₃)₂

b. (benzyl)—CH₂Br $\xrightarrow[\substack{[2]\ BuLi \\ [3]\ C_6H_5CH_2CH_2CHO}]{[1]Ph_3P}$ —CH=CHCH₂CH₂C₆H₅

c. (cyclopentane)—CH₂Cl $\xrightarrow[\substack{[2]\ BuLi \\ [3]\ CH_3CH_2CH_2CHO}]{[1]\ Ph_3P}$ —CH=CHCH₂CH₂CH₃

21.48

a. Ph₃P=CHCH₂CH₂CH₃ ⟹ BrCH₂CH₂CH₂CH₃

b. Ph₃P=C(CH₂CH₂CH₃)₂ ⟹ BrCH(CH₂CH₂CH₃)₂

c. Ph₃P=CHCH=CH₂ ⟹ BrCH₂CH=CH₂

21.49

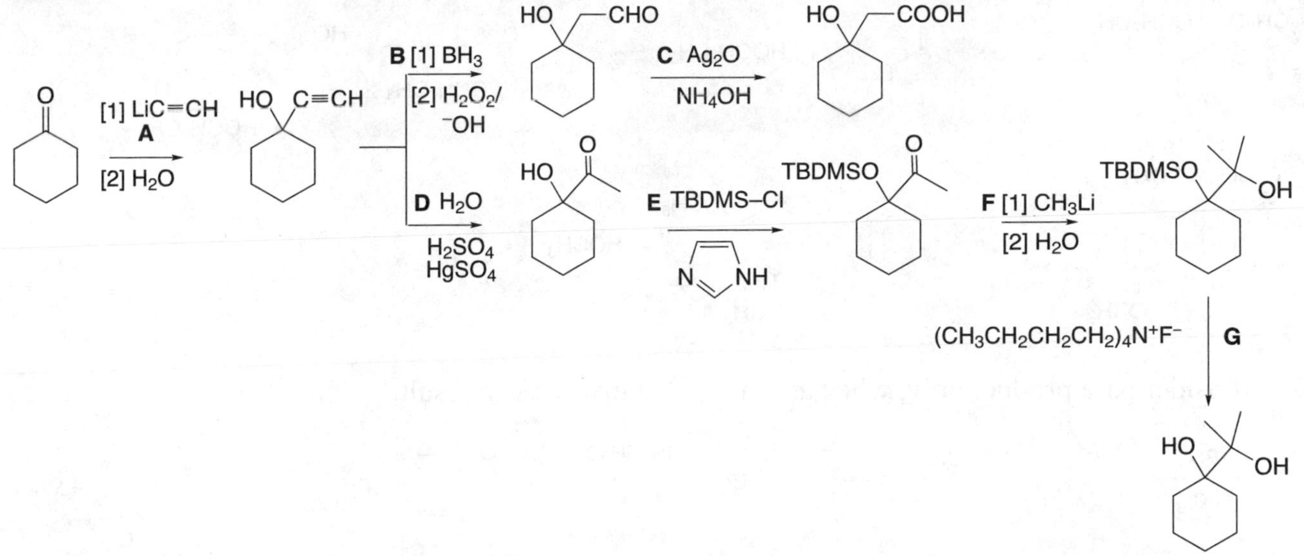

21.50

a. 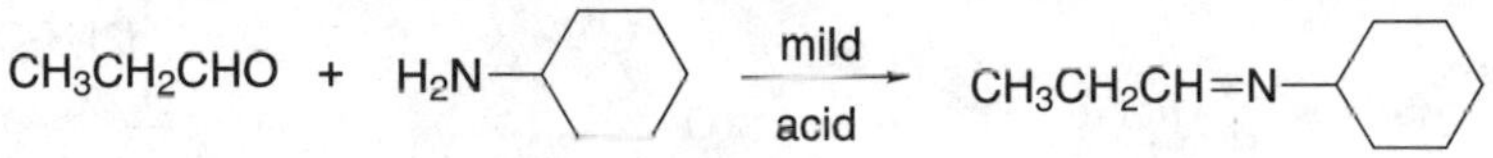 CH_3CH_2CHO + H_2N—⬡ $\xrightarrow[\text{acid}]{\text{mild}}$ $CH_3CH_2CH=N$—⬡

b.

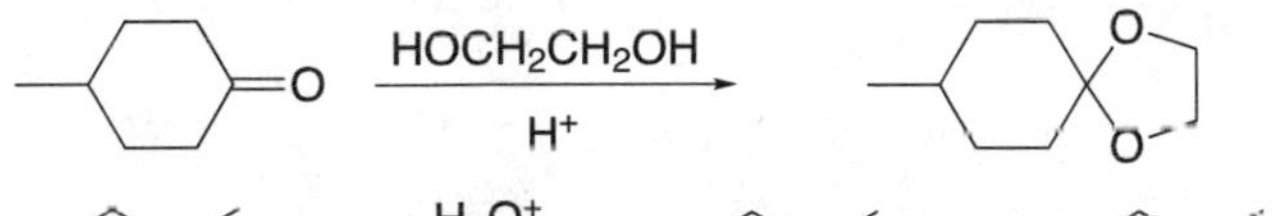

c.

d. 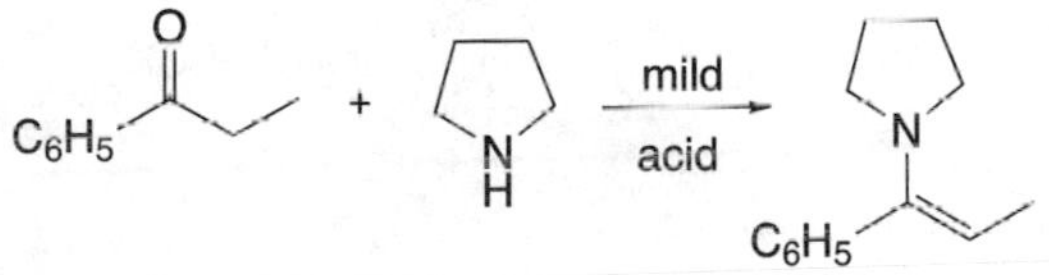 $(E$ and Z isomers)

e.

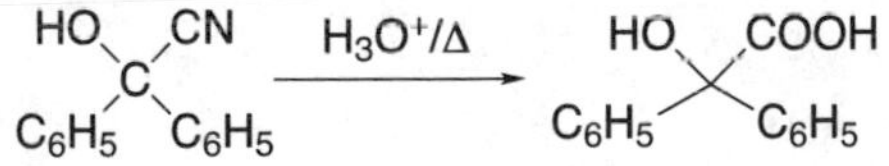

f.

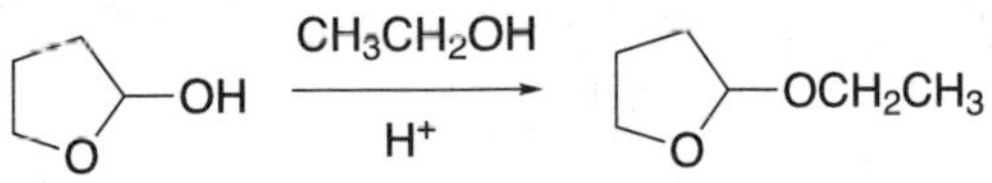

g.

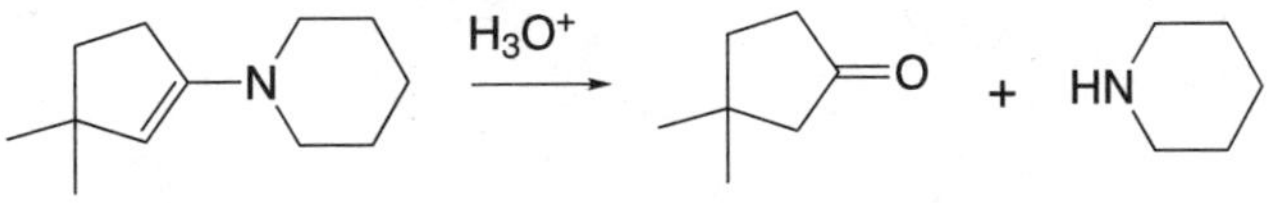

h. 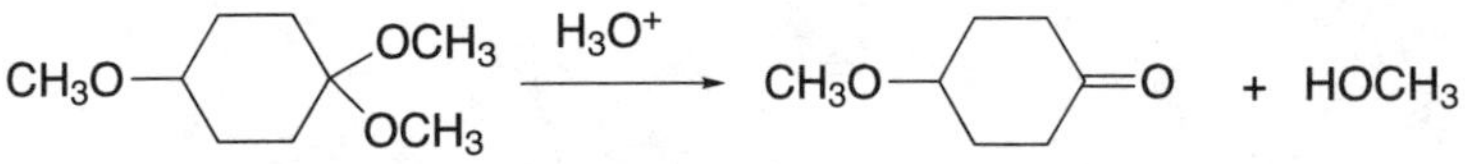 CH_3O—

21.51

a. 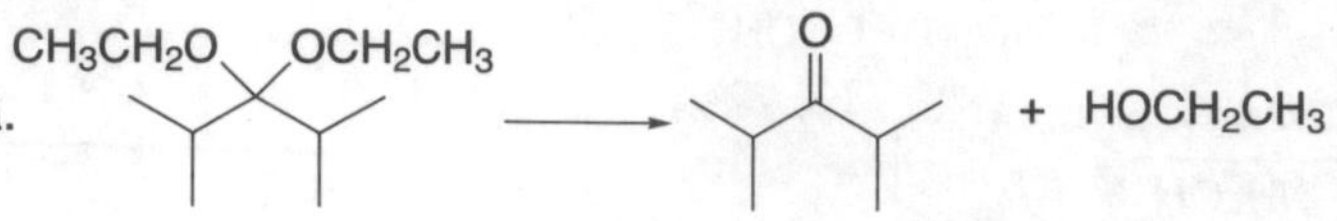

c. 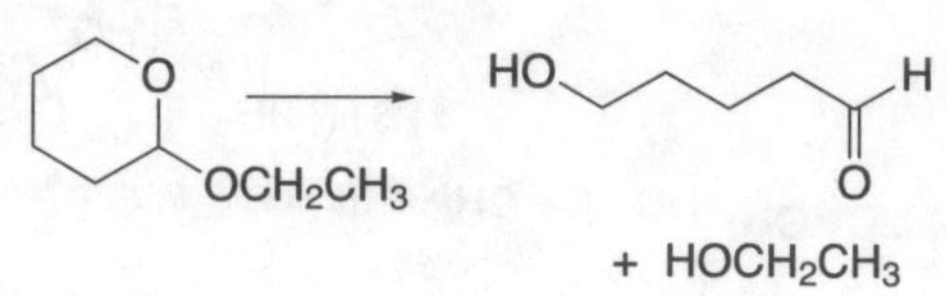

b. 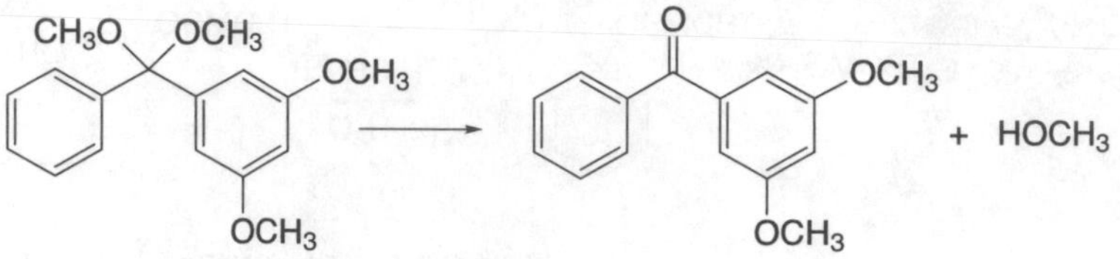

21.52 Consider para product only, when an ortho, para mixture can result.

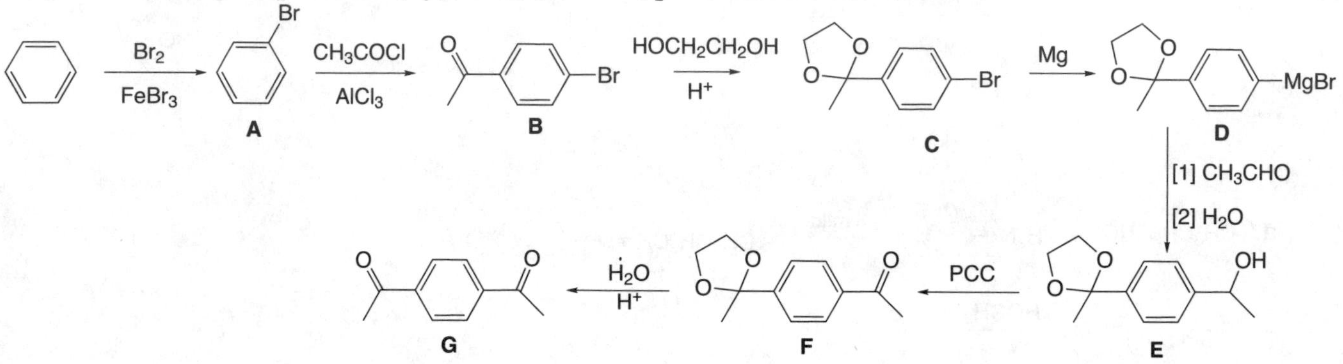

21.53

a. 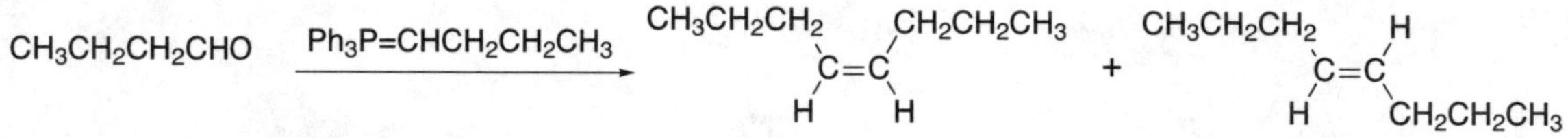

b. 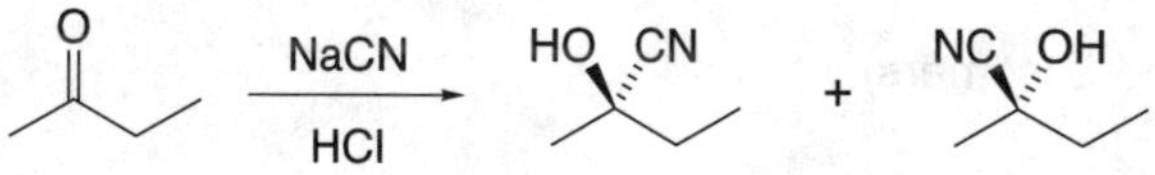

c. 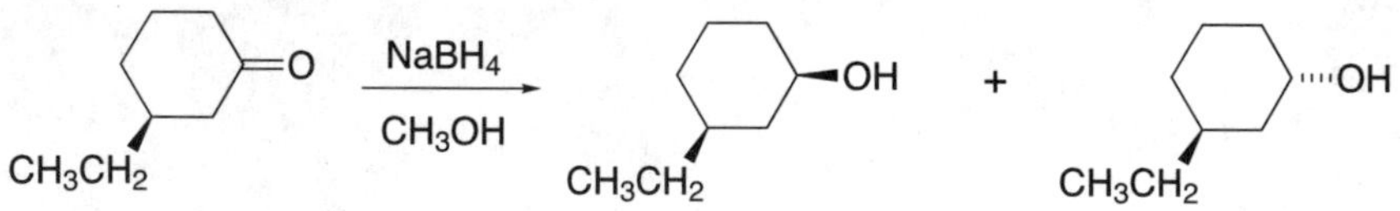

d. 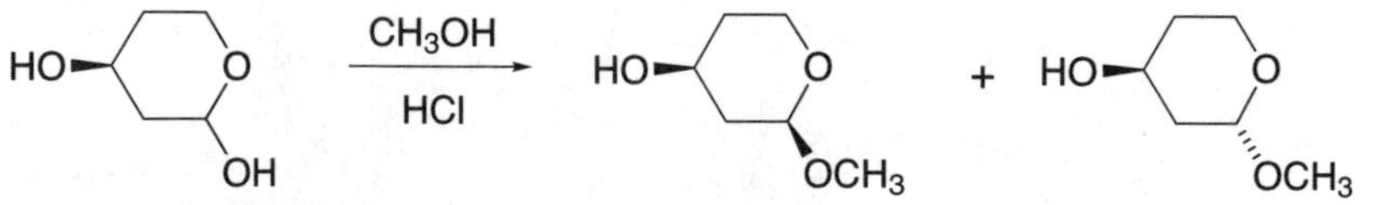

21.54

new stereogenic center

An equal mixture of enantiomers results, so the product is optically inactive.

A achiral

new stereogenic center

A mixture of diastereomers results. Both compounds are chiral and they are not enantiomers, so the mixture is optically active.

B chiral

21.55

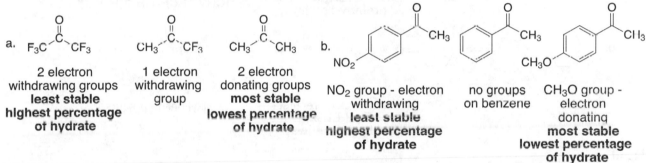

acetal **frontalin**

acetal **multistriatin**

21.56 As the number of R groups bonded to the carbonyl C increases, stability also increases due to electron donating effects. Then use the principles from Answer 21.29.

a.

F_3C—C—CF_3
2 electron withdrawing groups
least stable
highest percentage of hydrate

CH_3—C—CF_3
1 electron withdrawing group

CH_3—C—CH_3
2 electron donating groups
most stable
lowest percentage of hydrate

b.

NO_2 group - electron withdrawing
least stable
highest percentage of hydrate

no groups on benzene

CH_3O group - electron donating
most stable
lowest percentage of hydrate

21.57 Use the rule from Answer 21.1.

a.

increasing reactivity
decreasing steric hindrance

b.

increasing reactivity
decreasing steric hindrance

21.58

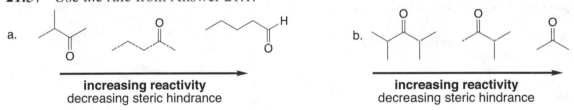

Less stable carbonyl compounds give a higher percentage of hydrate. Cyclopropanone is an unstable carbonyl compound because the bond angles around the carbonyl carbon deviate considerably from the desired angle. Since the carbonyl carbon is sp^2 hybridized, the optimum bond angle is 120°, but the three-membered ring makes the C–C–C bond angles only 60°. This destabilizes the ketone, giving a high concentration of hydrate when dissolved in H_2O.

21.59 Use the principles from Answer 21.21.

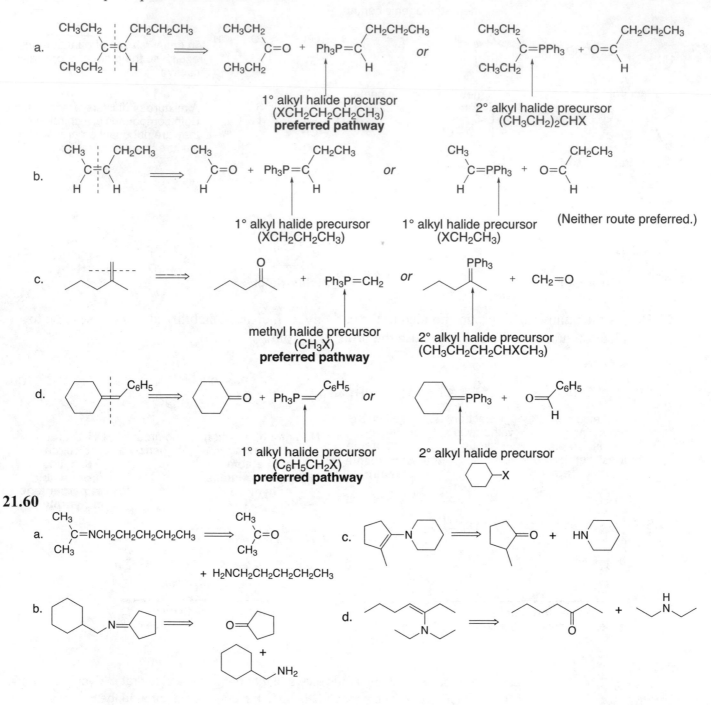

21.60

21.61

21.62

a. $C_6H_5-CH_2OH$ $\xrightarrow{PCC}$ C_6H_5-CHO

b. C_6H_5-COCl $\xrightarrow[\text{[2] }H_2O]{\text{[1] LiAlH[OC(CH}_3)_3]_3}}$ C_6H_5-CHO

c. $C_6H_5-COOCH_3$ $\xrightarrow[\text{[2] }H_2O]{\text{[1] DIBAL-H}}$ C_6H_5-CHO

d. C_6H_5-COOH $\xrightarrow[\text{[2] }H_2O]{\text{[1] LiAlH}_4}$ $C_6H_5-CH_2OH$ $\xrightarrow{PCC}$ C_6H_5-CHO

e. $C_6H_5-CH_3$ $\xrightarrow{KMnO_4}$ C_6H_5-COOH $\xrightarrow[\text{[2] }H_2O]{\text{[1] LiAlH}_4}$ $C_6H_5-CH_2OH$ $\xrightarrow{PCC}$ C_6H_5-CHO

f. $C_6H_5-CH=CH_2$ $\xrightarrow[\text{[2] Zn/}H_2O]{\text{[1] }O_3}$ C_6H_5-CHO

g. $C_6H_5-CH=NCH_2CH_2CH_3$ $\xrightarrow[H^+]{H_2O}$ C_6H_5-CHO

h. $C_6H_5-CH(OCH_2CH_3)_2$ $\xrightarrow[H^+]{H_2O}$ C_6H_5-CHO

21.63

a. $\xrightarrow{PCC}$

b. $\xrightarrow{(CH_3)_2CuLi}$

c. CH_3COCl $\xrightarrow{(CH_3CH_2)_2CuLi}$

d. $CH_3CH_2C\equiv CH$ $\xrightarrow[HgSO_4]{\begin{array}{c}H_2O\\H_2SO_4\end{array}}$

e. $CH_3C\equiv CCH_3$ $\xrightarrow[HgSO_4]{\begin{array}{c}H_2O\\H_2SO_4\end{array}}$

21.64

a. One-step sequence: $\xrightarrow{Ph_3P\diagup}$ preferred route only one product formed

or

Two-step sequence: $\xrightarrow[\text{[2] }H_2O]{\text{[1] CH}_3CH_2CH_2MgBr}$ $\xrightarrow{H_2SO_4}$ +

b. One-step sequence: $\xrightarrow{Ph_3P\diagup}$ preferred route only one product formed

or

Two-step sequence: $\xrightarrow[\text{[2] }H_2O]{\text{[1] (CH}_3)_2CHMgBr}$ $\xrightarrow{H_2SO_4}$ +

+ other alkenes that result from carbocation rearrangement

21.65

a.

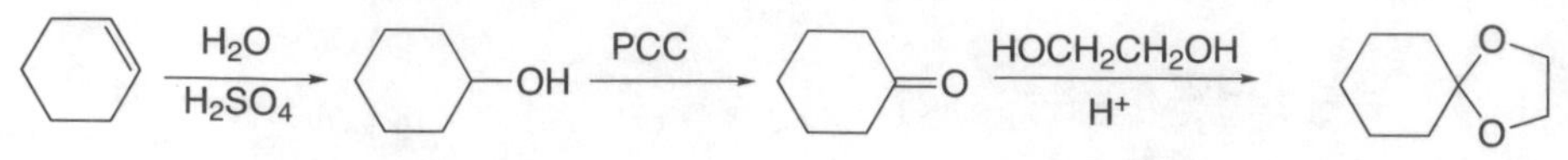

b.

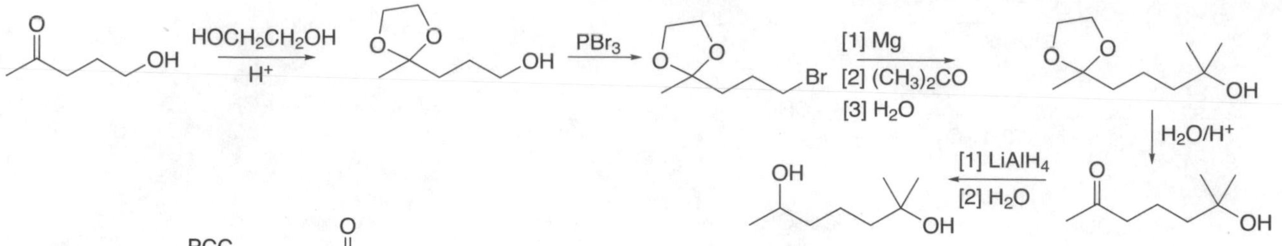

c.

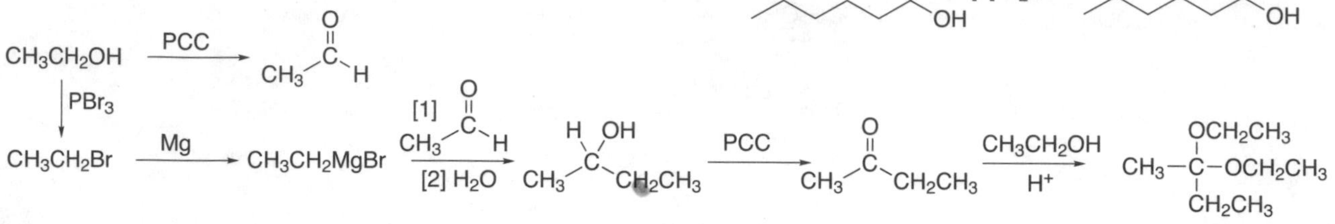

d.

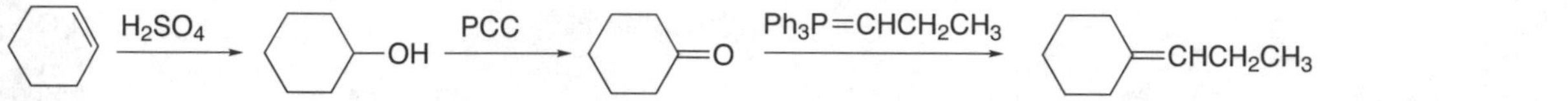

e.

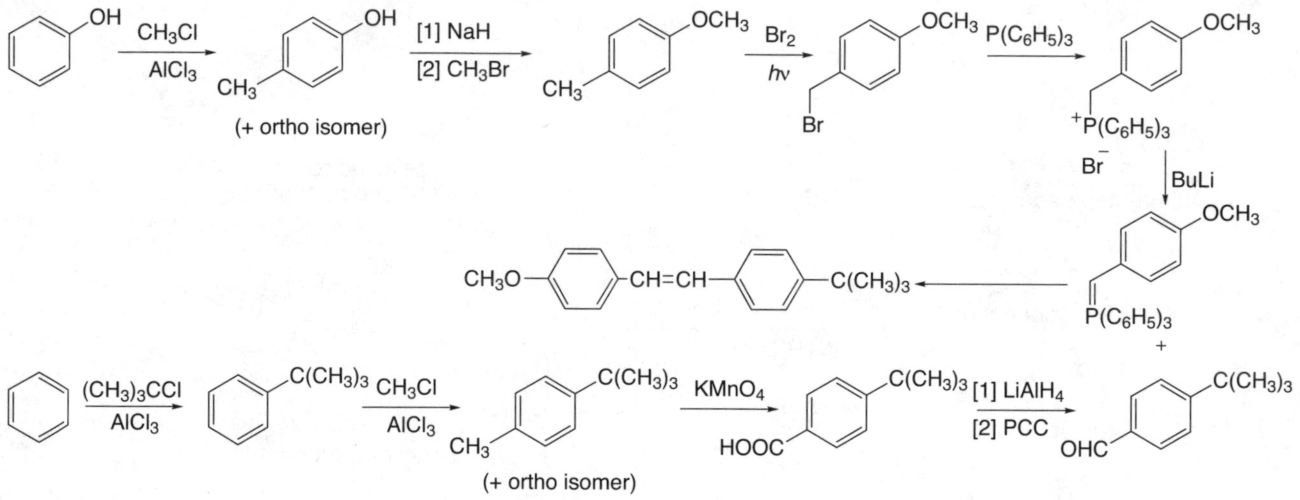

21.66

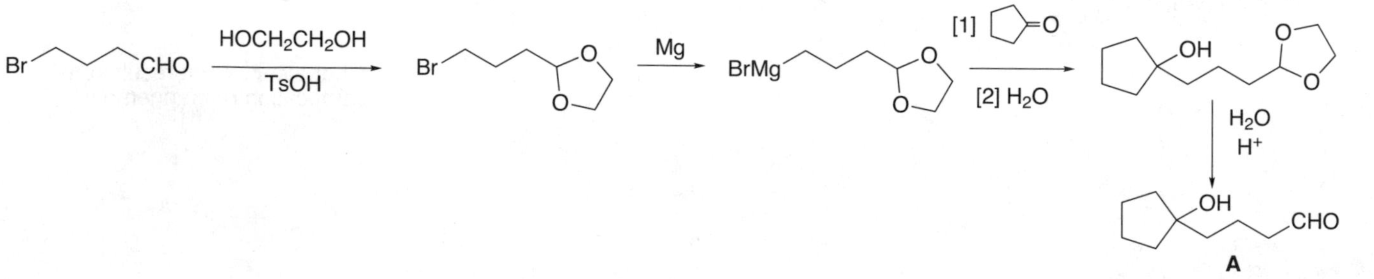

21.67

a. The α,β-unsaturated carbonyl has a δ^+ distributed over two carbons—the carbonyl carbon and the β carbon. As a result, there is less (+) charge on the carbonyl carbon, making it less electrophilic, and less reactive towards nucleophiles.

Three resonance structures:

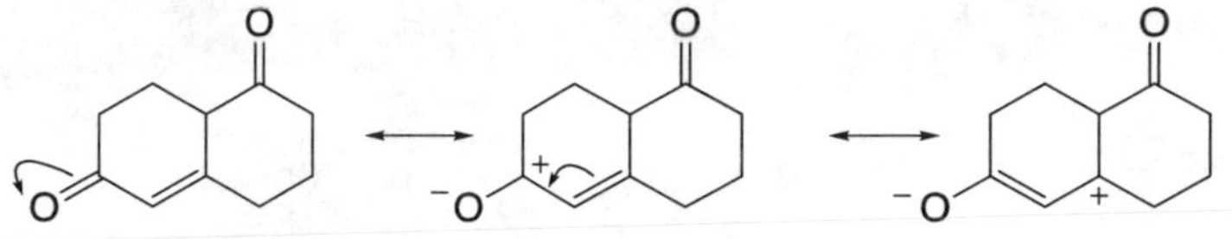

b.

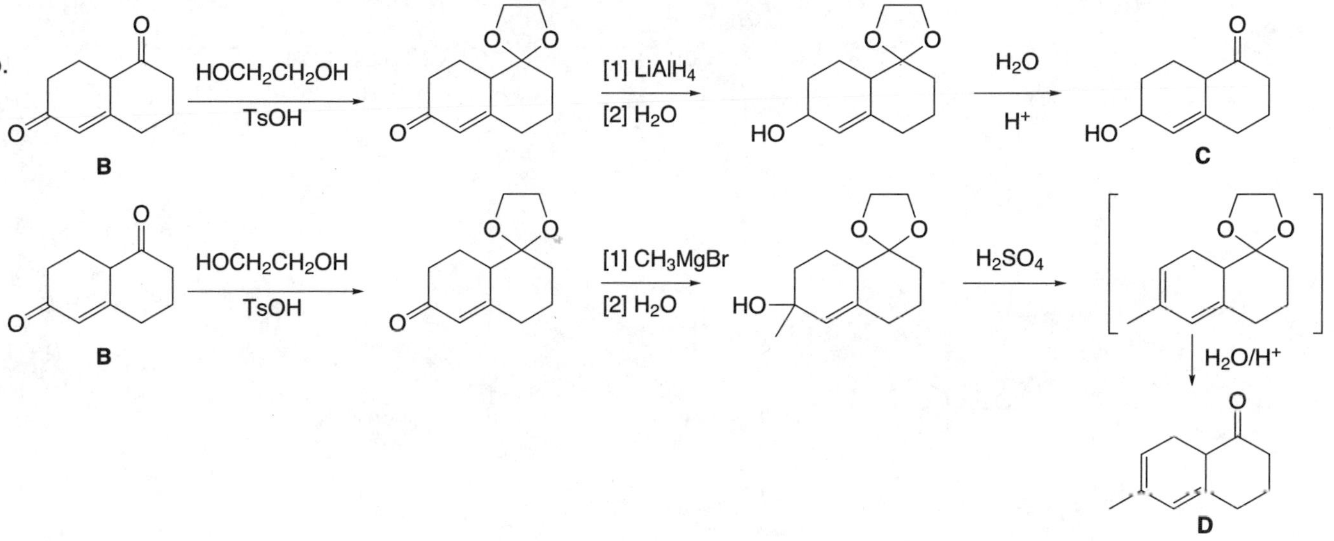

21.68

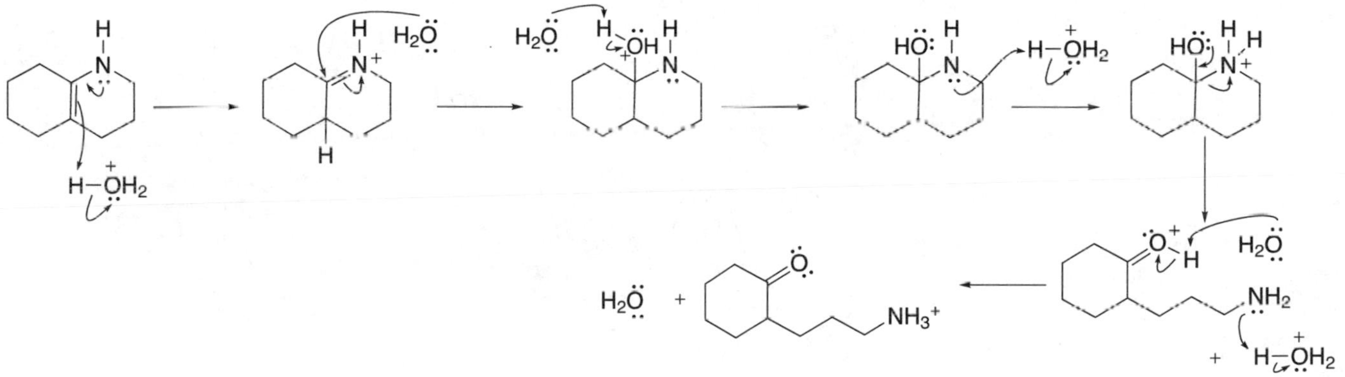

21.69

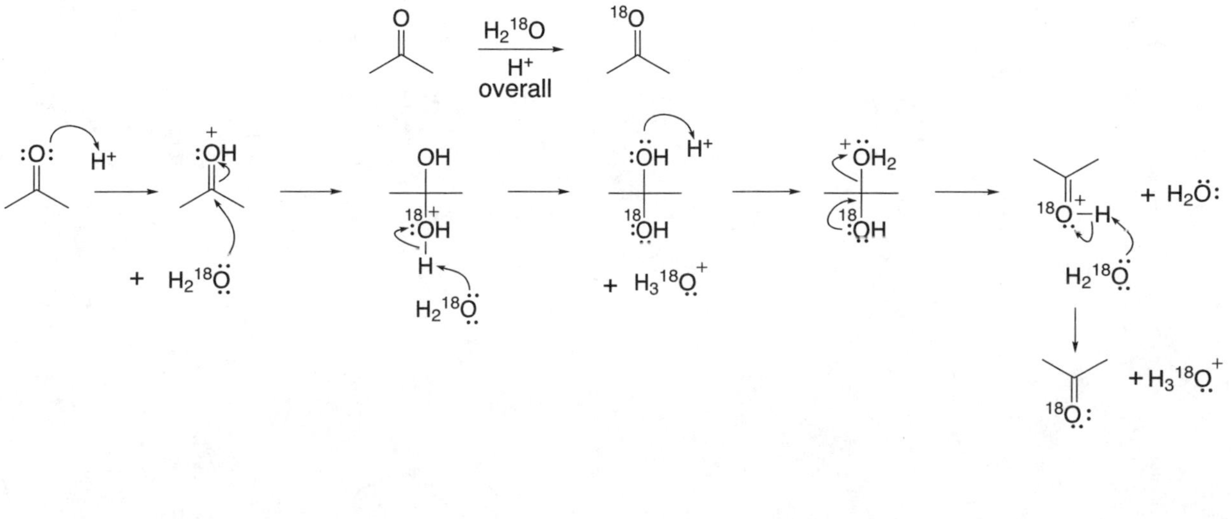

21.70

a.

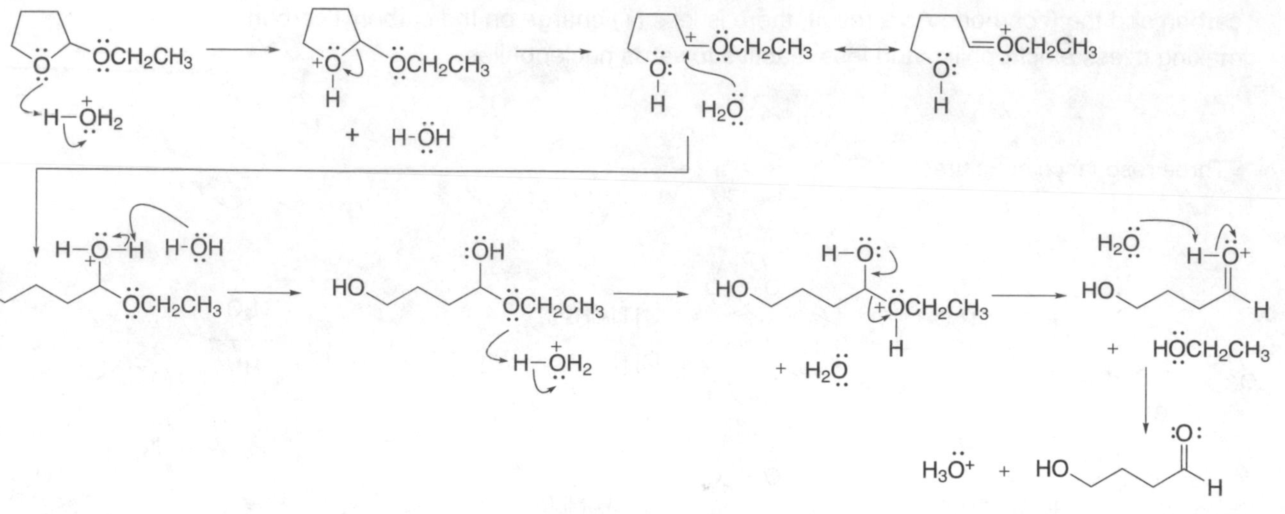

b.

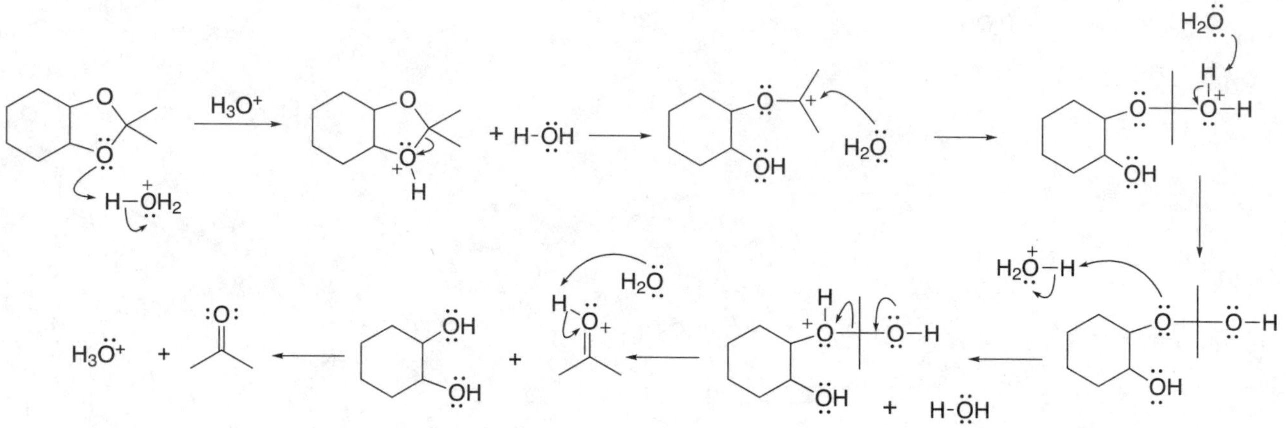

21.71

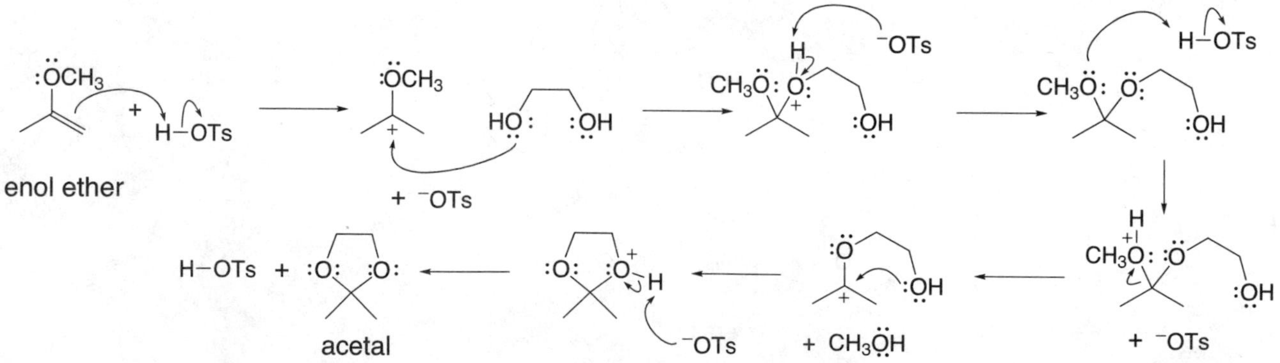

enol ether

acetal

21.72

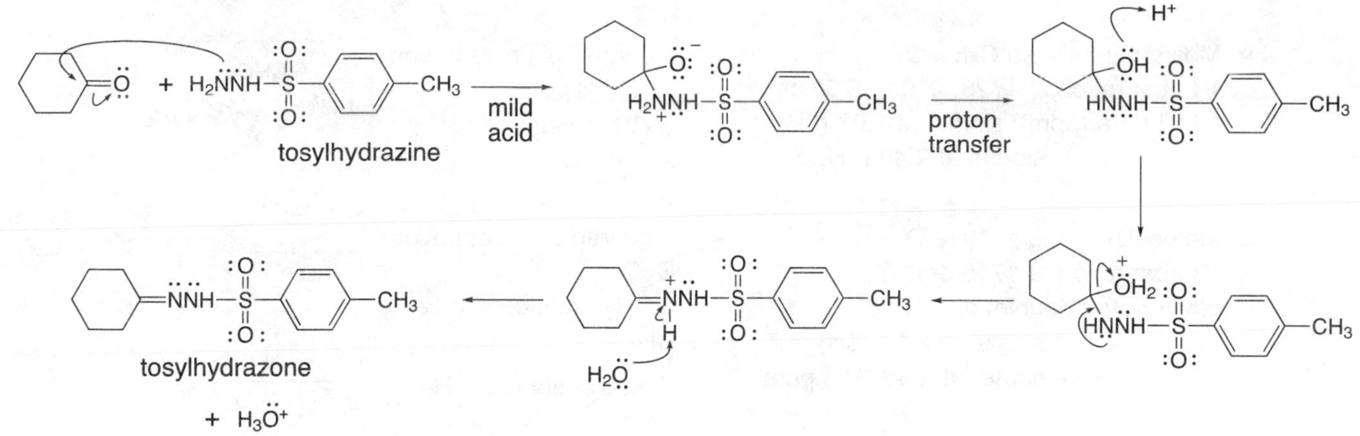

tosylhydrazine → mild acid → proton transfer

tosylhydrazone

+ $H_3\overset{..}{O}^+$

21.73

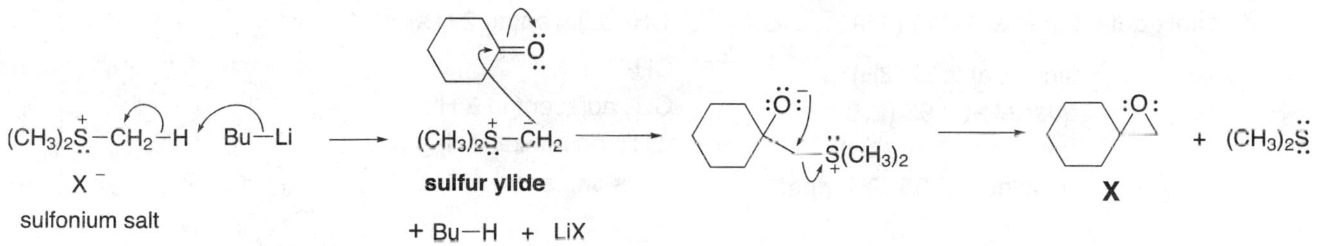

$(CH_3)_2\overset{+}{\overset{..}{S}}-CH_2-H$ Bu—Li → $(CH_3)_2\overset{+}{\overset{..}{S}}-CH_2$ →

X$^-$

sulfonium salt

sulfur ylide

+ Bu—H + LiX

X   + $(CH_3)_2\overset{..}{\overset{..}{S}}$

21.74 Hemiacetal **A** is in equilibrium with its acyclic hydroxy aldehyde. The aldehyde is susceptible to hydride reduction and this forms 1,4-butanediol.

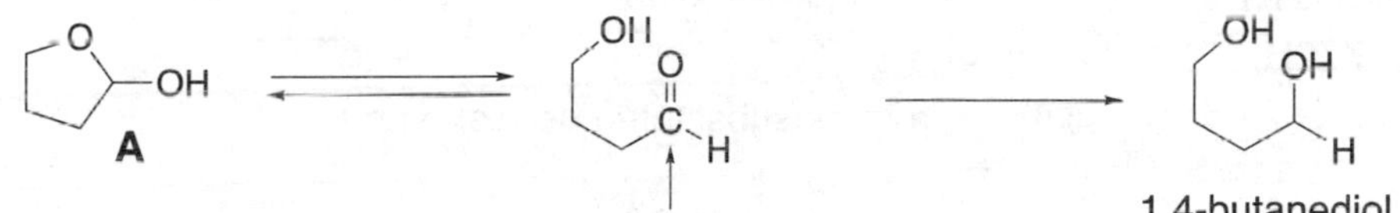

A

This can now be reduced with NaBH$_4$.

1,4-butanediol

21.75

a. 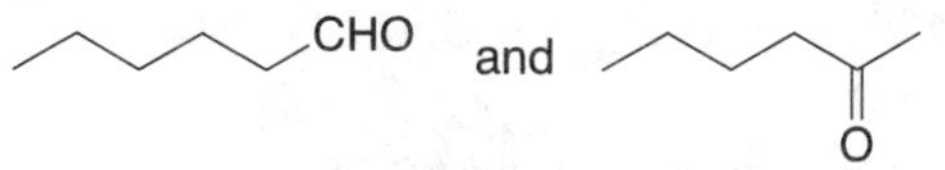 and

aldehyde   ketone

The sp^2 hybridized C–H bond of the aldehyde absorbs at 2700–2830 cm^{-1}.

c. and

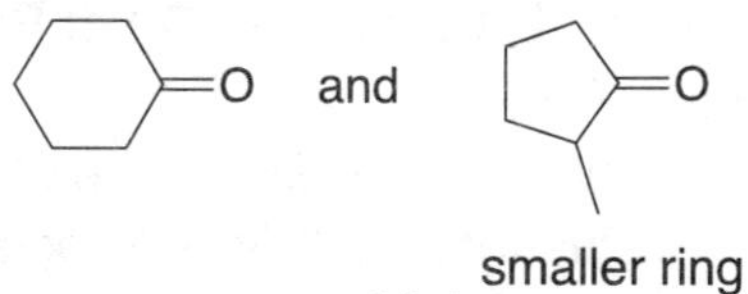

smaller ring
higher wavenumber for C=O

b. 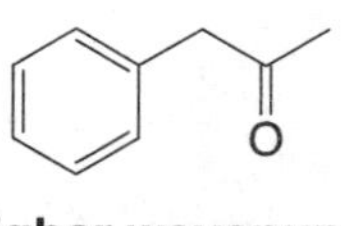 and

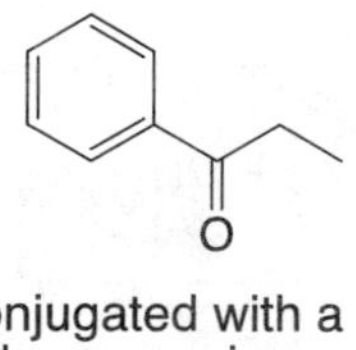

higher wavenumber for C=O

conjugated with a benzene ring
lower wavenumber

21.76

A. Molecular formula $C_5H_{10}O$ ⟶ 1 degree of unsaturation
    IR absorption at 1728, 2791, 2700 cm^{-1} ⟶ C=O, CHO
    NMR data (ppm): singlet at 1.08 (9H) ⟶ 3 CH$_3$ groups
          singlet at 9.48 (1H) ppm ⟶ CHO

B. Molecular formula $C_5H_{10}O$ ⟶ 1 degree of unsaturation
    IR absorption at 1718 cm^{-1} ⟶ C=O
    NMR data : doublet at 1.10 (6H) ⟶ 2 CH$_3$'s adjacent to H
          singlet at 2.14 (3H) ⟶ CH$_3$
          septet at 2.58 (1H) ppm ⟶ CH adjacent to 2 CH$_3$'s

C. Molecular formula $C_{10}H_{12}O$ ⟶ 5 degrees of unsaturation (4 due to a benzene ring)
    IR absorption at 1686 cm^{-1} ⟶ C=O
    NMR data: triplet at 1.21δ (3H) ⟶ CH$_3$ adjacent to 2 H's

          singlet at 2.39 (3H) ⟶ CH$_3$
          quartet at 2.95 (2H) ⟶ CH$_2$ adjacent to 3 H's
          doublet at 7.24 (2H) ⟶ 2 H's on benzene ring
          doublet at 7.85 (2H) ppm ⟶ 2 H's on benzene ring

D. Molecular formula $C_{10}H_{12}O$ ⟶ 5 degrees of unsaturation (4 due to a benzene ring)
    IR absorption at 1719 cm^{-1} ⟶ C=O
    NMR data: triplet at 1.02 (3H) ⟶ CH$_3$ adjacent 2 H's

          quartet at 2.45 (2H) ⟶ 2 H's adjacent to 3 H's
          singlet at 3.67 (2H) ⟶ CH$_2$
          multiplet at 7.06–7.48 (5H) ppm ⟶ a monosubstituted benzene ring

21.77

A. Molecular formula $C_9H_{10}O$
    IR absorption at 1700 cm^{-1} → C=O
    IR absorption at ~2700 cm^{-1} → CH of RCHO
    NMR data (ppm):
          triplet at 1.2 (2H's adjacent)
          quartet at 2.7 (3H's adjacent)
          doublet at 7.3 (2H's on benzene)
          doublet at 7.7 (2H's on benzene)
          singlet at 9.9 (CHO)

B. Molecular formula $C_9H_{10}O$
    IR absorption at 1720 cm^{-1} → C=O
    IR absorption at ~2700 cm^{-1} → CH of RCHO
    NMR data (ppm):
          2 triplets at 2.85 and 2.95
          multiplet at 7.2 (benzene H's)
          multiplet at 9.8 (CHO)

21.78

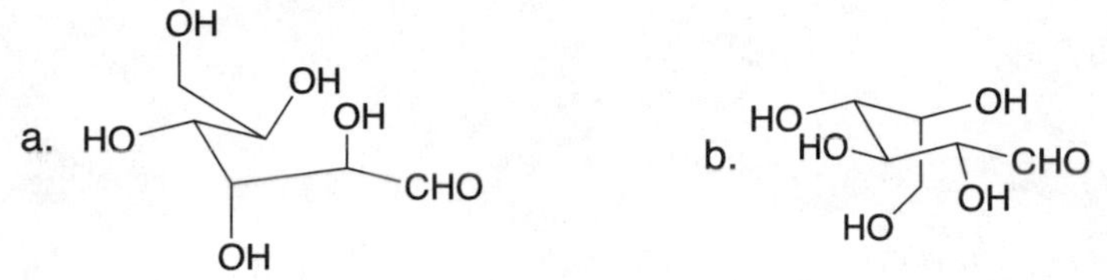

a.   [structure with OH, OH, OH, HO, OH, CHO, OH]

b.   [structure with HO, HO, OH, CHO, OH, HO]

21.79

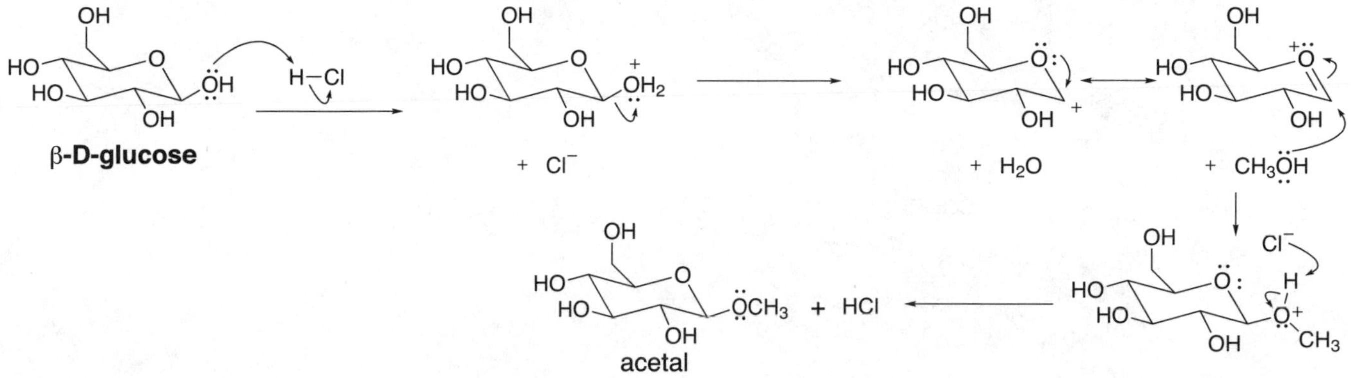

β-D-glucose

+ Cl⁻   + H₂O   + CH₃ÖH

acetal

HO ... ÖCH₃ + HCl

21.80

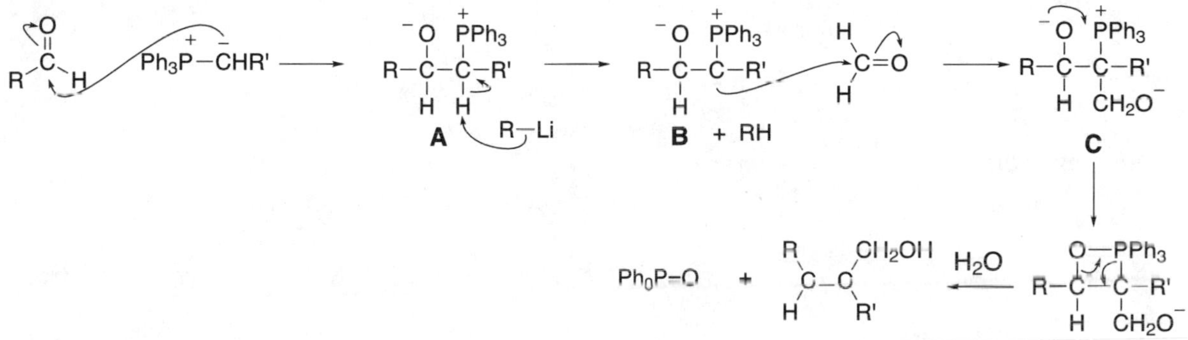

A   R–Li   B   + RH   C

Ph₃P=O   +   [structure R, CH₂OH, H, R']   H₂O

21.81

H⁺

H₂Ö:

+ H₃Ö⁺

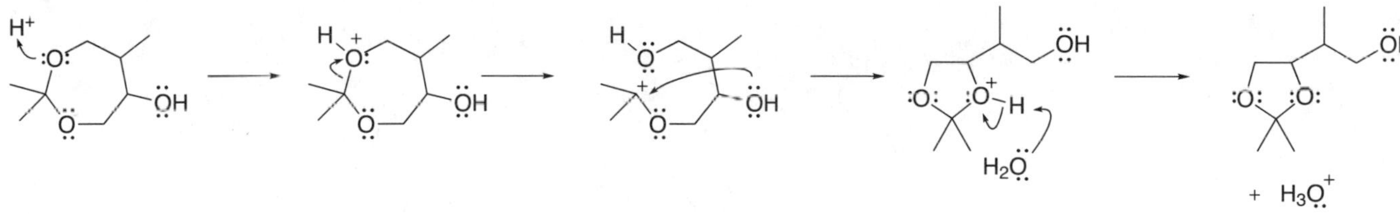

21.82

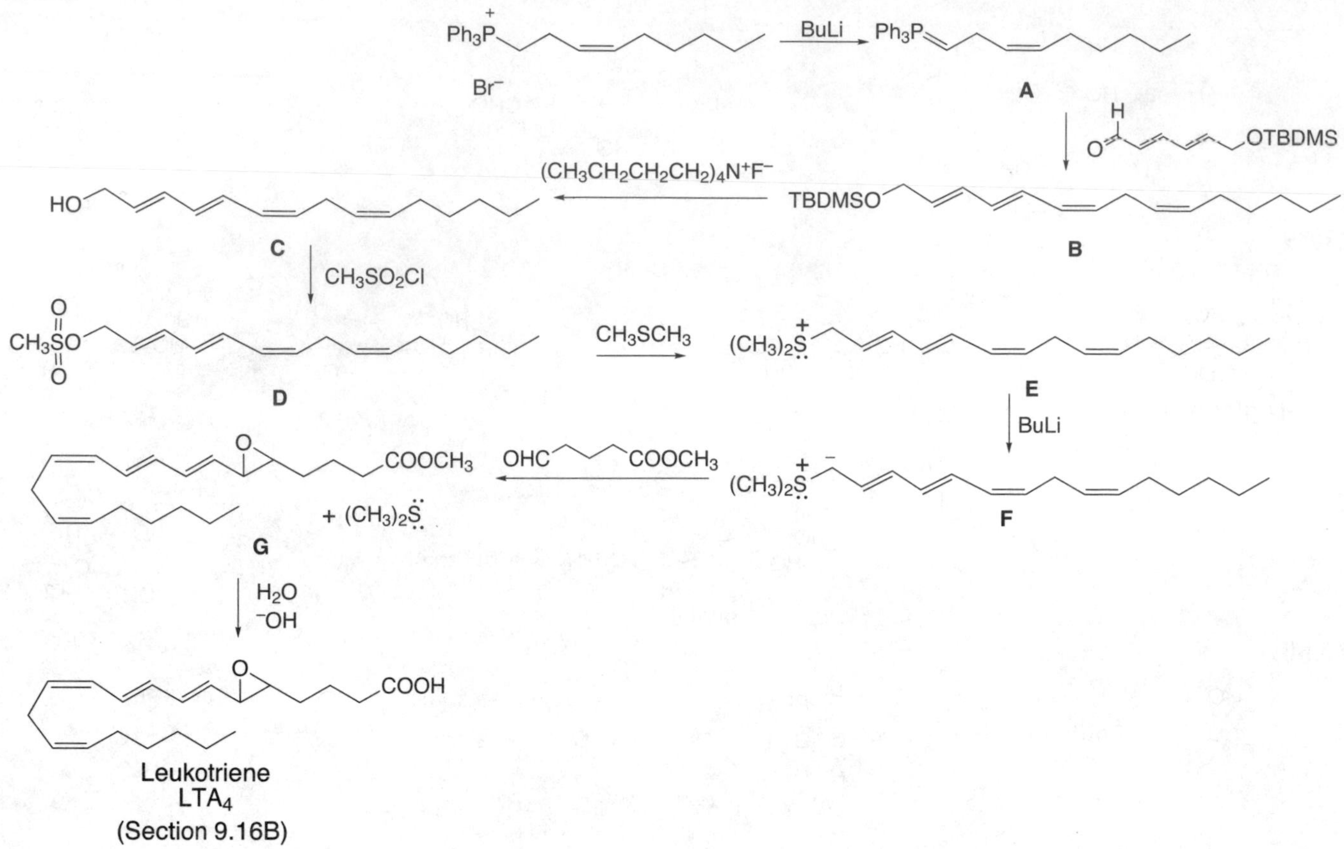

Ph$_3$P$^+$—\ \ \ \ \ \ \ \ \ \ \ \ \ \ \ Br$^-$ $\xrightarrow{\text{BuLi}}$ Ph$_3$P= \ \ \ \ \ \ \ \ \ **A**

$\downarrow$ O=CH—\ \ \ \ —OTBDMS

TBDMSO \ \ \ \ \ \ \ \ \ \ \ \ \ \ \ \ **B**

HO \ \ \ \ \ \ \ \ \ \ \ \ \ **C** $\xleftarrow{(CH_3CH_2CH_2CH_2)_4N^+F^-}$

$\downarrow$ CH$_3$SO$_2$Cl

D: CH$_3$SO$_2$O—\ \ \ \ \ \ \ \ \ \ \ \ **D** $\xrightarrow{CH_3SCH_3}$ (CH$_3$)$_2$S$^+$—\ \ \ \ \ \ \ \ **E**

$\downarrow$ BuLi

G: \ \ \ \ \ \ \ —COOCH$_3$ + (CH$_3$)$_2$S $\xleftarrow{\text{OHC}\ \ \ \text{COOCH}_3}$ (CH$_3$)$_2$S$^+$—\ \ \ \ \ \ \ \ **F**

G

$\downarrow$ H$_2$O, $^-$OH

\ \ \ \ \ \ \ —COOH

Leukotriene
LTA$_4$
(Section 9.16B)

Chapter 22: Carboxylic Acids and Their Derivatives—Nucleophilic Acyl Substitution

◆ Summary of spectroscopic absorptions of RCOZ (22.5)

IR absorptions	• All ROCZ compounds have a C=O absorption in the region 1600–1850 cm^{-1}.
	• RCOCl: 1800 cm^{-1}
	• $(RCO)_2O$: 1820 and 1760 cm^{-1} (two peaks)
	• RCOOR': 1735–1745 cm^{-1}
	• $RCONR'_2$: 1630–1680 cm^{-1}
	• Additional amide absorptions occur at 3200–3400 cm^{-1} (N–H stretch) and 1640 cm^{-1} (N–H bending).
	• Decreasing the ring size of a cyclic lactone, lactam, or anhydride increases the frequency of the C=O absorption.
	• Conjugation shifts the C=O to lower wave number.
^{1}H NMR absorptions	• C–H α to the C=O absorbs at 2–2.5 ppm.
	• N–H of an amide absorbs at 7.5–8.5 ppm.
^{13}C NMR absorption	• C=O absorbs at 160–180 ppm.

◆ ◆ Summary of spectroscopic absorptions of RCN (22.5)

IR absorption	• C≡N absorption at 2250 cm^{-1}
^{13}C NMR absorption	• C≡N absorbs at 115–120 ppm.

◆ Summary: The relationship between the basicity of Z⁻ and the properties of RCOZ

· **increasing basicity of the leaving group** (22.2)
· **increasing resonance stabilization** (22.2)

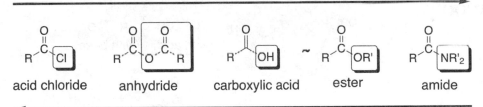

| acid chloride | anhydride | carboxylic acid | ester | amide |

· **increasing leaving group ability** (22.7B)
· **increasing reactivity** (22.7B)
· **increasing frequency of the C=O absorption in the IR** (22.5)

◆ General features of nucleophilic acyl substitution

- The characteristic reaction of compounds having the general structure RCOZ is nucleophilic acyl substitution (22.1).
- The mechanism consists of two basic steps (22.7A):
  [1] Addition of a nucleophile to form a tetrahedral intermediate
  [2] Elimination of a leaving group.
- More reactive acyl compounds can be used to prepare less reactive acyl compounds. The reverse is not necessarily true (22.7B).

♦ **Nucleophilic acyl substitution reactions**

[1] Reactions that synthesize acid chlorides (RCOCl)

[a] From RCOOH (22.10A):

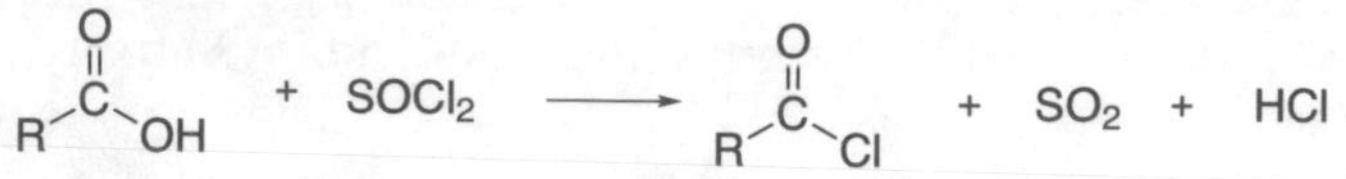

$$RCOOH + SOCl_2 \longrightarrow RCOCl + SO_2 + HCl$$

[2] Reactions that synthesize anhydrides [(RCO)₂O]

[a] From RCOCl (22.8):

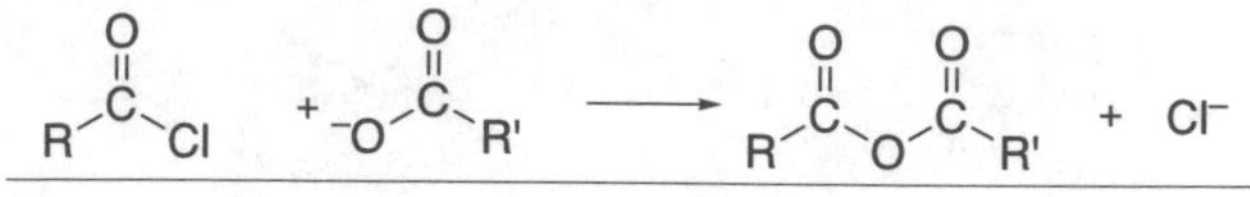

$$RCOCl + {}^-O-CO-R' \longrightarrow RCO-O-CO-R' + Cl^-$$

[b] From dicarboxylic acids (22.10B):

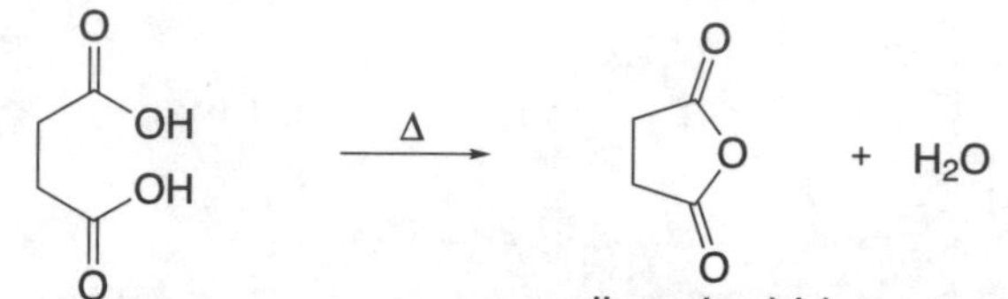

$$\xrightarrow{\Delta} \text{cyclic anhydride} + H_2O$$

[3] Reactions that synthesize carboxylic acids (RCOOH)

[a] From RCOCl (22.8):

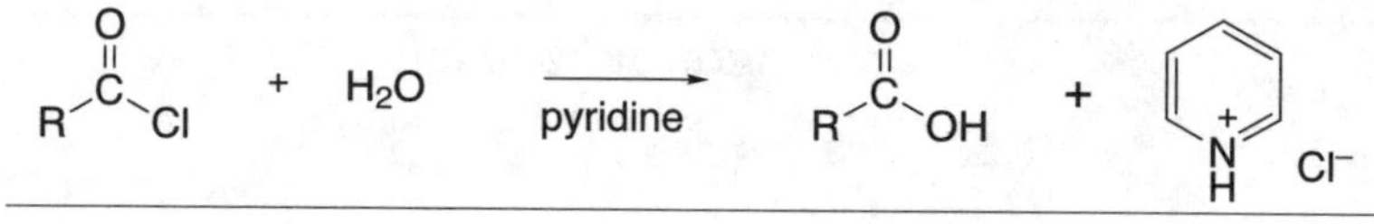

$$RCOCl + H_2O \xrightarrow{pyridine} RCOOH + \text{pyridinium}\ Cl^-$$

[b] From (RCO)₂O (22.9):

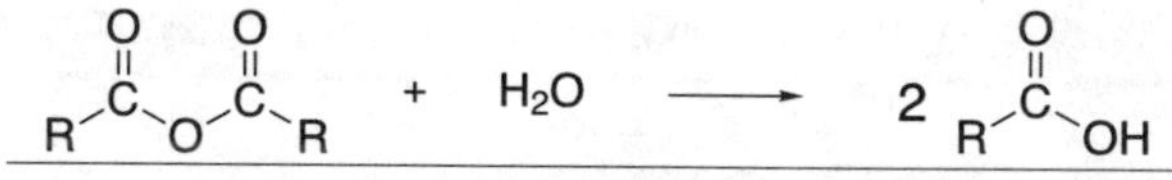

$$RCO-O-CO-R + H_2O \longrightarrow 2\ RCOOH$$

[c] From RCOOR′ (22.11):

$$RCOOR' + H_2O \xrightarrow{(H^+ \text{ or } {}^-OH)} RCOOH + RCOO^- + R'OH$$

(with acid) (with base)

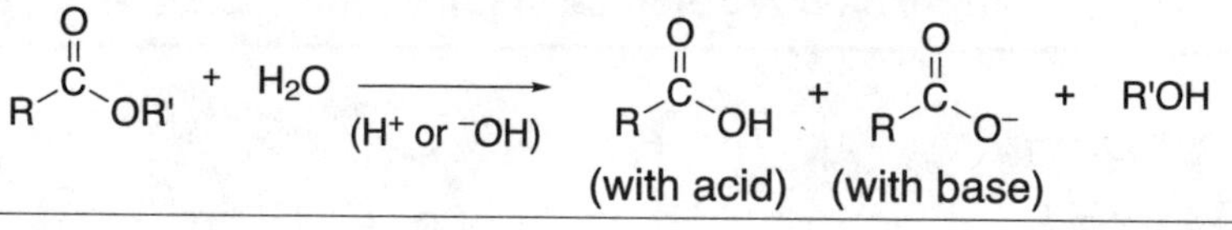

[d] From RCONR′₂ (R′ = H or alkyl, 22.13):

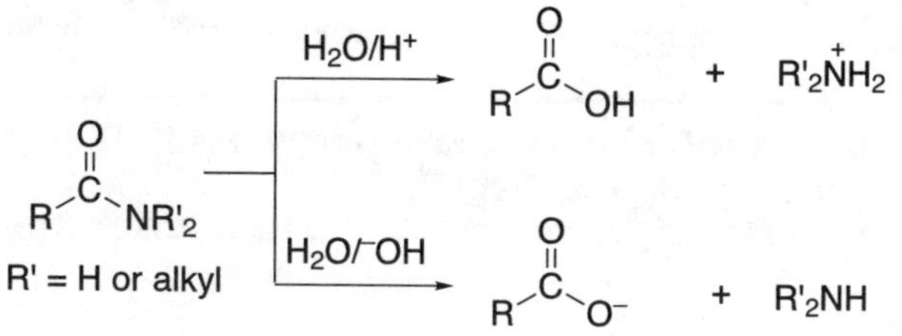

$$RCONR'_2 \quad (R' = H \text{ or alkyl})$$

$$\xrightarrow{H_2O/H^+} RCOOH + R'_2\overset{+}{N}H_2$$

$$\xrightarrow{H_2O/{}^-OH} RCOO^- + R'_2NH$$

[4] Reactions that synthesize esters (RCOOR′)

[a] From RCOCl (22.8):

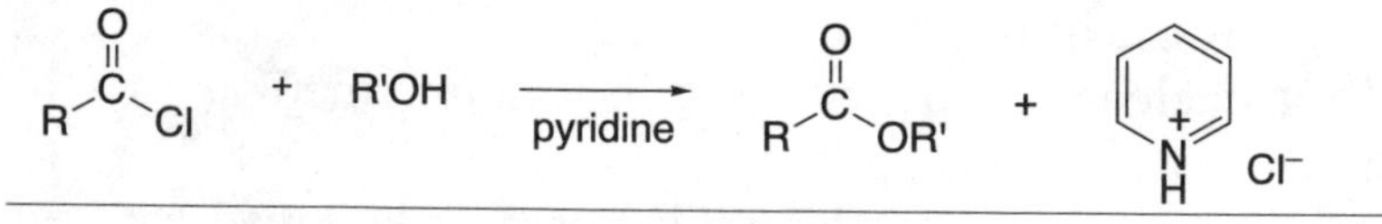

$$RCOCl + R'OH \xrightarrow{pyridine} RCOOR' + \text{pyridinium}\ Cl^-$$

[b] From (RCO)₂O (22.9):

$$R-\overset{\displaystyle O}{\underset{}{C}}-O-\overset{\displaystyle O}{\underset{}{C}}-R \;+\; R'OH \longrightarrow R-\overset{\displaystyle O}{\underset{}{C}}-OR' \;+\; RCOOH$$

[c] From RCOOH (22.10C):

$$R-\overset{\displaystyle O}{\underset{}{C}}-OH \;+\; R'OH \xrightarrow{\;H_2SO_4\;} R-\overset{\displaystyle O}{\underset{}{C}}-OR' \;+\; H_2O$$

[5] Reactions that synthesize amides (RCONH₂) [The reactions are written with NH₃ as nucleophile to form RCONH₂. Similar reactions occur with R'NH₂ to form RCONHR', and with R'₂NH to form RCONR'₂.]

[a] From RCOCl (22.8):

$$R-\overset{\displaystyle O}{\underset{}{C}}-Cl \;+\; \underset{(2\ equiv)}{NH_3} \longrightarrow R-\overset{\displaystyle O}{\underset{}{C}}-NH_2 \;+\; NH_4^+Cl^-$$

[b] From (RCO)₂O (22.9):

$$R-\overset{\displaystyle O}{\underset{}{C}}-O-\overset{\displaystyle O}{\underset{}{C}}-R \;+\; \underset{(2\ equiv)}{NH_3} \longrightarrow R-\overset{\displaystyle O}{\underset{}{C}}-NH_2 \;+\; RCOO^- NH_4^+$$

[c] From RCOOH (22.10D):

$$R-\overset{\displaystyle O}{\underset{}{C}}-OH \xrightarrow[{[2]\ \wedge}]{\;[1]\ NH_3\;} R-\overset{\displaystyle O}{\underset{}{C}}-NH_2 \;+\; H_2O$$

$$R-\overset{\displaystyle O}{\underset{}{C}}-OH \;+\; R'NH_2 \xrightarrow[{DCC}]{} R-\overset{\displaystyle O}{\underset{}{C}}-NHR' \;+\; H_2O$$

[d] From RCOOR' (22.11):

$$R-\overset{\displaystyle O}{\underset{}{C}}-OR' \;+\; NH_3 \longrightarrow R-\overset{\displaystyle O}{\underset{}{C}}-NH_2 \;+\; R'OH$$

♦ **Nitrile synthesis (22.18)**

Nitriles are prepared by S$_N$2 substitution using unhindered alkyl halides as starting materials.

$$R-X \;+\; {}^-CN \xrightarrow{\;S_N2\;} R-C\equiv N \;+\; X^-$$

$$R = CH_3,\ 1°$$

♦ **Reactions of nitriles**

[1] Hydrolysis (22.18A)

$$R-C\equiv N \xrightarrow[{(H^+\ or\ {}^-OH)}]{\;H_2O\;} \underset{(with\ acid)}{R-\overset{\displaystyle O}{\underset{}{C}}-OH} \;\; or \;\; \underset{(with\ base)}{R-\overset{\displaystyle O}{\underset{}{C}}-O^-}$$

[2] Reduction (22.18B)

$$R-C≡N \quad \xrightarrow[\text{[2] } H_2O]{\text{[1] LiAlH}_4} \quad R-CH_2NH_2$$

1° amine

$$R-C≡N \quad \xrightarrow[\text{[2] } H_2O]{\text{[1] DIBAL–H}} \quad \underset{\text{aldehyde}}{R-\overset{\overset{\displaystyle O}{\|}}{C}-H}$$

[3] Reaction with organometallic reagents (22.18C)

$$R-C≡N \quad \xrightarrow[\text{[2] } H_2O]{\text{[1] R'MgX or R'Li}} \quad \underset{\text{ketone}}{R-\overset{\overset{\displaystyle O}{\|}}{C}-R'}$$

22.1 The number of C–N bonds determines the classification as a 1°, 2°, or 3° amide.

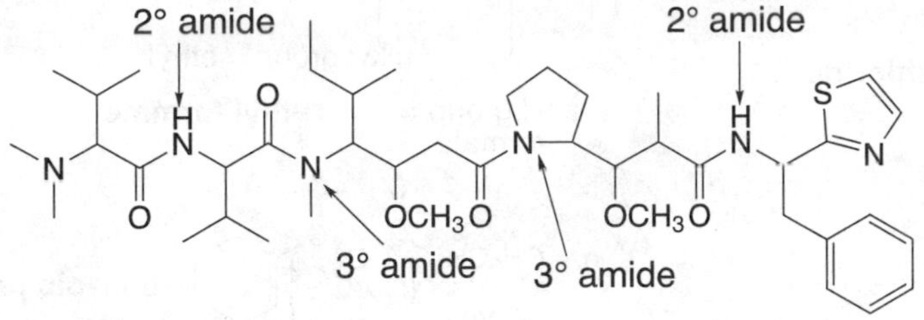

22.2 As the basicity of Z increases, the stability of RCOZ increases because of added resonance stabilization.

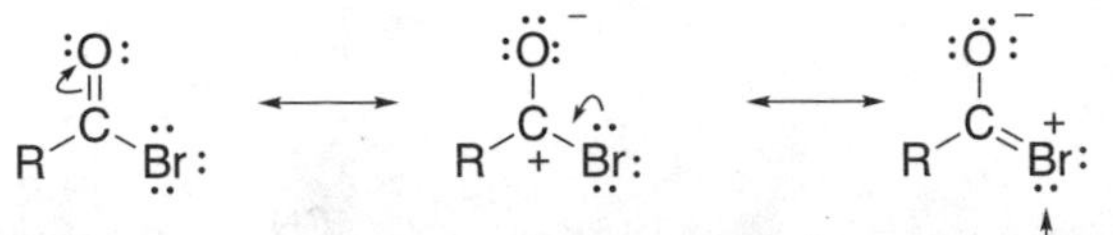

The **basicity of Z** determines how much
this structure contributes to the hybrid.
Br⁻ is less basic than ⁻OH, so RCOBr
is less stable than RCOOH.

22.3

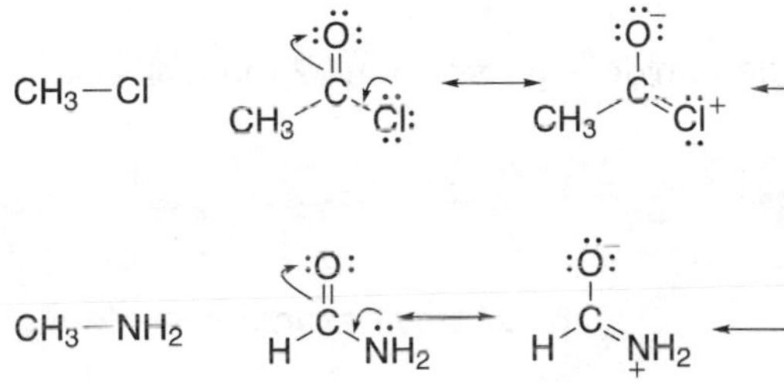

CH_3-Cl

This resonance structure contributes little to the hybrid
since Cl⁻ is a weak base. Thus, the C–Cl
bond has little double bond character, making it similar in
length to the C–Cl bond in CH_3Cl.

CH_3-NH_2

This resonance structure contributes more to the hybrid
since ⁻NH₂ is more basic. Thus, the C–N bond
in $HCONH_2$ has more double bond character, making
it shorter than the C–N bond in CH_3NH_2.

22.4

$CH_3-C\equiv N:$

2 groups
sp hybridized
lone pair in an *sp* hybrid orbital

22.5

a. (CH₃CH₂)₂CH—COCl

|redraw

2-ethylbutanoyl chloride
← 2-ethyl

b. C₆H₅COOCH₃

|redraw

methyl benzoate
↑ alkyl group = methyl
| acyl group = benzoate

c. CH₃CH₂CON(CH₃)CH₂CH₃

|redraw

← N-ethyl-N-methyl
↑ acyl group = propanamide
N-ethyl-N-methylpropanamide

d.

↑ alkyl group = ethyl
acyl group = formate
ethyl formate

e. CH₃CH₂

benzoic propanoic anhydride

acyl group = propanoic   acyl group = benzoic

f. 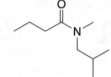 CN   **3-ethylhexanenitrile**

6 carbon chain = hexanenitrile

22.6

a. 5-methylheptanoyl chloride

b. isopropyl propanoate

c. acetic formic anhydride

d. N-isobutyl-N-methylbutanamide

e. 3-methylpentanenitrile

f. o-cyanobenzoic acid

g. sec-butyl 2-methylhexanoate

h. N-ethylhexanamide

22.7 CH₃CONH₂ has two H's bonded to N that can hydrogen bond. CH₃CON(CH₃)₂ does not have any H's capable of hydrogen bonding. This means CH₃CONH₂ has much stronger intermolecular forces, which leads to a higher boiling point.

22.8

a. 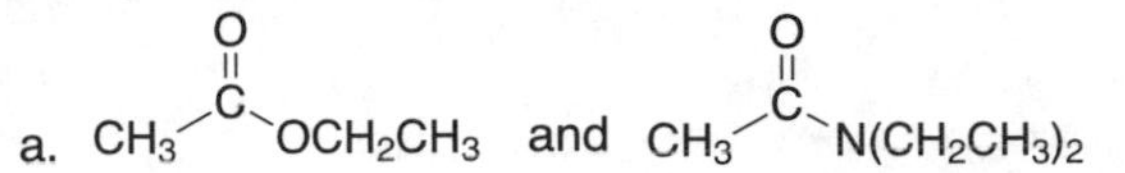 CH$_3$COCH$_2$CH$_3$ and CH$_3$CON(CH$_2$CH$_3$)$_2$

amide: C=O at
lower wave number

c. 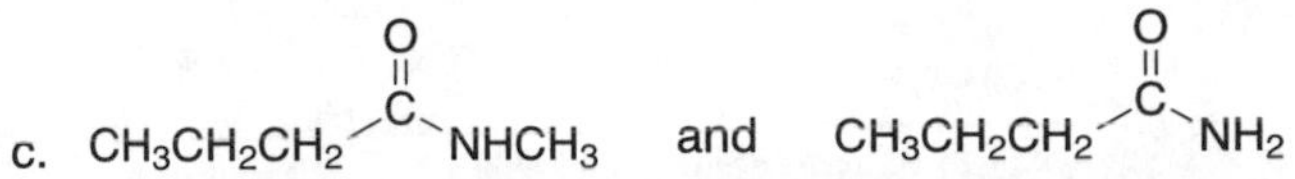 CH$_3$CH$_2$CH$_2$CONHCH$_3$ and CH$_3$CH$_2$CH$_2$CONH$_2$

2° amide: 1 N–H
absorption at
3200–3400 cm^{-1}

1° amide: 2
N–H absorptions

b. 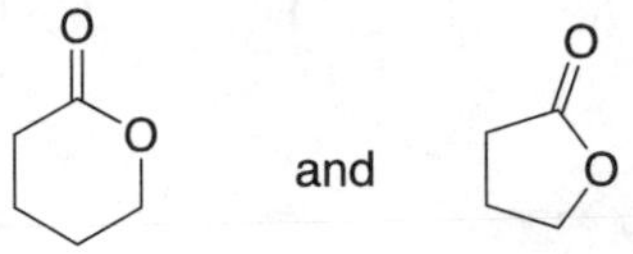 and

smaller ring:
C=O at a higher
wave number

d. 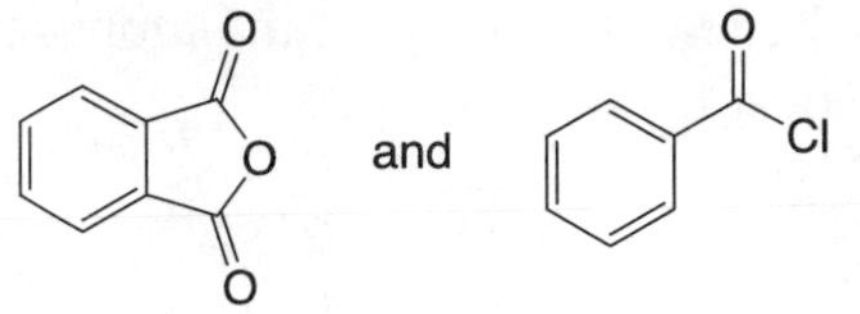 and

anhydride:
2 C=O peaks

22.9

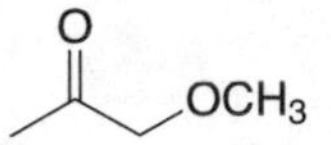

α-methoxyacetone
IR: C=O absorption at **slightly lower** wave numbers
^{1}H NMR: 3 singlets
^{13}C NMR: 4 peaks The C=O (a ketone) comes further
downfield than an ester (190–215 ppm vs 160–180 ppm)

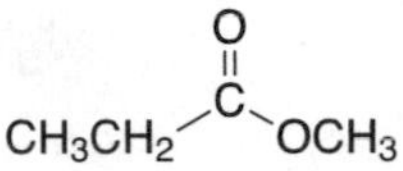

CH$_3$CH$_2$COCH$_3$

methyl propionate
IR: C=O absorption at **somewhat
higher** wave numbers
^{1}H NMR: 3 peaks – a triplet (CH$_3$), a quartet
(CH$_2$), and a singlet (OCH$_3$)
^{13}C NMR: 4 peaks

22.10

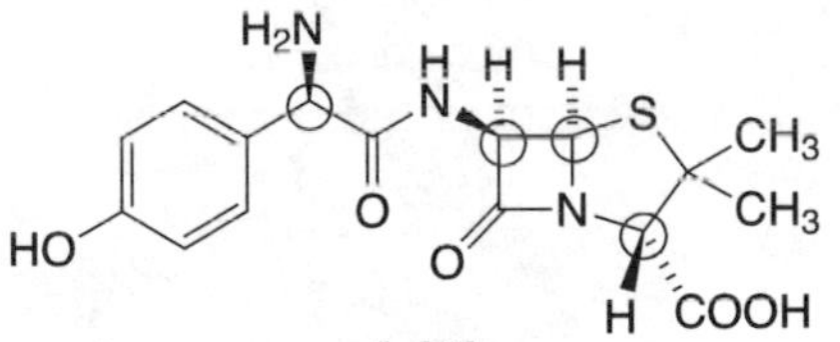

amoxicillin

a. 4 stereogenic centers
b. 2^4 = 16 possible stereoisomers
c. enantiomer

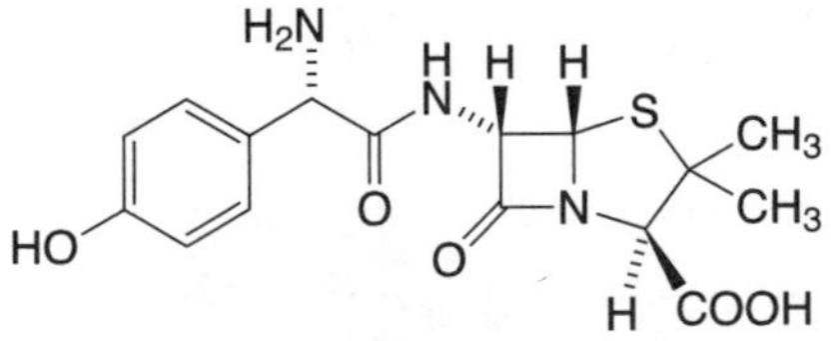

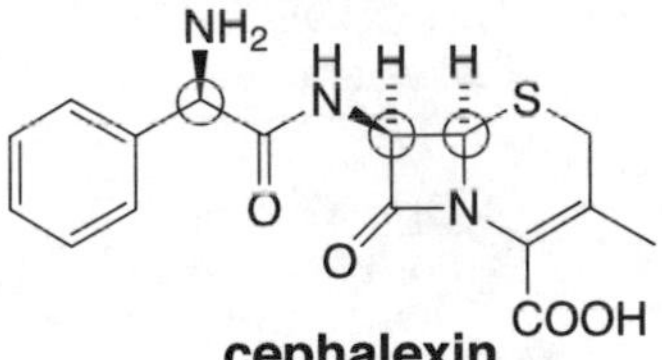

cephalexin
(Trade name: Keflex)

a. 3 stereogenic centers
b. 2^3 = 8 possible stereoisomers
c. enantiomer

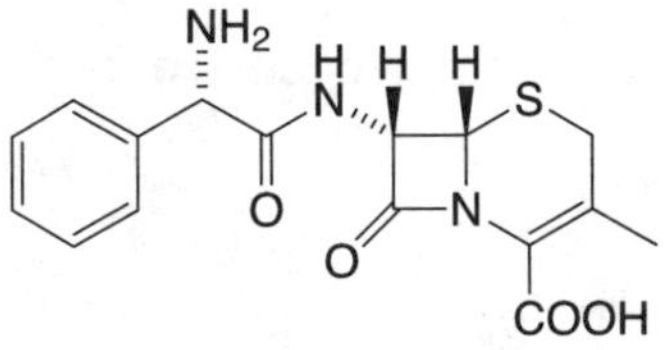

22.11 To draw the products of these nucleophilic acyl substitution reactions, find the nucleophile and the leaving group. Then replace the leaving group with the nucleophile and draw a neutral product.

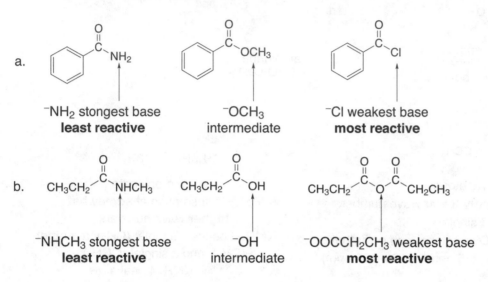

a. CH₃COCl + CH₃OH → CH₃COOCH₃ + HCl
leaving group → nucleophile

b. CH₃COOCH₂CH₃ + NH₃ → CH₃CONH₂ + HOCH₂CH₃
nucleophile → leaving group

22.12 The better the leaving group is, the more reactive the carboxylic acid derivative. The weakest base is the best leaving group.

a.

⁻NH₂ stongest base
least reactive

⁻OCH₃
intermediate

⁻Cl weakest base
most reactive

b. CH₃CH₂CONHCH₃ CH₃CH₂COOH CH₃CH₂CO-O-COCH₂CH₃

⁻NHCH₃ stongest base
least reactive

⁻OH
intermediate

⁻OOCCH₂CH₃ weakest base
most reactive

22.13 More reactive acyl compounds can be converted to less reactive acyl compounds.

a. CH₃COCl ——→ CH₃COOH
more reactive **YES** less reactive

c. CH₃COOCH₃ ——→ CH₃COCl
less reactive **NO** more reactive

b. CH₃CONHCH₃ ——→ CH₃COOCH₃
less reactive **NO** more reactive

22.14

CH₃CO-O-COCH₃
acetic anhydride

Cl₃C-CO-O-CO-CCl₃
trichloroacetic anhydride

The Cl atoms are electron withdrawing, which makes the conjugate base (the leaving group, CCl₃COO⁻) weaker and more stable.

22.15

a. H₂O / pyridine → benzoic acid (C₆H₅COOH) + pyridinium Cl⁻

b. CH₃COO⁻ → C₆H₅CO-O-COCH₃ + Cl⁻

c. NH₃ / excess → C₆H₅CONH₂ + NH₄⁺Cl⁻

d. (CH₃)₂NH / excess → C₆H₅CON(CH₃)₂ + (CH₃)₂NH₂⁺ Cl⁻

22.16 The mechanism has 3 steps: [1] nucleophilic attack by O; [2] proton transfer; [3] elimination of the Cl⁻ leaving group to form the product.

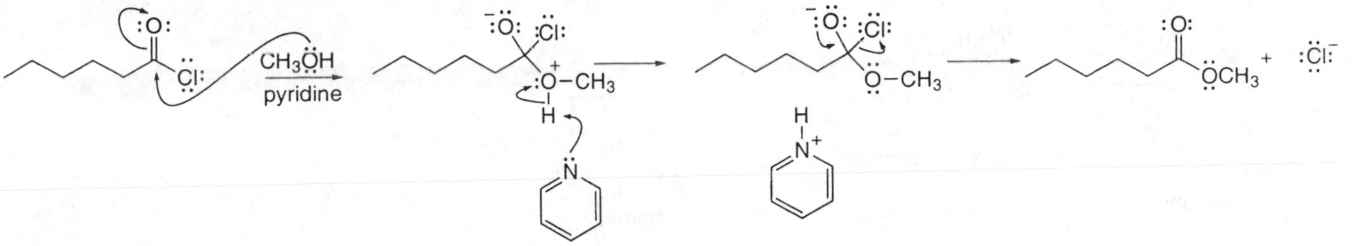

22.17

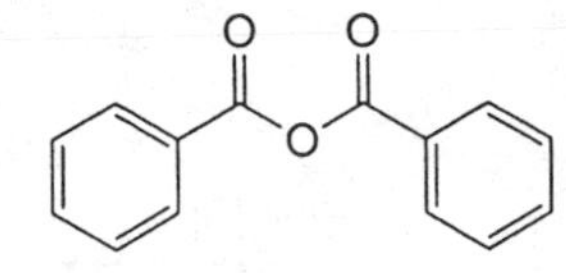

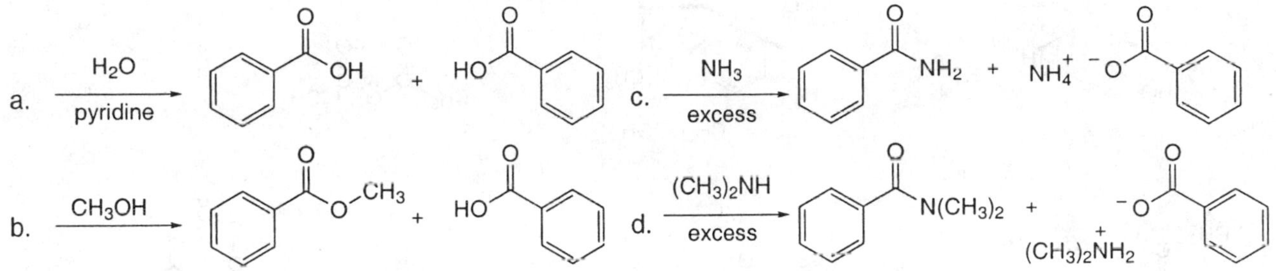

a. $\dfrac{H_2O}{\text{pyridine}}$

b. $\xrightarrow{CH_3OH}$

c. $\dfrac{NH_3}{\text{excess}}$

d. $\dfrac{(CH_3)_2NH}{\text{excess}}$

22.18

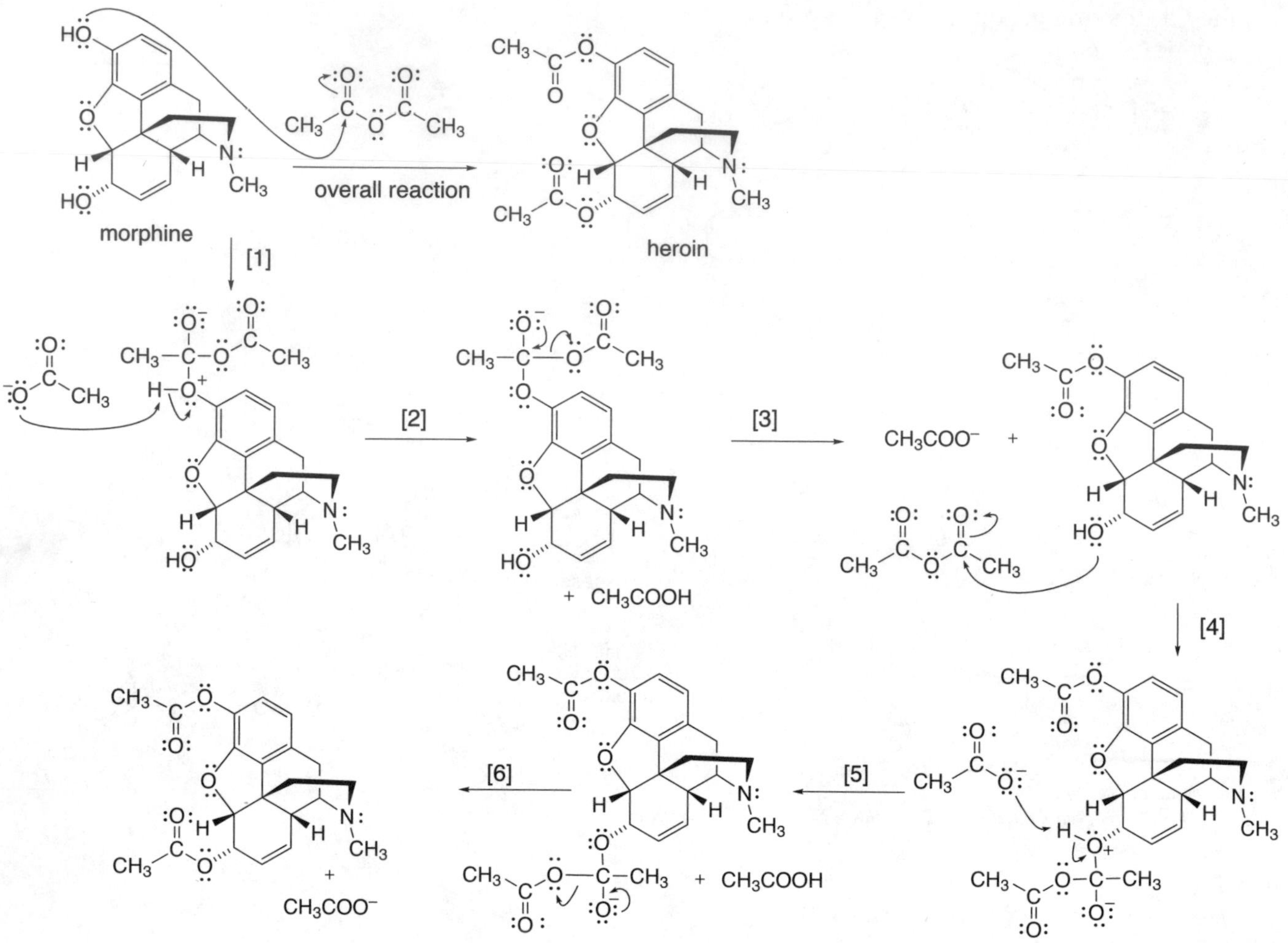

morphine → [1] → [2] → [3] → [4] → [5] → [6]

overall reaction

heroin

+ CH₃COOH

CH₃COO⁻ +

22.19 Reaction of a carboxylic acid with thionyl chloride converts it to an acid chloride.

a. 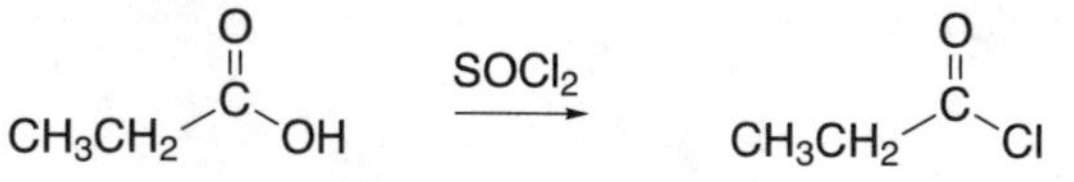 CH₃CH₂COOH →(SOCl₂)→ CH₃CH₂COCl

b. 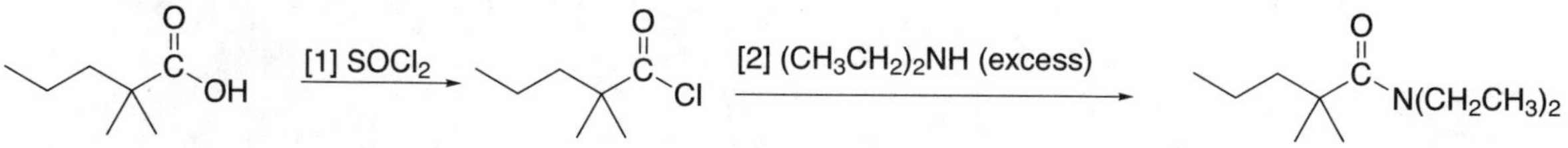 →[1] SOCl₂→ →[2] (CH₃CH₂)₂NH (excess)→ ...N(CH₂CH₃)₂

22.20

a. 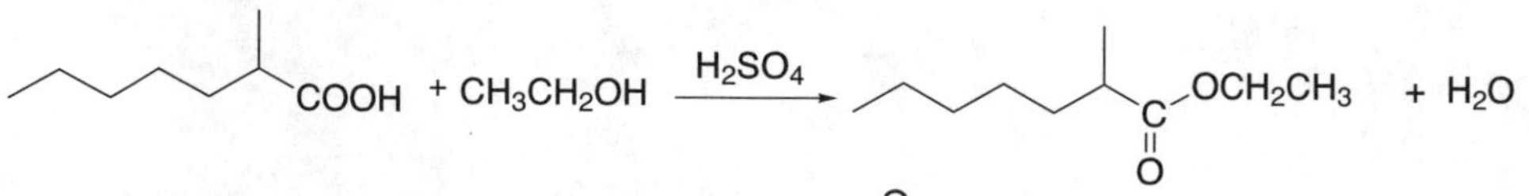 ...COOH + CH₃CH₂OH →(H₂SO₄)→ ...C-OCH₂CH₃ + H₂O

b. 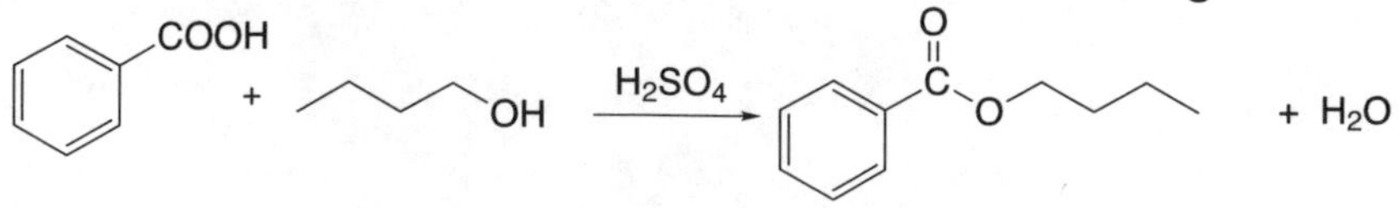 ...COOH + ...OH →(H₂SO₄)→ ... + H₂O

c.

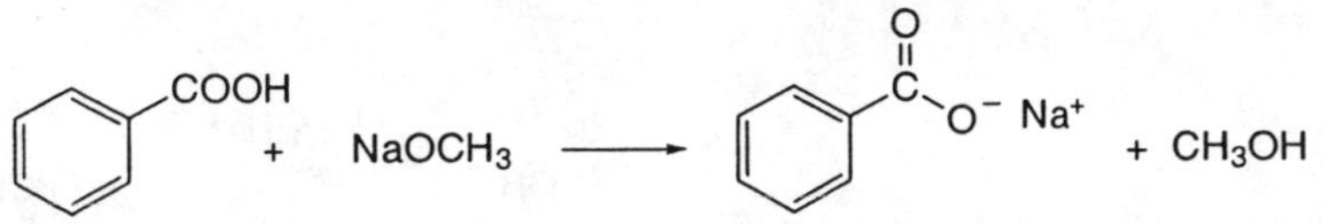

+ NaOCH₃ ⟶ [benzoate] O⁻ Na⁺ + CH₃OH

d. 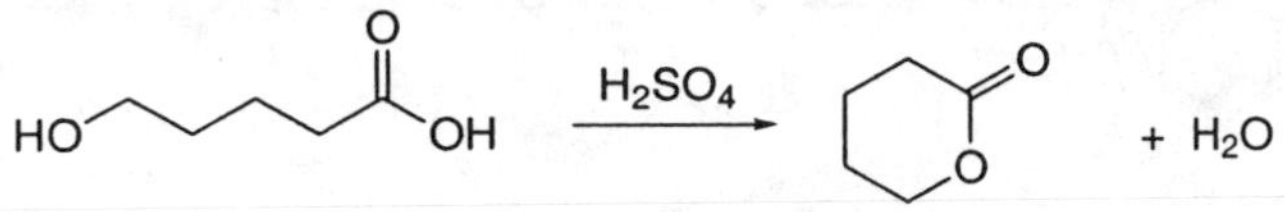 HO⟍⟍⟍COOH

$\xrightarrow{H_2SO_4}$

[δ-valerolactone] + H₂O

22.21

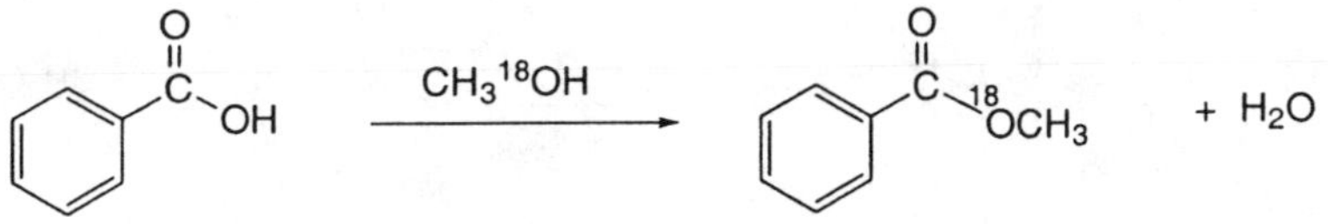

$\xrightarrow{CH_3{}^{18}OH}$

[benzoate ester with ¹⁸OCH₃] + H₂O

22.22

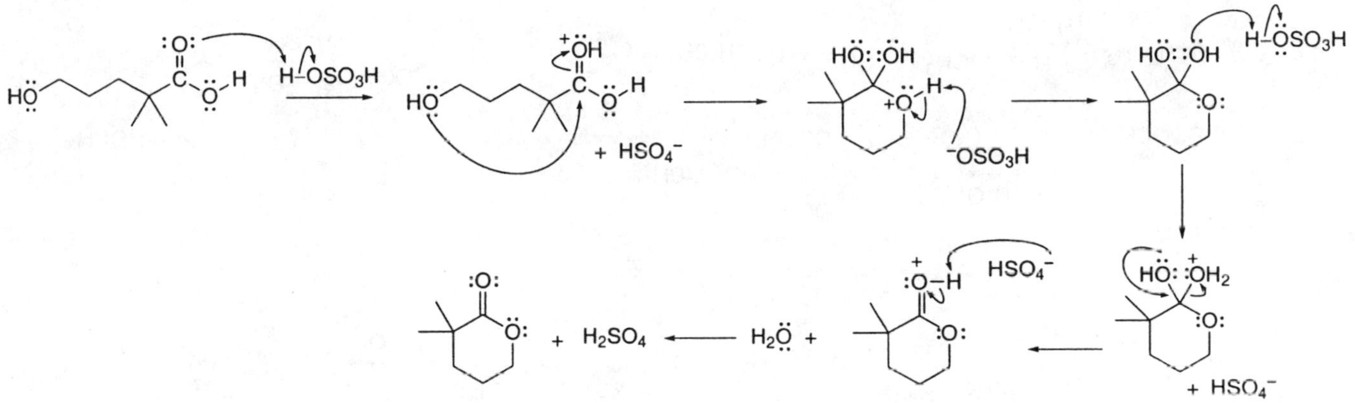

22.23

a.

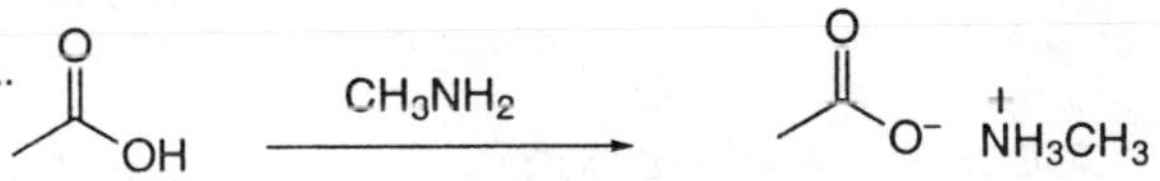

$\xrightarrow{CH_3NH_2}$ [acetate] O⁻ ⁺NH₃CH₃

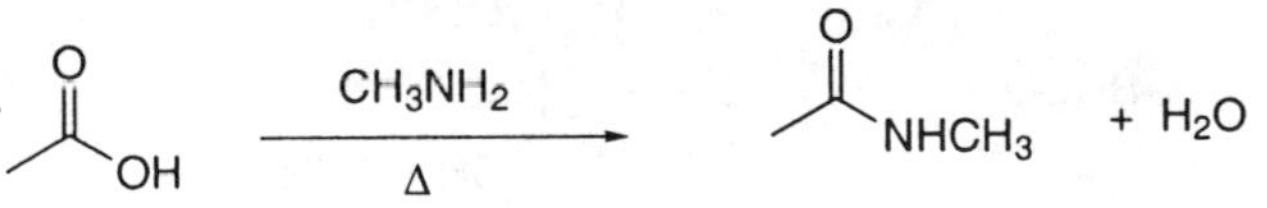

c.

$\xrightarrow[DCC]{CH_3NH_2}$ [N-methylacetamide] NHCH₃

b.

$\xrightarrow[\Delta]{CH_3NH_2}$ [N-methylacetamide] NHCH₃ + H₂O

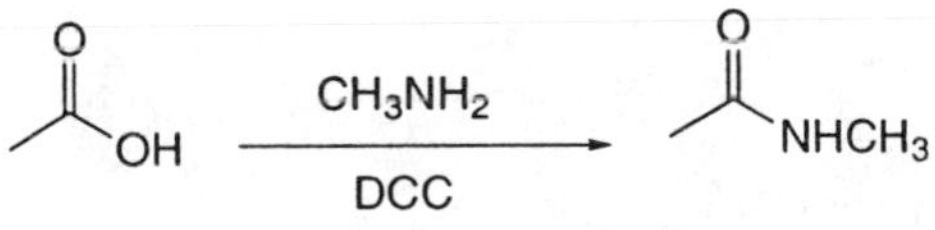

22.24

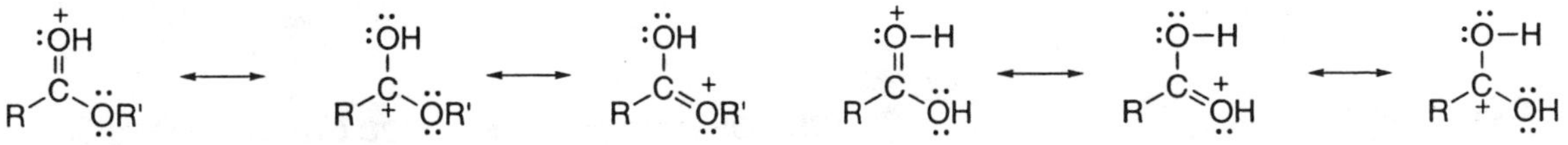

product of step [1]   product of step [5]

22.25

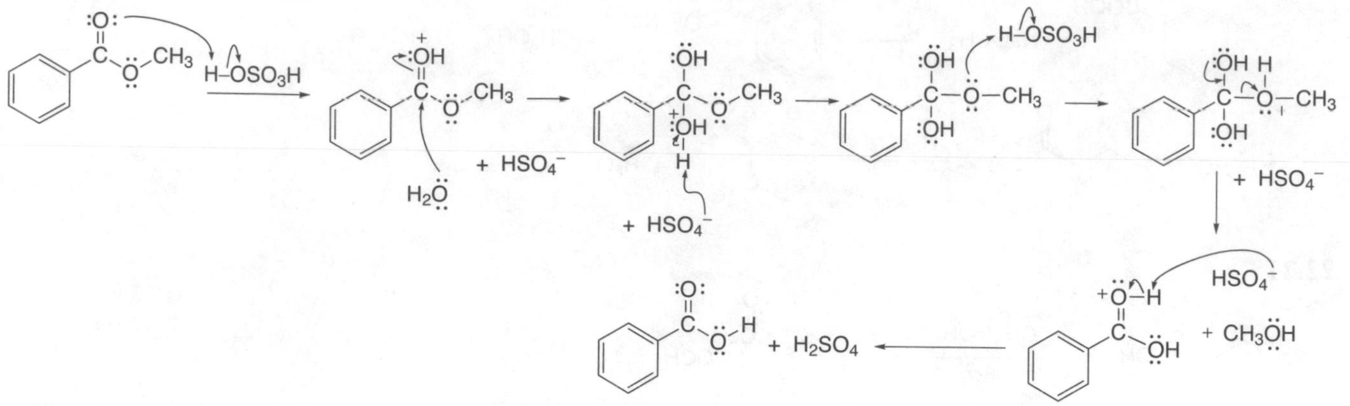

22.26 The ester is cleaved in these reactions to form a carboxylic acid and an alcohol.

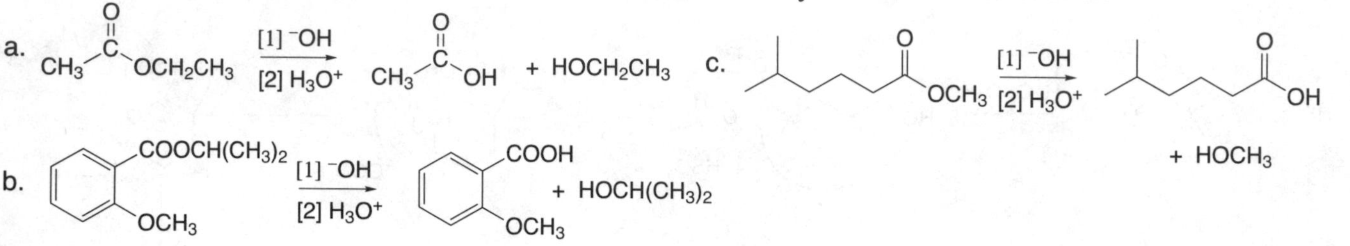

a.

b.

c.

22.27

a.

b.

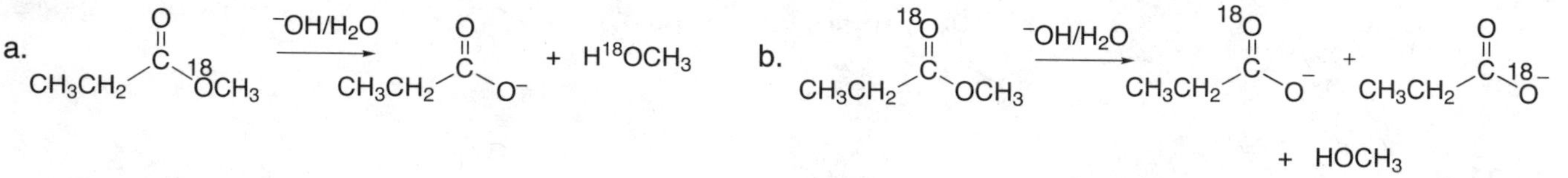

22.28

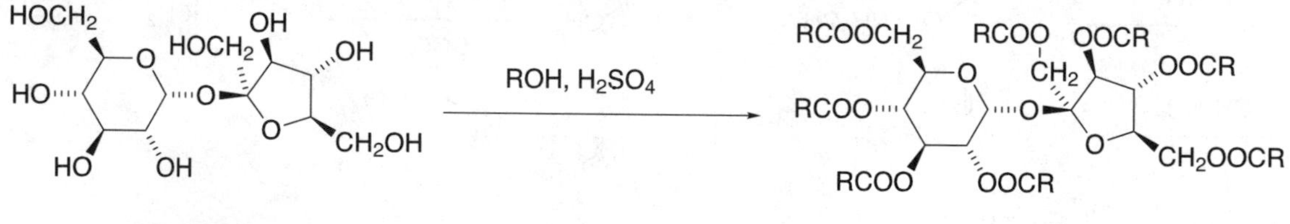

sucrose   olestra

22.29

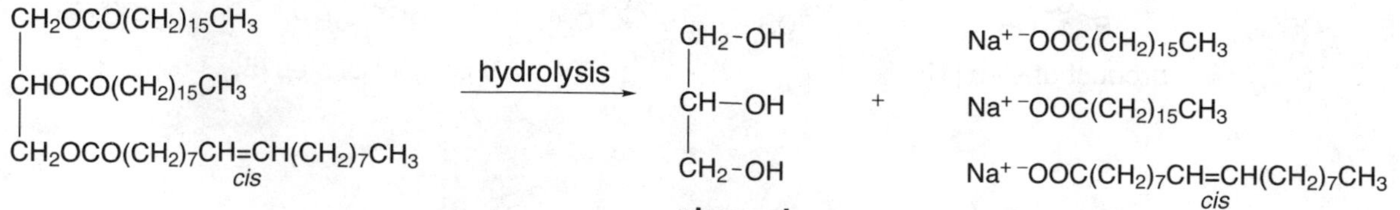

22.30

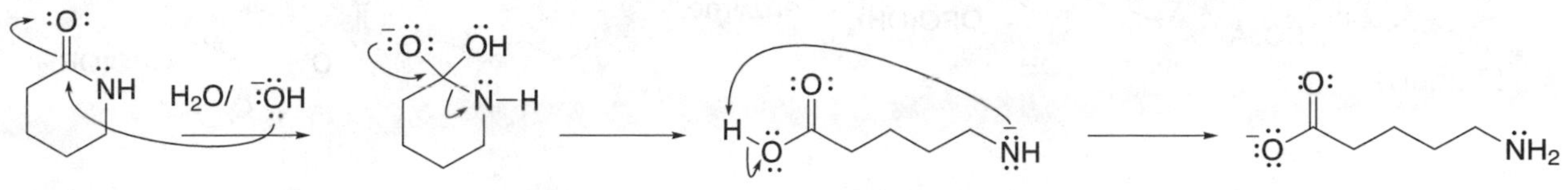

22.31

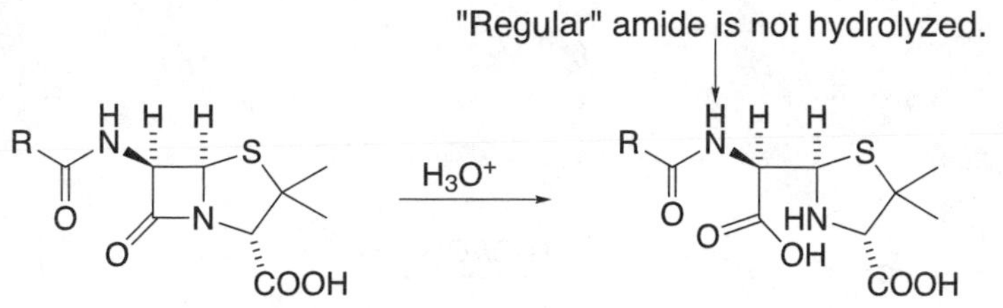

"Regular" amide is not hydrolyzed.

22.32

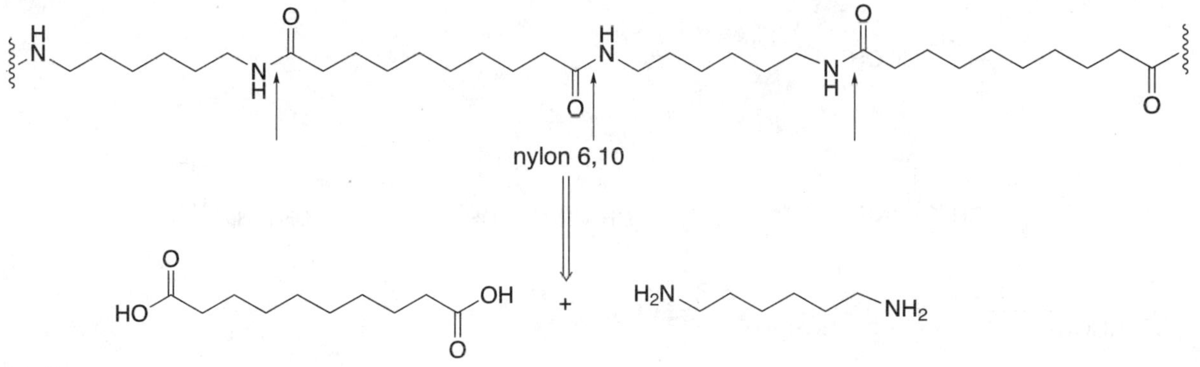

nylon 6,10

22.33

a.

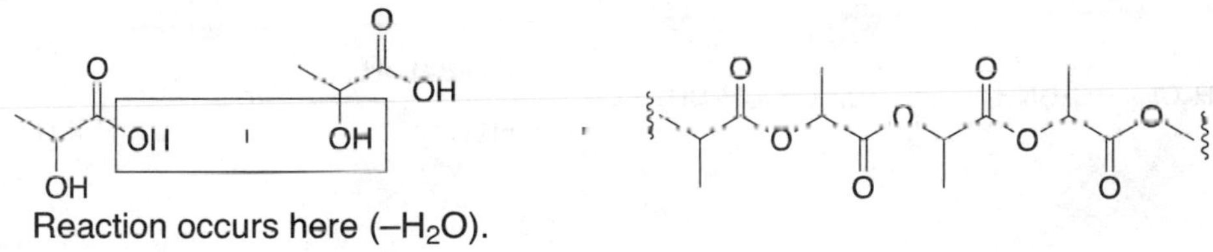

Reaction occurs here (–H₂O).

PLA
polylactic acid

b.

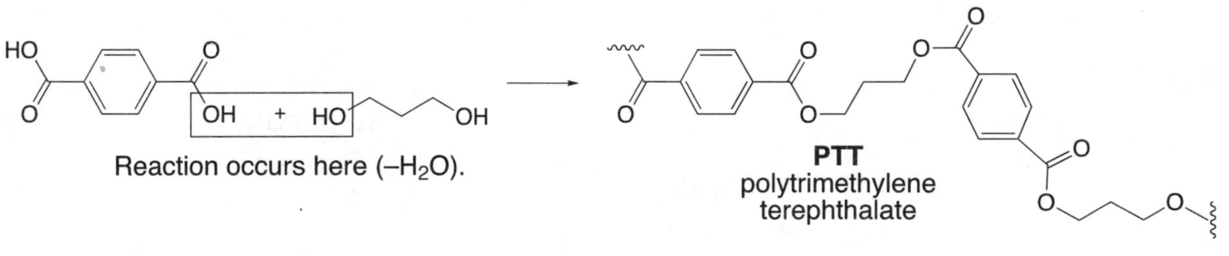

Reaction occurs here (–H₂O).

PTT
polytrimethylene
terephthalate

22.34

a.

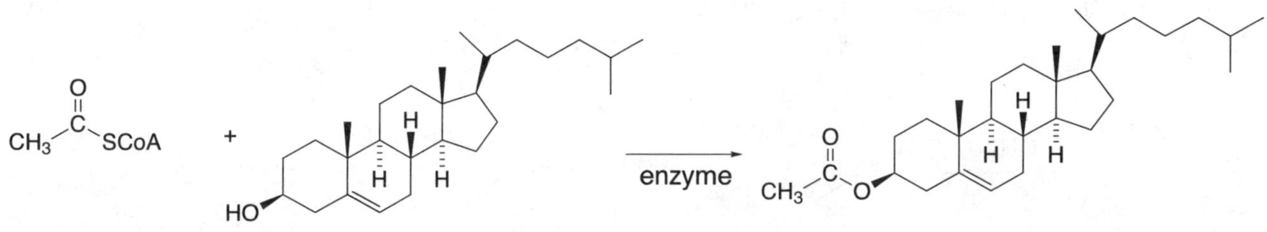

enzyme

b.

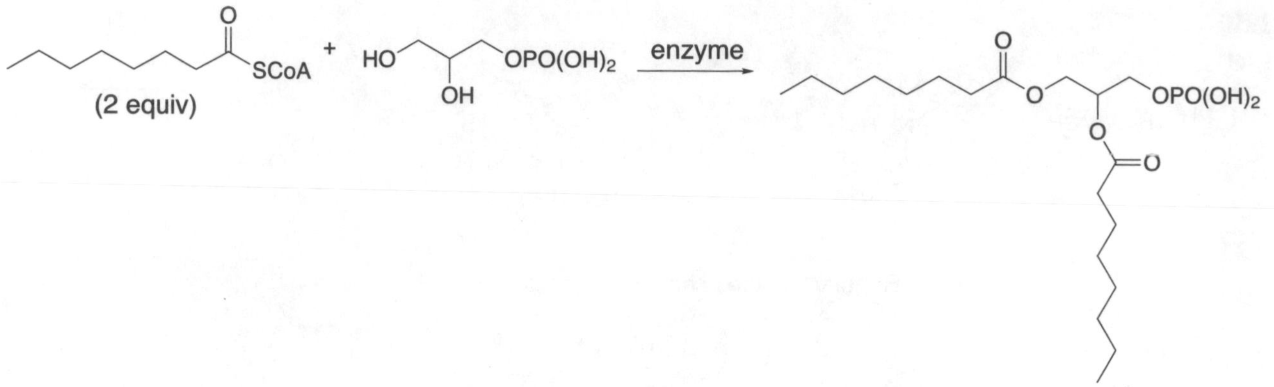

(2 equiv)

enzyme

22.35

a. 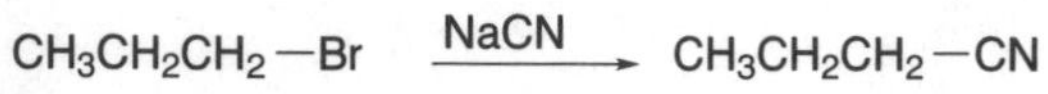 $CH_3CH_2CH_2-Br$ $\xrightarrow{NaCN}$ $CH_3CH_2CH_2-CN$

c.

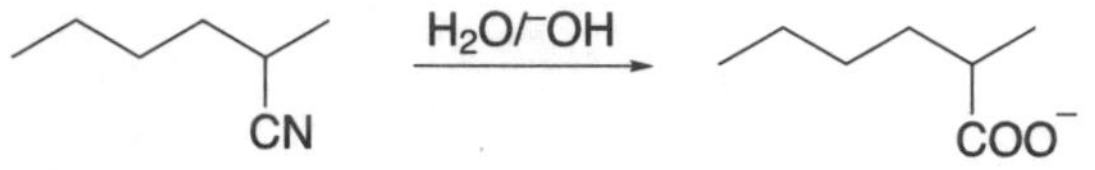

$\xrightarrow{H_2O/^-OH}$

b.

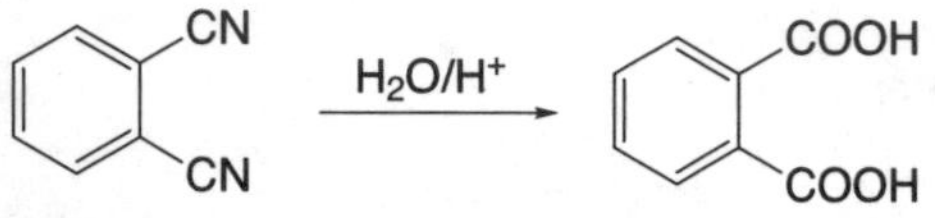

$\xrightarrow{H_2O/H^+}$

22.36

a.

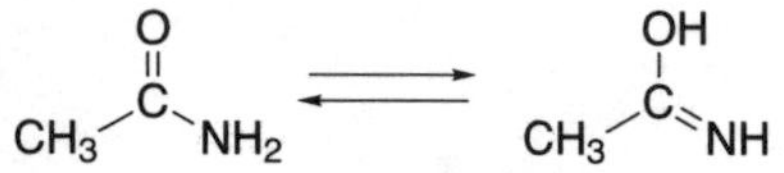

c.

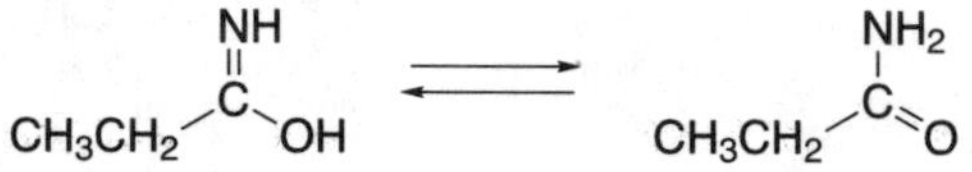

b.

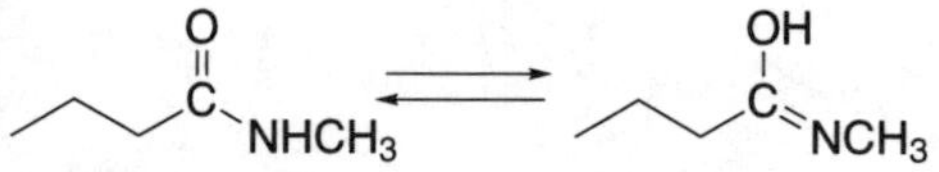

22.37

a. 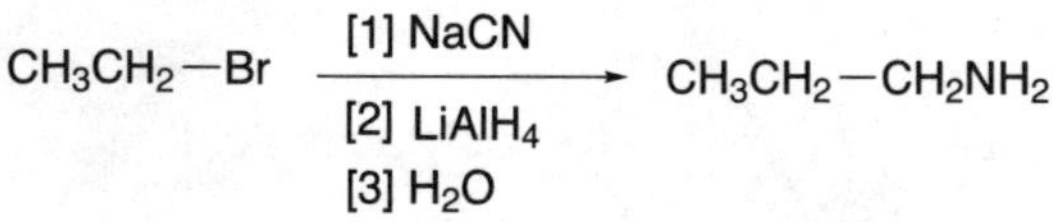 CH_3CH_2-Br $\xrightarrow[\substack{[2]\ LiAlH_4 \\ [3]\ H_2O}]{[1]\ NaCN}$ $CH_3CH_2-CH_2NH_2$

b. 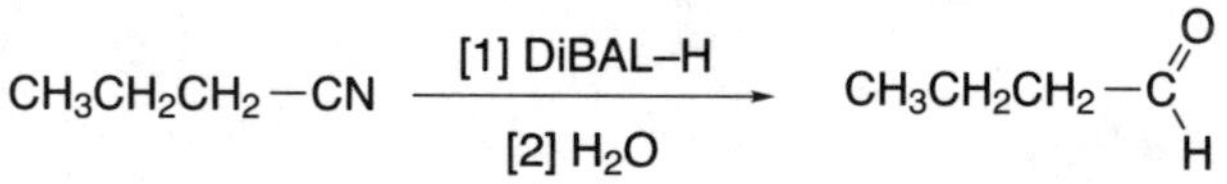 $CH_3CH_2CH_2-CN$ $\xrightarrow[\substack{[2]\ H_2O}]{[1]\ DiBAL-H}$ $CH_3CH_2CH_2-\overset{\overset{O}{\parallel}}{C}{-}H$

22.38

a.

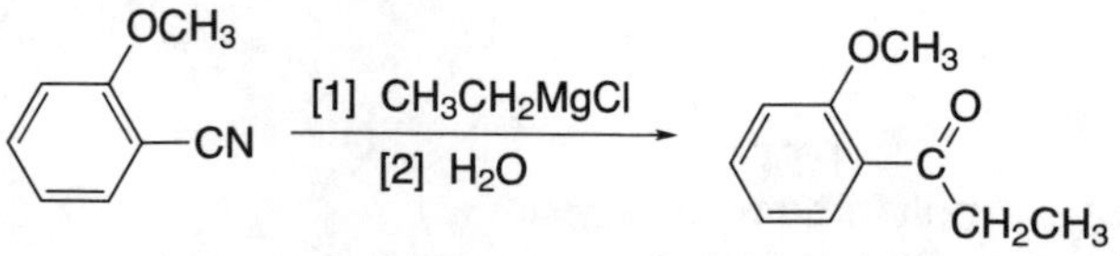

$\xrightarrow[\substack{[2]\ H_2O}]{[1]\ CH_3CH_2MgCl}$

b.

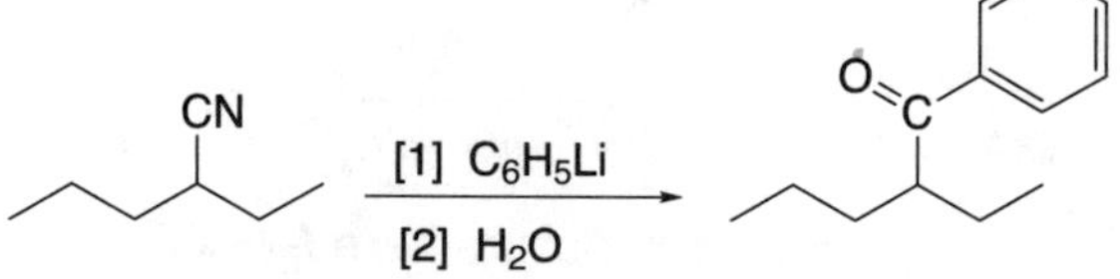

$\xrightarrow[\substack{[2]\ H_2O}]{[1]\ C_6H_5Li}$

22.39

a.

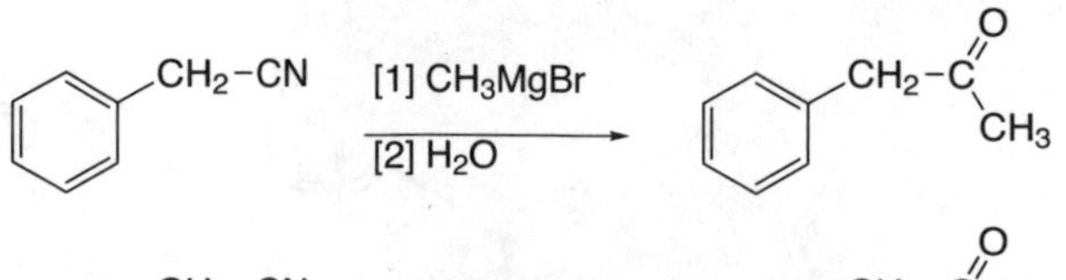

$\xrightarrow[\substack{[2]\ H_2O}]{[1]\ CH_3MgBr}$

c.

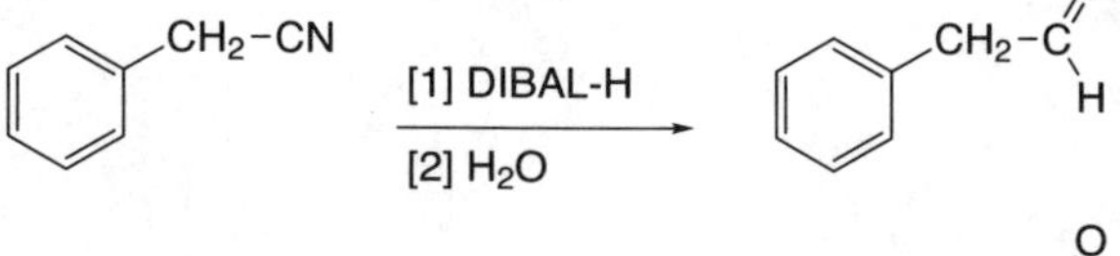

$\xrightarrow[\substack{[2]\ H_2O}]{[1]\ DiBAL-H}$

b.

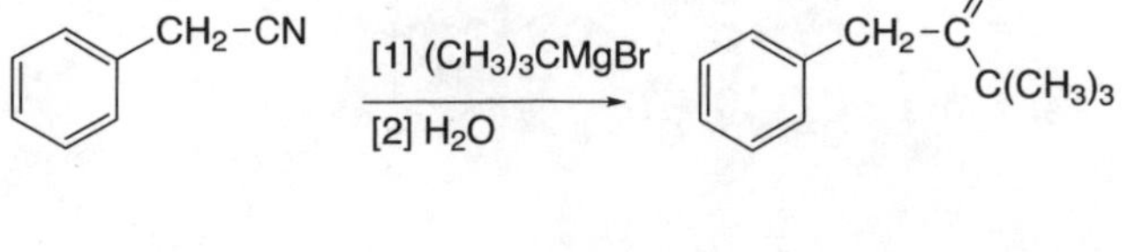

$\xrightarrow[\substack{[2]\ H_2O}]{[1]\ (CH_3)_3CMgBr}$

d.

$\xrightarrow{H_3O^+}$

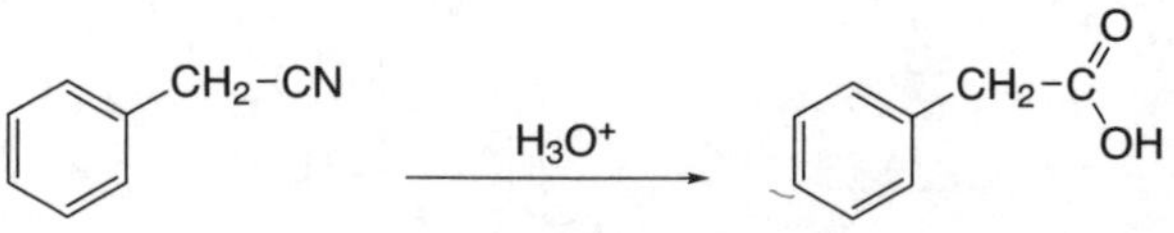

22.40

$CH_3-C\equiv N$ $\xrightarrow[\text{[2] } H_2O]{\text{[1] } CH_3CH_2MgBr}$

$CH_3CH_2-C\equiv N$ $\xrightarrow[\text{[2] } H_2O]{\text{[1] } CH_3MgBr}$

22.41

a. 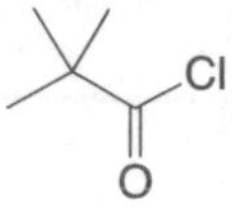 **2,2-dimethylpropanoyl chloride**

b. 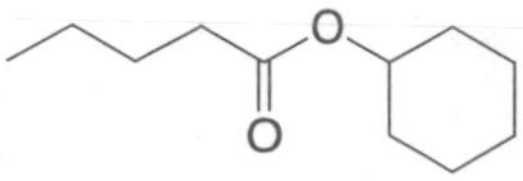 **cyclohexyl pentanoate**

c. 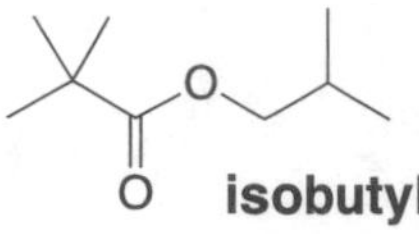 **isobutyl 2,2-dimethylpropanoate**

d. 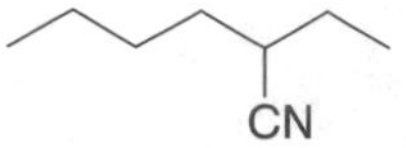 **2-ethylhexanenitrile**

e.

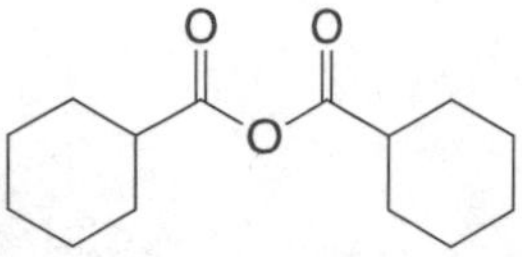

cyclohexanecarboxylic anhydride

f. 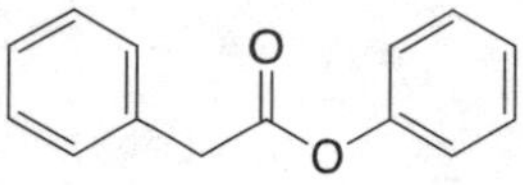 **phenyl phenylacetate**

g. 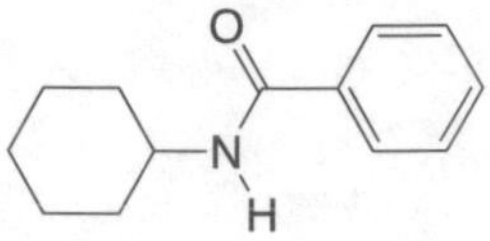 **N-cyclohexylbenzamide**

h. 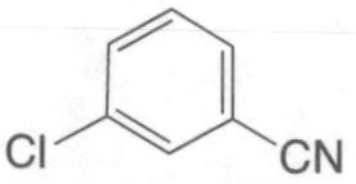 **m-chlorobenzonitrile**

i. 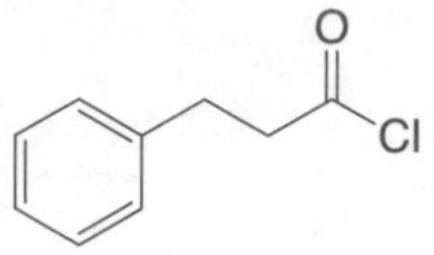 **3-phenylpropanoyl chloride**

j. 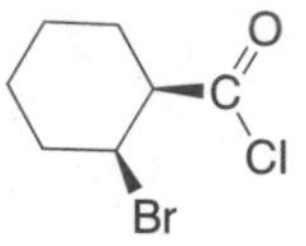 **cis-2-bromocyclohexane-carbonyl chloride**

k. 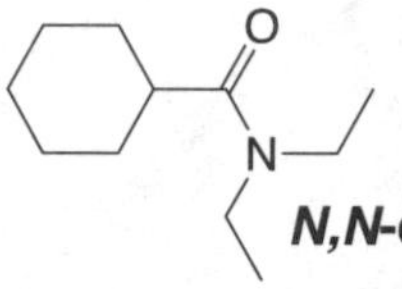 **N,N-diethylcyclohexanecarboxamide**

l. 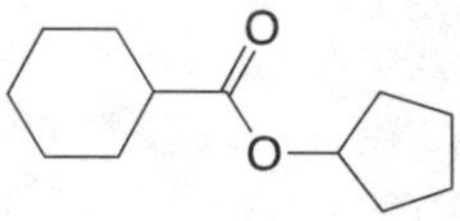 **cyclopentyl cyclohexanecarboxylate**

22.42

a. propanoic anhydride

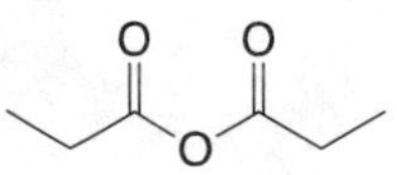

b. α-chlorobutyryl chloride

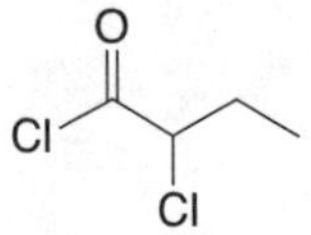

c. cyclohexyl propanoate

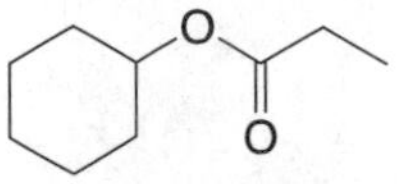

d. cyclohexanecarboxamide

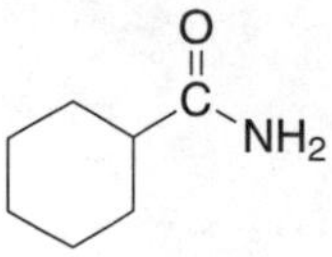

e. isopropyl formate

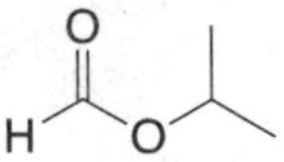

f. N-cyclopentylpentanamide

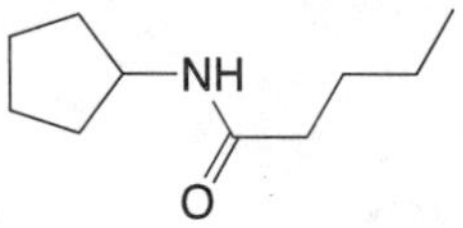

g. 4-methylheptanenitrile

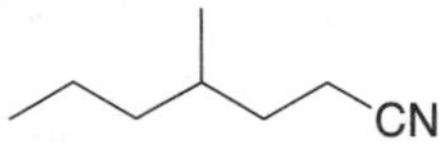

h. vinyl acetate

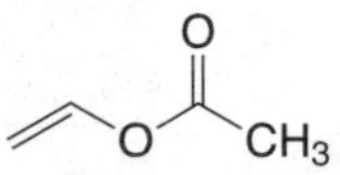

i. benzoic propanoic anhydride

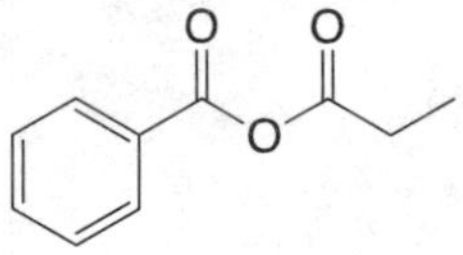

j. 3-methylhexanoyl chloride

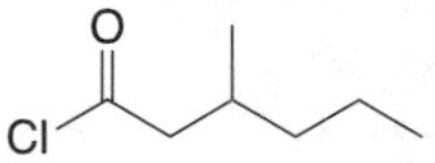

k. octyl butanoate

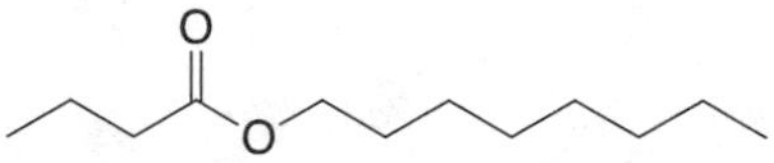

l. N,N-dibenzylformamide

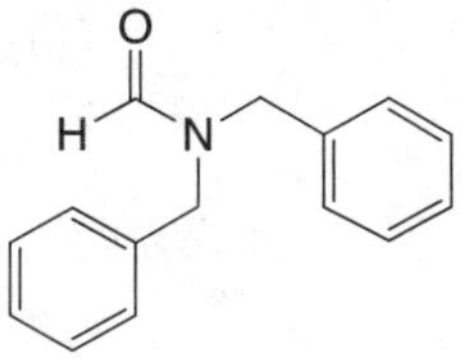

22.43 Rank the compounds using the rules from Answer 22.12.

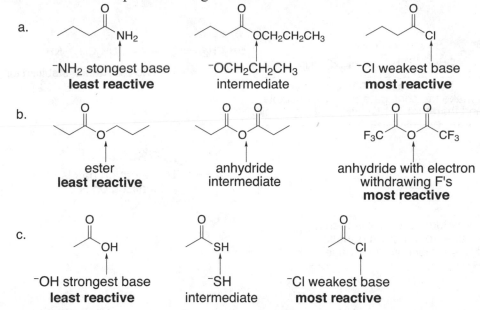

a.

$^-$NH$_2$ stongest base
least reactive

$^-$OCH$_2$CH$_2$CH$_3$
intermediate

$^-$Cl weakest base
most reactive

b.

ester
least reactive

anhydride
intermediate

anhydride with electron
withdrawing F's
most reactive

c.

$^-$OH strongest base
least reactive

$^-$SH
intermediate

$^-$Cl weakest base
most reactive

22.44

Better leaving groups make acyl compounds more reactive. **A** has an electron withdrawing NO$_2$ group, which stabilizes the negative charge of the leaving group, whereas **B** has an electron donating OCH$_3$ group, which destabilizes the leaving group.

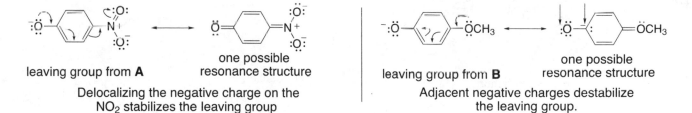

A

B

an electron withdrawing substituent an electron donating substituent

leaving group from **A**

one possible
resonance structure

leaving group from **B**

one possible
resonance structure

Delocalizing the negative charge on the
NO$_2$ stabilizes the leaving group
making **A** more reactive than **B**.

Adjacent negative charges destabilize
the leaving group.

22.45

resonance structures for the leaving group

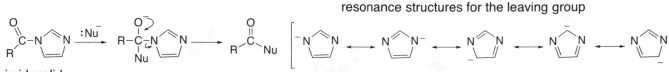

imidazolide

The leaving group is both resonance stabilized and aromatic
(6π electrons), making it a much better leaving group than
exists in a regular amide.

22.46

Reaction as an acid:

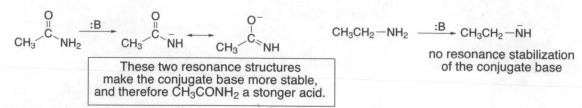

These two resonance structures make the conjugate base more stable, and therefore CH_3CONH_2 a stonger acid.

$CH_3CH_2-NH_2$ $\xrightarrow{:B}$ $CH_3CH_2-\overset{\bar{}}{N}H$

no resonance stabilization of the conjugate base

Reaction as a base:

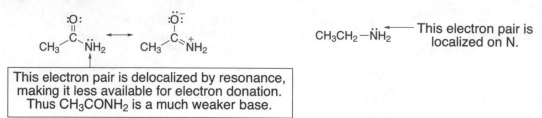

This electron pair is delocalized by resonance, making it less available for electron donation. Thus CH_3CONH_2 is a much weaker base.

$CH_3CH_2-\ddot{N}H_2$ ← This electron pair is localized on N.

22.47

$CH_3CH_2CH_2CH_2COCl$

a. $\xrightarrow[\text{pyridine}]{H_2O}$ $CH_3CH_2CH_2CH_2COOH$

b. $\xrightarrow[\text{pyridine}]{CH_3CH_2OH}$ $CH_3CH_2CH_2CH_2COOCH_2CH_3$

c. $\xrightarrow{CH_3COO^-}$ $CH_3CH_2CH_2CH_2\overset{O}{\underset{}{C}}-O-\overset{O}{\underset{}{C}}CH_3$

d. $\xrightarrow[\text{excess}]{NH_3}$ $CH_3CH_2CH_2CH_2CONH_2$

e. $\xrightarrow[\text{excess}]{(CH_3CH_2)_2NH}$ $CH_3CH_2CH_2CH_2CON(CH_2CH_3)_2$

f. $\xrightarrow[\text{excess}]{C_6H_5NH_2}$ $CH_3CH_2CH_2CH_2CONHC_6H_5$

22.48

a. $\xrightarrow{SOCl_2}$ no reaction

b. $\xrightarrow{H_2O}$ 2 (valeric acid, COOH)

c. $\xrightarrow{CH_3OH}$ (ester OCH_3) + (acid OH)

d. $\xrightarrow{NaCl}$ no reaction

e. $\xrightarrow[\text{excess}]{(CH_3CH_2)_2NH}$ (amide N(CH_2CH_3)_2) + (carboxylate O⁻ $H_2\overset{+}{N}(CH_2CH_3)_2$)

f. $\xrightarrow[\text{excess}]{CH_3CH_2NH_2}$ (amide NHCH_2CH_3) + (carboxylate O⁻ $H_3\overset{+}{N}CH_2CH_3$)

22.49

a. $\xrightarrow{\text{NaHCO}_3}$ → phenylacetate sodium salt + H_2CO_3

b. $\xrightarrow{\text{NaOH}}$ → phenylacetate sodium salt + H_2O

c. $\xrightarrow{\text{SOCl}_2}$ → acid chloride (Cl)

d. $\xrightarrow{\text{NaCl}}$ no reaction

e. $\xrightarrow[\text{(1 equiv)}]{\text{NH}_3}$ → phenylacetate ammonium salt ($O^- \ NH_4^+$)

f. $\xrightarrow[\Delta]{\text{NH}_3}$ → amide (NH_2)

g. $\xrightarrow[\text{H}_2\text{SO}_4]{\text{CH}_3\text{OH}}$ → methyl ester (OCH_3)

h. $\xrightarrow[\text{}^-\text{OH}]{\text{CH}_3\text{OH}}$ → carboxylate (O^-)

i. $\xrightarrow{\text{[1] NaOH} \ \text{[2] CH}_3\text{COCl}}$ → anhydride

j. $\xrightarrow[\text{DCC}]{\text{CH}_3\text{NH}_2}$ → amide ($NHCH_3$)

k. $\xrightarrow{\text{[1] SOCl}_2 \ \text{[2] CH}_3\text{CH}_2\text{CH}_2\text{NH}_2}$ → amide ($NHCH_2CH_2CH_3$)

l. $\xrightarrow{\text{[1] SOCl}_2 \ \text{[2] (CH}_3)_2\text{CHOH}}$ → isopropyl ester ($OCH(CH_3)_2$)

22.50

a. $\xrightarrow{\text{SOCl}_2}$ no reaction

b. $\xrightarrow{\text{H}_3\text{O}^+}$ → carboxylic acid (OH) + HO—ethyl

c. $\xrightarrow{\text{H}_2\text{O}/\text{}^-\text{OH}}$ → carboxylate (O^-) + HO—ethyl

d. $\xrightarrow{\text{NH}_3}$ → amide (NH_2) + HO—ethyl

e. $\xrightarrow{\text{CH}_3\text{CH}_2\text{NH}_2}$ → amide ($NHCH_2CH_3$) + HO—ethyl

22.51

a. $\xrightarrow{\text{H}_3\text{O}^+}$ → CH_2COOH

b. $\xrightarrow{\text{H}_2\text{O}/\text{}^-\text{OH}}$ → CH_2COO^-

22.52

a. $\xrightarrow{\text{H}_3\text{O}^+}$ → $COOH$

b. $\xrightarrow{\text{H}_2\text{O}/\text{}^-\text{OH}}$ → COO^-

c. $\xrightarrow{\text{[1] CH}_3\text{MgBr} \ \text{[2] H}_2\text{O}}$ → ketone (CH_3)

d. $\xrightarrow{\text{[1] CH}_3\text{CH}_2\text{Li} \ \text{[2] H}_2\text{O}}$ → ketone

e. $\xrightarrow{\text{[1] DIBAL-H} \ \text{[2] H}_2\text{O}}$ → aldehyde (CHO)

f. $\xrightarrow{\text{[1] LiAlH}_4 \ \text{[2] H}_2\text{O}}$ → CH_2NH_2

22.53

a.

b. C$_6$H$_5$COCl + (pyrrolidine, excess) → C$_6$H$_5$—C(=O)—N(pyrrolidine)

c. C$_6$H$_5$CN $\xrightarrow[\text{[2] H}_2\text{O}]{\text{[1] CH}_3\text{CH}_2\text{CH}_2\text{MgBr}}$ C$_6$H$_5$—C(=O)—CH$_2$CH$_2$CH$_3$

d. (CH$_3$)$_2$CHCOOH + CH$_3$CH$_2$CHOH(CH$_3$) $\xrightarrow{\text{H}_2\text{SO}_4}$ (CH$_3$)$_2$CH—C(=O)—OCH(CH$_3$)CH$_2$CH$_3$

e. (C$_6$H$_5$)—NHCOCH$_3$ $\xrightarrow[\text{$^-$OH}]{\text{H}_2\text{O}}$ CH$_3$—C(=O)—O$^-$ + (C$_6$H$_5$)—NH$_2$

f. $\xrightarrow{\text{H}_3\text{O}^+}$

g. CH$_3$CH$_2$CH$_2$CH$_2$Br $\xrightarrow[\text{[2] H}_2\text{O/}^-\text{OH}]{\text{[1] NaCN}}$ CH$_3$CH$_2$CH$_2$CH$_2$—C(=O)—O$^-$

h. C$_6$H$_5$CH$_2$COOH $\xrightarrow{\substack{\text{[1] SOCl}_2 \\ \text{[2] CH}_3\text{CH}_2\text{CH}_2\text{CH}_2\text{NH}_2 \\ \text{[3] LiAlH}_4 \\ \text{[4] H}_2\text{O}}}$ C$_6$H$_5$CH$_2$CH$_2$NHCH$_2$CH$_2$CH$_2$CH$_3$

i. C$_6$H$_5$CH$_2$CH$_2$CH$_2$CN $\xrightarrow{\text{H}_3\text{O}^+}$ C$_6$H$_5$CH$_2$CH$_2$CH$_2$COOH

j. HOOC—CH=CH—COOH $\xrightarrow{\Delta}$ (maleic anhydride)

k. (CH$_3$CO)$_2$O + (cyclohexyl)—NH$_2$ (excess) → (cyclohexyl)—NHCOCH$_3$ + CH$_3$COO$^-$ H$_3$N$^+$—(cyclohexyl)

l. C$_6$H$_5$CH$_2$CH$_2$COOCH$_2$CH$_3$ $\xrightarrow[\text{$^-$OH}]{\text{H}_2\text{O}}$ C$_6$H$_5$CH$_2$CH$_2$COO$^-$ + HOCH$_2$CH$_3$

22.54

22.55

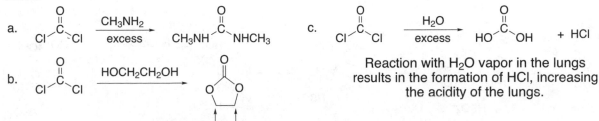

a.

b.

c.

d.

22.56

a. <chem>CI-C(=O)-CI</chem> →(CH₃NH₂, excess)→ <chem>CH₃NH-C(=O)-NHCH₃</chem>

b. <chem>CI-C(=O)-CI</chem> →(HOCH₂CH₂OH)→ cyclic carbonate

All H's on these C's are identical.

c. <chem>CI-C(=O)-CI</chem> →(H₂O, excess)→ <chem>HO-C(=O)-OH</chem> + HCl

Reaction with H₂O vapor in the lungs results in the formation of HCl, increasing the acidity of the lungs.

22.57

aspartame →(H₂O)→ + phenylalanine + CH₃OH

22.58

CH₃CH(Br)CH(COOH)— →(NaOH, E2 elimination)→ acrylic acid derivative

(*E* and *Z* isomers possible)
A = $C_4H_6O_2$

Base removes the elements of H and Br from the α and β carbons to form the E2 elimination product. IR data show the presence of an OH group (3500-2500 cm^{-1}), a C=O (1703 cm^{-1}), and a C=C (1656 cm^{-1}).

Br-CH₂CH₂CH₂-C(=O)-OH →(NaOH, proton transfer)→ →(SN2)→ five-membered lactone

B = $C_4H_6O_2$

B has a C=O at high wave number (1770 cm^{-1}), indicating the presence of the the five-membered ring lactone.

A similar reaction does not occur with $CH_3CH(Br)CH_2COOH$ because it would lead to a strained four-membered ring. Moreover, the 2° alkyl halide is more crowded, making nucleophilic substitution difficult.

CH₃CH(Br)CH₂-C(=O)-OH →(NaOH, proton transfer)→ 2° halide →×→ four-membered lactone

22.59

a.

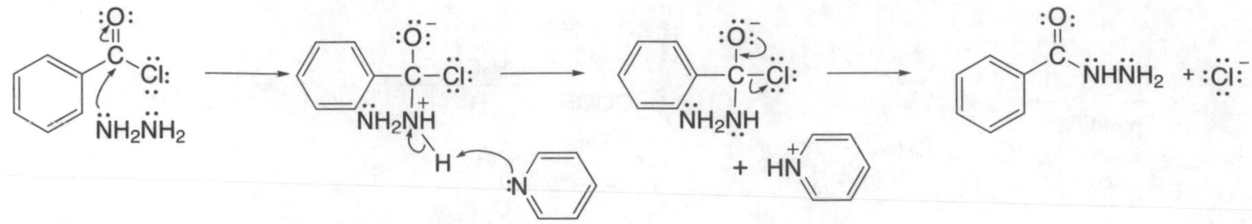

b.

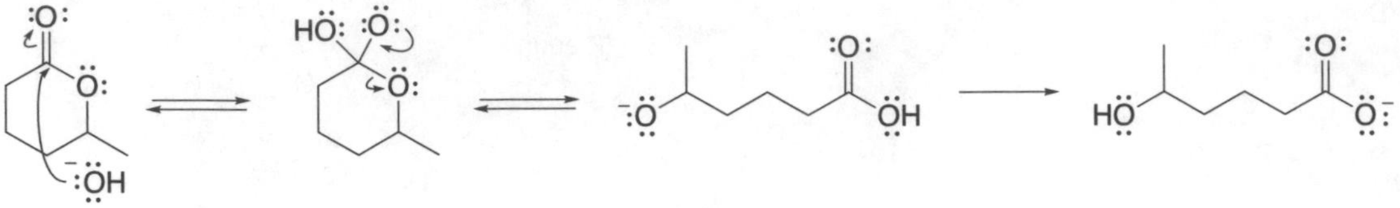

c.

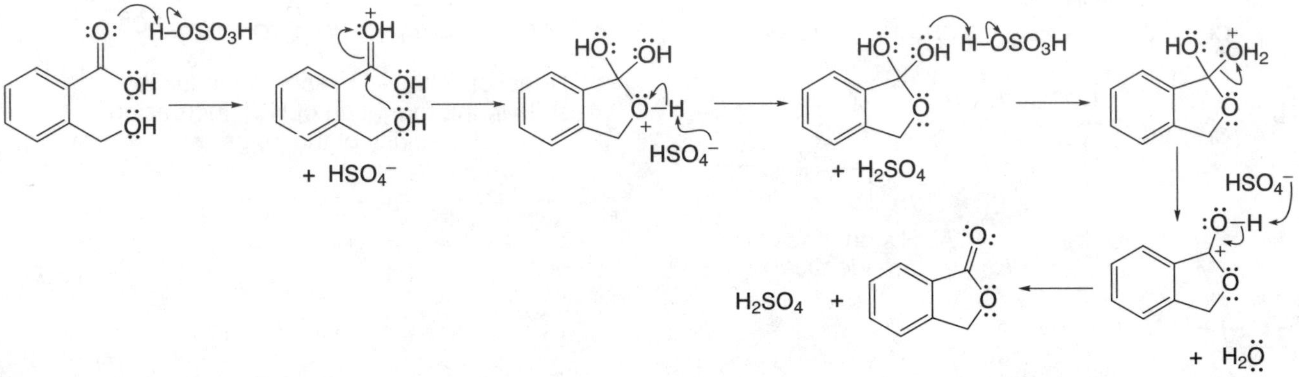

22.60

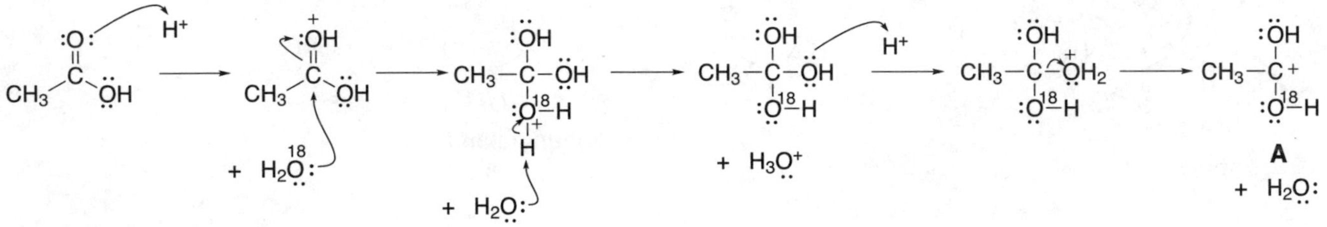

Two possibilities for **A**:

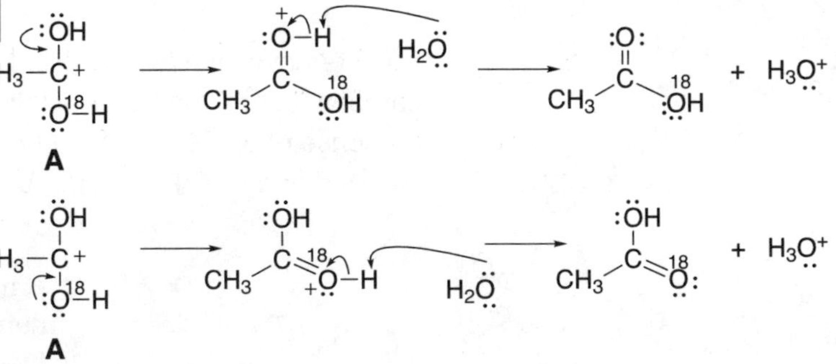

22.61

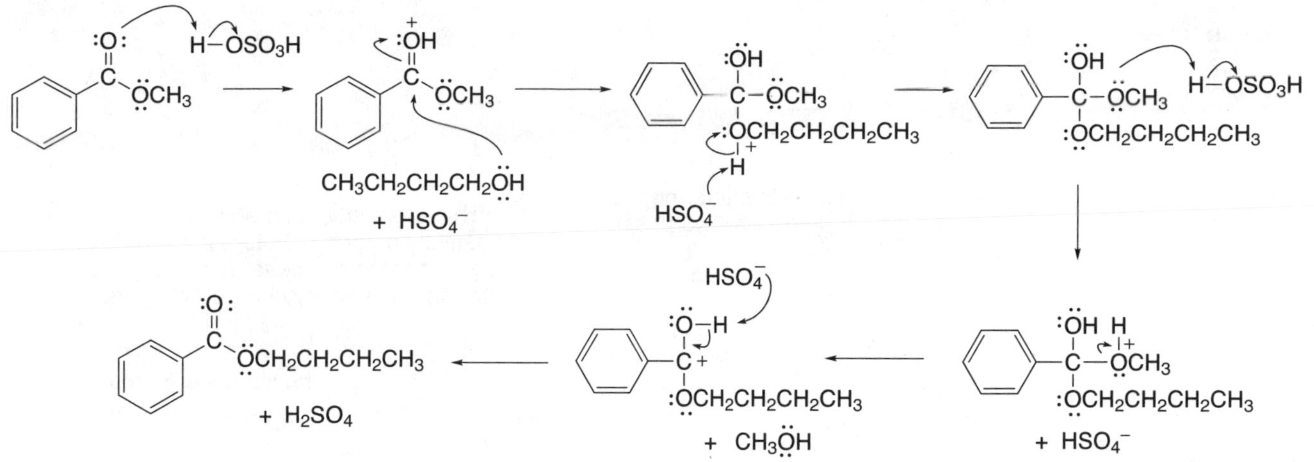

22.62

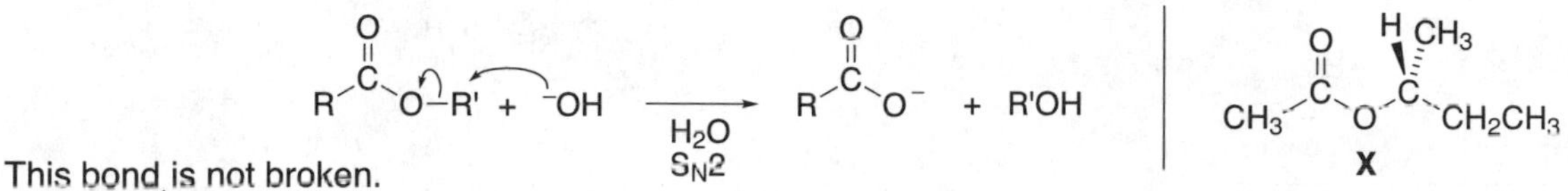

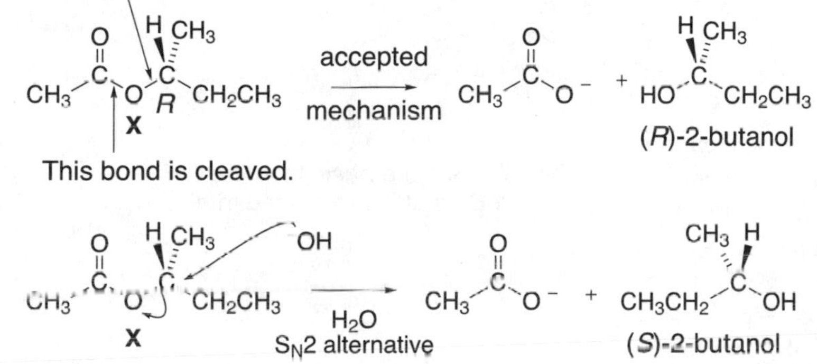

This bond is not broken.

This bond is cleaved.

accepted
mechanism

(R)-2-butanol

According to the accepted mechanism, the stereochemistry around the stereogenic center is retained in the product.

X

S_N2 alternative

(S)-2-butanol

This S_N2 mechanism would form the product of inversion leading to (S)-2-butanol. Since (R)-2-butanol is the only product formed, the S_N2 mechanism does not occur during ester hydrolysis.

22.63

Reaction in base:

Once this intermediate forms, there are two possibilities.

or

This product is favored since the negatively charged O atom is a weaker base and therefore better leaving group than the negatively charged N atom.

poorer leaving group

Reaction in acid:

Intermediated **X** can be re-converted to starting material **B** or go on to form **A** by a stepwise process.

Once this product forms it cannot revert to **B** since RNH₃⁺ no longer has a lone pair so it is not a nucleophile.

This NH₂ is more basic than the OH group in **B**, so it is protonated in acid.

22.64

diethyl carbonate

(any base)

22.65

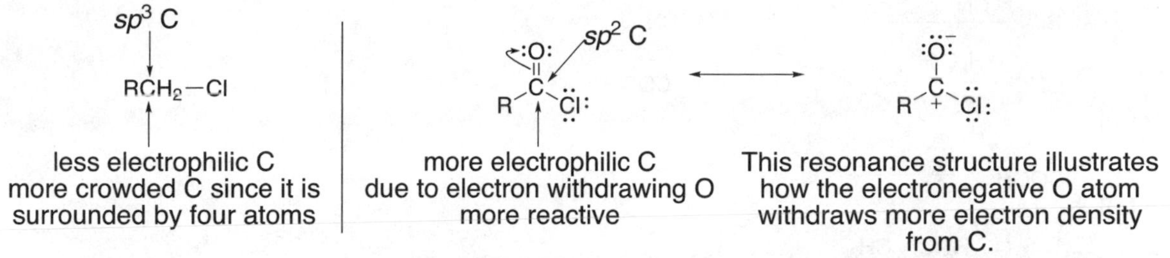
sp3 C
RCH2—Cl
less electrophilic C
more crowded C since it is
surrounded by four atoms
sp2 C
more electrophilic C
due to electron withdrawing O
more reactive
This resonance structure illustrates
how the electronegative O atom
withdraws more electron density
from C.

The sp^2 hybridized C of RCOCl is much less crowded, and this makes nucleophilic attack easier as well.

22.66 Fischer esterification is treatment of a carboxylic acid with an alcohol in the presence of an acid catalyst to form an ester.

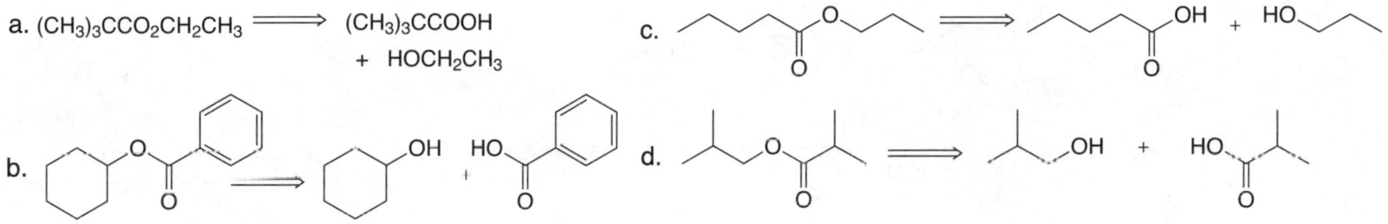
a. (CH3)3CCO2CH2CH3
(CH3)3CCOOH
+ HOCH2CH3
b.
OH
HO
c.
OH
HO
d.
OH
HO

22.67

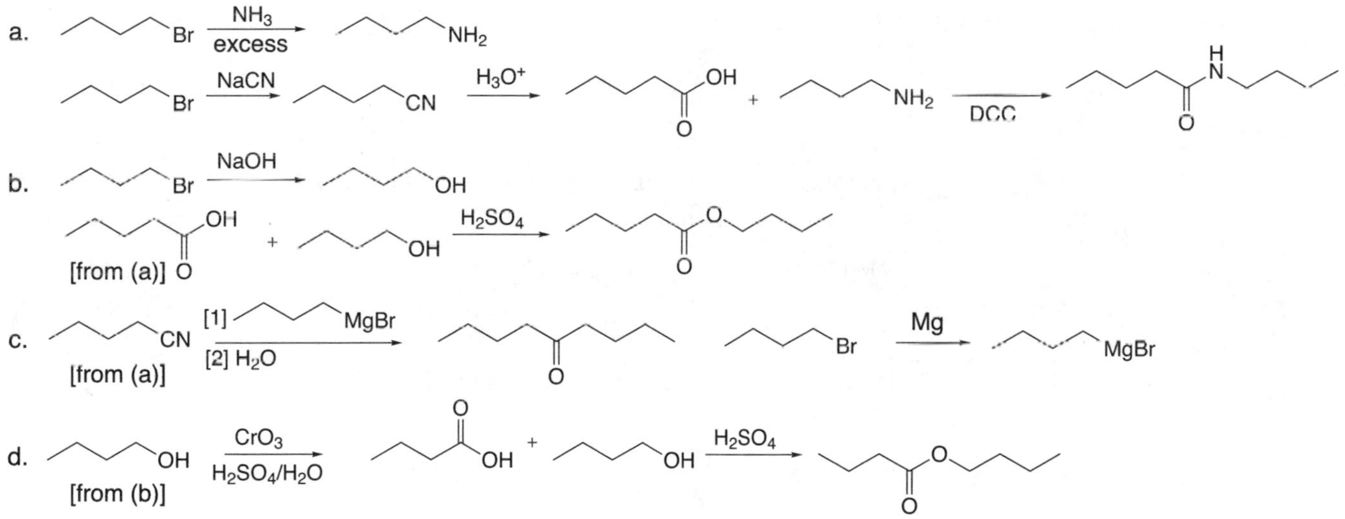
a.
Br
NH3
excess
NH2
NaCN
CN
H3O+
OH
NH2
DCC
b.
NaOH
OH
[from (a)]
H2SO4
c.
CN
[from (a)]
[1] MgBr
[2] H2O
Br
Mg
MgBr
d.
OH
[from (b)]
CrO3
H2SO4/H2O
OH
H2SO4

22.68

a. (structure) ~Br $\xrightarrow{\text{–CN}}$ (structure) ~CN

b. (structure) ~CN [from (a)] $\xrightarrow{\text{H}_2\text{O/H}^+}$ (structure) ~COOH

c. (structure) ~COOH [from (b)] $\xrightarrow{\text{SOCl}_2}$ (structure) ~COCl

d. (structure) ~COOH [from (b)] $\xrightarrow[\text{H}_2\text{SO}_4]{\text{HOCH}_2\text{CH}_3}$ (structure) ~CO$_2$CH$_2$CH$_3$

e. (structure) ~CN [from (a)] $\xrightarrow[\text{[2] H}_2\text{O}]{\text{[1] CH}_3\text{MgBr}}$ (structure) $\overset{\text{CH}_3}{\underset{\text{O}}{\text{C}}}$

f. (structure) ~CN [from (a)] $\xrightarrow[\text{[2] H}_2\text{O}]{\text{[1] DIBAL–H}}$ (structure) ~CHO

g. (structure) ~CN [from (a)] $\xrightarrow[\text{[2] H}_2\text{O}]{\text{[1] LiAlH}_4}$ (structure) ~CH$_2$NH$_2$

h. (structure) ~CH$_2$NH$_2$ [from (g)] $\xrightarrow[\text{DCC}]{\text{CH}_3\text{COOH}}$ (structure) ~CH$_2$NHCOCH$_3$

22.69

a. CH$_3$Cl + NaCN $\longrightarrow$ CH$_3$—CN $\xrightarrow{\text{H}_3\text{O}^+}$ CH$_3$—COOH

CH$_3$—Cl + Mg $\longrightarrow$ CH$_3$—MgCl $\xrightarrow[\text{[2] H}_3\text{O}^+]{\text{[1] CO}_2}$ CH$_3$—COOH

b. (benzene with Br, sp^2) + NaCN $\longrightarrow$ This method can't be used because an S$_N$2 reaction can't be done on an sp^2 hybridized C.

(benzene with Br) + Mg $\longrightarrow$ (benzene with MgBr) $\xrightarrow[\text{[2] H}_3\text{O}^+]{\text{[1] CO}_2}$ (benzene with COOH)

c. (CH$_3$)$_3$CCl + NaCN $\longrightarrow$ This method can't be used because an S$_N$2 reaction can't be done on a 3° C.

(CH$_3$)$_3$C—Cl + Mg $\longrightarrow$ (CH$_3$)$_3$C—MgCl $\xrightarrow[\text{[2] H}_3\text{O}^+]{\text{[1] CO}_2}$ (CH$_3$)$_3$C—COOH

d. HOCH$_2$CH$_2$CH$_2$CH$_2$Br + NaCN $\longrightarrow$ HOCH$_2$CH$_2$CH$_2$CH$_2$—CN $\xrightarrow{\text{H}_3\text{O}^+}$ HOCH$_2$CH$_2$CH$_2$CH$_2$—COOH

HOCH$_2$CH$_2$CH$_2$CH$_2$—Br + Mg $\longrightarrow$ This method can't be used because you can't make a Grignard reagent with an acidic OH group.

22.70

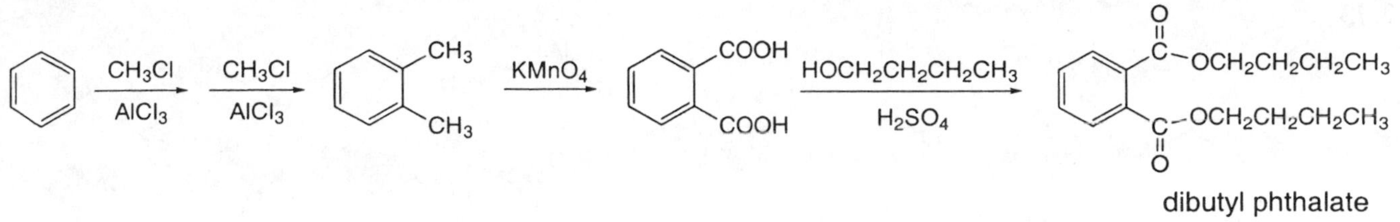

dibutyl phthalate

22.71

a.

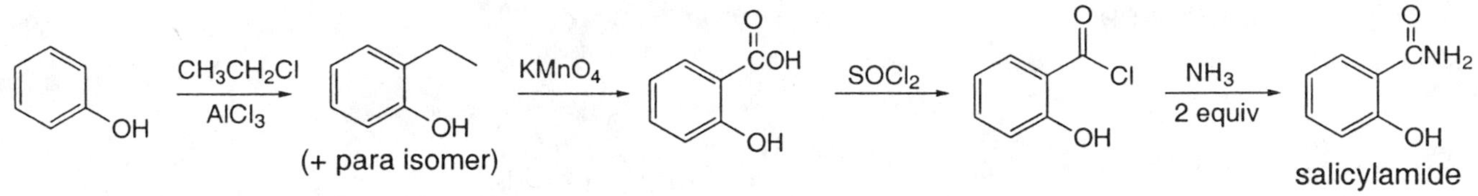

(+ para isomer)

salicylamide

b.

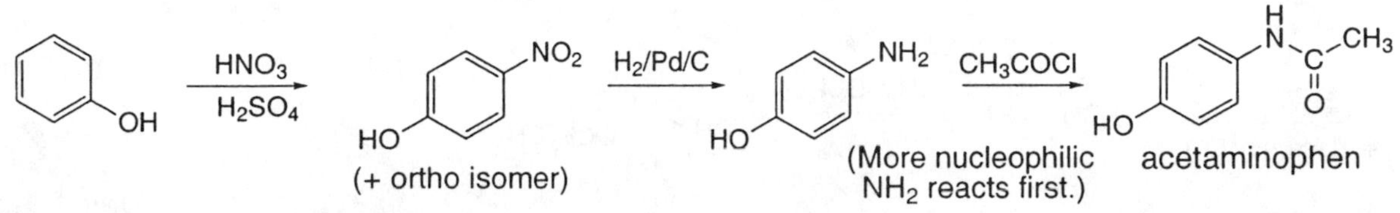

(+ ortho isomer)

(More nucleophilic NH₂ reacts first.)

acetaminophen

c.

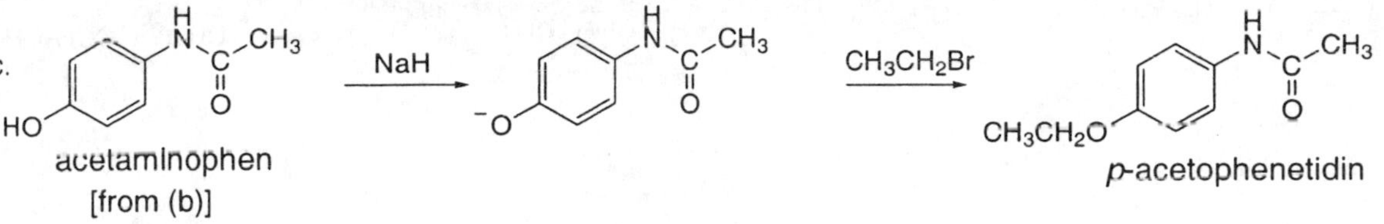

acetaminophen
[from (b)]

p-acetophenetidin

22.72

a

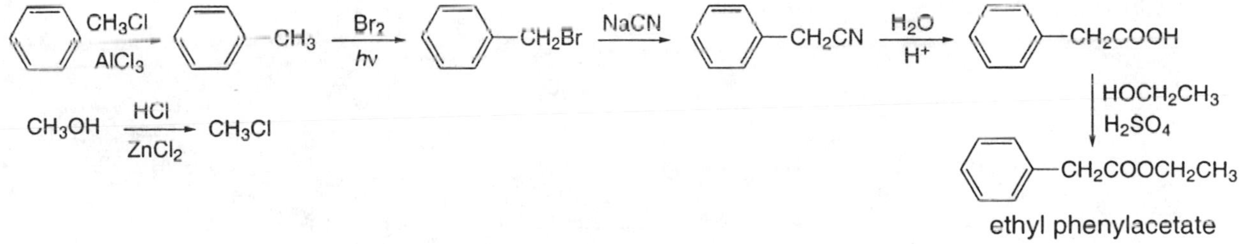

ethyl phenylacetate

$$CH_3OH \xrightarrow[ZnCl_2]{HCl} CH_3Cl$$

b.

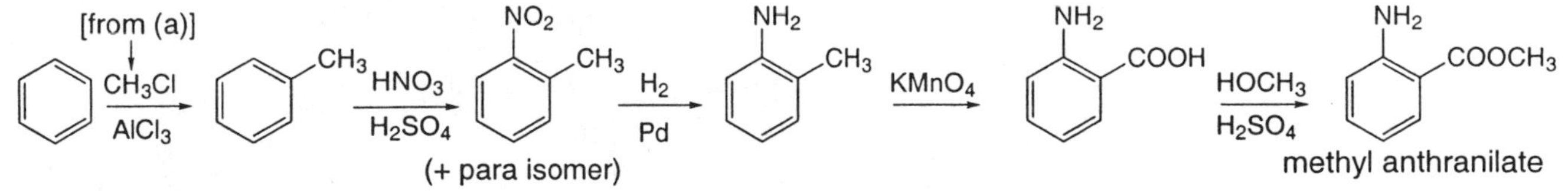

[from (a)]

(+ para isomer)

methyl anthranilate

c.

[from (a)]

benzyl acetate

$$CH_3CH_2OH \xrightarrow[H_2SO_4/H_2O]{CrO_3} CH_3COOH$$

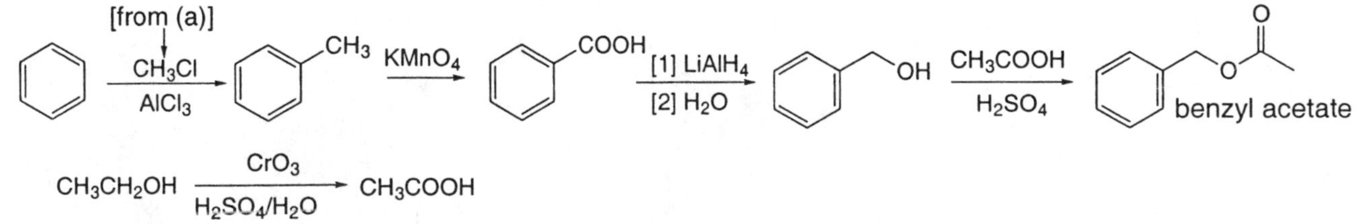

22.73

Cl~~~~Cl $\xrightarrow[\text{excess}]{^-\text{CN}}$ NC~~~~CN

$\xrightarrow{\text{[1] LiAlH}_4}$
$\xrightarrow{\text{[2] H}_2\text{O}}$ H_2N~~~~NH_2

$\xrightarrow[\text{H}_2\text{SO}_4]{\text{H}_2\text{O}}$ HO~~~~OH (diacid)

22.74

a. CH_3—COOH $\xrightarrow[\text{H}_2\text{SO}_4]{\text{CH}_3{}^{13}\text{CH}_2\text{OH}}$ CH_3—CO—O^{13}CH$_2$CH$_3$

b. $CH_3{}^{13}CH_2OH$ $\xrightarrow[\text{H}_2\text{SO}_4/\text{H}_2\text{O}]{\text{CrO}_3}$ CH_3—^{13}C(O)—OH $\xrightarrow[\text{H}_2\text{SO}_4]{\text{CH}_3\text{CH}_2\text{OH}}$ CH_3—^{13}C(O)—OCH$_2$CH$_3$

c. CH_3CH_2Br $\xrightarrow[\text{+ base (H}^{18}\text{O}^-)]{\text{H}_2{}^{18}\text{O}}$ $CH_3CH_2{}^{18}OH$ $\xrightarrow{\text{CH}_3\text{COCl}}$ CH_3—CO—^{18}OCH$_2$CH$_3$

d. $CH_3{}^{13}CH_2OH$ $\xrightarrow{\text{PBr}_3}$ $CH_3{}^{13}CH_2Br$ $\xrightarrow{^{18}\text{OH}}$ $CH_3{}^{13}CH_2{}^{18}OH$ $\xrightarrow[\text{H}_2\text{SO}_4/\text{H}_2{}^{18}\text{O}]{\text{CrO}_3}$ CH_3—^{13}C(^{18}O)—^{18}OH $\xrightarrow[\text{H}_2\text{SO}_4]{\text{CH}_3\text{CH}_2\text{OH}}$ CH_3—^{13}C(^{18}O)—OCH$_2$CH$_3$
+
$H_2{}^{18}O$

22.75

a. HO—⬡—OH and HO—CO—CH$_2$CH$_2$—CO—OH $\longrightarrow$ (polyester product)

b. ClOC—⬢—COCl and H_2N~~~~NH_2 $\longrightarrow$ (polyamide product)

22.76

a.

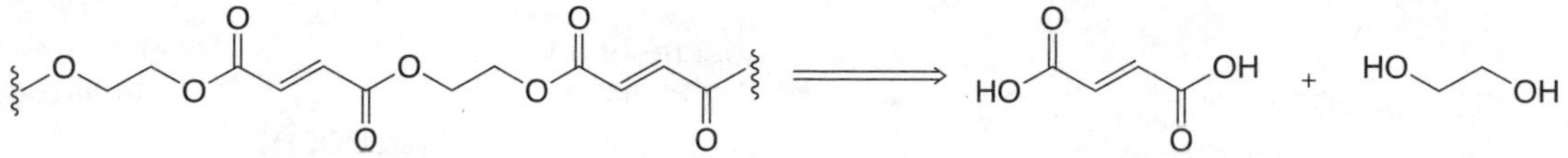

b.

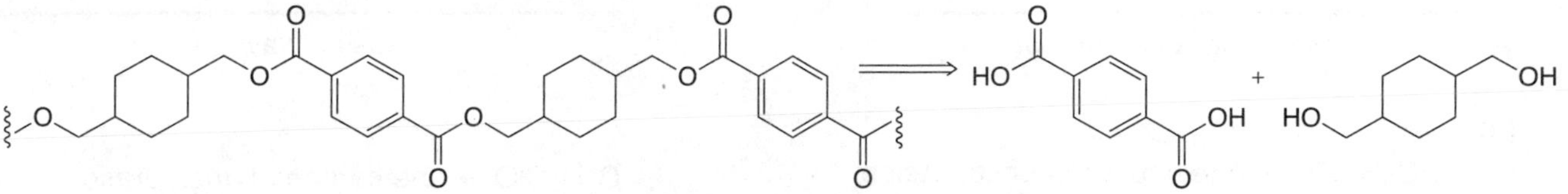

c.

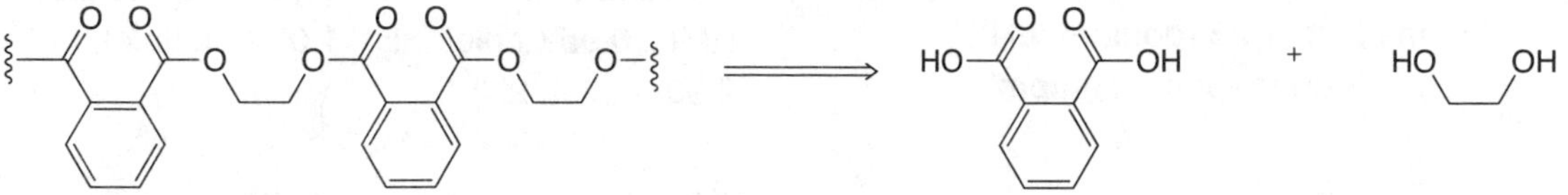

22.77

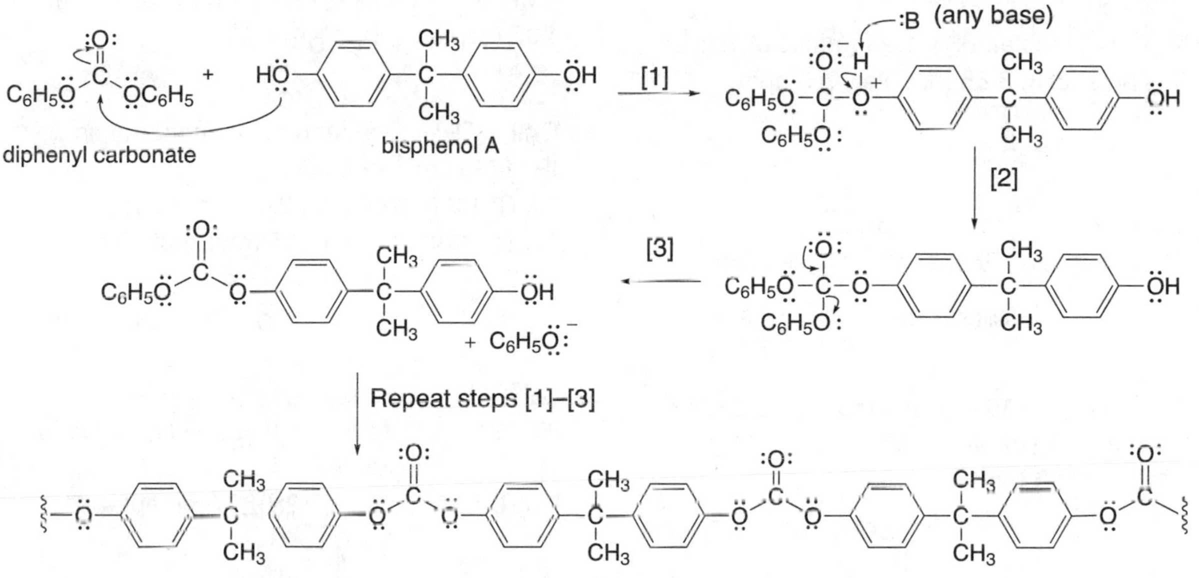

diphenyl carbonate   bisphenol A

Repeat steps [1]–[3]

Lexan

22.78

a.    and

contains a broad, strong OH
absorption at 3500–2500 cm^{-1}

b. 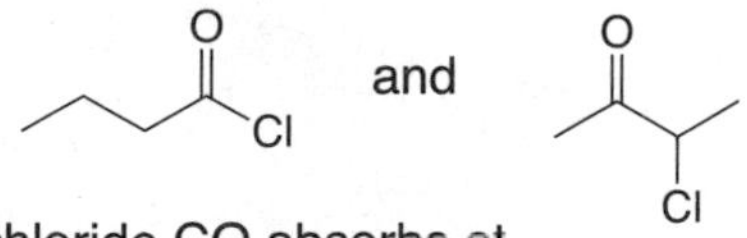   and

Acid chloride CO absorbs at
much higher wave number.   ketone

c.    and

C=O at <1700 cm^{-1} 2 NH absorptions at 3200–3400 cm^{-1}
due to the stabilized amide   C=O absorption higher frequency

d. 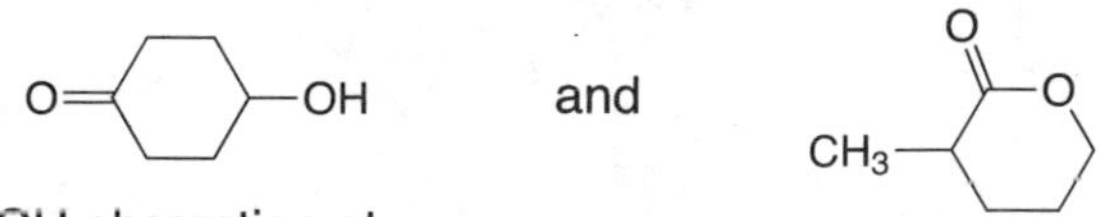   and

OH absorption at
3400–3200 cm^{-1} + C=O   only C=O

22.79

a. $C_6H_5COOCH_2CH_3$ $CH_3CH_2COOCH_2CH_3$

b. most resonance stabilized ——— least resonance stabilized

CH_3CONH_2 CH_3COOCH_3 CH_3COCl

increasing wavenumber →   increasing wavenumber →

22.80

a. $C_6H_{12}O_2 \rightarrow$ one degree of unsaturation
   IR: 1738 cm^{-1} → C=O
   NMR: 1.12 (triplet, 3H), 1.23 (doublet, 6H),
   2.28 (quartet, 2H), 5.00 (septet, 1H) ppm

b. $C_6H_{12}O_2 \rightarrow$ one degree of unsaturation
   IR: 1746 cm^{-1} → C=O
   NMR: 0.94 (doublet, 6H), 1.93 (multiplet,1H),
   2.05 (singlet,3H), 3.85 (doublet, 2H) ppm

c. C_4H_7N
   IR: 2250 cm^{-1} → triple bond
   NMR: 1.08 (triplet, 3H), 1.70 (multiplet, 2H),
   2.34 (triplet, 2H)
   $CH_3CH_2CH_2C{\equiv}N$

d. C_8H_9NO
   IR: 3328 (NH), 1639 (conjugated amide C=O) cm^{-1}
   NMR: 2.95 (singlet, 3H), 6.95 (singlet, 1H),
   7.3–7.7 (multiplet, 5H)

e. $C_4H_7ClO \rightarrow$ one degree of unsaturation
   IR: 1802 cm^{-1} → C=O (high wave number, RCOCl)
   NMR: 0.95 (triplet, 3H), 1.07 (multiplet, 2H),
   2.90 (triplet, 2H)

f. $C_5H_{10}O_2 \rightarrow$ one degree of unsaturation
   IR: 1750 cm^{-1} → C=O
   NMR: 1.20 (doublet, 6H), 2.00 (singlet, 3H),
   4.95 (septet, 1H)

g. $C_{10}H_{12}O_2 \rightarrow$ five degrees of unsaturation
   IR: 1740 cm^{-1} → C=O
   NMR: 1.2 (triplet, 3H), 2.4 (quartet, 2H),
   5.1 (singlet, 2H), 7.1–7.5 (multiplet, 5H)

h. $C_8H_{14}O_3 \rightarrow$ two degrees of unsaturation
   IR: 1810, 1770 cm^{-1} → 2 absorptions due to
      C=O (anhydride)
   NMR: 1.25 (doublet, 12H), 2.65 (septet, 2H)

22.81

A. Molecular formula $C_{10}H_{12}O_2 \rightarrow$ five degrees of unsaturation

IR absorption at 1718 cm$^{-1} \rightarrow$ C=O

NMR data (ppm):

triplet at 1.4 (CH$_3$ adjacent to 2H's)

singlet at 2.4 (CH$_3$)

quartet at 4.4 (CH$_2$ adjacent to 3H's)

doublet at 7.2 (2H's on benzene ring)

doublet at 7.9 (2H's on benzene ring)

B. IR absorption at 1740 cm$^{-1} \rightarrow$ C=O

NMR data (ppm):

singlet at 2.0 (CH$_3$)

triplet at 2.9 (CH$_2$ adjacent to CH$_2$)

triplet at 4.4 (CH$_2$ adjacent to CH$_2$)

multiplet at 7.3 (5H's, monosubstituted benzene)

22.82

Molecular formula $C_{10}H_{13}NO_2$

IR absorptions at 3300 (NH) and 1680 (C=O, amide or conjugated) cm^{-1}

NMR data (ppm):

triplet at 1.4 (CH$_3$ adjacent to 2H's)

singlet at 2.2 (CH$_3$C=O)

quartet at 3.9 (CH$_2$ adjacent to 3H's)

doublet at 6.8 (2H's on benzene ring)

singlet at 7.2 (NH)

doublet at 7.4 (2H's on benzene ring)

CH_3CH_2O

phenacetin

22.83

8.0 ppm

2 CH$_3$'s at
2.93 and 3.03 ppm

CH$_3$ (on the same side as the O atom)

CH$_3$ (on the same side as the H atom)

Resonance stabilization of the amide causes restricted rotation around the C–N amide bond, so the two methyl groups on N are in different environments. Thus they give two different NMR signals.

22.84

ethyl benzoate

Two OH groups are now equivalent and either can lose H$_2$O to form labeled or unlabeled ethyl benzoate.

Unlabeled starting material was recovered.

22.85

Chapter 23: Substitution Reactions of Carbonyl Compounds at the α Carbon

♦ Kinetic versus thermodynamic enolates (23.4)

kinetic enolate

Kinetic enolate
- The less substituted enolate
- Favored by strong base, polar aprotic solvent, low temperature: LDA, THF, –78 °C

thermodynamic enolate

Thermodynamic enolate
- The more substituted enolate
- Favored by strong base, protic solvent, higher temperature: $NaOCH_2CH_3$, CH_3CH_2OH, room temperature

♦ Halogenation at the α carbon

[1] Halogenation in acid (23.7A)

$$R-\overset{\overset{\displaystyle O}{\|}}{C}-\overset{\text{H}}{\underset{}{C}} \xrightarrow[CH_3COOH]{X_2} \boxed{\begin{array}{c} R-\overset{\overset{\displaystyle O}{\|}}{C}-\overset{X}{\underset{}{C}} \\ \text{α-halo aldehyde} \\ \text{or ketone} \end{array}}$$

$X_2 = Cl_2,\ Br_2,\ or\ I_2$

- The reaction occurs via enol intermediates.
- Monosubstitution of X for H occurs on the α carbon.

[2] Halogenation in base (23.7B)

$$R-\overset{\overset{\displaystyle O}{\|}}{C}-\overset{R}{\underset{\underset{\displaystyle H\ \ H}{}}{C}} \xrightarrow[^-OH]{X_2} \boxed{R-\overset{\overset{\displaystyle O}{\|}}{C}-\overset{R}{\underset{\underset{\displaystyle X\ \ X}{}}{C}}}$$

$X_2 = Cl_2,\ Br_2,\ or\ I_2$

- The reaction occurs via enolate intermediates.
- Polysubstitution of X for H occurs on the α carbon.

[3] Halogenation of *methyl* ketones in base—The haloform reaction (23.7B)

$$R-\overset{\overset{\displaystyle O}{\|}}{C}-CH_3 \xrightarrow[^-OH]{X_2\ (excess)} \boxed{R-\overset{\overset{\displaystyle O}{\|}}{C}-O^-}\ +\ HCX_3 \quad \text{haloform}$$

$X_2 = Cl_2,\ Br_2,\ or\ I_2$

- The reaction occurs with methyl ketones, and results in cleavage of a carbon–carbon σ bond.

◆ Alkylation reactions at the α carbon

[1] Direct alkylation at the α carbon (23.8)

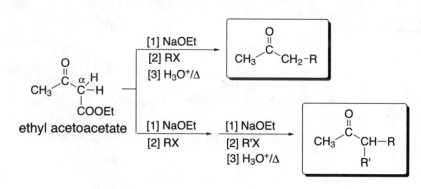

- The reaction forms a new C–C bond to the α carbon.
- LDA is a common base used to form an intermediate enolate.
- The alkylation in Step [2] follows an S_N2 mechanism.

[2] Malonic ester synthesis (23.9)

- The reaction is used to prepare α substituted carboxylic acids with one or two alkyl groups on the α carbon.
- The alkylation in Step [2] follows an S_N2 mechanism.

[3] Acetoacetic ester synthesis (23.10)

- The reaction is used to prepare α substituted ketones with one or two alkyl groups on the α carbon.
- The alkylation in Step [2] follows an S_N2 mechanism.

Chapter 23: Answers to Problems

23.1 • To convert a ketone to its enol tautomer, change the C=O to C–OH, make a new double bond to an α carbon, and remove a proton at the other end of the C=C.
• To convert an enol to its keto form, find the C=C bonded to an OH. Change the C–OH to a C=O, add a proton to the other end of the C=C, and delete the double bond.
[In cases where *E* and *Z* isomers are possible, only one stereoisomer is drawn.]

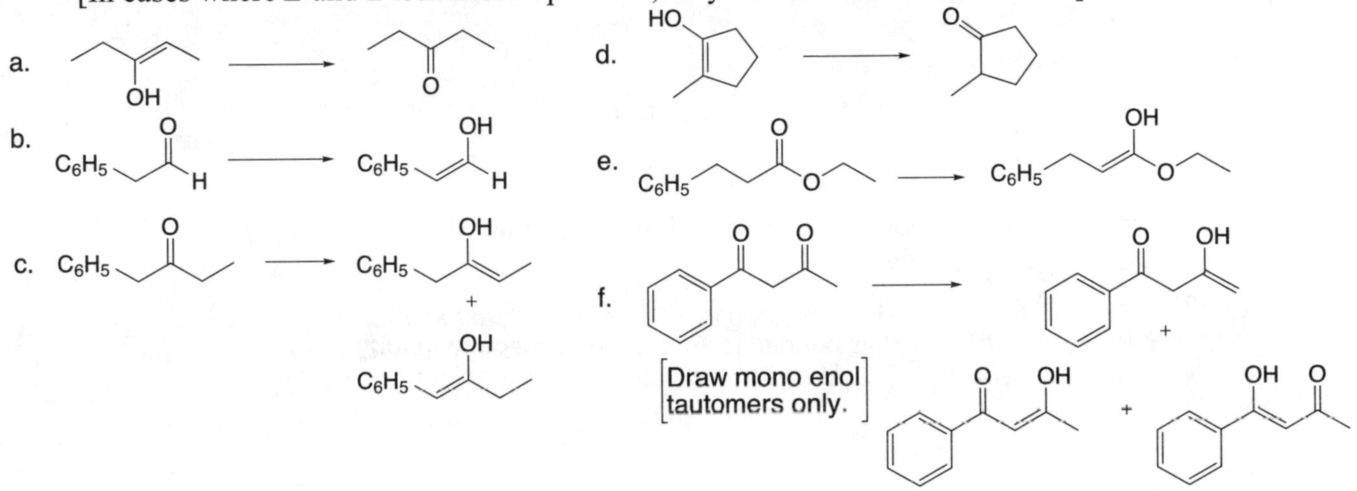
a.
b.
c.
d.
e.
f.
Draw mono enol tautomers only.

23.2

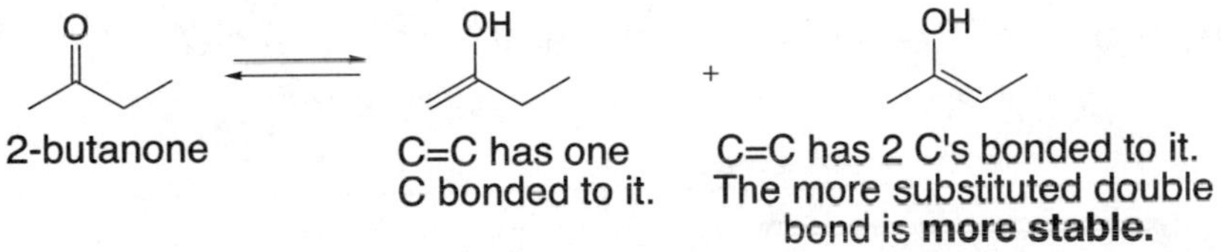
2-butanone
C=C has one C bonded to it.
C=C has 2 C's bonded to it. The more substituted double bond is **more stable.**

23.3

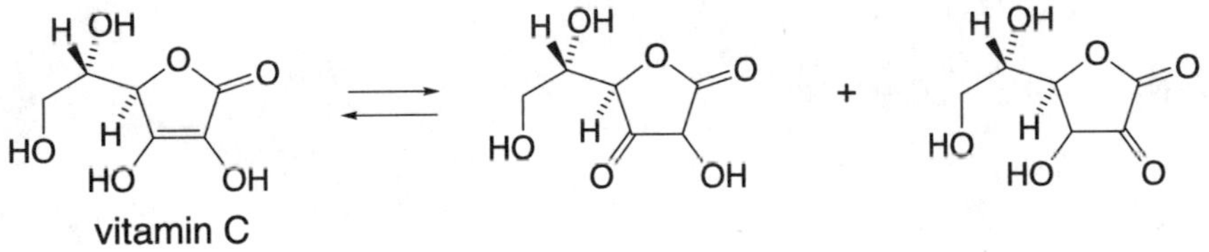
vitamin C

23.4 The mechanism has two steps: protonation followed by deprotonation.

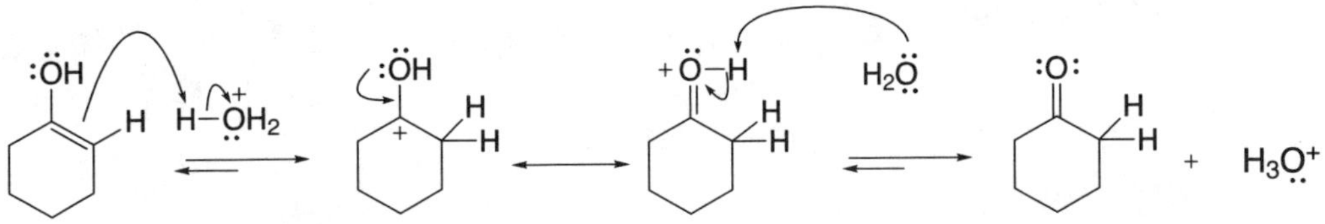

23.5

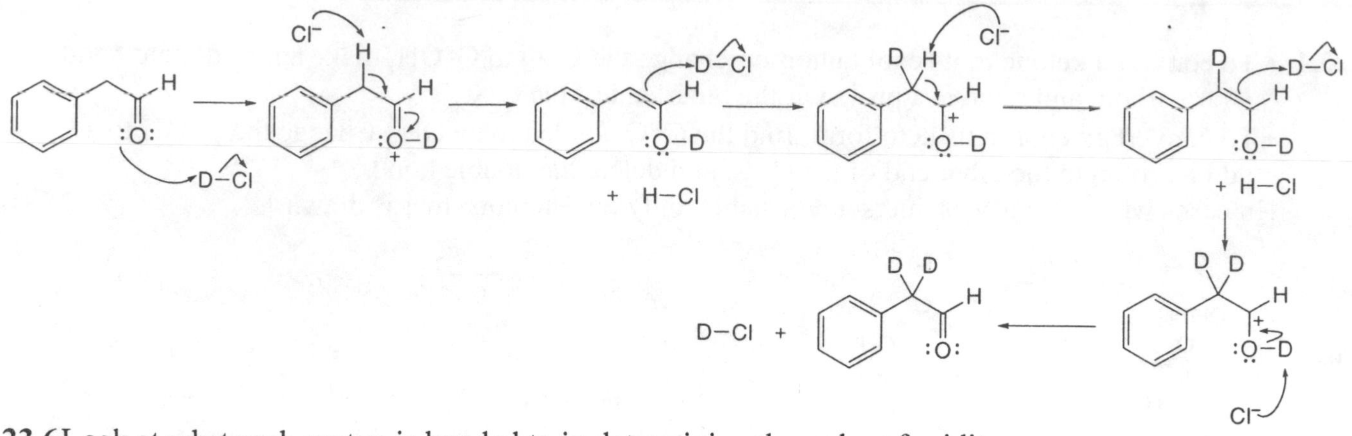

23.6 Look at what each proton is bonded to in determining the order of acidity.

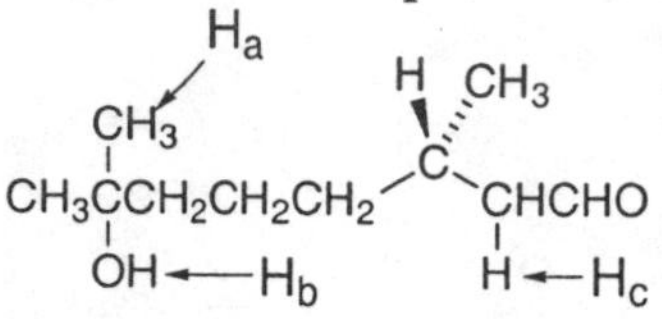

$H_a < H_c < H_b$
H_a is bonded to a CH_3 group = **least acidic**.
H_c is bonded to an α C = **intermediate acidity**.
H_b is bonded to an O = **most acidic**.

23.7

a.

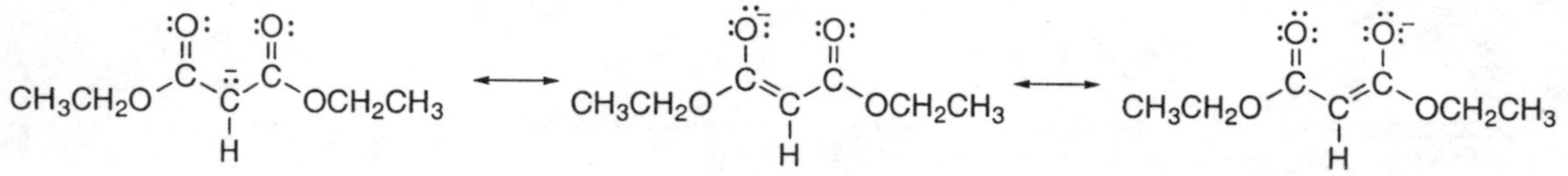

b.

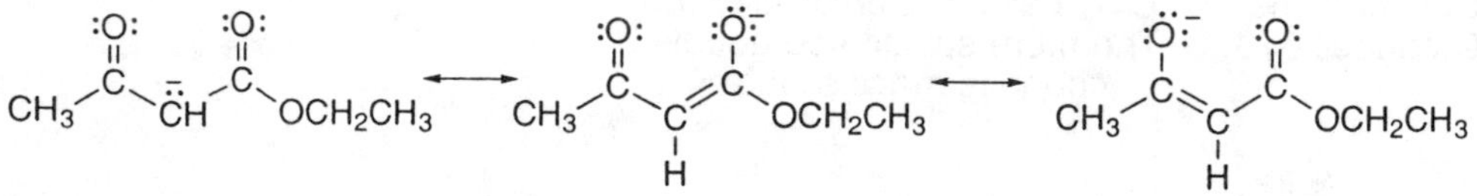

c.

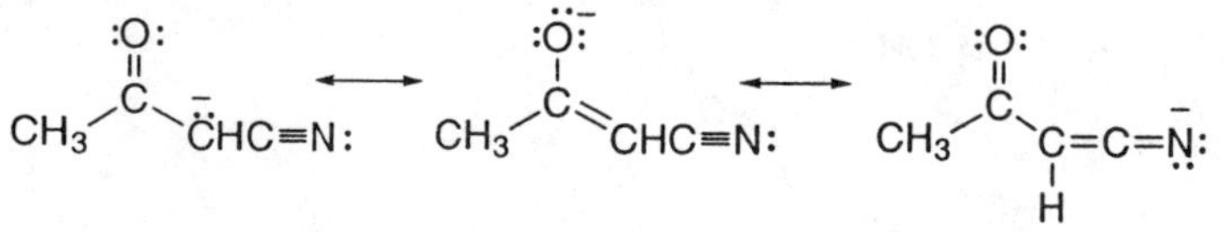

23.8

a.   b. $CH_3CH_2CH_2-CN$ c. 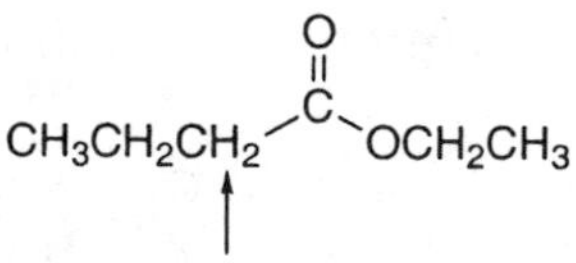   d.

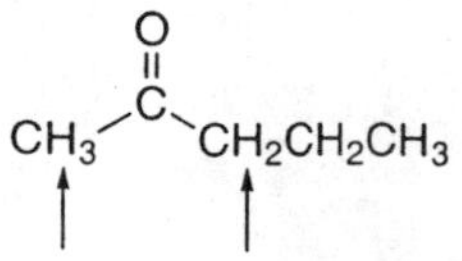

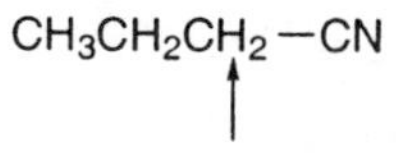

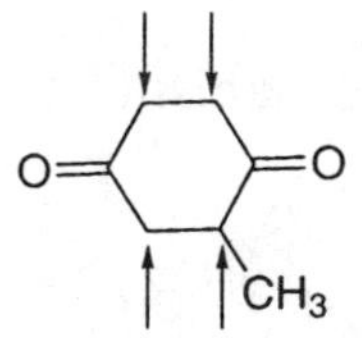

23.9

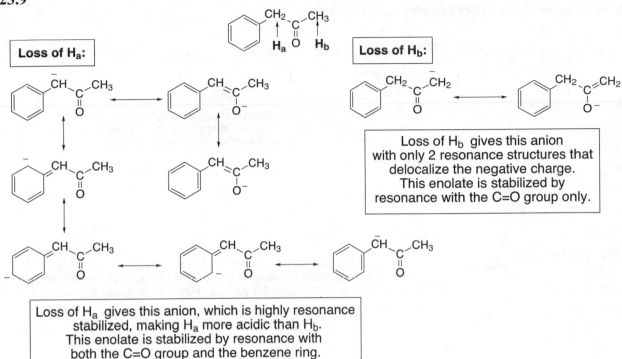

Loss of H$_a$:

Loss of H$_b$:

Loss of H$_b$ gives this anion with only 2 resonance structures that delocalize the negative charge. This enolate is stabilized by resonance with the C=O group only.

Loss of H$_a$ gives this anion, which is highly resonance stabilized, making H$_a$ more acidic than H$_b$. This enolate is stabilized by resonance with both the C=O group and the benzene ring.

23.10 In each of the reactions, the LDA pulls off the most acidic proton.

a.
$$\xrightarrow[\text{THF}]{\text{LDA}}$$

b.
$$\xrightarrow[\text{THF}]{\text{LDA}}$$

c.
$$CH_3-C \underset{OCH_2CH_3}{\overset{O}{\parallel}} \xrightarrow[\text{THF}]{\text{LDA}} \bar{C}H_2-C \underset{OCH_2CH_3}{\overset{O}{\parallel}}$$

d.
$$\xrightarrow[\text{THF}]{\text{LDA}}$$

23.11• LDA/THF treatment forms the kinetic enolate by removing a proton from the less substituted C.
    • Treatment with NaOCH$_3$/CH$_3$OH forms the thermodynamic enolate by removing a proton from the more substituted C.

a.
LDA/THF

NaOCH$_3$
CH$_3$OH

b.
LDA/THF

NaOCH$_3$
CH$_3$OH

c.
LDA/THF

NaOCH$_3$
CH$_3$OH

23.12

This acidic H is removed with base to form an achiral enolate.

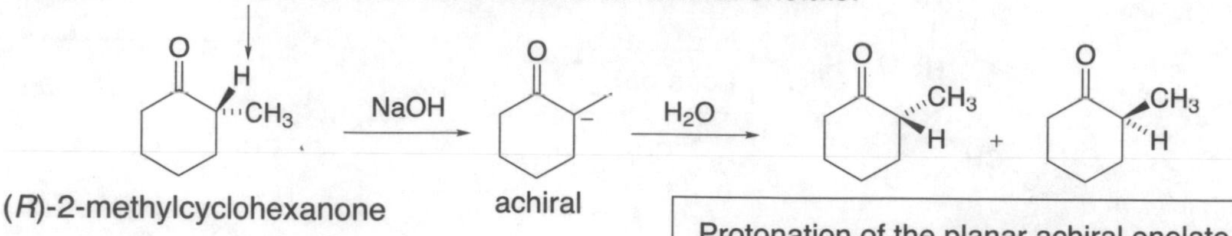

(R)-2-methylcyclohexanone   achiral

Protonation of the planar achiral enolate occurs with equal probability from two sides so a racemic mixture is formed. The racemic mixture is optically inactive.

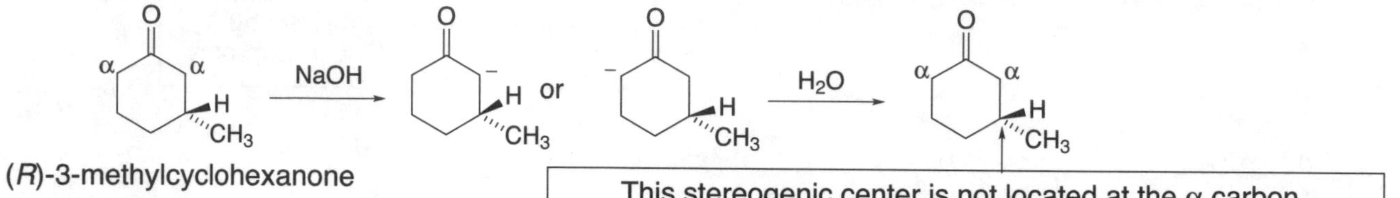

(R)-3-methylcyclohexanone

This stereogenic center is not located at the α carbon, so it is not deprotonated with base. Its configuration is retained in the product, and the product remains optically active.

23.13

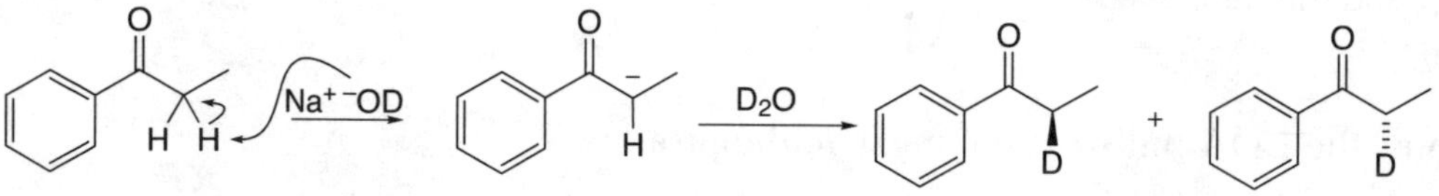

Removal of a proton from the α carbon forms an enolate, which can be protonated from either face to form two enantiomers in equal amounts.

23.14

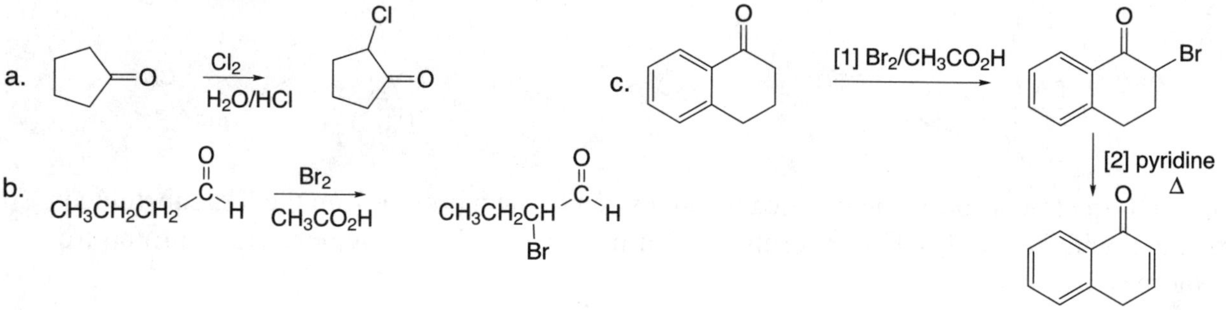

a.

b.

c.

23.15

a.

b.

c.

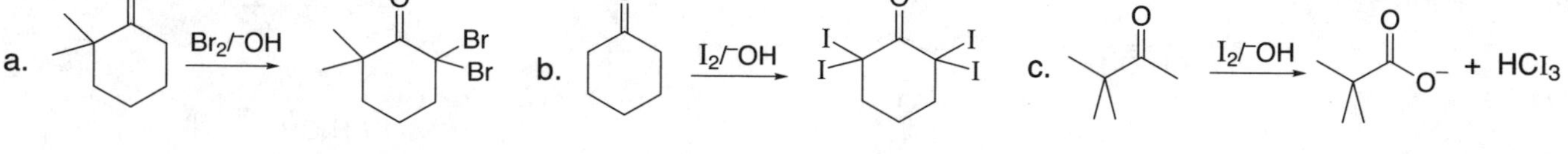

23.16

a. CH_3CH_2—C(=O)—CH_2CH_3 $\xrightarrow[\text{[2] CH}_3\text{CH}_2\text{I}]{\text{[1] LDA/THF}}$ CH_3CH_2—C(=O)—CHCH$_3$ (with CH$_2$CH$_3$ branch)

b. $\xrightarrow[\text{[2] CH}_3\text{CH}_2\text{I}]{\text{[1] LDA/THF}}$

c. $\xrightarrow[\text{[2] CH}_3\text{CH}_2\text{I}]{\text{[1] LDA/THF}}$

d. Ph—CH_2CH_2CN $\xrightarrow[\text{[2] CH}_3\text{CH}_2\text{I}]{\text{[1] LDA/THF}}$ Ph—CH_2CHCN (with CH$_2$CH$_3$ branch)

23.17

a. $\xrightarrow[\text{[2] CH}_3\text{I}]{\text{[1] LDA/THF}}$

b. $\xrightarrow[\text{[2] CH}_3\text{I}]{\text{[1] LDA/THF}}$ +

c. Ph—CH_2COOCH_3 $\xrightarrow[\text{[2] CH}_3\text{I}]{\text{[1] LDA/THF}}$ +

23.18

a. $\xrightarrow[\text{THF}]{\text{LDA}}$ $\xrightarrow{\text{CH}_3\text{CH}_2\text{Br}}$

b. $\xrightarrow[\text{CH}_3\text{CH}_2\text{OH}]{\text{NaOCH}_2\text{CH}_3}$ $\xrightarrow{\text{CH}_3\text{Br}}$ $\xrightarrow[\text{CH}_3\text{CH}_2\text{OH}]{\text{NaOCH}_2\text{CH}_3}$ $\xrightarrow{\text{CH}_3\text{Br}}$

c. $\xrightarrow[\text{THF}]{\text{LDA}}$ $\xrightarrow{\text{CH}_3\text{CH}_2\text{Br}}$ $\xrightarrow[\text{THF}]{\text{LDA}}$ $\xrightarrow{\text{CH}_3\text{I}}$

d. $\xrightarrow[\text{CH}_3\text{CH}_2\text{OH}]{\text{NaOCH}_2\text{CH}_3}$ $\xrightarrow{\text{CH}_3\text{Br}}$ $\xrightarrow[\text{THF}]{\text{LDA}}$ $\xrightarrow{\text{CH}_3\text{Br}}$

23.19

$\xrightarrow[\text{THF}]{\text{LDA}}$ $\xrightarrow{\text{CH}_3\text{I}}$ $\xrightarrow[\text{CH}_3\text{CO}_2\text{H}]{\text{Br}_2}$ $\xrightarrow[\Delta]{\text{pyridine}}$

A   B   C   α-methylene-γ-butyrolactone

23.20 Decarboxylation occurs only when a carboxyl group is bonded to the α C of another carbonyl group.

a.

YES

b.

NO

c.

YES

d.

NO

23.21

a. $CH_2(CO_2Et)_2$ $\xrightarrow[\text{[2]}]{\text{[1] NaOEt}}$ $\xrightarrow[\Delta]{H_3O^+}$ —CH_2–CH_2COOH

b. $CH_2(CO_2Et)_2$ $\xrightarrow[\text{[2] CH}_3\text{Br}]{\text{[1] NaOEt}}$ $\xrightarrow[\text{[2] CH}_3\text{Br}]{\text{[1] NaOEt}}$ $\xrightarrow[\Delta]{H_3O^+}$

23.22

a. $\longrightarrow$

b. $\longrightarrow$ —COOH

23.23 Locate the α C to the COOH group, and identify all of the alkyl groups bonded to it. These groups are from alkyl halides, and the remainder of the molecule is from diethyl malonate.

a. $\boxed{(CH_3)_2CHCH_2CH_2CH_2CH_2}$ —CH_2COOH
   |
   α

$CH_2(CO_2Et)_2$ $\xrightarrow[\text{[2] (CH}_3\text{)}_2\text{CHCH}_2\text{CH}_2\text{CH}_2\text{CH}_2\text{Br}]{\text{[1] NaOEt}}$ $\xrightarrow[\Delta]{H_3O^+}$ $(CH_3)_2CHCH_2CH_2CH_2CH_2CH_2COOH$

b.

$CH_2(CO_2Et)_2$ $\xrightarrow[\text{[2] CH}_3\text{CH}_2\text{CH}_2\text{Br}]{\text{[1] NaOEt}}$ $\xrightarrow[\text{[2] CH}_3\text{CH}_2\text{CH(CH}_3\text{)CH}_2\text{CH}_2\text{Br}]{\text{[1] NaOEt}}$ $\xrightarrow[\Delta]{H_3O^+}$

c. $\boxed{(CH_3CH_2CH_2)_2}CHCOOH$
   |
   α

$CH_2(CO_2Et)_2$ $\xrightarrow[\text{[2] CH}_3\text{CH}_2\text{CH}_2\text{Br}]{\text{[1] NaOEt}}$ $\xrightarrow[\text{[2] CH}_3\text{CH}_2\text{CH}_2\text{Br}]{\text{[1] NaOEt}}$ $\xrightarrow[\Delta]{H_3O^+}$ $(CH_3CH_2CH_2)_2CHCOOH$

23.24 The alkyl halide must be 1° or CH$_3$X since there is an S$_N$2 reaction.

a. (CH$_3$)$_3$C—CH$_2$COOH

(CH$_3$)$_3$CX
3° alkyl halide
(too crowded)

b.

aryl halide
(leaving group on an
sp^2 hybridized C)

Aryl halides are unreactive
in S$_N$2 reactions.

c. (CH$_3$)$_3$C—COOH (α)

This compound has 3 CH$_3$ groups on the α
carbon to the COOH. The malonic ester
synthesis can be used to prepare mono- and
disubstituted carboxylic acids only: RCH$_2$COOH
and R$_2$CHCOOH, but not R$_3$CCOOH.

23.25

a.

CH$_3$—C(=O)—CH$_2$CO$_2$Et

[1] NaOEt
[2] CH$_3$I
[3] H$_3$O$^+$/Δ

CH$_3$—C(=O)—CH$_2$CH$_3$

b.

CH$_3$—C(=O)—CH$_2$CO$_2$Et

[1] NaOEt
[2] CH$_3$CH$_2$CH$_2$Br
[3] NaOEt
[4] C$_6$H$_5$CH$_2$I
[5] H$_3$O$^+$/Δ

CH$_3$—C(=O)—CH—CH$_2$—(phenyl) with CH$_2$CH$_2$CH$_3$

23.26 Locate the α C. All alkyl groups on the α C come from alkyl halides, and the remainder of the
molecule comes from ethyl acetoacetate.

a. CH$_3$—C(=O)—CH$_2$CH$_2$CH$_3$ (α)

CH$_3$—C(=O)—CH$_2$—COOEt

[1] NaOEt
[2] CH$_3$CH$_2$Br

H$_3$O$^+$
Δ

CH$_3$—C(=O)—CH$_2$CH$_2$CH$_3$

b. CH$_3$—C(=O)—CH(CH$_2$CH$_3$)$_2$ (α)

CH$_3$—C(=O)—CH$_2$—COOEt

[1] NaOEt
[2] CH$_3$CH$_2$Br

[1] NaOEt
[2] CH$_3$CH$_2$Br

H$_3$O$^+$
Δ

CH$_3$—C(=O)—CH(CH$_2$CH$_3$)$_2$

c.

CH$_3$—C(=O)—CH$_2$—COOEt

[1] NaOEt
[2] CH$_3$CH$_2$Br

[1] NaOEt
[2] CH$_3$(CH$_2$)$_3$Br

H$_3$O$^+$
Δ

23.27

CH$_3$—C(=O)—CH$_2$CO$_2$Et + Br—CH$_2$CH$_2$—Br

NaOEt
(2 equiv)

CH$_3$—C(=O)—(cyclopropane with CO$_2$Et)   **X**

23.28

a.

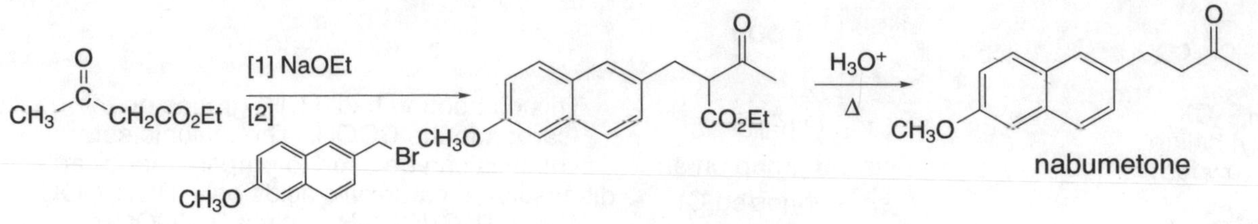

O
CH3
C
CH2CO2Et
[1] NaOEt
[2]
Br
CH3O
CH3O
O
CO2Et
H3O+
Δ
O
CH3O
nabumetone

b.

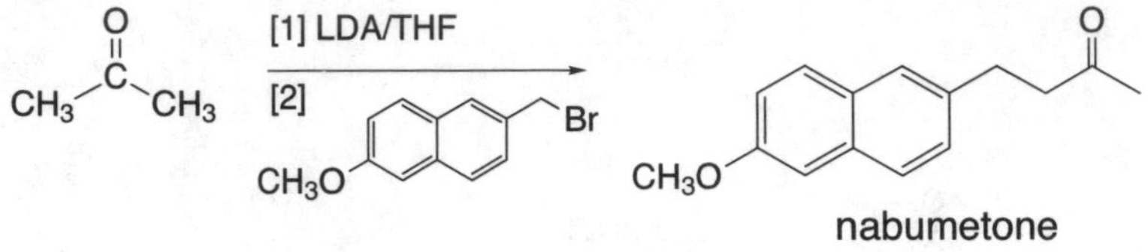

O
CH3
C
CH3
[1] LDA/THF
[2]
Br
CH3O
O
CH3O
nabumetone

23.29 Use the directions from Answer 23.1 to draw the enol tautomer(s). In cases where *E* and *Z* isomers can form, only one isomer is drawn.

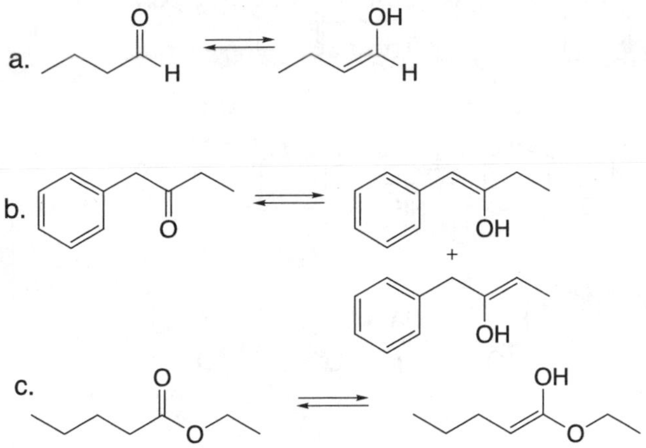

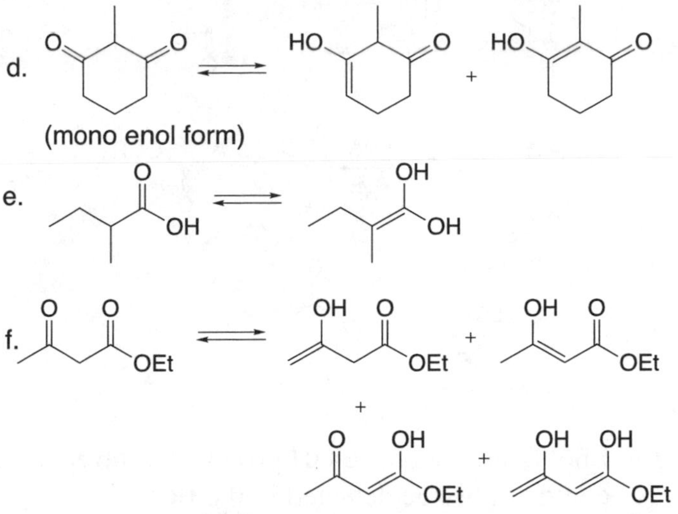

a.

b.

c.

d.

(mono enol form)

e.

f.

23.30

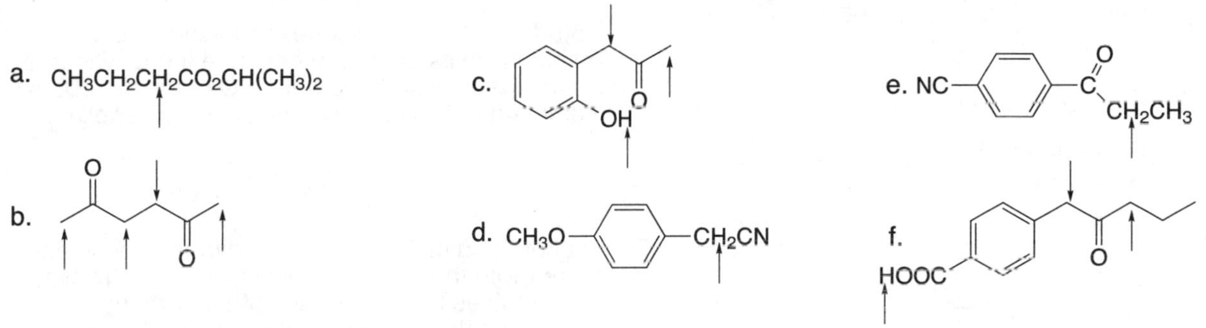

a. $CH_3CH_2CH_2CO_2CH(CH_3)_2$

b.

c.

d. CH_3O—⟨⟩—CH_2CN

e. NC—⟨⟩—$C(=O)CH_2CH_3$

f. $HOOC$—⟨⟩—

23.31

a. CH_3CH_2—$C(=O)OH$

H_a H_b H_c

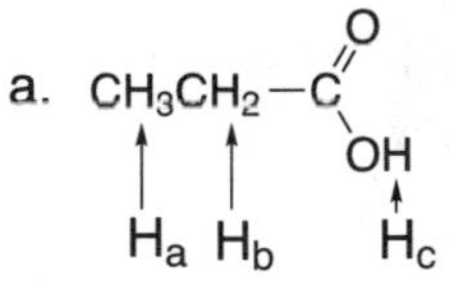

H_a is bonded to a CH_3 group = **least acidic**.
H_b is bonded to an α C = **intermediate acidity**.
H_c is bonded to O = **most acidic**.

c.

H_c
COOH
$H \leftarrow H_b$
$H \leftarrow H_a$

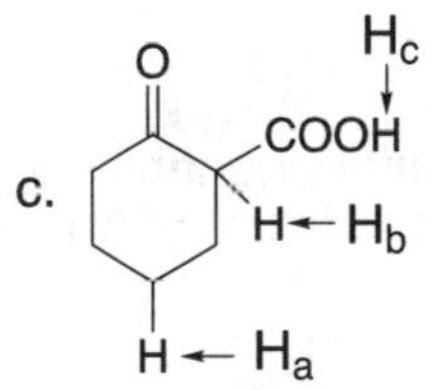

H_a is bonded to a CH_2 group = **!east acidic**.
H_b is bonded to an α C = **intermediate acidity**.
H_c is bonded to O = **most acidic**.

b.

H_a H_b H_c

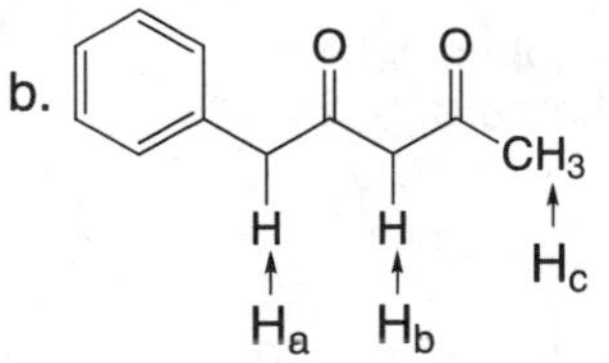

H_c is bonded to an α C = **least acidic**.
H_a is bonded to an α C , and is adjacent
    to a benzene ring = **intermediate acidity**.
H_b is bonded to an α C, between two
    C=O groups = **most acidic**.

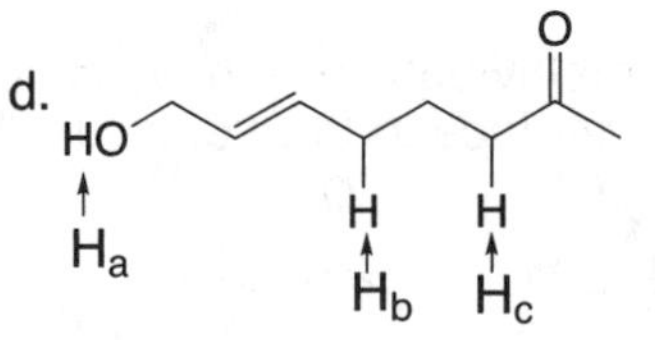

d.

HO
H_a H_b H_c

H_b is bonded to a CH_2 group = **least acidic**.
H_c is bonded to an α C = **intermediate acidity**.
H_a is bonded to O = **most acidic**.

23.32

a.

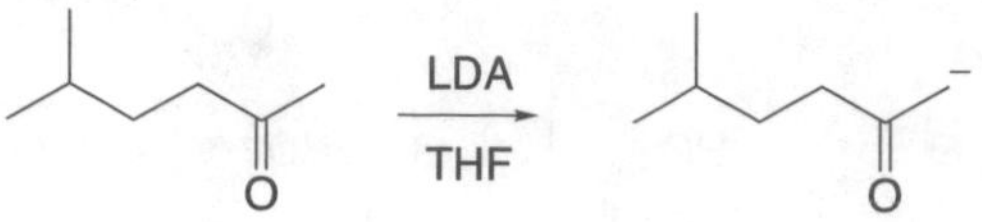

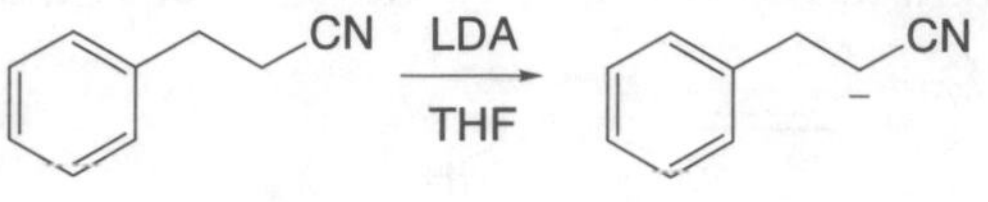

b.

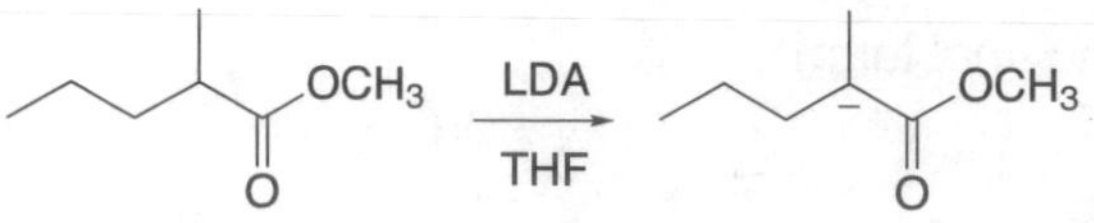

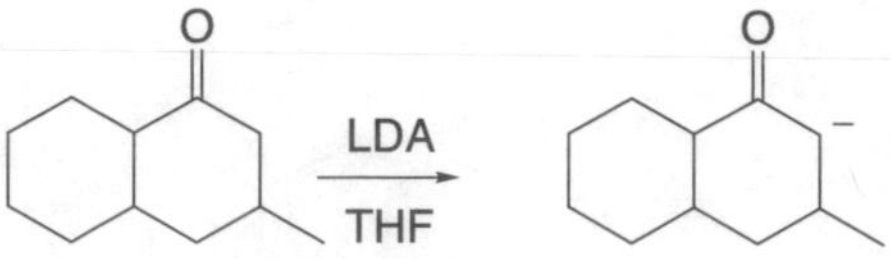

c.

d.

e.

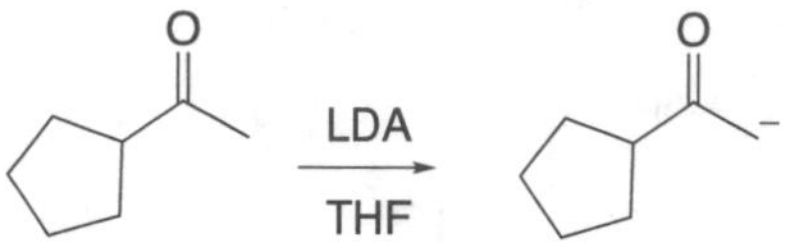

f.

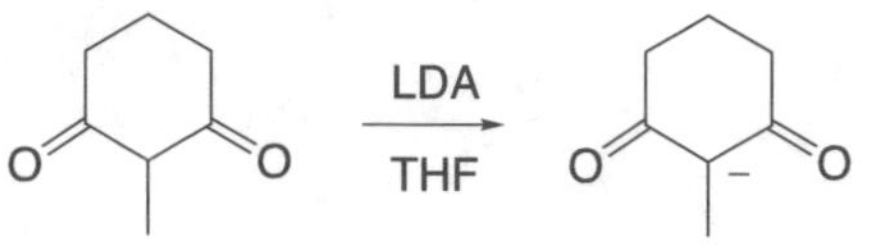

23.33 Enol tautomers have OH groups that give a broad OH absorption at 3600–3200 cm^{-1}, which could readily be detected in the IR.

23.34

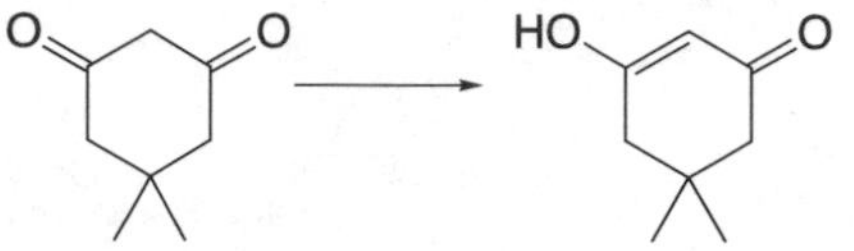

5,5-dimethyl-1,3-cyclohexanedione

5,5-Dimethyl-1,3-cyclohexanedione exists predominantly in its enol form because the C=C of the enol is conjugated with the other C=O of the dicarbonyl compound. Conjugation stabilizes this enol.

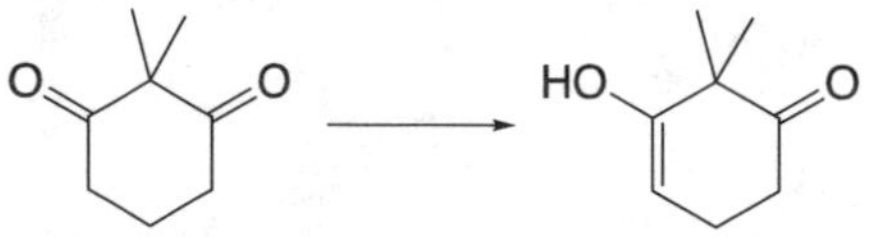

2,2-dimethyl-1,3-cyclohexanedione

The enol of 2,2-dimethyl-1,3-cyclohexanedione is not conjugated with the other carbonyl group. In this way it resembles the enol of any other carbonyl compound, and thus it is present in low concentration.

23.35

In the presence of acid, (R)-α-methylbutyrophenone enolizes to form an achiral enol. Protonation of the enol from either face forms an equal mixture of two enantiomers, making the solution optically inactive.

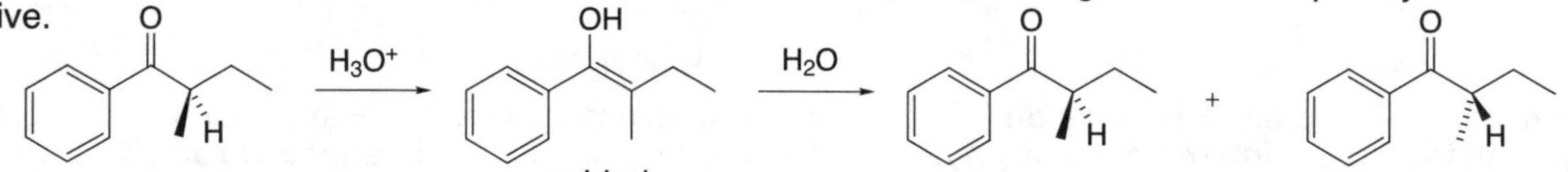

(R)-α-methylbutyrophenone

achiral

In the presence of base, (R)-α-methylbutyrophenone is deprotonated to form an achiral enolate, which can then be protonated from either face to form an optically inactive mixture of two enantiomers.

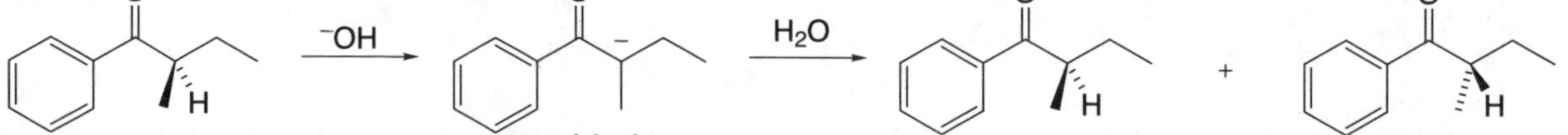

(R)-α-methylbutyrophenone

achiral

23.36

:O:
CH₃–C–O–CH₃ ⟶ :O: ⁻CH₂–C–O–CH₃ ⟶ :Ö⁻ CH₂=C–O–CH₃ ⟶ :Ö⁻ ⁻CH₂–C=O⁺–CH₃
ester

The O atom of the ester OR group donates electron density by a resonance effect. The resulting resonance structure keeps a negative charge on the less electronegative C end of the enolate. This destabilizes the resonance hybrid of the conjugate base, and makes the α H's of the ester less acidic.

:O:
CH₃–C–CH₃ ⟶ :O: ⁻CH₂–C–CH₃ ⟶ :Ö⁻ CH₂=C–CH₃
ketone

No additional resonance structures.

This structure, which places a negative charge on the O atom, is the major contributor to the hybrid, stabilizing it, and making the α H's of the ketone more acidic.

23.37

LDA reacts with the most acidic proton. If there is any H₂O present, it would immediately react with the base:

[(CH₃)₂CH]₂N̈: Li⁺ + H–ÖH ⟶ [(CH₃)₂CH]₂N̈H + Li⁺ ⁻OH

LDA

23.38

The most acidic proton in methyl acetate is on the α carbon. LDA removes this proton to form an enolate, which goes on to react with CH₃I to form CH₃CH₂COOCH₃.

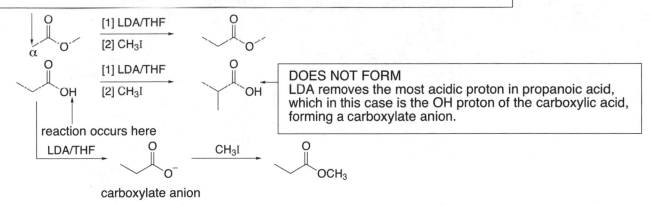

DOES NOT FORM
LDA removes the most acidic proton in propanoic acid, which in this case is the OH proton of the carboxylic acid, forming a carboxylate anion.

reaction occurs here

carboxylate anion

23.39 The mechanism of acid-catalyzed halogenation consists of two parts: **tautomerization** of the carbonyl compound to the enol form, and **reaction of the enol with halogen.**

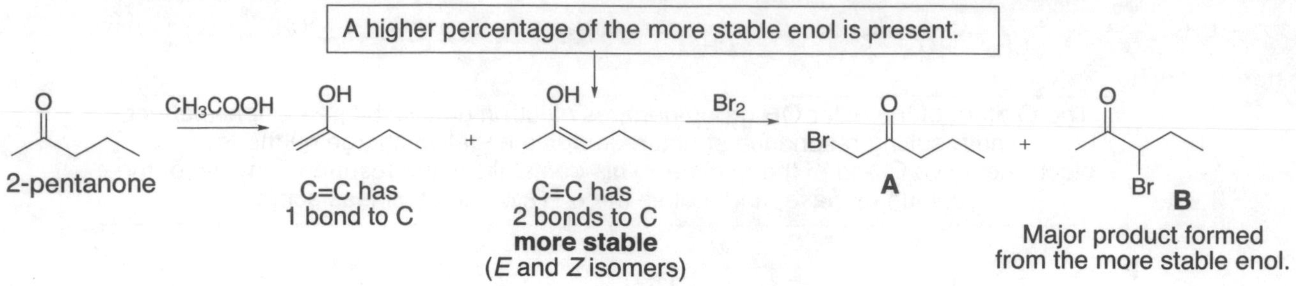
A higher percentage of the more stable enol is present.
CH3COOH
Br2
2-pentanone
C=C has
1 bond to C
C=C has
2 bonds to C
more stable
(E and Z isomers)
A
B
Major product formed
from the more stable enol.

23.40 • The mechanism of acid-catalyzed halogenation (Part a) consists of two parts: **tautomerization** of the carbonyl compound to the enol form, and **reaction of the enol with halogen**.
• In the haloform reaction (Part b), the three H's of the CH_3 group are successively replaced by X, to form an intermediate that is oxidatively cleaved with base.

a.

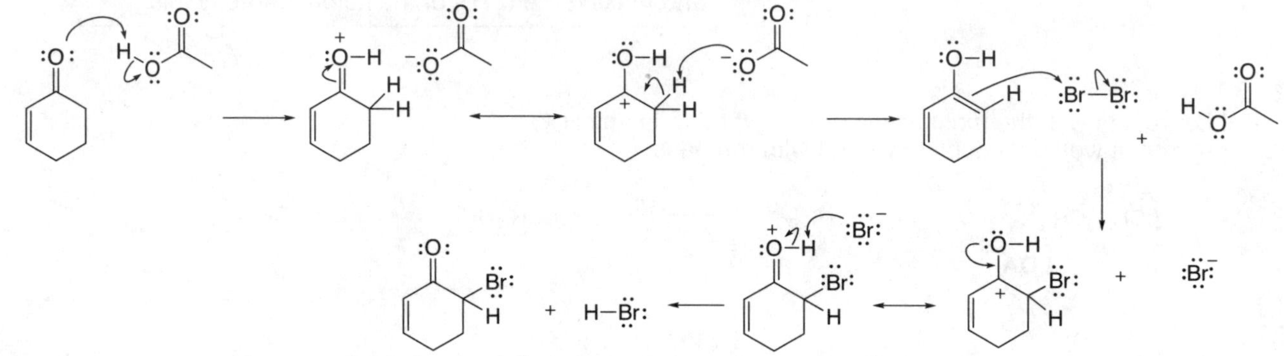

b.

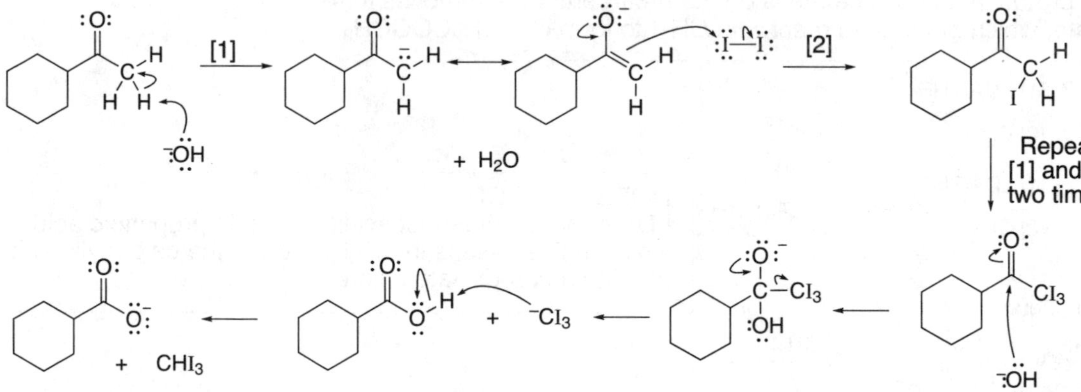
[1]
[2]
+ H2O
Repeat
[1] and [2]
two times.
+ CHI3

23.41 Use the directions from Answer 23.23.

a.

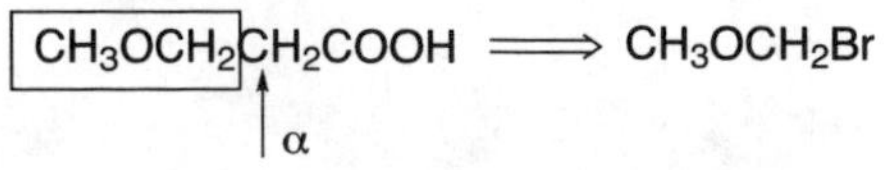
CH3OCH2CH2COOH
CH3OCH2Br
α

c.

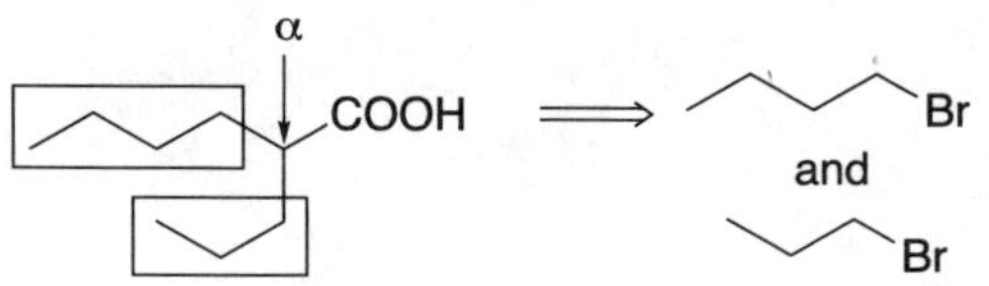
α
COOH
Br
and
Br

b.

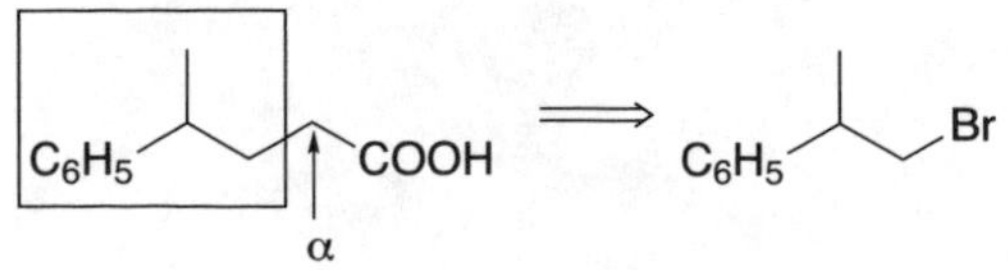
C6H5
COOH
α
C6H5
Br

23.42

a. $\boxed{CH_3CH_2CH_2CH_2CH_2}CH_2COOH$ $CH_2(CO_2Et)_2$ $\xrightarrow[\text{[2] } CH_3CH_2CH_2CH_2CH_2Br]{\text{[1] NaOEt}}$ $\xrightarrow[\Delta]{H_3O^+}$ $CH_3CH_2CH_2CH_2CH_2CH_2COOH$

α

b. $CH_2(CO_2Et)_2$ $\xrightarrow[\text{[2] } CH_3CH_2CH_2CH_2CH_2Br]{\text{[1] NaOEt}}$ $\xrightarrow[\text{[2] } CH_3Br]{\text{[1] NaOEt}}$ $\xrightarrow[\Delta]{H_3O^+}$

α

c. $CH_2(CO_2Et)_2$ $\xrightarrow[\text{[2] } (CH_3)_2CHCH_2CH_2CH_2CH_2Br]{\text{[1] NaOEt}}$ $\xrightarrow[\text{[2] } CH_3Br]{\text{[1] NaOEt}}$ $\xrightarrow[\Delta]{H_3O^+}$

α

23.43

a. $CH_2(CO_2Et)_2$ $\xrightarrow[\substack{\text{[2] } BrCH_2CH_2CH_2CH_2CH_2Br \\ \text{[3] NaOEt}}]{\text{[1] NaOEt}}$ $\xrightarrow[\Delta]{H_3O^+}$ —COOH

b. —COOH $\xrightarrow[\text{[2] } H_2O]{\text{[1] LiAlH}_4}$ —CH$_2$OH

[from (a)]

c. —COOH $\xrightarrow[H_2SO_4]{CH_3OH}$ —CO$_2$CH$_3$ $\xrightarrow[\text{[2] } H_2O]{\text{[1] } CH_3MgBr \text{ (2 equiv)}}$

[from (a)]

d. —COOH $\xrightarrow[H_2SO_4]{CH_3CH_2OH}$ —COOCH$_2$CH$_3$ $\xrightarrow[\text{[2] } CH_3I]{\text{[1] LDA}}$

[from (a)]

23.44

a. $\xrightarrow[\text{[2] } H_2O]{\text{[1] } Na^+ \, ^-CH(COOEt)_2}$ $CH_3\overset{\overset{\displaystyle HO}{|}}{C}HCH_2-CH(COOEt)_2$

nucleophilic attack here

b. $CH_2{=}O$ $\xrightarrow[\text{[2] } H_2O]{\text{[1] } Na^+ \, ^-CH(COOEt)_2}$ $HOCH_2CH(COOEt)_2$

c. $\xrightarrow[\text{[2] } H_2O]{\text{[1] } Na^+ \, ^-CH(COOEt)_2}$

d. $\xrightarrow[\text{[2] } H_2O]{\text{[1] } Na^+ \, ^-CH(COOEt)_2}$

     + CH_3COOH

23.45 Use the directions from Answer 23.26.

a. $CH_3\overset{\displaystyle O}{\underset{}{C}}CH_2COOEt$ $\xrightarrow[\text{[2] } Br\!\!\diagup\!\!\diagdown\!\!\diagup]{\text{[1] NaOEt}}$ $\xrightarrow[\Delta]{H_3O^+}$

b. $CH_3\overset{\displaystyle O}{\underset{}{C}}CH_2COOEt$ $\xrightarrow[\text{[2] } CH_3CH_2Br]{\text{[1] NaOEt}}$ $\xrightarrow[\text{[2] } Br\!\!\diagdown\!\!\bigcirc]{\text{[1] NaOEt}}$ $\xrightarrow[\Delta]{H_3O^+}$

c.

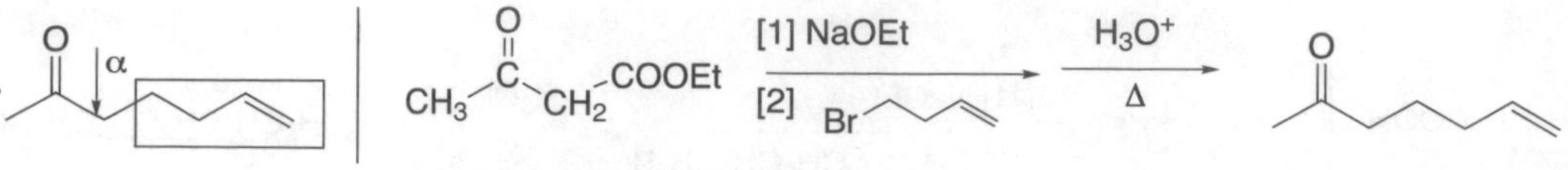

[1] NaOEt
[2] Br

H₃O⁺
Δ

d.

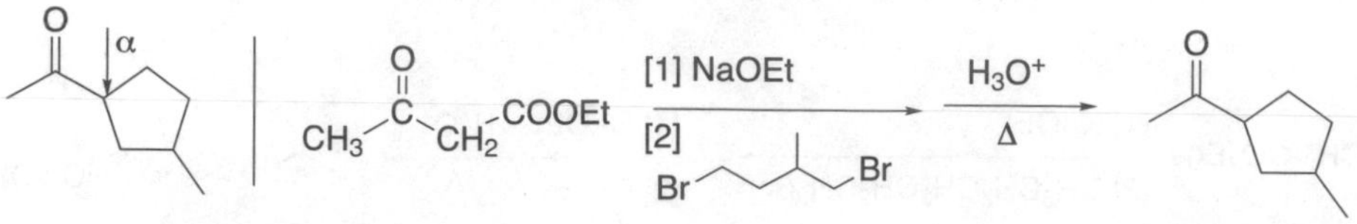

[1] NaOEt
[2]
Br   Br

H₃O⁺
Δ

23.46

a. 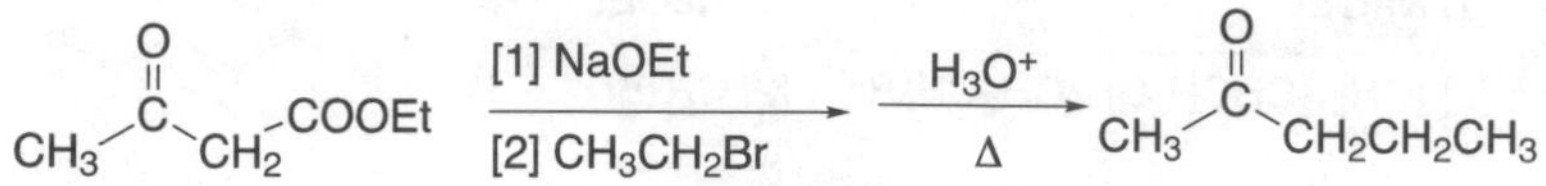   CH₃—C—CH₂—COOEt

[1] NaOEt
[2] CH₃CH₂Br

H₃O⁺
Δ

CH₃—C—CH₂CH₂CH₃

b. 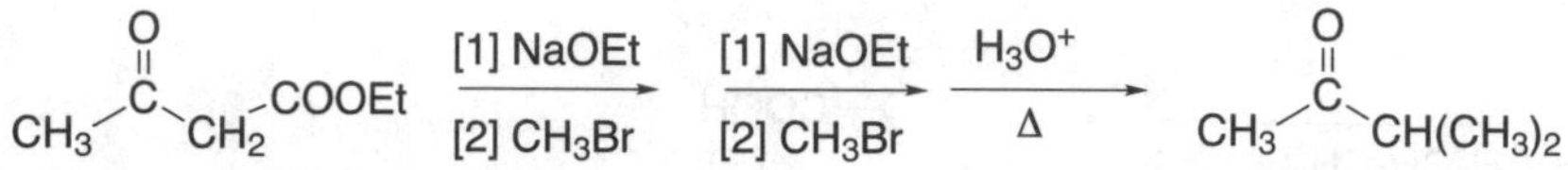   CH₃—C—CH₂—COOEt

[1] NaOEt
[2] CH₃Br

[1] NaOEt
[2] CH₃Br

H₃O⁺
Δ

CH₃—C—CH(CH₃)₂

c.   CH₃—C—CH(CH₃)₂
[from (b)]

LDA
THF

CH₂—C—CH(CH₃)₂

CH₃I

CH₃CH₂—C—CH(CH₃)₂

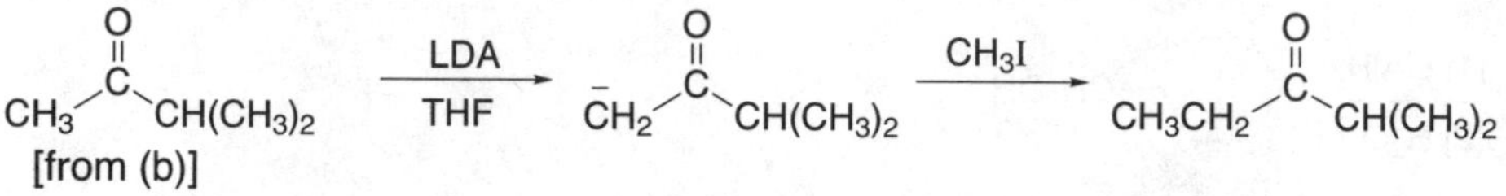

d. 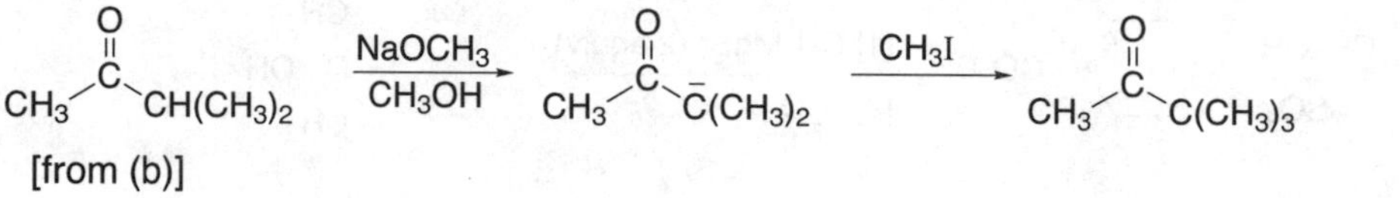   CH₃—C—CH(CH₃)₂
[from (b)]

NaOCH₃
CH₃OH

CH₃—C—C(CH₃)₂

CH₃I

CH₃—C—C(CH₃)₃

23.47

a.

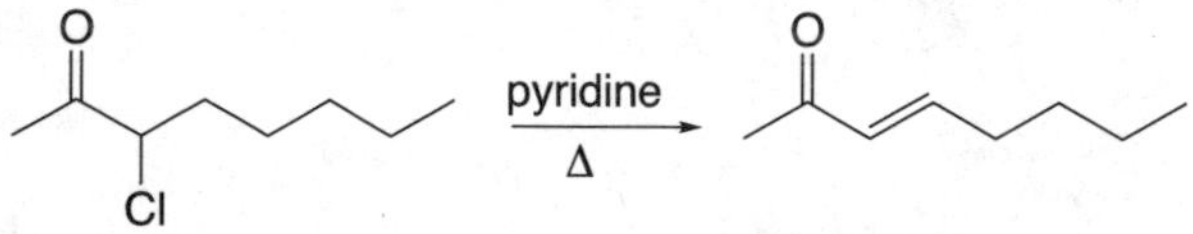

Cl

pyridine
Δ

b.

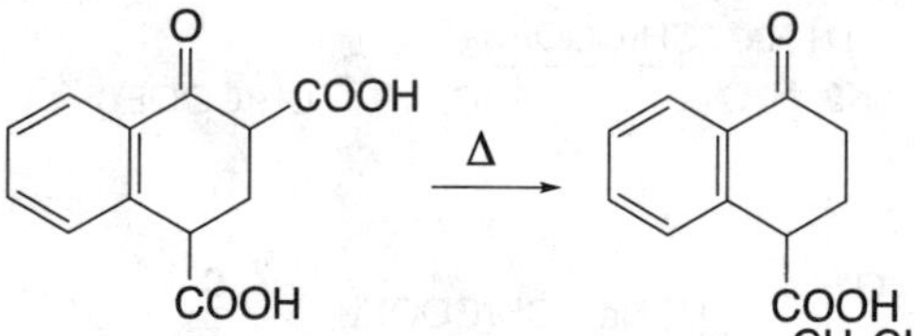

COOH

COOH

Δ

COOH

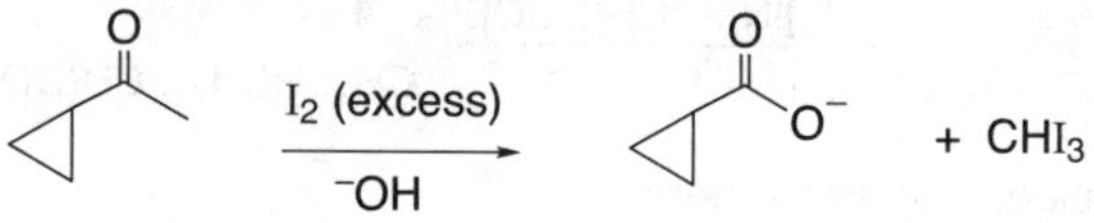

c. 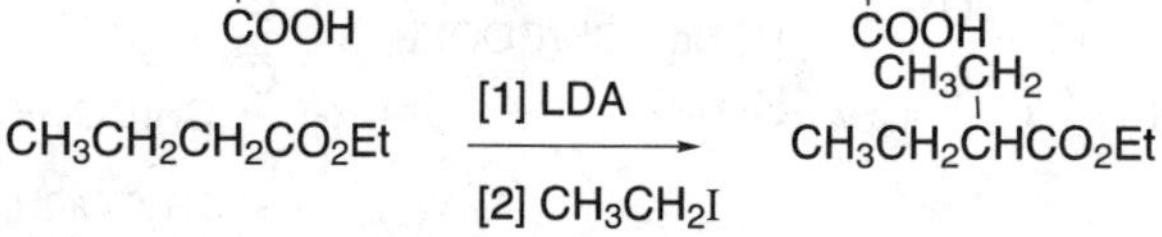 CH₃CH₂CH₂CO₂Et

[1] LDA
[2] CH₃CH₂I

CH₃CH₂
CH₃CH₂CHCO₂Et

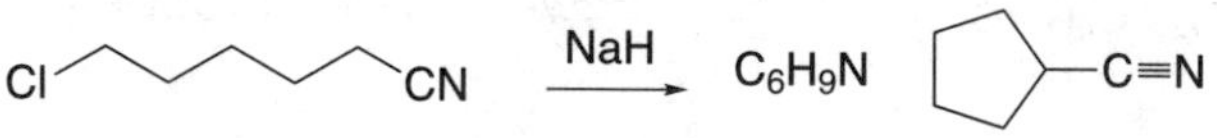

d.

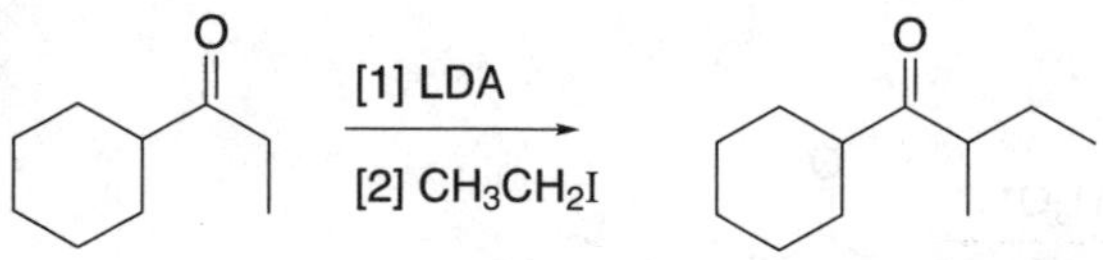

[1] LDA
[2] CH₃CH₂I

e.

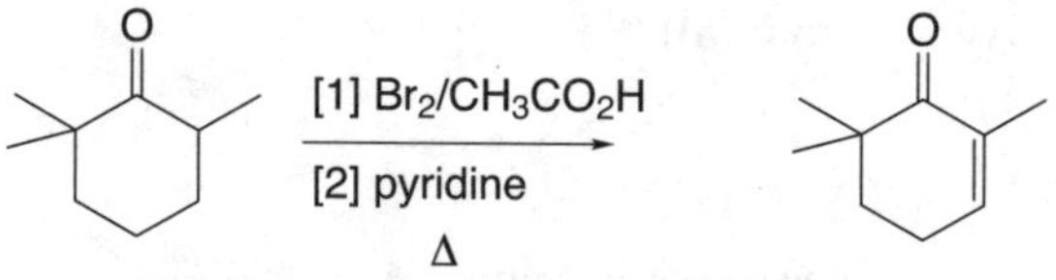

[1] Br₂/CH₃CO₂H
[2] pyridine
Δ

f.

I₂ (excess)
⁻OH

O⁻   + CHI₃

g. Cl   CN

NaH
C₆H₉N

C≡N

h.

Br₂ (excess)
⁻OH

Br Br

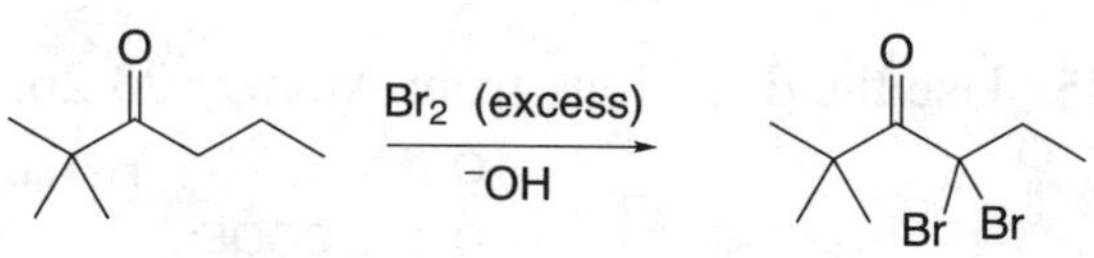

23.48

a.

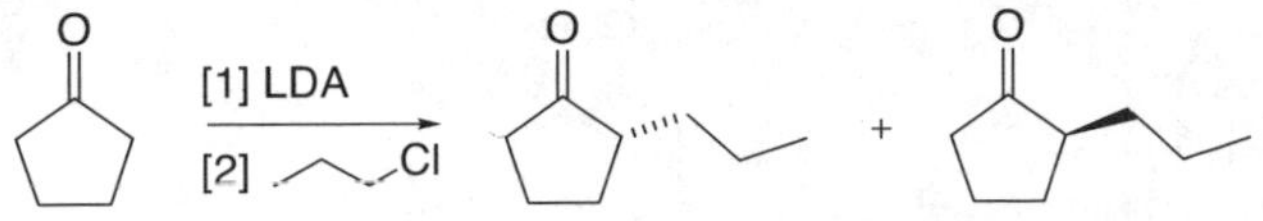

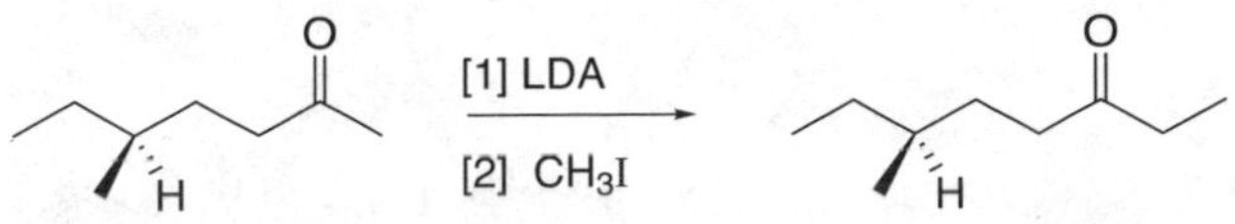

b.

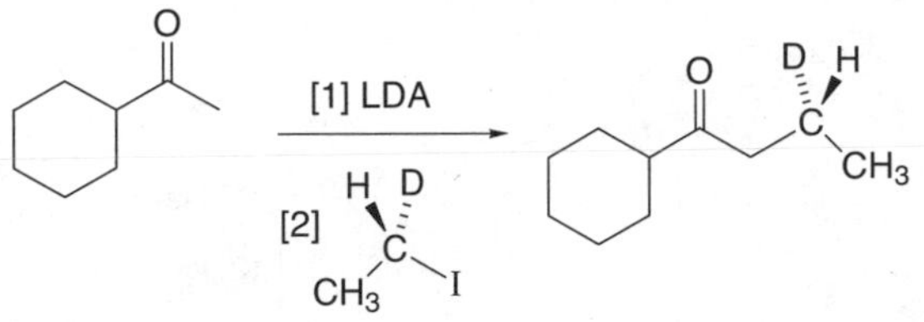

c.

23.49

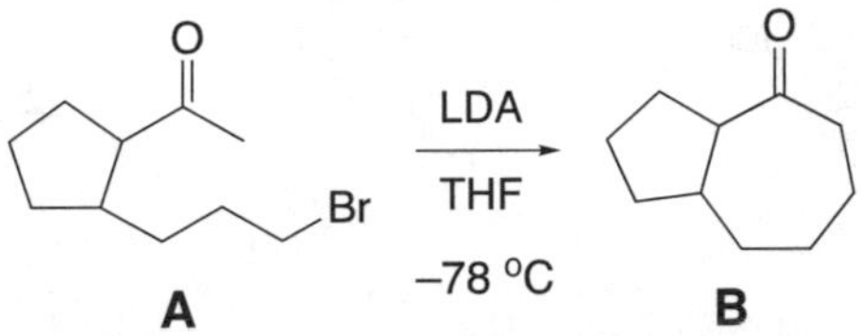

23.50

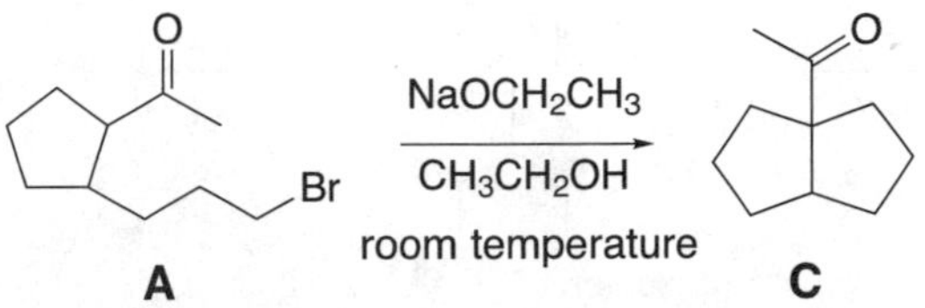

a.

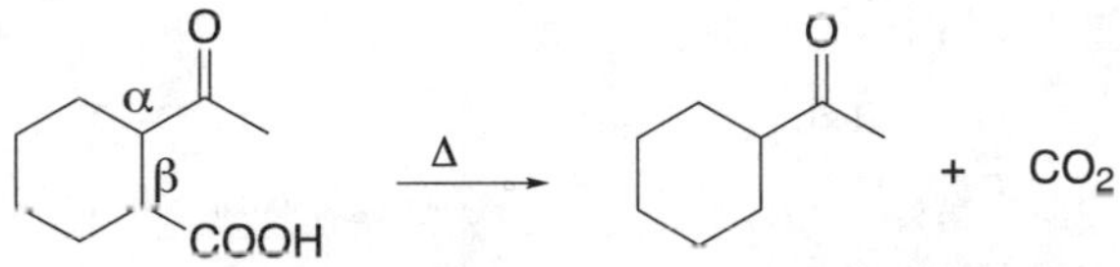

In order for decarboxylation to occur readily, the COOH group must be bonded to the α C of another carbonyl group. In this case, it is bonded to the β carbon.

b. $CH_2(CO_2Et)_2$ $\xrightarrow[\text{[2] } (CH_3CH_2)_3CBr]{\text{[1] NaOEt}}$ $(CH_3CH_2)_3CCH(CO_2Et)_2$

The 3° alkyl halide is too crowded to react with the strong nucleophile by an S$_N$2 mechanism.

c.

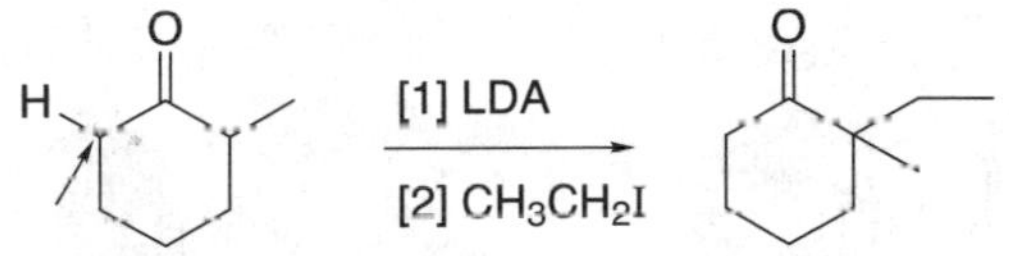

LDA removes a H from the less substituted C, forming the kinetic enolate. This product is from the thermodynamic enolate, which gives substitution on the more substituted α C.

23.51

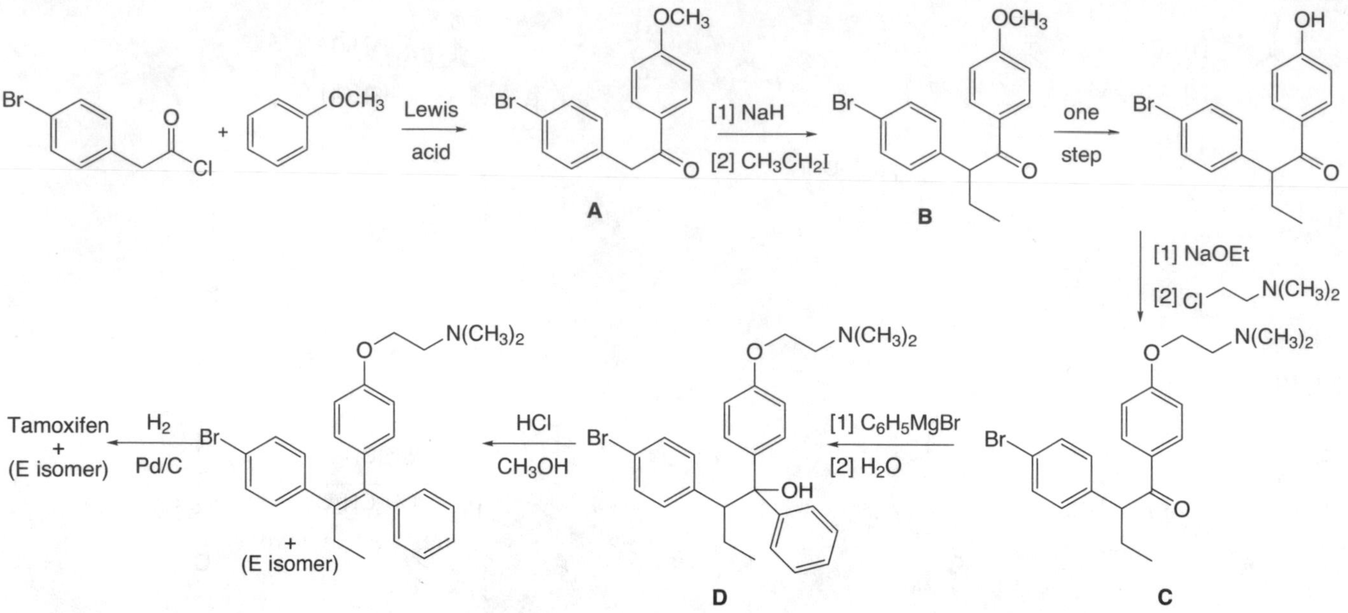

23.52

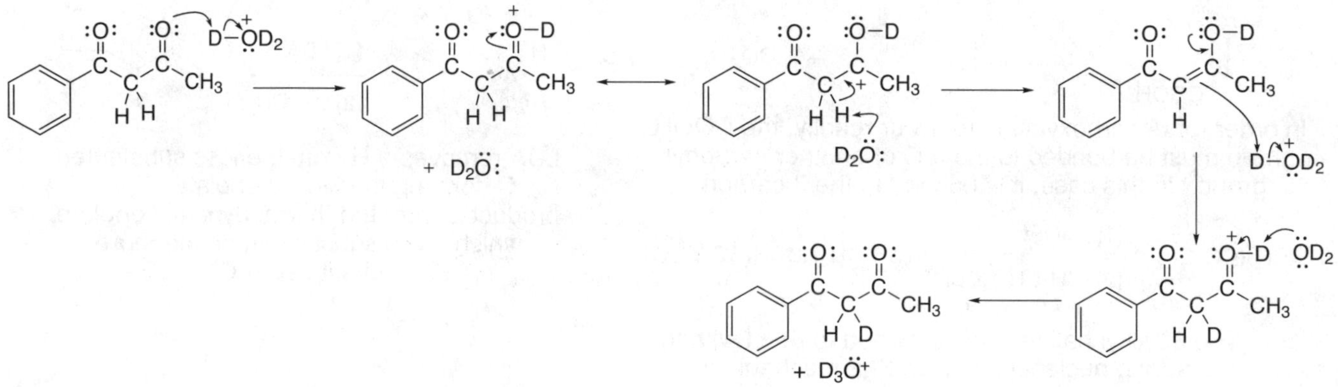

23.53

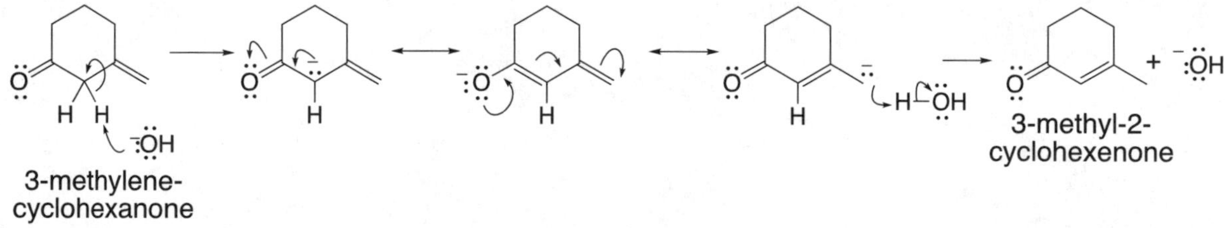

3-methylene-
cyclohexanone

3-methyl-2-
cyclohexenone

23.54

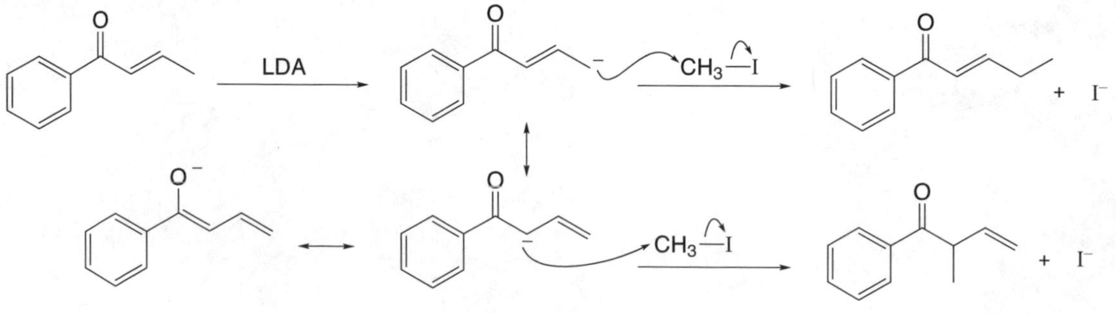

23.55

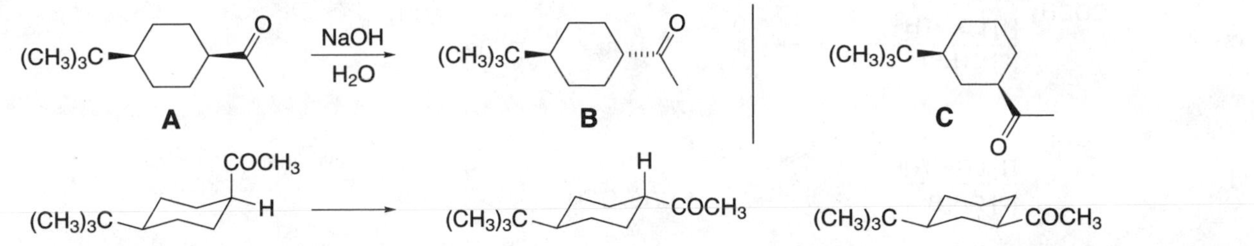

One axial and one equatorial group, **less stable**

Both groups are equatorial. **more stable**

Both groups are equatorial.

This isomerization will occur since it makes a more stable compound.

Compound **C** will not isomerize since it already has the more stable arrangement of substituents.

Isomerization occurs by way of an intermediate enolate, which can be protonated to either re-form **A**, or give **B**. Since **B** has two large groups equatorial, it is favored at equilibrium.

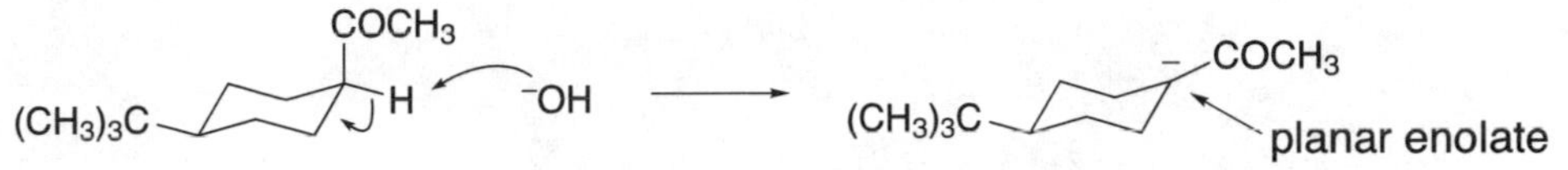

23.56

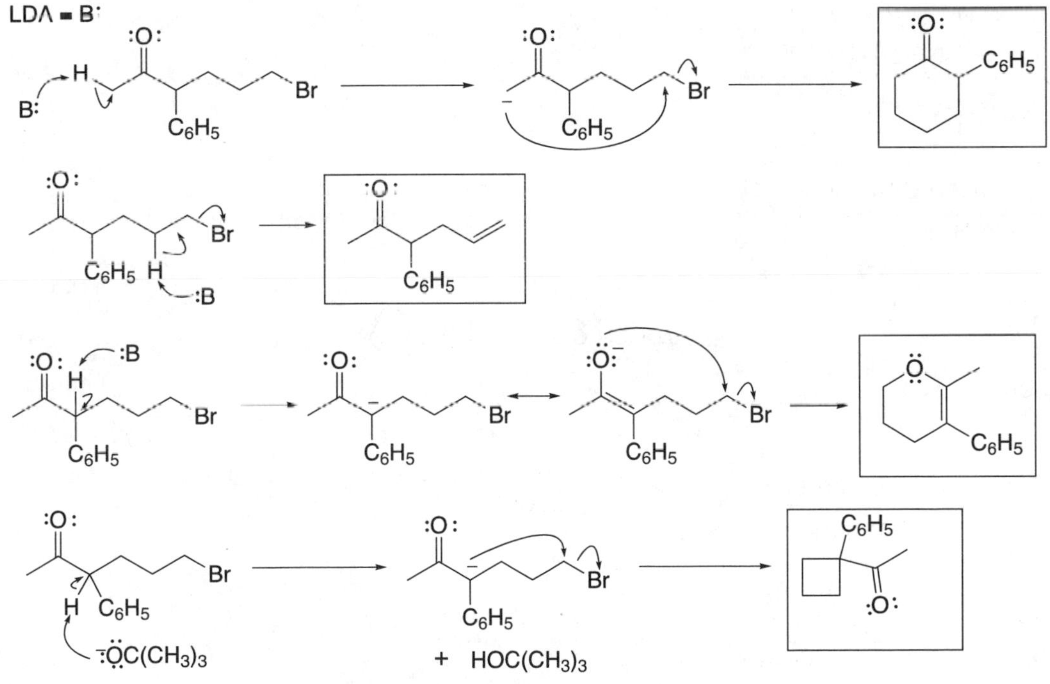

LDA = B:

This reaction occurs with both bases [LDA or KOC(CH₃)₃].

23.57

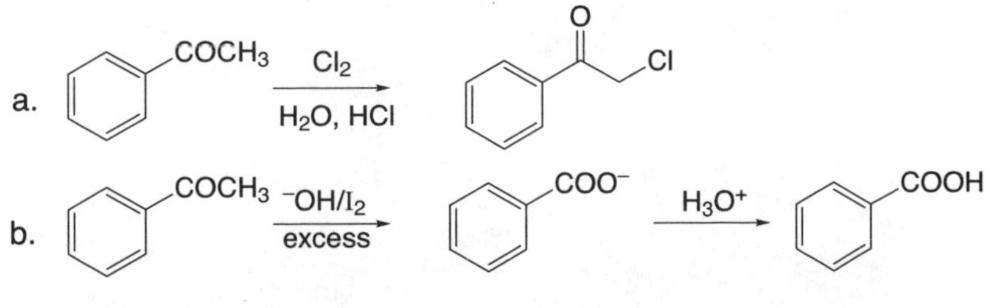

a.

b.

c.

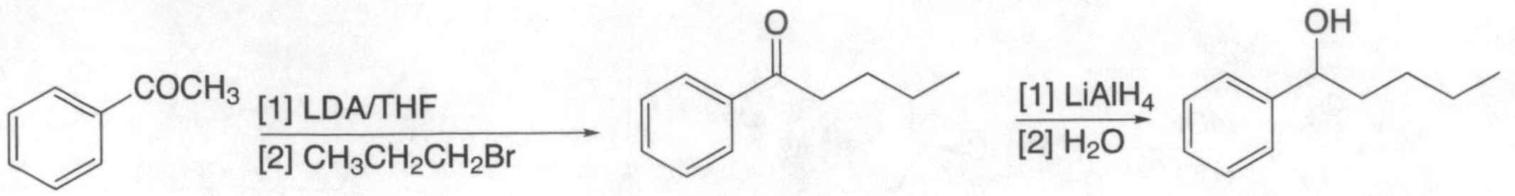

d.

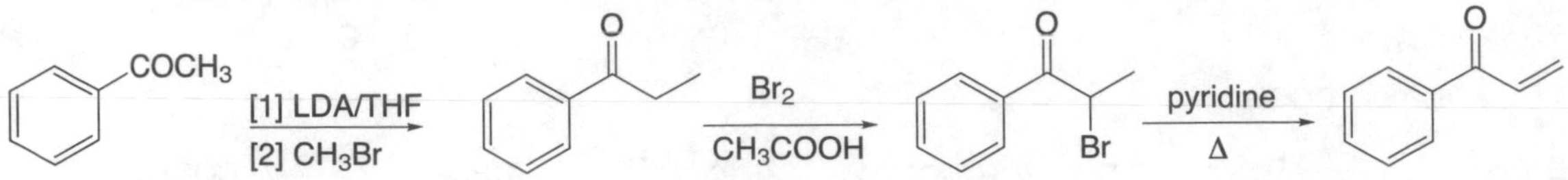

23.58

a.

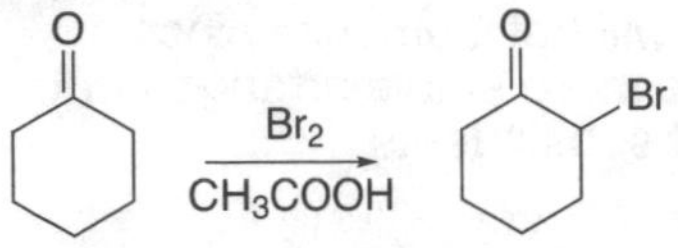

b.

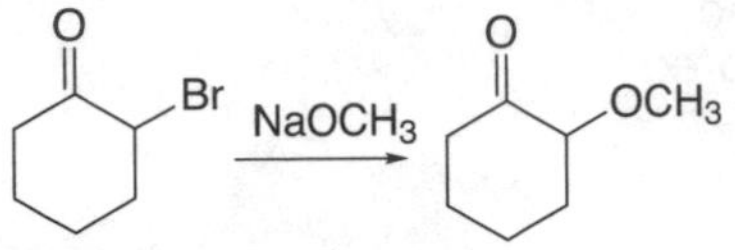

[from (a)]

c.

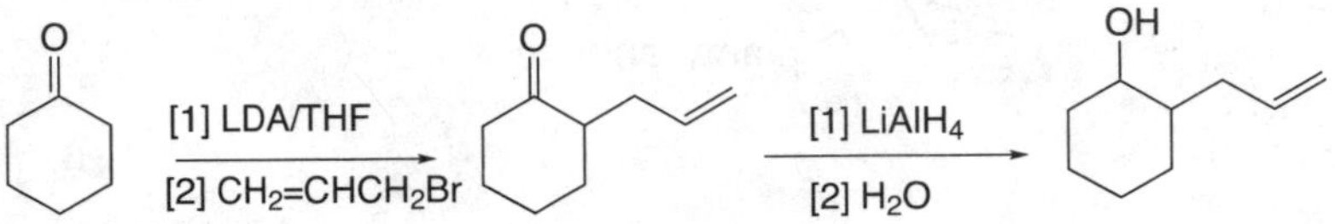

d.

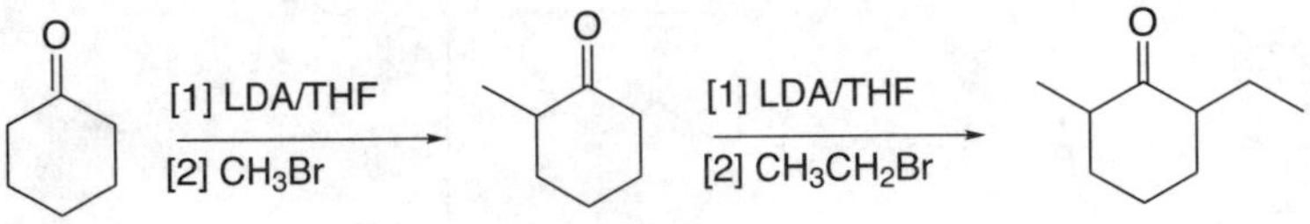

e.

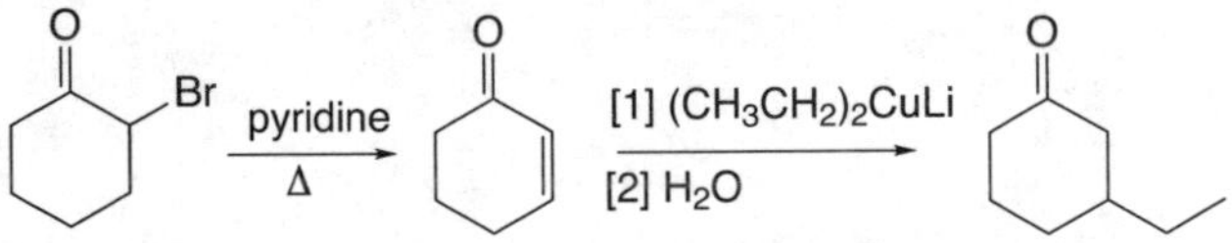

[from (a)]

f.

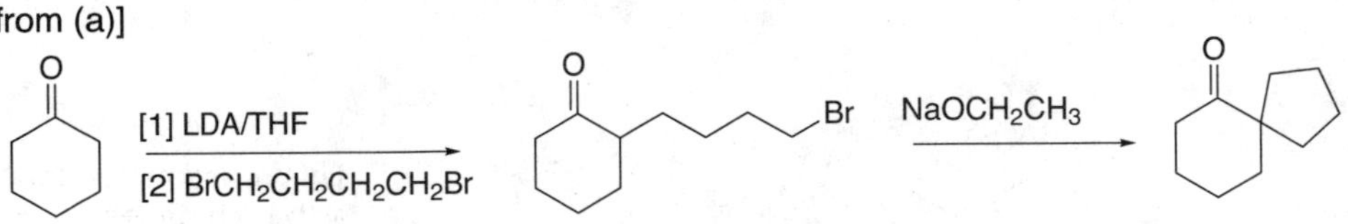

g.

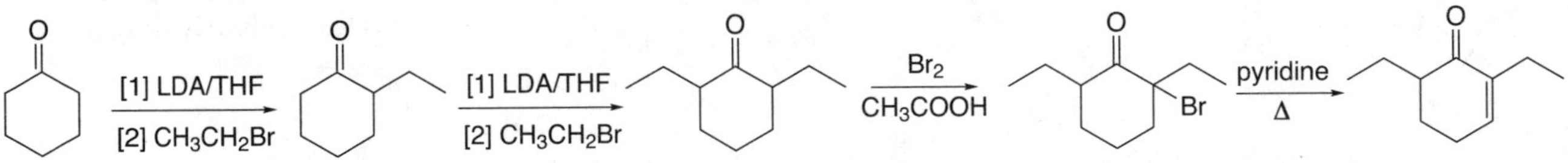

h.

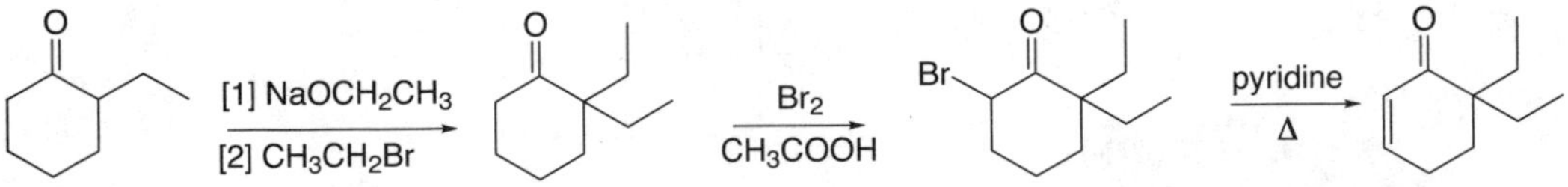

[from (g)]

23.59

a.

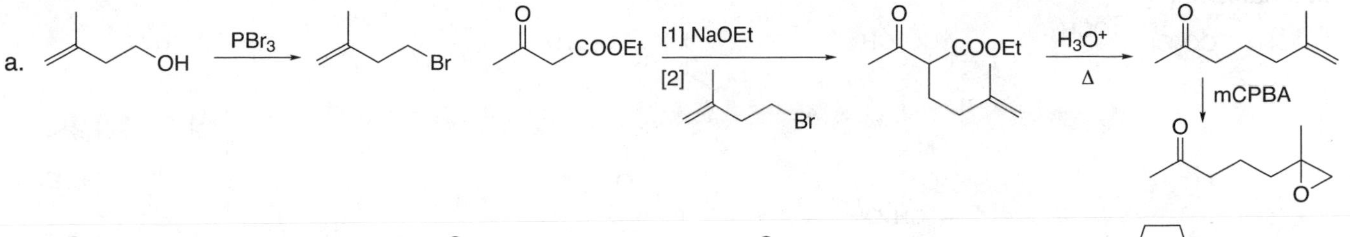

b.

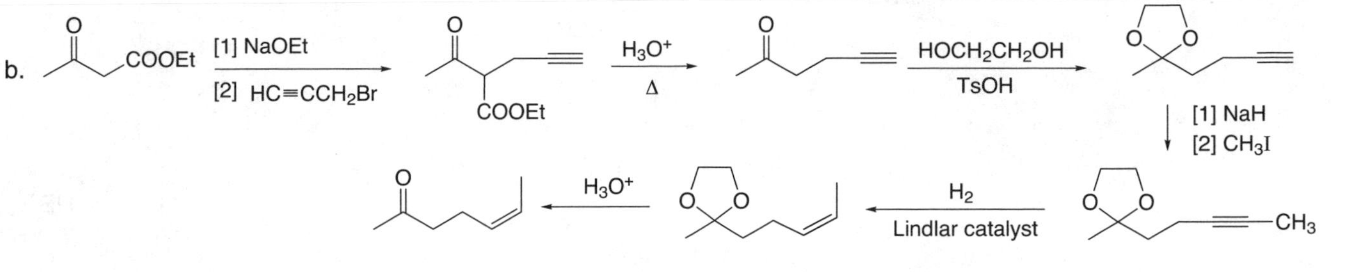

23.60

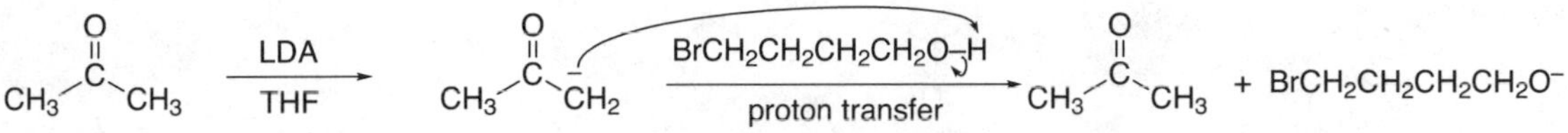

The OH group donates a proton to the enolate, which acts as a base.

To synthesize the desired product, a protecting group is needed:

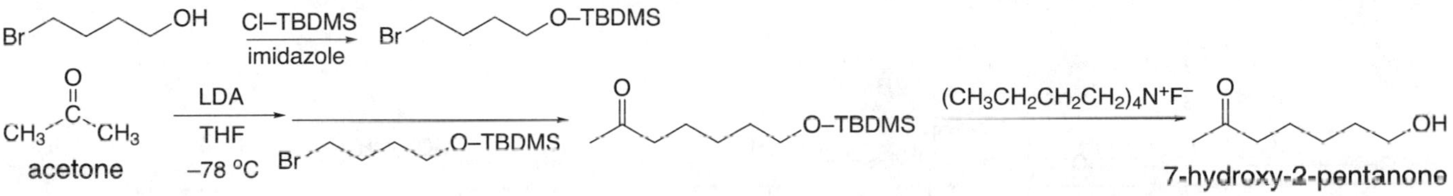

23.61

> Removal of H_a with base does not generate an anion that can delocalize onto the carbonyl O atom, whereas removal of H_b generates an enolate that is delocalized on O.

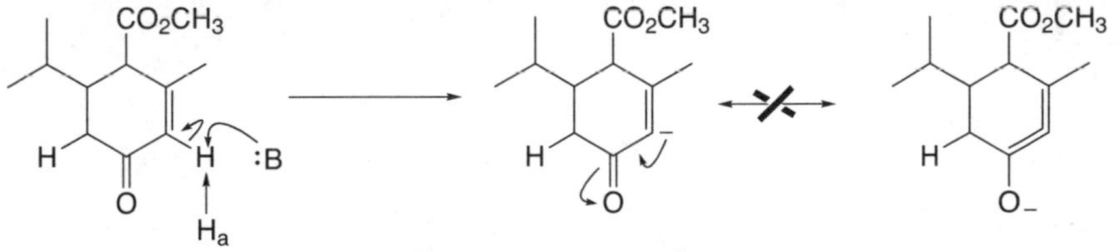

Delocalization of this sort can't occur by removal of H_a, making H_a less acidic.

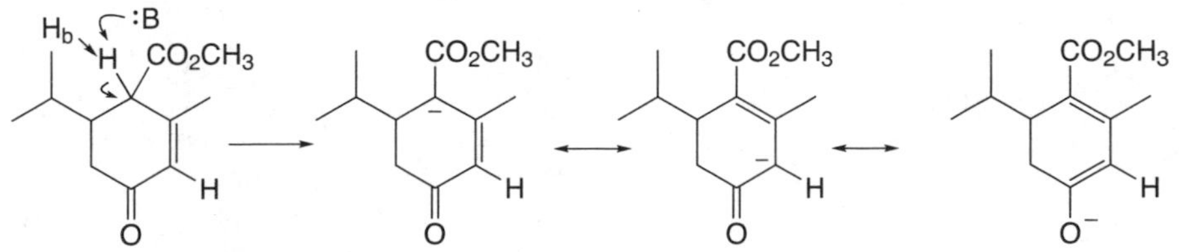

Removal of H_b gives an anion that is resonance stabilized so H_b is more acidic.

Mechanism:

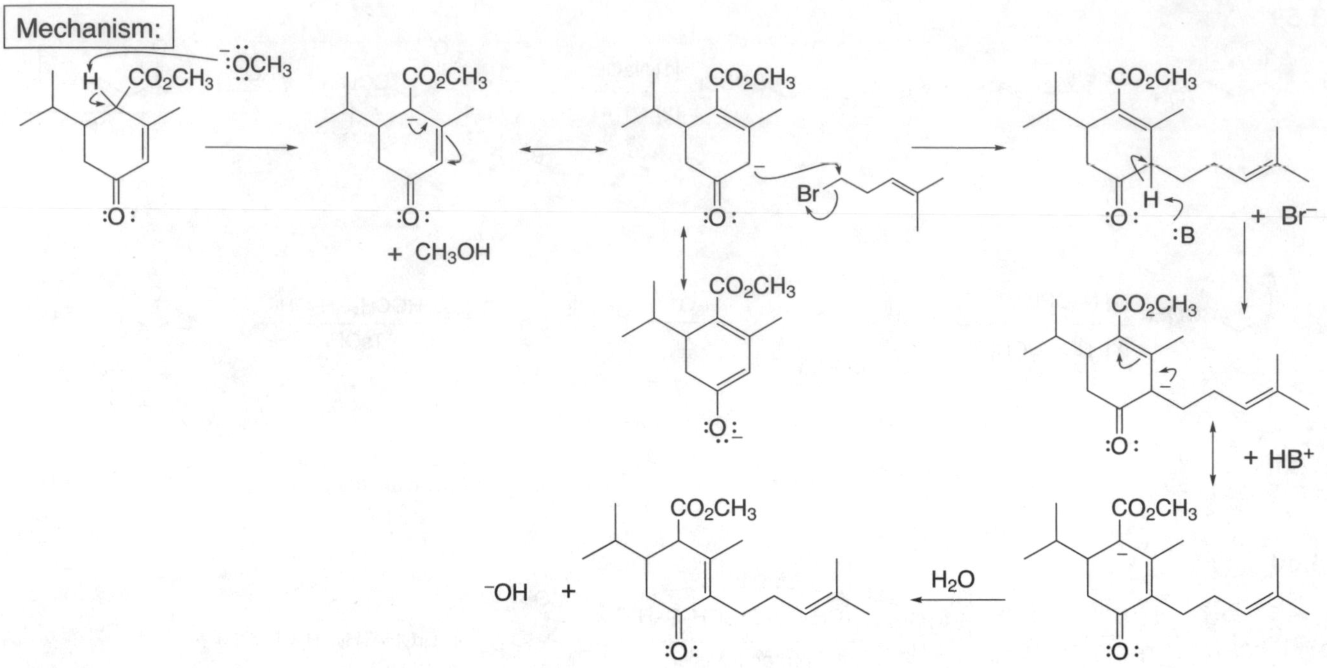

23.62

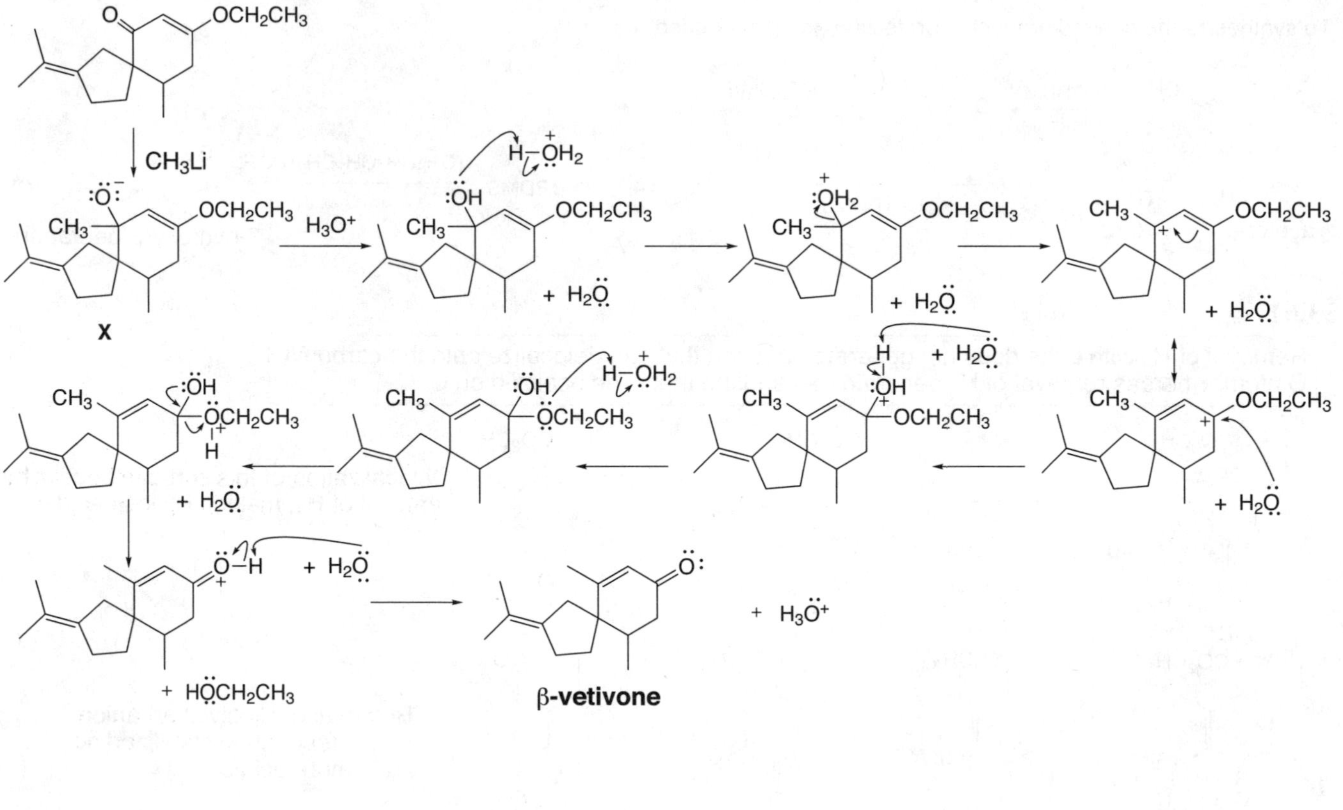

β-vetivone

Chapter 24: Carbonyl Condensation Reactions

♦ The four major carbonyl condensation reactions

Reaction type	Reaction
[1] Aldol reaction (24.1)	
[2] Claisen reaction (24.5)	
[3] Michael reaction (24.8)	
[4] Robinson annulation (24.9)	

♦ Useful variations

[1] Directed aldol reaction (24.3)

[2] Intramolecular aldol reaction (24.4)

[a] With 1,4-dicarbonyl compounds:

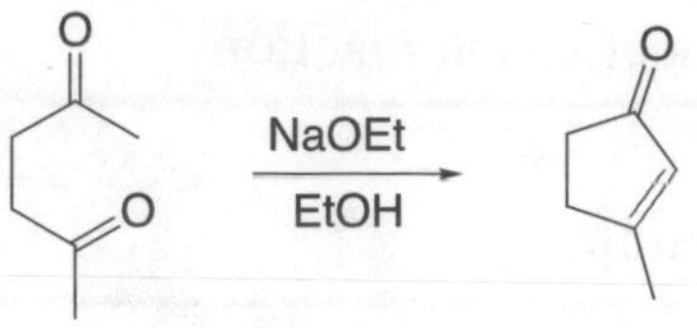

NaOEt / EtOH

[b] With 1,5-dicarbonyl compounds:

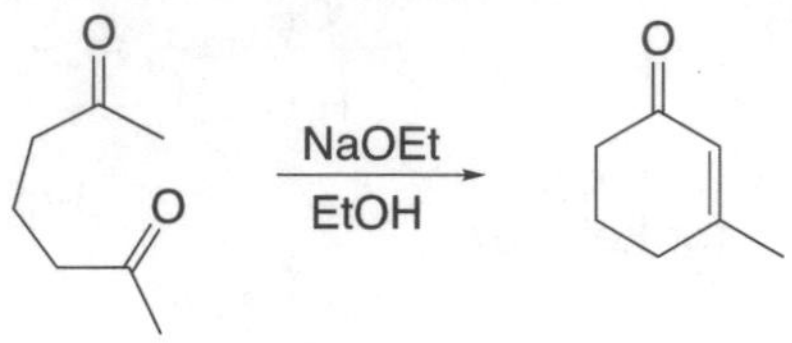

NaOEt / EtOH

[3] Dieckmann reaction (24.7)

[a] With 1,6-diesters:

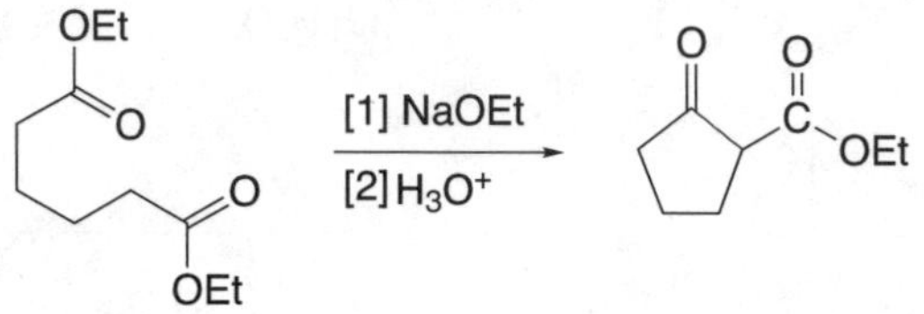

[1] NaOEt
[2] H$_3$O$^+$

[b] With 1,7-diesters:

[1] NaOEt
[2] H$_3$O$^+$

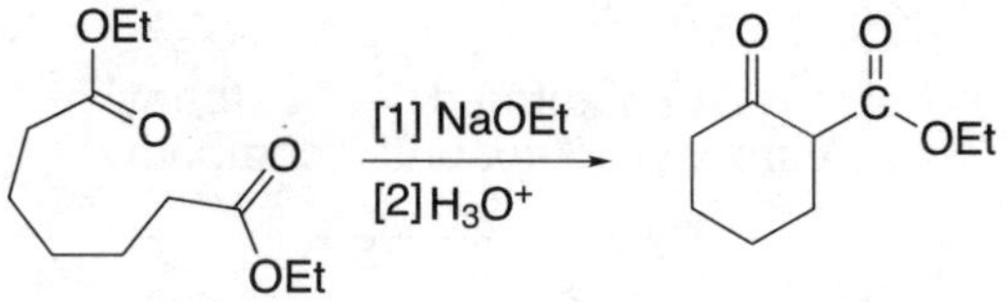

24.1

a.

b. (CH₃)₃CCH₂CHO ⟶ (CH₃)₃CCH₂C—C—CHO

c. CH₃—C—CH₃ ⟶ CH₃—C—CH₂—C—CH₃

d.

24.2

a.

b.

c.

24.3

Locate the α and β C's to the carbonyl group, and break the molecule into two halves at this bond. The α C and all of the atoms bonded to it belong to one carbonyl component. The β C and all the atoms bonded to it belong to the other carbonyl component.

a.

b.

c.

d.

24.4

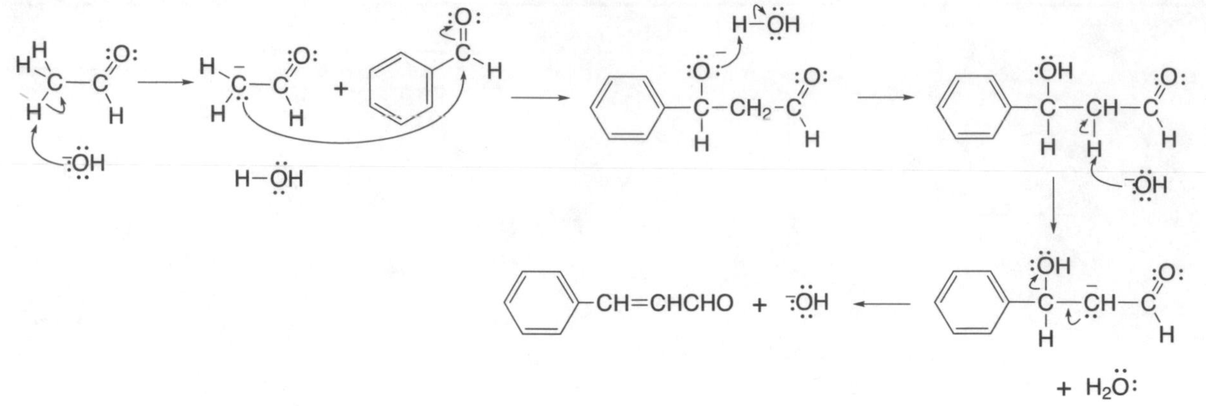

-CH=CHCHO + :ÖH
+ H₂Ö:

24.5

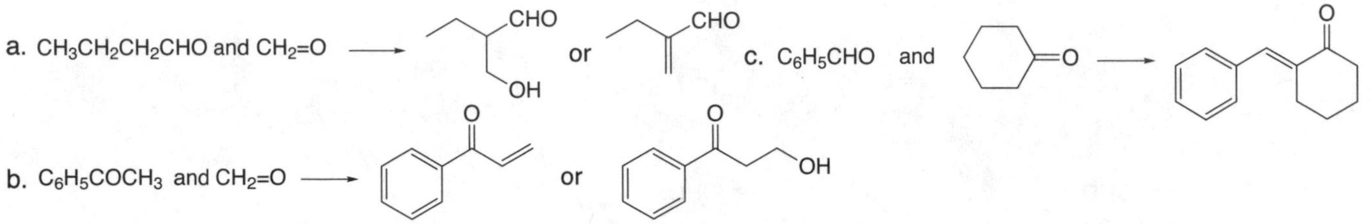

a. CH₃CH₂CH₂CHO and CH₂=O
CHO
OH
or
CHO
b. C₆H₅COCH₃ and CH₂=O
or
OH
c. C₆H₅CHO and

24.6

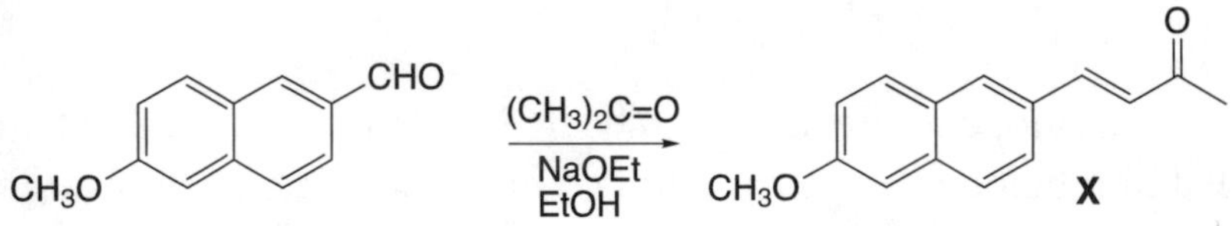

CHO
CH₃O
(CH₃)₂C=O
NaOEt
EtOH
CH₃O
X

24.7

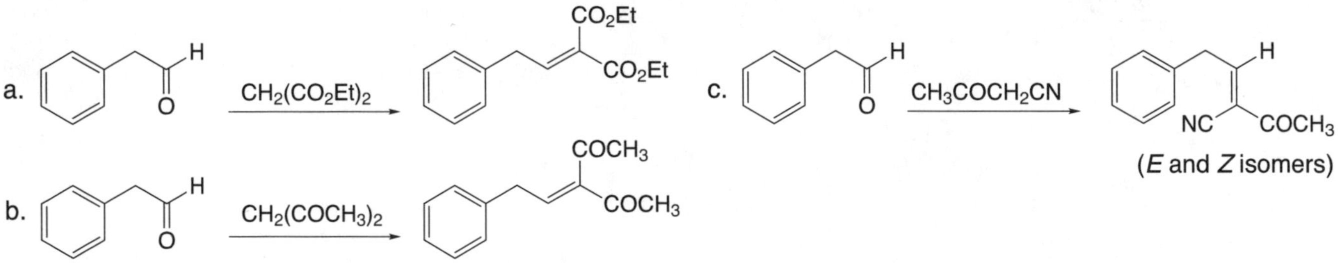

a.
CH₂(CO₂Et)₂
CO₂Et
CO₂Et
b.
CH₂(COCH₃)₂
COCH₃
COCH₃
c.
CH₃COCH₂CN
NC
COCH₃
(E and Z isomers)

24.8 Find the α and β C's to the carbonyl group and break the bond between them.

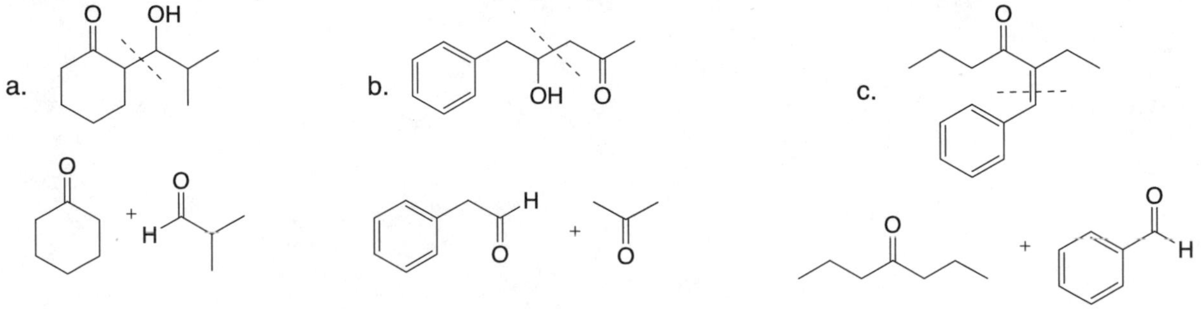

a.
OH
b.
OH
c.

24.9

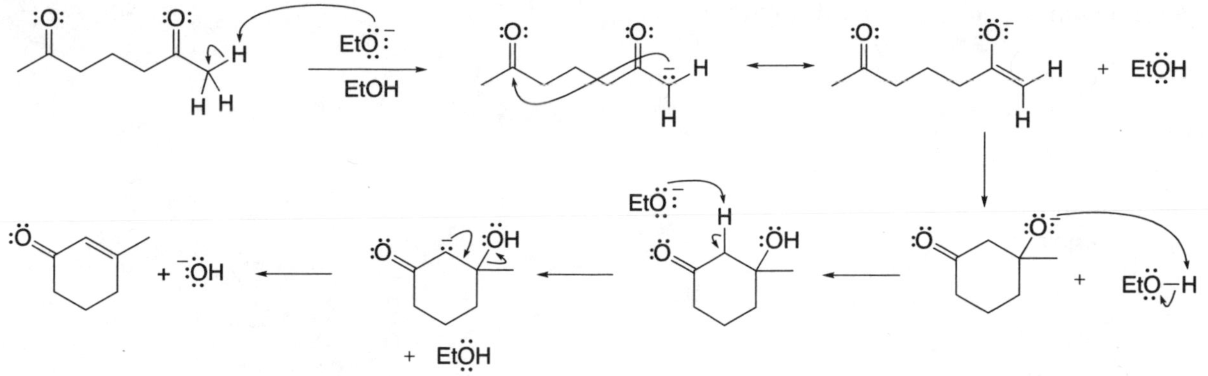

24.10

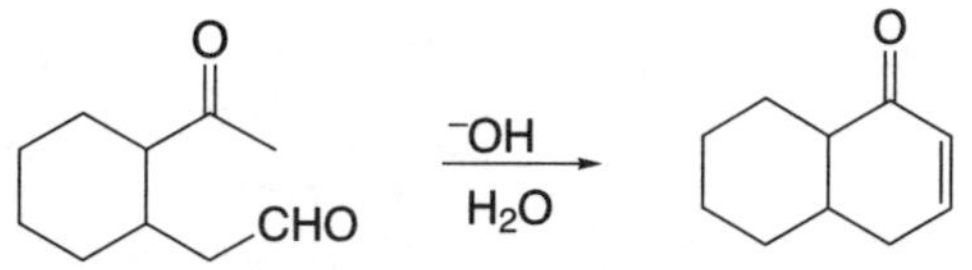

$$\xrightarrow[\text{H}_2\text{O}]{^-\text{OH}}$$

CHO

24.11

6-oxoheptanal $\xrightarrow[\text{H}_2\text{O}]{^-\text{OH}}$

CHO

1-acetylcyclopentene
major product

CHO

+

+

enolate precursors:

These two reacting functional groups are further away, making it harder for them to find each other to react. Also, the product contains a less stable 7-membered ring.

CH_3 CHO CH_3 CHO CH_2 CHO

A   **B**   **C**

most stable enolate

less hindered carbonyl

more hindered ketone carbonyl

less stable enolate

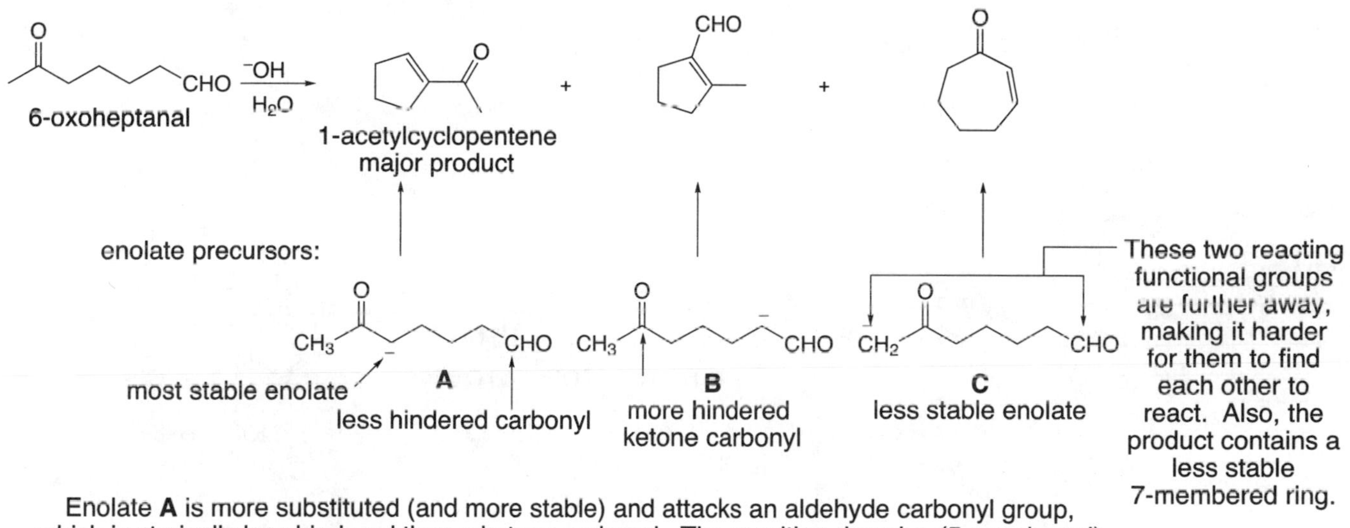

Enolate **A** is more substituted (and more stable) and attacks an aldehyde carbonyl group, which is sterically less hindered than a ketone carbonyl. The resulting ring size (5-membered) is also quite stable. That's why 1-acetylcyclopentene is the major product.

24.12 Join the α C of one ester to the carbonyl C of the other ester to form the β-keto ester.

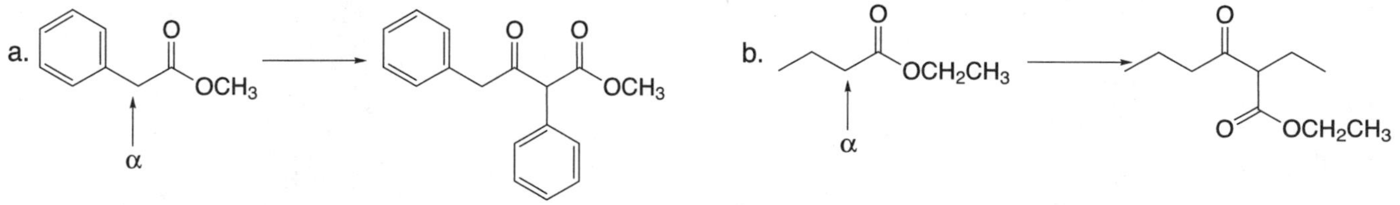

a.

OCH_3 $\longrightarrow$ OCH_3

α

b.

OCH_2CH_3 $\longrightarrow$

OCH_2CH_3

α

24.13

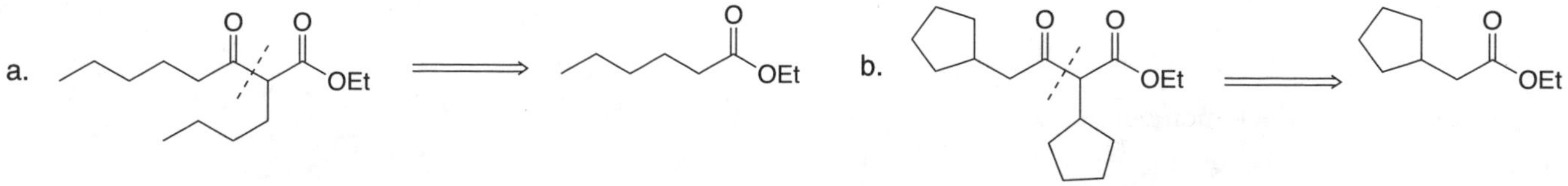

a.

OEt $\Longrightarrow$ OEt

b.

OEt $\Longrightarrow$ OEt

24.14 In a crossed Claisen reaction between an ester and a ketone, the enolate is always formed from the ketone, and the product is a β-dicarbonyl compound.

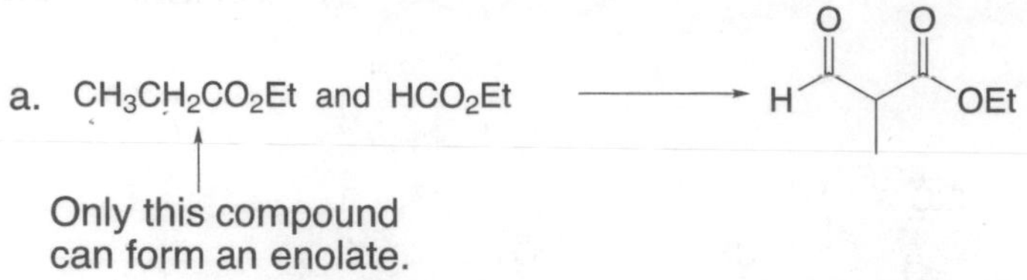
a. CH3CH2CO2Et and HCO2Et
H
OEt
Only this compound
can form an enolate.

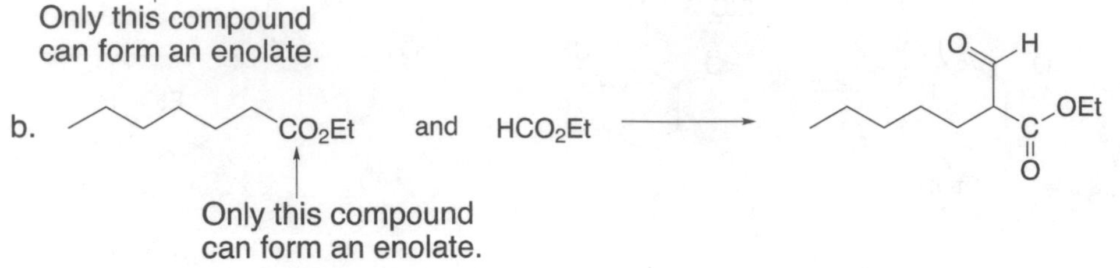
b.
CO2Et
and
HCO2Et
H
OEt
Only this compound
can form an enolate.

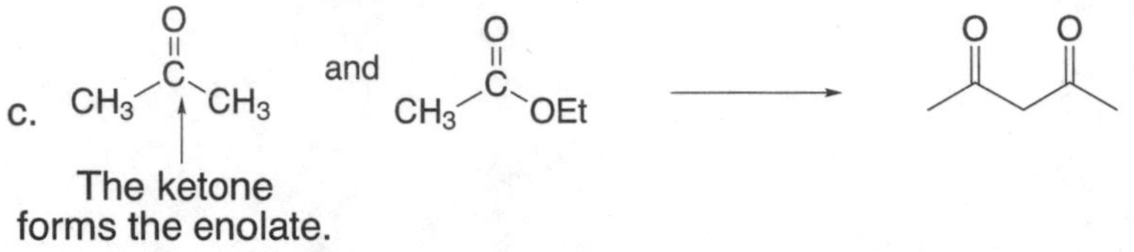
c. CH3
CH3
and
CH3
OEt
The ketone
forms the enolate.

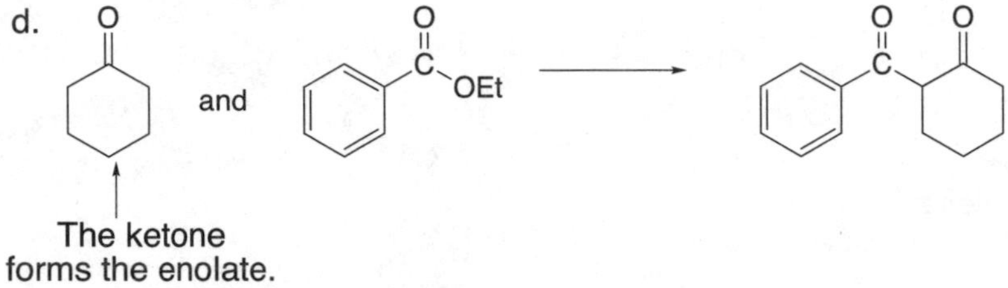
d.
and
OEt
The ketone
forms the enolate.

24.15

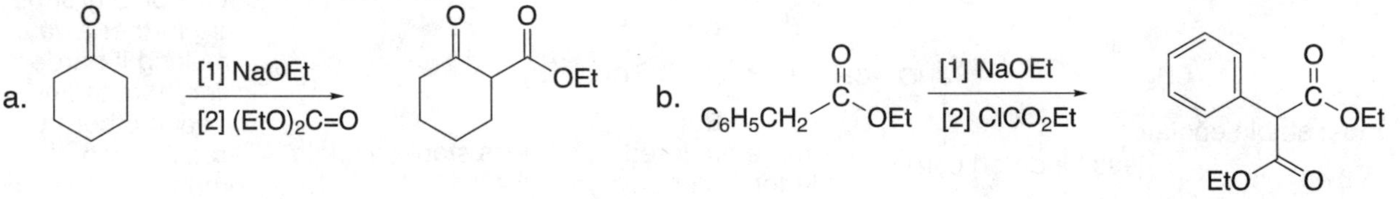
a.
[1] NaOEt
[2] (EtO)2C=O
OEt
b.
C6H5CH2
OEt
[1] NaOEt
[2] ClCO2Et
OEt
EtO

24.16

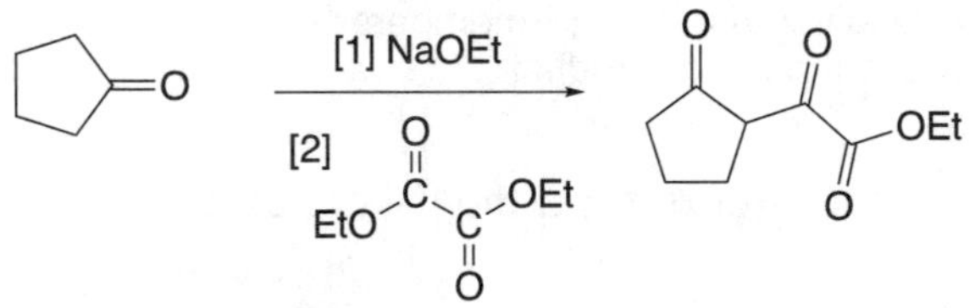
[1] NaOEt
[2]
EtO
OEt
OEt

24.17

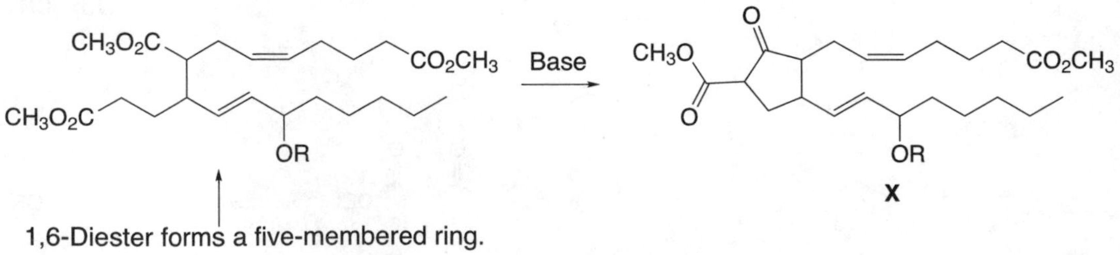
CH3O2C
CO2CH3
CH3O2C
OR
Base
CH3O
CO2CH3
OR
X
1,6-Diester forms a five-membered ring.

24.18

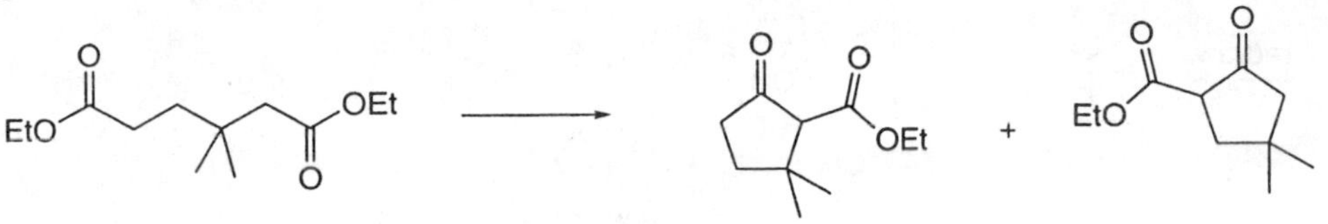

24.19

a. 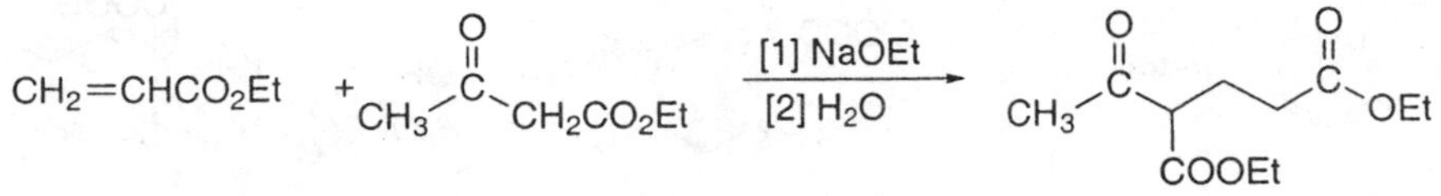 $CH_2{=}CHCO_2Et$ + $\xrightarrow[\text{[2] H}_2\text{O}]{\text{[1] NaOEt}}$

b. 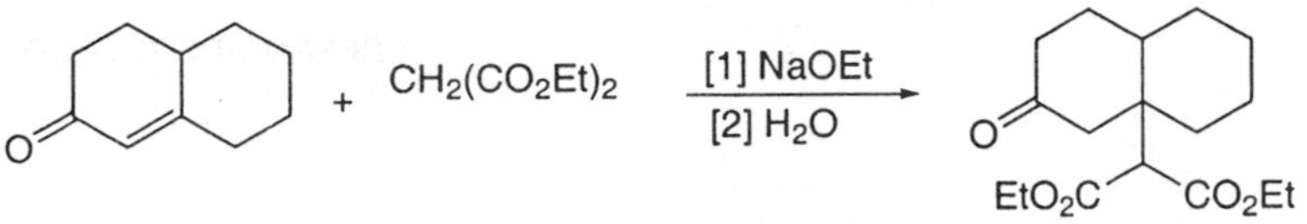   + $CH_2(CO_2Et)_2$ $\xrightarrow[\text{[2] H}_2\text{O}]{\text{[1] NaOEt}}$

c. 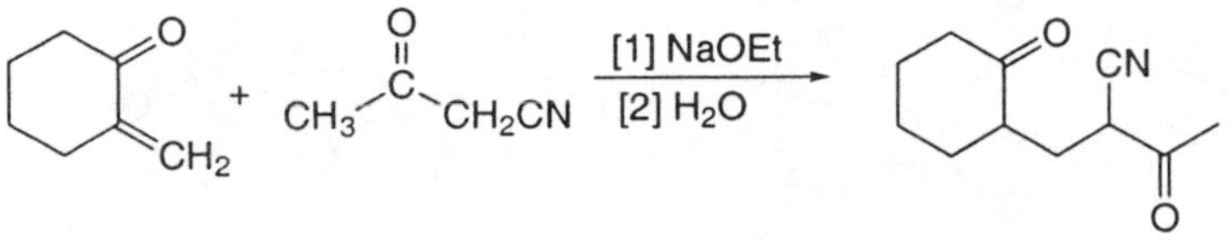   + $CH_3{-}\overset{\text{O}}{\underset{}{C}}{-}CH_2CN$ $\xrightarrow[\text{[2] H}_2\text{O}]{\text{[1] NaOEt}}$

24.20

a. 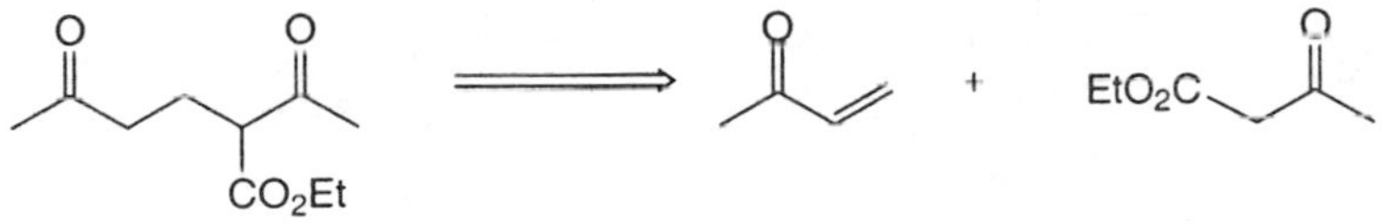 $\Longrightarrow$ + EtO_2C

b. 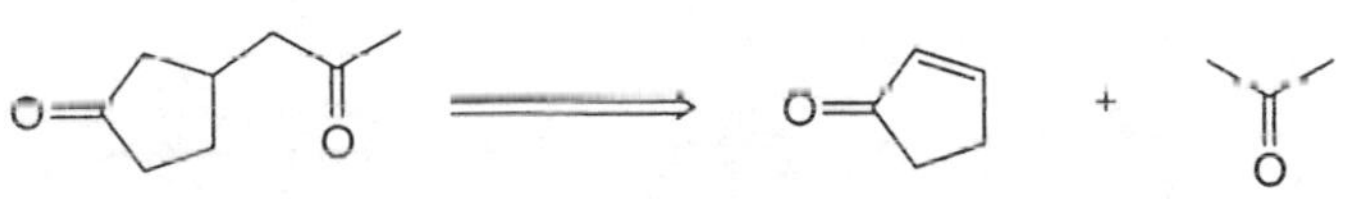 $\Longrightarrow$ +

24.21 The Robinson annulation forms a six-membered ring and three new carbon-carbon bonds: two σ bonds and one π bond.

a. 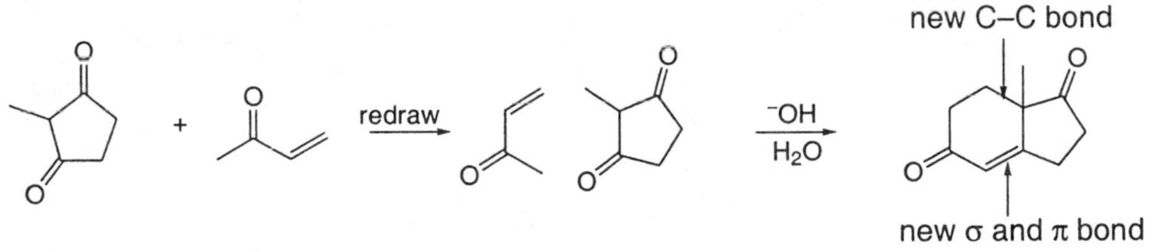   + $\xrightarrow{\text{redraw}}$ $\xrightarrow[\text{H}_2\text{O}]{{}^-\text{OH}}$

new C–C bond

new σ and π bond

b. 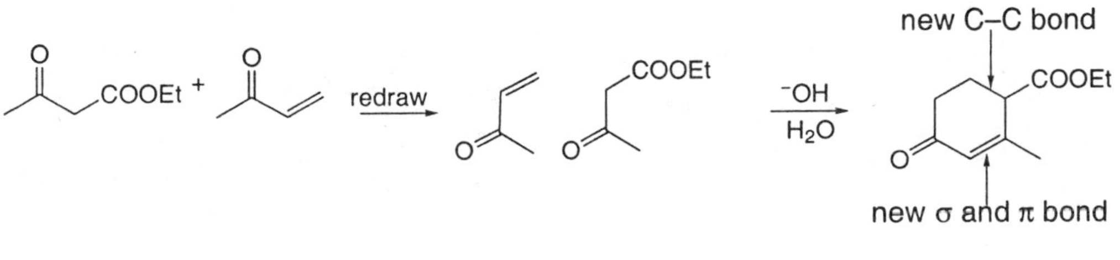   + $\xrightarrow{\text{redraw}}$ $\xrightarrow[\text{H}_2\text{O}]{{}^-\text{OH}}$

new C–C bond

new σ and π bond

c.

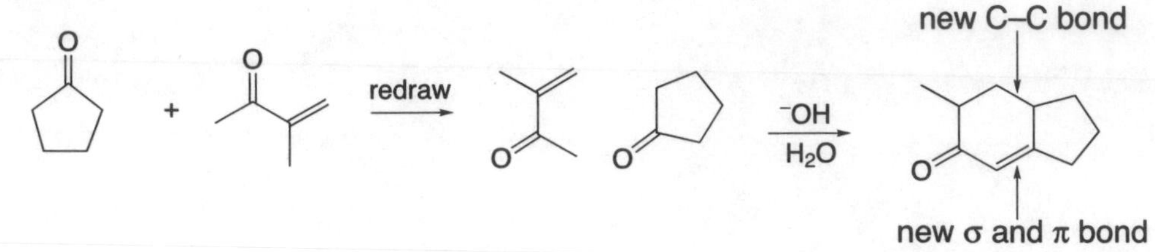
redraw
⁻OH
H₂O
new C–C bond
new σ and π bond

d.

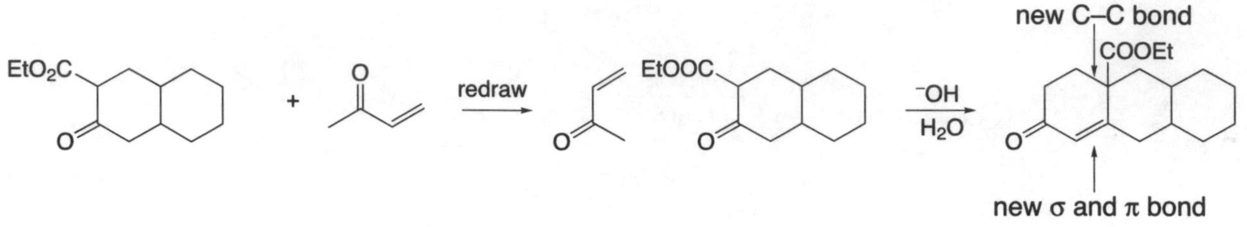
EtO₂C
redraw
EtOOC
⁻OH
H₂O
new C–C bond
COOEt
new σ and π bond

24.22

a.

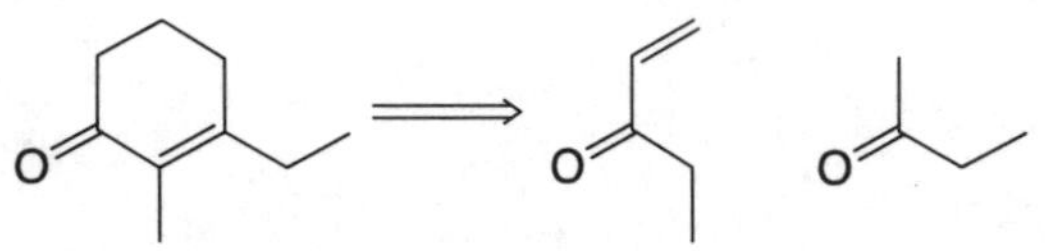

c.

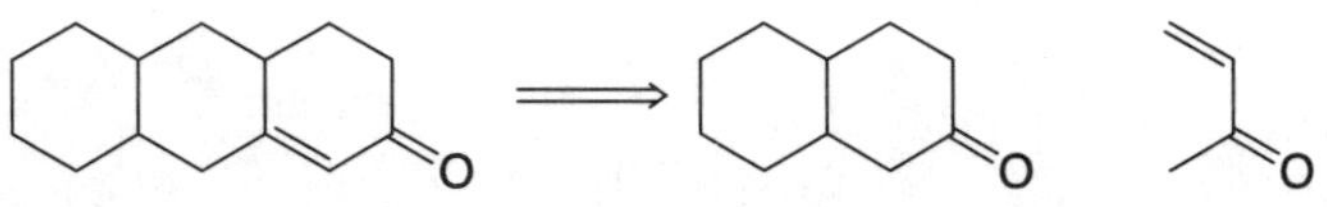

b.

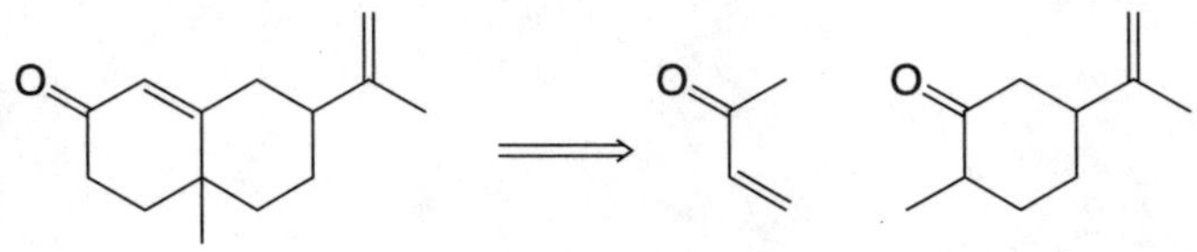

24.23 The product of an aldol reaction is a β-hydroxy carbonyl compound or an α,β-unsaturated carbonyl compound. The latter type of compound is drawn as product unless elimination of H$_2$O cannot form a conjugated system.

a. (CH$_3$)$_2$CHCHO only $\xrightarrow[\text{H}_2\text{O}]{^-\text{OH}}$ (CH$_3$)$_2$CHCHC(CH$_3$)$_2$ (with OH and CHO substituents)

d. (CH$_3$CH$_2$)$_2$C=O only $\xrightarrow[\text{H}_2\text{O}]{^-\text{OH}}$

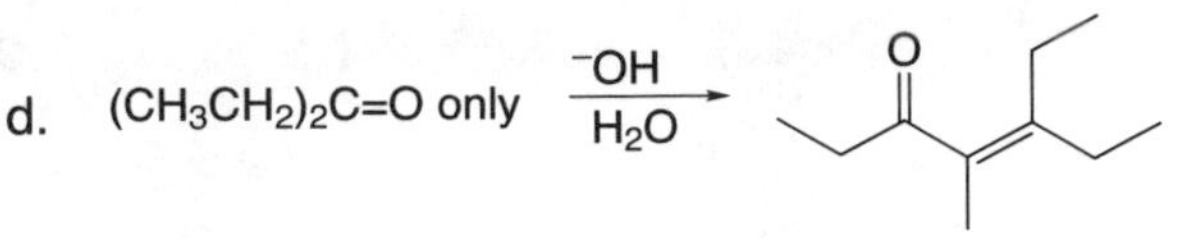

b. (CH$_3$)$_2$CHCHO + CH$_2$=O $\xrightarrow[\text{H}_2\text{O}]{^-\text{OH}}$ CH$_3$–C(CH$_3$)(CH$_2$OH)–CHO

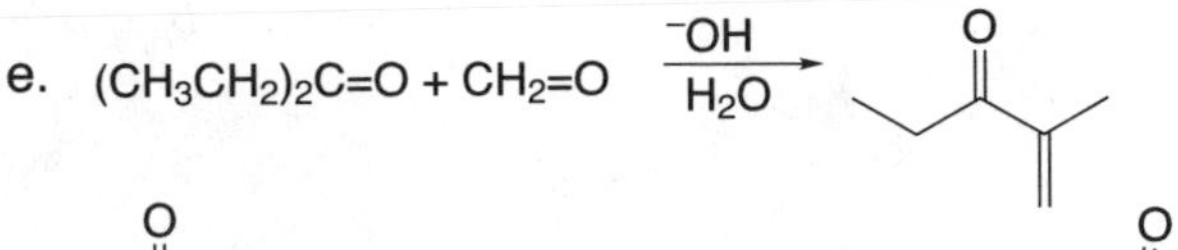

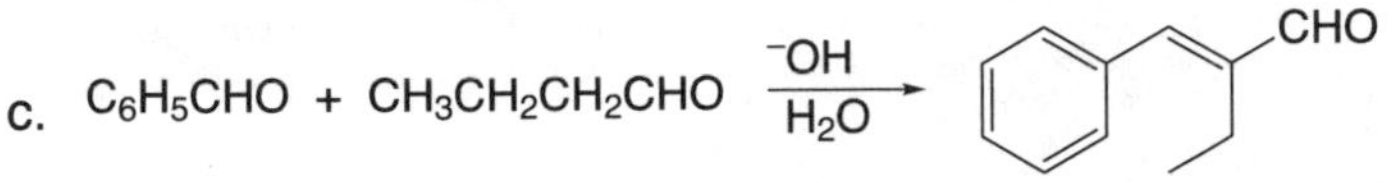

e. (CH$_3$CH$_2$)$_2$C=O + CH$_2$=O $\xrightarrow[\text{H}_2\text{O}]{^-\text{OH}}$

c. C$_6$H$_5$CHO + CH$_3$CH$_2$CH$_2$CHO $\xrightarrow[\text{H}_2\text{O}]{^-\text{OH}}$

(E and Z isomers)

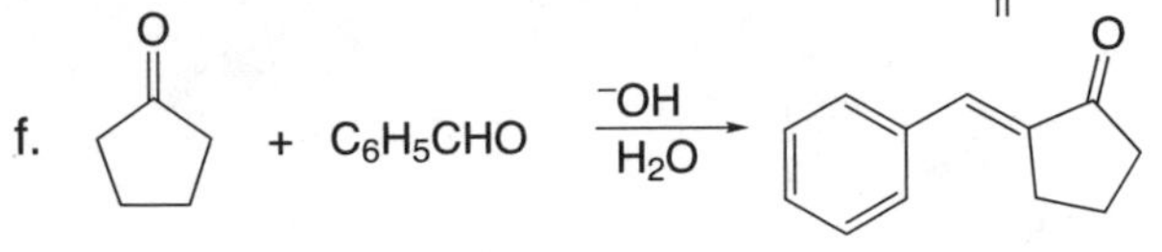

f. + C$_6$H$_5$CHO $\xrightarrow[\text{H}_2\text{O}]{^-\text{OH}}$

24.24

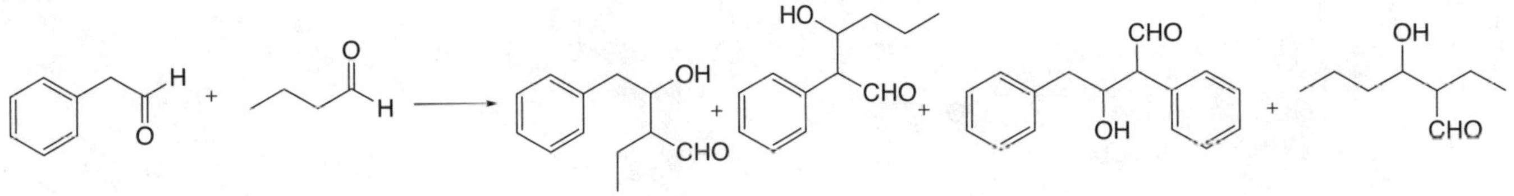

24.25

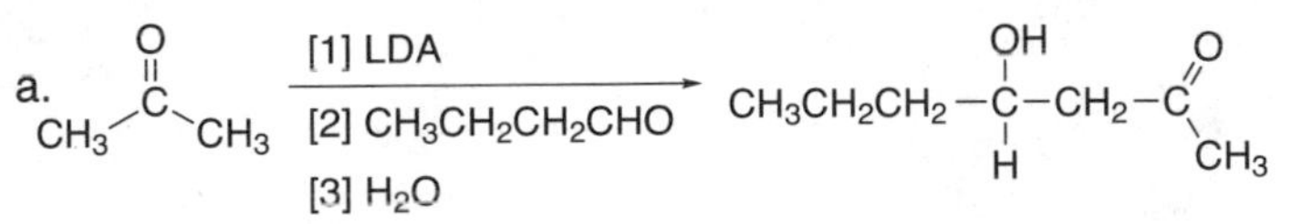

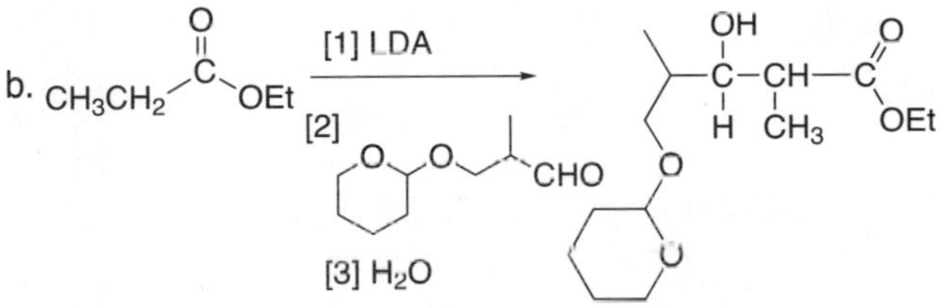

a.

b.

24.26

a.

b.

c. OHC⁓⁓⁓CHO ⟶

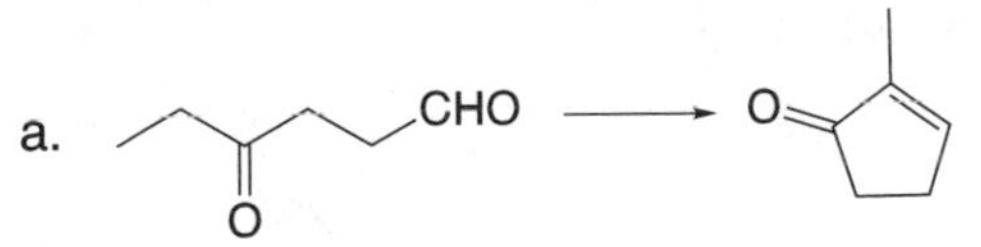

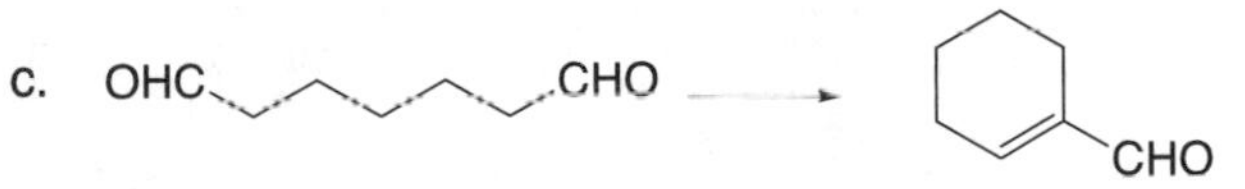

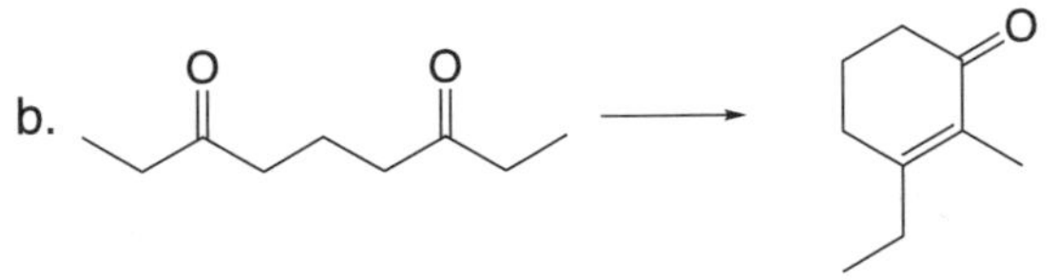

24.27 Locate the α and β C's to the carbonyl group, and break the molecule into two halves at this bond. The α C and all of the atoms bonded to it belong to one carbonyl component. The β C and all the atoms bonded to it belong to the other carbonyl component.

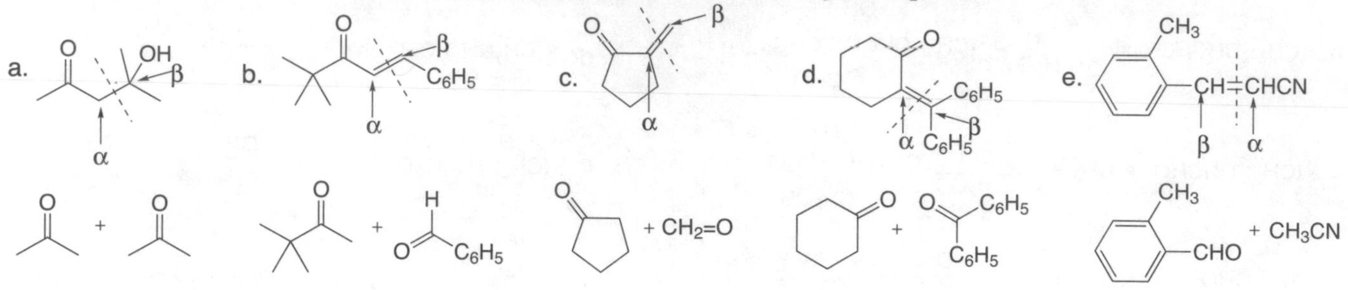

a.   b.   c.   d.   e.

24.28

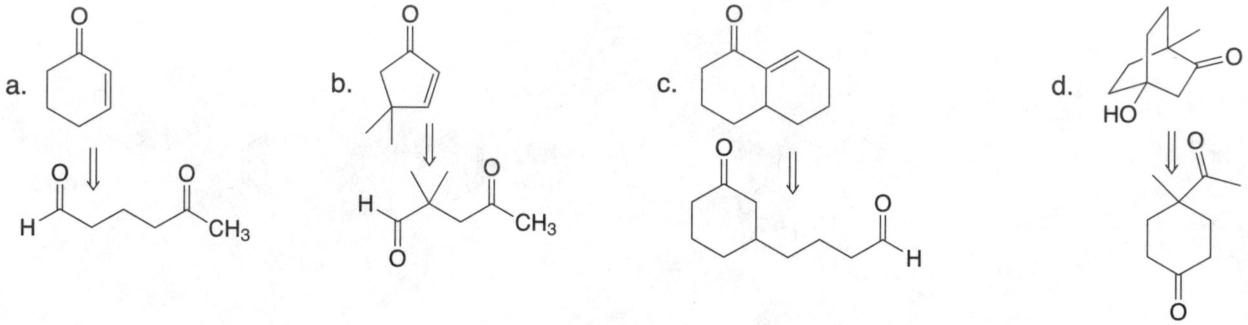

a.   b.   c.   d.

24.29

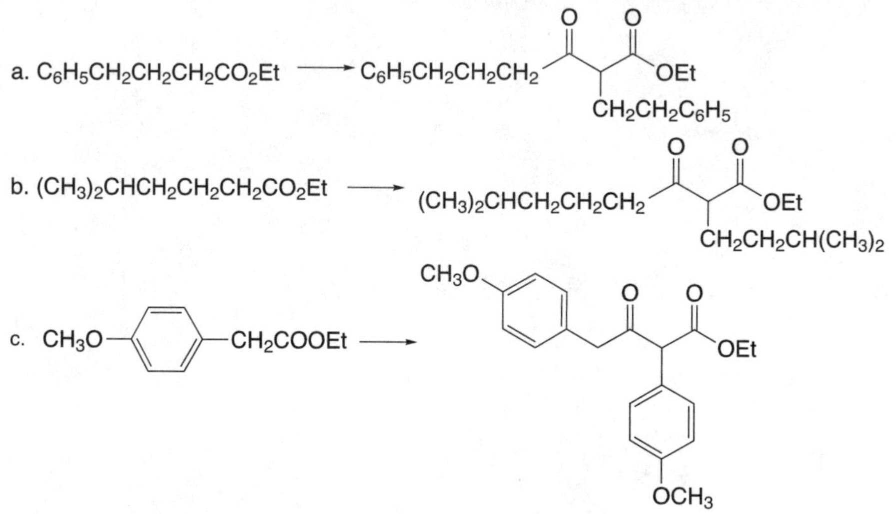

a. $C_6H_5CH_2CH_2CH_2CO_2Et \longrightarrow C_6H_5CH_2CH_2CH_2$ ... OEt, $CH_2CH_2C_6H_5$

b. $(CH_3)_2CHCH_2CH_2CH_2CO_2Et \longrightarrow (CH_3)_2CHCH_2CH_2CH_2$ ... OEt, $CH_2CH_2CH(CH_3)_2$

c. CH_3O-⟨⟩$-CH_2COOEt \longrightarrow$

24.30

$CH_3CH_2CH_2CH_2CO_2Et + CH_3CH_2CO_2Et \longrightarrow$ ... OEt + ... OEt + ... OEt + ... OEt

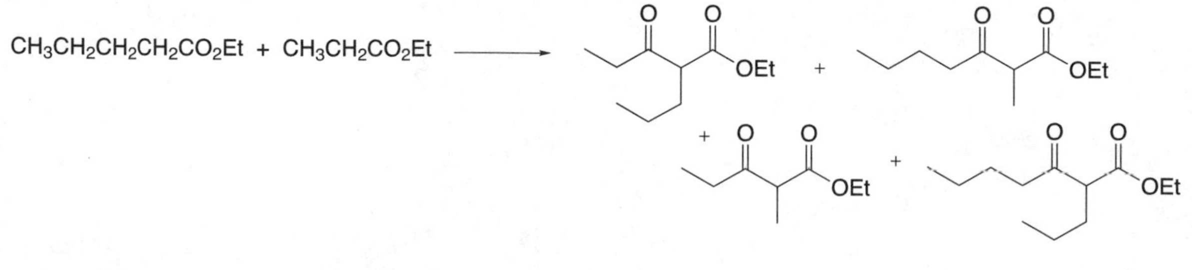

24.31

a. $CH_3CH_2CH_2CO_2Et$ only $\longrightarrow$

b. $CH_3CH_2CH_2CO_2Et + C_6H_5CO_2Et \longrightarrow$

c. $CH_3CH_2CH_2CO_2Et + (CH_3)_2C=O \longrightarrow$

d. $EtO_2CC(CH_3)_2CH_2CH_2CH_2CO_2Et \longrightarrow$

e. $C_6H_5COCH_2CH_3 + C_6H_5CO_2Et \longrightarrow$

f. $CH_3CH_2CO_2Et + (EtO)_2C=O \longrightarrow$

g. $+$ $HCO_2Et \longrightarrow$

h.

24.32

a.   or

b.

c.

d. $C_6H_5CH(COOEt)_2 \implies$

24.33

To form bond a:

To form bond b:

bond "a"   bond "b"

24.34

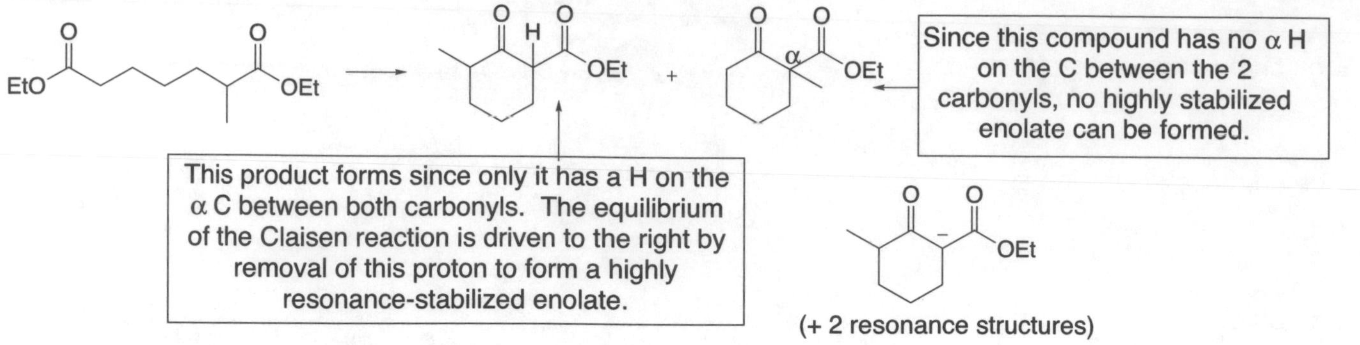
EtO
OEt
H
OEt
+
α
OEt
Since this compound has no α H on the C between the 2 carbonyls, no highly stabilized enolate can be formed.
This product forms since only it has a H on the α C between both carbonyls. The equilibrium of the Claisen reaction is driven to the right by removal of this proton to form a highly resonance-stabilized enolate.
OEt
(+ 2 resonance structures)

24.35

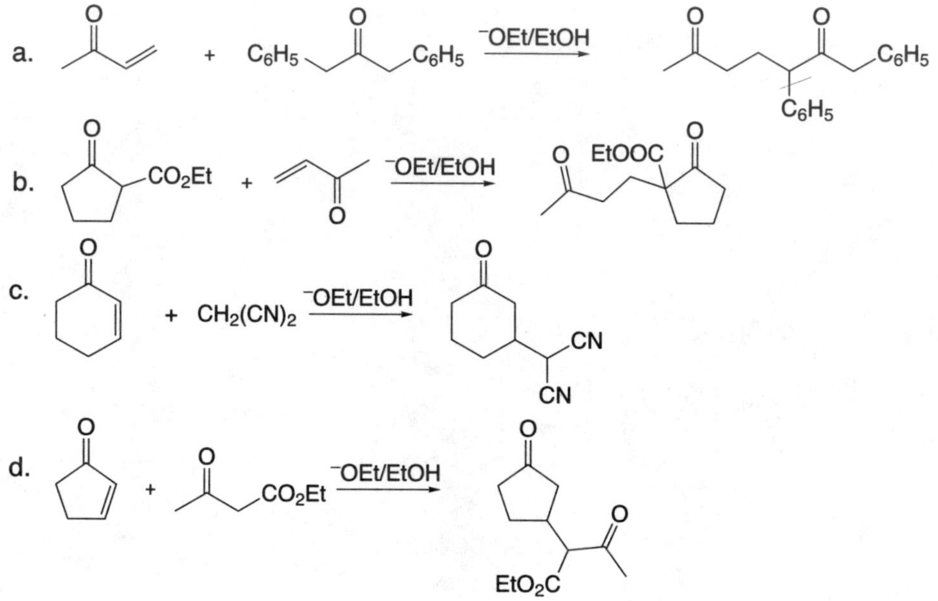
a.
C_6H_5
C_6H_5
$^-OEt/EtOH$
C_6H_5
C_6H_5
b.
CO_2Et
$^-OEt/EtOH$
EtOOC
c.
+ $CH_2(CN)_2$
$^-OEt/EtOH$
CN
CN
d.
CO_2Et
$^-OEt/EtOH$
EtO_2C

24.36

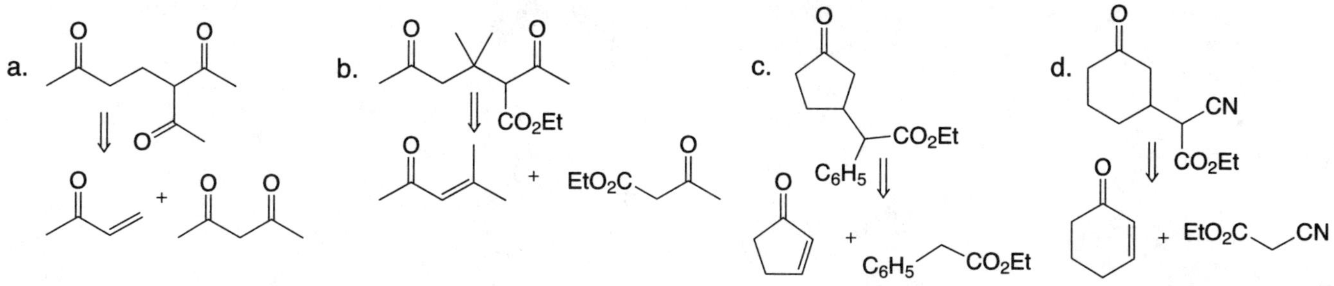
a.
b.
CO_2Et
EtO_2C
c.
CO_2Et
C_6H_5
C_6H_5
CO_2Et
d.
CN
CO_2Et
EtO_2C
CN

24.37

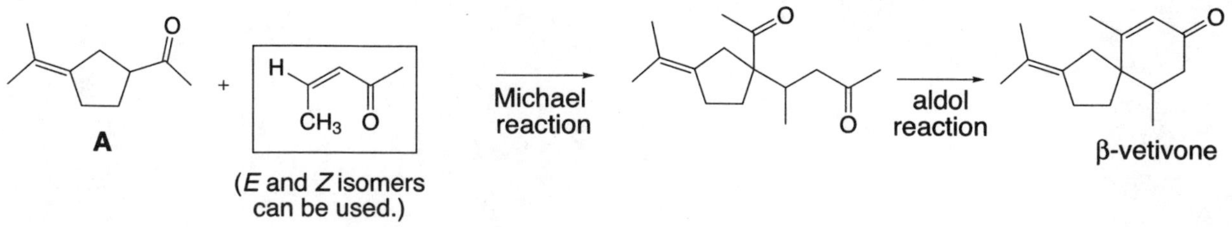
A
H
CH_3
(*E* and *Z* isomers can be used.)
Michael reaction
aldol reaction
β-vetivone

24.38

a.

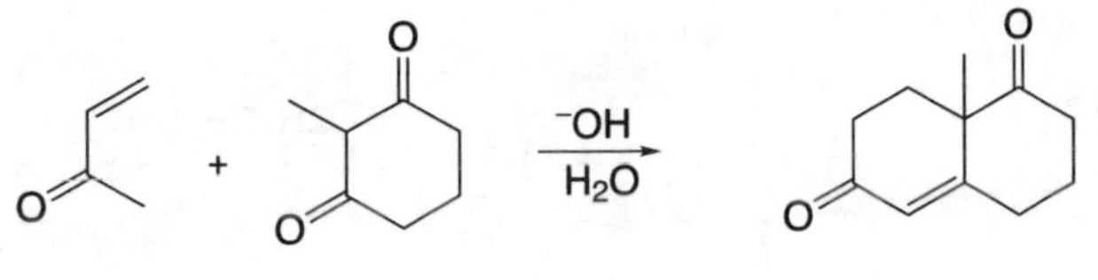

b.

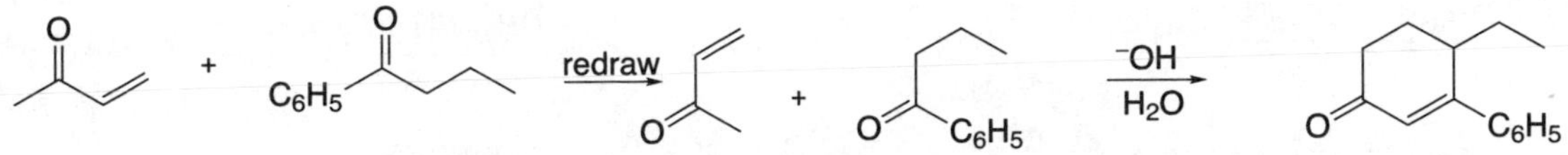

c.

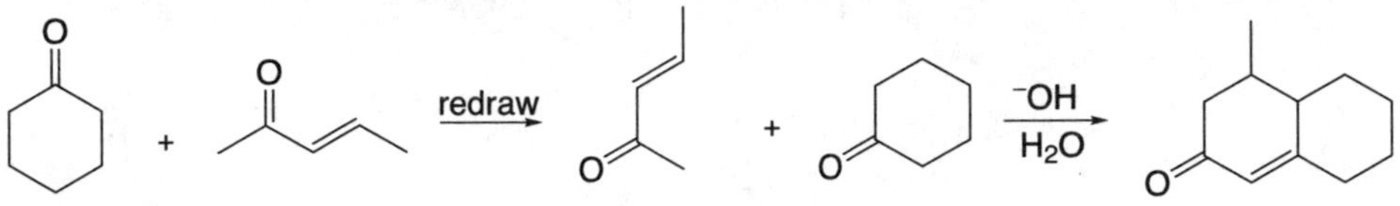

d.

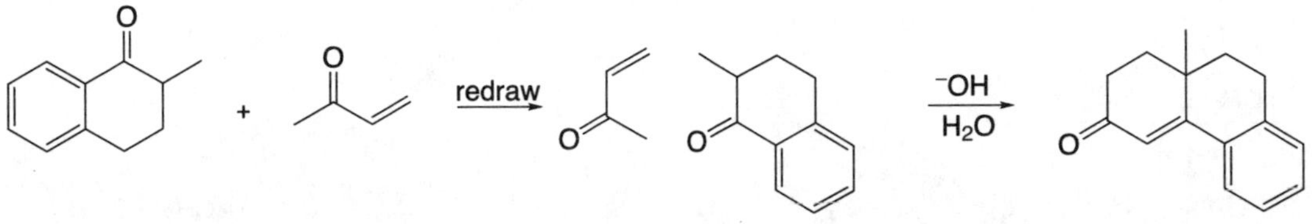

24.39

a.

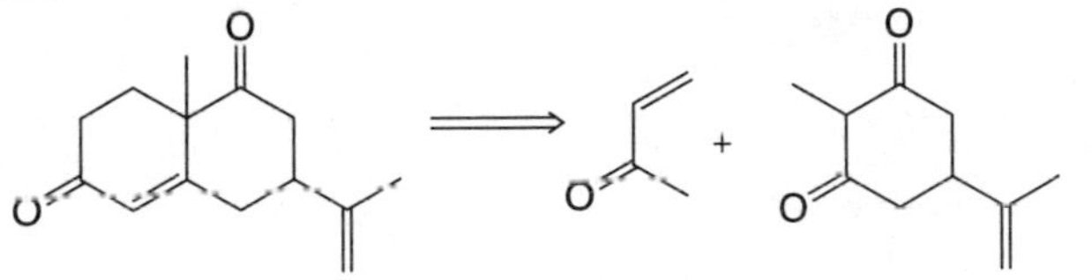

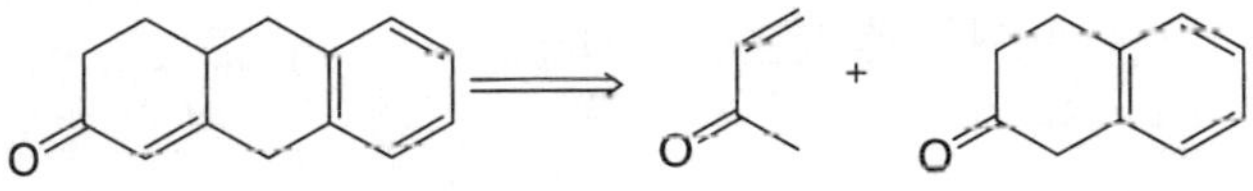

b.

c.

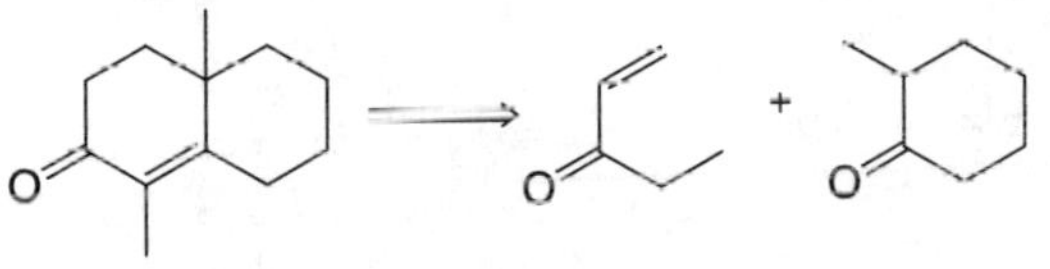

d.

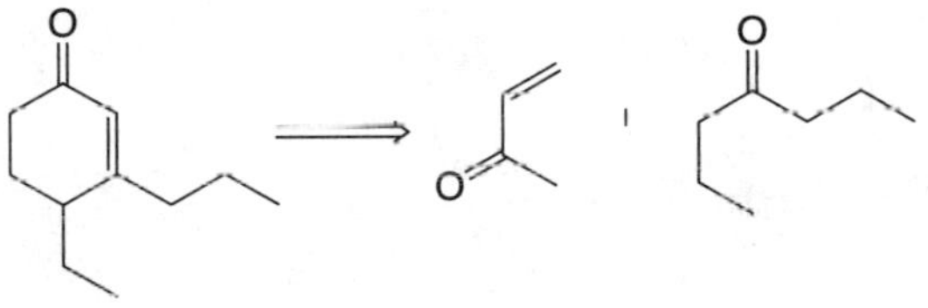

24.40

CH₃CH₂CH₂CHO

a. ⁻OH/H₂O → CH₃CH₂CH₂ / CHO / H / CH₂CH₃ (*E* and *Z*)

b. ⁻OH /CH₂=O/H₂O → CH₂=C / CHO / CH₂CH₃

c. [1] LDA; [2] CH₃CHO; [3] H₂O → CH₃ / CHO / H / CH₂CH₃ (*E* and *Z*)

d. CH₂(CO₂Et)₂/NaOEt/EtOH → CH₃CH₂CH₂ / H / EtO₂C / CO₂Et

e. [1] CH₃Li; [2] H₂O → OH structure

f. NaBH₄/CH₃OH → CH₃CH₂CH₂CH₂OH

g. H₂/Pd → CH₃CH₂CH₂CH₂OH

h. HOCH₂CH₂OH/TsOH → dioxolane with propyl, O O, H

i. CH₃NH₂, mild acid → CH₃CH₂CH₂ / CH₃ / C=N / H

j. (CH₃)₂NH, mild acid → CH₃CH₂CH=CHN(CH₃)₂ (*E* and *Z*)

k. CrO₃/H₂SO₄ CH₃CH₂CH₂COOH

l. Br₂/CH₃COOH → O / H / Br structure

m. Ph₃P=CH₂ → CH₃CH₂CH₂ / C=CH₂ / H

n. NaCN/HCl CH₃CH₂CH₂ / OH / C—H / CN

o. [1] LDA; [2] CH₃I → O / H structure

24.41

a. cyclopentanone ⁻OH / H₂O → product

e. CH₃—C(=O)—CH₃ [1] LDA [2] CH₃CH₂CHO [3] H₂O → (*E* and *Z*)

b. 1,3-cyclohexanedione NaOEt/EtOH / (CH₃)₂C=O → product

f. cyclohexanone + C₆H₅CH=CHC(=O)CH₃ NaOCH₃ / CH₃OH → C₆H₅ bicyclic product

c. NCCH₂CO₂Et NaOEt/EtOH cyclohexanone → CN / CO₂Et product

g. naphthalene-CHO + C(=O)C₆H₅ ⁻OH / H₂O → naphthalene-CH=CH-C(=O)-C₆H₅ (*E* and *Z*)

d. furan-CHO + CH₃—C(=O)—CH₃ ⁻OH / H₂O → furan-CH=CH-C(=O) (*E* and *Z*)

h. benzene-CH₂CO₂Et / CH₂CO₂Et [1] NaOEt/EtOH [2] H₃O⁺ → indanone product O, OEt

24.42

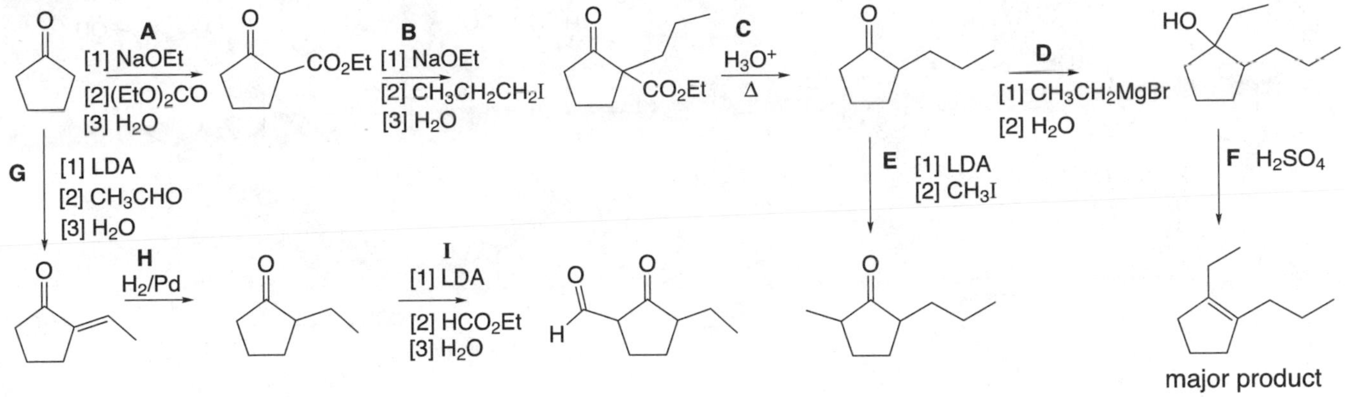

major product

24.43

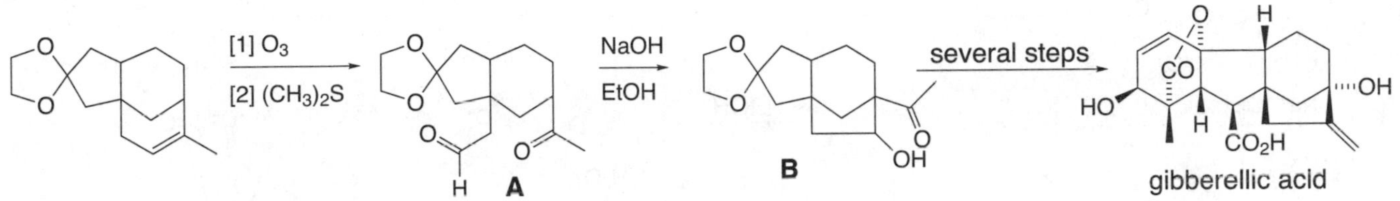

gibberellic acid

24.44

cis-jasmone

B
not formed

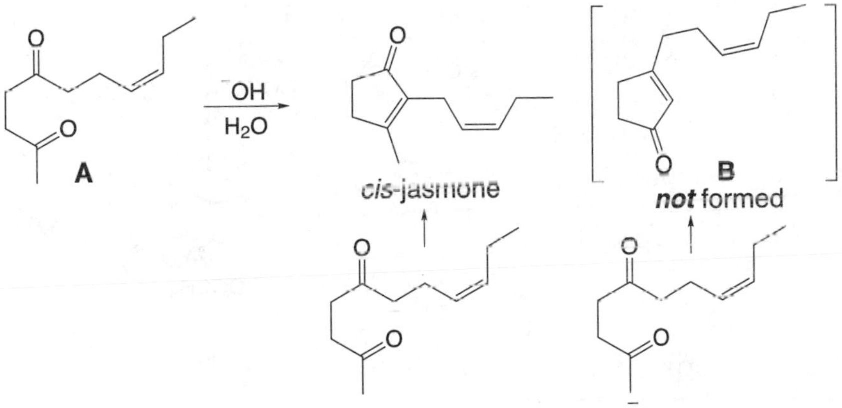

cis-Jasmone is formed from this more substituted and therefore more stable enolate. Under equilibrium conditions (⁻OH, H₂O) this thermodynamic enolate predominates.

This less substituted enolate is not favored by the base/solvent conditions chosen.

24.45

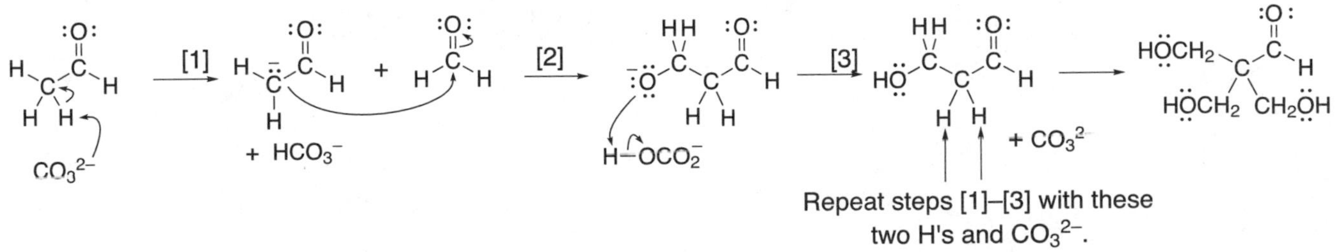

Repeat steps [1]–[3] with these two H's and CO₃²⁻.

24.46

a.

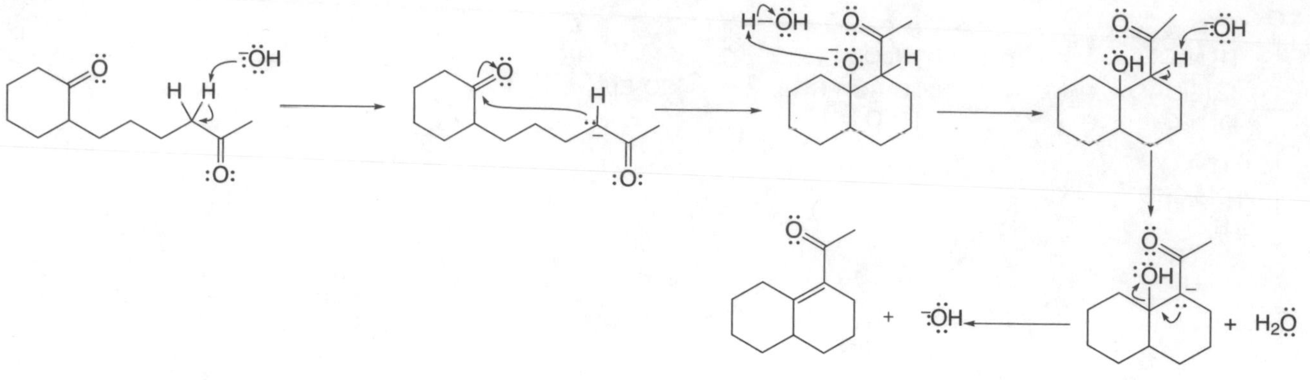

b.

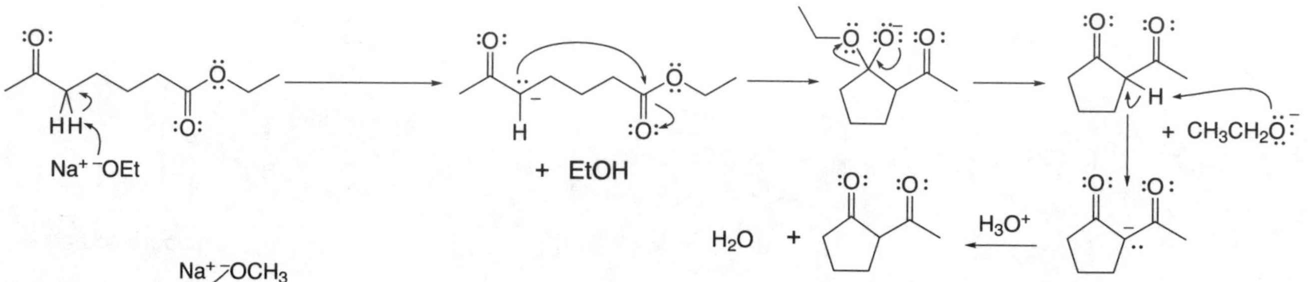

c.

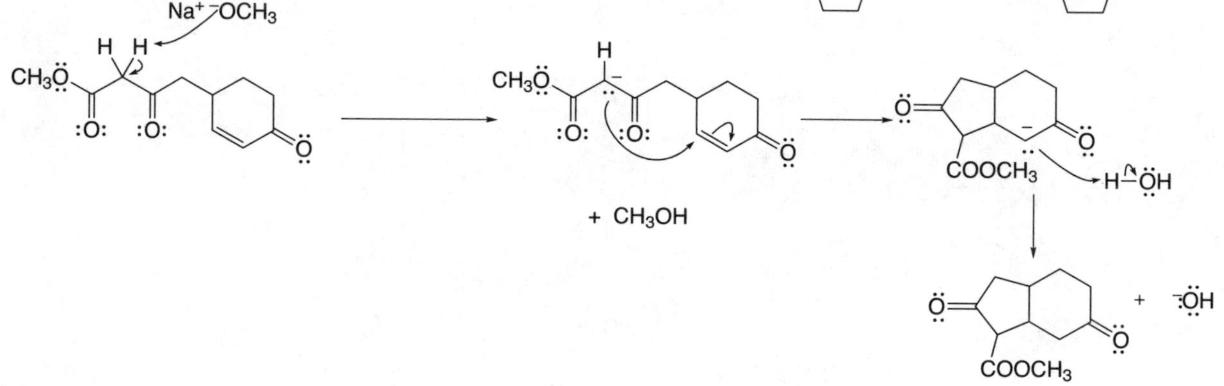

24.47

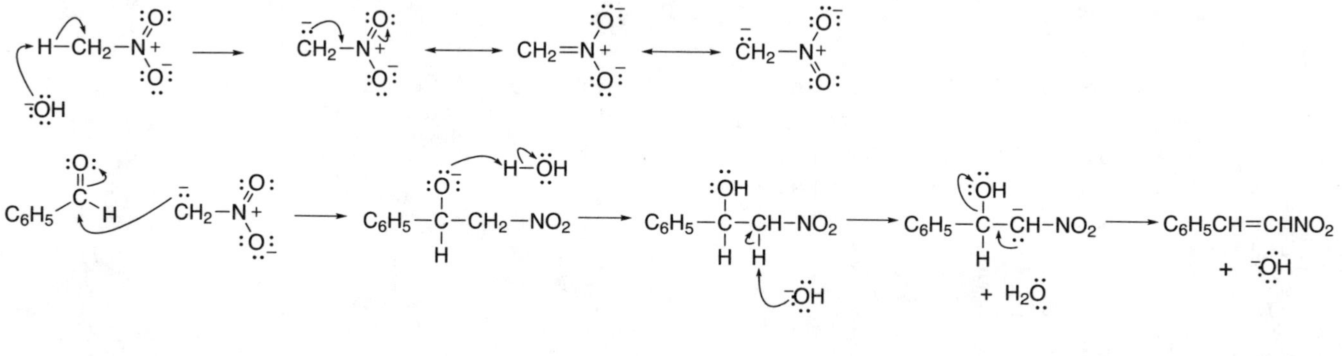

24.48

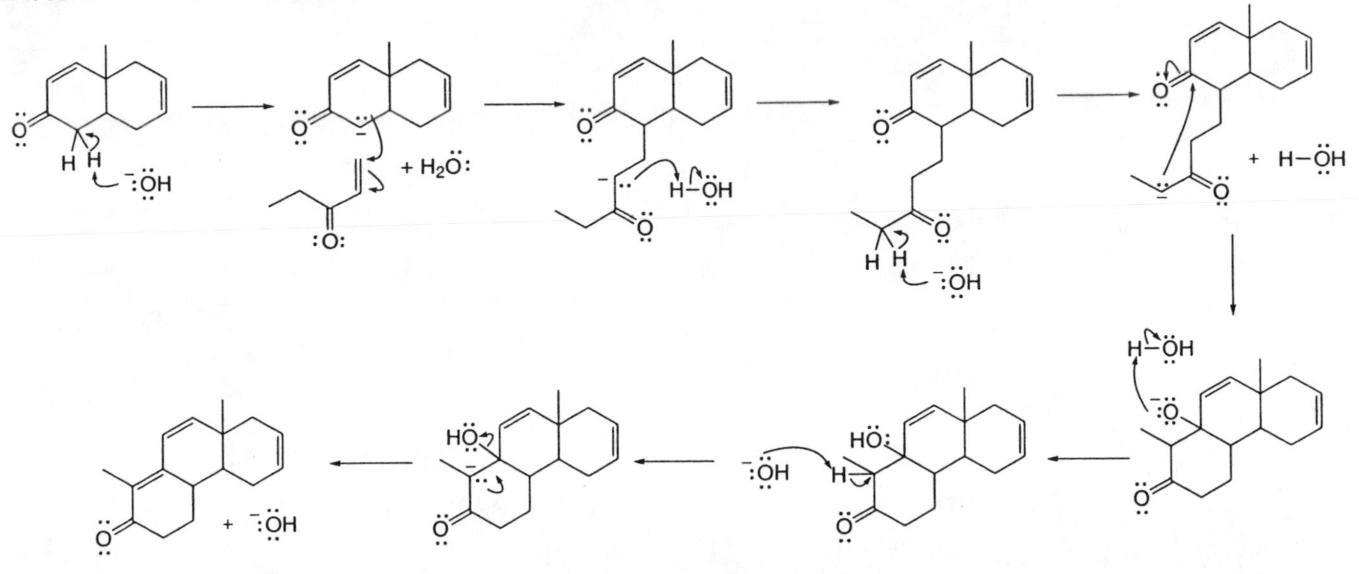

24.49

a.

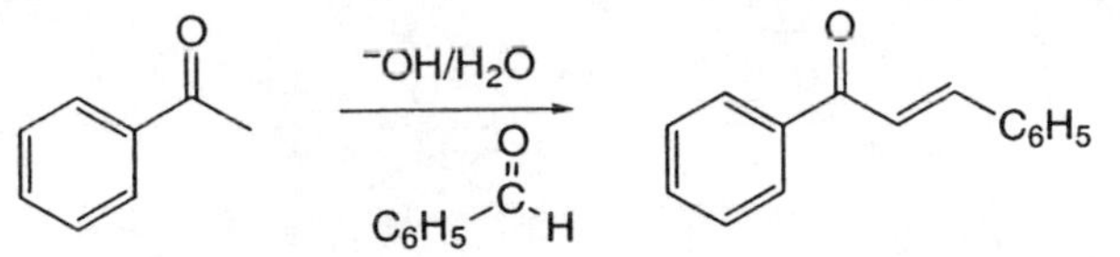

b.

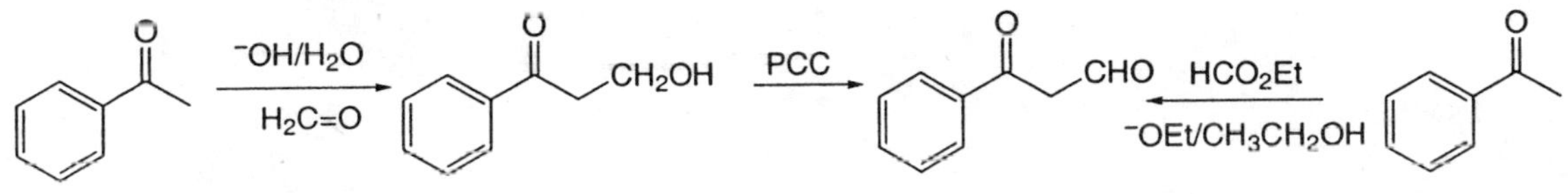

c.

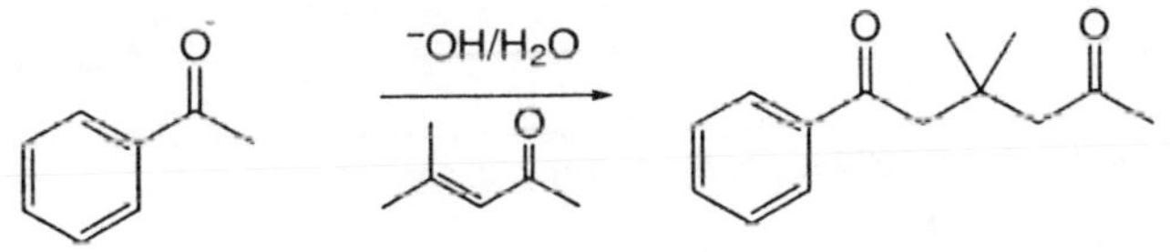

d.

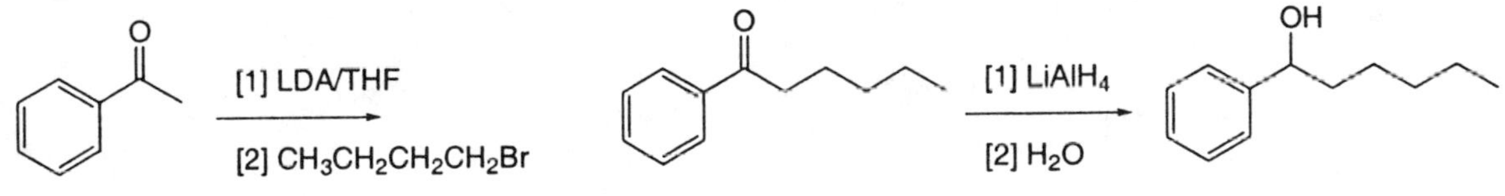

24.50

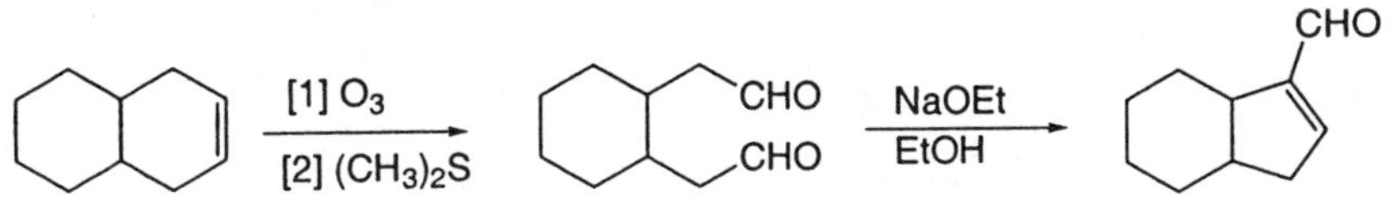

24.51

a.

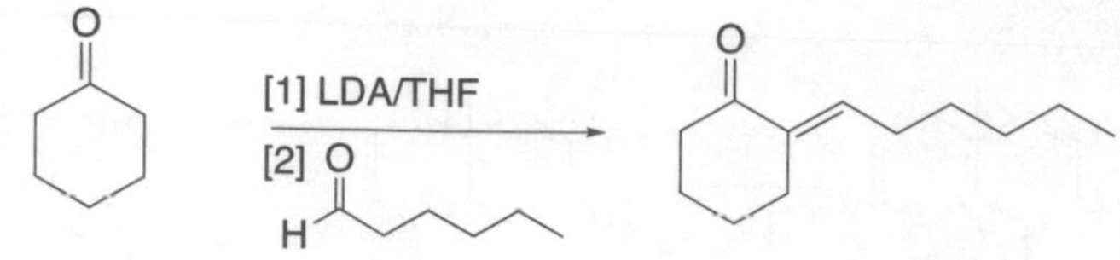

b.

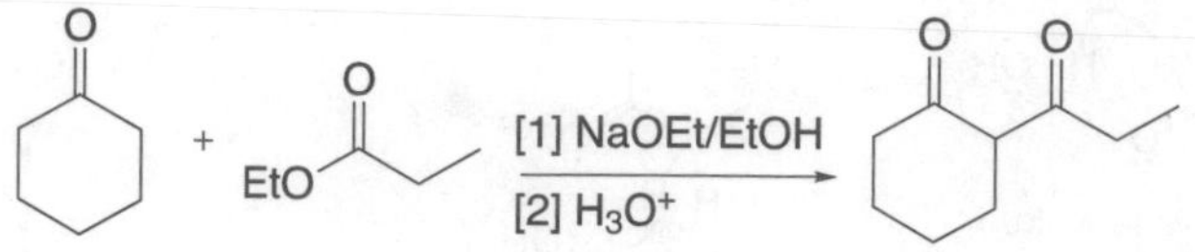

c.

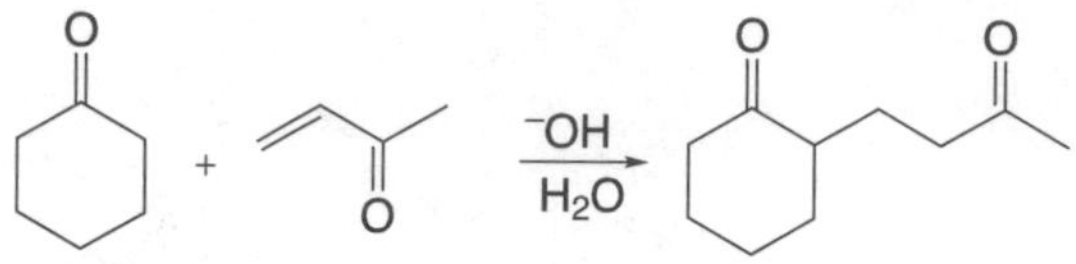

d.

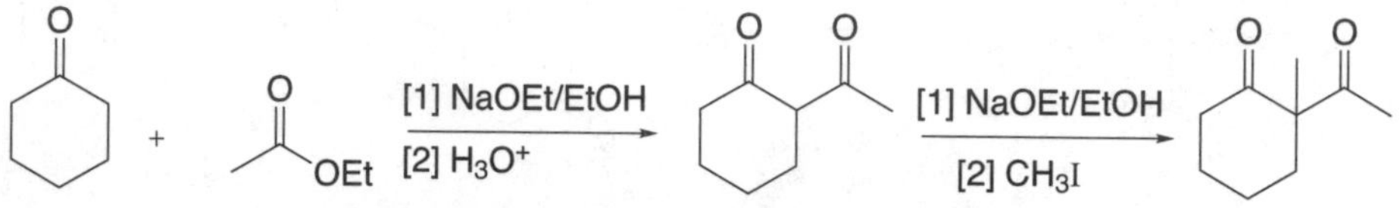

e.

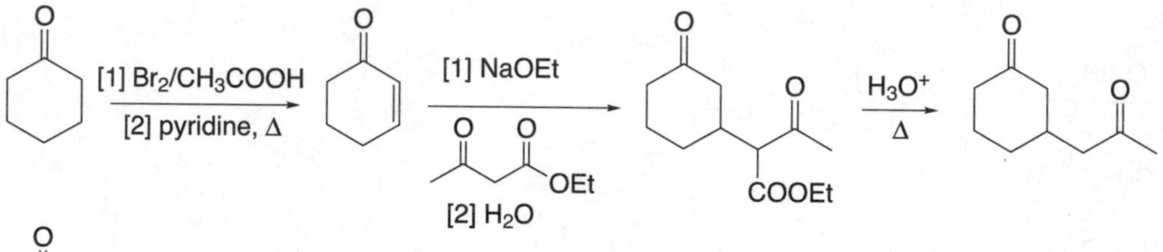

f.

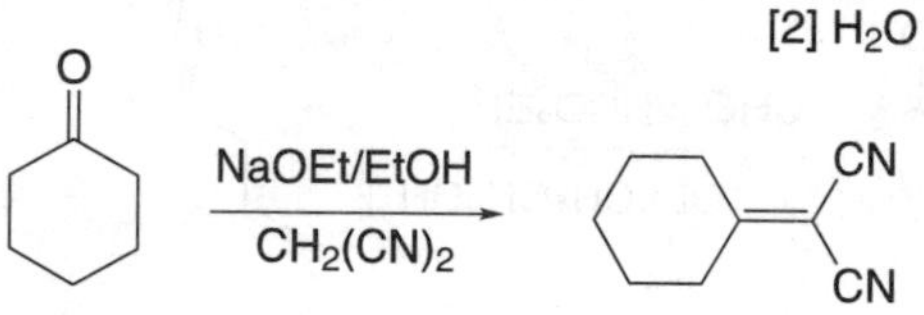

g.

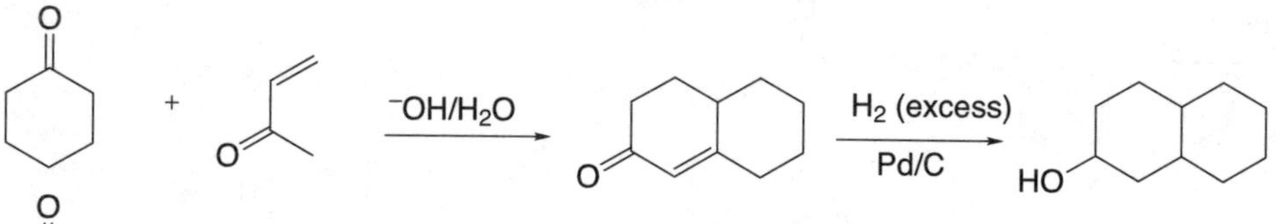

h.

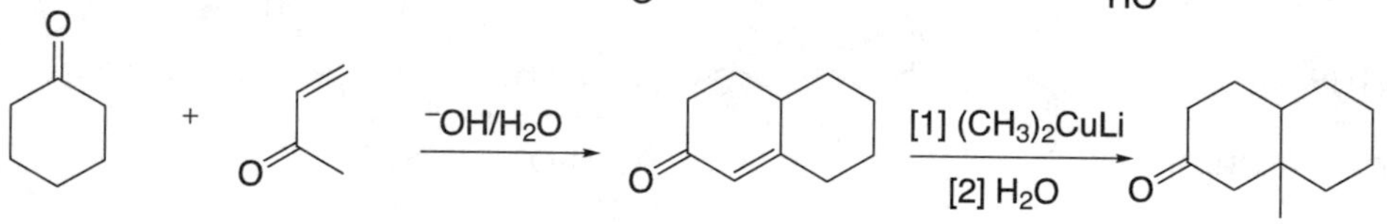

24.52

a.

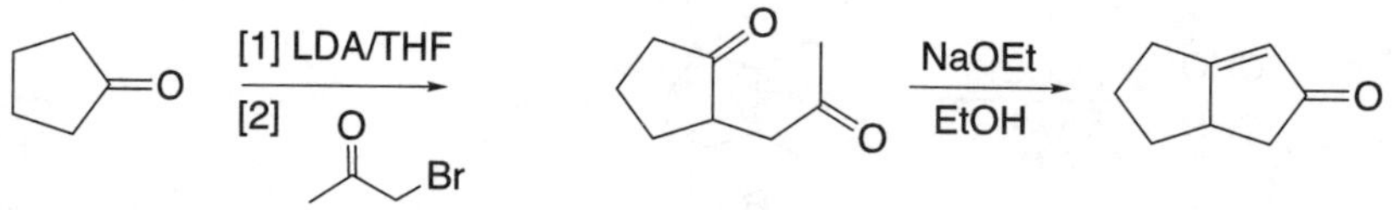

b.

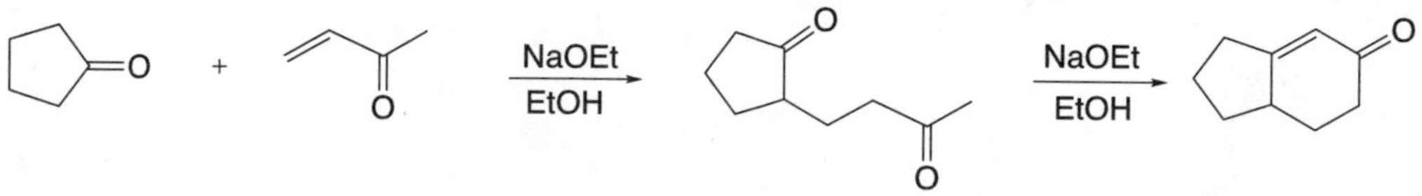

24.53

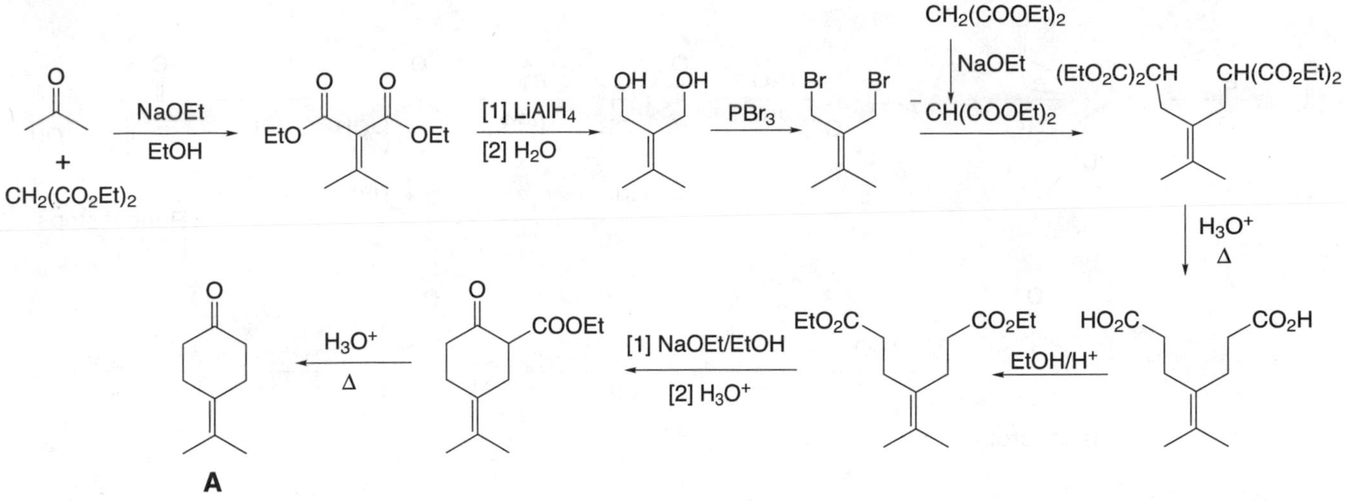

A

24.54

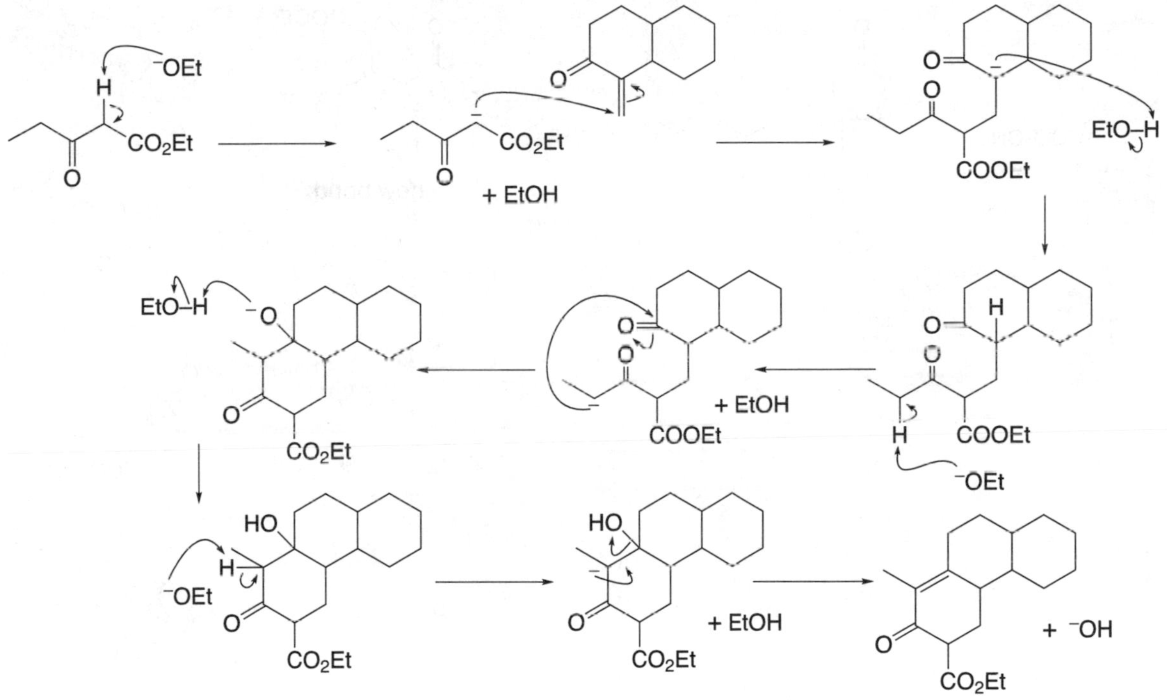

24.55

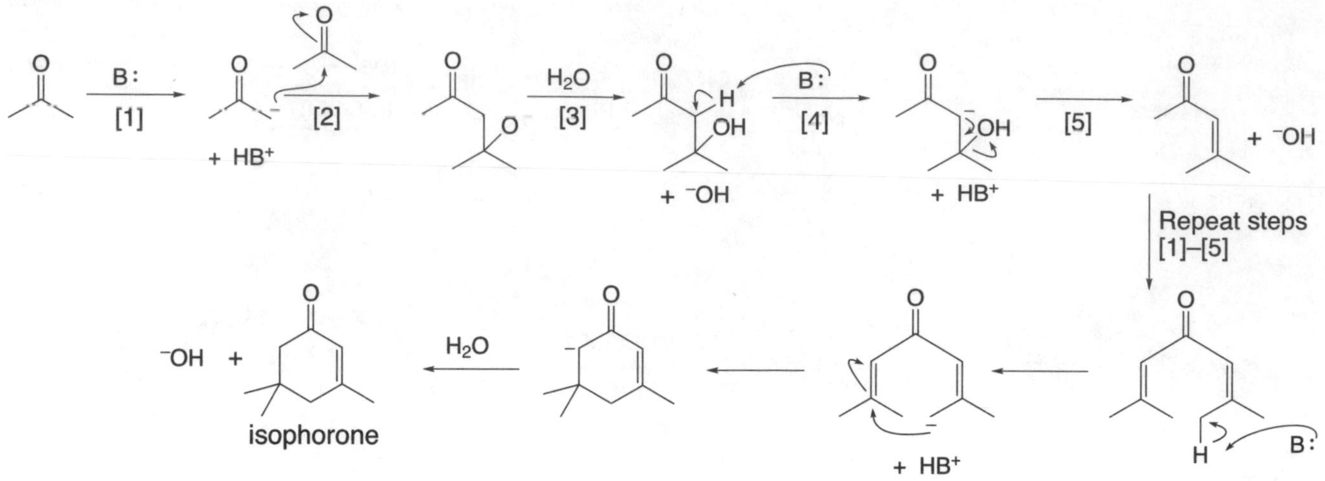

isophorone

24.56

new bond

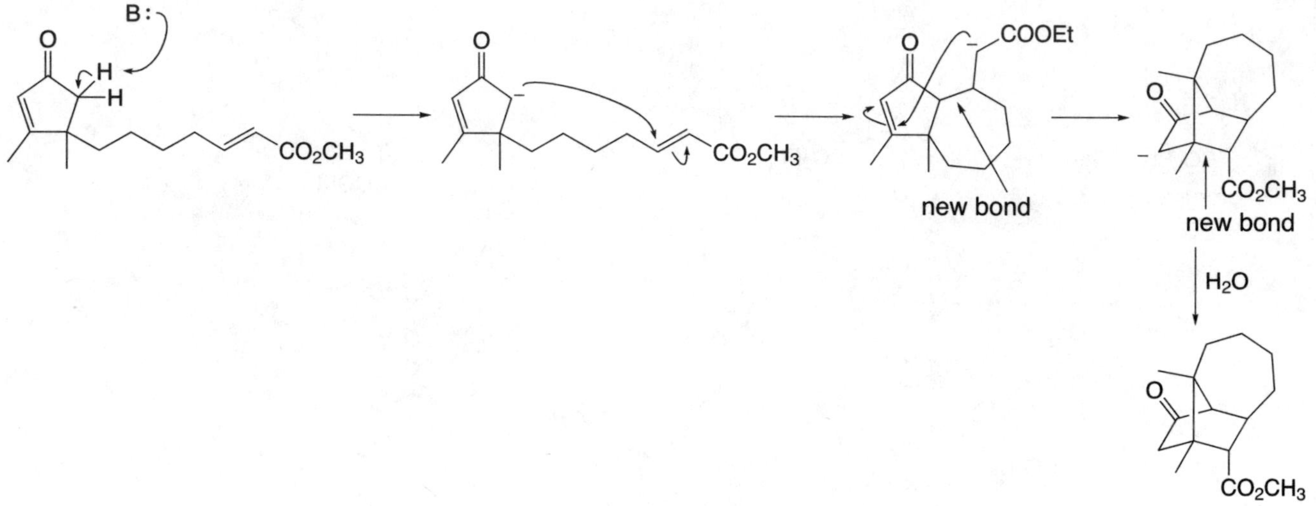

new bond

H_2O

Chapter 25: Amines

♦ General facts

- Amines are organic nitrogen compounds having the general structure RNH_2, R_2NH, or R_3N, with a lone pair of electrons on N (25.2).
- Amines are named using the suffix –*amine* (25.3).
- All amines have polar C–N bonds. $1°$ and $2°$ Amines have polar N–H bonds and are capable of intermolecular hydrogen bonding (25.4).
- The lone pair on N makes amines strong organic bases and nucleophiles (25.8)

♦ Summary of spectroscopic absorptions (25.5)

Mass spectra	Molecular ion	Amines with an odd number of N atoms give an odd molecular ion.
IR absorptions	N–H	$3300–3500$ cm^{-1} (two peaks for RNH_2, one peak for R_2NH)
^{1}H NMR absorptions	NH	$0.5–5$ ppm (no splitting with adjacent protons)
	CH–N	$2.3–3.0$ ppm (deshielded Csp^3–H)
^{13}C NMR absorption	C–N	$30–50$ ppm

♦ Comparing the basicity of amines and other compounds (25.10)

- Alkylamines (RNH_2, R_2NH, and R_3N) are more basic than NH_3 because of the electron donating R groups (25.10A).
- Alkylamines (RNH_2) are more basic than aryl amines ($C_6H_5NH_2$), which have a delocalized lone pair from the N atom (25.10B).
- Arylamines with electron donor groups are more basic than arylamines with electron withdrawing groups (25.10B).
- Alkylamines (RNH_2) are more basic than amides ($RCONH_2$), which have a delocalized lone pair from the N atom (25.10C).
- Aromatic heterocycles with a localized electron pair on N are more basic than those with a delocalized lone pair from the N atom (25.10D).
- Alkylamines with a lone pair in an sp^3 hybrid orbital are more basic than those with a lone pair in an sp^2 hybrid orbital (25.10E).

♦ Preparation of amines (25.7)

[1] Direct nucleophilic substitution with NH_3 and amines (25.7A)

- The mechanism is S_N2.
- The reaction works best for CH_3X or RCH_2X.
- The reaction works best to prepare $1°$ amines and quaternary ammonium salts.

[2] Gabriel synthesis (25.7A)

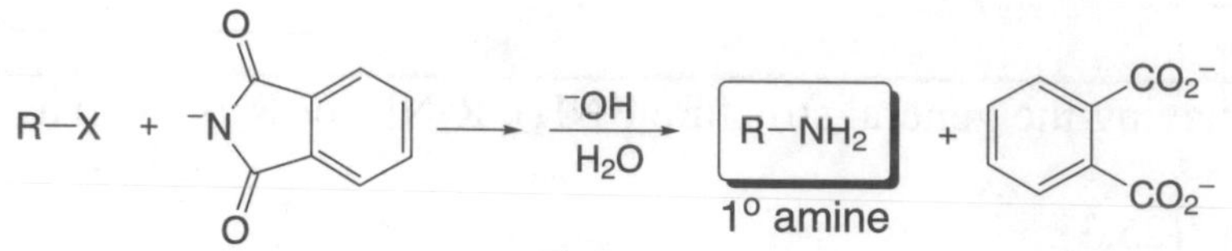

$R-X$ + (phthalimide anion) $\xrightarrow{}$ $\xrightarrow[H_2O]{^-OH}$ $R-NH_2$ **1° amine** + (benzene with CO_2^- and CO_2^-)

- The mechanism is S_N2.
- The reaction works best for CH_3X or RCH_2X.
- Only 1° amines can be prepared.

[3] Reduction methods (25.7B)

[a] From nitro compounds

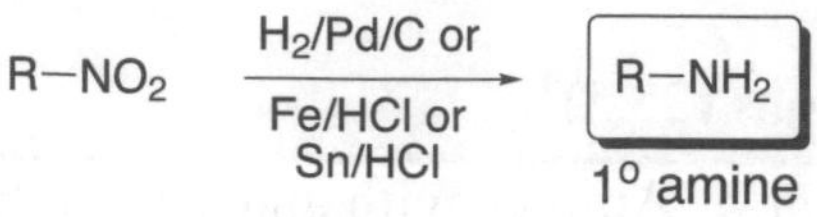

$R-NO_2$ $\xrightarrow[\text{Fe/HCl or}\atop\text{Sn/HCl}]{H_2/Pd/C \text{ or}}$ $R-NH_2$ **1° amine**

[b] From nitriles

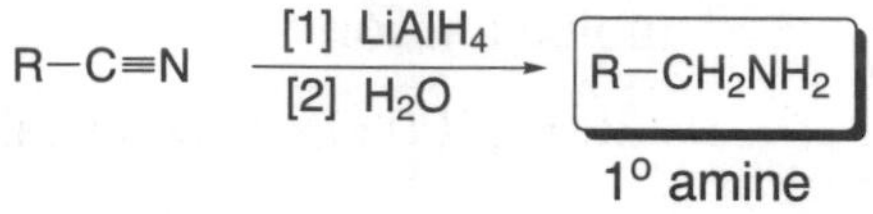

$R-C\equiv N$ $\xrightarrow[\text{[2] }H_2O]{\text{[1] LiAlH}_4}$ $R-CH_2NH_2$ **1° amine**

[c] From amides

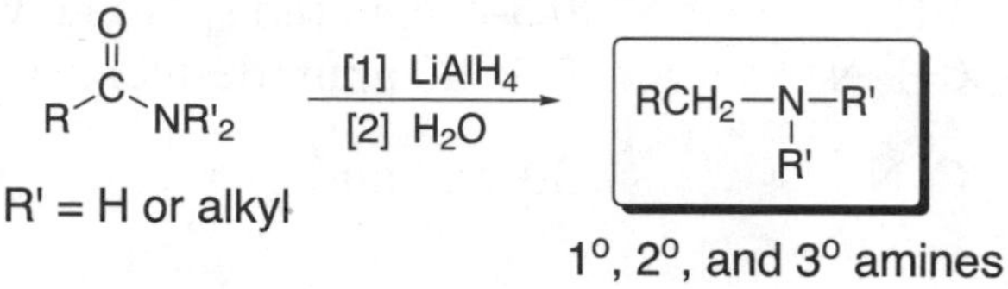

(amide: $R-\overset{O}{\overset{\|}{C}}-NR'_2$)

$R' = H$ or alkyl

$\xrightarrow[\text{[2] }H_2O]{\text{[1] LiAlH}_4}$ $RCH_2-\overset{R'}{\underset{R'}{N}}$

1°, 2°, and 3° amines

[4] Reductive amination (25.7C)

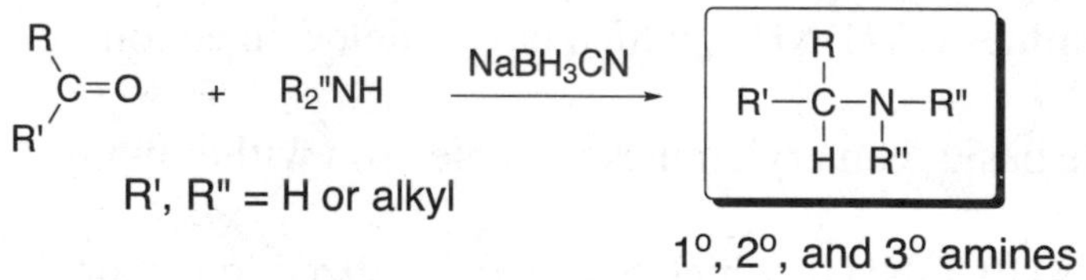

$\overset{R}{\underset{R'}{C}}=O$ + $R_2''NH$ $\xrightarrow{\text{NaBH}_3\text{CN}}$ $R'-\overset{R}{\underset{H}{\overset{|}{C}}}-\overset{|}{\underset{R''}{N}}-R''$

$R', R'' = H$ or alkyl

1°, 2°, and 3° amines

- Reductive amination adds one alkyl group (from an aldehyde or ketone) to a nitrogen nucleophile.
- 1°, 2°, and 3° Amines can be prepared.

♦ Reactions of amines

[1] Reaction as a base (25.9)

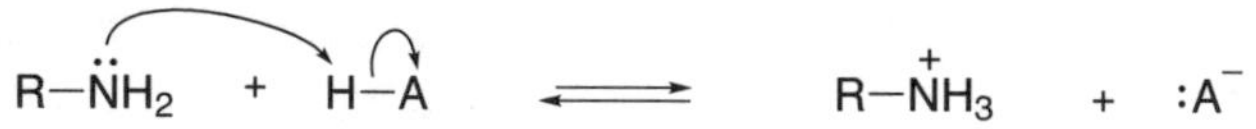

$R-\ddot{N}H_2$ + $H-A$ $\rightleftharpoons$ $R-\overset{+}{N}H_3$ + $:A^-$

[2] Nucleophilic addition to aldehydes and ketones (25.11)

With 1° amines:

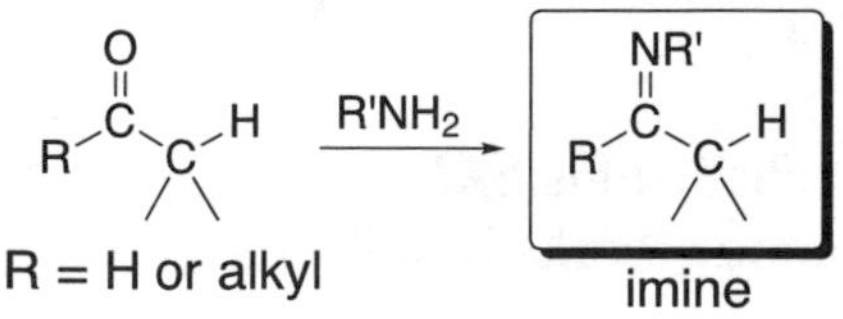

$R-\overset{O}{\overset{\|}{C}}-\underset{|}{\overset{|}{C}}-H$ $\xrightarrow{R'NH_2}$ $R-\overset{NR'}{\overset{\|}{C}}-\underset{|}{\overset{|}{C}}-H$

$R = H$ or alkyl   **imine**

With 2° amines:

$R-\overset{O}{\overset{\|}{C}}-\underset{|}{\overset{|}{C}}-H$ $\xrightarrow{R'_2NH}$ $R-\overset{NR'_2}{\overset{|}{C}}=\overset{|}{C}$

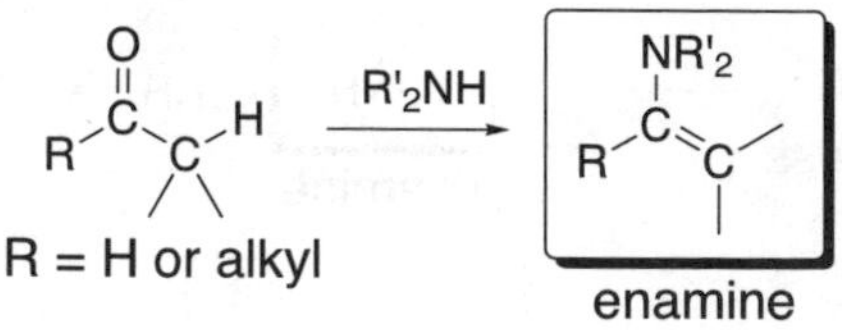

$R = H$ or alkyl   **enamine**

[3] Nucleophilic substitution with acid chlorides and anhydrides (25.11)

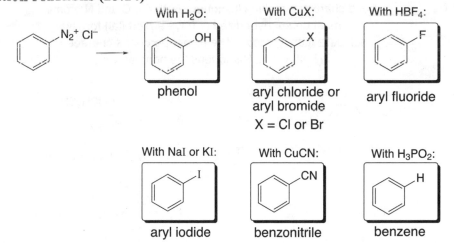

$$R\overset{\overset{\displaystyle O}{\parallel}}{C}Z \; + \; \underset{\text{(2 equiv)}}{R'_2NH} \longrightarrow \boxed{R\overset{\overset{\displaystyle O}{\parallel}}{C}NR'_2}$$

Z = Cl or OCOR
R' = H or alkyl

1°, 2°, and 3° amides

[4] Hofmann elimination (25.12)

$$-\underset{\underset{H}{|}}{C}-\underset{\underset{NH_2}{|}}{C}- \xrightarrow[\substack{[2]\; Ag_2O \\ [3]\; \Delta}]{[1]\; CH_3I \;(\text{excess})} \boxed{\overset{\diagup}{\underset{\diagup}{C}}=\overset{\diagdown}{\underset{\diagdown}{C}}}$$

alkene

• The less substituted alkene is the major product.

[5] Reaction with nitrous acid (25.13)

With 1° amines:

$$R-NH_2 \xrightarrow[HCl]{NaNO_2} \boxed{R-\overset{+}{N}\equiv N: \; Cl^-}$$

alkyl diazonium salt

With 2° amines:

$$R-\underset{\underset{R}{|}}{N}-H \xrightarrow[HCl]{NaNO_2} \boxed{R-\overset{..}{N}-\overset{..}{N}=\overset{..}{O}:}$$
$$\qquad\qquad\qquad\quad \underset{R}{|}$$

N-nitrosamine

◆ Reactions of diazonium salts

[1] Substitution reactions (25.14)

With H₂O:

phenol

With CuX:

aryl chloride or
aryl bromide
X = Cl or Br

With HBF₄:

aryl fluoride

With NaI or KI:

aryl iodide

With CuCN:

benzonitrile

With H₃PO₂:

benzene

[2] Coupling to form azo compounds (25.15)

azo compound + HCl

Y = NH₂, NHR, NR₂, OH
(a strong electron
donor group)

Chapter 25: Answers to Problems

25.1 Amines are classified as 1°, 2°, or 3° by the number of alkyl groups bonded to the *nitrogen* atom.

a.

b. CH_3CH_2O ... N—CH_3
3° amine

c.

25.2 The N atom of a quaternary ammonium salt is a stereogenic center when the N is surrounded by four different groups. All stereogenic centers are circled.

a.
$$CH_3-\overset{+}{\underset{CH_3}{N}}-CH_2CH_2-\overset{+}{N}-CH_2CH_3$$
N has 3 similar groups.

b.

c.

25.3

$$CH_3 \dashrightarrow \ddot{N}H_2$$
1.47 Å

The C–N bond is formed from two sp^3 hybridized atoms and the lone pair is localized on N.

+ 3 more resonance structures

1.40 Å

Because the lone pair on N can be delocalized on the benzene ring, the C–N bond has partial double bond character, making it shorter. Both the C and N atoms must be sp^2 hybridized (+ have a p orbital) for delocalization to occur. The higher percent s-character in both C and N shortens the bond as well.

partial double bond character

25.4

a. $CH_3CH_2CH(NH_2)CH_3$

2-butanamine
or
sec-butylamine

b. $(CH_3CH_2CH_2CH_2)_2NH$

dibutylamine

c. —$N(CH_3)_2$

N,N-dimethylcyclohexanamine

d.

2-methyl-5-nonanamine

e.

N-ethyl-3-hexanamine

f.

N-propyl-2-methylcyclopentanamine

25.5 Aromatic amines are named as derivatives of aniline.

a. *N*-methylaniline

b. *m*-ethylaniline

c. 3,5-diethylaniline

d. *N,N*-diethylaniline

25.6 An **NH₂** group named as a substituent is called an **amino group**.

a. 2,4-dimethyl-3-hexanamine

c. *N*-isopropyl-*p*-nitroaniline

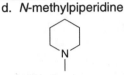

e. *N,N*-dimethylethylamine

g. 1-propylcyclohexylamine

b. *N*-methylpentylamine

d. *N*-methylpiperidine

f. 2-aminocyclohexanone

h. *p*-butyl-*N*-ethylaniline

25.7 1° and 2° Amines have higher bp's than similar compounds (like ethers) incapable of hydrogen bonding, but lower bp's than alcohols that have stronger intermolecular hydrogen bonds. 3° Amines have lower boiling points than 1° and 2° amines of comparable molecular weight because they have no N–H bonds.

a. $(CH_3)_2CHCH_2CH_3$

alkane
**lowest
boiling point**

$CH_3-\overset{O}{\overset{\|}{C}}-CH_2CH_3$

ketone
**intermediate
boiling point**

$(CH_3)_2CHCH_2NH_2$

amine
N–H can hydrogen bond
highest boiling point

b.

—CH_3

alkane
**lowest
boiling point**

—CH_3

ether
**intermediate
boiling point**

—NH_2

amine
N–H can hydrogen bond
highest boiling point

25.8 1° Amines show *two* N–H absorptions at 3300–3500 cm⁻¹. 2° Amines show *one* N–H absorption at 3300–3500 cm⁻¹.

molecular weight = 59
one IR peak = 2° amine

$CH_3-\underset{\underset{H}{|}}{N}-CH_2CH_3$

25.9

molecular weight = 87
$C_5H_{13}N$
two IR peaks = 1° amine

25.10 **The NH signal occurs between 0.5 and 5.0 ppm.** The protons on the carbon bonded to the amine nitrogen are deshielded and typically absorb at 2.3–3.0 ppm. The NH protons are not split.

molecular formula $C_6H_{15}N$
¹H NMR absorptions (ppm):
    0.9 (singlet, 1H) ⟶ NH
    1.10 (triplet, 3H) ⟶ CH_3 adjacent to CH_2
    1.15 (singlet, 9H) ⟶ $(CH_3)_3C$
    2.6 (quartet, 2H) ⟶ CH_2 adjacent to CH_3

25.11 The atoms of 2-phenylethylamine are in bold.

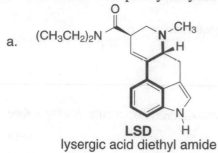

a.

LSD
lysergic acid diethyl amide

b.

codeine

25.12 S_N2 reaction of an alkyl halide with NH_3 or an amine forms an amine.

a. ~~~~~Cl $\xrightarrow[\text{excess}]{NH_3}$ ~~~~~NH_2

b. ⬡—$CH_2CH_2NH_2$ $\xrightarrow[\text{excess}]{CH_3CH_2Br}$ ⬡—$CH_2CH_2\overset{+}{N}(CH_2CH_3)_3$ Br^-

25.13

$\xrightarrow{KOH}$ A $\xrightarrow{C_6H_5CH_2Cl}$ B $\xrightarrow[\text{H}_2\text{O}]{^-OH}$ [COO⁻ / COO⁻] + H_2N—CH₂-phenyl

A   B   C

25.14 The Gabriel synthesis converts an alkyl halide into a $1°$ amine by a two-step process: nucleophilic substitution followed by hydrolysis.

a. ~~~~~NH_2
⇓
~~~~~Br

b. $(CH_3)_2CHCH_2CH_2NH_2$
⇓
$(CH_3)_2CHCH_2CH_2Br$

c. CH_3O—⬡—$CH_2CH_2NH_2$ ⇒ CH_3O—⬡—CH_2CH_2Br

25.15 **Nitriles are reduced to $1°$ amines with LiAlH$_4$. Nitro groups are reduced to $1°$ amines** using a variety of reducing agents. **$1°$, $2°$, and $3°$ Amides are reduced to $1°$, $2°$, and $3°$ amines** respectively, using LiAlH$_4$.

a. $\underset{CH_3}{CH_3CHCH_2NH_2}$ ⇒ $\underset{CH_3}{CH_3CHCH_2NO_2}$ $\underset{CH_3}{CH_3CHC\equiv N}$ $\underset{CH_3}{CH_3CH}\overset{O}{\overset{\|}{C}}NH_2$

b. ⬡—CH_2NH_2 ⇒ ⬡—CH_2NO_2 ⬡—$C\equiv N$ ⬡—$\overset{O}{\overset{\|}{C}}NH_2$

c. ~~~~~NH_2 ⇒ ~~~~~NO_2 ~~~~~$C\equiv N$ ~~~~~$\overset{O}{\overset{\|}{C}}NH_2$

25.16 **$1°$, $2°$, and $3°$ Amides are reduced to $1°$, $2°$, and $3°$ amines** respectively, using LiAlH$_4$.

a. ⬡—$CONH_2$ ⟶ ⬡—CH_2NH_2

b. ~~~$\overset{O}{\overset{\|}{C}}$—N(Et)~ ⟶ ~~~~N(Et)~

c. ~~~$\overset{O}{\overset{\|}{C}}$NHCH₃ ⟶ ~~~~NHCH₃

25.17

$(CH_3)_2CHNH_2$

isopropylamine

General reaction:

$$R-C\equiv N \xrightarrow{[H]} R\overset{\uparrow}{C}H_2NH_2$$

The amine needs 2 H's here.

The C bonded to the N must have 2 H's to be formed by reduction of a nitrile.

25.18 Reductive amination is a two-step method that converts aldehydes and ketones into 1°, 2°, and 3° amines. Reductive amination replaces a C=O by a C–H and C–N bond.

a.

$$\text{Ar–CHO} \xrightarrow[\text{NaBH}_3\text{CN}]{\text{CH}_3\text{NH}_2} \text{Ar–CH}_2\text{NHCH}_3$$

b.

$$\xrightarrow[\text{NaBH}_3\text{CN}]{\text{NH}_3}$$

c.

$$\xrightarrow[\text{NaBH}_3\text{CN}]{\text{(CH}_3\text{CH}_2)_2\text{NH}} \text{cyclohexyl–N(CH}_2\text{CH}_3)_2$$

25.19 In reductive amination, one alkyl group on N comes from the carbonyl compound. The remainder of the molecule comes from NH_3 or an amine.

a. cyclopentyl–NH_2 $\Longrightarrow$ cyclopentyl=O + NH_3

b. $CH_3CH_2-\overset{\underset{\displaystyle CH_3}{|}}{N}-CH_2CH_3 \Longrightarrow$

$$\underset{H}{\overset{O}{\underset{|}{\overset{||}{C}}}}\text{H} + HN(CH_2CH_3)_2$$

$$\underset{H}{\overset{O}{\overset{||}{C}}}CH_3 + CH_3CH_2NHCH_3$$

c. $H-\overset{\underset{\displaystyle CH_3}{|}}{N}-CH(CH_3)_2 \Longrightarrow$

$$\underset{H}{\overset{O}{\overset{||}{C}}}H + NH_2CH(CH_3)_2$$

or

$$CH_3\overset{O}{\overset{||}{C}}CH_3 + NH_2CH_3$$

25.20

$(CH_3)_3CNH_2$
tert-butylamine

Only amines that have a C bonded to a H and N atom can be made by reductive amination; that is, an amine must have the following structural feature:

$$-\overset{\underset{\displaystyle |}{|}}{\underset{H}{\overset{H}{C}}}-\overset{|}{N}-$$

The only alkyl group in *tert*-butylamine does not have a H on the C bonded to N, so it cannot be made by reductive amination.

25.21 The pK_a of many protonated amines is 10–11, the pK_a of the starting acid must be **less than 10** for equilibrium to favor the products. Amines are thus readily protonated by strong inorganic acids like HCl and H_2SO_4, and by carboxylic acids as well.

a. $CH_3CH_2CH_2CH_2-NH_2$ + HCl $\rightleftharpoons$ $CH_3CH_2CH_2CH_2-\overset{+}{N}H_3$ + Cl⁻

pK_a = –7   p$K_a \approx 10$
     weaker acid
     products favored

b. C_6H_5COOH + $(CH_3)_2NH$ $\rightleftharpoons$ $(CH_3)_2\overset{+}{N}H_2$ + $C_6H_5COO^-$

pK_a = 4.2   pK_a = 10.7
     weaker acid
     products favored

c.

+ H_2O $\rightleftharpoons$ + HO⁻

pK_a = 15.7   p$K_a \approx 10$
weaker acid
reactants favored

25.22 An amine can be separated from other organic compounds by converting it to a water-soluble ammonium salt by an acid–base reaction.

In each case, the extraction procedure would employ the following steps:
• Dissolve the amine and either **X** or **Y** in CH_2Cl_2.
• Add a solution of 10% HCl. The amine will be protonated and dissolve in the aqueous layer, while **X** or **Y** will remain in the organic layer as a neutral compound.
• Separate the layers.

a. [cyclohexyl]–NH_2 and [cyclohexyl]–CH_3 $\xrightarrow{\text{H–Cl}}$ [cyclohexyl]–$\overset{+}{N}H_3$ Cl^- + [cyclohexyl]–CH_3

   X

   • **soluble in H_2O**
   • **insoluble in CH_2Cl_2**

   X

   • insoluble in H_2O
   • soluble in CH_2Cl_2

b. $(CH_3CH_2CH_2CH_2)_3N$ and $(CH_3CH_2CH_2CH_2)_2O$ $\xrightarrow{\text{H–Cl}}$ $(CH_3CH_2CH_2CH_2)_3\overset{+}{N}H$ Cl^- + $(CH_3CH_2CH_2CH_2)_2O$

   Y

   • **soluble in H_2O**
   • **insoluble in CH_2Cl_2**

   Y

   • insoluble in H_2O
   • soluble in CH_2Cl_2

25.23 The weaker the conjugate acid, the higher its pK_a and the stronger the base. (pK_a values are for the conjugate acid of a given amine.)

a. CH_3NH_2 ($pK_a = 10.7$)   and $CH_3CH_2NH_2$ ($pK_a = 10.8$)

   stronger conjugate acid
   weaker base

   weaker conjugate acid
   stronger base

b. $(CH_3CH_2)_3N$ ($pK_a = 11.0$)   and $(CH_3)_3N$ ($pK_a = 9.8$)

   weaker conjugate acid
   stronger base

   stronger conjugate acid
   weaker base

25.24 1°, 2°, and 3° Alkylamines are more basic than NH_3 because of the electron donating inductive effect of the R groups.

   a. $(CH_3)_2NH$ and NH_3

   2° alkylamine
   CH_3 groups are electron donating.
   stronger base

   b. $CH_3CH_2NH_2$ and $ClCH_2CH_2NH_2$

   1° alkylamine
   stronger base

   1° alkylamine
   Cl is electron withdrawing.
   weaker base

25.25 Arylamines are less basic than alkylamines because the electron pair on N is delocalized. Electron donor groups add electron density to the benzene ring making the arylamine more basic than aniline. Electron withdrawing groups remove electron density from the benzene ring, making the arylamine less basic than aniline.

a. [structures with NH_2]

   CH_3O–[ring]–NH_2

   electron
   donating group
   most basic

   [ring]–NH_2

   arylamine
   **intermediate
   basicity**

   CH_3OOC–[ring]–NH_2

   electron
   withdrawing group
   least basic

b.

   O_2N–[ring]–NH_2

   electron
   withdrawing group
   least basic

   [ring]–NH_2

   arylamine
   **intermediate
   basicity**

   [cyclohexyl]–NH_2

   alkylamine
   most basic

25.26 Amides are much less basic than amines because the electron pair on N is highly delocalized.

amide	arylamine	alkylamine
least basic	**intermediate basicity**	**most basic**

25.27

a.

sp² hybridized
more basic

This N is also *sp²* hybridized so the electron pair can delocalize onto the aromatic ring. Delocalization makes this N less basic.

DMAP
4-(dimethylamino)pyridine

b.

— *sp³* hybridized N
stronger base

nicotine

sp² hybridized N
higher percent *s*-character
weaker base

25.28

a.

This electron pair is delocalized, making it a weaker base.

:NH₂

stronger base
This compound is similar to DMAP in Problem 25.27a.

b.

stronger base
sp³ hybridized N
25% *s*-character

sp² hybridized N
33% s-character

c.

sp² hybridized N
33% *s*-character

stronger base
sp³ hybridized N
25% *s*-character

25.29 Amines attack carbonyl groups to form products of nucleophilic addition or substitution.

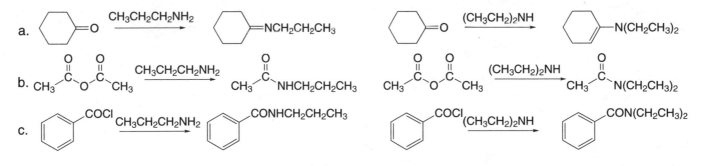

25.30 [1] Convert the amine (aniline) into an amide (acetanilide).
[2] **Carry out the Friedel-Crafts reaction.**
[3] **Hydrolyze the amide** to generate the free amino group.

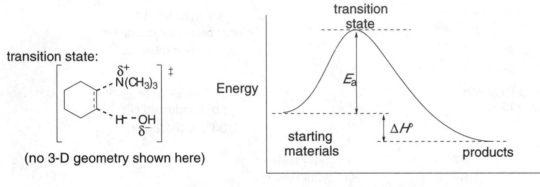

25.31

transition state:

(no 3-D geometry shown here)

transition
state

Energy

E_a

starting
materials

ΔH^o

products

reaction coordinate

25.32

a. $CH_3CH_2CH_2CH_2-NH_2$ $\xrightarrow[\text{[3] }\Delta]{\text{[1] CH}_3\text{I (excess)}}{\text{[2]Ag}_2\text{O}}$ $CH_3CH_2CH=CH_2$

c. ⬠—NH_2 $\xrightarrow[\text{[3] }\Delta]{\text{[1] CH}_3\text{I (excess)}}{\text{[2]Ag}_2\text{O}}$ ⬠

b. $(CH_3)_2CHNH_2$ $\xrightarrow[\text{[3] }\Delta]{\text{[1] CH}_3\text{I (excess)}}{\text{[2]Ag}_2\text{O}}$ $CH_3CH=CH_2$

25.33 In a Hofmann elimination, the base removes a proton from the less substituted, more accessible β carbon atom, because of the bulky leaving group on the nearby α carbon.

a. ⬡—CH_2CHCH_3 (with NH_2) $\xrightarrow[\text{[3] }\Delta]{\text{[1] CH}_3\text{I (excess)}}{\text{[2] Ag}_2\text{O}}$ ⬡—$CH=CHCH_3$ + ⬡—$CH_2CH=CH_2$
major product

b. [structure with H_2N and cyclohexane] $\xrightarrow[\text{[3] }\Delta]{\text{[1] CH}_3\text{I (excess)}}{\text{[2] Ag}_2\text{O}}$ [vinyl cyclohexane]
major product
+
[ethylidene cyclohexane]

c. β [piperidine ring with β, β, β labels] β
(3 β C's) $\xrightarrow[\text{[3] }\Delta]{\text{[1] CH}_3\text{I (excess)}}{\text{[2] Ag}_2\text{O}}$ $CH_3CH=CH(CH_2)_3N(CH_3)_2$
+
$CH_2=CH(CH_2)_2CHN(CH_3)_2$ (with CH_3)
+
$CH_2=CH(CH_2)_4N(CH_3)_2$
major product

25.34

a. 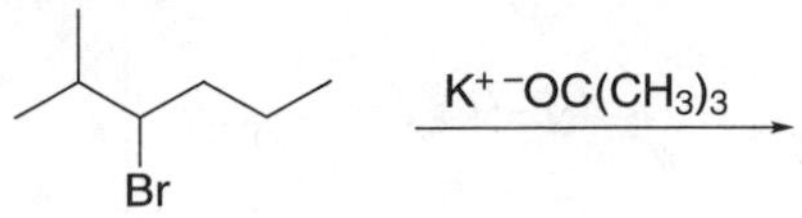
Br
K+ −OC(CH3)3

c. 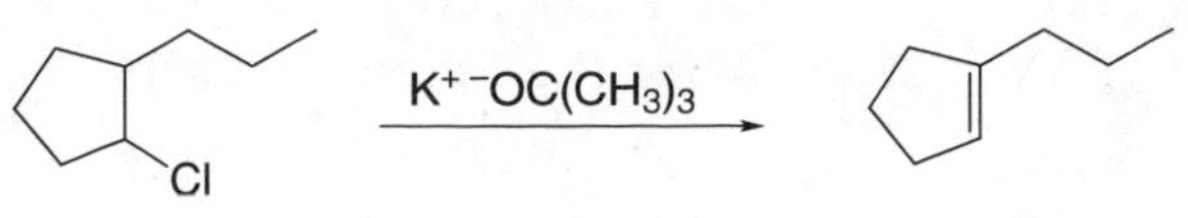
Cl
K+ −OC(CH3)3

b. 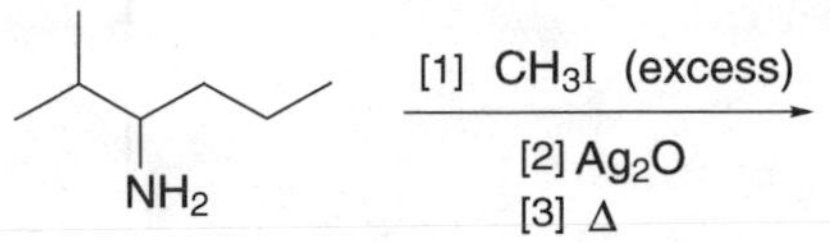
NH2
[1] CH3I (excess)
[2] Ag2O
[3] Δ
(E and Z)

d. 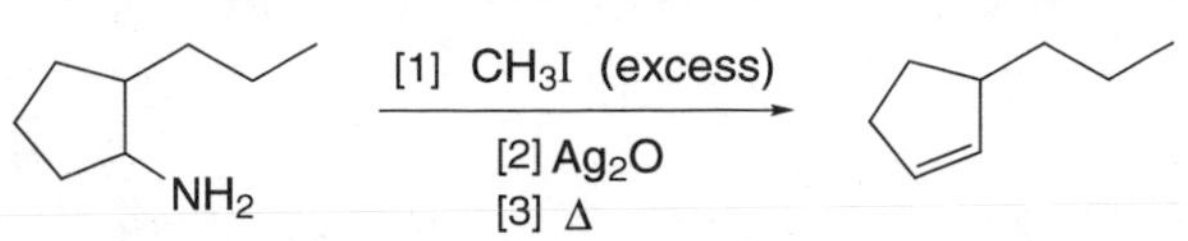
NH2
[1] CH3I (excess)
[2] Ag2O
[3] Δ

25.35

a. 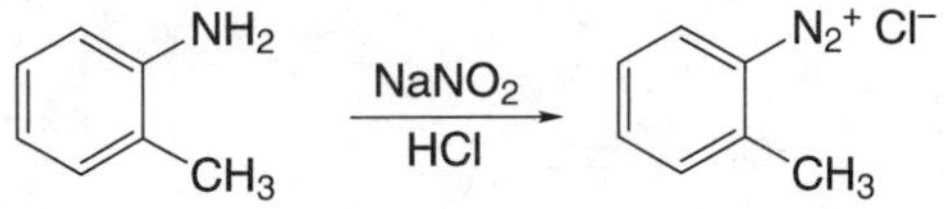
NH2
CH3
NaNO2
HCl
N2+ Cl−
CH3

c. 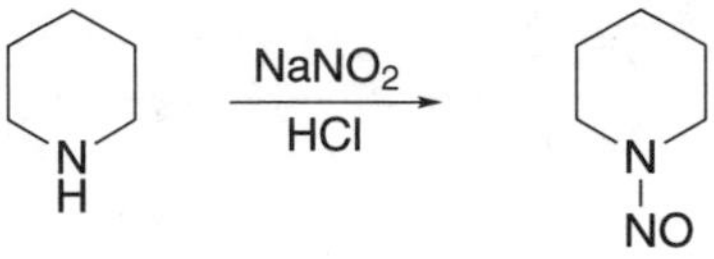
N
H
NaNO2
HCl
N
NO

b. 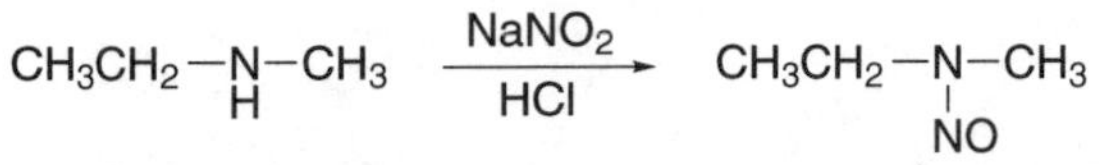
CH3CH2–N–CH3
H
NaNO2
HCl
CH3CH2–N–CH3
NO

d. 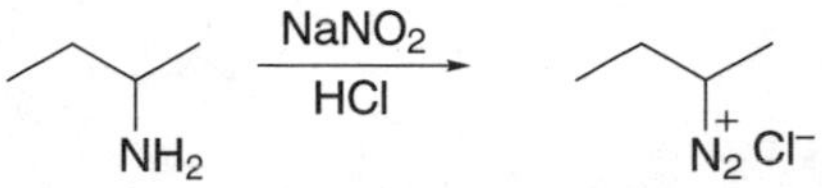
NH2
NaNO2
HCl
N2+ Cl−

25.36

a. 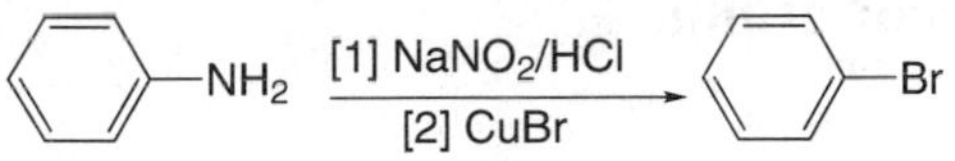
NH2
[1] NaNO2/HCl
[2] CuBr
Br

c. 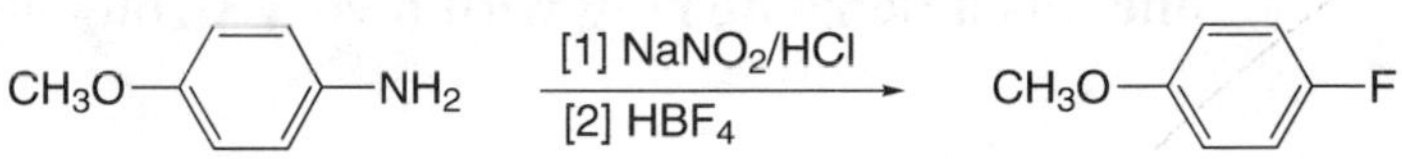
CH3O
NH2
[1] NaNO2/HCl
[2] HBF4
CH3O
F

b. 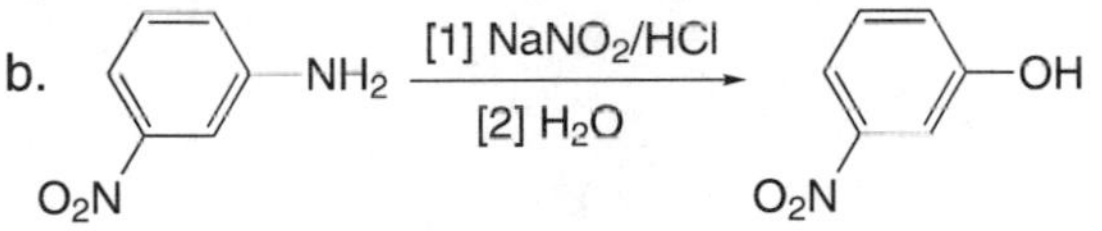
NH2
O2N
[1] NaNO2/HCl
[2] H2O
OH
O2N

d. 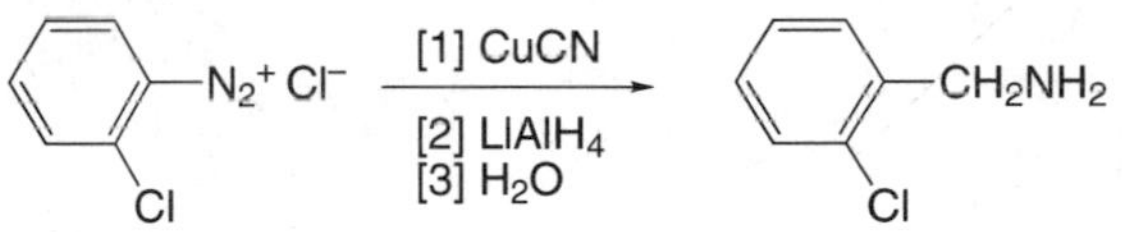
N2+ Cl−
Cl
[1] CuCN
[2] LiAlH4
[3] H2O
CH2NH2
Cl

25.37

a. 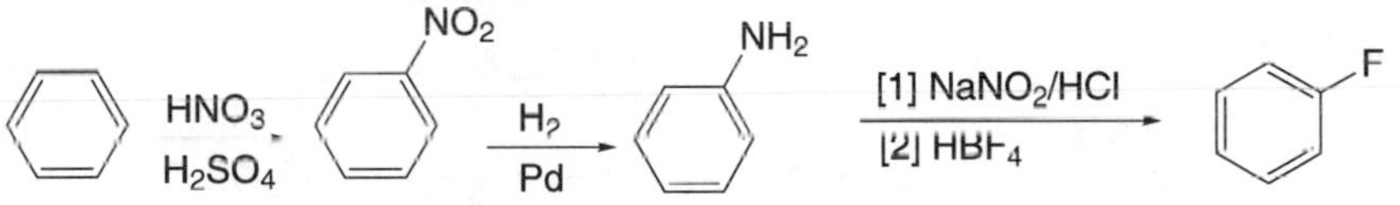
HNO3
H2SO4
NO2
H2
Pd
NH2
[1] NaNO2/HCl
[2] HBF4
F

b. 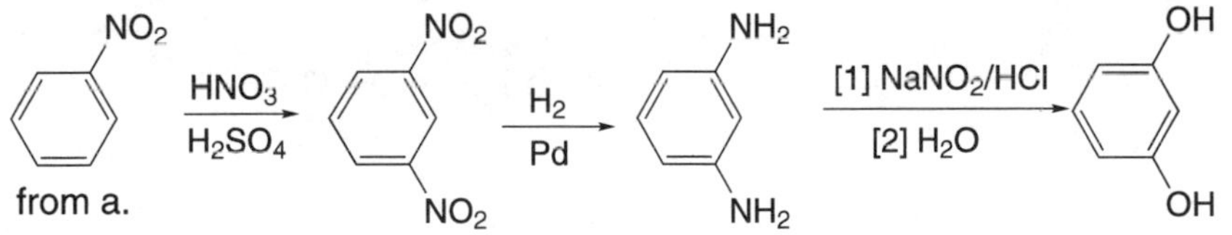
NO2
from a.
HNO3
H2SO4
NO2
NO2
H2
Pd
NH2
NH2
[1] NaNO2/HCl
[2] H2O
OH
OH

c. 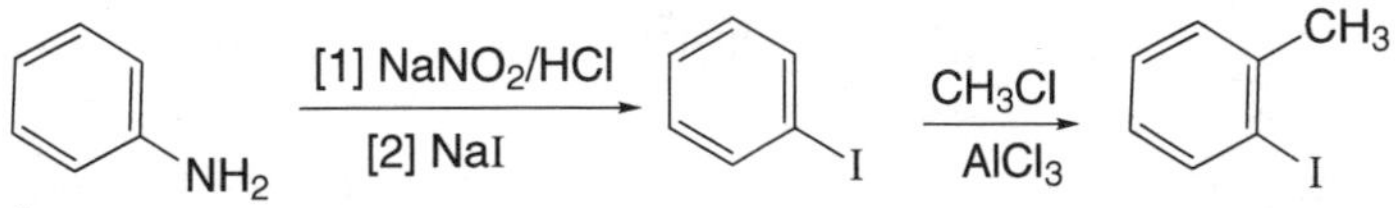
NH2
[1] NaNO2/HCl
[2] NaI
I
CH3Cl
AlCl3
CH3
I

(+ *para* isomer)

from a.

d. 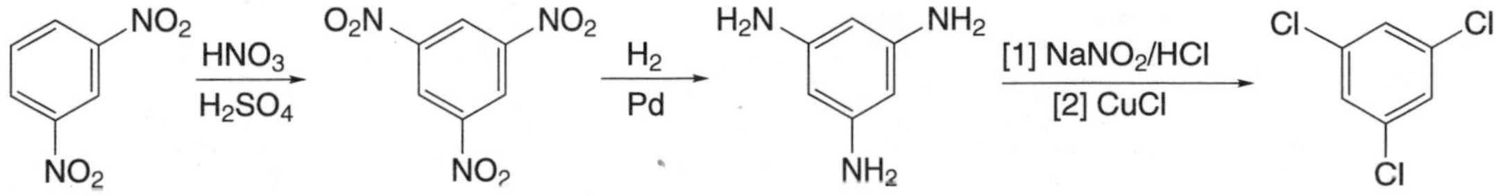
NO2
NO2
HNO3
H2SO4
O2N
NO2
NO2
H2
Pd
H2N
NH2
NH2
[1] NaNO2/HCl
[2] CuCl
Cl
Cl
Cl

from b.

25.38

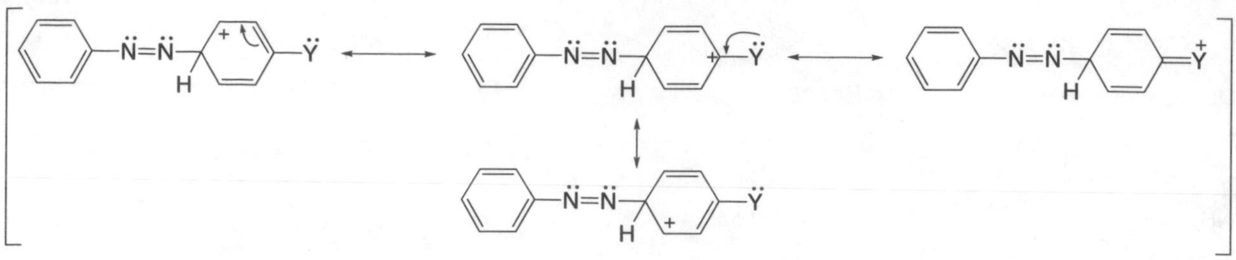

25.39

a.

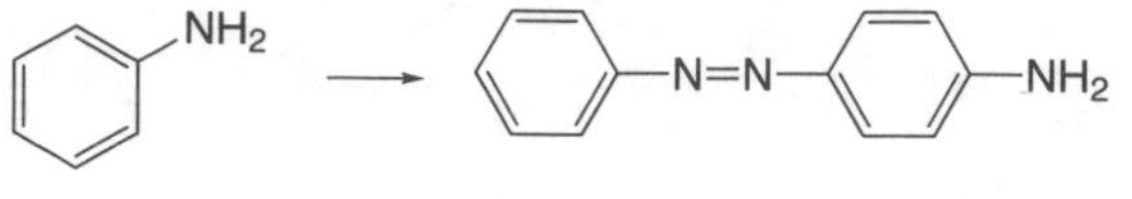

c.

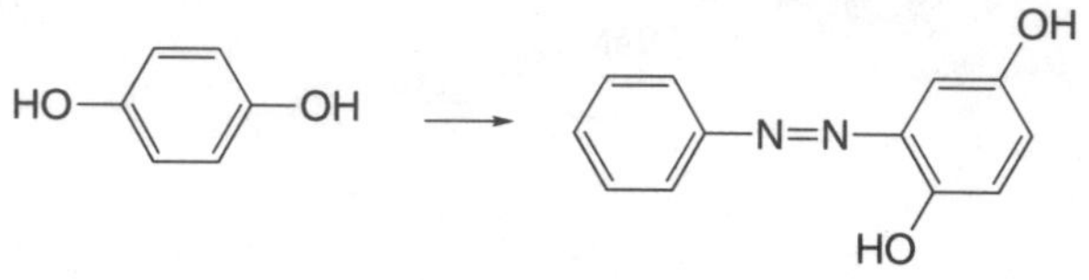

b.

25.40 To determine what starting materials are needed to synthesize a particular azo compound, always divide the molecule into two components: **one has a benzene ring with a diazonium ion, and one has a benzene ring with a very strong electron donor group.**

a.

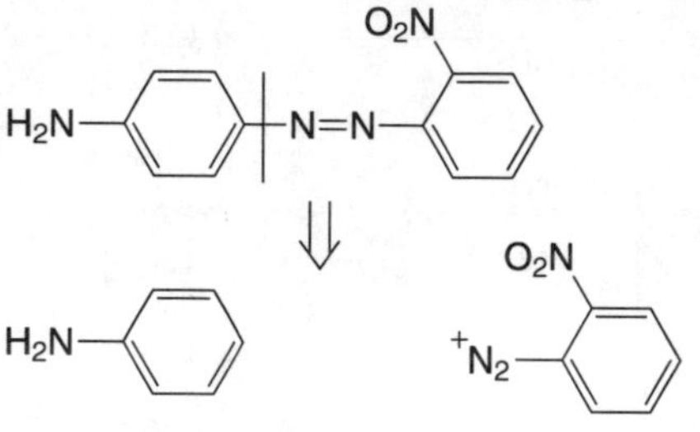

b.

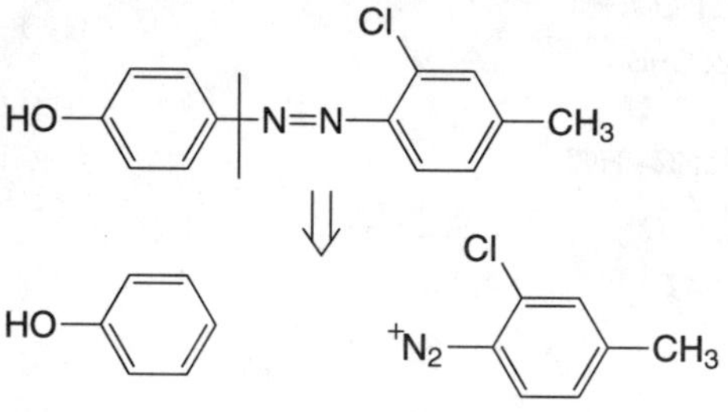

25.41

a.

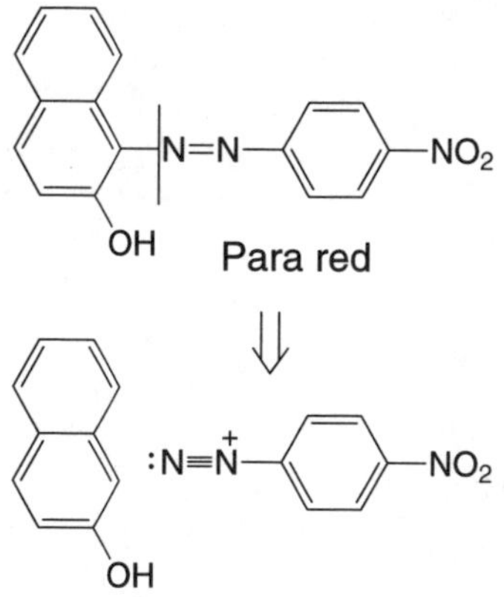

Para red

b.

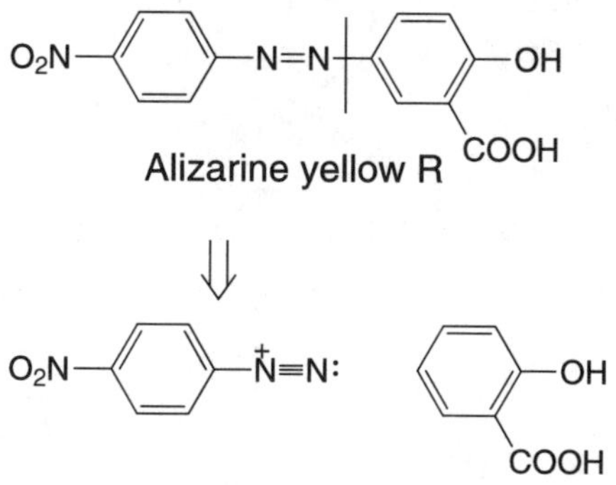

Alizarine yellow R

25.42

dacron

methyl orange

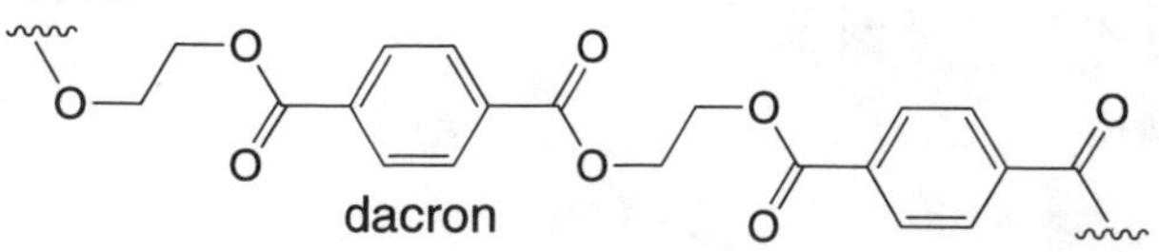

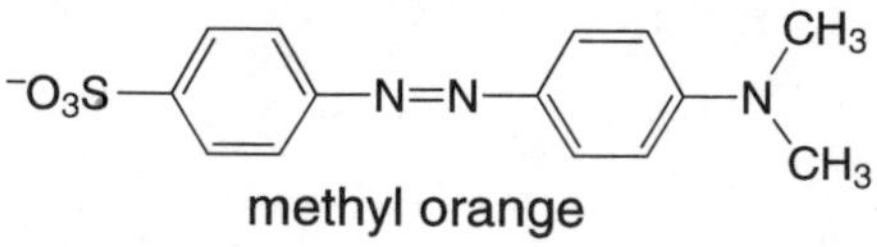

To bind to fabric, methyl orange (an anion) needs to interact with positively charged sites. Since dacron is a neutral compound with no cationic sites on the chain, it does not bind methyl orange.

25.43

a. CH$_3$NHCH$_2$CH$_2$CH$_2$CH$_3$

    N-methyl-1-butanamine
    (N-methylbutylamine)

b. [structure: CH$_2$ chain with NH$_2$]

    1-octanamine
    (octylamine)

c. [branched chain structure with NH$_2$]

    4,6-dimethyl-1-heptanamine

d. [cyclohexane with N(CH$_3$)(CH$_2$CH$_2$CH$_3$)]

    N-methyl-N-propylcyclohexanamine

e. (CH$_3$CH$_2$CH$_2$)$_3$N

    tripropylamine

f. (C$_6$H$_5$)$_2$NH

    diphenylamine

g. [phenyl-N-C(CH$_3$)$_3$ with CH$_2$CH$_3$]

    N-tert-butyl-N-ethylaniline

h. O=[cyclohexane]-NH$_2$

    4-aminocyclohexanone

i. [pyrrolidine ring with CH$_2$CH$_3$]

    2-ethylpyrrolidine

j. CH$_3$CH$_2$CH$_2$CH(NH$_2$)CH(CH$_3$)$_2$

    2-methyl-3-hexanamine

k. [cyclohexane with NH$_2$, ethyl, methyl substituents]

    3-ethyl-2-methylcyclohexanamine

l. [cycloheptane-N(CH$_2$CH$_3$)$_2$]

    N,N-diethylcycloheptanamine

25.44

a. cyclobutylamine

[cyclobutane-NH$_2$]

b. N-isobutylcyclopentylamine

[cyclopentane-N(H)-isobutyl]

c. tri-tert-butylamine

    N[C(CH$_3$)$_3$]$_3$

d. N,N-diethylaniline

[phenyl-N(CH$_2$CH$_3$)$_2$]

e. N-methylpyrrole

[pyrrole ring with N-CH$_3$]

f. N-methylcyclopentylamine

[cyclopentane-NHCH$_3$]

g. cis-2-aminocyclohexanol

[cyclohexane with NH$_2$ and OH]

h. 3-methyl-2-hexanamine

[structure with NH$_2$]

i. 2-sec-butylpiperidine

[piperidine ring with sec-butyl]

j. (S)-2-heptanamine

[structure with H, NH$_2$]

25.45 [* denotes a stereogenic center.]

a. [branched structure with N(CH$_3$)$_2$, stereocenter marked *]

    1 stereogenic center
    2 stereoisomers

[two stereoisomer structures with N(CH$_3$)$_2$, H wedge/dash]

b. CH$_3$CH$_2$*CHCH$_2$CH$_2$CH$_2$—*N$^+$—CH$_2$CH$_2$CH$_2$CH$_3$
with CH$_2$CH$_3$, CH$_3$ groups, Cl$^-$

    2 stereogenic centers
    4 stereoisomers

[four stereoisomer structures with N$^+$ and Cl$^-$]

25.46

a. $(CH_3CH_2)_2NH$ or

sp^3 hybridized N sp^2 hybridized N
stronger base   **weaker base**

b. $C_6H_5NHCH_3$ or $C_6H_5CH_2NH_2$

   arylamine   alkylamine
 weaker base   **stronger base**

c. $HCON(CH_3)_2$ or $(CH_3)_3N$

   amide   alkylamine
weaker base   **stronger base**

d. $(CH_3CH_2)_2NH$ or $(ClCH_2CH_2)_2NH$

                 2° alkylamine
2° alkylamine   Cl is electron withdrawing.
stronger base   **weaker base**

25.47

a.

arylamine   **intermediate**   alkylamine
least basic   **basicity**   **most basic**

b.

delocalized
electron pair on N sp^2 hybridized N sp^3 hybridized N
least basic   **intermediate**   **most basic**
               basicity

c.

   electron
withdrawing group   **intermediate**   electron
least basic   **basicity**   donating group
                              most basic

d. $(C_6H_5)_2NH$ $C_6H_5NH_2$

diarylamine   arylamine   alkylamine
least basic   **intermediate**   **most basic**
           basicity

25.48

a.

sp^3 hybridized N
stronger base

$(CH_3CH_2)_2NCH_2CH_2O$

delocalized
electron pair

b.

sp^3 hybridized N
stronger base

sp^2 hybridized N

25.49

a.

$N_b < N_a < N_c$

Order of basicity: $N_b < N_a < N_c$
N_b – The electron pair on this N atom is delocalized
     on the O atom; least basic.
N_a – The electron pair on this N atom is not
     delocalized, but is on an sp^2 hybridized atom.
N_c – The electron pair on this N atom is on an sp^3
     hybridized N; most basic.

b.

$N_b < N_a < N_c$

Order of basicity: $N_b < N_a < N_c$
N_b – The electron pair on this N atom is delocalized
     on the aromatic five-membered ring; least basic.
N_a – The electron pair on this N atom is not
     delocalized, but is on an sp^2 hybridized atom.
N_c – The electron pair on this N atom is on an sp^3
     hybridized N; most basic.

25.50

The para isomer is the weaker base because the electron pair on its NH_2 group can be delocalized onto the NO_2 group. In the meta isomer, no resonance structure places the electron pair on the NO_2 group, and fewer resonance structures can be drawn:

meta

para

25.51

A

pK_a of the conjugate acid = 5.2
stronger conjugate acid
weaker base
The electron pair of this arylamine
is delocalized on the benzene ring,
decreasing its basicity.

B

This two-carbon bridge makes it difficult for the lone pair on N to delocalize on the aromatic ring.

pK_a of the conjugate acid = 7.29
weaker conjugate acid
stronger base

Resonance structures that place a double bond between the N atom and the benzene ring are destabilized. Since the electron pair is more localized on N, compound **B** is more basic.

B

Geometry makes having a double bond here difficult.

25.52

a. $C_6H_5CH_2CH_2CH_2Br$ $\xrightarrow[\text{excess}]{NH_3}$ $C_6H_5CH_2CH_2CH_2NH_2$

b. $C_6H_5CH_2CH_2Br$ $\xrightarrow{NaCN}$ $C_6H_5CH_2CH_2CN$ $\xrightarrow[\text{[2] H}_2\text{O}]{\text{[1] LiAlH}_4}$ $C_6H_5CH_2CH_2CH_2NH_2$

c. $C_6H_5CH_2CH_2CH_2NO_2$ $\xrightarrow[\text{Pd/C}]{H_2}$ $C_6H_5CH_2CH_2CH_2NH_2$

d. $C_6H_5CH_2CH_2CONH_2$ $\xrightarrow[\text{[2] H}_2\text{O}]{\text{[1] LiAlH}_4}$ $C_6H_5CH_2CH_2CH_2NH_2$

e. $C_6H_5CH_2CH_2CHO$ $\xrightarrow[\text{NaBH}_3\text{CN}]{NH_3}$ $C_6H_5CH_2CH_2CH_2NH_2$

25.53

a. $(CH_3CH_2)_2NH$

$\Downarrow$

$CH_3-\overset{\overset{O}{\|}}{C}-NHCH_2CH_3$

b. [structure with NH_2]

$\Downarrow$

[structure with $\overset{\overset{O}{\|}}{}$ and NH_2]

c. [structure with $N(CH_3)_2$]

$\Downarrow$

[structure with $N(CH_3)_2$ and O]

or

[structure] $\overset{CH_3}{\underset{\overset{\|}{C}-H}{N}}$ with O

d. [structure with $\overset{H}{N}$]

$\Downarrow$

[structure with $\overset{H}{N}$ and O]

25.54 Use the directions from Answer 25.19.

a. [structure with NH_2] $\Longrightarrow$ [structure with H, O] $+$ NH_3

b. [structure with $\overset{H}{N}$, C_6H_5] $\Longrightarrow$ [structure H, O, C_6H_5] $+$ [structure NH_2] or [structure H, O] $+$ H_2N [structure C_6H_5]

c. $(CH_3CH_2CH_2)_2N(CH_2)_2CH(CH_3)_2$ $\Longrightarrow$ $(CH_3CH_2CH_2)_2NH$ $+$ [structure O, H] or $H-\overset{O}{\underset{CH_2CH_3}{C}}$ $+$ $CH_3CH_2CH_2\overset{}{\underset{H}{N}}$ [structure]

d. [structure with $\overset{}{\underset{H}{N}}$] $\Longrightarrow$ [structure H, O] $+$ [structure NH_2]

25.55

a. C_6H_5 [structure with O] $\xrightarrow[\text{NaBH}_3\text{CN}]{\text{[structure NH}_2]}$ C_6H_5 [structure with NH]

b. [ring with O] $\xrightarrow[\text{NaBH}_3\text{CN}]{(CH_3)_2NH}$ [ring with $N(CH_3)_2$]

c. C_6H_5 [structure CHO] $\xrightarrow[\text{NaBH}_3\text{CN}]{\text{NH}_3}$ C_6H_5 [structure CH_2NH_2]

d. [structure with O] $\xrightarrow[\text{NaBH}_3\text{CN}]{\text{[cyclohexyl-NH}_2]}$ [structure with NH and cyclohexyl]

25.56

a. [benzyl Br] $\xrightarrow[\text{excess}]{\text{NH}_3}$ [CH_2NH_2]

b. [benzene CN] $\xrightarrow[\text{[2] H}_2\text{O}]{\text{[1] LiAlH}_4}$ [CH_2NH_2]

c. [benzene $CONH_2$] $\xrightarrow[\text{[2] H}_2\text{O}]{\text{[1] LiAlH}_4}$ [CH_2NH_2]

d. [benzene CHO] $\xrightarrow[\text{NaBH}_3\text{CN}]{\text{NH}_3}$ [CH_2NH_2]

e.

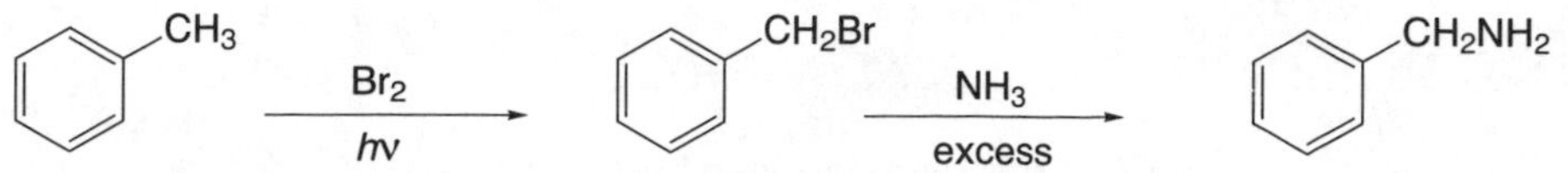

f.

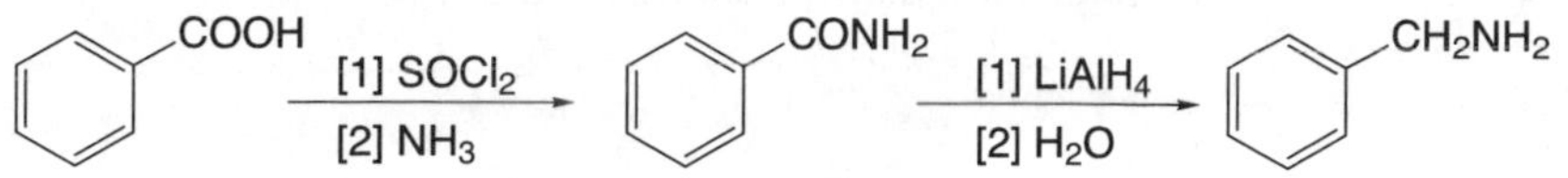

g.

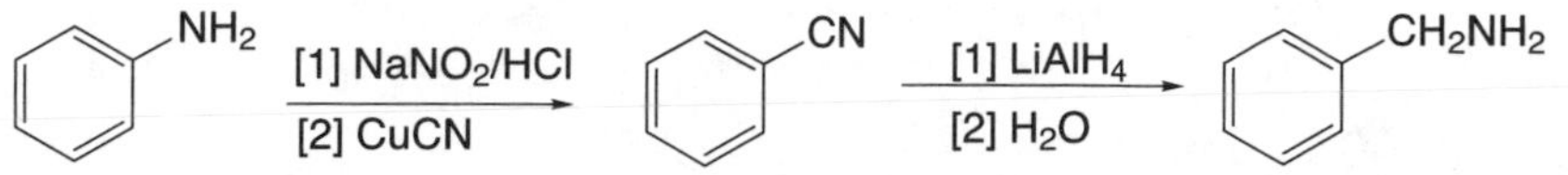

h.

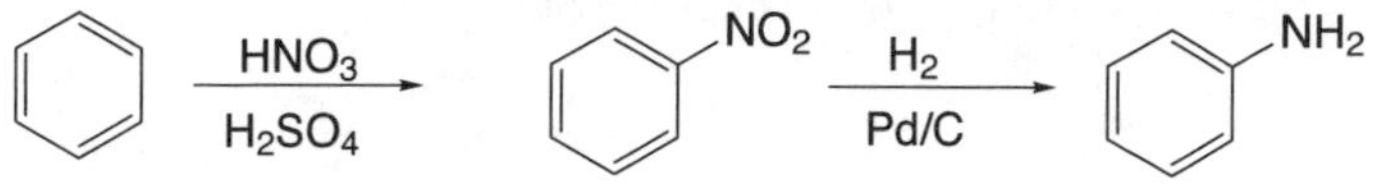

then as in (g).

25.57 Use the directions from Answer 25.22.

Separation can be achieved because benzoic acid reacts with aqueous base and aniline reacts with aqueous acid acording to the following equations:

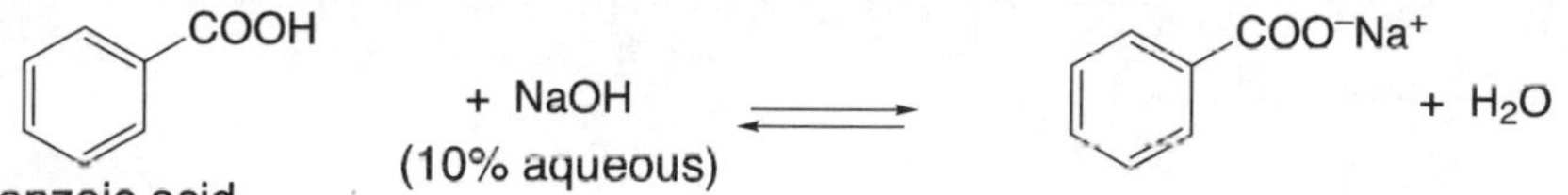

benzoic acid
• soluble in CH_2Cl_2
• insoluble in H_2O

+ NaOH
(10% aqueous)

• soluble in H_2O
• insoluble in CH_2Cl_2

+ H_2O

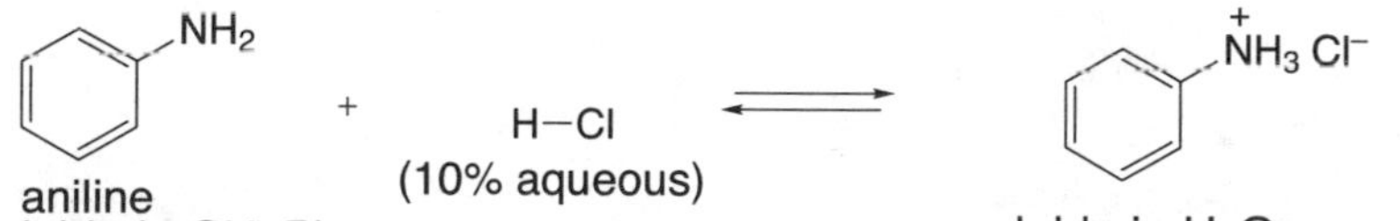

aniline
• soluble in CH_2Cl_2
• insoluble in H_2O

+ H—Cl
(10% aqueous)

• soluble in H_2O
• insoluble in CH_2Cl_2

Toluene ($C_6H_5CH_3$), on the other hand, is not protonated or deprotonated in aqueous solution so it is always soluble in CH_2Cl_2 and is insoluble in H_2O. The following flow chart illustrates the process.

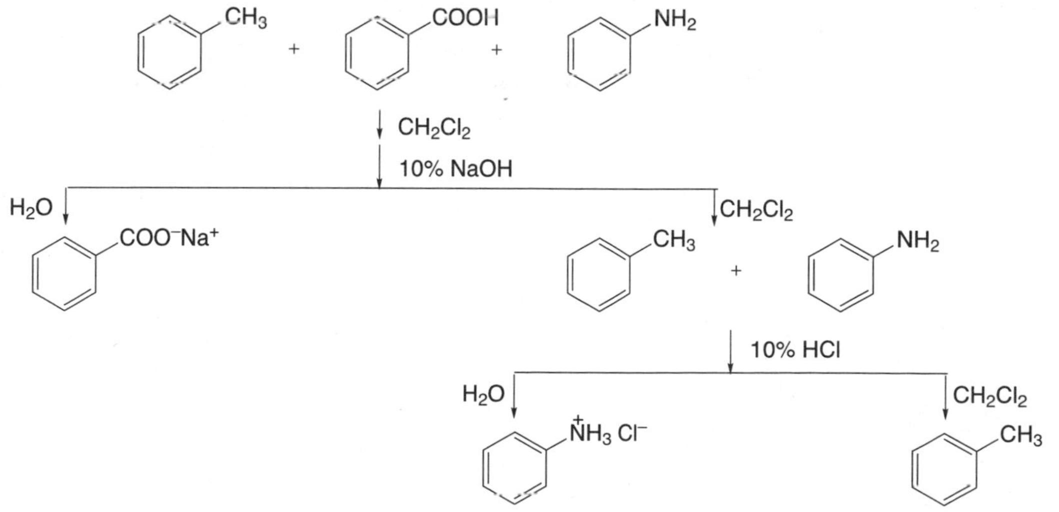

25.58

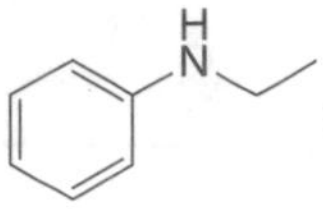

N-ethylaniline

a. HCl

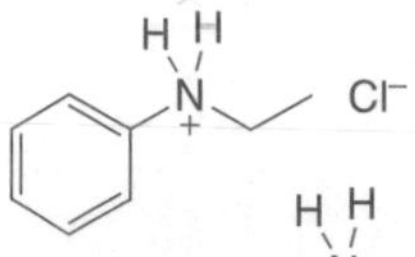

b. CH$_3$COOH

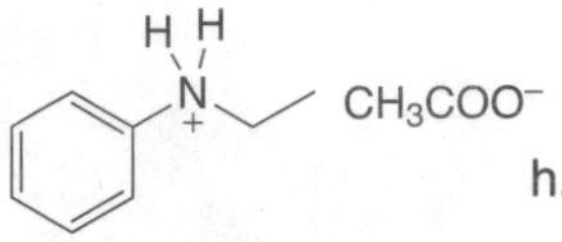

c. (CH$_3$)$_2$C=O

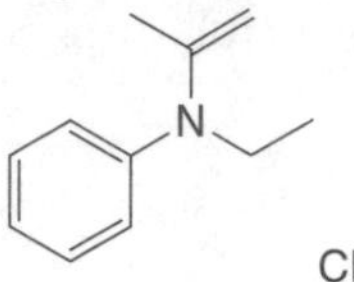

d. CH$_2$O/NaBH$_3$CN

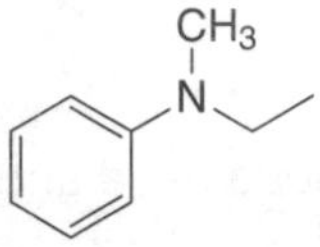

e. CH$_3$I (excess)

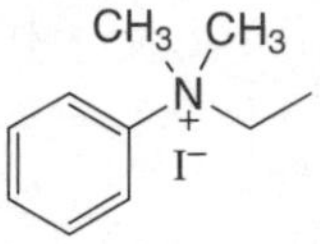

f. CH$_3$I (excess), followed by Ag$_2$O and Δ

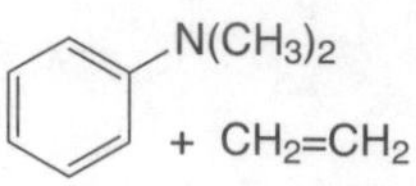

g. CH$_3$CH$_2$COCl

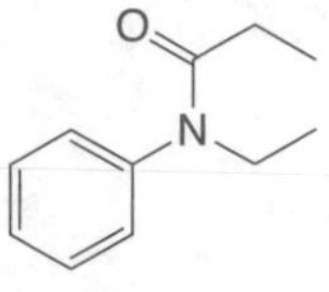

h. The product in (g), then HNO$_3$/H$_2$SO$_4$

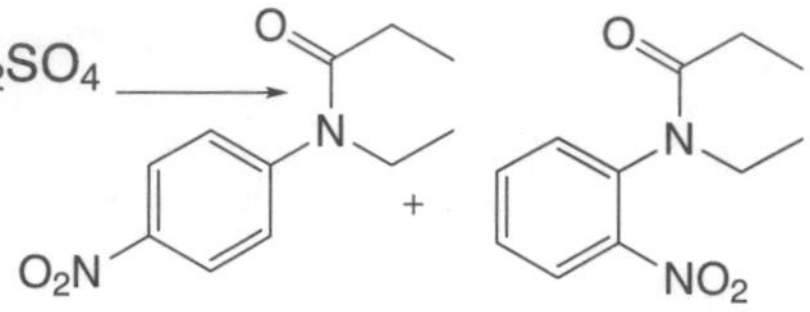

i. The product in (g), then [1] LiAlH$_4$; [2] H$_2$O

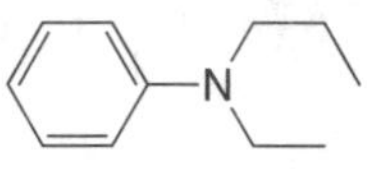

j. The product in (h), then H$_2$/Pd

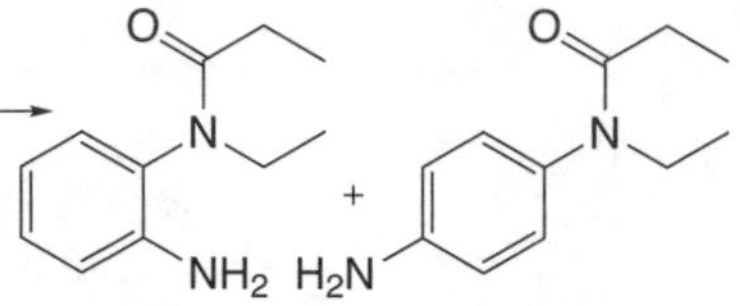

25.59

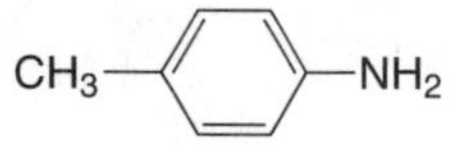

CH$_3$—⟨⟩—NH$_2$ p-methylaniline

a. HCl

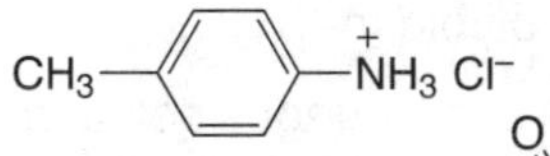

b. CH$_3$COCl

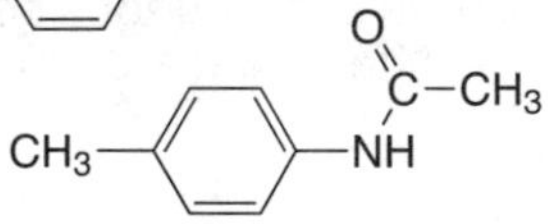

c. (CH$_3$CO)$_2$O

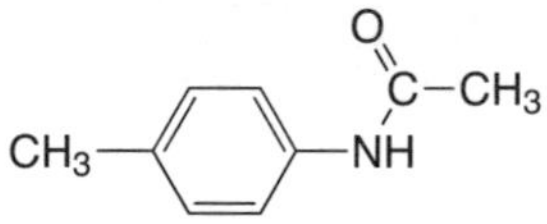

d. excess CH$_3$I

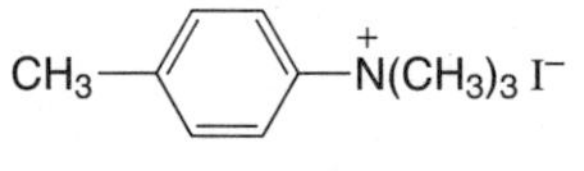

e. (CH$_3$)$_2$C=O

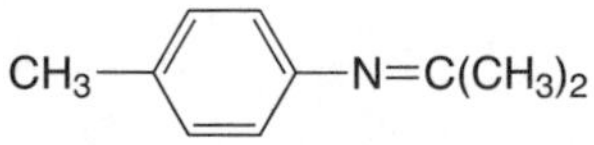

f. CH$_3$COCl/AlCl$_3$

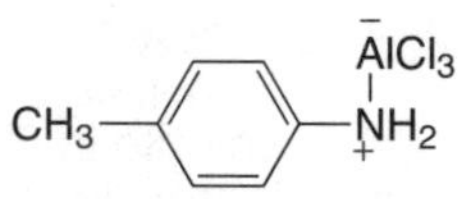

g. CH$_3$COOH

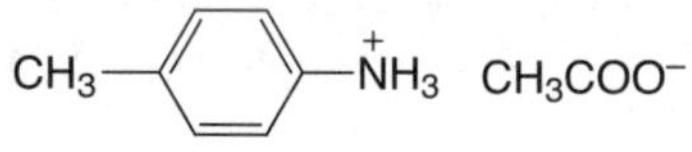

h. NaNO$_2$/HCl

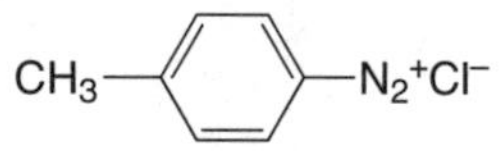

i. step (b), then CH$_3$COCl/AlCl$_3$

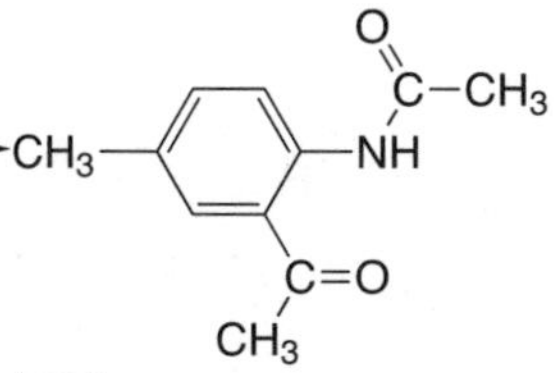

j. CH$_3$CHO/NaBH$_3$CN

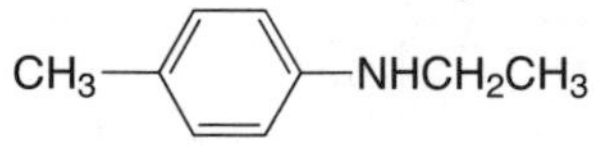

25.60

a. $CH_3CH_2CH_2CH_2NH_2 \xrightarrow{ClCOC_6H_5} CH_3CH_2CH_2CH_2NHCOC_6H_5$

b. $CH_3CH_2CH_2CH_2NH_2 \xrightarrow{O{=}C(CH_2CH_3)_2} CH_3CH_2CH_2CH_2N{=}C(CH_2CH_3)_2$

c. $CH_3CH_2CH_2CH_2NH_2 \xrightarrow[\text{[2] } Ag_2O \text{ [3] } \Delta]{\text{[1] } CH_3I \text{ (excess)}} CH_3CH_2CH{=}CH_2$

d. $CH_3CH_2CH_2CH_2NH_2 \xrightarrow{C_6H_5CH_2Br} CH_3CH_2CH_2CH_2NHCH_2C_6H_5$

e. $CH_3CH_2CH_2CH_2NH_2 \xrightarrow{CH_3CH_2Br} CH_3CH_2CH_2CH_2NHCH_2CH_3$

f. $CH_3CH_2CH_2CH_2NH_2 \xrightarrow[\text{excess}]{CH_3I} [CH_3CH_2CH_2CH_2N(CH_3)_3]^+I^-$

25.61

a. $CH_3(CH_2)_6NH_2 \xrightarrow[\text{[2] } Ag_2O \text{ [3] } \Delta]{\text{[1] } CH_3I \text{ (excess)}} CH_3(CH_2)_4CH{=}CH_2$

b.
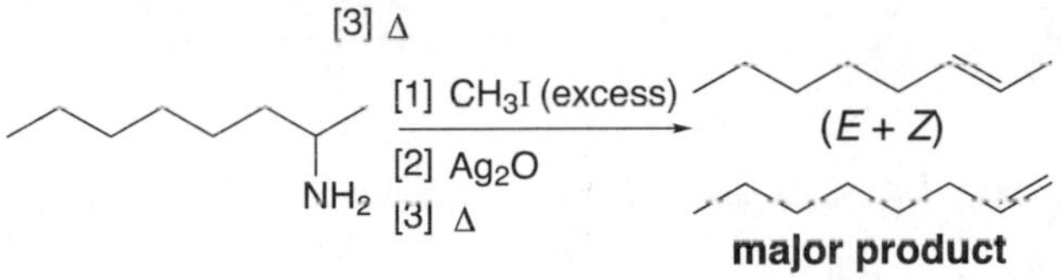

c.
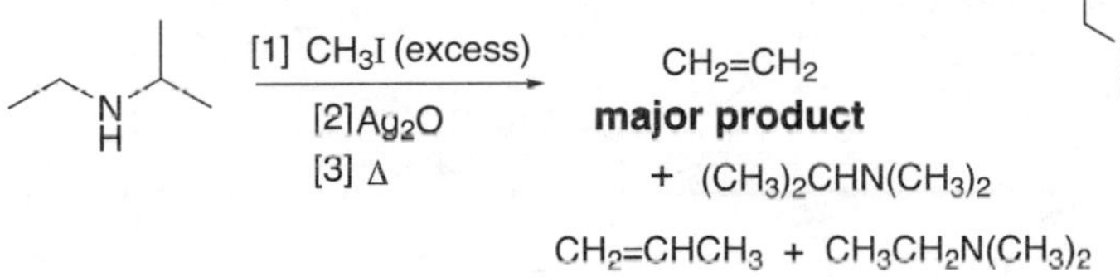

d.
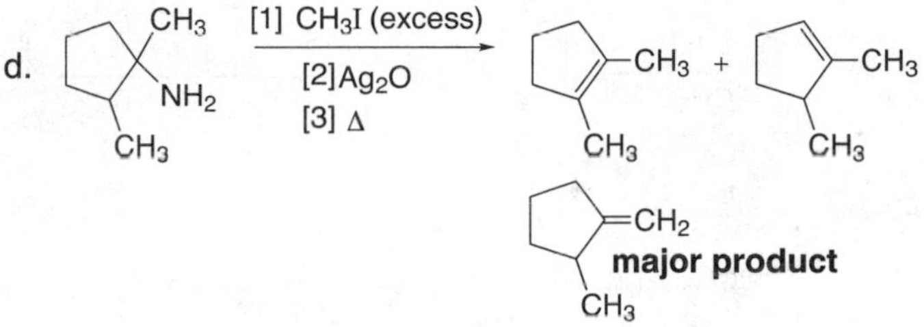

e.
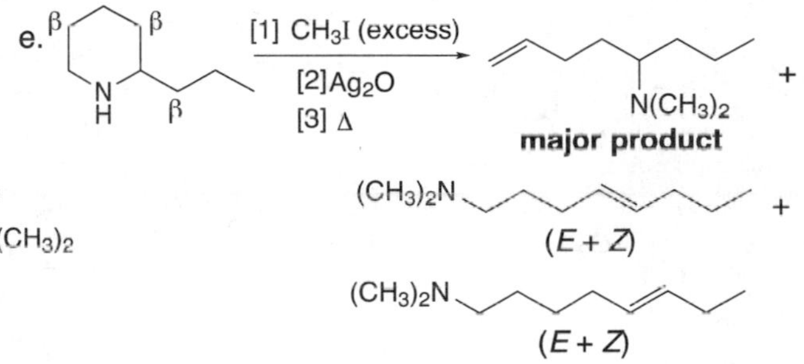

25.62

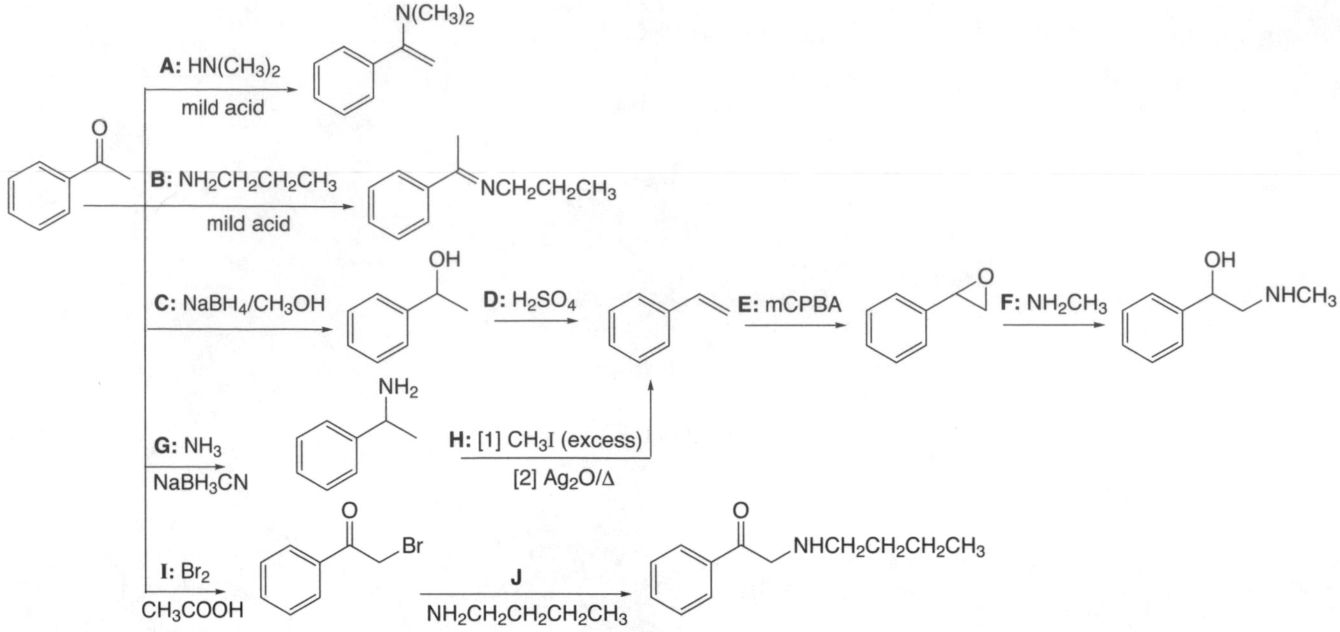

A: HN(CH₃)₂ mild acid

B: NH₂CH₂CH₂CH₃ mild acid

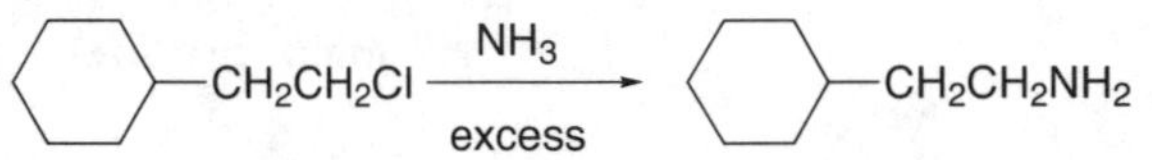

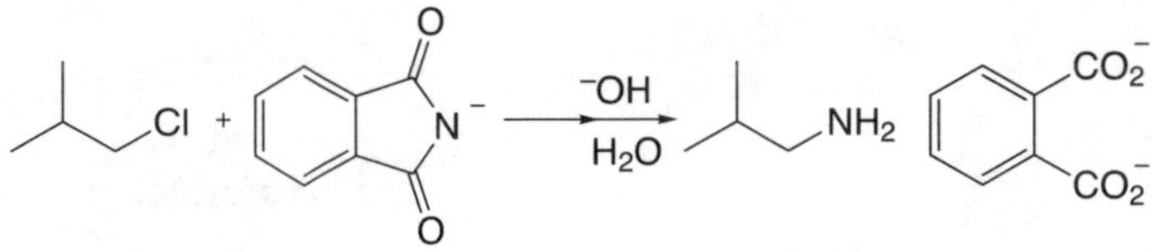

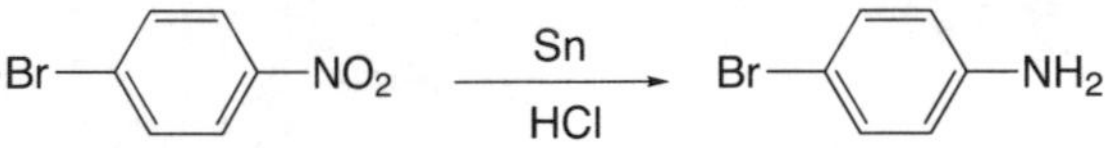

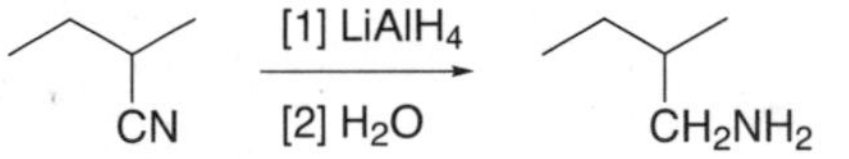

C: NaBH₄/CH₃OH

D: H₂SO₄

E: mCPBA

F: NH₂CH₃

G: NH₃ / NaBH₃CN

H: [1] CH₃I (excess) [2] Ag₂O/Δ

I: Br₂ / CH₃COOH

J: NH₂CH₂CH₂CH₂CH₃

25.63

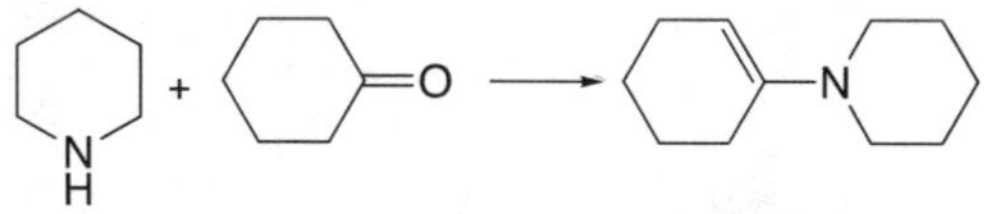

a.

b.

c.

d. [1] LiAlH₄ [2] H₂O

e. [1] LiAlH₄ [2] H₂O

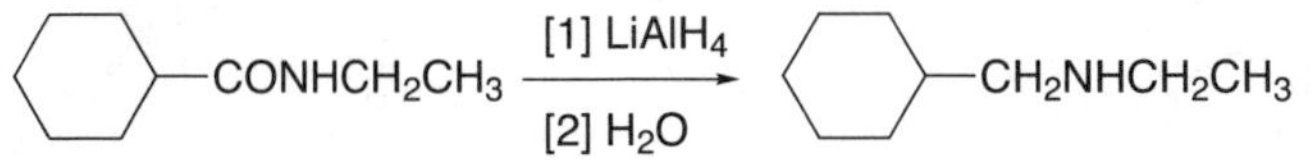

f.   C₆H₅CH₂CH₂NH₂ + (C₆H₅CO)₂O ⟶
C₆H₅CH₂CH₂NHCOC₆H₅

g.
NaNO₂ / HCl

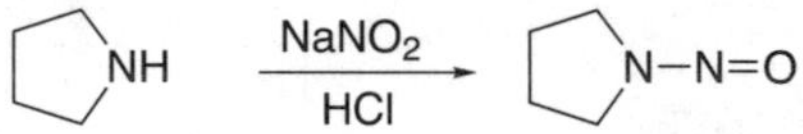

h.
+ C₆H₅CHO →(NaBH₃CN)→

i.

j.   CH₃CH₂CH₂—N(H)—CH(CH₃)₂   [1] CH₃I (excess) [2] Ag₂O [3] Δ
CH₃CH=CH₂ + (CH₃)₂NCH(CH₃)₂
CH₃CH₂CH₂N(CH₃)₂

25.64

a.
⁻OH

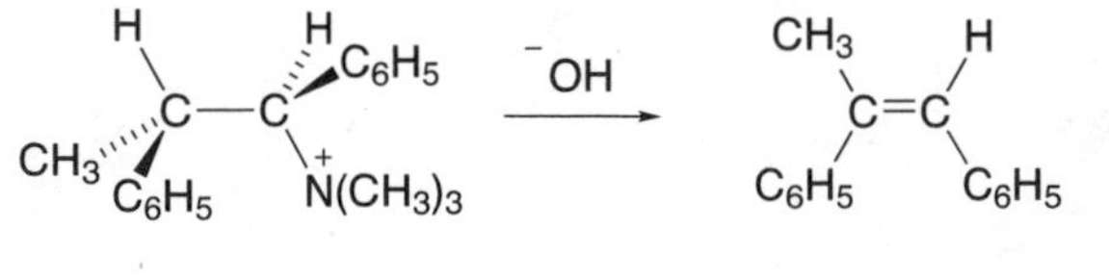

b.
⁻OH

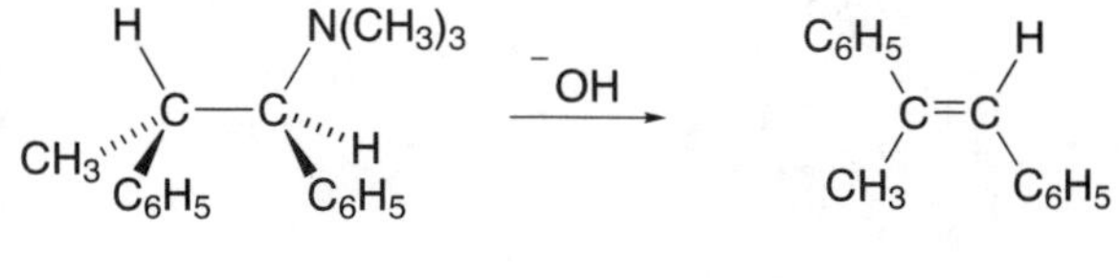

25.65

A (structure: benzene ring with $N_2^+ Cl^-$ and Cl substituents)

a. H_2O → (structure: benzene ring with HO and Cl)

b. H_3PO_2 → (structure: benzene ring with H and Cl)

c. CuCl → (structure: benzene ring with Cl and Cl)

d. CuBr☐ → (structure: benzene ring with Br and Cl)

e. CuCN → (structure: benzene ring with NC and Cl)

f. HBF_4 → (structure: benzene ring with F and Cl)

g. NaI → (structure: benzene ring with I and Cl)

h. $C_6H_5NH_2$ → (structure: benzene ring with Cl, –N=N–, ring with NH_2)

i. C_6H_5OH → (structure: benzene ring with Cl, –N=N–, ring with OH)

j. KI → (structure: benzene ring with I and Cl)

25.66

Under the acidic conditions of the reaction, aniline is first protonated to form an ammonium salt that has a positive charge on the atom bonded to the benzene ring. The $-NH_3^+$ is now an electron withdrawing meta director, so significant amounts of meta substitution occurs.

(reaction: benzene-$\ddot{N}H_2$ + $H–\ddot{O}SO_3H$ → benzene-$\overset{+}{N}H_3$ + HSO_4^-)

This group is now a *meta* director.

25.67

(structure A: $CH_3CH_2CH_2$–C(CH₃)(H)–NH_2, with R configuration)

[1] CH_3I (excess)
[2] Ag_2O
[3] Δ

→ (structure B: $CH_2=C(H)$ connected to $CH_2CH_2CH_3$, alkene)

[1] O_3
[2] CH_3SCH_3

→ (product: H–CHO (formaldehyde)) + (product: $O=C(H)CH_2CH_2CH_3$)

25.68

(resonance structures of aryl diazonium salt: benzene ring–$\overset{+}{N}\equiv\ddot{N}$ ↔ several resonance forms ↔ benzene ring–$\overset{+}{N}\equiv N$)

aryl diazonium salt

(alkyl diazonium salt: $R–\overset{+}{N}\equiv\ddot{N}$)

alkyl diazonium salt

The N_2^+ group on an aromatic ring is stabilized by resonance, whereas the alkyl diazonium salt is not.

25.69

a.

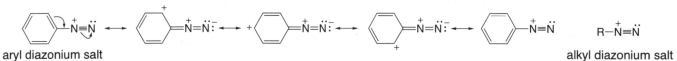

(reaction mechanism scheme: Br–(chain)–Br + $CH_3CH_2\ddot{N}H_2$ → Br–(chain)–$\overset{+}{N}$(H H)(CH₂CH₃) Br^-, with $Na^+ :\ddot{O}H^-$ → Br–(chain)–$\ddot{N}H$(CH₂CH₃) + Br^- + $H_2\ddot{O}$, $Na^+ :\ddot{O}H^-$ → pyrrolidine ring–$\overset{+}{N}$(+)(H CH₂CH₃) → pyrrolidine ring–N–CH_2CH_3 + $H_2\ddot{O}$ + NaBr)

b.

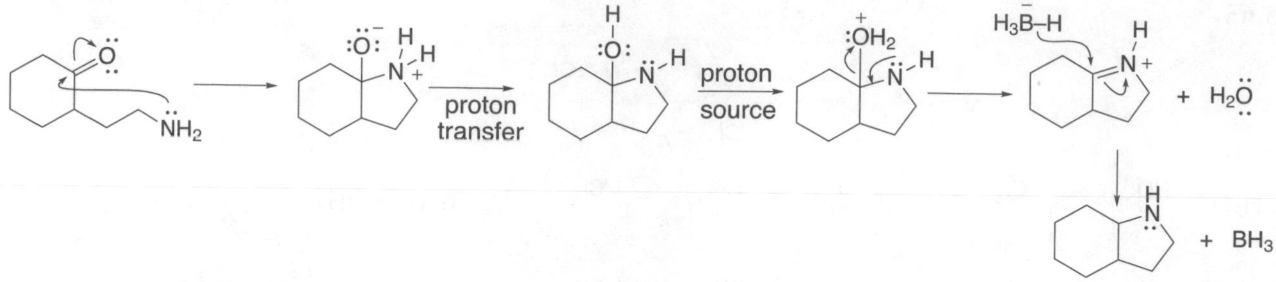

25.70

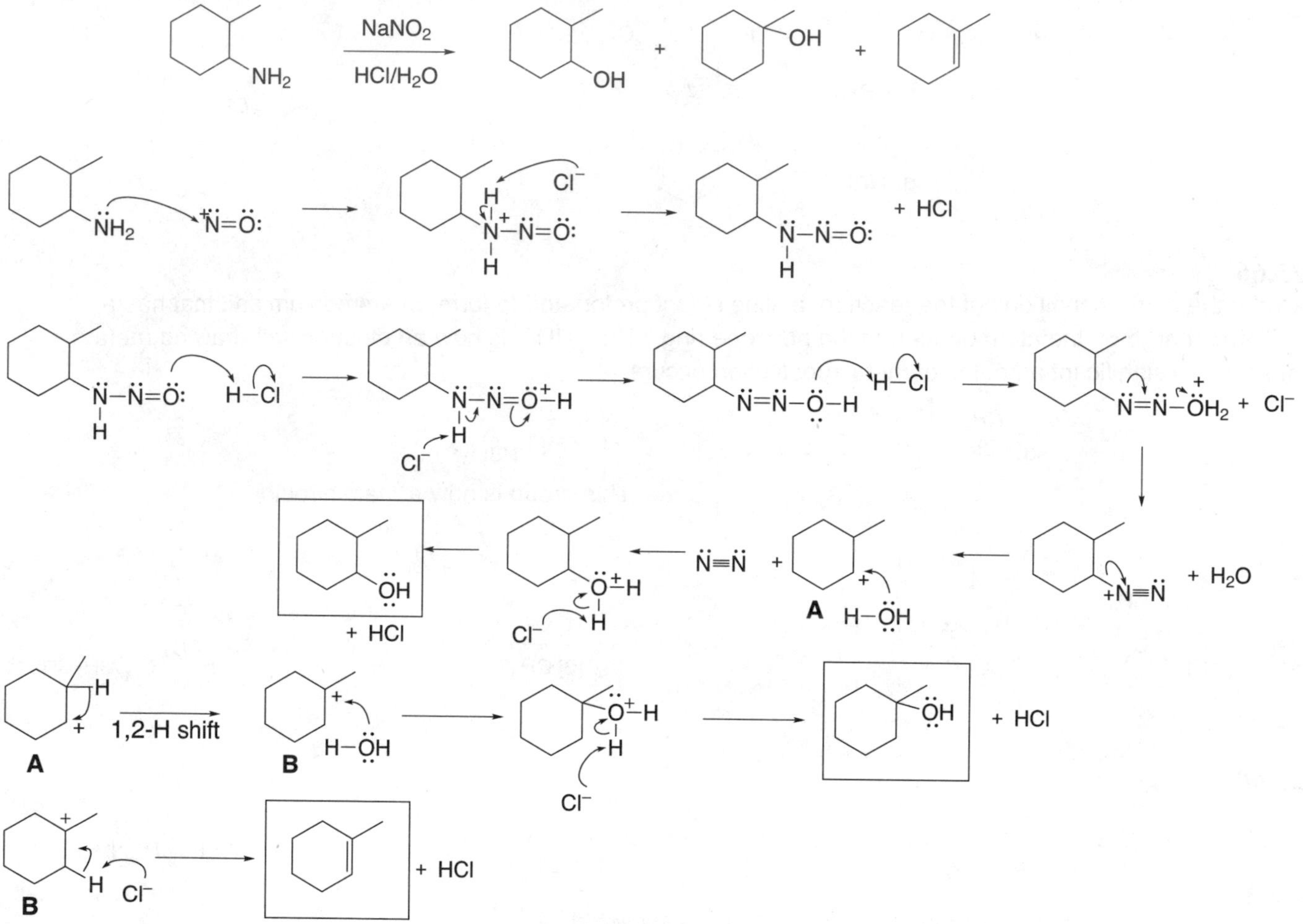

25.71

a.

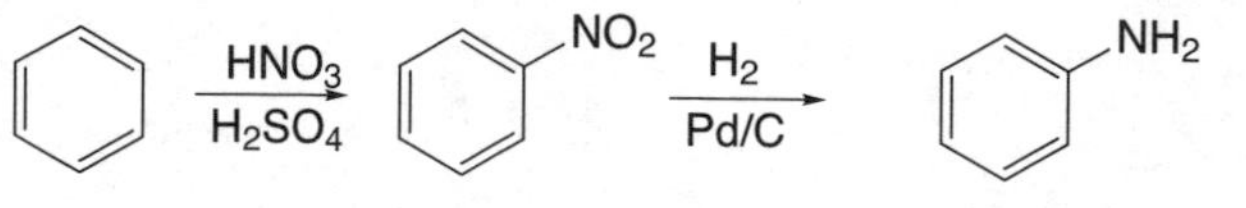

b.

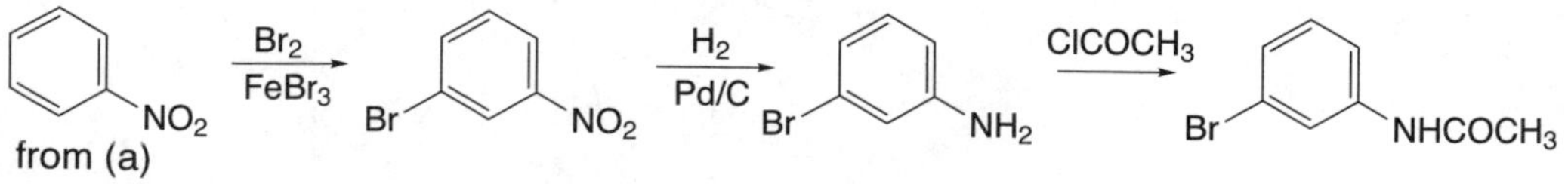

c.

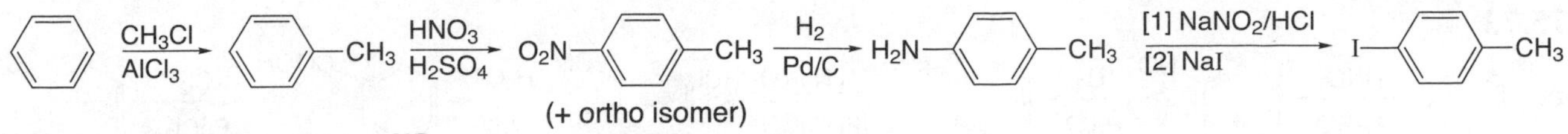

(+ ortho isomer)

d.

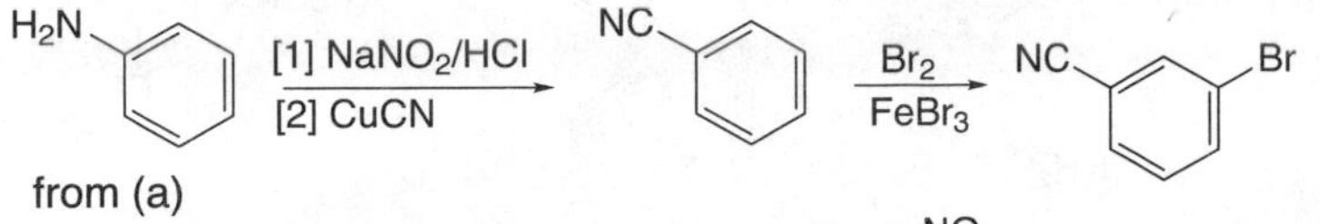

from (a)

e.

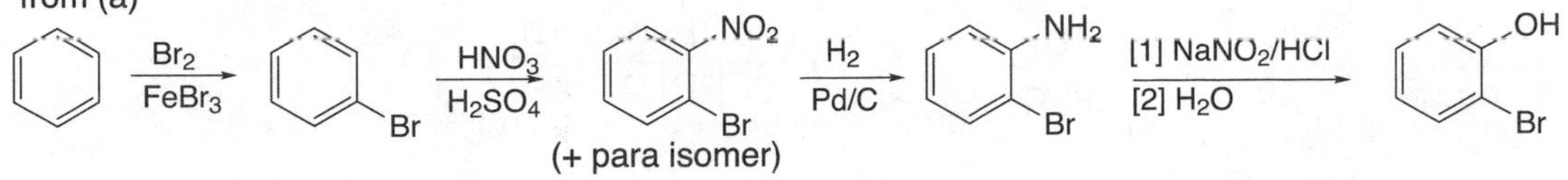

(+ para isomer)

f.

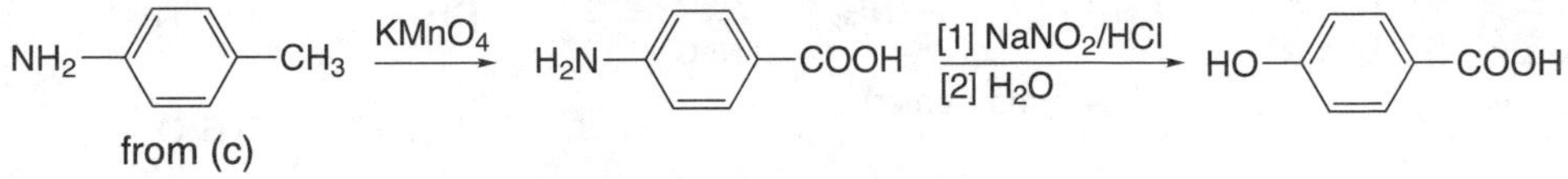

from (c)

g.

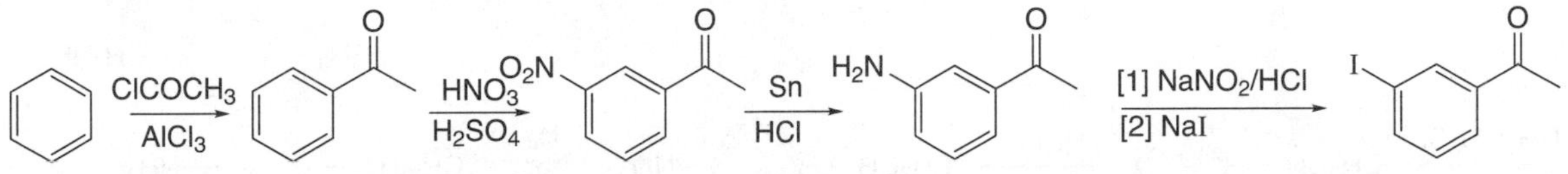

h.

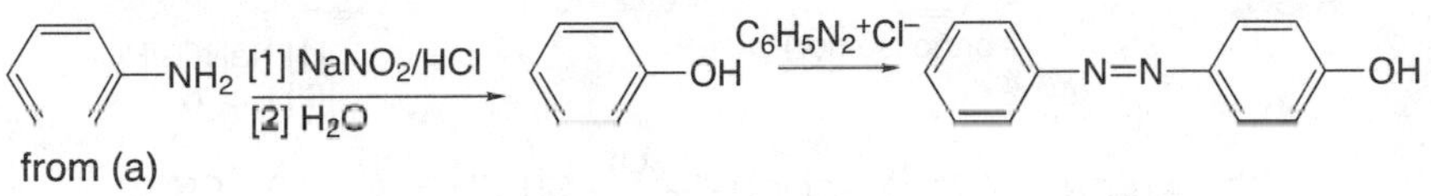

from (a)

25.72

a.

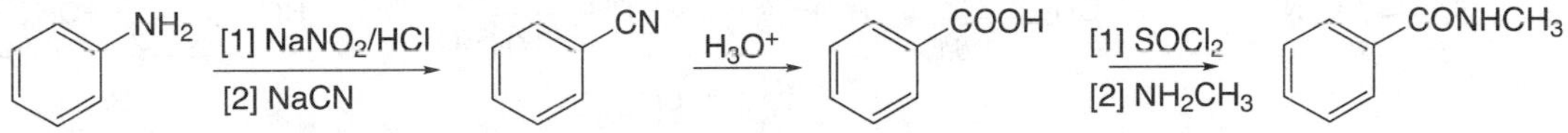

b.

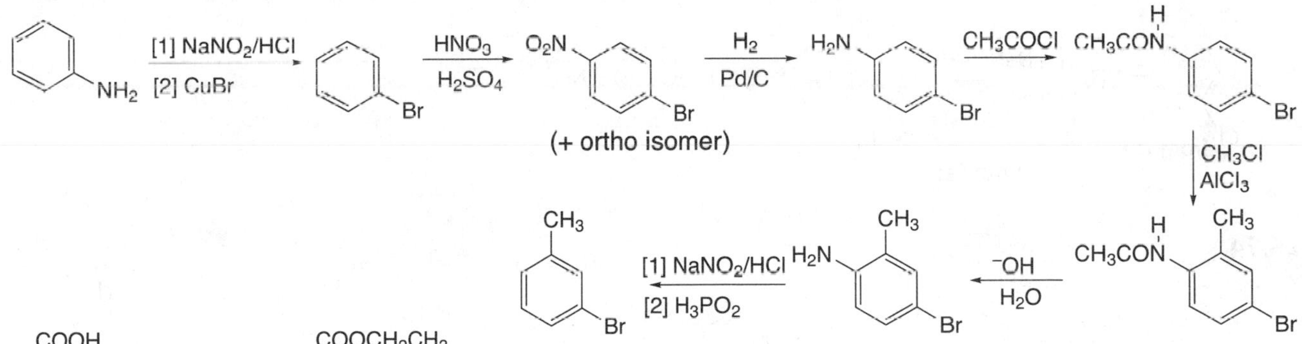

(+ ortho isomer)

c.

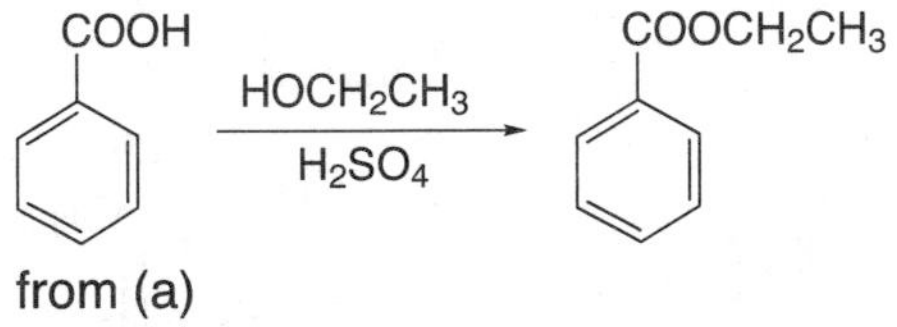

from (a)

d.

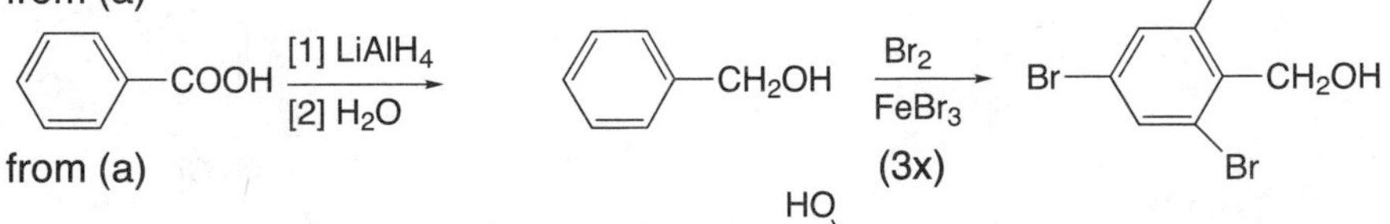

from (a)

(3x)

e.

(+ ortho isomer)

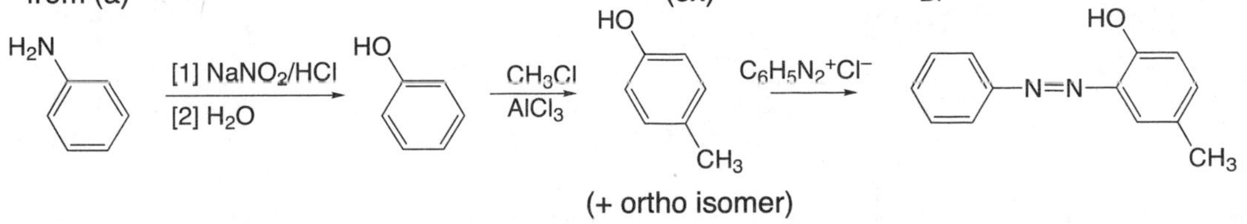

25.73

a.

b.

c.

d.

e.

f.

25.74

a.

b.

from (a)

[1] NaNO$_2$/HCl
[2] H$_2$O

HNO$_3$
H$_2$SO$_4$

(+ ortho isomer)

H$_2$
Pd/C

c.

ClCOCH$_2$CH$_3$
AlCl$_3$

Cl$_2$
H$_2$O/HCl

NH$_2$CH$_3$
Δ

NaBH$_4$
CH$_3$OH

25.75

[1] LiAlH$_4$
[2] H$_2$O

Compound A: C$_8$H$_7$N
IR absorption at 2230 cm^{-1} → triple bond
^{1}H NMR peaks at (ppm):
  2.4 (singlet, 3H) CH$_3$
  7.2 (2H) ⎤
  7.5 (2H) ⎦ disubstituted benzene ring

Compound B: C$_8$H$_{11}$N
IR absorption at 3370, 3290 cm^{-1}
  two IR peaks → 1° amine
^{1}H NMR peaks at (ppm):
  1.4 (singlet, 2H) CH$_2$ or NH$_2$
  2.3 (singlet, 3H) CH$_3$
  3.8 (singlet, 2H) CH$_2$
  7.0–7.3 (4H) disubstituted benzene ring

25.76

Compound A: C$_8$H$_{11}$N
IR absorption at 3400 cm^{-1} → amine
^{1}H NMR peaks at (ppm):
  1.3 (triplet, 3H) CH$_3$ adjacent to 2 H's
  3.1 (quartet, 2H) CH$_2$ adjacent to 3 H's
  3.6 (singlet, 1H) amine H
  6.8–7.2 (multiplet, 5H) benzene ring

Compound B: C$_8$H$_{11}$N
IR absorption at 3310 cm^{-1} → amine
^{1}H NMR peaks at (ppm):
  1.4 (singlet, 1H) amine H
  2.4 (singlet, 3H) CH$_3$
  3.8 (singlet, 2H) CH$_2$
  7.2 (multiplet, 5H) benzene ring

Compound C: C$_8$H$_{11}$N
IR absorption at 3430 and
  3350 cm^{-1} → amine
^{1}H NMR peaks at (ppm):
  1.3 (triplet, 3H) CH$_3$ near CH$_2$
  2.5 (quartet, 2H) CH$_2$ near CH$_3$
  3.6 (singlet, 2H) amine H's
  6.7 (doublet, 2H) ⎤ disubstituted
  7.0 (doublet, 2H) ⎦ benzene ring

25.77

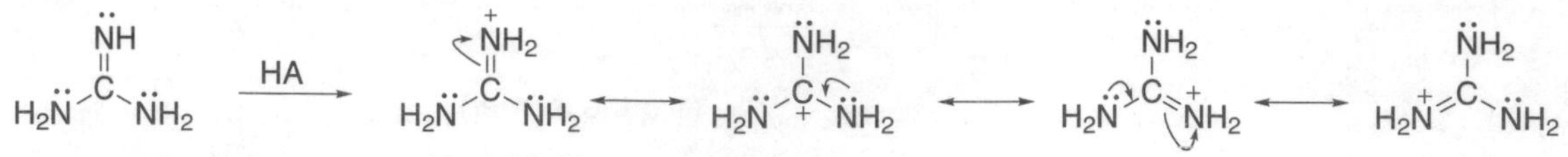

guanidine

$pK_a = 13.6$

Guanidine is a strong base because its conjugate acid is very stabilized by resonance. This resonance delocalization makes guanidine easily donate its electron pair; thus it's a strong base.

25.78

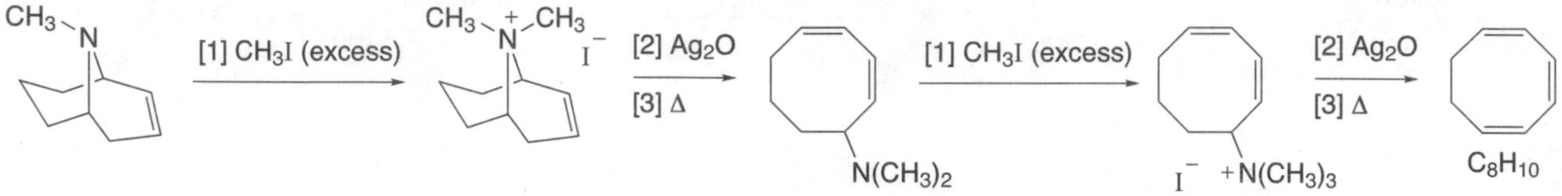

C_8H_{10}

Y

25.79

Possible products of Hofmann elimination. Look for H's on β carbons to the N atom.

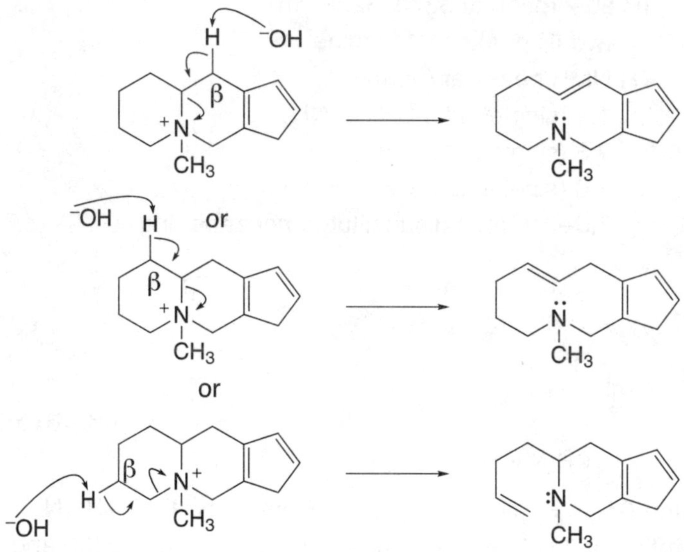

Only this product has a conjugated double bond so it is the preferred product.

25.80

One possibility:

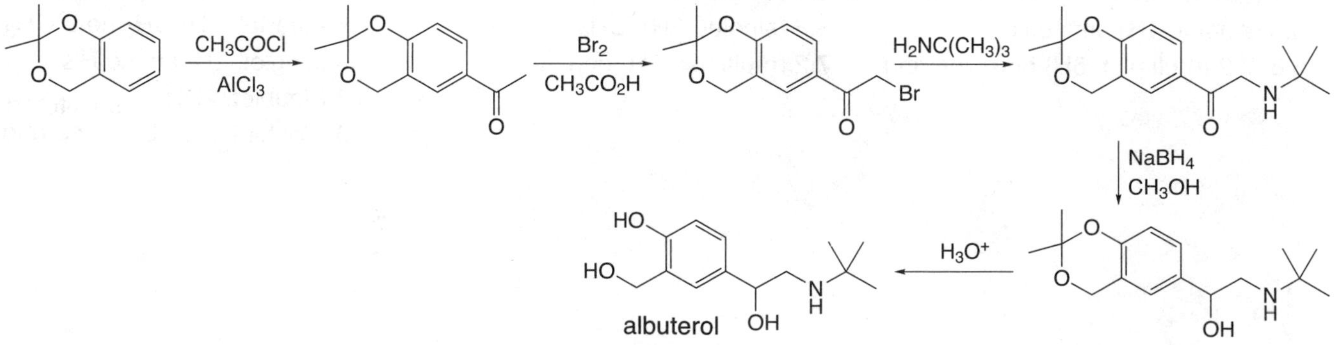

albuterol

Chapter 26: Lipids

◆ Hydrolyzable lipids

[1] **Waxes (26.2)**—Esters formed from a long chain alcohol and a long chain carboxylic acid.

R, R' = long chains of C's

[2] **Triacylglycerols (26.3)**—Triesters of glycerol with three fatty acids.

R, R', R" = alkyl groups with 11–19 C's

[3] **Phospholipids (26.4)**

[a] Phosphatidylethanolamine (cephalin)

R, R' = long carbon chain

[b] Phosphatidylcholine (lecithin)

R, R' = long carbon chain

[c] Sphingomyelin

R = long carbon chain
R' = H or CH$_3$

◆ Nonhydrolyzable lipids

[1] **Fat-soluble vitamins (26.5)**—Vitamins A, D, E, and K.

[2] **Eicosanoids** (26.6)—Compounds containing 20 carbons derived from arachidonic acid. There are four types: prostaglandins, thromboxanes, prostacyclins, and leukotrienes.

[3] **Terpenes** (26.7)—Lipids composed of repeating five-carbon units called isoprene units.

Isoprene unit	Types of terpenes			
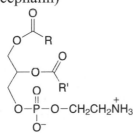	[1] monoterpene	10 C's	[4] sesterterpene	25 C's
	[2] sesquiterpene	15 C's	[5] triterpene	30 C's
	[3] diterpene	20 C's	[6] tetraterpene	40 C's

[4] **Steroids (26.8)**—Tetracyclic lipids composed of three six-membered and one five-membered ring.

Chapter 26: Answers to Problems

26.1 Waxes are esters (RCOOR′) formed from a high molecular weight alcohol (R′OH) and a fatty acid (RCOOH).

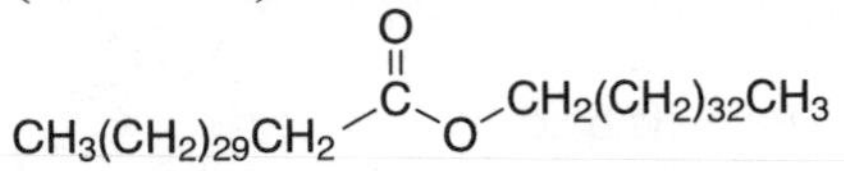

$CH_3(CH_2)_{29}CH_2$
O
C
O
$CH_2(CH_2)_{32}CH_3$

carnauba wax component

26.2 Eicosapentaenoic acid has 20 C's and 5 C=C's. Since an increasing number of double bonds decreases the melting point, eicosapentaenoic acid should have a melting point lower than arachidonic acid, that is < –49°C.

26.3

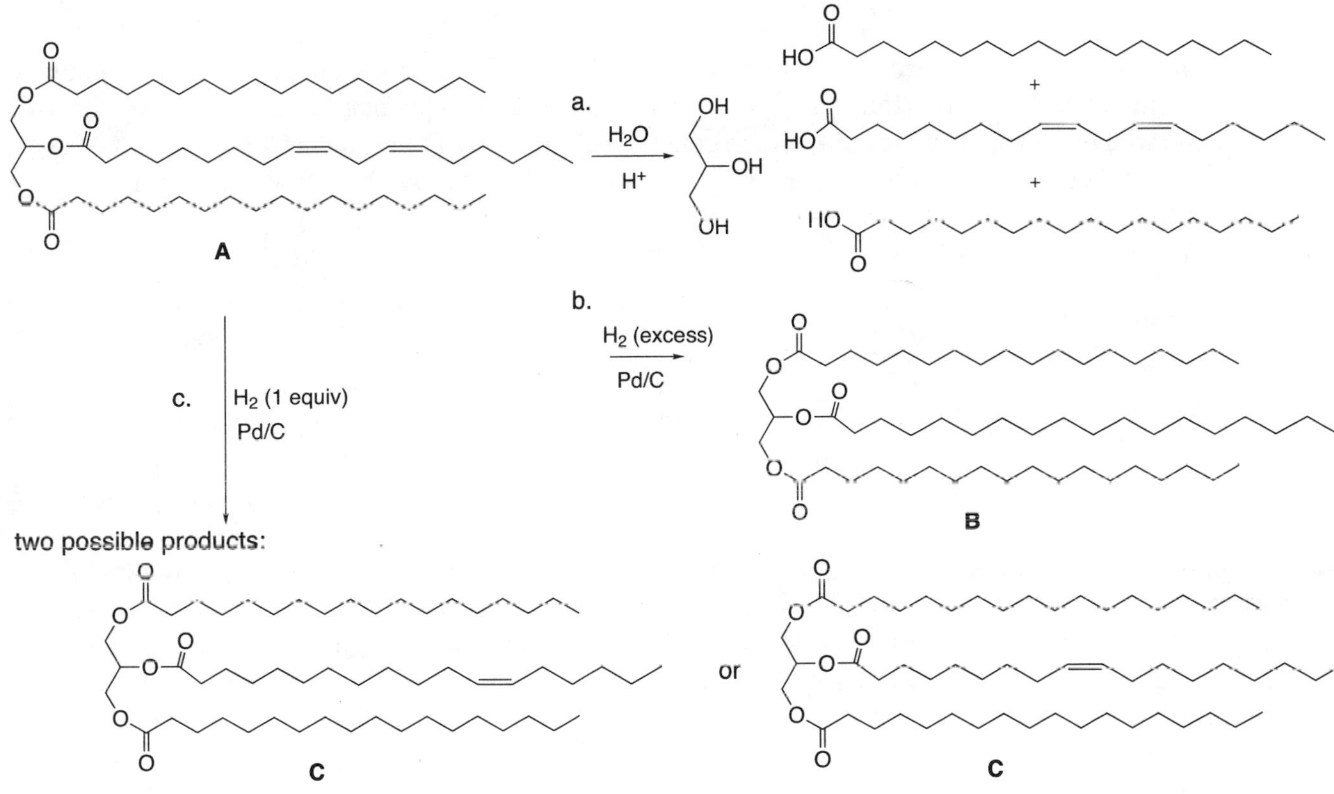

A
a.
H_2O
H^+
OH
OH
OH
HO
+
HO
+
HO
b.
H_2 (excess)
Pd/C
B
c.
H_2 (1 equiv)
Pd/C
two possible products:
C
or
C

26.4

A	<	C	<	B
2 double bonds		1 double bond		0 double bonds
lowest melting point		**intermediate melting point**		**highest melting point**

26.5 A lecithin is a type of phosphoacylglycerol. Two of the hydroxyl groups of glycerol are esterified with fatty acids. The third OH group is part of a phosphodiester, which is also bonded to another low molecular weight alcohol.

general structure
of a lecithin

26.6 Soaps and phosphoacylglycerols have hydrophilic and hydrophobic components. Both compounds have an ionic "head" that is attracted to polar solvents like H_2O. This head is small in size compared to the hydrophobic region, which consists of one or two long hydrocarbons chains. These nonpolar chains consist of only C–C and C–H bonds and exhibit only van der Waals forces.

26.7 Phospholipids have a polar (ionic) head and two nonpolar tails. These two regions, which exhibit very different forces of attraction, allow the phospholipids to form a bilayer with a central hydrophobic region that serves as a barrier to agents crossing a cell membrane, while still possessing an ionic head to interact with the aqueous environment inside and outside the cell. Two different regions are needed in the molecule. Triacyglycerols have three polar, uncharged ester groups, but they are not nearly as polar as phospholipids. They do not have an ionic head with nonpolar tails and so they do not form bilayers. They are largely nonpolar C–C and C–H bonds so they are not attracted to an aqueous medium, making them H_2O insoluble.

26.8 Fat-soluble vitamins are hydrophobic and therefore are readily stored in the fatty tissues of the body. Water-soluble vitamins, on the other hand, are readily excreted in the urine and large concentrations cannot build up in the body.

26.9

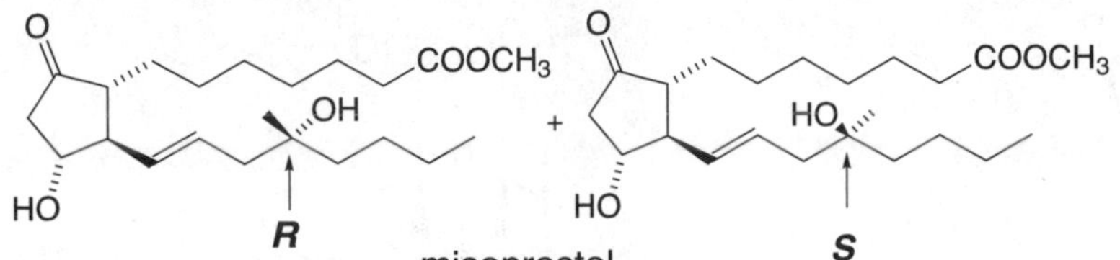
COOCH3
OH
+
HO
HO
R
S

misoprostol
diastereomers

Only one tetrahedral stereogenic center is different in these two compounds.

26.10

a. 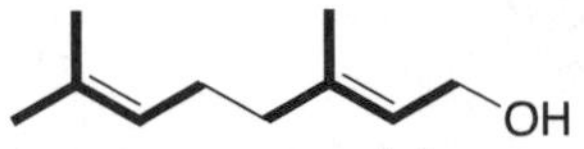

geraniol
(roses and geraniums)

c. 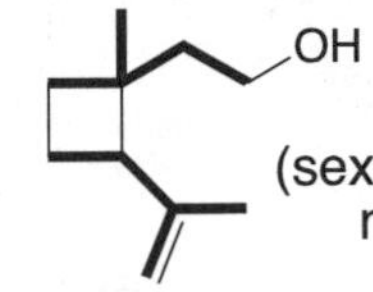

grandisol
(sex phoromone of the
male boll weevil)

b. 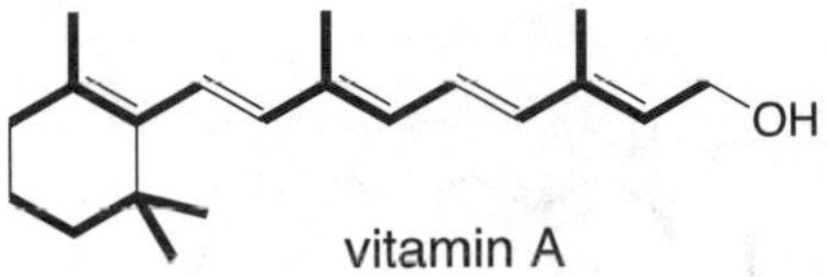

vitamin A

d. 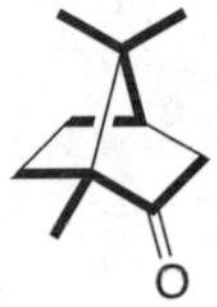

camphor

26.11

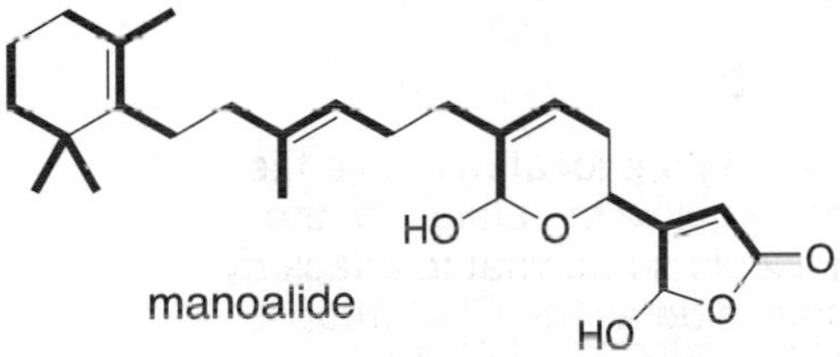

manoalide

26.12

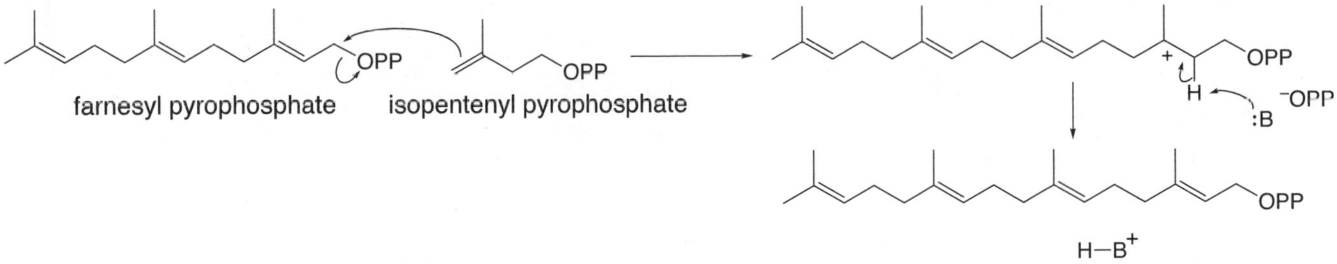
OPP
farnesyl pyrophosphate
OPP
isopentenyl pyrophosphate
OPP
H
:B
⁻OPP
OPP
H—B⁺

26.13

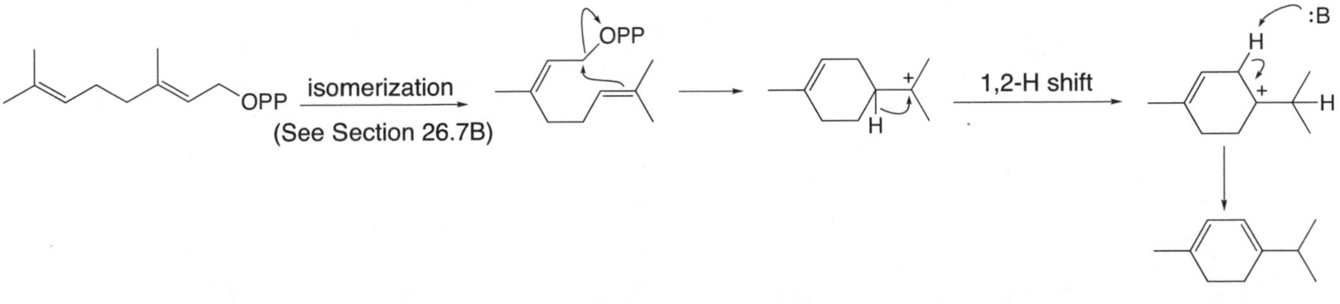
OPP
isomerization
(See Section 26.7B)
OPP
H
1,2-H shift
:B
H
H

α-terpinene

26.14

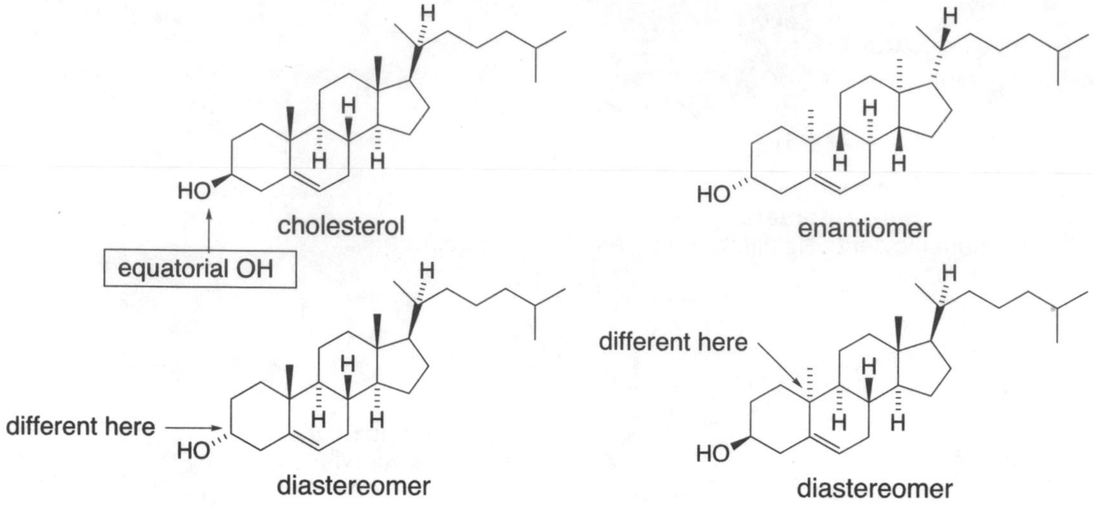

cholesterol

equatorial OH

enantiomer

different here

diastereomer

different here

diastereomer

26.15

A

B

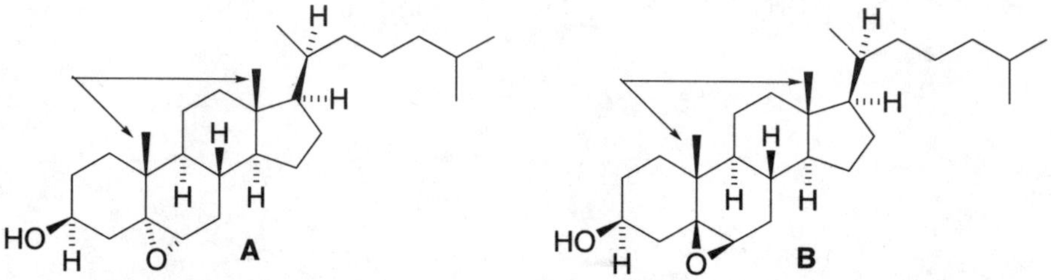

All four rings are in the same plane. The bulkyl CH$_3$ groups (arrow) are located above the plane. Epoxide **A** is favored, because it results from epoxidation below the plane, on the opposite side from the CH$_3$ groups that shield the top of the molecule somewhat to attack by reagents. In **B**, the epoxide ring is above the plane on the same side as the CH$_3$ groups. Formation of **B** would require epoxidation of the planar C=C from the less accessible, more sterically hindered side of the double bond. This path is thus disfavored.

26.16

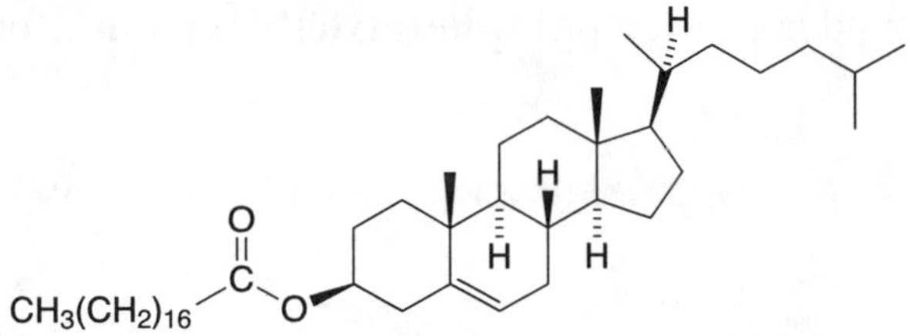
H
H
H
H
O
C
CH3(CH2)16
O

26.17 Each compound has one tetrahedral stereogenic center, so there are two stereoisomers (two enantiomers) possible.

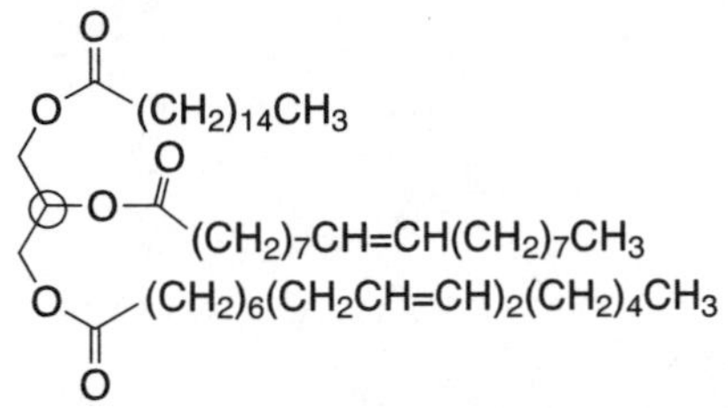
(CH2)14CH3
(CH2)7CH=CH(CH2)7CH3
(CH2)6(CH2CH=CH)2(CH2)4CH3

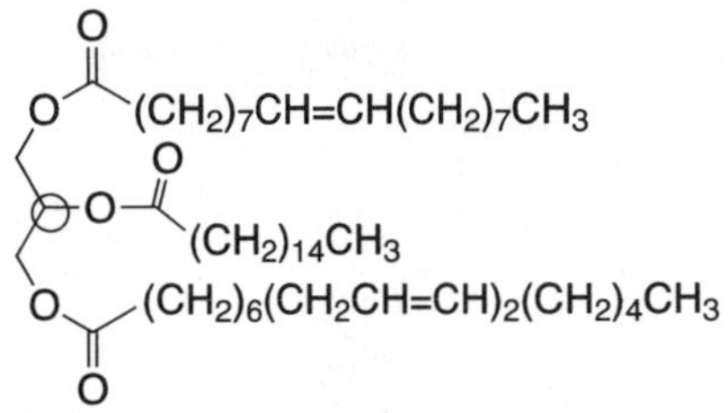
(CH2)7CH=CH(CH2)7CH3
(CH2)14CH3
(CH2)6(CH2CH=CH)2(CH2)4CH3

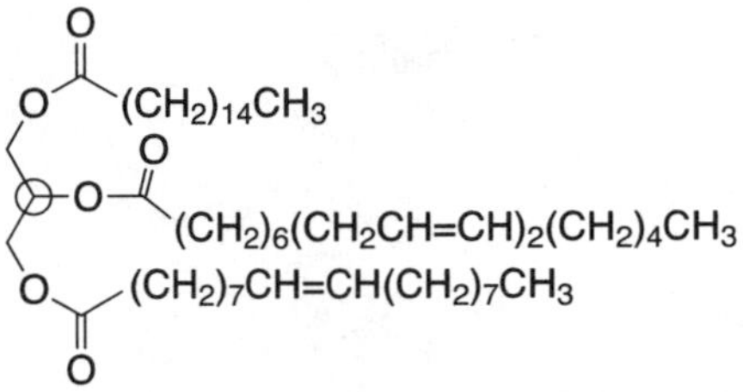
(CH2)14CH3
(CH2)6(CH2CH=CH)2(CH2)4CH3
(CH2)7CH=CH(CH2)7CH3

26.18

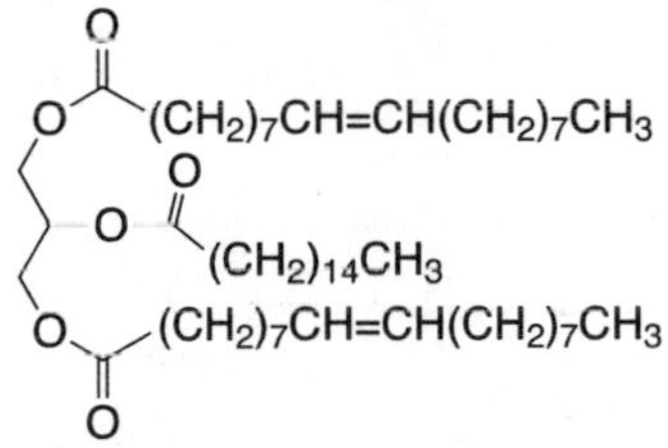
(CH2)7CH=CH(CH2)7CH3
(CH2)14CH3
(CH2)7CH=CH(CH2)7CH3

26.19

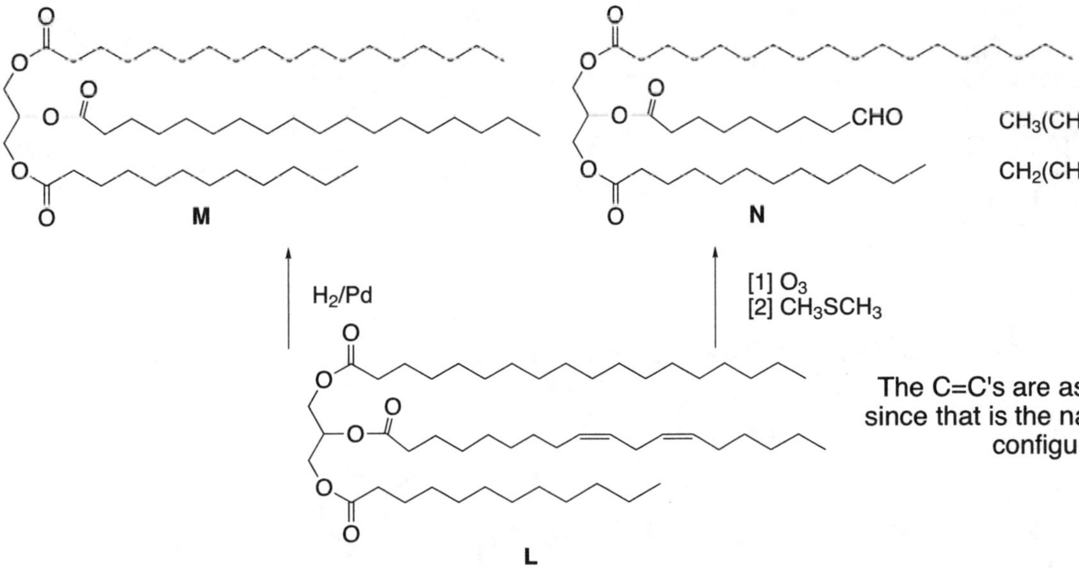
M
N
CHO
CH3(CH2)4CHO = O
CH2(CHO)2 = P
H2/Pd
[1] O3
[2] CH3SCH3
L

The C=C's are assumed to be *Z*, since that is the naturally occurring configuration.

26.20 When R″ = CH$_2$CH$_2$NH$_3$$^+$, the compound is called a **phosphatidylethanolamine** or **cephalin**.

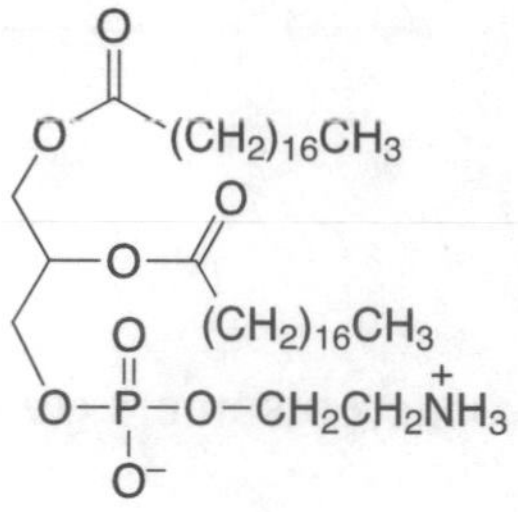

cephalin

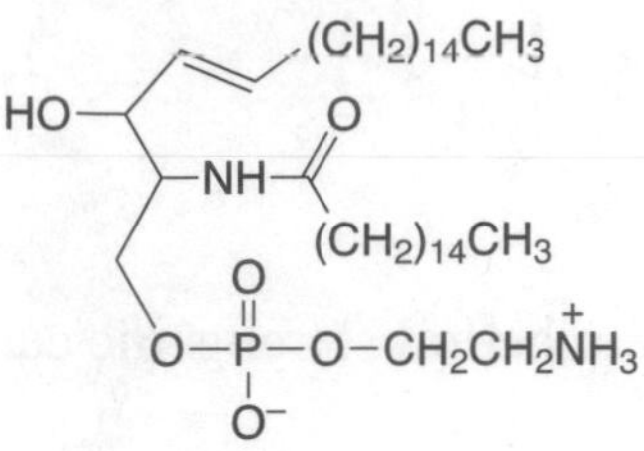

sphingomyelin

26.21

a.

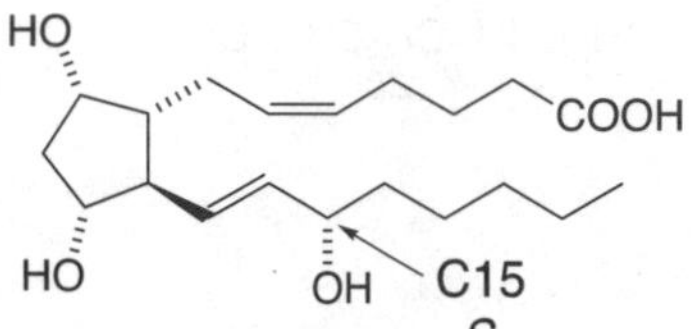

PGF$_{2\alpha}$

b.

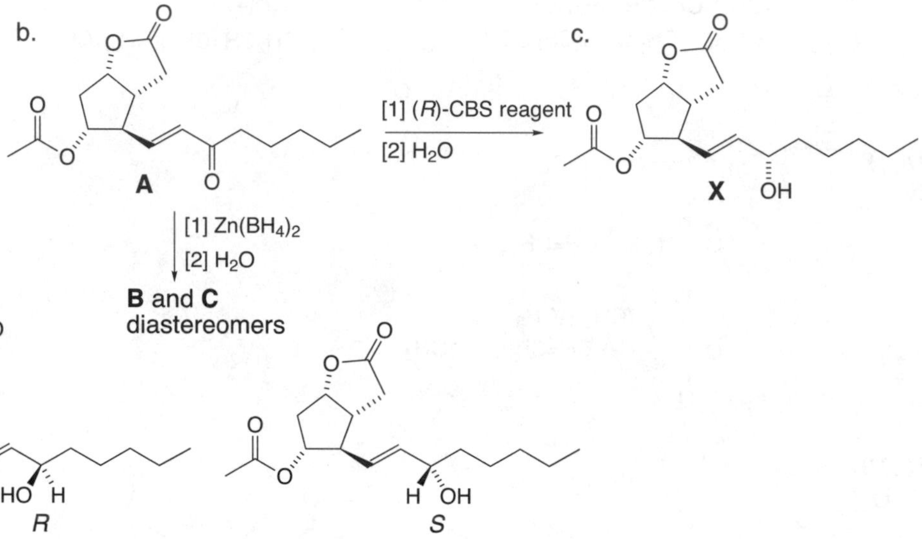

A

[1] (R)-CBS reagent

[2] H$_2$O

c.

X

[1] Zn(BH$_4$)$_2$

[2] H$_2$O

B and **C**
diastereomers

R

S

26.22

a.

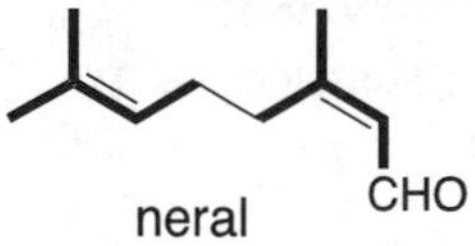

neral

b.

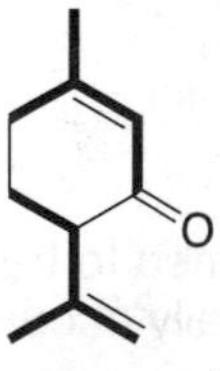

carvone

c.

α-pinene

d.

lycopene

e.

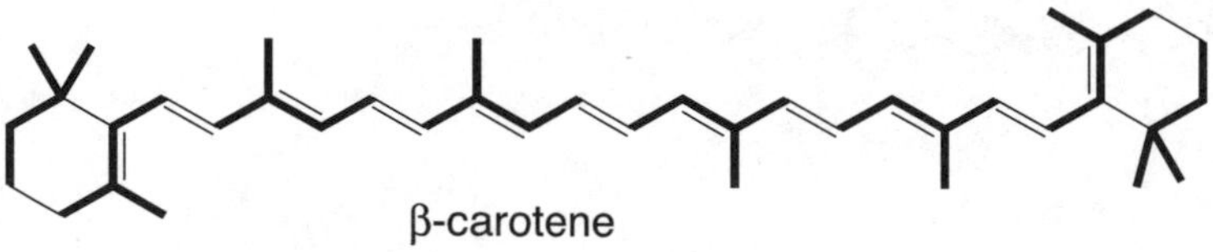

β-carotene

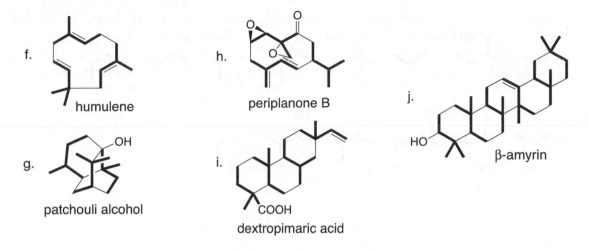

f. humulene

h. periplanone B

j. β-amyrin

g. patchouli alcohol

i. dextropimaric acid

26.23 A *monoterpene* **contains 10 carbons** and two isoprene units; a *sesquiterpene* **contains 15 carbons** and three isoprene units, etc. See Table 26.5.

a. monoterpene CHO

b. monoterpene

c. monoterpene

d. tetraterpene

e. tetraterpene

f. sesquiterpene

g. sesquiterpene

h. sesquiterpene

i. diterpene COOH

j. triterpene

26.24

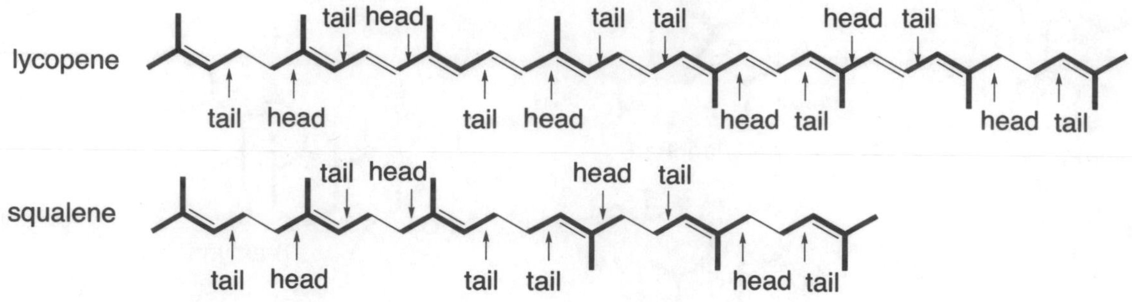

lycopene

squalene

26.25

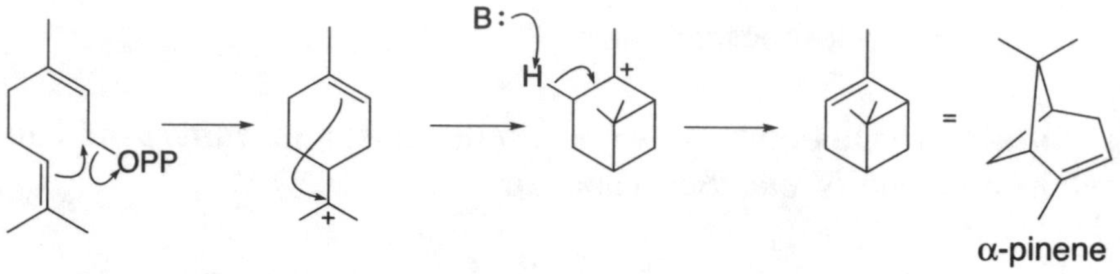

α-pinene

26.26

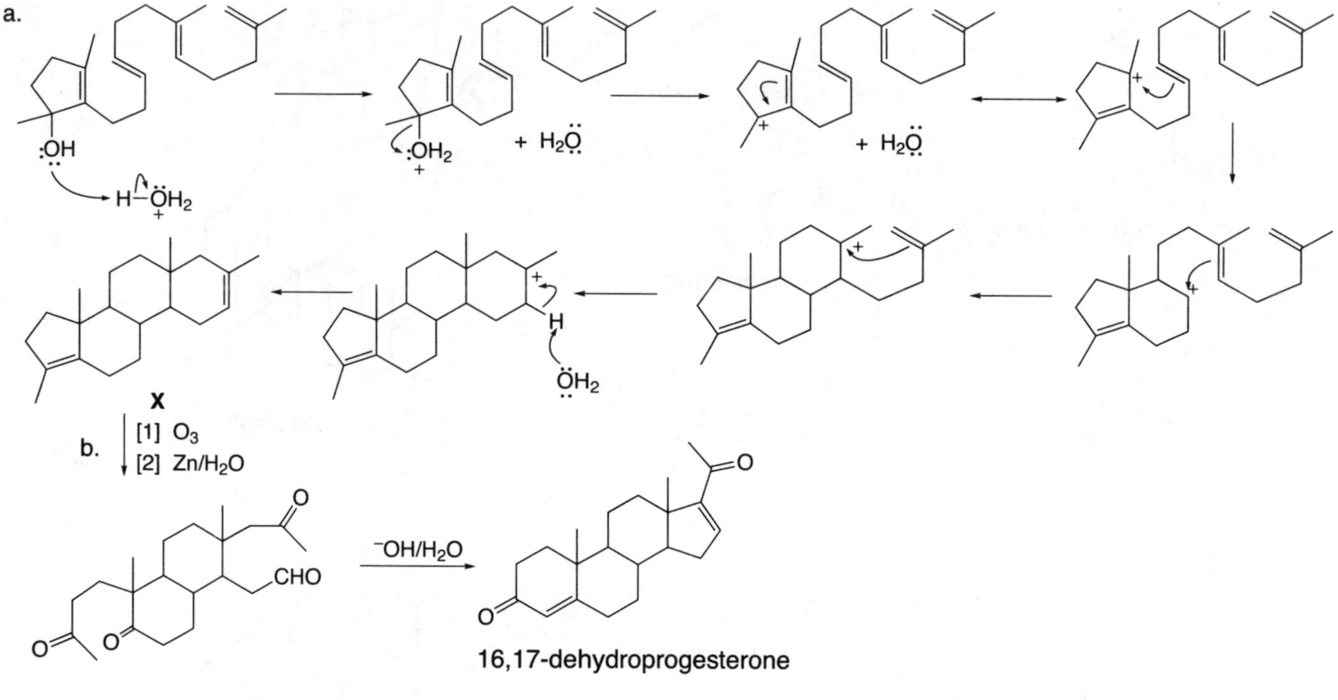

a.

X

b. [1] O₃
   [2] Zn/H₂O

16,17-dehydroprogesterone

26.27

a.

b.

26.28

a. HO equatorial OH

b. HO axial OH

c. HO axial OH

d. HO equatorial OH

26.29

a. HO OH (ax) (eq) Axial reacts faster.

b. HO OH (eq) OH (ax) Axial reacts faster.

26.30

HO = OH H

26.31

CH₃ groups make this face more sterically hindered

a. O = O

b. H₂/Pd → O H

The bottom face is more accessible so the H₂ is added from this face to form an equatorial OH.

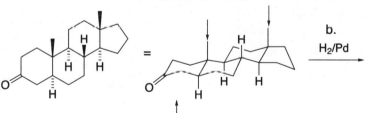

26.32

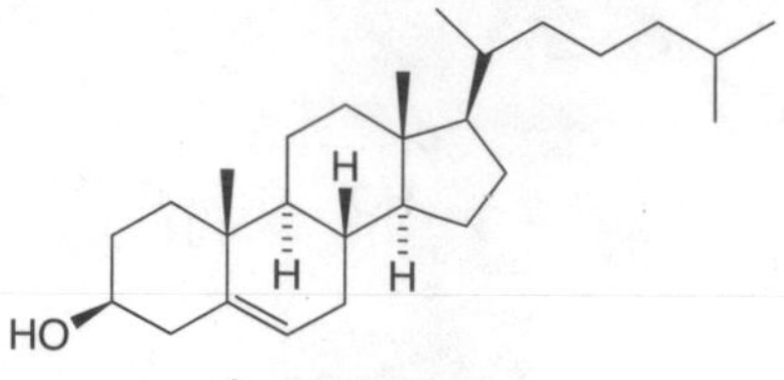

cholesterol

a. CH₃COCl

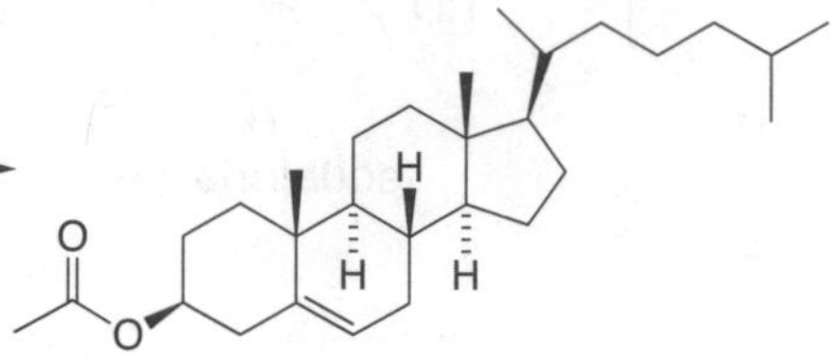

d. stearic acid/H⁺

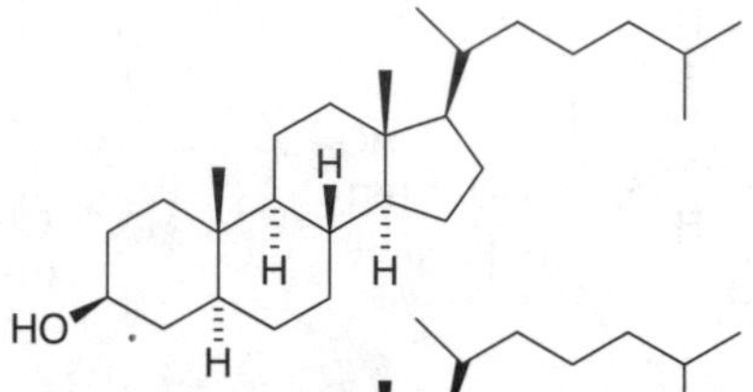

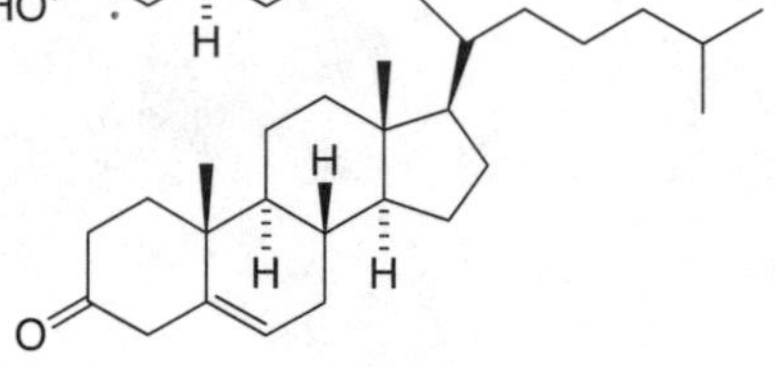

b. H₂/Pd/C

e. [1] BH₃/THF; [2] H₂O₂/⁻OH

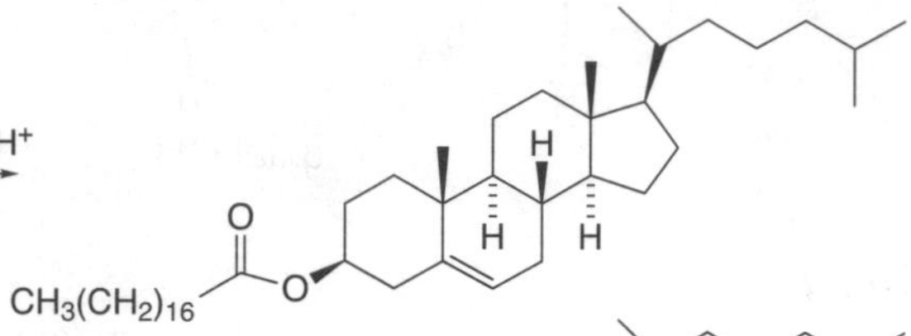

c. PCC

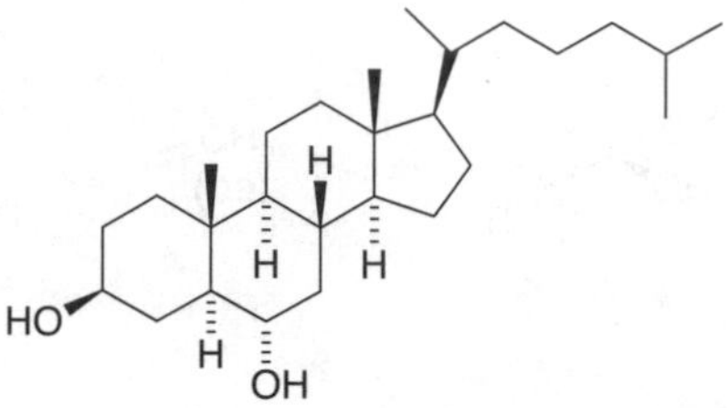

26.33

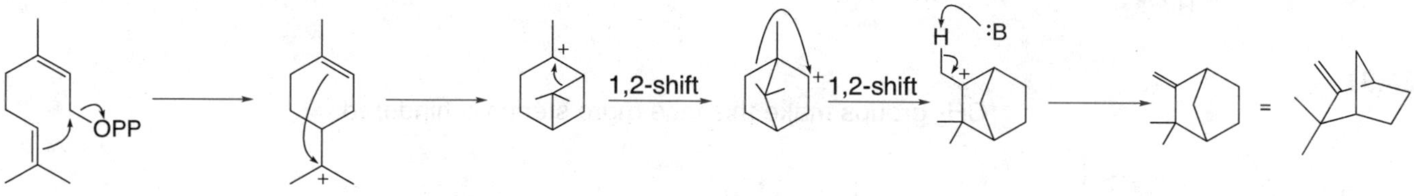

26.34 Re-draw the starting material in a conformation that suggests the structure of the product.

Chapter 27: Carbohydrates

◆ Important terms

- **Aldose** — A monosaccharide containing an aldehyde (27.2)
- **Ketose** — A monosaccharide containing a ketone (27.2)
- **D-Sugar** — A monosaccharide with the O bonded to the stereogenic center furthest from the carbonyl group drawn on the right in the Fischer projection (27.2C)
- **Epimers** — Two diastereomers that differ in configuration around one stereogenic center only (27.3)
- **Anomers** — Monosaccharides that differ in configuration at only the hemiacetal OH group (27.6)
- **Glycoside** — An acetal derived from a monosaccharide hemiacetal (27.7)

◆ Acyclic, Haworth, and 3-D representations for D-glucose (27.6)

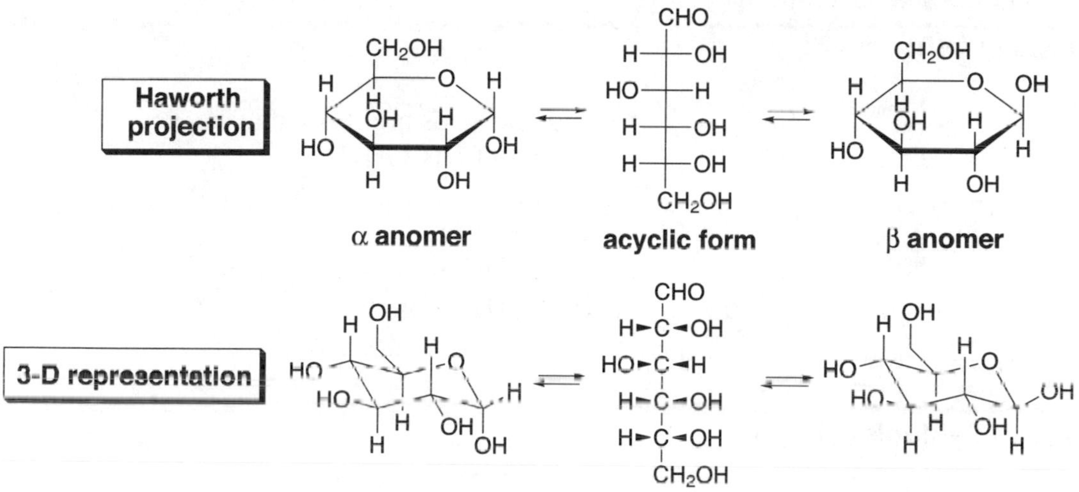

◆ Reactions of monosaccharides involving the hemiacetal

[1] Glycoside formation (27.7A)

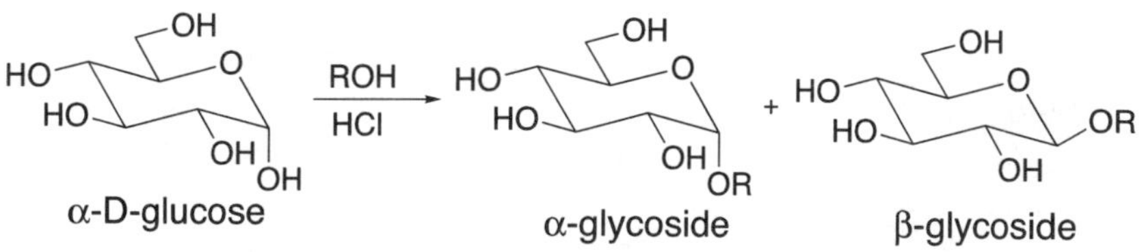

- Only the hemiacetal OH reacts.
- A mixture of α- and β-glycosides forms.

[2] Glycoside hydrolysis (27.7B)

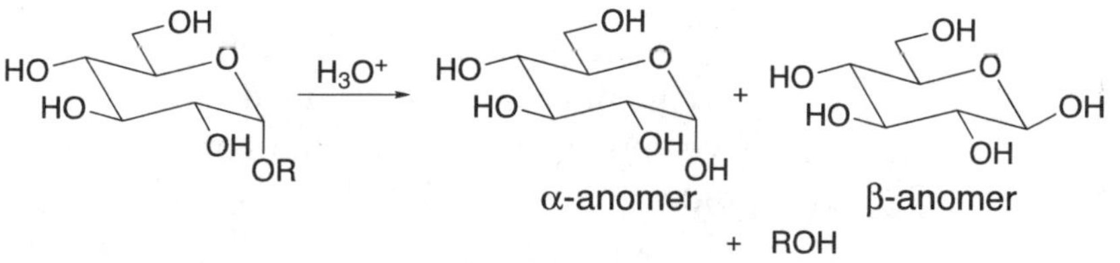

- A mixture of α and β anomers forms.

◆ Reactions of monosaccharides at the OH groups

[1] Ether formation (27.8)

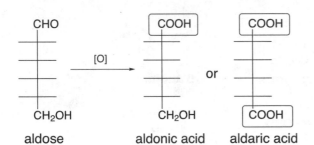

- All OH groups react.
- The stereochemistry at all stereogenic centers is retained.

[2] Ester formation (27.8)

- All OH groups react.
- The stereochemistry at all stereogenic centers is retained.

◆ Reactions of monosaccharides at the carbonyl group

[1] Oxidation of aldoses (27.9B)

- Aldonic acids are formed using:
  - Ag_2O/NH_4OH
  - Cu^{2+}
  - Br_2/H_2O
- Aldaric acids are formed with HNO_3/H_2O.

[2] Reduction of aldoses to alditols (27.9A)

[3] Wohl degradation (27.10A)

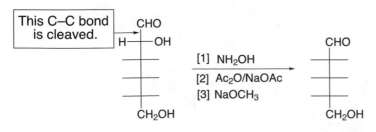

- The C1–C2 bond is cleaved to shorten an aldose chain by one carbon.
- The stereochemistry at all other stereogenic centers is retained.
- Two epimers at C2 form the same product.

[4] Kiliani-Fischer synthesis (27.10B)

- One carbon is added to the aldehyde end of an aldose.
- Two epimers at C2 are formed.

◆ **Other reactions**

[1] Hydrolysis of disaccharides (27.12)

This bond is cleaved.

A mixture of anomers is formed.

[2] Formation of N-glycosides (27.14B)

- Two anomers are formed.

Chapter 27: Answers to Problems

27.1 A *ketose* is a monosaccharide containing a ketone. An *aldose* is a monosaccharide containing an aldehyde. A monosaccharide is called: a *triose* if it has three C's; a *tetrose* if it has four C's; a *pentose* if it has five C's; a *hexose* if it has six C's, and so forth.

a. a ketotetrose

$$
\begin{array}{c}
CH_2OH \\
| \\
C=O \\
| \\
H-C-OH \\
| \\
CH_2OH
\end{array}
$$

b. an aldopentose

$$
\begin{array}{c}
CHO \\
| \\
H-C-OH \\
| \\
H-C-OH \\
| \\
H-C-OH \\
| \\
CH_2OH
\end{array}
$$

c. an aldotetrose

$$
\begin{array}{c}
CHO \\
| \\
H-C-OH \\
| \\
H-C-OH \\
| \\
CH_2OH
\end{array}
$$

27.2 Rotate and re-draw each molecule to place the horizontal bonds in front of the plane and the vertical bonds behind the plane. Then use a cross to represent the stereogenic center in a Fischer projection formula.

a. Br–C–Cl (CHO top, H bottom) = Br—Cl Fischer projection (CHO top, H bottom)

c. OHC, CH₂CH₃, C, HOCH₂ H — re-draw → H–C–CH₂OH (CHO top, CH₂CH₃ bottom) = H—CH₂OH Fischer (CHO top, CH₂CH₃ bottom)

b. HO, C, H, CHO, CH₃ — re-draw → HO–C–CH₃ (CHO top, H bottom) = HO—CH₃ Fischer (CHO top, H bottom)

d. OH, H–C–CHO, CH₃ — re-draw → HO–C–CH₃ (CHO top, H bottom) = HO—CH₃ Fischer (CHO top, H bottom)

27.3 For each molecule:
   [1] Convert the Fischer projection formula to a representation with wedges and dashes.
   [2] Assign priorities (Section 5.6).
   [3] Determine *R* or *S* in the usual manner. Reverse the answer if priority group [4] is oriented forward (on a wedge).

a. Cl—CH₂Br (CH₂NH₂ top, H bottom) → [1] → Cl–C–CH₂Br (CH₂NH₂ top, H bottom) → [2] → 1 Cl–C–CH₂Br 2 (CH₂NH₂ top = 3, H bottom = 4) → [3] → 1 Cl–C–CH₂Br 2 (CH₂NH₂ = 3, H bottom) **S configuration**

b. Cl—H (CHO top, CH₂NH₂ bottom) → [1] → Cl–C–H (CHO top, CH₂NH₂ bottom) → [2] → 1 Cl–C–H 4 (CHO top = 2, CH₂NH₂ bottom = 3) → [3] → 1 Cl–C–H (CHO = 2, CH₂NH₂ = 3) — H forward **S configuration**

c. Cl—H (CHO top, CH₂OH bottom) → [1] → Cl–C–H (CHO top, CH₂OH bottom) → [2] → 1 Cl–C–H 4 (CHO top = 2, CH₂OH bottom = 3) → [3] → 1 Cl–C–H (CHO = 2, CH₂OH = 3) — H forward **S configuration**

d. Cl—CH₂Br (COOH top, H bottom) → [1] → Cl–C–CH₂Br (COOH top, H bottom) → [2] → 1 Cl–C–CH₂Br 2 (COOH top = 3, H bottom = 4) → [3] → 1 Cl–C–CH₂Br 2 (COOH = 3, H bottom) **S configuration**

27.4

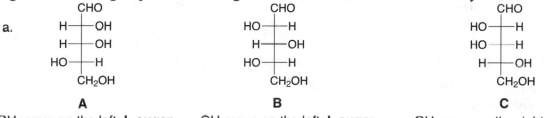

CHO   *R*
H►C◄OH *S*
HO►C◄H *R*
H►C◄OH *R*
H►C◄OH
CH₂OH

D-glucose

27.5

a. aldotetrose: 2 stereogenic centers   b. a ketohexose: 3 stereogenic centers

CHO
H—C—OH
H—C—OH
CH₂OH

CH₂OH
C=O
H—C—OH
H—C—OH
H—C—OH
CH₂OH

27.6 A D sugar has the OH group on the stereogenic center furthest from the carbonyl on the right. An L sugar has the OH group on the stereogenic center furthest from the carbonyl on the left.

a.

CHO
H——OH
H——OH
HO——H
CH₂OH

A

OH group on the left: **L sugar.**

CHO
HO——H
H——OH
HO——H
CH₂OH

B

OH group on the left: **L sugar.**

CHO
HO——H
HO——H
H——OH
CH₂OH

C

OH group on the right: **D sugar.**

b. A and B are diastereomers.
A and C are enantiomers.
B and C are diastereomers.

27.7

CHO
H——OH
H——OH
CH₂OH

D-erythrose

CHO
H——OH
HO——H
CH₂OH

L-threose

diastereomers

CHO
HO——H
HO——H
CH₂OH

L-erythrose

CHO
H——OH
HO——H
CH₂OH

L-threose

diastereomers

27.8 The D- notation signifies the position of the OH group on the stereogenic carbon furthest from the carbonyl group, and does not correlate with dextrorotatory or levorotatory. These terms describe a physical phenomenon, the direction of rotation of plane polarized light.

27.9 There are 32 aldoheptoses; 16 are D sugars.

27.10 *Epimers* are two diastereomers that differ in the configuration around only one stereogenic center.

epimers

CHO
H——OH
H——OH
CH₂OH

D-erythrose

CHO
HO——H
H——OH
CH₂OH

and

CHO
H——OH
HO——H
CH₂OH

27.11   a. D-allose and L-allose: **enantiomers**
   b. D-altrose and D-gulose: **diastereomers** but not epimers
   c. D-galactose and D-talose: **epimers**
   d. D-mannose and D-fructose: **constitutional isomers**
   e. D-fructose and D-sorbose: **diastereomers** but not epimers
   f. L-sorbose and L-tagatose: **epimers**

27.12

a.

CH₂OH
C=O
HO——H
H——OH
H——OH
CH₂OH

D-fructose

CH₂OH
C=O
H——OH
HO——H
HO——H
CH₂OH

L-fructose

enantiomers

b.

CH₂OH
C=O
HO——H
HO——H
H——OH
CH₂OH

D-tagatose

c.

CH₂OH
C=O
HO——H
H——OH
HO——H
CH₂OH

L-sorbose

27.13   Step [1]: Place the O atom in the upper right corner of a hexagon, and add the CH₂OH group on the first carbon counterclockwise from the O atom.
   Step [2]: Place the anomeric carbon on the first carbon clockwise from the O atom.
   Step [3]: Add the substituents at the three remaining stereogenic centers, clockwise around the ring.

a. Draw the α anomer of:

CHO
H——OH
H——OH
H——OH
H——OH ← furthest away C,
CH₂OH   OH on right = D sugar

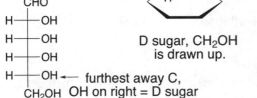

[1]

D sugar, CH₂OH
is drawn up.

[2]

α anomer
OH is down
for a D sugar

[3]

First 3 substituents are
on the right so they are
drawn down.

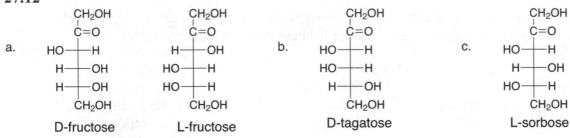

b. Draw the α anomer of:

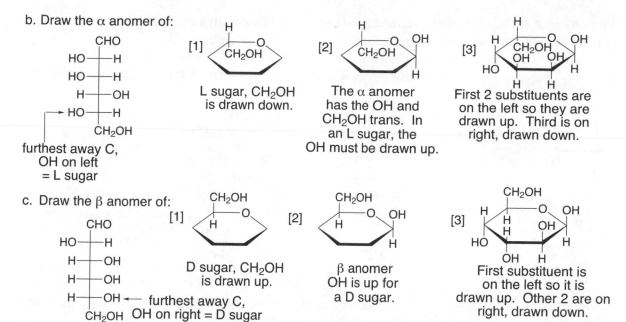

[1] L sugar, CH₂OH is drawn down.

[2] The α anomer has the OH and CH₂OH trans. In an L sugar, the OH must be drawn up.

[3] First 2 substituents are on the left so they are drawn up. Third is on right, drawn down.

furthest away C, OH on left = L sugar

c. Draw the β anomer of:

[1] D sugar, CH₂OH is drawn up.

furthest away C, OH on right = D sugar

[2] β anomer OH is up for a D sugar.

[3] First substituent is on the left so it is drawn up. Other 2 are on right, drawn down.

27.14 To convert each Haworth projection into its acyclic form:

[1] Draw the C skeleton with the CHO on the top and the CH₂OH on the bottom.

[2] Draw in the OH group furthest from the C=O.

A CH₂OH group drawn up means a D sugar; a CH₂OH group drawn down means an L sugar.

[3] Add the three other stereogenic centers, counterclockwise around the ring.

"Up" groups go on the left and "down" groups go on the right.

a.

"up" group on left

CH₂OH is up = D sugar

"down" groups on right

[1] CHO ... CH₂OH

[2] CHO ... H—OH ... CH₂OH

OH on right = D sugar

[3] CHO
H—OH
H—OH
HO—H
H—OH
CH₂OH

b.

CH₂OH is down = L sugar

"down" groups on right

"up" group on left

[1] CHO ... CH₂OH

[2] CHO ... HO—H ... CH₂OH

OH on left = L sugar

[3] CHO
HO—H
H—OH
H—OH
HO—H
CH₂OH

27.15 To convert a Haworth projection into a 3-D representation with a chair cyclohexane:
[1] Draw the pyranose ring as a chair with the O as an "up" atom.
[2] Add the substituents around the ring.

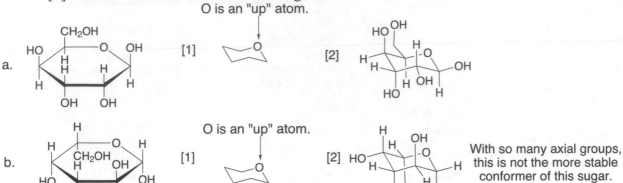

27.16 Cyclization always forms a new stereogenic center at the anomeric carbon, so two different anomers are possible.

Two anomers of D-erythrose:

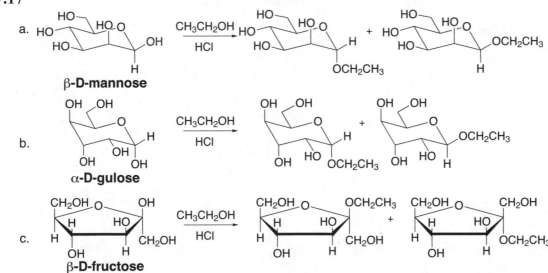

27.17

27.18

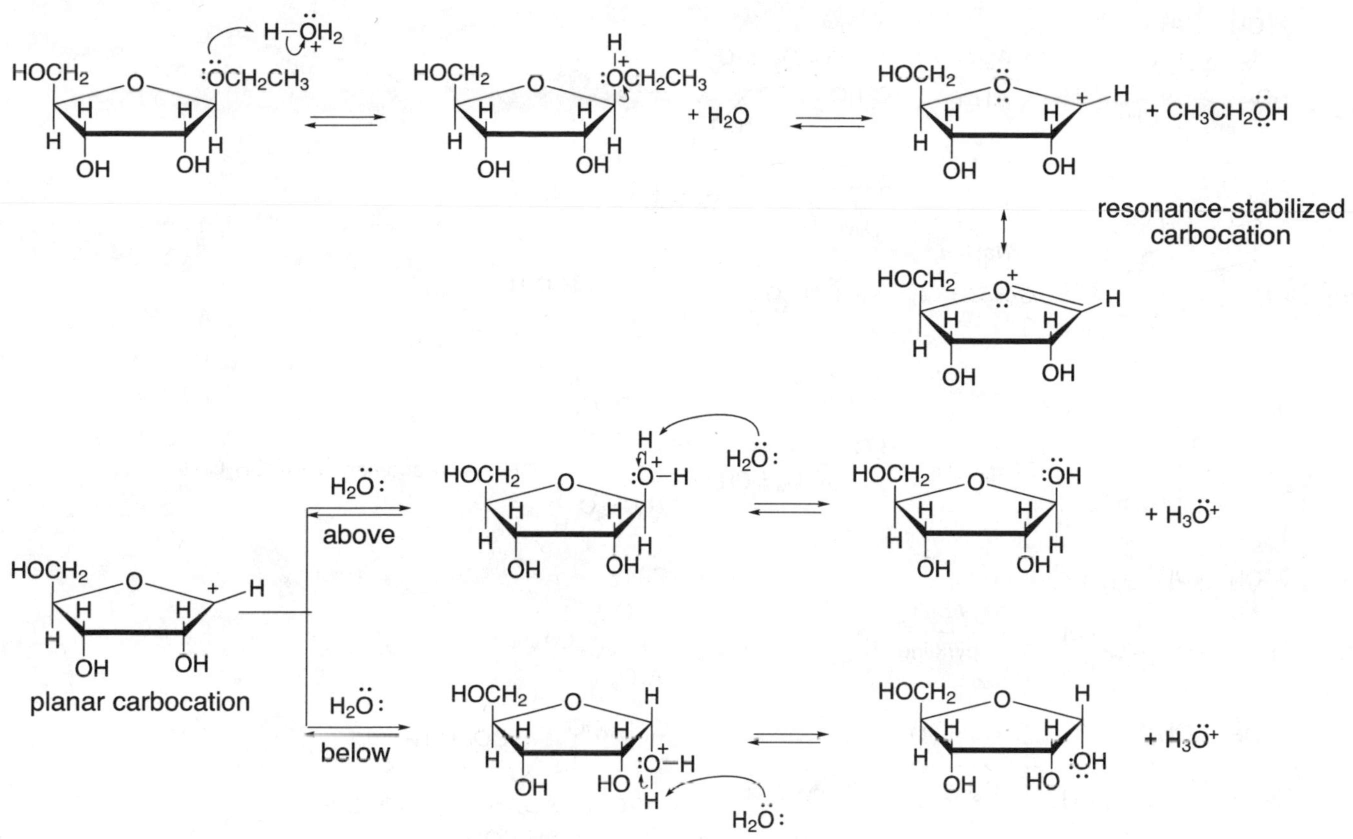
+ H_2O
+ CH_3CH_2OH
resonance-stabilized carbocation
planar carbocation
above
below
+ H_3O^+
+ H_3O^+

27.19

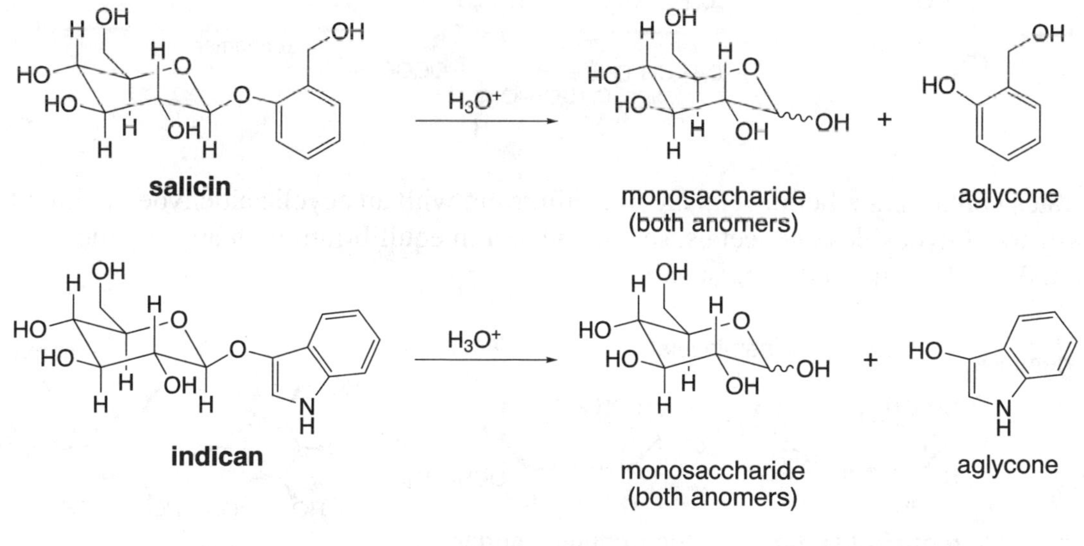
salicin
H_3O^+
monosaccharide (both anomers)
aglycone
indican
H_3O^+
monosaccharide (both anomers)
aglycone

27.20

a.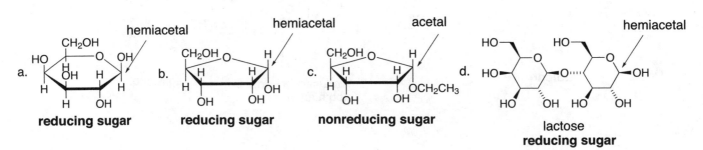

b.

c.

d.

e.

f.

27.21 Carbohydrates containing a hemiacetal are in equilibrium with an acyclic aldehyde, making them reducing sugars. Glycosides are acetals, so they are not in equilibrium with any acyclic aldehyde, making them nonreducing sugars.

a. hemiacetal
reducing sugar

b. hemiacetal
reducing sugar

c. acetal
nonreducing sugar

d. hemiacetal
lactose
reducing sugar

27.22

a.
$$\begin{array}{c} CHO \\ HO-\!\!-H \\ H-\!\!-OH \\ H-\!\!-OH \\ CH_2OH \end{array} \xrightarrow[NH_4OH]{Ag_2O} \begin{array}{c} COOH \\ HO-\!\!-H \\ H-\!\!-OH \\ H-\!\!-OH \\ CH_2OH \end{array}$$

c.
$$\begin{array}{c} CHO \\ HO-\!\!-H \\ H-\!\!-OH \\ H-\!\!-OH \\ CH_2OH \end{array} \xrightarrow[H_2O]{HNO_3} \begin{array}{c} COOH \\ HO-\!\!-H \\ H-\!\!-OH \\ H-\!\!-OH \\ COOH \end{array}$$

b.
$$\begin{array}{c} CHO \\ HO-\!\!-H \\ H-\!\!-OH \\ H-\!\!-OH \\ CH_2OH \end{array} \xrightarrow[H_2O]{Br_2} \begin{array}{c} COOH \\ HO-\!\!-H \\ H-\!\!-OH \\ H-\!\!-OH \\ CH_2OH \end{array}$$

27.23 Molecules with a plane of symmetry are optically inactive.

a.
$$\begin{array}{c} CHO \\ H-\!\!-OH \\ H-\!\!-OH \\ CH_2OH \end{array} \longrightarrow \begin{array}{c} COOH \\ H-\!\!-OH \\ H-\!\!-OH \\ COOH \end{array}$$

D-erythrose   **optically inactive**

c.
$$\begin{array}{c} CHO \\ H-\!\!-OH \\ HO-\!\!-H \\ HO-\!\!-H \\ H-\!\!-OH \\ CH_2OH \end{array} \longrightarrow \begin{array}{c} COOH \\ H-\!\!-OH \\ HO-\!\!-H \\ HO-\!\!-H \\ H-\!\!-OH \\ COOH \end{array}$$

D-galactose   **optically inactive**

b.
$$\begin{array}{c} CHO \\ HO-\!\!-H \\ HO-\!\!-H \\ H-\!\!-OH \\ CH_2OH \end{array} \longrightarrow \begin{array}{c} COOH \\ HO-\!\!-H \\ HO-\!\!-H \\ H-\!\!-OH \\ COOH \end{array}$$

D-lyxose   **optically active**

27.24

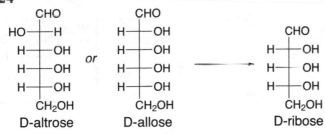

$$\begin{array}{c} CHO \\ HO-\!\!-H \\ H-\!\!-OH \\ H-\!\!-OH \\ H-\!\!-OH \\ CH_2OH \end{array}$$
D-altrose

or

$$\begin{array}{c} CHO \\ H-\!\!-OH \\ H-\!\!-OH \\ H-\!\!-OH \\ H-\!\!-OH \\ CH_2OH \end{array}$$
D-allose

$$\longrightarrow$$

$$\begin{array}{c} CHO \\ H-\!\!-OH \\ H-\!\!-OH \\ H-\!\!-OH \\ CH_2OH \end{array}$$
D-ribose

27.25

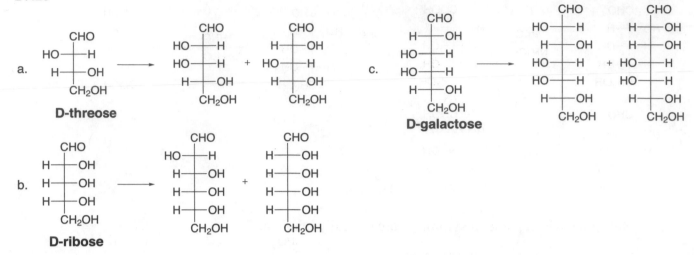

a. D-threose

b. D-ribose

c. D-galactose

27.26

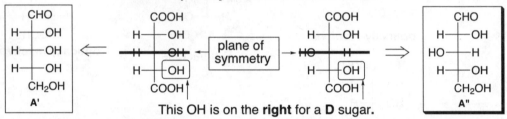

Possible optically inactive D-aldaric acids:

This OH is on the **right** for a **D** sugar.

There are two possible structures for the D-aldopentose (**A'** and **A"**), and the Wohl degradation determines which structure corresponds to **A**.

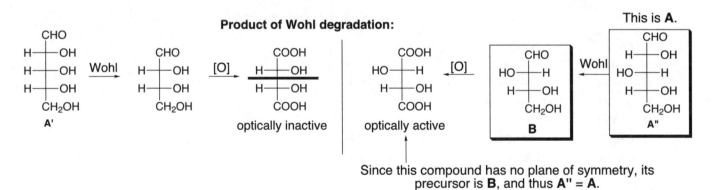

Product of Wohl degradation:

This is **A**.

optically inactive   optically active

Since this compound has no plane of symmetry, its precursor is **B**, and thus **A" = A**.

27.27

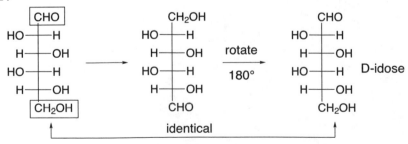

rotate 180°   D-idose

identical

27.28

Optically inactive alditols formed from NaBH₄ reduction of a D-aldohexose.

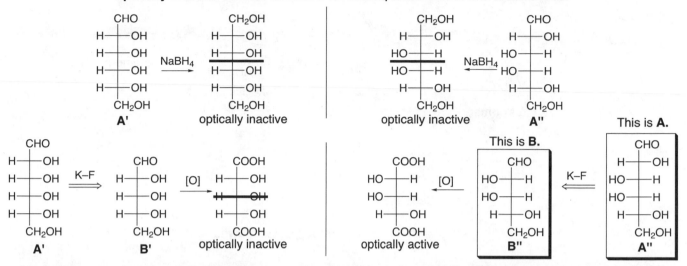

Two D-aldohexoses (**A'** and **A"**) give optically inactive alditols on reduction. **A"** is formed from **B"** by Kiliani–Fischer synthesis. Since **B"** affords an optically active aldaric acid on oxidation, **B"** is **B** and **A"** is **A**. The alternate possibility (**A'**) is formed from an aldopentose **B'** that gives an optically inactive aldaric acid on oxidation.

27.29

β-D-glucose

planar carbocation

above

β-D-glucose

+ H₃O⁺

below

α-D-glucose

27.30

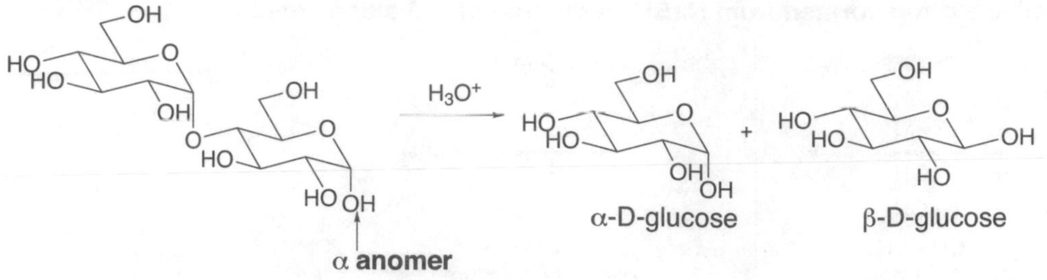

The same products are formed on hydrolysis of the α and β anomers of maltose.

α anomer

α-D-glucose + β-D-glucose

27.31

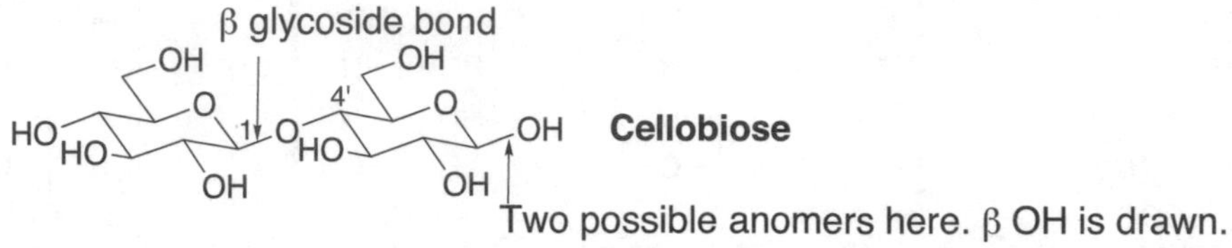

β glycoside bond

Cellobiose

Two possible anomers here. β OH is drawn.

27.32

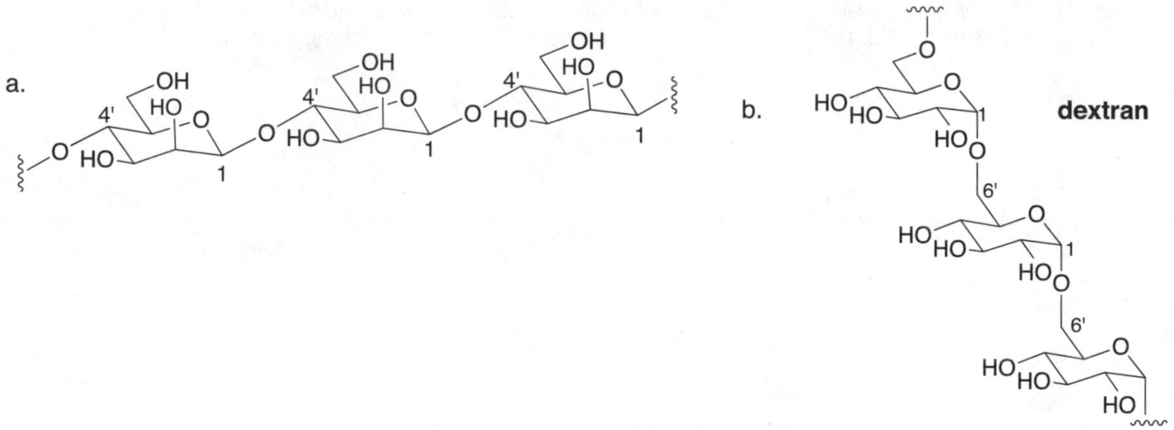

a.

b.

dextran

27.33

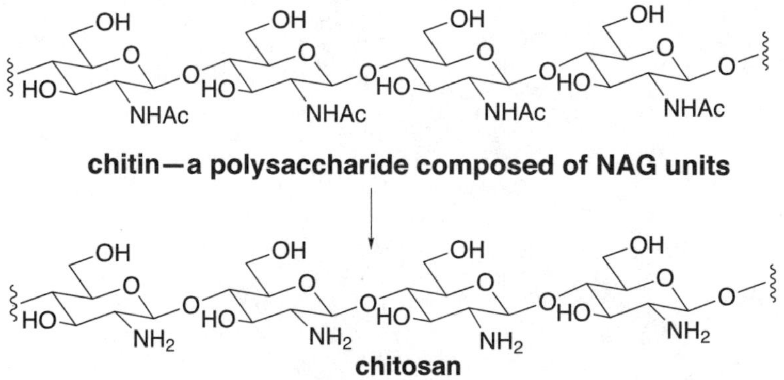

chitin—a polysaccharide composed of NAG units

chitosan

27.34

a

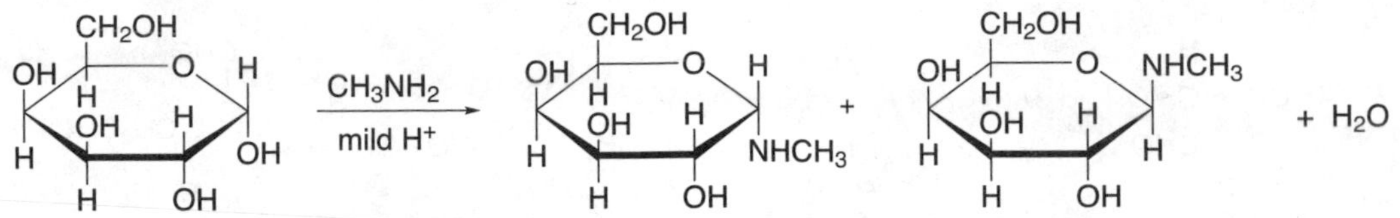

b.

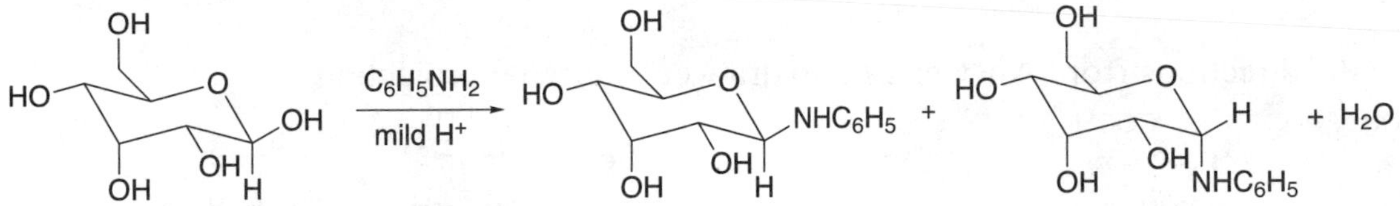

27.35

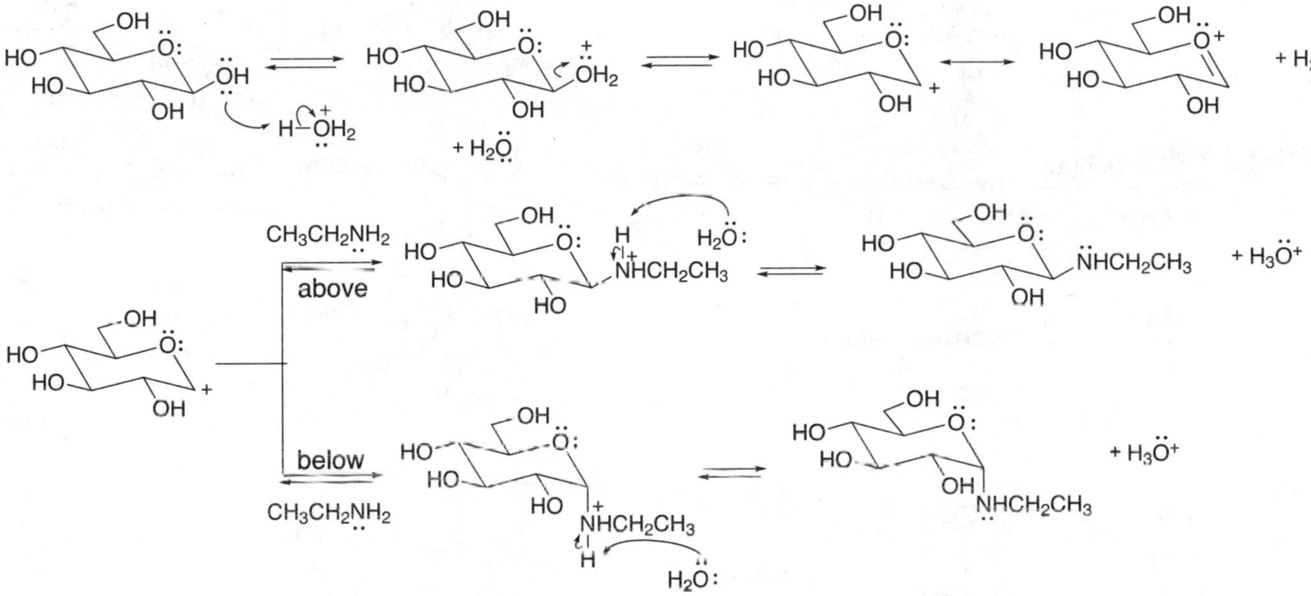

27.36

a.

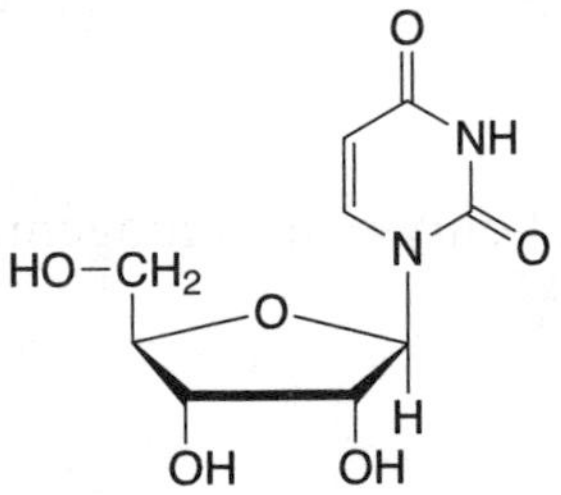

b.

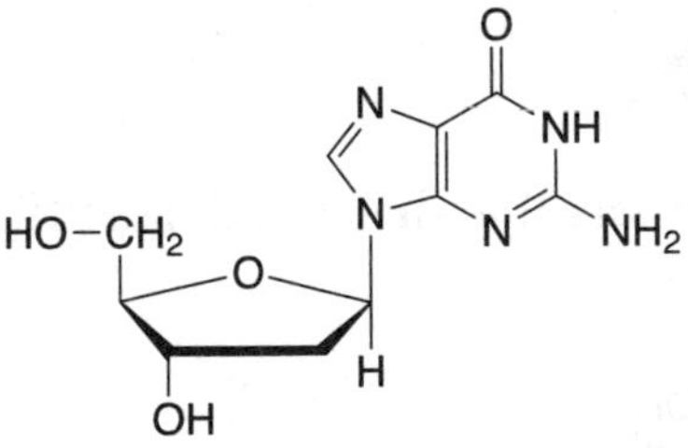

27.37 Label the compounds with *R* or *S* and then classify.

CHO
H——OH
CH₂CH₃

A
R

a. CH₃CH₂—C—OH (H top, CHO bottom)

R
identical

b. (CHO structure)

R
identical

c. CH₃CH₂—C—CHO (H, OH top)

S
enantiomer

d. HO—C—CH₂CH₃ (CHO top, H bottom)

S
enantiomer

27.38 Use the directions from Answer 27.2 to draw each Fischer projection.

a. CH₃—C—Br (COOH top, H bottom) = CH₃—**S**—Br (COOH top, H bottom)

b. CH₃—C—Cl (Br top, H bottom) —re-draw→ H—C—Br (CH₃ top, Cl bottom) = H—**S**—Br (CH₃ top, Cl bottom)

c. (CH₃O, OCH₂CH₃, CH₃, CH₂CH₃) —re-draw→ CH₃—C—CH₂CH₃ (OCH₃ top, OCH₂CH₃ bottom) = CH₃—**S**—CH₂CH₃ (OCH₃ top, OCH₂CH₃ bottom)

d. CH₃CH₂—C—H (Cl top, Br bottom) —re-draw→ CH₃CH₂—C—Cl (H top, Br bottom) = CH₃CH₂—**S**—Cl (H top, Br bottom)

e. H—C—Br; Cl—C—H (CH₃ top, CH₂CH₃ bottom) = H—**S**—Br; Cl—**S**—H (CH₃ top, CH₂CH₃ bottom)

f. (H, Cl, Br, Br, Cl, H with **R S**) —re-draw→ Br—C—H; Cl—C—Br (Cl top, H bottom) (**R**, **S**) = Br—**R**—H; Cl—**S**—Br

g. (CH₃, H, Br, H, Br, CH₃) —re-draw→ H—C—Br; Br—C—H (CH₃ top and bottom) = H—**S**—Br; Br—**S**—H (CH₃ top and bottom)

h. HO (structure) —re-draw→ HO—C—H; HO—C—H; H—C—OH (CHO top, CH₂OH bottom) = HO—**S**—H; HO—**S**—H; H—**R**—OH (CHO top, CH₂OH bottom)

27.39

CHO ——C2
H——OH
H——OH
H——OH
CH₂OH
D-ribose

CHO ——C2
H——OH
HO——H
H——OH
CH₂OH
D-xylose

R at C2.

27.40 *Epimers* are two diastereomers that differ in the configuration around only one stereogenic center.

CHO
H——OH
HO——H
H——OH
CH₂OH
D-xylose

CHO
H——OH
HO——H
HO——H
CH₂OH
L-arabinose

C4

27.41

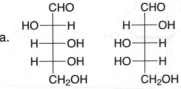

a.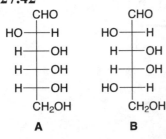

D-arabinose **enantiomer**

b.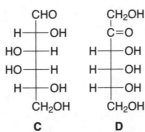

epimer

c.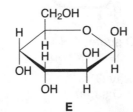

diastereomer
(but not epimer)

d.

constitutional
isomer

27.42

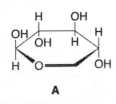 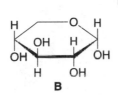 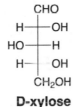

A   B   C   D   E

F

a. **A** and **B** **epimers**
b. **A** and **C** **diastereomers**

c. **B** and **C** **enantiomers**
d. **A** and **D** **constitutional isomers**

e. **E** and **F** **diastereomers**

27.43

a. anomers, epimers, diastereomers, reducing sugars
b. .

 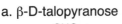 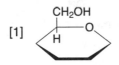 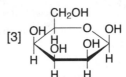

A   B

CHO

This is the acyclic form of
both **A** and **B**.

D-xylose

27.44 Use the directions from Answer 27.13.

a. β-D-talopyranose

CHO
HO——H
HO——H
HO——H
H——OH ← furthest away C,
CH₂OH OH on right = D sugar
D-talose

[1] CH₂OH ...

D sugar, CH₂OH
is drawn up.

[2] β anomer
OH is up for
a D sugar.

[3]

b. β-D-mannopyranose

CHO
HO——H
HO——H
H——OH
H——OH ← furthest away C,
CH₂OH OH on right = D sugar
D-mannose

[1] CH₂OH ...

D sugar, CH₂OH
is drawn up.

[2] β anomer
OH is up for
a D sugar.

[3]

c. α-D-galactopyranose

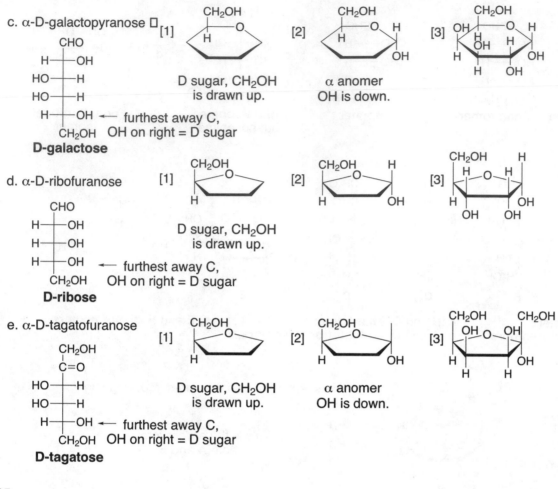

d. α-D-ribofuranose

e. α-D-tagatofuranose

27.45

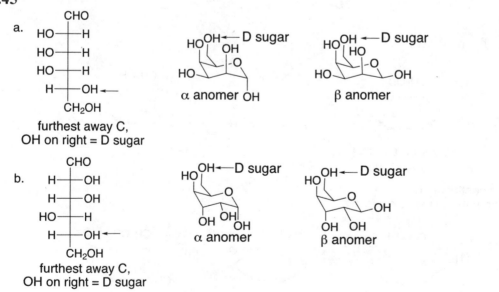

c.

CHO
HO——H
H——OH
HO——H
H——OH ←
CH₂OH

furthest away C,
OH on right = D sugar

α anomer

HO OH ← D sugar
OH
-O
OH   OH

β anomer

HO OH ← D sugar
HO
-O
OH
OH

27.46 Use the directions from Answer 27.14.

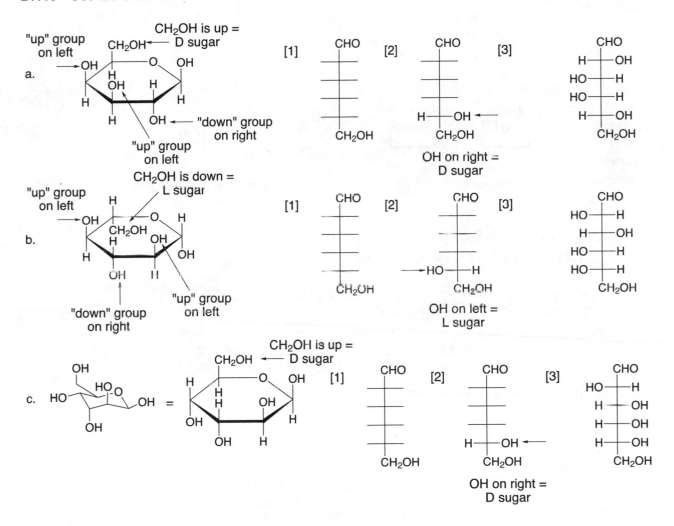

a.

"up" group
on left

CH₂OH is up =
CH₂OH ← D sugar
O   OH
OH
H
OH   H
H
H   OH ← "down" group
on right
"up" group
on left

[1]   CHO

CH₂OH

[2]   CHO

H——OH ←

CH₂OH

OH on right =
D sugar

[3]   CHO

H——OH
HO——H
HO——H
H——OH

CH₂OH

b.

"up" group
on left

CH₂OH is down =
L sugar
H
O   H
CH₂OH
H   OH
OH
H
OH   H

"down" group
on right

"up" group
on left

[1]   CHO

CH₂OH

[2]   CHO

HO——H ←

CH₂OH

OH on left =
L sugar

[3]   CHO

HO——H
H——OH
HO——H
HO——H

CH₂OH

c.

OH
HO   HO O
OH
OH

=

CH₂OH is up =
CH₂OH ← D sugar
H
O   OH
H
OH
H
OH   H

[1]   CHO

CH₂OH

[2]   CHO

H——OH ←

CH₂OH

OH on right =
D sugar

[3]   CHO

HO——H
H——OH
H——OH
H——OH

CH₂OH

d. **D sugar**

e.

f.

27.47

D-arabinose

a.
β anomer
α anomer

b.
two anomers in the pyranose form

27.48
Two anomers of D-idose, as well as two conformers of each anomer:

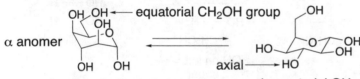

α anomer ⇌ equatorial CH₂OH group

4 axial substituents
4 equatorial OH groups
More stable conformer for the α anomer—the
CH₂OH is axial, but all other groups are equatorial.

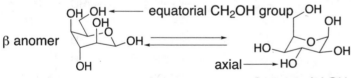

β anomer ⇌ equatorial CH₂OH group

3 axial substituents
3 equatorial OH groups
The more stable conformer for the β anomer—the
CH₂OH is axial, as is the anomeric OH, but three
other OH groups are equatorial.

27.49

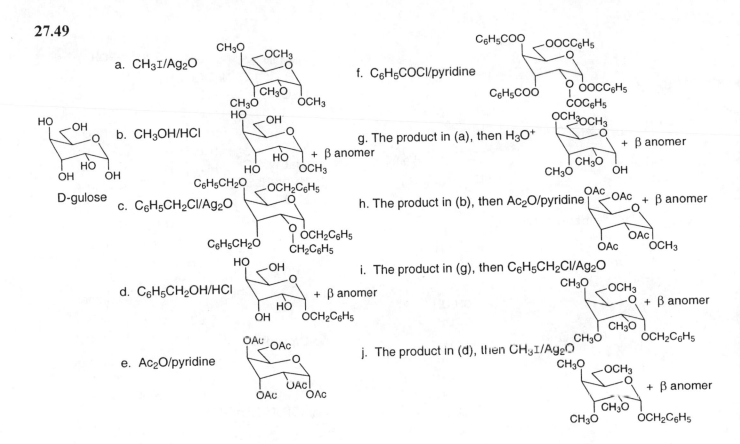

a. CH₃I/Ag₂O

b. CH₃OH/HCl

c. C₆H₅CH₂Cl/Ag₂O

d. C₆H₅CH₂OH/HCl

e. Ac₂O/pyridine

f. C₆H₅COCl/pyridine

g. The product in (a), then H₃O⁺

h. The product in (b), then Ac₂O/pyridine

i. The product in (g), then C₆H₅CH₂Cl/Ag₂O

j. The product in (d), then CH₃I/Ag₂O

D-gulose

27.50

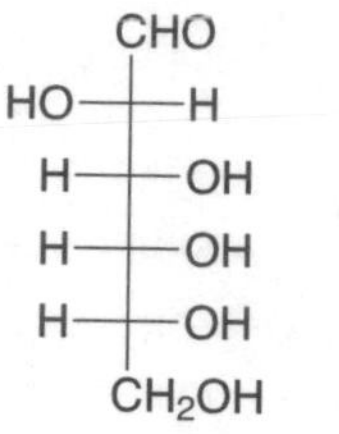

CHO
HO——H
H——OH
H——OH
H——OH
CH₂OH

D-altrose

a. CH₃OH/HCl

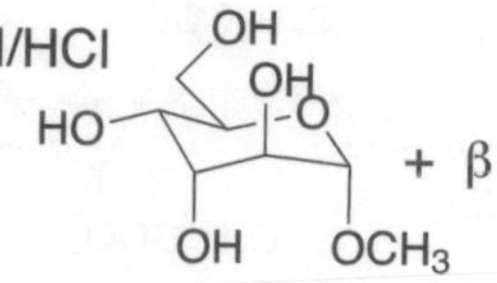

+ β anomer

b. (CH₃)₂CHOH/HCl

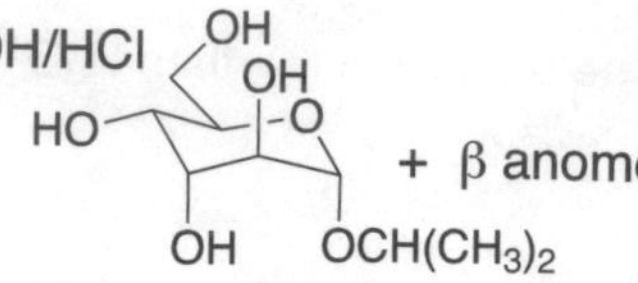

+ β anomer

c. NaBH₄/CH₃OH

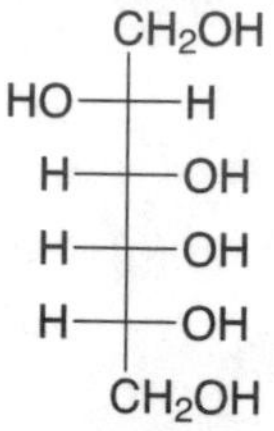

CH₂OH
HO——H
H——OH
H——OH
H——OH
CH₂OH

d. Br₂/H₂O

COOH
HO——H
H——OH
H——OH
H——OH
CH₂OH

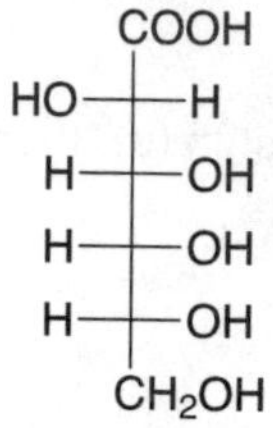

e. HNO₃/H₂O

COOH
HO——H
H——OH
H——OH
H——OH
COOH

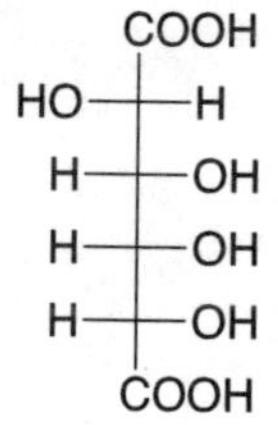

f. [1] NH₂OH
   [2] (CH₃CO)₂O, NaOCOCH₃
   [3] NaOCH₃

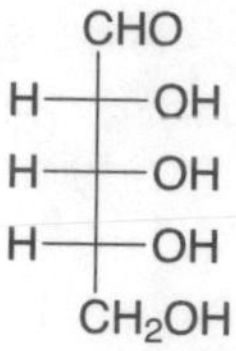

CHO
H——OH
H——OH
H——OH
CH₂OH

g. [1] NaCN, HCl
   [2] H₂/Pd/BaSO₄
   [3] H₃O⁺

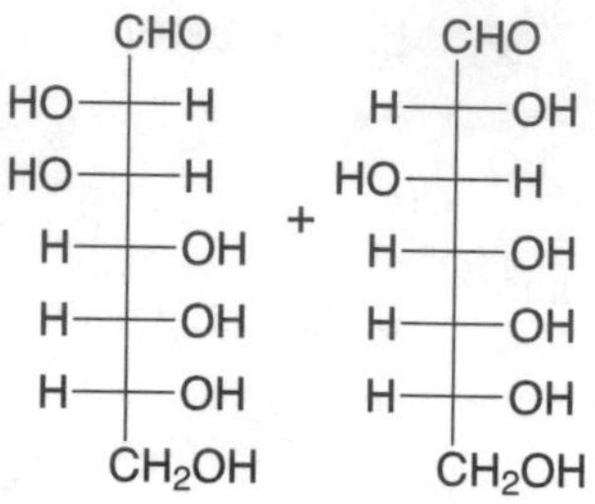

CHO
HO——H
HO——H
H——OH
H——OH
H——OH
CH₂OH

+

CHO
H——OH
HO——H
H——OH
H——OH
H——OH
CH₂OH

h. CH₃I/Ag₂O

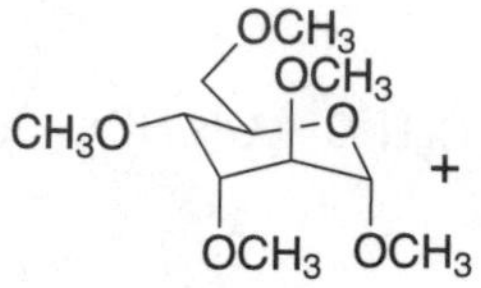

+ β anomer

i. Ac₂O/pyridine

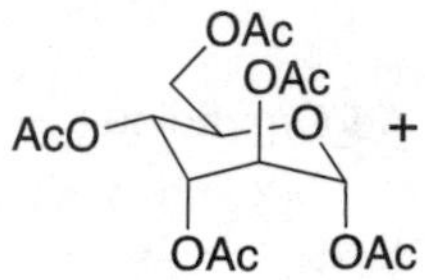

+ β anomer

j. C₆H₅CH₂NH₂/ mild H⁺

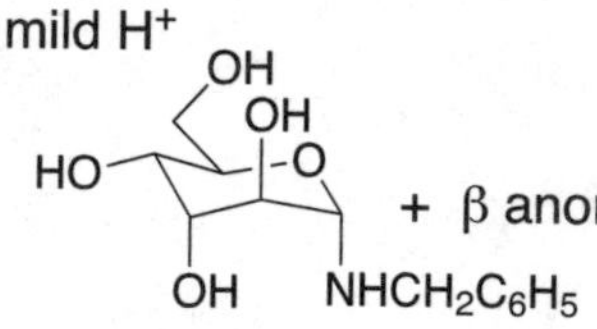

+ β anomer

27.51

a. CH$_3$OH/HCl

[structure: methyl furanoside] + β anomer

b. (CH$_3$)$_2$CHOH/HCl

[structure]

CHO
H—OH
HO—H
H—OH
CH$_2$OH
D-xylose

[structure] + β anomer

c. NaBH$_4$/CH$_3$OH

CH$_2$OH
H—OH
HO—H
H—OH
CH$_2$OH

d. Br$_2$/H$_2$O

COOH
H—OH
HO—H
H—OH
CH$_2$OH

e. HNO$_3$/H$_2$O

COOH
H—OH
HO—H
H—OH
COOH

f. [1] NH$_2$OH
   [2] (CH$_3$CO)$_2$O, NaOCOCH$_3$
   [3] NaOCH$_3$

CHO
HO—H
H—OH
CH$_2$OH

g. [1] NaCN, HCl
   [2] H$_2$/Pd/BaSO$_4$
   [3] H$_3$O$^+$

CHO   CHO
HO—H   H—OH
H—OH   H—OH
HO—H   +   HO—H
H—OH   H—OH
CH$_2$OH   CH$_2$OH

h. CH$_3$I/Ag$_2$O

[structure with CH$_2$OCH$_3$, OCH$_3$] + β anomer

i. Ac$_2$O/pyridine

[structure with CH$_2$OAc, OAc] + β anomer

j. C$_6$H$_5$CH$_2$NH$_2$/ mild H$^+$

[structure with CH$_2$OH, NHCH$_2$C$_6$H$_5$] + β anomer

27.52

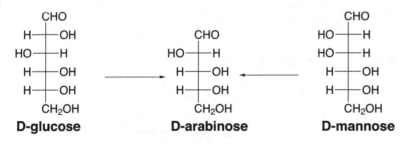

CHO
H—OH
HO—H
H—OH
H—OH
CH$_2$OH
D-glucose

CHO
HO—H
H—OH
H—OH
CH$_2$OH
D-arabinose

CHO
HO—H
HO—H
H—OH
H—OH
CH$_2$OH
D-mannose

27.53

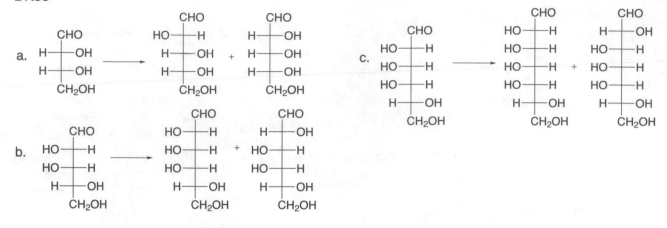

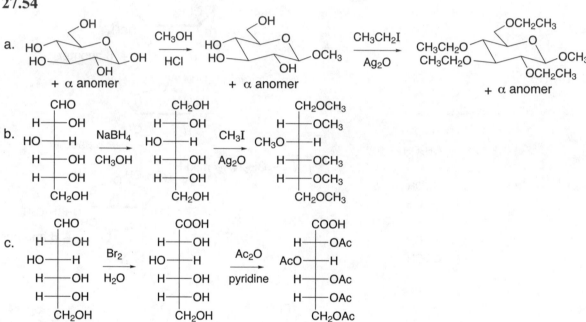

27.54

27.55 Molecules with a plane of symmetry are optically inactive.

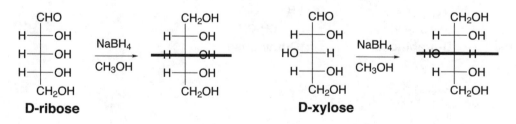

27.56

a.

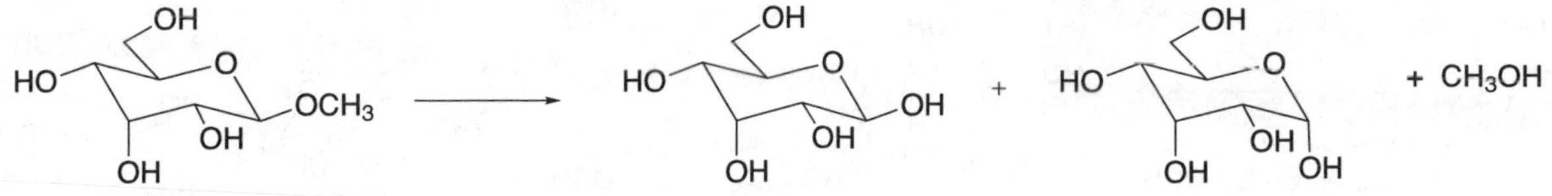

b.

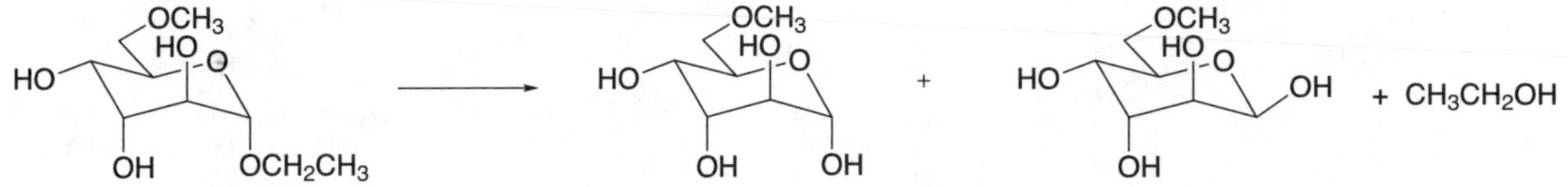

c.

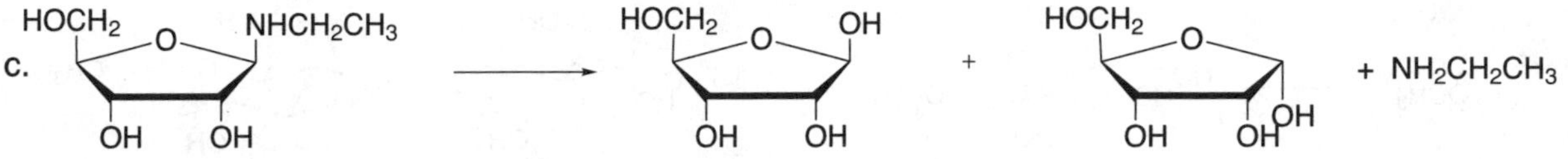

27.57

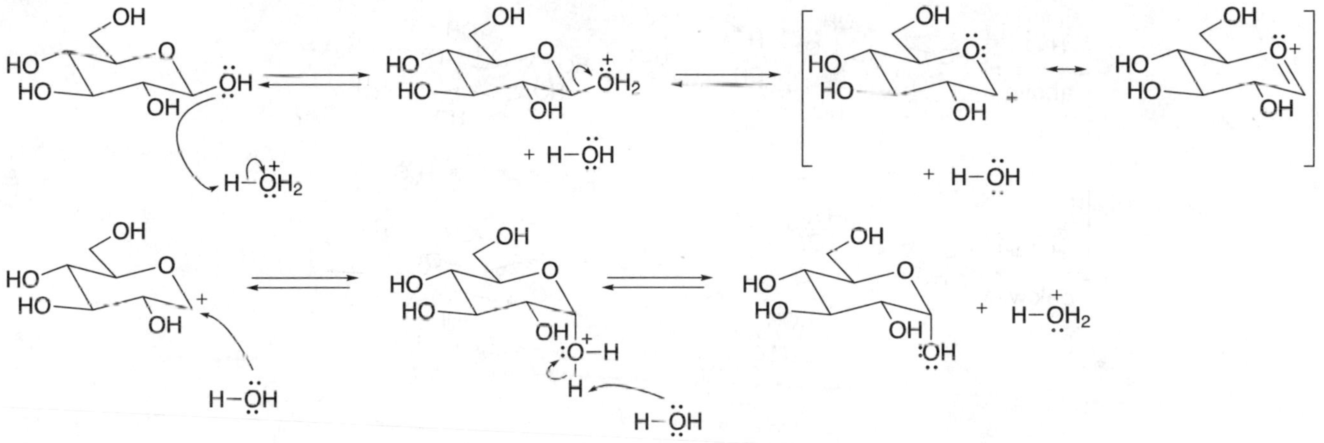

27.58

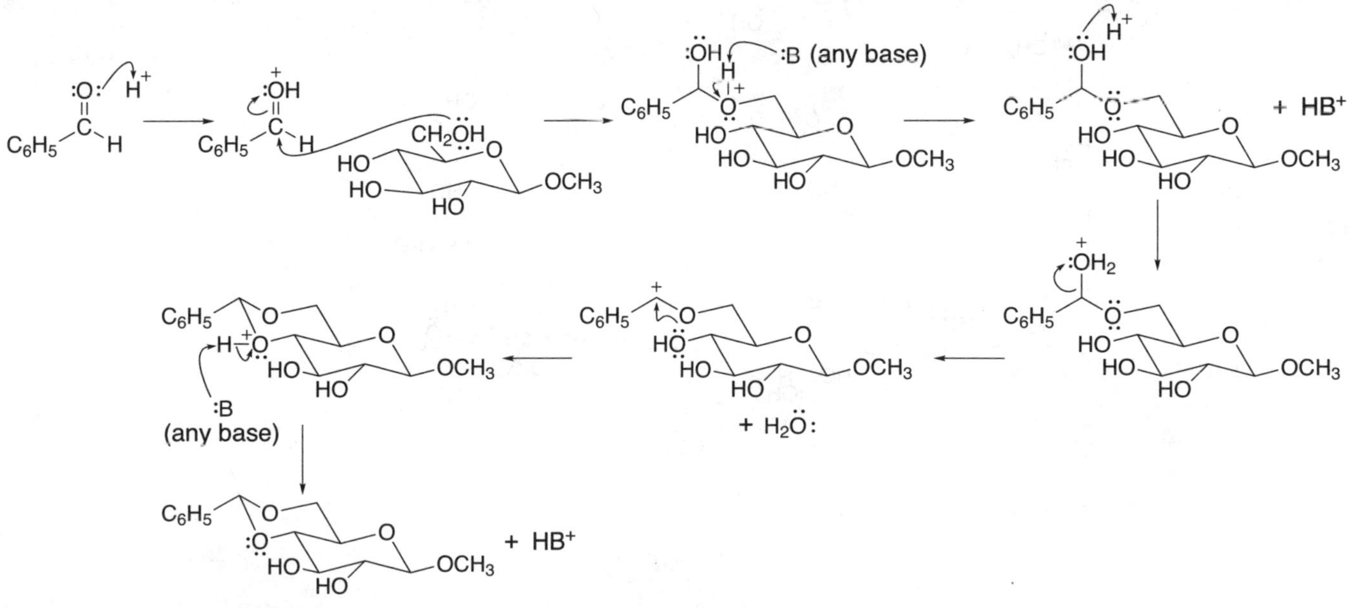

27.59

27.60

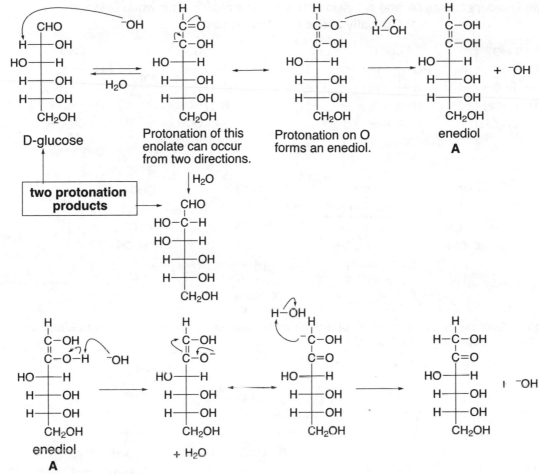

D-glucose

Protonation of this enolate can occur from two directions.

Protonation on O forms an enediol.

enediol
A

two protonation products

enediol
A

Deprotonation of one OH of the enediol forms a new enolate that goes on to form the ketohexose.

27.61

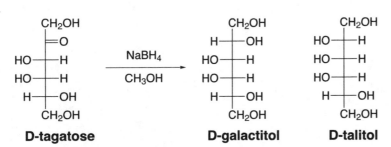

D-tagatose $\xrightarrow[\text{CH}_3\text{OH}]{\text{NaBH}_4}$ **D-galactitol**   **D-talitol**

27.62

Two D-aldopentoses (**A'** and **A"**) yield optically active aldaric acids when oxidized.

Optically active D-aldaric acids:

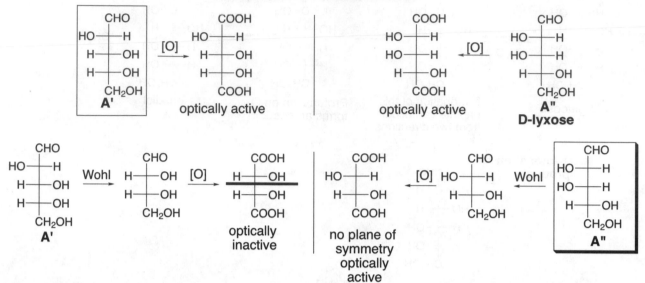

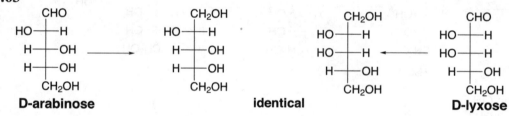

Only **A"** undergoes Wohl degradation to an aldotetrose that is oxidized to an optically active aldaric acid so **A"** is the structure of the D-aldopentose in question.

27.63

| CHO | CH₂OH | CH₂OH | CHO |

D-arabinose identical D-lyxose

27.64

Only two D-aldopentoses (**A'** and **A"**) yield optically inactive aldaric acids (**B'** and **B"**).

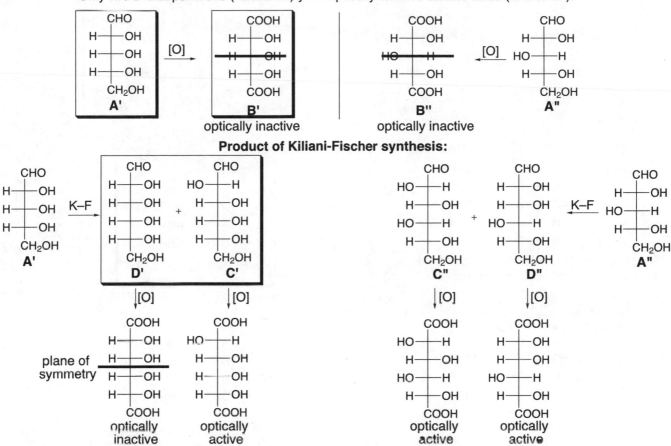

Product of Kiliani-Fischer synthesis:

Only **A'** fits the criteria. Kiliani–Fischer synthesis of **A'** forms **C'** and **D'** which are oxidized to one optically active and one optically inactive aldaric acid. A similar procedure with **A"** forms two optically active aldaric acids. Thus, the structures of **A–D**, correspond to the structures of **A'–D'**.

27.65

Only two D-aldopentoses (**A'** and **A"**) are reduced to optically active alditols.

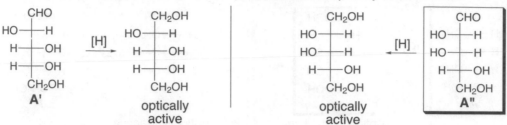

Product of Kiliani-Fischer synthesis:

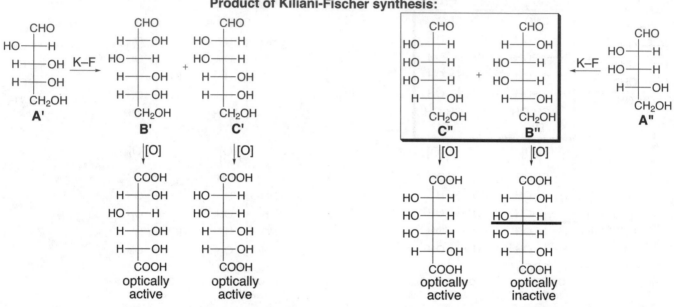

Only **A"** fits the remaining criteria. Kiliani–Fischer synthesis of **A"** forms **B"** and **C"**, which are oxidized to one optically inactive and one optically active diacid. A similar procedure with **A'** forms two optically active diacids. Thus, the structures of **A–C** correspond to **A"–C"**.

27.66

D-gulose

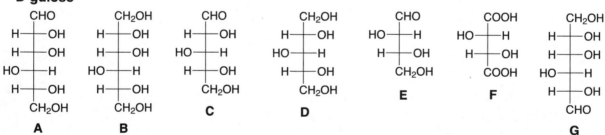

27.67

a.

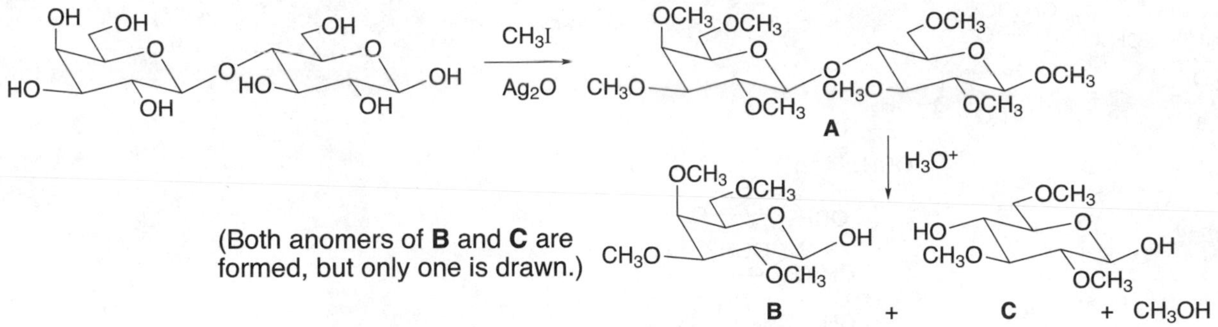

CH_3I / Ag_2O → **A**

H_3O^+ ↓

B + **C** + CH_3OH

(Both anomers of **B** and **C** are formed, but only one is drawn.)

b.

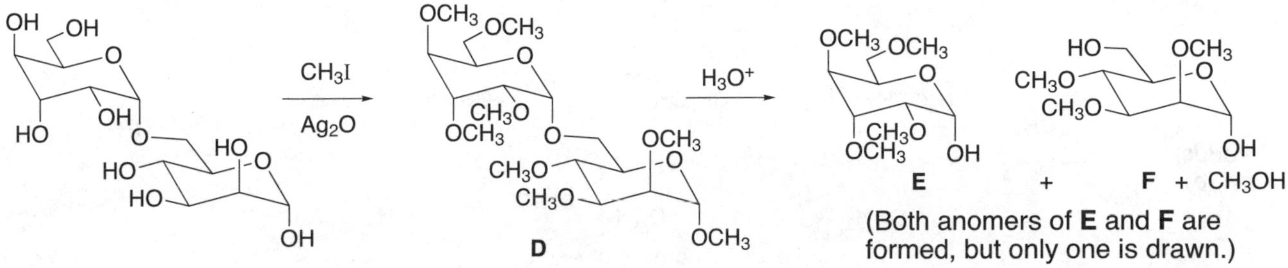

CH_3I / Ag_2O → **D**

H_3O^+ → **E** + **F** + CH_3OH

(Both anomers of **E** and **F** are formed, but only one is drawn.)

27.68

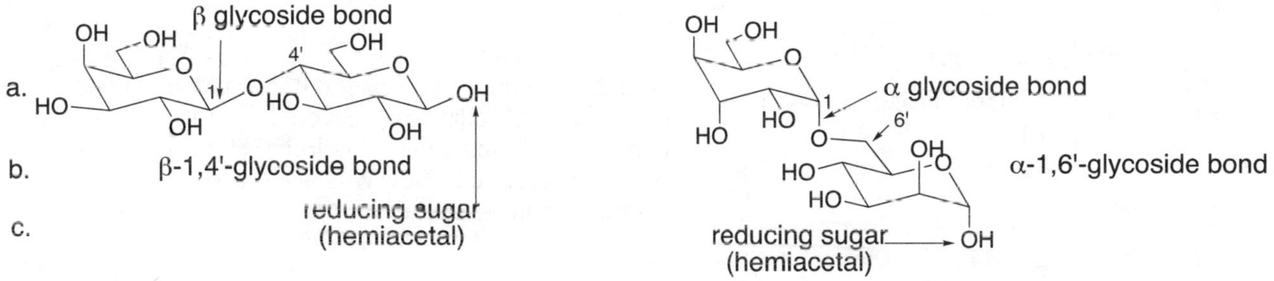

a.

β glycoside bond

reducing sugar (hemiacetal)

α glycoside bond

α-1,6'-glycoside bond

reducing sugar (hemiacetal)

b. β-1,4'-glycoside bond

c.

27.69

a and b.

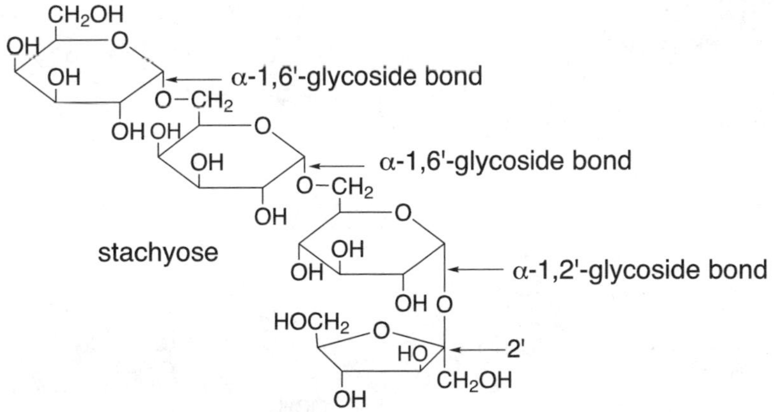

α-1,6'-glycoside bond

α-1,6'-glycoside bond

α-1,2'-glycoside bond

stachyose

c.
H_3O^+ →

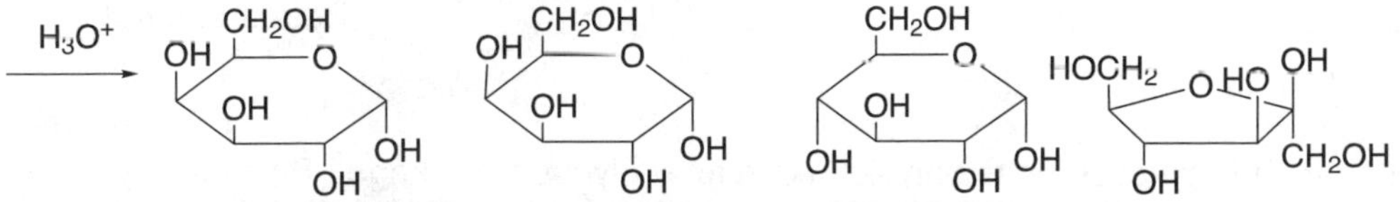

α and β Anomers of each monosaccharide are formed, but only one anomer is drawn.

d. Stachyose is not a reducing sugar since it contains no hemiacetal.

e.
$$\xrightarrow[\text{Ag}_2\text{O}]{\text{CH}_3\text{I}}$$

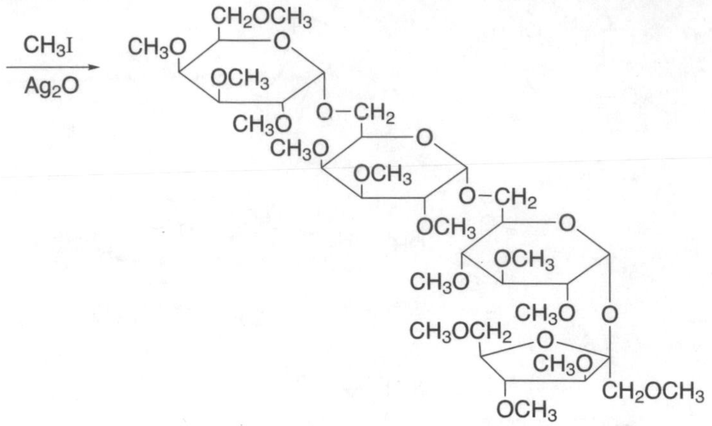

f. product in (e) $\xrightarrow{\text{H}_3\text{O}^+}$

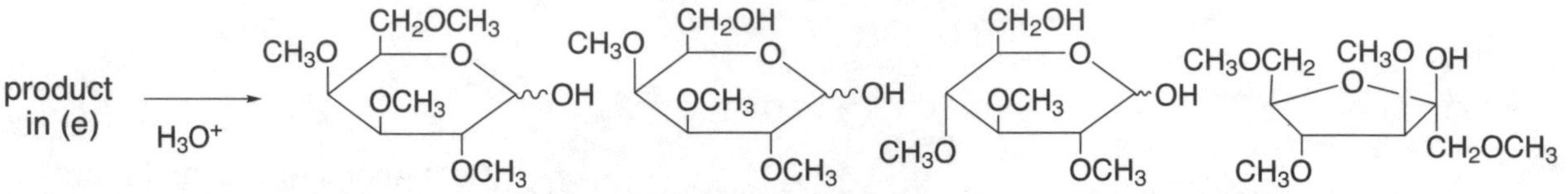

α and β Anomers of each monosaccharide are formed.

27.70

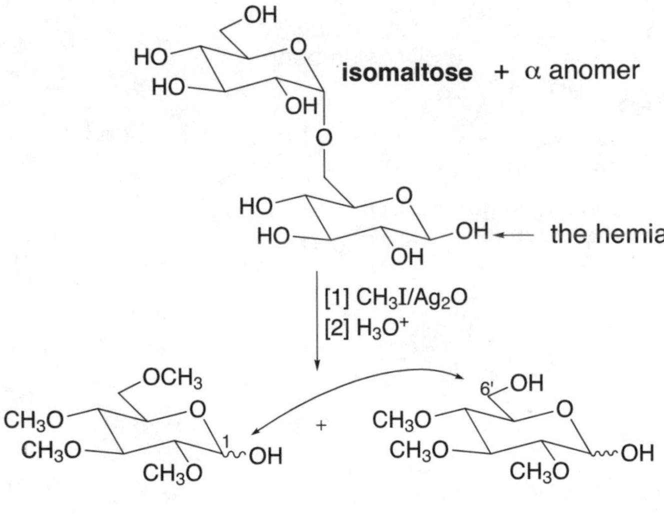

isomaltose + α anomer

Isomaltose must be composed of two glucose units in an α glycosidic linkage. Since it is a reducing sugar it contains a hemiacetal. The free OH groups in the hydrolysis products show where the two monosaccharides are joined.

the hemiacetal

[1] CH$_3$I/Ag$_2$O
[2] H$_3$O$^+$

(Both anomers are present.)

27.71

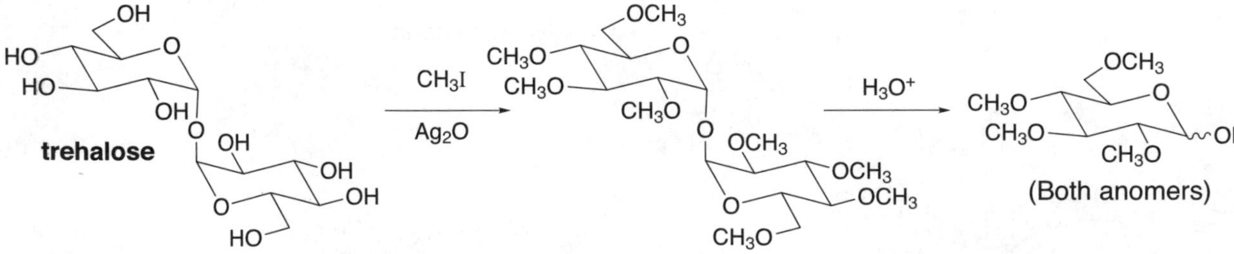

trehalose

$$\xrightarrow[\text{Ag}_2\text{O}]{\text{CH}_3\text{I}}$$

$$\xrightarrow{\text{H}_3\text{O}^+}$$

(Both anomers)

Trehalose must be composed of D-glucose units only, joined in an α glycosidic linkage. Since trehalose is nonreducing it contains no hemiacetal. Since there is only one product formed after methylation and hydrolysis, the two anomeric C's must be joined.

27.72

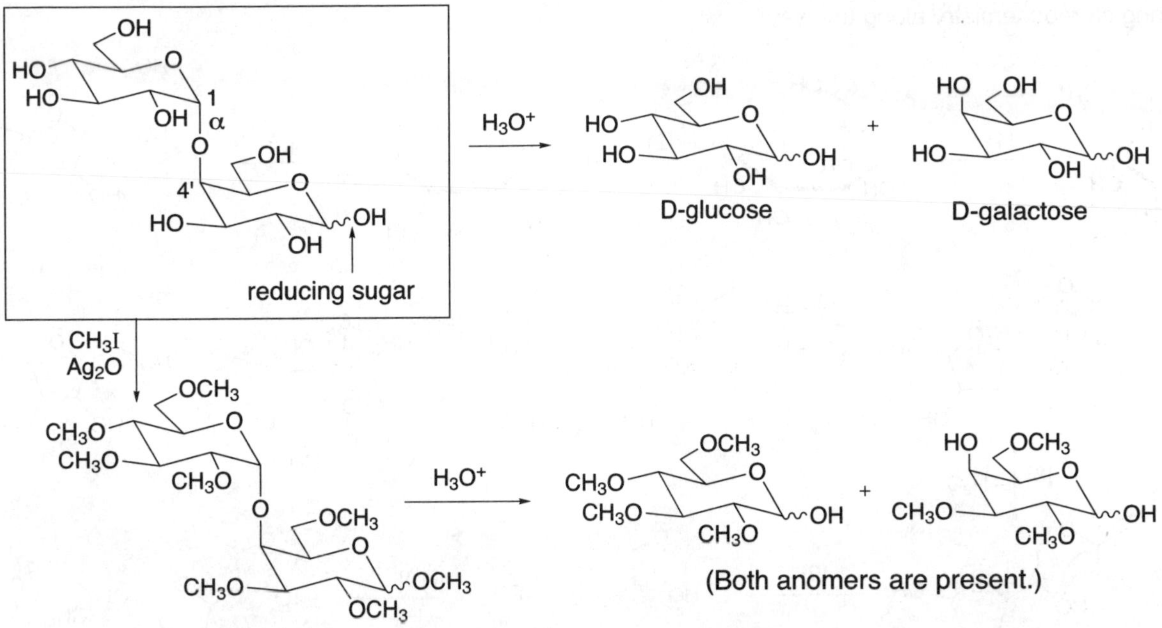

The disaccharide is composed of D-glucose and D-galactose joined in an α-glycosidic linkage. Since it is a reducing sugar it contains a hemiacetal. The free OH's in the two-step reaction show where the two monosaccharides were joined.

27.73

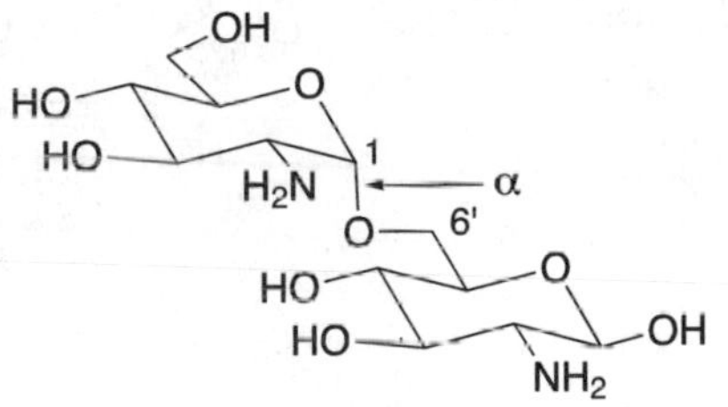

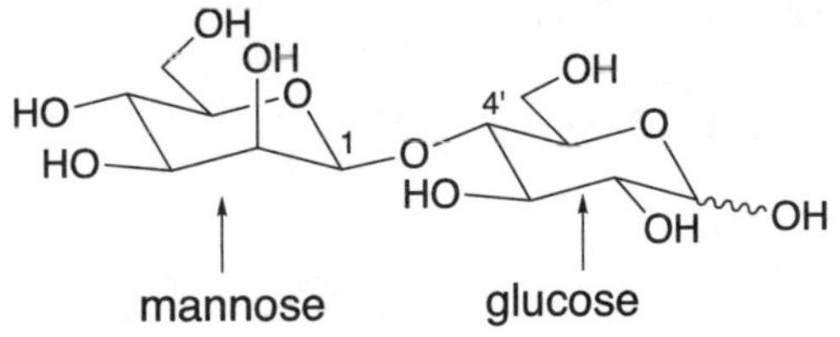

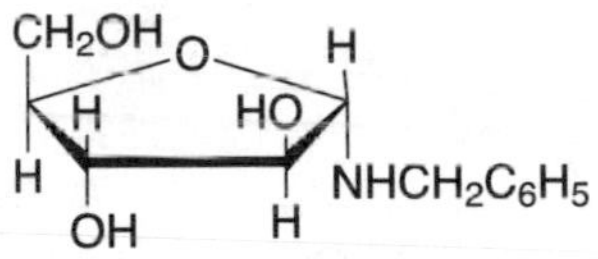

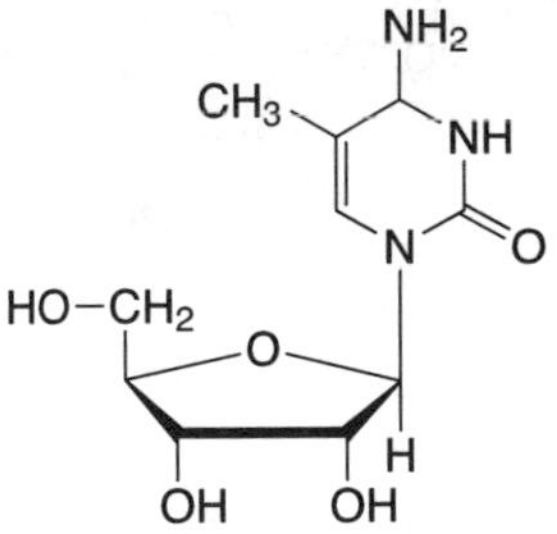

27.74

Ignoring stereochemistry along the way:

27.75

The hydrolysis data suggest that the trisaccharide has D-galactose on one end and D-fructose on the other. D-galactose must be joined to its adjacent sugar by a β glycosidic linkage. D-Fructose must be joined to its adjacent sugar by an α glycosidic linkage.

2,3,4,6-tetra-*O*-methyl-D-galactose   2,3,4-tri-*O*-methyl-D-glucose   1,3,6-tri-*O*-methyl-D-fructose

(Both anomers of each compound are formed.)

Chapter 28: Amino Acids and Proteins

◆ Synthesis of amino acids (28.2)

[1] From α-halo carboxylic acids by S_N2 reaction

$$R-\underset{\underset{Br}{|}}{CH}COOH \xrightarrow[\substack{\text{(large excess)}\\ \mathbf{S_N2}}]{NH_3} R-\underset{\underset{NH_3^+}{|}}{CH}COO^- + NH_4^+\,Br^-$$

[2] By alkylation of diethyl acetamidomalonate

$$\underset{CH_3}{\overset{O}{\underset{|}{C}}}-\underset{H}{\overset{H}{N}}-\underset{COOEt}{\overset{H}{\underset{|}{C}}}-COOEt \xrightarrow[\substack{[2]\ RX \\ [3]\ H_3O^+/\ \Delta}]{[1]\ NaOEt} H_2N-\underset{H}{\overset{R}{\underset{|}{C}}}-COOH$$

• Alkylation occurs with unhindered alkyl halides—that is, CH_3X and RCH_2X.

[3] Strecker synthesis

$$\underset{R}{\overset{O}{\underset{}{C}}}{}_{H} \xrightarrow[NaCN]{NH_4Cl} R-\underset{H}{\overset{NH_2}{\underset{|}{C}}}-CN \xrightarrow{H_3O^+} R-\underset{H}{\overset{NH_2}{\underset{|}{C}}}-COOH$$

α-amino nitrile

◆ Preparation of optically active amino acids

[1] Resolution of enantiomers by forming diastereomers (23.8A)

- Convert a racemic mixture of amino acids into a racemic mixture of N-acetyl amino acids [(S)- and (R)-CH₃CONHCH(R)COOH].
- React the enantiomers with a chiral amine to form a mixture of diastereomers.
- Separate the diastereomers.
- Regenerate the amino acids by protonation of the carboxylate salt and hydrolysis of the N-acetyl group.

[2] Kinetic resolution using enzymes (23.8B)

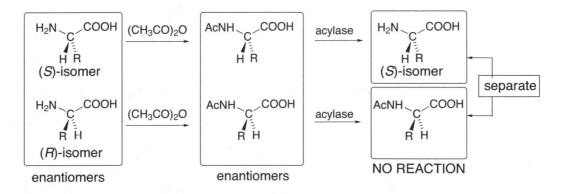

[3] By enantioselective hydrogenation (28.4)

Rh^* = chiral Rh hydrogenation catalyst

♦ **Adding and removing protecting groups for amino acids (28.6)**

[1] Protection of an amino group as a BOC derivative

[2] Deprotection of a BOC-protected amino acid

[3] Protection of a carboxyl group as an ester

methyl ester   benzyl ester

[4] Deprotection of an ester group

methyl ester   benzyl ester

♦ **Synthesis of dipeptides (28.6)**

[1] Amide formation with DCC

[2] Four steps are needed to synthesize a dipeptide:

  [a] **Protect** the amino group of one amino acid using a BOC group
  [b] **Protect** the carboxyl group of the second amino acid using an ester.
  [c] Form the amide bond with **DCC.**
  [d] **Remove both protecting groups** in one or two reactions.

♦ Summary of the Merrifield method of peptide synthesis (28.7)

[1] Attach a BOC-protected amino acid to a polymer derived from polystyrene.
[2] Remove the BOC-protecting group.
[3] Form the amide bond with a second BOC-protected amino acid using DCC.
[4] Repeat steps [2] and [3].
[5] Detach the peptide from the polymer and remove the protecting group.

Chapter 28: Answers to Problems

28.1

S H$_2$N–C–H
S CH$_3$–C–H
CH$_2$CH$_3$
L-isoleucine

R H–C–NH$_2$
S CH$_3$–C–H
CH$_2$CH$_3$

R H–C–NH$_2$
R H–C–CH$_3$
CH$_2$CH$_3$

S H$_2$N–C–H
R H–C–CH$_3$
CH$_2$CH$_3$

28.2

a.

$(CH_3)_2CH-\overset{NH_3^+}{\underset{H}{C}}-COO^-$

b.

$(CH_3)_2CHCH_2-\overset{NH_3^+}{\underset{H}{C}}-COO^-$

c.

d.

$HOOCCH_2CH_2-\overset{NH_3^+}{\underset{H}{C}}-COO^-$

28.3 In an amino acid, the electron withdrawing carboxyl group destabilizes the ammonium ion (–NH$_3^+$), making it more readily donate a proton; that is, it makes it a stronger acid. Also, the electron withdrawing carboxyl group removes electron density from the amino group (–NH$_2$) of the conjugate base, making it a weaker base than a 1° amine, which has no electron withdrawing group.

28.4 The most direct way to synthesize an α-amino acid is by **S$_N$2 reaction of an α-halo carboxylic acid with a large excess of NH$_3$.**

a. Br–CH–COOH $\xrightarrow[\text{(large excess)}]{NH_3}$ H$_2$N–CH–COOH
   (H below each CH)

c. Br–CH–COOH $\xrightarrow[\text{(large excess)}]{NH_3}$ H$_2$N–CH–COOH
   (CH$_2$–phenyl below each)

b. Br–CH–COOH $\xrightarrow[\text{(large excess)}]{NH_3}$ H$_2$N–CH–COOH
   CH–CH$_3$ CH–CH$_3$
   CH$_2$ CH$_2$
   CH$_3$ CH$_3$

28.5

a. $\xrightarrow{CH_3I}$ H$_2$N–$\overset{CH_3}{\underset{H}{C}}$–COOH   alanine

c. $\xrightarrow{CH_3CH_2CH(CH_3)Br}$ H$_2$N–$\overset{CH(CH_3)CH_2CH_3}{\underset{H}{C}}$–COOH   isoleucine

b. $\xrightarrow{(CH_3)_2CHCH_2Cl}$ H$_2$N–$\overset{CH_2CH(CH_3)_2}{\underset{H}{C}}$–COOH   leucine

28.6

$\xrightarrow[\substack{[2]\ CH_2=O \\ [3]\ H_3O^+/\ \Delta}]{[1]\ NaOEt}$ H$_2$N–$\overset{CH_2OH}{\underset{H}{C}}$–COOH

serine

28.7

a. H₂N−CHCOOH ⟹   (CH₃)₂CH−C(=O)−H
      CH−CH₃
      CH₃
   valine

b. H₂N−CHCOOH ⟹   (CH₃)₂CHCH₂−C(=O)−H
      CH₂
      CH−CH₃
      CH₃
   leucine

c. H₂N−CHCOOH ⟹   C₆H₅CH₂−C(=O)−H
      CH₂
      (phenyl ring)
   phenylalanine

28.8

a. $BrCH_2COOH \xrightarrow[\text{large excess}]{NH_3} NH_2CH_2COO^-\ NH_4^+$

c. $CH_3CH_2CH(CH_3)CHO \xrightarrow[\text{[2] }H_3O^+]{\text{[1] }NH_4Cl/NaCN} $ H₂N−CHCOOH
      CH(CH₃)CH₂CH₃

b. CH₃CONH−C(H)(COOEt)−COOEt $\xrightarrow[\text{[3] }H_3O^+/\Delta]{\substack{\text{[1] NaOEt} \\ \text{[2] (CH}_3)_2\text{CHCl}}}$ H₂N−C(H)(COOH)−CH(CH₃)₂

d. CH₃CONH−C(H)(COOEt)−COOEt $\xrightarrow[\text{[3] }H_3O^+/\Delta]{\substack{\text{[1] NaOEt} \\ \text{[2] BrCH}_2\text{CO}_2\text{Et}}}$ H₂N−C(H)(COOH)−CH₂CO₂H

28.9 A chiral amine must be used to resolve a racemic mixture of amino acids.

a. C₆H₅CH₂CH₂NH₂
   achiral

b.
achiral

c. CH₃CH₂−C(CH₃H)(NH₂)
   chiral
   (can be used)

d.
chiral
(can be used)

28.10

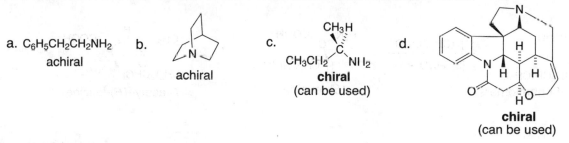

| To begin: |
| Convert the amino acids into N-acetyl amino acids (two enantiomers). |

NH₂−C(H)(COOH)−CH₂CH(CH₃)₂ **S**   +   NH₂−C(COOH)((CH₃)₂CHCH₂)(H) **R**   **enantiomers**

↓ Ac₂O

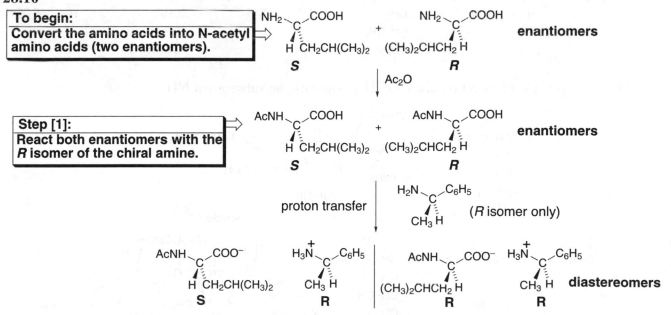

| Step [1]: |
| React both enantiomers with the **R** isomer of the chiral amine. |

AcNH−C(H)(COOH)−CH₂CH(CH₃)₂ **S**   +   AcNH−C(COOH)((CH₃)₂CHCH₂)(H) **R**   **enantiomers**

H₂N−C(C₆H₅)(CH₃H)   (*R* isomer only)

proton transfer

AcNH−C(H)(COO⁻)−CH₂CH(CH₃)₂ **S**   H₃N⁺−C(C₆H₅)(CH₃ H) **R** | AcNH−C(COO⁻)((CH₃)₂CHCH₂)(H) **R**   H₃N⁺−C(C₆H₅)(CH₃ H) **R**   **diastereomers**

These salts have the *same* configuration around one stereogenic center, but the *opposite* configuration about the other stereogenic center.

Step [2]:
Separate the diastereomers.

separate

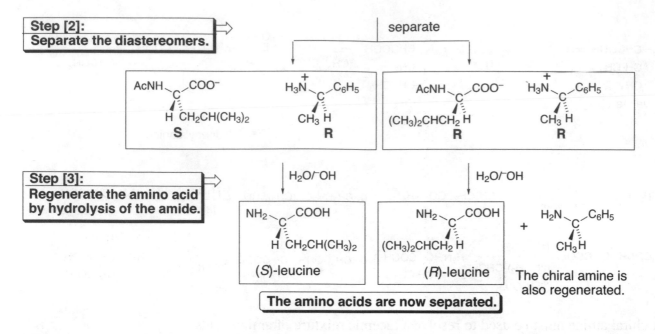

Step [3]:
Regenerate the amino acid by hydrolysis of the amide.

(S)-leucine

(R)-leucine

The chiral amine is also regenerated.

The amino acids are now separated.

28.11

COOH
H₂N—C—H
CH₂CH(CH₃)₂
(mixture of enantiomers)

[1] (CH₃CO)₂O
[2] acylase

H₂N   COOH
C
H CH₂CH(CH₃)₂
(S)-leucine

+

CH₃   N—H   COOH
C   C
O   H
(CH₃)₂CHCH₂
N-acetyl-(R)-leucine

28.12

a. H₂N—CHCOOH ⟹
        CH₃

H   NHAc
C=C
H   COOH

b. H₂N—CHCOOH ⟹
        CH₂
        CH-CH₃
        CH₃

(CH₃)₂CH   NHAc
C=C
H   COOH

c. H₂N—CHCOOH ⟹
        CH₂
        CH₂
        CONH₂

H₂NCOCH₂   NHAc
C=C
H   COOH

28.13 Draw the peptide by bonding one COOH group with the subsequent NH₂.

a.
O
H₂N—CH—C—OH
        CH-CH₃
        CH₃
Val

O
H₂N—CH—C—OH
        CH₂
        CH₂
        COOH
Glu

⟶

amide
(CH₃)₂CH H   O
        C   N   C
H₂N   C   OH
        O   H CH₂CH₂COOH
N-terminal   ← C-terminal
Val–Glu

b.
O
H₂N—CH—C—OH
        H
Gly

O
H₂N—CH—C—OH
        CH₂
        NH
        N
His

O
H₂N—CH—C—OH
        CH₂
        CH-CH₃
        CH₃
Leu

⟶

amide
                (CH₃)₂CHCH₂
H   H   O   H
    C   N   C   N   C
H₂N   C   C   OH
    O   H CH₂   O
N-terminal   amide   C-terminal
        CH₂
        NH
        N
Gly–His–Leu

c.

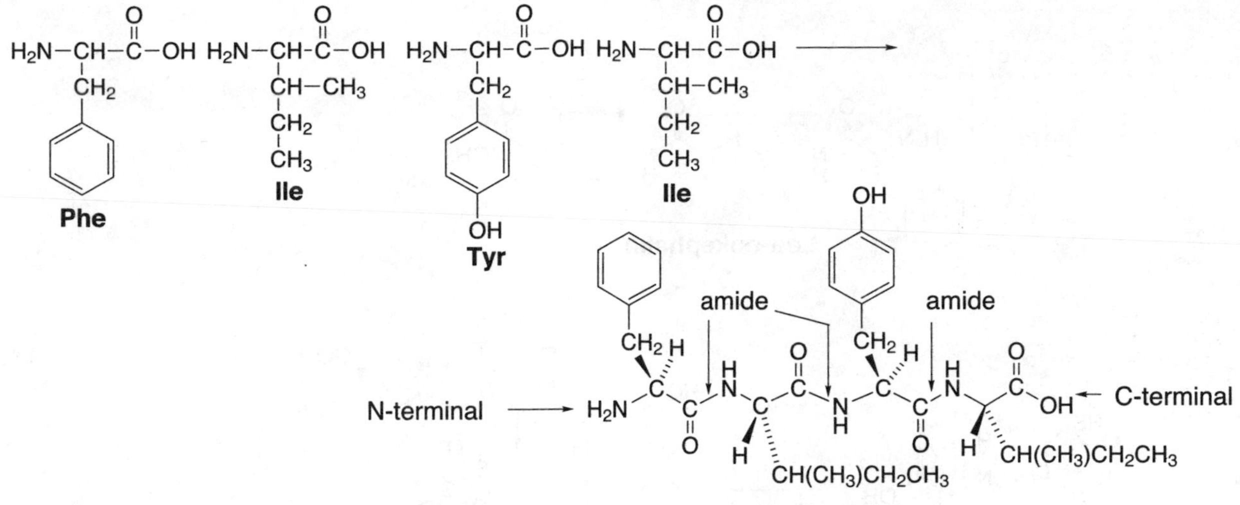

Phe   **Ile**   **Tyr**   **Ile**

Phe–Ile–Tyr–Ile

N-terminal   C-terminal

amide   amide

28.14

a. 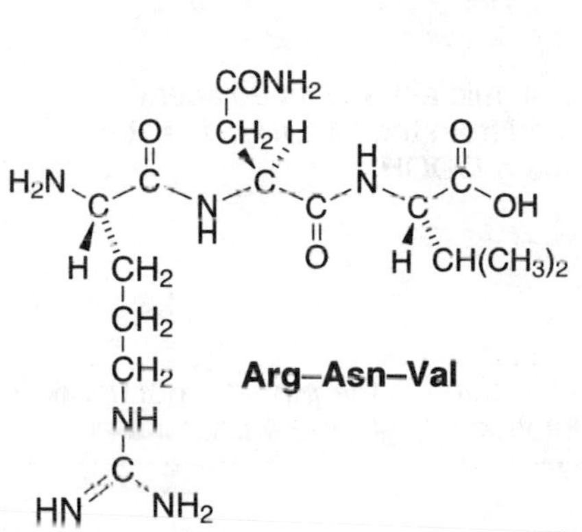   **Arg–Asn–Val**

b. 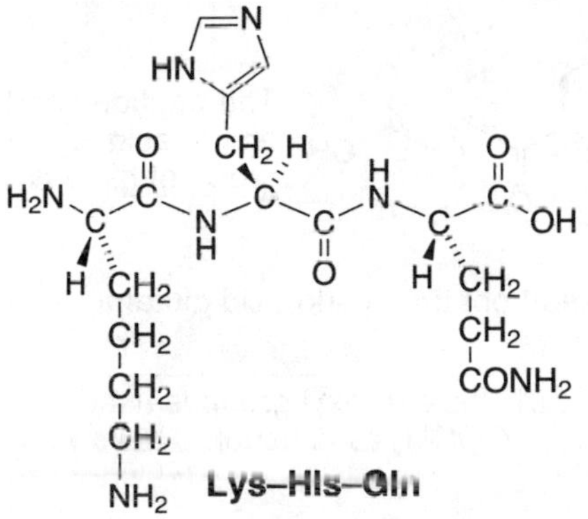   **Lys–His–Gln**

28.15 There are six different tripeptides that can be formed from three amino acids: A–B–C, A–C–B, B–A–C, B–C–A, C–A–B, C–B–A.

28.16 The *s*-trans conformation has the two R groups oriented on *opposite* sides of the C–N bond. The *s*-cis conformation has the two R groups oriented on the *same* side of the C–N bond.

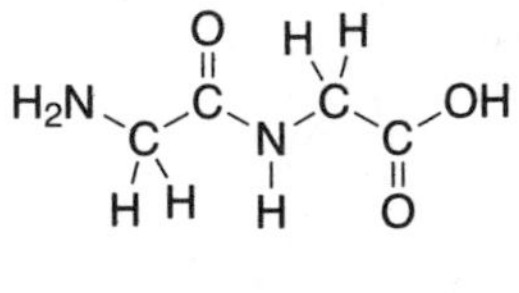

s-trans 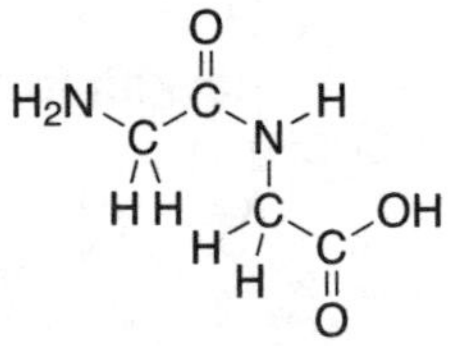   *s*-cis

28.17

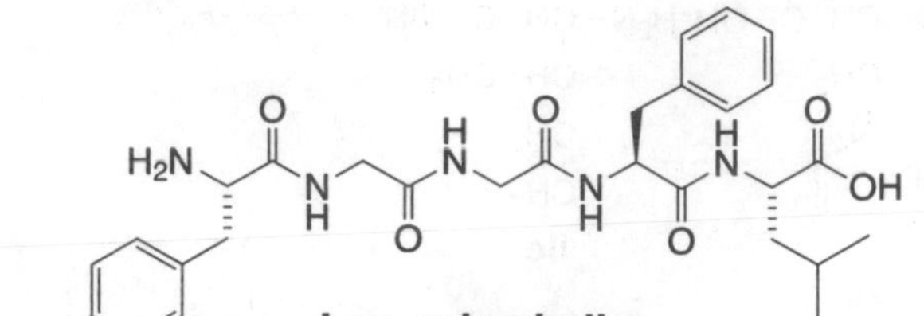

Leu-enkephalin

28.18

a

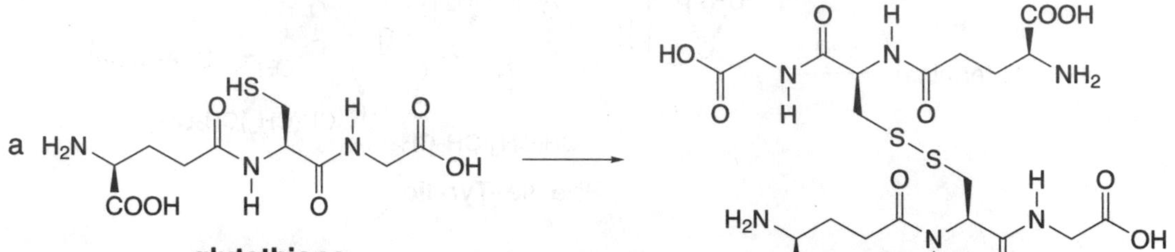

glutathione

b.

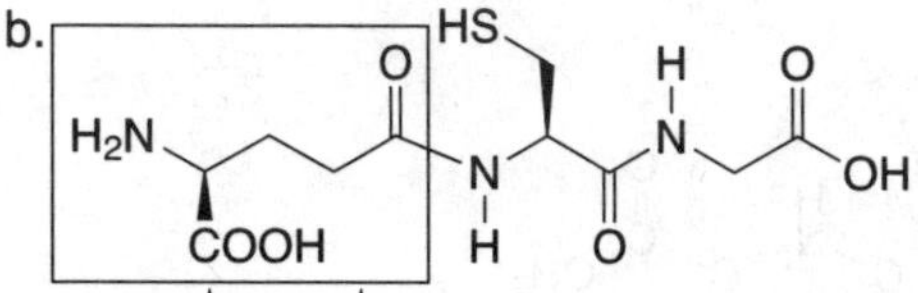

The peptide bond beween glutamic aid and its adjacent amino acid (cysteine) is formed from the COOH in the R group of glutamic acid, not the α COOH.

α COOH   This comes from the amino acid glutamic acid.

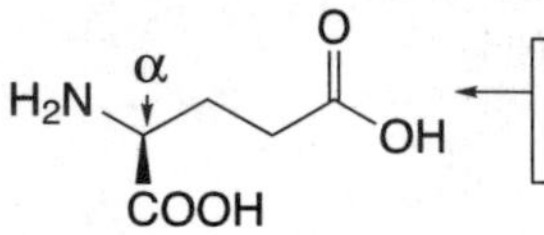

This carboxyl group is used to form the amide bond in the peptide, not the α COOH, as is usual. That's what makes glutathione's structure unusual.

glutamic acid

28.19

a.

Leu

Val

new amide bond

Leu–Val

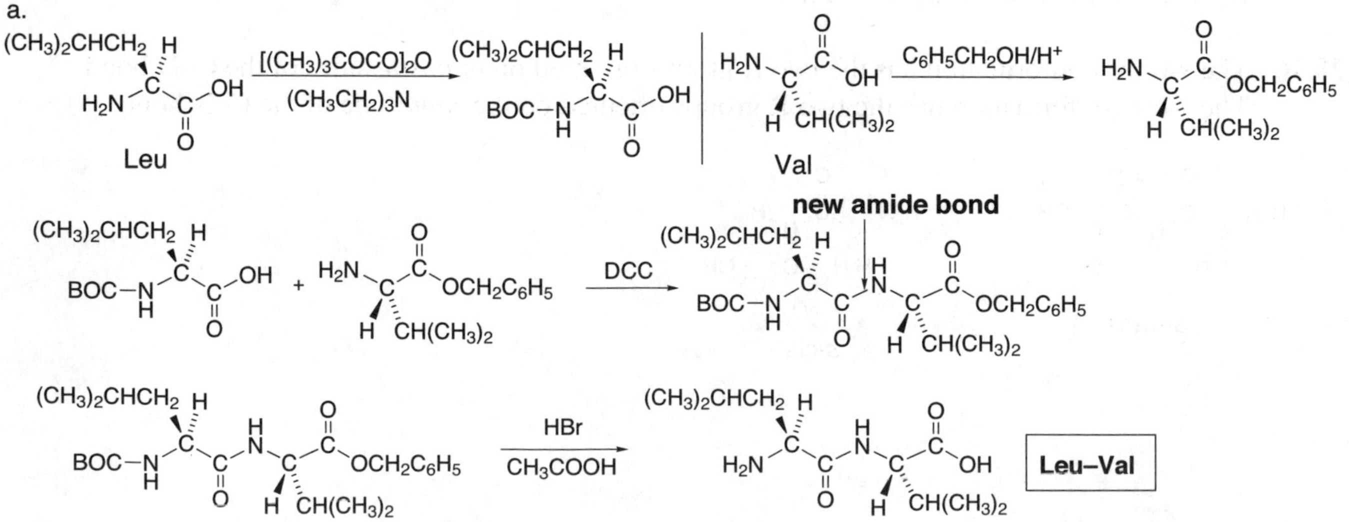

b.

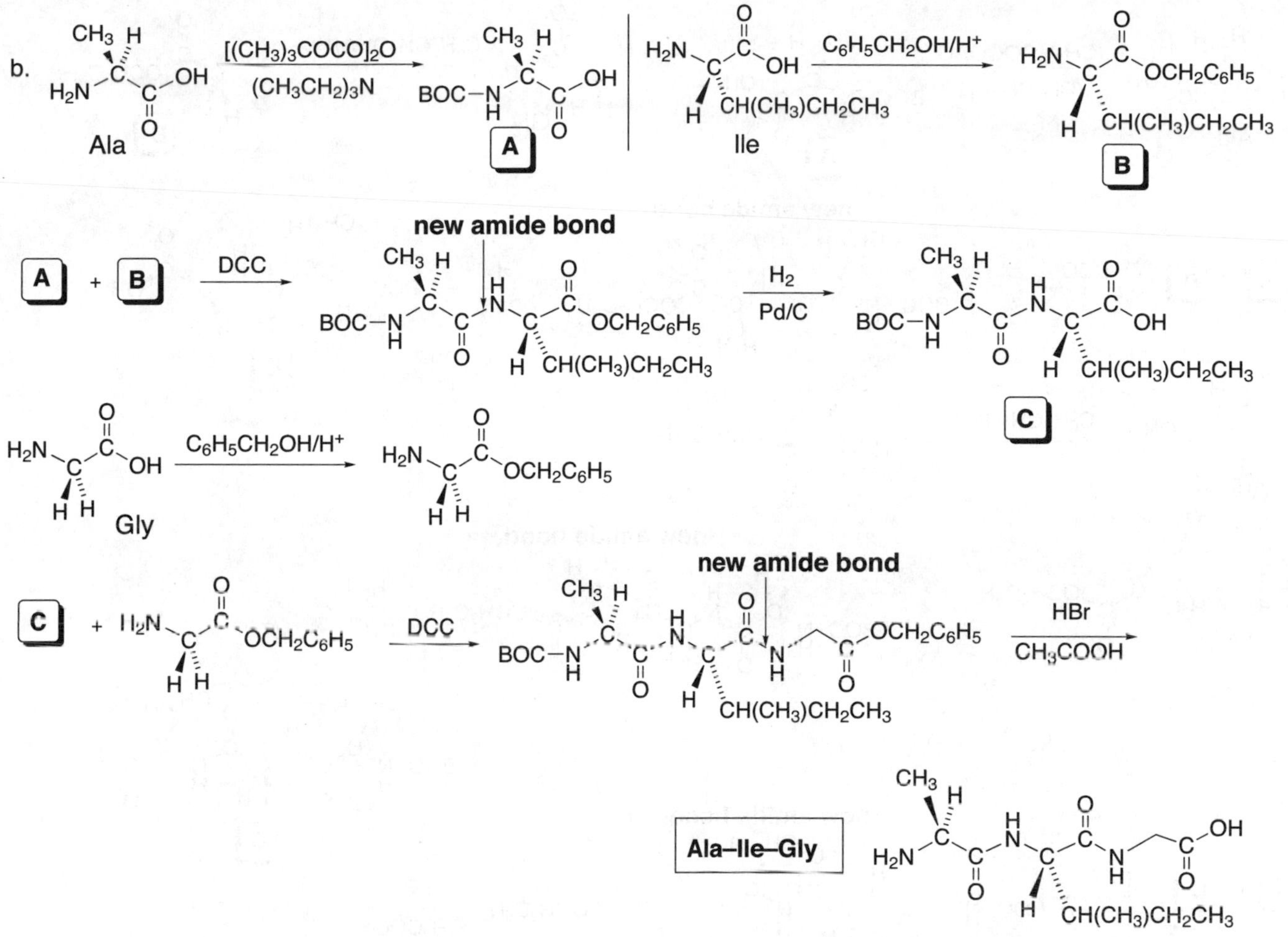

new amide bond

A + B $\xrightarrow{DCC}$

$\xrightarrow[\text{Pd/C}]{H_2}$

C

Gly $\xrightarrow{C_6H_5CH_2OH/H^+}$

new amide bond

C + H_2N—... $OCH_2C_6H_5$ $\xrightarrow{DCC}$

$\xrightarrow[\text{CH}_3\text{COOH}]{HBr}$

Ala–Ile–Gly

c.

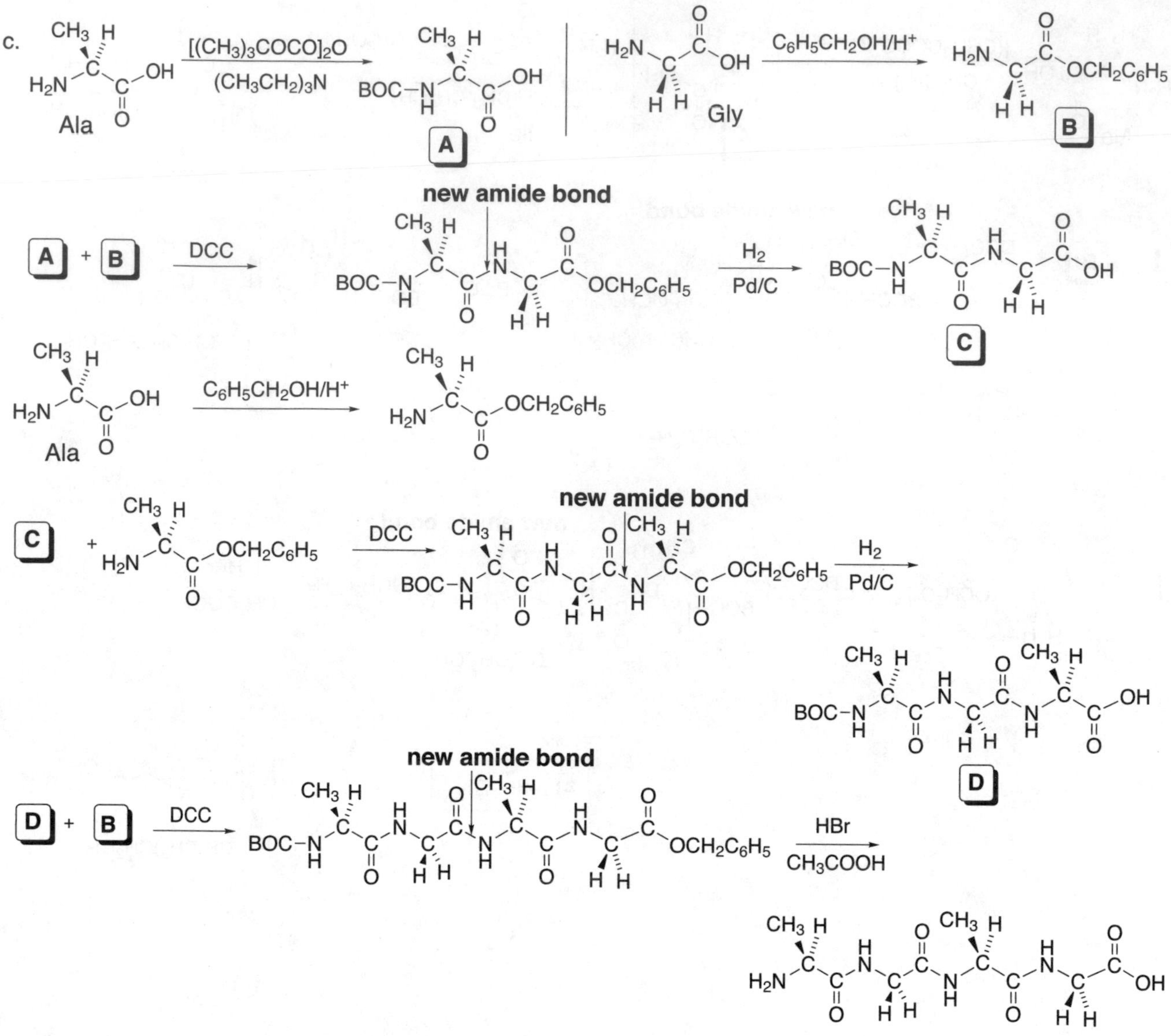

new amide bond

new amide bond

new amide bond

Ala–Gly–Ala–Gly

28.20

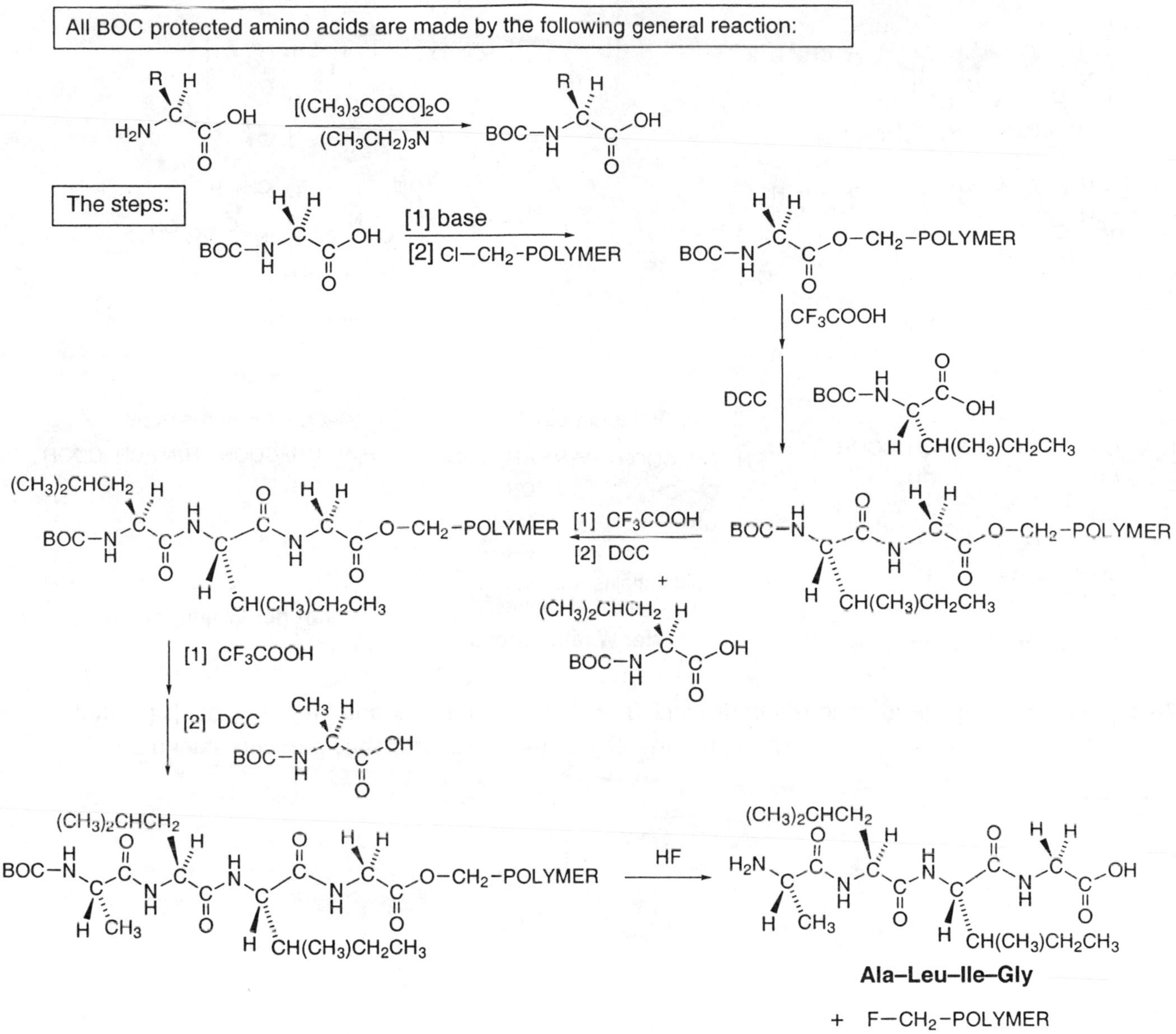
All BOC protected amino acids are made by the following general reaction:
[(CH3)3COCO]2O
(CH3CH2)3N
The steps:
[1] base
[2] Cl–CH2-POLYMER
O–CH2–POLYMER
CF3COOH
DCC
CH(CH3)CH2CH3
[1] CF3COOH
[2] DCC
(CH3)2CHCH2
HF
Ala–Leu–Ile–Gly
+ F–CH2–POLYMER

28.21 Antiparallel β-pleated sheets are more stable then parallel β-pleated sheets because of geometry. The N–H and C=O of one chain are directly aligned with the N–H and C=O of an adjacent chain in the antiparallel β-pleated sheet, whereas they are not in the parallel β-pleated sheet. This makes the latter set of hydrogen bonds weaker.

28.22 In a *parallel* β-pleated sheet, the strands run in the *same* direction from the N- to C-terminal amino acid. In an *antiparallel* β-pleated sheet, the strands run in the *opposite* direction.

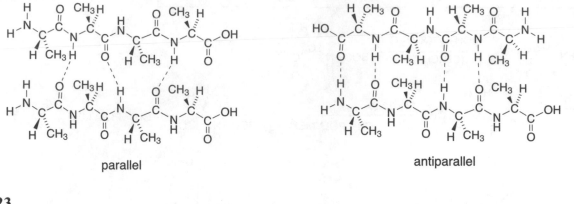

parallel   antiparallel

28.23

a. Ser and Tyr

H₂N−CH−COOH   H₂N−CH−COOH

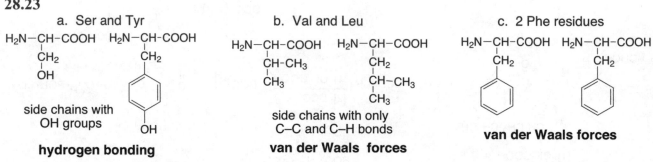

side chains with
OH groups

hydrogen bonding

b. Val and Leu

side chains with only
C–C and C–H bonds

van der Waals forces

c. 2 Phe residues

van der Waals forces

28.24 The R group for glycine is a hydrogen. The R groups must be small to allow the β-pleated sheets to stack on top of each other. With large R groups, steric hindrance prevents stacking.

28.25 All L-amino acids except cysteine have the **S configuration**. L-Cysteine has the *R* configuration because the R group contains a sulfur atom, which has higher priority.

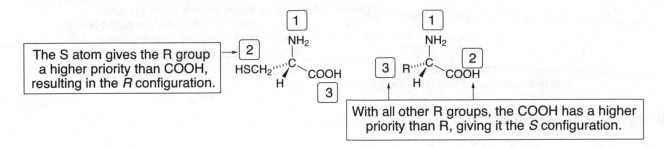

The S atom gives the R group a higher priority than COOH, resulting in the *R* configuration.

With all other R groups, the COOH has a higher priority than R, giving it the *S* configuration.

28.26

a.

(*R*)-penicillamine
L-penicillamine

(*S*)-penicillamine
D-penicillamine

b.

28.27 Amino acids are insoluble in diethyl ether because amino acids are highly polar; they exist as salts in their neutral form. Diethyl ether is weakly polar, so amino acids are not soluble in it. *N*-Acetyl amino acids are soluble because they are polar but not salts.

amino acid, a salt
H_2O soluble and ether insoluble

N-acetyl amino acid
Ether soluble

28.28

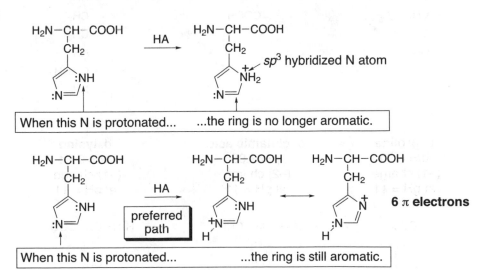

When this N is protonated... ...the ring is no longer aromatic.

When this N is protonated... ...the ring is still aromatic.

preferred path

6 π electrons

28.29

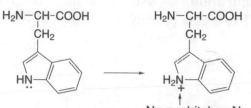

The ring structure on tryptophan is aromatic since each atom contains a *p* orbital. Protonation of the N atom would disrupt the aromaticity, making this a less favorable reaction.

No *p* orbital on N.
The five-membered ring is no longer aromatic.

28.30 At its isoelectric point, each amino acid is neutral.

a.
$$H_3\overset{+}{N}-\underset{\underset{CH_3}{|}}{\overset{\overset{COO^-}{|}}{C}}-H$$
alanine

b.
$$H_3\overset{+}{N}-\underset{\underset{CH_2CH_2SCH_3}{|}}{\overset{\overset{COO^-}{|}}{C}}-H$$
methionine

c.
$$H_3\overset{+}{N}-\underset{\underset{CH_2COOH}{|}}{\overset{\overset{COO^-}{|}}{C}}-H$$
aspartic acid

d.
$$H_2N-\underset{\underset{CH_2CH_2CH_2CH_2\overset{+}{N}H_3}{|}}{\overset{\overset{COO^-}{|}}{C}}-H$$
lysine

28.31

    a. [1] glutamic acid: use the pK_a's $2.10 + 4.07$
       [2] lysine: use the pK_a's $8.95 + 10.53$
       [3] arginine: use the pK_a's $9.04 + 12.48$
    b. In general the p*I* of an acidic amino acid is lower than that of a neutral amino acid.
    c. In general the p*I* of a basic amino acid is higher than that of a neutral amino acid.

28.32

a. **threonine**	b. **methionine**	c. **aspartic acid**	d. **arginine**
p*I* = 5.06	p*I* = 5.74	p*I* = 2.98	p*I* = 5.41
(+1) charge at pH = 1	(+1) charge at pH = 1	(+1) charge at pH = 1	(+2) charge at pH = 1

a.
$$H_3\overset{+}{N}-CH-COOH$$
$$\underset{\underset{CH_3}{|}}{\overset{|}{CH}}-OH$$

b.
$$H_3\overset{+}{N}-CH-COOH$$
$$\overset{|}{CH_2}$$
$$\overset{|}{CH_2}$$
$$\overset{|}{S}$$
$$\overset{|}{CH_3}$$

c.
$$H_3\overset{+}{N}-CH-COOH$$
$$\overset{|}{CH_2}$$
$$\overset{|}{COOH}$$

d.
$$H_3\overset{+}{N}-CH-COOH$$
$$\overset{|}{CH_2}$$
$$\overset{|}{CH_2}$$
$$\overset{|}{CH_2}$$
$$\overset{|}{NH}$$
$$\overset{|}{\underset{\overset{|}{NH_2}}{C}}=\overset{+}{N}H_2$$

28.33

a. **valine**	b. **proline**	c. **glutamic acid**	d. **lysine**
p*I* = 6.00	p*I* = 6.30	p*I* = 3.08	p*I* = 9.74
(−1) charge at pH = 11	(−1) charge at pH = 11	(−2) charge at pH = 11	(−1) charge at pH = 11

a.
$$H_2N-CH-COO^-$$
$$\underset{\underset{CH_3}{|}}{\overset{|}{CH}}-CH_3$$

b.
$$COO^-$$
(proline ring)
HN

c.
$$H_2N-CH-COO^-$$
$$\overset{|}{CH_2}$$
$$\overset{|}{CH_2}$$
$$\overset{|}{COO^-}$$

d.
$$H_2N-CH-COO^-$$
$$\overset{|}{CH_2}$$
$$\overset{|}{CH_2}$$
$$\overset{|}{CH_2}$$
$$\overset{|}{CH_2}$$
$$\overset{|}{NH_2}$$

28.34

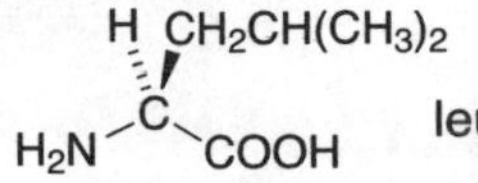

leucine

a. CH₃OH/H⁺

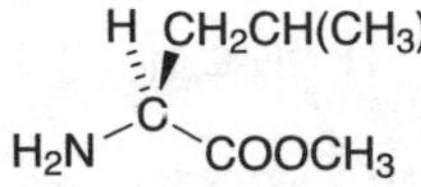

b. CH₃COCl/pyridine

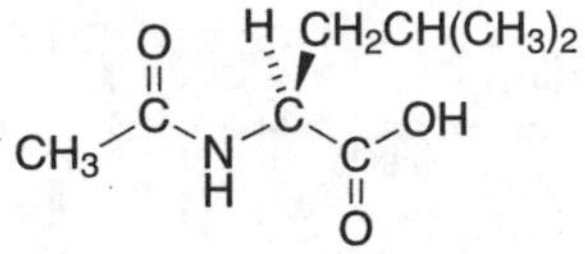

c. C₆H₅CH₂OH/H⁺

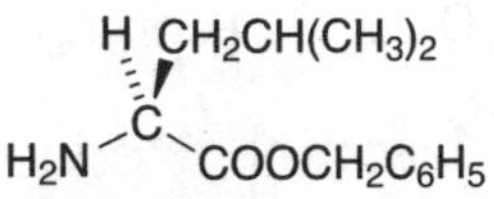

d. Ac₂O/pyridine

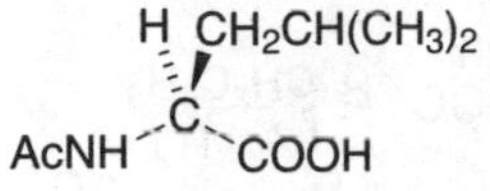

e. HCl (1 equiv)

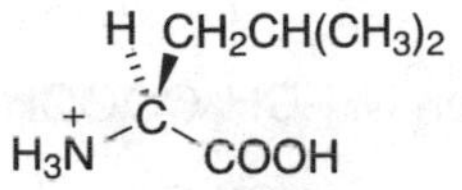

f. NaOH (1 equiv)

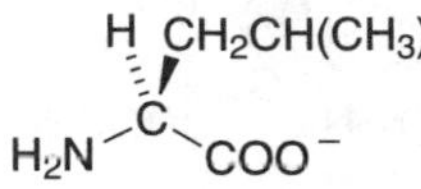

g. C₆H₅COCl/pyridine

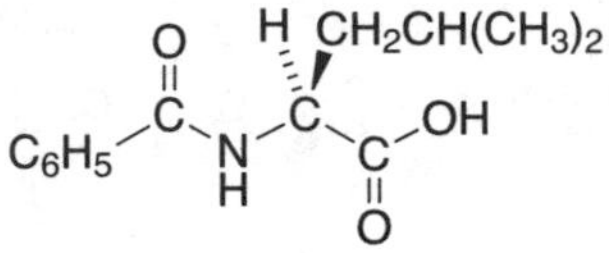

h. [(CH₃)₃COCO]₂O

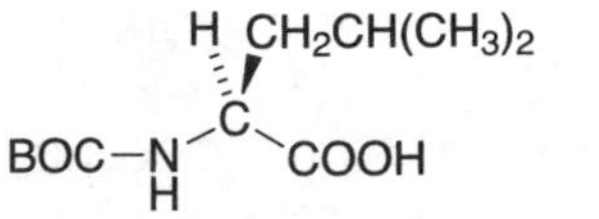

i. the product in (d), then
NH₂CH₂COOCH₃ + DCC

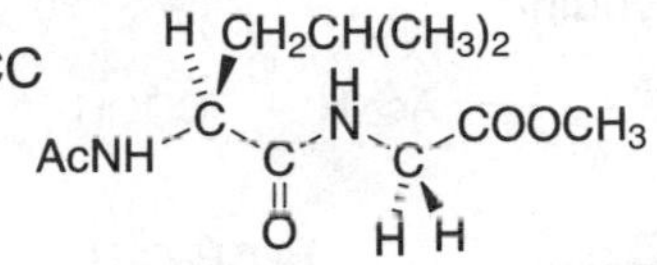

j. the product in (h), then NH₂CH₂COOCH₃ + DCC

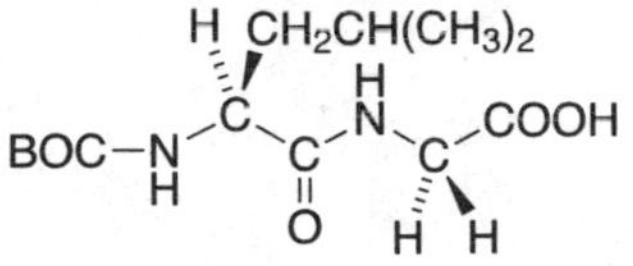

28.35

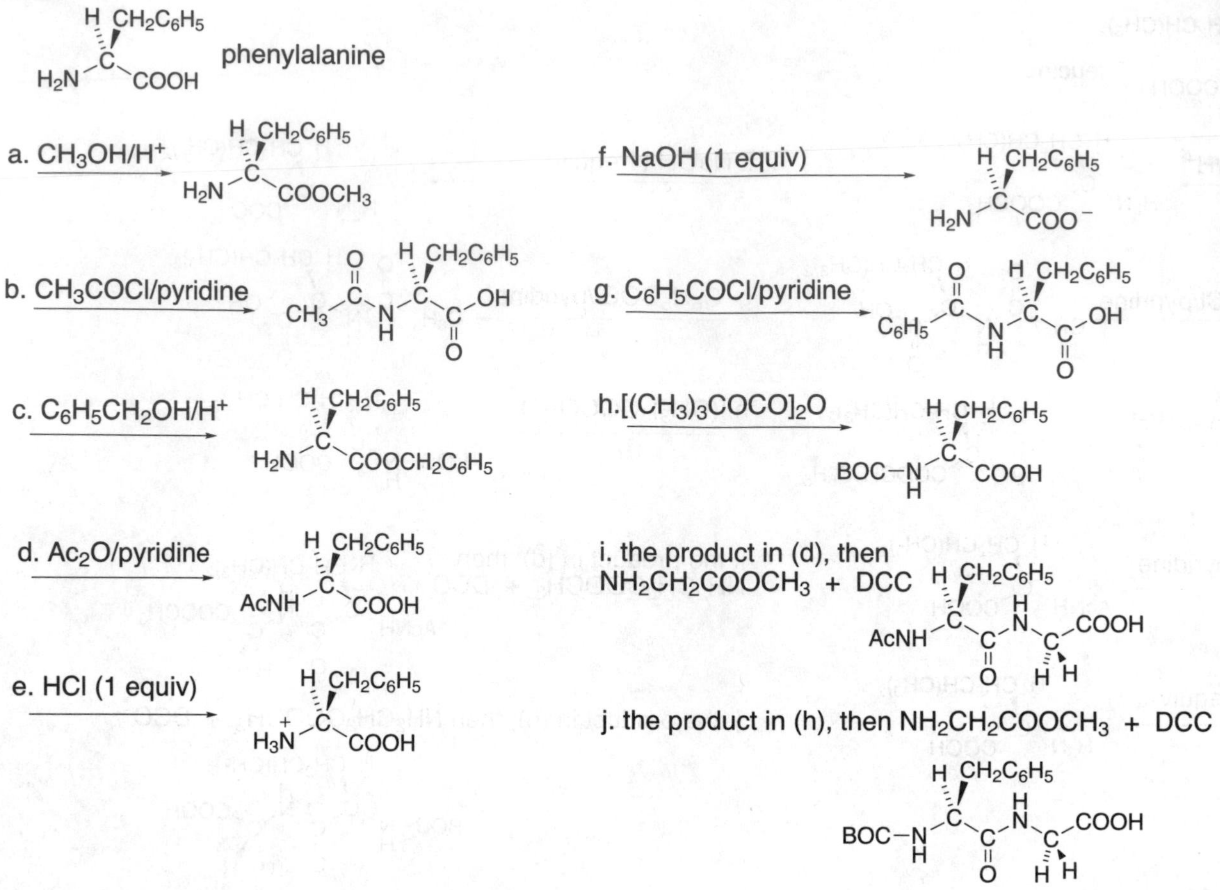

phenylalanine

a. CH₃OH/H⁺ →

b. CH₃COCl/pyridine →

c. C₆H₅CH₂OH/H⁺ →

d. Ac₂O/pyridine →

e. HCl (1 equiv) →

f. NaOH (1 equiv) →

g. C₆H₅COCl/pyridine →

h. [(CH₃)₃COCO]₂O →

i. the product in (d), then NH₂CH₂COOCH₃ + DCC

j. the product in (h), then NH₂CH₂COOCH₃ + DCC

28.36

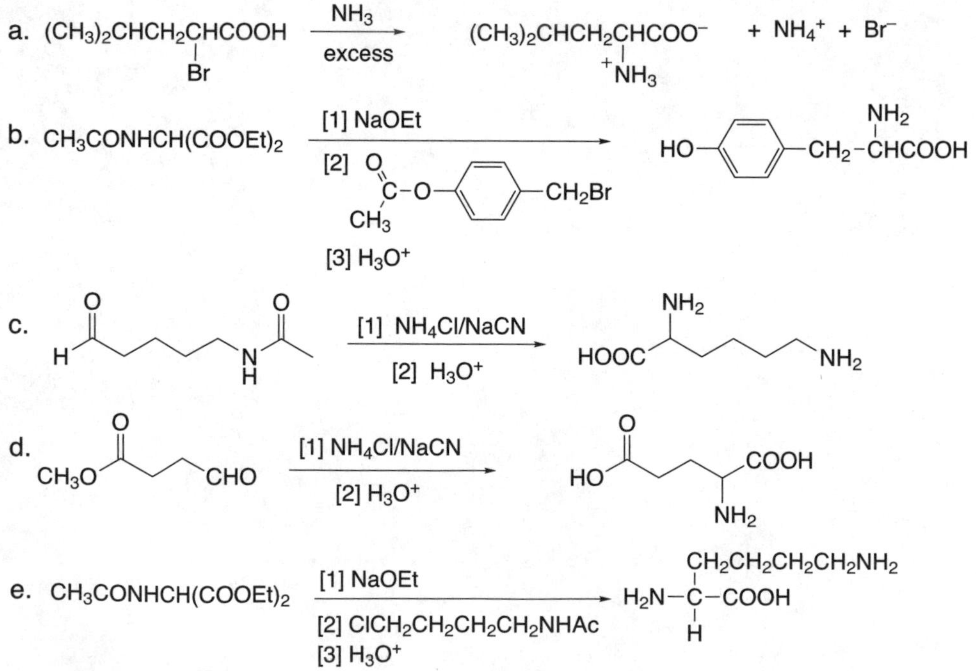

a. (CH₃)₂CHCH₂CHCOOH ──NH₃, excess──→ (CH₃)₂CHCH₂CHCOO⁻ + NH₄⁺ + Br⁻
          |Br   ⁺NH₃

b. CH₃CONHCH(COOEt)₂ ──[1] NaOEt; [2] ; [3] H₃O⁺──→

c. ──[1] NH₄Cl/NaCN; [2] H₃O⁺──→

d. ──[1] NH₄Cl/NaCN; [2] H₃O⁺──→

e. CH₃CONHCH(COOEt)₂ ──[1] NaOEt; [2] ClCH₂CH₂CH₂CH₂NHAc; [3] H₃O⁺──→

28.37

a. Asn

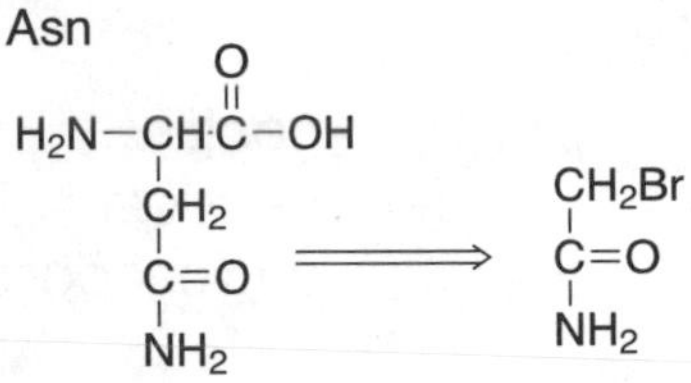

b. His

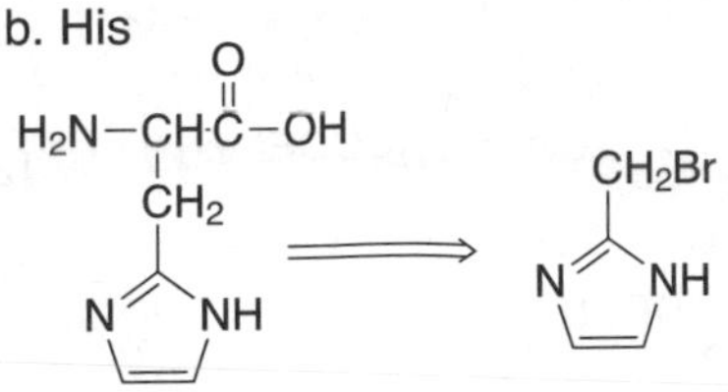

c. Trp

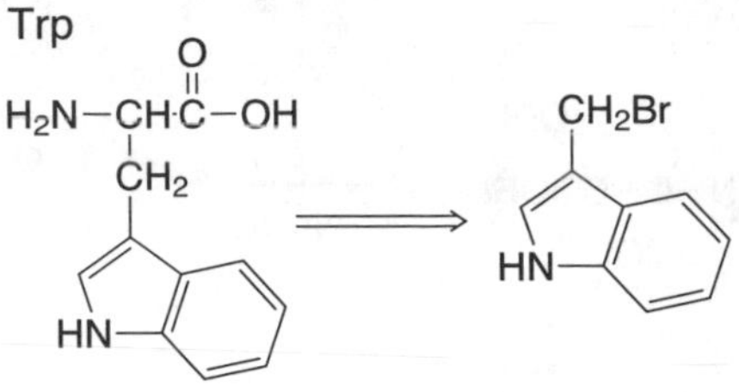

28.38

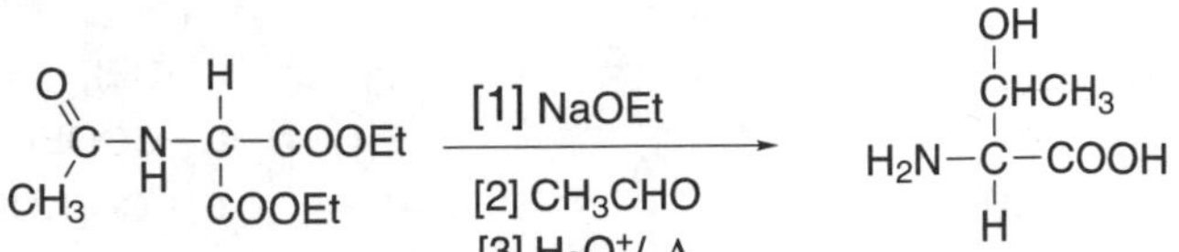

[1] NaOEt

[2] CH$_3$CHO

[3] H$_3$O$^+$/ Δ

threonine

28.39

a. CH$_3$CHO $\xrightarrow[\text{CH}_3\text{COOH}]{\text{Br}_2}$ BrCH$_2$CHO $\xrightarrow[\text{H}_2\text{SO}_4/\text{H}_2\text{O}]{\text{CrO}_3}$ BrCH$_2$COOH $\xrightarrow[\text{excess}]{\text{NH}_3}$ H$_3$$\overset{+}{\text{N}}CH_2COO^-$

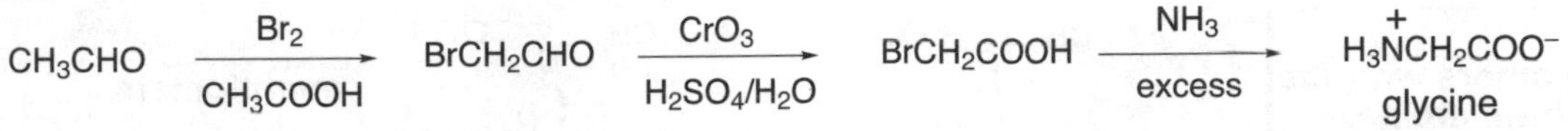

glycine

b.

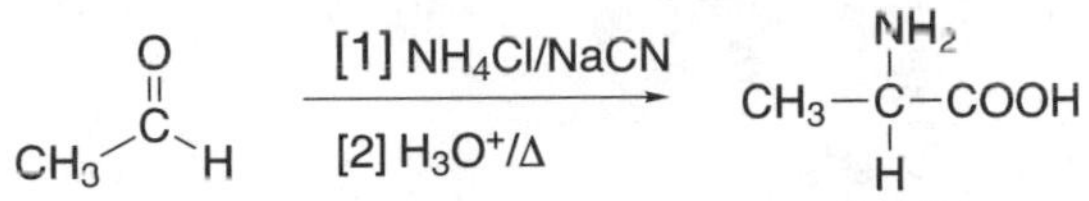

[1] NH$_4$Cl/NaCN

[2] H$_3$O$^+$/Δ

alanine

28.40

a. (CH$_3$)$_2$CHCH$_2$CHO $\xrightarrow[\text{CH}_3\text{COOH}]{\text{Br}_2}$ (CH$_3$)$_2$CHCHCHO $\xrightarrow[\text{H}_2\text{SO}_4/\text{H}_2\text{O}]{\text{CrO}_3}$ (CH$_3$)$_2$CHCHCOOH $\xrightarrow[\text{excess}]{\text{NH}_3}$ (CH$_3$)$_2$CHCHCOO$^-$

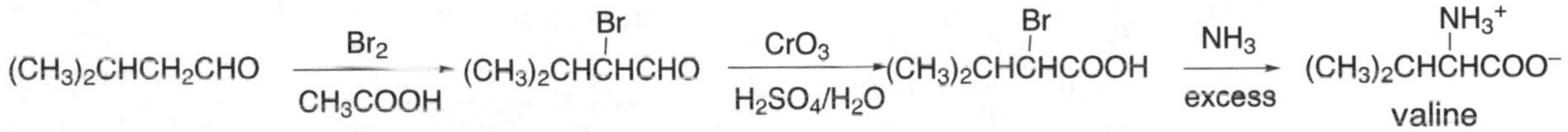

valine

b.

[1] NH$_4$Cl/NaCN

[2] H$_3$O$^+$/Δ

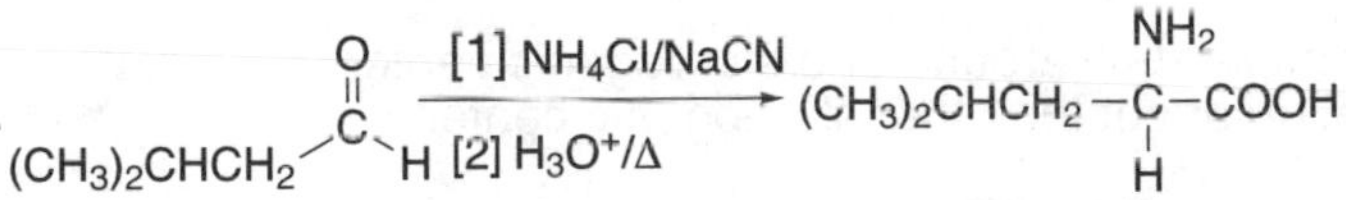

leucine

28.41

CH$_2$(COOEt)$_2$ $\xrightarrow[\text{CH}_3\text{COOH}]{\text{Br}_2}$ Br—CH(COOEt)$_2$

A

B

[1] NaOEt

[2] ClCH$_2$CH$_2$SCH$_3$

D

[1] NaOH/H$_2$O

[2] H$_3$O$^+$

C

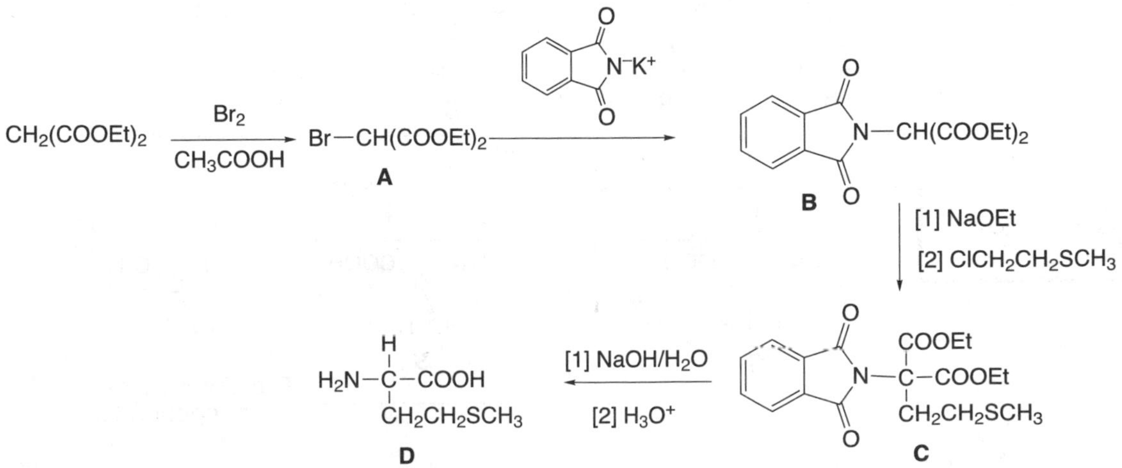

28.42

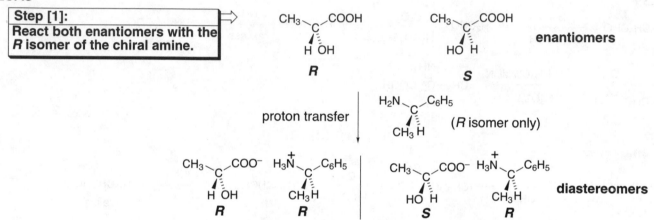

28.43

| Step [1]: |
| React both enantiomers with the **R** isomer of the chiral amine. |

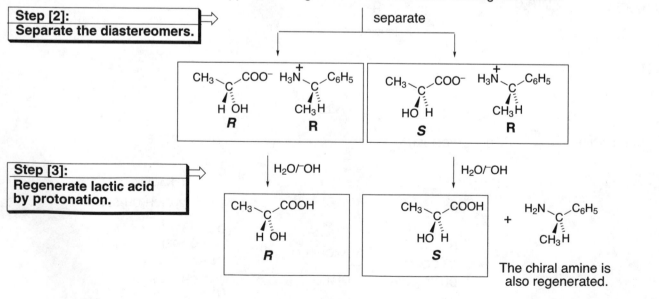

These salts have the *same* configuration around one stereogenic center, but the *opposite* configuration about the other stereogenic center.

| Step [2]: |
| Separate the diastereomers. |

| Step [3]: |
| Regenerate lactic acid by protonation. |

The chiral amine is also regenerated.

28.44

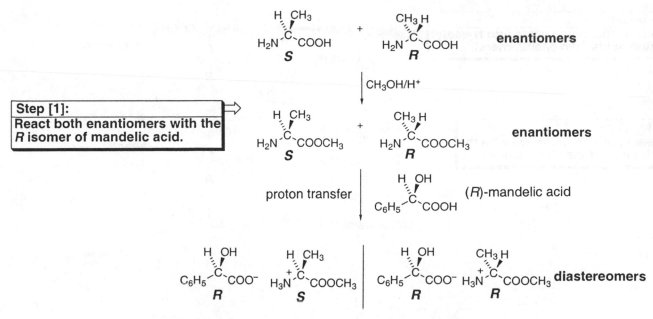

Step [1]:
React both enantiomers with the R isomer of mandelic acid.

proton transfer

(R)-mandelic acid

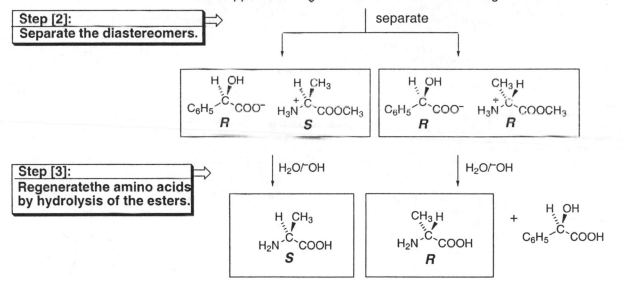

These salts have the *same* configuration around one stereogenic center,
but the *opposite* configuration about the other stereogenic center.

Step [2]:
Separate the diastereomers.

separate

Step [3]:
Regenerate the amino acids by hydrolysis of the esters.

28.45

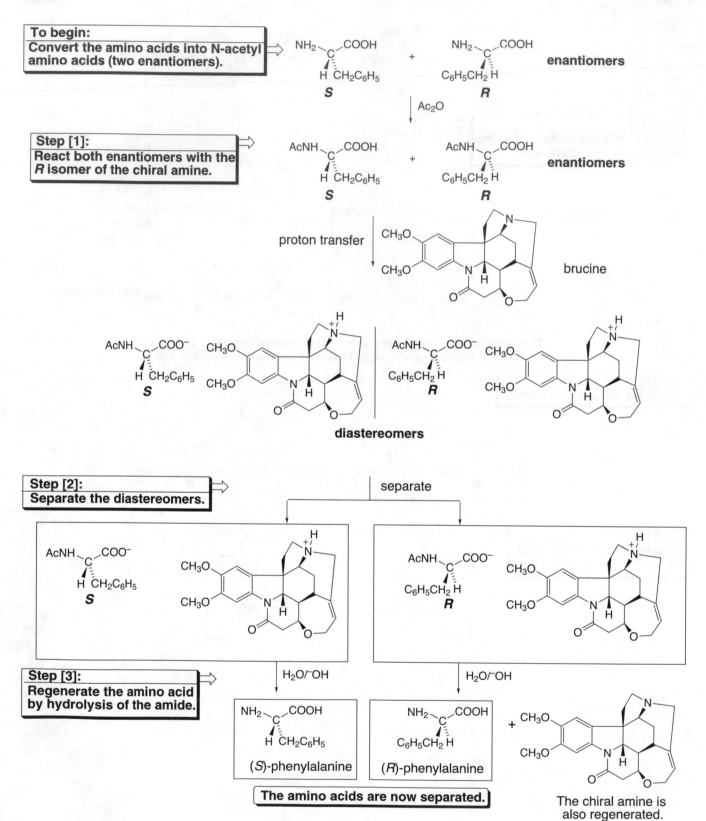

To begin:
Convert the amino acids into N-acetyl amino acids (two enantiomers).

enantiomers

S   R

Ac₂O

Step [1]:
React both enantiomers with the R isomer of the chiral amine.

enantiomers

S   R

proton transfer

brucine

diasteromers

Step [2]:
Separate the diastereomers.

separate

Step [3]:
Regenerate the amino acid by hydrolysis of the amide.

H₂O/⁻OH   H₂O/⁻OH

(S)-phenylalanine   (R)-phenylalanine

The amino acids are now separated.

The chiral amine is also regenerated.

28.46

a. (CH₃)₂CH–CH(NH₂)–COOH   racemic mixture

$(CH_3)_2CH-CH(NH_2)-COOH$ racemic mixture

$\xrightarrow{Ac_2O}$ $\xrightarrow{acylase}$

H₂N–C(H)(CH(CH₃)₂)–COOH + AcNH–C(H)(CH(CH₃)₂)–COOH

b.

CH₃CONH—...—NHCOCH₃ / COOH $\xrightarrow[\text{chiral Rh catalyst}]{H_2}$ $\xrightarrow[H_2O]{^-OH}$ H₂N–C(H)(CH₂CH₂CH₂CH₂NH₂)–COOH (S)-isomer

c.

(indole) COOH / NHAc $\xrightarrow[\text{chiral Rh catalyst}]{H_2}$ $\xrightarrow[H_2O]{^-OH}$ H₂N–C(H)(CH₂–indole)–COOH (S)-isomer

28.47

a. C₆H₅CH₂ H / H₂N–C– ... –C–OH   **Phe–Ala**

b. H H / H₂N–C– ... CH₂ / CH₂CONH₂   **Gly–Gln**

c. H₂NCH₂CH₂CH₂CH₂ ... **Lys–Gly**

d. NH₂C(NH)NHCH₂CH₂CH₂ ... CH₂ / imidazole **Arg–His**

28.48 Amide bonds are bold lines but not wedges.

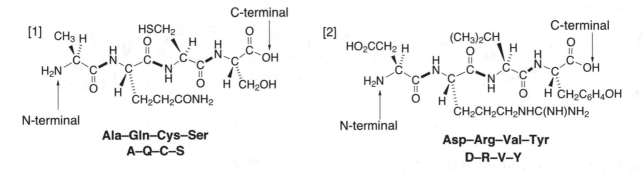

[1] CH₃ H HSCH₂ C-terminal
H₂N–C– ... –OH
N-terminal
CH₂CH₂CONH₂   H CH₂OH
Ala–Gln–Cys–Ser
A–Q–C–S

[2] HO₂CCH₂ H (CH₃)₂CH C-terminal
H₂N–C– ... –OH
N-terminal
CH₂CH₂CH₂NHC(NH)NH₂   H CH₂C₆H₄OH
Asp–Arg–Val–Tyr
D–R–V–Y

28.49 Name a peptide from the N-terminal to the C-terminal end.

a.

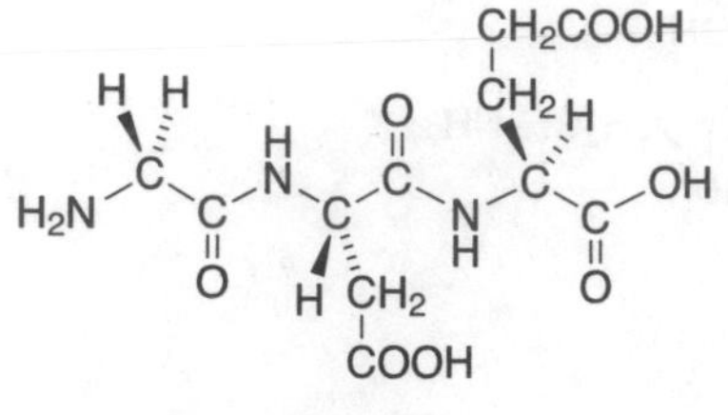

Gly–Asp–Glu
G–D–E

b.

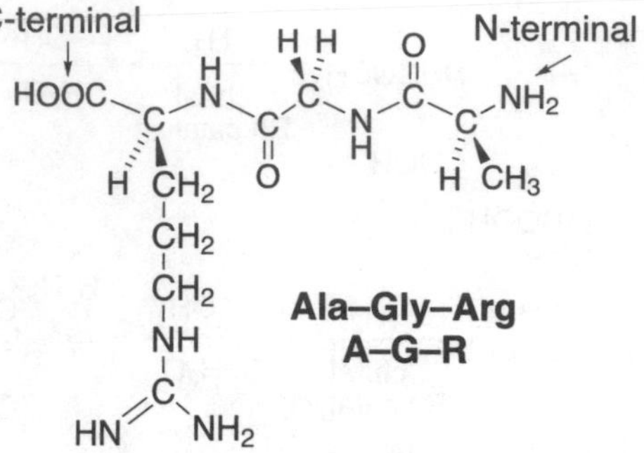

Ala–Gly–Arg
A–G–R

28.50 A peptide C–N bond is stronger than an ester C–O bond because the C–N bond has more double bond character due to resonance. Since N is more basic than O, an amide C–N bond is more stabilized by delocalization of the lone pair on N.

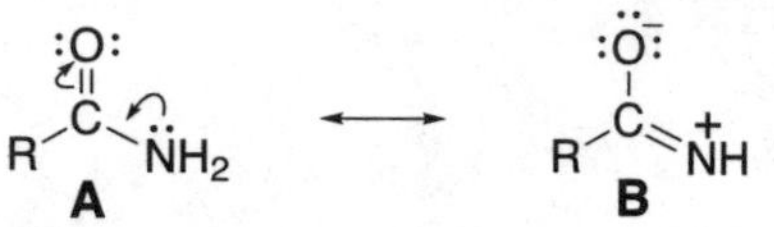

Structure **B** contributes greatly to the resonance hybrid and thus shortens and strengthens the C–N bond.

28.51 Use the principles from Answer 28.16.

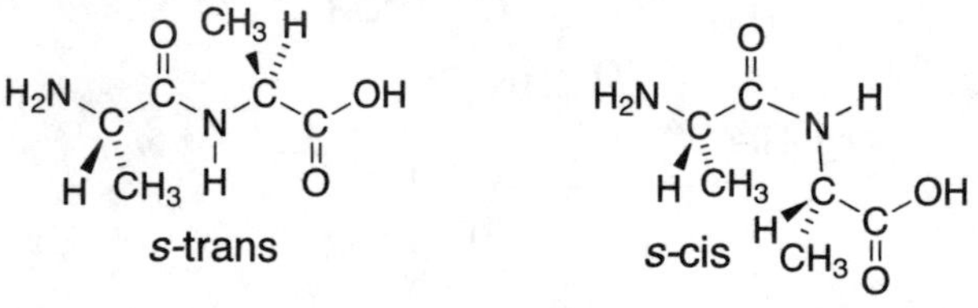

s-trans

s-cis

28.52 **A** and **B** can react to form an amide, or two molecules of **B** can form an amide.

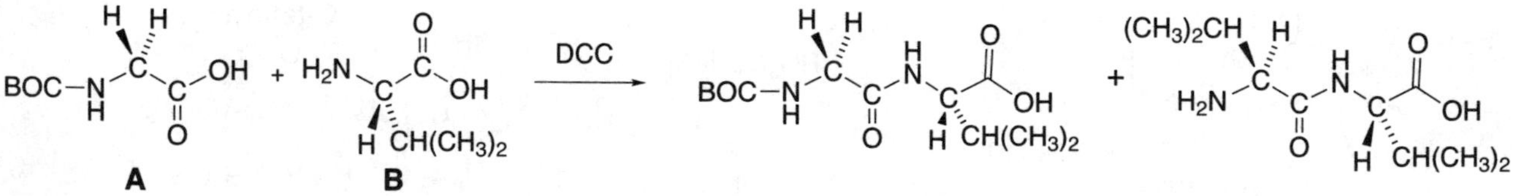

28.53

a.

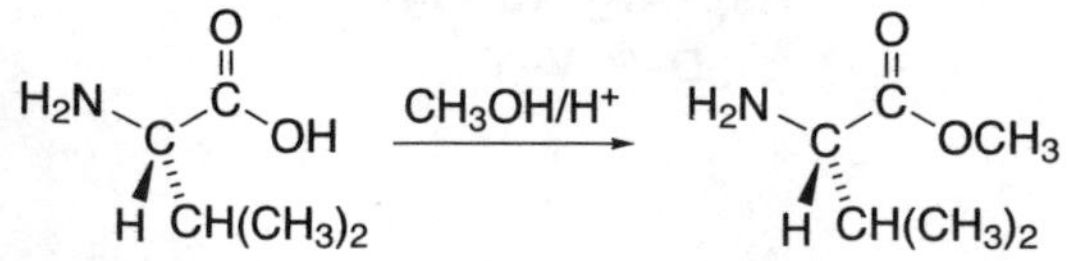

b.

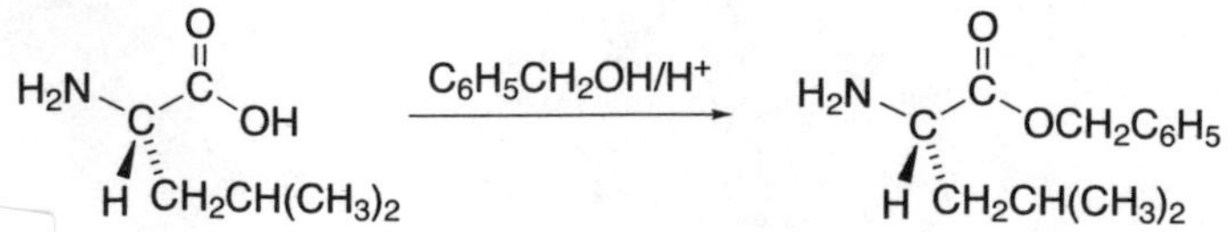

c. NH_2CH_2COOH $\xrightarrow[\text{(CH}_3\text{CH}_2)_3\text{N}]{\text{[(CH}_3)_3\text{COCO]}_2\text{O}}$ BOC—N—C—COOH (BOC-protected glycine)

d. product in (b) + product in (c) $\xrightarrow{\text{DCC}}$ BOC—N—C—C(=O)—N—C(—CH$_2$CH(CH$_3$)$_2$)—C(=O)OCH$_2$C$_6$H$_5$

e. $(CH_3)_3CO$—C(=O)—N(H)—C(H)(CH(CH$_3$)$_2$)—C(=O)OCH$_2$C$_6$H$_5$ $\xrightarrow[\text{Pd/C}]{\text{H}_2}$ $(CH_3)_3CO$—C(=O)—HN—C(H)(CH(CH$_3$)$_2$)—C(=O)OH

f. starting material in (e) $\xrightarrow[\text{CH}_3\text{COOH}]{\text{HBr}}$ H_2N—C(H)(CH(CH$_3$)$_2$)—C(=O)OH

g. product in (e) $\xrightarrow{\text{CF}_3\text{COOH}}$ H_2N—C(H)(CH(CH$_3$)$_2$)—C(=O)OH

28.54

R—$\overset{:\overset{..}{O}:}{C}$(—$\overset{..}{O}$—R)(—$\overset{..}{N}$(H)—R) $\longleftrightarrow$ R—$\overset{:\overset{..}{O}:^-}{C}$(=$\overset{+}{N}$(H)—R)(—$\overset{..}{O}$—R) $\longleftrightarrow$ R—$\overset{:\overset{..}{O}:^-}{C}$(=$\overset{+}{O}$—R)(—$\overset{..}{N}$(H)—R) $\longleftrightarrow$ R—$\overset{:\overset{..}{O}:^-}{C}$(—$\overset{..}{O}$—R)(—$\overset{+}{N}$(H)—R)

carbamate

28.55

a. Gly: H_2N—C(H)(H)—C(=O)OH $\xrightarrow[\text{(CH}_3\text{CH}_2)_3\text{N}]{\text{[(CH}_3)_3\text{COCO]}_2\text{O}}$ BOC—N(H)—C(H)(H)—C(=O)OH **A**

Ala: H_2N—C(H)(CH$_3$)—C(=O)OH $\xrightarrow{\text{C}_6\text{H}_5\text{CH}_2\text{OH/H}^+}$ H_2N—C(H)(CH$_3$)—C(=O)OCH$_2$C$_6$H$_5$ **B**

A + **B** $\xrightarrow{\text{DCC}}$ BOC—N(H)—C(H)(H)—C(=O)—N(H)—C(H)(CH$_3$)—C(=O)OCH$_2$C$_6$H$_5$ $\xrightarrow[\text{CH}_3\text{COOH}]{\text{HBr}}$ H_2N—C(H)(H)—C(=O)—N(H)—C(H)(CH$_3$)—C(=O)OH

Gly–Ala

b.

$C_6H_5CH_2$ ... H, H_2N, OH, O — Phe
$\xrightarrow{\substack{[(CH_3)_3COCO]_2O \\ (CH_3CH_2)_3N}}$
$C_6H_5CH_2$ ... H, BOC—N ... OH, O — **A**

H_2N, OH, O, H, $CH_2CH(CH_3)_2$ — Leu
$\xrightarrow{C_6H_5CH_2OH/H^+}$
H_2N, O, $OCH_2C_6H_5$, H, $CH_2CH(CH_3)_2$ — **B**

A + **B** $\xrightarrow{DCC}$
$C_6H_5CH_2$ ... H, BOC—N, H, N, O, $OCH_2C_6H_5$, H, $CH_2CH(CH_3)_2$
$\xrightarrow{\substack{HBr \\ CH_3COOH}}$
$C_6H_5CH_2$ ... H, H_2N, H, N, O, OH, H, $CH_2CH(CH_3)_2$

Phe–Leu

c.

CH_3CH_2CH, CH_3, H, H_2N, OH, O — Ile
$\xrightarrow{\substack{[(CH_3)_3COCO]_2O \\ (CH_3CH_2)_3N}}$
CH_3CH_2CH, CH_3, H, BOC—N, OH, O — **A**

H_2N, O, OH, H, CH_3 — Ala
$\xrightarrow{C_6H_5CH_2OH/H^+}$
H_2N, O, $OCH_2C_6H_5$, H, CH_3 — **B**

A + **B** $\xrightarrow{DCC}$
CH_3CH_2CH, CH_3, H, BOC—N, H, N, O, $OCH_2C_6H_5$, H, CH_3
$\xrightarrow{\substack{H_2 \\ Pd/C}}$
CH_3CH_2CH, CH_3, H, BOC—N, H, N, O, OH, H, CH_3 — **C**

$C_6H_5CH_2$ ... H, H_2N, OH, O — Phe
$\xrightarrow{C_6H_5CH_2OH/H^+}$
$C_6H_5CH_2$ ... H, H_2N, O, $OCH_2C_6H_5$
$\xrightarrow[DCC]{\boxed{C}}$
CH_3CH_2CH, CH_3, H, BOC—N, H, N, O, $C_6H_5CH_2$, H, $OCH_2C_6H_5$, H, CH_3, O

$\xrightarrow{\substack{HBr \\ CH_3COOH}}$

CH_3CH_2CH, CH_3, H, H_2N, H, N, O, $C_6H_5CH_2$, H, N, H, OH, O, H, CH_3

Ile–Ala–Phe

28.56 Make all the BOC derivatives as described in Problem 28.20.

a

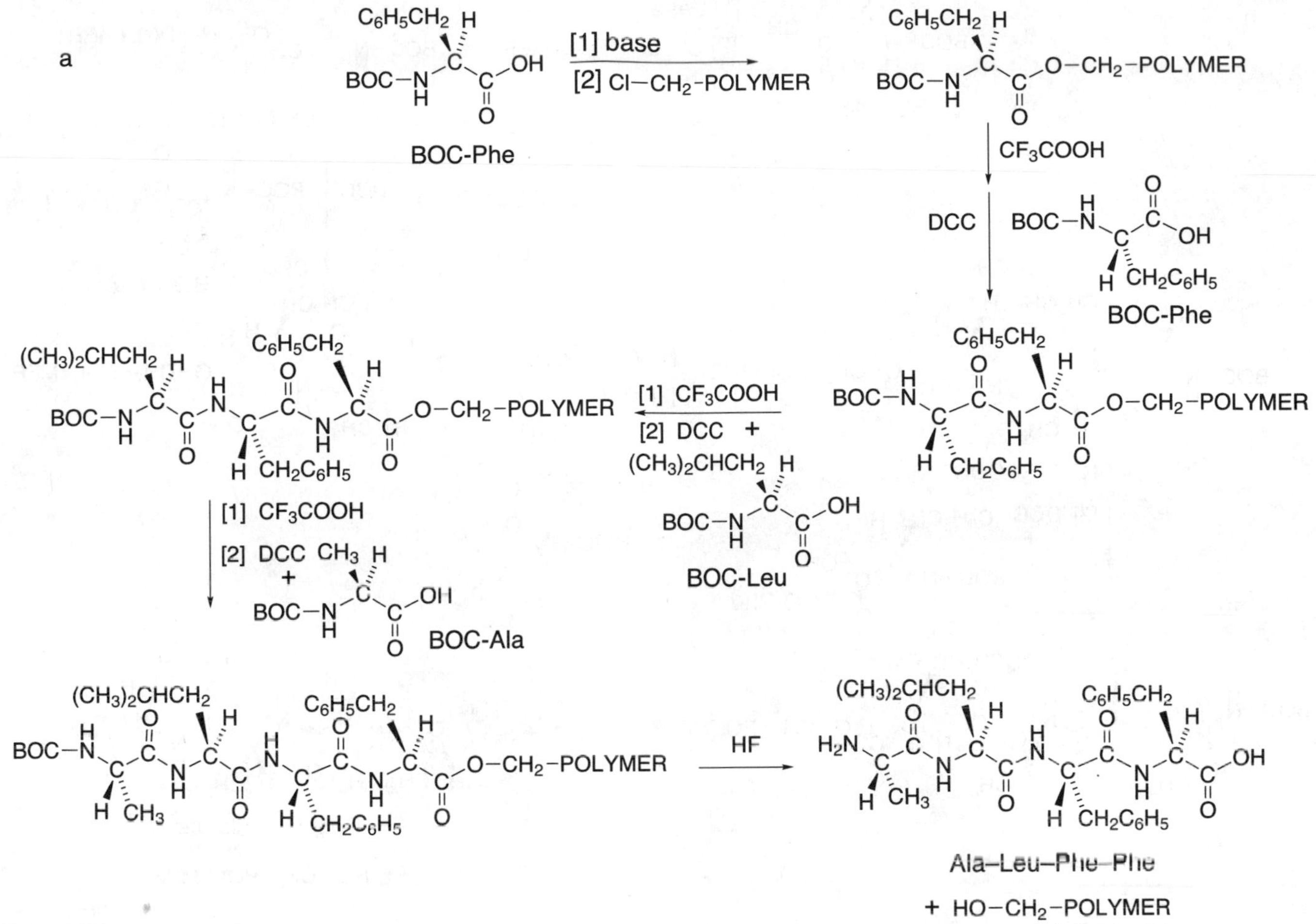

Ala–Leu–Phe–Phe

+ HO–CH₂–POLYMER

b.

28.57 An acetyl group on the NH₂ forms an amide. Although this amide does block an amino group from reaction, this amide is no different in reactivity than any of the peptide amide bonds. To remove the acetyl group after the peptide bond is formed would require harsh reaction conditions that would also cleave the amide bonds of the peptide.

N-acetyl amino acid

28.58

a. A *p*-nitrophenyl ester activates the carboxyl group of the first amino acid to amide formation by converting the OH group into a good leaving group, the *p*-nitrophenoxide group, which is highly resonance stabilized. In this case the electron withdrawing NO_2 group further stabilizes the leaving group.

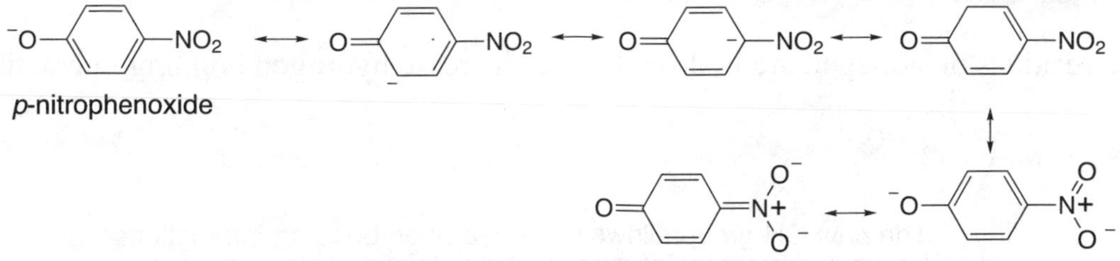

p-nitrophenoxide

The negative charge is delocalized on the O atom of the NO_2 group.

b. The *p*-methoxyphenyl ester contains an electron donating OCH_3 group, making $CH_3OC_6H_4O^-$ a poorer leaving group than $NO_2OC_6H_4O^-$, so this ester does not activate the amino acid to amide formation as much.

28.59

a.

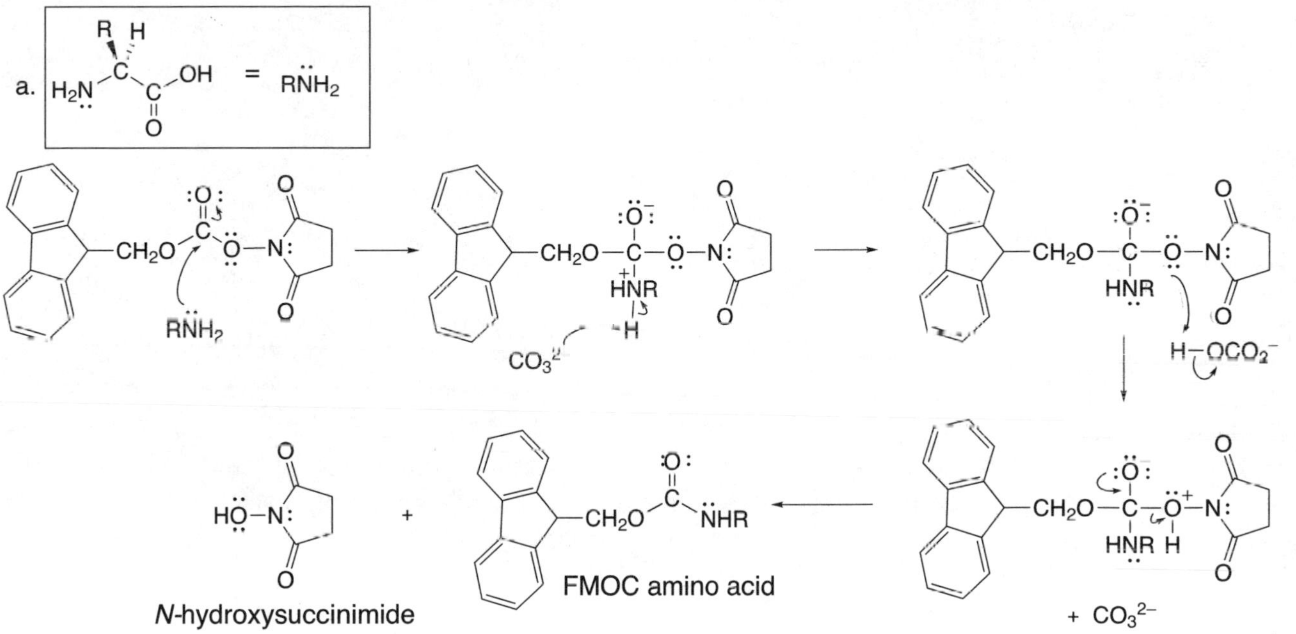

b.

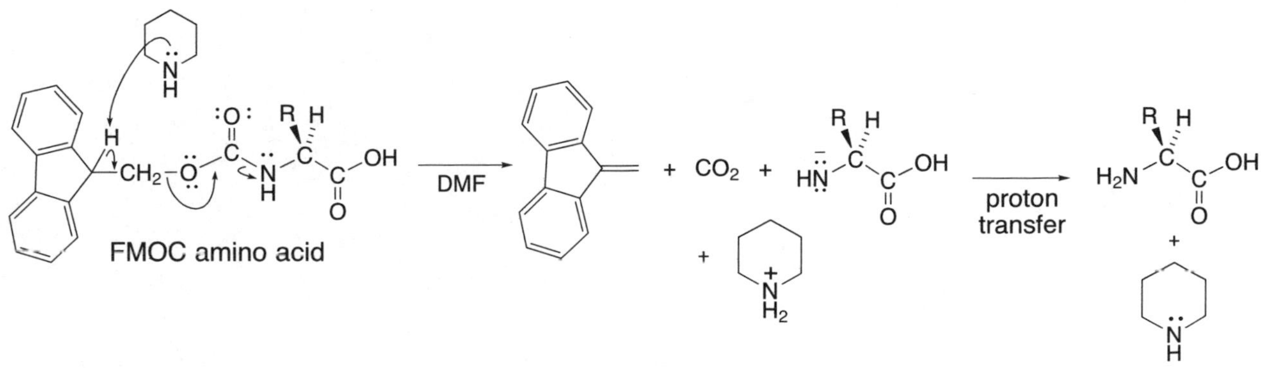

28.60 Amino acids commonly found on the interior of a globular protein have nonpolar or weakly polar side chains: isoleucine and phenylalanine. Amino acids commonly found on the surface have COOH, NH_2, and other groups that can hydrogen bond to water: aspartic acid, lysine, arginine, and glutamic acid.

28.61 The proline residues on collagen are hydroxylated to increase hydrogen bonding interactions.

The new OH group allows more hydrogen bonding interactions between the chains of the triple helix, thus stabilizing it.

Notes

Notes

Notes

Notes

Notes